Lecture Notes in Computer Science 12905

More information about this subseries at http://www.springer.com/series/7412

Marleen de Bruijne · Philippe C. Cattin ·
Stéphane Cotin · Nicolas Padoy ·
Stefanie Speidel · Yefeng Zheng ·
Caroline Essert (Eds.)

Medical Image Computing and Computer Assisted Intervention – MICCAI 2021

24th International Conference
Strasbourg, France, September 27 – October 1, 2021
Proceedings, Part V

Springer

Editors
Marleen de Bruijne (iD)
Erasmus MC - University Medical Center
Rotterdam
Rotterdam, The Netherlands

University of Copenhagen
Copenhagen, Denmark

Stéphane Cotin (iD)
Inria Nancy Grand Est
Villers-lès-Nancy, France

Stefanie Speidel (iD)
National Center for Tumor Diseases
(NCT/UCC)
Dresden, Germany

Caroline Essert (iD)
ICube, Université de Strasbourg, CNRS
Strasbourg, France

Philippe C. Cattin (iD)
University of Basel
Allschwil, Switzerland

Nicolas Padoy (iD)
ICube, Université de Strasbourg, CNRS
Strasbourg, France

Yefeng Zheng (iD)
Tencent Jarvis Lab
Shenzhen, China

ISSN 0302-9743   ISSN 1611-3349 (electronic)
Lecture Notes in Computer Science
ISBN 978-3-030-87239-7   ISBN 978-3-030-87240-3 (eBook)
https://doi.org/10.1007/978-3-030-87240-3

LNCS Sublibrary: SL6 – Image Processing, Computer Vision, Pattern Recognition, and Graphics

This Springer imprint is published by the registered company Springer Nature Switzerland AG
The registered company address is: Gewerbestrasse 11, 6330 Cham, Switzerland

Preface

The 24th edition of the International Conference on Medical Image Computing and Computer Assisted Intervention (MICCAI 2021) has for the second time been placed under the shadow of COVID-19. Complicated situations due to the pandemic and multiple lockdowns have affected our lives during the past year, sometimes perturbing the researchers work, but also motivating an extraordinary dedication from many of our colleagues, and significant scientific advances in the fight against the virus. After another difficult year, most of us were hoping to be able to travel and finally meet in person at MICCAI 2021, which was supposed to be held in Strasbourg, France. Unfortunately, due to the uncertainty of the global situation, MICCAI 2021 had to be moved again to a virtual event that was held over five days from September 27 to October 1, 2021. Taking advantage of the experience gained last year and of the fast-evolving platforms, the organizers of MICCAI 2021 redesigned the schedule and the format. To offer the attendees both a strong scientific content and an engaging experience, two virtual platforms were used: Pathable for the oral and plenary sessions and SpatialChat for lively poster sessions, industrial booths, and networking events in the form of interactive group video chats.

These proceedings of MICCAI 2021 showcase all 531 papers that were presented at the main conference, organized into eight volumes in the Lecture Notes in Computer Science (LNCS) series as follows:

- Part I, LNCS Volume 12901: Image Segmentation
- Part II, LNCS Volume 12902: Machine Learning 1
- Part III, LNCS Volume 12903: Machine Learning 2
- Part IV, LNCS Volume 12904: Image Registration and Computer Assisted Intervention
- Part V, LNCS Volume 12905: Computer Aided Diagnosis
- Part VI, LNCS Volume 12906: Image Reconstruction and Cardiovascular Imaging
- Part VII, LNCS Volume 12907: Clinical Applications
- Part VIII, LNCS Volume 12908: Microscopic, Ophthalmic, and Ultrasound Imaging

These papers were selected after a thorough double-blind peer review process. We followed the example set by past MICCAI meetings, using Microsoft's Conference Managing Toolkit (CMT) for paper submission and peer reviews, with support from the Toronto Paper Matching System (TPMS), to partially automate paper assignment to area chairs and reviewers, and from iThenticate to detect possible cases of plagiarism.

Following a broad call to the community we received 270 applications to become an area chair for MICCAI 2021. From this group, the program chairs selected a total of 96 area chairs, aiming for diversity — MIC versus CAI, gender, geographical region, and a mix of experienced and new area chairs. Reviewers were recruited also via an open call for volunteers from the community (288 applications, of which 149 were selected by the program chairs) as well as by re-inviting past reviewers, leading to a total of 1340 registered reviewers.

We received 1630 full paper submissions after an original 2667 intentions to submit. Four papers were rejected without review because of concerns of (self-)plagiarism and dual submission and one additional paper was rejected for not adhering to the MICCAI page restrictions; two further cases of dual submission were discovered and rejected during the review process. Five papers were withdrawn by the authors during review and after acceptance.

The review process kicked off with a reviewer tutorial and an area chair meeting to discuss the review process, criteria for MICCAI acceptance, how to write a good (meta-)review, and expectations for reviewers and area chairs. Each area chair was assigned 16–18 manuscripts for which they suggested potential reviewers using TPMS scores, self-declared research area(s), and the area chair's knowledge of the reviewers' expertise in relation to the paper, while conflicts of interest were automatically avoided by CMT. Reviewers were invited to bid for the papers for which they had been suggested by an area chair or which were close to their expertise according to TPMS. Final reviewer allocations via CMT took account of reviewer bidding, prioritization of area chairs, and TPMS scores, leading to on average four reviews performed per person by a total of 1217 reviewers.

Following the initial double-blind review phase, area chairs provided a meta-review summarizing key points of reviews and a recommendation for each paper. The program chairs then evaluated the reviews and their scores, along with the recommendation from the area chairs, to directly accept 208 papers (13%) and reject 793 papers (49%); the remainder of the papers were sent for rebuttal by the authors. During the rebuttal phase, two additional area chairs were assigned to each paper. The three area chairs then independently ranked their papers, wrote meta-reviews, and voted to accept or reject the paper, based on the reviews, rebuttal, and manuscript. The program chairs checked all meta-reviews, and in some cases where the difference between rankings was high or comments were conflicting, they also assessed the original reviews, rebuttal, and submission. In all other cases a majority voting scheme was used to make the final decision. This process resulted in the acceptance of a further 325 papers for an overall acceptance rate of 33%.

Acceptance rates were the same between medical image computing (MIC) and computer assisted interventions (CAI) papers, and slightly lower where authors classified their paper as both MIC and CAI. Distribution of the geographical region of the first author as indicated in the optional demographic survey was similar among submitted and accepted papers.

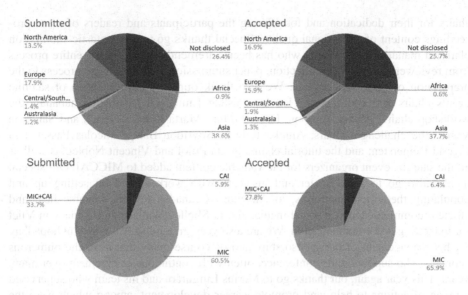

New this year, was the requirement to fill out a reproducibility checklist when submitting an intention to submit to MICCAI, in order to stimulate authors to think about what aspects of their method and experiments they should include to allow others to reproduce their results. Papers that included an anonymous code repository and/or indicated that the code would be made available were more likely to be accepted. From all accepted papers, 273 (51%) included a link to a code repository with the camera-ready submission.

Another novelty this year is that we decided to make the reviews, meta-reviews, and author responses for accepted papers available on the website. We hope the community will find this a useful resource.

The outstanding program of MICCAI 2021 was enriched by four exceptional keynote talks given by Alyson McGregor, Richard Satava, Fei-Fei Li, and Pierre Jannin, on hot topics such as gender bias in medical research, clinical translation to industry, intelligent medicine, and sustainable research. This year, as in previous years, high-quality satellite events completed the program of the main conference: 28 workshops, 23 challenges, and 14 tutorials; without forgetting the increasingly successful plenary events, such as the Women in MICCAI (WiM) meeting, the MICCAI Student Board (MSB) events, the 2nd Startup Village, the MICCAI-RSNA panel, and the first "Reinforcing Inclusiveness & diverSity and Empowering MICCAI" (or RISE-MICCAI) event.

MICCAI 2021 has also seen the first edition of CLINICCAI, the clinical day of MICCAI. Organized by Nicolas Padoy and Lee Swanstrom, this new event will hopefully help bring the scientific and clinical communities closer together, and foster collaborations and interaction. A common keynote connected the two events. We hope this effort will be pursued in the next editions.

We would like to thank everyone who has contributed to making MICCAI 2021 a success. First of all, we sincerely thank the authors, area chairs, reviewers, and session chairs for their dedication and for offering the participants and readers of these proceedings content of exceptional quality. Special thanks go to our fantastic submission platform manager Kitty Wong, who has been a tremendous help in the entire process from reviewer and area chair selection, paper submission, and the review process to the preparation of these proceedings. We also thank our very efficient team of satellite events chairs and coordinators, led by Cristian Linte and Matthieu Chabanas: the workshop chairs, Amber Simpson, Denis Fortun, Marta Kersten-Oertel, and Sandrine Voros; the challenges chairs, Annika Reinke, Spyridon Bakas, Nicolas Passat, and Ingerid Reinersten; and the tutorial chairs, Sonia Pujol and Vincent Noblet, as well as all the satellite event organizers for the valuable content added to MICCAI. Our special thanks also go to John Baxter and his team who worked hard on setting up and populating the virtual platforms, to Alejandro Granados for his valuable help and efficient communication on social media, and to Shelley Wallace and Anna Van Vliet for marketing and communication. We are also very grateful to Anirban Mukhopadhay for his management of the sponsorship, and of course many thanks to the numerous sponsors who supported the conference, often with continuous engagement over many years. This year again, our thanks go to Marius Linguraru and his team who supervised a range of actions to help, and promote, career development, among which were the mentorship program and the Startup Village. And last but not least, our wholehearted thanks go to Mehmet and the wonderful team at Dekon Congress and Tourism for their great professionalism and reactivity in the management of all logistical aspects of the event.

Finally, we thank the MICCAI society and the Board of Directors for their support throughout the years, starting with the first discussions about bringing MICCAI to Strasbourg in 2017.

We look forward to seeing you at MICCAI 2022.

September 2021

Marleen de Bruijne
Philippe Cattin
Stéphane Cotin
Nicolas Padoy
Stefanie Speidel
Yefeng Zheng
Caroline Essert

Organization

General Chair

Caroline Essert — Université de Strasbourg, CNRS, ICube, France

Program Chairs

Marleen de Bruijne — Erasmus MC Rotterdam, The Netherlands, and University of Copenhagen, Denmark
Philippe C. Cattin — University of Basel, Switzerland
Stéphane Cotin — Inria, France
Nicolas Padoy — Université de Strasbourg, CNRS, ICube, IHU, France
Stefanie Speidel — National Center for Tumor Diseases, Dresden, Germany
Yefeng Zheng — Tencent Jarvis Lab, China

Satellite Events Coordinators

Cristian Linte — Rochester Institute of Technology, USA
Matthieu Chabanas — Université Grenoble Alpes, France

Workshop Team

Amber Simpson — Queen's University, Canada
Denis Fortun — Université de Strasbourg, CNRS, ICube, France
Marta Kersten-Oertel — Concordia University, Canada
Sandrine Voros — TIMC-IMAG, INSERM, France

Challenges Team

Annika Reinke — German Cancer Research Center, Germany
Spyridon Bakas — University of Pennsylvania, USA
Nicolas Passat — Université de Reims Champagne-Ardenne, France
Ingerid Reinersten — SINTEF, NTNU, Norway

Tutorial Team

Vincent Noblet — Université de Strasbourg, CNRS, ICube, France
Sonia Pujol — Harvard Medical School, Brigham and Women's Hospital, USA

Clinical Day Chairs

Nicolas Padoy	Université de Strasbourg, CNRS, ICube, IHU, France
Lee Swanström	IHU Strasbourg, France

Sponsorship Chairs

Anirban Mukhopadhyay	Technische Universität Darmstadt, Germany
Yanwu Xu	Baidu Inc., China

Young Investigators and Early Career Development Program Chairs

Marius Linguraru	Children's National Institute, USA
Antonio Porras	Children's National Institute, USA
Daniel Racoceanu	Sorbonne Université/Brain Institute, France
Nicola Rieke	NVIDIA, Germany
Renee Yao	NVIDIA, USA

Social Media Chairs

Alejandro Granados Martinez	King's College London, UK
Shuwei Xing	Robarts Research Institute, Canada
Maxence Boels	King's College London, UK

Green Team

Pierre Jannin	INSERM, Université de Rennes 1, France
Étienne Baudrier	Université de Strasbourg, CNRS, ICube, France

Student Board Liaison

Éléonore Dufresne	Université de Strasbourg, CNRS, ICube, France
Étienne Le Quentrec	Université de Strasbourg, CNRS, ICube, France
Vinkle Srivastav	Université de Strasbourg, CNRS, ICube, France

Submission Platform Manager

Kitty Wong	The MICCAI Society, Canada

Virtual Platform Manager

John Baxter	INSERM, Université de Rennes 1, France

Program Committee

Ehsan Adeli	Stanford University, USA
Iman Aganj	Massachusetts General Hospital, Harvard Medical School, USA
Pablo Arbelaez	Universidad de los Andes, Colombia
John Ashburner	University College London, UK
Meritxell Bach Cuadra	University of Lausanne, Switzerland
Sophia Bano	University College London, UK
Adrien Bartoli	Université Clermont Auvergne, France
Christian Baumgartner	ETH Zürich, Switzerland
Hrvoje Bogunovic	Medical University of Vienna, Austria
Weidong Cai	University of Sydney, Australia
Gustavo Carneiro	University of Adelaide, Australia
Chao Chen	Stony Brook University, USA
Elvis Chen	Robarts Research Institute, Canada
Hao Chen	Hong Kong University of Science and Technology, Hong Kong SAR
Albert Chung	Hong Kong University of Science and Technology, Hong Kong SAR
Adrian Dalca	Massachusetts Institute of Technology, USA
Adrien Depeursinge	HES-SO Valais-Wallis, Switzerland
Jose Dolz	ÉTS Montréal, Canada
Ruogu Fang	University of Florida, USA
Dagan Feng	University of Sydney, Australia
Huazhu Fu	Inception Institute of Artificial Intelligence, United Arab Emirates
Mingchen Gao	University at Buffalo, The State University of New York, USA
Guido Gerig	New York University, USA
Orcun Goksel	Uppsala University, Sweden
Alberto Gomez	King's College London, UK
Ilker Hacihaliloglu	Rutgers University, USA
Adam Harrison	PAII Inc., USA
Mattias Heinrich	University of Lübeck, Germany
Yi Hong	Shanghai Jiao Tong University, China
Yipeng Hu	University College London, UK
Junzhou Huang	University of Texas at Arlington, USA
Xiaolei Huang	The Pennsylvania State University, USA
Jana Hutter	King's College London, UK
Madhura Ingalhalikar	Symbiosis Center for Medical Image Analysis, India
Shantanu Joshi	University of California, Los Angeles, USA
Samuel Kadoury	Polytechnique Montréal, Canada
Fahmi Khalifa	Mansoura University, Egypt
Hosung Kim	University of Southern California, USA
Minjeong Kim	University of North Carolina at Greensboro, USA

Ender Konukoglu	ETH Zürich, Switzerland
Bennett Landman	Vanderbilt University, USA
Ignacio Larrabide	CONICET, Argentina
Baiying Lei	Shenzhen University, China
Gang Li	University of North Carolina at Chapel Hill, USA
Mingxia Liu	University of North Carolina at Chapel Hill, USA
Herve Lombaert	ÉTS Montréal, Canada, and Inria, France
Marco Lorenzi	Inria, France
Le Lu	PAII Inc., USA
Xiongbiao Luo	Xiamen University, China
Dwarikanath Mahapatra	Inception Institute of Artificial Intelligence, United Arab Emirates
Andreas Maier	FAU Erlangen-Nuremberg, Germany
Erik Meijering	University of New South Wales, Australia
Hien Nguyen	University of Houston, USA
Marc Niethammer	University of North Carolina at Chapel Hill, USA
Tingying Peng	Technische Universität München, Germany
Caroline Petitjean	Université de Rouen, France
Dzung Pham	Henry M. Jackson Foundation, USA
Hedyeh Rafii-Tari	Auris Health Inc, USA
Islem Rekik	Istanbul Technical University, Turkey
Nicola Rieke	NVIDIA, Germany
Su Ruan	Laboratoire LITIS, France
Thomas Schultz	University of Bonn, Germany
Sharmishtaa Seshamani	Allen Institute, USA
Yonggang Shi	University of Southern California, USA
Darko Stern	Technical University of Graz, Austria
Carole Sudre	King's College London, UK
Heung-Il Suk	Korea University, South Korea
Jian Sun	Xi'an Jiaotong University, China
Raphael Sznitman	University of Bern, Switzerland
Amir Tahmasebi	Enlitic, USA
Qian Tao	Delft University of Technology, The Netherlands
Tolga Tasdizen	University of Utah, USA
Martin Urschler	University of Auckland, New Zealand
Archana Venkataraman	Johns Hopkins University, USA
Guotai Wang	University of Electronic Science and Technology of China, China
Hongzhi Wang	IBM Almaden Research Center, USA
Hua Wang	Colorado School of Mines, USA
Qian Wang	Shanghai Jiao Tong University, China
Yalin Wang	Arizona State University, USA
Fuyong Xing	University of Colorado Denver, USA
Daguang Xu	NVIDIA, USA
Yanwu Xu	Baidu, China
Ziyue Xu	NVIDIA, USA

Zhong Xue   Shanghai United Imaging Intelligence, China
Xin Yang   Huazhong University of Science and Technology,
                               China
Jianhua Yao   National Institutes of Health, USA
Zhaozheng Yin   Stony Brook University, USA
Yixuan Yuan   City University of Hong Kong, Hong Kong SAR
Liang Zhan   University of Pittsburgh, USA
Tuo Zhang   Northwestern Polytechnical University, China
Yitian Zhao   Chinese Academy of Sciences, China
Luping Zhou   University of Sydney, Australia
S. Kevin Zhou   Chinese Academy of Sciences, China
Dajiang Zhu   University of Texas at Arlington, USA
Xiahai Zhuang   Fudan University, China
Maria A. Zuluaga   EURECOM, France

Reviewers

Alaa Eldin Abdelaal   Chloé Audigier
Khalid Abdul Jabbar   Kamran Avanaki
Purang Abolmaesumi   Angelica Aviles-Rivero
Mazdak Abulnaga   Suyash Awate
Maryam Afzali   Dogu Baran Aydogan
Priya Aggarwal   Qinle Ba
Ola Ahmad   Morteza Babaie
Sahar Ahmad   Hyeon-Min Bae
Euijoon Ahn   Woong Bae
Alireza Akhondi-Asl   Junjie Bai
Saad Ullah Akram   Wenjia Bai
Dawood Al Chanti   Ujjwal Baid
Daniel Alexander   Spyridon Bakas
Sharib Ali   Yaël Balbastre
Lejla Alic   Marcin Balicki
Omar Al-Kadi   Fabian Balsiger
Maximilian Allan   Abhirup Banerjee
Pierre Ambrosini   Sreya Banerjee
Sameer Antani   Shunxing Bao
Michela Antonelli   Adrian Barbu
Jacob Antunes   Sumana Basu
Syed Anwar   Mathilde Bateson
Ignacio Arganda-Carreras   Deepti Bathula
Mohammad Ali Armin   John Baxter
Md Ashikuzzaman   Bahareh Behboodi
Mehdi Astaraki   Delaram Behnami
Angélica Atehortúa   Mikhail Belyaev
Gowtham Atluri   Aicha BenTaieb

Camilo Bermudez
Gabriel Bernardino
Hadrien Bertrand
Alaa Bessadok
Michael Beyeler
Indrani Bhattacharya
Chetan Bhole
Lei Bi
Gui-Bin Bian
Ryoma Bise
Stefano B. Blumberg
Ester Bonmati
Bhushan Borotikar
Jiri Borovec
Ilaria Boscolo Galazzo
Alexandre Bousse
Nicolas Boutry
Behzad Bozorgtabar
Nathaniel Braman
Nadia Brancati
Katharina Breininger
Christopher Bridge
Esther Bron
Rupert Brooks
Qirong Bu
Duc Toan Bui
Ninon Burgos
Nikolay Burlutskiy
Hendrik Burwinkel
Russell Butler
Michał Byra
Ryan Cabeen
Mariano Cabezas
Hongmin Cai
Jinzheng Cai
Yunliang Cai
Sema Candemir
Bing Cao
Qing Cao
Shilei Cao
Tian Cao
Weiguo Cao
Aaron Carass
M. Jorge Cardoso
Adrià Casamitjana
Matthieu Chabanas

Ahmad Chaddad
Jayasree Chakraborty
Sylvie Chambon
Yi Hao Chan
Ming-Ching Chang
Peng Chang
Violeta Chang
Sudhanya Chatterjee
Christos Chatzichristos
Antong Chen
Chang Chen
Cheng Chen
Dongdong Chen
Geng Chen
Hanbo Chen
Jianan Chen
Jianxu Chen
Jie Chen
Junxiang Chen
Lei Chen
Li Chen
Liangjun Chen
Min Chen
Pingjun Chen
Qiang Chen
Shuai Chen
Tianhua Chen
Tingting Chen
Xi Chen
Xiaoran Chen
Xin Chen
Xuejin Chen
Yuhua Chen
Yukun Chen
Zhaolin Chen
Zhineng Chen
Zhixiang Chen
Erkang Cheng
Jun Cheng
Li Cheng
Yuan Cheng
Farida Cheriet
Minqi Chong
Jaegul Choo
Aritra Chowdhury
Gary Christensen

Daan Christiaens
Stergios Christodoulidis
Ai Wern Chung
Pietro Antonio Cicalese
Özgün Çiçek
Celia Cintas
Matthew Clarkson
Jaume Coll-Font
Toby Collins
Olivier Commowick
Pierre-Henri Conze
Timothy Cootes
Luca Corinzia
Teresa Correia
Hadrien Courtecuisse
Jeffrey Craley
Hui Cui
Jianan Cui
Zhiming Cui
Kathleen Curran
Claire Cury
Tobias Czempiel
Vedrana Dahl
Haixing Dai
Rafat Damseh
Bilel Daoud
Neda Davoudi
Laura Daza
Sandro De Zanet
Charles Delahunt
Yang Deng
Cem Deniz
Felix Denzinger
Hrishikesh Deshpande
Christian Desrosiers
Blake Dewey
Neel Dey
Raunak Dey
Jwala Dhamala
Yashin Dicente Cid
Li Ding
Xinghao Ding
Zhipeng Ding
Konstantin Dmitriev
Ines Domingues
Liang Dong

Mengjin Dong
Nanqing Dong
Reuben Dorent
Sven Dorkenwald
Qi Dou
Simon Drouin
Niharika D'Souza
Lei Du
Hongyi Duanmu
Nicolas Duchateau
James Duncan
Luc Duong
Nicha Dvornek
Dmitry V. Dylov
Oleh Dzyubachyk
Roy Eagleson
Mehran Ebrahimi
Jan Egger
Alma Eguizabal
Gudmundur Einarsson
Ahmed Elazab
Mohammed S. M. Elbaz
Shireen Elhabian
Mohammed Elmogy
Amr Elsawy
Ahmed Eltanboly
Sandy Engelhardt
Ertunc Erdil
Marius Erdt
Floris Ernst
Boris Escalante-Ramírez
Maria Escobar
Mohammad Eslami
Nazila Esmaeili
Marco Esposito
Oscar Esteban
Théo Estienne
Ivan Ezhov
Deng-Ping Fan
Jingfan Fan
Xin Fan
Yonghui Fan
Xi Fang
Zhenghan Fang
Aly Farag
Mohsen Farzi

Lina Felsner
Jun Feng
Ruibin Feng
Xinyang Feng
Yuan Feng
Aaron Fenster
Aasa Feragen
Henrique Fernandes
Enzo Ferrante
Jean Feydy
Lukas Fischer
Peter Fischer
Antonio Foncubierta-Rodríguez
Germain Forestier
Nils Daniel Forkert
Jean-Rassaire Fouefack
Moti Freiman
Wolfgang Freysinger
Xueyang Fu
Yunguan Fu
Wolfgang Fuhl
Isabel Funke
Philipp Fürnstahl
Pedro Furtado
Ryo Furukawa
Jin Kyu Gahm
Laurent Gajny
Adrian Galdran
Yu Gan
Melanie Ganz
Cong Gao
Dongxu Gao
Linlin Gao
Siyuan Gao
Yixin Gao
Yue Gao
Zhifan Gao
Alfonso Gastelum-Strozzi
Srishti Gautam
Bao Ge
Rongjun Ge
Zongyuan Ge
Sairam Geethanath
Shiv Gehlot
Nils Gessert
Olivier Gevaert

Sandesh Ghimire
Ali Gholipour
Sayan Ghosal
Andrea Giovannini
Gabriel Girard
Ben Glocker
Arnold Gomez
Mingming Gong
Cristina González
German Gonzalez
Sharath Gopal
Karthik Gopinath
Pietro Gori
Michael Götz
Shuiping Gou
Maged Goubran
Sobhan Goudarzi
Dushyant Goyal
Mark Graham
Bertrand Granado
Alejandro Granados
Vicente Grau
Lin Gu
Shi Gu
Xianfeng Gu
Yun Gu
Zaiwang Gu
Hao Guan
Ricardo Guerrero
Houssem-Eddine Gueziri
Dazhou Guo
Hengtao Guo
Jixiang Guo
Pengfei Guo
Xiaoqing Guo
Yi Guo
Yulan Guo
Yuyu Guo
Krati Gupta
Vikash Gupta
Praveen Gurunath Bharathi
Boris Gutman
Prashnna Gyawali
Stathis Hadjidemetriou
Mohammad Hamghalam
Hu Han

Liang Han
Xiaoguang Han
Xu Han
Zhi Han
Zhongyi Han
Jonny Hancox
Xiaoke Hao
Nandinee Haq
Ali Hatamizadeh
Charles Hatt
Andreas Hauptmann
Mohammad Havaei
Kelei He
Nanjun He
Tiancheng He
Xuming He
Yuting He
Nicholas Heller
Alessa Hering
Monica Hernandez
Carlos Hernandez-Matas
Kilian Hett
Jacob Hinkle
David Ho
Nico Hoffmann
Matthew Holden
Sungmin Hong
Yoonmi Hong
Antal Horváth
Md Belayat Hossain
Benjamin Hou
William Hsu
Tai-Chiu Hsung
Kai Hu
Shi Hu
Shunbo Hu
Wenxing Hu
Xiaoling Hu
Xiaowei Hu
Yan Hu
Zhenhong Hu
Heng Huang
Qiaoying Huang
Yi-Jie Huang
Yixing Huang
Yongxiang Huang

Yue Huang
Yufang Huang
Arnaud Huaulmé
Henkjan Huisman
Yuankai Huo
Andreas Husch
Mohammad Hussain
Raabid Hussain
Sarfaraz Hussein
Khoi Huynh
Seong Jae Hwang
Emmanuel Iarussi
Kay Igwe
Abdullah-Al-Zubaer Imran
Ismail Irmakci
Mobarakol Islam
Mohammad Shafkat Islam
Vamsi Ithapu
Koichi Ito
Hayato Itoh
Oleksandra Ivashchenko
Yuji Iwahori
Shruti Jadon
Mohammad Jafari
Mostafa Jahanifar
Amir Jamaludin
Mirek Janatka
Won-Dong Jang
Uditha Jarayathne
Ronnachai Jaroensri
Golara Javadi
Rohit Jena
Rachid Jennane
Todd Jensen
Won-Ki Jeong
Yuanfeng Ji
Zhanghexuan Ji
Haozhe Jia
Jue Jiang
Tingting Jiang
Xiang Jiang
Jianbo Jiao
Zhicheng Jiao
Amelia Jiménez-Sánchez
Dakai Jin
Yueming Jin

Bin Jing
Anand Joshi
Yohan Jun
Kyu-Hwan Jung
Alain Jungo
Manjunath K N
Ali Kafaei Zad Tehrani
Bernhard Kainz
John Kalafut
Michael C. Kampffmeyer
Qingbo Kang
Po-Yu Kao
Neerav Karani
Turkay Kart
Satyananda Kashyap
Amin Katouzian
Alexander Katzmann
Prabhjot Kaur
Erwan Kerrien
Hoel Kervadec
Ashkan Khakzar
Nadieh Khalili
Siavash Khallaghi
Farzad Khalvati
Bishesh Khanal
Pulkit Khandelwal
Maksim Kholiavchenko
Naji Khosravan
Seyed Mostafa Kia
Daeseung Kim
Hak Gu Kim
Hyo-Eun Kim
Jae-Hun Kim
Jaeil Kim
Jinman Kim
Mansu Kim
Namkug Kim
Seong Tae Kim
Won Hwa Kim
Andrew King
Atilla Kiraly
Yoshiro Kitamura
Tobias Klinder
Bin Kong
Jun Kong
Tomasz Konopczynski

Bongjin Koo
Ivica Kopriva
Kivanc Kose
Mateusz Kozinski
Anna Kreshuk
Anithapriya Krishnan
Pavitra Krishnaswamy
Egor Krivov
Frithjof Kruggel
Alexander Krull
Elizabeth Krupinski
Serife Kucur
David Kügler
Hugo Kuijf
Abhay Kumar
Ashnil Kumar
Kuldeep Kumar
Nitin Kumar
Holger Kunze
Tahsin Kurc
Anvar Kurmukov
Yoshihiro Kuroda
Jin Tae Kwak
Yongchan Kwon
Francesco La Rosa
Aymen Laadhari
Dmitrii Lachinov
Alain Lalande
Tryphon Lambrou
Carole Lartizien
Bianca Lassen-Schmidt
Ngan Le
Leo Lebrat
Christian Ledig
Eung-Joo Lee
Hyekyoung Lee
Jong-Hwan Lee
Matthew Lee
Sangmin Lee
Soochahn Lee
Étienne Léger
Stefan Leger
Andreas Leibetseder
Rogers Jeffrey Leo John
Juan Leon
Bo Li

Chongyi Li
Fuhai Li
Hongming Li
Hongwei Li
Jian Li
Jianning Li
Jiayun Li
Junhua Li
Kang Li
Mengzhang Li
Ming Li
Qing Li
Shaohua Li
Shuyu Li
Weijian Li
Weikai Li
Wenqi Li
Wenyuan Li
Xiang Li
Xiaomeng Li
Xiaoxiao Li
Xin Li
Xiuli Li
Yang Li
Yi Li
Yuexiang Li
Zeju Li
Zhang Li
Zhiyuan Li
Zhjin Li
Gongbo Liang
Jianming Liang
Libin Liang
Yuan Liang
Haofu Liao
Ruizhi Liao
Wei Liao
Xiangyun Liao
Roxane Licandro
Gilbert Lim
Baihan Lin
Hongxiang Lin
Jianyu Lin
Yi Lin
Claudia Lindner
Geert Litjens

Bin Liu
Chi Liu
Daochang Liu
Dong Liu
Dongnan Liu
Feng Liu
Hangfan Liu
Hong Liu
Huafeng Liu
Jianfei Liu
Jingya Liu
Kai Liu
Kefei Liu
Lihao Liu
Mengting Liu
Peng Liu
Qin Liu
Quande Liu
Shengfeng Liu
Shenghua Liu
Shuangjun Liu
Sidong Liu
Siqi Liu
Tianrui Liu
Xiao Liu
Xinyang Liu
Xinyu Liu
Yan Liu
Yikang Liu
Yong Liu
Yuan Liu
Yue Liu
Yuhang Liu
Andrea Loddo
Nicolas Loménie
Daniel Lopes
Bin Lou
Jian Lou
Nicolas Loy Rodas
Donghuan Lu
Huanxiang Lu
Weijia Lu
Xiankai Lu
Yongyi Lu
Yueh-Hsun Lu
Yuhang Lu

Imanol Luengo
Jie Luo
Jiebo Luo
Luyang Luo
Ma Luo
Bin Lv
Jinglei Lv
Junyan Lyu
Qing Lyu
Yuanyuan Lyu
Andy J. Ma
Chunwei Ma
Da Ma
Hua Ma
Kai Ma
Lei Ma
Anderson Maciel
Amirreza Mahbod
S. Sara Mahdavi
Mohammed Mahmoud
Saïd Mahmoudi
Klaus H. Maier-Hein
Bilal Malik
Ilja Manakov
Matteo Mancini
Tommaso Mansi
Yunxiang Mao
Brett Marinelli
Pablo Márquez Neila
Carsten Marr
Yassine Marrakchi
Fabio Martinez
Andre Mastmeyer
Tejas Sudharshan Mathai
Dimitrios Mavroeidis
Jamie McClelland
Pau Medrano-Gracia
Raghav Mehta
Sachin Mehta
Raphael Meier
Qier Meng
Qingjie Meng
Yanda Meng
Martin Menten
Odyssée Merveille
Islem Mhiri

Liang Mi
Stijn Michielse
Abhishek Midya
Fausto Milletari
Hyun-Seok Min
Zhe Min
Tadashi Miyamoto
Sara Moccia
Hassan Mohy-ud-Din
Tony C. W. Mok
Rafael Molina
Mehdi Moradi
Rodrigo Moreno
Kensaku Mori
Lia Morra
Linda Moy
Mohammad Hamed Mozaffari
Sovanlal Mukherjee
Anirban Mukhopadhyay
Henning Müller
Balamurali Murugesan
Cosmas Mwikirize
Andriy Myronenko
Saad Nadeem
Vishwesh Nath
Rodrigo Nava
Fernando Navarro
Amin Nejatbakhsh
Dong Ni
Hannes Nickisch
Dong Nie
Jingxin Nie
Aditya Nigam
Lipeng Ning
Xia Ning
Tianye Niu
Jack Noble
Vincent Noblet
Alexey Novikov
Jorge Novo
Mohammad Obeid
Masahiro Oda
Benjamin Odry
Steffen Oeltze-Jafra
Hugo Oliveira
Sara Oliveira

Arnau Oliver
Emanuele Olivetti
Jimena Olveres
John Onofrey
Felipe Orihuela-Espina
José Orlando
Marcos Ortega
Yoshito Otake
Sebastian Otálora
Cheng Ouyang
Jiahong Ouyang
Xi Ouyang
Michal Ozery-Flato
Danielle Pace
Krittin Pachtrachai
J. Blas Pagador
Akshay Pai
Viswanath Pamulakanty Sudarshan
Jin Pan
Yongsheng Pan
Pankaj Pandey
Prashant Pandey
Egor Panfilov
Shumao Pang
Joao Papa
Constantin Pape
Bartlomiej Papiez
Hyunjin Park
Jongchan Park
Sanghyun Park
Seung-Jong Park
Seyoun Park
Magdalini Paschali
Diego Patiño Cortés
Angshuman Paul
Christian Payer
Yuru Pei
Chengtao Peng
Yige Peng
Antonio Pepe
Oscar Perdomo
Sérgio Pereira
Jose-Antonio Pérez-Carrasco
Fernando Pérez-García
Jorge Perez-Gonzalez
Skand Peri

Matthias Perkonigg
Mehran Pesteie
Jorg Peters
Jens Petersen
Kersten Petersen
Renzo Phellan Aro
Ashish Phophalia
Tomasz Pieciak
Antonio Pinheiro
Pramod Pisharady
Kilian Pohl
Sebastian Pölsterl
Iulia A. Popescu
Alison Pouch
Prateek Prasanna
Raphael Prevost
Juan Prieto
Sergi Pujades
Elodie Puybareau
Esther Puyol-Antón
Haikun Qi
Huan Qi
Buyue Qian
Yan Qiang
Yuchuan Qiao
Chen Qin
Wenjian Qin
Yulei Qin
Wu Qiu
Hui Qu
Liangqiong Qu
Kha Gia Quach
Prashanth R.
Pradeep Reddy Raamana
Mehdi Rahim
Jagath Rajapakse
Kashif Rajpoot
Jhonata Ramos
Lingyan Ran
Hatem Rashwan
Daniele Ravì
Keerthi Sravan Ravi
Nishant Ravikumar
Harish RaviPrakash
Samuel Remedios
Yinhao Ren

Yudan Ren
Mauricio Reyes
Constantino Reyes-Aldasoro
Jonas Richiardi
David Richmond
Anne-Marie Rickmann
Leticia Rittner
Dominik Rivoir
Emma Robinson
Jessica Rodgers
Rafael Rodrigues
Robert Rohling
Michal Rosen-Zvi
Lukasz Roszkowiak
Karsten Roth
José Rouco
Daniel Rueckert
Jaime S. Cardoso
Mohammad Sabokrou
Ario Sadafi
Monjoy Saha
Pramit Saha
Dushyant Sahoo
Pranjal Sahu
Maria Sainz de Cea
Olivier Salvado
Robin Sandkuehler
Gianmarco Santini
Duygu Sarikaya
Imari Sato
Olivier Saut
Dustin Scheinost
Nico Scherf
Markus Schirmer
Alexander Schlaefer
Jerome Schmid
Julia Schnabel
Klaus Schoeffmann
Andreas Schuh
Ernst Schwartz
Christina Schwarz-Gsaxner
Michaël Sdika
Suman Sedai
Anjany Sekuboyina
Raghavendra Selvan
Sourya Sengupta

Youngho Seo
Lama Seoud
Ana Sequeira
Maxime Sermesant
Carmen Serrano
Muhammad Shaban
Ahmed Shaffie
Sobhan Shafiei
Mohammad Abuzar Shaikh
Reuben Shamir
Shayan Shams
Hongming Shan
Harshita Sharma
Gregory Sharp
Mohamed Shehata
Haocheng Shen
Li Shen
Liyue Shen
Mali Shen
Yiqing Shen
Yiqiu Shen
Zhengyang Shen
Kuangyu Shi
Luyao Shi
Xiaoshuang Shi
Xueying Shi
Yemin Shi
Yiyu Shi
Yonghong Shi
Jitae Shin
Boris Shirokikh
Suprosanna Shit
Suzanne Shontz
Yucheng Shu
Alberto Signoroni
Wilson Silva
Margarida Silveira
Matthew Sinclair
Rohit Singla
Sumedha Singla
Ayushi Sinha
Kevin Smith
Rajath Soans
Ahmed Soliman
Stefan Sommer
Yang Song

Youyi Song
Aristeidis Sotiras
Arcot Sowmya
Rachel Sparks
William Speier
Ziga Spiclin
Dominik Spinczyk
Jon Sporring
Chetan Srinidhi
Anuroop Sriram
Vinkle Srivastav
Lawrence Staib
Marius Staring
Johannes Stegmaier
Joshua Stough
Robin Strand
Martin Styner
Hai Su
Yun-Hsuan Su
Vaishnavi Subramanian
Gérard Subsol
Yao Sui
Avan Suinesiaputra
Jeremias Sulam
Shipra Suman
Li Sun
Wenqing Sun
Chiranjib Sur
Yannick Suter
Tanveer Syeda-Mahmood
Fatemeh Taheri Dezaki
Roger Tam
José Tamez-Peña
Chaowei Tan
Hao Tang
Thomas Tang
Yucheng Tang
Zihao Tang
Mickael Tardy
Giacomo Tarroni
Jonas Teuwen
Paul Thienphrapa
Stephen Thompson
Jiang Tian
Yu Tian
Yun Tian

Aleksei Tiulpin
Hamid Tizhoosh
Matthew Toews
Oguzhan Topsakal
Antonio Torteya
Sylvie Treuillet
Jocelyne Troccaz
Roger Trullo
Chialing Tsai
Sudhakar Tummala
Verena Uslar
Hristina Uzunova
Régis Vaillant
Maria Vakalopoulou
Jeya Maria Jose Valanarasu
Tom van Sonsbeek
Gijs van Tulder
Marta Varela
Thomas Varsavsky
Francisco Vasconcelos
Liset Vazquez Romaguera
S. Swaroop Vedula
Sanketh Vedula
Harini Veeraraghavan
Miguel Vega
Gonzalo Vegas Sanchez-Ferrero
Anant Vemuri
Gopalkrishna Veni
Mitko Veta
Thomas Vetter
Pedro Vieira
Juan Pedro Vigueras Guillén
Barbara Villarini
Satish Viswanath
Athanasios Vlontzos
Wolf-Dieter Vogl
Bo Wang
Cheng Wang
Chengjia Wang
Chunliang Wang
Clinton Wang
Congcong Wang
Dadong Wang
Dongang Wang
Haifeng Wang
Hongyu Wang

Hu Wang
Huan Wang
Kun Wang
Li Wang
Liansheng Wang
Linwei Wang
Manning Wang
Renzhen Wang
Ruixuan Wang
Sheng Wang
Shujun Wang
Shuo Wang
Tianchen Wang
Tongxin Wang
Wenzhe Wang
Xi Wang
Xiaosong Wang
Yan Wang
Yaping Wang
Yi Wang
Yirui Wang
Zeyi Wang
Zhangyang Wang
Zihao Wang
Zuhui Wang
Simon Warfield
Jonathan Weber
Jürgen Weese
Dong Wei
Donglai Wei
Dongming Wei
Martin Weigert
Wolfgang Wein
Michael Wels
Cédric Wemmert
Junhao Wen
Travis Williams
Matthias Wilms
Stefan Winzeck
James Wiskin
Adam Wittek
Marek Wodzinski
Jelmer Wolterink
Ken C. L. Wong
Chongruo Wu
Guoqing Wu

Ji Wu
Jian Wu
Jie Ying Wu
Pengxiang Wu
Xiyin Wu
Ye Wu
Yicheng Wu
Yifan Wu
Tobias Wuerfl
Pengcheng Xi
James Xia
Siyu Xia
Wenfeng Xia
Yingda Xia
Yong Xia
Lei Xiang
Deqiang Xiao
Li Xiao
Yiming Xiao
Hongtao Xie
Lingxi Xie
Long Xie
Weidi Xie
Yiting Xie
Yutong Xie
Xiaohan Xing
Chang Xu
Chenchu Xu
Hongming Xu
Kele Xu
Min Xu
Rui Xu
Xiaowei Xu
Xuanang Xu
Yongchao Xu
Zhenghua Xu
Zhoubing Xu
Kai Xuan
Cheng Xue
Jie Xue
Wufeng Xue
Yuan Xue
Faridah Yahya
Ke Yan
Yuguang Yan
Zhennan Yan

Changchun Yang
Chao-Han Huck Yang
Dong Yang
Erkun Yang
Fan Yang
Ge Yang
Guang Yang
Guanyu Yang
Heran Yang
Hongxu Yang
Huijuan Yang
Jiancheng Yang
Jie Yang
Junlin Yang
Lin Yang
Peng Yang
Xin Yang
Yan Yang
Yujiu Yang
Dongren Yao
Jiawen Yao
Li Yao
Qingsong Yao
Chuyang Ye
Dong Hyc Ye
Menglong Ye
Xujiong Ye
Jingru Yi
Jirong Yi
Xin Yi
Youngjin Yoo
Chenyu You
Haichao Yu
Hanchao Yu
Lequan Yu
Qi Yu
Yang Yu
Pengyu Yuan
Fatemeh Zabihollahy
Ghada Zamzmi
Marco Zenati
Guodong Zeng
Rui Zeng
Oliver Zettinig
Zhiwei Zhai
Chaoyi Zhang

Daoqiang Zhang
Fan Zhang
Guangming Zhang
Hang Zhang
Huahong Zhang
Jianpeng Zhang
Jiong Zhang
Jun Zhang
Lei Zhang
Lichi Zhang
Lin Zhang
Ling Zhang
Lu Zhang
Miaomiao Zhang
Ning Zhang
Qiang Zhang
Rongzhao Zhang
Ru-Yuan Zhang
Shihao Zhang
Shu Zhang
Tong Zhang
Wei Zhang
Weiwei Zhang
Wen Zhang
Wenlu Zhang
Xin Zhang
Ya Zhang
Yanbo Zhang
Yanfu Zhang
Yi Zhang
Yishuo Zhang
Yong Zhang
Yongqin Zhang
You Zhang
Youshan Zhang
Yu Zhang
Yue Zhang
Yueyi Zhang
Yulun Zhang
Yunyan Zhang
Yuyao Zhang
Can Zhao
Changchen Zhao
Chongyue Zhao
Fenqiang Zhao
Gangming Zhao

He Zhao
Jun Zhao
Li Zhao
Qingyu Zhao
Rongchang Zhao
Shen Zhao
Shijie Zhao
Tengda Zhao
Tianyi Zhao
Wei Zhao
Xuandong Zhao
Yiyuan Zhao
Yuan-Xing Zhao
Yue Zhao
Zixu Zhao
Ziyuan Zhao
Xingjian Zhen
Guoyan Zheng
Hao Zheng
Jiannan Zheng
Kang Zheng
Shenhai Zheng
Yalin Zheng
Yinqiang Zheng
Yushan Zheng
Jia-Xing Zhong
Zichun Zhong

Bo Zhou
Haoyin Zhou
Hong-Yu Zhou
Kang Zhou
Sanping Zhou
Sihang Zhou
Tao Zhou
Xiao-Yun Zhou
Yanning Zhou
Yuyin Zhou
Zongwei Zhou
Dongxiao Zhu
Hancan Zhu
Lei Zhu
Qikui Zhu
Xinliang Zhu
Yuemin Zhu
Zhe Zhu
Zhuotun Zhu
Aneeq Zia
Veronika Zimmer
David Zimmerer
Lilla Zöllei
Yukai Zou
Lianrui Zuo
Gerald Zwettler
Reyer Zwiggelaar

Outstanding Reviewers

Neel Dey
Monica Hernandez
Ivica Kopriva
Sebastian Otálora

Danielle Pace
Sérgio Pereira
David Richmond
Rohit Singla
Yan Wang

New York University, USA
University of Zaragoza, Spain
Rudjer Boskovich Institute, Croatia
University of Applied Sciences and Arts Western
    Switzerland, Switzerland
Massachusetts General Hospital, USA
Lunit Inc., South Korea
IBM Watson Health, USA
University of British Columbia, Canada
Sichuan University, China

Honorable Mentions (Reviewers)

Mazdak Abulnaga	Massachusetts Institute of Technology, USA
Pierre Ambrosini	Erasmus University Medical Center, The Netherlands
Hyeon-Min Bae	Korea Advanced Institute of Science and Technology, South Korea
Mikhail Belyaev	Skolkovo Institute of Science and Technology, Russia
Bhushan Borotikar	Symbiosis International University, India
Katharina Breininger	Friedrich-Alexander-Universität Erlangen-Nürnberg, Germany
Ninon Burgos	CNRS, Paris Brain Institute, France
Mariano Cabezas	The University of Sydney, Australia
Aaron Carass	Johns Hopkins University, USA
Pierre-Henri Conze	IMT Atlantique, France
Christian Desrosiers	École de technologie supérieure, Canada
Reuben Dorent	King's College London, UK
Nicha Dvornek	Yale University, USA
Dmitry V. Dylov	Skolkovo Institute of Science and Technology, Russia
Marius Erdt	Fraunhofer Singapore, Singapore
Ruibin Feng	Stanford University, USA
Enzo Ferrante	CONICET/Universidad Nacional del Litoral, Argentina
Antonio Foncubierta-Rodríguez	IBM Research, Switzerland
Isabel Funke	National Center for Tumor Diseases Dresden, Germany
Adrian Galdran	University of Bournemouth, UK
Ben Glocker	Imperial College London, UK
Cristina González	Universidad de los Andes, Colombia
Maged Goubran	Sunnybrook Research Institute, Canada
Sobhan Goudarzi	Concordia University, Canada
Vicente Grau	University of Oxford, UK
Andreas Hauptmann	University of Oulu, Finland
Nico Hoffmann	Technische Universität Dresden, Germany
Sungmin Hong	Massachusetts General Hospital, Harvard Medical School, USA
Won-Dong Jang	Harvard University, USA
Zhanghexuan Ji	University at Buffalo, SUNY, USA
Neerav Karani	ETH Zurich, Switzerland
Alexander Katzmann	Siemens Healthineers, Germany
Erwan Kerrien	Inria, France
Anitha Priya Krishnan	Genentech, USA
Tahsin Kurc	Stony Brook University, USA
Francesco La Rosa	École polytechnique fédérale de Lausanne, Switzerland
Dmitrii Lachinov	Medical University of Vienna, Austria
Mengzhang Li	Peking University, China
Gilbert Lim	National University of Singapore, Singapore
Dongnan Liu	University of Sydney, Australia

Bin Lou	Siemens Healthineers, USA
Kai Ma	Tencent, China
Klaus H. Maier-Hein	German Cancer Research Center (DKFZ), Germany
Raphael Meier	University Hospital Bern, Switzerland
Tony C. W. Mok	Hong Kong University of Science and Technology, Hong Kong SAR
Lia Morra	Politecnico di Torino, Italy
Cosmas Mwikirize	Rutgers University, USA
Felipe Orihuela-Espina	Instituto Nacional de Astrofísica, Óptica y Electrónica, Mexico
Egor Panfilov	University of Oulu, Finland
Christian Payer	Graz University of Technology, Austria
Sebastian Pölsterl	Ludwig-Maximilians Universität, Germany
José Rouco	University of A Coruña, Spain
Daniel Rueckert	Imperial College London, UK
Julia Schnabel	King's College London, UK
Christina Schwarz-Gsaxner	Graz University of Technology, Austria
Boris Shirokikh	Skolkovo Institute of Science and Technology, Russia
Yang Song	University of New South Wales, Australia
Gérard Subsol	Université de Montpellier, France
Tanveer Syeda-Mahmood	IBM Research, USA
Mickael Tardy	Hera-MI, France
Paul Thienphrapa	Atlas5D, USA
Gijs van Tulder	Radboud University, The Netherlands
Tongxin Wang	Indiana University, USA
Yirui Wang	PAII Inc., USA
Jelmer Wolterink	University of Twente, The Netherlands
Lei Xiang	Subtle Medical Inc., USA
Fatemeh Zabihollahy	Johns Hopkins University, USA
Wei Zhang	University of Georgia, USA
Ya Zhang	Shanghai Jiao Tong University, China
Qingyu Zhao	Stanford University, China
Yushan Zheng	Beihang University, China

Mentorship Program (Mentors)

Shadi Albarqouni	Helmholtz AI, Helmholtz Center Munich, Germany
Hao Chen	Hong Kong University of Science and Technology, Hong Kong SAR
Nadim Daher	NVIDIA, France
Marleen de Bruijne	Erasmus MC/University of Copenhagen, The Netherlands
Qi Dou	The Chinese University of Hong Kong, Hong Kong SAR
Gabor Fichtinger	Queen's University, Canada
Jonny Hancox	NVIDIA, UK

Nobuhiko Hata · Harvard Medical School, USA
Sharon Xiaolei Huang · Pennsylvania State University, USA
Jana Hutter · King's College London, UK
Dakai Jin · PAII Inc., China
Samuel Kadoury · Polytechnique Montréal, Canada
Minjeong Kim · University of North Carolina at Greensboro, USA
Hans Lamecker · 1000shapes GmbH, Germany
Andrea Lara · Galileo University, Guatemala
Ngan Le · University of Arkansas, USA
Baiying Lei · Shenzhen University, China
Karim Lekadir · Universitat de Barcelona, Spain
Marius George Linguraru · Children's National Health System/George
Washington University, USA

Herve Lombaert · ETS Montreal, Canada
Marco Lorenzi · Inria, France
Le Lu · PAII Inc., China
Xiongbiao Luo · Xiamen University, China
Dzung Pham · Henry M. Jackson Foundation/Uniformed Services
University/National Institutes of Health/Johns
Hopkins University, USA

Josien Pluim · Eindhoven University of Technology/University
Medical Center Utrecht, The Netherlands
Antonio Porras · University of Colorado Anschutz Medical
Campus/Children's Hospital Colorado, USA
Islem Rekik · Istanbul Technical University, Turkey
Nicola Rieke · NVIDIA, Germany
Julia Schnabel · TU Munich/Helmholtz Center Munich, Germany,
and King's College London, UK

Debdoot Sheet · Indian Institute of Technology Kharagpur, India
Pallavi Tiwari · Case Western Reserve University, USA
Jocelyne Troccaz · CNRS, TIMC, Grenoble Alpes University, France
Sandrine Voros · TIMC-IMAG, INSERM, France
Linwei Wang · Rochester Institute of Technology, USA
Yalin Wang · Arizona State University, USA
Zhong Xue · United Imaging Intelligence Co. Ltd, USA
Renee Yao · NVIDIA, USA
Mohammad Yaqub · Mohamed Bin Zayed University of Artificial
Intelligence, United Arab Emirates, and University
of Oxford, UK

S. Kevin Zhou · University of Science and Technology of China, China
Lilla Zollei · Massachusetts General Hospital, Harvard Medical
School, USA
Maria A. Zuluaga · EURECOM, France

Contents – Part V

Outcome/Disease Prediction

Computer Aided Diagnosis

DeepStationing: Thoracic Lymph Node Station Parsing in CT Scans Using Anatomical Context Encoding and Key Organ Auto-Search

Dazhou Guo[1](✉), Xianghua Ye[2](✉), Jia Ge[2], Xing Di[3], Le Lu[1], Lingyun Huang[4], Guotong Xie[4], Jing Xiao[4], Zhongjie Lu[2], Ling Peng[5], Senxiang Yan[2](✉), and Dakai Jin[1](✉)

[1] PAII Inc., Bethesda, MD, USA
{guodazhou999,jindakai376}@paii-labs.com
[2] The First Affiliated Hospital Zhejiang University, Hangzhou, China
{hye1982,yansenxiang}@zju.edu.cn
[3] Johns Hopkins University, Baltimore, USA
[4] Ping An Insurance Company of China, Shenzhen, China
[5] Zhejiang Provincial People's Hospital, Hangzhou, China

Abstract. Lymph node station (LNS) delineation from computed tomography (CT) scans is an indispensable step in radiation oncology workflow. High inter-user variabilities across oncologists and prohibitive laboring costs motivated the automated approach. Previous works exploit anatomical priors to infer LNS based on predefined ad-hoc margins. However, without the voxel-level supervision, the performance is severely limited. LNS is highly context-dependent—LNS boundaries are constrained by anatomical organs—we formulate it as a deep spatial and contextual parsing problem via encoded anatomical organs. This permits the deep network to better learn from both CT appearance and organ context. We develop a stratified referencing organ segmentation protocol that divides the organs into anchor and non-anchor categories and uses the former's predictions to guide the later segmentation. We further develop an auto-search module to identify the key organs that opt for the optimal LNS parsing performance. Extensive four-fold cross-validation experiments on a dataset of 98 esophageal cancer patients (with the most comprehensive set of *12 LNSs + 22 organs* in thoracic region to date) are conducted. Our LNS parsing model produces significant performance improvements, with an average Dice score of 81.1%±6.1%, which is 5.0% and 19.2% higher over the pure CT-based deep model and the previous representative approach, respectively.

1 Introduction

Cancers in thoracic region are the most common cancers worldwide [17] and significant proportions of patients are diagnosed at late stages involved with lymph

D. Guo, X. Ye—Equal contribution.

© Springer Nature Switzerland AG 2021
M. de Bruijne et al. (Eds.): MICCAI 2021, LNCS 12905, pp. 3–12, 2021.
https://doi.org/10.1007/978-3-030-87240-3_1

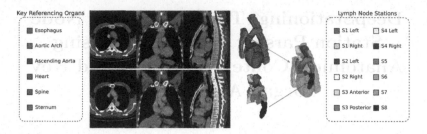

Fig. 1. An illustration of LNS and key referencing organs. The top row illustrates the auto-searched top-6 key referencing organs; the bottom row depicts the 12 LNSs.

node (LN) metastasis. The treatment protocol is a sophisticated combination of surgical resection and chemotherapy and/or radiotherapy [5]. Assessment of involved LNs [1,21] and accurate labeling their corresponding stations are essential for the treatment selection and planning. For example, in radiation therapy, the delineation accuracy of gross tumor volume (GTV) and clinical target volume (CTV) are the two most critical factors impacting the patient outcome. For CTV delineation, areas containing metastasis lymph nodes (LNs) should be included to sufficiently cover the sub-clinical disease regions [2]. One strategy to outline the sub-clinical disease region is to include the lymph node station (LNS) that containing the metastasized LNs [14,19]. Thoracic LNS is determined according to the text definitions of International Association for the Study of Lung Cancer (IASLC) [15]. The delineation of LNS in the current clinical workflow is predominantly a manual process using computed tomography (CT) images. Visual assessment and manual delineation is a challenging and time-consuming task even for experienced physicians, since converting text definitions of IASLC to precise 3D voxel-wise annotations can be error prone leading to large intra- and inter-user variability [2].

Deep convolutional neural networks (CNNs) have made remarkable progress in segmenting organs and tumors in medical imaging [4,7–9,18,20]. Only a handful of non-deep learning studies have tackled the automated LNS segmentation [3,11,13,16]. A LNS atlas was established using deformable registration [3]. Predefined margins from manually selected organs, such as the aorta, trachea, and vessels, were applied to infer LNSs [11], which was not able to accurately adapt to individual subject. Other methods [13,16] built fuzzy models to directly parse the LNS or learn the relative positions between LNS and some referencing organs. Average location errors ranging from 6.9mm to 34.2mm were reported using 22 test cases in [13], while an average Dice score (DSC) of 66.0% for 10 LNSs in 5 patients was observed in [16].

In this work, we propose the DeepStationing – an anatomical context encoded deep LNS parsing framework with key organ auto-search. We first segment a comprehensive set of 22 chest organs related to the description of LNS according to IASLC guideline. As inspired by [4], the 22 organs are stratified into the anchor or non-anchor categories. The predictions of the former category are exploited

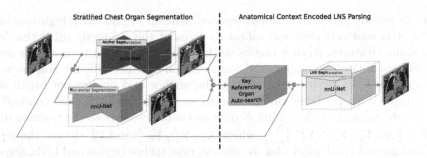

Fig. 2. Overall workflow of our DeepStationing, which consists of stratified chest organ segmentation and anatomical context encoded LNS parsing with key organ auto-search.

to guide and boost the segmentation performance of the later category. Next, CT image and referencing organ predictions are combined as different input channels to the LNS parsing module. The 22 referencing organs are identified by human experts. However, relevant but different from the human process, CNN may require a particular set of referencing organs (key organs) that can opt for optimal performance. Therefore, we automatically search for the key organs by applying a channel-weighting to the input organ prediction channels based on differentiable neural search [10]. The auto-searched final top-6 key organs, i.e., esophagus, aortic arch, ascending aorta, heart, spine and sternum (shown in Fig. 1), facilitate our DeepStationing method to achieve high LNS parsing accuracy. We adopt 3D nnU-Net [6] as our segmentation and parsing backbone. Extensive 4-fold cross-validation is conducted using a dataset of 98 CT images with 12 LNS + 22 Organ labels each, as *the first of its kind* to date. Experimental results demonstrate that deep model encoded with the spatial context of auto-searched key organs significantly improves the LNS paring performance, resulting in an average Dice score (DSC) of 81.1% ± 6.1%, which is 5.0% and 19.2% higher over the pure CT-based deep model and the most recent relevant work [11] (from our re-implementations), respectively.

2 Method

Figure 2 depicts the overview of our DeepStationing framework, consisting of two major modularized components: (1) stratified chest organ segmentation; (2) context encoded LNS parsing with key organ auto-search.

2.1 Stratified Chest Organ Segmentation

To provide the spatial context for LNS parsing, we first segment a comprehensive set of 22 chest organs related to the description of LNS. Simultaneously segmenting a large number of organs increase optimization difficulty leading to sub-optimal performance. Motivated by [4], we stratify 22 chest organs into the anchor and non-anchor categories. Anchor organs have high contrast, hence, it is relatively easy and robust to segment them directly using the deep appearance features. Anchor organs are first segmented, and their results serve as ideal candidates to support the segmentation of other difficult non-anchors. We use two CNN branches to stratify the anchor and non-anchor organ segmentation. With predicted anchor organs as additional input, the non-anchor organs are segmented. Assuming N data instances, we denote the training data as $\mathbb{S} = \left\{ X_n, Y_n^A, Y_n^{\neg A}, Y_n^L, \right\}_{n=1}^N$, where X_n, Y_n^A, $Y_n^{\neg A}$ and Y_n^L denote the input CT and ground-truth masks for the anchor, non-anchor organs and LNS, respectively. Assuming there are C_A and $C_{\neg A}$ classes for anchor and non-anchor organs and dropping n for clarity, our organ segmentation module generate the anchor and non-anchor organ predictions at every voxel location, j, and every output class, c:

$$\hat{Y}_c^A(j) = p^A \left(Y^A(j) = c \,|\, X; \mathbf{W}^A \right), \quad \hat{\mathbf{Y}}^A = \left[\hat{Y}_1^A \dots \hat{Y}_{C_A}^A \right], \quad (1)$$

$$\hat{Y}_c^{\neg A}(j) = p^{\neg A} \left(Y^{\neg A}(j) = c \,|\, X, \hat{\mathbf{Y}}^A; \mathbf{W}^{\neg A} \right), \quad \hat{\mathbf{Y}}^{\neg A} = \left[\hat{Y}_1^{\neg A} \dots \hat{Y}_{C_{\neg A}}^{\neg A} \right], \quad (2)$$

where $p^{(*)}(.)$ denotes the CNN functions and $\hat{Y}_c^{(*)}$ for the output segmentation maps. Here, we combine both anchor and non-anchor organ predictions into an overall prediction map $\hat{\mathbf{Y}}^{\mathfrak{A}} = \hat{\mathbf{Y}}^A \cup \hat{\mathbf{Y}}^{\neg A}$. Predictions are vector valued 3D masks as they provide a pseudo-probability for every class. $\mathbf{W}^{(*)}$ represents the corresponding CNN parameters.

2.2   Anatomical Context Encoded LNS Parsing

Segmenting LNS by only CT appearance can be error prone, since LNS highly relies on the spatial context of adjacent anatomical structures. Emulating the clinical practice of IASLC guidelines, we incorporate the referencing organs into the training process of LNS parsing. Given C_L classes of the LNSs, as illustrated in Fig. 2, we combine the above organ predictions with CT images to create a multi-channel input: $\left[X, \hat{\mathbf{Y}}^{\mathfrak{A}} \right]$:

$$\hat{Y}_c^L(j) = p^L \left(Y^L(j) = c \,|\, X, \hat{\mathbf{Y}}^{\mathfrak{A}}; \mathbf{W}^L \right), \quad \hat{\mathbf{Y}}^L = \left[\hat{Y}_1^L \dots \hat{Y}_{C_L}^L \right]. \quad (3)$$

Thereupon, the LNS parsing module leverages both the CT appearance and the predicted anatomical structures, implicitly encoding the spatial distributions of referencing organs during training. Similar to Eq. (1), we have the LNS prediction in its vector-valued form as $\hat{\mathbf{Y}}^L$.

Key Organ Auto-Search. The 22 referencing organs are previously selected according to the IASLC guideline. Nevertheless for deep learning based LNS model training, those manually selected organs might not lead to the optimal performance. Considering the potential variations in organ location and size distributions, and differences in automated organ segmentation accuracy, we hypothesize that the deep LNS parsing model would benefit from an automated reference organ selection process that are tailored to this purpose. Hence, we use the differentiable neural search [4] to search the key organs by applying a channel-weighting strategy to input organ masks. We make the search space continuous by relaxing the selection of the referencing organs to a Softmax function over the channel weights of the one-hot organ predictions $\hat{\mathbf{Y}}^{\mathfrak{A}}$. For C_L classes, we define a set of C_L learn-able logits for each channel, denoted as $\alpha_c, \forall c \in [1 \cdots C_L]$. The channel weight ϕ_c for a referencing organ is defined as:

$$\phi_c = \frac{\exp(\alpha_c)}{\sum_{m=1}^{C_L} \exp(\alpha_m)}, \quad \Phi = [\phi_1 \cdots \phi_{C_L}], \tag{4}$$

$$F(\hat{Y}_c^{\mathfrak{A}}, \phi_c) = \phi_c \cdot \hat{Y}_c^{\mathfrak{A}}, \quad F(\hat{\mathbf{Y}}^{\mathfrak{A}}, \Phi) = \left[F(\hat{Y}_1^{\mathfrak{A}}, \phi_1) \cdots F(\hat{Y}_{C_L}^{\mathfrak{A}}, \phi_{C_L}) \right] \tag{5}$$

where Φ denotes the set of channel weights and $F(\phi_c, \hat{Y}_c^{\mathfrak{A}})$ denotes the channel-wise multiplication between the scalar ϕ_c and the organ prediction $\hat{Y}_c^{\mathfrak{A}}$. The input of LNS parsing model becomes $\left[X, F(\hat{\mathbf{Y}}^{\mathfrak{A}}, \Phi) \right]$. As the results of the key organ auto-search, we select the organs with the top-n weights to be the searched n key organs. In this paper, we heuristically select the $n = 6$ based on the experimental results. Last, we train the LNS parsing model using the combination of original CT images and the auto-selected top-6 key organs' segmentation predictions.

3   Experimental Results

Dataset. We collected 98 contrast-enhanced venous-phase CT images of patients with esophageal cancers underwent surgery and/or radiotherapy treatments. A board-certified radiation oncologist with 15 years of experience annotated each patient with 3D masks of 12 LNSs, involved LNs (if any), and 22 referencing organs related to LNS according to IASLC guideline. The 12 annotated LN stations are: S1 *(left + right)*, S2 *(left + right)*, S3 *(anterior + posterior)*, S4 *(left + right)*, S5, S6, S7, S8. The average CT image size is $512 \times 512 \times 80$ voxels with an average resolution of $0.7 \times 0.7 \times 5.0$ mm. Extensive four-fold cross-validation (CV), separated at the patient level, was conducted. We report the segmentation performance using DSC in percentage, Hausdor distance (HD) and average surface distance (ASD) in mm.

Implementation Details. We adopt the nnU-Net [6] with DSC+CE losses as our backbone for all experiments due to its high accuracy on many medical image segmentation tasks. The nnU-Net has been proposed to automatically adapt different preprocessing strategies (i.e., the training image patch size, resolution, and learning rate) to a given 3D medical imaging dataset. We use the default nnU-Net settings for our model training. The total training epochs is 1000. For the organ auto-search parameter α_c, we first fix the α_c for 200 epochs and alternatively update the α_c and the network weights for another 800 epochs.

Table 1. Mean DSCs, HDs, and ASDs, and their standard deviations of LNS parsing performance using: (1) only CT appearance; (2) CT+all 22 referencing organ ground-truth masks; (3) CT+all 22 referencing organ predicted masks; (4) CT+auto-searched 6 referencing organ predicted masks. The best performance scores are shown in **bold**.

LNS	CT Only	+22 Organ GT	+22 Organ Pred	+6 Searched Organ Pred
DSC				
S1 Left	78.1 ± 6.8	84.3 ± 4.5	82.3 ± 4.6	**85.1 ± 4.0**
S1 Right	76.8 ± 5.0	84.3 ± 3.4	82.2 ± 3.4	**85.0 ± 4.1**
S2 Left	66.9 ± 11.4	75.8 ± 9.0	73.7 ± 8.9	**76.1 ± 8.2**
S2 Right	70.7 ± 8.5	74.8 ± 7.6	72.8 ± 7.6	**77.5 ± 6.4**
S3 Anterior	77.4 ± 4.9	79.8 ± 5.6	79.7 ± 5.6	**81.5 ± 4.9**
S3 Posterior	84.6 ± 3.1	87.9 ± 2.8	87.8 ± 2.9	**88.6 ± 2.7**
S4 Left	74.1 ± 8.2	77.0 ± 8.9	76.9 ± 8.9	**77.9 ± 9.4**
S4 Right	73.8 ± 8.9	74.9 ± 9.3	74.9 ± 9.4	**76.7 ± 8.3**
S5	72.6 ± 6.7	73.2 ± 7.4	73.2 ± 7.4	**77.9 ± 8.0**
S6	72.4 ± 5.7	74.9 ± 4.4	74.8 ± 4.5	**75.7 ± 4.3**
S7	85.0 ± 5.1	86.6 ± 5.8	86.6 ± 5.8	**88.0 ± 6.1**
S8	80.9 ± 6.1	84.0 ± 5.9	82.0 ± 5.9	**84.3 ± 6.3**
Average	76.1 ± 6.7	79.8 ± 6.2	78.9 ± 6.3	**81.1 ± 6.1**
HD				
S1 Left	11.9 ± 3.2	12.3 ± 6.0	27.6 ± 38.8	**10.3 ± 4.1**
S1 Right	18.0 ± 29.3	10.6 ± 2.6	61.1 ± 97.6	**9.7 ± 1.8**
S2 Left	13.3 ± 9.2	9.7 ± 3.1	35.6 ± 76.9	**9.2 ± 3.1**
S2 Right	36.3 ± 61.7	10.8 ± 3.0	10.8 ± 3.0	**9.5 ± 3.2**
S3 Anterior	41.7 ± 62.4	13.5 ± 4.9	50.4 ± 79.1	**12.2 ± 4.3**
S3 Posterior	9.1 ± 3.3	8.0 ± 2.0	18.0 ± 30.9	**7.6 ± 1.9**
S4 Left	11.5 ± 4.9	14.7 ± 22.2	14.5 ± 22.2	**9.8 ± 3.8**
S4 Right	32.8 ± 69.7	**9.8 ± 3.5**	16.2 ± 21.5	9.8 ± 3.6
S5	36.4 ± 56.4	20.5 ± 35.2	38.1 ± 60.3	**10.9 ± 4.0**
S6	19.2 ± 30.6	8.6 ± 2.5	52.5 ± 85.3	**8.5 ± 2.7**
S7	26.3 ± 42.6	9.6 ± 3.7	9.6 ± 3.7	**9.5 ± 3.5**
S8	14.5 ± 6.0	13.6 ± 5.7	13.1 ± 5.8	**12.2 ± 6.2**
Average	22.6 ± 31.6	11.8 ± 7.9	28.9 ± 43.8	**9.9 ± 3.5**
ASD				
S1 Left	1.6 ± 0.8	1.3 ± 0.6	1.4 ± 1.0	**0.9 ± 0.5**
S1 Right	1.8 ± 0.8	1.2 ± 0.5	1.6 ± 1.1	**0.9 ± 0.5**
S2 Left	1.4 ± 0.8	1.0 ± 0.6	1.3 ± 0.8	**0.8 ± 0.6**
S2 Right	1.5 ± 0.8	1.3 ± 0.7	1.3 ± 0.7	**1.0 ± 0.7**
S3 Anterior	1.0 ± 0.8	0.7 ± 0.4	0.9 ± 0.9	**0.6 ± 0.4**
S3 Posterior	0.9 ± 0.5	**0.6 ± 0.3**	0.8 ± 1.1	**0.6 ± 0.4**
S4 Left	1.0 ± 0.6	1.4 ± 2.7	1.2 ± 1.6	**0.8 ± 0.6**
S4 Right	1.5 ± 1.0	1.4 ± 1.0	1.5 ± 1.0	**1.3 ± 1.0**
S5	1.3 ± 0.6	1.9 ± 3.4	1.6 ± 1.8	**1.0 ± 0.5**
S6	0.8 ± 0.4	0.7 ± 0.3	1.0 ± 1.1	**0.6 ± 0.3**
S7	0.9 ± 0.7	0.8 ± 0.6	0.8 ± 0.6	**0.7 ± 0.6**
S8	1.7 ± 1.2	1.6 ± 1.1	1.6 ± 1.1	**1.3 ± 1.3**
Average	1.3 ± 0.7	1.1 ± 1.0	1.3 ± 1.1	**0.9 ± 0.6**

The rest settings are the same as the default nnU-Net setup. We implemented our DeepStationing method in PyTorch, and an NVIDIA Quadro RTX 8000 was used for training. The average training/inference time is 2.5 GPU days or 3 mins.

Quantitative Results. We first evaluate the performance of our stratified referencing organ segmentation. The average DSC, HD and ASD for anchor and nonanchor organs are $90.0 \pm 4.3\%$, 16.0 ± 18.0 mm, 1.2 ± 1.1 mm, and $82.1 \pm 6.0\%$, 19.4 ± 15.0 mm, 1.2 ± 1.4 mm, respectively. We also train a model by segmenting all organs using only one nnUNet. The average DSCs of the anchor, non-anchor, and all organs are $86.4 \pm 5.1\%$, $72.7 \pm 8.7\%$, and $80.8 \pm 7.06\%$, which are 3.6%, 9.4%, and 5.7% less than the stratified version, respectively. The stratified organ segmentation demonstrates high accuracy, which provides robust organ predictions for the subsequent LNS parsing model.

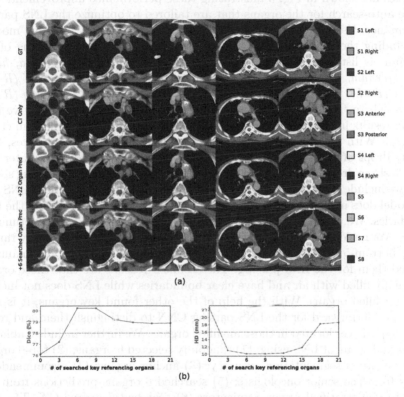

(a)

(b)

Fig. 3. (a) Examples of LNS parsing results using different setups. For better comparison, red arrows are used to depict visual improvements. (b) The bottom charts demonstrate the performance using different numbers of searched referencing organs.

Table 1 outlines the quantitative comparisons on different deep LNS parsing setups. Columns 1 to 3 show the results using: 1) only CT images, 2) CT + all 22 ground-truth organ masks, and 3) CT + all 22 predicted organ masks. Using only CT images, LNS parsing exhibits lowest performance with an average DSC of 76.1% and HD of 22.6 mm. E.g., distant false predictions is observed in the first image 2nd row of Fig. 3 and false-positive S3 posterior is predicted (in pink) between the S1 and S2. When adding 22 ground-truth organ masks as spatial context, both DSC and HD show remarked improvements: from 76.1% to 79.8% in DSC and 22.6mm to 11.8mm in HD. This verifies the importance and effectiveness of referencing organs in inferring LNS boundaries. However, when predicted masks of the 22 organs are used (the real testing condition), it has a significant increase in HD from 11.8 mm to 28.9 mm as compared to that using ground truth organ masks. This shows the necessity to select the key organs suited for the deep parsing model. Finally, using the top-6 auto-searched referencing organs, our DeepStationing model achieves the best performance reaching **81.1 ± 6.1% DSC, 9.9 ± 3.5 mm HD and 0.9 ± 0.6 mm** ASD. Qualitative examples are shown in Fig. 3 illustrating these performance improvements.

We auto-search for the organs that are tailored to optimize the LNS parsing performance. Using an interval of 3, we train 7 additional LNS parsing models, by including the top-3 up to top-21 organs. The auto-searched ranking of the 22 organs is listed as follows: *esophagus, aortic arch, ascending aorta, heart, spine, sternum, V.BCV (R+L), V.pulmonary, descending aorta, V.IJV (R+L), A.CCA (R+L), V.SVC, A.pulmonary, V.azygos, bronchus (R+L), lung (R+L), trachea*, where '*A*' and '*V*' denote the *Artery* and *Vein*. The quantitative LNS parsing results in selecting the top-n organs are illustrated in the bottom charts of Fig. 3. With more organs included gradually, the DSC first improves, then slightly drops after having more than top-6 organs. The performance later witnesses a sharp drop after including more than top-9 organs, then becoming steady when we include more than top-15 organs. This demonstrates that deep LNS paring model does not need a complete set of referencing organs to capture the LNS boundaries. We choose the top-6 as our final key organs based on experimental results. We notice that the trachea, lungs, and bronchus are surprisingly ranked in the bottom-5 of the auto-search, although previous works [11,12] manually selected them for the LNS parsing. The assumed reasons are that those organs are usually filled with air and have clear boundaries while LNS does not include air or air-filled organs. With the help of the other found key organs, it is relatively straightforward for the LNS parsing CNN to distinguish them and reject the false-positives located in those air-filled organs. We further include 6 ablation studies and segment LNS using: (1) randomly selected 6 organs; (2) top-6 organs with best organ segmentation accuracy; (3) anchor organs; (4) recommended 6 organs from the senior oncologists; (5) searched 6 organs predictions from less accurate non-stratified organ segmentor; (6) searched 6 organs GT. The randomly selected 6 organs are: *V.BCV (L), V.pulmonary, V.IJV (R), heart, spine, trachea*; The 6 organs with the best segmentation accuracy are: *lungs (R+L), descending aorta, heart, trachea, spine*; Oncologists recommended 6 organs are: *trachea, aortic arch, spine, lungs (R+L), descending aorta*; The DSCs for setups (1–6) are 77.2%, 78.2%, 78.6%, 79.0%, 80.2%, 81.7%; the HDs are 19.3 mm, 11.8 mm, 12.4 mm, 11.0 mm, 10.1 mm, 8.6 mm, respectively. In comparison to the LNS predictions using only CT images, the ablation studies demonstrate that using the referencing organ for LNS segmentation is the key contributor for the performance gain, and the selection and the quality of supporting organs are the main factors for the performance boost, e.g., our main results of the setups

(5) and (6) show that better searched-organ delineation can help get superior LNS segmentation performance.

Comparison to Previous Work. We compare the DeepStationing to the previous most relevant approach [11] that exploits heuristically pre-defined spatial margins for LNS inference. The DeepStationing outperforms [11] by 19.2% in DSC, 30.2 mm in HD, and 5.2 mm in ASD. For the ease of comparison, similar to [11], we also merge our LNSs into four LN zones, i.e., *supraclavicular* (S1), *superior* (S2, S3, and S4), *aortic* (S5 and S6) and *inferior* (S7 and S8) zones, and calculate the accuracy of LN instances that are correctly located in the predicted zones. DeepStationing achieves an average accuracy of 96.5%, or 13.3% absolutely superior than [11] in LN instance counting accuracy. We tested additionally 2 backbone networks: 3D PHNN (3D UNet with a light-weighted decoding path) and 2D UNet. The DSCs of 3D PHNN and 2D UNet are 79.5% and 78.8%, respectively. The assumed reason for the performance drop might be the loss of the boundary precision/3D information.

4   Conclusion

In this paper, we propose DeepStationing as a novel framework that performs key organ auto-search based LNS parsing on contrasted CT images. Emulating the clinical practices, we segment the referencing organs in thoracic region and use the segmentation results to guide LNS parsing. Different from employing the key organs directly suggested by oncologists, we search for the key organs automatically as a neural architecture search problem that can opt for optimal performance. Evaluated using a most comprehensive LNS dataset, DeepStationing method outperforms previous most relevant approach by a significant quantitative margin of 19.2% in DSC, and is coherent to clinical explanation. This work is an important step towards reliable and automated LNS segmentation.

References

1. Chao, C.-H., et al.: Lymph node gross tumor volume detection in oncology imaging via relationship learning using graph neural network. In: Martel, A.L., et al. (eds.) MICCAI 2020. LNCS, vol. 12267, pp. 772–782. Springer, Cham (2020). https://doi.org/10.1007/978-3-030-59728-3_75
2. Chapet, O., et al.: CT-based definition of thoracic lymph node stations: an atlas from the University of Michigan. Int. J. Radiat. Oncol.* Biol.* Phys. **63**(1), 170–178 (2005)
3. Feuerstein, M., Glocker, B., Kitasaka, T., Nakamura, Y., Iwano, S., Mori, K.: Mediastinal atlas creation from 3-D chest computed tomography images: application to automated detection and station mapping of lymph nodes. Med. Image Anal. **16**(1), 63–74 (2012)
4. Guo, D., et al.: Organ at risk segmentation for head and neck cancer using stratified learning and neural architecture search. In: Proceedings of the IEEE/CVF Conference on Computer Vision and Pattern Recognition, pp. 4223–4232 (2020)

5. Hirsch, F.R., et al.: Lung cancer: current therapies and new targeted treatments. Lancet **389**(10066), 299–311 (2017)
6. Isensee, F., Jaeger, P.F., Kohl, S.A., Petersen, J., Maier-Hein, K.H.: nnU-Net: a self-configuring method for deep learning-based biomedical image segmentation. Nat. Methods **18**(2), 203–211 (2020)
7. Jin, D., et al.: Accurate esophageal gross tumor volume segmentation in PET/CT using two-stream chained 3D deep network fusion. In: Shen, D., et al. (eds.) MICCAI 2019. LNCS, vol. 11765, pp. 182–191. Springer, Cham (2019). https://doi.org/10.1007/978-3-030-32245-8_21
8. Jin, D., et al.: Deep esophageal clinical target volume delineation using encoded 3D spatial context of tumors, lymph nodes, and organs at risk. In: Shen, D., et al. (eds.) MICCAI 2019. LNCS, vol. 11769, pp. 603–612. Springer, Cham (2019). https://doi.org/10.1007/978-3-030-32226-7_67
9. Jin, D., et al.: DeepTarget: gross tumor and clinical target volume segmentation in esophageal cancer radiotherapy. Med. Image Anal. **68**, 101909 (2020)
10. Liu, H., Simonyan, K., Yang, Y.: DARTS: differentiable architecture search. arXiv preprint arXiv:1806.09055 (2018)
11. Liu, J., et al.: Mediastinal lymph node detection and station mapping on chest CT using spatial priors and random forest. Med. Phys. **43**(7), 4362–4374 (2016)
12. Lu, K., Taeprasartsit, P., Bascom, R., Mahraj, R.P., Higgins, W.E.: Automatic definition of the central-chest lymph-node stations. Int. J. Comput. Assist. Radiol. Surg. **6**(4), 539–555 (2011)
13. Matsumoto, M.M., Beig, N.G., Udupa, J.K., Archer, S., Torigian, D.A.: Automatic localization of iaslc-defined mediastinal lymph node stations on CT images using fuzzy models. In: Medical Imaging 2014: Computer-Aided Diagnosis, vol. 9035, p. 90350J. International Society for Optics and Photonics (2014)
14. Pignon, J.P., et al.: A meta-analysis of thoracic radiotherapy for small-cell lung cancer. N. Engl. J. Med. **327**(23), 1618–1624 (1992)
15. Rusch, V.W., Asamura, H., Watanabe, H., Giroux, D.J., Rami-Porta, R., Goldstraw, P.: The IASLC lung cancer staging project: a proposal for a new international lymph node map in the forthcoming seventh edition of the TNM classification for lung cancer. J. Thoracic Oncol. **4**(5), 568–577 (2009)
16. Sarrut, D., et al.: Learning directional relative positions between mediastinal lymph node stations and organs. Med. Phys. **41**(6Part1), 061905 (2014)
17. Sung, H., et al.: Global cancer statistics 2020: GLOBOCAN estimates of incidence and mortality worldwide for 36 cancers in 185 countries. CA Cancer J. Clin. **71**, 209–249 (2021)
18. Tang, H., et al.: Clinically applicable deep learning framework for organs at risk delineation in CT images. Nat. Mach. Intell. **1**(10), 480–491 (2019)
19. Yuan, Y., et al.: Lymph node station-based nodal staging system for esophageal squamous cell carcinoma: a large-scale multicenter study. Ann. Surg. Oncol. **26**(12), 4045–4052 (2019)
20. Zhang, L., et al.: Robust pancreatic ductal adenocarcinoma segmentation with multi-institutional multi-phase partially-annotated CT scans. In: Martel, A.L., et al. (eds.) MICCAI 2020. LNCS, vol. 12264, pp. 491–500. Springer, Cham (2020). https://doi.org/10.1007/978-3-030-59719-1_48
21. Zhu, Z., et al.: Lymph node gross tumor volume detection and segmentation via distance-based gating using 3D CT/PET imaging in radiotherapy. In: Martel, A.L., et al. (eds.) MICCAI 2020. LNCS, vol. 12267, pp. 753–762. Springer, Cham (2020). https://doi.org/10.1007/978-3-030-59728-3_73

Hepatocellular Carcinoma Segmentation from Digital Subtraction Angiography Videos Using Learnable Temporal Difference

Wenting Jiang[1,2], Yicheng Jiang[1,2], Lu Zhang[3], Changmiao Wang[2],
Xiaoguang Han[1,2(✉)], Shuixing Zhang[3(✉)], Xiang Wan[2], and Shuguang Cui[1,2,4]

[1] School of Science and Engineering, The Chinese University of Hong Kong,
Shenzhen, Shenzhen, China
hanxiaoguang@cuhk.edu.cn
[2] Shenzhen Research Institute of Big Data, Shenzhen, China
[3] Department of Radiology, The First Affiliated Hospital of Jinan University,
Guangzhou, China
[4] Future Network of Intelligence Institute, Shenzhen, China

Abstract. Automatic segmentation of hepatocellular carcinoma (HCC) in Digital Subtraction Angiography (DSA) videos can assist radiologists in efficient diagnosis of HCC and accurate evaluation of tumors in clinical practice. Few studies have investigated HCC segmentation from DSA videos. It shows great challenging due to motion artifacts in filming, ambiguous boundaries of tumor regions and high similarity in imaging to other anatomical tissues. In this paper, we raise the problem of HCC segmentation in DSA videos, and build our own DSA dataset. We also propose a novel segmentation network called DSA-LTDNet, including a segmentation sub-network, a temporal difference learning (TDL) module and a liver region segmentation (LRS) sub-network for providing additional guidance. DSA-LTDNet is preferable for learning the latent motion information from DSA videos proactively and boosting segmentation performance. All of experiments are conducted on our self-collected dataset. Experimental results show that DSA-LTDNet increases the DICE score by nearly 4% compared to the U-Net baseline.

1 Introduction

Digital Subtraction Angiography (DSA) is good at displaying vascular lesions, thus it is widely used in clinical detection of blood flow, stenosis, thrombosis, etc.

The work was supported in part by Key-Area Research and Development Program of Guangdong Province [2020B0101350001], in part by the National Key R&D Program of China with grant No. 2018YFB1800800, by Shenzhen Outstanding Talents Training Fund, and by Guangdong Research Project No. 2017ZT07X152. It was also supported by NFSC 61931024 and National Natural Science Foundation of China (81871323). W. Jiang, Y. Jiang, L. Zhang—Contribute equally to this work.

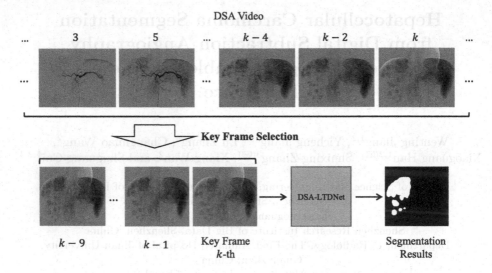

Fig. 1. Samples of DSA video frames and the pipeline of our methodology.

DSA has been one of the mainstream medical imaging tools for evaluating blood supply of tumors in Transcatheter Arterial Chemoembolization (TACE) for Hepatocellular Carcinoma (HCC) [1]. It is also essential for quantitative analysis of tumor blood supply in clinical diagnosis, which provides important information and guidance for operative drug planning and postoperative prognostic. In particular, radiologists are usually required to accurately identify multiple tumors of various sizes, healthy anatomical tissues and motion artifacts before evaluating tumors when the imaging of tumors becomes stable in videos. Automatically segmenting HCC from DSA videos can assist radiologists in locating and delineating tumor regions quickly, which is of great value to clinical diagnosis.

It is regretful that few studies have investigated segmentation of DSA videos before. Segmentation of DSA Videos is similar to the video object segmentation (VOS) task. There are several methods of the VOS task using traditional object detection solutions, such as optical flow, frame difference and background subtraction [2–4]. These solutions focus on learning objects' displacement to improve object detection performance. Apparent temporal motion information of tumors also exists in DSA videos. However, different from the VOS task, the HCC segmentation aims to segment tumors on one key frame (Fig. 1) without inputting segmentation masks. The temporal motion information of tumors in DSA video reflects in regional expansion, color tarnish and edge sharpness over time, which is distinct from the information in the VOS task of natural scenes as well. Thus, the methods of VOS cannot be used directly in our task.

Although the HCC segmentation can be solved with classic methods of medical image segmentation, like U-Net [5], the results based on these methods are not satisfactory (Sect. 4.3). Two key issues in our problem remain unsolved: (1) Inevitable and irregular motion artifacts appearing throughout DSA videos especially those near the liver regions, slash the contrast of tumors and blur the boundary of tumors, which impairs the segmentation performance greatly. (2) Other anatomical structures (like gastrointestinal tissue filled with contrast agent) have similar shape, contrast and intensity to tumors in DSA videos leading to segmentation inaccuracy. Therefore, precise and robust segmentation is highly desired and difficult for DSA videos at the same time.

In this paper, we propose a novel segmentation network called DSA-LTDNet to address the aforementioned issues by learning the distinct motion information of tumors from DSA videos and integrating the prior anatomical knowledge of liver regions as a supplemental information. All experiments are performed on our self-collected dataset. The experimental results of our method are improved by nearly 4% of DICE coefficient compared to the baseline. Our main contributions are:

1) We raise the problem of HCC segmentation in DSA videos for the first time, which is significant and challenging in clinical practice. We also build our self-collected DSA dataset of 488 samples, with careful annotations;
2) The creative proposed temporal difference learning network can efficiently capture the distinct motion information of tumors from DSA videos, which is helpful for delineating tumor regions;
3) Prior anatomical knowledge of liver regions integrated in our network can support locating tumor regions.

2   Methodology

The pipeline of the proposed DSA-LTDNet is plotted in Fig. 2, including a temporal difference learning (TDL) module, a liver region segmentation (LRS) subnetwork and a final fusion segmentation (FFS) sub-network. We also develop a simple approach for selecting key frames. The proposed TDL with multi-frame inputs is adopted for auto-learning temporal difference under the supervision of frame difference images. Our method also integrates prior anatomical knowledge of liver regions as guidance. The 3 inputs of FFS are the key frame, liver region masks predicted by LRS and the learned temporal difference by TDL.

2.1   Key Frame Selection and Segmentation

Selection. In clinical practice, radiologists choose the frame where the imaging of tumors is the clearest, the most stable and the most complete in the DSA video, for diagnosis and measurement of tumor parameters. Thus, we define such frames as the key frames and pick out them from DSA videos as a simulation of diagnosis process. We design a simple method of automatic selection for this task. Based on an observed regular pattern, we firstly calculate the difference image between 2 adjacent frames in the last 15 frames of DSA videos and sum to gain the total pixel values. To judge whether a frame is the key frame, we average the pixel values of the 2 difference images before and after the frame.

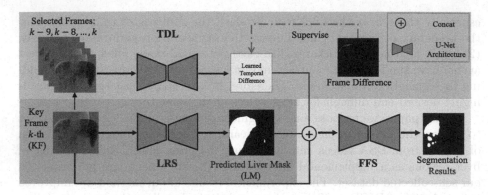

Fig. 2. Architecture of DSA-LTDNet

We select the frame with the minimum average pixel value as the key frame because the imaging in the key frame is the most stable imaging, representing the minimum average pixel value.

Segmentation. Several classical medical image segmentation networks using the single key frame as the input are tested. Their segmentation results on the key frames (Sect. 4.2) are similar to each other. Their overall performances are barely satisfactory due to the challenging work. We finally choose the basic U-Net as the baseline because it is the easiest one to implement and modify.

2.2 Temporal Difference Learning

Motion Information. The result of the U-Net baseline on the key frames reflect many problems, including multiple false positive and negative regions in the segmentation mask (Sect. 4.3). However, the motion information in the frames, such as the pixel-based frame difference (Fig. 1), can help delineate the tumor regions better. Consequently, we propose to add motion information across frames into our network for superior segmentation performance. We attempted 3 classic VOS methods of extracting motion information, which are the optical flow [6], frame difference [7] and background subtraction difference [8]. Our experiments (Sect. 4.2 and Table 2) show that FSS with inputs, key frames(KF) concating frame differences(FD) performs best (71.73%). FDs are calculated between the $(k-9)-th$ and the $k-th$ for the purpose of acquiring the most representative difference from the $k-th$ key frame (Fig. 2). We choose the parameter, 9, based on our experimental results with different parameters: 5, 7, 9, 11, 13. The model with k-9 has the best result due to its bigger difference than that of 5, 7 and less noise than that of 11, 13 according to our analysis.

Learnable Difference. Such motion information as FD is simple and direct, while humans can obtain more complex motion information. Thus, we also propose a Temporal Difference Learning (TDL) module, to better learn the latent temporal difference automatically. TDL module concatenates the 10 consecutive frames (Fig. 2) as its input and uses U-Net to learn the latent motion information from the input. We believe that this module has the ability to dig out more useful motion information than using the frame difference directly. However, the results in Sect. 4.2 show that TDL cannot perform well without any supervision because it is difficult to converge. Therefore, we creatively use the frame difference as ancillary information to supervise TDL. Specifically, we creatively initialize TDL guided by FD, and finetunc it with a relatively small loss in DSA-LTDNet. Experimental results (Sect. 4.2) show that the learned temporal difference of TDL can effectively boost the segmentation performance quantitatively and qualitatively.

2.3 Liver Region Guidance

Our experimental results show that the basic models on the key frame sometimes segment the tumor mask outside the liver region (Fig. 3), while such mistakes do not occur in clinical practice due to their medical knowledge that tumors are always located in or near the liver regions which generally appear at the top left of DSA videos (Fig. 1). Such mistakes happen because some anatomical structures (especially gastrointestinal tissues and healthy liver tissues filled with contrast agent) have highly similar shape, contrast and intensity to tumors in DSA videos. Consequently, the LRS sub-network based on U-Net, is co-trained on the key frames to predict the liver regions. The predicted liver region masks (LM) produced by LRS are plugged the spatial prior knowledge of tumor regions into our network, which successfully guides our model to locate tumors accurately.

2.4 Architecture and Loss Function

We integrate the aforementioned module and sub-network into the whole architecture (Fig. 2). Key frames $(256 \times 256 \times 1)$, along with temporal difference $(256 \times 256 \times 1)$ learned by LTD and the predicted liver region masks $(256 \times 256 \times 1)$ by LRS, are concatenated and fed into the final fusion segmentation(FFS) network. We co-train the 3 networks simultaneously. The followings are the loss functions:

$$L_{LTD} = |I_{LTD} - I_{FD}|_{L1} \tag{1}$$

$$L_{LRS} = a * |I_{LRS} - I_{LM}|_{BCE} + (1 - a) * |I_{LRS} - I_{LM}|_{DICE} \tag{2}$$

$$L_{seg} = a * |I_{seg} - I_{GT}|_{BCE} + (1 - a) * |I_{seg} - I_{GT}|_{DICE} \tag{3}$$

$$L_{Total} = \lambda_0 * L_{LTD} + \lambda_1 * L_{LRS} + L_{seg} \tag{4}$$

where L_{LTD} denotes the loss of LTD; L_{LRS} denotes the loss of LRS; L_{seg} denotes the segmentation loss of tumors; L_{Total} is the total loss of the whole network); a, λ_0 and λ_1 are 0.5, 0.1 and 1 respectively.

3 Dataset Construction

Collection. The DSA data is acquired from 486 HCC patients in the TACE procedure (3 DSA images are acquired form one patient, giving 488 samples in total) with 362 for training and 124 for testing at Hospital A. All of patients have been diagnosed with HCC in BCLC stage B [9] among January 2010 and December 2019, which was confirmed by biopsy or radiological imaging studies according to the guidelines [10]. The included criteria are as followings: 1) over 18 years of age; 2) without any prior treatment or surgery for HCC and any known active malignancy in the past 3 years; 3) Child-Pugh class A or B [11]; 4) Eastern Cooperative Oncology Group (ECOG) [12] performance scores are 2 or below. The following situations are excluded in the selection: 1) concurrent ischemic heart disease or heart failure; 2) history of acute tumor rupture with hemoperitoneum; 3) Child-Pugh class C cirrhosis; 4) history of hepatic encephalopathy, intractable ascites, biliary obstruction, and variceal bleeding; 5) extrahepatic metastasis, and tumor invasion of the portal vein or branch.

Annotation. Each DSA video contains 20–30 frames with 1021×788 pixel-wise resolution. We invited experienced radiologists to annotate the ground truth of the HCC and the corresponding liver regions. We augment the data by selecting 2 more consecutive frames after the key frame for the training, which is similar to key frame because the imaging of tumors keeps stable shortly after the key frame. All of annotations are performed with Labelme platform [13]. There are 760 annotated tumors in total with 212 tumors of poor blood, and 548 tumors of rich blood in our DSA dataset of 488 samples. Note that each patient has up to 10 tumors. The range of tumor size is [6.5, 217] and the average is 69.37 in pixels.

4 Experiments and Results

4.1 Implementation Details

All the models are trained from scratch based on PyTorch, using a NVIDIA GTX 8119 MiB GPU. We do not apply the popular data augmentation (e.g., rotation) due to the special anatomical knowledge of HCC and liver area. In the training settings, the batch size is set to 8 and our proposed network consisted of 3 models is co-trained for 150 epochs. We used the Adam optimizer and the CosineAnnealingLR scheduler to update the network parameters. The initial learning rate is set to 0.001. In experimenting U-Net, Attention U-Net [14] U-Net++ [15] and nnU-Net [16] are trained separately to segment tumors on the key frame at first. At the second stage, we concat the key frame, with one of the followings respectively, which are optical flow, frame difference, background subtraction, learned temporal difference and predicted liver region masks and trained the segmentation networks for comparison. Eventually, we combine the TDL module and the LRS sub-network with the FFS sub-network, and co-trained the DSA-LTDNet.

Table 1. Dice scores (%) for basic models and proposed methods

Method	Dice
U-Net (baseline)	70.75
Attention U-Net	71.03
U-Net++	70.73
nnU-Net	71.89

Method	Dice
FFS + LRS	72.01 (+ 1.26)
FFS + TDL	72.32 (+ 1.57)
DSA-LTDNet (Our Method)	**73.68 (+ 2.97)**

Table 2. Dice scores (%) for TDL comparison experiments

Method	Inputs (concat)	Dice
FFS	KF + OF	68.35 (−2.40)
FFS	KF + FD	71.73 (+0.98)
FFS	KF + BS	68.89 (−1.86)
FFS	KF + TDL w/o supervision	69.90 (−0.85)
FFS	KF + TDL w/ supervision	**72.32 (+1.57)**

4.2 Quantitative Analysis

The results of basic models and our proposed networks are in Table 1. The left part of Table 1 displays that 4 basic models, U-Net, Attention U-Net U-Net++ and nnU-Net have similar segmentation capabilities on the key frames. That tumors of poor blood supply have low contrast and blurry boundaries in DSA videos, along with other mentioned problems, brings great difficulty to HCC segmentation of basic models. Considering models' complexity of implementation and modification, we choose the easiest U-Net as the backbone of DSA-LTDNet. U-Net's dice score is 70.75%, which is regarded as our baseline. In the right part, it is worth noting that our proposed auxiliary models substantially improve the performance of HCC segmentation in DSA videos. With LRS and TDL added separately, the results are improved by 1.26% and 1.57%. Note that, the proposed method with both TDL and LRS added, achieves the best result, which is higher than U-Net by 2.97%. These numbers indicate that our methods take remarkable effects. The combination of LRS and TDL is successful as well.

At the second stage of our experiments, we tried to use traditional object detection methods as auxiliary information to support the network learn the motion information from DSA videos and segment tumors. The results of experiments in this stage are listed in Table 2. The traditional methods did not work well. The results of optical flow and background subtraction are even reduced by 2.40% and 1.9%. It is because that the motion information of other anatomical tissues displays similarly in optical flow and background subtraction as well, which increases the difficulty of final segmentation instead. Afterwards, we used a CNN network to automatically learn the latent temporal motion without any supervision but it did not succeed as expected, whose result is 69.90%.

According to our analysis, the failure is caused by the directionless learning without supervision, which might confuse the TDL network and feedback some noise information to the FFS network. We creatively propose to use frame difference, which increases the result by 0.98%, to guide the TDL network. Eventually, this architecture improves the results by 1.57% compared to the baseline, which satisfies our expectations.

4.3   Qualitative Results

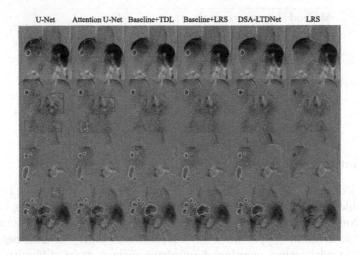

Fig. 3. Visualization of the segmentation results by different methods on the testing data. Ground truth and predicted mask of tumors are labeled in yellow and cyan-blue. Predicted masks of liver regions by LRS are labeled in green in the last column. False positive regions and missed tumor regions in predicted mask are marked in red and blue. (Color figure online)

Figure 3 illustrates a visual comparison of the ground truth, 2 basic models, baseline with TDL, baseline with LRS and our DSA-LTDNet on the key frame. It displays that basic models successfully produce accurate segmentation on tumors, but tumors' locations and shapes are not as accurate as our proposed methods. Such observations are particularly apparent in the first 3 rows of Fig. 3. Moreover, our proposed methods can obtain more accurate results by removing false positive regions under the guidance of liver regions and reducing the negative effect of motion artifacts, compared with basic models in the first 3 rows. It is noteworthy that DSA-LTDNet's performance on multiple small tumors marked blue in the last row, is significantly better than that of basic models due to the effective learning of temporal difference.

5 Discussions and Conclusions

We raise the problem of the HCC segmentation in DSA videos, which is proven significant and challenging in clinical practice. The novel DSA-LTDNet with a dominant TDL module, a supplementary LRS sub-network and a FFS sub-network is proposed. The TDL module can efficiently capture the distinct motion information of changing tumors for delineating tumor regions, while the prior anatomical knowledge of liver regions produced by LRS can support locating tumor regions. We also collect and annotate a relatively large DSA dataset, where we tested our networks. Experimental results show that our networks can effectively learn the latent motion information and achieve the state-of-the-art performance (DICE coefficient: 0.7368) on commonly used evaluation metrics. In the future, in order to improve the robustness, we plan to apply our networks to other DSA datasets from multiple hospitals, which are being collected. We will also facilitate the translation of CNN to clinical practice, especially in the treatment of HCC in the future.

References

1. Liapi, E., Georgiades, C.C., Hong, K., Geschwind, J.-F.H.: Transcatheter arterial chemoembolization: current technique and future promise. Tech. Vasc. Interv. Radiol. **10**(1), 2–11 (2007)
2. Cheng, J., Tsai, Y.-H., Wang, S., Yang, M.-H.: SegFlow: joint learning for video object segmentation and optical flow. In: Proceedings of the IEEE International Conference on computer Vision, pp. 686–695 (2017)
3. Guo, J., Wang, J., Bai, R., Zhang, Y., Li, Y.: A new moving object detection method based on frame-difference and background subtraction. In: IOP Conference Series: Materials Science and Engineering, vol. 242, p. 012115. IOP Publishing (2017)
4. Tsai, Y.-H., Yang, M.-H., Black, M.J.: Video segmentation via object flow. In: Proceedings of the IEEE Conference on Computer Vision and Pattern Recognition, pp. 3899–3908 (2016)
5. Ronneberger, O., Fischer, P., Brox, T.: U-Net: convolutional networks for biomedical image segmentation. In: Navab, N., Hornegger, J., Wells, W.M., Frangi, A.F. (eds.) MICCAI 2015. LNCS, vol. 9351, pp. 234–241. Springer, Cham (2015). https://doi.org/10.1007/978-3-319-24574-4_28
6. Shafie, A.A., Hafiz, F., Ali, M., et al.: Motion detection techniques using optical flow. World Acad. Sci. Eng. Technol. **56**, 559–561 (2009)
7. Singla, N.: Motion detection based on frame difference method. Int. J. Inf. Comput. Technol. 4(15), 1559–1565 (2014)
8. Kavitha, K., Tejaswini, A.: Vibe: background detection and subtraction for image sequences in video. Int. J. Comput. Sci. Inf. Technol. 3(5), 5223–5226 (2012)
9. Forner, A., Reig, M.E., de Lope, C.R., Bruix, J.: Current strategy for staging and treatment: the BCLC update and future prospects. In: Seminars in Liver Disease, vol. 30, pp. 061–074. Thieme Medical Publishers (2010)
10. Bruix, J., Sherman, M.: Management of hepatocellular carcinoma. Hepatology **42**(5), 1208–1236 (2005)

11. Durand, F., Valla, D.: Assessment of the prognosis of cirrhosis: Child-Pugh versus MELD. J. Hepatol. **42**(1), S100–S107 (2005)
12. Oken, M.M., et al.: Toxicity and response criteria of the eastern cooperative oncology group. Am. J. Clin. Oncol. **5**(6), 649–656 (1982)
13. Russell, B.C., Torralba, A., Murphy, K.P., Freeman, W.T.: LabelMe: a database and web-based tool for image annotation. Int. J. Comput. Vis. **77**(1–3), 157–173 (2008)
14. Oktay, O., et al.: Attention U-Net: learning where to look for the pancreas. arXiv preprint arXiv:1804.03999 (2018)
15. Zhou, Z., Rahman Siddiquee, M.M., Tajbakhsh, N., Liang, J.: UNet++: a nested U-Net architecture for medical image segmentation. In: DLMIA/ML-CDS -2018. LNCS, vol. 11045, pp. 3–11. Springer, Cham (2018). https://doi.org/10.1007/978-3-030-00889-5_1
16. Isensee, F., et al.: nnU-Net: self-adapting framework for U-Net-based medical image segmentation. arXiv preprint arXiv:1809.10486 (2018)

CA-Net: Leveraging Contextual Features for Lung Cancer Prediction

Mingzhou Liu[1,3], Fandong Zhang[3], Xinwei Sun[2(✉)], Yizhou Yu[3,4], and Yizhou Wang[1]

[1] Department of Computer Science and Technology,
Peking University, Beijing, China
[2] Peking University, Beijing, China
sxwxiaoxiaohehe@pku.edu.cn
[3] Deepwise AI Lab, Beijing, China
[4] University of Hong Kong, Pokfulam, Hong Kong

Abstract. In the early diagnosis of lung cancer, an important step is classifying malignancy/benignity for each lung nodule. For this classification, the nodule's features (e.g., shape, margin) have traditionally been the main focus. Recently, the contextual features attract increasing attention, due to the complementary information they provide. Clinically, such contextual features refer to the features of nodule's surrounding structures, such that (together with nodule's features) they can expose discriminate patterns for the malignant/benign, such as vascular convergence and fissural attachment. To leverage such contextual features, we propose a **C**ontext **A**ttention **Net**work (CA-Net) which extracts both nodule's and contextual features and then effectively fuses them during malignancy/benignity classification. To accurately identify the contextual features that contain structures distorted/attached by the nodule, we take the nodule's features as a reference via an attention mechanism. Further, we propose a feature fusion module that can adaptively adjust the weights of nodule's and contextual features across nodules. The utility of our proposed method is demonstrated by a noticeable margin over the 1st place on Data Science Bowl 2017 dataset in Kaggle's competition.

Keywords: Lung cancer prediction · Context · Domain knowledge

1 Introduction

Early diagnosis of lung cancer, as commonly adopted clinically to improve survival rate, contains an important step of predicting malignancy/benignity for each nodule. Compared with benign ones, the malignant nodules can expose irregular patterns or textures, providing as clues that traditional malignancy prediction methods have been based on.

Electronic supplementary material The online version of this chapter (https://doi.org/10.1007/978-3-030-87240-3_3) contains supplementary material, which is available to authorized users.

© Springer Nature Switzerland AG 2021
M. de Bruijne et al. (Eds.): MICCAI 2021, LNCS 12905, pp. 23–32, 2021.
https://doi.org/10.1007/978-3-030-87240-3_3

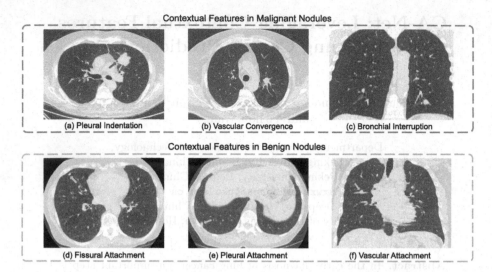

Fig. 1. Illustration of contextual features. The top figure shows contextual features in malignant nodules, including (a) pleural indentation: pulling of the pleura towards the nodule; (b) vascular convergence: vessels converge towards the nodule; and (c) bronchial interruption: the bronchus penetrates into the nodule with tapered narrowing and interruption. The bottom figure shows contextual features in the benign class: (d) fissural attachment; (e) pleural attachment; and (f) vascular attachment.

In addition to these nodule-related features, the features of nodule's surrounding structures, such as the distortion or the presence/absence of attachment, can provide complementary information for malignancy/benignity classification. For example, it has been clinically pointed out that the formation of malignant nodules often involves distortion of surrounding structures, such as pleural indentation [22]; vascular convergence [6]; bronchial interruption [14], as respectively illustrated by Fig. 1(a, b, c). Besides, it is more likely to be benign for the nodules with attachment to particular structures [19], such as fissures, the pleura and vessels (Fig. 1(d), (e), (f)). In this paper, we call such features related to nodule's surrounding structures that are discriminate in malignancy/benignity classification as "contextual features", which should be leveraged but ignored by most existing works in the literature.

To leverage these contextual features into lung cancer diagnosis, we propose a novel **C**ontext **A**ttention **Net**work (CA-Net), which extracts both nodule's and contextual features and effectively fuses them during nodule malignancy classification. To accurately capture the contextual features that are distorted/attached by the nodule, we incorporate the nodule's information as a reference via an attention mechanism. Besides, the effect of contextual features on classification can be varied across nodules. For examples, nodules with larger diameters may show more contextual features, while those with small diameters may show less. To model such a prior, we propose a feature fusion module, which adaptively learn the weights of nodule's and contextual features nodule by nodule.

To verify the utility of our CA-Net, we evaluate it on the Data Science Bowl (DSB) 2017, a dataset for lung cancer prediction in Kaggle's competition [1]. It yields that the CA-Net can outperform the 1st place by 2.5% on AUC. Besides, the visualization results qualitatively show that our method can well capture the contextual features with reasonable weights.

Our contributions can be summarized as follows: i) **Ideologically**, we propose to leverage contextual features in lung cancer prediction; ii) **Methodologically**, we propose a novel CA-Net which can accurately capture contextual features by utilizing the nodule's information; iii) **Experimentally**, our CA-Net outperforms the 1st place by a noticeable margin on Kaggle DSB2017 dataset.

1.1 Related Work

There is a large literature in malignancy/benignity classification for lung nodule, among which most of them [3,8–10,13,16,18] ignored the contextual information. Another branch of works that are more related to us, proposed to enlarge the receptive fields, in order to include contextual features during classification, such as [17] with a multi-crop pooling mechanism and [11] using a multi-scale features fusion operation. However, these methods can select irrelevant features empirically, as they fail to utilize the prior that the contextual features only contain those surrounding structures that are both nodule and disease-related, such as distortion or attachment by the nodule. In contrast, our CA-Net incorporates the nodule's information as a reference, leading to better ability to identify the contextual features.

2 Methodology

Problem Setting. Our dataset contains $\{I^i, Y^i\}_{i \in \{1,...,n\}}$, in which I denotes the patient's CT image and $Y \in \{0, 1\}$ denotes the lung cancer status (with 1 denoting the disease). During test stage, our goal is to predict Y given a new CT image I.

The whole pipeline of our method, namely **C**ontext **A**ttention **Net**work (CA-Net), is illustrated in Fig. 2. As shown, it is the sequential of three stages: **(i)** *Nodule Detection* that detects all nodules from the CT image; **(ii)** *Nodule Malignancy Classification* that predicts the malignancy probability for each nodule; **(iii)** *Cancer Prediction* that predicts the final label Y by assembling all nodules' malignancy probabilities. In the following, we will respectively introduce above three stages in Sect. 2.1, 2.2 and 2.3.

2.1 Nodule Detection

Given a whole CT image I, we implement a state-of-the-art nodule detector [9] to annotate all nodules $N_1, \ldots, N_K$. The impacts of false positive detection are minimized by a strict confidence threshold. For each nodule N_k, we obtain its radius and a cropped image I_{N_k} at the center. The size of for each I_{N_k} is large enough, so that the nodule's and contextual information are contained.

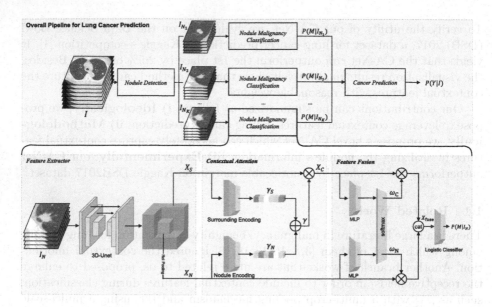

Fig. 2. The schematic overview of CA-Net. The top figure shows the procedure of lung cancer prediction, namely *Nodule Detection, Nodule Malignancy Classification*, and *Cancer Prediction*. The bottom figure shows details of our nodule malignancy classification, where we first extract nodule's and contextual features (*Feature Extractor* and *Contextual Attention*), then effectively fuse them into classification (*Feature Fusion*).

2.2 Nodule Malignancy Classification

After obtaining all nodules' images $\{I_{N_1}, \ldots, I_{N_K}\}$ from I, we predict the malignancy for each image I_{N_k}. For simplicity, we omit the index k in the following. As illustrated by the bottom figure in Fig. 2, such a malignancy classification contains three modules: *Feature Extractor*, *Contextual Attention* and *Feature Fusion*.

Feature Extractor. For each I_N, we obtain the featuremaps X of I_N by feeding I_N into the feature extractor with 3D-Unet as the backbone.

Contextual Attention. Given featuremaps X, this module extracts nodule's and contextual features, namely X_N and X_C, respectively. For X_N, as the nodule's radius can be calculated by the *Nodule Detector*, we can obtain the nodule's features X_N via the Region-Of-Interest pooling on X.

Denote $X_S := X \backslash X_N$ as the complementary set of X_N in X, which represents all features other than the nodule's. To capture X_C that refers to elements in X_S such that they are discriminate in the malignant/benign (*e.g.*, structures distortion for the malignant or attachment for the benign), we adopt an attention mechanism $X_C = \gamma \circ X_S{}^1$, in which the attention vector $\gamma := \gamma_N + \gamma_S$

---

[1] The $a \circ b$ denotes the element-wise product between a, b.

incorporates the information of both the nodule (as a reference) via γ_N and the surroundings (disease-related features in X_S) via γ_S.

Specifically, we obtain γ_N and γ_S by respectively feeding X_N, X_S into a nodule encoding block $F_{\theta_N^{enc}}(\cdot)$ (parameterized by θ_N^{enc}) and a surrounding encoding block $F_{\theta_S^{enc}}(\cdot)$ (parameterized by θ_S^{enc}):

$$\gamma_N = F_{\theta_N^{enc}}(X_N), \quad \gamma_S = F_{\theta_S^{enc}}(X_S) \tag{1}$$

Feature Fusion. To account for varied effect on classification across nodules, we for each nodule adaptively learn ω_N and ω_C subjected to $\omega_N + \omega_C = 1$, which respectively denote fusing weights in X_{fuse}:

$$X_{\text{fuse}} = \text{Concat}(\omega_N \times X_N, \omega_C \times X_C). \tag{2}$$

Specifically, we respectively calculate the ω_N and ω_C (with $\omega_N + \omega_C = 1$) via $\omega_N = F_{\theta_N^{fuse}}(X_N)$ and $\omega_N = F_{\theta_C^{fuse}}(X_C)$, with F denoting a two-layer perceptions followed by a softmax layer. After getting X_{fuse}, the prediction of the malignancy $P(M|I_N)$ is given by the sigmoid function parameterized by θ_{cls}:

$$P(M|I_N) = \text{sigmoid}_{\theta_{\text{cls}}}(X_{\text{fuse}}).$$

2.3 Cancer Prediction

After obtaining $P(M|I_{N_1}), \ldots, P(M|I_{N_K})$, we predict the cancer status Y by assembling the malignancy probabilities of all nodules via a leaky noisy-OR model proposed in [9]:

$$P(Y|I) = 1 - (1 - p_l) \prod_{k=1}^{K} (1 - P(M|I_{N_k})) \tag{3}$$

where the $p_l \in (0, 1)$ denotes the learned leaky parameter.

3   Experiments

We conduct experiments on DSB2017 dataset[2]. This dataset provides pathologically proven lung cancer label for each patient. There are 1397, 198, and 506 patients in the training, validation, and test set, respectively.

3.1   Experimental Results

Evaluation Metrics. We adopt the following metrics for evaluation: the Accuracy (Acc) under the threshold of 0.5, the Area Under the Curve (AUC) and the Log Loss[3].

---

[2] LIDC-IDRI [12] is not considered for the lack of biopsy confirmed lung cancer labels.
[3] Official evaluation metric of the competition, lower value means better performance.

Implementation Details. The image crop size is set to $96 \times 96 \times 96$ mm to contain all contextual features. Our features extractor is a modified 3D-Unet [9]. For the contextual attention module, we adopt a Squeeze-and-Excitation layer [7] as our nodule encoding block, and an Encoding Layer [20] as our surrounding encoding block since it has been empirically proven to capture global statistics. The training takes 130 epochs. We implement the SGD optimizer, with the learning rate set to 0.01 and decays at rate 0.1 in the 20, 70, and 110 epochs; the momentum set to 0.9; and weight decay set to 0.0001.

Compared Baselines. We compare our CA-Net with top 5 from the Kaggle competition leader board and other baselines. Among these baselines, 3D-ResNet18 [4] and MV-KBC [18] have ever achieved competitive results on nodule malignancy classification in LIDC-IDRI dataset; MV-KBC proposed a 2D multi-view network that jointly learned different nodule's features (i.e. texture, internal characteristics, and margin); Ozdemir *et al.* proposed a 3D probabilistic model to explore the coupling between nodule detection and malignancy classification. Liao *et al.* shared the nodule detection and cancer prediction stages with us, but only incorporated nodule's features into classification by a center 2×2 pooling.

Table 1. Evaluations on DSB2017 dataset.

	Method	Acc (%) ↑	AUC (%) ↑	LogLoss ↓
Leader board (top 5)	grt123	–	–	.3998
	Julian de Wit & Daniel Hammack	–	–	.4012
	Aidence	–	–	.4013
	qfpxfd	–	–	.4018
	Pierre Fillard	–	–	.4041
Other baselines	3D-ResNet18 [4]	76.28	82.76	.4748
	MV-KBC [18]	78.57	84.67	.4887
	Ozdemir *et al.* [13]	–	86.90	–
	Liao *et al.* [9]	81.42	87.00	.3989
	CA-Net (ours)	**83.79**	**89.52**	**.3725**

Results. As shown from Table 1, our proposed CA-Net achieves state-of-the-art performance in DSB2017 dataset with a noticeable margin (2.4% on Acc, 2.5% on AUC, and 2.6% on Log Loss), which demonstrates the effectiveness of our method in lung cancer prediction. Compared with other baselines that only take nodule's features into consideration, our superiority mainly comes from the effective incorporation of contextual features.

3.2 Ablation Study

To verify the effectiveness of each module in our CA-Net during nodule malignancy classification, we conduct ablation study as shown in Table 2. As we can see, our contextual attention module brings a major improvement of 0.8% on AUC and 1.6% on LogLoss, which demonstrates its ability to capture contextual features. The reason behind this can be contributed to the incorporation of nodule's information as a reference into identifying contextual features from nodule's surroundings. Another improvement comes from our feature fusion module (1.0% on AUC and 0.9% on LogLoss), due to our adaptively learned weights that can account for the varied effect of contextual features across nodules.

Table 2. Ablation study on different modules. 'cat' denotes native concatenation, $F_{\theta_S^{enc}}(\cdot)$ and $F_{\theta_N^{enc}}(\cdot)$ respectively denote the surrounding and nodule encoding block.

X_N	X_S	Features fusion		Contextual attention		Acc (%) ↑	AUC (%) ↑	LogLoss ↓
		Cat	Ours	$F_{\theta_S^{enc}}(\cdot)$	$F_{\theta_N^{enc}}(\cdot)$			
√						83.00	87.78	.3975
√	√	√				83.20	88.31	.3901
√	√		√			83.40	88.70	.3886
√	√		√	√		83.20	89.18	.3811
√	√		√	√	√	**83.79**	**89.52**	**.3725**

To further validate the effectiveness of our contextual attention module, we compare it with other existing candidates: Spatial Pyramid Pooling (SPP) [5], Multi-Crop pooling [17], Atrous Spatial Pyramid Pooling (A-SPP) [2], and Pyramid Scene Parsing module (PSP) [21]. As shown from Tab. 3, our attention module outperforms all of these methods. The reference of nodule's information via an attention mechanism may be the main reason for its superiority.

Table 3. Comparison of different contextual features caption methods.

SPP [5]	Multi-crop [17]	A-SPP [2]	PSP [21]	Ours	Acc (%) ↑	AUC(%) ↑	LogLoss ↓
√					83.00	88.72	.3849
	√				83.20	88.83	.3846
		√			83.79	88.67	.3914
			√		83.00	89.02	.3834
				√	**83.79**	**89.52**	**.3725**

3.3 Visualization

We show visualization results to qualitatively demonstrate that our method can better identify the contextual features (Fig. 3) and model their varied effects on classification (Fig. 4). More examples are provided in the supplementary due to space limit.

We visualize selected contextual features of different methods by gradient class activation map (Grad-CAM) [15]. The results are shown in Fig. 3. The top row nodule presents pleural indentation while the bottom one has vascular distortion. As we can see, due to the incorporation of nodule's information, our CA-Net can better capture the structures distorted by the nodule.

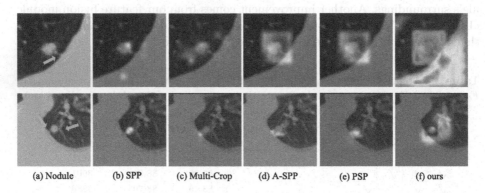

    (a) Nodule   (b) SPP   (c) Multi-Crop   (d) A-SPP   (e) PSP   (f) ours

Fig. 3. Visualization of contextual features captured by different methods. The top row shows a nodule with pleural indentation, and the bottom row shows a nodule with both pleural indentation and vascular distortion.

Figure 4 shows learned fusion weights (ω_N, ω_C) across nodules. Nodules in the top row present contextual features (arrows), leading to high contextual weights. Contextual features are absent in bottom row nodules, resulting in low contextual weights. This demonstrates that the learned weights of contextual features can well reflect the presence/absence of contextual features across nodules.

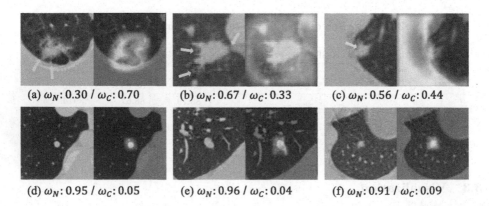

  (a) ω_N: 0.30 / ω_C: 0.70   (b) ω_N: 0.67 / ω_C: 0.33   (c) ω_N: 0.56 / ω_C: 0.44

  (d) ω_N: 0.95 / ω_C: 0.05   (e) ω_N: 0.96 / ω_C: 0.04   (f) ω_N: 0.91 / ω_C: 0.09

Fig. 4. Visualization of learned weights (ω_N, ω_C) in our feature fusion module. The top row nodules respectively show pleural indentation (a, b), bronchial interruption (b), and pleural attachment (c). The bottom row is nodules without any contextual features.

4 Conclusions

In this paper, a novel Context Attention Network (CA-Net) is proposed to leverage contextual features in lung cancer prediction. By adopting the attention mechanism, the CA-Net can incorporate the nodule's information for better identification of contextual features. By using the adaptive weighting strategy, our feature fusion module can effectively fuse the nodule's and contextual features for all nodules. With experiments conducted on Kaggle DSB2017 dataset, we have shown that our method can accurately captures the contextual features and hence achieves better prediction power than others.

Acknowledgements. This work was supported by MOST-2018AAA0102004, NSFC-61625201, and the Beijing Municipal Science and Technology Planning Project (Grant No. Z201100005620008).

References

1. Data science bowl 2017 (2017). https://www.kaggle.com/c/data-science-bowl-2017
2. Chen, L.C., Papandreou, G., Schroff, F., Adam, H.: Rethinking atrous convolution for semantic image segmentation. arXiv preprint arXiv:abs/1706.05587 (2017)
3. Chen, S., et al.: Automatic scoring of multiple semantic attributes with multi-task feature leverage: a study on pulmonary nodules in CT images. IEEE Trans. Med. Imaging 36(3), 802–814 (2017). https://doi.org/10.1109/TMI.2016.2629462
4. He, K., Zhang, X., Ren, S., Sun, J.: Deep residual learning for image recognition. In: 2016 IEEE Conference on Computer Vision and Pattern Recognition (CVPR), pp. 770–778 (2016). https://doi.org/10.1109/CVPR.2016.90
5. He, K., Zhang, X., Ren, S., Sun, J.: Spatial pyramid pooling in deep convolutional networks for visual recognition. In: Fleet, D., Pajdla, T., Schiele, B., Tuytelaars, T. (eds.) ECCV 2014. LNCS, vol. 8691, pp. 346–361. Springer, Cham (2014). https://doi.org/10.1007/978-3-319-10578-9_23
6. Hu, H., Wang, Q., Tang, H., Xiong, L., Lin, Q.: Multi-slice computed tomography characteristics of solitary pulmonary ground-glass nodules: differences between malignant and benign. Thoracic Cancer 7(1), 80–87 (2016). https://doi.org/10.1111/1759-7714.12280
7. Hu, J., Shen, L., Sun, G.: Squeeze-and-excitation networks. In: 2018 IEEE/CVF Conference on Computer Vision and Pattern Recognition, pp. 7132–7141 (2018). https://doi.org/10.1109/CVPR.2018.00745
8. Lei, Y., Tian, Y., Shan, H., Zhang, J., Wang, G., Kalra, M.K.: Shape and margin-aware lung nodule classification in low-dose CT images via soft activation mapping. Med. Image Anal. 60, 101628 (2020). https://doi.org/10.1016/j.media.2019.101628
9. Liao, F., Liang, M., Li, Z., Hu, X., Song, S.: Evaluate the malignancy of pulmonary nodules using the 3-d deep leaky noisy-or network. IEEE Trans. Neural Netw. Learn. Syst. 30(11), 3484–3495 (2019). https://doi.org/10.1109/TNNLS.2019.2892409
10. Liu, L., Dou, Q., Chen, H., Qin, J., Heng, P.A.: Multi-task deep model with margin ranking loss for lung nodule analysis. IEEE Trans. Med. Imaging 39(3), 718–728 (2020). https://doi.org/10.1109/TMI.2019.2934577

11. Liu, X., Hou, F., Qin, H., Hao, A.: Multi-view multi-scale CNNs for lung nodule type classification from CT images. Pattern Recogn. **77**, 262–275 (2018). https://doi.org/10.1016/j.patcog.2017.12.022

12. McNitt-Gray, M.F., et al.: The lung image database consortium (LIDC) data collection process for nodule detection and annotation. Acad. Radiol. **14**(12), 1464–1474 (2007). https://doi.org/10.1016/j.acra.2007.07.021

13. Ozdemir, O., Russell, R.L., Berlin, A.A.: A 3D probabilistic deep learning system for detection and diagnosis of lung cancer using low-dose CT scans. IEEE Trans. Med. Imaging **39**(5), 1419–1429 (2020). https://doi.org/10.1109/TMI.2019.2947595

14. Qiang, J., et al.: The relationship between solitary pulmonary nodules and bronchi: multi-slice CT-pathological correlation. Clin. Radiol. **59**(12), 1121–1127 (2004). https://doi.org/10.1016/j.crad.2004.02.018

15. Selvaraju, R.R., Cogswell, M., Das, A., Vedantam, R., Parikh, D., Batra, D.: Grad-CAM: visual explanations from deep networks via gradient-based localization. In: 2017 IEEE International Conference on Computer Vision (ICCV), pp. 618–626 (2017). https://doi.org/10.1109/ICCV.2017.74

16. Shen, W., et al.: Learning from experts: developing transferable deep features for patient-level lung cancer prediction. In: Ourselin, S., Joskowicz, L., Sabuncu, M.R., Unal, G., Wells, W. (eds.) MICCAI 2016. LNCS, vol. 9901, pp. 124–131. Springer, Cham (2016). https://doi.org/10.1007/978-3-319-46723-8_15

17. Shen, W., et al.: Multi-crop convolutional neural networks for lung nodule malignancy suspiciousness classification. Pattern Recogn. **61**, 663–673 (2017). https://doi.org/10.1016/j.patcog.2016.05.029

18. Xie, Y., et al.: Knowledge-based collaborative deep learning for benign-malignant lung nodule classification on chest CT. IEEE Trans. Med. Imaging **38**(4), 991–1004 (2019). https://doi.org/10.1109/TMI.2018.2876510

19. Xu, D.M., et al.: Smooth or attached solid indeterminate nodules detected at baseline CT screening in the Nelson study: cancer risk during 1 year of follow-up. Radiology **250**(1), 264–272 (2009). https://doi.org/10.1148/radiol.2493070847

20. Zhang, H., Xue, J., Dana, K.: Deep TEN: texture encoding network. In: 2017 IEEE Conference on Computer Vision and Pattern Recognition (CVPR), pp. 2896–2905 (2017). https://doi.org/10.1109/CVPR.2017.309

21. Zhao, H., Shi, J., Qi, X., Wang, X., Jia, J.: Pyramid scene parsing network. In: 2017 IEEE Conference on Computer Vision and Pattern Recognition (CVPR), pp. 6230–6239 (2017). https://doi.org/10.1109/CVPR.2017.660

22. Zwirewich, C.V., Vedal, S., Miller, R.R., Müller, N.L.: Solitary pulmonary nodule: high-resolution CT and radiologic-pathologic correlation. Radiology **179**(2), 469–476 (1991). https://doi.org/10.1148/radiology.179.2.2014294

Semi-supervised Learning for Bone Mineral Density Estimation in Hip X-Ray Images

Kang Zheng[1]([✉]), Yirui Wang[1], Xiao-Yun Zhou[1], Fakai Wang[1,2], Le Lu[1], Chihung Lin[3], Lingyun Huang[4], Guotong Xie[4], Jing Xiao[4], Chang-Fu Kuo[3], and Shun Miao[1]

[1] PAII Inc., Bethesda, MD, USA
[2] University of Maryland, College Park, MD, USA
[3] Chang Gung Memorial Hospital, Linkou, Taoyuan City, Taiwan, Republic of China
[4] Ping An Technology, Shenzhen, China

Abstract. Bone mineral density (BMD) is a clinically critical indicator of osteoporosis, usually measured by dual-energy X-ray absorptiometry (DEXA). Due to the limited accessibility of DEXA machines and examinations, osteoporosis is often under-diagnosed and under-treated, leading to increased fragility fracture risks. Thus it is highly desirable to obtain BMDs with alternative cost-effective and more accessible medical imaging examinations such as X-ray plain films. In this work, we formulate the BMD estimation from plain hip X-ray images as a regression problem. Specifically, we propose a new semi-supervised self-training algorithm to train a BMD regression model using images coupled with DEXA measured BMDs and unlabeled images with pseudo BMDs. Pseudo BMDs are generated and refined iteratively for unlabeled images during self-training. We also present a novel adaptive triplet loss to improve the model's regression accuracy. On an in-house dataset of 1,090 images (819 unique patients), our BMD estimation method achieves a high Pearson correlation coefficient of 0.8805 to ground-truth BMDs. It offers good feasibility to use the more accessible and cheaper X-ray imaging for opportunistic osteoporosis screening.

Keywords: Bone mineral density estimation · Hip X-ray · Semi-supervised learning

1 Introduction

Osteoporosis is a common skeletal disorder characterized by decreased bone mineral density (BMD) and bone strength deterioration, leading to an increased risk of fragility fracture. All types of fragility fractures affect the elderly with multiple morbidities, reduced life quality, increased dependence, and mortality. FRAX is a clinical tool for assessing bone fracture risks by integrating clinical risk factors and BMD. While some clinical risk factors such as age, gender, and body mass

© Springer Nature Switzerland AG 2021
M. de Bruijne et al. (Eds.): MICCAI 2021, LNCS 12905, pp. 33–42, 2021.
https://doi.org/10.1007/978-3-030-87240-3_4

index can be obtained from electronic medical records, the current gold standard to measure BMD is dual-energy X-ray absorptiometry (DEXA). However, due to the limited availability of DEXA devices, especially in developing countries, osteoporosis is often under-diagnosed and under-treated. Therefore, alternative lower-cost BMD evaluation protocols and methods using more accessible medical imaging examinations, e.g., X-ray plain films, are highly desirable.

Previous methods aiming to use imaging obtained for other indications to estimate BMD or classify osteoporosis have been proposed. The clinical value of this "opportunistic screening" approach seems apparent with low additional cost, high speed to obtain essential patient information, and potentially greater accuracy over traditional prediction models. For example, previous studies [1,2,6,9,11,12] have attempted to estimate BMD through CT scans for osteoporosis screening. However, the undesired prediction performance, high cost, long acquisition time, and particularly high radiation dose of CT imaging compared to DEXA and plain film X-ray are barriers to their wide clinical adoption. In contrast, plain film X-ray is a more accessible and lower-cost imaging tool than CT for opportunistic screening. Existing works also classify osteoporosis from hip or spine X-rays [8,15,16]. However, they either perform poorly or require patient information, such as age, sex, and fracture history. In this work, we attempt to estimate BMD from plain film hip X-ray images for osteoporosis screening.

Our work is based on the assumption that hip X-ray images contain sufficient information on visual cues for BMD estimation. We use a convolutional neural network (CNN) to regress BMD from hip X-ray images. Paired hip X-ray image and DEXA measured BMD taken within six months apart (as ground-truth) are collected as labeled data for supervised regression learning. To improve regression accuracy, we also propose a novel adaptive triplet loss, enabling the model to better discriminate samples with dissimilar BMDs in the feature space. Since it is often difficult to obtain a large amount of hip X-ray images paired with DEXA measured BMDs, we propose a new semi-supervised self-training algorithm to exploit the usefulness of large-scale hip X-ray images without ground-truth BMDs, which can be efficiently collected at scale. Our method is evaluated on an in-house dataset of 1,090 images from 819 patients and achieves a Pearson correlation coefficient of 0.8805 to ground-truth BMDs.

Our main contributions are three-fold: 1) This work is the first to estimate bone mineral density from the hip plain films, which offers a potential computer-aided diagnosis (CAD) application to opportunistic osteoporosis screening with more accessibility and reduced costs. 2) We propose a novel adaptive triplet loss to improve the model's regression accuracy, applicable to other regression tasks as well. 3) We present a new semi-supervised self-training algorithm to boost the BMD estimation accuracy via exploiting unlabeled hip X-ray images.

2 Method

Given a hip X-ray image, we first crop a region-of-interest (ROI) around the femoral neck[1] as input to the convolutional neural network. Denoting the hip

---

[1] This automated ROI localization module is achieved by re-implementing the deep adaptive graph (DAG) network [10].

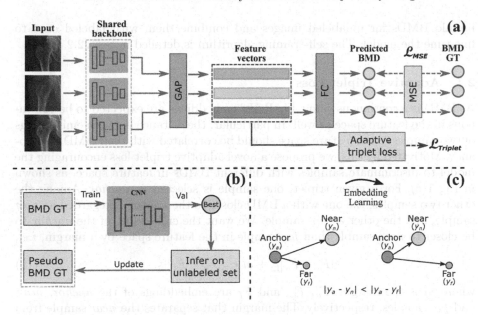

Fig. 1. An overview of our proposed framework, consisting of (a) supervised pre-training stage and (b) semi-supervised self-training stage. In the supervised pre-training stage, we train the model on labeled images using the MSE loss and (c) a novel adaptive triplet loss, which encourages the distance between feature embeddings of samples correlated to their BMD difference. In the self-training stage, we fine-tune the model on labeled data and pseudo-labeled data. We update the pseudo labels when the model achieves higher performance on the validation set.

ROI as I and its associated ground-truth (GT) BMD as y, our goal is to provide an estimation y' close to y. We formulate the problem as a regression problem since BMD values are continuous.

Figure 1 illustrates the overall framework of our proposed method, which consists of two stages: supervised pre-training and semi-supervised self-training. In the first stage, we pre-train a model on a small amount of labeled images. We adopt VGG-11 [13] as the backbone with its last two fully-connected (FC) layers replaced by a global average pooling (GAP) layer to obtain a 512-dimensional embedding feature vector. Then, an FC layer is applied to regress BMD from the feature vector. We train the model using Mean Square Error (MSE) loss and a novel adaptive triplet loss, as shown in Fig. 1(a). The MSE loss is defined between the predicted BMD y' and the GT BMD y, written as:

$$\mathcal{L}_{mse} = (y - y')^2. \tag{1}$$

With the proposed adaptive triplet loss, the model learns more discriminative feature embeddings for images with different BMDs, thus improving its regression accuracy. We elaborate on the adaptive triplet loss in Sect. 2.1. In the second stage, we fine-tune the model with a novel semi-supervised self-training algorithm using both labeled and unlabeled images. We iteratively predict the pseudo BMDs for unlabeled images and combine them with labeled ones to fine-tune the model. The self-training algorithm is detailed in Sect. 2.2.

2.1 Adaptive Triplet Loss

As BMD is a continuous value, hip ROIs' embeddings are expected to be continuous in the feature space as well. In particular, the distance between embeddings of two samples in the feature space should be correlated with their BMD discrepancy. Motivated by this, we propose a novel adaptive triplet loss encouraging the model to discriminate samples with different BMDs in feature space, as shown in Fig. 1(c). For a given triplet, one sample is selected as *anchor*. Among the other two samples, the one with a BMD closer to the *anchor* is regarded as *near* sample, and the other as *far* sample. We want the embedding of the *anchor* to be closer to *near* sample than *far* sample in the feature space by a margin, *i.e.*,

$$||F_a - F_n||_2^2 + m < ||F_a - F_f||_2^2, \tag{2}$$

where m is the margin. F_a, F_n, and F_f are embeddings of the *anchor*, *near*, and *far* samples, respectively. The margin that separates the *near* sample from the *far* sample should account for their relative BMD differences. Therefore, we define the adaptive triplet loss as follows:

$$\mathcal{L}_{triplet} = \left[||F_a - F_n||_2^2 - ||F_a - F_f||_2^2 + \alpha m\right]_+. \tag{3}$$

α is the adaptive coefficient based on the BMD differences, defined as:

$$\alpha = ||y_a - y_f||_2^2 - ||y_a - y_n||_2^2, \tag{4}$$

where y_a, y_n, and y_f are the GT BMD values of the *anchor*, *near*, and *far* samples, respectively. For network training, we combine the MSE loss with adaptive triplet loss as follows:

$$\mathcal{L} = \mathcal{L}_{mse} + \lambda \mathcal{L}_{triplet}, \tag{5}$$

where λ is the weight for adaptive triplet loss. We set λ to 0.5 for all experiments. The adaptive triplet loss is calculated for triplets randomly constructed within each mini-batch. Specifically, each sample is used as an anchor with two other samples randomly selected as the near/far samples depending on their BMD distances to the anchor.

2.2 Semi-supervised Self-training

Given limited images coupled with GT BMDs, the model can easily overfit the training data and yield poor performance on unseen test data. To overcome this issue, we propose a semi-supervised self-training algorithm to leverage both labeled and unlabeled data. An overview of the proposed self-training algorithm is illustrated in Fig. 1(b). We first use the pre-trained model to predict pseudo GT BMDs on unlabeled images to obtain additional supervisions. The unlabeled images with pseudo GT BMDs are subsequently combined with labeled

---

Algorithm 1. Semi-supervised self-training algorithm

1: Initialize the best R-value $\hat{\eta} := 0$ and MSE $\hat{\epsilon} := \infty$
2: Initialize training epoch $e := 0$ and set total training epoch E
3: Initialize model with pre-trained weights
4: **while** $e < E$ **do**
5:      Evaluate model performance R-value η and MSE ϵ on the validation set
6:      **if** $\eta > \hat{\eta}$ and $\epsilon < \hat{\epsilon}$ **then**
7:          $\hat{\eta} := \eta$
8:          $\hat{\epsilon} := \epsilon$
9:          Generate pseudo BMDs for unlabeled images
10:     Fine-tune model on labeled images and unlabeled images with pseudo BMDs
11:     $e := e + 1$

---

images to fine-tune the model. To improve the quality of estimated pseudo GT BMDs, we propose to refine them using the better fine-tuned models during self-training. We assume that if a fine-tuned model achieves higher performance on the validation set, it can also produce more accurate/reliable pseudo GT BMDs for unlabeled images. Specifically, after each self-training epoch, we evaluate the model performance on the validation set using Pearson correlation coefficient and mean square error. If the model indeed achieves higher correlation coefficient and lower mean square error than previous models at the same time, we use it to re-generate pseudo GT BMDs for the unlabeled images for self-training. This process is repeated until the total self-training epoch is reached. We summarize this semi-supervised self-training algorithm in Algorithm 1.

In addition, we augment each image twice and employ consistency constraints between their features and between predicted BMDs, which regularize the model to avoid being misled by inaccurate pseudo labels. Denoting two augmentations of the same image as I_1 and I_2, their features F_1 and F_2, and their predicted BMDs y_1 and y_2, we define the consistency loss as:

$$\mathcal{L}_c = ||F_1 - F_2||_2^2 + ||y_1 - y_2||_2^2. \tag{6}$$

For self-training, the total loss is:

$$\mathcal{L} = \mathcal{L}_{mse} + \lambda \mathcal{L}_{triplet} + \lambda' \mathcal{L}_c, \tag{7}$$

where λ' is the consistency loss weight set to 1.0.

3   Experiment Results

Dataset and Evaluation Metrics. We collected 1,090 hip X-ray images with associated DEXA-measured BMD values from 819 patients. The X-ray images are taken within six months of the BMD measurement. We randomly split the images into training, validation, and test sets of 440, 150, and 500 images based on patient identities. For semi-supervised learning, 8,219 unlabeled hip X-ray images are collected. To extract hip ROIs around the femoral neck,

Table 1. Comparison of baseline methods using different backbones.

Backbone	VGG-11	VGG-13	VGG-16	ResNet-18	ResNet-34	ResNet-50
R-value	**0.8520**	0.8501	0.8335	0.8398	0.8445	0.8448
RMSE	**0.0831**	0.0855	0.1158	0.0883	0.0946	0.1047

we train a localization model with the deep adaptive graph (DAG) network (re-implemented from [10]) using about 100 images with manually annotated anatomical landmarks. We adopt two metrics, Pearson correlation coefficient (R-value) and Root Mean Square Error (RMSE), to evaluate the proposed method and all compared methods.

Implementation Details. The ROIs are resized to 512×512 pixels as model input. Random affine transformations, color jittering, and horizontal flipping are applied to each ROI during training. We adopt VGG-11 with batch normalization [5] and squeeze-and-excitation (SE) layer [4] as the backbone, since it outperforms other VGG networks [13] and ResNets [3] as demonstrated later. The margin for adaptive triplet loss is set to 0.5. We use Adam optimizer with an initial learning rate of 10^{-4} and a weight decay of 4×10^{-4} to train the network on labeled images for 200 epochs. The learning rate is decayed to 10^{-5} after 100 epochs. For semi-supervised learning, we fix the learning rate at 10^{-5} to fine-tune for $E = 100$ epochs. After each training/fine-tuning epoch, we evaluate the model on the validation set and select the one with the highest Pearson correlation coefficient for testing. All models are implemented using PyTorch 1.7.1 and trained on a workstation with an Intel(R) Xeon(R) CPU, 128G RAM, and a 12G NVIDIA Titan V GPU. We set the batch size to 16.

Backbone Selection. Here we study how different backbones affect the baseline performance without adaptive triplet loss or self-training. The compared backbones include VGG-11, VGG-13, VGG-16, ResNet-18, ResNet-34, and ResNet-50. As shown in Table 1, VGG-11 achieves the best R-value of 0.8520 and RMSE of 0.0831. The lower performance of other VGG networks and ResNets may be attributed to overfitting from more learnable parameters. In light of this, we conjecture there might be a more appropriate network backbone for BMD regression, which is outside this paper's scope. Therefore, we adopt VGG-11 with batch normalization and SE layer as our network backbone.

Quantitative Comparison Results. We compare our method with three existing semi-supervised learning (SSL) methods: Π-model [7], temporal ensembling [7], and mean teacher [14]. Regression MSE loss between predicted and GT BMDs is used on labeled images for all semi-supervised learning methods. Π-model is trained to encourage consistent network output between two augmentations of the same input image. Temporal ensembling produces pseudo labels

Table 2. Comparison with semi-supervised learning methods. (Temp. Ensemble: temporal ensembling)

Method	Π-model	Temp. ensemble	Mean teacher	Proposed
R-value	0.8637	0.8722	0.8600	**0.8805**
RMSE	0.0828	0.0832	0.0817	**0.0758**

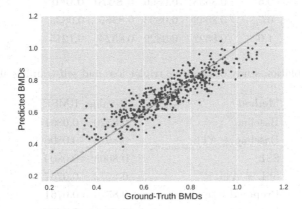

Fig. 2. Predicted BMDs against ground-truth BMDs on the test set.

via calculating the exponential moving average of predictions after every training epoch. Pseudo labels are then combined with labeled images to train the model. Instead of directly ensembling predictions, mean teacher uses an exponential moving average of model weights to produce pseudo labels for unlabeled images. For the proposed method and compared SSL methods, all models are fine-tuned from the weights pre-trained on labeled images.

As shown in Table 2, the proposed method achieves the best R-value of 0.8805 and RMSE of 0.0758. Π-model outperforms the baseline by enforcing output consistency as a regularization. While both temporal ensembling and mean teacher obtain improvements with the additional pseudo label supervision, averaging labels or weights can accumulate more errors over time. In contrast, our method only updates pseudo labels when the model performs better on the validation set, proving to be an effective strategy. Figure 2 shows the predicted BMDs of our model against GT BMDs. Overall, the model has a larger prediction error for lower or higher BMD cases. This situation is expected because lower/higher BMD cases are less common, and the model tends to predict moderate values.

Ablation Study. To assess the importance of various components in our proposed method, we conduct experiments to remove some components or vary some hyper-parameters while keeping others fixed. Firstly, we analyze the proposed adaptive triplet loss (ATL) by comparing it to its counterpart without the adaptive coefficient in Eq. 4 and varying the preset margin. Results in Table 3 show

Table 3. Ablation study of adaptive triplet loss.

Margin	Triplet loss		ATL	
	R-value	RMSE	R-value	RMSE
0.1	0.8515	0.0917	0.8563	0.1013
0.3	0.8459	0.1000	0.8657	0.0814
0.5	0.8538	0.0866	**0.8670**	**0.0806**
0.7	0.8549	0.0823	0.8565	0.0836
1.0	0.8522	0.0829	0.8524	0.1215

Table 4. Ablation study of adaptive triplet loss and self-training algorithm.

Method	R-value	RMSE
Baseline	0.8520	0.0831
Baseline + ATL	0.8670	0.0806
SSL	0.8605	0.0809
SSL + ATL	0.8772	0.0767
Proposed w/o Consistency	0.8776	0.0761
Proposed	**0.8805**	**0.0758**

that the non-adaptive counterpart deteriorates the model's regression accuracy, which proves the necessity of adaptive coefficient. Since BMD differences vary for different triplets, it is unreasonable to use a fixed margin to uniformly separate samples with dissimilar BMDs. We also observe that using ATL achieves higher R-values than the baseline regardless of the margin value. Specifically, $m = 0.5$ produces the best R-value of 0.8670 and RMSE of 0.0806. It demonstrates the effectiveness of our proposed adaptive triplet loss.

To better understand the benefits of our proposed semi-supervised self-training algorithm, we compare it with a straightforward semi-supervised learning (SSL) approach. We use the pre-trained model to predict pseudo BMDs for unlabeled images and combine them with labeled ones for further fine-tuning. We apply this approach to fine-tune models pre-trained using only MSE loss and the combination of MSE and ATL losses, denoted as SSL and SSL+ATL, respectively. Finally, we also analyze the contribution of the consistency loss in Eq. 6 by removing it during the self-training stage. The comparison results are shown in Table 4.

Employing the straightforward SSL strategy is effective as expected. It increases the baseline R-value to 0.8605 and reduces the RMSE to 0.0809. The model pre-trained using MSE and ATL losses is also improved. The results prove the effectiveness of using pseudo labels of unlabeled images. Our proposed self-training algorithm can obtain further improvement by updating pseudo labels during fine-tuning. However, the results show that the improvement becomes marginal from 0.8772 to 0.8776 in R-value without the consistency loss, even if pseudo labels are updated a few more times. The consistency loss can regularize model training by encouraging consistent output and features. Without it, the model is prone to overfitting to inaccurate pseudo labels and deteriorates. Combining ATL, self-training algorithm, and consistency loss, our proposed method achieves the best R-value of 0.8805 and RMSE of 0.0758. Compared to the baseline, the R-value is improved by 3.35% and the RMSE is reduced by 8.78%.

4   Conclusion

In this work, we investigate the feasibility of obtaining bone mineral density (BMD) from hip X-ray images instead of DEXA measurement. It is highly desirable for various practical reasons. We employ a convolutional neural network to estimate BMDs from preprocessed hip ROIs. Besides the MSE loss, a novel adaptive triplet loss is proposed to train the network on hip X-ray images with paired ground-truth BMDs. We further present a semi-supervised self-training algorithm to improve our model using large-scale unlabeled hip X-ray images. Experiment results on an in-house dataset of 1,090 images from 819 patients show that our method effectively achieves a high Pearson correlation coefficient of 0.8805. It implies the strong feasibility of X-ray based BMD estimation, which can provide potential opportunistic osteoporosis screening with more accessibility and at reduced cost.

References

1. Alacreu, E., Moratal, D., Arana, E.: Opportunistic screening for osteoporosis by routine ct in southern europe. Osteoporosis Int. **28**(3), 983–990 (2017)
2. Dagan, N., et al.: Automated opportunistic osteoporotic fracture risk assessment using computed tomography scans to aid in frax underutilization. Nat. Med. **26**(1), 77–82 (2020)
3. He, K., Zhang, X., Ren, S., Sun, J.: Deep residual learning for image recognition. In: Proceedings of the IEEE Conference on Computer Vision and Pattern Recognition, pp. 770–778 (2016)
4. Hu, J., Shen, L., Sun, G.: Squeeze-and-excitation networks. In: Proceedings of the IEEE Conference on Computer Vision and Pattern Recognition, pp. 7132–7141 (2018)
5. Ioffe, S., Szegedy, C.: Batch normalization: accelerating deep network training by reducing internal covariate shift. In: International Conference on Machine Learning, pp. 448–456. PMLR (2015)
6. Jang, S., Graffy, P.M., Ziemlewicz, T.J., Lee, S.J., Summers, R.M., Pickhardt, P.J.: Opportunistic osteoporosis screening at routine abdominal and thoracic ct: normative l1 trabecular attenuation values in more than 20 000 adults. Radiology **291**(2), 360–367 (2019)
7. Laine, S., Aila, T.: Temporal ensembling for semi-supervised learning. arXiv preprint arXiv:1610.02242 (2016)
8. Lee, S., Choe, E.K., Kang, H.Y., Yoon, J.W., Kim, H.S.: The exploration of feature extraction and machine learning for predicting bone density from simple spine x-ray images in a korean population. Skeletal Radiol. **49**(4), 613–618 (2020)

9. Lee, S.J., Anderson, P.A., Pickhardt, P.J.: Predicting future hip fractures on routine abdominal ct using opportunistic osteoporosis screening measures: a matched case-control study. Am. J. Roentgenol **209**(2), 395–402 (2017)

10. Li, W., et al.: Structured landmark detection via topology-adapting deep graph learning. arXiv preprint arXiv:2004.08190 (2020)

11. Mookiah, M.R.K., et al.: Feasibility of opportunistic osteoporosis screening in routine contrast-enhanced multi detector computed tomography (mdct) using texture analysis. Osteoporosis Int. **29**(4), 825–835 (2018)

12. Pan, Y., et al.: Automatic opportunistic osteoporosis screening using low-dose chest computed tomography scans obtained for lung cancer screening. Eur. Radiol. **30**, 1–10 (2020)

13. Simonyan, K., Zisserman, A.: Very deep convolutional networks for large-scale image recognition. arXiv preprint arXiv:1409.1556 (2014)

14. Tarvainen, A., Valpola, H.: Mean teachers are better role models: Weight-averaged consistency targets improve semi-supervised deep learning results. arXiv preprint arXiv:1703.01780 (2017)

15. Yamamoto, N., et al.: Deep learning for osteoporosis classification using hip radiographs and patient clinical covariates. Biomolecules **10**(11), 1534 (2020)

16. Zhang, B., et al.: Deep learning of lumbar spine x-ray for osteopenia and osteoporosis screening: a multicenter retrospective cohort study. Bone **140**, 115561 (2020)

DAE-GCN: Identifying Disease-Related Features for Disease Prediction

Churan Wang[1,4], Xinwei Sun[2(✉)], Fandong Zhang[4], Yizhou Yu[4,5],
and Yizhou Wang[3]

[1] Center for Data Science, Peking University, Beijing, China
[2] Peking University, Beijing, China
sxwxiaoxiaohehe@pku.edu.cn
[3] Department of Computer Science and Technology, Peking University,
Beijing, China
[4] Deepwise AI Lab, Beijing, China
[5] The University of Hong Kong, Pokfulam, Hong Kong

Abstract. Learning disease-related representations plays a critical role
in image-based cancer diagnosis, due to its trustworthy, interpretable
and good generalization power. A good representation should not only
be disentangled from the disease-irrelevant features, but also incorporate the information of lesion's attributes (*e.g.*, shape, margin) that
are often identified first during cancer diagnosis clinically. To learn
such a representation, we propose a **D**isentangle **A**uto-**E**ncoder with
Graph **C**onvolutional **N**etwork (DAE-GCN), which adopts a *disentangling mechanism* with the guidance of a GCN model in the AE-based
framework. Specifically, we explicitly separate the encoded features into
disease-related features and others. Among such features that all participate in image reconstruction, we only employ the disease-related features
for disease prediction. Besides, to account for lesions' attributes, we propose to leverage the attributes and adopt the GCN to learn them during
training. Take mammogram mass benign/malignant classification as an
example, our DAE-GCN helps improve the performance and the interpretability of cancer prediction, which can be verified by state-of-the-art
performance on one public dataset DDSM and three in-house datasets.

Keywords: Disease prediction · Disentangle · GCN

1 Introduction

For image-based disease benign/malignant diagnosis, it is crucial to learn the
disease-related representation for prediction, due to the necessity of trustworthy (to patients), explainable (to clinicians) and good generalization ability in
healthcare. A good representation, should not only be disentangled from the

Electronic supplementary material The online version of this chapter (https://
doi.org/10.1007/978-3-030-87240-3_5) contains supplementary material, which is available to authorized users.

© Springer Nature Switzerland AG 2021
M. de Bruijne et al. (Eds.): MICCAI 2021, LNCS 12905, pp. 43–52, 2021.
https://doi.org/10.1007/978-3-030-87240-3_5

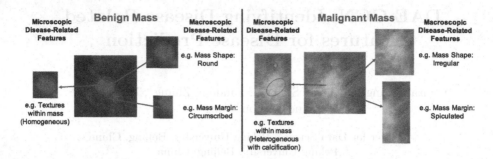

Fig. 1. Illustrations of macroscopic/microscopic disease-related features in the mass. The left figure shows the benign case while the right shows the malignant case. Macroscopic-related features (*e.g.* margins, shapes) perform regular or clear in a benign mass while behave spiculated and irregular in a malignant mass. Microscopic-related features (*e.g.* the textures within the mass) perform differently between benign/malignant cases by taking a closer inspection. Textures within a malignant mass are heterogeneous (with calcification sometimes) but the textures in a benign mass are homogeneous.

disease-irrelevant features, but should also extract both macroscopic attributes and microscopic features of diseases. The macroscopic attributes mainly refer to the morphological attributes(*e.g.* shape, margin) [14] as summarized in American College of Radiology (ACR) [13], which are commonly adopted by doctors in benign/malignant diagnosis of many kinds of diseases. As all these attributes share the disease's information, they are correlated among each other [13]. In addition to these attributes, one can also extract some disease-related and microscopic features from the image, such as textures, curvatures of contour [5], *etc.*, which are helpful for prediction but may be beyond the observing ability of clinicians. Both examples of disease-related features, *i.e.*, macroscopic and microscopic ones are illustrated in Fig. 3.

Typically, in deep supervised learning, such a representation is regarded as the final hidden layer of the neural network [8,11]. However, it has been criticized [7] that these hidden layers can learn non-interpretable/semantic features, due to lack of prior knowledge which refers to clinical attributes in our scenario. Other representation learning works although leverage attributes [4,9] into learning, lack consideration of the microscopic features and empirical assurance to disentangle disease-related features from others. Besides, they do not model the correlation among attributes, which can make learning inefficient (Fig. 1).

For better representation learning, the disentanglement mechanism has been proved to be an effective way [1,3,12], since such a mechanism prompts different independent latent units to encode different independent ground truth generation factors that vary in the data [1]. Based on the above, to capture the disease-related features without mixing other irrelevant information, in this paper we propose a **D**isentangle **A**uto-**E**ncoder with **G**raph **C**onvolutional **N**etwork (DAE-GCN), which incorporates a *disentangling mechanism* into an AE framework, equipped with attribution data during training stage (the attributes are not provided during the test). Specifically, in our encoder network, we explicitly encode the image into three hidden factors: h_{ma}, h_{mi}, h_i which respectively correspond to macroscopic-related, microscopic-related and disease-irrelevant features. To achieve disentanglement, these hidden factors are fed into different constrains during the training phase. In details, among all h_{ma}, h_{mi}, h_i that participate in reconstruction of the whole image, we only use h_{ma}, h_{mi} in disease prediction and only h_{ma} to predict attributes, enforcing the disentanglement of h_{ma}, h_{mi}, h_i. To further leverage the correlation among attributes, we implement the GCN to facilitate learning.

To verify the utility of our method for diagnosing cancer-based on learning disease-related representations, we apply the DAE-GCN on Digital Database for Screening Mammography (DDSM) [2]) and three in-house datasets in mammogram mass benign/malignant diagnosis. It yields that our method successfully learns disease-related features and leads to a large classification improvement (4% AUC) over others on all datasets.

To summarize, our contributions are mainly three-fold: **a)** We propose a novel and general DAE-GCN framework, which helps disentangle the disease-related features from others to prompt image-based diagnosis; **b)** We leverage the GCN which accounts for correlations among attributes to facilitate learning; **c)** Our model can achieve state-of-the-art prediction performance for mass benign/malignant diagnosis on both the public and in-house datasets.

2 Methodology

Problem Setup. Our dataset contains $\{x_i, A_i, y_i\}_{i \in \{1,\dots,n\}}$, in which x, A, y respectively denote the patch-level mass image, attributes (*e.g.*, circumscribed-margin, round-shape, irregular-shape), and the binary disease label. During test stage, only the image data is provided for feature extraction and prediction.

Figure 2 outlines the overall pipeline of our method, namely **D**isentangle **A**uto-**E**ncoder with **G**raph **C**onvolutional **N**etwork(DAE-GCN). As shown, our method is based on the auto-encoder framework, which has been empirically validated to extract effective features [10]. In the encoder, we separate encoded factors into three parts: macroscopic attributes h_{ma}, microscopic features h_{mi} and disease-irrelevant features h_i. During disentangle phase, we design the training with three constrains, which respectively reconstruct x, A, y via *image reconstruction*, *GCN learning* and *disease prediction*. In the following, we will introduce our encoder and disentangle training in Sect. 2.1, 2.2.

2.1 Encoder

We encode the original image into hidden factors h via an encoding network $f_{\theta_{\text{enc}}}$ parameterized by θ_{enc}: $h = f_{\theta_{\text{enc}}}^{\theta}(x)$. Such a hidden factor can contain many variations, which as a whole can be roughly categorized as three types: macroscopic disease-related attributes, microscopic disease-related features and other disease-irrelevant features. To only capture disease-related features into prediction, we explicitly separate h into three parts: h_{ma}, h_{mi}, h_i that aims to

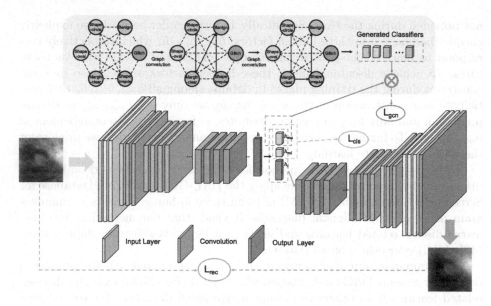

Fig. 2. Overview of the DAE-GCN, which is based on the auto-encoder framework. The encoded features h are separated into h_{ma}, h_{mi}, h_i marked by green, pink and blue. The decoder leads to three loss functions: the $\mathcal{L}_{\mathrm{rec}}$ for reconstructing the original image using h as input; the $\mathcal{L}_{\mathrm{gcn}}$ for learning attributes with h_{ma} as input; the $\mathcal{L}_{\mathrm{cls}}$ for disease prediction with h_{ma}, h_{mi} as inputs. Specifically, we implement GCN for attribute learning to leverage the correlation among attributes. (Color figure online)

respectively capture the above three types of features. Accordingly, we append a three-constrain disentangle training, as elaborated in the subsequent section.

2.2   Disentangle Training

To achieve disentanglement of h_{ma}, h_{mi}, h_i, in particular the disease-related features (h_{ma}, h_{mi}) from others, we design a three-constrain training: *image reconstruction, GCN learning* and *disease prediction* with utilizing different factors.

Image Reconstruction. Since h_{ma}, h_{mi}, h_i describe different contents, they all participate in reconstructing the original image, via an decoding network $f_{\theta_{\mathrm{dec}}}$: $\hat{x} = f_{\theta_{\mathrm{dec}}}(f_{\theta_{\mathrm{enc}}}(x))$. The reconstruction loss $\mathcal{L}_{\mathrm{rec}}$ is defined as:

$$\mathcal{L}_{\mathrm{rec}}(\theta_{\mathrm{enc}}, \theta_{\mathrm{dec}}) := \|x - \hat{x}(\theta_{\mathrm{enc}}, \theta_{\mathrm{dec}})\|_1 \qquad (1)$$

Graph Convolutional Network Learning. To leverage correlation among attributes A into learning h_{ma}, we propose to implement the graph convolutional network (GCN), in which the computation rule is based on the graph $G := (V, E)$. Each node in V denotes the word embeddings of each attribute (details in supplementary). Each edge connecting V_i and V_j in E denotes the inner relevance between attributes. Specifically, we adopt [4] to define E as follows:

$$E_{i,j} = \begin{cases} p/\sum_{j=1}^{c} \tilde{E}_{i,j}, & \text{if } i \neq j \\ 1 - p, & \text{if } i = j, \end{cases} \qquad (2)$$

where $\tilde{E}_{i,j} = 0$ if $M_{i,j}/T_i < \tau$ and $= 1$ otherwise for some threshold hyperparameter $\tau \in (0,1)$, with $M_{i,j}$ denoting the number of concurrences of the i-th attribute and the j-th attribute, T_i denoting the number of occurrences of the i-th attribute. The $p \in (0,1)$ denotes the weight hyperparameter to avoid the over-smoothed problem, $i.e.$, the node from different clusters (shape-related vs margin-related) are indistinguishable [4]. Denote $H_l \in \mathbb{R}^{c \times d_l}$ as the output of the l-th layer, with c denoting the number of nodes. With $H_{l+1} = \sigma(EH_lW_l)$ for nonlinear activation σ and $W_l \in \mathbb{R}^{d_l \times d_{l+1}}$, each layer's output processes to the next, the last of which ($H_L \in \mathbb{R}^{c \times d_L}$ with $d_L = \dim(h_{ma})$) is fed into a softmax classifier for multi-label classification:

$$\hat{y}_i := \frac{\exp(\langle H_{i,L}, h_a \rangle)}{\left(\sum_{j \in \text{clus}(i)} \exp(\langle H_{j,L}, h_a \rangle)\right)}$$

where the clus(i) denoting the index set of attributes that belong to the same cluster ($e.g.$, shape, margin) with the i-th attribute. Denote θ_{gcn} as the whole set of parameters, $i.e.$, $\theta_{\text{gcn}} := \{W_l\}_{l \in \{0,\dots,L\}}$. The multi-label loss function for training such θ_{gcn} (and also θ_{enc}) $\mathcal{L}_{\text{gcn}}$ is denoted as the summation of cross-entropy loss for each cluster k:

$$\mathcal{L}_{\text{gcn}}(\theta_{\text{enc}}, \theta_{\text{gcn}}) := \frac{1}{N} \sum_n \sum_k \sum_{i \in \text{clus}_k} \left(\mathbb{1}(y_i = k) \log \hat{y}_i(H_0^n, \theta_{\text{gcn}}) \right). \qquad (3)$$

Disease Prediction. We feed disease-related features h_{ma}, h_{mi} into disease prediction via the classification network $f_{\theta_{\text{cls}}}$, with the θ_{cls} trained via the binary cross-entropy loss $\mathcal{L}_{\text{cls}}(\theta_{\text{enc}}, \theta_{\text{cls}})$. Combining $\mathcal{L}_{\text{cls}}$ with the $\mathcal{L}_{\text{rec}}, \mathcal{L}_{\text{gcn}}$ defined in Eq. (1), (3), the overall loss function $\mathcal{L}$ is:

$$\mathcal{L}(\theta_{\text{enc}}, \theta_{\text{dec}}, \theta_{\text{gcn}}, \theta_{\text{cls}}) = \mathcal{L}_{\text{rec}}(\theta_{\text{enc}}, \theta_{\text{dec}}) + \mathcal{L}_{\text{gcn}}(\theta_{\text{enc}}, \theta_{\text{gcn}}) + \mathcal{L}_{\text{cls}}(\theta_{\text{enc}}, \theta_{\text{cls}}).$$

3 Experiments

To evaluate the effectiveness of our DAE-GCN, we verify it on the patch-level mammogram mass benign/malignant classification. We consider both the public dataset DDSM [2] and three in-house datasets: Inhouse1, Inhouse2 and Inhouse3. For each dataset, the region of interests (ROIs) (malignant/benign masses) are cropped based on the annotations of radiologists the same as [9][1]. For all datasets, we randomly[2] divide the whole set into training, validation and testing

---

[1] We leave the number of ROIs and patients of each dataset and the description about the selection of attributes for DDSM in supplementary.

[2] Existing works about DDSM do not publish their splitting way and mention smaller count number of ROIs in DDSM compared with our statistics.

Table 1. AUC evaluation on public DDSM [2] and three in-house datasets.

Methodology	Inhouse1	Inhouse2	Inhouse3	DDSM [2]
Vanilla [8]	0.888	0.847	0.776	0.847
Chen et al. [4]	0.924	0.878	0.827	0.871
Guided-VAE [6]	0.921	0.867	0.809	0.869
ICADx [9]	0.911	0.871	0.816	0.879
Li et al. [11]	0.908	0.859	0.828	0.875
DAE-GCN (Ours)	**0.963**	**0.901**	**0.857**	**0.919**

as 8:1:1 in patient-wise. To provide convenience for latter works, we publish our spitted test set of DDSM [2] in supplementary.

3.1 Results

Compared Baselines. We compare our DAE-GCN the following methods: a) Li et al. [11] propose to reconstruct benign and malignant images separately via adversarial training; b) ICADx [9] also adopts the adversarial learning method and additionally introduces the attributes for reconstruction; c) Vanilla [8] directly trains the classifier via Resnet34; d) Guided-VAE [6] also implements disentangle network but lack the medical knowledge during learning; e) Chen et al. [4] only implements GCN to learning disease label and attributes.

Implementation Details. We implement Adam to train our model. For a fair comparison, all methods are conducted under the same setting; besides, they share the network structure of ResNet34 [8] as the encoder backbone. Area Under the Curve (AUC) is used as evaluation metrics in image-wise. More details are shown in the supplementary. For implementation of compared baselines, we directly load the published trained model of Vanilla [8], Chen et al. [4] during test; while for Guided-VAE [6], ICADx [9] and Li et al. [11] that without published source codes, we re-implement their methods.

Results and Analysis. As shown in Table 1, our methods can achieve state-of-the-art results on all datasets. Taking a closer look, the advantage of Guided-VAE [6] over Vanilla [8] may due to the disentangle learning in the former method. With further exploration of attributes via GCN, our method can outperform Guided VAE [6]. Although ICADx [9] incorporats the attributes during learning, they fail to model correlations among attributes, which limits their performance. Compared to Chen et al. [4] that also implement GCN to learn attributes, we additionally use disentangled learning which can help to identify disease-related features during prediction.

Table 2. Ablation Studies on (a) Inhouse1; (b) Inhouse2; (c) Inhouse3; (d) pubilic dataset DDSM [2]

Disentangle	Attribute Learning	$\mathbf{L_{rec}}$	h_i	AUC(a)	AUC(b)	AUC(c)	AUC(d)
×	×	×	×	0.888	0.847	0.776	0.847
×	Multi-task	×	×	0.890	0.857	0.809	0.863
×	$\mathcal{L}_{gcn}$	×	×	0.924	0.878	0.827	0.871
×	×	✓	×	0.899	0.863	0.795	0.859
✓	×	✓	×	0.920	0.877	0.828	0.864
✓	×	✓	✓	0.941	0.882	0.835	0.876
✓	$\mathcal{L}_{gcn}$	✓	✓	**0.963**	**0.901**	**0.857**	**0.919**

Table 3. Overall prediction accuracy of multi attributes (mass shape,mass margin) on (a) Inhouse1; (b) Inhouse2; (c) Inhouse3; (d) pubilic dataset DDSM [2].

Methodology	ACC (Inhouse1)	ACC (Inhouse2)	ACC (Inhouse3)	ACC (DDSM [2])
Vanilla-multitask	0.715	0.625	0.630	0.736
Chen et al. [4]	0.855	0.829	0.784	0.875
ICADX [9]	0.810	0.699	0.671	0.796
Proposed Method	**0.916**	**0.880**	**0.862**	**0.937**

3.2 Ablation Study

To verify the effectiveness of each component in our model, we evaluate some variant models: *Disentangle* that denotes whether implement disentangled learning during reconstructing phase; and *multi-task* denotes using the multi-task model to learn attributes. The results are shown in Table 2.

As shown, deleting or changing any of the four components would lead to a descent of the classification performance. To be worthy of attention, using naive GCN also leads to a boosting of around 3%. Such a result can validate that the attributes data is quite helpful for the guidance of disease-related features learning. Meanwhile, the model with disentangle learning outperforms the one without it by a noticeable margin, which may be due to that the disease-related features can be identified without mixing information from others via disentangle learning. Moreover, with the guidance of exploring attributes, disease-related features can be disentangled better.

3.3 Interpretability

Attributes Prediction. To verify the prediction power of our learned representation h_{ma}, we report the accuracy of multi-label classification. Table 3 shows the results. As shown, our DAE-GCN outperforms other considered methods, which demonstrates that the learned h_{ma} can well capture the information to predict attributes.

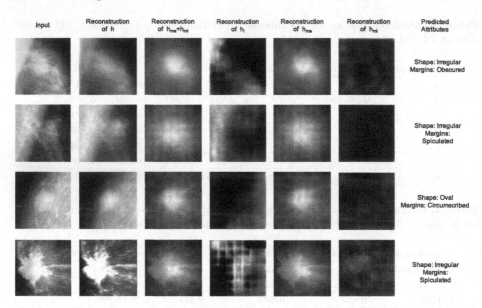

Fig. 3. Visualization of the disentanglement via reconstruction. From the left column to the right column: input; reconstruction using h $(h_{ma} + h_{mi} + h_i)$; diseased-related features $(h_{ma} + h_{mi})$; disease-irrelevant features h_i; macroscopic attributes h_{ma}; microscopic features h_{mi} and predicted results of attributes.

Visualization. To further evaluate our learned representation, we visualize different parts in Fig. 3 via reconstruction effect. As we can see, the disease-related features $(h_{mi} + h_{ma})$ mainly reflect the lesion-related information since they only reconstruct the lesion regions without mixing others. The disease-irrelevant features h_i mainly learn features such as the contour of the breasts, pectoralis and other irrelevant grands without lesion information. The macroscopic attributes h_{ma} capture macroscopic attributes of the lesions; while the microscopic features h_{mi} learn features like global context, density or other invisible features but related to classification. For better validation of the success of disentangling, we use disease-related features $(h_{mi} + h_{ma})$ and disease-irrelevant features h_i encoded from our model while testing to train an SVM classifier respectively. The AUC of disease-related features $(h_{mi} + h_{ma})$ and disease-irrelevant features h_i for benign/malignant classification is separately 0.90 and 0.57 in DDSM [2]. This result further indicates the effectiveness and interpretability of our proposed DAE-GCN.

4 Conclusions and Disscusions

In this paper, we propose a novel approach called **D**isentangle **A**uto-**E**ncoder with **G**raph **C**onvolutional **N**etwork (DAE-GCN) to improve the mammogram classification performance. The proposed method performs disentangle learning by exploiting the attribute prior effectively. The promising results achieved on mammogram classification shows the potential of our method in benefiting the diagnosis of other types of diseases, *e.g.*, lung cancer, liver cancer and pancreatic cancer, which are left as our future work.

Acknowledgement. This work was supported by MOST-2018AAA0102004, NSFC-61625201 and ZheJiang Province Key Research & Development Program (No. 2020C03073).

References

1. Bengio, Y., Courville, A., Vincent, P.: Representation learning: a review and new perspectives. IEEE Trans. Pattern Anal. Mach. Intell. **35**(8), 1798–1828 (2013). https://doi.org/10.1109/TPAMI.2013.50
2. Bowyer, K., et al.: The digital database for screening mammography. In: Third International Workshop on Digital Mammography, vol. 58, p. 27 (1996)
3. Burgess, C.P., et al.: Understanding disentangling in β-vae. arXiv preprint arXiv:1804.03599 (2018)
4. Chen, Z.M., Wei, X.S., Wang, P., Guo, Y.: Multi-label image recognition with graph convolutional networks. In: Proceedings of the IEEE/CVF Conference on Computer Vision and Pattern Recognition, pp. 5177–5186 (2019)
5. Ding, J., et al.: Optimizing the peritumoral region size in radiomics analysis for sentinel lymph node status prediction in breast cancer. Acad. Radiol. (2020). https://doi.org/10.1016/j.acra.2020.10.015
6. Ding, Z., et al.: Guided variational autoencoder for disentanglement learning. In: Proceedings of the IEEE/CVF Conference on Computer Vision and Pattern Recognition, pp. 7920–7929 (2020)
7. Geirhos, R., Rubisch, P., Michaelis, C., Bethge, M., Wichmann, F.A., Brendel, W.: Imagenet-trained CNNs are biased towards texture; increasing shape bias improves accuracy and robustness. In: International Conference on Learning Representations (2019). https://openreview.net/forum?id=Bygh9j09KX
8. He, K., Zhang, X., Ren, S., Sun, J.: Deep residual learning for image recognition. In: Proceedings of the IEEE Conference on Computer Vision and Pattern Recognition, pp. 770–778 (2016)
9. Kim, S.T., Lee, H., Kim, H.G., Ro, Y.M.: Icadx: interpretable computer aided diagnosis of breast masses. In: Medical Imaging 2018: Computer-Aided Diagnosis, vol. 10575, p. 1057522. International Society for Optics and Photonics (2018). https://doi.org/10.1117/12.2293570
10. Klingler, S., Wampfler, R., Käser, T., Solenthaler, B., Gross, M.: Efficient feature embeddings for student classification with variational auto-encoders. International Educational Data Mining Society (2017)
11. Li, H., Chen, D., Nailon, W.H., Davies, M.E., Laurenson, D.I.: Signed laplacian deep learning with adversarial augmentation for improved mammography diagnosis. In: Shen, D., et al. (eds.) MICCAI 2019. LNCS, vol. 11769, pp. 486–494. Springer, Cham (2019). https://doi.org/10.1007/978-3-030-32226-7_54
12. Ridgeway, K.: A survey of inductive biases for factorial representation-learning. arXiv preprint arXiv:1612.05299 (2016)

13. Sickles, E., D'Orsi, C., Bassett, L.: Acr bi-rads® mammography. acr bi-rads® atlas, breast imaging reporting and data system (2013)
14. Surendiran, B., Vadivel, A.: Mammogram mass classification using various geometric shape and margin features for early detection of breast cancer. Int. J. Med. Eng. Inf. 4(1), 36–54 (2012). https://doi.org/10.1504/IJMEI.2012.045302

Enhanced Breast Lesion Classification via Knowledge Guided Cross-Modal and Semantic Data Augmentation

Kun Chen[1,2], Yuanfan Guo[1,2], Canqian Yang[1,2], Yi Xu[1,2(✉)], Rui Zhang[1], Chunxiao Li[3], and Rong Wu[3]

[1] Shanghai Jiao Tong University, Shanghai, China
xuyi@sjtu.edu.cn
[2] MoE Key Lab of Artificial Intelligence, AI Institute,
Shanghai Jiao Tong University, Shanghai, China
[3] Shanghai General Hospital, Shanghai, China

Abstract. Ultrasound (US) imaging is a fundamental modality for detecting and diagnosing breast lesions, while shear-wave elastography (SWE) serves as a crucial complementary counterpart. Although an automated breast lesion classification system is desired, training of such a system is constrained by data scarcity and modality imbalance problems due to the lack of SWE devices in rural hospitals. To enhance the diagnosis with only US available, in this work, we propose a knowledge-guided data augmentation framework, which consists of a modal translater and a semantic inverter, achieving cross-modal and semantic data augmentation simultaneously. Extensive experiments show a significant improvement of AUC from 84.36% to 86.71% using the same ResNet18 classifier after leveraging our augmentation framework, further improving classification performance based on conventional data augmentation methods and GAN-based data augmentation methods without knowledge guidance.

Keywords: Breast lesion classification · B-mode ultrasound · Shear-wave elastography · Generative adversarial network · Data augmentation

1 Introduction

Breast cancer, the most commonly diagnosed cancer, is the fifth leading cause of cancer death all over the world [24]. Currently, ultrasound (US) imaging has

K. Chen and Y. Guo—Contributed equally to this article.

Electronic supplementary material The online version of this chapter (https://doi.org/10.1007/978-3-030-87240-3_6) contains supplementary material, which is available to authorized users.

© Springer Nature Switzerland AG 2021
M. de Bruijne et al. (Eds.): MICCAI 2021, LNCS 12905, pp. 53–63, 2021.
https://doi.org/10.1007/978-3-030-87240-3_6

shown great potential in detecting and diagnosing breast lesions [4]. The most widely used B-mode US only provides diagnostic information related to the lesion structure and internal echogenicity, which limits the performance of computer-aided diagnosis (CAD) systems to some extent [5, 21, 22]. It has been investigated that breast cancer tissue tends to be harder than normal breast tissue and benign tumor tissue [1], indicating that stiffness information is discriminative for breast cancer diagnosis. Hence, shear-wave elastography (SWE) is developed to utilize the stiffness information [2, 23]. It has been proven that the use of SWE features, singly or in combination, as an adjunct to conventional B-mode US features, would improve the accuracy of assessing a lesion's probability of malignancy [8, 12, 27, 28].

Considering the shortage of radiologists and SWE devices in rural hospitals, an automated deep learning based breast lesion classification system has great potential to help treatment planning for radiologists with US only. However, without a sufficient number of annotated US-SWE image pairs, deep learning based classification networks are highly likely to suffer from the over-fitting problem. One standard solution to over-fitting is image manipulation, including various simple transformations of the data, such as translation, flipping, scaling, and shearing. However, these transformations cannot introduce new knowledge but only the variants of original images, bringing limited performance improvement [3]. Instead, synthetic data augmentation methods based on image-to-image Generative Adversarial Networks (GANs) [6, 11, 29] are able to generate sophisticated types of transformed images based on original images, providing more variability to enrich the training data.

Recently, image-to-image GANs have shown promising performance for data augmentation in medical image analysis [7, 16, 17, 19, 25]. Liu *et al.* [16] utilized both shape information and intensity distributions for input annotations to generate more realistic images, significantly improving the performance on cell detection and segmentation tasks. Gupta *et al.* [7] proposed a GAN-based image-to-image translation approach to generate multiple corresponding virtual samples of different modalities from a given pathological image, achieving an improvement on downstream cell segmentation task. Liu *et al.* [17] explored object detection in the small data regime from a generative modeling perspective by learning to generate chest X-ray images with inserted nodules as semantic data augmentation, and using the expanded dataset for training an object detector. While previous works have shown the promising performance of cross-modal and semantic data augmentation respectively, few efforts have been made to integrate both augmentations into a unified framework to improve downstream tasks' performance further.

In this paper, we propose a knowledge-guided data augmentation framework for ultrasound images, where both cross-modal and semantic knowledge are considered for data augmentation. For cross-modal data augmentation, we propose a knowledge distillation guided modal translater, which synthesizes SWE based on US, achieving modality enrichment for US images. Semantic knowledge guided data augmentation is achieved by a semantic inverter which maps a US image to the opposite category. Further experiments prove that our framework is able to synthesize virtual US and SWE images with US images only, significantly enhancing the breast lesion classification network based on conventional data

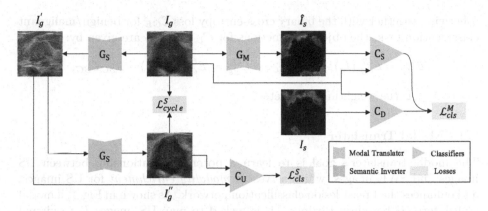

Fig. 1. Overview of the proposed augmentation framework. G_M is based on Pix2pix [11] and G_S is based on CycleGAN [29]. Original loss of [11] and adversarial loss of [29] are not shown for simplicity. Please refer to the main text for the definition of notations.

augmentation methods and outperforming conventional GANs without knowledge guidance.

2   Methodology

An overview of our proposed knowledge-guided data augmentation framework is shown in Fig. 1. The framework mainly consists of: (1) two baseline classification networks C_U and C_D pre-trained on US images and US-SWE image pairs respectively to provide knowledge guidance for semantic inverter and modal translator, (2) a knowledge distillation [10] guided modal translator G_M to synthesize SWE images based on US images, (3) a semantic inverter G_S to synthesize US images with opposite category and (4) a student classification network C_S conditioned on real and virtual US images and corresponding virtual SWE images, which serves as our final breast lesion classification network.

2.1   Baseline Classification Networks

We first pre-train two ResNet18 [9] based classification networks C_U and C_D with uni-modal real US images I_g and dual-modal real US-SWE image pairs (I_g, I_s) respectively. C_U serves as the baseline model and is utilized to provide semantic knowledge for training the semantic inverter G_S. C_D has full knowledge of both US and SWE images; hence serves as the oracle model, and is utilized to guide the training of modal translator G_M and student classification network C_S. Both classification networks are trained with conventional data augmentation, including random translation, flipping, scaling, and shearing. During our experiments, we find using Supervised Contrastive Loss [13] improves the performance of both C_U and C_D. Let $\mathcal{L}_{SC}$ be the supervised contrastive loss, which takes samples of the same class within a batch as positive pairs, and negative pairs otherwise, together with the binary cross-entropy loss $\mathcal{L}_{CE}$ for benign/malignant classification task, the objective functions for C_U and C_D are given by:

$$\mathcal{L}_{cls}^{C_U} = \mathcal{L}_{CE}^{C_U}(I_g) + \lambda_{SC}\mathcal{L}_{SC}, \quad \mathcal{L}_{cls}^{C_D} = \mathcal{L}_{CE}^{C_D}(I_g, I_s) + \lambda_{SC}\mathcal{L}_{SC}, \tag{1}$$

where λ_{SC} is the weighting parameter.

2.2 Modal Translater

The modal translater's goal is to learn a potential relationship between US images and SWE images, which serves as *modality enrichment* for US images, and enhances the breast lesion classification network. As shown in Fig. 1, a modal translater G_M based on Pix2pix [11] is tasked to map US images I_g to virtual SWE images $\hat{I}_s$:

$$\hat{I}_s = G_M(I_g). \tag{2}$$

Both I_g and $\hat{I}_s$ are utilized to train the student classification network C_S.

While Pix2pix is optimized for image realism, it does not guarantee good feature quality for the downstream semantic classification task. Intuitively, we can directly employ the pre-trained classification network C_D to introduce the *semantic guidance* on G_M by supervision of the ground truth labels. However, in our early attempts, we find that such 'hard' supervision increases the difficulty of training, and thus the model fails to achieve a promising convergence, which is consistent with the theoretical analysis in [10].

To this end, we propose to achieve a milder *semantic guidance* by distilling knowledge from C_D to train G_M and C_S in a joint optimization manner. The involvement of C_S benefits G_M from two aspects. On the one hand, C_S distills knowledge embedded in C_D and propagates it to G_M as a reasonable *semantic guidance*. On the other hand, C_S can adjust itself to provide extra tolerance to the inferior performance of G_M in the early training stage, which stabilizes the training and leads to better convergence. Besides, the cooperation of modality variability introduced by G_M and 'soft' supervision guided by C_D also serves as a reasonable pre-training, which is supposed to facilitate the performance of C_S further.

Hence, the objective functions of C_S and G_M are given by:

$$\mathcal{L}_{C_S} = \mathcal{L}_{CE}^{C_S}(I_g, \hat{I}_s) + \lambda_{SC}\mathcal{L}_{SC} + \lambda_{dist}\|C_D(I_g, I_s) - C_S(I_g, \hat{I}_s)\|_2^2, \tag{3}$$

$$\mathcal{L}_{G_M} = \mathcal{L}_{P2p} + \mathcal{L}_{cls}^M, \tag{4}$$

where λ_{dist} is the weighting parameter. The third term in Eq. 3 distills semantic knowledge by matching the output logits $C_D(I_g, I_s)$ and $C_S(I_g, \hat{I}_s)$ in a mean squared error manner. For G_M, we integrate the loss of Pix2pix $\mathcal{L}_{P2p}$ [11] and the loss $\mathcal{L}_{C_S}$ as semantic guidance, i.e., $\mathcal{L}_{cls}^M = \mathcal{L}_{C_S}$.

2.3 Semantic Inverter

With a limited number of US images available for training our classification networks, a knowledge-guided semantic inverter is proposed to increase US images' semantic diversity. As shown in Fig. 1, US images I_g are processed twice by a semantic inverter G_S to synthesize virtual US images I'_g and I''_g with opposite and same category respectively:

$$I'_g = G_S(I_g), I''_g = G_S(I'_g). \tag{5}$$

Instead of training two category-specified networks as done in CycleGAN [29], we share G_S among benign and malignant US images and force it to invert the category of US images without category prior to the input. This manner not only saves network parameters, but also enables the employment on the test set for visualization and further cooperation with G_M where semantic labels are unavailable.

Based on an unsupervised image translation scenario providing no benign-malignant US image pair that refers to the same lesion, there is no ground truth image for I'_g. Only a cycle-consistency loss [29] is employed to provide *morphological supervision*:

$$\mathcal{L}^S_{cycle} = \|I''_g - I_g\|_1. \tag{6}$$

For an input US image I_g, G_S aims to convert it into the opposite category and further convert it back. Traditional GANs optimized for image realism provide no guarantee for such semantic conversion; hence we utilize the pre-trained classification network C_U to provide *semantic supervision*. Specifically, I'_g and I''_g should be considered as the opposite category and same category w.r.t I_g by C_U respectively:

$$\mathcal{L}^S_{cls} = \mathcal{L}^{C_U}_{CE}(I'_g) + \mathcal{L}^{C_U}_{CE}(I''_g). \tag{7}$$

Together with the established GAN loss $\mathcal{L}_{GAN}$ [29], the full objective function of G_S is given by:

$$\mathcal{L}_{G_S} = \mathcal{L}_{GAN} + \lambda^S_{cycle}\mathcal{L}^S_{cycle} + \lambda^S_{cls}\mathcal{L}^S_{cls}, \tag{8}$$

where λ^S_{cycle} and λ^S_{cls} are the weighting parameters.

2.4 Final Classification Network

After the training of modal translater G_M and semantic inverter G_S, both G_M and G_S are utilized to fine-tune the student classification network C_S pre-trained with G_M. Specifically, G_S is utilized to synthesize virtual US images I'_g and I''_g from I_g, while G_M is employed to synthesize corresponding virtual SWE images $\hat{I}_s$, $\hat{I}'_s$ and $\hat{I}''_s$. All virtual US-SWE image pairs are utilized to fine-tune G_M and C_S. Hence, the objective function of C_S at the fine-tuning stage is given by:

$$\mathcal{L}^F_{C_S} = \mathcal{L}_{C_S} + \mathcal{L}^{C_S}_{CE}(I'_g, \hat{I}'_s) + \mathcal{L}^{C_S}_{CE}(I''_g, \hat{I}''_s). \tag{9}$$

After fine-tuning, C_S serves as our final classification network.

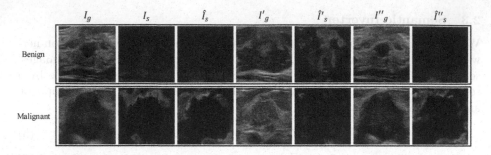

$$I_g \qquad I_s \qquad \hat{I}_s \qquad I'_g \qquad \hat{I}'_s \qquad I''_g \qquad \hat{I}''_s$$

Fig. 2. Examples of real and virtual US and SWE images.

3   Experiments

3.1   Dataset Preparation

From September 2020 to January 2021, a total of 2,008 images of benign lesions and 1,466 images of malignant lesions from 593 patients were collected as the dataset used in this paper. The Super Linear SL-15-4 probe of ultrafast ultrasound device Aixplorer (Super Sonic Imagine, Aix-en-Provence, France) was used for imaging data collection. The maximum stiffness scale of SWE images was selected as 180 Kilopascal (kPa). We divided the dataset into five subsets with no patient overlapping, each containing roughly the same number of US-SWE image pairs and the same proportion of benign and malignant images. Examples of US and SWE images can be found in the first two columns of Fig. 2.

3.2   Implementation Details

We employ U-Net [20] for modal translater and semantic inverter, and Patch-GAN [15] for discriminators. We adopt ResNet18 as backbone for all classification networks. For training, conventional data augmentation (translation, flipping, scaling and shearing) is applied, then the images are resized to 256×256 pixels before fed into models. All generators and classification networks are trained with batchsize 4 and Adam [14] optimizer. For training, we keep the same learning rate of 2e-4 for the first 100 epochs, and linearly decay it to 0 over the next 100 epochs. When fine-tuning, an initial learning rate of 2e-5 is adopted and linearly decays to 0 over the next 20 epochs. For the weighting parameters, we use $\lambda_{SC} = 0.2$, $\lambda_{dist} = 0.01$, $\lambda_{cycle}^S = 3$, $\lambda_{cls}^S = 0.5$. The settings of L_{P2p} and L_{GAN} follows those in [11,29]. *Code will be released soon.*

3.3   Experimental Results

Evaluation of Synthesized Images: Examples of synthesized US and SWE images through modal translater G_M and semantic inverter G_S compared to real images are shown in Fig. 2. We can see that the generated SWE images $\hat{I}_s$ roughly capture the lesions' structure. We employ three metrics to measure the

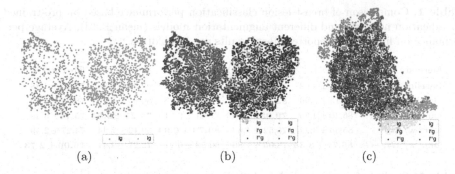

(a)   (b)   (c)

Fig. 3. t-SNE visualization of US images in training sets. Benign images are shown in blue, and malignant images are shown in red. Converted images are shown in darker colors. (a) Original training set; (b) Training set expanded by G_S; (c) Training set expanded by G_S without semantic supervision. (Color figure online)

quality of $\hat{I}_s$ generated by G_M and Pix2pix, i.e., G_M without semantic guidance, including (1) mean absolute error (MAE), (2) peak signal-to-noise ratio (PSNR), and (3) structural similarity (SSIM) [26]. As shown in Table 2, G_M yields better results than Pix2pix, suggesting that the semantic guidance helps synthesize images with better visual quality.

Evaluation of Semantic Inversion: To ensure the virtual US images synthesized by semantic inverter G_S have meaningful features for the classification task, we use t-SNE [18] dimension reduction algorithm for visualization. As shown in Fig. 3 (a)–(c), we employ the pre-trained classification network C_U as the feature extractor for US images on the original and two variant training sets, namely the version expanded by G_S, and that expanded by G_S without semantic supervision. We can see that G_S effectively increases the semantic diversity of training samples while not disturbing the original distribution too much.

Evaluation of Breast Lesion Classification: We employ five metrics for performance evaluation of the baseline classification model C_U, oracle model C_D, and the final model C_S, including AUC, accuracy, recall, precision, and F1-score. The average performance on five folds is reported in Table 1. For each fold of the five-fold cross-validation, four subsets are used for training and one subset for testing; no validation sets are used. As can be seen, either G_M or G_S leads to improvements in AUC, and the joint utilization of G_M and G_S further enhances the performance, proving the effectiveness of our knowledge-guided data augmentation framework.

Table 1. Comparison of breast lesion classification performance based on pre-trained classification networks and different augmentation models (mean ± std). Average performance on five folds is reported. Best results are in boldface.

Method	AUC(%)	Accuracy(%)	Recall(%)	Precision(%)	F1-score(%)
Baseline	84.36 ± 3.62	76.29 ± 2.30	78.70 ± 9.44	69.29 ± 0.98	73.43 ± 4.38
Oracle	89.13 ± 2.99	82.74 ± 2.69	77.58 ± 7.75	80.86 ± 2.46	78.96 ± 4.09
Ours w/ G_M	86.26 ± 4.65	**79.03 ± 3.52**	77.41 ± 9.91	**74.21 ± 3.81**	75.44 ± 5.46
Ours w/ G_S	85.50 ± 3.22	76.52 ± 1.69	**80.74 ± 6.94**	69.12 ± 2.44	74.27 ± 2.59
Ours w/ G_M+G_S	**86.71 ± 3.35**	78.65 ± 1.86	80.53 ± 6.21	72.27 ± 2.41	**76.00 ± 2.73**

Table 2. Quality of synthesized virtual SWE images and classification performance of Pix2pix and modal translator G_M (mean ± std). Average performance on five folds is reported. Best results are in boldface.

Method	MAE	PSNR	SSIM	AUC(%)	Recall(%)
Pix2pix	16.52 ± 7.52	18.89 ± 2.76	0.5977 ± 0.1200	84.66 ± 2.93	**77.88 ± 8.84**
G_M	**15.75 ± 7.55**	**19.29 ± 2.89**	**0.6073 ± 0.1212**	**86.26 ± 4.65**	77.41 ± 9.91

3.4   Ablation Study

Modal Translater: We investigate the effectiveness of the semantic guidance for G_M on improving classification performance. The results are shown in Table 2. As can be seen, G_M outperforms Pix2pix in terms of both image quality and classification performance, indicating that semantic guidance for G_M plays a vital role in improving the performance of virtual SWE synthesis and breast lesion classification.

Semantic Inverter: To investigate the importance of semantic supervision for training G_S, we provide an ablation study on $\mathcal{L}_{cls}^S$ and explore the choices of which virtual US images to be added into the training set. As shown in Table 3, for G_S with $\mathcal{L}_{cls}^S$, adding either I_g' or I_g'' into training set leads to improved AUC, while joint utilization of I_g' and I_g'' further improves the AUC. For G_S without $\mathcal{L}_{cls}^S$, adding I_g' degrades the AUC, while adding I_g'' achieves similar improvement to G_S with $\mathcal{L}_{cls}^S$. Two observations can be found in Table 3. *First*, $\mathcal{L}_{cls}^S$ is necessary for G_S to perform semantic conversion. Otherwise, G_S will produce virtual US images with high realism but low feature quality for classification. *Second*, cycle-consistency loss provides a regularization effect for G_S with and without $\mathcal{L}_{cls}^S$, which makes I_g'' a similar variant of I_g and suitable for data augmentation.

Table 3. Comparison of different augmentation choices and the importance of semantic supervision for G_S (mean $\pm$ std). Average performance on five folds is reported. Best results are in boldface.

I_g	I_g'	I_g''	$\mathcal{L}_{cls}^S$	AUC(%)	Recall(%)
✓				84.36 ± 3.62	78.70 ± 9.44
✓	✓			84.02 ± 4.19	76.94 ± 10.97
✓		✓		85.03 ± 4.11	79.72 ± 7.90
✓	✓	✓		84.62 ± 4.21	78.43 ± 9.87
✓	✓		✓	85.27 ± 3.34	80.73 ± 6.55
✓		✓	✓	85.04 ± 3.26	80.13 ± 7.78
✓	✓	✓	✓	$\mathbf{85.50 \pm 3.22}$	$\mathbf{80.74 \pm 6.94}$

4   Conclusion

In this paper, we propose a knowledge-guided data augmentation framework consisting of a modal translater and a semantic inverter for joint breast US-to-SWE cross-modal augmentation and US-to-US semantic augmentation. Experiments show that our knowledge-guided framework is able to synthesize US and SWE images with better image quality and feature quality than the conventional image-to-image translation GANs. A baseline ResNet18 based classification network trained with the expanded dataset exhibits significantly improved performance based on conventional data augmentation methods and outperforms conventional GANs without knowledge guidance. We believe this framework can be extended to other imaging modalities to help alleviate data scarcity and modality imbalance problems.

Acknowledgment. This work was supported in part by Shanghai Municipal Science and Technology Major Project (2021SHZDZX0102), 111 project (BP0719010), Shanghai Science and Technology Committee (18DZ2270700) and Shanghai Jiao Tong University Science and Technology Innovation Special Fund (ZH2018ZDA17).

References

1. Barr, R., et al.: Wfumb guidelines and recommendations for clinical use of ultrasound elastography: Part 2: breast. Ultrasound Med. Biol. **41**(5), 1148–60 (2015)
2. Berg, W.A., et al.: Shear-wave elastography improves the specificity of breast us: the bel multinational study of 939 masses. Radiology **262**(2), 435–449 (2012)
3. Chen, X., Li, Y., Yao, L., Adeli, E., Zhang, Y.: Generative adversarial u-net for domain-free medical image augmentation. ArXiv abs/2101.04793 (2021)
4. Cheng, H., Shan, J., Ju, W., Guo, Y., Zhang, L.: Automated breast cancer detection and classification using ultrasound images: a survey. Pattern Recogn. **43**, 299–317 (2010)
5. Fujioka, T., et al.: Classification of breast masses on ultrasound shear wave elastography using convolutional neural networks. Ultrasonic Imaging **42**(4–5), 213–220 (2020)

6. Goodfellow, I., et al.: Generative adversarial nets. In: Advances in Neural Information Processing Systems, pp. 2672–2680 (2014)
7. Gupta, L., Klinkhammer, B.M., Boor, P., Merhof, D., Gadermayr, M.: GAN-based image enrichment in digital pathology boosts segmentation accuracy. In: Shen, D., et al. (eds.) MICCAI 2019. LNCS, vol. 11764, pp. 631–639. Springer, Cham (2019). https://doi.org/10.1007/978-3-030-32239-7_70
8. Han, X., Wang, J., Zhou, W., Chang, C., Ying, S., Shi, J.: Deep doubly supervised transfer network for diagnosis of breast cancer with imbalanced ultrasound imaging modalities. In: Martel, A.L., et al. (eds.) MICCAI 2020. LNCS, vol. 12266, pp. 141–149. Springer, Cham (2020). https://doi.org/10.1007/978-3-030-59725-2_14
9. He, K., Zhang, X., Ren, S., Sun, J.: Deep residual learning for image recognition. In: Computer Vision and Pattern Recognition, pp. 770–778 (2016)
10. Hinton, G.E., Vinyals, O., Dean, J.: Distilling the knowledge in a neural network. ArXiv abs/1503.02531 (2015)
11. Isola, P., Zhu, J., Zhou, T., Efros, A.A.: Image-to-image translation with conditional adversarial networks. In: Computer Vision and Pattern Recognition, pp. 5967–5976 (2017)
12. Jian, W., et al.: Auto-weighting for breast cancer classification in multimodal ultrasound. arXiv: Image and Video Processing (2020)
13. Khosla, P., et al.: Supervised contrastive learning. ArXiv abs/2004.11362 (2020)
14. Kingma, D.P., Ba, J.: Adam: a method for stochastic optimization. CoRR abs/1412.6980 (2015)
15. Li, C., Wand, M.: Precomputed real-time texture synthesis with markovian generative adversarial networks. In: Leibe, B., Matas, J., Sebe, N., Welling, M. (eds.) ECCV 2016. LNCS, vol. 9907, pp. 702–716. Springer, Cham (2016). https://doi.org/10.1007/978-3-319-46487-9_43
16. Liu, J., Shen, C., Liu, T., Aguilera, N., Tam, J.: Active appearance model induced generative adversarial network for controlled data augmentation. In: Shen, D., et al. (eds.) MICCAI 2019. LNCS, vol. 11764, pp. 201–208. Springer, Cham (2019). https://doi.org/10.1007/978-3-030-32239-7_23
17. Liu, L., Muelly, M., Deng, J., Pfister, T., Li, L.J.: Generative modeling for small-data object detection. In: ICCV, pp. 6072–6080 (2019)
18. Maaten, L.V.D., Hinton, G.E.: Visualizing data using t-sne. J. Mach. Learn. Res. **9**, 2579–2605 (2008)
19. Pan, Y., Liu, M., Lian, C., Xia, Y., Shen, D.: Disease-image specific generative adversarial network for brain disease diagnosis with incomplete multi-modal neuroimages. In: Shen, D., et al. (eds.) MICCAI 2019. LNCS, vol. 11766, pp. 137–145. Springer, Cham (2019). https://doi.org/10.1007/978-3-030-32248-9_16
20. Ronneberger, O., Fischer, P., Brox, T.: U-Net: convolutional networks for biomedical image segmentation. In: Navab, N., Hornegger, J., Wells, W.M., Frangi, A.F. (eds.) MICCAI 2015. LNCS, vol. 9351, pp. 234–241. Springer, Cham (2015). https://doi.org/10.1007/978-3-319-24574-4_28
21. Schmidt, T., et al.: Diagnostic accuracy of phase-inversion tissue harmonic imaging versus fundamental b-mode sonography in the evaluation of focal lesions of the kidney. Am. J. Roentgenol **180**(6), 1639–1647 (2003)
22. Sehgal, C.M., Weinstein, S.P., Arger, P.H., Conant, E.F.: A review of breast ultrasound. J. Mammary Gland Biol. Neoplasia **11**(2), 113–123 (2006)
23. Sigrist, R.M., Liau, J., El Kaffas, A., Chammas, M.C., Willmann, J.K.: Ultrasound elastography: review of techniques and clinical applications. Theranostics **7**(5), 1303 (2017)

24. Sung, H., et al.: Global cancer statistics 2020: globocan estimates of incidence and mortality worldwide for 36 cancers in 185 countries. CA Cancer J. Clin. **71**, 209–249 (2021)
25. Tang, Y., Tang, Y., Sandfort, V., Xiao, J., Summers, R.: Tuna-net: task-oriented unsupervised adversarial network for disease recognition in cross-domain chest x-rays. ArXiv abs/1908.07926 (2019)
26. Wang, Z., Bovik, A., Sheikh, H.R., Simoncelli, E.P.: Image quality assessment: from error visibility to structural similarity. IEEE Trans. Image Process. **13**, 600–612 (2004)
27. Zhang, Q., et al.: Deep learning based classification of breast tumors with shear-wave elastography. Ultrasonics **72**, 150–157 (2016)
28. Zhou, Y., et al.: A radiomics approach with cnn for shear-wave elastography breast tumor classification. IEEE Trans. Biomed. Eng. **65**(9), 1935–1942 (2018)
29. Zhu, J.Y., Park, T., Isola, P., Efros, A.A.: Unpaired image-to-image translation using cycle-consistent adversarial networks. In: ICCV, pp. 2242–2251 (2017)

Multiple Meta-model Quantifying for Medical Visual Question Answering

Tuong Do[1]($\boxtimes$), Binh X. Nguyen[1], Erman Tjiputra[1], Minh Tran[1],
Quang D. Tran[1], and Anh Nguyen[2]

[1] AIOZ, Singapore, Singapore
{tuong.khanh-long.do,binh.xuan.nguyen,erman.tjiputra,
minh.quang.tran,quang.tran}@aioz.io
[2] University of Liverpool, Liverpool, UK
anh.nguyen@liverpool.ac.uk

Abstract. Transfer learning is an important step to extract meaningful features and overcome the data limitation in the medical Visual Question Answering (VQA) task. However, most of the existing medical VQA methods rely on external data for transfer learning, while the meta-data within the dataset is not fully utilized. In this paper, we present a new multiple meta-model quantifying method that effectively learns meta-annotation and leverages meaningful features to the medical VQA task. Our proposed method is designed to increase meta-data by auto-annotation, deal with noisy labels, and output meta-models which provide robust features for medical VQA tasks. Extensively experimental results on two public medical VQA datasets show that our approach achieves superior accuracy in comparison with other state-of-the-art methods, while does not require external data to train meta-models. Source code available at: https://github.com/aioz-ai/MICCAI21_MMQ.

Keywords: Visual Question Answering · Meta learning

1 Introduction

A medical Visual Question Answering (VQA) system can provide meaningful references for both doctors and patients during the treatment process. Extracting image features is one of the most important steps in a medical VQA framework which outputs essential information to predict answers. Transfer learning, in which the pretrained deep learning models [9,12,13,24,36] that are trained on the large scale labeled dataset such as ImageNet [32], is a popular way to initialize the feature extraction process. However, due to the difference in visual concepts between ImageNet images and medical images, finetuning process is not sufficient

Electronic supplementary material The online version of this chapter (https://doi.org/10.1007/978-3-030-87240-3_7) contains supplementary material, which is available to authorized users.

[26]. Recently, Model Agnostic Meta-Learning [6] (MAML) has been introduced to overcome the aforementioned problem by learning meta-weights that quickly adapt to visual concepts. However, MAML is heavily impacted by the meta-annotation phase for all images in the medical dataset [26]. Different from normal images, transfer learning in medical images is more challenging due to: *(i)* noisy labels may occur when labeling images in an unsupervised manner; *(ii)* high-level semantic labels cause uncertainty during learning; and *(iii)* difficulty in scaling up the process to all unlabeled images in medical datasets.

In this paper, we introduce a new Multiple Meta-model Quantifying (MMQ) process to address these aforementioned problems in MAML. Intuitively MMQ is designed to: *(i)* effectively increase meta-data by auto-annotation; *(ii)* deal with the noisy labels in the training phase by leveraging the uncertainty of predicted scores during the meta-agnostic process; and *(iii)* output meta-models which contain robust features for down-stream medical VQA task. Note that, compared with the recent approach for meta-learning in medical VQA [26], our proposed MMQ does not take advantage of additional out-of-dataset images, while achieves superior accuracy in two challenging medical VQA datasets.

2   Literature Review

Medical Visual Question Answering. Based on the development of VQA in general images, the medical VQA task inherits similar techniques and achieves certain achievements [1, 2, 18, 19, 28, 45]. Specifically, the attention mechanisms such as MCB [7], SAN [43], BAN [15], or CTI [5] are applied in [1, 26, 28, 41, 45] to learn joint representation between medical visual information and questions. Additionally, in [17, 18, 28, 45], the authors take advantage of transfer learning for extracting medical image features. Recently, approaches which directly solve different aspects of medical VQA are introduced, including reasoning [17, 44], diagnose model behavior [40], multi-modal fusion [35], dedicated framework design [8, 20], and generative model for dealing with abnormality questions [31].

Meta-learning. Traditional machine learning algorithms, specifically deep learning-based approaches, require a large-scale labeled training set [3, 4, 21, 23, 25]. Therefore, meta-learning [11, 14, 34, 42], which targets to deal with the problem of data limitation when learning new tasks, is applied broadly. There are three common approaches to meta-learning, namely model-based [22, 33], metric-based [16, 37–39], and optimization-based [6, 27, 30]. A notable optimization-based work, MAML [6], helps to learn a meta-model then quickly adapt it to other tasks. The authors in [26] used MAML to overcome the data limitation problem in medical VQA. However, their work required the use of external data during the training.

3   Methodology

3.1   Method Overview

Our approach comprises two parts: our proposed multiple meta-model quantifying (MMQ - Fig. 1) and a VQA framework for integrating meta-models outputted from MMQ (Fig. 2). MMQ addresses the meta-annotation problem by outputting multiple meta-models. These models are expected to robust to each other and have high accuracy during the inference phase of model-agnostic tasks. The VQA framework aims to leverage different features extracted from candidate meta-models and then generates predicted answers.

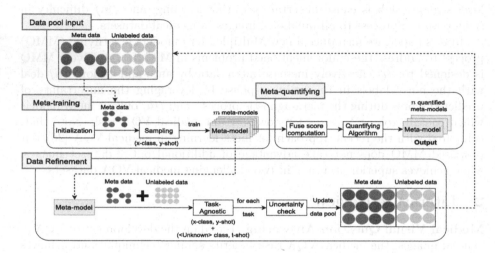

Fig. 1. Multiple meta-model quantifying in medical VQA. Dotted lines denote looping steps, the number of loop equals to m required meta-models.

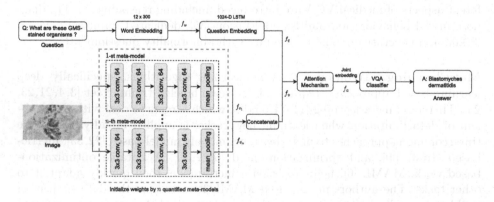

Fig. 2. Our VQA framework is designed to integrate robust image features extracted from multiple meta-models outputted from MMQ.

3.2 Multiple Meta-model Quantifying

Multiple meta-model quantifying (Fig. 1) contains three modules: *(i)* **Meta-training** which trains a specific meta-model for extracting image features used in medical VQA task by following MAML [6]; *(ii)* **Data refinement** which increases the training data by auto-annotation and deal with the noisy label by leveraging the uncertainty of predicted scores; and *(iii)* **Meta-quantifying** which selects meta-models whose robust to each others and have high accuracy during inference phase of model-agnostic tasks.

Algorithm 1: Model-Agnostic for data refinement

Input: $\rho(\mathcal{T})$ distribution over tasks; data pool $\mathcal{D}$; meta-model weights θ
Output: Updated data pool $\mathcal{D}'$

1   Sample batch of tasks $\mathcal{T}_i \sim \rho(\mathcal{T})$

2   Establish list A with contains list of (Score S, Label L) of each sample in data pool D. (S, L) is from the predicted process of Classifier $\mathcal{C}$ of each task $\mathcal{T}_i$.

3   Set α and β be uncertainty checking threshold.

4   **For all task $\mathcal{T}_i$ do**

5   **For all image I_k in $\mathcal{T}_i$ batch do**

6       $(S_k^i, L_k^i) \leftarrow \mathcal{C}_i(\mathcal{T}_i, \theta, I_k)$. Where $\mathcal{C}_i$ is the i-th classifier of task $\mathcal{T}_i$.

7   Append (S_k^i, L_k^i) into $A[I_k]$.

8   Establish new version of Meta data split $\mathcal{M}'$ and new version of Unlabeled data split $\mathcal{U}'$ of $\mathcal{D}$

9   **For all element $A[I_j]$ in list A do**

10   **If $A[I_j]$ in Meta data split $\mathcal{M}$ of $\mathcal{D}$**

11   **If $\exists A[I_j]\{S\} < \alpha$ and $A[I_j]\{L\}$ is $A[I_j]\{GT_j\}$.** Where GT_j is the ground-truth label of I_j.

12   Append $(I_j, A[I_j]\{L\})$ into $\mathcal{U}'$

13   Remove $(I_j, A[I_j]\{L\})$ from $\mathcal{M}$

14   **If $A[I_j]$ in Unlabeled data split $\mathcal{U}$ of $\mathcal{D}$**

15   **If $\exists A[I_j]\{S\} > \beta$**

16   Append $argmax_{A[I_j]\{S\}}(I_j, A[I_j]\{L\})$ into $\mathcal{M}'$

17   $\mathcal{U}_f = \mathcal{U} - \mathcal{M}' + \mathcal{U}'$. Where $\mathcal{U}_f$ is the updated Unlabeled data split of $\mathcal{D}'$.

18   $\mathcal{M}_f = \mathcal{M} - \mathcal{U}' + \mathcal{M}'$. Where $\mathcal{M}_f$ is the updated Meta data split of $\mathcal{D}'$.

19   **return** $\mathcal{M}_f, \mathcal{U}_f$ of $\mathcal{D}'$

Meta-training. We generally follow MAML [6] to do meta-training. Let f_θ be the classification meta-model. Hence, θ represents the parameters of f_θ while $\{\theta'_0, \theta'_1, ... \theta'_x\}$ is the adapting parameters list of classification models for x given tasks $\mathcal{T}_i$ and their associated dataset $\{\mathcal{D}_i^{tr}, \mathcal{D}_i^{val}\}$. Specifically, for each iteration, x tasks are sampled with y examples of each task. Then we calculate the gradient descent $\nabla_\theta L_{\mathcal{T}_i}(f_\theta(\mathcal{D}_i^{tr}))$ of the classification loss $L_{\mathcal{T}_i}$ and update the corresponding adapting parameters as follow.

$$\theta'_i = \theta - \alpha \nabla_\theta L_{\mathcal{T}_i}(f_\theta(\mathcal{D}_i^{tr})) \tag{1}$$

At the end of each iteration, the meta-model parameters θ are updated throughout validation sets of all tasks sampled to learn the generalized features as:

$$\theta \leftarrow \theta - \beta \nabla_\theta \sum_{T_i} L_{T_i}(f_{\theta_i'}(\mathcal{D}_i^{val})) \tag{2}$$

Unlike MAML [6] where only one meta-model is selected, we develop the following refinement and meta-quantifying steps to select high-quality meta-models for transfer learning to the medical VQA framework later.

Data Refinement. After finishing the meta-training phase, the weights of the meta-models are used for refining the dataset. The module aims to expand the meta-data pool for meta-training and removes samples that are expected to be hard-to-learn or have noisy labels (See Algorithm 1 for more details).

Meta-quantifying. This module aims to identify candidate meta-models that are useful for the medical VQA task. A candidate model θ should achieve high performance during the validating process and its features distinct from other features from other candidate models.

To achieve these goals, we design a fuse score S_F as described in (3).

$$S_F = \gamma S_P + (1 - \gamma) \sum_{t=1}^{m} 1 - Cosine\,(F_c, F_t)\,\forall F_c \neq F_t \tag{3}$$

where S_P is the predicted score of the current meta-model over ground-truth label; F_c is the feature extracted from the aforementioned meta-model that needs to compute the score; F_t is the feature extracted from t-th model of the list of meta-model Θ; Cosine is using for similarity checking between two features.

Since the predicted score S_P at the ground-truth label and diverse score are co-variables, therefore the fuse score S_F is also covariate with both aforementioned scores. This means that the larger S_F is, the higher chance of the model to be selected for the VQA task. Algorithm 2 describes our meta-quantifying algorithm in details.

3.3   Integrate Quantified Meta-models to Medical VQA Framework

To leverage robust features extracted from quantified meta-models, we introduce a VQA framework as in Fig. 2. Specifically, each input question is trimmed to a 12-word sentence and then zero-padded if its length is less than 12. Each word is represented by a 300-D GloVe word embedding [29]. The word embedding is fed into a 1024-D LSTM to produce the question embedding f_q.

Each input image is passed through n quantified meta-models got from the meta-quantifying module, which produce n vectors. These vectors are concatenated to form an enhanced image feature, denoted as f_v in Fig. 2. Since this vector contains multiple features extracted from different high-performed meta-models and each model has different views, the VQA framework is expected to

---

Algorithm 2: Meta-quantifying algorithm

Input: Data pool $\mathcal{D}_T$; list of meta-model $\Theta \in [\theta_0, \theta_1, ..., \theta_m]$ where m denotes the number of candidate meta-models; number of quantified model n.

Output: List of Quantified meta-models $\Theta_n \in [\theta_0, \theta_1, ..., \theta_n]$. $n < m$.

1 For all n meta-models, sample batch of tasks $\mathcal{T}_i \sim \rho(\mathcal{T})$
2 Establish list A with contains list of (Score S_P, Feature F) of each sample in quantify data pool $\mathcal{D}_T$. (S_P, F) is got from the predicted process of Classifier $\mathcal{C}$ of each task $\mathcal{T}_i$. S_P is the predicted score at ground-truth label.
3 Set γ be effectiveness - robustness balancing hyper-parameter.
4 Establish Fuse Score list $\mathcal{L}_{S_F}$ for all meta-model in Θ.
5 **For all** task $\mathcal{T}_i$ **do**
6   **For all** image I_k in $\mathcal{T}_i$ batch **do**
7   **For all** meta-model Θ_t in Θ **do**
8       $(S_k^i, F_k^i)^{\Theta_t} \leftarrow \mathcal{C}_i(\mathcal{T}_i, \theta, I_k)$. Where $\mathcal{C}_i$ is the i-th classifier of task $\mathcal{T}_i$.
9   Append $(S_k^i, F_k^i)^{\Theta_t}$ into $A[I_k]$.
10   **For all** meta-model Θ_t in Θ **do**
11   **For all** $A[I_k]$ **do**
12       $S_F^{\Theta_t} \leftarrow$ Compute fuse score using Equation (3).
13       $\mathcal{L}_{S_F}^{\Theta_t} += S_F^{\Theta_t}$.
14 $\mathcal{L}_{S_F} \leftarrow$ Sort $\mathcal{L}_{S_F}$ decreasingly along with corresponding θ.
15 **return** $\Theta_n \leftarrow n$-first meta-models selected from $\mathcal{L}_{S_F}$.

---

be less affected by the bias problem. Image feature f_v and question embedding f_q are fed into an attention mechanism (BAN [15] or SAN [43]) to produce a joint representation f_a. This feature f_a is used as input for a multi-class classifier (over the set of predefined answer classes [18]). To train the proposed model, we use a Cross Entropy loss for the answer classification task. The whole VQA framework is then fine-tuned in an end-to-end manner.

4 Experiments

4.1 Dataset

We use the VQA-RAD [18] and PathVQA [10] in our experiments. The VQA-RAD [18] dataset contains 315 images and 3,515 corresponding questions. Each image is associated with more than one question. The PathVQA [10] dataset consists of 32,799 question-answer pairs generated from 1,670 pathology images collected from two pathology textbooks, and 3,328 pathology images collected from the PEIR digital library.

4.2 Experimental Details

Meta-training. Similar to [26], we first create the meta-annotation for training MAML. For the VQA-RAD dataset, we re-use the meta-annotation created by [26]. Note that we do not use their extra collected data in our experiment.

For the PathVQA dataset, we create the meta-annotation by categorizing all training images into 31 classes based on body parts, types of images, and organs.

For every iteration of MAML training, 5 tasks are sampled per iteration in RAD-VQA while in PathVQA, this value is 4 instead. For each task, in RAD-VQA, we randomly select 3 classes from 9 classes while in PathVQA we select 5 classes from 31 aforementioned classes. For each class, in RAD-VQA, we randomly select 6 images in which 3 images are used for updating task models and the remaining 3 images are used for updating meta-model. In PathVQA, the same process is applied with 20 random selected images, 5 of them are used for updating task models and the remains are used for updating meta-model.

Data Refinement. The meta-model outputted from the meta-training step is then used for updating the data pool through the algorithm described in Sect. 3.2. The refined data pool is then leveraged as the input for the meta-training step to output another meta-model. This loop is applied by a maximum of 7 times to output up to 7 different meta-models.

Meta-quantifying. All meta-models got from the previous step are passed through the Algorithm 2 to quantify their effectiveness. A maximum of 4 models which have high performance is applied to VQA training.

VQA Training. After selecting candidate meta-models from the meta quantifying module, we use their trained weights to initialize the image feature extraction component in the VQA framework. We then finetune the whole VQA model using the VQA training set. The output vector of each meta-model is set to 32-D in PathVQA and 64-D in VQA-RAD dataset. We use 50% of meta-annotated images for training meta-models. The effect of meta-annotated images can be found in our supplementary material.

Baselines. We compare our MMQ results with recent methods in medical VQA: MAML [6], MEVF [26], stacked attention network (StAN) with VGG-16 [10], and bilinear attention network (BiAN) with Faster-RCNN [10]. Two attentions methods SAN [43] and BAN [15] are used in MAML, MEVF, and MMQ. Note that, StAN and BiAN only use pretrained models from the ImageNet dataset, MEVF [26] uses extra collected data to train their meta-model, while our MMQ relies solely on the images from the dataset. For the question feature extraction, all baselines and our method use the same pretrained models (i.e., Glove [29]) and then finetuning on VQA-RAD or PathVQA dataset.

4.3   Results

Table 1 presents comparative results between different methods. The results show that our MMQ significantly outperforms other meta-learning methods by a large margin. Besides, the gain in performance of MMQ is stable with different attention mechanisms (BAN [15] or SAN [43]) in the VQA task. It worth noting that, compared with the most recent state-of-the-art method MEVF [26], we outperform 5.3% in free-form questions of the PathVQA dataset and 9.8% in the Open-ended questions of the VQA-RAD dataset, respectively. Moreover, no out-of-dataset images are used in MMQ for learning meta-models. The results imply that our proposed MMQ learns essential representative information from the input images and leverage effectively the features from meta-models to deal with challenging questions in medical VQA datasets.

Table 1. Performance comparison on VQA-RAD and Path-VQA test set. (*) indicates methods used pre-trained model on ImageNet dataset. We refine data 5 times ($m = 5$) and use 3 meta-models ($n = 3$) in our MMQ.

Reference methods	Attention method	PathVQA			VQA-RAD		
		Free-form	Yes/No	Over-all	Open-ended	Close-ended	Over-all
StAN [10] (*)	SAN	1.6	59.4	30.5	24.2	57.2	44.2
BiAN [10] (*)	BAN	2.9	68.2	35.6	28.4	67.9	52.3
MAML [6]	SAN	5.4	75.3	40.5	38.2	69.7	57.1
	BAN	5.9	79.5	42.9	40.1	72.4	59.6
MEVF [26]	SAN	6.0	81.0	43.6	40.7	74.1	60.7
	BAN	8.1	81.4	44.8	43.9	75.1	62.7
MMQ (ours)	SAN	11.2	82.7	47.1	46.3	75.7	64.0
	BAN	**13.4**	**84.0**	**48.8**	**53.7**	**75.8**	**67.0**

4.4 Ablation Study

Table 2 presents our MMQ accuracy in PathVQA dataset when applying m times refining data and n quantified meta-models. The results show that, by using only 1 quantified meta-model outputted from our MMQ, we significantly outperform both MAML and MEVF baselines. This confirms the effectiveness of the proposed MMQ for dealing with the limitation of meta-annotation in medical VQA, i.e., noisy labels and scalability. Besides, leveraging more quantified meta-models also further improves the overall performance.

We note that the improvements of our MMQ are more significant on free-form questions over yes/no questions. This observation implies that the free-form questions/answers which are more challenging and need more information from input images benefits more from our proposed method.

Table 2 also shows that increasing the number of refinement steps and the number of quantified meta-models can improve the overall result, but the gain is smaller after each loop. The training time also increases when the number of meta-models is set higher. However, our testing time and the total number of parameters are only slightly higher than MAML [6] and MEVF [26]. Based on the empirical results, we recommend applying 5 times refinement with a maximum of 3 quantified meta-models to balance the trade-off between the accuracy performance and the computational cost.

Table 2. The effectiveness of our MMQ under m times refining data and n quantified meta-models on PathVQA test set. BAN is used as the attention method.

Methods	m	n	Free-form	Yes/No	Over-all	Train time (hours)	Test time (s/sample)	#Paras (M)
MAML [6]	-	-	5.9	79.5	42.9	2.1	0.007	27.2
MEVF [26]	-	-	8.1	81.4	44.8	2.5	0.008	27.9
MMQ (ours)	3	1	10.1	82.1	46.2	5.8	0.008	27.8
	4	2	12.0	83.0	47.6	7.3	0.009	28.1
	5	3	13.4	84.0	48.8	8.9	0.010	28.3
	7	4	13.6	84.0	48.8	12.1	0.011	28.5

5 Conclusion

In this paper, we proposed a new multiple meta-model quantifying method to effectively leverage meta-annotation and deal with noisy labels in the medical VQA task. The extensively experimental results show that our proposed method outperforms the recent state-of-the-art meta-learning based methods by a large margin in both PathVQA and VQA-RAD datasets. Our implementation and trained models will be released for reproducibility.

References

1. Abacha, A.B., Gayen, S., Lau, J.J., Rajaraman, S., Demner-Fushman, D.: NLM at ImageCLEF 2018 visual question answering in the medical domain. In: CEUR Workshop Proceedings (2018)
2. Abacha, A.B., Hasan, S.A., Datla, V.V., Liu, J., Demner-Fushman, D., Müller, H.: VQA-Med: overview of the medical visual question answering task at ImageCLEF 2019. In: CLEF (Working Notes) (2019)
3. Bar, Y., Diamant, I., Wolf, L., Greenspan, H.: Deep learning with non-medical training used for chest pathology identification. In: Medical Imaging: Computer-Aided Diagnosis (2015)
4. Chi, W., et al.: Collaborative robot-assisted endovascular catheterization with generative adversarial imitation learning. In: ICRA (2020)
5. Do, T., Do, T.T., Tran, H., Tjiputra, E., Tran, Q.D.: Compact trilinear interaction for visual question answering. In: ICCV (2019)
6. Finn, C., Abbeel, P., Levine, S.: Model-agnostic meta-learning for fast adaptation of deep networks. In: ICML (2017)
7. Fukui, A., Park, D.H., Yang, D., Rohrbach, A., Darrell, T., Rohrbach, M.: Multimodal compact bilinear pooling for visual question answering and visual grounding. In: EMNLP (2016)
8. Gupta, D., Suman, S., Ekbal, A.: Hierarchical deep multi-modal network for medical visual question answering. Expert Syst. Appl. (2021)
9. He, K., Zhang, X., Ren, S., Sun, J.: Deep residual learning for image recognition. In: CVPR (2016)
10. He, X., Zhang, Y., Mou, L., Xing, E., Xie, P.: PathVQA: 30000+ questions for medical visual question answering. arXiv preprint arXiv:2003.10286 (2020)
11. Hsu, K., Levine, S., Finn, C.: Unsupervised learning via meta-learning. In: ICLR (2019)

12. Huang, B., et al.: Tracking and visualization of the sensing area for a tethered laparoscopic gamma probe. Int. J. Comput. Assist. Radiol. Surg. **15**(8), 1389–1397 (2020). https://doi.org/10.1007/s11548-020-02205-z
13. Huang, B., et al.: Self-supervised generative adversarial network for depth estimation in laparoscopic images. In: MICCAI (2021)
14. Khodadadeh, S., Bölöni, L., Shah, M.: Unsupervised meta-learning for few-shot image classification. In: NIPS (2019)
15. Kim, J.H., Jun, J., Zhang, B.T.: Bilinear attention networks. In: NIPS (2018)
16. Koch, G., Zemel, R., Salakhutdinov, R.: Siamese neural networks for one-shot image recognition. In: ICML Deep Learning Workshop (2015)
17. Kornuta, T., Rajan, D., Shivade, C., Asseman, A., Ozcan, A.S.: Leveraging medical visual question answering with supporting facts. arXiv:1905.12008 (2019)
18. Lau, J.J., Gayen, S., Abacha, A.B., Demner-Fushman, D.: A dataset of clinically generated visual questions and answers about radiology images. Nature (2018)
19. Liu, S., Ding, H., Zhou, X.: Shengyan at VQA-Med 2020: an encoder-decoder model for medical domain visual question answering task. CLEF (2020)
20. Lubna, A., Kalady, S., Lijiya, A.: MoBVQA: a modality based medical image visual question answering system. In: TENCON (2019)
21. Maicas, G., Bradley, A.P., Nascimento, J.C., Reid, I., Carneiro, G.: Training medical image analysis systems like radiologists. In: Frangi, A.F., Schnabel, J.A., Davatzikos, C., Alberola-López, C., Fichtinger, G. (eds.) MICCAI 2018. LNCS, vol. 11070, pp. 546–554. Springer, Cham (2018). https://doi.org/10.1007/978-3-030-00928-1_62
22. Munkhdalai, T., Yu, H.: Meta networks. In: ICML (2017)
23. Nguyen, A.: Scene understanding for autonomous manipulation with deep learning. arXiv preprint arXiv:1903.09761 (2019)
24. Nguyen, A., et al.: End-to-end real-time catheter segmentation with optical flow-guided warping during endovascular intervention. In: ICRA (2020)
25. Nguyen, A., Nguyen, N., Tran, K., Tjiputra, E., Tran, Q.: Autonomous navigation in complex environments with deep multimodal fusion network. In: IROS (2020)
26. Nguyen, B.D., Do, T.-T., Nguyen, B.X., Do, T., Tjiputra, E., Tran, Q.D.: Overcoming data limitation in medical visual question answering. In: Shen, D., et al. (eds.) MICCAI 2019. LNCS, vol. 11767, pp. 522–530. Springer, Cham (2019). https://doi.org/10.1007/978-3-030-32251-9_57
27. Nichol, A., Achiam, J., Schulman, J.: On first-order meta-learning algorithms. arXiv preprint arXiv:1803.02999 (2018)
28. Peng, Y., Liu, F., Rosen, M.P.: UMass at ImageCLEF medical visual question answering (Med-VQA) 2018 task. In: CEUR Workshop Proceedings (2018)
29. Pennington, J., Socher, R., Manning, C.D.: GloVe: global vectors for word representation. In: EMNLP (2014)
30. Ravi, S., Larochelle, H.: Optimization as a model for few-shot learning. In: ICLR (2017)
31. Ren, F., Zhou, Y.: CGMVQA: a new classification and generative model for medical visual question answering. IEEE Access **8**, 50626–50636 (2020)
32. Russakovsky, O., et al.: ImageNet large scale visual recognition challenge. Int. J. Comput. Vis. **115**(3), 211–252 (2015). https://doi.org/10.1007/s11263-015-0816-y
33. Santoro, A., Bartunov, S., Botvinick, M., Wierstra, D., Lillicrap, T.: Meta-learning with memory-augmented neural networks. In: ICML (2016)
34. Schmidhuber, J.: Evolutionary principles in self-referential learning (1987)
35. Shi, L., Liu, F., Rosen, M.P.: Deep multimodal learning for medical visual question answering. In: CLEF (Working Notes) (2019)

36. Simonyan, K., Zisserman, A.: Very deep convolutional networks for large-scale image recognition. In: ICLR (2015)
37. Snell, J., Swersky, K., Zemel, R.S.: Prototypical networks for few-shot learning. In: NIPS (2017)
38. Sung, F., Yang, Y., Zhang, L., Xiang, T., Torr, P.H., Hospedales, T.M.: Learning to compare: relation network for few-shot learning. In: CVPR (2018)
39. Vinyals, O., Blundell, C., Lillicrap, T., Kavukcuoglu, K., Wierstra, D.: Matching networks for one shot learning. In: NIPS (2016)
40. Vu, M.H., Löfstedt, T., Nyholm, T., Sznitman, R.: A question-centric model for visual question answering in medical imaging. IEEE TMI **39**, 2856–2868 (2020)
41. Vu, M., Sznitman, R., Nyholm, T., Löfstedt, T.: Ensemble of streamlined bilinear visual question answering models for the ImageCLEF 2019 challenge in the medical domain. In: Conference and Labs of the Evaluation Forum (2019)
42. Wang, Y.X., Hebert, M.: Learning from small sample sets by combining unsupervised meta-training with CNNs. In: NIPS (2016)
43. Yang, Z., He, X., Gao, J., Deng, L., Smola, A.J.: Stacked attention networks for image question answering. In: CVPR (2016)
44. Zhan, L.M., Liu, B., Fan, L., Chen, J., Wu, X.M.: Medical visual question answering via conditional reasoning. In: ACM International Conference on Multimedia (2020)
45. Zhou, Y., Kang, X., Ren, F.: Employing Inception-Resnet-v2 and Bi-LSTM for medical domain visual question answering. In: CEUR Workshop Proceedings (2018)

mfTrans-Net: Quantitative Measurement of Hepatocellular Carcinoma via Multi-Function Transformer Regression Network

Jianfeng Zhao[1,5], Xiaojiao Xiao[2], Dengwang Li[3(✉)], Jaron Chong[1],
Zahra Kassam[4], Bo Chen[1], and Shuo Li[1,5(✉)]

[1] Western University, London, ON, Canada
sli287@uwo.ca
[2] Taiyuan University of Technology, Shanxi 030000, China
[3] Shandong Key Laboratory of Medical Physics and Image Processing, Shandong
Institute of Industrial Technology for Health Sciences and Precision Medicine, School
of Physics and Electronics, Shandong Normal University, Jinan 250358, China
dengwang@sdnu.edu.cn
[4] London Health Sciences Centre, London, ON, Canada
[5] Digital Imaging Group of London, London, ON, Canada

Abstract. Quantitative measurement of hepatocellular carcinoma
(HCC) on multi-phase contrast-enhanced magnetic resonance imaging
(CEMRI) is one of the key processes for HCC treatment and prognosis.
However, direct automated quantitative measurement using the CNN-
based network a still challenging task due to: (1) The lack of ability for
capturing long-range dependencies of multi-anatomy in the whole medical image; (2) The lack of mechanism for fusing and selecting multi phase
CEMRI information. In this study, we propose a multi-function Transformer regression network (mfTrans-Net) for HCC quantitative measurement. Specifically, we first design three parallel CNN-based encoders for
multi-phase CEMRI feature extraction and dimension reducing. Next,
the non-local Transformer makes our mfTrans-Net self-attention for capturing the long-range dependencies of multi-anatomy. At the same time,
a phase-aware Transformer captures the relevance between multi-phase
CEMRI for multi-phase CEMRI information fusion and selection. Lastly,
we proposed a multi-level training strategy, which enables an enhanced
loss function to improve the quantification task. The mfTrans-Net is validated on multi-phase CEMRI of 138 HCC subjects. Our mfTrans-Net
achieves high performance of multi-index quantification that the mean
absolute error of center point, max-diameter, circumference, and area
is down to $2.35\,mm$, $2.38\,mm$, $8.28\,mm$, and $116.15\,mm^2$, respectively.
The results show that mfTrans-Net has great potential for small lesions
quantification in medical images and clinical application value.

Keywords: Quantitative measurement · Non-local transformer ·
Phase-aware transformer

© Springer Nature Switzerland AG 2021
M. de Bruijne et al. (Eds.): MICCAI 2021, LNCS 12905, pp. 75–84, 2021.
https://doi.org/10.1007/978-3-030-87240-3_8

1   Introduction

Liver cancer is the second cancer-related death globally, in which hepatocellular carcinoma (HCC) represents about 90% of primary liver cancers [13]. The quantitative measurement of HCC is one of the key processes for HCC treatment and prognosis [5,8,12,19]. As shown in Fig. 1(b), in clinic, the quantification of HCC is usually completed by physicians observing the multi-phase contrast-enhanced magnetic resonance imaging (CEMRI). This is because multi-phase CEMRI shows more sensitive than other modalities [21]. However, it still suffers from manpower-consuming, time-consuming, subjective, and changeable due to performed manually by physicians [22]. And there is also a problem of intra- and inter-observer variability [10]. Therefore, if the quantification of HCC can be achieved automatically, it will overcome the shortcomings in the clinic, which as shown in Fig. 1(a).

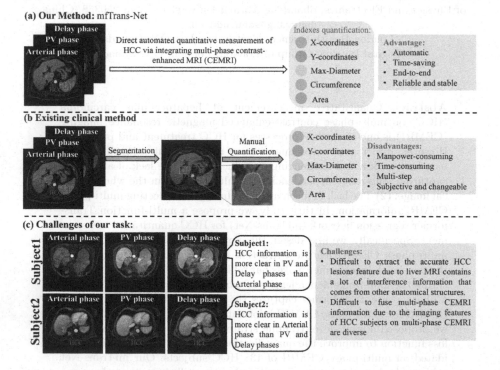

Fig. 1. Advantages and challenges of our method to tackle the quantification of HCC via using multi-phase CEMRI information. From (a) to (c), they are our method, existing clinical method, and challenges of our tasks.

As shown in Fig. 1(c), the automatic quantitative measurement of HCC on multi-phase CEMRI (i.e. arterial phase, portal venous (PV) phase, and delay phase) is exceedingly intractable task due to the following challenges: (1) Difficult to extract the accurate HCC lesions feature due to liver MRI contains a lot of interference information that comes from other anatomical structures. It means that it is necessary to explicitly capture non-local dependencies of multi-anatomy feature representations regardless of locations [21]. (2) Difficult to fuse and select the multi-phase CEMRI information due to the imaging characteristics of multi-phase CEMRI are diverse among different HCC subjects [1], which as the two subjects shown in Fig. 1(c).

In recent years, an increasing number of works focus on quantitative measurement in medical image [3,4,11,14,15,20]. Although these works achieved promising performance in the quantification of the renal, spine, or cardiac image. They are incapable of direct measuring the multi-index of HCC on multi-phase CEMRI because they are limits to the following shortcomings. (1) The lack of ability for capturing long-range dependencies of multi-anatomy in the whole medical image. Because traditional CNN-based networks are good at capturing local information and ignoring the dependence on global information [9,18]. (2) The lack of mechanism for fusing and selecting multi-phase CEMRI information. Because there is a non-linear relationship between the imaging characteristics of multi-phase CEMRI and different HCC subjects [1], which cause the information fusion and selection of multi-phase CEMRI difficult.

In this study, we propose a multi-function Transformer regression network (mfTrans-Net) for HCC multi-index quantitative measurement (i.e. center point $(O(X,Y))$, max-diameter (MDIA), circumference(CIR), and Area). Our basic assumption is that enhance the non-local feature extraction in single-phase CEMRI and capture the relevance between multi-phase CEMRI among HCC subjects for HCC quantitative measurement. Specifically, the mfTrans-Net first utilizes three parallel CNN-based encoders for multi-phase CEMRI feature extraction and dimension reducing. Next, to enhance the non-local feature extraction in single-phase CEMRI for improving capturing long-range dependencies of multi-anatomy, three non-local Transformer modules are embedded behind the parallel encoder to make the mfTrans-Net self-attention. At the same time, to capture the relevance between multi-phase CEMRI for multi-phase CEMRI information fusion and selection, a phase-aware Transformer is performed on the feature maps. Lastly, we proposed a multi-level training strategy that the multi-index quantification is not only executed after the multi-function Transformer (i.e. non-local Transformer and phase-aware Transformer) but also each non-local Transformer. It enables an enhanced loss function to constrain the quantification task.

The contributions of this work mainly contain: (1) To the best of our knowledge, it is the first time to achieve automated multi-index quantification of HCC via integrating multi-phase CEMRI, which will provide a time-saving, reliable, and stable tool for the clinical diagnosis of HCC. (2) Our mfTrans-Net provided a novel architecture to combine CNN and Transformer, which provided a solution for the medical tasks when using multi-modality images. (3) In which the newly designed multi-function Transformer (mf-Trans) provided a solution to capture the non-local information and multi-phase CEMRI relevance. (4) We proposed a multi-level constraint strategy, which enables an enhanced loss function to constrain the mfTrans-Net for improving the quantification performance.

2 Method

The mfTrans-Net integrates multi-phase CEMRI information for HCC quantification. Specifically, as shown in Fig. 2, the mfTrans-Net is fed with the sequences of arterial phase CEMRI ($\mathcal{X}^A \in \mathbb{R}^{H \times W \times N}$), PV phase CEMRI ($\mathcal{X}^P \in \mathbb{R}^{H \times W \times N}$), and delay phase CEMRI ($\mathcal{X}^D \in \mathbb{R}^{H \times W \times N}$) to three parallel CNN-based encoders (i.e. Encoder-A, Encoder-P, and Encoder-D), respectively. And the mfTrans-Net outputs the final results of HCC multi-index quantification $\mathcal{Y}_Q$. The mfTrans-Net is performed via the following three stages: (1) Using the three parallel encoders for feature extraction. (2) Using the newly designed mf-Trans for non-local feature extraction and capture the relevance between multi-phase CEMRI. (3) Using an enhanced loss function for task constraint.

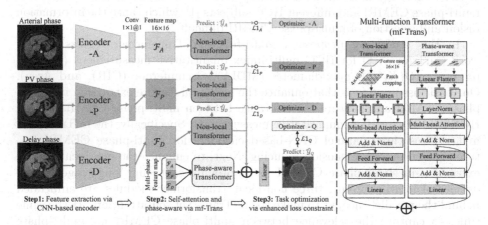

Fig. 2. Overview of the proposed mfTrans-Net. There are three main steps in the regression of HCC multi-index: (1) Feature extraction via using three parallel CNN-based encoder. (2) Feature fusion via using newly designed mf-Trans. (3) Task constraint via using enhanced loss function.

2.1 CNN-based Encoder for Feature Extraction and Dimension Reducing

As step1 shown in Fig. 2, all three parallel CNN-based encoders have the same network architecture, which is the typical shallow convolutional network with four convolution blocks. And each convolution block consists of a convolution layer, batch normalization layer, ReLU layer, and max-pooling layer. At the end of each encoder, one convolution layer with 1×1 kernel size and a single channel is used for changing the channel of feature maps. The reasons why we design the encoder in this way are as follows: (1) Using the CNN-based encoder rather than fully Transformer-based architecture, which is conducive to the linear flatten in mf-Trans because it reduces the dimension of the medical image. Thus, it greatly reduces the amount of parameter calculation that comes from linear flatten. (2) Using three parallel encoders instead of a shared encoder is to extract multi-phase CEMRI information separately, which is conducive to the position encoding of multi-phase CEMRI in the phase-aware Transformer.

2.2 mf-Trans for Self-attention and Phase-Aware

The Transformer is proposed by Vaswani et al. [17], and has since become the state-of-the-art method in many NLP tasks. Recently, Dosovitskiy et al. [2] proposed the vision transformer achieved promising performance in the field of computer vision. Inspired by these works, we design the mf-Trans for enhancing the non-local feature extraction and capture the relevance between multi-phase CEMRI. As step2 shown in Fig. 2, the mf-Trans contains the non-local Transformer and phase-aware Transformer. And both of non-local Transformer and phase-aware Transformer are designed based on the typical Transformer [17].

Non-local Transformer Makes mfTrans-Net Self-attention for Enhancing Non-local Feature Extraction. For the non-local Transformer, it first fed with the feature maps of $\mathcal{F}^A \in \mathbb{R}^{H \times W}$, $\mathcal{F}^P \in \mathbb{R}^{H \times W}$, and $\mathcal{F}^D \in \mathbb{R}^{H \times W}$ those comes from CNN separately. To describe the calculation clear, we assume the input is $\mathcal{F}^A$ here. After feeding feature map $\mathcal{F}^A$, it performs the operation of 2D patch cropping on the $\mathcal{F}^A$. To handle 2D patches, we reshape them into the sequence of $\mathcal{X}^A \in \mathbb{R}^{N \times P^2}$. Where $H \times W$ represents the resolution of the original feature map $\mathcal{F}^A$, (P, P) represents the resolution of each image patch, and $N = HW/P^2$ represents the number of patches. To make our mfTrans-Net self-attention for enhancing non-local feature extraction, a learnable positional encoding of the same dimensions is added to $\mathcal{X}^A$. The output embeddings $\mathcal{Z}^A \in \mathbb{R}^{N \times P^2}$ can serve as the effective input sequence length for the multi-head attention layer. Thus, the computation of self-attention (SA) matrix and the multi-head attention are identical to those defined in Transformer [17]:

$$SA : Attention(\mathcal{Q}, \mathcal{K}, \mathcal{V}) = softmax(\frac{\mathcal{Q}\mathcal{K}^T}{\sqrt{d_k}})(\mathcal{V}) \tag{1}$$

$$MultiHead(\mathcal{Q}, \mathcal{K}, \mathcal{V}) = Concat(head_1, ..., head_h)\mathcal{W}^O \tag{2}$$

$$head_i = Attention(\mathcal{Q}\mathcal{W}_i^{\mathcal{Q}}, \mathcal{K}\mathcal{W}_i^{\mathcal{O}}, \mathcal{V}\mathcal{W}_i^{\mathcal{V}}) \tag{3}$$

where query $\mathcal{Q}$, key $\mathcal{K}$, and value $\mathcal{V}$ are all vectors for flattened feature map $\mathcal{X}^A$. $\mathcal{W}_i^{\mathcal{O}}$ is the projection matrix. And the $\frac{1}{\sqrt{d_k}}$ is the scaling factor.

Phase-Aware Transformer Captures the Relevance of Multi-phase CEMRI for Multi-phase CEMRI Information Fusion and Selection. Different from the non-local Transformer, the phase-aware Transformer is fed with the feature maps of $\mathcal{F}^A$, $\mathcal{F}^P$, and $\mathcal{F}^D$ together. And there is no operation of 2D patch cropping. Thus, it enables to embed a learnable positional encoding among the multi-phase CEMRI. Besides, the feature maps come from three different encoders here, so that adding a normalization layer (i.e. LayerNorm shown in Fig. 2) before the multi-head attention layer is necessary. The specific detailed computations of the self-attention and the multi-head attention are similar to the non-local Transformer.

At the end of the step2, the concatenation operation (i.e. $\oplus$ shown in Fig. 2) is performed among three non-local Transformers and the phase-aware Transformer. And mfTrans-Net outputs the final quantification results by performing a linear layer to the concatenation.

2.3 Multi-level Constraint Strategy of Enhanced Loss Function

The basic loss function used to constrain our mfTrans-Net is the $\mathcal{L}1$ loss (i.e. mean absolute error (MAE)), which calculates the deviation of quantified multi-index value. In this paper, we adopts a multi-level constraint strategy. Specifically, before the final quantified multi-index of $\hat{\mathcal{Y}}_Q$, the mfTrans-Net predicts the quantified multi-index of $\hat{\mathcal{Y}}_A$, $\hat{\mathcal{Y}}_P$, and $\hat{\mathcal{Y}}_D$ after each non-local Transformer branch. And then the $\mathcal{L}1$ is performed to $\hat{\mathcal{Y}}_A$, $\hat{\mathcal{Y}}_P$, $\hat{\mathcal{Y}}_D$, and $\hat{\mathcal{Y}}_Q$ respectively. In this situation, the $\mathcal{L}1$ not only constrains the whole mfTrans-Net but also constrains the three parallel branches separately, which is beneficial to improve the network performance. The enhanced loss function can be defined as:

$$\mathcal{L}_{quan} = \mathcal{L}1_A(\hat{\mathcal{Y}}_A, \mathcal{Y}_A) + \mathcal{L}1_P(\hat{\mathcal{Y}}_P, \mathcal{Y}_P) + \mathcal{L}1_D(\hat{\mathcal{Y}}_D, \mathcal{Y}_D) + \lambda_q \mathcal{L}1_Q(\hat{\mathcal{Y}}_Q, \mathcal{Y}_Q) \quad (4)$$

where the $\mathcal{L}1_k$ can be formulated as:

$$\mathcal{L}1_k(\hat{\mathcal{Y}}_i^m, \mathcal{Y}_i^m) = \frac{1}{N} \sum_{i=1}^{N} |\mathcal{Y}_i^m - \hat{\mathcal{Y}}_i^m| \quad (5)$$

where $\hat{\mathcal{Y}}_i^m$ represents the predicted multi-index value, $\mathcal{Y}_i^m$ represents the ground truth of multi-index value, $m = \{O(X,Y), MDIA, CIR, Area\}$ represents the type of multi-index. $k = \{A, P, D, Q\}$ represents the corresponding loss function. And N represents the images number in the sequences of multi-phase CEMRI.

3 Experiment and Results

Experimental results show that mfTrans-Net successfully measures the multi-index value of HCC. Quantitative results of the measurement are evaluated by MAE. The MAE of the quantification of $O(X,Y)$, MDIA, CIR, and Area is down to 2.35 mm, 2.38 mm, 8.28 mm and 116.15 mm^2, respectively.

Table 1. The quantitative evaluation of the HCC quantification. The criteria of MAE evaluated the performance of our mfTrans-Net and other three other methods.

Method	$O(X, Y)$ (mm)	MDIA (mm)	CIR (mm)	Area (mm^2)
VGG-16 [16]	3.51 ± 0.54	9.91 ± 3.66	32.94 ± 10.55	383.21 ± 101.31
ResNet-50 [6]	2.92 ± 0.35	6.59 ± 2.35	21.19 ± 6.41	268.27 ± 83.16
DenseNet [7]	2.76 ± 0.31	6.25 ± 2.12	18.68 ± 6.34	240.74 ± 75.81
mfTrans-Net	$\mathbf{2.35 \pm 0.25}$	$\mathbf{2.38 \pm 1.53}$	$\mathbf{8.28 \pm 4.82}$	$\mathbf{116.15 \pm 64.24}$

Dataset. The effectiveness of the mfTrans-Net is validated in the datasets of 4000 images that come from 138 HCC subjects. All subjects underwent standard clinical liver MRI examinations. And every subject has corresponding arterial phase CEMRI (256×256 pixel), PV phase CEMRI (256×256 pixel), and delay phase CEMRI (256×256 pixel). The multi-phase CEMRI was obtained using the protocol of gadobutrol 0.1 mmol/kg on a 3T MRI scanner. Two radiologists with more than ten years of experience manually quantified HCC. The annotation rule was discussed and agreed upon for the main variability.

Configuration. The mfTrans-Net was trained on one 5-fold cross-validation test for performance evaluation and comparison. Specifically,the dataset of 4000 images is divided into 5 groups. In which four groups are employed to train the mfTrans-Net, and the last group is used for testing. This procedure was executed five times in a loop until each subject is used as a training object and test object. The mfTrans-Net was trained for 100 epochs with the Adam optimizer, the learning of 1e-4, and batech size of 1. And an enhanced loss function of four items, as shown in equation (4), is employed for mfTrans-Net training. In which the hyper-parameter λ_q is set to 2 in all experiments for enhancing the weight of multi-phase CEMRI fusion information. Inspired by work Vaswani et al. [17], the scaling factor d_k set to 64 in equation (1). The mfTrans-Net was implemented on Ubuntu 18.04 platform, Python v3.6, Pytorch v0.4.0, and CUDA v9.0 library, and running on Intel(R) Core(TM) i9-9900K CPU @ 3.60 GHz, RAM 16 GB, and GeForce GTX 1080Ti 11 GB.

3.1   Multi-index Quantification Comparison

The mfTrans-Net has been validated by comparing with three state-of-the-art methods (i.e. VGG-16 [16], ResNet-50 [6], and DenseNet [7] with 5-layer dense block). In order to ensure that the input of the networks is the same as mfTrans-Net, we set the channel number of the first convolution layer in the comparison method to 3, which corresponding to the arterial phase CEMRI, PV phase CEMRI, and delay phase CEMRI respectively. The quantitative analysis results of HCC multi-index quantification are shown in Table 1. It is clear that our mfTrans-Net outperforms the other methods. Especially for the quantitative value of MDIA, CIR, and Area.

3.2   Performance in Different Size of HCC

To verify the performance of mfTrans-Net in different HCC sizes, we divided the quantitative results into two groups (i.e. MDIA<30mm of 85 subjects and MDIA≥30mm of 53 subects) according to the HCC size to evaluate mfTrans-Net. The quantitative analysis results of HCC multi-index quantification are shown in Table 2. It is clear our mfTrans-Net still performs well even when the tumor size is less than 30 mm, which demonstrates that the performance of mfTrans-Net is robust to small lesions.

Table 2. Performance of mfTrans-Net in terms of MAE under different HCC sizes.

HCC size	$O(X,Y)$ (mm)	MDIA (mm)	CIR (mm)	Area (mm^2)
MDIA<30 mm	2.39 ± 0.29	2.51 ± 1.88	9.45 ± 5.71	123.46 ± 67.90
MDIA≥30 mm	2.29 ± 0.19	2.09 ± 0.97	6.40 ± 3.39	104.43 ± 58.37
Total	**2.35 ± 0.25**	**2.38 ± 1.53**	**8.28 ± 4.82**	**116.15 ± 64.24**

3.3   Ablation Study

In order to prove the contributions of the non-local Transformer and the phase-aware Transformer. We removed the mf-Trans and using three parallel encoders only as of the baseline method (i.e. No mf-Trans). And we also removed the non-local Transformer (i.e. No NL-Trans) and phase-aware Transformer (i.e. No PA-Trans) for comparison, respectively. Besides, in order to prove the contribution of the enhanced loss function, performed the comparison between mfTrans-Net and mfTrans-Net with $\mathcal{L}1_Q$ only (i.e. No enhanced loss). The results of ablation studies demonstrate that every part of the mfTrans-Net contributes to multi-index quantification of HCC. And it showed that the mf-Trans made the greatest contribution, which should be attributed to fact that mf-Trans improves the ability of non-local feature extraction and phase perception (Table 3).

Table 3. Performance of mfTrans-Net in terms of MAE under different configurations for O(X, Y), MDIA, CIR, and Area.

Method	$O(X,Y)$ (mm)	MDIA (mm)	CIR (mm)	Area (mm^2)
No mf-Trans	2.75 ± 0.31	6.13 ± 2.05	18.44 ± 6.19	220.5 ± 74.33
No NL-Trans	2.46 ± 0.28	4.18 ± 1.75	12.05 ± 5.25	160.37 ± 68.51
No PA-Trans	2.57 ± 0.29	5.53 ± 1.83	14.56 ± 5.77	182.11 ± 70.27
No enhanced loss	2.41 ± 0.26	3.02 ± 1.60	9.89 ± 5.06	135.05 ± 66.39
mfTrans-Net	**2.35 ± 0.25**	**2.38 ± 1.53**	**8.28 ± 4.82**	**116.15 ± 64.24**

4 Conclusion

Our proposed mfTrans-Net achieved the multi-index quantification of HCC via integrating multi-phase CEMRI information. It provided a novel architecture to combine CNN and Transformer, which provided a solution for the medical tasks when using multi-modality images. In which the newly designed mf-Trans makes the mfTrans-Net non-local and phase-aware, which improved the feature extraction and fusion. Moreover, the multi-level training strategy enables an enhanced loss function for task constrain. All these contributions are conducive to improve the performance of HCC quantification. Thus, the mfTrans-Net has great potential for small lesions quantification in medical images and clinical application value. For future works, mfTrans-Net will be further tested on other medical tasks via using multi-modality medical images.

Acknowledgements. This work is partly supported by the China Scholarship Council (No.202008370191).

References

1. Cereser, L., et al.: Comparison of portal venous and delayed phases of gadolinium-enhanced magnetic resonance imaging study of cirrhotic liver for the detection of contrast washout of hypervascular hepatocellular carcinoma. J. Comput. Assist. Tomog **34**(5), 706–711 (2010)
2. Dosovitskiy, A., et al.: An image is worth 16x16 words: Transformers for image recognition at scale. arXiv preprint arXiv:2010.11929 (2020)
3. Ge, R., et al.: K-net: integrate left ventricle segmentation and direct quantification of paired echo sequence. IEEE Trans. Med. Imaging **39**(5), 1690–1702 (2019)
4. Ge, R., et al.: Pv-lvnet: direct left ventricle multitype indices estimation from 2d echocardiograms of paired apical views with deep neural networks. Med. Image Anal. **58**, 101554 (2019)
5. Goh, B.K.: Importance of tumor size as a prognostic factor after partial liver resection for solitary hepatocellular carcinoma: Implications on the current ajcc staging system. J. Surg. Oncol **113**(1), 89–93 (2016)
6. He, K., Zhang, X., Ren, S., Sun, J.: Deep residual learning for image recognition. In: Proceedings of the IEEE Conference on Computer Vision and Pattern Recognition, pp. 770–778 (2016)
7. Huang, G., Liu, Z., Van Der Maaten, L., Weinberger, K.Q.: Densely connected convolutional networks. In: Proceedings of the IEEE Conference on Computer Vision and Pattern Recognition, pp. 4700–4708 (2017)
8. Hwang, S., et al.: The impact of tumor size on long-term survival outcomes after resection of solitary hepatocellular carcinoma: single-institution experience with 2558 patients. J. Gastrointest. Surg. **19**(7), 1281–1290 (2015)
9. Jaderberg, M., Simonyan, K., Zisserman, A., Kavukcuoglu, K.: Spatial transformer networks. arXiv preprint arXiv:1506.02025 (2015)
10. Kim, Y.S., et al.: Strahlentherapie und Onkologie **192**(10), 714–721 (2016). https://doi.org/10.1007/s00066-016-1028-2
11. Lin, L., et al.: Multiple axial spine indices estimation via dense enhancing network with cross-space distance-preserving regularization. IEEE J. Biomed. Health Inf. **24**(11), 3248–3257 (2020)

12. Liu, H., et al.: Reclassification of tumor size for solitary hbv-related hepatocellular carcinoma by minimum p value method: a large retrospective study. World J. Surg. Oncol. **18**(1), 1–10 (2020)
13. Liver, E.A.F.T.S.O.T., et al.: Easl clinical practice guidelines: management of hepatocellular carcinoma. J. Hepatol. **69**(1), 182–236 (2018)
14. Pang, Shumao, Leung, Stephanie, Ben Nachum, Ilanit, Feng, Qianjin, Li, Shuo: Direct automated quantitative measurement of spine via cascade amplifier regression network. In: Frangi, Alejandro F.., Schnabel, Julia A.., Davatzikos, Christos, Alberola-López, Carlos, Fichtinger, Gabor (eds.) MICCAI 2018. LNCS, vol. 11071, pp. 940–948. Springer, Cham (2018). https://doi.org/10.1007/978-3-030-00934-2_104
15. Ruan, Y., et al.: Mb-fsgan: joint segmentation and quantification of kidney tumor on ct by the multi-branch feature sharing generative adversarial network. Med. Image Anal. **64**, 101721 (2020)
16. Simonyan, K., Zisserman, A.: Very deep convolutional networks for large-scale image recognition. arXiv preprint arXiv:1409.1556 (2014)
17. Vaswani, A., et al.: Attention is all you need. arXiv preprint arXiv:1706.03762 (2017)
18. Wang, X., Girshick, R., Gupta, A., He, K.: Non-local neural networks. In: Proceedings of the IEEE Conference on Computer Vision and Pattern Recognition, pp. 7794–7803 (2018)
19. Wu, G., Wu, J., Wang, B., Zhu, X., Shi, X., Ding, Y.: Importance of tumor size at diagnosis as a prognostic factor for hepatocellular carcinoma survival: a population-based study. Cancer Manag. Res. **10**, 4401 (2018)
20. Xu, C., Howey, J., Ohorodnyk, P., Roth, M., Zhang, H., Li, S.: Segmentation and quantification of infarction without contrast agents via spatiotemporal generative adversarial learning. Med. Image Anal. **59**, 101568 (2020)
21. Zhao, J., et al.: Tripartite-gan: synthesizing liver contrast-enhanced mri to improve tumor detection. Med. Image Anal. **63**, 101667 (2020)
22. Zhao, J., Li, D., Xiao, X., Chong, J., Chen, B., Li, S.: United adversarial learning for liver tumor segmentation and detection of multi-modality non-contrast mri. Med. Image Anal., 102154 (2021)

You only Learn Once: Universal Anatomical Landmark Detection

Heqin Zhu[1], Qingsong Yao[1], Li Xiao[1], and S. Kevin Zhou[1,2(✉)]

[1] Key Lab of Intelligent Information Processing of Chinese Academy of Sciences (CAS), Institute of Computing Technology, CAS, Beijing 100190, China
[2] Medical Imaging, Robotics, and Analytic Computing Laboratory and Engineering (MIRACLE) School of Biomedical Engineering and Suzhou Institute for Advanced Research, University of Science and Technology of China, Suzhou 215123, China

Abstract. Detecting anatomical landmarks in medical images plays an essential role in understanding the anatomy and planning automated processing. In recent years, a variety of deep neural network methods have been developed to detect landmarks automatically. However, all of those methods are unary in the sense that a highly specialized network is trained for a single task say associated with a particular anatomical region. In this work, for the first time, we investigate the idea of "You Only Learn Once (YOLO)" and develop a universal anatomical landmark detection model to realize multiple landmark detection tasks with end-to-end training based on mixed datasets. The model consists of a local network and a global network: The local network is built upon the idea of universal U-Net to learn multi-domain local features and the global network is a parallelly-duplicated sequential of dilated convolutions that extract global features to further disambiguate the landmark locations. It is worth mentioning that the new model design requires much fewer parameters than models with standard convolutions to train. We evaluate our YOLO model on three X-ray datasets of 1,588 images on the head, hand, and chest, collectively contributing 62 landmarks. The experimental results show that our proposed universal model behaves largely better than any previous models trained on multiple datasets. It even beats the performance of the model that is trained separately for every single dataset. Our code is available at https://github.com/ICT-MIRACLE-lab/YOLO_Universal_Anatomical_Landmark_Detection.

Keywords: Landmark detection · Multi-domain learning

1 Introduction

Landmark detection plays an important role in varieties of medical image analysis tasks [25,26]. For instance, landmarks of vertebrae are helpful for surgery planning [2], which determines the positions of implants. Furthermore, the landmark locations can be used for segmentation [5] and registration [9] of medical images.

© Springer Nature Switzerland AG 2021
M. de Bruijne et al. (Eds.): MICCAI 2021, LNCS 12905, pp. 85–95, 2021.
https://doi.org/10.1007/978-3-030-87240-3_9

Because it is time-consuming and labor-intensive to annotate landmarks manually in medical images, many computer-assisted landmark detection methods are developed in the past years. These methods can be categorized into two types: traditional and deep learning based methods. Traditional methods aim at designing image filters and extracting invariant features, such as SIFT [15]. Liu et al. [14] present a submodular optimization framework to utilize the spatial relationships between landmarks for detecting them. Lindner et al. [13] propose a landmark detection algorithm in the use of supervised random forest regression. These methods are less accurate and less robust in comparison to deep neural network methods. Yang et al. [23] make use of a deep neural network and propose a deep image-to-image network built up with an encoder-decoder architecture for initializing vertebra locations, which are evolved with another ConvLSTM model and refined by a shape-based network. Payer et al. [16] propose a novel CNN-based neural network which integrates spatial configuration into the heatmap and demonstrate that, for landmark detection, local features are accurate but potentially ambiguous, while global features eliminate ambiguities but are less accurate [10,27]. Recently, Lian et al. [12] develop a multi-task dynamic transformer network for bone segmentation and large-scale landmark localization with dental CBCT, which also makes use of global features when detecting landmarks.

However, all of those methods are unary in the sense that a highly specialized network is trained for a single task say associated with a particular anatomical region (such as head, hand, or spine), often based on a single dataset and not robust enough [24]. It's promising and desirable to develop a model which is learned once and works for all the tasks[4,11], that is "You Only Learn Once". We, for the first time in the literature, develop a powerful, **universal model** for detecting the landmarks associated with different anatomies, each exemplified by a dataset. Our approach attempts to *unleash the potential of "bigger data"* as it utilizes the aggregate of all training images and builds a model that outperforms the models that are individually trained. We believe that *there are common knowledge among the seemingly different anatomical regions*, observing that the local features of landmarks from different datasets share some characters (such as likely locating at corners, endpoints, extrema of curves or surfaces, etc.); after all, they are all landmarks. We attempt to design a model that is able to capture these common knowledge to gain more effectiveness while taking into account the differences among various tasks. To the best of our knowledge, this marks *first such attempt* for landmark detection.

Our model, named as Global Universal U-Net (**GU2Net**), is inspired by the universal design of Huang et al. [4] and the local-global design of Payer et al. [16], reaping the benefits of both worlds. As shown in Fig. 1, it has a local network and a global network. Motivated by our observation, the local network is designed to be similar to a universal U-Net, which is a U-Net [17] with each convolution replaced with *separable convolution*. The separable convolution is composed of *channel-wise convolution* and *point-wise convolution* [4], which model task-shared and task-specific knowledge, respectively, and hence has fewer parameters than a normal convolution. Differently, instead of outputting a segmentation mask in [4], we output a landmark heatmap. The local network extracts local features which are mostly accurate but still possibly ambiguous for landmarks. The global network is designed to extract global features and further reduce ambiguities when detecting landmarks. Different from Payer et al. [16], our global network takes downsampled image and local features as input and uses dilated convolutions for enlarging receptive fields.

In sum, we make the following contributions:

- The *first attempt* in the literature, to the best of our knowledge, to develop a universal landmark detection model that works for multiple datasets and different anatomical regions, unleashing the potential of "bigger data";
- *State-of-the-art performances* of detecting a total of 62 landmarks based on three X-ray datasets of head, hand, and chest, totaling 1,588 images but using only one model which needs fewer parameters.

2   Method

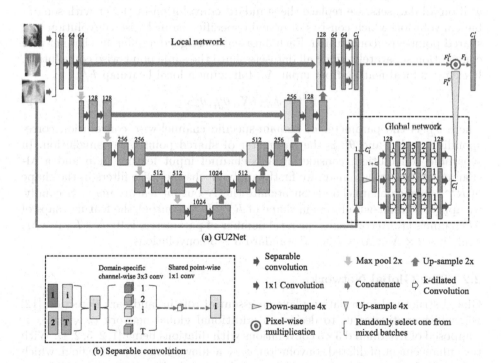

(a) GU2Net

(b) Separable convolution

Fig. 1. (a) The structure of the GU2Net model, consisting of two parts, namely the local network and the global network. The local network is a U-Net structure with each convolution replaced with separable convolution which consists of channel-wise convolution and point-wise convolution. The global network is parallel-duplicated sequential of five dilated small-kernel-size convolutions. (b) Separable convolution. Each 3x3 convolution is followed by batch-normalization and Leaky ReLU activation. The global network takes four times downsampled image and local heatmap as input and outputs four times upsampled heatmap.

Problem Definition: Let $\{D_1, D_2, \ldots, D_T\}$ be a set of T datasets, which are potentially from different anatomical regions. Giving an image $X_i \in R^{C_i \times H_i \times W_i}$ from Dataset D_i along with corresponding landmarks $\{(x_{iC_i'}, y_{iC_i'})\}$, we obtain the kth ($k \in [1, 2, \ldots, C_i']$) landmark' heatmap $Y_{ik} \in R^{C_i' \times H_i \times W_i}$ by using Gaussian function:

$$Y_{ik} = \frac{1}{\sqrt{2\pi}\sigma} \exp(-\frac{(x - x_{ik})^2 + (y - y_{ik})^2}{2\sigma^2}), \tag{1}$$

where C_i is the number of channels on the input image (i.e. $C_i = 1$ for a X-ray image); C_i' is the number of channels on output heatmap, namely the number of landmarks; H_i is the height of image X_i and W_i is the width of image X_i.

2.1   The Local Network

The local network ϕ_{LN} aims at extracting local features and generating a local heatmap that is used to determine the accurate location of a landmark. To work well on all datasets, we replace the standard convolution in U-Net with separable convolution which consists of domain-specific channel-wise convolution and shared point-wise convolution. Each data set is assigned a different channel-wise convolution separately, while all datasets share the same point-wise convolution. It extracts local feature from input X_i, following a local heatmap $\tilde{F}_i^L$:

$$\tilde{F}_i^L = \phi_{LN}(X_i; \theta_{di}^L, \theta_s^L) \tag{2}$$

where θ_{di}^L is the parameter of domain-specific channel-wise convolution corresponding to D_i; and θ_s^L is the parameter of shared point-wise convolution. In separable convolution, considering a N-channel input feature map and a M-channel output feature map, we firstly apply N channel-wise filters in the shape of $R^{3 \times 3}$ to each channel and concatenate the N output feature maps. Secondly, we apply M point-wise filters in shape of $R^{1 \times 1 \times N}$ to output the feature maps of M channels [4]. Accordingly, the total number of parameters is $9 \times N \times T + N \times M$, while it's $9 \times N \times M \times T$ for T standard 3×3 convolutions.

2.2   The Global Network

Global structural information plays an essential role in landmark detection [12, 16], which motivates us to design an additional global network ϕ_{GN}. ϕ_{GN} is composed of five dilated 3×3 convolutions with dilations being $[1, 2, 5, 2, 1]$. With the enhancement of dilated convolution, ϕ_{GN} achieves large receptive field, which is benefit for capturing the important global information [22]. Since different anatomical regions vary a lot in appearance, we duplicate our global network for each dataset (see Fig. 1, resulting in domain-specific parameters θ_{di}^G. As shown in Fig. 1(a)), ϕ_{GN} takes image X_i and local feature $\tilde{F}_i^L$ as input and aggregates the global information at a coarse-grained scale, flowing global heatmap $\tilde{F}_i^G$:

$$\tilde{F}_i^G = \phi_{GN}(X_i, \tilde{F}_i^L; \theta_{di}^G) \tag{3}$$

2.3 Loss Function

As shown in Fig. 1(a), we combine the local and global information by point-wise multiplying the local heatmap $\tilde{F}_i^L$ and global heatmap $\tilde{F}_i^G$, resulting in final heatmap $\tilde{F}_i = \tilde{F}_i^G \odot \tilde{F}_i^L$, where $\odot$ is the pixel-wise multiplication. In the training stage, we penalize the final heatmap $\tilde{F}_i$ and the ground truth Y_i (defined in Eq. 1):

$$L_i = \sum_{y \in Y_i, f \in \tilde{F}_i} -y \log f - (1-y) \log (1-f) \tag{4}$$

In the inference stage, the k-th landmark is obtained after finding maximum location of the k-th channel in output heatmaps $\tilde{F}_i$.

$$\text{landmark}_k = argmax(\tilde{F}_{ik}) \tag{5}$$

3 Experiments

In this section, we qualitatively and quantitatively evaluate our universal model and compare it with the state-of-the-art methods on three public X-ray datasets of head, hand, and chest. Except as otherwise noted, models are learned only once on the three datasets which are randomly mixed by batch. Evaluation is carried out on a single dataset separably for each model. Furthermore, we conduct an ablation study to demonstrate how different components help to improve the performance of our universal model.

3.1 Settings

Our deep networks are implemented in Pytorch 1.3.0 and run on a TITAN RTX GPU with CUDA version being 11. Each 3×3 convolution is followed by batch normalization [7] and Leaky ReLU activation [3]. We convert landmarks to a Gaussian heatmap which retains the probablility distribution of landmark in each pixel, with σ set to 3.

For data augmentation, we rotate the input image by 2 degrees with 0.1 probability and do the translation by 10 pixels in each direction with 0.1 probability. When training networks, we set batch-size to 4 and learning-rate to [1e–4,1e–2]. The binary cross-entropy (BCE) loss and an Adam optimizer are used to train the network up to 100 epochs and a cyclic scheduler [18] is used to decrease learning rate from 1e–2 to 1e–4 dynamically. For evaluation, the inference model is chosen as the one with minimum validation loss and evaluated on two metrics: mean radial error (MRE) and successful detection rates (SDR).

3.2 Dataset

Head. The head dataset is an open-source dataset that contains 400 cephalometric X-ray images [21]. We choose the first 150 images for training and the other 250 images for testing. Each image is of size 2400×1935 with a resolution of 0.1 mm × 0.1 mm. When training networks on the head dataset, we resize the original image to the size of 512 × 416 to keep the ratio of width and height. There are 19 landmarks manually labeled by two medical experts and we use the average labels like [16].

Hand. The hand dataset is also a public dataset[1] which contains 909 X-ray images. The first 609 images are used for training and the other 300 images are used for testing. The sizes of these images are not all the same, so we resize images to the shape of 512 × 368. To calculate the physical distance and compare it with other methods, we assume that the width of the wrist is 50 mm, following Payer et al. [16]. With the two endpoints of the wrist being p, q, the physical distance can be calculated as the multiplication of the pixel distance and $\frac{50}{\|p-q\|_2}$. A total of 37 landmarks have been manually labeled by Payer et al. [16] and the two endpoints of the wrist can be obtained from the first and fifth points, respectively.

Chest. We also adopt a public chest dataset[2] which contains two subset: the China set and the Montgomery set [1,8]. We select the China set and exclude cases that are labeled as abnormal lungs(diseased lungs) to form our experimental dataset. Our chest dataset has 279 X-ray images. The first 229 images are used for training and the last 50 images are used for testing. Since the chest dataset has no landmark labels, we manually annotate six landmarks in each image. The left three landmarks are the top, the bottom, and the right boundary point of the right lung. It's the same with the right three landmarks. In the rough, the six landmarks determine the boundary of the lung (see Fig. 2). We resize the input image to the shape of 512 × 512. Besides, since the chest dataset only contains png images and the physical spacing is not known, we use pixel distance to measure the model's performance.

3.3   Evaluation

Table 1 shows the experimental results in comparison of different methods on the head, hand, and chest datasets, with different SDR thresholds. Images are resized during training and testing.

On the head dataset, our GU2Net achieves the best accuracy within all thresholds (2 mm, 2.5 mm, 3 mm, 4 mm) and obtains a MRE of 1.54 ± 2.37 mm, behaving much better than U-Net which is also learned on the mixed multiple datasets. GU2Net even beats all models marked with * which are learned on a single dataset. Within 2 mm, GU2Net achieves the best SDR of 77.79%, outperforming the previous state-of-the-art method   [16] by 4.46%. Such an improvement is consistent among SDRs at other distances.

---

[1] https://ipilab.usc.edu/research/baaweb.
[2] https://www.kaggle.com/nikhilpandey360/chest-xray-masks-and-labels.

Table 1. Quality metrics of different models on head, hand, and chest datasets. * represents the performances copied from the original paper. + represents the model is learned on the mixed three datasets. - represents that no experimental results can be found in the original paper. The best results are in **bold** and the second best results are underlined.

Models	MRE (mm)	Head SDR(%)				MRE (mm)	Hand SDR(%)			MRE (px)	Chest SDR(%)		
		2 mm	2.5 mm	3 mm	4 mm		2 mm	4 mm	10 mm		3px	6px	9px
Ibragimov et al. [6]*	1.84	68.13	74.63	79.77	86.87	–	–	–	–	–	–	–	–
Štern et al. [19]*	–	–	-	-	-	0.80	92.20	98.45	99.83	–	–	–	–
Lindner et al. [13]*	1.67	70.65	76.93	82.17	89.85	0.85	93.68	98.95	99.94	–	–	–	–
Urschler et al. [20]*	-	70.21	76.95	82.08	89.01	0.80	92.19	98.46	99.95	–	–	–	–
Payer et al. [16]*	–	73.33	78.76	83.24	89.75	0.66	94.99	99.27	99.99	–	–	–	–
U-Net [17]+	12.45	52.08	60.04	66.54	73.68	6.14	81.16	92.46	93.76	5.61	51.67	82.33	90.67
GU2Net (Ours) +	1.54	77.79	84.65	89.41	94.93	0.84	95.40	99.35	99.75	5.57	57.33	82.67	89.33

On the hand dataset, our GU2Net also reaches the best accuracy of 95.40% within 2 mm and 99.35% within 4 mm which is far ahead of other models learned on the single hand dataset. GU2Net also performs better than U-Net which is learned on the mixed multiple datasets by a margin of 14.24% within 2 mm. For SDR within 10 mm, the performance of GU2Net is not the best but very close to the best.

On the chest dataset, our GU2Net obtains an MRE of 5.57 ± 20.54 px, behaving better than U-Net within 3px and 6px, but a little worse than U-Net within 9px. This is probably due to that for the head and hand datasets, the landmarks are mostly with bone structures, while here the landmarks are with the lung, which concerns soft-tissue contrast. This motivates us in the future to design a new universal model to better capture the nuances between bone and soft tissue.

In summary, our proposed GU2Net generally outperforms any other state-of-the-art methods learned on a single dataset or mixed multiple datasets, especially under high-precision conditions, which is evident from the SDR values within say 2 mm, 4 mm, 3px, and 6px.

3.4   Ablation Study

In order to demonstrate the effectiveness of our local network ϕ_{LN} and global network ϕ_{GN}, we perform ablation study on the mixed dataset by merging the three datasets together. There are total $250 + 300 + 50 = 600$ images for testing. The average MRE and SDR on the mixed dataset are adopted as metrics. We evaluate the performance on U-Net, Tri-UNet, ϕ_{GN}, ϕ_{LN}(with local network only) and ϕ_{GN}(with global network only), and GU2net. Here Tri-UNet is a U-Net with each convolution duplicated 3 times, and with ReLU replaced with leaky ReLU.

Table 2. Ablation study of our universal model with local network and global network. θ_d denotes domain-specific parameters while θ_s denotes shared parameters.

Models	Parameter		MRE±STD	SDR(%)		
	Number	Type	(px)	2px	4px	6px
U-Net [17]	17.3M	θ_s	7.99±26.14	71.67	86.40	89.26
Tri-UNet	51.92M	θ_d	1.16±5.21	88.19	97.46	99.14
ϕ_{GN}	1.35M	θ_d	2.31±5.64	67.39	94.22	97.62
ϕ_{LN}	2.12M	θ_d, θ_s	1.20±4.98	88.68	97.29	99.12
GU2Net($\phi_{LN} \odot \phi_{GN}$)	4.44M	θ_d, θ_s	1.14±3.94	89.09	97.49	99.52

As shown in Table 2, when comparing ϕ_{LN} with U-Net, it is evident that separable convolution in the local network improves the model's performance by a large margin. Thus, the architecture of local network is more capable for multi-domain learning. By comparing the results of GU2Net with ϕ_{LN} and ϕ_{GN}, we observe much improvement of detection accuracy, which demonstrates the effectiveness of fusing local information and global information. Thus global information and local information are equally important for accurate localization

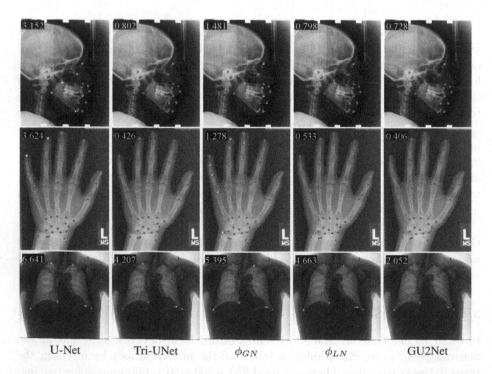

| U-Net | Tri-UNet | ϕ_{GN} | ϕ_{LN} | GU2Net |

Fig. 2. Qualitative comparison of different models on head, hand, and chest datasets. All images are randomly selected. The red points • are the learned landmarks while the green points • are the ground truth labels. The MRE value is displayed on the top left corner of the image for reference. (Color figure online)

of anatomical landmarks. Since U-Net only has shared parameters, its performance is the worst among all models and falls behind others by a huge gap. U-Net is even inferior to our global network ϕ_{GN} which consists of small amount of domain-specific parameters. By comparing GU2Net with Tri-UNet and U-Net, our GU2Net behaves better than them under all the measurements even though the number of parameters of GU2Net is about 3 times less than that of U-Net and 10 times less than Tri-UNet's, which demonstrates the superiority of our architecture and the indispensability of both shared and domain-specific parameters.

To qualitatively show the superiority of our universal model, we further visualize the learned landmarks in Fig. 2. The MRE value is displayed on the top left of the image for reference. The red points are the learned landmarks while the green points are the ground truth labels. It's clear that our model's results have more overlap regions of red and green points than other models', which also can be verified according to the MRE values.

4   Conclusions

To build a universal model, we propose our global universal U-Net to integrate the global and local features, both of which are beneficial for accurately localizing landmarks. Our universal model makes the first attempt in the literature to train a single network on multiple datasets for landmark detection. Using the separable convolution makes it possible to learn multi-domain information, even with a reduced number of parameters. Experimental results qualitatively and quantitatively show that our proposed model performs better than other models trained on multiple datasets and even better than models trained on a single dataset separately. Future works include designing different network architectures and exploring more datasets to further improve the performances on all datasets simultaneously.

References

1. Candemir, S., et al.: Lung segmentation in chest radiographs using anatomical atlases with nonrigid registration. IEEE Trans. Med. Imaging **33**(2), 577–590 (2013)
2. Chiras, J., Depriester, C., Weill, A., Sola-Martinez, M., Deramond, H.: Percutaneous vertebral surgery. technics and indications. J. Neuroradiol.= Journal de neuroradiologie **24**(1), 45–59 (1997)
3. Glorot, X., Bordes, A., Bengio, Y.: Deep sparse rectifier neural networks. In: Proceedings of the Fourteenth International Conference on Artificial Intelligence and Statistics, pp. 315–323. JMLR Workshop and Conference Proceedings (2011)
4. Huang, C., Han, H., Yao, Q., Zhu, S., Zhou, S.K.: 3D U²-Net: a 3D universal u-net for multi-domain medical image segmentation. In: Shen, D., et al. (eds.) MICCAI 2019. LNCS, vol. 11765, pp. 291–299. Springer, Cham (2019). https://doi.org/10.1007/978-3-030-32245-8_33

5. Ibragimov, B., Korez, R., Likar, B., Pernuš, F., Xing, L., Vrtovec, T.: Segmentation of pathological structures by landmark-assisted deformable models. IEEE Trans. Med. Imaging **36**(7), 1457–1469 (2017)
6. Ibragimov, B., Likar, B., Pernuš, F., Vrtovec, T.: Shape representation for efficient landmark-based segmentation in 3-d. IEEE Trans. Med. Imaging **33**(4), 861–874 (2014)
7. Ioffe, S., Szegedy, C.: Batch normalization: accelerating deep network training by reducing internal covariate shift. In: International Conference on Machine Learning, pp. 448–456. PMLR (2015)
8. Jaeger, S., et al.: Automatic tuberculosis screening using chest radiographs. IEEE Trans. Med. Imaging **33**(2), 233–245 (2013)
9. Lange, T., et al.: 3d ultrasound-ct registration of the liver using combined landmark-intensity information. Int. J. Comput. Assist. Radiol. Surg. **4**(1), 79–88 (2009)
10. Lay, N., Birkbeck, N., Zhang, J., Zhou, S.K.: Rapid multi-organ segmentation using context integration and discriminative models. In: Gee, J.C., Joshi, S., Pohl, K.M., Wells, W.M., Zöllei, L. (eds.) IPMI 2013. LNCS, vol. 7917, pp. 450–462. Springer, Heidelberg (2013). https://doi.org/10.1007/978-3-642-38868-2_38
11. Li, H., Han, H., Zhou, S.K.: Bounding maps for universal lesion detection. In: Martel, A.L., et al. (eds.) MICCAI 2020. LNCS, vol. 12264, pp. 417–428. Springer, Cham (2020). https://doi.org/10.1007/978-3-030-59719-1_41
12. Lian, C., et al.: Multi-task dynamic transformer network for concurrent bone segmentation and large-scale landmark localization with dental CBCT. In: Martel, A.L., et al. (eds.) MICCAI 2020. LNCS, vol. 12264, pp. 807–816. Springer, Cham (2020). https://doi.org/10.1007/978-3-030-59719-1_78
13. Lindner, C., Bromiley, P.A., Ionita, M.C., Cootes, T.F.: Robust and accurate shape model matching using random forest regression-voting. IEEE Trans. Pattern Anal. Mach. Intell. **37**(9), 1862–1874 (2014)
14. Liu, D., Zhou, S.K., Bernhardt, D., Comaniciu, D.: Search strategies for multiple landmark detection by submodular maximization. In: 2010 IEEE Conference on Computer Vision and Pattern Recognition (CVPR), pp. 2831–2838. IEEE (2010)
15. Lowe, D.G.: Object recognition from local scale-invariant features. In: Proceedings of the Seventh IEEE International Conference on Computer Vision, vol. 2, pp. 1150–1157. IEEE (1999)
16. Payer, C., Štern, D., Bischof, H., Urschler, M.: Integrating spatial configuration into heatmap regression based cnns for landmark localization. Med. Image Anal. **54**, 207–219 (2019)
17. Ronneberger, O., Fischer, P., Brox, T.: U-Net: convolutional networks for biomedical image segmentation. In: Navab, N., Hornegger, J., Wells, W.M., Frangi, A.F. (eds.) MICCAI 2015. LNCS, vol. 9351, pp. 234–241. Springer, Cham (2015). https://doi.org/10.1007/978-3-319-24574-4_28
18. Smith, L.N.: Cyclical learning rates for training neural networks. In: 2017 IEEE Winter Conference on Applications of Computer Vision (WACV), pp. 464–472. IEEE (2017)
19. Štern, D., Ebner, T., Urschler, M.: From local to global random regression forests: exploring anatomical landmark localization. In: Ourselin, S., Joskowicz, L., Sabuncu, M.R., Unal, G., Wells, W. (eds.) MICCAI 2016. LNCS, vol. 9901, pp. 221–229. Springer, Cham (2016). https://doi.org/10.1007/978-3-319-46723-8_26
20. Urschler, M., Ebner, T., Štern, D.: Integrating geometric configuration and appearance information into a unified framework for anatomical landmark localization. Med. Image Anal. **43**, 23–36 (2018)

21. Wang, C.W., et al.: A benchmark for comparison of dental radiography analysis algorithms. Med. Image Anal. **31**, 63–76 (2016)
22. Wang, P., et al.: Understanding convolution for semantic segmentation. In: 2018 IEEE Winter Conference on Applications of Computer Vision (WACV), pp. 1451–1460. IEEE (2018)
23. Yang, D., et al.: Deep image-to-image recurrent network with shape basis learning for automatic vertebra labeling in large-scale 3d ct volumes. In: Descoteaux, M., Maier-Hein, L., Franz, A., Jannin, P., Collins, D.L., Duchesne, S. (eds.) MICCAI 2017. LNCS, vol. 10435, pp. 498–506. Springer, Cham (2017). https://doi.org/10.1007/978-3-319-66179-7_57
24. Yao, Q., He, Z., Han, H., Zhou, S.K.: Miss the point: targeted adversarial attack on multiple landmark detection. In: Martel, A.L., et al. (eds.) MICCAI 2020. LNCS, vol. 12264, pp. 692–702. Springer, Cham (2020). https://doi.org/10.1007/978-3-030-59719-1_67
25. Zhou, S.K., et al.: A review of deep learning in medical imaging: imaging traits, technology trends, case studies with progress highlights, and future promises. In: Proceedings of the IEEE (2021)
26. Zhou, S.K., Rueckert, D., Fichtinger, G.: Handbook of Medical Image Computing and Computer Assisted Intervention. Academic Press, Cambridge (2019)
27. Zhou, S.K.: Shape regression machine and efficient segmentation of left ventricle endocardium from 2d b-mode echocardiogram. Med. Image Anal. **14**(4), 563–581 (2010)

A Coherent Cooperative Learning Framework Based on Transfer Learning for Unsupervised Cross-Domain Classification

Xinxin Shan, Ying Wen[✉], Qingli Li, Yue Lu, and Haibin Cai

Shanghai Key Laboratory of Multidimensional Information Processing,
School of Communication and Electronic Engineering,
East China Normal University, Shanghai, China
ywen@cs.ecnu.edu.cn

Abstract. In the practical application of medical image analysis, due to the different data distributions of source domain and target domain and the lack of the labels of target domain, domain adaptation for unsupervised cross-domain classification attracts widespread attention. However, current methods take knowledge transfer model and classification model as two separate training stages, which inadequately considers and utilizes the intrinsic information interaction between modules. In this paper, we propose a coherent cooperative learning framework based on transfer learning for unsupervised cross-domain classification. The proposed framework is constructed by two classifiers trained by transfer learning, which can respectively classify images of source domain and target domain, and a Wasserstein CycleGAN for image translation and data augmentation. In the coherent process, all modules are updated in turn, and the data is transferred between different modules to realize the knowledge transfer and collaborative training. The final prediction is obtained by a voting result of two classifiers. Experimental results on three pneumonia databases demonstrate the effectiveness of our framework with diverse backbones.

Keywords: Unsupervised cross-domain classification · Transfer learning · Collaborative training · Wasserstein CycleGAN

1 Introduction

Many of the basic assumptions in machine learning are based on the fact that the data of source domain (training dataset) and the target domain (testing dataset) is independent and identically distributed. When the data distributions of source domain and target domain are different, domain adaptation for

Electronic supplementary material The online version of this chapter (https://doi.org/10.1007/978-3-030-87240-3_10) contains supplementary material, which is available to authorized users.

cross-domain classification becomes an effective solution [9]. Generally, similarity [19], data global structure [16,31] and feature alignment [18] are advisable solutions to cross-domain classification. In addition, many measurement criteria such as maximum mean discrepancy (MMD) [27] and multiple kernel MMD [15] are generally applied in domain adaptation and have inspired many recognition models [6,8,28,35]. Fine-tuning pre-trained models [20,23] and domain adversarial neural network [5,32] are also two efficient techniques for cross-domain classification. Thus, domain adaptation is critical to promote the generalization ability of neural network [7].

When lacking the labels of target domain in cross-domain classification, unsupervised domain adaptation has more profound significance [2]. For instance, CycleGAN [33] and its variants [14,17,29] can translate images from target to source [13,34] to make the distribution approximate for classification [24]. However, there are still some issues that need further consideration. (i) Class-imbalanced datasets [11]: commonly exist in medical datasets, which may result in over-fitting during model training process and degrade the performance. (ii) Lack of labeled medical images [13]: leads to poor use of supervised learning. Fortunately, transfer learning [7,15] and unsupervised learning [2,18,26] are employed to deal with such challenges. (iii) Separate training of different modules: current methods separately train knowledge transfer model and classification model from beginning to end [4]. That is, the training process is divided into two separate stages, which ignores knowledge transfer and information interaction between modules.

To address the above issues, we propose a coherent cooperative learning (CCL) framework based on transfer learning for unsupervised cross-domain classification, which is constructed by a proposed Wasserstein CycleGAN (WCycle-GAN) for image translation and two classifiers for prediction. First, by training the WCycleGAN with the original images from the source domain and the target domain, we obtain a class-balanced dataset used to fine-tune two classifiers that are convolutional neural networks (CNNs) pre-trained on ImageNet. Specially, the classifier of the target domain uses the proposed cooperative mechanism and MMD criterion to achieve unsupervised cross-domain classification. Finally, we input both the probe image and its translated image generated by WCycleGAN into two classifiers, and get the final prediction by a voting strategy.

There are three contributions in the proposed method. (i) The proposed WCycleGAN and two classifiers are iteratively updated in a united process. (ii) The proposed collaborative training makes different modules complement each other. (iii) Knowledge transfer is reflected from three aspects in CCL: image translation by WCycleGAN, transfer learning by fine-tuning the CNN pre-trained on ImageNet to identify pneumonia, and the parameter passing in fine-tuning.

2   Method

The proposed CCL is constructed by a WCycleGAN G and two classifiers C_t, C_s, as shown in Fig. 1. The details will be introduced in the following section.

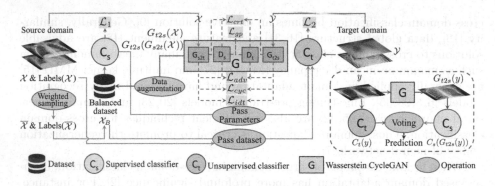

Fig. 1. The training (a) and testing (b) diagrams of the proposed method.

2.1 Wasserstein CycleGAN in the Proposed Method

WCycleGAN is designed for data augmentation and domain translation. We are inspired by Wasserstein GAN [1], Wasserstein GAN with gradient penalty [10] and CycleGAN [33] to develop the proposed WCycleGAN. The inputs of the source domain and the target domain are respectively described as labeled $\mathcal{X} = \{x_i\}_{i=1}^n$ and unlabeled $\mathcal{Y} = \{y_i\}_{i=1}^n$, where n is the number of images in a batch.

To solve the issue of class-imbalanced dataset, we construct an extreme class-balanced dataset $\mathcal{X}_B = \{x_i\}_{i=1}^m$ by G, where m is the total number of images and the number of images per class is the same. As shown in Fig. 1 (a), in order to construct $\mathcal{X}_B$, we first obtain a balanced dataset $\overline{\mathcal{X}}$ by performing weighted sampling on $\mathcal{X}$. Then we supplement $\overline{\mathcal{X}}$ with some intermediate images, which are generated by G during the calculation of losses. In this way, some images may be resampled for $\mathcal{X}_B$. Next, we will elaborate on the acquisition of the intermediate images during training G.

We respectively denote $P_\mathcal{X}$ and $P_\mathcal{Y}$ as the data distributions of the source domain and the target domain. There are four components in WCycleGAN: the two discriminators D_s and D_t can distinguish the domain of images; the two generators are able to translate images, i.e. $G_{s2t}: \mathcal{X} \to \mathcal{Y}$ and $G_{t2s}: \mathcal{Y} \to \mathcal{X}$.

The adversarial loss constrains the generators and discriminators:

$$\mathcal{L}_{adv} = \mathop{\mathbb{E}}_{x \sim P_\mathcal{X}} [-D_s(G_{s2t}(x))] + \mathop{\mathbb{E}}_{y \sim P_\mathcal{Y}} [-D_t(G_{t2s}(y))] \tag{1}$$

In order to enforce G_{s2t} and G_{t2s} to be cycle-consistent with each other, we calculate the cycle consistency loss:

$$\mathcal{L}_{cyc} = \mathop{\mathbb{E}}_{x \sim P_\mathcal{X}} [\|G_{t2s}(G_{s2t}(x)) - x\|_1] + \mathop{\mathbb{E}}_{y \sim P_\mathcal{Y}} [\|G_{t2s}(G_{s2t}(y)) - y\|_1] \tag{2}$$

To further ensure the generation capacity of the generator, we also introduce the identity mapping loss [26]:

$$\mathcal{L}_{idt} = \mathop{\mathbb{E}}_{x \sim P_\mathcal{X}} [\|G_{t2s}(x) - x\|_1] + \mathop{\mathbb{E}}_{y \sim P_\mathcal{Y}} [\|G_{s2t}(y) - y\|_1] \tag{3}$$

It is noted that G_{t2s} and G_{s2t} are generators shown in Fig. 1 (a), from which we can get $G_{t2s}(G_{t2s}(x))$ of Eq. (2) and $G_{t2s}(x)$ of Eq. (3) as the intermediate images for data augmentation. The overall loss of generators is:

$$\mathcal{L}_{gen} = \mathcal{L}_{adv} + \mathcal{L}_{cyc} + \mathcal{L}_{idt} \tag{4}$$

Then, we take D_s for example to illustrate the optimal objective of discriminators, and its critic loss is:

$$\mathcal{L}_{crt}(D_s, G_{t2s}, \mathcal{X}, \mathcal{Y}) = \mathop{\mathbb{E}}_{y \sim P_y}[D_s(G_{t2s}(y))] - \mathop{\mathbb{E}}_{x \sim P_{\mathcal{X}}}[D_s(x)] \tag{5}$$

On this basis, the gradient penalty term is added to constrain the gradient norm of the outputs of discriminator:

$$\mathcal{L}_{gp}(D_s, \mathcal{X}) = \mathop{\mathbb{E}}_{\hat{x} \sim P_{\hat{x}}}[\|(\nabla_{\hat{x}} D_s(\hat{x})\|_2 - 1)^2] \tag{6}$$

where $\hat{x} = \epsilon x + (1 - \epsilon)G_{s2t}(x)$ represents sampling from source data x and generating data $G_{s2t}(x)$, and $\epsilon \in U([0, 1])$ is a random number. Therefore, the overall loss of D_s is:

$$\mathcal{L}_{dis}(D_s, G_{t2s}, \mathcal{X}, \mathcal{Y}) = \lambda_1 \mathcal{L}_{crt}(D_s, G_{t2s}, \mathcal{X}, \mathcal{Y}) + \lambda_2 \mathcal{L}_{gp}(D_s, \mathcal{X}) \tag{7}$$

where λ_1 and λ_2 are two hyper-parameters. Similarly, $\mathcal{L}_{dis}(D_t, G_{s2t}, \mathcal{Y}, \mathcal{X})$ is the loss for D_t. Finally, the optimization task of WCycleGAN is formulated by:

$$\mathcal{L}_{total} = \mathcal{L}_{gen} + \mathcal{L}_{dis}(D_s, G_{t2s}, \mathcal{X}, \mathcal{Y}) + \mathcal{L}_{dis}(D_t, G_{s2t}, \mathcal{Y}, \mathcal{X}) \tag{8}$$

The WCycleGAN G is an important part of CCL: $\mathcal{X}_B$ is formed with the help of G, and G is able to translate images to a given domain.

2.2   Cooperative Training of Classifiers Based on Transfer Learning

The classifiers C_s and C_t in a collaborative way are trained with a batch during each iterative update, and they are respectively designed for the classification tasks of the source domain and the target domain.

Supervised Classifier C_s is obtained by fine-tuning a CNN that pre-trained on ImageNet. It can be used to classify images of the source domain and conditionally assist the training of C_t.

In the current iteration, after constructing $\mathcal{X}_B$ by G and feeding $\mathcal{X}_B$ into C_s, we can get the feature maps $\mathcal{F} = \{f_i\}_{i=1}^m \in \mathbb{R}^{m \times (D \times K)}$ before the final softmax layer, where D is the dimension of images and K is the number of classes. Each $x_i \in \mathcal{X}_B$ in the source domain is labeled as $Label(x_i)$, so $f_{i,Label(x_i)}$ is a D-dimensional feature map of x_i. The objective function of C_s is a combination of the cross entropy loss $\mathcal{L}_S$ and the center loss $\mathcal{L}_C$ [30] balanced by a hyper-parameter λ_3:

$$\mathcal{L}_1 = \mathcal{L}_S + \lambda_3 \mathcal{L}_C = -\sum_{i=1}^{m} log \left(\frac{e^{f_{i,Label(x_i)}}}{\sum_{k=1}^{K} e^{f_{i,k}}} \right) + \frac{\lambda_3}{2} \sum_{i=1}^{m} \left\| f_{i,Label(x_i)} - c_{l_i} \right\|_2^2 \quad (9)$$

where c_{l_i} is the feature center of the l_i class, and l_i is the label of the i-th images.

Unsupervised Classifier C_t is also obtained by fine-tuning the CNN that pretrained on ImageNet, which is an unsupervised classifier and can classify images of the target domain. During each iterative training of C_t, we exploit labeled $\mathcal{X}$, $\mathcal{X}_B$ and unlabeled $\mathcal{Y}$.

The unsupervised classifier C_t needs to receive the parameters and images from the labeled source domain, so as to realize unsupervised domain adaptation well. During the iterative training process, as shown in Fig. 1 (a), we design a cooperative mechanism to control C_s to pass parameters to C_t under two conditions: (i) the training/verification accuracy of C_s does not reach the threshold τ; (ii) the predictions of C_t per batch all belong to a certain class (over-fitting).

When the above conditions are not satisfied in the current iteration, C_t will train by itself to update, in which MMD [27] criterion is utilized to minimize the distribution distance between $\mathcal{X}$ and $\mathcal{Y}$. The squared MMD distance is:

$$d(\mathcal{X}, \mathcal{Y})^2 = \left\| \frac{1}{n} \left(\sum_{x_i \in \mathcal{X}} \phi(x_i) - \sum_{y_i \in \mathcal{Y}} \phi(y_i) \right) \right\|_H^2 \quad (10)$$

where $\phi(\cdot)$ is the mapping function corresponding to the Gaussian kernel, and the subscript H means the distance is measured by using $\phi(\cdot)$ to map the data into reproducing kernel Hilbert space.

By combining Eq. (10) with Eq. (9) using a hyper-parameter λ_4, the optimized object of C_t is:

$$\mathcal{L}_2 = \mathcal{L}_1 + \lambda_4 d(\mathcal{X}, \mathcal{Y})^2 \quad (11)$$

2.3   Prediction

After the above iterative training, C_s and C_t are able to classify images of the corresponding domain, and G can translate the original images to a given domain. The output of each classifier is the predicted probability for each class. For obtaining more reliable prediction, we utilize a voting strategy of Fig. 1 (b) to fuse predictions.

For a probe y from the target domain, we can obtain $y' = G_{t2s}(y)$ translated by G, whose distribution is as consistent as possible with $\mathcal{X}$. Then we use both y and y' for prediction. The label of y is expressed as $Label(y) = l_k$, where l_k represents the label of the k-th class. The indicator k is worked out by:

$$k = \begin{cases} mode(\arg\max_k \{C_s^k(y'), C_t^k(y)\}), & \text{if it is unique} \\ \arg\max_k (C_s^k(y') + C_t^k(y)) & \text{else} \end{cases} \quad (12)$$

where the '$mode$' of Eq. (12) chooses a value that appears the most frequently.

3 Experiments

3.1 Databases and Settings

Databases. We use three databases in the experiments, and their information is shown in Table 1. The Chest X-Ray[1] is divided into training dataset and testing dataset [12]. Single lesion[2] and Multiple lesions[3] are the training datasets of two open lesion recognition competitions. We name them according to the number of lesions: each image consists of at most/least one lesion.

Table 1. The number of images and examples of (a) Chest X-Ray database; (b) Single lesion database; (c) Multiple lesions database. Their data distributions vary due to different imaging equipment/standards.

Name	Normal	Pneumonia	Sum	Class-imbalanced
Chest X-Ray (for training)	1349	3883	5232	✓
Chest X-Ray (for testing)	234	390	624	✓
Single lesion	20025	5659	25684	✓
Multiple lesions	15504	4509	20013	✓

Comparative Methods. Kermany et al. [12], Ayan et al. [3] and Gu et al. [7] (a two-step progressive transfer learning technique) are all the transfer learning methods for medical image classification. Additionally, we introduce the domain-adversarial training of neural networks (DANN) [5] and [7]-GAN (an adversarial learning technique with CycleGAN) for cross-domain classification.

Settings. All the experiments are carried out on Intel® Xeon® Gold 6230, and CCL is implemented with Pytorch 1.6.0. We set $\lambda_1 = 0.5$ [33] and $\lambda_2 = 10.0$ [10] in Eq. (7), $\lambda_3 = 1.0$ [30] in Eq. (9) and $\lambda_4 = 2.0$ [27] in Eq. (11). During training, the initial learning rate for SGD optimizer is 0.001, the weight decay is 0.0005 and the threshold τ of cooperative mechanism is 0.8 (see the Supplement Sect. 1 for the related experiments). The number per batch $n = 32$, and one batch constitutes n mini-batches including 1 images in training WCycleGAN. We can load a simple pre-trained WCycleGAN to speed up the convergence, and the training can be completed within 10 epochs with the optimal model stored

---

[1] https://data.mendeley.com/datasets/rscbjbr9sj/2/files/.
[2] https://www.kaggle.com/c/rsna-pneumonia-detection-challenge/data.
[3] https://god.yanxishe.com/18.

after every 50 batches. The visualization of the training process for verifying the performance of CCL can be seen from the Supplement Sect. 2.

Evaluation Metrics. The evaluation metrics are accuracy, precision, recall and F1 score. We take the normal diagnosis as the positive in this paper.

3.2 Performance of the Supervised Classifier

Table 2. Results of supervised classification on Chest X-Ray.

Method	Accuracy	Precision	Recall	F1 score	Backbone
Kermany *et al.* [12]	83.97%	65.81%	88.51%	0.75	Inception v3
Ayan *et al.* [3]	82.00%	73.08%	85.00%	0.75	Xception
Ayan *et al.* [3]	87.00%	78.63%	85.19%	0.82	Vgg16
Gu *et al.* [7]	88.00%	47.86%	91.80%	0.63	ResNet 152
CCL	89.42%	79.91%	90.78%	0.85	MobileNet v2
CCL	88.94%	76.92%	92.31%	0.84	Inception v3
CCL	**91.35%**	**80.34%**	**95.92%**	**0.87**	Vgg16

We first verify the performance of the supervised classifier C_s on Chest X-Ray. Consider the experimental results with diverse backbones, we choose MobileNetV2 (fine-tune the last 6 layers) [21], InceptionV3 (fine-tune the classification layer) [25] and Vgg16 (fine-tune the last 5 layers) [22] as backbones. From Table 2, C_s based on diverse backbones is always superior compared with other methods, which indicates that C_s is capable of classification in the same domain very well.

3.3 Evaluation of Unsupervised Cross-Domain Classification

We follow the common leave-one-domain-out strategy as [18], and use the three databases in pairs to test the performance of CCL on cross-domain classification. We get the average result of 3 runs. To ensure the comparison and repeatability of experimental results, we use the existed divided datasets for training/testing/validation, instead of using cross-validation that will change the composition of datasets. All samples of the source domain are used for training, and Chest X-Ray (for training) and Chest X-Ray (for testing) in the target domain are respectively used for validation and testing to ascertain the iterations when convergence. Then we can apply the iterations as stop criteria to test other datasets in the target domain.

From the results in Table 3, CCL has the best accuracy and F1 score for all tasks. For different classification task, diverse backbones have their own advantages: (i) when training with the large dataset and the limited device memory, the lightweight network such as MobileNetV2 is a priority; (ii) complex CNNs have better ability to prevent over-fitting to some extent.

Table 3. Results of unsupervised cross-domain classification

Training on Chest X-Ray. Testing on (a) Single lesion; (b) Multiple lesions.

IDX	Method	Acc	Precision	Recall	F1 score	IDX	Acc	Precision	Recall	F1 score	Backbone
(a)	DANN [5]	0.57	0.51	**0.89**	0.65	(b)	0.54	0.44	**0.92**	0.60	–
	[12]	0.24	0.95	0.03	0.05		0.25	0.95	0.03	0.05	InceptionV3
	[3]	0.60	0.55	0.89	0.68		0.53	0.87	0.46	0.60	Vgg16
	[7]	0.77	**0.99**	0.77	0.87		0.71	0.91	0.76	0.83	ResNet152
	[7]-GAN	0.33	0.18	0.83	0.30		0.23	0.17	0.66	0.00	ResNet152
	CCL	0.73	0.81	0.83	0.82		0.62	0.60	0.87	0.71	As DANN
	CCL	**0.78**	**0.99**	0.78	**0.88**		0.74	0.89	0.80	0.84	MobileNetV2
	CCL	0.50	0.56	0.73	0.63		**0.77**	**0.99**	0.77	**0.87**	InceptionV3
	CCL	0.76	0.89	0.82	0.85		0.76	0.92	0.80	0.86	Vgg16

Training on Single lesion. Testing on (a) Chest X-Ray; (b) Multiple lesions.

IDX	Method	Acc	Precision	Recall	F1 score	IDX	Acc	Precision	Recall	F1 score	Backbone
(a)	DANN [5]	0.77	0.66	0.73	0.70	(b)	0.72	0.73	0.89	0.80	–
	[12]	0.74	0.62	0.76	0.68		0.73	0.93	0.71	0.80	InceptionV3
	[3]	0.77	0.73	0.70	0.71		0.77	0.76	**0.93**	0.84	Vgg16
	[7]	0.65	0.66	0.53	0.59		0.23	0.02	0.56	0.04	ResNet152
	[7]-GAN	0.38	**0.97**	0.38	0.54		0.73	0.93	0.77	0.84	ResNet152
	CCL	0.78	0.45	0.98	0.62		0.72	0.70	0.92	0.80	As DANN
	CCL	0.71	0.36	**0.77**	0.49		**0.79**	0.89	0.85	0.87	MobileNetV2
	CCL	0.67	0.43	0.60	0.50		0.78	**0.99**	0.79	**0.88**	InceptionV3
	CCL	**0.79**	0.81	0.69	**0.75**		0.72	0.70	0.91	0.79	Vgg16

Training on Multiple lesions. Testing on (a) Chest X-Ray; (b) Single lesion.

IDX	Method	Acc	Precision	Recall	F1 score	IDX	Acc	Precision	Recall	F1 score	Backbone
(a)	DANN [5]	0.75	0.81	0.64	0.72	(b)	0.77	0.76	0.92	0.84	–
	[12]	0.69	0.55	**0.91**	0.69		0.77	0.94	0.75	0.83	InceptionV3
	[3]	0.73	0.76	0.62	0.68		0.77	0.76	**0.93**	0.84	Vgg16
	[7]	0.59	0.42	0.45	0.43		0.55	0.66	0.73	0.69	ResNet152
	[7]-GAN	0.38	**0.98**	0.38	0.55		0.58	0.65	0.77	0.70	ResNet152
	CCL	0.77	0.61	0.76	0.68		0.77	0.83	0.87	0.85	As DANN
	CCL	0.71	0.26	0.79	0.39		**0.79**	0.97	0.80	**0.88**	MobileNetV2
	CCL	0.68	0.27	0.71	0.39		**0.79**	**0.98**	0.80	**0.88**	InceptionV3
	CCL	**0.80**	0.72	0.74	**0.73**		0.77	0.80	0.89	0.84	Vgg16

3.4 Visualization and Ablation Experiments

Figure 2 is the visualization of translating original images. Figure 2 (b) and (c) are respectively translated to another domain by CycleGAN and WCycleGAN, from which it is clear that WCycleGAN has the better generation capacity (e.g. the edges of ribs are as clear as the images of the source domain) than CycleGAN. Figure 2 (d) and (e) are respectively the intermediates in Eq. (2) and Eq. (3) and also used for data augmentation, which are similar to Fig. 2 (a) but not identical.

We take Vgg16 as the backbone to do the ablation experiments of method[#]1 wo passing balanced dataset, method[#]2 wo passing parameters and method[#]3 wo generating balanced dataset, which are the operations of Fig. 1. From Fig. 3, it is obvious that method[#]3 causes severe over-fitting, and method[#]1 and [#]2 are also greatly affected. Hence, these operations in CCL are all essential and beneficial.

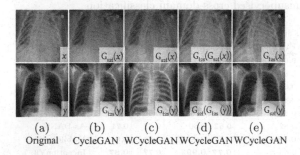

(a)   (b)   (c)   (d)   (e)
Original   CycleGAN WCycleGAN WCycleGAN WCycleGAN

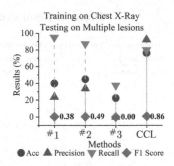

Fig. 2. The visual image translation between Chest X-Ray (row1) and Multiple lesions (row2).

Fig. 3. The results of ablation experiments.

4   Discussion and Conclusion

In this paper, we present an effective framework named CCL based on transfer learning for unsupervised cross-domain classification. The class-balanced dataset of CCL contributes to avoiding over-fitting. Besides, the proposed method can overcome the problem of insufficient labels in medical data by combining transfer learning and unsupervised learning. During the training and testing process, WCycleGAN and two classifiers complement each other by cooperative learning, whose backbones can be flexibly modified to obtain competitive results. Experiments on three pneumonia databases indicate that the propose method achieves promising performance in unsupervised cross-domain classification.

Acknowledgements. This work was supported in part by 2030 National Key Research and Development Program of China (2018AAA0100500), the National Nature Science Foundation of China (no. 61773166), Projects of International Cooperation of Shanghai Municipal Science and Technology Committee (14DZ2260800), the Fundamental Research Funds for the Central Universities, and the ECNU Academic Innovation Promotion Program for Excellent Doctoral Students (YBNLTS2021-040).

References

1. Arjovsky, M., Chintala, S., Bottou, L.: Wasserstein GaN. In: ICML, pp. 1–18 (2017)
2. Armanious, K., Jiang, C., Abdulatif, S., Küstner, T., Gatidis, S., Yang, B.: Unsupervised medical image translation using Cycle-MeDGAN. In: EUSIPCO, pp. 1–5 (2019)
3. Ayan, E., Ünver, H.M.: Diagnosis of pneumonia from chest X-ray images using deep learning. In: EBBT, pp. 7–11 (2019)
4. Chiou, E., Giganti, F., Punwani, S., Kokkinos, I., Panagiotaki, E.: Harnessing uncertainty in domain adaptation for MRI prostate lesion segmentation. In: Martel, A.L., et al. (eds.) MICCAI 2020. LNCS, vol. 12261, pp. 510–520. Springer, Cham (2020). https://doi.org/10.1007/978-3-030-59710-8_50
5. Ganin, Y., et al.: Domain-adversarial training of neural networks. J. Mach. Learn. Res. **17**, 189–209 (2016)

6. Ghifary, M., Kleijn, W.B., Zhang, M.: Domain adaptive neural networks for object recognition. In: Pham, D.-N., Park, S.-B. (eds.) PRICAI 2014. LNCS (LNAI), vol. 8862, pp. 898–904. Springer, Cham (2014). https://doi.org/10.1007/978-3-319-13560-1_76

7. Gu, Y., Ge, Z., Bonnington, C.P., Zhou, J.: Progressive transfer learning and adversarial domain adaptation for cross-domain skin disease classification. IEEE J. Biomed. Health Inform. 24(5), 1379–1393 (2020)

8. Han, X., Wang, J., Zhou, W., Chang, C., Ying, S., Shi, J.: Deep doubly supervised transfer network for diagnosis of breast cancer with imbalanced ultrasound imaging modalities. In: Martel, A.L., et al. (eds.) MICCAI 2020. LNCS, vol. 12266, pp. 141–149. Springer, Cham (2020). https://doi.org/10.1007/978-3-030-59725-2_14

9. He, Y., Carass, A., Zuo, L., Dewey, B.E., Prince, J.L.: Self domain adapted network. In: Martel, A.L., et al. (eds.) MICCAI 2020. LNCS, vol. 12261, pp. 437–446. Springer, Cham (2020). https://doi.org/10.1007/978-3-030-59710-8_43

10. Ishaan G., Faruk A., Martin A., Vincent D., Aaron, C.: Improved training of Wasserstein GANs. In: NeurIPS, pp. 1–11 (2017)

11. Jamal, M.A., Brown, M., Yang, M.H., Wang, L., Gong, B.: Rethinking class-balanced methods for long-tailed visual recognition from a domain adaptation perspective. In: CVPR, pp. 7610–7619 (2020)

12. Kermany, D.S., et al.: Identifying medical diagnoses and treatable diseases by image-based deep learning. Cell 172(5), 1122.e9–1131.e9 (2018)

13. Liu, J., Li, J., Liu, T., Tam, J.: Graded image generation using stratified Cycle-GAN. In: Martel, A.L., et al. (eds.) MICCAI 2020. LNCS, vol. 12262, pp. 760–769. Springer, Cham (2020). https://doi.org/10.1007/978-3-030-59713-9_73

14. Liu, J., et al.: Illumination-invariant flotation froth color measuring via Wasserstein distance-based CycleGAN with structure-preserving constraint. IEEE Trans. Cybern. 51, 839–852 (2020)

15. Long, M., Cao, Y., Wang, J., Jordan, M.I.: Learning transferable features with deep adaptation networks. In: ICML, pp. 97–105 (2015)

16. Ma, Y., Xu, Y., Liu, F.: Multi-perspective dynamic features for cross-database face presentation attack detection. IEEE Access 8, 26505–26516 (2020)

17. McDermott, M.B., et al.: Semi-supervised biomedical translation with cycle Wasserstein regression GaNs. In: AAAI, pp. 2363–2370 (2018)

18. Meng, Q., Rueckert, D., Kainz, B.: Unsupervised cross-domain image classification by distance metric guided feature alignment. In: Hu, Y., et al. (eds.) ASMUS/PIPPI -2020. LNCS, vol. 12437, pp. 146–157. Springer, Cham (2020). https://doi.org/10.1007/978-3-030-60334-2_15

19. Ali, N., Neagu, D., Trundle, P.: Classification of heterogeneous data based on data type impact on similarity. In: Lotfi, A., Bouchachia, H., Gegov, A., Langensiepen, C., McGinnity, M. (eds.) UKCI 2018. AISC, vol. 840, pp. 252–263. Springer, Cham (2019). https://doi.org/10.1007/978-3-319-97982-3_21

20. Othman, E., Werner, P., Saxen, F., Al-Hamadi, A., Walter, S.: Cross-database evaluation of pain recognition from facial video. In: International Symposium on Image and Signal Processing and Analysis, pp. 181–186 (2019)

21. Sandler, M., Howard, A., Zhu, M., Zhmoginov, A., Chen, L.C.: MobileNetV2: inverted residuals and linear bottlenecks. In: CVPR, pp. 4510–4520 (2018)

22. Simonyan, K., Zisserman, A.: Very deep convolutional networks for large-scale image recognition. In: ICLR, pp. 1–14 (2015)

23. Su, Y., Fu, Y., Tian, Q., Gao, X.: Cross-database age estimation based on transfer learning. In: ICASSP, pp. 1270–1273 (2010)

24. Sun, Y., Yang, G., DIng, D., Cheng, G., Xu, J., Li, X.: A GAN-based domain adaptation method for glaucoma diagnosis. In: IJCNN, pp. 1–8 (2020)
25. Szegedy, C., Vanhoucke, V., Ioffe, S., Shlens, J., Wojna, Z.: Rethinking the inception architecture for computer vision. In: CVPR, pp. 2818–2826 (2016)
26. Taigman, Y., Polyak, A., Wolf, L.: Unsupervised cross-domain image generation. In: ICLR, pp. 1–14 (2017)
27. Tzeng, E., Hoffman, J., Zhang, N., Saenko, K., Darrell, T.: Deep domain confusion: maximizing for domain invariance. arXiv preprint arXiv:1412.3474, pp. 1–9 (2014)
28. Venkateswara, H., Eusebio, J., Chakraborty, S., Panchanathan, S.: Deep hashing network for unsupervised domain adaptation. In: CVPR, pp. 5385–5394 (2017)
29. Wang, C., Zhang, F., Yu, Y., Wang, Y.: BR-GAN: bilateral residual generating adversarial network for mammogram classification. In: Martel, A.L., et al. (eds.) MICCAI 2020. LNCS, vol. 12262, pp. 657–666. Springer, Cham (2020). https://doi.org/10.1007/978-3-030-59713-9_63
30. Wen, Y., Zhang, K., Li, Z., Qiao, Yu.: A discriminative feature learning approach for deep face recognition. In: Leibe, B., Matas, J., Sebe, N., Welling, M. (eds.) ECCV 2016. LNCS, vol. 9911, pp. 499–515. Springer, Cham (2016). https://doi.org/10.1007/978-3-319-46478-7_31
31. Xia, K., Ni, T., Yin, H., Chen, B.: Cross-domain classification model with knowledge utilization maximization for recognition of epileptic EEG signals. IEEE/ACM Trans. Comput. Biol. Bioinf. 18, 53–61 (2020)
32. Yang, F.E., Chang, J.C., Tsai, C.C., Wang, Y.C.F.: A Multi-domain and multimodal representation disentangler for cross-domain image manipulation and classification. IEEE Trans. Image Process. 29, 2795–2807 (2020)
33. Zhu, J.Y., Park, T., Isola, P., Efros, A.A.: Unpaired image-to-image translation using cycle-consistent adversarial networks. In: ICCV, pp. 2242–2251 (2017)
34. Zhu, Y., et al.: Cross-domain medical image translation by shared latent Gaussian mixture model. In: Martel, A.L., et al. (eds.) MICCAI 2020. LNCS, vol. 12262, pp. 379–389. Springer, Cham (2020). https://doi.org/10.1007/978-3-030-59713-9_37
35. Zhu, Y., et al.: Multi-representation adaptation network for cross-domain image classification. Neural Netw. 119, 214–221 (2019)

Towards a Non-invasive Diagnosis of Portal Hypertension Based on an Eulerian CFD Model with Diffuse Boundary Conditions

Lixin Ren[1,3], Shang Wan[2], Yi Wei[2], Xiaowei He[1,3(✉)], Bin Song[2(✉)], and Enhua Wu[1,3,4]

[1] SKLCS, Institute of Software, Chinese Academy of Sciences, Beijing, China
xiaowei@iscas.ac.cn
[2] Department of Radiology, West China Hospital, Sichuan University, Chengdu, China
songlab_radiology@163.com
[3] University of Chinese Academy of Sciences, Beijing, China
[4] Faculty of Science and Technology, University of Macau, Macao, China

Abstract. Portal hypertension is one of the major complications in patients with chronic liver diseases (CLD) which induces the increase in portal vein gradient pressure. At advanced stage, it can cause the esophageal varices and variceal hemorrhage. Therefore, portal hypertension has been the leading cause of mortality in CLD patients. To diagnose portal hypertension, the invasive hepatic venous pressure gradient (HVPG) measurement is still the only validated technique to accurately evaluate changes in portal pressure and regarded as the standard reference. However, it entails the limitation of invasive procedure and have the risk of further bleeding and inflammation. In this paper we propose an Eulerian computational fluid dynamics (CFD) model to facilitate hemodynamics analysis. To enable consistent simulation results with different boundary conditions, a diffuse boundary handling technique was proposed to impose smooth boundary conditions for both the pressure and velocity fields. We also propose a computational workflow for quantifying patient-specific hemodynamics in portal vein systems non-invasively. The simulation is performed on patient-specific PV models reconstructed from CT angiographic images. Experiments show that pressure changes in the PV of patients with portal hypertension due to blockage of the RPV are significantly lower than that of normal subjects.

Keywords: Diffuse boundary conditions · Pressure change · Non-invasive diagnosis · Portal hypertension · CFD

L. Ren and S. Wan—These authors contributed equally to this work.

Electronic supplementary material The online version of this chapter (https://doi.org/10.1007/978-3-030-87240-3_11) contains supplementary material, which is available to authorized users.

1 Introduction

Portal hypertension is the hemodynamics abnormality associated with the most severe complications of cirrhosis (including ascites, hepatic encephalopathy and bleeding from gastroesophageal varices), and has emerged as the leading cause of mortality in cirrhotic patients [3]. While liver cirrhosis arise from various causes including viruses, toxins, and genetics, a common theory indicates that portal hypertension mainly arise from the increased blood pressure in portal vein, possibly due to an increased resistance to blood flow through the portal system [15]. According to the anatomy, the portal vein system is joined by the superior mesenteric vein (SMV) and splenic vein (SV), and then divides into the left portal vein (LPV) and the right portal vein (RPV) branches which entering the left and right liver lobes, respectively [2]. At advanced stage of liver cirrhosis, patients with terminal hepatic failure (THF) were observed with atrophy in the right liver lobe while hypertrophy in the left liver lobe [17]. This phenomenon indicates the resistance to blood flow in the right portal vein may be increased, causing increased blood pressure in the veins of the portal system. However, the direct measurement of blood pressure inside the portal vein is difficult. The invasive hepatic venous pressure gradient (HVPG) measurement is still the only validated technique to accurately evaluate changes in portal pressure.

In recent years, computational fluid dynamics (CFD) has shown great potential in hemodynamic analysis, together with noninvasive and invasive imaging techniques [18]. Since CFD itself has a variety of different methods and each has its own advantages and shortcomings, it should be rather careful when adopting a numerical method to study the hepatic flow. For example, the accuracy of the finite element method (FEM) is largely influenced by the temporal/spatial resolutions, numerical solvers and boundary handing techniques [8]. In hemodynamics analysis, the choice of boundary conditions is in fact of particular importance to better reproduce *in vivo* conditions because only a small part of the PV system will be retained for simulation [11]. Despite the recognized importance [5,16], previous works on how to find a CFD-based identifier that can help diagnose portal hypertension noninvasively are rather limited [14] and only a few studies have devoted efforts to addressing the intrinsic issues in computational models to better understanding hemodynamics in PV systems. Most of them only apply the FEM solver to solve hemodynamics in the PV system, yet pay no attention on the numerical problems [12].

In this paper, we propose an Eulerian CFD model to facilitate hemodynamics analysis. Our method is based on the assumption that the increased blood pressure in the portal vein system is caused by the blockage in the blood flow through the cirrhotic liver tissue. For patients with portal hypertension, the large veins are subsequently developed to get around the blockage. Therefore, if the same boundary conditions are applied and a blockage is applied on right portal vein, the pressure change in PV systems of patients with portal hypertension is expected to be smaller than those without portal hypertension. To verify this idea, the following contributions were made: (1) an Eulerian computational fluid dynamics model for quantifying patient-specific hemodynamics in portal vein systems; (2) a diffuse boundary handling technique that is able to impose smooth boundary conditions for both the pressure and velocity fields; (3) a workflow for quantifying pressure changes in PV that helps discriminate patients with portal hypertension from normal subjects.

2   Methodology

2.1   Blood Flow Model

In 3D Eulerian space, the general model for the hemodynamics in portal vein (PV) systems is given by the Navier–Stokes equation

$$\rho\left(\frac{\partial \mathbf{v}}{\partial t} + \mathbf{v} \cdot \nabla \mathbf{v}\right) = -\nabla p + \mu\nabla \cdot \tau + \mathbf{b}, \tag{1}$$

where ρ is the fluid density, $\mathbf{v} = (u, v, w)$ is the velocity vector, p is the pressure, $\tau = \nabla \mathbf{v} + \nabla \mathbf{v}^T$ is the stress tensor, μ is the dynamic viscosity coefficient, $\mathbf{b}$ is the external force per unit volume and ∇ represents the gradient operator. For simplicity, we assume the blood behaves as an incompressible Newtonian fluid and the vessel walls are rigid. We also assume that there are no body forces acting on the blood flow. Besides, since the PV is far from the heart, the influence of the cardiac cycle can be neglected [6]. The momentum Eq. (1) can then be simplified into

$$-\nabla p + \mu\Delta \mathbf{v} = 0, \tag{2}$$

where Δ is the Laplacian operator that is equal to $\nabla \cdot \nabla$. Since there are no sources or sinks of blood inside a vessel, the velocity field should satisfy the following continuity equation as well

$$\nabla \cdot \mathbf{v} - 0, \tag{3}$$

which indicates the blood flowing in the vessel from all inlets equals to the total amount of blood flowing out through outlets.

2.2   Diffuse Boundary Conditions

To solve Eq. (2) and (3), we propose to discretize all physical quantities on a uniform marker and cell (MAC) grid [7], as shown in Fig. 1. It is important to note that all three components of $\mathbf{v}$ must be defined on centers of cell faces (two components in a 2D space), staggered by half a grid spacing with respect cell center on which pressure values are defined. This is to avoid non-physical wiggle solutions for the pressure and velocity fields [10]. To solve accurate hemody-namics, cells inside the vessel should be identified first. It is a common way to use a binary mask to isolate the fluid domain of interest from other tissues [9]. However, the problem with a binary threshold is that it will introduce stair-step grid artifacts into the simulation results [1].

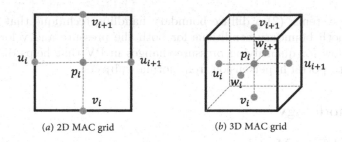

(a) 2D MAC grid   (b) 3D MAC grid

Fig. 1. An illustration of the MAC grid used to discretize the computational domain.

To remove the stair-step grid artifacts, we propose to impose diffuse boundary conditions for Eq. (2) and (3). We first introduce a signed distance function $\phi(\mathbf{x})$ to represent the blood vessel implicitly (Information on the practical implementation of how reconstruct ϕ is given below under Sect. 2.3). For an arbitrary point $\mathbf{x} \in \mathbb{R}^3$, $|\phi(\mathbf{x})|$ tells us its distance to the vessel boundary, and $sign(\mathbf{x}) > 0$ means $\mathbf{x}$ is located inside a blood vessel while $sign(\mathbf{x}) < 0$ means the opposite case. In the following discussion, we only demonstrate how to impose diffuse boundary conditions in a one-dimensional space. Consider a cell inside the vessel that is far away from the boundary (i.e., u_{i+1}, u_{i-1} and u_i are all located inside the computational domain), the Laplacian of u can be discretized into

$$\mathcal{L}(u_i) = \frac{u_{i+1} - 2u_i + u_{i-1}}{d^2}, \tag{4}$$

where d represents the grid spacing. Assume u_i is now located inside the boundary and u_{i+1} is located outside of the boundary, as shown in Fig. 2(a). Note that no velocity sampling point is located just on the boundary. To impose a Dirichlet velocity boundary condition of u_b, we assume u_{i-1}, u_{i+1} and u_b satisfy the following linear relationship

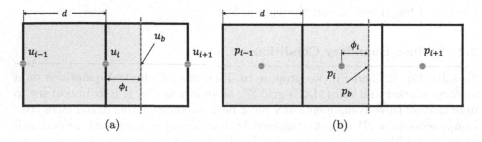

(a)   (b)

Fig. 2. Illustration of boundary cells. (a) velocity samples; (b) pressure samples.

$$\frac{u_b - u_{i-1}}{d + \phi_i^u} = \frac{u_{i+1} - u_b}{d - \phi_i^u}, \tag{5}$$

where ϕ_i^u is the signed distance of u_i. Since u_{i+1} is outside of the boundary, we are able to reformulate $\mathcal{L}(u_i)$ as

$$\mathcal{L}(u_i) = \frac{A_i u_b + (2 - A_i)\, u_{i-1} - 2u_i}{d^2}, \quad A_i = \frac{2d}{d + \phi_i^u} \tag{6}$$

after substituting Eq. 5 into Eq. 4. By clamping the value of ϕ_i^u into a range of $(0, d]$, A_i becomes a constant ranging from 1 to 2. Note if u_i is located on the boundary, i.e., $\phi_i = 0$, $\mathcal{L}(u_i)$ equals to $\frac{2u_b - 2u_i}{d^2}$. This indicate the velocity boundary condition is directly imposed on the velocity sampling point of u_i. Otherwise, if $\phi_i > d$, Eq. 6 is just simplified into the form of Eq. 4 and the velocity boundary condition is imposed on the velocity sampling point u_{i+1}. In a similar way, the Laplacian of p (see an illustration in Fig. 2(b)) can be formulated as

$$\mathcal{L}(p_i) = \frac{B_i p_b + (2 - B_i)\, p_{i-1} - 2p_i}{d^2}, \quad B_i = \frac{2d}{d + \phi_i^p} \tag{7}$$

where B_i is a constant ranging from 1 to 2, and ϕ_i^p represents the clamped signed distance of p_i.

Finally, let us consider a standard discretization of the divergence of u

$$\mathcal{D}(u_i) = \frac{u_{i+1} - u_i}{d}. \tag{8}$$

By imposing a Dirichlet boundary condition, we assume u_{i+1}, u_i and u_b satisfy the following linear relationship

$$\frac{u_b - u_i}{\frac{d}{2} + \phi_i^p} = \frac{u_{i+1} - u_i}{d}, \tag{9}$$

where ϕ_i^p is the signed distance of p_i. Substituting Eq. 9 into Eq. 8, the divergence of u can be reformulated as

$$\mathcal{D}(u_i) = C_i \frac{u_b - u_i}{d}, \quad C_i = \frac{2d}{d + 2\phi_i^p}, \tag{10}$$

where C_i is a constant ranging from 1 to 2 since ϕ_i^p is in a range of $[0,\, d/2]$.

2.3   Data Pre-processing and Numerical Implementation

To validate the proposed boundary handling technique, both model analysis and patient-specific analysis were conducted. For the patient-specific analysis, all analyses were conducted in accordance with the principles of West China Hospital and met the requirements of medical ethics. Patients' approval and informed consent were waived as our study was purely observational and retrospective in nature. All patients underwent CT angiography, and the invasive transjugular HVPG measurement. Normal subjects only underwent CT angiography. Figure 3 outlines a pipeline showing all data-processing steps and CFD model to estimate pressure change in PV caused by blood blockage in the RPV.

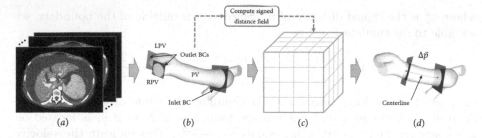

Fig. 3. A pipeline demonstration of all data pre-processing steps and CFD model to estimate pressure change in PV caused by blood blockage in the RPV.

Transjugular HVPG Measurement. The transjugular HVPG measurement (reference standard) was performed by following the established standards [3]. Pressure measurements were conducted by using a balloon catheter (Edwards Lifesciences, Irvine, California) with a pressure transducer at the tip. A zero measurement with transducer open to air was needed before the transjugular catheterization. Free hepatic venous pressure was measured in the right hepatic vein. As the balloon was inflated for total occlusion of the right hepatic vein, the wedged hepatic venous pressure was measured. Continuous recording was necessary until the pressure reached a plateau. All measurements were taken in triplicate and then averaged. HVPG was calculated by subtracting the free venous hepatic pressure from the wedged hepatic pressure.

CT Image Acquisition. Datas from all the subjects were collected through computed tomography systems (Sensation 64 CT (Siemens), or Sensation 16 CT (Siemens)) in West China Hospital, affiliated with Sichuan University. The triple-phase CT examinations including non-enhanced, arterial, and portal vein phase were obtained, in which potal vein phase was used to reconstruct geometric models. The scout of abdomen was acquired from the lung bases to the iliac crests including the entire liver. Arterial phase of the same region started about $20-30s$ after the contrast agent administration, and followed with portal venous phase $(30-40s)$. The reconstruction was performed on advantage 3D workstation and the reconstitution thickness was set at $1-2mm$.

Segmentation of Portal Vein Systems. The open-source software 3D Slicer [4] was used to reconstruct the geometry of the PV system models for CFD simulation. To facilitate the hemodynamic calculations, some parts of the reconstructed surface mesh of the PV system were pruned, retaining only the PV and its main branch vessels including the left portal vein (LPV), right portal vein (RPV), superior mesenteric vein (SMV) and splenic vein (SV). Vascular centerline was extracted by the computational geometry algorithm library (CGAL) to facilitate the measurement of the distance between two probing planes (see Fig. 3(d)).

Estimation of Pressure Change in PV. Signed distance fields of blood vessels were first reconstructed with the Fast Sweeping Method [19]. Other parts were implemented with C++ and the SIMPLE algorithm [10] was adopted to solve the discretized governing equations (2) and (3) (see supplementary materials for more details). Motived by virtual HVPG [12], two simulations were taken for each patient-specific data. In the first simulation, Neumann velocity boundary conditions were imposed on the outlets of both LPV and RPV. In the second simulation, we virtually blocked the outlet of RPV but left LPV open to mimic a situation when patients with cirrhosis suffered from an increased resistance to blood flow through the portal system. A constant flat profile of $\bar{u} = 0.1364$ m/s was imposed as the inlet velocity on portal vein for all models [13], including both patients and normal subjects. Finally, the pressure change before and after RPV blockage was be calculated.

3   Experiments and Results

3.1   In Silico Study on a Steady Poiseuille Flow

An analytic laminar steady flow case was first studied to verify the accuracy of our method, as sketched in Fig. 4. Since the flow is caused by a pressure gradient, the analytical solution to the Poiseuille flow in a circular pipe along axis x can be written as

$$\delta p = \frac{4\mu L u_{max}}{R^2} \qquad (11)$$

where δp represents the pressure drop over a length L of the pipe, R is the pipe radius and u_{max} is the maximum velocity. Since Ansys Fluent has been commonly used to analyze hemodynamics, the steady Poiseuille flow was first solved with a finite element method embedded in ANSYS 2020 R2. To verify the accuracy under different boundary conditions, two velocity profiles were imposed at the inlet cross-sections, including a constant flat profile of $\bar{u} = 0.2$ m/s and the corresponding exact profile $u(r) = 2\bar{u}\left(1 - \frac{r^2}{R^2}\right)$ m/s with $R = 5.4 \times 10^{-3}$ m. Besides, a zero-pressure boundary condition was imposed at the outlet cross-sections and a non-slip boundary condition was imposed at the vessel walls for all models. It can be noted from Fig. 4(a) that if the exact velocity profile is given, the simulation results show good convergence to the exact solution for all three different spatial resolutions. Our method does not outperform FEM at a spatial resolution of $d = 2$ mm, but shows a good performance as the spatial resolution is increased to $d = 0.5$ mm. Otherwise, if the constant velocity profile is given, the simulation with FEM fails to converge as grid resolution is increased, the relative error of pressure with the finniest spatial resolution is still over 104%. In contrast, our method does not suffer from the sensitivity problem caused by boundary conditions and is able to reproduce consistent simulation results for both cases. This feature makes our method more applicable for further clinical applications because the exact velocity profile is usually unattainable in real situations and only an approximate average velocity can be measured (e.g., with Doppler US).

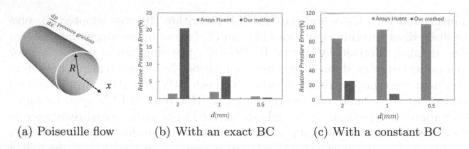

(a) Poiseuille flow   (b) With an exact BC   (c) With a constant BC

Fig. 4. In silico study of a steady Poiseuille flow. (a) an illustration of the steady Poiseuille flow; (b) a comparison of simulation results under an exact velocity profile; (c) a comparison of simulation results under a constant velocity profile.

3.2 In-Vivo Validation with Invasive HVPG

After in silico validation with our method, we applied it on real models constructed from CT angiographic images. We enrolled consecutive CLD patients who underwent contrast-enhanced CT within 2 weeks of HPVG measurement. Patients were excluded according to the following criteria: (a) a prior variceal treatment (i.e., band ligation and endoscopic varices ligation (EVL)) before admission; (b) patients with histopathologically confirmed as hepatocellular carcinoma (HCC); (c) a history of splenectomy, hepatectomy or portal-azygous disconnection. A total of three patients with liver cirrhosis and three subjects with normal livers are studied. The pressure difference δp before and after RPV was blocked is colored mapped and demonstrated in Fig. 5. In addition, an average pressure change in PV is calculate as

$$\delta \bar{p}_{PV} = \frac{\int_{V_{PV}} \delta p \, dV}{l},\tag{12}$$

where V_{PV} represents the total volume enclosed by the two probing planes as shown in Fig. 3(d) and l is the length of the centerline. Note the pressure changes in PV of patients with portal hypertension (see the top row in Fig. 5) are significantly lower than that of normal subjects (see the bottom row in Fig. 5).

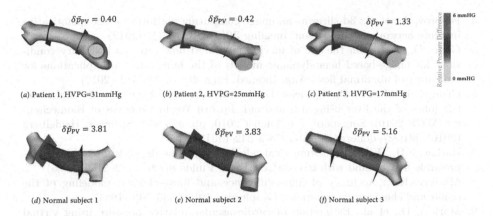

$\delta \bar{p}_{PV} = 0.40$ $\delta \bar{p}_{PV} = 0.42$ $\delta \bar{p}_{PV} = 1.33$

(a) Patient 1, HVPG=31mmHg   (b) Patient 2, HVPG=25mmHg   (c) Patient 3, HVPG=17mmHg

$\delta \bar{p}_{PV} = 3.81$ $\delta \bar{p}_{PV} = 3.83$ $\delta \bar{p}_{PV} = 5.16$

(d) Normal subject 1   (e) Normal subject 2   (f) Normal subject 3

Fig. 5. In-vivo validation with invasive HVPG. Experiments show that pressure changes in the PV of patients with portal hypertension due to blockage of the RPV is lower than that of normal subjects.

4 Conclusion

This paper proposed an Eulerian CFD model for quantifying patient-specific hemodynamics in portal vein systems non-invasively. To address boundary problems within traditional models, diffuse boundary conditions were proposed to impose smooth boundary conditions on both the pressure and velocity fields. Experiments show that our approach is less sensitive to boundary conditions and is able to reproduce consistent simulation results under two commonly used boundary conditions. The approach is also performed on patient-specific PV models reconstructed from CT angiographic images, additional experiments show that our method is able to capture pressure changes in the PV that show a significant difference between patients with portal hypertension and normal subjects.

Acknowledgements. This research was supported by the National Natural Science Foundation of China (No.61872345, No.62072449, No.61632003), Science and Technology Support Program of Sichuan Province(No.2021YFS0144, No.2021YFS0021), Post-Doctor Research Project, West China Hospital, Sichuan University (No.2020HXBH130), Youth Innovation Promotion Association, CAS (No.2019109).

References

1. Batty, C., Bertails, F., Bridson, R.: A fast variational framework for accurate solid-fluid coupling. ACM Trans. Graph. (TOG) **26**(3), 100-es (2007)
2. Bosch, J., Pizcueta, P., Feu, F., Fernández, M., García-Pagán, J.C.: Pathophysiology of portal hypertension. Gastroenterol. Clin. North Am. **21**(1), 1–14 (1992)
3. De Franchis, R.: Expanding consensus in portal hypertension: report of the baveno vi consensus workshop: stratifying risk and individualizing care for portal hypertension. J. Hepatol. **63**(3), 743–752 (2015)

4. Fedorov, A., et al.: 3d slicer as an image computing platform for the quantitative imaging network. Magn. Reson. Imaging **30**(9), 1323–1341 (2012)
5. Gallo, D., et al.: On the use of in vivo measured flow rates as boundary conditions for image-based hemodynamic models of the human aorta: implications for indicators of abnormal flow. Ann. Biomed. Eng. **40**(3), 729–741 (2012)
6. George, S., Martin, D., Giddens, D.: Portal vein contribution to the right and left lobes of the liver using mri and cfd. In: 6th World Congress of Biomechanics (WCB 2010), Singapore, 1–6 August 2010, pp. 473–476. Springer, Heidelberg (2010). https://doi.org/10.1007/978-3-642-14515-5_121
7. Harlow, F.H., Welch, J.E.: Numerical calculation of time-dependent viscous incompressible flow of fluid with free surface. Phys. Fluids **8**(12), 2182–2189 (1965)
8. Marfurt, K.J.: Accuracy of finite-difference and finite-element modeling of the scalar and elastic wave equations. Geophysics **49**(5), 533–549 (1984)
9. Marlevi, D., et al.: Estimation of cardiovascular relative pressure using virtual work-energy. Sci. Rep. **9**(1), 1–16 (2019)
10. Patankar, S.: Numerical heat transfer and fluid flow. Taylor & Francis (2018)
11. Pirola, S., Cheng, Z., Jarral, O., O'Regan, D., Pepper, J., Athanasiou, T., Xu, X.: On the choice of outlet boundary conditions for patient-specific analysis of aortic flow using computational fluid dynamics. J. Biomech. **60**, 15–21 (2017)
12. Qi, X., et al.: Virtual hepatic venous pressure gradient with ct angiography (chess 1601): a prospective multicenter study for the noninvasive diagnosis of portal hypertension. Radiology **290**(2), 370–377 (2019)
13. Qin, Y.J., Yao, Y.Z.: Characteristic of hemodynamics on liver cirrhosis resulting from chronic hepatitis b. J. Pathogen Biol. **4** (2007)
14. Souguir, S.B.C., Fathallah, R., Abdallah, A.B., Hmida, B., Bedoui, M.H.: Numerical simulation of venous system blood flow for hepatic donor patient and portal hypertension's proposed measurement. In: Chaari, L. (ed.) Digital Health in Focus of Predictive, Preventive and Personalised Medicine. APPPM, vol. 12, pp. 33–38. Springer, Cham (2020). https://doi.org/10.1007/978-3-030-49815-3_5
15. Toubia, N., Sanyal, A.J.: Portal hypertension and variceal hemorrhage. Med. Clin. North Am. **92**(3), 551–574 (2008)
16. Vignon-Clementel, I.E., Figueroa, C.A., Jansen, K.E., Taylor, C.A.: Outflow boundary conditions for three-dimensional finite element modeling of blood flow and pressure in arteries. Comput. Methods Appl. Mech. Eng. **195**(29–32), 3776–3796 (2006)
17. Yamagishi, Y., et al.: Value of computed tomography-derived estimated liver volume/standard liver volume ratio for predicting the prognosis of adult fulminant hepatic failure in japan. J. Gastroenterol. Hepatol. **20**(12), 1843–1849 (2005)
18. Zhang, J.M., et al.: Perspective on cfd studies of coronary artery disease lesions and hemodynamics: a review. Int. J. Numer. Methods Biomed. Eng. **30**(6), 659–680 (2014)
19. Zhao, H.: A fast sweeping method for eikonal equations. Math. Comput. **74**(250), 603–627 (2005)

A Segmentation-Assisted Model for Universal Lesion Detection with Partial Labels

Fei Lyu[1], Baoyao Yang[2], Andy J. Ma[3], and Pong C. Yuen[1(✉)]

[1] Department of Computer Science, Hong Kong Baptist University,
Kowloon Tong, Hong Kong
{feilyu,pcyuen}@comp.hkbu.edu.hk
[2] School of Computers, Guangdong University of Technology, Guangzhou, China
[3] School of Computer Science and Engineering, Sun Yat-Sen University,
Guangzhou, China

Abstract. Developing a Universal Lesion Detector (ULD) that can detect various types of lesions from the whole body is of great importance for early diagnosis and timely treatment. Recently, deep neural networks have been applied for the ULD task, and existing methods assume that all the training samples are well-annotated. However, the partial label problem is unavoidable when curating large-scale datasets, where only a part of instances are annotated in each image. To address this issue, we propose a novel segmentation-assisted model, where an additional semantic segmentation branch with superpixel-guided selective loss is introduced to assist the conventional detection branch. The segmentation branch and the detection branch help each other to find unlabeled lesions with a mutual-mining strategy, and then the mined suspicious lesions are ignored for fine-tuning to reduce their negative impacts. Evaluation experiments on the DeepLesion dataset demonstrate that our proposed method allows the baseline detector to boost its average precision by 13%, outperforming the previous state-of-the-art methods.

Keywords: Universal lesion detection · Partial labels

1 Introduction

Developing a Universal Lesion Detector (ULD) that can discover various types of lesions could provide great clinical value [18]. For training an effective ULD with deep neural networks, a large-scale and high-quality labeled dataset is required [12,14]. In clinical practice, it is common that radiologists only mark the partial lesions that are most representative for disease diagnosis and ignore those trivial lesions [5], which brings the partial label problem. The fatal drawback caused by the partial label problem is that a detector will suffer from incorrect supervision due to the missing annotations, *i.e.*, the true object area (positive) with missing annotations are treated as background (negative), which will confuse the model during training and make the detector bias to the background [9,13].

© Springer Nature Switzerland AG 2021
M. de Bruijne et al. (Eds.): MICCAI 2021, LNCS 12905, pp. 117–127, 2021.
https://doi.org/10.1007/978-3-030-87240-3_12

The availability of large-scale labeled datasets has greatly promoted the development in universal lesion detection [11,20,22]. For example, DeepLesion [21] is a recent public CT dataset with over 32K annotated lesions, it also suffers from the partial label problem and around 50% of the lesions are left unlabeled. Some methods have been developed to tackle the partial label problem for universal lesion detection. Wang et al. [17] proposed to mine unlabeled lesions by considering the continuity of multi-slice axial CT images. Yan et al. [19] leveraged the knowledge learned from several fully-labeled single-type lesion datasets to find unlabeled lesions. Cai et al. [2] proposed an effective framework called Lesion-harvester, where a lesion proposal generator and a lesion proposal classifier are chained together to iteratively mine unlabeled lesions and hard negatives for re-training. Nevertheless, these methods either fail to find missing annotations which do not match with any existing annotations, or rely on extra fully-labeled datasets which are usually unavailable in the real-world scenarios.

In this paper, we propose a segmentation-assisted model to address the partial label problem in the universal lesion detection task. Inspired by the advances in Co-teaching [6], we propose to add a peer network for effectively mining unlabeled lesions, and reducing their negative impacts on model training. Specifically, we add an additional semantic segmentation branch as the peer network, which can leverage the spatial context and provide complementary information for the detection task [3,10]. To handle the segmentation task with partial labels, we propose a superpixel-guided selective loss where only the pixels within the selective masks contribute to the loss and the selective masks are generated based on superpixels [1]. Compared to conventional methods using selective masks based on bounding boxes, superpixels can generate perceptually meaningful selective masks by utilizing local contextual information. In this way, more informative pixels can be involved for calculating the loss, and thus benefit for training the segmentation network. The detection and segmentation branches have different learning abilities and can generate two different sets of lesion detection results, therefore we introduce a mutual-mining strategy, where the unlabeled lesions are mined from one branch and then fed back to the other branch to alleviate the potential errors accumulated issue.

Our main contributions are summarized as follows: (1) We introduce a novel segmentation-assisted model to handle the partial label problem in universal lesion detection, where a semantic segmentation branch with superpixel-guided selective loss is added. (2) We propose a mutual-mining strategy to effectively find unlabeled lesions for reducing their negative impacts. (3) The ULD performance on DeepLesion is improved over previous state-of-the-art methods.

2   Method

Figure 1 illustrates the proposed segmentation-assisted model. The detection branch is based on Mask R-CNN [7], which is shown in red. We employ the SLIC algorithm [1] to produce superpixels, and add an additional branch with superpixel-guided selective loss to predict the pixel-wise segmentation map for the whole image, which is highlighted in green. These two branches can

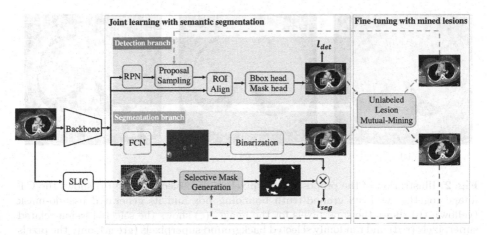

Fig. 1. Overview of our proposed segmentation-assisted model for universal lesion detection with partial labels. Best viewed in color. (Color figure online)

generate two different sets of lesion detection results. After going through the unlabeled lesion mutual-mining module, the mined suspicious lesions from one branch are fed back to the other branch for fine-tuning. Note that the model has two branches during training and only the detection branch during inference.

2.1   Joint Learning with Semantic Segmentation

Segmentation with Superpixel-Guided Selective Loss. Training a segmentation model requires supervision in the form of dense pixel-wise annotation of the full-image. However, we only have the pseudo-masks generated by RECIST measurement [16], and the remaining regions can not be simply regarded as background because they may contain unlabeled lesions. While selective pixel loss [15] can be used to handle segmentation task with partial labels, their selective masks are only based on the bounding boxes without considering the contextual information in the image, and many lesion-surrounding pixels that may also include some lesion level information [8] are ignored. Moreover, a significant portion of the unlabeled background pixels do not contribute to the loss, bringing a high false positive rate for the segmentation result. To overcome the aforementioned shortcomings, we introduce a superpixel-guided selective loss as illustrated in Fig. 2. Superpixels can group the original image pixels into locally homogeneous blocks, and the selective masks based on these meaningful atomic regions can involve more informative pixels for training, and thus benefit for the segmentation performance.

N superpixels that are overlapped with or neighbouring to the lesion areas are first selected as lesion-related superpixels. For better utilizing the unlabeled pixels which mainly belong to the background, another N superpixels are randomly sampled from the remaining superpixels. Though the selected superpixels may encounter unannotated lesion pixels, the possibility is greatly reduced due

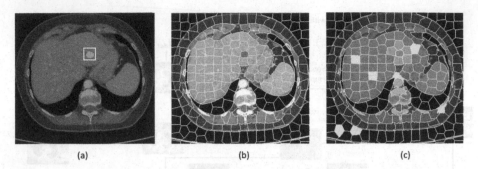

(a)   (b)   (c)

Fig. 2. Illustration of the proposed superpixel-guided selective loss. (a) shows the CT image together with its ground-truth bounding box and its generated pseudo-mask (yellow). (b) shows the superpixels for the image. (c) shows the selected lesion-related superpixels (red) and randomly selected background superpixels (green), only the pixels within the selected superpixels SP have nonzero coefficients on the selective mask and contribute to the loss. Best viewed in color. (Color figure online)

to random sampling. We generate a selective mask W where only pixels within the selected superpixels have nonzero values. The superpixel-guided selective loss $\mathcal{L}_{seg}$ is formulated as a per-pixel weighted binary cross-entropy loss:

$$\mathcal{L}_{seg} = -\frac{1}{n} \sum_i \sum_j [w_{ij} \cdot y_{ij} \cdot \log \hat{y}_{ij} + w_{ij} \cdot (1 - y_{ij}) \cdot \log(1 - \hat{y}_{ij})] \qquad (1)$$

where $\hat{y}_{ij}$ is the predicted lesion probability on pixel (i, j) of the segmentation map and y_{ij} is the pixel label. w_{ij} is the coefficient on pixel (i, j) of the selective mask W, where $w_{ij} = 1$ for the pixels within the selected superpixels while $w_{ij} = 0$ for the others. n is the number of pixels which have nonzero coefficients.

Joint Learning. The segmentation branch is jointly trained with the detection branch, so the overall loss formulation $\mathcal{L}$ of the model is:

$$\mathcal{L} = \mathcal{L}_{det} + \beta \cdot \mathcal{L}_{seg}$$
$$\mathcal{L}_{det} = \mathcal{L}_{cls} + \mathcal{L}_{reg} + \mathcal{L}_{mask} \qquad (2)$$

where $\mathcal{L}_{det}$ is the loss from the detection branch, which follows the same definition as in Mask R-CNN [7]. It combines three terms $\mathcal{L}_{cls}$, $\mathcal{L}_{reg}$ and $\mathcal{L}_{mask}$, respectively for classification, bounding box regression and mask prediction. The coefficient β is used to balance the contributions of detection and segmentation.

2.2   Fine-Tuning with Mined Lesions

Unlabeled Lesion Mutual-Mining. In this subsection, we aim to mine unlabeled lesions for reducing their ill-effects. The detection branch predicts bounding boxes with classification scores. From the segmentation branch, we can generate bounding boxes from the segmentation probability map by applying image binarization, and calculate mean confidence score for each bounding box. We first remove the bounding boxes that are overlapped with existing annotations or with lower confidence scores below τ_{low}. We denote remaining bounding boxes from the detection branch and the segmentation branch as D and S, respectively. We can further divide them into the intersected ones $D \cap S$ and the respective ones, $D \backslash (D \cap S)$ and $S \backslash (D \cap S)$. As for $D \cap S$, since two peer networks with different learning abilities agree they are lesions, we have reason to believe that they are highly likely to be true lesions. As for the respective ones, they are uncertain lesions and may contain false positives, and the errors would be accumulated if we directly feed back to itself. Therefore, we propose a mutual-mining strategy to alleviate this issue. Specifically, we select α (*i.e.* selection ratio) samples with higher scores from the respective bounding boxes of one branch together with all the intersected ones as suspicious lesions and feed back to the other branch.

Fine-Tuning. Each mined lesion can either be an unlabeled lesion or a false positive, and it will bring more noise if we directly treat them as true lesions, thus we choose to ignore them. During fine-tuning, b_{det} mined from the detection branch are sent back to the segmentation branch and b_{seg} from the segmentation branch are sent back to the detection branch.

Fine-Tuning Detection Branch. After Region Proposal Network (RPN), labels are assigned to each generated proposal p_i based on their maximum Intersection over Union (IoU) with ground truth (GT). Moreover, we also regard those proposals that have a maximum IoU with the mined suspicious lesions over τ_{sus} as ignored samples. In this way, the regions of interest (RoIs) sampled from the positive and negative proposals are less likely to be assigned with wrong labels.

$$Label(p_i) = \begin{cases} -1, & \text{if } I_m(p_i, GT) \in [\tau_{neg}, \tau_{pos}] \text{ or } I_m(p_i, b_{seg}) > \tau_{sus} \\ 1, & \text{if } I_m(p_i, GT) > \tau_{pos} \text{ and } I_m(p_i, b_{seg}) < \tau_{sus} \\ 0, & \text{if } I_m(p_i, GT) < \tau_{neg} \text{ and } I_m(p_i, b_{seg}) < \tau_{sus} \end{cases} \quad (3)$$

where $I_m(a, b)$ denotes the maximum IoU between a and b.

Fine-tuning Segmentation Branch. Coefficients of pixels within the mined lesions are set to zero on the selective mask W generated by selected superpixels SP. Because they are likely to be lesion pixels but are falsely regarded as background pixels, they will not propagate incorrect gradients if we exclude them in the loss. The coefficient w_{ij} in W is then formulated as:

$$w_{ij} = \begin{cases} 1, & \text{if } (i,j) \in SP \text{ and } (i,j) \notin b_{det} \\ 0, & otherwise \end{cases} \quad (4)$$

3 Experiments

3.1 Experiment Setup

Dataset and Metric. DeepLesion [21] is currently the largest dataset for universal lesion detection, but with around 50% of unlabeled lesions. We use the official training set for training. The official testing set may also contain missing labels, which will cause unreliable evaluation results. Cai *et al.* [2] completely annotate and publicly release 1915 of the DeepLesion subvolumes. Of these, 844 subvolumes are selected from the original training set for supporting lesion harvesting. The other 1,071 subvolumes are selected from the original testing set for evaluation. They also introduce a new pseudo 3D (P3D) evaluation metric, which can serve a better measurement of 3D detection performance than current practices. The detected 2D bounding boxes are first stacked to generate 3D bounding boxes, and a predicted 3D bounding box is regarded as correct if any 2D bounding box within the 3D bounding box volume has an intersection over union (IoU) with a ground-truth box which is larger than 0.5. We study sensitivities at operating points varying from false positive (FP) rates of 0.125 to 8 per volume and the average precision for evaluating the detection performance.

Table 1. Sensitivity at various FPs per image on the completely annotated testing set of DeepLesion, we denote the official annotated lesions, mined suspicious lesions and mined hard negative samples as R, P^+ and P^- respectively.

Method	R	P^+	P^-	Sensitivity (%) at different FPs per sub-volume							AP (%)
				0.125	0.25	0.5	1	2	4	8	
MULAN [20]	✓			11.43	18.69	26.98	38.99	50.15	60.38	69.71	41.8
Lesion Harvester [2]	✓			11.92	18.42	27.54	38.91	50.15	60.76	69.82	43.0
Mask R-CNN [7]	✓			13.64	19.95	27.48	36.73	47.01	57.41	66.38	41.4
Ours	✓			14.84	20.08	27.93	37.14	47.59	57.65	66.47	42.1
Wang *et al.* [17]	✓	✓		15.88	20.98	29.40	38.36	47.89	58.14	67.89	43.2
Lesion Harvester [2]	✓	✓		13.40	19.16	27.34	37.54	49.33	60.52	70.18	43.6
Ours	✓	✓		20.00	25.65	33.44	42.81	52.39	60.90	68.79	47.5
Lesion Harvester [2]	✓	✓	✓	19.86	27.11	36.21	46.82	56.89	**66.82**	**74.73**	51.9
Ours	✓	✓	✓	**30.89**	**37.28**	**43.94**	**51.05**	**57.41**	63.68	69.81	**54.7**

Implementation Details. Similar to [2,20], we use a 2.5D truncated DenseNet-121 as the backbone, and each input includes 9 adjacent slices. We adopt a Fully Convolution Network (FCN) with three layers for semantic segmentation. In the pre-processing step, the CT slices are resampled to 2 mm thickness in the z-axis and resized to 512×512 pixels in the xy-plane. We rescale the 12-bit CT intensity range to [0,255] using a single windowing ($-1024 \sim 3071$ HU). We employ the SLIC algorithm [1] to produce superpixels, and the total number of superpixel is 150. The loss coefficient β is set to 1, and the selection ratio α for mutual-mining is 50%. For the threshold parameters, we remove the detected bounding boxes with confidence scores lower than τ_{low}, we assign label -1 to those proposals from RPN which have a maximum IoU with the mined lesions higher than τ_{sus}, and those proposals with label -1 will not be sampled during fine-tuning. τ_{neg} and τ_{pos} are used to assign labels to each proposal as lesion or background. These thresholds τ_{low}, τ_{neg}, τ_{pos}, τ_{sus} are empirically set to 0.8, 0.4, 0.5 and 0.5. Unless otherwise noted, all hyperparameters are as in [4] for Mask R-CNN. We adopt Stochastic gradient descent (SGD) optimizer to train the model for 5 epochs with the base learning rate 0.008 in the joint training stage. The learning rate is then reduced by a factor of 10 after the 5-th and 7-th epoch with another 3 epochs, for fine-tuning. Following [19], we stack the predicted 2D boxes to 3D ones if the IoU of two 2D boxes in consecutive slices is greater than 0.5.

3.2   Comparison with State-of-the-Arts

Table 1 displays the comparison between our proposed method and previous state-of-the-art methods on the new testing set of DeepLesion. We train the detection model with and without mined samples. Since MULAN [20] requires tags, which are not available for the mined samples, we only report the results using the official annotated lesions. Lesion-harvester [2] has achieved the state-of-the-art performance by iteratively mining suspicious lesions and hard negative samples for re-training. Unlike [2], our mined suspicious lesions are generated by a mutual-mining module during training without additional annotations. For a fair comparison with [2], we apply our trained detector on their 844 completely annotated training subvolumes, and select the hard negative samples from the false positive detection results with scores greater than 0.8 for retraining. As Table 1 demonstrates, using mined suspicious lesions and hard negatives can greatly improve the detection result. Remarkably, our method that is trained with the mined suspicious lesions and hard negative samples allows a Mask R-CNN based detector to boost its average precision (AP) by 13%, and achieves the best performance at 54.7% AP and higher sensitivity at lower tolerated FPs. Compared with [2], our method is slightly inferior when FP is greater than 2. The possible reason is that our mined hard negative samples are less than theirs, and no extra lesion classifiers are used for refinement.

3.3   Analysis of Our Method

Segmentation with Superpixel-Guided Selective Loss. To validate the effectiveness of our proposed superpixel-guided selective loss, we conduct experiments with different selective masks for calculating the loss, and show the detection results after fine-tuning with the mined suspicious lesions. 'whole image' means all pixels from the whole image have nonzero coefficients on the selective mask, while 'bounding box' means that only the pixels within the ground truth bounding box have nonzero coefficients. Figure 3a shows that the superpixel-guided selective loss outperforms the others. The reason is that more lesion-surrounding pixels and background pixels can contribute to the loss, the false positive rate is reduced and the segmentation performance is improved, and thus the lesion detection results from segmentation branch are more reliable for finding unlabeled lesions.

Unlabeled Lesion Mutual-Mining. To analyze the contribution of unlabeled lesion mutual-mining, we explore different strategies for finding suspicious unlabeled lesions. 'detection' mines the suspicious lesions from the detection branch only, while 'segmentation' from the segmentation branch only. 'intersection' takes the intersection of two branches as suspicious lesions. Figure 3b shows our mutual-mining strategy achieves better detection performance than the others.

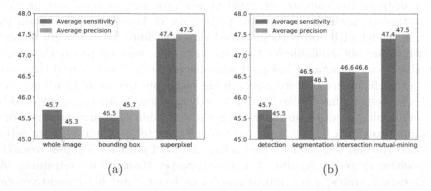

Fig. 3. (a) Detection performance of loss functions with different selective masks. (b) Detection performance of different unlabeled lesion mining strategies.

We also perform a case study to analyze our mutual-mining strategy. Figure 4 shows the suspicious lesions from the detection branch and segmentation branch during training. If these two branches generate similar lesion proposals, they are likely to be true lesions. Those only appear in one branch are uncertain lesions and may contain false positives, and the proposed mutual-mining strategy can help alleviate the errors accumulated issue by exchanging the uncertain lesions.

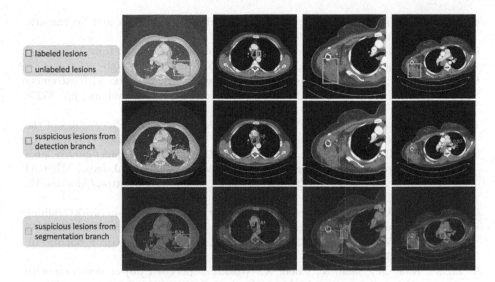

Fig. 4. Visualization of two different sets of suspicious lesions generated by the detection branch and segmentation branch during training. Best viewed in color. (Color figure online)

4   Conclusion

In this paper, we propose a segmentation-assisted model to address the partial label problem in the universal lesion detection task. We add an additional semantic segmentation branch with superpixel-guided selective loss, which is complementary to the detection task by exploring more spatial context. Moreover, we propose to mine unlabeled lesions through a mutual-mining strategy, and then ignore them for fine-tuning the detector. In this way, incorrect information is not propagated and the ill-effects caused by missing annotations can be reduced. Experiment results have demonstrated the advantages of our new approach over the state-of-the-art alternatives.

Acknowledgments. This work was supported by the Health and Medical Research Fund Project under Grant 07180216.

References

1. Achanta, R., Shaji, A., Smith, K., Lucchi, A., Fua, P., Süsstrunk, S.: Slic superpixels compared to state-of-the-art superpixel methods. IEEE Trans. Pattern Anal. Mach. Intell. **34**(11), 2274–2282 (2012)
2. Cai, J., et al.: Lesion-harvester: iteratively mining unlabeled lesions and hard-negative examples at scale. IEEE Trans. Med. Imaging (2020)
3. Chen, K., et al.: Hybrid task cascade for instance segmentation. In: Proceedings of the IEEE/CVF Conference on Computer Vision and Pattern Recognition pp. 4974–4983 (2019)

4. Chen, K., et al.: Mmdetection: Open mmlab detection toolbox and benchmark. arXiv preprint arXiv:1906.07155 (2019)

5. Eisenhauer, E.A., et al.: New response evaluation criteria in solid tumours: revised recist guideline (version 1.1). Eur. J. Cancer **45**(2), 228–247 (2009)

6. Han, B., et al.: Co-teaching: robust training of deep neural networks with extremely noisy labels. In: Advances in Neural Information Processing Systems, pp. 8527–8537 (2018)

7. He, K., Gkioxari, G., Dollár, P., Girshick, R.: Mask r-cnn. In: Proceedings of the IEEE International Conference on Computer Vision, pp. 2961–2969 (2017)

8. Li, H., Wei, D., Cao, S., Ma, K., Wang, L., Zheng, Y.: Superpixel-guided label softening for medical image segmentation. In: Martel, A.L., et al. (eds.) MICCAI 2020. LNCS, vol. 12264, pp. 227–237. Springer, Cham (2020). https://doi.org/10.1007/978-3-030-59719-1_23

9. Li, H., et al.: A novel loss calibration strategy for object detection networks training on sparsely annotated pathological datasets. In: Martel, A.L., et al. (eds.) MICCAI 2020. LNCS, vol. 12265, pp. 320–329. Springer, Cham (2020). https://doi.org/10.1007/978-3-030-59722-1_31

10. Li, X., Kan, M., Shan, S., Chen, X.: Weakly supervised object detection with segmentation collaboration. In: Proceedings of the IEEE/CVF International Conference on Computer Vision, pp. 9735–9744 (2019)

11. Li, Z., Zhang, S., Zhang, J., Huang, K., Wang, Y., Yu, Y.: MVP-net: multi-view FPN with position-aware attention for deep universal lesion detection. In: Shen, D., Liu, T., Peters, T.M., Staib, L.H., Essert, C., Zhou, S., Yap, P.-T., Khan, A. (eds.) MICCAI 2019. LNCS, vol. 11769, pp. 13–21. Springer, Cham (2019). https://doi.org/10.1007/978-3-030-32226-7_2

12. Litjens, G., et al.: A survey on deep learning in medical image analysis. Med. Image Anal. **42**, 60–88 (2017)

13. Niitani, Y., Akiba, T., Kerola, T., Ogawa, T., Sano, S., Suzuki, S.: Sampling techniques for large-scale object detection from sparsely annotated objects. In: Proceedings of the IEEE/CVF Conference on Computer Vision and Pattern Recognition, pp. 6510–6518 (2019)

14. Sahiner, B., et al.: Deep learning in medical imaging and radiation therapy. Med. Phys. **46**(1), e1–e36 (2019)

15. Tajbakhsh, N., Jeyaseelan, L., Li, Q., Chiang, J.N., Wu, Z., Ding, X.: Embracing imperfect datasets: a review of deep learning solutions for medical image segmentation. Med. Image Anal. **63**, 101693 (2020)

16. Tang, Y.B., Yan, K., Tang, Y.X., Liu, J., Xiao, J., Summers, R.M.: Uldor: a universal lesion detector for ct scans with pseudo masks and hard negative example mining. In: 2019 IEEE 16th International Symposium on Biomedical Imaging (ISBI 2019), pp. 833–836. IEEE (2019)

17. Wang, Z., Li, Z., Zhang, S., Zhang, J., Huang, K.: Semi-supervised lesion detection with reliable label propagation and missing label mining. In: Lin, Z., Wang, L., Yang, J., Shi, G., Tan, T., Zheng, N., Chen, X., Zhang, Y. (eds.) PRCV 2019. LNCS, vol. 11858, pp. 291–302. Springer, Cham (2019). https://doi.org/10.1007/978-3-030-31723-2_25

18. Yan, K., Bagheri, M., Summers, R.M.: 3D context enhanced region-based convolutional neural network for end-to-end lesion detection. In: Frangi, A.F., Schnabel, J.A., Davatzikos, C., Alberola-López, C., Fichtinger, G. (eds.) MICCAI 2018. LNCS, vol. 11070, pp. 511–519. Springer, Cham (2018). https://doi.org/10.1007/978-3-030-00928-1_58

19. Yan, K., et al.: Learning from multiple datasets with heterogeneous and partial labels for universal lesion detection in ct. IEEE Trans. Med. Imaging (2020)
20. Yan, K., et al.: MULAN: multitask universal lesion analysis network for joint lesion detection, tagging, and segmentation. In: Shen, D., et al. (eds.) MICCAI 2019. LNCS, vol. 11769, pp. 194–202. Springer, Cham (2019). https://doi.org/10.1007/978-3-030-32226-7_22
21. Yan, K., et al.: Deep lesion graphs in the wild: relationship learning and organization of significant radiology image findings in a diverse large-scale lesion database. In: Proceedings of the IEEE Conference on Computer Vision and Pattern Recognition, pp. 9261–9270 (2018)
22. Zlocha, M., Dou, Q., Glocker, B.: Improving RetinaNet for CT lesion detection with dense masks from weak RECIST labels. In: Shen, D., et al. (eds.) MICCAI 2019. LNCS, vol. 11769, pp. 402–410. Springer, Cham (2019). https://doi.org/10.1007/978-3-030-32226-7_45

Constrained Contrastive Distribution Learning for Unsupervised Anomaly Detection and Localisation in Medical Images

Yu Tian[1,3]([✉]), Guansong Pang[1], Fengbei Liu[1], Yuanhong Chen[1],
Seon Ho Shin[2], Johan W. Verjans[1,2,3], Rajvinder Singh[2],
and Gustavo Carneiro[1]

[1] Australian Institute for Machine Learning, University of Adelaide, Adelaide,
Australia
yu.tian01@adelaide.edu.au
[2] Faculty of Health and Medical Sciences, University of Adelaide, Adelaide, Australia
[3] South Australian Health and Medical Research Institute, Adelaide, Australia

Abstract. Unsupervised anomaly detection (UAD) learns one-class classifiers exclusively with normal (i.e., healthy) images to detect any abnormal (i.e., unhealthy) samples that do not conform to the expected normal patterns. UAD has two main advantages over its fully supervised counterpart. Firstly, it is able to directly leverage large datasets available from health screening programs that contain mostly normal image samples, avoiding the costly manual labelling of abnormal samples and the subsequent issues involved in training with extremely class-imbalanced data. Further, UAD approaches can potentially detect and localise any type of lesions that deviate from the normal patterns. One significant challenge faced by UAD methods is how to learn effective low-dimensional image representations to detect and localise subtle abnormalities, generally consisting of small lesions. To address this challenge, we propose a novel self-supervised representation learning method, called Constrained Contrastive Distribution learning for anomaly detection (CCD), which learns fine-grained feature representations by simultaneously predicting the distribution of augmented data and image contexts using contrastive learning with pretext constraints. The learned representations can be leveraged to train more anomaly-sensitive detection models. Extensive experiment results show that our method outperforms current state-of-the-art UAD approaches on three different colonoscopy and fundus screening datasets. Our code is available at https://github.com/tianyu0207/CCD.

Keywords: Anomaly detection · Unsupervised learning · Lesion detection and segmentation · Self-supervised pre-training · Colonoscopy

M. de Bruijne et al. (Eds.): MICCAI 2021, LNCS 12905, pp. 128–140, 2021.
https://doi.org/10.1007/978-3-030-87240-3_13

1   Introduction

Classifying and localising malignant tissues have been vastly investigated in medical imaging [1,11,22–24,26,29,42,43]. Such systems are useful in health screening programs that require radiologists to analyse large quantities of images [35,41], where the majority contain normal (or healthy) cases, and a small minority have abnormal (or unhealthy) cases that can be regarded as anomalies. Hence, to avoid the difficulty of learning from such class-imbalanced training sets and the prohibitive cost of collecting large sets of manually labelled abnormal cases, several papers investigate anomaly detection (AD) with a few or no labels as an alternative to traditional fully supervised imbalanced learning [1,26,28,32,33,37,38,43–45]. UAD methods typically train a one-class classifier using data from the normal class only, and anomalies (or abnormal cases) are detected based on the extent the images deviate from the normal class.

Current anomaly detection approaches [7,8,14,27,37,43,46] train deep generative models (e.g., auto-encoder [19], GAN [15]) to reconstruct normal images, and anomalies are detected from the reconstruction error [33]. These approaches rely on a low-dimensional image representation that must be effective at reconstructing normal images, where the main challenge is to detect anomalies that show subtle deviations from normal images, such as with small lesions [43]. Recently, self-supervised methods that learn auxiliary pretext tasks [2,6,13,17,18,25] have been shown to learn effective representations for UAD in general computer vision tasks [2,13,18], so it is important to investigate if self-supervision can also improve UAD for medical images.

The main challenge for the design of UAD methods for medical imaging resides in how to devise effective pretext tasks. Self-supervised pretext tasks consist of predicting geometric or brightness transformations [2,13,18], or contrastive learning [6,17]. These pretext tasks have been designed to work for downstream classification problems that are not related to anomaly detection, so they may degrade the detection performance of UAD methods [47]. Sohn et al. [40] tackle this issue by using smaller batch sizes than in [6,17] and a new data augmentation method. However, the use of self-supervised learning in UAD for medical images has not been investigated, to the best of our knowledge. Further, although transformation prediction and contrastive learning show great success in self-supervised feature learning, there are no studies on how to properly combine these two approaches to learn more effective features for UAD.

In this paper, we propose Constrained Contrastive Distribution learning (CCD), a new self-supervised representation learning designed specifically to learn normality information from exclusively normal training images. The contributions of CCD are: a) contrastive distribution learning, and b)two pretext learning constraints, both of which are customised for anomaly detection (AD). Unlike modern self-supervised learning (SSL) [6,17] that focuses on learning generic semantic representations for enabling diverse downstream tasks, CCD instead contrasts the distributions of strongly augmented images (e.g., random permutations). The strongly augmented images resemble some types of abnormal images, so CCD is enforced to learn discriminative normality representations

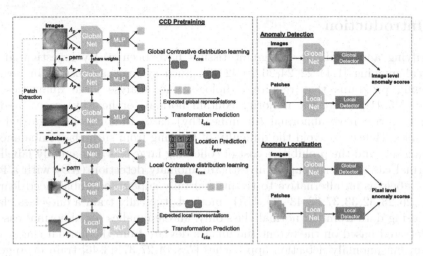

Fig. 1. Our proposed CCD framework. **Left** shows the proposed pre-training method that unifies a contrastive distribution learning and pretext learning on both global and local perspectives (Sect. 2.1), **Right** shows the inference for detection and localisation (Sect. 2.2).

by its contrastive distribution learning. The two pretext learning constraints on augmentation and location prediction are added to learn fine-grained normality representations for the detection of subtle abnormalities. These two unique components result in significantly improved self-supervised AD-oriented representation learning, substantially outperforming previous general-purpose SOTA SSL approaches [2,6,13,18]. Another important contribution of CCD is that it is agnostic to downstream anomaly classifiers. We empirically show that our CCD improves the performance of three diverse anomaly detectors (f-anogan [37], IGD [8], MS-SSIM) [48]). Inspired by IGD [8], we adapt our proposed CCD pre-training on global images and local patches, respectively. Extensive experimental results on three different health screening medical imaging benchmarks, namely, colonoscopy images from two datasets [4,27], and fundus images for glaucoma detection [21], show that our proposed self-supervised approach enables the production of SOTA anomaly detection and localisation in medical images.

2   Method

In this section, we introduce the proposed approach, depicted in the diagram of Fig. 1. Specifically, given a training medical image dataset $\mathcal{D} = \{\mathbf{x}_i\}_{i=1}^{|\mathcal{D}|}$, with all images assumed to be from the normal class and $\mathbf{x} \in \mathcal{X} \subset \mathbb{R}^{H \times W \times C}$, our approach aims to learn anomaly detection and localisation using three modules: 1) a self-supervised constrained contrastive feature learner that pre-trains an encoding network $f_\theta : \mathcal{X} \rightarrow \mathcal{Z}$ (with $\mathcal{Z} \subset \mathbb{R}^{d_z}$) tailored for anomaly detection, 2) an anomaly classification model $h_\psi : \mathcal{Z} \rightarrow [0,1]$ that is built upon the pre-trained network, and 3) an anomaly localiser that leverages the classifier $h_\psi(f_\theta(\mathbf{x}_\omega))$ to localise an abnormal image region $\mathbf{x}_\omega \in \mathbb{R}^{\hat{H} \times \hat{W} \times C}$, centred at $\omega \in \Omega$ (Ω is the image lattice) with height $\hat{H} << H$ and width $\hat{W} << W$. The approach is evaluated on a testing set $\mathcal{T} = \{(\mathbf{x}, y, \mathbf{m})_i\}_{i=1}^{|\mathcal{T}|}$, where $y \in \mathcal{Y} = \{\text{normal}, \text{abnormal}\}$, and $\mathbf{m} \in \mathcal{M} \subset \{0,1\}^{H \times W \times C}$ denotes the segmentation mask of the lesion in the image $\mathbf{x}$. For adapting our CCD pretraining on patch representations, we simply crop the training images into patches before applying our method.

2.1 Constrained Contrastive Distribution Learning

Contrastive learning has been used by self-supervised learning methods to pre-train encoders with data augmentation [6,17,47] and contrastive learning loss [39]. The idea is to sample functions from a data augmentation distribution (e.g., geometric and brightness transformations), and assume that the same image, under separate augmentations, form one class to be distinguished against all other images in the batch [2,13]. Another form of pre-training is based on a pretext task, such as solving jigsaw puzzle and predicting geometric and brightness transformations [6,17]. These self-supervised learning approaches are useful to pre-train classification [6,17] and segmentation models [31,49]. Only recently, self-supervised learning using contrastive learning [40] and pretext learning [2,13] have been shown to be effective in anomaly detection. However, these two approaches are explored separately. In this paper, we aim at harnessing the power of both approaches to learn more expressive pre-trained features specifically for UAD. To this end, we propose the novel Constrained Contrastive Distribution learning method (CCD).

Contrastive distribution learning is designed to enforce a non-uniform distribution of the representations in the space $\mathcal{Z}$, which has been associated with more effective anomaly detection performance [40]. Our CCD method constrains the constrastive distribution learning with two pretext learning tasks, with the goal of enforcing further the non-uniform distribution of the representations. The CCD loss is defined as

$$\ell_{CCD}(\mathcal{D}; \theta, \beta, \gamma) = \ell_{con}(\mathcal{D}; \theta) + \ell_{cla}(\mathcal{D}; \beta) + \ell_{pos}(\mathcal{D}; \gamma), \tag{1}$$

where $\ell_{con}(\cdot)$ is the contrastive distribution loss, ℓ_{cla} and ℓ_{pos} are two pretext learning tasks added to constrain the optimisation; and θ, β and γ are trainable parameters. The contrastive distribution learning uses a dataset of **weak data augmentations** $\mathcal{A}_p = \{a_l : \mathcal{X} \to \mathcal{X}\}_{l=1}^{|\mathcal{A}_p|}$ and **strong data augmentations** $\mathcal{A}_n = \{a_l : \mathcal{X} \to \mathcal{X}\}_{l=1}^{|\mathcal{A}_n|}$, where $a_l(\mathbf{x})$ denotes a particular data augmentation applied to $\mathbf{x}$, and the loss is defined as

$$\ell_{con}(\mathcal{D}; \theta) =$$

$$- \mathbb{E}\left[\log \frac{\exp\left[\frac{1}{\tau} f_\theta(a(\tilde{\mathbf{x}}^j))^\top f_\theta(a'(\tilde{\mathbf{x}}^j))\right]}{\exp\left[\frac{1}{\tau} f_\theta(a(\tilde{\mathbf{x}}^j))^\top f_\theta(a'(\tilde{\mathbf{x}}^j))\right] + \sum_{i=1}^{M} \exp\left[\frac{1}{\tau} f_\theta(a(\tilde{\mathbf{x}}^j))^\top f_\theta(a'(\tilde{\mathbf{x}}_i^j))\right]}\right],$$

$$\tag{2}$$

where the expectation is over $\mathbf{x} \in \mathcal{D}$, $\{\mathbf{x}_i\}_{i=1}^M \subset \mathcal{D} \setminus \{\mathbf{x}\}$, $a(.), a'(.) \in \mathcal{A}_p$, $\tilde{\mathbf{x}}^j = a_j(\mathbf{x})$, $\tilde{\mathbf{x}}_i^j = a_j(\mathbf{x}_i)$, and $a_j(.) \in \mathcal{A}_n$. The images augmented with the functions from the strong set $\mathcal{A}_n$ carry some 'abnormality' compared to the original images, which is helpful to learn a non-uniform distribution in the representation space $\mathcal{Z}$.

We can then constrain further the training to learn more non-uniform representations with a self-supervised classification constraint $\ell_{cla}(\cdot)$ that enforces the model to achieve accurate classification of the strong augmentation function:

$$\ell_{cla}(\mathcal{D}; \beta) = -\mathbb{E}_{\mathbf{x} \in \mathcal{D}, a(.) \in \mathcal{A}_n} \left[\log \mathbf{a}^\top f_\beta(f_\theta(a(\mathbf{x}))) \right], \tag{3}$$

where $f_\beta : \mathcal{Z} \to [0,1]^{|\mathcal{A}_n|}$ is a fully-connected (FC) layer, and $\mathbf{a} \in \{0,1\}^{|\mathcal{A}_n|}$ is a one-hot vector representing the strong augmentation $a(.) \in \mathcal{A}_n$.

The second constraint is based on the relative patch location from the centre of the training image – this positional information is important for segmentation tasks [20, 31]. This constraint is added to learn fine-grained features and achieve more accurate anomaly localisation. Inspired by [10], the positional constraint predicts the relative position of the paired image patches, with its loss defined as

$$\ell_{pos}(\mathcal{D}; \gamma) = -\mathbb{E}_{\{\mathbf{x}_{\omega_1}, \mathbf{x}_{\omega_2}\} \sim \mathbf{x} \in \mathcal{D}} \left[\log \mathbf{p}^\top f_\gamma(f_\theta(\mathbf{x}_{\omega_1}), f_\theta(\mathbf{x}_{\omega_2})) \right], \tag{4}$$

where $\mathbf{x}_{\omega_1}$ is a randomly selected fixed-size image patch from $\mathbf{x}$, $\mathbf{x}_{\omega_2}$ is another image patch from one of its eight neighbouring patches (as shown in 'patch location prediction' in Fig. 1), $f_\gamma : \mathcal{Z} \times \mathcal{Z} \to [0,1]^8$, and $\mathbf{p} = \{0,1\}^8$ is a one-hot encoding of the synthetic class label.

Overall, the constraints in (3) and (4) to the contrastive distribution loss in (2) are designed to increase the non-uniform representation distribution and to improve the representation discriminability between normal and abnormal samples, compared with [40].

2.2 Anomaly Detection and Localisation

Building upon the pre-trained encoder $f_\theta(\cdot)$ using the loss in (1), we fine-tune two state-of-the-art UAD methods, IGD [8] and F-anoGAN [37], and a baseline method, multi-scale structural similarity index measure (MS-SSIM)-based auto-encoder [48]. All UAD methods use the same training set $\mathcal{D}$ that contains only normal image samples.

IGD [8] combines three loss functions: 1) two reconstruction losses based on local and global multi-scale structural similarity index measure (MS-SSIM) [48] and mean absolute error (MAE) to train the encoder $f_\theta(\cdot)$ and decoder $g_\phi(\cdot)$, 2) a regularisation loss to train adversarial interpolations from the encoder [3], and 3) an anomaly classification loss to train $h_\psi(\cdot)$. The anomaly detection score of image $\mathbf{x}$ is

$$s_{IGD}(\mathbf{x}) = \xi \ell_{rec}(\mathbf{x}, \tilde{\mathbf{x}}) + (1 - \xi)(1 - h_\psi(f_\theta(\mathbf{x}))), \tag{5}$$

where $\tilde{\mathbf{x}} = g_\phi(f_\theta(\mathbf{x}))$, $h_\psi(f_\theta(\mathbf{x})) \in [0,1]$ returns the likelihood that $\mathbf{x}$ belongs to the normal class, $\xi \in [0,1]$ is a hyper-parameter, and

$$\ell_{rec}(\mathbf{x}, \tilde{\mathbf{x}}) = \rho \|\mathbf{x} - \tilde{\mathbf{x}}\|_1 + (1 - \rho)(1 - (\nu m_G(\mathbf{x}, \tilde{\mathbf{x}}) + (1 - \nu) m_L(\mathbf{x}, \tilde{\mathbf{x}}))), \tag{6}$$

with $\rho, \nu \in [0, 1]$, $m_G(\cdot)$ and $m_L(\cdot)$ denoting the global and local MS-SSIM scores [8]. Anomaly localisation uses (5) to compute $s_{IGD}(\mathbf{x}_\omega)$, $\forall \omega \in \Omega$, where $\mathbf{x}_\omega \in \mathbb{R}^{\hat{H} \times \hat{W} \times C}$ is an image region–this forms a heatmap, where large values denote anomalous regions.

F-anoGAN [37] combines generative adversarial networks (GAN) and auto-encoder models to detect anomalies. Training involves the minimisation of reconstruction losses in both the original image and representation spaces to model $f_\theta(\cdot)$ and $g_\phi(\cdot)$. It also uses a GAN loss [15] to model $g_\phi(\cdot)$ and $h_\psi(\cdot)$. Anomaly detection for image $\mathbf{x}$ is

$$s_{FAN}(\mathbf{x}) = \|\mathbf{x} - g_\phi(f_\theta(\mathbf{x}))\| + \kappa\|f_\theta(\mathbf{x}) - f_\theta(g_\phi(f_\theta(\mathbf{x})))\|. \qquad (7)$$

Anomaly localisation at $\mathbf{x}_\omega \in \mathbb{R}^{\hat{H} \times \hat{W} \times C}$ is achieved by $\|\mathbf{x}_\omega - g_\phi(f_\theta(\mathbf{x}_\omega))\|$, $\forall \omega \in \Omega$.

For the MS-SSIM auto-encoder [48], we train it with the MS-SSIM loss for reconstructing the training images. Anomaly detection for $\mathbf{x}$ is based on $s_{MSI}(\mathbf{x}) = 1 - (\nu m_G(\mathbf{x}, \tilde{\mathbf{x}}) + (1 - \nu)m_L(\mathbf{x}, \tilde{\mathbf{x}}))$, with $\tilde{\mathbf{x}}$ as defined in (5). Anomaly localisation is performed with $s_{MSI}(\mathbf{x}_\omega)$ at image regions $\mathbf{x}_\omega \in \mathbb{R}^{\hat{H} \times \hat{W} \times C}$, $\forall \omega \in \Omega$. Inspired by IGD [8], we also pretrain a local model using our CCD pretraining approach based on the local patches for F-anogan [37] and MS-SSIM autoencoder [48], respectively.

3 Experiments

3.1 Dataset

We test our framework on three health screening datasets. We test both anomaly detection and localisation on the colonoscopy images of Hyper-Kvasir dataset [4]. On the glaucoma datasets using fundus images [21] and colonoscopy dataset [27] that do not have lesion masks, we test anomaly detection only. Detection is assessed with area under the ROC curve (AUC). Localisation is measured with intersection over union (ioU).

Hyper-Kvasir is a large multi-class public gastrointestinal dataset. The data was collected from the gastroscopy and colonoscopy procedures from Baerum Hospital in Norway. All labels were produced by experienced radiologists. The dataset contains 110,079 images from abnormal (i.e., unhealthy) and normal (i.e., healthy) patients, with 10,662 labelled. We use part of the clean images from the dataset to train our UAD methods. Specifically, 2,100 images from 'cecum', 'ileum' and 'bbps-2-3' are selected as normal, from which we use 1,600 for training and 500 for testing. We also take 1,000 abnormal images and their segmentation masks and stored them in the testing set.

LAG is a large scale fundus image dataset for glaucoma detection [21], containing 4,854 fundus images with 1,711 positive glaucoma scans and 3,143 negative glaucoma scans. We reorganised this dataset for training the UAD methods, with 2,343 normal (negative glaucoma) images for training, and 800 normal images and 1,711 abnormal images with positive glaucoma for testing.

Liu et al.'s colonoscopy dataset is a colonoscopy image dataset for UAD using 18 colonocopy videos from 15 patients [27]. The training set contains 13,250 normal (healthy) images without any polyps, and the testing set contains 967 images, having 290 abnormal images with polyps and 677 normal (healthy) images without polyps.

3.2   Implementation Details

For pre-training, we use Resnet18 [16] as the backbone architecture for the encoder $f_\theta(\mathbf{x})$, and similarly to previous works [6,40], we add an MLP to this backbone as the projection head for the contrastive learning. All images from the Hyper-Kvasir [4] and LAG [21] datasets are resized to 256 × 256 pixels. For the Liu et al.'s colonoscopy dataset, images are resized to 64 × 64 pixels. The batch size is set to 32 and learning rate to 0.01 for the self-supervised pre-training. We investigate the impact of different strong augmentations in $\mathcal{A}_n$ such as rotation, permutation, cutout and Gaussian noise. All weak augmentations in $\mathcal{A}_p$ are the same as SimCLR [6] (i.e., colour jittering, random grey scale, crop, resize, and Gaussian blur). The model is trained using SGD optimiser with temperature 0.2. The encoder $f_\theta(\cdot)$ outputs a 128 dimensional feature in $\mathcal{Z}$. All datasets are pre-trained for 2,000 epochs.

For the training of IGD [8], F-anoGAN [37] and MS-SSIM auto-encoder [8], we use the hyper-parameters suggested by the respective papers. For localisation, we compute the heatmap based on the localised anomaly scores from IGD, where the final map is obtained by summing the global and local maps. In our experiments, the local map is obtained by considering each 32 × 32 image patch as a instance and apply our proposed self-supervised learning to it. The global map is computed based on the whole image sized as 256 × 256. For F-anoGAN and MS-SSIM auto-encoder, we use the same setup as the IGD, where models based the 256 × 256 whole image and the 32 × 32 patches are trained, respectively. Code will be made publicly available upon paper acceptance.

3.3   Ablation Study

In Fig. 2 (right), we explore the influence of strong augmentation strategies, represented by rotation, permutation, cutout and Gaussian noise, on the AUC results on Hyper-Kvasir dataset, based on our self-supervised pre-training with IGD as anomaly detector. The experiment indicates that the use of random permutations as strong augmentations yields the best AUC results. We also explore the relation between batch size and AUC results in Fig. 2 (left). The results suggest that small batch size (equal to 16) leads to a relatively low AUC, which increases for batch size 32, and then decreases for larger batch sizes. Given these results, we use permutation as the strong augmentation for colonoscopy images and training batch size is set to 32. For the LAG dataset, we omit the results, but we use rotation as the strong augmentation because it produced the largest AUC. We also used batch size of 32 for the LAG dataset.

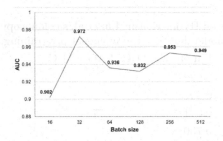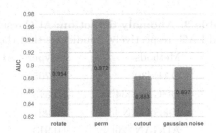

Fig. 2. Left: Anomaly detection performance results based on different batch sizes of self-supervised pre-training. **Right**: Anomaly detection performance in terms of different types of strong augmentations. Both results are on Hyper-Kvasir test set using IGD as anomaly detector.

Table 1. Ablation study of the loss terms in (1) **on Hyper-Kvasir, using IGD as anomaly detector.**

ℓ_{con}[6,17]	ℓ_{con}	ℓ_{pre}	ℓ_{pat}	AUC - Hyper-Kvasir
✓				0.913
	✓			0.937
	✓	✓		0.964
	✓	✓	✓	**0.972**

Table 2. Anomaly localisation: Mean IoU results on Hyper-Kvasir on 5 different groups of 100 images with ground truth masks. * indicates that we pretrained the geometric transformation-based anomaly detection [13] using IGD [8] as the UAD method.

Supervision	Methods	Localisation - IoU
Supervised	U-Net [36]	0.746
	U-Net++ [50]	0.743
	ResUNet [9]	**0.793**
	SFA [12]	0.611
Unsupervised	RotNet [13]+IGD [8]*	0.276
	CAVGA-R_u [46]	0.349
	Ours - IGD	**0.372**

We also present an ablation study that shows the influence of each loss term in (1) in Table 1, again on Hyper-Kvasir dataset, based on our self-supervised pre-training with IGD. The vanilla contrastive learning in [6,17] only achieves 91.3% of AUC. After replacing it with our distribution contrastive loss from (2), the performance increases by 2.4% AUC. Adding distribution classification and patch position prediction losses boosts the performance by another 2.7% and 0.8% AUC, respectively.

3.4 Comparison to SOTA Models

In Table 3, we show the results of anomaly detection on Hyper-Kvasir, Liu et al.'s colonoscopy dataset and LAG datasets. The IGD, F-anoGAN and MS-SSIM methods improve their baselines (without our self-supervision method) from 3.3% to 5.1% of AUC on Hyper-Kvasir, from −0.3% to 12.2% on Liu et al.'s dataset, and from 0.9% to 7.8% on LAG. The IGD with our pre-trained

Table 3. Anomaly detection: AUC results on Hyper-Kvasir, Liu et al.'s colonocopy and LAG, respectively. * indicates that the model does not use imagenet pre-training.

Methods	Hyper - AUC	Liu et al. - AUC	LAG - AUC
DAE [30]	0.705	0.629 *	–
OCGAN [34]	0.813	0.592 *	–
F-anoGAN [37]	0.907	0.691 *	0.778
ADGAN [26]	0.913	0.730 *	–
CAVGA-R_u [46]	0.928	–	–
MS-SSIM [8]	0.917	0.799	0.823
IGD [8]	0.939	0.787	0.796
RotNet [13]+IGD [8]	0.905	–	–
Ours - MS-SSIM	0.945	0.796	0.839
Ours - F-anoGAN	0.958	0.813	0.787
Ours - IGD	**0.972**	**0.837**	**0.874**

features achieves SOTA anomaly detection AUC on all three datasets. Such results suggest that our self-supervised pre-training can effectively produce good representations for various types of anomaly detectors and datasets. OCGAN [34] constrained the latent space based on two discriminators to force the latent representations of normal data to fall at a bounded area. CAVGA-R_u [46] is a recently proposed approach for anomaly detection and localisation that uses an attention expansion loss to encourage the model to focus on normal object regions in the images. These two methods achieve 81.3% and 92.8% AUC on Hyper-Kvasir, respectively, which are well behind our self-supervised pre-training with IGD of 97.2% AUC.

We also investigate the anomaly localisation performance on Hyper-Kvasir in Table 2. Compared to the SOTA UAD localisation method, CAVGA-R_u [46], our approach with IGD is more than 3% better in terms of IoU. We also compare our results to **fully supervised methods** [9,12,36,50] to assess how much performance is lost by suppressing supervision from abnormal data. The fully supervised baselines [9,12,36,50] use 80% of the annotated 1,000 colonoscopy images containing polyps during training, and 10% for validation and 10% for testing. We validate our approach using the same number of testing samples, but without using abnormal samples for training. The localisation results are post processed by the Connected Component Analysis (CCA) [5]. Notice on Table 2 that we lose between 0.3 and 0.4 IoU for not using abnormal samples for training.

We present visual anomaly localisation results of our IGD with self-supervised pre-training on the abnormal images from Hyper Kvasir [4] test set in Fig. 3. Notice how our model can accurately localise polyps with various size and textures.

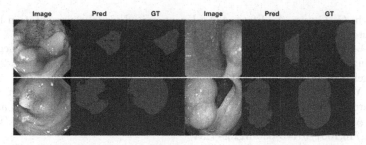

Fig. 3. Qualitative results of our localisation network based on IGD with self-supervised pre-training on the abnormal images from Hyper Kvasir [4] test set.

4 Conclusion

To conclude, we proposed a self-supervised pre-training for UAD named as constrained contrastive distribution learning for anomaly detection. Our approach enforces non-uniform representation distribution by constraining contrastive distribution learning with two pretext tasks. We validate our approach on three medical imaging benchmarks and achieve SOTA anomaly detection and localisation results using three UAD methods. In future work, we will investigate more choices of pretext tasks for UAD.

References

1. Baur, C., Wiestler, B., Albarqouni, S., Navab, N.: Scale-space autoencoders for unsupervised anomaly segmentation in brain MRI. In: Martel, A.L., et al. (eds.) MICCAI 2020. LNCS, vol. 12264, pp. 552–561. Springer, Cham (2020). https://doi.org/10.1007/978-3-030-59719-1_54
2. Bergman, L., Hoshen, Y.: Classification-based anomaly detection for general data. arXiv preprint arXiv:2005.02359 (2020)
3. Berthelot, D., Raffel, C., Roy, A., Goodfellow, I.: Understanding and improving interpolation in autoencoders via an adversarial regularizer. arXiv preprint arXiv:1807.07543 (2018)
4. Borgli, H., et al.: Hyperkvasir, a comprehensive multi-class image and video dataset for gastrointestinal endoscopy. Sci. Data **7**(1), 1–14 (2020)
5. Chai, B.B., Vass, J., Zhuang, X.: Significance-linked connected component analysis for wavelet image coding. IEEE Trans. Image Process. **8**(6), 774–784 (1999)
6. Chen, T., Kornblith, S., Norouzi, M., Hinton, G.: A simple framework for contrastive learning of visual representations. In: ICML, pp. 1597–1607. PMLR (2020)
7. Chen, X., You, S., Tezcan, K.C., Konukoglu, E.: Unsupervised lesion detection via image restoration with a normative prior. Med. Image Anal. **64**, 101713 (2020)
8. Chen, Y., Tian, Y., Pang, G., Carneiro, G.: Unsupervised anomaly detection and localisation with multi-scale interpolated gaussian descriptors. arXiv preprint arXiv:2101.10043 (2021)
9. Diakogiannis, F.I., et al.: Resunet-a: a deep learning framework for semantic segmentation of remotely sensed data. ISPRS J. Photogrammetry Remote. Sens. **162**, 94–114 (2020)

10. Doersch, C., Gupta, A., Efros, A.A.: Unsupervised visual representation learning by context prediction. In: ICCV, pp. 1422–1430 (2015)
11. Fan, D.-P., et al.: PraNet: parallel reverse attention network for polyp segmentation. In: Martel, A.L., et al. (eds.) MICCAI 2020. LNCS, vol. 12266, pp. 263–273. Springer, Cham (2020). https://doi.org/10.1007/978-3-030-59725-2_26
12. Fang, Y., Chen, C., Yuan, Y., Tong, K.: Selective feature aggregation network with area-boundary constraints for polyp segmentation. In: Shen, D., et al. (eds.) MICCAI 2019. LNCS, vol. 11764, pp. 302–310. Springer, Cham (2019). https://doi.org/10.1007/978-3-030-32239-7_34
13. Golan, I., El-Yaniv, R.: Deep anomaly detection using geometric transformations. arXiv preprint arXiv:1805.10917 (2018)
14. Gong, D., et al.: Memorizing normality to detect anomaly: Memory-augmented deep autoencoder for unsupervised anomaly detection. In: ICCV, pp. 1705–1714 (2019)
15. Goodfellow, I.: Generative adversarial nets. In: Advances in Neural Information Processing Systems, pp. 2672–2680 (2014)
16. He, K., et al.: Deep residual learning for image recognition. In: CVPR, pp. 770–778 (2016)
17. He, K., et al.: Momentum contrast for unsupervised visual representation learning. In: CVPR, pp. 9729–9738 (2020)
18. Hendrycks, D., et al.: Using self-supervised learning can improve model robustness and uncertainty. arXiv preprint arXiv:1906.12340 (2019)
19. Kingma, D.P., Welling, M.: Auto-encoding variational bayes. arXiv preprint arXiv:1312.6114 (2013)
20. Kolesnikov, A., Zhai, X., Beyer, L.: Revisiting self-supervised visual representation learning. CoRR abs/1901.09005 (2019)
21. Li, L., et al.: Attention based glaucoma detection: a large-scale database and CNN model. In: CVPR, pp. 10571–10580 (2019)
22. Litjens, G., et al.: A survey on deep learning in medical image analysis. Med. Image Anal. **42**, 60–88 (2017)
23. Liu, F., Tian, Y., Cordeiro, F.R., Belagiannis, V., Reid, I., Carneiro, G.: Noisy label learning for large-scale medical image classification. arXiv preprint arXiv:2103.04053 (2021)
24. Liu, F., Tian, Y., et al.: Self-supervised mean teacher for semi-supervised chest x-ray classification. arXiv preprint arXiv:2103.03629 (2021)
25. Liu, F., Jonmohamadi, Y., Maicas, G., Pandey, A.K., Carneiro, G.: Self-supervised depth estimation to regularise semantic segmentation in knee arthroscopy. In: Martel, A.L., et al. (eds.) MICCAI 2020. LNCS, vol. 12261, pp. 594–603. Springer, Cham (2020). https://doi.org/10.1007/978-3-030-59710-8_58
26. Liu, Y., et al.: Photoshopping colonoscopy video frames. In: ISBI, pp. 1–5 (2020). https://doi.org/10.1109/ISBI45749.2020.9098406
27. Liu, Y., et al.: Photoshopping colonoscopy video frames. In: ISBI, pp. 1–5 (2020)
28. Luo, W., Gu, Z., Liu, J., Gao, S.: Encoding structure-texture relation with p-net for anomaly detection in retinal images
29. LZ, C.T.P., et al.: Computer-aided diagnosis for characterisation of colorectal lesions: a comprehensive software including serrated lesions. Gastrointest. Endosc. **92**(4), 891–899 (2020)
30. Masci, J., Meier, U., Cireşan, D., Schmidhuber, J.: Stacked convolutional auto-encoders for hierarchical feature extraction. In: Honkela, T., Duch, W., Girolami, M., Kaski, S. (eds.) ICANN 2011. LNCS, vol. 6791, pp. 52–59. Springer, Heidelberg (2011). https://doi.org/10.1007/978-3-642-21735-7_7

31. Noroozi, M., Favaro, P.: Unsupervised learning of visual representations by solving jigsaw puzzles. In: Leibe, B., Matas, J., Sebe, N., Welling, M. (eds.) ECCV 2016. LNCS, vol. 9910, pp. 69–84. Springer, Cham (2016). https://doi.org/10.1007/978-3-319-46466-4_5

32. Ouardini, K., et al.: Towards practical unsupervised anomaly detection on retinal images. In: Wang, Q., et al. (eds.) DART/MIL3ID -2019. LNCS, vol. 11795, pp. 225–234. Springer, Cham (2019). https://doi.org/10.1007/978-3-030-33391-1_26

33. Pang, G., Shen, C., Cao, L., Hengel, A.V.D.: Deep learning for anomaly detection: a review. ACM Comput. Surv. (CSUR) 54(2), 1–38 (2021)

34. Perera, P., Nallapati, R., Xiang, B.: Ocgan: one-class novelty detection using gans with constrained latent representations. In: CVPR, pp. 2898–2906 (2019)

35. Pu, L., Tao, Z.C., et al.: Prospective study assessing a comprehensive computer-aided diagnosis for characterization of colorectal lesions: results from different centers and imaging technologies. In: Journal of Gastroenterology and Hepatology, vol. 34, pp. 25–26. WILEY 111 RIVER ST, HOBOKEN 07030–5774, NJ USA (2019)

36. Ronneberger, O., Fischer, P., Brox, T.: U-Net: convolutional networks for biomedical image segmentation. In: Navab, N., Hornegger, J., Wells, W.M., Frangi, A.F. (eds.) MICCAI 2015. LNCS, vol. 9351, pp. 234–241. Springer, Cham (2015). https://doi.org/10.1007/978-3-319-24574-4_28

37. Schlegl, T., et al.: f-anogan: fast unsupervised anomaly detection with generative adversarial networks. Med. Image Anal. 54, 30–44 (2019)

38. Seeböck, P., et al.: Exploiting epistemic uncertainty of anatomy segmentation for anomaly detection in retinal oct. IEEE Trans. Med. Imaging 39(1), 87–98 (2019)

39. Sohn, K.: Improved deep metric learning with multi-class n-pair loss objective. In: Proceedings of the 30th International Conference on Neural Information Processing Systems, pp. 1857–1865 (2016)

40. Sohn, K., Li, C.L., Yoon, J., Jin, M., Pfister, T.: Learning and evaluating representations for deep one-class classification. arXiv preprint arXiv:2011.02578 (2020)

41. Tian, Y., otherss: Detecting, localising and classifying polyps from colonoscopy videos using deep learning. arXiv preprint arXiv:2101.03285 (2021)

42. Tian, Y., et al.: One-stage five-class polyp detection and classification. In: 2019 IEEE 16th International Symposium on Biomedical Imaging (ISBI 2019), pp. 70–73. IEEE (2019)

43. Tian, Yu., Maicas, G., Pu, L.Z.C.T., Singh, R., Verjans, J.W., Carneiro, G.: Few-shot anomaly detection for polyp frames from colonoscopy. In: Martel, A.L., et al. (eds.) MICCAI 2020. LNCS, vol. 12266, pp. 274–284. Springer, Cham (2020). https://doi.org/10.1007/978-3-030-59725-2_27

44. Tian, Y., et al.: Weakly-supervised video anomaly detection with robust temporal feature magnitude learning. arXiv preprint arXiv:2101.10030 (2021)

45. Uzunova, H., Schultz, S., Handels, H., Ehrhardt, J.: Unsupervised pathology detection in medical images using conditional variational autoencoders. Int. J. Comput. Assist. Radiol. Surg. 14(3), 451–461 (2018). https://doi.org/10.1007/s11548-018-1898-0

46. Venkataramanan, S., Peng, K.-C., Singh, R.V., Mahalanobis, A.: Attention guided anomaly localization in images. In: Vedaldi, A., Bischof, H., Brox, T., Frahm, J.-M. (eds.) ECCV 2020. LNCS, vol. 12362, pp. 485–503. Springer, Cham (2020). https://doi.org/10.1007/978-3-030-58520-4_29

47. Wang, T., Isola, P.: Understanding contrastive representation learning through alignment and uniformity on the hypersphere. In: ICML, pp. 9929–9939. PMLR (2020)

48. Wang, Z., et al.: Multiscale structural similarity for image quality assessment. In: The Thrity-Seventh Asilomar Conference on Signals, Systems & Computers, 2003, vol. 2, pp. 1398–1402. IEEE (2003)
49. Yi, J., Yoon, S.: Patch svdd: patch-level svdd for anomaly detection and segmentation. In: ACCV (2020)
50. Zhou, Z., Rahman Siddiquee, M.M., Tajbakhsh, N., Liang, J.: UNet++: a nested u-net architecture for medical image segmentation. In: Stoyanov, D., et al. (eds.) DLMIA/ML-CDS -2018. LNCS, vol. 11045, pp. 3–11. Springer, Cham (2018). https://doi.org/10.1007/978-3-030-00889-5_1

Conditional Training with Bounding Map for Universal Lesion Detection

Han Li[1,2], Long Chen[2,3], Hu Han[2(✉)], Ying Chi[4], and S. Kevin Zhou[1,2(✉)]

[1] Medical Imaging, Robotics, Analytic Computing Laboratory/Engineering (MIRACLE), School of Biomedical Engineering & Suzhou Institute for Advanced Research, University of Science and Technology of China, Suzhou, China
[2] Key Laboratory of Intelligent Information Processing of Chinese Academy of Sciences (CAS), Institute of Computing Technology, CAS, Beijing, China
{han.li,long.chen}@miracle.ict.ac.cn, hanhu@ict.ac.cn
[3] School of Electronic, Electrical and Communication Engineering, University of the Chinese Academy of Science, Beijing, China
[4] Healthcare Intelligence, AIC, DAMO Academy, Alibaba Group, Hangzhou, China
xinyi.cy@alibaba-inc.com

Abstract. Universal Lesion Detection (ULD) in computed tomography plays an essential role in computer-aided diagnosis. Promising ULD results have been reported by coarse-to-fine two-stage detection approaches, but such two-stage ULD methods still suffer from issues like imbalance of positive v.s. negative anchors during object proposal and insufficient supervision problem during localization regression and classification of the region of interest (RoI) proposals. While leveraging pseudo segmentation masks such as bounding map (BM) can reduce the above issues to some degree, it is still an open problem to effectively handle the diverse lesion shapes and sizes in ULD. In this paper we propose a BM-based conditional training for two-stage ULD, which can (i) reduce positive vs. negative anchor imbalance via a BM-based conditioning (BMC) mechanism for anchor sampling instead of traditional IoU-based rule; and (ii) adaptively compute size-adaptive BM (ABM) from lesion bounding-box, which is used for improving lesion localization accuracy via ABM-supervised segmentation. Experiments with four state-of-the-art methods show that the proposed approach can bring an almost free detection accuracy improvement without requiring expensive lesion mask annotations.

Keywords: Universal lesion detection · Adaptive bounding map · Conditional training

This research was supported in part by the Natural Science Foundation of China (grant 61732004), Youth Innovation Promotion Association CAS (grant 2018135) and Alibaba Group through Alibaba Innovative Research Program.

Electronic supplementary material The online version of this chapter (https://doi.org/10.1007/978-3-030-87240-3_14) contains supplementary material, which is available to authorized users.

1   Introduction

Universal Lesion Detection (ULD) in computed tomography (CT) [1–14], which aims to localize different types of lesions instead of identifying lesion types [15–26], plays an essential role in computer-aided diagnosis (CAD) [27,28]. ULD is a challenging task because different lesions have very diverse shapes and sizes, easily leading to false positive and false negative detections. Most existing ULD methods are mainly inspired by the successful deep models in object detection from natural images. These ULD approaches adapt the Mask-RCNN [29] framework by constructing a pseudo segmentation mask for lesion regions as the extra supervision [1,2,5,8,9,13] or extract more 3D context information from multi CT slice [3,4,6–10,12] to assist the ULD training.

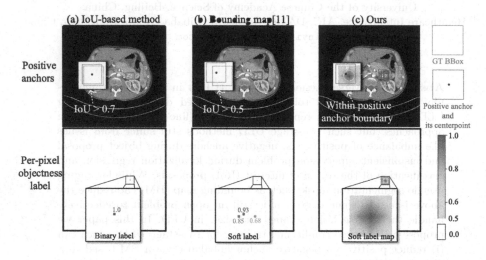

Fig. 1. The IoU-based method (a) produces very few positive anchors thus biases the RPN training. The BM method [11] (b) relieves this issue to some degree via a soft-label map BM and a lower threshold but the imbalance issue remains severe. Our BMC mechanism (c) predicts a per-pixel objectness map and any anchor within the BM with a value larger than a threshold can be a positive anchor. Thus, the positive vs. negative anchor imbalance issue can be effectively relieved.

Most of the above approaches proposed [2,4–9,12,13] for ULD are designed based on a two-stage, anchor-based detection framework, i.e., proposal generation followed by classification and regression as in Faster R-CNN [30]. While achieving good success, such a framework has inherent limitations: (i) *Anchor imbalance in stage-1* [31]. In the first stage, anchor-based methods first find out the positive (lesion) anchors and use them as the region of interest (RoI) proposals according to the IoU between anchors and ground-truth (GT) BBoxs. An anchor is considered positive if its IoU with any GT BBox is greater than the IoU threshold and negative otherwise. This idea helps natural images to get enough positive anchors because they may have a lot of GT BBoxs per image [31], but it isn't suitable for ULD. Most CT slices only have one or two GT lesion BBox(s), so the amount of positive anchors of lesions is rather limited. This limitation can cause severe RoI imbalance (our empirical statistics show the positive vs. negative proposal imbalance can be as large as 1:200 which is showed in Suppl. Material), thereby leading to difficulty in network convergence. (ii) *Insufficient supervision in stage-2.* In the second stage, each RoI proposal from the first stage gets a classification score indicating its probability of lesion vs. non-lesion. The RoI proposals with high classification scores are chosen to obtain the final lesion BBox predictions. Since various lesions may have similar appearances with the other tissues; the RoI proposals from non-lesion regions can also get very high lesion classification scores. Hence, a single classification score cannot provide sufficient supervision information to remove the false detections in the second stage.

To tackle the above two challenges, recently Li et al. [11] propose a bounding map (BM) based two-stage ULD method, which uses a lower IoU threshold to determine anchor categories based on IoU mechanism, and uses the value of BM as the soft labels. In addition, the BMs are also used as pseudo lesion masks for a newly introduced segmentation branch in stage-2. However, such an approach still suffers from several issues: (i) generating BMs from BBox in a fixed manner without considering lesion size and shape differences leads to sub-optimum BM representation for small lesions; and (ii) using IoU to determine anchor categories may lead to incorrect classification for lesions with irregular shapes.

To address the above issues, while exploiting the advantages of the BM-based two-stage ULD method, we propose a novel training mechanism for ULD to effectively reduce positive vs. negative anchor imbalance via a BM-based conditioning (BMC) mechanism in stage-1 and improve lesion localization accuracy by leveraging a size-adaptive BM (ABM) for supervising the segmentation branch in stage-2. Different from the traditional IoU-based positive and negative anchors selection mechanism, we use two independent anchor classification and regression tasks by directly predicting a **per-pixel objectness map** and selecting positive anchors for regression based on BMs [11]. Specifically as shown in Fig. 1, we use the BM_{xy} [11] as objectness GT map and select anchors whose value is greater than a threshold B in BM_{xy} to the region proposal network (RPN) for pixel-wise objectness map regression. We further randomly mask out some background pixels in BM_{xy} during objectness training to keep the number of background pixels no more than two times of the foreground pixel number. In addition, we extend BM into size-adaptive BMs (ABMs) to handle diverse lesions, in which the small lesions will be enhanced to compensate its limited pixel number while the big lesions are weakened. We use an ABM-supervised segmentation branch in stage-2 to improve the lesion localization accuracy.

Our method works perfectly with two-stage ULD methods, so it can be integrated with the state-of-the-art (SOTA) two-stage ULD methods. We conduct extensive experiments on the DeepLesion dataset [32] with three SOTA ULD methods to validate the effectiveness of our method.

2   Method

As shown in Fig. 2, we utilize the BMC mechanism ((a) & (b) in Fig. 2) to reduce positive vs. negative anchor imbalance in stage-1 and improve lesion localization accuracy by adding a supervised segmentation branch based on size-adaptive BM (ABM) ((c) in Fig. 2) in stage-2. Section 2.1 details the BM generation process [11]; Sect. 2.2 introduces the BMC mechanism, and Sect. 2.3 explains the newly introduced ABM-supervised segmentation branch.

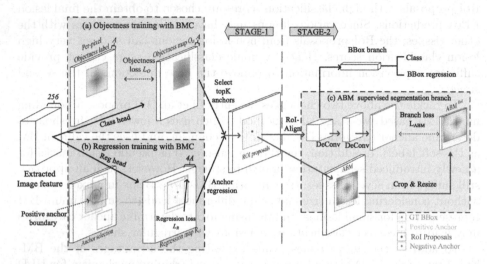

Fig. 2. The network architecture of the proposed approach, in which, BMC mechanism (a&b) and ABM branch (c) are respectively used in the two stages of the ULD method.

2.1   Bounding Map Generation

For the nth GT BBox in an image I_{ct}, the BMs $BM_x^{(n)}$ and $BM_y^{(n)}$ are computed based on an all-zero map, in which the element values within the BBox at the same location as the GT BBox are assigned to values linearly interpolated from 1 to 0.5. The all-zero map and the final BMs have the same size as the image feature map output by the backbone network. Specifically,

$$BM_x^{(n)}(x,y) = \begin{cases} 0 & (x,y) \notin S_{BBox}^{(n)} \\ 1 - k_x^{(n)} \left| x^{(n)} - x_{ctr}^{(n)} \right| & (x,y) \in S_{BBox}^{(n)} \end{cases}, \quad (1)$$

$$BM_y^{(n)}(x,y) = \begin{cases} 0 & (x,y) \notin S_{BBox}^{(n)} \\ 1 - k_y^{(n)} \left| y^{(n)} - y_{ctr}^{(n)} \right| & (x,y) \in S_{BBox}^{(n)} \end{cases}, \quad (2)$$

where $S_{BBox}^{(n)}$ denotes the nth GT BBox $(x_1^{(n)}, y_1^{(n)}, x_2^{(n)}, y_2^{(n)})$, whose center is at $(x_{ctr}^{(n)}, y_{ctr}^{(n)})$. The slope $k^{(n)}$ calculation is defined as:

$$k_x^{(n)} = \frac{1}{x_2^{(n)} - x_1^{(n)}}, \quad k_y^{(n)} = \frac{1}{y_2^{(n)} - y_1^{(n)}}, \quad (3)$$

and the final BM_x (BM_y) is generated by aggregating all the $BM_x^{(n)}$s ($BM_y^{(n)}$s):

$$BM_x = \min\left[\sum_{i=1}^{I} BM_x^{(n)}, 1\right], BM_y = \min\left[\sum_{i=1}^{I} BM_y^{(n)}, 1\right], \tag{4}$$

where N is the number of GT BBoxs of a CT image. Finally, the BM_{xy} is formed by element-wise multiplication between BM_x and BM_y [11]:

$$BM_{xy} = \sqrt[2]{BM_x \odot BM_y}, \tag{5}$$

Different from [11], we use only BM_{xy} and we will call it BM below.

2.2 BM-based Conditioning Mechanism

The RPN is trained to produce object bounding map $R_{ct} \in \mathcal{R}^{\frac{W}{R} \times \frac{H}{R} \times 4A}$ via regression and objectness score map $O_{ct} \in \mathcal{R}^{\frac{W}{R} \times \frac{H}{R} \times A}$ via classification at each position (or an anchor's centerpoint) in an input image $I_{ct} \in \mathcal{R}^{W \times H}$, where R and A are the network output stride and the number of anchor classes, respectively. In previous ULDs, anchors are selected for training objectness and regression in conventional RPN; here we develop two independent BM-based processes for RoI proposal: (i) objectness map prediction with BMC and (ii) BBox regression with BMC.

Objectness Map Prediction with BMC. Conventionally, the objectness GT labels of positive, negative and otherwise anchors are set as 1, 0, and -1, respectively, and only the positive and negative anchors are used for loss calculation. Motivated by FCOS [33], we first resize $BM \in \mathcal{R}^{W \times H \times A}$ into $BM^r \in \mathcal{R}^{\frac{W}{R} \times \frac{H}{R} \times A}$ in a linear interpolation manner, then directly use BM^r as the objectness GT label and train RPN to predict the objectness map in a per-pixel manner as showed in Fig. 2 (a). Considering the lesion sizes variation and foreground-background pixel imbalance, we further randomly ignore some background pixels in BM^r to keep the number of background pixels no more than two times of the foreground pixel number N_f. Specifically, we first find out the foreground and background pixels set S_f and S_b based on BM^r:

$$S_f = \{(x_f, y_f)|BM_{xy}^r(x_f, y_f) \geq 0.5\}, \ S_b = \{(x_b, y_b)|BM_{xy}^r(x_b, y_b) < 0.5\}. \tag{6}$$

Then the number of foreground pixels $N_f = Card(S_f)$ can be counted. After that, we randomly sample $2N_f$ background pixels from the background pixels set S_b as the training set S_b^t. Finally, we train the RPN for predicting objectness map $O_{ct} \in \mathcal{R}^{\frac{W}{R} \times \frac{H}{R} \times A}$ with a binary cross entropy (BCE) loss:

$$\mathcal{L}_o = \sum_{j=1}^{J} \mathcal{L}_{BCE}(O_{ct}(x_j, y_j), BM_{xy}^r(x_j, y_j)) \quad (x_j, y_j) \in S_b^t \cup S_f, \tag{7}$$

where J is the total number of pixels in $S_b^t \cup S_f$.

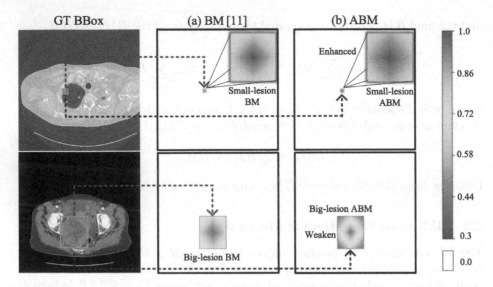

Fig. 3. Different from the BM (a) that uses a fixed generation method [11], the proposed ABM (b) can enhance the small lesions while weaken the big lesions.

BBox Regression with BMC. In our method, the loss computation during BBox regression is still applied to positive anchors, but we propose for BBox regression a new positive anchor selection method based on BM (see Fig. 2 (b)). We first define a positive anchor boundary value B and set the anchors whose corresponding BM_{xy}^r value is greater than B as positive anchors:

$$Loc_p^{(n)} = \begin{cases} \{x_p^{(n)}, y_p^{(n)} | BM_{xy}^{r(i)}(x_p^{(n)}, y_p^{(n)}) \geq B\} & A_{BBox}^{(n)} \geq 16 \\ \{x_p^{(n)}, y_p^{(n)} | BM_{xy}^{r(i)}(x_p^{(n)}, y_p^{(n)}) \geq 0.5\} & else \end{cases}, \quad (8)$$

where $Loc_p^{(n)}$ and $A_{BBox}^{(n)}$ is the positive anchor's centerpoint location set for the nth GT BBox and the area of the nth GT BBox. Finally, we union all $Loc_p^{(n)}$ and get Loc_p, the anchor centerpoint location set for the image.

Finally the BBox regression loss can be written as:

$$\mathcal{L}_r = \sum_{m=1}^M \mathcal{L}_{reg}(x_m, y_m) \quad (x_m, y_m) \in Loc_p, \quad (9)$$

where $\mathcal{L}_{reg}$ is the original regression loss function and M is the number of anchors' centerpoint locations in Loc_p.

2.3   Pseudo Lesion Segmentation via ABM Supervision

The top-K objectness anchors regressed by the regression are used as RoI proposals and the input of the second-stage classification and regression. However, as we analyzed in Sect. 1, the BBox branch does not provide enough supervision for the complex ULD task. We introduce an extra branch, i.e., ABM supervised pseudo lesion segmentation branch.

Size Adaptive Bounding Maps (ABM). As shown in Sect. 2.1, BM uses a fixed generating manner without considering lesion size and shape differences leads to sub-optimum representations for small lesions. As shown in Fig. 3, we extend BM to make it adaptive to lesion size, so that a small lesion is enhanced to compensate its limited area. Specifically, we multiply the two slopes $k_x^{(n)}$ and $k_y^{(n)}$ by a ratio α when generating the $ABM_x^{(n)}$ and $ABM_y^{(n)}$:

$$
ABM_x^{(n)}(x,y) = \begin{cases} 0 & (x,y) \notin S_{BBox}^{(n)} \\ 1 - \alpha k_x^{(n)} \left| x^{(n)} - x_{ctr}^{(n)} \right| & (x,y) \in S_{BBox}^{(n)} \end{cases} , \quad (10)
$$

$$
ABM_y^{(n)}(x,y) = \begin{cases} 0 & (x,y) \notin S_{BBox}^{(n)} \\ 1 - \alpha k_y^{(n)} \left| y^{(n)} - y_{ctr}^{(n)} \right| & (x,y) \in S_{BBox}^{(n)} \end{cases} , \quad (11)
$$

$$
\alpha = \begin{cases} 0 & A_{BBox}^{(n)} < A_s \\ 1 & A_s \le A_{BBox}^{(n)} < A_m \\ 1.4 & A_{BBox}^{(n)} \ge A_m \end{cases} , \quad (12)
$$

where the A_s, A_m are the size thresholds for small and medium lesions. The final ABM_{xy} generation is the same as the BM_{xy}.

Segmentation with ABM Supervision. As shown in Fig. 2 (c), the ABM supervised segmentation branch is the same as the mask branch in Mask R-CNN [29]. It is parallel to the BBox classification and regression branch. The ABM is first cropped based on the RoI BBox and resized to the size of the output to obtain $ABM_x^{RoI} \in \mathcal{R}^{W_b \times H_b \times 1}$, where W_b and H_b are the output $A\hat{B}M^{RoI}$'s width and height of ABM-supervised segmentation branch, respectively. Then the loss function of ABM branch for each RoI can be defined as a norm-2 loss:

$$
\mathcal{L}_{ABM} = \mathcal{L}_2(A\hat{B}M^{RoI}, ABM^{RoI}). \quad (13)
$$

3 Experiments

3.1 Dataset and Setting

We conduct experiments on the DeepLesion dataset [32]. The dataset contains 32,735 lesions on 32,120 axial slices from 10,594 CT studies of 4,427 unique patients. Most existing datasets typically focus on one type of lesion, while DeepLesion contains a variety of lesions with large diameters ranges (from 0.21 to 342.5 mm). The 12-bit intensity CT is rescaled to [0,255] with different window ranges settings used in different frameworks. Also, every CT slice is resized and interpolated according to the detection frameworks' setting. We follow the

Table 1. Sensitivity (%) at various FPPI on the testing set of DeepLesion [32].

Methods	slices	@0.5	@1	@2	@4	Avg.[0.5,1,2,4]
Faster R-CNN [30]	3	57.17	68.82	74.97	82.43	70.85
Faster R-CNN+BM [11]	3	63.96 (6.79↑)	74.43 (5.61↑)	79.80 (4.83↑)	86.28 (3.85↑)	76.12(5.27↑)
Faster R-CNN+Ours	3	65.37 (**8.20↑**)	76.31 (**7.49↑**)	81.03 (**6.06↑**)	87.98 (**5.55↑**)	77.67(**6.82↑**)
3DCE [6]	9	59.32	70.68	79.09	84.34	73.36
3DCE+BM [11]	9	64.38 (5.06↑)	75.55 (4.87↑)	82.74 (3.65↑)	87.78 (3.44↑)	77.62(4.26↑)
3DCE+Ours	9	66.98 (7.66↑)	77.25 (6.57↑)	83.64 (4.55↑)	88.41 (4.07↑)	79.07(5.71↑)
MVP-Net [7]	9	70.07	78.77	84.91	87.33	80.27
MVP-Net+BM [11]	9	72.12 (2.05↑)	80.51 (1.74↑)	86.08 (1.17↑)	88.41 (1.08↑)	81.78(1.51↑)
MVP-Net+Ours	9	73.05 (2.98↑)	81.41 (2.64↑)	87.22 (2.31↑)	89.37 (2.04↑)	82.76(2.49↑)
MULAN w/o (SRL&RECIST) [8]	9	72.34	80.17	85.21	89.41	81.78
MULAN w/o (SRL&RECIST) +BM [11]	9	74.71 (2.37↑)	81.62 (1.45↑)	86.44 (1.23↑)	89.82 (0.41↑)	83.15(1.37↑)
MULAN w/o (SRL&RECIST) +Ours	9	75.97 (3.63↑)	82.81 (2.64↑)	86.85 (1.64↑)	90.00(0.59↑)	83.91(2.13↑)
MULAN w/o SRL [8]	9	73.85	81.02	85.98	90.01	82.71
MULAN w/o SRL +Ours	9	76.28 (2.43↑)	83.13 (2.11↑)	87.30 (1.32↑)	90.49 (0.48↑)	84.30(1.59↑)
AlignShift w/o RECIST [9]	7	76.31	84.07	88.08	91.62	85.02
AlignShift w/o RECIST+BM [11]	7	77.51 (1.20↑)	84.95 (0.88↑)	88.63 (0.55↑)	92.08 (0.46↑)	85.79(0.77↑)
AlignShift w/o RECIST+Ours	7	79.03 (2.72↑)	85.63 (1.56↑)	89.53 (1.45↑)	92.43 (0.81↑)	86.66(1.64↑)
AlignShift [9]	7	77.20	84.38	89.03	92.31	85.73
AlignShift +Ours	7	**79.17** (1.97↑)	**85.71** (1.33↑)	**89.80** (0.77↑)	**92.65** (0.34↑)	**86.83**(1.10↑)
FCOS (anchor-free) [34]	3	37.78	54.84	64.12	77.84	58.65
Objects as points (anchor-free) [33]	3	34.87	43.58	52.41	64.01	48.72
Deformable detr (anchor-free) [35]	3	57.62	65.64	70.65	75.58	67.37

official split, i.e., 70% for training, 15% for validation and 15% for testing. The number of false positives per image (FPPI) is used as the evaluation metric. We set the positive anchor boundary value B in Eq. 8 as 0.25, size thresholds S_s and S_m in Eq. 12 as 250 and 1000 respectively. For training, we use the original network architecture and settings, and initialize the network using pretrained ULD models on DeepLesion dataset [32].

3.2   Lesion Detection Performance

We apply our method with four SOTA ULD approaches [6–9] to evaluate the effectiveness. We also compare with three SOTA anchor-free [33–35] and one two-stage anchor-based [30] detection methods. As shown in Table 1, our method brings promising detection performance improvements for all baselines. The improvements of Faster R-CNN [30], 9-slice 3DCE, and MVP-Net are more pronounced than those of MULAN w/o SRL [8] and AlignShift [9]. This is because MULAN and AlignShift introduce extra weakly segmentation mask generated from radiologist-annotated RECIST labels. The anchor-free methods get unsatisfactory results mainly because they are lack of the initialize advantage of anchors and the coarse-to-fine training advantage of the two-stage mechanism. We also provide a case to show our method's effectiveness in Suppl. Material.

3.3   Ablation Study

Ablation study is provided to evaluate the importance of the three key components of the proposed method: (i) Objectness map prediction with BMC, (ii) Box regression with BMC, and (iii) *ABM* supervised segmentation. As shown in Table 2, using objectness map prediction with BMC, we obtain a 1.79% improvement over the baseline. Further adding BBox regression training with BMC accounts for another 0.18% improvement and gives the best performance. Without using extra RECIST label, *ABM* and *BM* increases the performance by 1.08% and 0.02%, respectively. In addition, the use of our method brings a minor increase (less than 10%) in the inference time on a Titan RTX GPU.

Table 2. Ablation study of our method at various FPs per image (FPPI).

AlignShift [9]	RECIST label	BMC objectness training	BMC regression training	ABM branch	BM branch [11]	$FPPI=0.5$	$FPPI=1$	Inference (s/img)
✓	✓					77.20	84.38	0.1758
✓	✓	✓				78.99	85.65	0.1747
✓	✓	✓	✓			**79.17**	**85.71**	0.1833
✓		✓	✓			77.95	84.99	0.1784
✓		✓	✓	✓		**79.03**	85.63	0.1814
✓		✓	✓		✓	77.97	85.04	0.1755

4   Conclusion

In this paper, we try to overcome the two intrinsic limitations of two-stage ULD methods: anchor imbalance in stage-1 and insufficient supervision in stage-2. To relieve these, we propose a new BMC mechanism in stage-1 and an ABM supervised segmentation branch in stage-2. Extensive experiments using several SOTA baselines on the DeepLesion dataset show that our approach can effectively boost the ULD performance with almost no additional computational cost.

References

1. Zlocha, M., Dou, Q., Glocker, B.: Improving RetinaNet for CT lesion detection with dense masks from weak RECIST labels. In: Shen, D., et al. (eds.) MICCAI 2019. LNCS, vol. 11769, pp. 402–410. Springer, Cham (2019). https://doi.org/10. 1007/978-3-030-32226-7_45

2. Tao, Q., Ge, Z., Cai, J., Yin, J., See, S.: Improving deep lesion detection using 3D contextual and spatial attention. In: Shen, D., et al. (eds.) MICCAI 2019. LNCS, vol. 11769, pp. 185–193. Springer, Cham (2019). https://doi.org/10.1007/978-3-030-32226-7_21

3. Zhang, N., et al.: 3d anchor-free lesion detector on computed tomography scans. arXiv:1908.11324 (2019)

4. Zhang, N., Cao, Y., Liu, B., Luo, Y.: 3d aggregated faster R-CNN for general lesion detection. arXiv:2001.11071 (2020)

5. Tang, Y., Yan, K., Tang, Y., Liu, J., Xiao, J., Summers, R.M.: Uldor: a universal lesion detector for ct scans with pseudo masks and hard negative example mining. In: IEEE ISBI, pp. 833–836 (2019)

6. Yan, K., Bagheri, M., Summers, R.M.: 3D context enhanced region-based convolutional neural network for end-to-end lesion detection. In: Frangi, A.F., Schnabel, J.A., Davatzikos, C., Alberola-López, C., Fichtinger, G. (eds.) MICCAI 2018. LNCS, vol. 11070, pp. 511–519. Springer, Cham (2018). https://doi.org/10.1007/978-3-030-00928-1_58

7. Li, Z., Zhang, S., Zhang, J., Huang, K., Wang, Y., Yu, Y.: MVP-Net: multi-view FPN with position-aware attention for deep universal lesion detection. In: Shen, D., et al. (eds.) MICCAI 2019. LNCS, vol. 11769, pp. 13–21. Springer, Cham (2019). https://doi.org/10.1007/978-3-030-32226-7_2

8. Yan, K., et al.: MULAN: multitask universal lesion analysis network for joint lesion detection, tagging, and segmentation. In: Shen, D., et al. (eds.) MICCAI 2019. LNCS, vol. 11769, pp. 194–202. Springer, Cham (2019). https://doi.org/10.1007/978-3-030-32226-7_22

9. Yang, J., et al.: *AlignShift*: bridging the gap of imaging thickness in 3D anisotropic volumes. In: Martel, A.L., et al. (eds.) MICCAI 2020. LNCS, vol. 12264, pp. 562–572. Springer, Cham (2020). https://doi.org/10.1007/978-3-030-59719-1_55

10. Cai, J., et al.: Deep volumetric universal lesion detection using light-weight pseudo 3D convolution and surface point regression. In: Martel, A.L., et al. (eds.) MICCAI 2020. LNCS, vol. 12264, pp. 3–13. Springer, Cham (2020). https://doi.org/10.1007/978-3-030-59719-1_1

11. Li, H., Han, H., Zhou, S.K.: Bounding maps for universal lesion detection. In: Martel, A.L., et al. (eds.) MICCAI 2020. LNCS, vol. 12264, pp. 417–428. Springer, Cham (2020). https://doi.org/10.1007/978-3-030-59719-1_41

12. Zhang, S., et al.: Revisiting 3D context modeling with supervised pre-training for universal lesion detection in CT slices. In: Martel, A.L., et al. (eds.) MICCAI 2020. LNCS, vol. 12264, pp. 542–551. Springer, Cham (2020). https://doi.org/10.1007/978-3-030-59719-1_53

13. Yan, K., et al.: Learning from multiple datasets with heterogeneous and partial labels for universal lesion detection in CT. IEEE Trans. Med. Imaging (2020)

14. Li, H., et al.: High-resolution chest x-ray bone suppression using unpaired CT structural priors. IEEE Trans. Med. Imaging 39(10), 3053–3063 (2020)

15. Liao, F., Liang, M., Li, Z., Hu, X., Song, S.: Evaluate the malignancy of pulmonary nodules using the 3-d deep leaky noisy-or network. IEEE Trans. Neural Netw. Learn. Syst. 30(11), 3484–3495 (2019)

16. Lin, Y., et al.: Automated pulmonary embolism detection from CTPA images using an end-to-end convolutional neural network. In: Shen, D., et al. (eds.) MICCAI 2019. LNCS, vol. 11767, pp. 280–288. Springer, Cham (2019). https://doi.org/10.1007/978-3-030-32251-9_31

17. Wang, X., Han, S., Chen, Y., Gao, D., Vasconcelos, N.: Volumetric attention for 3D medical image segmentation and detection. In: Shen, D., et al. (eds.) MICCAI 2019. LNCS, vol. 11769, pp. 175–184. Springer, Cham (2019). https://doi.org/10.1007/978-3-030-32226-7_20

18. Astaraki, M., Toma-Dasu, I., Smedby, Ö., Wang, C.: Normal appearance autoencoder for lung cancer detection and segmentation. In: Shen, D., et al. (eds.) MICCAI 2019. LNCS, vol. 11769, pp. 249–256. Springer, Cham (2019). https://doi.org/10.1007/978-3-030-32226-7_28

19. Tang, H., Zhang, C., Xie, X.: NoduleNet: decoupled false positive reduction for pulmonary nodule detection and segmentation. In: Shen, D., et al. (eds.) MICCAI 2019. LNCS, vol. 11769, pp. 266–274. Springer, Cham (2019). https://doi.org/10.1007/978-3-030-32226-7_30

20. Shao, Q., Gong, L., Ma, K., Liu, H., Zheng, Y.: Attentive CT lesion detection using deep pyramid inference with multi-scale booster. In: Shen, D., et al. (eds.) MICCAI 2019. LNCS, vol. 11769, pp. 301–309. Springer, Cham (2019). https://doi.org/10.1007/978-3-030-32226-7_34

21. Liu, J., Cao, L., Akin, O., Tian, Y.: 3DFPN-HS^2: 3D feature pyramid network based high sensitivity and specificity pulmonary nodule detection. In: Shen, D., et al. (eds.) MICCAI 2019. LNCS, vol. 11769, pp. 513–521. Springer, Cham (2019). https://doi.org/10.1007/978-3-030-32226-7_57

22. Boot, T., Irshad, H.: Diagnostic assessment of deep learning algorithms for detection and segmentation of lesion in mammographic images. In: Martel, A.L., et al. (eds.) MICCAI 2020. LNCS, vol. 12264, pp. 56–65. Springer, Cham (2020). https://doi.org/10.1007/978-3-030-59719-1_6

23. Yu, X., et al.: Deep attentive panoptic model for prostate cancer detection using biparametric MRI scans. In: Martel, A.L., et al. (eds.) MICCAI 2020. LNCS, vol. 12264, pp. 594–604. Springer, Cham (2020). https://doi.org/10.1007/978-3-030-59719-1_58

24. Yao, Q., Xiao, L., Liu, P., Zhou, S.K.: Label-free segmentation of covid-19 lesions in lung CT. IEEE Trans. Med. Imaging (2021)

25. Yao, Q., He, Z., Han, H., Zhou, S.K.: Miss the point: targeted adversarial attack on multiple landmark detection. In: Martel, A.L., et al. (eds.) MICCAI 2020. LNCS, vol. 12264, pp. 692–702. Springer, Cham (2020). https://doi.org/10.1007/978-3-030-59719-1_67

26. Yao, Q., Quan, Q., Xiao, L., Zhou, S.K.: One-shot medical landmark detection. arXiv preprint arXiv:2103.04527 (2021)

27. Zhou, S.K., et al.: A review of deep learning in medical imaging: imaging traits, technology trends, case studies with progress highlights, and future promises. In: Proceedings of the IEEE (2021)

28. Zhou, S.K., Rueckert, D., Fichtinger, G.:Handbook of Medical Image Computing and Computer Assisted Intervention. Academic Press, London (2019)

29. He, K., Gkioxari, G., Dollár, P., Girshick, R.: MASK R-CNN. In: IEEE ICCV, pp. 2961–2969 (2017)

30. Ren, S., He, K., Girshick, R., Sun, J.: Faster R-CNN: towards real-time object detection with region proposal networks. In: NIPS, pp. 91–99 (2015)

31. Oksuz, K., Cam, B.C., Kalkan, S., Akbas, E.: Imbalance problems in object detection: a review. Trans. Pattern Anal. Mach. Intell. 43(10), 3388–3415 (2020)

32. Yan, K., et al.: Deep lesion graphs in the wild: relationship learning and organization of significant radiology image findings in a diverse large-scale lesion database. In: IEEE CVPR, pp. 9261–9270 (2018)

33. Zhou, X., Wang, D., Krähenbühl, P.: Objects as points. arXiv:1904.07850 (2019)
34. Tian, Z., Shen, C., Chen, H., He, T.: Fcos: fully convolutional one-stage object detection. In: IEEE ICCV, pp. 9627–9636 (2019)
35. Zhu, X., Su, W., Lu, L., Li, B., Wang, X., Dai, J.: Deformable detr: Deformable transformers for end-to-end object detection. arXiv preprint arXiv:2010.04159 (2020)

Focusing on Clinically Interpretable Features: Selective Attention Regularization for Liver Biopsy Image Classification

Chong Yin, Siqi Liu, Rui Shao, and Pong C. Yuen[(⊠)]

Department of Computer Science, Hong Kong Baptist University,
kowloon, Hong Kong
{chongyin,siqiliu,ruishao,pcyuen}@comp.hkbu.edu.hk

Abstract. Liver biopsy image analysis is the gold standard for early diagnosis of non-alcoholic fatty liver disease (NAFLD) worldwide. Deep neural networks offer an effective tool for image analysis. However, when applying deep learning methods to smaller histological image datasets, the model may be distracted by dominant normal tissues and ignore critical tissue alterations that pathologists focus on. In this paper, we propose a selective attention regularization module (SAttenReg) to mimic the diagnosis process of pathologists. Specifically, to explicitly encourage the model to focus on clinically interpretable features (e.g., nuclei and fat droplets), SAttenReg learns the attention map with the regularization of clinically interpretable features. Furthermore, with the different contributions of histological features, the model can selectively focus on different histological features based on the distribution of nuclei in each instance. Experiments conducted on the in-house Liver-NAS and public Biopsy4Grading biopsy image datasets show that our method achieves superior classification performance with promising localization results.

Keywords: Liver biopsy images · Selective attention regularization

1 Introduction

Nonalcoholic fatty liver disease (NAFLD) is the most common cause of liver disease worldwide [21,22]. It is estimated that the prevalence of NAFLD is between 25% and 45%, which has become an important public health concern [4,20]. Liver biopsy images are the gold standard for early diagnosis of NAFLD. It mainly involves the quantitative analysis of histological patterns, including fibrosis and three other NAS-related components (ballooning degeneration, lobular inflammation, and steatosis) [11]. Due to the heavy burden of diagnosis, it is important to develop an automatic image analysis model to assist pathologists.

Recently, deep learning methods have achieved promising results in many computer vision tasks including image classification [26] and segmentation [10].

© Springer Nature Switzerland AG 2021
M. de Bruijne et al. (Eds.): MICCAI 2021, LNCS 12905, pp. 153–162, 2021.
https://doi.org/10.1007/978-3-030-87240-3_15

There are also some deep learning-based works on gigapixel whole slide images [5,12,23]. For liver biopsy images, due to differences in resolution and staining methods, fibrosis analysis is always separated from NAS-related components [11]. [6,23] apply a convolutional neural network for fibrosis recognition and get a promising result. The analysis of NAS-related components is more challenging and rarely studied.

When applying existing deep learning methods to this specific task, the model may suffer from low reliability due to attention distraction. The model may shift the focus to dominant normal tissues in the biopsy image. As a result, histological features of clinical concern (e.g., nuclei and fat droplets) [2,16] are ignored, leading to mistaken diagnosis. Attention-based methods [8,9,13,24] can alleviate this problem to a certain extent. However, due to the lack of explicit supervision of attention, it is difficult to determine exactly where the learning model should pay its attention on the complex tissue structures.

In this paper, we introduce a selective attention regularization method to clearly utilize clinically interpretable features for the analysis of NAS-related components. Specifically, based on a conventional classification network, we propose a selective attention regularization module (SAttenReg). SAttenReg mimics the process by which pathologists usually look at clinically interpretable features (e.g., nuclei and fat droplets) during diagnosis. The areas of the nuclei and fat droplets are first extracted to guide and regularize the attention maps to have a large overlap with the areas of clinically interpretable features. By adding explicit regularization constraints to the attention map, the model can iteratively focus on clinically interpretable features, thereby improving the ability to identify NAS-related components. Furthermore, clinically interpretable features have different contributions to the recognition of NAS-related components. The nuclei density is a key feature that distinguishes between normal tissues (presenting uniformly distributed nuclei) and NAS-related components (presenting high or low nuclei density) [5,18]. SAttReg regresses the nuclei density with the Gaussian distribution, and assigns Gaussian distribution-induced weights to adjust the supervision of attention regularization. Gaussian distribution-induced weights allow the model to selectively focus on nuclei and fat droplets, corresponding to the identification of lobular inflammation and steatosis or ballooning degeneration, respectively. Our contributions are:

- We propose a selective attention regularization (SAttenReg) method to analyze NAS-related components in liver biopsy images. It explicitly drives the model to focus on clinically interpretable features (e.g., nuclei and fat droplets) to improve the interpretability and reliability of the model.
- SAttenReg introduces Gaussian distribution-induced weights based on nuclei density to adjust the attention regularization supervision, thus the model can selectively focus on clinically interpretable features, corresponding to its different contributions to the recognition of NAS-related components.
- Experiments show that our method achieves superior classification performance with a promising localization ability.

2   Proposed Method

As shown in Fig. 1, given an input biopsy image, we feed it into a conventional classification network to learn the feature maps. To show the region that the model focuses on, the attention map is calculated by the weighted sum of feature maps. On top of that, we propose a novel selective attention regularization module (SAttenReg) to explicitly drive the model to focus on clinically interpretable features. To mimic the diagnosis process of pathologists, we take the nuclei and fat droplets into consideration. The clinically interpretable features extracted by the histological features extractor are used to guide and regularize attention maps to have a large overlap with the masks of clinically interpretable features. Considering the different contributions of clinically interpretable features to diagnosis, different regularization weights induced by nuclei density distribution are assigned to each sample to adjust the attention regularization.

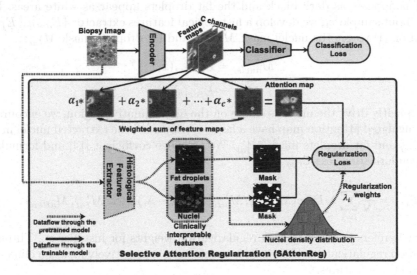

Fig. 1. Illustration of our proposed method.

2.1   Image Classification

Given a collection of labeled samples $x_i \in \mathcal{D}$ and corresponding labels y_i, we feed them into the conventional classification network which consists of a feature encoder g and a classifier f. The corresponding classification loss $\mathcal{L}_{cls}$ is:

$$\mathcal{L}_{cls} = \frac{1}{N} \sum_{i=1}^{N} y_i log \hat{y}_i \qquad (1)$$

where $\hat{y}_i = g \circ f(x_i)$ refers the predicated label.

In order to mimic the diagnosis process of pathologists, we first calculate attention maps, extract the nuclei and fat droplet regions, and then extend them to a selective attention regularization model.

2.2 Selective Attention Regularization

Inspired by the class activation map (CAM) [25], the weighted sum of feature maps reflects the class-specific image regions used by the model for prediction:

$$M_{atten}^i = \sum_{c=1}^C \alpha_c f_c(x_i) \tag{2}$$

where α_c are the parameters of the classifier. f_c refers the feature map at channel c.

In clinical practice, nuclei and fat droplets play a key role in identifying NAS-related components. The spectrum of NAFLD starts with the accumulation of fat in liver cells [3]. The accumulation of fat droplets may lead to microcystic steatosis with preserved cellular architecture, and macrovesicular steatosis where the nucleus is displaced [19]. When stained with hematoxylin-eosin (H&E), the nucleus appears as dark black and the fat droplets appear as white areas. For each input sample x_i, we develop a histological features extractor $\{E_{nuclei}, E_{fat}\}$ based on [1] to get the nuclei mask M_{nuclei}^i and fat droplets mask M_{fat}^i:

$$M_{nuclei}^i = E_{nuclei}(x_i) \tag{3}$$

$$M_{fat}^i = E_{fat}(x_i) \tag{4}$$

To explicitly drive the model to focus on the discriminative region, we encourage the calculated attention map have a large overlap with the extracted nuclei mask M_{nuclei}^i and fat droplets mask M_{fat}^i. We use Dice coefficient [14] and formulate the regularization loss $\mathcal{L}_{reg}$:

$$\mathcal{L}_{reg} = \frac{1}{N} \sum_{i=1}^N \lambda_i Dice(M_{nuclei}^i, M_{atten}^i) + (1 - \lambda_i) Dice(M_{fat}^i, M_{atten}^i) \tag{5}$$

where λ_i refers to the attention regularization weights for instance x_i. Through attention regularization, the model is encouraged to iteratively focus on clinically interpretable features.

The nuclei and fat droplets contribute differently to the identification of three NAS-related components. First, the cell nuclei location distribution is different. Different from the normal tissue, the location distribution of nuclei is homogeneous, presenting a similar distance. In contrast, inflammation areas present a high density of nuclei density, whereas the nuclei density is very low in fatty regions (e.g., ballooning degeneration and steatosis). Second, the accumulation of fat droplets is different. The uneven distribution of the nucleus is more likely to present the accumulation of fat droplets. Inspired by this, we make the nuclei density conform to the Gaussian distribution:

$$||M_{nuclei}^i||_1 \sim \mathcal{N}(\mu, \sigma^2) \tag{6}$$

where $||.||_1$ denotes the norm of nuclei mask, and $\mathcal{N}$ refers to the Gaussian distribution with mean μ and standard variance σ. μ and σ are learned from the Gaussian regression on all training samples.

The attention regularization weights λ_i is drawn from the Gaussian distribution:

$$\lambda_i = \begin{cases} 1 - \frac{1}{\sigma\sqrt{2\pi}} exp^{-\frac{1}{2}(\frac{||M^i_{nuclei}||_1 - \mu}{\sigma})^2} & ||M^i_{nuclei}||_1 \geq \mu \\ \frac{1}{\sigma\sqrt{2\pi}} exp^{-\frac{1}{2}(\frac{||M^i_{nuclei}||_1 - \mu}{\sigma})^2} & ||M^i_{nuclei}||_1 < \mu \end{cases} \quad (7)$$

This formulation encourages the model to selectively focus on nuclei and fat droplets when learning the discriminative features for NAS-related components recognition.

So, the final objective function $\mathcal{L}$ can be formulated as the combination of two loss terms:

$$\mathcal{L} = \mathcal{L}_{cls} + \mathcal{L}_{reg} \quad (8)$$

3 Experiments

3.1 Experiment Settings

Dataset. We conduct experiments on two liver biopsy image datasets. **Biopsy4Grading** [7] is a public liver section dataset collected from animals studies. Each liver tiles (299×299 pixels) were assigned to discrete pathologist-like sub-scores for quantifying NAS-related components of ballooning degeneration (0–2), lobular inflammation (0–3), steatosis (0–3) and fibrosis (0–4), corresponding to the Kleiner score system [11]. **Liver-NAS** is a private dataset of liver biopsy images collecting from 9 patients. Image tiles were generated from whole slide images ($\sim 106259 \times 306939$ pixels) with an area of 224×224 pixels, which can guarantee the pathologist to sufficiently identify the relevant histological features within the tile. The **Liver-NAS** dataset has liver biopsy image tiles with steatosis (N = 3838), ballooning degeneration (N = 298), lobular inflammation (N = 69) and others (N = 1659). On both datasets, we randomly split the data and report the results using 5-fold cross-validation. For each cross-validated fold, the dataset is randomly partitioned into a training set (70%), a validation set (10%) and test set (20%). The performance on the validation set is monitored during training and used for model selection.

Evaluation Metrics. For NAS-related components identification and quantification, we choose sensitivity, specificity and F1 score for evaluation. Beyond the classification ability, we also investigate the model performance in localizing the supporting evidence. To achieve this, we further invite the pathologist to annotate liver biopsy images to indicate the location of NAS-related components on the test dataset. The Dice metric [17] is adopted for measuring localization performance which reflects the overlap between the binary attention map ($threshold = 0.5$) and the annotated region.

Training Details. The model is trained using the Adam optimizer with weight decay $\alpha = 1e^{-4}$, $\beta_1 = 0.9$, $\beta_2 = 0.999$. The initial learning rate is set to 0.1. We train the model with 100 epochs and reduce the learning rate by 10 times for every 30 epochs. The batch size is set to 64. The nuclei density follows the Gaussian distribution with mean value $\mu \in [0.060, 0.065]$, and the standard variance $\sigma \in [0.069, 0.071]$.

3.2   Quantitative Analysis

We choose attention-based approach SCNet [13] as the baseline network, and compare it with the state-of-art method [7].

NAS-Related Components Identification. Table 1 shows results on Liver-NAS dataset for NAS-related components identification. Comparing with the attention-based approach SCNet, the introduction of attention regularization model ($\lambda_i = \frac{1}{2}$) improves lobular inflammation (LI) from 37.60 to 43.38, ballooning degeneration (BD) from 76.86 to 79.12 with respect to $F1$. When adopting different weights ($\lambda_i = \mathcal{N}$), the performance can further be improved 8.22 and 2.90, respectively.

NAS-related Components Localization. As shown in Table 2, the introduction of selective attention regularization (With SAttenReg) consistently performs better in localizing NAS-related components as compared to the baseline model (W/o SAttenReg). SAttenReg enforces that the attention map has a large overlap with the area of clinically interpretable features, and improves the interpretability.

NAS-related Components Quantification. Table 3 shows results on Biopsy4Grading for quantifying NAS-related components. Our method shows a better overall performance. Through the attention regularization, the ability to recognize the NAS-components is improved after iteratively focusing on the

Table 1. Quantitative comparisons on the private Liver-NAS dataset for NAS-related components classification. ($\lambda_i = \mathcal{N}$ refers to Gaussian distribution-induced weights)

Methods		Others	LI	BD	Steatosis
Fabian et al. [7]	Sensitivity	91.14 ± 3.01	**43.82 ± 21.67**	78.16 ± 15.97	97.06 ± 3.50
	Specificity	95.12 ± 2.52	99.38 ± 0.65	99.26 ± 7.81	95.86 ± 1.78
	F1	89.66 ± 4.09	45.46 ± 23.49	80.62 ± 6.61	97.38 ± 1.42
SCNet [13]	Sensitivity	93.34 ± 3.41	28.50 ± 19.07	75.90 ± 21.58	97.28 ± 2.18
	Specificity	95.10 ± 1.71	99.72 ± 0.43	99.08 ± 0.96	**96.74 ± 2.51**
	F1	90.74 ± 3.59	37.60 ± 24.44	76.86 ± 12.52	97.72 ± 0.84
Ours ($\lambda_i = \frac{1}{2}$)	Sensitivity	93.54 ± 2.75	35.44 ± 15.87	76.68 ± 17.72	97.78 ± 1.82
	Specificity	95.80 ± 1.39	99.66 ± 0.24	99.22 ± 0.81	96.56 ± 1.99
	F1	91.64 ± 2.52	43.38 ± 17.97	79.12 ± 12.81	97.96 ± 0.58
Ours ($\lambda_i = \mathcal{N}$)	Sensitivity	**93.92 ± 1.17**	43.48 ± 20.38	**78.58 ± 14.59**	**97.94 ± 1.43**
	Specificity	**95.96 ± 1.19**	**99.74 ± 0.21**	**99.42 ± 0.64**	96.60 ± 1.53
	F1	**92.02 ± 1.89**	**51.60 ± 21.93**	**82.02 ± 6.35**	**98.06 ± 0.46**

Table 2. Quantitative comparisons for weakly supervised localization ability with or without SAttenReg on private Liver-NAS dataset.

Method	Dice		
	LI	BD	Steatosis
W/o SAttenReg	0.4299 ± 0.1065	0.3713 ± 0.1030	0.2262 ± 0.1128
With SAttenReg	**0.4699 ± 0.1034**	**0.5072 ± 0.0640**	**0.6144 ± 0.0375**

Table 3. Quantitative comparisons on quantifying the NAS-related components on the Biopsy4Grading [7] dataset.

Methods		LI	BD	Steatosis
Fabian et al. [7]	Sensitivity	79.36 ± 2.27	85.83 ± 1.35	94.20 ± 1.50
	Specificity	94.60 ± 0.60	93.94 ± 0.78	98.73 ± 0.38
	F1	80.16 ± 1.63	87.61 ± 0.90	93.78 ± 1.16
SCNet [13]	Sensitivity	78.52 ± 2.68	85.63 ± 2.44	94.13 ± 1.38
	Specificity	94.31 ± 0.81	93.55 ± 1.04	98.68 ± 0.28
	F1	78.94 ± 1.83	87.11 ± 1.38	93.85 ± 0.98
Ours ($\lambda_i = \frac{1}{2}$)	Sensitivity	78.53 ± 2.72	86.30 ± 1.08	94.17 ± 1.08
	Specificity	94.39 ± 2.72	93.91 ± 0.57	98.70 ± 0.30
	F1	79.24 ± 2.49	87.90 ± 0.98	93.90 ± 0.79
Ours ($\lambda_i = \mathcal{N}$)	Sensitivity	**80.42** ± 2.17	**86.46** ± 2.93	**94.68** ± 1.12
	Specificity	**94.92** ± 0.99	**94.13** ± 1.15	**98.78** ± 0.29
	F1	**81.03** ± 0.99	**88.01** ± 1.84	94.42 ± 0.81

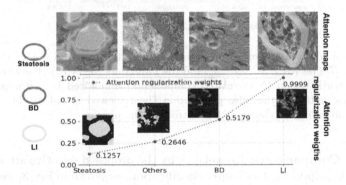

Fig. 2. Visual attention maps (upper row) and corresponding attention regularization weights (lower row) based on the nuclei density. The circle denotes the ground truth region (NAS-related components) annotated by pathologists (best viewed in color).

discriminative tissue areas. It allows the better discovery of evidence to support the quantification and improve the reliability of the model.

3.3   Qualitative Analysis

Selective Attention Regularization Weights: To better understand our model, we first visualize the visual attention maps generated by Grad-CAM [15], and draw the corresponding attention regularization weights λ_i based on the nuclei density of sample x_i. We simply upsample the attention map to image size (224 × 224) using bilinear interpolation.

As shown in Fig. 2, the extracted clinically interpretable features show a strong correlation with the discriminative region visualized by the attention map. The model attends to the fat droplets more when identifying the steatosis and ballooning degeneration. The low attention regularization weights are 0.1257 and 0.2646 respectively, which can indicate this choice. In contrast, for the lobular inflammation identification, the model attends to nucleus regions more with larger attention regularization weights 0.999.

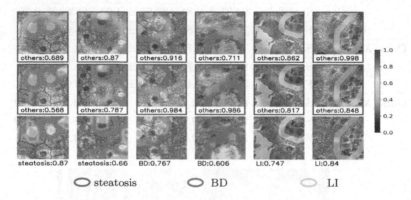

Fig. 3. Visual comparisons of attention maps generated by Grad-CAM [15] between Fabian et al. [7] (1st row), SCNet [13] (2nd row) and ours (3rd row). The circle denotes the ground truth region (NAS-related components) annotated by pathologists. Red bounding box indicates the false predication (best viewed in color). The predicted class and probability are shown at the bottom of the picture.

Attention Comparisons: To explain why the proposed selective attention regularization is helpful for liver biopsy classification, as shown in Fig. 3, we visualize the attention map generated by different methods. It can be clearly seen that our attention can more accurately locate the NAS-related components (highlighted by a circle) and produce better predictions (red border indicates incorrect predications). Without explicit attention regularization, the model may shift the focus to the frequently presented normal tissues in the biopsy image. In contrast, the introduction of attention regularization can calibrate the attention and enforce the model to focus on clinically interpretable features.

4   Conclusion

In this paper, we present a selective attention regularization (SAttenReg) method for liver biopsy image analysis. SAttenReg mimics pathologists to encourage the model to focus on clinically interpretable features using attention regularization. Gaussian distribution-induced weights are assigned to the attention regularization to allow the model to selectively focus on clinically interpretable features in each instance. Experiments conducted on the in-house Liver-NAS and public

Biopsy4Grading biopsy image datasets show that our method yields superior classification performance with promising localization results. Future work will focus on integrating the local patch information for whole-slide pathology image understanding.

Acknowledgement. This work was supported by the Health and Medical Research Fund Project under Grant 07180216. We acknowledge insightful discussion with Anthony W.H. CHAN. We also thank Vincent WS WONG, Grace LH WONG, and Howard H.W. LEUNG from the Chinese University of Hong Kong for help with data preparation.

References

1. Angulo, J., Jeulin, D.: Stochastic watershed segmentation. In: ISMM (1), pp. 265–276 (2007)
2. Boyd, A., Cain, O., Chauhan, A., Webb, G.J.: Medical liver biopsy: background, indications, procedure and histopathology. Frontline Gastroenterol. **11**(1), 40–47 (2020)
3. Brunt, E.M.: Pathology of nonalcoholic fatty liver disease. Nat. Rev. Gastroenterol. Hepatol. **7**(4), 195–203 (2010)
4. Chalasan, N., et al.: The diagnosis and management of non-alcoholic fatty liver disease: Practice guideline by the American association for the study of liver diseases, American college of gastroenterology, and the American gastroenterological association. Hepatology **55**(6), 2005–2023 (2012)
5. Forlano, R., et al.: High-throughput, machine learning-based quantification of steatosis, inflammation, ballooning, and fibrosis in biopsies from patients with nonalcoholic fatty liver disease. Clin. Gastroenterol. Hep. **18**(9), 2081–2090 (2020)
6. Fu, X., Liu, T., Xiong, Z., Smaill, B.H., Stiles, M.K., Zhao, J.: Segmentation of histological images and fibrosis identification with a convolutional neural network. Comput. Biol. Med. **98**, 147–158 (2018)
7. Heinemann, F., Birk, G., Stierstorfer, B.: Deep learning enables pathologist-like scoring of nash models. Sci. Rep. **9**(1), 1–10 (2019)
8. Hu, J., Shen, L., Sun, G.: Squeeze-and-excitation networks. In: Proceedings of the IEEE Conference on Computer Vision and Pattern Recognition, pp. 7132–7141 (2018)
9. Huang, Y., Chung, A.C.S.: Evidence localization for pathology images using weakly supervised learning. In: Shen, D., et al. (eds.) MICCAI 2019. LNCS, vol. 11764, pp. 613–621. Springer, Cham (2019). https://doi.org/10.1007/978-3-030-32239-7_68
10. Ji, Y., Zhang, R., Li, Z., Ren, J., Zhang, S., Luo, P.: UXNet: searching multi-level feature aggregation for 3D medical image segmentation. In: Martel, A.L., et al. (eds.) MICCAI 2020. LNCS, vol. 12261, pp. 346–356. Springer, Cham (2020). https://doi.org/10.1007/978-3-030-59710-8_34
11. Kleiner, D.E., et al.: Design and validation of a histological scoring system for nonalcoholic fatty liver disease. Hepatology **41**(6), 1313–1321 (2005)
12. Lerousseau, M., et al.: Weakly supervised multiple instance learning histopathological tumor segmentation. In: Martel, A.L., et al. (eds.) MICCAI 2020. LNCS, vol. 12265, pp. 470–479. Springer, Cham (2020). https://doi.org/10.1007/978-3-030-59722-1_45

13. Liu, J.J., Hou, Q., Cheng, M.M., Wang, C., Feng, J.: Improving convolutional networks with self-calibrated convolutions. In: Proceedings of the IEEE/CVF Conference on Computer Vision and Pattern Recognition, pp. 10096–10105 (2020)
14. Milletari, F., Navab, N., Ahmadi, S.A.: V-net: Fully convolutional neural networks for volumetric medical image segmentation. In: 2016 Fourth International Conference on 3D Vision (3DV), pp. 565–571. IEEE (2016)
15. Selvaraju, R.R., Cogswell, M., Das, A., Vedantam, R., Parikh, D., Batra, D.: Gradcam: visual explanations from deep networks via gradient-based localization. In: Proceedings of the IEEE International Conference on Computer Vision, pp. 618–626 (2017)
16. Sethunath, D.: Detection of histological features in liver biopsy images to help identify Non-Alcoholic Fatty Liver Disease. Ph.D. thesis (2018)
17. Srinidhi, C.L., Ciga, O., Martel, A.L.: Deep neural network models for computational histopathology: a survey. Med. Image Anal. **67**, 101813 (2020)
18. Takahashi, Y., Fukusato, T.: Histopathology of nonalcoholic fatty liver disease/nonalcoholic steatohepatitis. World J. Gastroenterol. WJG **20**(42), 15539 (2014)
19. Wei, Y., Rector, R.S., Thyfault, J.P., Ibdah, J.A.: Nonalcoholic fatty liver disease and mitochondrial dysfunction. World J. Gastroenterol. WJG **14**(2), 193 (2008)
20. Williams, C.D., et al.: Prevalence of nonalcoholic fatty liver disease and nonalcoholic steatohepatitis among a largely middle-aged population utilizing ultrasound and liver biopsy: a prospective study. Gastroenterology **140**(1), 124–131 (2011)
21. Wong, R.J., et al.: Nonalcoholic steatohepatitis is the second leading etiology of liver disease among adults awaiting liver transplantation in the united states. Gastroenterology **148**(3), 547–555 (2015)
22. Younossi, Z., et al.: Global perspectives on nonalcoholic fatty liver disease and nonalcoholic steatohepatitis. Hepatology **69**(6), 2672–2682 (2019)
23. Yu, Y., et al.: Deep learning enables automated scoring of liver fibrosis stages. Sci. Rep. **8**(1), 1–10 (2018)
24. Zagoruyko, S., Komodakis, N.: Paying more attention to attention: Improving the performance of convolutional neural networks via attention transfer. arXiv preprint arXiv:1612.03928 (2016)
25. Zhou, B., Khosla, A., Lapedriza, A., Oliva, A., Torralba, A.: Learning deep features for discriminative localization. In: Proceedings of the IEEE Conference on Computer Vision and Pattern Recognition, pp. 2921–2929 (2016)
26. Zhuang, J., Cai, J., Wang, R., Zhang, J., Zheng, W.-S.: Deep kNN for medical image classification. In: Martel, A.L., et al. (eds.) MICCAI 2020. LNCS, vol. 12261, pp. 127–136. Springer, Cham (2020). https://doi.org/10.1007/978-3-030-59710-8_13

Categorical Relation-Preserving Contrastive Knowledge Distillation for Medical Image Classification

Xiaohan Xing[1], Yuenan Hou[2], Hang Li[3], Yixuan Yuan[4(✉)], Hongsheng Li[1], and Max Q.-H. Meng[1,5(✉)]

[1] Department of Electronic Engineering, The Chinese University of Hong Kong, Shatin, Hong Kong, China
[2] Department of Information Engineering, The Chinese University of Hong Kong, Shatin, Hong Kong, China
[3] School of Informatics, Xiamen University, Xiamen, China
[4] Department of Electrical Engineering, City University of Hong Kong, Kowloon, Hong Kong, China
yxyuan.ee@cityu.edu.hk
[5] Department of Electronic and Electrical Engineering, Southern University of Science and Technology, Shenzhen, China
max.meng@sustech.edu.cn

Abstract. The amount of medical images for training deep classification models is typically very scarce, making these deep models prone to overfit the training data. Studies showed that knowledge distillation (KD), especially the mean-teacher framework which is more robust to perturbations, can help mitigate the over-fitting effect. However, directly transferring KD from computer vision to medical image classification yields inferior performance as medical images suffer from higher intra-class variance and class imbalance. To address these issues, we propose a novel Categorical Relation-preserving Contrastive Knowledge Distillation (CRCKD) algorithm, which takes the commonly used mean-teacher model as the supervisor. Specifically, we propose a novel Class-guided Contrastive Distillation (CCD) module to pull closer positive image pairs from the same class in the teacher and student models, while pushing apart negative image pairs from different classes. With this regularization, the feature distribution of the student model shows higher intra-class similarity and inter-class variance. Besides, we propose a Categorical Relation Preserving (CRP) loss to distill the teacher's relational knowledge in a robust and class-balanced manner. With the contribution of the CCD and CRP, our CRCKD algorithm can distill the relational knowledge more comprehensively. Extensive experiments on the HAM10000 and APTOS datasets demonstrate the superiority of the proposed CRCKD method. The source code is available at https://github.com/hathawayxxh/CRCKD.

Electronic supplementary material The online version of this chapter (https://doi.org/10.1007/978-3-030-87240-3_16) contains supplementary material, which is available to authorized users.

Keywords: Medical image classification · Knowledge distillation · Contrastive learning

1   Introduction

With the recent progress of deep learning techniques, computer-aided diagnosis has shown human-level performance for some diseases and reduced the workload of human screening [1]. However, the amount of training data for most diseases are limited, making the deep models prone to overfit the training data [2,3]. To tackle the over-fitting issue, many learning schemes have been proposed, such as transfer learning [4,5], dropout [6], and label-smoothing regularization [7,8]. Another effective solution is knowledge distillation (KD), where a trained teacher model provides soft labels that supply secondary information to the student model, thus relieving the over-fitting problem [9].

Among existing KD frameworks, the self-ensembling mean-teacher [10] is widely studied in medical image classification. Updated by the temporal moving average of the student model, the mean-teacher produces feature distribution and predictions that are robust to different perturbations, thus showing higher generalizability even with the limited amount of data. Therefore, to train a student with high accuracy and generalizability, it is crucial to maximally distill knowledge from the mean-teacher. Some researchers distilled the individual sample knowledge from the teacher, such as output logits [11] and feature maps [12]. Recently, Liu et al. [13] took the relation among mini-batch samples as distilling targets and demonstrated its superiority over the individual KD counterparts.

However, most of the existing KD methods [11–17] are directly transferred from the computer vision field, without fully considering the following challenges in the medical domain. First, the intra-class variation and inter-class similarity in medical datasets are more severe than those in the natural domain. In specific, two types of diseases may exhibit extremely similar color, shape, and texture, making them less distinguishable than two classes of natural images (dogs vs. cats). Second, medical image datasets usually suffer from severe class imbalance since some diseases are common while others are rare. Due to this, the knowledge distilled by current KD may be biased towards the majority class and has insufficient representation for the minority classes.

To tackle the above-mentioned challenges, we propose a novel distillation approach, termed Categorical Relation-preserving Contrastive Knowledge Distillation (CRCKD), for medical image classification. Built upon the mean-teacher framework, we propose two novel KD paradigms, i.e., *Class-guided Contrastive Distillation* (CCD) and *Categorical Relation Preserving* (CRP), to distill the rich structural knowledge from the mean-teacher model. The main contributions are summarized as: (1) We propose the CCD module to pull closer positive image pairs from the same class in the teacher and student models, while pushing apart negative image pairs from different classes. With this regularization, the feature distribution of the student model exhibits higher intra-class similarity and inter-class variance. (2) To distill more robust and fine-grained relational knowledge, we propose the CRP loss that utilizes category centroids as anchors to regulate each sample's relation with different categories. Compared with previous relational KD [13,18] that adopts images in a mini-batch as anchors, the category centroids in our method serve as more reliable anchors and naturally mitigate the class imbalance problem. (3) Experimental results on HAM10000 and APTOS datasets demonstrated the efficacy of our proposed CRCKD method, as well as the superiority of the CCD and CRP over existing relational KD paradigms.

2 Method

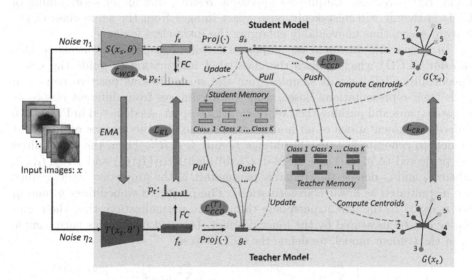

Fig. 1. Overview of the proposed Categorical Relation-preserving Contrastive Knowledge Distillation (CRCKD) framework. The student model is supervised by the weighted cross-entropy loss $\mathcal{L}_{WCE}$ and knowledge distillation losses ($\mathcal{L}_{KL}$, $\mathcal{L}_{CCD}^{(S)}$, and $\mathcal{L}_{CRP}$). The dashed orange lines indicate the back-propagation paths of the gradients. [Best viewed in color] (Color figure online)

Figure 1 illustrates our proposed CRCKD framework. It consists of a student model and a mean-teacher model. The student model is optimized by stochastic gradient descent while the teacher weights θ' are updated by the exponential moving average (EMA) of the student weights θ. Given an image x, it is augmented twice by adding different perturbations (i.e., random flipping and affine transformation) and produces two different images x_s and x_t. Taking the corresponding augmented image as an input, the student model and teacher model extract feature representations f_s and f_t, and predict output probabilities p_s and p_t, respectively. The student's prediction p_s is supervised by the weighted cross-entropy loss $\mathcal{L}_{WCE}$ and the KL divergence $\mathcal{L}_{KL}$ with p_t. To constrain the consistency between the student's and teacher's structural information, we propose the $\mathcal{L}_{CCD}$ loss that pulls the positive pairs (f_s, f_t) from the same class while pushing way the negative feature pairs from different classes. Furthermore, in each model, we construct a novel relation graph $G(x_s)$ $(G(x_t))$ between the sample feature and the category centroids. The $\mathcal{L}_{CRP}$ loss is proposed to regularize the consistency between the relation graphs of the teacher and student models.

2.1   Class-Guided Contrastive Distillation (CCD)

Recently, *Contrastive Representation Distillation* (CRD) has achieved impressive distillation performance via incorporating contrastive learning into the conventional KD paradigm. Despite its appealing results, one major shortcoming of CRD is that it will mistakenly push apart images from the same class in the feature space, thus unavoidably enlarging the intra-class variance.

To tackle this dilemma, we propose a novel *Class-guided Contrastive Distillation* (CCD) which utilizes the class-label information to guide the CRD. Specifically, CCD regards samples from the same class as positive pairs and pulls their representations closer, while taking images from different classes as negative pairs and pushing their representations apart. As depicted in Fig. 1, two different augmentations of an image are processed by the teacher and student models to generate feature embeddings f_s and $f_t \in R^D$. Then, the embeddings are projected to g_s and $g_t \in R^d (d < D)$ through $Proj(f_s; \zeta_s)$ and $Proj(f_t; \zeta_t)$, where ζ_s and ζ_t denote the trainable parameters in the projection layers (which are instantiated as linear transformation). The projected embeddings g_s and g_t are mapped to the unit hypersphere through L_2 normalization, thus their similarity can be measured by the inner product. Inspired by [19], for each sample g_s in the student model, we define the CCD loss as

$$\mathcal{L}_{CCD}^{(S)}(\theta, \zeta_s) = -\frac{1}{k_P} \sum_{i=1}^{k_P} (\log \frac{e^{(g_s \cdot g_{t,i}/\tau)}}{e^{(g_s \cdot g_{t,i}/\tau)} + \frac{k_N}{M}} + \sum_{j=1}^{k_N} \log(1 - \frac{e^{(g_s \cdot g_{t,j}/\tau)}}{e^{(g_s \cdot g_{t,j}/\tau)} + \frac{k_N}{M}})), \tag{1}$$

where τ is the temperature that controls the concentration level, k_P and k_N denote the number of positive samples and negative samples, respectively. M is the cardinality of the dataset. By minimizing $L_{CCD}^{(S)}$, the student model is optimized to produce feature representations that are more similar with the k_P positive pairs while differing from the k_N negative samples in the teacher model. Similarly, the CCD loss for the teacher model is defined as

$$\mathcal{L}_{CCD}^{(T)}(\zeta_t) = -\frac{1}{k_P} \sum_{i=1}^{k_P} (\log \frac{e^{(g_t \cdot g_{s,i}/\tau)}}{e^{(g_t \cdot g_{s,i}/\tau)} + \frac{k_N}{M}} + \sum_{j=1}^{k_N} \log(1 - \frac{e^{(g_t \cdot g_{s,j}/\tau)}}{e^{(g_t \cdot g_{s,j}/\tau)} + \frac{k_N}{M}})). \tag{2}$$

It is noteworthy that the $\mathcal{L}_{CCD}^{(T)}$ loss merely updates the projection head of the teacher model. The CCD loss regularizes the consistency of teacher and student's inter-sample structural knowledge by enlarging intra-class similarity and inter-class divergence between these two models, thus yielding performance gains.

As suggested in [19,20], a large number of negative samples is required to ensure the performance of contrastive learning. To get access to a large number of negative samples and avoid large batch size, we follow Wu et al. [21] to construct a memory bank $M \in R^{N \times d}$ that stores the d-dimensional embeddings of all N training images. We denote the memory bank for the student (teacher) model as M_s (M_t). As shown in Fig. 1, in each forward propagation, only the features of the query samples in the mini-batch are updated while all other samples retain their embeddings at the last step. For each query sample x_s in the student model, the k_P positive and k_N negative samples in Eq. 1 are randomly selected from the teacher's memory M_t. Similarly, the Eq. 2 is computed in a similar manner.

2.2 Categorical Relation Preserving (CRP)

Although the proposed CCD can regularize the structural consistency of the teacher and student, the regularization is relatively coarse since each sample pushes apart negative image pairs from different classes without differentiation. However, some categories of diseases are much more similar than other categories, thus their distributions should be closer in the embedding space. To capture fine-grained relational knowledge, [13,18,22] proposed to distill the pairwise relations between data samples in a mini-batch. However, for the dataset with severe class imbalance, most samples in a mini-batch belong to the majority class, thus the constructed relation graphs may suffer from class bias.

To settle the above issues and capture class-balanced relational knowledge, we propose a novel *Categorical Relation Preserving* (CRP) loss that utilizes category centroids to construct the relation graph. Specifically, in the student (teacher) model, we compute the centroid of the i-th category by averaging the features of all samples in the i-th class (retrieved from the memory bank M_s (M_t)):

$$C_i^S = \frac{1}{|C_i|} \sum_{m_s \in C(i)} m_s, \quad C_i^T = \frac{1}{|C_i|} \sum_{m_t \in C(i)} m_t, \tag{3}$$

where $|C_i|$ denotes the number of samples in the i-th class. Then, for each query sample x_s (x_t) in the mini-batch, we compute its cosine similarity with all category centroids in the student (teacher) model. After softmax over all classes, we obtain the categorical relation between the sample x_s (x_t) and the i-th category:

$$R(x_s, C_i^S) = \frac{e^{g_s \cdot C_i^S}}{\sum_{i=1}^{K} e^{g_s \cdot C_i^S}}, \quad R(x_t, C_i^T) = \frac{e^{g_t \cdot C_i^T}}{\sum_{i=1}^{K} e^{g_t \cdot C_i^T}}, \tag{4}$$

where K is the total number of classes in the dataset, g_s and g_t denote the sample's representations extracted by the student and teacher, respectively. Then, we propose the CRP loss to minimize the KL divergence between the teacher and student's categorical relation graphs:

$$\mathcal{L}_{CRP} = \sum_{x_s, x_t \in T} \sum_{i=1}^{K} R(x_t, C_i^T) \log \frac{R(x_t, C_i^T)}{R(x_s, C_i^S)}. \tag{5}$$

Compared with existing relational KD [13,18,22], using category centroids as anchors in our methods has two advantages: 1) One anchor (class centroid) is utilized to represent each class (regardless of the number of images in that class), thus the constructed relation graphs naturally mitigate the bias caused by class imbalance. 2) The anchors in the CRP retrieved from the memory bank (momentum aggregation of temporal steps) are more robust than the anchors that rely on the current step. To this end, the CRP loss regularizes the reliable categorical relation graphs built with more robust and representative anchors, thus is expected to better mimic the relational knowledge in the teacher model.

2.3 Training and Testing

In the proposed framework, the weights θ of the student model and ζ_s, ζ_t of the projection layers are optimized by the loss function defined as:

$$\mathcal{L} = \mathcal{L}_{WCE} + \lambda_1 \cdot \mathcal{L}_{KL} + \lambda_2 \cdot \mathcal{L}_{CCD} + \lambda_3 \cdot \mathcal{L}_{CRP}, \tag{6}$$

where $\mathcal{L}_{WCE}$ denotes the weighted cross-entropy loss supervised by the ground-truth labels. The weight for each class in $\mathcal{L}_{WCE}$ is inverse proportional to the number of images in that class. $\mathcal{L}_{KL}, \mathcal{L}_{CCD}, \mathcal{L}_{CRP}$ are used to distill the individual and structural knowledge from the teacher model. In the first T epochs, the trade-off weights λ_1 and λ_3 would gradually ramp-up from 0 to 1 according to a Gaussian warming up function $\lambda(t) = 1 * e^{(-5(1-t/T)^2)}$, while λ_2 is set as 0.1. After T epochs, we fix the value of λ_1 and λ_3 as 1, and set λ_2 as 0.01. The teacher weights θ' are updated as the EMA of the student weights.

At the testing stage, we discard the mean teacher and the projection heads, so the inference time is the same as the vanilla student model.

3   Experiments

3.1   Dataset and Implementation Details

Dataset: We evaluated our proposed CRCKD framework on the HAM10000 [23,24] and APTOS datasets [25]. The HAM10000 consists of 10015 dermoscopy images labeled by 7 types of skin lesions. In APTOS, there are 3662 fundus images for grading diabetic retinopathy into five categories. These two datasets both suffer from severe class imbalance. A detailed description of these two datasets is provided in the supplementary material. For both datasets, we performed five-fold cross-validation and reported the average testing performance over the five folds. We evaluated the classification performance by overall accuracy (ACC), average precision (AP), balanced multi-class accuracy (BMA), and $F1$ score. Due to the class imbalance, BMA is considered the most important metric in this task.

Table 1. Five-fold cross-validation results on HAM10000 and APTOS datasets. The highest rankings are highlighted in **bold**. Our method: B2 + CCD + CRP. Detailed performance on the mean and standard deviation of each algorithm is provided in the supplementary material.

Methods	HAM10000				APTOS			
	ACC	AP	BMA	F1	ACC	AP	BMA	F1
DenseNet121 (B1)	84.30	74.16	72.19	72.53	83.83	71.85	67.51	69.14
B1 + MT (B2)	85.01	74.19	76.07	74.38	83.77	71.66	68.79	69.89
B2 + CCD	85.52	74.87	77.64	75.45	84.42	72.79	70.42	71.23
B2 + CRP	85.32	75.06	77.06	75.37	84.47	73.07	69.87	71.07
Our method	**85.66**	**76.35**	**78.07**	**76.45**	**84.87**	**73.18**	**71.90**	**72.22**
B2 + CRD [19]	85.33	74.41	76.44	74.77	84.09	71.82	69.38	70.27
B2 + SP [18]	85.13	74.92	76.06	74.48	83.16	70.51	69.15	69.54
B2 + FitNet [27]	84.13	72.85	76.38	73.98	83.76	71.97	69.88	70.52

Implementation: Our method was implemented in Python with the Pytorch library. We employed the pre-trained DenseNet121 [26] as the backbone of the teacher and student model. The network was trained with two P40 GPUs in parallel and the batch size was set to 64. Adam with $\beta_1 = 0.5$ and $\beta_2 = 0.999$ was used for network optimization. We trained the network for 80 epochs with ramp-up epoch T set as 30. The initial learning rate was set to 0.0001 and decayed by the one-cycle schedule. The temperature τ in Eq. 1 and Eq. 2 is set as 0.07. For each query sample, the number of positive pairs k_P and negative pairs k_N was empirically set as 20 and 4096, respectively.

3.2   Experimental Results

Quantitative Results: Table 1 summarizes the performance of our CRCKD and baseline algorithms on the HAM10000 and APTOS datasets. Compared with the vanilla student model "DenseNet121 (B1)", "B1 + MT (B2)" achieves much better classification performance, showing the superiority of introducing the mean-teacher guidance. What's more, Table 1 (*row 3–4*) indicates the effectiveness of the proposed CCD and CRP, because involving either of them leads to relatively better performance than "B2". We conjecture that performance gains brought by CCD and CRP are attributed to the distillation of structural knowledge. Further, the combination of CCD and CRP in our method achieves the best performance with BMA of 78.07% (2.00% higher than "B2") on the HAM10000 dataset and BMA of 71.90% (3.11% higher than "B2") on the APTOS dataset, demonstrating the efficacy of the proposed method.

To further validate the effectiveness of the proposed CCD and CRP, we performed a comparison with the other three KD paradigms (see the last three rows in Table 1). "B2 + CCD" achieves better performance than "B2 + CRD" [19],

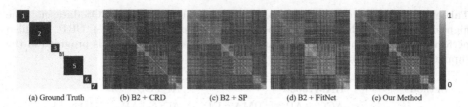

(a) Ground Truth   (b) B2 + CRD   (c) B2 + SP   (d) B2 + FitNet   (e) Our Method

Fig. 2. Visualization of the relation matrices between mini-batch samples (batch size = 256) in the HAM10000 dataset. Input samples are grouped by ground truth class along each axis. The color from dark blue to light green indicates increased similarity. (Color figure online)

Table 2. Comparison with state-of-the-art methods on the HAM10000 dataset.

Methods	ACC (%)	AP (%)	BMA (%)	F1 (%)
Yan et al. [28]	77.16 ± 0.80	62.71 ± 1.66	73.37 ± 1.24	66.70 ± 1.33
Zhang et al. [29]	82.34 ± 0.73	72.32 ± 1.38	74.01 ± 1.76	72.28 ± 1.04
Zhang et al. [30]	81.61 ± 0.94	71.44 ± 1.22	73.34 ± 1.86	71.65 ± 0.85
Liu et al. [13]	84.73 ± 1.00	73.88 ± 1.24	76.55 ± 1.32	74.63 ± 0.99
Our method	$\mathbf{85.66 \pm 0.97}$	$\mathbf{76.35 \pm 0.99}$	$\mathbf{78.07 \pm 1.28}$	$\mathbf{76.45 \pm 0.66}$

indicating the contribution of introducing the class-label guidance, which is consistent with our analysis in Sect. 2.1. Besides, the superiority over "B2 + SP" [18] suggests that our proposed CRP can better distill relational knowledge by utilizing more robust and representative class centroids as anchors. Finally, both "B2 + CCD" and "B2 + CRP" outperform "B2 + FitNet" [27] (a classical KD method that distills the intermediate features of individual samples), suggesting the necessity and superiority of relation-preserving KD studied in this work.

Qualitative Analysis: Taking the HAM10000 dataset as an example, we visualize inter-sample relations produced by the features of different methods in Fig. 2. The diagonal blue blocks in Fig. 2 (a) denote intra-class similarity while other parts represent inter-class relation. As shown in Fig. 2 (b), "B2 + CRD" exhibits low intra-class similarity (especially for the $1st$, $2nd$, and $5th$ block). In contrast, "B2 + SP" and "B2 + FitNet" (Fig. 2 (c, d)) yield high inter-class relations. With the regularization of the proposed CCD and CRP, our method (Fig. 2 (e)) exhibits lower inter-class similarity and higher intra-class similarity. To quantitatively compare the relation matrices, we resort to $\mathcal{R}_d = \overline{\mathcal{R}}_{intra}/\overline{\mathcal{R}}_{inter}$, where $\overline{\mathcal{R}}_{intra}$ denotes the average pair-wise similarity between samples from the same class, $\overline{\mathcal{R}}_{inter}$ is the average sample similarity between different classes. The larger value of $\mathcal{R}_d$ indicates higher intra-class similarity and lower inter-class similarity. The $\mathcal{R}_d$ value of our method is 1.41, outperforming the "B2 + CRD" (1.36), "B2 + SP" (1.35), and "B2 + FitNet" (1.22). These results demonstrate that our proposed method can effectively alleviate the issue of high intra-class variance and inter-class similarity in the medical domain.

Comparison with Contemporary Methods: On the HAM10000 dataset, we further evaluated the performance of our proposed method with four state-of-the-art methods for skin lesion classification: attention-based methods [28,29], synergic deep learning [30], and KD method based on sample relations [13]. For a fair comparison, we changed [13] to full supervision, used the implementations of [13,28–30] suggested by the authors, and evaluated the performance using the same dataset as our method. As shown in Table 2, the proposed method outperforms existing methods with an improvement of 4.70%, 4.06%, 4.73%, 1.52% in BMA, further validating the effectiveness of our proposed method.

4   Conclusion

In this paper, we present a novel Categorical Relation-preserving Contrastive Knowledge Distillation (CRCKD) framework for medical image classification. Against the unique challenges of high inter-class similarity and class-imbalance in the medical domain, we propose two novel KD paradigms, i.e., CCD and CRP, to distill rich structural knowledge from the mean-teacher model. Experimental results on the HAM10000 and APTOS datasets demonstrate the effectiveness of the proposed CCD and CRP over other KD paradigms. On the HAM10000 dataset, experiments show that our CRCKD method outperforms many state-of-the-art methods.

Acknowledgements. The work described in this paper was supported by National Key R&D program of China with Grant No. 2019YFB1312400, Hong Kong RGC CRF grant C4063-18G, and Hong Kong RGC GRF grant #14211420.

References

1. Litjens, G., et al.: A survey on deep learning in medical image analysis. Med. Image Anal. **42**, 60–88 (2017)
2. Yang, C., Xie, L., Su, C., Yuille, A.L.: Snapshot distillation: teacher-student optimization in one generation. In: Proceedings of the CVPR, pp. 2859–2868 (2019)
3. Zhuang, J., Cai, J., Wang, R., Zhang, J., Zheng, W.-S.: Deep kNN for medical image classification. In: Martel, A.L., et al. (eds.) MICCAI 2020. LNCS, vol. 12261, pp. 127–136. Springer, Cham (2020). https://doi.org/10.1007/978-3-030-59710-8_13
4. Cheplygina, V., de Bruijne, M., Pluim, J.P.: Not-so-supervised: a survey of semi-supervised, multi-instance, and transfer learning in medical image analysis. Med. Image Anal. **54**, 280–296 (2019)
5. Shang, H., et al.: Leveraging other datasets for medical imaging classification: evaluation of transfer, multi-task and semi-supervised learning. In: Shen, D., et al. (eds.) MICCAI 2019. LNCS, vol. 11768, pp. 431–439. Springer, Cham (2019). https://doi.org/10.1007/978-3-030-32254-0_48

6. Srivastava, N., Hinton, G., Krizhevsky, A., Sutskever, I., Salakhutdinov, R.: Dropout: a simple way to prevent neural networks from overfitting. J. Mach. Learn. Res. **15**(1), 1929–1958 (2014)

7. Szegedy, C., Vanhoucke, V., Ioffe, S., Shlens, J., Wojna, Z.: Rethinking the inception architecture for computer vision. In: Proceedings of the CVPR, pp. 2818–2826 (2016)

8. Müller, R., Kornblith, S., Hinton, G.: When does label smoothing help? arXiv preprint arXiv:1906.02629 (2019)

9. Hinton, G., Vinyals, O., Dean, J.: Distilling the knowledge in a neural network. arXiv preprint arXiv:1503.02531 (2015)

10. Tarvainen, A., Valpola, H.: Mean teachers are better role models: Weight-averaged consistency targets improve semi-supervised deep learning results. In: Advances in Neural Information Processing Systems, pp. 1195–1204 (2017)

11. Thiagarajan, J.J., Kashyap, S., Karargyris, A.: Distill-to-label: weakly supervised instance labeling using knowledge distillation. In: 2019 18th IEEE International Conference on Machine Learning And Applications (ICMLA), pp. 902–907. IEEE (2019)

12. Wu, J., et al.: Leveraging undiagnosed data for glaucoma classification with teacher-student learning. In: Martel, A.L., et al. (eds.) MICCAI 2020. LNCS, vol. 12261, pp. 731–740. Springer, Cham (2020). https://doi.org/10.1007/978-3-030-59710-8_71

13. Liu, Q., Yu, L., Luo, L., Dou, Q., Heng, P.A.: Semi-supervised medical image classification with relation-driven self-ensembling model. IEEE Trans. Med. Imaging **39**, 3429–3440 (2020)

14. Unnikrishnan, B., Nguyen, C.M., Balaram, S., Foo, C.S., Krishnaswamy, P.: Semi-supervised classification of diagnostic radiographs with noteacher: a teacher that is not mean. In: Martel, A.L., et al. (eds.) MICCAI 2020. LNCS, vol. 12261, pp. 624–634. Springer, Cham (2020). https://doi.org/10.1007/978-3-030-59710-8_61

15. Abbasi, S., et al.: Classification of diabetic retinopathy using unlabeled data and knowledge distillation. arXiv preprint arXiv:2009.00982 (2020)

16. Patra, A., et al.: Efficient ultrasound image analysis models with sonographer gaze assisted distillation. In: Shen, D., et al. (eds.) MICCAI 2019. LNCS, vol. 11767, pp. 394–402. Springer, Cham (2019). https://doi.org/10.1007/978-3-030-32251-9_43

17. Hou, Y., Ma, Z., Liu, C., Loy, C.C.: Learning lightweight lane detection CNNs by self attention distillation. In: Proceedings of the ICCV, pp. 1013–1021 (2019)

18. Tung, F., Mori, G.: Similarity-preserving knowledge distillation. In: Proceedings of the ICCV, pp. 1365–1374 (2019)

19. Tian, Y., Krishnan, D., Isola, P.: Contrastive representation distillation. arXiv preprint arXiv:1910.10699 (2019)

20. Saunshi, N., Plevrakis, O., Arora, S., Khodak, M., Khandeparkar, H.: A theoretical analysis of contrastive unsupervised representation learning. In: International Conference on Machine Learning, pp. 5628–5637 (2019)

21. Wu, Z., Xiong, Y., Yu, S.X., Lin, D.: Unsupervised feature learning via non-parametric instance discrimination. In: Proceedings of the CVPR, pp. 3733–3742 (2018)

22. Park, W., Kim, D., Lu, Y., Cho, M.: Relational knowledge distillation. In: Proceedings of the CVPR, pp. 3967–3976 (2019)

23. Tschandl, P., Rosendahl, C., Kittler, H.: The HAN10000 dataset, a large collection of multi-source dermatoscopic images of common pigmented skin lesions. Sci. Data **5**, 180161 (2018)

24. Codella, N.C., et al.: Skin lesion analysis toward melanoma detection: a challenge at the 2017 international symposium on biomedical imaging (ISBI), hosted by the international skin imaging collaboration (ISIC). In: Proceedings of the ISBI, pp. 168–172. IEEE (2018)
25. Aptos 2019 blindness detection. https://www.kaggle.com/c/aptos2019-blindness-detection/data
26. Huang, G., Liu, Z., Van Der Maaten, L., Weinberger, K.Q.: Densely connected convolutional networks. In: Proceedings of the CVPR, pp. 4700–4708 (2017)
27. Yim, J., Joo, D., Bae, J., Kim, J.: A gift from knowledge distillation: fast optimization, network minimization and transfer learning. In: Proceedings of the CVPR, pp. 4133–4141 (2017)
28. Yan, Y., Kawahara, J., Hamarneh, G.: Melanoma recognition via visual attention. In: Chung, A.C.S., Gee, J.C., Yushkevich, P.A., Bao, S. (eds.) IPMI 2019. LNCS, vol. 11492, pp. 793–804. Springer, Cham (2019). https://doi.org/10.1007/978-3-030-20351-1_62
29. Zhang, J., Xie, Y., Wu, Q., Xia, Y.: Skin lesion classification in dermoscopy images using synergic deep learning. In: Frangi, A.F., Schnabel, J.A., Davatzikos, C., Alberola-López, C., Fichtinger, G. (eds.) MICCAI 2018. LNCS, vol. 11071, pp. 12–20. Springer, Cham (2018). https://doi.org/10.1007/978-3-030-00934-2_2
30. Zhang, J., Xie, Y., Xia, Y., Shen, C.: Attention residual learning for skin lesion classification. IEEE Trans. Med. Imaging 38(9), 2092–2103 (2019)

Tensor-Based Multi-index Representation Learning for Major Depression Disorder Detection with Resting-State fMRI

Dongren Yao[1,2], Erkun Yang[1], Hao Guan[1], Jing Sui[3], Zhizhong Zhang[4(✉)], and Mingxia Liu[1(✉)]

[1] Department of Radiology and BRIC, University of North Carolina at Chapel Hill, Chapel Hill, NC 27599, USA
mxliu@med.unc.edu
[2] Brainnetome Center & National Laboratory of Pattern Recognition, Institute of Automation, Chinese Academy of Sciences, Beijing 100190, China
[3] State Key Laboratory of Cognitive Neuroscience and Learning, Beijing Normal University, Beijing 100678, China
[4] School of Computer Science and Technology, East China Normal University, Shanghai 200241, China
zzzhang@cs.ecnu.edu.cn

Abstract. Major depressive disorder (MDD) is a common and costly mental illness whose pathophysiology is difficult to clarify. Resting-state functional MRI (rs-fMRI) provides a non-invasive solution for the study of functional brain network abnormalities in MDD patients. Existing studies have shown that multiple indexes derived from rs-fMRI, such as fractional amplitude of low-frequency fluctuations (fALFF), voxel-mirrored homotopic connectivity (VMHC), and degree centrality (DC) help depict functional mechanisms of brain disorders from different perspectives. However, previous methods generally treat these indexes independently, without considering their potentially complementary relationship. Moreover, it is usually very challenging to effectively fuse multi-index representations for disease analysis, due to the significant heterogeneity among indexes in the feature distribution. In this paper, we propose a tensor-based multi-index representation learning (TMRL) framework for fMRI-based MDD detection. In TMRL, we first generate multi-index representations (i.e., fALFF, VMHC and DC) for each subject, followed by patch selection via group comparison for each index. We further develop a tensor-based multi-task learning model (with a tensor-based regularizer) to align multi-index representations into a common latent space, followed by MDD prediction. Experimental results on 533 subjects with rs-fMRI data demonstrate that the TMRL outperforms several state-of-the-art methods in MDD identification.

Keywords: Major depressive disorder · rs-fMRI · Diagnosis

Electronic supplementary material The online version of this chapter (https://doi.org/10.1007/978-3-030-87240-3_17) contains supplementary material, which is available to authorized users.

1 Introduction

Major depression disorder (MDD) is one of the most prevalent disabling disorder, characterized by depressed mood, loss of interest or pleasure in nearly all activities. This mental illness has a high mortality rate due to the suicidal behavior of MDD patients, while the high cost of treatment troubles patients, their family members, and society [1,2]. Even though many efforts have been made in clinical neuroscience and psychiatric research, the unknown etiology and pathological mechanism still prevent us from fully understanding the disease.

Resting-state functional MRI (rs-fMRI) has become an essential non-invasive tool for assessing the brain substrates underlying mental disorders. [3–5]. Recent studies report that MDD is not only related to regional deficits, but also related to abnormal functional integration of distributed brain regions [6–8]. To measure spontaneous neural activities in rs-fMRI, multiple indexes have been designed, such as fractional amplitude of low-frequency fluctuation (fALFF) [9], voxel-mirrored homotopic connectivity (VMHC) [10], and degree centrality (DC) [11]. These indexes have been shown to help reveal the functional mechanisms of brain disorders [12–14]. However, existing methods generally treat multiple indexes independently, without considering their potentially complementary relationship. Also, it is usually very challenging to effectively fuse multi-index representations because of the significant between-index heterogeneity in feature distribution.

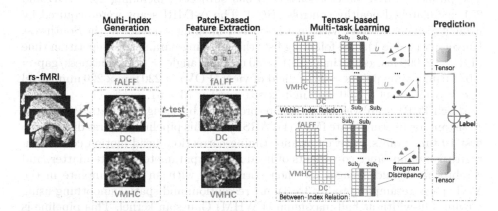

Fig. 1. Framework of the tensor-based multi-index representation learning (TMRL) for MDD identification with rs-fMRI, including (1) multi-index generation, (2) patch-based feature extraction, (3) tensor-based multi-task learning, and (4) prediction.

To address these issues, we propose a tensor-based multi-index representation learning (**TMRL**) framework for MDD detection with rs-fMRI. As illustrated in Fig. 1, we first generate three indexes (i.e., fALFF, VMHC, and DC) from pre-processed fMRI. For each index, we perform patch-wise group comparison (via

Table 1. Demographic information of studied subjects. M: Male; F: Female; Y: Yes; N: No; D: Lack of record; Mean ± Standard.

Category	Gender	Age	Education year	First period	On medication	Illness time
MDD	99M/183F	38.7 ± 13.6	10.8 ± 3.6	209(Y)/49(N) 24(D)	124(Y)/125(N) 33(D)	50.0 ± 65.9 35(D)
HC	87M/164F	39.6 ± 15.8	13.0 ± 3.9	–	–	–

t-test) between MDD and healthy controls (HCs) to select the most discriminative patches based on training images. We hypothesize that *patches at the same location for the same subject from different indexes contain similar neural activity.* Accordingly, we propose a tensor-regularized multi-task learning model (with a tensor-based regularizer) to align multi-index features into a common latent space. This helps mitigate distribution differences and capture potential relationships between indexes. We finally perform prediction using the learned new features. Experiments on 533 subjects with rs-fMRI suggest the efficacy of our method in MDD detection compared with previous state-of-the-arts.

2   Materials and Method

2.1   Subjects and Image Pre-processing

The public rs-fMRI dataset consists of 533 subjects[1], including 282 MDD and 251 age-matched healthy controls (HCs). The rs-fMRI scans were acquired by using a Siemens scanner with an echo-planar imaging sequence at the Southwest University (Table 1). The following lists the scanning parameters: repetition time (TR) = $2,000$ ms, echo time (TE) = 30 ms, flip angle = $90°$, thickness/gap = $3.0/1.0$ mm, time points = $242\,s$, field of view (FOV) = 220 mm × 220 mm, voxel size = $3.44 \times 3.44 \times 4.00$, and matrix size = $61 \times 73 \times 61$.

Each fMRI scan was basically pre-processed by using the Data Processing Assistant for Resting-State fMRI (DPARSF). In this pipeline, we first discard the first 10 time points, followed by slice timing correction, head motion correction, regression of nuisance co-variants of head motion parameters, white matter, and cerebrospinal fluid. Images are then normalized with an EPI template in the MNI space, resampling to $3 \times 3 \times 3$ mm^3 resolution, and spatial smoothing using a 6 mm full-width at half-maximum (FWHM) Gaussian kernel. This pipeline is slightly changed to generate multi-index representations (see below).

2.2   Methodology

Multi-index Generation. For each subject with rs-fMRI scan, we first extract multi-index representations, i.e., fALFF, VMHC, and DC (size: $61 \times 73 \times 61$). (1) The fALFF index is employed to depict the relationship between field potential activity and cognitive-emotional processing [9]. Fast Fourier Transform is

---

[1] http://rfmri.org/REST-meta-MDD.

used to transform the time series of each voxel to the frequency domain. And the mean square root is obtained across $0.01 - 0.1Hz$ band for each voxel. The fALFF as a ratio is standardized by the mean fALFF of the global brain for all voxels. No temporal bandpass filtering $(0.01 - 0.1Hz)$ is required to extract the fALFF index. (2) The VMHC index is used to measure the functional connection between two voxels across the contralateral hemispheres [10]. In specific, it represents the Pearson's correlation coefficient between the time series of each voxel and that of its counterpart voxel at the same location in the opposite hemisphere. (3) The DC index is a graph-based measurement of brain networks, which captures the number of instantaneous functional connections between a voxel and the rest voxels in the entire brain network. We calculate the weighted sum of positive correlations by keeping inter-voxel connectivities with Pearson's correlation coefficient greater than a threshold $r = 0.25$, leading to the DC matrix.

Patch-Based Feature Extraction. Three activation maps (i.e., indexes) are calculated at the voxel level, but some voxels may not provide informative information for MDD diagnosis. To this end, we employ a patch-based feature extraction strategy by selecting the most discriminative patches in each index. We first partition each index into multiple non-overlapping $3 \times 3 \times 3$ patches. We then perform *patch-wise group comparison* between MDD and HC populations to select the most discriminative patches in one index based on training subjects. For each location in a specific index, we vectorize the corresponding patches of MDD and HC subjects, yielding two 27-dimensional feature vectors. Each element in the feature vector denotes the intensity or degree value of a specific voxel within a patch. We then leverage the t-test algorithm to compare these two feature vectors. If a patch contains more than 9 elements whose $p \leq 0.05$, we treat it as a discriminative patch. To make features between different indexes consistent and comparable, we perform a union operation on patches selected within three indexes and obtain a total of 308 patches. The feature vector of each of three indexes is $8,316$-dimensional to represent each subject.

Tensor-Regularized Multi-task Learning. To alleviate the inter-index heterogeneity and capture the underlying relationship among indexes, we develop a tensor-regularized multi-task learning model to fuse multi-index representations. Let $\mathbf{X}^{(v)} = [\boldsymbol{x}_1^{(v)}, \boldsymbol{x}_2^{(v)}, \cdots, \boldsymbol{x}_n^{(v)}] \in \mathbb{R}^{d \times n}$ denotes the v-th $(v = 1, \cdots, V)$ index feature matrix, where each column is a d-dimensional feature vector corresponding to the v-th index and n is the number of training samples. To explore the consistent cues from various functional characteristics, we further assume that there exist V projection matrices $\{\mathbf{U}^{(v)}\}_{v=1}^{V}$ to transform multi-index representations $\{\mathbf{X}^{(v)}\}_{v=1}^{V}$ into a latent space. Therefore, an ideal latent space is expected in which discrepancy and redundancy of multi-index representations is eliminated, and the useful complementary information is easily captured.

To this end, we formulate multi-index representation fusion as a tensor-regularized multi-task learning problem, with each task corresponding to a specific index. Denote C ($C = 2$ in this work) as the number of categories and $\mathbf{Y}^{(v)} \in \mathbb{R}^{C \times n}$ as the label matrix for training samples in the v-th index space. Let $\mathbf{U}^{(v)} \in \mathbb{R}^{d \times C}$ denotes the v-th projection matrix corresponding to the v-th index/task. Denote $\| \cdot \|_F$ as the Frobenius norm of a matrix. Our tensor-regularized multi-task learning model is formulated as:

$$\min_{\{\mathbf{U}^{(v)}\}_{v=1}^{V}} \sum_{v=1}^{V} \|\mathbf{U}^{(v)\top} \mathbf{X}^{(v)} - \mathbf{Y}^{(v)}\|_F^2 + \alpha \|\mathcal{U}\|_\circledast + \beta \sum_{1 \leq i < j \leq V} \|\mathbf{U}^{(i)} - \mathbf{U}^{(j)}\|_F^2, \quad (1)$$

where the 1st term is the empirical loss on training samples, the 2nd term is a tensor-based regularizer, while the last term is the Bregman divergence [15].

In Eq. (1), $\mathcal{U} = \Phi(\mathbf{U}^{(1)}, \mathbf{U}^{(2)}, \cdots, \mathbf{U}^{(V)}) \in \mathbb{R}^{d \times C \times V}$ is a tensor by merging different $\mathbf{U}^{(v)}$ to a 3-order tensor along the third dimension, where each frontal slice of $\mathcal{U}$ is our task-specific projection matrix (i.e., $\mathcal{U}(:,:,v) = \mathbf{U}^{(v)}$). The high-order tensor low-rank norm $\|\mathcal{U}\|_\circledast$ measures the rank of a block circulant matrix constructed by all projection matrices, where $\| \cdot \|_\circledast$ is the tensor nuclear norm [16]. By comparing each column and each row of frontal slices (i.e., $\mathbf{U}^{(v)}$) via a low-rank constraint, this tensor-based regularizer helps explore shared and complementary information among multi-index representations. Also, all task-specific classifiers are correlated via this regularizer, so that knowledge can be flexibly transferred between tasks/indexes. In addition, the Bregman divergence is used here to encourage that the discrepancy between multiple transformations should be small, which helps reduce the heterogeneity among different indexes. That is, with the last two terms in Eq. (1), the proposed model helps mitigate distribution differences and capture the underlying relationship between indexes.

Due to the use of tensor nuclear norm and Bregman divergence regularizers, the objective function defined in Eq. (1) is difficult to optimize. We develop a unique optimization algorithm to solve this problem. To facilitate optimization, we first construct two block matrices as follows:

$$\tilde{\mathbf{U}} = [\tilde{\mathbf{U}}^{(1)}; \tilde{\mathbf{U}}^{(2)}; \cdots; \tilde{\mathbf{U}}^{(V)}], \quad \tilde{\mathbf{Y}} = [\mathbf{Y}^{(1)}, \mathbf{Y}^{(2)}, \cdots, \mathbf{Y}^{(V)}]. \quad (2)$$

Denote $\tilde{\mathbf{X}}$ as a diagonal matrix, where its v-th diagonal element is $\mathbf{X}^{(v)}$ and others are 0, and $\mathbf{I} \in \mathbb{R}^{d \times d}$ as an identity matrix. We also define a matrix $\mathbf{M}$ as:

$$\mathbf{M} = \begin{bmatrix} (V-1)\mathbf{I} & -\mathbf{I} & \cdots & -\mathbf{I} \\ -\mathbf{I} & (V-1)\mathbf{I} & \cdots & -\mathbf{I} \\ \vdots & \vdots & \ddots & \vdots \\ -\mathbf{I} & -\mathbf{I} & \cdots & (V-1)\mathbf{I} \end{bmatrix}. \quad (3)$$

Then, Eq. (1) can be rewritten as follows:

$$\min_{\tilde{\mathbf{U}}} \|\tilde{\mathbf{U}}^\top \tilde{\mathbf{X}} - \tilde{\mathbf{Y}}\|_F^2 + \alpha \|\mathcal{U}\|_\circledast + \beta tr(\tilde{\mathbf{U}}^\top \mathbf{M} \tilde{\mathbf{U}}), \quad (4)$$

which can be solved by using the Augmented Lagrange Multiplier (ALM) [17]. Specifically, by introducing an auxiliary tensor variable $\mathcal{G}$, the problem in Eq. (3) can be solved by minimizing the following problem:

$$\min_{\tilde{\mathbf{U}}, \mathcal{G}} \|\tilde{\mathbf{U}}^\top \tilde{\mathbf{X}} - \tilde{\mathbf{Y}}\|_F^2 + \alpha \|\mathcal{G}\|_\circledast + \beta tr(\tilde{\mathbf{U}}^\top \mathbf{M} \tilde{\mathbf{U}}) + \langle \mathcal{W}, \mathcal{U} - \mathcal{G} \rangle + \frac{\rho}{2} \|\mathcal{U} - \mathcal{G}\|_F^2, \quad (5)$$

where the tensor $\mathcal{W}$ is the Lagrange multiplier and ρ is the penalty parameter. We employ an alternating scheme to optimize Eq. (5), with details given below.

   (1) Ũ-subproblem: When the tensor $\mathcal{G}$ is fixed, since $\mathbf{G}^{(v)} = \varPhi_v^{-1}(\mathcal{G})$ and $\mathbf{W}^{(v)} = \varPhi_v^{-1}(\mathcal{W})$, where $\varPhi_v^{-1}$ is the inverse operation w.r.t $\varPhi$ by clipping v-th frontal slice of the tensor, the optimization task can be rewritten as:

$$\min_{\tilde{\mathbf{U}}} ||\tilde{\mathbf{U}}^\top \tilde{\mathbf{X}} - \tilde{\mathbf{Y}}||_F^2 + \beta tr(\tilde{\mathbf{U}}^\top \mathbf{M}\tilde{\mathbf{U}}) + <\tilde{\mathbf{W}}, \tilde{\mathbf{U}} - \tilde{\mathbf{G}}> + \frac{\rho}{2}||\tilde{\mathbf{U}} - \tilde{\mathbf{G}}||_F^2, \quad (6)$$

where $\tilde{\mathbf{W}}$ and $\tilde{\mathbf{G}}$ are the block matrices constructed by $[\mathbf{W}^{(1)}; \mathbf{W}^{(2)}; \cdots ; \mathbf{W}^{(V)}]$ and $[\mathbf{G}^{(1)}; \mathbf{G}^{(2)}; \cdots ; \mathbf{G}^{(V)}]$, respectively.

   (2) $\mathcal{G}$-subproblem: When $\tilde{\mathbf{U}}$ is fixed, the objective function in Eq. (5) can be solved through Theorem 1 in the *Supplementary Materials*.

   The above two steps are alternately repeated until the convergence condition is satisfied. Based on [18], we prove the convergence of our optimization algorithm through Theorem 2 in the *Supplementary Materials*. The code will be freely released to the public via GitHub.

Prediction with Metric Learning. With the learned $\mathbf{U}$, we can calculate the distance between two subjects $\mathbf{x}_i$ and $\mathbf{x}_j$ with their new representations as:

$$d(\mathbf{x}_i, \mathbf{x}_j) = \sqrt{(\mathbf{x}_i - \mathbf{x}_j)\mathbf{M}(\mathbf{x}_i - \mathbf{x}_j)^\top} = \left\|\mathbf{U}^\top \mathbf{x}_i - \mathbf{U}^\top \mathbf{x}_j\right\|_2, \quad (7)$$

where $\mathbf{M} = \mathbf{U}^\top \mathbf{U}$. Given a new unseen test subject $\mathbf{z}$ with rs-fMRI, we first extract its multi-index representations and generate the new feature vector via $\hat{\mathbf{z}} = \mathbf{U}^\top \mathbf{z}$. A metric learning method [19] is used for prediction at the test stage. We randomly select m MDD and m HC subjects from training data and calculate the average distance from the test subject to the m sample within each group (i.e., MDD or HC). We set $m = 5$ to avoid the bias caused by random selection. The class label of the group with the smaller distance will be assigned to $\mathbf{z}$.

3   Experiment and Results

Experimental Settings. We evaluate the effectiveness of the proposed model using a 5-fold cross-validation (CV) strategy. To avoid bias introduced by a random partition, this 5-fold CV procedure is repeated 5 times. The performance of MDD identification from age-matched HCs is measured by four metrics, including accuracy (ACC), sensitivity (SEN), specificity (SPE), and F1-score (F1).

Competing Methods. We compare the proposed TMRL with two traditional methods, i.e., (1) **Baseline** that concatenates all features from different views into one vector, (2) **MKLpy** [20] with a multi-kernel learning technique, as well as three state-of-the-art multi-view learning methods, i.e., (3) **SNMF** [21] that uses a shallow non-negative matrix factorization method, (4) **DMF** [22] that employs a deep matrix factorization technique, and (5) **McDR** [23] that uses a multi-view feature reduction method. The parameters α and β in our TMRL model are chosen from $\{0.01, 0.02, \cdots, 1\}$ via an inner cross-validation strategy based on only the training data. The parameters for five competing methods are set according to the original papers, and they use a linear support vector machine (SVM) as the classifier to detect MDD patients from HC subjects. In these methods, the parameter C for SVM is selected from the range of $\{0.01, 0.05, 0.1, 0.15, \cdots, 10\}$ via inner cross-validation on the training samples.

Table 2. Classification results of different methods in MDD detection.

Method	ACC	SEN	SPE	F1
SVM	0.561 ± 0.012	0.599 ± 0.018	0.518 ± 0.021	0.591 ± 0.014
MKLpy	0.585 ± 0.026	0.589 ± 0.023	0.579 ± 0.019	0.606 ± 0.029
NMF	0.579 ± 0.006	0.596 ± 0.014	0.559 ± 0.016	0.602 ± 0.011
DMF	0.588 ± 0.033	0.614 ± 0.027	0.562 ± 0.039	0.614 ± 0.020
McDR	0.594 ± 0.021	0.621 ± 0.015	0.577 ± 0.022	0.626 ± 0.019
TMRL (Ours)	**0.642 ± 0.027**	**0.643 ± 0.013**	**0.639 ± 0.016**	**0.654 ± 0.028**

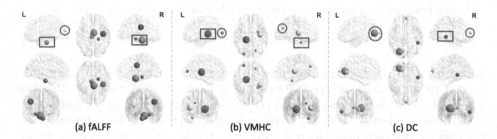

(a) fALFF   (b) VMHC   (c) DC

Fig. 2. Top 5 discriminative patches identified by our TMRL in three indexes. (Color figure online)

Results of MDD Detection. We first report the results of six different methods in MDD vs. HC classification in Table 2. It can be seen from this table that the proposed TMRL produces the best performance in MDD detection among the six methods. Compared with five competing methods, the TMRL achieves at least 4% improvement in terms of ACC and SPE values and 2% improvement in terms of SEN and F1 metrics. This implies that our method is able to learn effective representations from three indexes and boost the detection performance. Besides, compared with the simple feature concatenation method (i.e., SVM), the multi-kernel method (i.e., MKLpy) and the multi-view learning methods (i.e., NMF, DMF, and McDR) generally yield better results. This further validates the necessity of modeling the potential relationships of multiple indexes (as we do in this work) in order to improve the detection performance of MDD.

Visualization of Discriminating Regions. To display the most discriminative features identified by the proposed TMRL, we visually show the top 5 clusters of our selected patches within each of three indexes in Fig. 2. Since many patches (size: $3 \times 3 \times 3$ without overlap) are very close, we show their clusters here to facilitate visualization. Specifically, if the distance between two patches along a single axis is less than 9, we treat them as the same cluster.

As can be observed from Fig. 2, the temporal regions (marked by blue rectangles) are simultaneously identified by TMRL in the three indexes. Similarly, the occipital lobe (marked by red circles) also appears in three indexes and is regarded as the discriminative region by the proposed TMRL. This is consistent with previous studies that both temporal and occipital lobe regions are associated with MDD [24–26]. These visual results validate the reliability of TMRL in identifying MDD-affected brain regions with multi-index representations.

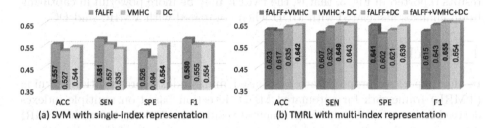

Fig. 3. Results of SVM with single-index representation and our TMRL that uses different combinations of three indexes in MDD vs. HC classification.

Table 3. Comparison with state-of-the-art methods for rs-fMRI based MDD detection.

Method	Model	Index	ACC (%)	SEN (%)	SPE (%)	F1 (%)
Gu *et al.* [27]	Kendall+SVM	DC	0.533	0.578	0.482	0.567
Li *et al.* [28]	SVM	DC	0.544	0.535	0.554	0.554
Jie *et al.* [29]	SVM-FoBa	fALFF	0.561	0.622	0.491	0.605
Guo *et al.* [30]	sROC	VMHC	0.509	0.553	0.462	0.544
Ours	TMRL	DC+fALLF+VMHC	**0.642**	**0.643**	**0.639**	**0.654**

Multi-Index Vs. Single-Index Representation. We further compare the results achieved by SVM with single-index representation and our TMRL with different combinations of three indexes, with results shown in Fig. 3. Figure 3 (a) suggests that compared with the VMHC and DC indexes, the fALFF with SVM shows the best performance. It can be seen from Fig. 3 (b) that our TMRL using multi-index representation (regardless of two or three indexes) consistently outperforms SVM that uses only one index in terms of four metrics. This implies that multi-index representation may provide complementary information for MDD detection, and using only one index cannot produce good performance.

Comparison with State-of-the-Arts. Even though many studies propose to use different indexes to study the brain changes associated with MDD [27–32], only a few studies directly employ these indexes to distinguish MDD from HCs. In Table 3, we briefly summarize several state-of-the-art (SOTA) studies for MDD vs. HC classification based on three indexes derived from rs-fMRI data. For a fair comparison, we reproduced their algorithms, performed experiments using the same data set as this work, and reported the classification results.

As can be seen from Table 3, the proposed TMRL yields the best performance in terms of four evaluation metrics, compared with four SOTA methods. This may be due to the use of multi-index representation in TMRL, while four SOTA methods use only a single index. On the other hand, the SVM-FoBa model proposed in [29] with the fALFF index produces the overall better performance (e.g., with higher ACC, SEN and F1 values), compared with the other three SOTA methods that use the DC or VMHC index. This is consistent with the results reported in Fig. 3, that is, the fALFF may be more powerful in capturing brain changes associated with MDD when compared with VMHC and DC.

4   Conclusion

In this paper, we proposed a tensor-based multi-index representation learning (TMRL) framework for automated MDD detection based on multiple indexes derived from rs-fMRI data. Experimental results on 533 subjects with rs-fMRI demonstrate that our method is superior to several state-of-the-art methods in MDD detection. In the current work, we only employ a multi-index representation derived from rs-fMRI for MDD detection. Considering the potential complementary property of different imaging techniques, we think it is interesting to explore multi-modality data such as DTI and structural MRI to further study the neurobiological mechanisms of MDD, which will be our future work.

Acknowledgements. This work was finished when D. Yao was visiting the University of North Carolina at Chapel Hill. D. Yao and M. Liu was partly supported by NIH grant (No. AG041721). Z. Zhang was partly supported by the National Key Research and Development Program of China (No. 2016YFD0700100).

References

1. Otte, C., et al.: Major depressive disorder. Nat. Rev. Dis. Primers **2**(1), 1–20 (2016)
2. Gray, J.P., Müller, V.I., Eickhoff, S.B., Fox, P.T.: Multimodal abnormalities of brain structure and function in major depressive disorder: a meta-analysis of neuroimaging studies. Am. J. Psychiatry **177**(5), 422–434 (2020)
3. Yao, D., et al.: A mutual multi-scale triplet graph convolutional network for classification of brain disorders using functional or structural connectivity. IEEE Trans. Med. Imaging **40**(4), 1279–1289 (2021)
4. Wang, M., Lian, C., Yao, D., Zhang, D., Liu, M., Shen, D.: Spatial-temporal dependency modeling and network hub detection for functional MRI analysis via convolutional-recurrent network. IEEE Trans. Biomed. Eng. **67**(8), 2241–2252 (2019)

5. Guo, X., et al.: Shared and distinct resting functional connectivity in children and adults with attention-deficit/hyperactivity disorder. Transl. Psychiatry **10**(1), 1–12 (2020)
6. Zhi, D., et al.: Aberrant dynamic functional network connectivity and graph properties in major depressive disorder. Front. Psychiatry **9**, 339 (2018)
7. Wang, M., Zhang, D., Huang, J., Yap, P.T., Shen, D., Liu, M.: Identifying autism spectrum disorder with multi-site fMRI via low-rank domain adaptation. IEEE Trans. Med. Imaging **39**(3), 644–655 (2019)
8. Yao, D., Sui, J., Yang, E., Yap, P.-T., Shen, D., Liu, M.: Temporal-adaptive graph convolutional network for automated identification of major depressive disorder using resting-state fMRI. In: Liu, M., Yan, P., Lian, C., Cao, X. (eds.) MLMI 2020. LNCS, vol. 12436, pp. 1–10. Springer, Cham (2020). https://doi.org/10.1007/978-3-030-59861-7_1
9. Zou, Q.H., et al.: An improved approach to detection of amplitude of low-frequency fluctuation (ALFF) for resting-state fMRI: Fractional ALFF. J. Neurosci. Methods **172**(1), 137–141 (2008)
10. Zuo, X.N., et al.: Growing together and growing apart: regional and sex differences in the lifespan developmental trajectories of functional homotopy. J. Neurosci. **30**(45), 15034–15043 (2010)
11. Bonacich, P.: Factoring and weighting approaches to status scores and clique identification. J. Math. Soc. **2**(1), 113–120 (1972)
12. Wei, J., et al.: Voxel-mirrored homotopic connectivity of resting-state functional magnetic resonance imaging in blepharospasm. Front. Psychol. **9**, 1620 (2018)
13. Sun, H., et al.: Regional homogeneity and functional connectivity patterns in major depressive disorder, cognitive vulnerability to depression and healthy subjects. J. Affect. Disord. **235**, 229–235 (2018)
14. Liu, W., et al.: Abnormal degree centrality of functional hubs associated with negative coping in older Chinese adults who lost their only child. Biol. Psychol. **112**, 46–55 (2015)
15. Banerjee, A., Merugu, S., Dhillon, I.S., Ghosh, J., Lafferty, J.: Clustering with Bregman divergences. J. Mach. Learn. Res. **6**(10), 1705–1749 (2005)
16. Lu, C., Feng, J., Chen, Y., Liu, W., Lin, Z., Yan, S.: Tensor robust principal component analysis with a new tensor nuclear norm. IEEE Trans. Pattern Anal. Mach. Intell. **42**(4), 925–938 (2019)
17. Lin, Z., Chen, M., Ma, Y.: The augmented lagrange multiplier method for exact recovery of corrupted low-rank matrices. arXiv preprint arXiv:1009.5055 (2010)
18. Boyd, S., Parikh, N., Chu, E.: Distributed Optimization and Statistical Learning via the Alternating Direction Method of Multipliers. Now Publishers Inc, Boston (2011)
19. Yang, L., Jin, R.: Distance metric learning: a comprehensive survey. Mich. State Univ. **2**(2), 4 (2006)
20. Lauriola, I., Aiolli, F.: MKLpy: A python-based framework for multiple kernel learning. arXiv preprint arXiv:2007.09982 (2020)
21. Févotte, C., Idier, J.: Algorithms for nonnegative matrix factorization with the β-divergence. Neural Comput. **23**(9), 2421–2456 (2011)
22. Zhao, H., Ding, Z., Fu, Y.: Multi-view clustering via deep matrix factorization. In: Proceedings of the AAAI Conference on Artificial Intelligence, vol. 31 (2017)
23. Zhang, C., Fu, H., Hu, Q., Zhu, P., Cao, X.: Flexible multi-view dimensionality co-reduction. IEEE Trans. Image Process. **26**(2), 648–659 (2016)

24. Beneyto, M., Kristiansen, L.V., Oni-Orisan, A., McCullumsmith, R.E., Meador-Woodruff, J.H.: Abnormal glutamate receptor expression in the medial temporal lobe in schizophrenia and mood disorders. Neuropsychopharmacology **32**(9), 1888–1902 (2007)
25. Caetano, S.C., et al.: Medial temporal lobe abnormalities in pediatric unipolar depression. Neurosci. Lett. **427**(3), 142–147 (2007)
26. Liao, Y., et al.: Is depression a disconnection syndrome? Meta-analysis of diffusion tensor imaging studies in patients with MDD. J. Psychiatry Neurosci. JPN **38**(1), 49 (2013)
27. Gu, L., Huang, L., Yin, F., Cheng, Y.: Classification of depressive disorder based on rs-fMRI using multivariate pattern analysis with multiple features. In: 2017 4th IAPR Asian Conference on Pattern Recognition (ACPR), pp. 61–66. IEEE (2017)
28. Li, M., et al.: Clinical utility of a short resting-state MRI scan in differentiating bipolar from unipolar depression. Acta Psychiatrica Scandinavica **136**(3), 288–299 (2017)
29. Jie, N.F., et al.: Discriminating bipolar disorder from major depression based on SVM-FoBa: efficient feature selection with multimodal brain imaging data. IEEE Trans. Auton. Ment. Dev. **7**(4), 320–331 (2015)
30. Guo, W., et al.: Decreased interhemispheric resting-state functional connectivity in first-episode, drug-naive major depressive disorder. Prog. Neuropsychopharmacol. Biol. Psychiatry **41**, 24–29 (2013)
31. Zhou, M., et al.: Intrinsic cerebral activity at resting state in adults with major depressive disorder: a meta-analysis. Prog. Neuropsychopharmacol. Biol. Psychiatry **75**, 157–164 (2017)
32. Fan, H., Yang, X., Zhang, J., Chen, Y., Li, T., Ma, X.: Analysis of voxel-mirrored homotopic connectivity in medication-free, current major depressive disorder. J. Affect. Disord. **240**, 171–176 (2018)

Region Ensemble Network for MCI Conversion Prediction with a Relation Regularized Loss

Yuan-Xing Zhao[1,2], Yan-Ming Zhang[2], Ming Song[2,3],
and Cheng-Lin Liu[1,2,4(✉)]

[1] School of Artificial Intelligence, University of Chinese Academy of Sciences,
Beijing 100149, China
{yuanxing.zhao,liucl}@nlpr.ia.ac.cn
[2] NLPR, Institute of Automation, Chinese Academy of Sciences,
Beijing 100190, China
{ymzhang,msong}@nlpr.ia.ac.cn
[3] Brainnetome Center, Institute of Automation, Chinese Academy of Sciences,
Beijing 100190, China
[4] CAS Center for Excellence of Brain Science and Intelligence Technology,
Beijing 100190, China

Abstract. Despite many recent advances, computer-aided mild cognitive impairment (MCI) conversion prediction is still a very challenging task due to: 1) the abnormal areas are subtle compared to the size of the whole brain, 2) the features' dimension is much larger than the number of samples. To tackle these problems, we propose a region ensemble model using a divide and conquer strategy to capture the disease's finer representation. Specifically, the features are independently extracted from non-overlapping regions and then fused to describe the subject according to the attention scores. Moreover, we design a novel loss that models the relationship between different stages of the disease to regularize the training process explicitly. Experiments on public data sets for MCI conversion prediction demonstrate that our method has achieved state-of-the-art performance. Specifically, the area under the receiver operating characteristic curve (AUC) is improved from 79.3% to 85.4%. Beyond that, each region's contribution can be assessed quantitatively, using the proposed method.

Keywords: Alzheimer's disease · Mild cognitive impairment · Region ensemble network · Relation regularized loss

1 Introduction

Mild cognitive impairment (MCI) is the prodromal stage of Alzheimer's disease (AD). A systematic review found that 32% of individuals with MCI would convert to AD within five years' follow-up. Hence, identifying which individuals with MCI are more likely to develop AD is a primary goal of current

© Springer Nature Switzerland AG 2021
M. de Bruijne et al. (Eds.): MICCAI 2021, LNCS 12905, pp. 185–194, 2021.
https://doi.org/10.1007/978-3-030-87240-3_18

research [1]. Before some noticeable symptoms of the disease, several subtle structural changes have already happened in the brain. As an essential computer-aided diagnosis technique, structural magnetic resonance imaging (sMRI) can non-invasively capture such changes. Therefore many machine learning or deep learning-based methods have been applied to AD diagnosis and MCI conversion prediction based on sMRI [2,3] and have reported remarkable success.

In general, all of these methods can be grouped into detection-dependent approaches and detection-free approaches, depending on whether they need a separate model to detect regions of interest (ROI) or not. Detection-dependent methods [4–9] first locate ROI based on prior domain knowledge using an independent detection model. It then constructs a diagnosis model based on the ROI's feature. These methods reduce the feature's dimension using the whole brain's sub-areas but may miss some critical regions in practice. Thus, it is hard for them to achieve high performance. To tackle this limitation, current state-of-the-art methods adopt the detection-free approach. Taking the whole brain as the input, the methods in [10–12] locate abnormal areas and predict the result simultaneously. In this way, they can extract the critical areas and discard useless regions in a data-driven and target-consistent way. While these methods are more powerful and flexible in principle, they tend to suffer from severe over-fitting due to the limited number of training samples. One way to alleviate this problem is to use auxiliary data. For example, the works [10] and [12] both pre-train an AD diagnosis model first and then fine-tune the model for the MCI conversion prediction task.

This paper proposes a region ensemble model together with a relation regularized loss for the MCI conversion prediction task. The model's core idea is to divide the brain into non-overlapping regions and learn a region-based diagnosis sub-network for each region. In this way, sub-networks overcome the curse of dimensionality by focusing on small regions of the brain. Finally, we construct an ensemble model by weighted fusion of the regional features with attention scores. Additionally, we propose a relation regularized loss based on an assumption of the disease to regularize the training process. More importantly, this loss allows our method to incorporate auxiliary samples to improve the performance.

In summary, the main contributions of this paper are three-fold. First, we propose a novel region ensemble model that uses a divide-and-conquer strategy and attention mechanism to extract the discriminative features and locate abnormal areas. Second, we propose a relation regularized loss to regularize the model's training process through additional samples. Third, on public data sets (ADNI-1 and ADNI-2), our method outperforms competing methods with a large margin and achieves the state-of-the-art performance.

2   Method

2.1   Region Ensemble Network

As shown in Fig. 1, our diagnosis model consists of three sequential components: 1) feature extraction sub-network, 2) region-based diagnosis sub-network, and

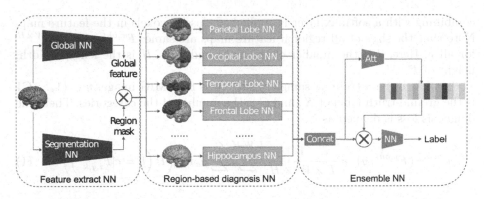

Fig. 1. The architecture of the region ensemble model. 15 region-based diagnosis subnetworks are adopted in our model.

3) ensemble sub-network. We first extract the global feature map from the whole brain through the feature extraction sub-network. This feature map then voxelwisely multiplies with the region masks, produced by a segmentation model, to create a raw feature map for each brain region. After that, region-based diagnosis sub-networks generate discriminative features from the regional feature maps. Finally, the ensemble sub-network fuses the regional representations with an attention mechanism to produce the final representation and make a classification.

In this framework, the segmentation network is trained separately using the dataset in [13], while the other parts are trained in an end-to-end manner.

Feature Extract Sub-Network. In this stage, a sub-network is used to extract a raw feature map for each region-based diagnosis sub-network. Concretely, we first extract the whole brain's feature map F^g using three feature extraction blocks, the first and the third blocks followed by a max-pooling layer. A block is composed by stacking a convolution layer, a batch normalization (BN) [14], a parametric rectified linear units (PReLU) [15], and a convolutional block attention module (CBAM) [16]. Meanwhile, we perform 3D whole brain segmentation using an auxiliary segmentation network and get one region mask M^r for each region r. After that, the input feature map $F^{r,in}$ for the r-th region diagnosis sub-network is calculated by $F^{r,in} = M^r \otimes F^g$, where $\otimes$ denote an element-wise multiplication.

Region-Based Diagnosis Sub-Network. For each region, we use an independent diagnosis sub-network to extract the discriminative features from $F^{r,in}$. We first adopt a convolution layer followed by a max-pooling layer to reduce the feature map scale. Next, 14 feature extraction blocks, each composed by stacking a convolution layer, a BN, a PReLU, and a CBAM, are used to obtain the final regional feature map $F^{r,out}$. Finally, a convolution layer with kernel size $1 \times 1 \times 1$

combining with a softmax layer is applied to classify voxels in the feature map. Note that the sizes of all regional feature maps are same: $\boldsymbol{F}^{r,out} \in \mathbb{R}^{d \times L \times W \times H}$ for all r. Here, d is the number of channels, and L, W, H is the length, width, height of $\boldsymbol{F}^{r,out}$.

Let $(\boldsymbol{X}, y)$ be a training sample. Here, $\boldsymbol{X}$ is the sMRI image, $y \in \{1, ..., C\}$ is the ground-truth label of $\boldsymbol{X}$, and C is the number of the categories. Then, the diagnosis loss is defined as

$$L^{voxel}\left(\boldsymbol{F}^{r,out}, y\right) = \frac{1}{L \times W \times H} \sum_{i,j,k=1}^{L,W,H} \sum_{c=1}^{C} y_c \log\left(P\left(\hat{y} = c | F_{i,j,k}^{r,out}\right)\right). \quad (1)$$

y_c is the binary indicator of the ground-truth label, which equals to 1 if $\boldsymbol{X}$ belong to class c and 0 otherwise. $P\left(\hat{y} = c | F_{i,j,k}^{r,out}\right)$ is the predicted probability for class c of voxel (i, j, k). It is noted that different from the common-used strategy, which first performs a global pooling to reduce $\boldsymbol{F}^{r,out}$ to a feature vector and then optimizes the loss defined on that vector, we optimize the loss defined on each voxel to prevent losing critical details.

Ensemble Sub-Network. We design an ensemble sub-network to automatically identify discriminative regions in the whole brain and perform classification. The structure of the ensemble sub-network is shown in Fig. 1. It includes two parts: an attention module and a classifier.

Because voxels have different discriminative abilities, we assign an attention score to each region's voxel independently. For simplicity, we only introduce the computation of attention score for one voxel. Given voxel (i, j, k)'s feature $\boldsymbol{f}^r = \boldsymbol{F}_{i,j,k}^{r,out} \in \mathbb{R}^d$ generated by the r-th region-based diagnosis sub-network, the attention module first transforms $\boldsymbol{f}^r$ into a scalar $f^r \in \mathbb{R}$ by

$$f^r = \delta\left(\boldsymbol{W}_2^r \delta\left(\boldsymbol{W}_1^r \boldsymbol{f}^r\right)\right), \quad (2)$$

where δ refers to the PReLU function, $\boldsymbol{W}_1^r \in \mathbb{R}^{3d \times d}$ and $\boldsymbol{W}_2^r \in \mathbb{R}^{1 \times 3d}$ are learnable parameters. To consider the relationship among regions, we combine the values at the same location (i, j, k) in different regional feature maps into $\boldsymbol{f} = [f^1, f^2, ..., f^R]$ and get the final attention score by

$$\boldsymbol{a} = \sigma(\boldsymbol{W}_2^a \delta(\boldsymbol{W}_1^a \boldsymbol{f})) \in \mathbb{R}^R, \quad (3)$$

where σ refers to the sigmoid function, $\boldsymbol{W}_1^a \in \mathbb{R}^{\frac{R}{3} \times R}$, and $\boldsymbol{W}_2^a \in \mathbb{R}^{R \times \frac{R}{3}}$. R is the number of regions. After computing the scores for each voxel, we can get the attention score $\boldsymbol{A}_r \in \mathbb{R}^{L \times W \times H}$ for each regional feature map $\boldsymbol{F}^{r,out}$ and then the ensemble feature map is computed as

$$\boldsymbol{F}^e = \sum_{r=1}^{R} \boldsymbol{F}^{r,out} \otimes \boldsymbol{A}_r. \quad (4)$$

Finally, a convolution layer with kernel size $1 \times 1 \times 1$ combining with a softmax layer is applied to $\boldsymbol{F}^e$ to classify all voxels. The probability of a given sMRI $\boldsymbol{X}$ to be predicted as class c is calculated by

$$P(\hat{y} = c | \boldsymbol{X}) = \frac{1}{L \times W \times H} \sum_{i,j,k=1}^{L,W,H} P\left(\hat{y} = c | F_{i,j,k}^e\right). \tag{5}$$

2.2 Relation Regularized Loss

The main challenge in MCI conversion prediction is the lack of training samples. To alleviate the problem, works [10] and [12] use AD and normal controls (NC) samples to pre-train a model and then fine-tune it on stable MCI (sMCI) and progressive MCI (pMCI) samples. In this work, we utilize AD/NC samples more sophisticatedly by introducing a novel ranking loss.

NC/sMCI/pMCI/AD labels are intrinsically ordered because MCI is a pro-dromal stage of AD, and its structural changes are between AD and NC [17]. Hence, we make an assumption as follows. We defined $P^c = P(\hat{y} = c | \boldsymbol{X})$ as the predicted probability of $\boldsymbol{X}$ belonging to class c, For a training sample $(\boldsymbol{X}, y)$,

1) if $y = \text{NC}$, we have $P^{\text{NC}} > P^{\text{sMCI}} > P^{\text{pMCI}} > P^{\text{AD}}$.
2) if $y = \text{sMCI}$, we have $P^{\text{sMCI}} > P^{\text{NC}}$ and $P^{\text{sMCI}} > P^{\text{pMCI}} > P^{\text{AD}}$.
3) if $y = \text{pMCI}$, we have $P^{\text{pMCI}} > P^{\text{AD}}$ and $P^{\text{pMCI}} > P^{\text{sMCI}} > P^{\text{NC}}$.
4) if $y = \text{AD}$, we have $P^{\text{AD}} > P^{\text{pMCI}} > P^{\text{sMCI}} > P^{\text{NC}}$.

In order to enforce such relationship, we define a ranking loss as

$$L^{rank}\left(P^{c_1}, P^{c_2}, z\right) = z \exp\left(-z\left(P^{c_1} - P^{c_2}\right)\right), \tag{6}$$

where $z \in \{0, 1\}$. For $z = 0$, L^{rank} equals to 0 and plays no role in learning. For $z = 1$, minimizing L^{rank} constrains the model to obey the relation $P^{c_1} > P^{c_2}$. It is noted that we only optimize the relation of $P^{c_1} > P^{c_2}$, because $P^{c_1} > P^{c_2}$ and $P^{c_1} < P^{c_2}$ are equivalent. To represent the pairwise relation between the predicted probabilities, a difference matrix $\boldsymbol{D} \in \mathbb{R}^{4 \times 4}$ is defined as $D_{ij} = P^{c_i} - P^{c_j}$. For each label c, a relation matrix $\boldsymbol{Z}^c \in \{0, 1\}^{4 \times 4}$ is defined as $Z_{ij}^c = 1$ for class c, $P^{c_i} > P^{c_j}$, according to relations explained above. Specifically, The rows and columns of the matrix are set in order of NC, sMCI, pMCI, and AD, (e.g. $Z^{NC}[2, 3]$ denotes the relation between the p^{sMCI} and p^{pMCI}) we have the following $\boldsymbol{Z}^c$

$$\boldsymbol{Z}^{\text{NC}} = \begin{bmatrix} 0 & 1 & 1 & 1 \\ 0 & 0 & 1 & 1 \\ 0 & 0 & 0 & 1 \\ 0 & 0 & 0 & 0 \end{bmatrix}, \boldsymbol{Z}^{\text{sMCI}} = \begin{bmatrix} 0 & 0 & 0 & 0 \\ 1 & 0 & 1 & 1 \\ 0 & 0 & 0 & 1 \\ 0 & 0 & 0 & 0 \end{bmatrix}, \boldsymbol{Z}^{\text{pMCI}} = \begin{bmatrix} 0 & 0 & 0 & 0 \\ 1 & 0 & 0 & 0 \\ 1 & 1 & 0 & 1 \\ 0 & 0 & 0 & 0 \end{bmatrix}, \boldsymbol{Z}^{\text{AD}} = \begin{bmatrix} 0 & 0 & 0 & 0 \\ 1 & 0 & 0 & 0 \\ 1 & 1 & 0 & 0 \\ 1 & 1 & 1 & 0 \end{bmatrix}.$$

Finally, the relation regularized loss can be defined as

$$L^{rank}(\boldsymbol{D}, \boldsymbol{Z}^c) = \frac{1}{nonzero(\boldsymbol{Z}^c)} \sum_{i,j=1}^{4} Z_{ij}^c \exp\left(-Z_{ij}^c D_{ij}\right), \tag{7}$$

where $nonzero(Y)$ is the number of non-zero elements in Y.

The overall loss optimized in our method is defined as

$$Loss(\boldsymbol{X}, y) = L^{voxel}(\boldsymbol{F}^e, y) + \frac{\lambda_1}{R} \sum_{r=1}^{R} L^{voxel}(\boldsymbol{F}^{r,out}, y) + \lambda_2 L^{rank}(\boldsymbol{D}, \boldsymbol{Z}^y), \quad (8)$$

where λ_1, λ_2 are hyperparameters to control the influences of L^{rank} and L^{voxel}.

3   Experiments

3.1   Dataset and Evaluation Metrics

We perform experiments on the public Alzheimer's Disease Neuroimaging Initiative (ADNI) dataset [18]. Following [9–12], in all experiments, we treat ADNI-1 as the training set and leave ADNI-2 for testing to make an easier comparison. The training and testing set contains 226 sMCI vs. 167 pMCI and 239 sMCI vs. 38 pMCI, respectively. We also collect 199 AD and 229 NC samples in ADNI-1 as the additional samples to optimize the proposed relation regularized loss. Diagnostic performance is assessed using four metrics: classification accuracy (ACC), sensitivity (SEN), specificity (SPE), and AUC.

3.2   Implementation Details

Since our method needs voxel-level annotation to extract brain regions, we use the dataset in [13] and the method in [19] to train a segmentation model which segments the whole brain into 134 regions. Then, we apply this model to sMRI images in ADNI and obtain the initial region annotations. Since small regions may result in unstable results, we merge the 134 regions into 15 according to anatomy knowledge. The resulting regions are shown in Fig. 3.

sMRI images are processed following a standard pipeline. Specifically, we use the segmentation model mentioned above to simultaneously strip the skull and split the brain region. All subjects were aligned by affine registration to Colin27 template [20] to remove the global linear difference. After that, voxels were resampled to an identical spatial resolution ($1 \times 1 \times 1\,\mathrm{mm}^3$), using SimpleITK [21]. To handle sMRI with different sizes, we crop or pad them (for large or small sMRI) into $160 \times 196 \times 152$ for both the training and testing phases. In the feature extract sub-network, the channel in the first building block is 16, and is increased by 16 after each block. In the region-based diagnosis sub-network, the channel in the first block is 24, and is increased by 14 after each block.

Stochastic gradient descent with momentum is used as the optimizer, and the learning rate is set to 0.05 initially and is decreased during the training process. The dropout is set to 0 initially and is increased to 0.1 until the 25th epochs. The batch size is 4 for each GPU. The method is implemented by PyTorch [22], and all experiments are conducted with two TITAN GPUs with 12 GB RAM.

3.3 Ablation Studies

To better understand our method, we conduct ablation experiments to examine how each proposed component affects performance. 1) The baseline model adopts neither region partition nor relation regularized loss and is trained with traditional cross-entropy loss. For a fair comparison, we expand the baseline model's channels so that the model's size is nearly the same as the proposed model. 2) The region ensemble model has 15 region diagnosis sub-networks based on the non-overlapped brain regions. 3) The baseline model is trained by the proposed relation regularized loss. 4) The region ensemble model is trained by the proposed relation regularized loss.

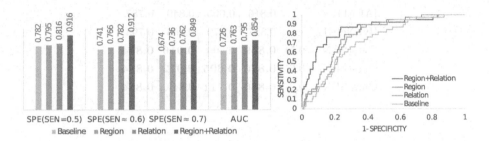

Fig. 2. Contribution of the proposed components in MCI conversion prediction

The results are shown in Fig. 2. The four models are denoted as Baseline, Region, Relation, and Region+Relation, separately. We have the following observations. First, our method consistently improves with each component's addition. Second, the region ensemble model outperforms the whole brain classification model. It implies that the region ensemble model can capture more helpful information. Third, the model trained with the relation regularized loss is more accurate than the baseline model and the region ensemble model. It indicates that using more training samples and exploiting the task's intrinsic structure are the critical factors for obtaining high performance.

3.4 Comparing with SOTA Methods

We compare our method with several approaches for MCI conversion prediction. We trained the model on ADNI-1 (using 10% of subjects for validation) and then used it to diagnose the subjects from ADNI-2. The classification results are summarized in Tab. 1. The results of compared methods are referred from [9–12]. All methods have used the same testing data. Furthermore, works in [10] and [12] also use the same AD and NC sample as ours to pre-train the model. For a more thorough and comprehensive evaluation, we compared the model's performance at different SEN values. From Table 1, we can see that our approach yields better results, demonstrating the advantage of our proposed strategies, i.e., the region ensemble network and relation regularized loss. Due to the limitations of sample quantity, we cannot get the SEN's value at 0.6 and 0.7 exactly, hence, we set the SEN near 0.6 and 0.7. Additionally, we also trained and tested on the ADNI-1, using 5-folder cross-validation, obtaining a AUC of 0.82.

Table 1. Results for MCI conversion prediction on ADNI-2.

Methods	ACC	SEN	SPE	AUC
ROI [4]	0.661	0.474	0.690	0.638
VBM [5]	0.643	0.368	0.686	0.593
DMIL [9]	0.769	0.421	0.824	0.776
H-FCN [10]	0.809	0.526	0.854	0.781
IAF [11]	0.816	0.605	0.849	0.787
HybNet [12]	0.827	0.579	0.866	0.793
Ours(SEN = 0.5)	**0.859**	0.500	**0.916**	**0.854**
Ours(SEN ≈ 0.6)	**0.870**	**0.605**	**0.912**	**0.854**
Ours(SEN ≈ 0.7)	**0.830**	**0.711**	0.849	**0.854**

3.5   Interpreting the Model's Prediction

We try to provide insights into the MCI conversion prediction problem by analyzing our model's intermediate result. First, we examine how different brain

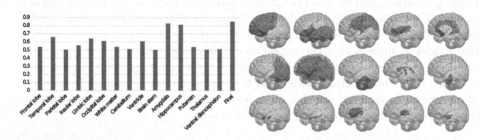

Fig. 3. The AUC of each brain region on ADNI-2 and the brain regions used to train the region diagnosis sub-networks, with the same order as the bins in the histogram, from left to right, top to bottom.

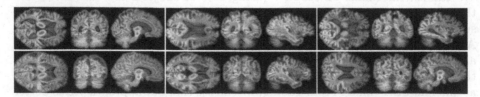

Fig. 4. Attention maps of sMCI subjects (top) and pMCI subjects (bottom).

regions are related to sMCI vs. pMCI classification. The AUC values for 15 region-based diagnosis sub-networks are shown in Fig. 3. We can see that Hippocampus and Amygdala are much more informative than other regions and can get similar AUC results as the ensemble model. On the other hand, Cerebellum, Parietal lobe, Thalamus, Ventral diencephalon, and Brain stem seemed valueless for the prediction.

In Fig. 4 we multiply each region mask M^r with the ensemble model's attention score A_r as each region's weight and visualize the weights at the individual level. As shown in the figure, our model can localize different subjects' abnormalities, which is valuable in clinical diagnosis.

4   Conclusion

MCI conversion prediction is a fundamental problem in the computer-aided diagnosis of Alzheimer's disease. This paper introduces a region ensemble model to predict the disease and identify the disease's critical brain regions. Additionally, we propose a relation regularized loss using the disease's intrinsic structure and AD/NC samples. Extensive experiments on public datasets show the superiority of our method. However, the critical brain region assessed by our method is relatively coarse because of the limitation of the GPU memory. In the future, we will investigate methods for evaluating the more delicate brain regions.

Acknowledgments. This work has been supported by the National Key Research and Development Program Grant 2018AAA0100400, the National Natural Science Foundation of China (NSFC) grants 61773376, 61836014, 61721004 and 31870984.

References

1. Association, A., et al.: 2020 Alzheimer's disease facts and figures. Alzheimer's Dement. **16**, 391–460 (2020)
2. Rathore, S., Habes, M., Iftikhar, M.A., Shacklett, A., Davatzikos, C.: A review on neuroimaging-based classification studies and associated feature extraction methods for alzheimer's disease and its prodromal stages. Neuroimage **155**, 530 (2017)
3. Leandrou, S., Petroudi, S., Reyes-Aldasoro, C.C., Kyriacou, P.A., Pattichis, C.S.: Quantitative MRI brain studies in mild cognitive impairment and alzheimer's disease: a methodological review. IEEE Rev. Biomed. Eng. **11**, 97–111 (2018)
4. Zhang, D., Wang, Y., Zhou, L., Yuan, H., Shen, D.: Multimodal classification of Alzheimer's disease and mild cognitive impairment. Neuroimage **55**, 856–867 (2011)
5. Ashburner, J., Friston, K.J.: Voxel-based morphometry–the methods. Neuroimage **11**(6), 805–821 (2000)
6. Zhang, J., Gao, Y., Gao, Y., Munsell, B.C., Shen, D.: Detecting anatomical landmarks for fast alzheimer's disease diagnosis. IEEE Trans. Med. Imaging **35**, 2524–2533 (2016)
7. Lei, B., Yang, P., Wang, T., Chen, S., Ni, D.: Relational-regularized discriminative sparse learning for alzheimer's disease diagnosis. IEEE Trans. Cybern. **47**, 1102–1113 (2017)

8. Cheng, B., Liu, M., Zhang, D., Shen, D., Initiative, A.D.N., et al.: Robust multi-label transfer feature learning for early diagnosis of alzheimer's disease. Brain Imaging Behav. **13**, 138–153 (2019)
9. Liu, M., Zhang, J., Adeli, E., Shen, D.: Landmark-based deep multi-instance learning for brain disease diagnosis. Med. Image Anal. **43**, 157–168 (2018)
10. Lian, C., Liu, M., Zhang, J., Shen, D.: Hierarchical fully convolutional network for joint atrophy localization and alzheimer's disease diagnosis using structural MRI. IEEE Trans. Pattern Anal. Mach. Intell. **42**, 880–893 (2018)
11. Li, Q., et al.: Novel iterative attention focusing strategy for joint pathology localization and prediction of mci progression. In: International Conference on Medical Image Computing and Computer-Assisted Intervention, pp. 307–315 (2019)
12. Lian, C., Liu, M., Pan, Y., Shen, D.: Attention-guided hybrid network for dementia diagnosis with structural mr images. IEEE Trans. Cybern. 1–12 (2020, early access)
13. Landman, B., Warfield, S.: Miccai 2012 workshop on multi-atlas labeling. In: Medical Image Computing and Computer Assisted Intervention Conference (2012)
14. Ioffe, S., Szegedy, C.: Batch normalization: accelerating deep network training by reducing internal covariate shift. In: International Conference on Machine Learning, pp. 448–456 (2015)
15. He, K., Zhang, X., Ren, S., Sun, J.: Delving deep into rectifiers: Surpassing human-level performance on imagenet classification. In: Proceedings of the IEEE International Conference on Computer Vision, pp. 1026–1034 (2015)
16. Woo, S., Park, J., Lee, J.-Y., So Kweon, I.: Cbam: Convolutional block attention module. In: Proceedings of the European Conference on Computer Vision (ECCV), pp. 3–19 (2018)
17. Coupé, P., Manjón, J.V., Lanuza, E., Catheline, G.: Lifespan changes of the human brain in alzheimer's disease. Sci. Rep. **9**(1), 1–12 (2019)
18. Jack, C.R., Bernstein, M.A., Fox, N.C., Thompson, P., Weiner, M.W.: The alzheimer's disease neuroimaging initiative (adni): Mri methods. J. Magn. Reson. Imaging **27**, 685–691 (2010). http://adni.loni.usc.edu
19. Zhao, Y.-X., Zhang, Y.-M., Song, M., Liu, C.-L.: Multi-view semi-supervised 3d whole brain segmentation with a self-ensemble network. In: International Conference on Medical Image Computing and Computer-Assisted Intervention, pp. 256–265 (2019)
20. Holmes, C.J., Hoge, R., Collins, L., Woods, R., Evans, A.C.: Enhancement of MR images using registration for signal averaging. J. Comput. Assist. Tomogr. **3**, 324–333 (1998)
21. Lowekamp, B.C., Chen, D.T., Ibáez, L., Blezek, D.: The design of simpleitk. Front. Neuroinformatics **7**, 45 (2013)
22. Paszke, A., et al.: Pytorch: an imperative style, high-performance deep learning library. Adv. Neural Inf. Process. Syst. **32**, 8026–8037 (2019)

Airway Anomaly Detection by Prototype-Based Graph Neural Network

Tianyi Zhao and Zhaozheng Yin[✉]

Stony Brook University, Stony Brook, NY, USA
zyin@cs.stonybrook.edu

Abstract. Detecting the airway anomaly can be an essential part to aid the lung disease diagnosis. Since normal human airways share an anatomical structure, we design a graph prototype whose structure follows the normal airway anatomy. Then, we learn the prototype and a graph neural network from a weakly-supervised airway dataset, i.e., only the holistic label is available, indicating if the airway has anomaly or not, but which bronchus node has the anomaly is unknown. During inference, the graph neural network predicts the anomaly score at both the holistic level and node-level of an airway. We initialize the airway anomaly detection problem by creating a large airway dataset with 2589 samples, and our prototype-based graph neural network shows high sensitivity and specificity on this new benchmark dataset. The code is available at https://github.com/tznbz/Airway-Anomaly-Detection-by-Prototype-based-Graph-Neural-Network.

Keywords: Anomaly detection · Bronchus classification · Graph neural network

1 Introduction

Computer-aided diagnosis (CAD) becomes more and more important to assist doctors in the interpretation of medical images, especially in the pandemic situation. Many lung-related image analysis tasks have been investigated such as lung segmentation [1], nodule detection [2] and airway segmentation [3–7]. A complete lung-related CAD system should not only detect if the disease exists, but also provide detailed analysis on the disease including localizing which lobe the region-of-interest (ROI) belongs to. The clinical definition of different lobes are directly related to the bronchus hierarchy. Recently, a classification algorithm is proposed to label the bronchus, which can help segment the lobe [7]. However, this algorithm has an assumption that the bronchi to be classified follow the anatomical structure. It is unaware of the anomaly appearing in the airway structure. In fact, detecting anomaly in an airway structure can aid the lung disease diagnosis. If a CAD system can provide an anomaly score for every bronchus, this could be considered as a new digital bio-marker (i.e., the anomaly bronchi deserve special attention or treatment from the doctor).

© Springer Nature Switzerland AG 2021
M. de Bruijne et al. (Eds.): MICCAI 2021, LNCS 12905, pp. 195–204, 2021.
https://doi.org/10.1007/978-3-030-87240-3_19

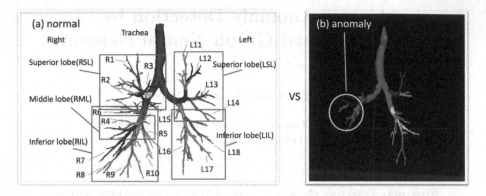

Fig. 1. (a) Normal human airway follows a common anatomical structure. (b) An airway with anomaly exhibits large airway tree variations compared to the anatomical structure, indicating the possible disease area or related surgery.

A normal human airway follows an anatomical structure that has 5 lobes and 18 segments with their specific characteristics, as shown in Fig. 1(a). An anomaly airway does not fully follow the pre-defined structure due to its disease or surgery (e.g., Fig. 1(b)). Inspired by these two observations, we propose a prototype-based airway anomaly detection algorithm, where the prototype is a learned graph representation of the normal airway and a graph neural network is learned to estimate the anomaly score for each bronchus node of an airway.

Though detecting airway anomaly is valuable to aid lung disease diagnosis, unfortunately there is no dataset with related labels in the community yet. Particularly, labeling the bronchus as anomaly or not one-by-one in CT image stacks is very tedious. For example, the three widely-used public datasets [8–10] only have segmentation labels on lung regions or the disease ROIs in CT images, without the detailed disease/anomaly/normal label on individual bronchus. To initialize the airway anomaly detection problem in our community, we explore a weakly-supervised anomaly detection approach using the existing datasets where only the holistic label is given, indicating if the airway contains anomaly or not, but which bronchus node in the airway has anomaly is unknown. We deploy the multi-instance setting, in which if all the bronchus nodes are normal, then the airway is normal, otherwise, if at least one node is abnormal, the airway has anomaly. In summary, this paper has three-fold contributions:

- We propose an airway anomaly detection algorithm to predict the anomaly score for the holistic airway and for each node of the airway tree. The anomaly score is calculated by a graph convolutional neural network (GCN).
- We propose a virtual prototype whose graph structure follows the anatomical structure of normal airways. The GCN and prototype are alternatively trained such that the GCN generates low anomaly scores on normal airways and the prototype and high anomaly scores on anomaly samples.
- We initialize the airway anomaly detection problem in our lung disease diagnose community by creating a large airway dataset and formulating it in a

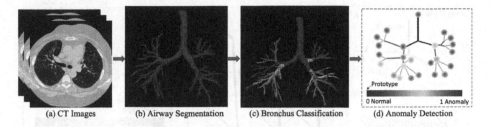

(a) CT Images   (b) Airway Segmentation   (c) Bronchus Classification   (d) Anomaly Detection

Fig. 2. A typical process of the airway-related CAD system. The airway is first segmented from the CT images. Then the bronchus are classified according to the anatomical structure. Finally, our proposed anomaly detection algorithm processes the graph-based classification result and estimates the anomaly at each node.

weakly-supervised way, i.e., given only the holistic level label for training, we train a graph neural network able to predict the anomaly score of an airway at both the holistic and node levels during inference.

2   Related Work

2.1   Airway-Related CAD System

The airway tree can provide valuable information for airway disease diagnosis (e.g., Chronic Obstructive Pulmonary Disease (COPD)). A typical process of the airway-related CAD system is shown in Fig. 2. Several Convolutional Neural Network (CNN) based methods have been proposed to segment the airway tree volume from the lung CT images, such as 3D U-Net based methods [3–5], 2.5D CNN [6], and a 2-stage 2D+3D CNN [7]. Based on the airway segmentation, a classification algorithm is proposed in [7] to classify each airway segment to one of the 5 lobes (and one of the 18 segmental bronchus). However, the bronchus classification method is unaware of the anomaly.

2.2   Anomaly Detection

Anomaly detection or novelty detection, is to detect an incoming sample with a new unknown class. Novelty detection is also related to the domain adaption problem, where the known class belongs to the source domain, and the novel class belongs to the new target domain. The novelty detection method can be distribution based [11,12], or probabilistic based [13]. For example, a softmax and threshold baseline is proposed in [11] for out-of-distribution detection. A novel margin-based loss term by self-supervised leave-out classifier is proposed in [12]. In this paper, we invent a prototype-based algorithm, where the anomaly of an airway including its nodes is predicted by a graph neural network.

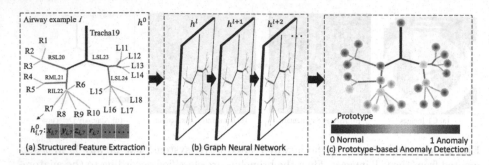

Fig. 3. The overview of our anomaly detection algorithm with three components: feature extraction, graph neural network and prototype-based anomaly prediction.

2.3 Graph Neural Network

Graph cnvolutioal neural network (GCN) is widely used for graph-structure data processing, such as the social network and biological data. The graph construction and layer-wise propagation rules have been described in [14,15] Recently, GCN has been enhanced in many ways such as Graph Attention Networks [16] with self-attentional layers; Graph-ResNet [17] with the residual block; Graph-SAGE [18] replacing the embedding vector of each graph node with a set of aggregator functions, and hierarchical deep representations [19,20] to learn the graph-level representation. In this paper, we integrate an anatomical-based graph module to GCN for the multi-level anomaly detection.

3   Method

The overview of our airway anomaly detection is shown in Fig. 3 with three components: (1) The feature extraction component extracts airway features from the airway segmentation results for every node, which composes the input of the graph neural network; (2) then, the graph neural network process the initial graph input; and (3) a prototype based anomaly detection algorithm computes the anomaly score for each node and the entire graph.

3.1   Structured Feature Extraction

Given the segmented and classified airway results shared by [7] (Fig. 2(b–c)), we encode the anatomical structure of the airways, as well as their image properties and graph properties into feature vectors. The feature vectors can be viewed as an anatomical representation of the airway, which will be used in the graph neural network for anomaly detection.

Formally, we define $h_{i,k}^0 \in R^D$ as the initial feature vector for node k of the ith airway sample (e.g., $h_{i,7}^0$ in Fig. 3(a) represents the initial feature vector of the R7 bronchus node in the right lobe of the ith sample). Based on the anatomical airway structure, we design 7 attributes in the feature representation:

(1) **The coordinates** of node k of sample i is represented as $(x_{i,k}, y_{i,k}, z_{i,k})$. The coordinates are centralized by the coordinates of the trachea and normalized by the spacing of the CT scan and the bounding box of the airway tree.

(2) **The direction** of node k of sample i is represented as $(r_{i,k})$ in 3D dimension. The direction is calculated from the coordinates of the current node to the average coordinates of the leaf nodes of the current node.

(3) **The level** of the node. The level of the root node (the trachea node) is 0. If the level of a node is l, then the children of this node are at level $l + 1$.

(4) **The bottom-up level** of the node. The level of the leaf node is 0. If the level of a node is l, then the parent of this node is at level $l + 1$.

(5) **The length** of the node is calculated from the coordinates of the current node to the coordinates of the furthest leaf node of the current node.

(6) **The number of the descendants** of the node.

(7) **The image feature** is extracted from the last layer of segmentation network based on the coordinates of the node. The dimension of the image feature is 64. The dimension of the entire feature vector is $D = 74$.

3.2 Graph Neural Network

After obtaining feature vectors $h^0_{i,k}$ for all individual nodes, we feed them into a graph convolutional neural network (GCN) modeling the dependencies of neighboring nodes in the tree graph. We use a shallow graph neural network which contains six graph convolutional layers and performs the anomaly detection task under multiple levels (i.e., the holistic level and the node level). Each graph-based convolutional layer is defined as:

$$h^{l+1}_{i,k} = w_0 h^l_{i,k} + \sum_{k' \subset N(k)} w_1 h^l_{i,k'}, \tag{1}$$

where $h^l_{i,k}$ is the feature vector of node k of sample i at layer l of the GCN. $N(k)$ represents the neighbor set of the node k. w_0 and w_1 are learned parameters of the graph-based convolutional kernels, with w_1 being shared by all edges. The graph convolutional layer accounts for the way in which nodes neighboring a given node regularize the node-to-neighbor exchange of information.

3.3 Prototype-Based Anomaly Detection

Our graph network (GCN) generates an anomaly score ($\hat{y}_{i,k} \in [0,1]$) for every node k of airway sample i (i.e., $G(h^0_i) : R^{K \times D} \to [0,1]^K$, where G represents the GCN function, h^0_i is the initial segmented airway graph of sample i, and K is the number of graph nodes, $K = 24$). In our study, we only have the holistic label, indicating if the airway is normal or anomaly, but we do not have the node-level label. During the training, we label all the nodes based on the holistic label, i.e., $y_{i,k} = y_i$. During the testing, if at least one node is anomaly, the airway is anomaly (i.e., if any $\hat{y}_{i,k} = 1$, then $\hat{y}_i = 1$), otherwise, the airway is normal.

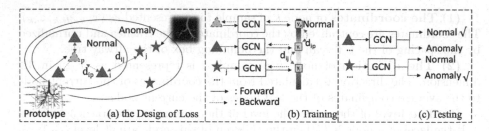

Fig. 4. (a) The design of loss to learn the prototype and GCN. (b) The training process on GCN and prototype. (c) Prediction during testing.

We define the prototype p of normal airways by its node feature matrix $h_p \in R^{K \times D}$. Our objective is to learn both the graph neural network function G and the prototype h_p, but G and h_p are intertwined together, so we propose to learn them in an alternative way:

First, suppose an initial prototype h_p^0 is given and the anomaly score for each node of the prototype is 0. The GCN function G is expected to generate a low anomaly score close to 0 for every node of a normal sample i (i.e., $G(h_p^0) \approx G(h_i^0)$). Meanwhile, G is expected to generate a large anomaly score for anomaly sample j. Motivated by these expectations, illustrated in Fig. 4(a) as well, we define the following loss function to train the GCN:

$$L_{GCN} = \sum_{i, y_i = 0} d_{ip}(\bar{G}(h_p^0), G(h_i^0)) - \sum_{i, y_i = 0} \sum_{j, y_j = 1} d_{ij}(G(h_i^0), G(h_j^0)), \qquad (2)$$

where d_{ip} is the distance between the prototype and the normal sample i. By minimizing d_{ip}, the normal training sample i is close to the prototype, leading to a low anomaly score on sample i. d_{ij} is the distance between a normal sample i and an anomaly sample j. By maximizing d_{ij}, we maintain the discrimination between normal and anomaly samples. We use the Euclidean distance for d_{ip} and d_{ij}. Note that, the GCN output on the prototype is represented by $\bar{G}(h_p^0)$, which means the prototype is fixed when training G.

Secondly, we assume G is given, represented by $\bar{G}$. Then, we learn the prototype h_p by minimizing the following loss function:

$$L_{prototype} = \sum_{k}^{K} f(\bar{G}(h_p)_k, 0), \qquad (3)$$

where $f(\cdot)$ is the binary cross entropy function. The alternative GCN and prototype learning is summarized in Algorithm 1 below.

Algorithm 1: The training procedure for GCN and prototype

Initialization: $h_p^0 = \frac{1}{N} \sum_i^N h_i^0$ on N samples, G is randomly initialized;

for $t = 1$:max iteration **do**

$\quad$ $G^{(t)} \leftarrow argmin_G$ Eq.2;

$\quad$ $\bar{G} \leftarrow G^{(t)}$;

$\quad$ $h_p^{(t)} \leftarrow argmin_{h_p}$ Eq.3;

$\quad$ $h_p^0 \leftarrow h_p^{(t)}$;

end

Result: G, h_p // The GCN function and the prototype

---

4   Experiments

Dataset: We collected datasets from 3 resources: The Lung Image Database Consortium (LIDC) [8], The Lung Tissue Research Consortium (LTRC) [9], and The National Lung Screening Trial (NLST) [10], leading to 62 normal samples and 23 anomaly samples in total. The available airway data in our community is quite limited and it is even costly to label node-level anomaly. Meanwhile, we observe three most common anomaly patterns: switch, shift and cut. Thus, we propose three augmentation methods to generate more anomaly samples from the collected dataset: (1) **Switch**: randomly switch two bronchus within the same lobe; (2) **Shift**: a bronchus is randomly shifted to one of its descendants; and (3) **Cut**: cut an airway branch and re-label the airway by the bronchus classification algorithm [7]. After the augmentation, we have 2396 training samples with 40 of them being normal samples, and 193 testing samples with 22 samples being normal samples. In total, a dataset of 2589 samples is established.

Evaluation Metric: We use two evaluation metrics in the experiment: the sensitivity that is the proportions of the correctly predicted anomaly data, and the specificity that is the proportions of the correctly predicted normal data.

Ablation Study: We evaluate the effectiveness of GCN by comparing it with CNN and evaluate the effectiveness of prototype by comparing networks with or without it. The ablation study is summarized in Table 1. Since our dataset has more anomaly samples than normal samples, the CNN w/o prototype method gets a high sensitivity but a low specificity, which means the method tends to predict the anomaly. With the prototype, specificity is increased, since the prototype represents and emphasizes the normal samples during training. The GCN w/o prototype method has a better average score than the CNN w/o prototype method. Compared to the CNN model, GCN can exchange information between neighboring nodes. In the anomaly case, only a small portion of the nodes are anomaly nodes. The GCN can make the prediction of each node more robust. Overall, our GCN with the prototype method has the best performance.

Table 1. The ablation study results.

Method	Sensitivity	Specificity	Average
CNN w/o prototype	**0.964**	0.681	0.822
CNN w prototype	0.935	0.772	0.853
GCN w/o prototype	0.923	0.818	0.871
GCN w prototype	0.912	**0.954**	**0.933**

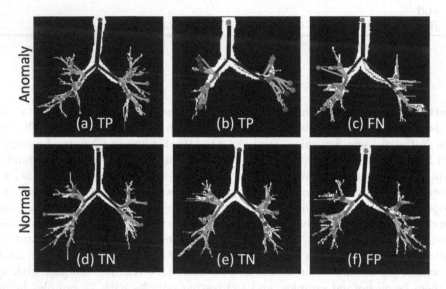

Fig. 5. Anomaly score of each node overlaid on the airway via heatmap color scheme. (Color figure online)

Qualitative Analysis: Some qualitative results are provided in Fig. 5. The segmented airway tree is represented by white pixels. The anomaly score is represented by color dot at each node (blue: low anomaly score; red: high anomaly score). In Fig. 5(a), one bronchus is correctly predicted as the anomaly and in Fig. 5(b), multiple bronchi are correctly predicted as the anomaly. A false negative example is given in Fig. 5(c). The anomaly in this example is very minor, where some bronchi are too thin to detect. Three normal samples are given in Fig. 5(d–f). Most of the bronchi are correctly predicted as normal, except some bronchi in the left inferior lobe of Fig. 5(f). In this case, multiple segmental bronchi are very close to each other, causing the confusing clutter.

5   Conclusion

We propose a prototype-based anomaly detection algorithm to predict the anomaly score for each node of an airway through a graph neural network, which can aid the lung disease diagnosis. We initialize the airway anomaly detection problem, and the newly created dataset along with our source codes will be released to the community.

References

1. Zhao, T., Gao, D., Wang, J., Yin, Z.: Lung segmentation in CT images using a fully convolutional neural network with multi-instance and conditional adversary loss. In: ISBI, pp. 505–509 (2018)
2. Ding, J., Li, A., Hu, Z., Wang, L.: Accurate pulmonary nodule detection in computed tomography images using deep convolutional neural networks. In: Descoteaux, M., Maier-Hein, L., Franz, A., Jannin, P., Collins, D.L., Duchesne, S. (eds.) MICCAI 2017. LNCS, vol. 10435, pp. 559–567. Springer, Cham (2017). https://doi.org/10.1007/978-3-319-66179-7_64
3. Jin, D., Xu, Z., Harrison, A.P., George, K., Mollura, D.J.: 3D convolutional neural networks with graph refinement for airway segmentation using incomplete data labels. In: Wang, Q., Shi, Y., Suk, H.-I., Suzuki, K. (eds.) MLMI 2017. LNCS, vol. 10541, pp. 141–149. Springer, Cham (2017). https://doi.org/10.1007/978-3-319-67389-9_17
4. Garcia-Uceda Juarez, A., Tiddens, H.A.W.M., de Bruijne, M.: Automatic airway segmentation in chest CT using convolutional neural networks. In: Stoyanov, D., et al. (eds.) RAMBO/BIA/TIA -2018. LNCS, vol. 11040, pp. 238–250. Springer, Cham (2018). https://doi.org/10.1007/978-3-030-00946-5_24
5. Meng, Q., Roth, H.R., Kitasaka, T., Oda, M., Ueno, J., Mori, K.: Tracking and segmentation of the airways in chest CT using a fully convolutional network. In: Descoteaux, M., Maier-Hein, L., Franz, A., Jannin, P., Collins, D.L., Duchesne, S. (eds.) MICCAI 2017. LNCS, vol. 10434, pp. 198–207. Springer, Cham (2017). https://doi.org/10.1007/978-3-319-66185-8_23
6. Yun, J., et al.: Improvement of fully automated airway segmentation on volumetric computed tomographic images using a 2.5 dimensional convolutional neural net. Med. Image Anal. (MedIA) 51, 13–20 (2019)
7. Zhao, T., Yin, Z., Wang, J., Gao, D., Chen, Y., Mao, Y.: Bronchus segmentation and classification by neural networks and linear programming. In: Shen, D., et al. (eds.) MICCAI 2019. LNCS, vol. 11769, pp. 230–239. Springer, Cham (2019). https://doi.org/10.1007/978-3-030-32226-7_26
8. Armato, S.G., III., et al.: The lung image database consortium (LIDC) and image database resource initiative (IDRI): a completed reference database of lung nodules on CT scans. Med. Phys. 38, 915–931 (2011)
9. Bartholmai, B., Karwoski, R., Zavaletta, V., Robb, R., Holmes, D.R.I.: The Lung Tissue Research Consortium: an extensive open database containing histological, clinical, and radiological data to study chronic lung disease. In: MICCAI Open Science Workshop (2006)
10. Clark, K., et al.: The Cancer Imaging Archive (TCIA): maintaining and operating a public information repository. J. Digit. Imaging 26(6), 1045–1057 (2013)
11. Vyas, A., Jammalamadaka, N., Zhu, X., Das, D., Kaul, B., Willke, T.L.: Out-of-distribution detection using an ensemble of self supervised leave-out classifiers. In: Ferrari, V., Hebert, M., Sminchisescu, C., Weiss, Y. (eds.) ECCV 2018. LNCS, vol. 11212, pp. 560–574. Springer, Cham (2018). https://doi.org/10.1007/978-3-030-01237-3_34

12. Hendrycks, D., Gimpel, K.: A baseline for detecting misclassified and out-of-distribution examples in neural networks. In: ICLR (2017)
13. Eskin, E.: Anomaly detection over noisy data using learned probability distributions. In: ICML (2000)
14. Bruna, J., Zaremba, W., Szlam, A., LeCun, Y.: Spectral networks and locally connected networks on graphs. arXiv preprint arXiv:1312.6203 (2013)
15. Bruna, J., Zaremba, W., Szlam, A., LeCun, Y.: Semi-supervised classification with graph convolutional networks. arXiv preprint arXiv:1609.02907 (2016)
16. Veličković, P., Cucurull, G., Casanova, A., Romero, A., Lio, P., Bengio, Y.: Graph attention networks. arXiv preprint arXiv:1710.10903 (2017)
17. Wang, N., Zhang, Y., Li, Z., Fu, Y., Liu, W., Jiang, Y.-G.: Pixel2Mesh: generating 3D mesh models from single RGB images. In: Ferrari, V., Hebert, M., Sminchisescu, C., Weiss, Y. (eds.) ECCV 2018. LNCS, vol. 11215, pp. 55–71. Springer, Cham (2018). https://doi.org/10.1007/978-3-030-01252-6_4
18. Hamilton, W.L., Ying, R., Leskovec, J.: Inductive representation learning on large graphs. arXiv preprint arXiv:1706.02216 (2017)
19. Ying, R., You, J., Morris, C., Ren, X., Hamilton, W.L., Leskovec, J.: Hierarchical graph representation learning with differentiable pooling. arXiv preprint arXiv:1806.08804 (2018)
20. Huang, J., Li, Z., Li, N., Liu, S., Li, G.: AttPool: towards hierarchical feature representation in graph convolutional networks via attention mechanism. In: ICCV, pp. 6480–6489 (2019)

Energy-Based Supervised Hashing for Multimorbidity Image Retrieval

Peng Huang[1], Xiuzhuang Zhou[1]([✉]), Zeqiang Wei[1], and Guodong Guo[2]

[1] School of Artificial Intelligence, Beijing University of Posts and Telecommunications, Beijing, China
xiuzhuang.zhou@bupt.edu.cn
[2] Institute of Deep Learning, Baidu Research, Beijing, China

Abstract. Content-based image retrieval (CBIR) has attracted increasing attention in the field of computer-aided diagnosis, for which learning-based hashing approaches represent the most prominent techniques for large-scale image retrieval. In this work, we propose a Supervised Hashing method with Energy-Based Modeling (SH-EBM) for scalable multi-label image retrieval, where concurrence of multiple symptoms with subtle differences in visual feature makes the search problem quite challenging, even for sophisticated hashing models built upon modern deep architectures. In addition to similarity-preserving ranking loss, multi-label classification loss is often employed in existing supervised hashing to further improve the expressiveness of hash codes, by optimizing a normalized probabilistic objective with tractable likelihood (e.g., multi-label cross-entropy). On the contrary, we present a multi-label EBM loss without restriction on the tractability of the log-likelihood, which is more flexible to parameterize and can model a more expressive probability distribution over multimorbidity image data. We further develop a multi-label Noise Contrastive Estimation (ml-NCE) algorithm for discriminative training of the proposed hashing network. On a multimorbidity dataset constructed by the NIH Chest X-ray, our SH-EBM outperforms most supervised hashing methods by a significant margin, implying its effectiveness in facilitating multilevel similarity preservation for scalable image retrieval.

Keywords: Multimorbidity medical image retrieval · Supervised hashing · Energy-based model · Noise contrastive estimation

1 Introduction

Over the past decade, the amount of radiology images has significantly increased due to the accessibility to modern imaging technologies (e.g., X-ray, CT and MR) in daily radiology workflow. However, analysis of these medical images are often labor-intensive and expertise-dependent. Unlike domain expert analysis, content-based image retrieval (CBIR) [1,2] can be more efficient in an automatic way, by

© Springer Nature Switzerland AG 2021
M. de Bruijne et al. (Eds.): MICCAI 2021, LNCS 12905, pp. 205–214, 2021.
https://doi.org/10.1007/978-3-030-87240-3_20

which visually similar images to the query being analyzed can be automatically retrieved in an existing images database for supporting a final diagnosis.

While massive amounts of medical images can provide rich cues for computer-aided analysis in daily radiology workflow, algorithms and techniques for efficient and accurate retrieval in large-scale medical image databases remain challenging. Image representation (i.e., feature extraction) and feature indexing are two key ingredients of a large-scale CBIR. Generally, image representation can be either hand-crafted or learning-based, while feature indexing can be implemented by specific indexing structure or feature compressing for computational efficiency. Deep representation learning [3] represents the most popular technologies for effective feature extraction that facilitates accurate medical image retrieval, by training deep convolutionary neural networks (DCNNs) on large-scale medical image databases [4]. As for the feature indexing, hashing technologies have been extensively studied and employed for large-scale image retrieval [5], due to their efficiency through mapping the high-dimensional data into compact binary codes with similarity-preserving.

The focus of this paper is on learning to hash for multi-label medical image retrieval. To date, learning-based hashing methods can generally be categorized into shallow learning-based [6–8] and deep learning-based [12–18], depending on the learning architectures they are built upon. Typically, the former employs hand-crafted image descriptors to learn mapping functions, in a unsupervised (e.g., Spectral Hashing [6] and Iterative Quantization [7]), supervised (e.g., Supervised Discrete Hashing [9]) or semi-supervised (e.g., Semi-supervised Hashing [8]) manner. Unlike this two-stage learning scheme that may be suboptimal, deep learning-based hashing represents a promising solution for large-scale retrieval applications, in which feature representation and indexing are jointly optimized in a end-to-end manner to output compact hash codes with similarity-preserving. Earlier works in this line include Deep Hashing [10], Deep Supervised Hashing [11], and etc. For preservation of semantic similarity, input data to the loss functions for hash learning can be pointwise [19], pairwise [5], or tripletwise [15], depending on the supervision level of the methods.

In the field of medical image retrieval, learning to hash with similarity preservation has also been investigated in the past few years. Seminal works in this research line include the residual hashing for chest X-ray images [14] and multiple instance hashing for tumor analysis [13]. To date, however, very few attempt has been made for hashing-based multimorbidity image retrieval, although hashing in the multi-label setting has been well investigated in computer vision [20]. Despite recent success of generic multi-label image retrieval, effective hashing for multi-label medical image retrieval is still a challenging topic. On one hand, visual features with different symptoms in multimorbidity medical images can be very subtle, and hence powerful image representation for more fine-grained analysis is required. On the other hand, how to model and learn multilevel semantic similarity inherent in multimorbidity images for hashing remains open. Recently, Chen et al. [15] proposed an order-sensitive deep hashing for multimorbidity images retrieval, by introducing a tripletwise multi-label ranking loss to model multilevel semantic similarity. In multimorbidity images, however, the concurrence of multiple symptoms with subtle differences in visual feature makes the discriminative representation learning very challenging. We argue that expressiveness of the multi-label cross-entropy they employed would be insufficient to learn a discriminative representation for multimorbidity images with complex visual patterns, resulting in suboptimal hashing performance even with a well-defined tripletwise multi-label ranking loss.

Most existing supervised hashing employed multi-label classification loss to improve the expressiveness of hash codes, by optimizing a normalized probabilistic objective with tractable likelihood (e.g., multi-label cross-entropy). Taking inspiration from energy-based model [21], in this paper we propose a multi-label energy-based loss for supervised hashing, without restriction on the tractability of the log-likelihood, which is more flexible to parameterize and can model a more expressive joint distribution over multimorbidity image-label space. We term our method as Supervised Hashing with Energy-Based Modeling (SH-EBM), as illustrated in Fig. 1. We further develop a multi-label Noise Contrastive Estimation (ml-NCE) algorithm for discriminative training of the proposed hashing network. The superiority of our formulation for scalable image retrieval is validated on a large multimorbidity dataset built on the NIH Chest X-ray [4].

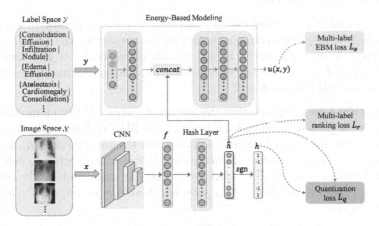

Fig. 1. Overview of the proposed SH-EBM method.

2   Methodology

Let $\mathcal{X} \equiv \mathcal{R}^d$ and $\mathcal{Y} \equiv \{1, 2, ..., C\}$ respectively denote the sample (image) space and its associated label space. Given a training set $X = \{x_i\}_{i=1}^N$ of N samples, each of which x_i is associated with a subset of labels $y_i \subseteq \mathcal{Y}$, the task of supervised hashing for multimorbidity image retrieval is to learn a mapping function to quantize each input sample into compact binary codes, such that multilevel semantic structure of the multimorbidity image is preserved in Hamming space.

2.1 Network Architecture

We design a deep supervised hashing model to jointly learn the multimorbidity image representation and feature hashing to generate compact binary codes in an end-to-end manner. As illustrated in Fig. 1, our two-stream network consists of three key modules: the CNN backbone for extraction of image feature, the hash layer for feature quantization (finally into binary codes), and the energy-based model (EBM). One stream receives a multimorbidity image x as input, which is first processed by a CNN backbone. The backbone can be any popular deep architectures used for representation learning (in this work we use AlexNet [3] for fair comparison with state-of-the-art methods). The extracted image feature f is then fed into the hash layer to quantize into a feature vector $\hat{h}$. Another stream (e.g., EBM) receives multi-label vector y (associated to x) as input, and feeds it into a deep neural network (DNN) to generate a transformed vector f_y, which is concatenated with feature vector $\hat{h}$ and then fed into another DNN. The EBM finally outputs a scalar value $u(x, y)$ termed *energy*. With well-defined loss objectives, we will show that the proposed SH-EBM network can facilitate learning to hash for effective multimorbidity image retrieval.

2.2 Learning Objectives

Ideally, multi-label supervised hashing methods learn to map and quantize each multimorbidity image into compact binary codes, preserving the semantic structure inherent in each sample in terms of both visual contents and multi-label cues. To this end, we introduce two loss functions as the learning objectives for training of our SH-EBM netowrk.

Quantization Loss. Suppose that the feature f_i extracted from a multimorbidity image x_i is passed through hash layer of the SH-EBM network to generate a feature vector $\hat{h}_i$, which is then quantized into the binary codes h_i using the $sgn(\cdot)$ function. To reduce the quantization error, feature vector $\hat{h}_i$ can be fed into an activation function (e.g., $\texttt{tanh}$) such that the magnitude of the output is within $[-1, 1]$ before the binarization process. Therefore, we impose the following quantization loss:

$$L_q = \frac{1}{n} \sum_{i=1}^{n} \|\hat{h}_i - h_i\|_2^2 \tag{1}$$

where $h_i = sgn(\hat{h}_i)$, $sgn(u) = 1$ if $u > 0$, and $sgn(u) = -1$ otherwise. To circumvent the non-smooth problem in optimization, we regard the binary code h_i as a fixed variable, back-propagation is only performed on $\hat{h}_i$.

Multi-label EBM Loss. To preserve the multilevel semantic structure inherent in multimorbidity images for supervised hashing, we propose a new loss objective to guide the training of the SH-EBM network. Typically, multi-label classification loss employed in existing supervised hashing methods aim to optimize a normalized probabilistic objective with tractable likelihood (e.g., multi-label cross-entropy). On the contrary, we construct a multi-label energy-based loss without restriction on the tractability of the log-likelihood, which is more flexible to parameterize and can model a more expressive probability distribution over multimorbidity image data. To this end, we draw inspiration from the EBM formulation [21] and rely on a DNN (parameterized by θ) that directly outputs a scalar value $u_\theta(x, y)$ for any data pair (x, y) over the image-label space $\mathcal{X} \times \mathcal{Y}$. In this way, our model of the conditional density $p(y|x)$ can then be defined by

$$p(y|x, \theta) = \frac{e^{u_\theta(x,y)}}{Z(x, \theta)} \tag{2}$$

where $Z(x, \theta) = \int e^{u_\theta(x, \tilde{y})} d\tilde{y}$ is the intractable normalizing constant (also known as the partition function). Based on this energy-based model, we can define our multi-label EBM loss L_e as the following negative log-likelihood:

$$L_e = \frac{1}{n} \sum_{i=1}^{n} - \log p(y_i|x_i, \theta) = \frac{1}{n} \sum_{i=1}^{n} \log \left(\int e^{u_\theta(x_i, y)} dy \right) - u_\theta(x_i, y_i) \tag{3}$$

While the introduced multi-label classification objective can be more expressive in modeling multimorbidity image data, in which the concurrence of multiple symptoms with subtle differences in visual feature requires more discriminative representation learning. However, training the multi-label EBM loss with intractable normalizing constant is quite difficult. In next section we will derive an effective training algorithm based on the noise contrastive estimation [22].

Overall Loss. The learning objective for our SH-EBM method mainly consists of the two loss functions discussed above. However, combined with most existing multi-label ranking loss functions, either pairwise [20] or tripletwise [15], can be helpful to further improve the overall retrieval performance, although this improvement is less significant when it is compared to our proposed multi-label EBM loss, as shown in the experiments. The overall loss can be written as:

$$L = L_e + \lambda_q L_q + \lambda_r L_r \tag{4}$$

where the ranking loss [15] for triplet input (q, i, j) is defined by $L_r = \sum_i \sum_{j:r_j < r_i} \frac{2^{r_i} - 2^{r_j}}{Z} \max(0, D(h_q, h_i) - D(h_q, h_j) + \rho)$, and λ_q and λ_r are two trade-off parameters. Here, D denotes the Hamming distance and r_i denotes the similarity level measured by the number of shared labels between the data point i and query q, ρ and Z are the margin and normalizing constant, respectively.

2.3 Training Algorithm

Our proposed SH-EBM model is trained by optimize the learning objective (4) based on the deep learning framework PyTorch [24]. As discussed earlier, the difficulty during optimization stems from the intractable partition function (2). One common solution to address this issue is to utilize Monte Carlo importance sampling to approximate the intractable normalizing constant in training. However, as pointed out by existing work on training EBMs [23], this simple sampling strategy may not capture the data distribution faithfully. To this end, we develop a training algorithm based on the Noise Contrastive Estimation [22] in the multi-label setting. Different from the simple Monte Carlo importance sampling, basic idea behind the NCE is that one can train an EBM by contrasting it with another known (noisy) distribution. More specifically, the multi-label EBM loss can now be defined as:

$$L_e = \frac{1}{n} \sum_{i=1}^{n} -\log \frac{e^{u_\theta(x_i, y^{(i,0)}) - \log p_N(y^{(i,0)}|y_i)}}{\sum_{m=0}^{M} e^{u_\theta(x_i, y^{(i,m)}) - \log p_N(y^{(i,m)}|y_i)}} \tag{5}$$

where $p_N(y|y_i)$ is a noisy multi-label proposal distribution used for drawing M noise samples $\{y^{(i,m)} \subseteq \mathcal{Y}\}_{m=1}^{M}$, with $y^{(i,0)} = y_i$. We use a mixture of K isotropic Gaussians centered at the ground-truth label y_i as the multi-label proposal, and we observed that it consistently performs well in our experiments in terms of different retrieval metrics. Specifically, the proposal distribution is defined by

$$p_N(y|y_i) = \frac{1}{K} \sum_{k=1}^{K} \mathcal{N}(\cdot|y_i, \sigma_k I) \tag{6}$$

where σ_k is the standard deviation for k-th Gaussian and I is the identity matrix.

3 Experiments and Results

3.1 Dataset

NIH Chest X-ray Dataset [4] is a large-scale chest X-ray database publicly available, built by the National Institutes of Health in 2017. It consists of 112,120 frontal-view X-ray images from 30,805 different patients, mined from the associated radiological reports. Each X-ray image can have multiple labels of multimorbidity. Following the data protocol suggested by the OSDH [15], we also choose 13,000 images of 13 thoracic pathologies with high incidence rate to constitute our dataset (including Atelectasis, Infiltration, Consolidation, Pneumothorax, Edema, Fibrosis, Emphysema, Pleural thickening, Effusion, Mass, Pneumonia, Cardiomegaly and Nodule), and the dataset is further divided into the training set (80%) and testing set (20%) with non-overlapping splits in both patient- and pathology-level.

3.2 Experiment Settings

We adopt AlexNet [3] as the backbone architecture for training of SH-EBM based on the PyTorch framework [24]. Our SH-EBM is trained from scratch using the Adam optimizer with the parameters β_1, β_2 and ϵ of 0.9, 0.999 and 10^{-8}, respectively. The batch size is set as 64, and the base learning rate is set as 2×10^{-5}. As for the multi-label ranking loss, we follow the parameters setting in [15]. For the multi-label EBM loss, we use $K = 10$ isotropic Gaussians with standard deviations $\sigma_k = 0.2k$ as the GMM proposal, and we sample $M = 10,240$ multimorbidity labels during training from the proposal (6). For the total objective function, the trade-off parameters λ_r and λ_q are empirically set as 0.1 and 0.1, respectively.

For the EBM architecture of our SH-EBM, taking the code length of 32 bits for example, the multimorbidity label y is first processed by a fully-connected layer ($13 \rightarrow 32$), generating $f_y \in \mathcal{R}^{32}$. The two feature vectors f_y and $\hat{h}$ are then concatenated and fed into three fully-connected layers ($64 \rightarrow 32$, $32 \rightarrow 32$, $32 \rightarrow 1$), outputting the energy $u(x, y) \in \mathcal{R}$. As for the DNN structure of the hash layer, it consists of a fully-connected layer ($4096 \rightarrow 32$), outputting 32-bits codes.

In the experiments, we use three widely used metrics [20] to evaluate the performance of different retrieval methods. These evaluation metrics are: average cumulative gain (ACG), normalized discounted cumulative gain (NDCG), and weighted mean average precision (mAP$_w$).

Table 1. Comparison of NDCG@100 with different hashing algorithms.

Methods	16 bits	32 bits	48 bits	64 bits
DH [10]	0.1233	0.1364	0.1384	0.1374
ITQ [7]	0.1545	0.1568	0.1569	0.1565
SSH [8]	0.1337	0.1403	0.1472	0.1495
SDH [10]	0.1868	0.1874	0.1923	0.1937
DSH [11]	0.1701	0.1645	0.1624	0.1670
OSDH [15]	0.2366	0.2396	0.2390	0.2422
DLBHC [16]	0.1294	0.3367	0.3180	0.2957
SH-EBM	**0.3482**	**0.3810**	**0.3548**	**0.3340**

Table 2. Impact of different components in SH-EBM on the retrieval performance.

Methods	NDCG	ACG	mAP$_w$
SH-EBM-e	0.2945	0.6001	0.4974
SH-EBM-q	0.3290	0.5990	0.5104
SH-EBM-r	0.3388	0.6207	0.5161
SH-EBM	0.3810	0.6842	0.5695

3.3   Results and Analysis

In this section, we first compare the proposed SH-EBM with the state-of-the-art hashing algorithms for multimorbidity image retrieval. These algorithms being compared are ITQ [7], SSH [8] based on shallow architectures, and DH [10], SDH [10], DSH [11], OSDH [15], DLBHC [16] based on deep networks. We reproduce their reported results based on the codes provided by the authors.

Table 1 summarizes the retrieval performance in terms of NDCG@100 for different hashing algorithms under different code lengths. The performance comparison in terms of more metrics (NDCG@100, ACG@100 and mAP$_w$) under varying code lengths for different hashing algorithms are shown in Fig. 2. From Table 1 we can observe that our SH-EBM consistently outperforms the compared hashing algorithms that built upon either shallow learning or deep learning architectures by a significant margin. While supervised hashing solutions

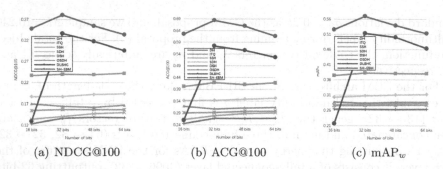

(a) NDCG@100   (b) ACG@100   (c) mAP$_w$

Fig. 2. Comparison of different hashing methods w.r.t. NDCG@100, ACG@100 and mAPw.

typically benefit from the order-preserving ranking loss, our proposed multi-label EBM formulation can bring more significant performance gains in terms of various evaluation metrics, as illustrated in Fig. 2. More specifically, compared to the extended cross-entropy (as the multi-label classification loss) adopted by OSDH [15], our energy-based modeling of multi-label classification over hash codes achieves significant performance improvement, indicating the superiority of EBM in mining and exploiting multi-label information for better supervised hashing. In Fig. 3, we visually present sampled ranking lists (top-10 retrieved images for each query) by our SH-EBM, where green boxes mean the highest semantic similarity level (i.e., total number of the shared labels between query image q and the i-th test image $|y_q \cap y_i| > 2$), boxes in pale green mean $|y_q \cap y_i| = 2$, and blue boxes mean $|y_q \cap y_i| = 1$. We can see that most images sharing more labels of pathologies with the query are appropriately ranked in the lists. This also demonstrates the effectiveness of our SH-EBM in modeling the multilevel semantic similarity for multimorbidity image retrieval. Last but not least, we perform ablation study for our method by removing different components in our objective function (4). More specifically, by removing L_e, L_r and L_q from L, respectively, we obtain three variants of our SH-EBM: SH-EBM-e, SH-EBM-r, and SH-EBM-q. We compare SH-EBM with its three variants in terms of three different evaluation metrics, and the performance comparison under 32 code bits are outlined in Table 2. From the table we can see that multi-label

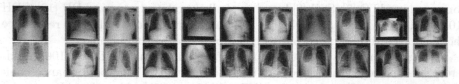

Fig. 3. Retrieved list by SH-EBM for the query with multimorbidity **Ed/Ef/In** (Top) and multimorbidity **At/Ef/In** (Bottom). Images with green, pale green, and blue boxes are respectively with $r > 2$, $r = 2$ and $r = 1$, where r is the semantic similarity level. (Color figure online)

EBM loss plays a most important role in improving the performance of multimorbidity image retrieval when compared to other two terms. By combining the three loss terms, our SH-EBM achieves better performance than other individuals. This indicates that the energy-based modeling of supervised hashing can better mine and exploit multi-label signal in multimorbidity images for modeling of multilevel semantic similarity.

4 Conclusion

In this paper, we have introduced a new supervised hashing method for scalable multimorbidity image retrieval. We proposed a multi-label EBM loss without restriction on the tractability of the log-likelihood, which is more flexible to parameterize and can model a more expressive probability distribution over multimorbidity image data. We conducted extensive experiments on a large multimorbidity image dataset to validate the superiority of our method when compared to several state-of-the-art alternatives. Incorporation of generative training in our energy-based supervised hashing can be helpful to further improve the retrieval performance. This appears to be an interesting direction of future work.

Acknowledgements. This work was supported in part by the National Natural Science Foundation of China under grants 61972046, and in part by the Beijing Natural Science Foundation under grants 4202051.

References

1. Li, Z., Zhang, X., Müller, H., Zhang, S.: Large-scale retrieval for medical image analytics: a comprehensive review. Med. Image Anal. **43**, 66–84 (2018)
2. Müller, H., Michoux, N., Bandon, D., Geissbuhler, A.: A review of content-based image retrieval systems in medical applications-clinical benefits and future directions. Int. J. Med. Informat. **73**(1), 1–23 (2004)
3. Krizhevsky, A., Sutskever, I., Hinton, G.E.: ImageNet classification with deep convolutional neural networks. In: NIPS, pp. 1106–1114 (2012)
4. Wang, X., Peng, Y., Lu, L., Lu, Z., Bagheri, M., Summers, R.M.: ChestX-ray8: hospital-scale chest X-ray database and benchmarks on weakly-supervised classification and localization of common thorax diseases. In: CVPR, pp. 3462–3471 (2017)
5. Wang, J., Zhang, T., Sebe, N., Shen, H.T.: A survey on learning to hash. TPAMI **40**(4), 769–790 (2017)
6. Weiss, Y., Torralba, A., Fergus, R.: Spectral hashing. In: NIPS, pp. 1753–1760 (2008)
7. Gong, Y., Lazebnik, S., Gordo, A., Perronnin, F.: Iterative quantization: a procrustean approach to learning binary codes for large-scale image retrieval. TPAMI **35**(12), 2916–2929 (2013)
8. Wang, J., Kumar, S., Chang, S.: Semi-supervised hashing for large-scale search. TPAMI **34**(12), 2393–2406 (2012)

9. Shen, F., Shen, C., Liu, W., Shen, H.T.: Supervised Discrete Hashing. In: CVPR, pp. 37–45 (2015)
10. Liong, V.E., Lu, J., Wang, G., Moulin, P., Zhou, J.: Deep hashing for compact binary codes learning. In: CVPR, pp. 2475–2483 (2015)
11. Liu, H., Wang, R., Shan, S., Chen, X.: Deep supervised hashing for fast image retrieval. In: CVPR, pp. 2064–2072 (2016)
12. Cao, Z., Long, M., Wang, J., Yu, P.S.: HashNet: deep learning to hash by continuation. In: ICCV, pp. 5608–5617 (2017)
13. Conjeti, S., Paschali, M., Katouzian, A., Navab, N.: Deep multiple instance hashing for scalable medical image retrieval. In: MICCAI, pp. 550–558 (2017)
14. Conjeti, S., Roy, A.G., Katouzian, A., Navab, N.: Hashing with residual networks for image retrieval. In: MICCAI, pp. 541–549 (2017)
15. Chen, Z., Cai, R., Lu, J., Lu, J., Feng, J., Zhou, J.: Order-sensitive deep hashing for multimorbidity medical image retrieval. In: MICCAI, pp. 620–628 (2018)
16. Lin, K., Yang, H.F., Hsiao, J.H., Chen, C.S.: Deep learning of binary hash codes for fast image retrieval. In: Proceedings of the IEEE Conference on Computer Vision and Pattern Recognition (CVPR) Workshops, pp. 27–35 (2015)
17. Li, Q., Sun, Z., He, R., Tan, T.: Deep supervised discrete hashing. In: NIPS, pp. 2479–2488 (2017)
18. Yang, H., Lin, K., Chen, C.: Supervised learning of semantics-preserving hash via deep convolutional neural networks. TPAMI 40(2), 437–451 (2018)
19. Liu, L., Rahimpour, A., Taalimi. A, Qi, H.: End-to-end binary representation learning via direct binary embedding. In: ICIP, pp. 1257–1261 (2017)
20. Rodrigues, J., Cristo, M., Colonna, J.G.: Deep hashing for multi-label image retrieval: a survey. Artif. Intell. Rev. 53(7), 5261–5307 (2020). https://doi.org/10.1007/s10462-020-09820-x
21. LeCun, Y., Chopra, S., Hadsell, R., Ranzato, M., Huang, F.: A tutorial on energy-based learning. Predicting Struc. Data 1 (2006)
22. Gutmann, M., Hyvärinen, A.: Noise-contrastive estimation: a new estimation principle for unnormalized statistical models. In: AISTATS, pp. 297–304 (2010)
23. Song, Y., Kingma, D.P.: How to train your energy-based models. arXiv preprint arXiv:2101.03288 (2021)
24. Paszke, A., et al.: PyTorch: an imperative style, high-performance deep learning library. In: NIPS, pp. 8024–8035 (2019)

Stochastic 4D Flow Vector-Field Signatures: A New Approach for Comprehensive 4D Flow MRI Quantification

Mohammed S. M. Elbaz[1]($\boxtimes$), Chris Malaisrie[2], Patrick McCarthy[2], and Michael Markl[1,3]

[1] Departments of Radiology, Northwestern University Feinberg School of Medicine, Chicago, IL, USA
mohammed.elbaz@northwestern.edu
[2] Cardiac Surgery, Northwestern University Feinberg School of Medicine, Chicago, IL, USA
[3] Department of Biomedical Engineering, Northwestern University, Evanston, IL, USA

Abstract. 4D Flow MRI has emerged as a new imaging technique for assessing 3D flow dynamics in the heart and great arteries (e.g., Aorta) in vivo. 4D Flow MRI provides in vivo voxel-wise mapping of 3D time-resolved three-directional velocity vector-field information. However, current techniques underutilize such comprehensive vector-field information by reducing it to aggregate or derivative scalar-field. Here we propose a new data-driven stochastic methodological approach to derive the unique 4D vector-field signature of the 3D flow dynamics. Our technique is based on stochastically encoding the profile of the underlying pairwise vector-field associations comprising the entire 3D flow-field dynamics. The proposed technique consists of two stages: 1) The 4D Flow vector-field signature profile is constructed by stochastically encoding the probability density function of the co-associations of millions of pair-wise vectors over the entire 4D Flow MRI domain. 2) The Hemodynamic Signature Index (HSI) is computed as a measure of the degree of alteration in the 4D Flow signature between patients. The proposed technique was extensively evaluated in three in vivo 4D Flow MRI datasets of 106 scans, including 34 healthy controls, 57 bicuspid aortic valve (BAV) patients and 15 Rescan subjects. Results demonstrate our technique's excellent robustness, reproducibility, and ability to quantify distinct signatures in BAV patients.

Keywords: 4D Flow MRI · Vector-field · Blood flow quantification

1 Introduction

4D Flow MRI has recently emerged as a novel in vivo imaging technique to visualize and evaluate 3D blood flow in the heart and great arteries (e.g., Aorta) in health and disease [1, 2]. 4D Flow MRI provides a comprehensive voxel-wise mapping of three-directional velocity vector-field $\mathbf{V} = (v_x, v_y, v_z)$, over the three-spatial dimensions and time-resolved over the cardiac cycle (Fig. 1). Recent 4D Flow MRI studies provide growing evidence that congenital bicuspid aortic valve (BAV) disease is associated with changes in aortic

© Springer Nature Switzerland AG 2021
M. de Bruijne et al. (Eds.): MICCAI 2021, LNCS 12905, pp. 215–224, 2021.
https://doi.org/10.1007/978-3-030-87240-3_21

3D hemodynamics that mediate abnormal aortic dilation and aortic wall degeneration on tissue histology [3–5]. These studies suggest that 4D flow MRI has the potential to unravel the association between aortic valve disease, aberrant aortic flow dynamics, and development of life-threatening complications such as aortic aneurysm or dissection [6]. Thus, may permit more individualized metrics to help diagnose and predict disease progression for tailored patient management. Nevertheless, regional and global dynamic 3D aortic blood flow changes can be complex including a mixture of flow jets (e.g., regurgitation) [7], vortical [8], and helical patterns [6, 9], for which a visual assessment is challenging with limited reproducibility. Existing quantitative 4D Flow MRI metrics, mainly based on a fluid dynamics operator, e.g., wall shear stress [10], kinetic energy [11–13], vorticity/vortex [11, 14, 15], energy loss [12, 16, 17] are useful, but suffer from limitations: 1) each metric quantifies only partial components of the overall complex flow changes (e.g., vorticity would capture vortical flow changes but not jets and kinetic energy would capture jets but not vortical flow); 2) unable to directly quantify the native 4D Flow 3-directional velocity vector-field, which is the primary/basis flow field, that embeds the full spectrum of measurable velocity/flow dynamics. Instead, current methods mainly require a prior conversion of such primary vector-field into a secondary derivative scalar field (i.e. computed as a function of the underlying velocity vector field). Hence, losing the potentially-diagnostic changes embedded in the velocity vector-field dynamics and directionality components. 3) Underutilizing the spatiotemporal voxel-wise distribution with the use of aggregated quantities (e.g., peak, average, sum).

In this work, we propose a new data-driven methodological stochastic signature approach that efficiently (linear complexity) quantifies directly and comprehensively the entire native 4D Flow MRI three-directional velocity vector-field information, both in 3D and over time. Our data-driven technique is based on stochastic encoding of the probability distribution of millions of pair-wise co-associations over the entire 4D velocity vector-field to derive a unique profile of the patient's complex 3D flow dynamics: 4D vector-field signature. Our proposed technique consists of two stages. First, a unique signature profile of the patient's complex 3D flow-field dynamics over time is constructed. Second, a Hemodynamic Signature Index (HSI) is computed as a measure of the degree of 4D flow signature alteration. Our technique was extensively evaluated in three datasets of a total of 106 4D Flow MRI scans from 91 subjects including 34 healthy controls, 57 BAV patients and 15 Rescan subjects. The experiments showed the techniques' convergence, high robustness to segmentation errors, excellent in vivo scan-rescan day-to-day reproducibility. The technique identified consistent signatures in healthy controls and distinctly altered signatures in BAV patients reflected in significantly different HSI.

2   Methodology

Our proposed 4D Flow vector-field signature is based on stochastically encoding the distribution of multi-million pair-wise vector-field co-associations of the in vivo measured 4D flow MRI data to derive the unique signature of the entire mixture of complex 3D flow vector-field dynamics. The proposed technique consists of two stages: signature profile construction and hemodynamic signature index computation.

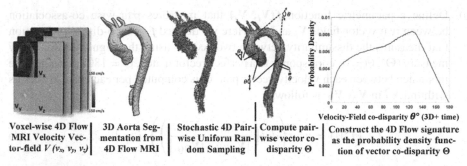

Fig. 1. Illustrative Summary of the stochastic 4D Flow Vector-Field Signature Construction Steps

2.1 Stage I: Stochastic 4D Flow Vector-Field Signature Construction

Given in vivo voxel-wise 4D Flow MRI velocity vector-field $\mathbf{V}(v_x, v_y, v_z)$, the following steps are performed (illustrated in Fig. 1):

1) Segment the region/organ of interest for analysis (i.e., 3D aorta) from 4D Flow MRI data.
2) Define a range of time points [t0, T] in the cardiac cycle to derive the signature. Here we studied aortic 4D flow overall systole time points; average 9-time points per subject.
3) Define $\mathbf{X}$ as a long 1D vector that includes the indices of all the voxels within the 3D segmentation (i.e., 3D aorta) over the entire time points as follows:

$$X = \left(X_1^{t0}...X_N^{t0}, X_1^{t0+1}...X_N^{t0+1}, ..., X_1^T...X_N^T \right) \tag{1}$$

with N as the total number of voxels in the 3D segmented aorta at a single time point.

4) Perform stochastic discrete uniform random sampling of η point (voxel) pairs over the domain X into two 1D indices vectors $\mathbf{A}$, $\mathbf{B}$ as follows:

$$A = U(X, \eta), B = U(X, \eta) \tag{2}$$

with U as a discrete uniform random sampler and η as the number of pair-wise random samples. To enable systematic patient-specific density sampling for consistent patient comparison, we define η as a function of each subject's 3D segmented volume (i.e., aorta size) over t_0 to T. If L is the length of the vector X (i.e., $L = N \times (T - t_0)$), then η is computed as:

$$\eta = \alpha \times L; \text{ such that } 0 < \alpha \leq L - 1 \text{ is the sampling density factor} \tag{3}$$

5) Define $\mathbf{V_A}$, $\mathbf{V_B}$ as two long velocity vector-field vectors, each with size $\eta \times 3$. Each element i in $\mathbf{V_A}$, $\mathbf{V_B}$ holds the 4D Flow MRI 3-directional velocity vector-field $\mathbf{V}_i$ (v_x, v_y, v_z) of the index given by the element i in the vector $\mathbf{A}$, and vector $\mathbf{B}$ from Eq. 2.

6) Define a parametric function $f(V_i, V_j)$ that measures pair-wise co-association between two vector-fields V_i and V_j. Here we defined f as a co-disparity function that measures the dissimilarity between two vectors using the angular dissimilarity measure (Θ). $\Theta = 0°$ corresponds to a perfect vector match, $\Theta = 180°$ as a complete mismatch between each velocity vector pair. Θ is computed per each pair of vectors with index i in V_A, V_B as follows:

$$f : \Theta°(i) = acos\left(dot\left(\frac{V_{A(i)}}{|V_{A(i)}|}, \frac{V_{B(i)}}{|V_{B(i)}|}\right)\right), i = 1, \ldots, \eta \qquad (4)$$

The output from step (6) is a scalar co-disparity 1D vector Θ with length η. Note that the angular dissimilarity measure is scale-invariant (i.e. invariant to vector magnitude).

7) For each subject k, compute the standardized signature S as the probability density function (pdf) of Θ as estimated by a frequency-normalized histogram of B bins:

$$S_k = pdf(\theta, B); \quad B = 180 \text{ equally}-\text{spaced bins was used here} \qquad (5)$$

2.2 Stage II: Hemodynamic Signature Index (HSI)

To enable consistent comparison of the derived 4D vector-field signature profile changes between patients and controls, we introduce this index to quantitatively assess the degree of alteration in each patient's signature compared to controls. A smaller HSI value means more similarity of the individual's signature to controls, while a larger value corresponds to more flow velocity-field alteration. To compute HSI for a specific patient, we first calculate the dissimilarity value (d) between the patient's 4D Flow signature profile (pdf) against each of the individual controls' signature profiles using a probability distribution distance measure. Here we used the Earth Mover's Distance (EMD) between discrete distributions [18] as the signature distance d:

$$d_{p,q} = EMD(S_p, S_q); \quad p = 1, \ldots, F; \quad q = 1, \ldots, Cn \qquad (6)$$

Such that F is the number of patients and Cn is the number of controls. Then the HSI of patients (HSI_{pat}) is computed as:

$$HSI_{pat\#p} = 95^{th} \text{ percentile } (d_{p,q=1}, d_{p,q=2}, \ldots, d_{p,q=Cn});$$
$$p = 1, \ldots, F; \quad q = 1, \ldots, Cn \qquad (7)$$

The output of this step is a vector (HSI_{pat}) containing a single index (HSI) per patient.

Controls' Inter-Individual Signature Variations (HSIcntrl). To capture and account for the degree of "normal" inter-individual physiologic variations in the 4D Flow signatures in controls when performing patient comparisons, we compute baseline HSI in controls. First, the signature distance (d) between individual controls is computed as:

$$d_{i,q} = EMD(S_i, S_q); \quad i \neq q; i = 1, \ldots, Cn; q = 1, \ldots, Cn \qquad (8)$$

Then the HSI of controls ($\mathbf{HSI_{cntrl}}$) is computed as:

$$\mathbf{HSI_{cntrl\#i}} = 95^{\text{th}} \text{ percentile } \left(\mathbf{d}_{i,q=1}, \mathbf{d}_{i,q=2}, \ldots, \mathbf{d}_{i,q=Cn}\right);$$

$$i = 1, \ldots, Cn; \quad q = 1, \ldots, Cn \tag{9}$$

The output of this step is a vector ($\mathbf{HSI_{cntrl}}$) containing a single index (HSI) per control.

3 Datasets and Preprocessing

We evaluated the technique in an IRB-approved retrospective database of three datasets: a dataset of 34 healthy controls (age, 44 ± 14 years; 12 women) and a dataset of 57 BAV patients (age, 45 years $\pm$ 12; 18 women). To test the reproducibility, another dataset of 15 healthy subjects (a subset of the controls) underwent a rescan 2–3 weeks post the first 4D Flow MRI scan. All subjects underwent in vivo aortic 4D Flow MRI scanning at 1.5 T or 3.0 T (Siemens) covering the 3D aorta with a spatial resolution of 1.66–2.81×1.66–2.81×2.20–$3.50 \, \text{mm}^3$ and a temporal resolution of 37–43 ms. Following standard 4D flow MRI preprocessing, data were corrected for phase offset errors, aliasing, and Maxwell terms and used to compute a static 3D phase-contrast angiography (PCMRA) [1]. The 3D aorta was manually segmented from the 3D PCMRA using commercial software (Mimics; Materialise) as previously described [1, 19].

4 Convergence Analysis of Optimal Sampling Density Factor (α)

To systematically identify an optimal stochastic sampling scheme that allows unbiased patient comparison, we define η (Eq. 3) as a function of each patient's 3D segmented volume over time (L); and a density sampling factor α. Since L is patient-specific and readily derived from the number of 3D segmentation voxels, we only need to identify optimal α that provides consistently stable 4D Flow signature computation in patients. To identify the optimal α, for each subject we started with $\alpha = 1$ and incremented α by 1 iteratively until convergence. Optimal α, i.e., convergence, was defined as the largest α value overall patients and controls at which the signature profile between two consecutive iterations does not change more than 1% (for 10 iterations) as defined by the coefficient of variation (CV) $\leq 1\%$. To avoid local minima and ensure generalization, we required for convergence that CV $\leq 1\%$ for 10 consecutive α iterations. Hence, the optimal α is defined as the α value at the 10th iteration of a consecutive CV $\leq 1\%$. The CV between signatures of two consecutive α iterations ($S_\alpha, S_{\alpha+1}$) was computed using the root mean square error method of duplicate measurements [20] as: $CV(\%) = 100 \times \sqrt{\sum(d/m)^2/2B}$, with $d = S_{\alpha+1} - S_\alpha$ and $m = S_{\alpha+1} + S_\alpha$.

5 Signature Sensitivity Analysis to Segmentation Errors

The premise of our proposed technique's robustness to potential segmentation errors is that the 4D flow field signature's definition as a probability density function by definition indicates that relatively small density errors (relative to the size and prevalence of flow dynamics in the aorta) would appear in the tail of the distribution in small densities. Hence, would have a minor impact on the overall signature profile (*pdf*). To assess the sensitivity of the proposed signature to potential segmentation errors, we simulated potential segmentation errors as randomized dilation, erosion or no change to the original 3D segmentation. To achieve this, we used two uniform random generators. The first random generator outputs a random value of 'Dilate', 'Erode', or 'No change' for each individual subject/patient. The second uniform random generator outputs the random width value in the range [1, 3] voxels for the cubic structure element for each subject (i.e. in case of 'Dilate' or 'Erode' choice was generated) used for dilation or erosion all around the 3D aorta segmentation. To mitigate bias to the random generator, we repeated this process 10 times overall controls and patients – at each iteration each subject receives new combination from the two random error generators. This randomized error approach was used to mitigate bias towards simple consistent errors and push the algorithm to mimic more realistic challenging variable segmentation errors.

6   In Vivo Scan-Rescan Reproducibility Study

Our proposed 4D Flow signature technique is aimed at evaluating in vivo 4D Flow MRI data of patients and healthy controls for which day-to-day physiologic variabilities between scans can exist. Hence, we aimed to assess the robustness of our proposed signature analysis to realistic day-to-day scan-rescan variabilities and uncertainties in an in vivo clinical setup. Therefore, we studied 15 healthy subjects who each underwent 4D Flow MRI scanning twice on different days with 2–3 weeks. We assessed the reproducibility between the scan and rescan 4D Flow signatures using Bland-Altman plots, coefficient of variation (CV) [20] and Intra-class correlation (ICC).

7   Signature Comparison of BAV Patients and Healthy Controls

Ultimately, the proposed 4D Flow signature technique aims to help assess the degree of complex changes in 3D blood flow dynamics in patients (i.e. versus healthy) from in vivo 4D Flow MRI. Therefore, we aimed to evaluate the potential of the proposed signature technique in identifying distinct flow signature alterations in BAV patients compared to the healthy controls. Here, we analyzed the 4D flow signature of the 3D aortic flow over the entire systolic time frames (9 ± 2 time frames identified from flow time-curves in all healthy controls and BAV patients). We computed and compared both the signature profile and the HSI measure. Statistical comparison was performed using Mann-Whitney U-test with p value <0.05 as significant.

8   Results

In all 106 analyzed 4D Flow MRI scans (34 healthy controls, 57 BAV patients, 15 Rescans), the proposed 4D Flow-field signature analysis was successful. Detailed results of the sampling density (α) convergence is given in Fig. 2 (left panel). Convergence

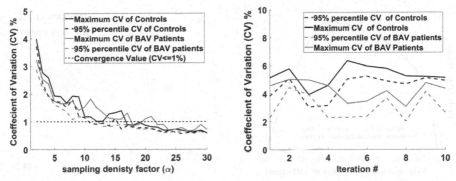

Fig. 2. (Left panel) Convergence analysis of the sampling density factor α. (Right panel) Sensitivity analysis for randomly-generated segmentation errors up to 3-voxels (dilation or erosion) over 10 iterations. Each point corresponds to the maximum (solid) or 95% percentile Coefficient of Variation (CV) % overall 34 healthy controls (black lines) and 57 patients (red lines) (Color figure online).

analysis identified the optimal density sampling $\alpha = 30$ overall healthy controls and patients. Given that α is used to derive the patient-specific number of pair-wise samples η (Eq. 3), using the optimal $\alpha = 30$ resulted in $\eta = 5.5 \pm 2$ million (95%CI: [4.8, 6.2] million; CI: Confidence Interval) pair-wise samples in healthy controls and $\eta = 6.2 \pm 1.8$ million (95%CI = [5.7, 6.7] million) pair-wise samples in BAV patients. The optimal $\alpha = 30$ was then used for all experiments in this work. The sensitivity analysis results to potential segmentation errors over the 10 iterations are shown in Fig. 2 (right panel). Compared to the signature generated from original segmentation, the results indicate excellent robustness in BAV patients with CV = 0.95% ± 0.91% (95%CI = [0.29%, 3.9%]) and in healthy controls CV = 1.3% ± 1.4% (95%CI = [0.3%, 5.2%]). That is, the random segmentation errors (up to 3 voxels) all around the 3D aorta, induced on average, a very low variability of only ~1% (overall <5% in the worst iterations).

Collectively, these results indicate the strong robustness of our proposed technique to typical segmentation errors. Notably, the proposed signature showed excellent in vivo Scan-Rescan reproducibility with CV = 1.8% ± 1.2% (95%CI = [0.4%, 4.7%]) and ICC = 0.99 ± 0.007 (95%CI = [0.97, 0.99]). The excellent Signature profile reproducibility between Scan-Rescans is shown in Fig. 3 (left panel). The Bland-Altman plot showed negligible bias and narrow limits of agreement (Fig. 3 right panel). Importantly, the 4D Flow signature analysis identified significant differences in the signature profile between BAV patients and healthy controls (Fig. 4). The controls' 4D Flow signatures were consistent, while BAV patient signatures were distinctly altered (Fig. 4 – left panel). BAV signature profiles showed higher velocity vector-field disparities indicated by higher probability density of the higher Θ values (i.e., mismatching pair-wise vectors) and lower probability density of low Θ values (matching pair-wise vectors). These differences were reflected in a higher HSI in BAV patients (Fig. 4 – right panel, p < 0.0001).

Efficiency and Processing Time. The vectorized pair-wise computation nature of the algorithm (Eq. 1–5) makes it efficient for processing with linear complexity $O(\eta)$ with

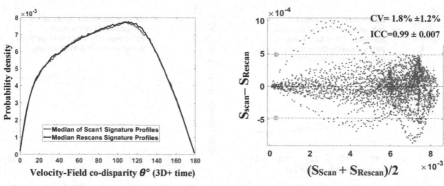

Fig. 3. Scan-Rescan Signature reproducibility results in 15 healthy subjects showing (Left panel) Comparison of the median signature profiles of all baseline scans i.e. scan 1 (in black) versus the median of all Rescans (in red). (Right panel) Bland-Altman plot of all subjects signature profiles (i.e., 180 probability density bins of Θ values per subject) of the baseline scan (S_{scan}) and Repeat scan i.e. Rescan (S_{Rescan}). ICC: Intraclass correlation; CV: coefficient of variation. (Color figure online)

potential for further efficiency by parallelization. The proposed algorithm was implemented in Matlab R2018a (Mathworks, Inc.) on an average DELL desktop computer with Processor Intel(R) Core(TM) i7-8700K CPU @ 3.70 GHz, 3696 MHz with 64 GB RAM. The analysis was performed using serial processing (no parallel programming). In this work, signature computation time was 2.8 ± 0.9 s (95%CI = [2.4, 3.0] s) in controls and 3.1 ± 0.8 s (95%CI = [2.9, 3.3] s) in BAV patients.

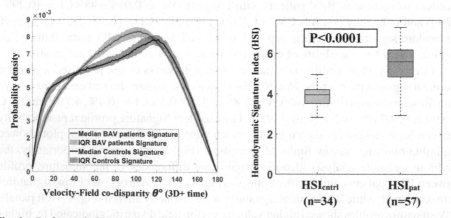

Fig. 4. (Left panel) The median and interquartile range (IQR) of 4D Flow signature profiles of all 34 healthy controls (black) and all 57 BAV patients (red). (Right panel) Box plot of the derived Hemodynamic Signature Index (HSI) values for 34 healthy controls and 57 BAV patients. (Color figure online)

9 Discussion and Conclusions

This paper introduces a new data-driven methodological stochastic approach for quantifying 4D flow vector-field data from 4D Flow MRI with application to aortic valve disease. The proposed quantitative signature stochastically encodes the individual's unique profile of the overall complex 3D flow dynamics. The proposed signature is derived directly from the entire 3D time-resolved three-directional 4D Flow MRI velocity vector-field, enabling a comprehensive utilization and highly automated analysis. Our stochastic technique does not require any spatial or temporal derivative/gradient computations. Therefore, this makes our technique inherently more robust to spatiotemporal resolutions than existing gradient based flow analysis methods that are intrinsically resolution-dependent and can be error-prone [1]. The reported results showed the technique's high potential in detecting significantly altered signatures of the complex 3D aorta flow dynamics in BAV patients versus healthy controls. Notably, the signature definition as a probability density function makes it 1) intrinsically standardized (area under the curve = 1), 2) volume-normalized (independent of aorta size) and, 3) does not require pre-registration for patient comparison, 4) robust to small-frequency errors as these would appear in the tail of the distribution. This was evidenced by the demonstrated excellent robustness to potential segmentation errors. The reported excellent Scan-Rescan reproducibility may indicate the reliability of the proposed signature towards day-to-day scan variability that would typically be part of a clinical routine, e.g., follow-up scanning. Notably, the patient-specific definition of the stochastic sampling density η as a function of the aorta size and the density factor α (Eq. 3) ensures equivalent sampling density even in subjects with different aorta sizes for unbiased signature comparison. While comprehensive, the algorithm is linear in complexity yielding near real-time results within average mere 3 s and can easily be further parallelized. Here, we assessed aortic flow dynamics over the systole. In the future, we will extend to the entire cardiac cycle. In principle, our technique is generic and could be applied to quantify other vector-field data beyond velocity and to be explored. We will also explore different parametric co-association functions to capture different level of vector-field details and will assess the impact of resolution and noise. Given the encouraging results, we will assess the technique in a large cohort of BAV patients and other cardiovascular diseases beyond the aorta (e.g. left atrium and left ventricle).

Acknowledgements. The first author's research is supported in part by Transformational Project Award AHA 20TPA35490311 from the American Heart Association (AHA), and grant R21 HL150498 from the National Heart, Lung, and Blood Institute of the National Institutes of Health (NIH-NHLBI).

References

1. Dyverfeldt, P., et al.: 4D flow cardiovascular magnetic resonance consensus statement. J. Cardiovasc. Magn. Reson. **17**, 72 (2015)
2. Markl, M., Frydrychowicz, A., Kozerke, S., Hope, M., Wieben, O.: 4D flow MRI. J. Magn. Reson. Imaging **36**, 1015–1036 (2012)

3. Mahadevia, R., et al.: Bicuspid aortic cusp fusion morphology alters aortic three-dimensional outflow patterns, wall shear stress, and expression of aortopathy. Circulation **129**, 673–682 (2014)
4. Bissell, M.M., et al.: Aortic dilation in bicuspid aortic valve disease: flow pattern is a major contributor and differs with valve fusion type. Circulat. Cardiovasc. Imaging **6**(4), 499–507 (2013). https://doi.org/10.1161/CIRCIMAGING.113.000528
5. Guzzardi, D.G., et al.: Valve-related hemodynamics mediate human bicuspid aortopathy: insights from wall shear stress mapping. J. Am. Coll. Cardiol. **66**, 892–900 (2015)
6. Garcia, J., Barker, A.J., Markl, M.: The role of imaging of flow patterns by 4D flow MRI in aortic stenosis. JACC: Cardiovasc. Imag. **12**(2), 252–266 (2019)
7. Hope, M.D., et al.: 4D flow CMR in assessment of valve-related ascending aortic disease. JACC: Cardiovasc. Imag. **4**(7), 781–787 (2011)
8. Hope, M.D., et al.: Bicuspid aortic valve: four-dimensional MR evaluation of ascending aortic systolic flow patterns. Radiology **255**, 53–61 (2010)
9. Garcia, J., Markl, M., Collins, J., Carr, J., Barker, A.J.: Volumetric quantification of localized normalized helicity in patients with bicuspid valve and aortic dilation. In: Proceedings of the International Society for Magnetic Resonance in Medicine, p. 2716.
10. Barker, A.J., et al.: Bicuspid aortic valve is associated with altered wall shear stress in the ascending aortaclinical perspective. Circul. Cardiovasc. Imag. **5**, 457–466 (2012)
11. Elbaz, M.S.M., et al.: Four-dimensional virtual catheter: noninvasive assessment of intra-aortic hemodynamics in bicuspid aortic valve disease. Radiology 190411
12. Dyverfeldt, P., Hope, M.D., Tseng, E.E., Saloner, D.: Magnetic resonance measurement of turbulent kinetic energy for the estimation of irreversible pressure loss in aortic stenosis. JACC: Cardiovasc. Imag. **6**(1), 64–71 (2013). https://doi.org/10.1016/j.jcmg.2012.07.017
13. Carlsson, M., Heiberg, E., Toger, J., Arheden, H.: Quantification of left and right ventricular kinetic energy using four-dimensional intracardiac magnetic resonance imaging flow measurements. Am. J. Physiol. Heart Circul. Physiol. **302**, H893–H900 (2012)
14. Elbaz, M.S., Calkoen, E.E., Westenberg, J.J., Lelieveldt, B.P., Roest, A.A., van der Geest, R.J.: Vortex flow during early and late left ventricular filling in normal subjects: quantitative characterization using retrospectively-gated 4D flow cardiovascular magnetic resonance and three-dimensional vortex core analysis. J. Cardiovasc. Magn. Reson. **16**, 78 (2014)
15. von Spiczak, J., Crelier, G., Giese, D., Kozerke, S., Maintz, D., Bunck, A.C.: Quantitative analysis of vortical blood flow in the thoracic aorta using 4D phase contrast MRI. PLoS ONE **10**, e0139025 (2015)
16. Barker, A.J., et al.: Viscous energy loss in the presence of abnormal aortic flow. Magn. Reson. Med. **72**, 620–628 (2014)
17. Elbaz, M.S., et al.: Assessment of viscous energy loss and the association with three-dimensional vortex ring formation in left ventricular inflow: in vivo evaluation using four-dimensional flow MRI. Magn. Reson. Med. **77**, 794–805 (2017)
18. Rubner, Y., Tomasi, C., Guibas, L.J.: The earth mover's distance as a metric for image retrieval. Int. J. Comput. Vision **40**, 99–121 (2000)
19. Bock, J., Kreher, B., Hennig, J., Markl, M.: Optimized pre-processing of time-resolved 2D and 3D phase contrast MRI data. In: Proceedings of the 15th Annual Meeting of ISMRM, Berlin, Germany, vol. 3138 (2007)
20. Hyslop, N.P., White, W.H.: Estimating precision using duplicate measurements. J. Air Waste Manag. Assoc. **59**, 1032–1039 (2009)

Source-Free Domain Adaptive Fundus Image Segmentation with Denoised Pseudo-Labeling

Cheng Chen$^{(\boxtimes)}$, Quande Liu, Yueming Jin, Qi Dou, and Pheng-Ann Heng

Department of Computer Science and Engineering,
The Chinese University of Hong Kong, Hong Kong, China
cchen@cse.cuhk.edu.hk

Abstract. Domain adaptation typically requires to access source domain data to utilize their distribution information for domain alignment with the target data. However, in many real-world scenarios, the source data may not be accessible during the model adaptation in the target domain due to privacy issue. This paper studies the practical yet challenging source-free unsupervised domain adaptation problem, in which only an existing source model and the unlabeled target data are available for model adaptation. We present a novel denoised pseudo-labeling method for this problem, which effectively makes use of the source model and unlabeled target data to promote model self-adaptation from pseudo labels. Importantly, considering that the pseudo labels generated from source model are inevitably noisy due to domain shift, we further introduce two complementary pixel-level and class-level denoising schemes with uncertainty estimation and prototype estimation to reduce noisy pseudo labels and select reliable ones to enhance the pseudo-labeling efficacy. Experimental results on cross-domain fundus image segmentation show that without using any source images or altering source training, our approach achieves comparable or even higher performance than state-of-the-art source-dependent unsupervised domain adaptation methods (Code is available at https://github.com/cchen-cc/SFDA-DPL).

Keywords: Source-free domain adaptation · Self-training · Pseudo-labeling

1 Introduction

Deep neural networks remain notoriously vulnerable to the domain shift between the training (source) and testing (target) data acquired under different conditions [6,7,12]. To address the domain shift problem, unsupervised domain adaptation (UDA) has been an active research topic. Existing UDA methods typically require to simultaneously access the source and target data for distribution alignment [2,10,11,16,20,23–27]. However, in real-world scenarios, the source data are often inaccessible during model adaptation in the target domain, because

© Springer Nature Switzerland AG 2021
M. de Bruijne et al. (Eds.): MICCAI 2021, LNCS 12905, pp. 225–235, 2021.
https://doi.org/10.1007/978-3-030-87240-3_22

medical data are strictly regulated and prohibitive to be shared before taking complex ethical procedures. An interesting yet less explored problem is: Can a model adapt to target domains without using the source data? This problem has high practical value as often only a learned model is given instead of the source training data due to data privacy and security issues.

We are concerned with a more challenging *source-free unsupervised domain adaptation* setting, in which only a well-trained source model and unlabeled target domain data are provided for model adaptation. Such a setting has wider applicability than vanilla UDA, since data transmission is not required and the model adaptation is only performed in the target domain. Previous UDA methods relying on distribution alignment are not suitable in this scenario as the source distribution information are not accessible. For example, the adversarial learning based methods [12,21] need to align the source and target features within a single model, which is infeasible without source data. A few very recent studies have started to explore the possibility of adapting a model in the absence of source data. Bateson et al. [1] introduce an auxiliary branch in source model to estimate class-ratio prior and propose a prior-regularized entropy minimization strategy for source-relaxed adaption. Stan et al. [19] encode the source samples into a prototypical distribution, which is transferred to the target domain for distribution alignment. In [13] and [9], auto-encoders are pre-trained in the source domain for denoising or image reconstruction and are leveraged in target domain for adaptation. However, all these methods need to alter the source-domain training with auxiliary branches or additional training tasks, hence cannot adapt the existing models that already been well-trained on the source domains.

Given only an existing model trained on the source domain and unlabeled target data, the challenge lies in how to realize the model self-adaptation using the knowledge purely from the target domain. Our insight is to make the best use of the source model and target data via self-training [14], that is generating pseudo labels in the target domain with the model's own predictions and re-training the model based on the generated pseudo labels. In real-world model deployment, the most common situation is the model adaptation cross different clinical centers, where the encountered domain shift caused by the scanner difference is generally minor compared with cross-modality discrepancy (e.g., CT and MRI). Considering this practical situation, a model's own predictions on target data can be reasonable and provide a meaningful supervision base for model self-training. However, some inevitable noise, that is the incorrect predictions caused by domain shift, shall harm pseudo-labeling for model self-adaptation. How to reduce the noisy pseudo labels is crucial for yielding its best efficacy.

In this paper, we present a novel denoised pseudo-labeling (DPL) method for source-free unsupervised domain adaptation. Our method roots in the pseudo-labeling strategy, which enables effective self-training in the target domain with only a pre-trained source model and unannotated target data, without using any source-domain data or altering source-domain training. Importantly, to reduce the training noise in pseudo-labeling, we propose two complementary pixel-level and class-level denoising schemes with uncertainty estimation and prototype estimation, to provide more discriminative and less noisy supervision for model adaptation. First, we estimate the model's pixel-level prediction uncertainty, and identify potential unreliable pseudo labels with high uncertainty. Second, we estimate the class-level prototypes motivated by the prototypical networks [18] and calculate relative feature distance, to suggest noisy pseudo labels which lie further away from their corresponding class prototypes. We evaluate our method on two public fundus image datasets for the multi-label optic disc and cup segmentation, which are popular benchmarks for UDA tasks. Without requiring any source data, our method achieves comparable or even higher performance than state-of-the-art source-dependent UDA methods.

2   Method

Figure 1 illustrates our source-free UDA framework via denoised pseudo-labeling. In this section, we first present the pseudo-label learning for source-free UDA. Next, we propose two complementary *pixel-level* and *class-level* pseudo label denoising schemes. The training procedures are finally described.

2.1   Pseudo-Labeling for Source-Free UDA

In source-free UDA setting, we are given a model $f^s : \mathcal{X}^s \to \mathcal{Y}^s$ trained from unknown source domain $\mathcal{D}^s = (\mathcal{X}^s, \mathcal{Y}^s)$, and an unlabeled dataset $\{x_i^t\}_{i=1}^{N^t}$ from target domain $\mathcal{D}^t$, where $x_i^t \in \mathcal{X}^t$. The goal of source-free UDA is to adapt the model f^s with only $\{x_i^t\}_{i=1}^{N^t}$, such that the obtained model $f^{s \to t}$ can perform well on the target domain distribution. We consider fundus image segmentation, which is a multi-label segmentation problem with $x_i \in \mathbb{R}^{H \times W \times 3}$ and $y_i \in \{0, 1\}^{H \times W \times C}$ (C denotes the class number).

Ideal model adaptation towards target distribution would require the ground truth labels at target domain, which are not available in the source-free UDA setting. To address this problem, we devise an effective approach to assign pseudo labels to the target samples so that the model could be adapted via supervised learning on the target domains. Specifically, given an unlabeled image from the target domain, we denote p_v as its prediction probability from the source model on v-th pixel. Then the corresponding pseudo label can be generated as:

$$\hat{y}_v^t = \mathbb{1}[\, p_v \geq \gamma \,], \tag{1}$$

where $\mathbb{1}(\cdot)$ is the indicator function, $\gamma \in (0, 1)$ is a probability threshold to generate binary pseudo labels for the multi-label segmentation task. With the generated pseudo labels, we can then adapt the source model towards target data distribution with supervised learning:

$$\mathcal{L}_{ce} = -\sum_v [\hat{y}_v^t \cdot log(p_v) + (1 - \hat{y}_v^t) \cdot log(1 - p_v)]. \tag{2}$$

The target function $f^{s \to t}$ is obtained by updating the source model over all target domain data $\{x_i^t\}_{i=1}^{N^t}$ with generated pseudo labels $\{\hat{y}_i^t\}_{i=1}^{N^t}$.

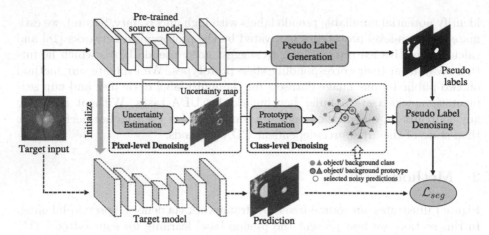

Fig. 1. Overview of our proposed *denoised pseudo-labeling* (DPL) for source-free UDA. Pseudo labels are generated from source model's predictions to guide the self-training in target domain. Two complementary denoising schemes with pixel-level uncertainty estimation and class-level prototype estimation are developed to reduce the noisy pseudo labels and preserve reliable ones to provide more helpful and less noisy supervision.

2.2   Pixel-Level Denoising via Uncertainty Estimation

Due to the discrepancy between source and target domains, the pseudo labels assigned from source model would unavoidably contain noisy and incorrect predictions. Adapting the model with false pseudo labels is harmful to further improve the model's discrimination capacity in target domain due to the accumulated segmentation errors. To this end, we aim to carefully filter out the unreliable pseudo labels and only preserve the accurate ones to enhance the model adaptation in target domain. The conventional way for pseudo label selection is directly based on the network output confidence probability to select high-confidence pseudo labels. However, the model's predictions under domain shift are prone to be over-confident and incorrect predictions can also have high confidence scores [8]. Motivated by the observation in previous supervised segmentation work that uncertainty measures correlate with incorrect predictions, we leverage the pixel-level uncertainty information from the model's predictions to better indicate the unreliability of pseudo label on each pixel of the segmentation map.

We estimate uncertainty with Monte Carlo Dropout [5] for Bayesian approximation. Specifically, given a target image x^t, we enable dropout at inference and perform K stochastic forward passes through the source model to obtain a set of predictive outputs, i.e., $p_k = f^s(x^t), k = 1, ..., K$. The uncertainty map $u^{H \times W \times C}$ is then estimated as the standard deviation of the K outputs, i.e., $u = std(p_1, ..., p_K)$. To utilize the uncertainty map to filter noisy pseudo labels, we define a label selection mask with a binary variable $m \subseteq \{0, 1\}^{H \times W \times C}$, whose value at pixel v is determined by the uncertainty measure u_v with:

$$m_v = \mathbb{1}[\, u_v < \eta \,], \tag{3}$$

where η is the uncertainty threshold. If uncertainty is sufficiently low, we have $m_v = 1$ to select the corresponding pseudo label $\hat{y}_v^t$; otherwise, the pseudo label $\hat{y}_v^t$ is regarded as a noisy one and excluded from the loss calculation in Eq. 2.

2.3 Class-Level Denoising via Prototype Estimation

The pixel-level denoising scheme estimates the prediction uncertainty for each pixel separately. However, the segmentation problem has structural characteristics and the segmentation regions of the same category are often highly correlated. In this regard, we also propose a new class-level denoising scheme by leveraging prototypes, i.e., class-wise feature centroids, to provide guidance for identifying potential unreliable pseudo labels. This is based on the assumption that for a reliable pseudo label, its feature should lie closer to its corresponding class prototype, otherwise a potential noisy pseudo label is implied.

Specifically, we exploit the relative feature distances to the class prototypes to determine whether a pseudo label is reliable. Given a target image x^t and its generated pseudo label $\hat{y}^t$, we denote its feature map from the layer before the last convolution as $e_l \in \mathbb{R}^{H_l \times W_l \times L}$ (L denotes the channel number), which is then bilinearly interpolated to $e \in \mathbb{R}^{H \times W \times L}$ to keep consistency with the dimensions of $\hat{y}^t \in \mathbb{R}^{H \times W \times C}$. To extract the class-wise prototypes, we utilize the proposed uncertainty-guided pseudo labels to indicate the category and collect corresponding feature vectors for each object (foreground) class and background class, given that the ground truth of target images is unavailable. Herein, the binary object mask $b^{obj} \in \mathbb{R}^{H \times W \times C}$ can be obtained from $\hat{y}^t$ as $b^{obj} = \mathbb{1}[\hat{y}^t = 1]\mathbb{1}[u < \eta]$, and the binary background mask $b^{bg} \in \mathbb{R}^{H \times W \times C}$ can be obtained as $b^{bg} = \mathbb{1}[\hat{y}^t = 0]\mathbb{1}[u < \eta]$. Then, the class-wise prototypes z^{obj} and z^{bg}, and the relative feature distance d_v^{obj} and d_v^{bg} between each feature vector for pixel v to the two prototypes are calculated as:

$$z^{obj} = \frac{\sum_v e_v \cdot b_v^{obj} \cdot p_v}{\sum_v b_v^{obj} \cdot p_v}, \quad z^{bg} = \frac{\sum_v e_v \cdot b_v^{bg} \cdot (1 - p_v)}{\sum_v b_v^{bg} \cdot (1 - p_v)}, \tag{4}$$

$$d_v^{obj} = ||e_v - z^{obj}||_2, \quad d_v^{bg} = ||e_v - z^{bg}||_2, \tag{5}$$

in which we also incorporate the network output probability p_v to weigh the contribution of each pixel to the prototype calculation. Based on the estimated prototypes and feature distance information, if $\hat{y}_v^t$ classifies pixel v as an object, but its feature vector e_v is further away from the object prototype z^{obj} than the background prototype z^{bg}, it is likely that the pseudo label is unreliable. We then remove it from the selection mask to denoise the pseudo labels. Based on this denoising scheme, the label selection mask in Eq. 3 is updated as:

$$m_v = \mathbb{1}[u_v < \eta]\mathbb{1}[\hat{y}_v^t = 1]\mathbb{1}[d_v^{obj} < d_v^{bg}] + \mathbb{1}[u_v < \eta]\mathbb{1}[\hat{y}_v^t = 0]\mathbb{1}[d_v^{obj} > d_v^{bg}]. \tag{6}$$

By combining the uncertainty estimation and prototype estimation in Eq. 6, a pseudo label $\hat{y}_v^t = 1$ ($\hat{y}_v^t = 0$) is selected when network's uncertainty on prediction is low and encoded feature lies closer to the object (background) prototype than the background (object) prototype. The segmentation loss with pseudo labels in Eq. 2 is updated with label selection mask as $\mathcal{L}_{seg} = \sum_v m_v * L_{ce,v}$.

2.4   Training Procedures and Implementation Details

Given the target images $\{x_i^t\}_{i=1}^{N^t}$ and the source model f^s, we first apply f^s on all the target images to generate predictions. Pseudo labels are obtained from these predictions with Eq. 1 and denoised with Eq. 6. We then initialize the target model $f^{s \rightarrow t}$ with f^s and optimize $f^{s \rightarrow t}$ with the denoised pseudo labels. We employ a MobileNetV2 adapted DeepLabv3+ [3] as network backbone, following the previous work [22] without the boundary branch. For uncertainty estimation with Monte Carlo Dropout, the dropout rate is set to 0.5 and 10 stochastic forward passes are performed to obtain the standard deviation of the output probabilities. The threshold γ for multi-label task is set as 0.75 referring to [22] and the uncertainty threshold η is set as 0.05 by visually inspecting the uncertainty map. We use weak augmentations including Gaussian noise, contrast adjustment, and random erasing to slightly disturb the inputs when training $f^{s \rightarrow t}$, in order to make the predictions deviate from pseudo labels. The model is trained using Adam optimizer with momentum of 0.9 and 0.99, and learning rate of 2e−3. We train the target model with batch size 8 and 2 epochs. The framework is implemented with Pytorch 0.4.1 using one NVIDIA TitanXp GPU.

3   Experiments

Dataset and Evaluation Metrics. We validate our method on optic disc and cup segmentation for retinal fundus images, using datasets collected from different clinical centers with distribution shifts. Specifically, we employ the training set of the REFUGE challenge [15] as the source domain, and the public RIM-ONE-r3 [4] and Drishti-GS [17] datasets as two different target domains. The source domain includes 400 annotated training images, and the two target domain data are split to 99/60 and 50/51 images for training/testing respectively, following the same experimental setup in [22]. Each image is pre-processed by cropping a 512×512 disc region as the network input. For evaluation, we employ two commonly-used metrics, including the Dice coefficient for pixel-wise accuracy measure and the Average Surface Distance (ASD) for boundary agreement assessment. The higher Dice and lower ASD indicate a better performance.

Comparison with State-of-the-Arts. We compare our method with recent state-of-the-art domain adaptation methods, including **BEAL** [22]: an UDA method with adversarial learning between *source* and target data with boundary prediction, which is the best reported UDA model on cross-domain fundus image segmentation; **AdvEnt** [21]: a popular UDA benchmark approach that encourages entropy consistency between the *source* and target domains; **SRDA** [1]: a domain adaptation approach that utilizes the task prior trained from *source* *domain* for model adaptation in the target domain; **DAE** [13]: a domain adaptation approach which utilizes a denoising auto-encoder learned from *source domain* to gradually refine the segmentation mask in target domain. Note that these methods can serve as strong baselines in our source-free UDA setting since they utilize more source domain information by either accessing source domain data or altering source domain training, while our method is completely free of the source domain. Since we follow the same network backbone and data split as BEAL, we referenced the performance reported in their paper for comparison.

Table 1. Quantitative comparison of different methods on the target domain datasets. (Note: S. denotes source domain, – means the results are not reported by that methods.)

Methods	Access/Alter		Optic disc segmentation		Optic cup segmentation	
	S. Data	S. Train	Dice [%]	ASD [pixel]	Dice [%]	ASD [pixel]
RIM-ONE-r3						
W/o adaptation			83.18 ± 6.46	24.15 ± 15.58	74.51 ± 16.40	14.44 ± 11.27
Oracle [22]			96.80	–	85.60	–
BEAL [22]	✓		89.80	–	**81.00**	–
AdvEnt [21]	✓		89.73 ± 3.66	9.84 ± 3.86	77.99 ± 21.08	**7.57 ± 4.24**
SRDA [1]		✓	89.37 ± 2.70	9.91 ± 2.45	77.61 ± 13.58	10.15 ± 5.75
DAE [13]		✓	89.08 ± 3.32	11.63 ± 6.84	79.01 ± 12.82	10.31 ± 8.45
DPL (ours)			**90.13 ± 3.06**	**9.43 ± 3.46**	79.78 ± 11.05	9.01 ± 5.59
Drishti-GS						
W/o adaptation			93.84 ± 2.91	9.05 ± 7.50	83.36 ± 11.95	11.39 ± 6.30
Oracle [22]			97.40	–	90.10	–
BEAL [22]	✓		96.10	–	**86.20**	–
AdvEnt [21]	✓		96.16 ± 1.65	4.36 ± 1.83	82.75 ± 11.08	**11.36 ± 7.22**
SRDA [1]		✓	96.22 ± 1.30	4.88 ± 3.47	80.67 ± 11.78	13.12 ± 6.48
DAE [13]		✓	94.04 ± 2.85	8.79 ± 7.45	83.11 ± 11.89	11.56 ± 6.32
DPL (ours)			**96.39 ± 1.33**	**4.08 ± 1.49**	83.53 ± 17.80	11.39 ± 10.18

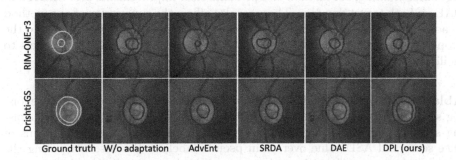

Ground truth   W/o adaptation   AdvEnt   SRDA   DAE   DPL (ours)

Fig. 2. Qualitative comparison of adaptation performance with different methods.

The quantitative comparison results are presented in Table 1, where we also include the "**W/o adaptation**" lower bound and the supervised training

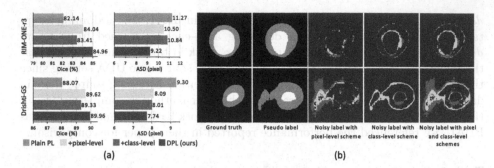

(a)   (b)

Fig. 3. (a) Ablation results with different denoising schemes in our method; (b) Examples of noisy pseudo labels identified with different denoising schemes. The correctly and falsely identified noisy pseudo labels are indicated in yellow and blue colors. (Color figure online)

upper bound in target domain (referred as "**Oracle**" with results from [22]). As observed, different domain adaptation methods can generally improve the performance over baseline, with BEAL which is the current state-of-the-art UDA method in fundus image segmentation performing the best. This benefits from their joint utilization of source and target domain data to perform adversarial alignment in both boundary and predicted entropy. Notably, without using any source data during adaptation, our method achieves 90.13% and 96.39% Dice scores for optic disc segmentation on the two target datasets respectively, which are even higher than the BEAL method. This indicates that source-free UDA would not necessarily underperform the vanilla UDA. A possible reason is that vanilla UDA heavily relies on finding an invariant latent space between source and target distributions which could be difficult, while our source-free method directly adapts the model towards target domain hence can capture more discriminative representations at the target distribution. In addition, without altering source training, our method also presents clear improvements over SRDA and DAE on both target domains. These results affirm the superiority of our method to adapt the model without using any source domain information, thanks to the proposed denoising schemes to explicitly filter out unreliable pseudo labels to facilitate the discriminative learning from target samples.

Ablation Study. We conduct ablation analysis to investigate the two denoising schemes. Figure 3(a) shows that adding the pixel-level or class-level denoising scheme can both clearly improve the adaptation performance in terms of Dice score and ASD value over plain pseudo labeling. Besides, integrating the two complementary denoising schemes completes our method and yields further improvements on both two target domains. Moreover, Fig. 3(b) shows that each denoising scheme could identify different subsets of noisy pseudo labels and combining them could complement each other. Precisely identifying all the noisy pseudo labels is indeed challenging and some misidentification could exist

Table 2. Comparison of pseudo-label selection accuracy with different methods.

Methods	RIM-ONE-r3		Drishti-GS	
	$PL\ Acc._{disc}$ [%]	$PL\ Acc._{cup}$ [%]	$PL\ Acc._{disc}$ [%]	$PL\ Acc._{cup}$ [%]
Plain PL	71.39	68.09	90.60	95.39
+pixel-level	74.19	70.20	90.88	95.60
+class-level	80.80	70.92	93.75	95.48
DPL (ours)	**81.35**	**72.26**	**93.88**	**95.69**

as observed in Fig. 3(b). We consider this is tolerable as it is more critical to ensure the remaining pseudo labels are reliable for model self-training, thus the sensitivity of finding noisy pseudo labels is more important than the specificity.

Table 2 shows the accuracy of the selected pseudo labels for quantitative comparison. We see that both two denoising schemes improve the pseudo label accuracy over the plain pseudo-labeling strategy, and their combination could further increase the accuracy on both optic disc and cup in the two target domains. It is worth noting that the proposed denoising schemes work much more effective when being used to denoise the pseudo labels than being used as post-processing for the segmentation results. We tried to apply the denoising schemes to post-process the segmentation results and found that the performance was even lower than the w/o adaptation model. The reason could be that denoising pseudo labels has a certain tolerance of noisy label misidentification, while such errors would directly decrease the performance if taking as post-processing.

4   Conclusion

We present a new method for the challenging source-free UDA problem. Without accessing source data or altering source training, our method achieves comparable or even better performance than source data-dependent approaches on cross-domain fundus image segmentation. The proposed pseudo label denoising schemes are also applicable to other scenarios such as semi-supervised learning. One limitation of the work is that we consider the generally minor domain shift under which the pseudo labels can provide a meaningful supervision base. Actually such kind of domain shift commonly exits in datasets collected with different scanners, to which our method provides a practical solution. For the more severe cross-modality domain shift, the source network may largely under-segment objects on the target images, which may make the limited number of pseudo labels for positive classes difficult to achieve desired model self-adaptation. In future studies, we will further explore the source domain distribution information embedded in the source model or knowledge prior about the shape to jointly work with the pseudo-labeling for handling more severe domain shift.

Acknowledgements. This work was supported in part by Key-Area Research and Development Program of Guangdong Province, China (2020B010165004), National

Natural Science Foundation of China with Project No. U1813204, and Hong Kong Innovation and Technology Fund (Project No. ITS/311/18FP and GHP/110/19SZ).

References

1. Bateson, M., Kervadec, H., Dolz, J., Lombaert, H., Ben Ayed, I.: Source-relaxed domain adaptation for image segmentation. In: Martel, A.L., et al. (eds.) MICCAI 2020. LNCS, vol. 12261, pp. 490–499. Springer, Cham (2020). https://doi.org/10.1007/978-3-030-59710-8_48

2. Chen, C., Dou, Q., Chen, H., Qin, J., Heng, P.A.: Unsupervised bidirectional cross-modality adaptation via deeply synergistic image and feature alignment for medical image segmentation. IEEE Trans. Med. Imaging **39**(7), 2494–2505 (2020)

3. Chen, L.-C., Zhu, Y., Papandreou, G., Schroff, F., Adam, H.: Encoder-decoder with atrous separable convolution for semantic image segmentation. In: Ferrari, V., Hebert, M., Sminchisescu, C., Weiss, Y. (eds.) ECCV 2018. LNCS, vol. 11211, pp. 833–851. Springer, Cham (2018). https://doi.org/10.1007/978-3-030-01234-2_49

4. Fumero, F., Alayón, S., Sanchez, J.L., Sigut, J., Gonzalez-Hernandez, M.: RIM-ONE: an open retinal image database for optic nerve evaluation. In: International Symposium on Computer-Based Medical Systems, pp. 1–6. IEEE (2011)

5. Gal, Y., Ghahramani, Z.: Dropout as a Bayesian approximation: representing model uncertainty in deep learning. In: ICML, pp. 1050–1059. PMLR (2016)

6. Ghafoorian, M., et al.: Transfer learning for domain adaptation in MRI: application in brain lesion segmentation. In: Descoteaux, M., Maier-Hein, L., Franz, A., Jannin, P., Collins, D.L., Duchesne, S. (eds.) MICCAI 2017. LNCS, vol. 10435, pp. 516–524. Springer, Cham (2017). https://doi.org/10.1007/978-3-319-66179-7_59

7. Gibson, E., et al.: Inter-site variability in prostate segmentation accuracy using deep learning. In: Frangi, A.F., Schnabel, J.A., Davatzikos, C., Alberola-López, C., Fichtinger, G. (eds.) MICCAI 2018. LNCS, vol. 11073, pp. 506–514. Springer, Cham (2018). https://doi.org/10.1007/978-3-030-00937-3_58

8. Guo, C., Pleiss, G., Sun, Y., Weinberger, K.Q.: On calibration of modern neural networks. In: ICML, pp. 1321–1330. PMLR (2017)

9. He, Y., Carass, A., Zuo, L., Dewey, B.E., Prince, J.L.: Self domain adapted network. In: Martel, A.L., et al. (eds.) MICCAI 2020. LNCS, vol. 12261, pp. 437–446. Springer, Cham (2020). https://doi.org/10.1007/978-3-030-59710-8_43

10. Huo, Y., et al.: SynSeg-Net: synthetic segmentation without target modality ground truth. IEEE TMI **38**(4), 1016–1025 (2018)

11. Ju, L., Wang, X., Zhao, X., Bonnington, P., Drummond, T., Ge, Z.: Leveraging regular fundus images for training UWF fundus diagnosis models via adversarial learning and pseudo-labeling. arXiv preprint arXiv:2011.13816 (2020)

12. Kamnitsas, K., et al.: Unsupervised domain adaptation in brain lesion segmentation with adversarial networks. In: Niethammer, M., et al. (eds.) IPMI 2017. LNCS, vol. 10265, pp. 597–609. Springer, Cham (2017). https://doi.org/10.1007/978-3-319-59050-9_47

13. Karani, N., Erdil, E., Chaitanya, K., Konukoglu, E.: Test-time adaptable neural networks for robust medical image segmentation. MIA **68**, 101907 (2021)

14. Lee, D.H.: Pseudo-label: the simple and efficient semi-supervised learning method for deep neural networks. In: Workshop on Challenges in Representation Learning, ICML, vol. 3 (2013)

15. Orlando, J.I., Fu, H., Breda, J.B., van Keer, K., Bathula, D.R., et al.: Refuge challenge: a unified framework for evaluating automated methods for glaucoma assessment from fundus photographs. Med. Image Anal. **59**, 101570 (2020)
16. Perone, C.S., Ballester, P., Barros, R.C., Cohen-Adad, J.: Unsupervised domain adaptation for medical imaging segmentation with self-ensembling. NeuroImage **194**, 1–11 (2019)
17. Sivaswamy, J., Krishnadas, S., Chakravarty, A., Joshi, G., Tabish, A.S., et al.: A comprehensive retinal image dataset for the assessment of glaucoma from the optic nerve head analysis. JSM Biomed. Imaging Data Pap. **2**(1), 1004 (2015)
18. Snell, J., Swersky, K., Zemel, R.S.: Prototypical networks for few-shot learning. arXiv preprint arXiv:1703.05175 (2017)
19. Stan, S., Rostami, M.: Privacy preserving domain adaptation for semantic segmentation of medical images. arXiv preprint arXiv:2101.00522 (2021)
20. Varsavsky, T., Orbes-Arteaga, M., Sudre, C.H., Graham, M.S., Nachev, P., Cardoso, M.J.: Test-time unsupervised domain adaptation. In: Martel, A.L., et al. (eds.) MICCAI 2020. LNCS, vol. 12261, pp. 428–436. Springer, Cham (2020). https://doi.org/10.1007/978-3-030-59710-8_42
21. Vu, T.H., Jain, H., Bucher, M., et al.: ADVENT: adversarial entropy minimization for domain adaptation in semantic segmentation. In: CVPR, pp. 2517–2526 (2019)
22. Wang, S., Yu, L., Li, K., Yang, X., Fu, C.-W., Heng, P.-A.: Boundary and entropy-driven adversarial learning for fundus image segmentation. In: Shen, D., et al. (eds.) MICCAI 2019. LNCS, vol. 11764, pp. 102–110. Springer, Cham (2019). https://doi.org/10.1007/978-3-030-32239-7_12
23. Xing, F., Bennett, T., Ghosh, D.: Adversarial domain adaptation and pseudo-labeling for cross-modality microscopy image quantification. In: Shen, D., et al. (eds.) MICCAI 2019. LNCS, vol. 11764, pp. 740–749. Springer, Cham (2019). https://doi.org/10.1007/978-3-030-32239-7_82
24. Zhang, L., Pereañez, M., Piechnik, S.K., Neubauer, S., Petersen, S.E., Frangi, A.F.: Multi-input and dataset-invariant adversarial learning (MDAL) for left and right-ventricular coverage estimation in cardiac MRI. In: Frangi, A.F., Schnabel, J.A., Davatzikos, C., Alberola-López, C., Fichtinger, G. (eds.) MICCAI 2018. LNCS, vol. 11071, pp. 481–489. Springer, Cham (2018). https://doi.org/10.1007/978-3-030-00934-2_54
25. Zhang, Y., Miao, S., Mansi, T., Liao, R.: Task driven generative modeling for unsupervised domain adaptation: application to X-ray image segmentation. In: Frangi, A.F., Schnabel, J.A., Davatzikos, C., Alberola-López, C., Fichtinger, G. (eds.) MICCAI 2018. LNCS, vol. 11071, pp. 599–607. Springer, Cham (2018). https://doi.org/10.1007/978-3-030-00934-2_67
26. Zhang, Z., Yang, L., Zheng, Y.: Translating and segmenting multimodal medical volumes with cycle-and shape-consistency generative adversarial network. In: CVPR, pp. 9242–9251 (2018)
27. Zhao, H., Li, H., Maurer-Stroh, S., Guo, Y., Deng, Q., Cheng, L.: Supervised segmentation of un-annotated retinal fundus images by synthesis. IEEE Trans. Med. Imaging **38**(1), 46–56 (2018)

ASC-Net: Adversarial-Based Selective Network for Unsupervised Anomaly Segmentation

Raunak Dey[1] and Yi Hong[2(✉)]

[1] Department of Computer Science, University of Georgia, Athens, USA
[2] Department of Computer Science and Engineering, Shanghai Jiao Tong University, Shanghai, China
yi.hong@sjtu.edu.cn

Abstract. We introduce a neural network framework, utilizing adversarial learning to partition an image into two cuts, with one cut falling into a reference distribution provided by the user. This concept tackles the task of unsupervised anomaly segmentation, which has attracted increasing attention in recent years due to their broad applications in tasks with unlabelled data. This Adversarial-based Selective Cutting network (ASC-Net) bridges the two domains of cluster-based deep learning methods and adversarial-based anomaly/novelty detection algorithms. We evaluate this unsupervised learning model on BraTS brain tumor segmentation, LiTS liver lesion segmentation, and MS-SEG2015 segmentation tasks. Compared to existing methods like the AnoGAN family, our model demonstrates tremendous performance gains in unsupervised anomaly segmentation tasks. Although there is still room to further improve performance compared to supervised learning algorithms, the promising experimental results shed light on building an unsupervised learning algorithm using user-defined knowledge.

1 Introduction

In computer vision and medical image analysis, unsupervised image segmentation has been an active research topic for decades [14,17,19,20,26], due to its potential of applying to many applications without requiring the data to be manually labelled. Recently, advances in GANs [15] have given rise to a class of anomaly detection algorithms, which are inspired by AnoGAN [24] to identify abnormal events, behaviors, or regions in images or videos [10,13,25]. The AnoGAN learns a manifold of normal images by mapping from image space to a latent space based on GANs. To detect the anomaly, AnoGAN needs iterative search in the latent space to find the closest corresponding images for a query image. The AnoGAN family, including f-AnoGAN [23] and other works [4,5,16,27,28], focus on the reconstruction of the corresponding normal images for a query image, but not directly working on the anomaly detection. As a result, their reconstruction quality heavily affects the performance of anomaly detection.

© Springer Nature Switzerland AG 2021
M. de Bruijne et al. (Eds.): MICCAI 2021, LNCS 12905, pp. 236–247, 2021.
https://doi.org/10.1007/978-3-030-87240-3_23

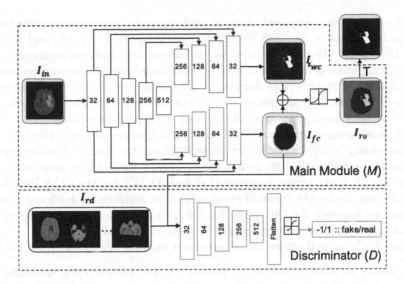

Fig. 1. Overview of our proposed ASC-Net for unsupervised anomaly segmentation. (Color figure online)

To center the focus on the anomaly without needing faithful reconstruction, we propose an adversarial-based selective cutting neural network (ASC-Net)[1], shown in Fig. 1. This network aims to decompose an image into two selective cuts based on a reference image distribution. Typically, the reference distribution is defined by a set of images provided by users or experts who have vague knowledge and expectation of normal cases. In this way, one cut will fall into the reference distribution, while other image content outside of the reference image distribution will group into the other cut. These two cuts allow to reconstruct the original input image semantically and perform a simple intensity thresholding to cluster normal and abnormal regions. To consider these two cuts simultaneously, we extend U-Net [21] with two upsampling branches, as used in CompNet [11], a supervised image segmentation approach. Meanwhile, one branch connects to a GAN's discriminator network, which allows introducing the knowledge contained in the reference image distribution. With the discriminator component aiding, the network can separate images into softly disjoint regions; that is, the generation of our selective cuts is under the constraint of the reference image distribution. As a result, we obtain a joint estimation of anomaly and the corresponding normal image, thus bypassing the need for perfect reconstruction. Furthermore, under the constraints of the GAN discriminator and the reconstruction of the original input, our ASC-Net becomes an unsupervised solution for anomaly detection, since we do not have any labels for the anomaly, with only a collection of normal images in the reference distribution.

---

[1] Our source code is available on Github: https://github.com/raun1/ASC-NET.

We evaluate our proposed unsupervised anomaly segmentation network on three public datasets, i.e., MS-SEG2015 [7], BraTS-2019 [1,2,18], and LiTS [6] datasets. For the MS-SEG2015 dataset, an exhaustive study on comparing multiple existing autoencoder-based models, variational-autoencoder-based models, and GAN-based models is performed in [3]. Compared to the best Dice scores reported in [3], we have significant gains in performance, which are increased by 23.24% without post-processing and 20.40% with post-processing[2]. For BraTS dataset, our experiments show that f-AnoGAN, the one performs the best after post-processing in [3], has difficulty reconstructing the normal images required for anomaly segmentation. By contrast, we obtain a mean Dice score of 63.67% for the BraTS brain tumor segmentation and 32.24% for the LiTS liver lesion segmentation, under the two-fold cross-validation settings for both datasets. In addition, we improve the Dice score for the liver lesion segmentation to 50.23% using a simple post-processing scheme of open and closed sets.

Overall, the contributions of our proposed method are summarized below:

- Proposing an adversarial based framework for unsupervised anomaly segmentation, which bypasses the normal image reconstruction and works on anomaly detection directly. This framework presents a general clustering strategy to generate two selective cuts based on a reference image set with human knowledge.
- To the best of our knowledge, our work is the first one to apply an unsupervised segmentation algorithm to the BraTS 2019 and LiTS liver lesion public datasets. Besides, our method outperforms the AnoGAN family and other popular methods presented in [3] on the publicly available MS-SEG2015 dataset.

2   Adversarial-Based Selective Cutting Network (ASC-Net)

2.1   Network Framework

Figure 1 describes the framework of our proposed ASC-Net, which includes two components, i.e., the main module M and the discriminator D, and one simple clustering step T based on thresholding. Overall, the main module includes normal and anomaly branches to semantically reconstruct the original image for clustering, while the discriminator brings user-defined knowledge into the normal branch in the main module.

Main Module M. The main module aims to generate two selective cuts, which guide a follow-up simple reconstruction of an input image to cluster image pixels based on intensity thresholding. The M follows an encoder-decoder architecture like the U-Net, including one encoder and two decoders. The encoder E extracts features of an input image I_{in}, which could be an image located within or outside

---

[2] Different from that in [3], we use a simple open-and-closed operation for post-processing.

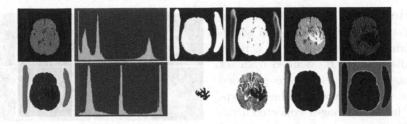

Fig. 2. Visualization of the "disjoincy" between images I_{fc} (top) and I_{wc} (bottom) generated by two cuts of ASC-Net. From left to right: the generated image, its histogram, and the following four columns representing the histogram equalized images of the thresholded peaks with the first peak being the first image, etc. The first peak of I_{fc} is disjoint with the last peak of I_{wc}, etc.

of the reference distribution $\{I_{rd}\}$, a collection of normal images. One decoder in green (the second branch) is designed to generate a "fence" cut C_f that is defined by an image fence formed by $\{I_{rd}\}$. The C_f aims to generate an image I_{fc} and tries to fool the discriminator D. The other decoder in blue (the first branch) is designed to generate another "wild" cut C_w, which captures leftover image content that is not included in I_{fc}. As a result, the C_w produces another images I_{wc} to complement the fence-cut output I_{fc}. The complementary relation between these two cuts C_f and C_w is enforced by a positive Dice loss discussed later. Figure 2 demonstrates the "disjoincy" of I_{fc} and I_{wc}, like their complementary histogram distribution and different thresholded images at different peaks.

The reconstructor R consists of a 1×1 convolution layer with the Sigmoid as the activation function, which is applied on the concatenation of the two-cut outputs I_{fc} and I_{wc} to regenerate the input image I_{in} back. This reconstructor R ensures that the C_f does not generate an image I_{fc} far from the input image I_{in} and also ensures that the C_w does not generate an empty image I_{wc} if the anomaly or novelty exists. Figure 3 shows the histogram separation of the reconstructed images, compared to the original input images which present complex histogram peaks and have difficulty in separating the brain tumor from background and other tissues via a simple thresholding. The discontinuous histogram distribution of I_{ro} is inherited from the two generated sub-images I_{fc} and I_{wc} through a simple weighted combination. As a result, the segmentation task becomes relatively easy to be done on the reconstructed image I_{ro}.

Discriminator D. The GAN discriminator tries to distinguish the generated image I_{fc}, according to a reference distribution R_d defined by a set of images $\{I_{rd}\}$, which are provided by the user or experts. The R_d typically includes images collected from the same group, for instance, normal brain scans, which share similar structures and lie on a manifold. Introducing D allows us to incorporate our vague prior knowledge about a task into a deep neural network. Typically, it is non-trivial to explicitly formulate such prior knowledge; however, it could be implicitly represented by a selected image set. The R_d is an essential component that makes our ASC-Net possible to generate selective cuts according to the user's input, without requiring other supervisions.

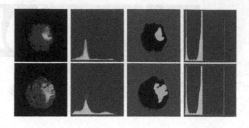

Fig. 3. Histogram comparison of two sample images. From left to right: the input image, its histogram, its reconstructed image using ASC-Net, and the histogram of the reconstructed image. The histograms of the input images vary greatly, while the ones of their reconstructions show peaks at similar ranges, which enables a thresholding based pixel-level separation.

Thresholding T. To cluster the reconstructed image I_{ro} into two groups at the pixel level, we choose the thresholding approach with the threshold values obtained using the histogram of I_{ro}. We observed that for an anomaly that is often brighter than the surrounding tissues like the BraTS brain tumor, the intensity value at the rightmost peak of the histogram is a desired threshold; while an opposite case like darker LiTS liver lesions, the value at the leftmost peak would be the threshold. We also observed that the histograms of the reconstructed images for different inputs reflect the same cut-off point for the left or right peaks, which allows using one threshold for an entire dataset.

Loss Functions. The main module M includes three loss functions: (i) the image generation loss for C_f ($Loss_{C_f}$), (ii) the "disjoincy" loss between C_f and C_w ($Loss_{C_w}$), and (iii) the reconstruction loss ($Loss_R$). In particular, the C_f tries to generate an image I_{fc} that fools the discriminator D by minimizing $Loss_{C_f} = \frac{1}{n} \sum_{i=1}^{n} |D(I_{fc}^{(i)}) - 1|$. Here, n is the number of samples in the training batch. The C_w tries to generate an image I_{wc} that is complement to I_{fc} by minimizing the positive Dice score $Loss_{C_w} = \frac{2|I_{fc} \cap I_{wc}|}{|I_{fc}| + |I_{wc}|}$. The last reconstruction takes an Mean-Squared-Error (MSE) loss between the input image I_{in} and the reconstructed image I_{ro}: $Loss_R = \frac{1}{n} \sum_{i=1}^{n} \|I_{in}^{(i)} - I_{ro}^{(i)}\|_2^2$. The discriminator D tries to reject the C_f output I_{fc} but accept the images from the reference distribution R_d, by minimizing the following loss function: $Loss_D = \frac{1}{n+m} \left(\sum_{i=1}^{n} |D(I_{fc}^{(i)}) - (-1)| + \sum_{i=1}^{m} |D(I_{R_d}^{(i)}) - 1| \right)$. Here, m is the number of the images in R_d. Even though D and C_{fc} are tied in an adversarial setup, here we do not use the Earth Mover distance [22] in the loss function, since we would like D to identify both positive samples and negative samples with equal precision. Therefore, we use Mean Absolute Error (MAE) instead.

2.2 Architecture Details and Training Scheme

We use the same network architecture for all of our experiments as shown in Fig. 1. The encoder E consists of four blocks of two convolution layers with a filter size of $(3, 3)$ followed by a max pooling layer with a filter size of $(2, 2)$ and batch normalization after every convolution layer. After every pooling layer we also introduce a dropout of 0.3. The number of feature maps in each of the convolution layer of a block are 32, 64, 128, and 256. Following these blocks is a transition layer of two convolution layers with feature maps of size 512 followed by batch normalization layers. The C_{fc} and C_{wc} decoders are connected to the E and mirror the layers with the pooling layers replaced with 2D transposed convolutional layers, which have the same number of feature maps as the blocks mirror those in the encoder. Similar to a U-Net, we also introduce skip connections across similar levels in the encoder and decoders. The reconstructor R is simply a Sigmoid layer applied to the concatenation of I_{fc} and I_{wc}, resulting in a simplified CompNet [11]. The Discriminator D mimics the architecture of the E, except for the last layer where a dense layer is used for classification. All the intermediate layers have ReLU activation function and the final output layers have the Sigmoid activation. The only exception is the output of the discriminator D, which has a Tanh activation function to separate I_{fc} and images from the R_d to the maximum extent.

We use Keras with Tensorflow backend and Adam optimizer with a learning rate of $5e-5$ to implement our architecture. We follow two distinct training stages:

- In the *first stage*, we train D and M in cycles. We start training D with $\{R_d\}$ with True labels and $\{I_{fc}\}$ with False labels. These training samples are shuffled randomly. Following D, we train M with $\{I_{in}\}$ as input and the weights of D frozen while preserving the connection between $\{I_{fc}\}$ and D. The objective of the M is to morph the appearance of $\{I_{in}\}$ into $\{I_{fc}\}$ to fool D with the frozen weights. We call these two steps one cycle, and in each step there may be more than one epochs of training for M or D.
- In the *second stage*, M and D continue to be trained alternatively; however, the input images to D are changed, since the training purpose at this stage is to focus on the differences between the $\{R_d\}$ and $\{I_{in}\}$, while ignoring the noisy biases created by the M in transforming $\{I_{in}\}$ to $\{I_{fc}\}$. To achieve this, we augment the reference distribution $\{R_d\}$ with its generated images via M, i.e., $\{I_{fc}(R_d)\}$. We treat them as true images, and the union set $\{R_d \cup I_{fc}(R_d)\}$ is used to update D.

Runtime Analysis. We use two Nvidia TitanX GPUs and on average a discriminator cycle takes 2.5 ms to process a single 2D image slice with size of 240 $\times$ 240, while the main module cycle takes 15.5 ms to process a single 2D image slice during training.

3 Applications

We evaluate our model on three unsupervised anomaly segmentation tasks: MS lesion segmentation, brain tumor segmentation, and liver lesion segmentation. We use the MS-SEG2015 [7] training set, BraTS [1,2,18], and LiTS [6] datasets in these tasks.

MS-SEG2015. The training set consists of 21 scans from 5 subjects with each scan dimensions of $181 \times 217 \times 181$. We resize the axial slices to 160×160, so that we can share the same network design as the rest of the experiments.

BraTS 2019. This dataset consists of 335 T1-w MRI brain scans collected from 259 subjects with high grade Glioma and 76 subjects with low grade Gliomas in the training set. The 3D dimensions of the images are $240 \times 240 \times 155$.

LiTS. The training set of LiTS consists of 130 abdomen CT scans of patients with liver lesions, collected from multiple institutions. Each scan has a varying number of slices with dimensions of 512×512. We resize these CT slices to 240×240 to share the same network architecture with other tasks.

For all experiments, the image intensity is normalized to $[0, 1]$ over the 3D volume; however, we perform the 3D segmentation task in the slice-by-slice manner using axial slices. To balance the sample size in I_{in} and R_d, we randomly sample and duplicate the number difference to the respective set.

MS Lesion Segmentation. In this task, we randomly sample 2870 non-tumor, non-zero, Brats-2019 training set slices to make our reference distribution R_d as in [3], while they use their own privately annotated healthy dataset. Meanwhile, the 2870 non zero 2D slices of the MS-SEG2015 training set are used in the main module M. We train this network using three cycles in the first stage and one cycle in the second stage and take the threshold at 254 intensity based on the right most peak of the image histogram.

We obtain an average Dice score of 32.94% without any post processing. By using a simple post-processing with erosion and dilation[3] with 5×5 filters, this number improves to 48.20% Dice score. In comparison, a similar study conducted by [3] consisting of a multitude of algorithms including AnoVAEGAN [4] and f-AnoGANS, obtained a best mean score of 27.8% Dice after post processing by f-AnoGANS. Before post processing the best method was Constrained AutoEncoder [8] with a score of 9.7% Dice. Sample images of our method are included in Fig. 4

Brain Tumor Segmentation. In this task, we perform patient-wise two-fold cross-validation on the Brats-2019 training set. In each training fold, we use a 90/10 split after removing empty slices. The 2D slices from the 90% split without tumors are used to make our reference distribution R_d; while the 2D slices with tumors from the 90% split and all the slices from the 10% split are used for training our model. As a result, the sample size of R_d for fold one and

---

[3] We use this operator to improve the connectivity of the generated anomaly mask.

two amounts to 11,745 and 12,407 respectively, while the size of I_{in} amounts to 11,364 and 10,786, respectively. We train this network using two cycles in the first stage and one cycle in the second stage.

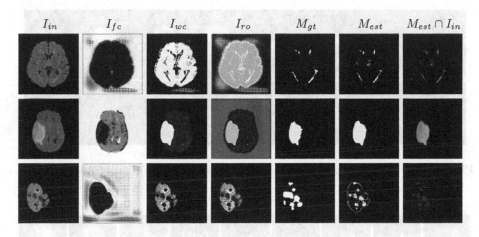

Fig. 4. Sample results of MS-SEG2015, Brats-2019 and LiTS (top to bottom) obtained from the various branches of the network. The I_{fc} in the second row is contrast enhanced to present the content contained in the brain region. None of these include any of the post processed images.

We obtain an average Dice score of 63.67% for the brain tumor segmentation. Figure 4 shows samples generated by our ASC-Net. Figure 5 shows our attempt to apply f-AnoGANs [23] by following their online instructions. The failure of AnoGANs in the reconstruction brings to light the issue with the regeneration based methods and the complexity and stability of GAN based image reconstruction.

Liver Lesion Segmentation. To generate the image data for this task, we remove the non-liver region by using the liver mask generated by CompNet [11] and take all non-zero images. We have 11,926 2D slices without liver lesions used in the reference distribution R_d. The remaining 6,991 images are then used for training the model. We perform slice-by-slice two-fold cross-validation and train the network using two cycles in both first and second stages. To extract the liver lesions, we first mask out the noises in the non-liver region of the reconstructed image I_{ro} and then invert the image to take a threshold value at 242, the rightmost peak of the inverted image.

We obtain an average Dice score of 32.24% for this liver lesion segmentation, which improves to 50.23% by using a simple post processing scheme of erosion and dilation with 5×5 filter. Sampled results are shown in Fig. 4. In comparison, a recent study [12] reports a cross-validation result of 67.3% under a supervised setting. Note that the annotation in the LiTS lesion dataset is imperfect with missing small lesions [9,12]. Since we use the imperfect annotation to select images for the reference distribution, some slices with small lesions may be included and treated as normal examples.

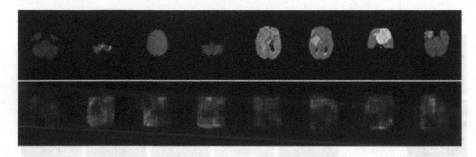

Fig. 5. Query images (top) and their reconstructions (bottom) using f-AnoGANs [23].

Fig. 6. Stability: The first image is the input image, the second is the ground truth. The rest of images are reconstruction from various re-runs of the framework with variable training cycles and stage. All runs are able to isolate the anomaly in question.

4   Discussion and Future Work

In this paper we have presented a framework that performs two-cut split in an unsupervised fashion guided by an reference distribution using GANs. Unlike the methods in the AnoGAN family which operate as a reconstruction-based method and needs faithful reconstruction of normal images to function properly, we treat the anomaly segmentation as a constrained two-cut problem that requires a semantical and reduced reconstruction for clustering. Our ASC-Net focuses on the anomaly detection with the normal image reconstruction as a byproduct, thus still producing competitive results where reconstruction dependent methods such as f-AnoGAN fails to work on. The current version of our ASC-Net aims to solve the two-cut problem, which will be tasked to handle more than two selective cuts in the future. Theoretical understanding of the proposed network is also required, which is left as a future work.

Limitations and Opportunities. One reason for our low Dice scores could be that we had to select non-tumor or normal slices as our reference distribution, which does not account for other co-morbidities. This affects the performance of the framework as it has no other guidance and would consider co-morbidities as an anomaly as well. However, this provides possibility of bringing other anomalies into the users' attention.

Termination and Stability. The termination point of this network training is periodic. The general guideline is that the peaks should be well separated and we terminate our algorithm at three or four peak separation. However, continuing to train further may not always result in the improvement for the purpose of segmentation due to accumulation of holes as shown in Fig. 7, even though visually the anomaly is captured in more intricate detail. We however encourage training longer as it reduces false positive and provide detailed anomaly reconstruction, though the Dice metric might not account for it. In our experiments, we specify the number of cycles in each stage. However, due to the random nature of the algorithm and the lack of a particular purpose and guidance, the

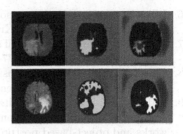

Fig. 7. Termination of network training affects the reconstruction result. Left to right columns in each row: the input image, the image reconstructed via two cycles in the first stage and one in the second stage, and the image reconstructed via adding one cycle in the second stage.

peak separation may occur much earlier, then training should be stopped accordingly. The reported network in our Brats-2019 experiments has an average Dice score of 6% over the network trained longer as shown in Fig. 7. Regarding the stability, Fig. 6 demonstrates an anomaly estimated by different networks that are trained with different number of training cycles. We observe that while the appearance of I_{ro} changes, we still obtain the anomaly as a separate cut since our framework works without depending on the quality of reconstruction.

Acknowledgements. This work was supported by NSF 1755970 and Shanghai Municipal Science and Technology Major Project 2021SHZDZX0102.

References

1. Bakas, S., et al.: Advancing the cancer genome atlas glioma MRI collections with expert segmentation labels and radiomic features. Sci. Data **4**, 170117 (2017)
2. Bakas, S., et al.: Identifying the best machine learning algorithms for brain tumor segmentation, progression assessment, and overall survival prediction in the brats challenge. arXiv preprint arXiv:1811.02629 (2018)
3. Baur, C., Denner, S., Wiestler, B., Navab, N., Albarqouni, S.: Autoencoders for unsupervised anomaly segmentation in brain MR images: a comparative study. Med. Image Anal. **69**, 101952 (2021)
4. Baur, C., Wiestler, B., Albarqouni, S., Navab, N.: Deep autoencoding models for unsupervised anomaly segmentation in brain MR images. In: Crimi, A., Bakas, S., Kuijf, H., Keyvan, F., Reyes, M., van Walsum, T. (eds.) BrainLes 2018. LNCS, vol. 11383, pp. 161–169. Springer, Cham (2019). https://doi.org/10.1007/978-3-030-11723-8_16
5. Berg, A., Ahlberg, J., Felsberg, M.: Unsupervised learning of anomaly detection from contaminated image data using simultaneous encoder training. arXiv preprint arXiv:1905.11034 (2019)

6. Bilic, P., Christ, P.F., Vorontsov, E., Chlebus, G., et al.: The liver tumor segmentation benchmark (LiTS). arXiv:1901.04056 (2019)
7. Carass, A., et al.: Longitudinal multiple sclerosis lesion segmentation: resource and challenge. NeuroImage **148**, 77–102 (2017)
8. Chen, X., Konukoglu, E.: Unsupervised detection of lesions in brain MRI using constrained adversarial auto-encoders. arXiv preprint arXiv:1806.04972 (2018)
9. Chlebus, G., Schenk, A., Moltz, J.H., van Ginneken, B., Hahn, H.K., Meine, H.: Automatic liver tumor segmentation in CT with fully convolutional neural networks and object-based postprocessing. Sci. Rep. **8**(1), 1–7 (2018)
10. Del Giorno, A., Bagnell, J.A., Hebert, M.: A discriminative framework for anomaly detection in large videos. In: Leibe, B., Matas, J., Sebe, N., Welling, M. (eds.) ECCV 2016. LNCS, vol. 9909, pp. 334–349. Springer, Cham (2016). https://doi.org/10.1007/978-3-319-46454-1_21
11. Dey, R., Hong, Y.: CompNet: complementary segmentation network for brain MRI extraction. In: Frangi, A.F., Schnabel, J.A., Davatzikos, C., Alberola-López, C., Fichtinger, G. (eds.) MICCAI 2018. LNCS, vol. 11072, pp. 628–636. Springer, Cham (2018). https://doi.org/10.1007/978-3-030-00931-1_72
12. Dey, R., Hong, Y.: Hybrid cascaded neural network for liver lesion segmentation. In: 2020 IEEE 17th International Symposium on Biomedical Imaging (ISBI), pp. 1173–1177. IEEE (2020)
13. Erfani, S.M., Rajasegarar, S., Karunasekera, S., Leckie, C.: High-dimensional and large-scale anomaly detection using a linear one-class SVM with deep learning. Pattern Recogn. **58**, 121–134 (2016)
14. Giordana, N., Pieczynski, W.: Estimation of generalized multisensor hidden Markov chains and unsupervised image segmentation. IEEE Trans. Pattern Anal. Mach. Intell. **19**(5), 465–475 (1997)
15. Goodfellow, I., et al.: Generative adversarial nets. In: Advances in Neural Information Processing Systems, pp. 2672–2680 (2014)
16. Kimura, M., Yanagihara, T.: Anomaly detection using GANs for visual inspection in noisy training data. In: Carneiro, G., You, S. (eds.) ACCV 2018. LNCS, vol. 11367, pp. 373–385. Springer, Cham (2019). https://doi.org/10.1007/978-3-030-21074-8_31
17. Lee, T.-W., Lewicki, M.S.: Unsupervised image classification, segmentation, and enhancement using ICA mixture models. IEEE Trans. Image Process. **11**(3), 270–279 (2002)
18. Menze, B.H., et al.: The multimodal brain tumor image segmentation benchmark (BRATS). IEEE Trans. Med. Imaging **34**(10), 1993–2024 (2014)
19. O'Callaghan, R.J., Bull, D.R.: Combined morphological-spectral unsupervised image segmentation. IEEE Trans. Image Process. **14**(1), 49–62 (2004)
20. Puzicha, J., Hofmann, T., Buhmann, J.M.: Histogram clustering for unsupervised image segmentation. In: Proceedings, 1999 IEEE Computer Society Conference on Computer Vision and Pattern Recognition (Cat. No PR00149), vol. 2, pp. 602–608. IEEE (1999)
21. Ronneberger, O., Fischer, P., Brox, T.: U-Net: convolutional networks for biomedical image segmentation. In: Navab, N., Hornegger, J., Wells, W.M., Frangi, A.F. (eds.) MICCAI 2015. LNCS, vol. 9351, pp. 234–241. Springer, Cham (2015). https://doi.org/10.1007/978-3-319-24574-4_28
22. Rubner, Y., Tomasi, C., Guibas, L.J.: The earth mover's distance as a metric for image retrieval. Int. J. Comput. Vis. **40**(2), 99–121 (2000)

23. Schlegl, T., Seeböck, P., Waldstein, S.M., Langs, G., Schmidt-Erfurth, U.: f-AnoGAN: fast unsupervised anomaly detection with generative adversarial networks. Med. Image Anal. **54**, 30–44 (2019)
24. Schlegl, T., Seeböck, P., Waldstein, S.M., Schmidt-Erfurth, U., Langs, G.: Unsupervised anomaly detection with generative adversarial networks to guide marker discovery. In: Niethammer, M., et al. (eds.) IPMI 2017. LNCS, vol. 10265, pp. 146–157. Springer, Cham (2017). https://doi.org/10.1007/978-3-319-59050-9_12
25. Seeböck, P., et al.: Identifying and categorizing anomalies in retinal imaging data. arXiv preprint arXiv:1612.00686 (2016)
26. Shi, J., Malik, J.: Normalized cuts and image segmentation. IEEE Trans. Pattern Anal. Mach. Intell. **22**(8), 888–905 (2000)
27. Zenati, H., Foo, C.S., Lecouat, B., Manek, G., Chandrasekhar, V.R.: Efficient GAN-based anomaly detection. arXiv preprint arXiv:1802.06222 (2018)
28. Zenati, H., Romain, M., Foo, C.-S., Lecouat, B., Chandrasekhar, V.: Adversarially learned anomaly detection. In: 2018 IEEE International Conference on Data Mining (ICDM), pp. 727–736. IEEE (2018)

Cost-Sensitive Meta-learning for Progress Prediction of Subjective Cognitive Decline with Brain Structural MRI

Hao Guan[1], Yunbi Liu[1,2], Shifu Xiao[3], Ling Yue[3(✉)], and Mingxia Liu[1(✉)]

[1] Department of Radiology and BRIC, University of North Carolina at Chapel Hill,
Chapel Hill, NC 27599, USA
mxliu@med.unc.edu

[2] School of Biomedical Engineering, Southern Medical University,
Guangzhou 510515, China

[3] Department of Geriatric Psychiatry, Shanghai Mental Health Center, Shanghai
Jiao Tong University School of Medicine, Shanghai 200030, China

Abstract. Subjective cognitive decline (SCD) is a preclinical phase of Alzheimer's disease (AD) which occurs before the deficits could be detected by cognitive tests. It is highly desired to predict the progress of SCD for possible intervention of AD-related cognitive decline. Many neuroimaging-based methods have been developed for AD diagnosis, but there are few studies devoted to automated progress prediction of SCD due to the limited number of SCD subjects. Even though some studies proposed to transfer models (trained on AD/MCI) to SCD analysis, the significant domain shift between their data distributions may degrade the prediction performance. To this end, this paper tackles the problem of learning a model from the source data for which can directly generalize to an unseen target domain for SCD prediction. We propose a cost-sensitive meta-learning scheme to simultaneously improve the model generalization and its sensitivity in MRI-based SCD detection. During training, the source domain is divided into virtual meta-train and meta-test sets to explicitly simulate the scenario for early-stage detection of AD. Considering the importance of sensitivity for progressive status detection, we further introduce cost-sensitive learning to enhance the meta-optimization process by encouraging the model to gain higher sensitivity for SCD detection with simulated domain shift. Experiments conducted on the large-scale ADNI dataset and a small-scale SCD dataset have demonstrated the effectiveness of the proposed method.

Keywords: Subjective cognitive decline · Meta-learning · Brain MRI

1 Introduction

Alzheimer's disease (AD) is a chronic neurodegenerative disease that usually starts slowly and worsens over time. As illustrated in Fig. 1, the spectrum of AD has been extended to an earlier stage even before its prodromal stage

M. de Bruijne et al. (Eds.): MICCAI 2021, LNCS 12905, pp. 248–258, 2021.
https://doi.org/10.1007/978-3-030-87240-3_24

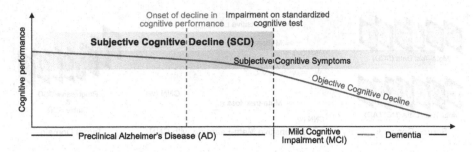

Fig. 1. Progress of disease pathology and clinical status of AD.

(*i.e.*, mild cognitive impairment, MCI), called subjective cognitive decline (SCD) or subjective memory complaint (SMC) that occurs before the deficits could be detected by cognitive tests [1,2]. In the literature, increasing evidence has shown that subjects with SCD have an increased risk of underlying AD pathology [3,4]. Therefore, predicting the future progress of subjects with SCD is fundamental for possible intervention of AD-related cognitive declines.

Neuroimaging-based methods have been widely used for AD/MCI diagnosis [5–8]. But only a few studies are devoted to SCD progress prediction, due to several challenges. (**i**) The number of SCD subjects is usually very limited (*e.g.*, tens), making it difficult to train a robust model with good generalization ability. Previous studies [9,10] proposed to augment data samples for model training, but usually failed to deal with domain shift between different domains/datasets. (**ii**) Even though some domain adaptation methods were designed to enhance the transferability of a learning model [11–16], they often require a part of labeled/unlabeled target samples to facilitate adaptation which cannot be satisfied in real applications. (**iii**) SCD appears at the preclinical stage of AD even without significant objective impairment in the brain. These challenges make it difficult to design a robust model for reliable SCD progress prediction.

To address these issues, we propose a cost-sensitive meta-learning (CSML) framework for structural MRI-based progress prediction of SCD, as illustrated in Fig. 2. During training, the source domain is divided into virtual meta-train and meta-test sets to explicitly simulate the scenario for early-stage detection of AD (*e.g.*, AD+MCI→ SCD). Considering the importance of sensitivity for detecting progressive status, we further introduce a cost-sensitive learning technique to enhance the meta-optimization process, by encouraging the model to gain higher sensitivity for progressive SCD (pSCD) detection with simulated domain shift. To the best of our knowledge, this is among the first attempts to perform SCD detection by explicitly considering cross-site domain shift and requiring no target data for model training. The proposed CSML is expected to simultaneously improve the model generalization and sensitivity in SCD progress prediction. Experimental results on a small-scale SCD dataset and the large-scale ADNI dataset demonstrate the effectiveness of the proposed method.

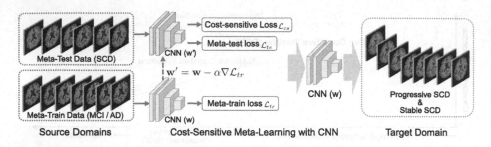

Fig. 2. Illustration of the cost-sensitive meta-learning (CSML) framework for SCD progress prediction with brain MRIs. CNN: Convolutional neural network.

2 Materials and Methodology

Datasets and MR Image Pre-processing. A total of $1,393$ T1-weighted structural MRIs from the publicly available Alzheimer's Disease Neuroimaging Initiative (ADNI) dataset[1] are used in this work to train a prediction model. Specifically, the ADNI contains 367 AD, 357 normal control (NC), 653 MCI subjects (253 pMCI, 400 sMCI), and 16 SCD subjects (11 pSCD and 5 sSCD). Among MCI subjects, 253 progressive MCI (pMCI) subjects would convert to AD within 36 months after baseline time, and the remaining 400 stable MCI (sMCI) would not. Also, 11 progressive SCD (pSCD) would convert to MCI within 36 months and the status of the 5 SCD subjects keep stable (called stable SCD, sSCD). We also collected 113 SCD subjects (with T1-weighted structural MRIs) from a local hospital as the target domain to evaluate the performance of the proposed model, including 40 pSCD subjects and 73 sSCD subjects. Note that we only used baseline MRIs in this work, without using longitudinal MRIs.

Following [17], all brain MRIs go through a standard pre-processing pipeline, including (i) skull stripping, (ii) intensity correction, (iii) re-sampling to the same resolution of $1 \times 1 \times 1$ mm^3 and (iv) spatial normalization to the Automated Anatomical Labeling (AAL) template. We employ the SPM software package[2] as the main tool to facilitate the MR image pre-processing.

Problem Formulation. In this work, we study the problem of learning a model based on N source domains, and apply it to precisely predict the future progress of SCD subjects in an unseen target domain. We treat different brain disorder categories (*e.g.*, AD, MCI, SCD) as different domains in this paper. Suppose there are N labeled source domains (*i.e.*, $\mathcal{D}_S = \{\mathcal{D}_1, \mathcal{D}_2, \cdots, \mathcal{D}_N\}$), as well as a to-be-analyzed target domain (*i.e.*, $\mathcal{D}_T$). All source and target domains share the same input MR image space $\mathcal{X}$ (with possibly different marginal distributions), but have different category labels. Our aim is to *utilize samples from source domains to train a convolutional neural network (CNN) that can be well*

---

[1] https://ida.loni.usc.edu.
[2] https://www.fil.ion.ucl.ac.uk/spm.

generalized to an unseen target domain with high sensitivity to SCD progress prediction.

Meta-learning Scheme. Since no target samples (either labeled or unlabeled ones) are available for model training or fine-tuning, we develop a meta-learning scheme to simulate the real-world scenario for cross-domain classification [18–20]. Different from conventional meta-learning schemes that only simulate domain shift with the assumption that the source and target domains share the same class labels, we also simulate domain shift by using different but related subjects (*e.g.*, MCI) for SCD prediction in the proposed meta-learning strategy.

As illustrated in Fig. 2, the source domains $\mathcal{D}_S$ are divided into a *virtual meta-train set* and a *virtual meta-test set*. To simulate SCD prediction on an independent target domain, the virtual test set consists of SCD samples whereas the virtual training set involves other types of samples (*e.g.*, MCI). Denote $\mathbf{w}$ as the parameters (weights) of a CNN model. In the meta-learning process, $\mathbf{w}$ goes through a two-step update and optimization. In the *first* step, a meta-train classification loss $\mathcal{L}_{tr} = \mathcal{F}(\mathcal{D}_{tr}; \mathbf{w})$ is computed using samples in the meta-train set, and the gradient with respect to $\mathbf{w}$ can be calculated and updated as:

$$\mathbf{w}' = \mathbf{w} - \lambda \nabla \mathcal{L}_{tr}(\mathcal{D}_{tr}; \mathbf{w}), \tag{1}$$

where λ is an update rate for the gradient of the CNN model. In the *second* step, the model with $\mathbf{w}'$ is applied to the meta-test set, yielding a meta-test loss $\mathcal{L}_{te} = \mathcal{G}(\mathcal{D}_{te}; \mathbf{w}')$. In Fig. 2, the CNNs in different colors merely indicate different update stages. By jointly optimizing the meta-test and meta-train losses, the CNN is expected to avoid bias towards the training set and learn some domain-invariant features that can be generalized to unseen target domains.

Cost-Sensitive Learning. Previous learning models for disease diagnosis or domain adaptation often treat misclassification errors equally for both positive or negative samples. Actually, incorrectly predicting a positive subject (unhealthy) as a negative case (healthy) is much more costly, because it will make subjects miss the best time for early intervention and treatment. To this end, we employ cost-sensitive learning [21–23] strategy to reduce the false negative rate and increase the sensitivity of our prediction model to possible patients. Specifically, we introduce a cost-sensitive loss $\mathcal{L}_{cs}$ into the meta-learning process of the proposed CNN. We impose this loss on the meta-test set to encourage the model to pursue high prediction accuracy of progressive SCD on unseen target domain.

Let $y \in \{0, 1\}$ denote the ground-truth label for a subject, where 0 represents sSCD and 1 is for pSCD, and the corresponding prediction by the model is $\hat{y}$. We design a cost matrix as follows:

$$\mathbf{M} = \begin{bmatrix} C_{11} & C_{12} \\ C_{21} & C_{22} \end{bmatrix}$$

where C_{ij} indicates the cost for a model when it predicts a subject belonging to the i-th category into the j-th category. For a specific prediction result $\hat{y}$, the

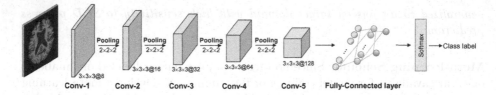

Fig. 3. Structure of the backbone CNN for MRI-based classification. The term $3 \times 3 \times 3@N$ indicates N convolution filters with the kernel size of $3 \times 3 \times 3$.

cost-sensitive loss can be calculated by the inner product of $\hat{y}$ and the row of $\mathbf{M}$ corresponding to its label y, $i.e.$, $\mathcal{L}_{cs} = \langle \hat{y}, M(y+1,\cdot) \rangle$. For example, for a pSCD subject ($y = 1$), the prediction of the learning model is: $\hat{y} = [1,0]$. Then, the cost is $\langle [1,0], [C_{21}, C_{22}] \rangle = C_{21}$. In this work, we empirically set C_{21} and C_{12} in $\mathbf{M}$ to η ($\eta=2$) and 1, respectively, whereas the other elements to 0.

Based on meta-learning and cost-sensitive learning schemes, the total loss for the CSML can be formulated as follows:

$$\mathcal{L}_{total} = \mathcal{L}_{tr} + \alpha \mathcal{L}_{te} + \beta \mathcal{L}_{cs}, \tag{2}$$

where α and β are parameters to control contributions of three losses. Our model is optimized via a two-step update strategy. First, the CNN is trained with the meta-train loss on meta-train set, and the weight $\mathbf{w}$ is updated according to Eq. (1). Then, the updated CNN is fed with the meta-test set, and the meta-test loss and cost-sensitive loss are utilized to further update the weight. The training on the meta-test set can be understood as a regularization term which enables the network to gain strong generalizability.

Network Architecture. We utilize an AlexNet-like 3D CNN as the backbone of our CSML model, with its architecture illustrated in Fig. 3. Specifically, this CNN contains 5 convolution layers and 3 fully-connected layers (with 128, 64 and 2 units, respectively). Each convolution layer consists of a 3D convolution filter (kernel size: $3 \times 3 \times 3$), batch normalization operation and ReLU activation function. We also add a $2 \times 2 \times 2$ max-pooling operation (stride: $2 \times 2 \times 2$) after each convolution layer. A softmax layer is used as the classification layer to outputs the probability of an input MRI belonging to each category.

Implementation Details. We implement the proposed CSML model using PyTorch[3]. The Adam algorithm is employed as the optimizer with a learning rate of 0.0001. The cross-entropy loss is used as the loss function for both $\mathcal{L}_{tr}$ and $\mathcal{L}_{te}$ for classification. Dropout with a probability of 0.5 is used during the training process to avoid overfitting. The gradient update rate λ in Eq. (1) is set to 0.0001. For simplicity, the parameters α and β in Eq. (2) are set to 1. The network is trained for 50 epochs with the batch size of 2.

---

[3] https://pytorch.org.

Table 1. Results of SCD progress prediction achieved by five methods.

Method	AUC (%)	ACC (%)	BAC (%)	SEN (%)	SPE (%)
Baseline-1	56.66 ± 0.87	56.63 ± 1.05	45.92 ± 1.52	21.50 ± 3.45	70.34 ± 1.27
Baseline-2	57.26 ± 1.03	59.12 ± 2.12	53.23 ± 1.66	35.00 ± 4.26	**71.45 ± 2.81**
ROI+SVM	58.18 ± 0.21	54.66 ± 0.01	52.07 ± 0.01	46.50 ± 0.02	61.64 ± 0.01
VoxCNN	56.30 ± 1.05	58.75 ± 2.81	53.42 ± 1.96	36.50 ± 4.58	69.67 ± 2.54
CSML (ours)	**65.36 ± 0.63**	**59.66 ± 2.53**	**60.51 ± 1.65**	**66.40 ± 3.01**	54.62 ± 1.96

3   Experiment

Experimental Setup. Five metrics are used for performance evaluation, *i.e.*, area under receiver operating characteristic curve (AUC), classification accuracy (ACC), balanced accuracy (BAC), sensitivity (SEN) and specificity (SPE).

In the experiments, we use AD/NC, MCI, SCD samples in the ADNI as three source domains, denoted as $\mathcal{A}$ (with AD and NC), $\mathcal{M}$ (with pMCI and sMCI) and $\mathcal{S}$ (with pSCD and sSCD), respectively. The proposed CSML model is first trained on these three source domains, and then applied to the target domain (with 113 SCD subjects collected from a local hospital). In the training phase, we use the domain $\mathcal{S}$ as the meta-test set. And we treat the domain $\mathcal{M}$ as the meta-train set, considering the closer relationship between SCD and MCI when compared with SCD and AD (see Fig. 1).

Competing Methods. We compare our CSML with four methods, including (i) **Baseline-1** that trains the backbone CNN with AD and NC samples and directly transfer it to the target domain; (ii) **Baseline-2** that trains the backbone CNN with pMCI and sMCI samples and directly transfers it to the target domain; (iii) **ROI+SVM** that uses pSCD+MCI subjects as positive samples and sSCD+NC as negative samples. The normalized gray matter volumes of 90 ROIs in AAL are used as MRI features to train a linear support vector machine (SVM) which is transferred to the target domain; and (iv) **VoxCNN** that is a state-of-the-art deep learning method deliberately designed for dementia classification [24] and contains 10 convolution layers and two fully-connected layers.

Prediction Results. The results of SCD progress prediction of different methods are shown in Table 1. From this table, we can derive the following observations. (i) Compared with four competing methods, our CSML can produce the overall best performance, especially with the highest sensitivity. This implies that the proposed meta-learning and cost-sensitive learning strategies in CSML help boost the prediction performance. (ii) Two baseline CNNs (i.e., Baseline-1 and Baseline-2) are weak in SCD progress prediction with a bias towards negative samples (i.e., stable SCD). This indicates that simply using AD or MCI samples to train a deep network and applying it to SCD analysis cannot achieve satisfactory performance. (iii) Baseline-2 that uses MCI samples for training can

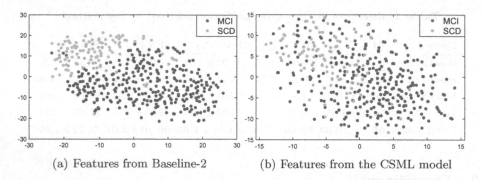

(a) Features from Baseline-2   (b) Features from the CSML model

Fig. 4. Visualization of feature distributions of MCI (source domain) and SCD subjects (target domain) from the baseline CNN and the proposed CSML model.

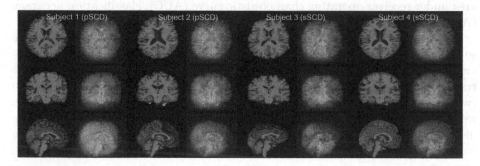

Fig. 5. Saliency maps generated by the proposed CSML model.

achieve better performance than Baseline-1 trained on AD/NC samples. This implies that a network trained on MCI conversion prediction can be reliably applied to SCD progress prediction, compared with that trained on AD/NC subjects. This may be because SCD is semantically closer to MCI in the AD spectrum.

Visualization Analysis. In Fig. 4, we visualize the feature distributions of MCI subjects in the source domain and SCD subjects in the target domain via t-SNE [25], where MRI features are generated from Baseline-2 and CSML, respectively. From Fig. 4(a), we can see that the feature distributions of MCI and SCD subjects generated from Baseline-2 (without meta-learning) have significant difference. As shown in Fig. 4(b), after the meta-training through our CSML model, the distribution gap between these two domains has been greatly reduced. This may indicate that our CSML can learn some domain invariant features, which can partly explain its superiority in cross-domain prediction.

We further visualize the saliency maps [26] generated by the proposed CSML for four SCD subjects in the target domain, as shown in Fig. 5. We compute the gradients of the output of the network w.r.t. each input MRI. Figure 5 shows

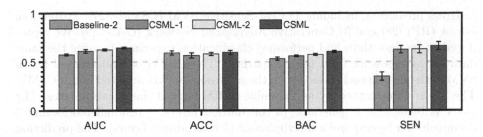

Fig. 6. Comparison between CSML and its variants in SCD progress prediction.

Table 2. Comparison with state-of-the-art methods in SCD progress prediction.

Method	Model	AUC (%)	ACC (%)	SEN (%)	SPE (%)
Yue et al. [27]	Cost-Sensitive SVM	59.8	56.6	55.0	**57.5**
Felpete et al. [28]	Random Forest	58.2	52.2	50.0	53.4
Liu et al. [9]	GAN	60.2	57.5	**70.0**	50.7
CSML (Ours)	Cost-Sensitive CNN	**65.0**	**59.3**	67.5	54.8

that even if it is not very prominent in the saliency maps, SCD may be related to several brain regions, such as the ventricle and hippocampus.

Ablation Study. Compared with baseline CNNs that are trained on the source domains and directly transferred to the target domain, our CSML model consists of meta-learning and cost-sensitive learning on virtual meta-train and meta-test data. To study their effects, we compare CSML with its two variants, including (i) **CSML-1** that uses $\mathcal{A}$ as the meta-train data without cost-sensitive learning, and (ii) **CSML-2** that uses $\mathcal{M}$ as the meta-train data without cost-sensitive learning. These two variants are compared with CSML and the Baseline-2 (trained on $\mathcal{M}$ without meta-learning), with results shown in Fig. 6. It can be seen from Fig. 6 that three meta-learning based methods (i.e., CSML, CSML-1 and CSML-2) generally outperform the Baseline-2 without meta-learning. Besides, our CSML achieves better SEN performance, compared with the three competing methods without cost-sensitive learning. We evaluate the results of CSML and its 2 variants (i.e., CSML-1 and CSML-2) through pair-wise t-test. The p-value for results of CSML vs. CSML-1 is 0.029, while that for CSML vs. CSML-2 is 0.033. This suggests that there is significant difference ($p < 0.05$) between CSML and CSML-1/CSML-2. These results verify the efficacy of the proposed meta-learning and cost-sensitive learning strategies. Also, CSML-2 (with MCI as meta-train data) usually outperforms CSML-1 (with AD/NC as meta-train data), which is consistent with the results of two baseline CNNs in Table 1.

Comparison with State-of-the-Arts. We further compare the proposed method with several state-of-the-art (SOTA) methods for MRI-based SCD progress prediction, including 1) Cost-Sensitive SVM (CSVM) [27], 2) Random Forest (RF) [28] and 3) Generative Adversarial Network (GAN) [9]. We reproduced these algorithms and performed classification experiments using the same data set as this work for a fair comparison. The CSVM is implemented by an SVM with linear kernel trained with the same cost matrix adopted in our CSML. The RF is implemented by an ensemble of 100 bagged classification trees. The GAN is built with a generator (3 convolution layers, 3 residual blocks and 2 deconvolution layers) and a discriminator (5 convolution layers). The prediction results of four different methods are reported in Table 2. It can be seen from Table 2 that, compared with the three SOTA methods, our CSML can produce better AUC and ACC results and comparable performance in terms of SEN and SPE. This may be attributed to the meta-learning scheme that can enhance the generalization ability of the prediction model and the cost-sensitive learning scheme that can produce competing prediction accuracy for both positive and negative samples.

4   Conclusion

In this paper, we present a cost-sensitive meta-learning (CSML) framework for SCD progress prediction based on brain structural MRIs. We train the CSML model on a relatively large-scale source domain and apply it to an unseen target domain with small-scale data for cross-domain SCD progress prediction. Specifically, the source domain (training set) is divided into a virtual meta-train set and a virtual meta-test set to simulate the domain shift among different domains. We also introduce cost-sensitive learning into the meta-training process which can further enhance the sensitivity for identifying progressive SCD subjects. Experimental results on the public ADNI database (with 1,393 subjects) and a private SCD dataset (with 113 subjects) demonstrate the effectiveness of the proposed method over previous state-of-the-arts. As a future work, we intend to leverage multi-modality data (such as MRI, PET and demographic information) for model training to further improve the performance of SCD progress prediction.

Acknowledgements. H. Guan and M. Liu were partly supported by NIH grant (No. AG041721).

References

1. Jessen, F., et al.: A conceptual framework for research on subjective cognitive decline in preclinical Alzheimer's disease. Alzheimer's Dementia **10**(6), 844–852 (2014)
2. Rabin, L.A., Smart, C.M., Amariglio, R.E.: Subjective cognitive decline in preclinical Alzheimer's disease. Annu. Rev. Clin. Psychol. **13**, 369–396 (2017)
3. Amariglio, R.E., et al.: Subjective cognitive complaints and amyloid burden in cognitively normal older individuals. Neuropsychologia **50**(12), 2880–2886 (2012)
4. Jessen, F., et al.: AD dementia risk in late MCI, in early MCI, and in subjective memory impairment. Alzheimer's Dementia **10**(1), 76–83 (2014)

5. Jack, C.R., Jr., et al.: The Alzheimer's disease neuroimaging initiative (ADNI): MRI methods. J. Magn. Resonan. Imaging Official J. Int. Soc. Magn. Resonan. Med. **27**(4), 685–691 (2008)
6. Liu, M., Zhang, J., Adeli, E., Shen, D.: Landmark-based deep multi-instance learning for brain disease diagnosis. Med. Image Anal. **43**, 157–168 (2018)
7. Wang, M., Zhang, D., Huang, J., Yap, P.T., Shen, D., Liu, M.: Identifying autism spectrum disorder with multi-site fMRI via low-rank domain adaptation. IEEE Trans. Med. Imaging **39**(3), 644–655 (2019)
8. Yao, D., Calhoun, V.D., Fu, Z., Du, Y., Sui, J.: An ensemble learning system for a 4-way classification of Alzheimer's disease and mild cognitive impairment. J. Neurosci. Methods **302**, 75–81 (2018)
9. Liu, Y., et al.: Joint neuroimage synthesis and representation learning for conversion prediction of subjective cognitive decline. In: Martel, A.L., et al. (eds.) MICCAI 2020. LNCS, vol. 12267, pp. 583–592. Springer, Cham (2020). https://doi.org/10.1007/978-3-030-59728-3_57
10. Cheng, N., et al.: Self-weighted multi-task learning for subjective cognitive decline diagnosis. In: Martel, A.L., et al. (eds.) MICCAI 2020. LNCS, vol. 12267, pp. 104–113. Springer, Cham (2020). https://doi.org/10.1007/978-3-030-59728-3_11
11. Wachinger, C., Reuter, M.: Domain adaptation for Alzheimer's disease diagnostics. NeuroImage **139**, 470–479 (2016)
12. Cheng, B., Liu, M., Zhang, D., Munsell, B.C., Shen, D.: Domain transfer learning for MCI conversion prediction. IEEE Trans. Biomed. Eng. **62**(7), 1805–1817 (2015)
13. Hosseini-Asl, E., Keynton, R., El-Baz, A.: Alzheimer's disease diagnostics by adaptation of 3D convolutional network. In: 2016 IEEE International Conference on Image Processing (ICIP), pp. 126–130. IEEE (2016)
14. Ghafoorian, M., et al.: Transfer learning for domain adaptation in MRI: application in brain lesion segmentation. In: Descoteaux, M., Maier-Hein, L., Franz, A., Jannin, P., Collins, D.L., Duchesne, S. (eds.) MICCAI 2017. LNCS, vol. 10435, pp. 516–524. Springer, Cham (2017). https://doi.org/10.1007/978-3-319-66179-7_59
15. Orbes-Arteaga, M., et al.: Multi-domain adaptation in brain MRI through paired consistency and adversarial learning. In: Wang, Q., et al. (eds.) DART/MIL3ID-2019. LNCS, vol. 11795, pp. 54–62. Springer, Cham (2019). https://doi.org/10.1007/978-3-030-33391-1_7
16. Li, W., Zhao, Y., Chen, X., Xiao, Y., Qin, Y.: Detecting Alzheimer's disease on small dataset: a knowledge transfer perspective. IEEE J. Biomed. Health Inform. **23**(3), 1234–1242 (2018)
17. Guan, H., Liu, Y., Yang, E., Yap, P.T., Shen, D., Liu, M.: Multi-site MRI harmonization via attention-guided deep domain adaptation for brain disorder identification. Med. Image Anal. **71**, 102076 (2021)
18. Finn, C., Abbeel, P., Levine, S.: Model-agnostic meta-learning for fast adaptation of deep networks. In: International Conference on Machine Learning, pp. 1126–1135. PMLR (2017)
19. Li, D., Yang, Y., Song, Y.Z., Hospedales, T.: Learning to generalize: Meta-learning for domain generalization. In: Proceedings of the AAAI Conference on Artificial Intelligence (2018)
20. Andrychowicz, M., et al.: Learning to learn by gradient descent by gradient descent. In: Advances in Neural Information Processing Systems (NeurIPS), pp. 3981–3989 (2016)
21. Ling, C.X., Sheng, V.S.: Cost-sensitive learning and the class imbalance problem. Encyclopedia Mach. Learn. **2008**, 231–235 (2011)

22. Kuo, W., Häne, C., Yuh, E., Mukherjee, P., Malik, J.: Cost-sensitive active learning for intracranial hemorrhage detection. In: Frangi, A.F., Schnabel, J.A., Davatzikos, C., Alberola-López, C., Fichtinger, G. (eds.) MICCAI 2018. LNCS, vol. 11072, pp. 715–723. Springer, Cham (2018). https://doi.org/10.1007/978-3-030-00931-1_82
23. Galdran, A., Dolz, J., Chakor, H., Lombaert, H., Ben Ayed, I.: Cost-sensitive regularization for diabetic retinopathy grading from eye fundus images. In: Martel, A.L., et al. (eds.) MICCAI 2020. LNCS, vol. 12265, pp. 665–674. Springer, Cham (2020). https://doi.org/10.1007/978-3-030-59722-1_64
24. Korolev, S., Safiullin, A., Belyaev, M., Dodonova, Y.: Residual and plain convolutional neural networks for 3D brain MRI classification. In: ISBI, pp. 835–838 (2017)
25. Maaten, L.V.D., Hinton, G.: Visualizing data using t-SNE. J. Mach. Learn. Res. 9(11), 2579–2605 (2008)
26. Simonyan, K., Vedaldi, A., Zisserman, A.: Deep inside convolutional networks: visualising image classification models and saliency maps. In: ICLR (2014)
27. Yue, L., et al.: Prediction of 7-year's conversion from subjective cognitive decline to mild cognitive impairment. Hum. Brain Mapp. 42(1), 192–203 (2021)
28. Felpete, A., et al.: Predicting progression in subjective cognitive decline (SCD) using a machine learning (ML) approach: the role of the complaint's severity. Alzheimer's Dementia 16, e043492 (2020)

Effective Pancreatic Cancer Screening on Non-contrast CT Scans via Anatomy-Aware Transformers

Yingda Xia[1], Jiawen Yao[2], Le Lu[2], Lingyun Huang[3], Guotong Xie[3], Jing Xiao[3], Alan Yuille[1], Kai Cao[4(✉)], and Ling Zhang[2]

[1] Johns Hopkins University, Baltimore, USA
[2] PAII Inc., Bethesda, USA
[3] PingAn Technology, Shenzhen, China
[4] Changhai Hospital, Shanghai, China

Abstract. Pancreatic cancer is a relatively uncommon but most deadly cancer. Screening the general asymptomatic population is not recommended due to the risk that a significant number of false positive individuals may undergo unnecessary imaging tests (e.g., multi-phase contrast-enhanced CT scans) and follow-ups, adding health care costs greatly and no clear patient benefits. In this work, we investigate the feasibility of using a single-phase non-contrast CT scan, a cheaper, simpler, and safer substituent, to detect resectable pancreatic mass and classify the detection as pancreatic ductal adenocarcinoma (PDAC) or other abnormalities (nonPDAC) or normal pancreas. This task is usually poorly performed by general radiologists or even pancreatic specialists. With pathology-confirmed mass types and knowledge transfer from contrast-enhanced CT to non-contrast CT scans as supervision, we propose a novel deep classification model with an anatomy-guided transformer. After training on a large-scale dataset including 1321 patients: 450 PDACs, 394 nonPDACs, and 477 normal, our model achieves a sensitivity of 95.2% and a specificity of 95.8% for the detection of abnormalities on the hold-out testing set with 306 patients. The mean sensitivity and specificity of 11 radiologists are 79.7% and 87.6%. For the 3-class classification task, our model outperforms the mean radiologists by absolute margins of 25%, 22%, and 8% for PDAC, nonPDAC, and normal, respectively. Our work sheds light on a potential new tool for large-scale (opportunistic or designed) pancreatic cancer screening, with significantly improved accuracy, lower test risk, and cost savings.

Keywords: Pancreatic cancer · Large-scale cancer screening · Transformers · Non-contrast CT

1 Introduction

Pancreatic cancer is the third leading cause of death among all cancers in the United States, with a 5-year overall survival rate of ∼10% [14]. Surgical resection

Y. Xia—Work done during an internship at PAII Inc.

M. de Bruijne et al. (Eds.): MICCAI 2021, LNCS 12905, pp. 259–269, 2021.
https://doi.org/10.1007/978-3-030-87240-3_25

by now remains the only treatment that offers curative potential [10], but more than 80% of patients with pancreatic cancer have already lost the opportunity of surgery at the first diagnosis. Thus, screening pancreatic cancer is very important to provide early diagnosis and patient risk monitoring. The most widely used imaging modality for the initial evaluation of suspected pancreatic cancer is the contrast-enhanced CT scan (CECT). The benefit of CECT for early pancreatic cancer detection includes high sensitivity and specificity, general standardization and availability and relatively easy interpretation [16]. However, CECT exposes patients to radiation and requires iodine contrast, which can cause reaction and potential risks in patients [16], making it hard to be recognized as a general protocol to screen for pancreatic cancer.

In this work, we investigate the possibility of using non-contrast CT scans (NCCT) to screen for pancreatic cancer with deep learning. Compared to CECT, NCCT is cheaper and safer, because it does not require iodine contrast and exposes patients to less radiation. NCCT has been generally applied in screening for lung nodules [11] which can possibly be reused for opportunistic pancreatic cancer screening as well. Nevertheless, due to the low contrast in NCCT pancreatic region, the difficulty of tumor detection rises significantly for radiologists without contrast enhancement. Deep networks, on the other hand, are particularly good at discovering local texture and shape geometry changes, which give us an opportunity to detect pancreatic cancer even without contrast enhancement on NCCT. Those miss detections by human eyes due to low visual contrast do not necessarily become the false negatives by deep learning (DL) detectors.

One major challenge of training deep learning models on NCCT is the difficulty of obtaining expert annotations. Even experienced radiologists could miss masses due to the low contrast on NCCT. This problem is tackled in the process of data collection from the following two aspects. (1) We obtain the pathology-confirmed mass type as classification ground-truth for patients with pancreatic ductal adenocarcinoma (PDAC) or non-PDAC. (2) For the pixel-level labeling of pancreatic tumor mass, the radiologist first annotates on the contrast-enhanced CT; we then transfer the segmentation mask from contrast-enhanced CT to non-contrast CT by performing volumetric registration on the same patient. The combined classification ground-truth labels and segmentation masks serve as the supervisions of our deep learning model with the input of non-contrast CT scans only. The pathology-confirmed mass type and knowledge transfer from CECT to NCCT are the two important pre-requisites for our model to surpass the human expert performance on detecting pancreatic cancer via NCCT.

In terms of the design of deep models, we extend the previous "Segmentation for Classification" [29] paradigm by building a deep classification on top of a segmentation model with transformers [18]. Given the fact that local texture could be insufficient to detect tumors in NCCT, we adopt Transformers to model the pancreas anatomy structure for better classification, which can capture the global context with multi-head attention. This is also in line with the practical diagnosis experience of the radiologists, where sometimes abnormality is discovered by the secondary-sign, such as swelling pancreas head/tail or pancreatic duct dilation, without actually seeing the tumor.

To validate the feasibility of the proposed solution, we collect a large-scale dataset, which covers 1627 patients: 558 PDACs, 474 nonPDACs, and 595 normal. Our model achieves a sensitivity of 95.2% and a specificity of 95.8% on the holdout test set, in terms of abnormality detection. In contrast, the average performance of 11 expert radiologists is 79.7% and 87.6%. This result illustrates the superiority of our designed deep learning-based framework in this specialized task of detecting pancreatic cancer in NCCT. This work sheds light on a potential viable and safe protocol to screen pancreatic cancers on general population.

The main contributions of this paper are summarized as follows.

- For the first time, non-contrast CT (NCCT) is proposed and validated as an effective imaging modality for full-spectrum taxonomy of pancreatic mass/disease screening using deep learning. This sheds light on new computing tools for large-scale opportunistic or designated pancreatic cancer screening of improved accuracy, lower test risk, and cost savings.
- We utilize the pathology-confirmed mass labels, and transfer the imaging information and knowledge from CECT to NCCT as supervision, which is a prerequisite to surpass human expert performance in this task.
- We propose a new framework, named Anatomy-aware Hybrid Transformers, outperforming the mainstream "Segmentation for Classification" paradigm.
- We achieve a sensitivity of 95.2% and a specificity of 95.8% on a large-scale dataset with 1627 patients, demonstrating the good potential of using more convenient non-contrast CT scans for pancreatic cancer screening.

1.1 Related Work

Automated Pancreatic Tumor Detection. Recent advances in deep learning have lead to tremendous improvement in pancreas segmentation [1,9,12,13, 19,23,27,28], an important pre-requisite step for pancreatic tumor detection. Researchers have started to explore the task of automated pancreatic tumor detection using contrast-enhanced CT scans with deep networks [4,20,24,25,29] and radiomics [3]; as well as the task of cancer prognosis prediction [22]. Different from previous work, our framework is designed for non-contrast CT scans, which is beneficial for general asymptomatic patients yet much more challenging.

Vision Transformers. Transformer [18] utilizes attention mechanism originally designed for language tasks. It has recently been applied into vision task, e.g., object detection [2], image recognition [18] and semantic segmentation [26], and achieved comparable or better performance than CNN based approaches.

2 Methodology

Problem Statement. We formulate the task of pancreatic cancer detection in non-contrast CT scans as a three-class classification problem. We denote $\mathcal{L} = \{0, 1, 2\}$ for the three patient classes, i.e., normal, PDAC and non-PDAC. The reasons of having these three classes are: (1) PDAC is a unique group with the most dismal prognosis; (2) any pancreas CT findings with a influence on patient management options. Screening for pancreatic cancer is much more difficult than lung nodules or mammography screening due to the challenge and visual ambiguity of soft-tissue tumor detection without CT contrast enhancement. A key part in our processing pipeline is the availability of knowledge transfer from contrast-enhanced CT by incorporating pathology-confirmed mass type as classification labels and segmentation labels (tumor/pancreas) used for intermediate supervision, as shown in Fig. 1. Denote the training set by $S = \{(\mathbf{X_i}, \mathbf{Y_i}, \mathbf{Z_i}) | i = 1, 2, ..M\}$, where $\mathbf{X_i} \in \mathbb{R}^{H_i \times W_i \times D_i}$, is the 3D volume representing the non-contrast CT scans of the i-th patient. $\mathbf{Y_i}$ is the voxel-wise annotated label map with the same spatial size as $\mathbf{X_i}$. $\mathbf{Z_i} \in \mathcal{L}$ is the class label of the image, confirmed by pathology, radiology, or clinical records. In the testing phase, only $\mathbf{X_i}$ is given, and our goal is to predict a class label for $\mathbf{X_i}$.

Knowledge Transfer from Contrast-Enhanced to Non-contrast CT. Considering the difficulties of mass annotation on non-contrast CT scans (e.g., tumors are barely visible), radiologists first annotate the voxel-wise mass mask on the contrast-enhanced CT scan with the same patient. We then perform image registration using DEEDS [8] from CECT to NCCT and apply the registration field on the manually segmented mass mask. In this way, we can obtain a relatively coarse, but the most reliable mass mask $\mathbf{Y_i}$ on the NCCT image.

2.1 Anatomy-Aware Classification with Transformers

Segmentation for classification is the most straightforward and adopted representation of the task of pancreatic tumor detection. We train a localization UNet [5] to segment pancreas and mass supervised by the transferred masks generated as above. This localization UNet is also used for cropping out the pancreas ROI region as shown in the test process in Fig. 1.

Given the superiority of the attention mechanism in modelling the global context, we build a hybrid Vision Transformer [6] on top of the UNet segmentation model (see Fig. 1). Since the transformer takes the input of the feature map of a segmentation network, we term it as Anatomy-ware Hybrid Transformer. We denote the pancreas ROI region by $\mathbf{X}$, and $\mathbf{X} \in \mathbb{R}^{H \times W \times D}$. We then forward the image $\mathbf{X}$ into a UNet, which consists of a feature extractor $\mathcal{F}$ and an output layer $\mathcal{G}$. This UNet has an intermediate supervision of the mask transferred from contrast-enhanced CT scan of the same patient where the human annotation is available. Therefore, the intermediate output segmentation $\mathbf{P}_s$ can be obtained by $\mathbf{P}_s = \mathcal{G}(\mathcal{F}(\mathbf{X}))$.

The input of the Vision Transformer $\mathcal{H}$ is the final feature map of the UNet right before the output layer, denoted as $\mathcal{F}(\mathbf{X})$, which has a spatial dimension of (C, W, H, D) and C is the number of channels of the feature map. We first use two consecutive 3D convolution layers with a kernel size of $k_1 \times k_2 \times 1$ and $1 \times 1 \times k_3$ to extract $\frac{W}{k_1} \times \frac{H}{k_2} \times \frac{D}{k_3}$ feature patches with C' dimensions each, where C' is also the dimension of the input sequences of the Transformer. Note that previous work [6] directly use one single convolution layer to extract patch features, while

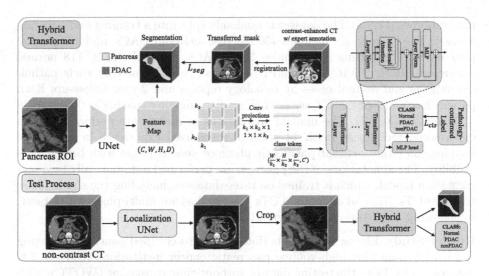

Fig. 1. A visual illustration of our whole framework. Top: we train our hybrid Vision Transformer on non-contrast CT via two supervisions: (i) class label of normal/PDAC/nonPDAC obtained by pathology-confirmed mass type, and (ii) coarse tumor segmentation label transferred from contrast-enhanced CT by registration. Bottom: in the testing phase, we first crop out the pancreas ROI with a localization UNet (separately trained) and output the class and segmentation prediction with the hybrid transformer given non-contrast CT scans.

we decompose it into two layers to reduce the number of parameters in our 3D settings. Learnable positional embeddings are then added to each patch. These patch features are forwarded through multiple transformer blocks with multi-head attention. Following ViT [6], we also use a class token for classification. The output embedding of the class token is used as the classification prediction after a MLP (multilayer perceptron). Our overall training objective is formulated as follows:

$$\mathcal{L} = L_{seg}(\mathbf{P}_s, \mathbf{Y}) + L_{cls}(\mathbf{P}_c, \mathbf{Z}), \tag{1}$$

where $\mathbf{P}_s = \mathcal{G}(\mathcal{F}(\mathbf{X}))$ and $\mathbf{P}_c = \mathcal{H}(\mathcal{F}(X))$ are the output segmentation of UNet and the final classification prediction of the Transformer, respectively. The loss function for classification is cross-entropy loss.

3 Experiments

Dataset and Ground Truth. Our dataset of CT scans of 1627 patients, is consecutively collected in the years of 2016–2018 from a high-volume pancreatic cancer institution. PDAC is of the highest priority among all pancreatic abnormalities with a 5-year survival rate of approximately 10% and is the most common type (about 90% of all pancreatic cancers). This is the main reason that we group all abnormalities into two classes of PDAC and nonPDAC (including nine subtypes [17,25]).The dataset is randomly split into a training and a testing dataset. The training set includes 450 PDACs, 394 nonPDACs, and 477 normal pancreases. The testing set includes 108 PDACs, 80 nonPDACs, 118 normal pancreases. Both PDAC and nonPDAC cases are confirmed by their pathology reports and normal cases by radiology reports and 2-year follow-up. Each patient has multi-phase CT scans. The median imaging spacing is 0.68 × 0.68 × 3.0 mm in [X,Y,Z]. The manual annotations of masses are performed by an experienced radiologist (with 14 years of specialized experience in pancreatic imaging) on either arterial/pancreatic phase or venous phase with better mass visibility. The annotations of the pancreas are performed automatically by a segmentation model, which is trained on three datasets, including the single-phase pancreas CTs [15] and abdominal CTs [7] as well as our multi-phase CT dataset, by following a self-training strategy [21,24].

Reader Study. Eleven radiologists (four are board-certified pancreatic imaging specialists) from four high-volume pancreatic cancer institutions read the 306 non-contrast CTs in the testing dataset without time constraint (WOTC), with a three-class decision by each reader: PDAC, nonPDAC, or normal.

Implementation Details. Each CT volume is firstly resampled into 0.68 × 0.68 × 3.0 mm spacing and normalized into zero mean and unit variance. In the training phase, we crop the foreground 3D bounding box of the pancreas region, randomly pad a small margin on each dimension, and resize the bounding box into a volume of shape (256, 256, 64). The input of the Vision Transformer is the final feature map of the UNet right before the output layer, which has a shape of (32, 256, 256, 64). The two consecutive 3D convolution layers have the kernel size of 32 × 32 × 1 and 1 × 1 × 8 which leads to 512 feature patches with 256 dimensions each. The transformer contains 12 consecutive 8-head attention blocks. We train our hybrid model in an end-to-end fashion with SGD optimizer. The initial learning rate is set to 1×10^{-3} and decays with a cosine learning rate schedule. In addition to the hybrid UNet-Transformer model, we also trained a standard UNet on the whole image for the localization of pancreas. In the inference phase, we first localize the bounding box of the pancreas region with aforementioned UNet, resize the pancreas region into (256, 256, 64) volume and then classify the pancreas region with our hybrid UNet-Transformer model.

Evaluation Methods and Metrics. We randomly split the training dataset into a training set (80% data) and a validation set (20% data). Since the primary goal of non-contrast CT screening is to distinguish between abnormal (PDAC+nonPDAC) and normal, a cutoff point (i.e., threshold) is used to dichotomize model's output probabilities into binary predictions. The cutoff point is predefined on the validation set by maximizing the value of (sensitivity + specificity) before model evaluation on the testing set. To further classify the abnormal as PDAC or nonPDAC, the one with a larger output probability is selected as the prediction. We first report the result of the 2-class classification (PDAC + nonPDAC vs. normal). The evaluation metrics include AUC (area under the ROC curve), sensitivity ($\frac{TP}{TP+FN}$) and specificity ($\frac{TN}{TN+FP}$). We also

Table 1. Results on two-class classification (PDAC+nonPDAC vs. normal) and three-class classification (PDAC vs. nonPDAC vs. normal). WOTC: without time constraint.

	2-class			3-class		
Method	AUC	Sensitivity	Specificity	PDAC	nonPDAC	Normal
S4C with UNet [29]	95.98	91.48	95.76	75.00	73.75	95.76
Hybrid CNN	98.25	94.68	94.91	76.85	80.00	94.91
Hybrid transformer	**98.37**	**95.21**	**95.76**	**78.70**	**80.00**	**95.76**
Mean radiologists WOTC	–	79.66	87.58	53.63	57.96	87.58

report the result of the 3-class classification (PDAC vs. nonPDAC vs. normal), measured by class accuracy. In addition, the mass detection rate by our model is assessed. A detection is considered successful if the intersection (between the ground truth and segmentation mask) over the ground truth is >0% – a coarse localization of mass would be useful in this application scenario.

Compared Methods. We compare our method to two baseline approaches. One is "segmentation for classification" [29] full-filled by a standard UNet where we classify a case as positive if the detected PDAC or nonPDAC tumor volume is larger than a certain threshold, which maximize the value of (sensitivity+specificity) on the validation set. The other is a hybrid CNN classifier built on the UNet feature map trained in an end-to-end fashion. Specifically, we integrate a classification head into the segmentation model. We extract multiple level of the UNet feature map, apply global max pooling on each feature map, concatenate them and forward into a single-layer perceptron for classification. Quantitative results are shown in Table 1. The 2-class ROC curve and a case study are shown in Fig. 2.

Anatomy-Aware Transformer Outperforms Baselines. Compared to two baselines, i.e., segmentation for classification (S4C) and hybrid CNN classifier, our hybrid transformer shows the best performance in all metrics (Table 1), with a relative low STD (about 1%) on sensitivity and specificity. Most medical segmentation models focus on local texture changes and lacks the ability to model the global context. In contrast, our anatomy-aware Hybrid Transformer is built on the locally discriminative features of the UNet, and captures the structural relationship over the whole pancreas region with multi-head attention. Our model is capable of improving the global decision process.

AI Models Outperforms Expert Radiologists on Non-contrast CT Scan. The performance of all 11 radiologists (WOTC) is below our model's ROC curve (Fig. 2). Our model has a mass detection rate of 87.76%, and its sensitivity in abnormality prediction (95%) outperforms the mean human performance (80%) by a large margin and also surpasses the best performing radiologist (R2: 91%) and specialist (S3: 89%), which is the main goal of pancreatic cancer screening using non-contrast CTs. More surprisingly, for the 3-class

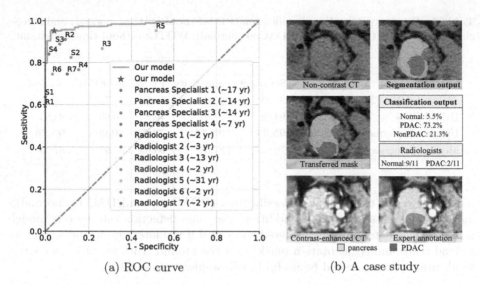

(a) ROC curve   (b) A case study

Fig. 2. (a) ROC diagram for our model result versus all other experts' referrals on the test set of n = 306 patients for 2-class classification. The asterisk denotes the performance of our model. Filled markers denote 11 experts' performances using the same non-contrast CT only. S1: Pancreas Specialist 1, R1: Radiologist 1. (b) A case study in the test set. This PDAC case is extremely challenging for radiologists (only 2/11 are correct) given the limited intensity contrast in non-contrast CT scans whereas our model can successfully locate the mass and predicts the class label.

classification task (Table 1), our model outperforms the mean radiologists (WOTC) by absolute margins of 25%, 22%, and 8% for PDAC, nonPDAC, and normal, respectively.

Human vision system requires adequate visual intensity contrast to distinguish mass from pancreas tissue, which is why contrast-enhanced CT scans are necessary for the diagnosis purpose. Given the surprising performance of DL models on non-contrast CT, we empirically hypothesize that machine vision is better at magnifying the local contrast changes to locate masses. Another crucial reason why computerized model substantially outperforming human performance on non-contrast CT is that we transfer the expert findings from contrast-enhanced CT. Most models are restricted to the performance upper bound of the human annotators. With annotations transfer from CECT (a more "doctor-friendly" modality), and pathology-confirmed labels, DL models are equipped with the essential information/knowledge to break the limit of human observers.

Impact and Future Work. From the reference of computerized performance using contrast-enhanced CT, (sensitivity, specificity) of PDAC vs. Normal is recorded as (92.7%, 99%) [29] and (97.1%, 96%) [20]. This work involves dealing with nonPDAC masses and is generally harder for deep learning [25]. Our performance on non-contrast CT scans (95.2%, 95.8%) is approaching those methods using contrast-enhanced CT. This finding sheds light on the opportunity to use automated methods to screen pancreatic cancers via non-contrast CT imaging. This may be very beneficial for patients, because non-contrast CT is much cheaper, simpler, and safer than its contrast-enhanced counterpart. We plan to conduct multi-institutional studies to validate the generalizability of our system.

4 Conclusion

In this paper, we explore detecting pancreatic cancer from non-contrast CT scans, as a relatively cheap, convenient, simple and safe imaging modality. We propose a hybrid transformer model which is trained by the supervision of pathology-confirmed mass types and the segmentation masks transferred from contrast-enhanced CT scans. We achieve high sensitivity and specificity on a large-scale dataset and outperform the mean radiologists by large margins. Our work suggests the good feasibility of using non-contrast CT scans as a promising clinical tool for large-scale pancreatic cancer screening.

References

1. Cai, J., Lu, L., Zhang, Z., Xing, F., Yang, L., Yin, Q.: Pancreas segmentation in MRI using graph-based decision fusion on convolutional neural networks. In: Ourselin, S., Joskowicz, L., Sabuncu, M.R., Unal, G., Wells, W. (eds.) MICCAI 2016. LNCS, vol. 9901, pp. 442–450. Springer, Cham (2016). https://doi.org/10.1007/978-3-319-46723-8_51
2. Carion, N., Massa, F., Synnaeve, G., Usunier, N., Kirillov, A., Zagoruyko, S.: End-to-end object detection with transformers. In: Vedaldi, A., Bischof, H., Brox, T., Frahm, J.-M. (eds.) ECCV 2020. LNCS, vol. 12346, pp. 213–229. Springer, Cham (2020). https://doi.org/10.1007/978-3-030-58452-8_13
3. Chu, L.C., et al.: Utility of CT radiomics features in differentiation of pancreatic ductal adenocarcinoma from normal pancreatic tissue. Am. J. Roentgenol. 213(2), 349–357 (2019)
4. Chu, L.C., et al.: Application of deep learning to pancreatic cancer detection: lessons learned from our initial experience. J. Am. Coll. Radiol. 16(9), 1338–1342 (2019)
5. Çiçek, Ö., Abdulkadir, A., Lienkamp, S.S., Brox, T., Ronneberger, O.: 3D U-Net: learning dense volumetric segmentation from sparse annotation. In: Ourselin, S., Joskowicz, L., Sabuncu, M.R., Unal, G., Wells, W. (eds.) MICCAI 2016. LNCS, vol. 9901, pp. 424–432. Springer, Cham (2016). https://doi.org/10.1007/978-3-319-46723-8_49
6. Dosovitskiy, A., et al.: An image is worth 16 × 16 words: transformers for image recognition at scale. ICLR (2021)
7. Gibson, E., et al.: Automatic multi-organ segmentation on abdominal CT with dense V-networks. IEEE Trans. Med. Imaging 37(8), 1822–1834 (2018)
8. Heinrich, M.P., Jenkinson, M., Brady, M., Schnabel, J.A.: MRF-based deformable registration and ventilation estimation of lung CT. IEEE Trans. Med. Imaging 32(7), 1239–1248 (2013)
9. Man, Y., Huang, Y., Feng, J., Li, X., Wu, F.: Deep Q learning driven CT pancreas segmentation with geometry-aware U-Net. IEEE Trans. Med. Imaging 38(8), 1971–1980 (2019)

10. Mizrahi, J.D., Surana, R., Valle, J.W., Shroff, R.T.: Pancreatic cancer. Lancet **395**(10242), 2008–2020 (2020)
11. Oudkerk, M., Liu, S., Heuvelmans, M.A., Walter, J.E., Field, J.K.: Lung cancer LDCT screening and mortality reduction–evidence, pitfalls and future perspectives. Nat. Rev. Clin. Oncol. **18**, 1–17 (2020)
12. Roth, H.R., et al.: DeepOrgan: multi-level deep convolutional networks for automated pancreas segmentation. In: Navab, N., Hornegger, J., Wells, W.M., Frangi, A.F. (eds.) MICCAI 2015. LNCS, vol. 9349, pp. 556–564. Springer, Cham (2015). https://doi.org/10.1007/978-3-319-24553-9_68
13. Roth, H.R., Lu, L., Farag, A., Sohn, A., Summers, R.M.: Spatial aggregation of holistically-nested networks for automated pancreas segmentation. In: Ourselin, S., Joskowicz, L., Sabuncu, M.R., Unal, G., Wells, W. (eds.) MICCAI 2016. LNCS, vol. 9901, pp. 451–459. Springer, Cham (2016). https://doi.org/10.1007/978-3-319-46723-8_52
14. Siegel, R.L., Miller, K.D., Jemal, A.: Cancer statistics, 2020. CA: Cancer J. Clin. **70**(1), 7–30 (2020). https://doi.org/10.3322/caac.21590
15. Simpson, A.L., et al.: A large annotated medical image dataset for the development and evaluation of segmentation algorithms. arXiv preprint arXiv:1902.09063 (2019)
16. Singhi, A.D., Koay, E.J., Chari, S.T., Maitra, A.: Early detection of pancreatic cancer: opportunities and challenges. Gastroenterology **156**(7), 2024–2040 (2019)
17. Springer, S., et al.: A multimodality test to guide the management of patients with a pancreatic cyst. Sci. Transl. Med. **11**(501) (2019)
18. Vaswani, A., et al.: Attention is all you need. arXiv preprint arXiv:1706.03762 (2017)
19. Xia, Y., Xie, L., Liu, F., Zhu, Z., Fishman, E.K., Yuille, A.L.: Bridging the gap between 2D and 3D organ segmentation with volumetric fusion net. In: Frangi, A.F., Schnabel, J.A., Davatzikos, C., Alberola-López, C., Fichtinger, G. (eds.) MICCAI 2018. LNCS, vol. 11073, pp. 445–453. Springer, Cham (2018). https://doi.org/10.1007/978-3-030-00937-3_51
20. Xia, Y., Yu, Q., Shen, W., Zhou, Y., Fishman, E.K., Yuille, A.L.: Detecting pancreatic ductal adenocarcinoma in multi-phase CT scans via alignment ensemble. In: Martel, A.L., et al. (eds.) MICCAI 2020. LNCS, vol. 12263, pp. 285–295. Springer, Cham (2020). https://doi.org/10.1007/978-3-030-59716-0_28
21. Xie, Q., Luong, M.T., Hovy, E., Le, Q.V.: Self-training with noisy student improves ImageNet classification. In: Proceedings of the IEEE/CVF Conference on Computer Vision and Pattern Recognition, pp. 10687–10698 (2020)
22. Yao, J., Shi, Yu., Lu, L., Xiao, J., Zhang, L.: DeepPrognosis: preoperative prediction of pancreatic cancer survival and surgical margin via contrast-enhanced CT imaging. In: Martel, A.L., et al. (eds.) MICCAI 2020. LNCS, vol. 12262, pp. 272–282. Springer, Cham (2020). https://doi.org/10.1007/978-3-030-59713-9_27
23. Yu, Q., Xie, L., Wang, Y., Zhou, Y., Fishman, E.K., Yuille, A.L.: Recurrent saliency transformation network: incorporating multi-stage visual cues for small organ segmentation. In: Proceedings of the IEEE Conference on Computer Vision and Pattern Recognition, pp. 8280–8289 (2018)
24. Zhang, L., et al.: Robust pancreatic ductal adenocarcinoma segmentation with multi-institutional multi-phase partially-annotated CT scans. In: Martel, A.L., et al. (eds.) MICCAI 2020. LNCS, vol. 12264, pp. 491–500. Springer, Cham (2020). https://doi.org/10.1007/978-3-030-59719-1_48
25. Zhao, T., et al.: 3D graph anatomy geometry-integrated network for pancreatic mass segmentation, diagnosis, and quantitative patient management. arXiv preprint arXiv:2012.04701 (2020)

26. Zheng, S., et al.: Rethinking semantic segmentation from a sequence-to-sequence perspective with transformers. arXiv preprint arXiv:2012.15840 (2020)
27. Zhou, Y., Xie, L., Shen, W., Wang, Y., Fishman, E.K., Yuille, A.L.: A fixed-point model for pancreas segmentation in abdominal CT scans. In: Descoteaux, M., Maier-Hein, L., Franz, A., Jannin, P., Collins, D.L., Duchesne, S. (eds.) MICCAI 2017. LNCS, vol. 10433, pp. 693–701. Springer, Cham (2017). https://doi.org/10.1007/978-3-319-66182-7_79
28. Zhu, Z., Xia, Y., Shen, W., Fishman, E., Yuille, A.: A 3D coarse-to-fine framework for volumetric medical image segmentation. In: 2018 International Conference on 3D Vision (3DV), pp. 682–690. IEEE (2018)
29. Zhu, Z., Xia, Y., Xie, L., Fishman, E.K., Yuille, A.L.: Multi-scale coarse-to-fine segmentation for screening pancreatic ductal adenocarcinoma. In: Shen, D., et al. (eds.) MICCAI 2019. LNCS, vol. 11769, pp. 3–12. Springer, Cham (2019). https://doi.org/10.1007/978-3-030-32226-7_1

Learning from Subjective Ratings Using Auto-Decoded Deep Latent Embeddings

Bowen Li[1]([✉]), Xinping Ren[2], Ke Yan[1], Le Lu[1], Lingyun Huang[4], Guotong Xie[4], Jing Xiao[4], Dar-In Tai[3], and Adam P. Harrison[1]

[1] PAII Inc., Bethesda, MD 20817, USA
[2] Ruijin Hospital, Shanghai, China
[3] Chang Gung Memorial Hospital, Linkou, Taoyuan, Taiwan ROC
[4] PingAn Technology, Shenzhen, China

Abstract. Depending on the application, radiological diagnoses can be associated with high inter- and intra-rater variabilities. Most computer-aided diagnosis (CAD) solutions treat such data as incontrovertible, exposing learning algorithms to considerable and possibly contradictory label noise and biases. Thus, managing subjectivity in labels is a fundamental problem in medical imaging analysis. To address this challenge, we introduce auto-decoded deep latent embeddings (ADDLE), which explicitly models the tendencies of each rater using an auto-decoder framework. After a simple linear transformation, the latent variables can be injected into any backbone at any and multiple points, allowing the model to account for rater-specific effects on the diagnosis. Importantly, ADDLE does not expect multiple raters per image in training, meaning it can readily learn from data mined from hospital archives. Moreover, the complexity of training ADDLE does not increase as more raters are added. During inference each rater can be simulated and a "mean" or "greedy" virtual rating can be produced. We test ADDLE on the problem of liver steatosis diagnosis from 2D ultrasound (US) by collecting 36 602 studies along with clinical US diagnoses originating from 65 different raters. We evaluated diagnostic performance using a separate dataset with gold-standard *biopsy* diagnoses. ADDLE can improve the partial areas under the curve (AUCs) for diagnosing severe steatosis by 10.5% over standard classifiers while outperforming other annotator-noise approaches, including those requiring 65 times the parameters.

Keywords: Subjective labels · Latent embedding · Liver steatosis · Ultrasound

1 Introduction

Deep learning has been widely applied to computer-aided diagnosis (CAD) tasks [3,18]. In order to train effective deep neural networks, large-scale labeled

Electronic supplementary material The online version of this chapter (https://doi.org/10.1007/978-3-030-87240-3_26) contains supplementary material, which is available to authorized users.

datasets are needed [8,22]. Yet, the labels within large-scale data, such as those found within hospital picture archiving and communication systems (PACSs), are typically image-based diagnoses, which are not always considered "ground-truth" [22]. Take for instance liver steatosis (fatty liver) diagnosis. Liver biopsy is considered the gold standard diagnosis for liver steatosis, but the risks and costs associated with such procedures hinder the collection of large-scale data. On the other hand, ultrasound (US) is the most common tool for assessing liver steatosis [12] and image/US-diagnosis pairs can be readily mined from PACSs. Unfortunately, US diagnoses are considered subjective and suffer from considerable inter- and intra-rater variability [12]. Thus, CAD solutions are faced with the quandary of needing to produce quantitative diagnostic scores when only subjectively labelled training data is available.

The problem of annotator noise has historically been addressed using expectation-maximization (EM) approaches [5,14,21], which typically attempt to estimate true labels and a conditional model that explains rater labels. However, apart from Khetan et al. [14], these require multiple ratings per sample, which is not a realistic requirement for large-scale data, especially for clinical PACS data. Another approach for annotator noise is multi-rater consensus modeling (MRCM) [25], which relies on a consensus loss that is only meaningful in a multiple rating per sample scenario. STAPLE is another popular EM approach that is used to impute ground truth from noisy labels [20], but it has the same multiple-rater requirements. In our problem setting and most clinical scenarios, there is only one rater annotating one image. Our solution is able to work with multi-rater data, but multi-rater algorithms cannot work with our data. In terms of methods that do not assume multi-ratings per training sample, Tanno et al. proposed an approach to estimate annotator confusion matrices that avoids the complexity of EM [19], but they assume that any annotator noise is dependent only on the true label and independent of the image itself, which will likely be violated. Another common limitation of the above approaches is that they are only applicable for categorical classification. More akin to our approach, some recent investigations have explicitly modelled each individual rater. For instance Chou et al. train separate models for each individual or group of raters [4], but this is very computationally expensive. Guan et al. train a separate classification head for each rater [9], but this is limited in modelling capability and their final solution requires a weighted average that again relies on multiple raters per sample, which are needed to calculate weights.

To fill these gaps, we introduce auto-decoded deep latent embeddings (ADDLE). In contrast to these above approaches, ADDLE models individual raters using a rich and expressive latent embedding that is probabilistically motivated. After a linear transformation to match dimensions, the encodings in rater-specific latent vectors can be added to convolutional or global features. The latent vector values, along with shared backbone weights, that best predict each rater's labels are then learned. ADDLE avoids the complexity of EM approaches and can be readily trained using typical gradient descent procedures. Additionally, although it admits multiple raters per sample, ADDLE naturally

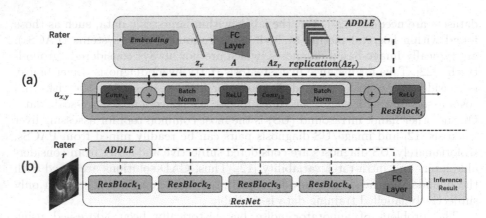

Fig. 1. Algorithmic workflow of ADDLE. (a): ADDLE and its integration into a Res-Block. (b): Potential injection points of ADDLE into a ResNet. ADDLE can also be used with any other classifcation, detection, or segmentation backbone.

works with single ratings per sample, addressing the most critical use-case within CAD. Moreover, ADDLE correctly models subjective ratings as conditional on both the image itself and the true label. Finally, ADDLE imposes no restrictions on what loss can be used and can be readily applied to any medical imaging task, including classification, detection, or segmentation.

We validate ADDLE on the task of liver steatosis diagnosis using conventional liver US images. Liver steatosis affects 20–30% of the global population and is associated with serious risk factors, such as liver fibrosis and hepatocellular carcinoma [24]. We trained ADDLE on a large-scale multi-scanner and multi-etiology PACS-mined dataset of image/US diagnosis pairs (3 790 patients, 312 848 images, and 36 602 studies) that were labelled by 65 different clinicians during clinical care. Apart from our methodological contributions, this alone represents a significant contribution in its own right. Deep learning has previously been applied to liver US, such as for diagnosing steatosis [1,2,10,17], diagnosing fibrosis [15], and detecting lesions [23]. For liver steatosis, datasets are small, with less than 200 patients and less than 1000 images [1,2,10,17] with only single scanners and etiologies, so method generalizability is under-tested.

In terms of methodological contributions, we show that ADDLE can learn from and model subjective US diagnoses. We evaluated ADDLE on a separate dataset of 148 patients with biopsy-proven diagnoses and show that the "mean" and "greedy" ADDLE virtual raters can outperform both standard classification baselines and also leading annotator noise approaches [4,9] in discriminating between healthy, mild, moderate, and severe steatosis. These results demonstrate that ADDLE provides a flexible, straightforward, and effective approach to manage subjectivity in medical imaging labels.

2   Methods

Figure 1 depicts the workings of ADDLE. In short, we assume we are given a dataset of images, noisy/subjective labels, and rater indices $\mathcal{X} = \{\mathbf{x}_i, y_i, r\}_{i=1}^N$ where $r = 1, \ldots R$ indexes which rater generated the label. Here for simplicity we assume a single rater per image and image-level labels, but ADDLE can easily admit multiple raters and other label types. We describe the training and inference process below.

Training. Park *et al.* [16] introduced an auto-decoder-based method, which learns the latent space of shapes for shape representation. ADDLE uses a similar approach to create rater-specific latent embeddings. We associate a set of latent codes, $\mathcal{Z} = \{\mathbf{z}_r\}_{r=1}^R$, to the raters. In practical terms, each $\mathbf{z}_r$ is a simple vector. We model the mapping from images and latent embeddings to a subjective rating using a deep network, $f_\theta(\mathbf{x}_i, \mathbf{z}_r)$, that accepts both the image and the corresponding latent code as input. The deep network can be any popular or well-known network that would normally accept the $\mathbf{x}_i$ input. We can inject the latent codes to any convolutational or global feature map along the network's forward propagation path via simple addition. In this way, the shared neural net weights encode common features that can be influenced by the latent embeddings to model any rater-specific tendencies. To retain any pretrained weights, we only inject the latent codes as part of an existing convolutional or fully-connected layer. For instance, if the latent code dimensionality is M and we wish to inject the latent codes in conjunction with a convolution with C outputs operating on a feature map, $\mathbf{a}_{x,y}$, we simply apply a linear transformation and replication operation:

$$\tilde{\mathbf{a}}_{x,y} = \text{conv}(\mathbf{a}_{x,y}) + \text{rep}(\mathbf{A}\mathbf{z}_r), \tag{1}$$

where $\tilde{\mathbf{a}}_{x,y}$ is the output feature map, conv(.) is the convolutional operator, $\mathbf{A}$ is an $C \times M$ matrix, and rep(.) replicates a vector across the convolutional spatial dimensions. Due to the linearity of convolution, (1) operates as if a large convolution was applied to a concatenated feature representation of $\mathbf{a}_{x,y}$ and $\mathbf{z}_r$. Importantly, any original pretrained weights in conv(.) can be kept. If injected to a global feature, the rep(.) operator can be forgone. Finally, if beneficial ADDLE allows latent codes to be injected at multiple points.

To learn the latent codes, we formulate a posterior composed of a product between a prior distribution and the likelihood of the observed labels:

$$p_\theta(\mathcal{Z}|\mathcal{X}) = \prod_r p(\mathbf{z}_r) \prod_{\{\mathbf{x}_i, y_i\} \in \mathcal{X}_r} p_\theta(y_i|\mathbf{x}_i, \mathbf{z}_r), \tag{2}$$

where $\mathcal{X}_r$ selects all samples labelled by the rth rater and θ parameterizes the likelihood. We assume that the prior, $p(\mathbf{z}_r)$, follows an isotropic zero-mean Gaussian distribution with a covariance of $\sigma^2\mathbf{I}$ (which is also how we initialize the latent embeddings). Given any loss, $\mathcal{L}(.,.)$, the likelihood can be expressed using the deep neural network described above:

$$p_\theta(y_i|\mathbf{x}_i, \mathbf{z}_r) = \exp(-\mathcal{L}(f_\theta(\mathbf{x}_i, \mathbf{z}_r), y_i)). \tag{3}$$

Loss functions are not constrained to be classification-based—indeed our own experiments employ an ordinal regression formulation. During training, (2) is maximized by minimizing the following sum with respect to the rater codes, $\mathcal{Z}$, and the shared network parameters, θ:

$$\arg\min_{\theta, \mathcal{Z}} \sum_{\{\mathbf{x}_i, y_i, r\} \in \mathcal{X}} (\mathcal{L}(f_\theta(\mathbf{x}_i, \mathbf{z}_r), y_i)) + \sum_{r=1}^{R} \left(\frac{1}{\sigma^2} \|\mathbf{z}_r\|_2^2\right), \tag{4}$$

which can be trained using gradient descent. The auto-decoding nature of ADDLE comes from the fact there is no encoding function for the latent embeddings. Instead the latent embedding values are learned solely based on the loss and prior formulation in (4). One downside of (4) is that if there are unbalanced numbers of samples labeled by the raters, then poorly represented raters may not be well optimized. To deal with this, after a solution to (4) is found the shared weights can be frozen and each rater's latent embedding can be individually fine-tuned:

$$\arg\min_{\mathbf{z}_r} \sum_{\{\mathbf{x}_i, y_i\} \in \mathcal{X}_r} (\mathcal{L}(f_\theta(\mathbf{x}_i, \mathbf{z}_r), y_i)) + \frac{1}{\sigma^2} \|\mathbf{z}_r\|_2^2, \tag{5}$$

where only the rater's samples, $\mathcal{X}_r$, are selected. Optimizing (5) is quick, since only each rater's latent vector needs to be fine-tuned, which are small in dimension.

Inference. After training, ADDLE should model how each rater would label a new image. But, emulating rater subjectivity alone is not necessarily useful in inference. Because ADDLE can provide simulated predictions for each rater, if there is *a priori* information on which raters are more experienced or trustworthy, ADDLE could simply simulate those ones. More generally, this is not available. One option is to greedily choose the "best" raters to average using a validation set with gold-standard diagnoses:

$$\bar{y} = \frac{1}{R} \sum_{r \in \mathcal{R}_{\text{greedy}}} f_\theta(\mathbf{x}_i, \mathbf{z}_r), \tag{6}$$

where $\mathcal{R}_{\text{greedy}}$ is a set of raters that are greedily chosen until validation performance tops out. When such validation sets are not available ADDLE can output a mean or majority rating:

$$\bar{y} = \frac{1}{R} \sum_{r=1}^{R} f_\theta(\mathbf{x}_i, \mathbf{z}_r). \tag{7}$$

3   Experiments

Datasets. We test ADDLE on the problem of liver steatosis diagnosis from US. Because US is the most common modality for clinically assessing liver steatosis, it is possible to collect large-scale datasets for algorithmic training. To this end, we collected a big-data **(BD)** cohort consisting of 3 790 patients, 312 848 US images, and 36 602 US studies from the PACS of *Anonymized*. Through the course of clinical care, each study was given a four-class ordinal assessment of either healthy, mild, moderate, or severe steatosis from one of 65 clinicians. We used 3 405 patients for training and left the rest as a stopping criteria validation set. We evaluated whether the ADDLE greedy and mean raters of (6) and (7), respectively, can provide a better quantitative score to differentiate more objective histopathology-derived steatosis severities. To do this we collected a separate histopathology **(HP)** dataset of 218 patients and US studies, all with accompanying biopsy-proven diagnoses within 3 months of the scan date. Histopathological diagnoses follow the same ordinal configuration from healthy to severe steatosis as the US ones. We used 70 of the patients for validation (**HP**-V), but only for model selection and greedy rater selection, and *not* as a validation set for stopping criteria. This left 148 patients as a test set (**HP**-T), with 25, 35, 36, and 52 diagnosed with healthy, mild, moderate, and severe steatosis, respectively. Patients with hepatitis B, hepatitis C, and non-alcoholic fatty liver disease are all represented in **BD** and **HP** studies originated from three different scanners with images corresponding to the six different viewpoints described by Li *et al.* [15].

Setup. The well-known binary decomposition loss of Frank and Hall for ordinal regression [6,7] was used for training on the ordered US severity labels. We used a ResNet-18 [11] backbone, which proved the most optimal, choosing a latent embedding dimensionality of 10, injecting the latent codes at the beginning of the second residual block, and using a σ^2 value of 1.0. An analysis of hyperparameter sensitivity, found in the supplementary, indicates that performance was not sensitive to the choice of σ^2 and was stable across most injection points and latent code sizes. Once trained, a simple summation of the Frank and Hall outputs can produce a single score [7]. In training we treat each US image in a study as an independent sample, but in inference we take the mean score across all images in a study to produce a study-wise score. Analysis on **HP**-V indicated that the top-two raters should be used for the greedy ADDLE variant of (6).

As baselines, we compare ADDLE to both a standard ResNet-18 network trained on **BD**'s US labels (denoted ResNet-18-**BD**) and a ResNet-18 trained only on the small-scale **HP**-T dataset using five-fold cross validation (denoted ResNet-18-**HP**). Comparisons with these baselines respectively reveal the impact of modelling rater tendencies and the importance of using large-scale data for training, even should it be subjectively labelled. We also compared against annotator noise methods that permit ordinal regression. The first comparison uses Chou *et al.*'s JLSL approach of training a separate model for each rater [4]. Note,

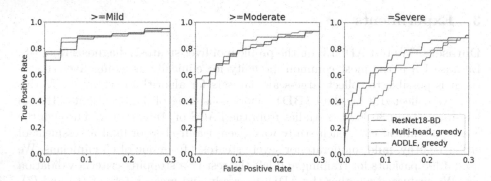

Fig. 2. Partial ROC curves (FPRs ≤ 0.30) on the biopsy-proven test set, excluding JLSL which requires 65 models.

Table 1. Diagnostic performance on the biopsy-proven test set. Note, ResNet18-**HP** is cross-validated on the **HP**-T test set itself, whereas all others are trained on the large-scale **BD** dataset with subjective US labels.

Model	JT	>= Mild		>= Moderate		= Severe		Params
		AUC_0	AUC_{0P}	AUC_1	AUC_{1P}	AUC_2	AUC_{2P}	(millions)
ResNet-18-**BD**	0.882	0.952	0.876	0.923	0.770	0.869	0.615	11.00
ResNet-18-**HP**	0.816	0.799	0.509	0.872	0.642	0.892	0.676	11.00
JLSL [4] (mean)	0.892	0.956	0.869	0.916	0.736	0.893	0.697	714.95
JLSL [4] (greedy)	0.878	0.944	0.860	0.897	0.722	0.883	0.694	714.95
Multi-head [9] (mean)	0.888	0.956	**0.875**	0.918	0.746	0.879	0.658	11.03
Multi-head [9] (greedy)	0.889	0.955	0.869	0.924	0.764	0.873	0.650	11.03
ADDLE (mean)	0.890	**0.957**	0.873	0.922	0.762	0.886	0.672	11.00
ADDLE (greedy)	**0.898**	0.955	0.864	**0.931**	**0.786**	**0.899**	**0.720**	11.00

Chou *et al.* also used additional components, but these rely on multiple raters per sample or the availability of gold-standard labels during training (which we do not assume), so we only evaluate the idea of a single model per rater. We also evaluate Guan *et al.*'s multi-head approach of using a separate classification layer for each rater [9], but forego their weighting process that also assumes multiple raters per sample. In short, JLSL [4] trains a separate model for each rater and multi-head [9] trains a separate classification head. Because we test on the same backbone, their implementation details are identical to ADDLE's apart from the above critical differences. For fairness, we finetune multi-head using an analogous version of (5). Like ADDLE, for each we evaluate their "mean" and "greedy" rater performance.

Evaluation Protocols. Because the problem is to discriminate between four ordered histopathological grades, standard binary receiver operating characteristic (ROC) analysis cannot be performed. Following standard practice, we use

ROC analysis on three different and ordered binary *cutoffs*, specifically >= mild, >= moderate, or = severe levels of steatosis. Associated AUC summary statistics for these cutoffs are denoted AUC_0, AUC_1, and AUC_2, respectively. Like was done by Li *et al.* for liver fibrosis [15], we primarily focus on partial ROC curves and AUCs only within the range of false positive rates that are <= 30%. The reasoning being that lower specificities are not clinically useful operating points to investigate. Partial AUCs are normalized to be within a range of 0 to 1. Additionally, we also report Jonckheere-Terpstra (JT) index values [13], which is a multi-partite generalization of the AUC [7] that also ranges from 0 to 1, with 1 corresponding to perfect discrimination between the four histopathological grades.

Results. Table 1 outlines the performance of all tested models on **HP-T**, and Fig. 2 depicts selected partial ROC curves of methods that only require one model. As can be seen, the standard ResNet-18 trained on US labels can post a good JT score and good AUCs. However, it struggles to identify severe steatosis, posting quite poor partial AUCs. As Fig. 2(c) demonstrates, this corresponds to only a sensitivity of 40% at a false positive rate (FPR) of 10%. Overall, ResNet-18-**HP** performs much poorer, even though it was cross-validated on **HP-T** itself and trained with actual histopatholgoical labels. This highlights the importance of training on large-scale data even if rater noise is present. Moving on to the competitor models, they are mostly able to boost the JT scores. For the most part, any gains are seen in the AUC_2 scores, with JLSL outperforming multi-head. However, for JLSL these gains require training 65 separate models. Moreover, only greedy multi-head is able to match the baseline's AUC_1 scores. Finally, JLSL's greedy rater performance is considerably worse than its mean rater variant, suggesting an overfitting problem that prevents rater performances generalizing from **HP-V** to **HP-T**.

When ADDLE's result are examined, it can be seen that it can boost the AUC_2 by even greater margins, with the greedy rater performing best (partial AUC_2 increases from 0.615 to 0.720). Compared to ResNet-18-**BD**, this is a boost of sensitivity from 40% to 65% at a 10% FPR. Greedy ADDLE can also boost the AUC_1 results (partial AUC increase from 0.770 to 0.786). Importantly, despite only requiring one model and incurring no practical computational cost over the baseline model, greedy ADDLE can still outperform the much costlier greedy JLSL. When greedy rater selection is not available, ADDLE and JLSL are more comparable, but mean ADDLE achieves this performance at 1/65 of the training cost.

The ADDLE latent space can itself be analyzed, which is an avenue of analysis not available elsewhere. For instance, as Fig. 3(a) demonstrates, there is considerable difference between the worst and best performing virtual-rater performance. Interpolating between the corresponding latent vectors produces a smooth performance curve. Additionally, as Fig. 3(b) illustrates, when conducting principal components analysis (PCA), varying the first principle component will produce variations in performance, suggesting that the latent embedding is indeed capturing differences in rater abilities. The first 6 principle components explain 75.2% variance, and the last principle component still explains 4.5% variance. These evidences support our choosing of latent embedding dimensionality 10, which is not likely to be further compressed based on the explained variance ratio. The norms of all latent vectors, z, range from 0.40 to 1.86, which is a reasonable range for our choice of $\sigma^2 = 1.0$. However, one interesting fact we observed is that 2-D PCA and 2-D t-SNE plots show no clustering of latent vectors with high performance, which could be a future research direction, and further investigation may provide additional insights into the latent space properties.

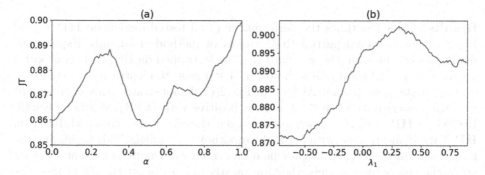

Fig. 3. The latent space and the effects on performance. (a): the test-set JT scores are measured as the latent vector is interploated between the virtual raters with the worst (z_0) and best (z_1) performance, *i.e.*, $z = z_0 + \alpha \cdot (z_1 - z_0)$. (b): the performance when using the first principle component of the PCA basis, scaled using $\lambda_1 = -0.72, \ldots 0.90$ (range calculated by projecting virtual raters to the first principle component). Graphs of other principle components can be found in the supplementary.

4   Conclusion

We introduced ADDLE as an effective approach to deal with subjective ratings using auto-decoded latent variables. ADDLE is highly *expressive*, *i.e.*, modelling rater labels as conditional on the image, *flexible*, *i.e.*, admitting any loss and number of ratings per sample, and *efficient*, *i.e.*, incurring no additional computational cost in training over standard models. We train ADDLE on 36 602 subjectively labelled US studies for liver steatosis, by far the largest such dataset used to date. When evaluated on a separate biopsy-proven dataset, ADDLE outperforms standard classifiers as well as leading annotator noise competitors [4, 9]. These results indicate that ADDLE can better learn from and exploit subjective labels to produce quantitative steatosis assessments. Given the prevalence of subjective labels in CAD training data, future work should validate ADDLE in other end applications, including for detection and segmentation tasks.

References

1. Biswas, M., Kuppili, V., Edla, D.R., Suri, H.S., Saba, L., Marinhoe, R.T., Sanches, J.M., Suri, J.S.: Symtosis: a liver ultrasound tissue characterization and risk stratification in optimized deep learning paradigm. Comput. Methods Programs Biomed. **155**, 165–177 (2018)
2. Byra, M., et al.: Transfer learning with deep convolutional neural network for liver steatosis assessment in ultrasound images. Int. J. Comput. Assist. Radiol. Surg. **13**(12), 1895–1903 (2018). https://doi.org/10.1007/s11548-018-1843-2
3. Cheng, C.T., et al.: A scalable physician-level deep learning algorithm detects universal trauma on pelvic radiographs. Nat. Commun. **12**(1), 1–10 (2021)
4. Chou, H., Lee, C.: Every rating matters: joint learning of subjective labels and individual annotators for speech emotion classification. In: ICASSP 2019–2019 IEEE International Conference on Acoustics, Speech and Signal Processing (ICASSP), pp. 5886–5890 (2019)
5. Dawid, A.P., Skene, A.M.: Maximum likelihood estimation of observer error-rates using the EM algorithm. J. Royal Stat. Soc. Ser. C (Appl. Stat.) **28**(1), 20–28 (1979)
6. Frank, E., Hall, M.: A simple approach to ordinal classification. In: De Raedt, L., Flach, P. (eds.) ECML 2001. LNCS (LNAI), vol. 2167, pp. 145–156. Springer, Heidelberg (2001). https://doi.org/10.1007/3-540-44795-4_13
7. Fürnkranz, J., Hüllermeier, E., Vanderlooy, S.: Binary decomposition methods for multipartite ranking. In: Buntine, W., Grobelnik, M., Mladenić, D., Shawe-Taylor, J. (eds.) ECML PKDD 2009. LNCS (LNAI), vol. 5781, pp. 359–374. Springer, Heidelberg (2009). https://doi.org/10.1007/978-3-642-04180-8_41
8. Greenspan, H., Van Ginneken, B., Summers, R.M.: Guest editorial deep learning in medical imaging: overview and future promise of an exciting new technique. IEEE Trans. Med. Imaging **35**(5), 1153–1159 (2016)
9. Guan, M., Gulshan, V., Dai, A., Hinton, G.: Who said what: modeling individual labelers improves classification (2018)
10. Gummadi, S., et al.: Automated machine learning in the sonographic diagnosis of non-alcoholic fatty liver disease. Adv. Ultrasound Diagnosis Therapy **4**(3), 176–182 (2020)
11. He, K., Zhang, X., Ren, S., Sun, J.: Deep residual learning for image recognition. In: Proceedings of the IEEE Conference on Computer Vision and Pattern Recognition, pp. 770–778 (2016)
12. Hernaez, R., et al.: Diagnostic accuracy and reliability of ultrasonography for the detection of fatty liver: a meta-analysis. Hepatology **54**(3), 1082–1090 (2011)
13. Jonckheere, A.R.: A distribution-free k-sample test against ordered alternatives. Biometrika **41**(1/2), 133–145 (1954)
14. Khetan, A., Lipton, Z.C., Anandkumar, A.: Learning from noisy singly-labeled data. In: International Conference on Learning Representations (2018)
15. Li, B., et al.: Reliable liver fibrosis assessment from ultrasound using global heteroimage fusion and view-specific parameterization. In: Martel, A.L., et al. (eds.) MICCAI 2020. LNCS, vol. 12263, pp. 606–615. Springer, Cham (2020). https://doi.org/10.1007/978-3-030-59716-0_58
16. Park, J.J., Florence, P., Straub, J., Newcombe, R., Lovegrove, S.: DeepSDF: learning continuous signed distance functions for shape representation. In: Proceedings of the IEEE Conference on Computer Vision and Pattern Recognition, pp. 165–174 (2019)

17. Reddy, D.S., Bharath, R., Rajalakshmi, P.: A novel computer-aided diagnosis framework using deep learning for classification of fatty liver disease in ultrasound imaging. In: 2018 IEEE 20th International Conference on e-Health Networking, Applications and Services (Healthcom), pp. 1–5. IEEE (2018)
18. Suzuki, K.: Overview of deep learning in medical imaging. Radiol. Phys. Technol. 10(3), 257–273 (2017). https://doi.org/10.1007/s12194-017-0406-5
19. Tanno, R., Saeedi, A., Sankaranarayanan, S., Alexander, D.C., Silberman, N.: Learning from noisy labels by regularized estimation of annotator confusion. In: 2019 IEEE/CVF Conference on Computer Vision and Pattern Recognition (CVPR), pp. 11236–11245 (2019)
20. Warfield, S.K., Zou, K.H., Wells, W.M.: Simultaneous truth and performance level estimation (staple): an algorithm for the validation of image segmentation. IEEE Trans. Med. Imaging 23(7), 903–921 (2004)
21. Welinder, P., Branson, S., Perona, P., Belongie, S.: The multidimensional wisdom of crowds. In: Lafferty, J., Williams, C., Shawe-Taylor, J., Zemel, R., Culotta, A. (eds.) Advances in Neural Information Processing Systems, vol. 23. Curran Associates, Inc. (2010)
22. Willemink, M.J., et al.: Preparing medical imaging data for machine learning. Radiology 295(1), 4–15 (2020)
23. Wu, K., Chen, X., Ding, M.: Deep learning based classification of focal liver lesions with contrast-enhanced ultrasound. Optik 125(15), 4057–4063 (2014)
24. Younossi, Z., et al.: Global burden of NAFLD and NASH: trends, predictions, risk factors and prevention. Nat. Rev. Gastroenterol. Hepatol. 15(1), 11–20 (2018). number: 1 Publisher: Nature Publishing Group
25. Yu, S., et al.: Difficulty-aware glaucoma classification with multi-rater consensus modeling. In: Martel, A.L., et al. (eds.) MICCAI 2020. LNCS, vol. 12261, pp. 741–750. Springer, Cham (2020). https://doi.org/10.1007/978-3-030-59710-8_72

VertNet: Accurate Vertebra Localization and Identification Network from CT Images

Zhiming Cui[1,2,5], Changjian Li[3], Lei Yang[2], Chunfeng Lian[4], Feng Shi[5], Wenping Wang[2], Dijia Wu[5], and Dinggang Shen[1,5(✉)]

[1] School of Biomedical Engineering, ShanghaiTech University, Shanghai, China
dgshen@shanghaitech.edu.cn
[2] Department of Computer Science, The University of Hong Kong, Hong Kong, China
[3] Department of Computer Science, University College London, London, UK
[4] School of Mathematics and Statistics, Xi'an Jiaotong University, Xi'an, China
[5] Shanghai United Imaging Intelligence Co. Ltd., Shanghai, China

Abstract. Accurate localization and identification of vertebrae from CT images is a fundamental step in clinical spine diagnosis and treatment. Previous methods have made various attempts in this task; however, they fail to robustly localize the vertebrae with challenging appearance or identify vertebra labels from CT images with a limited field of view. In this paper, we propose a novel two-stage framework, *VertNet*, for accurate and robust vertebra localization and identification from CT images. Our method first detects all vertebra centers by a weighted voting-based localization network. Then, an identification network is designed to identify the label of each detected vertebra in leveraging the synergy of global and local information. Specifically, a bidirectional relation module is designed to learn the global correlation among vertebrae along the upward and downward directions, and a continuous label map with dense annotation is employed to enhance the feature learning in local vertebra patches. Extensive experiments on a large dataset collected from real-world clinics show that our framework can accurately localize and identify vertebrae in various challenging cases and outperforms the state-of-the-art methods.

1 Introduction

Localizing and identifying each vertebra from CT images are two essential steps for clinical practice such as surgical planning [7,8], pathological diagnosis [5] and post-operative assessment [9], as the shape of the spine can serve as an important anatomical reference for other organs in these practices. To this end, doctors usually need to manually localize and identify the vertebrae in CT images, which is laborious and time-consuming. In this regard, a fully automatic method with high precision is practically demanded.

Automating these steps have long been the goal for medical imaging researches (e.g., [6,15,19]) but it remains a challenging task. This is because

© Springer Nature Switzerland AG 2021
M. de Bruijne et al. (Eds.): MICCAI 2021, LNCS 12905, pp. 281–290, 2021.
https://doi.org/10.1007/978-3-030-87240-3_27

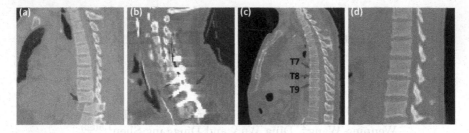

Fig. 1. Typical challenging cases: (a) Spine with pathological fracture; (b) Image with metal artifacts; (c) Adjacent vertebrae having similar shape appearance; (d) Image with a limited field of view.

many spines could be pathological (Fig. 1(a)) or with metal artifacts in the CT images (Fig. 1(b)). More recently, deep learning algorithms that can exploit the large-scale data have shown promising results for these two tasks. One line of the previous works [1,17] employ a one-stage framework to directly segment each of the vertebrae with the corresponding label. Yet, these methods are prone to produce segmentation artifacts in challenging cases, especially where adjacent vertebrae are similar in appearance (e.g., the 7th to the 9th thoracic vertebrae as shown in Fig. 1(c)). To improve the performance, a two-stage approach [4,10,13,16,18] has been proposed to first localize the vertebrae and then identify the label of each vertebra. These methods utilize the Recurrent Neural Network (RNN) or Long Short-Term Memory (LSTM) network for modeling the relationship of neighboring vertebrae. However, such a model may not fully capture the global dependency among all vertebrae and usually limit to local regions, which is important for handling vertebra identification from CT images with a limited field of view (Fig. 1(d)).

To address the aforementioned issues in challenging cases, we present a novel two-stage learning framework for automatic vertebra localization and labeling. Firstly, in the vertebra localization stage, instead of only utilizing the Gaussian-like heatmap to represent the vertebra centers, we combine it with 3D vertebra center offsets to generate more reliable vertebra positions with the supervision of the Chamfer distance. Then, in the second stage of vertebra identification, with the guidance of the detected vertebra centers, vertebra proposals are generated and fed to the identification network for vertebra labeling. As both global and local information are essential for accurate vertebra identification, we propose a bidirectional relation module to capture the global relationships among vertebrae using a self-attention mechanism. Moreover, we also introduce a continuous label map to parameterize the sequence of discrete vertebrae labels, and formulate the prediction of the continuous label map as an additional task for learning fine-grained features in local proposal patches. Our framework was extensively evaluated on a large dataset collected from clinics, which includes many challenging cases. Compared with the state-of-the-art performance, our proposed approach achieved superior results qualitatively and quantitatively, giving the high usability in real-world clinical practices.

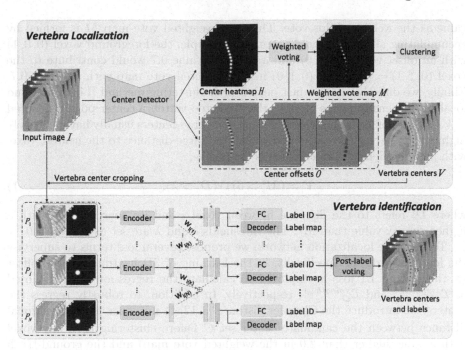

Fig. 2. An overview of the proposed *VertNet* for vertebra localization (Sect. 2.1) and identification (Sect. 2.2) from input CT image.

2   Method

An overview of our *VertNet* for vertebra localization and identification in CT images is shown in Fig. 2, which consists of the localization and identification sub-networks. We elaborate the framework in this section.

2.1   Vertebrae Localization

To localize each vertebra in CT images accurately, we formulate it as a vertebra center point prediction problem. An intuitive solution is to directly regress the 3D vertebra heatmap representing the centers, but it is prone to fail, especially around the cervical vertebrae that pack tightly and are hard to be separated.

Formally, as shown in Fig. 2, the localization network with two output branches takes as input a 3D CT image to predict a one channel Gaussian-like 3D heatmap H and a three channel offset map O, respectively. The former is centered at the vertebra center points with a small standard deviation $\delta=3$ voxel-size, while the later indicates the 3D offset vectors of each voxel pointing to its nearest vertebra center.

To localize each vertebra, we first obtain all foreground voxels F from the 3D heatmap H ($H > 0.2$). Then, for each foreground voxel $f^i \in F$, we consider its 3D offset as a vote to the vertebra center and treat the associated 3D heatmap value as the weight of this vote. Thus, the weighted vote map M is formed by accumulating all the weighted votes. For example, the foreground voxel $(0, 0, 1)$ with an offset vector $(0, 1, 0)$ and a heatmap value 0.7 would contribute to the voxel $(0, 1, 1) = (0, 0, 1) + (0, 1, 0)$ in the weighted vote map with a weight 0.7. Finally, we directly adopt a fast peak search clustering method [14] to find and localize the density peaks in M as the predicted vertebra center points, denoted as V. The rationale is that the clustering vertebrae centers usually have relatively high density values (i.e., weighted votes) and large distance to the nearest voxel with a higher density value, which is defined as:

$$V = (M^i > \delta) \cap (D^i > \lambda), \tag{1}$$

where D^i refers to the distance between voxel i and its nearest voxel with a higher density value than M^i. The thresholds δ and λ are set as 2.0 and 5.0.

To train the localization network, we propose several loss terms to supervise the learning process. Specifically, in the learning of 3D heatmap H and offset O, the smooth L1 loss is employed to calculate the regression error, denoted as $\mathcal{L}_H^{smoothL1}$ and $\mathcal{L}_O^{smoothL1}$, respectively. In addition, to robustly regress the centers, we introduce the Chamfer distance [3,12] to minimize the bidirectional distance between the candidate center set $\widehat{C}$ before clustering (i.e., any voxel with a value higher than 2.0 in the weighted vote map) and the ground-truth center set C, defined as:

$$\mathcal{L}_{CD} = \sum_{\hat{c}_i \in \hat{C}} \min_{c_k \in C} ||\hat{c}_i - c_k||_2^2 + \sum_{c_k \in C} \min_{\hat{c}_i \in \hat{C}} ||c_k - \hat{c}_i||_2^2. \tag{2}$$

Finally, the total loss $\mathcal{L}_{loc}$ of the localization network is formulated as $\mathcal{L}_{loc} = \mathcal{L}_H^{smoothL1} + \mathcal{L}_O^{smoothL1} + \beta \mathcal{L}_{CD}$, where β is the balancing weight and empirically set to 0.2 for all experiments.

2.2   Vertebrae Identification

As shown in the bottom part of Fig. 2, with all the detected vertebra centers, we further assign the label of each vertebra using an identification network. Considering both global and local information are important for accurate vertebra identification, we model the inter-vertebra relationships via a bidirectional relation module at the global scale, and further introduce a novel continuous vertebra label map to enhance the feature learning in each local vertebra patch.

Vertebra Proposal Generation. We generate vertebra proposals guided by the detected vertebra centers. First, we select one vertebra point and crop an image patch from the input 3D CT image I. We then generate an equal-sized center patch using a Gaussian filter centered at the selected vertebra point with a small standard deviation $\delta = 3$ voxel-size, which serves as a guidance signal. Finally, each image patch is concatenated with the corresponding center patch, yielding a two-channel proposal for the identification network.

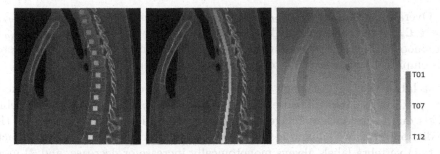

Fig. 3. The process of continuous label map generation, including (a) vertebra centers with labels, (b) B-spine curve with labels, and (c) continuous label map.

Bidirectional Relation Module. For each two-channel vertebra proposal $p_i \in P$, we utilize an shared encoder to extract the feature f_i. Since vertebrae are sequentially connected together from top to bottom and neighboring vertebrae share similar appearance, the contextual clues of the neighboring vertebrae are quite important for accurate identification. Thus, given a proposal p_i, we utilize the self-attention mechanism to obtain the correlation features r_i^{up} in upward direction and r_i^{down} in downward directions to encode the relationship with the other vertebrae in corresponding directions. We take the upward direction as an example to show the calculation of feature r_i^{up} of proposal p_i, and the same procedure is applied to computing feature r_i^{down}. Here, r_i^{up} is a weighted sum of features extracted from upward vertebra proposals defined as $r_i^{up} = \sum_{k=1}^{i-1} w_{i(k)}^{spatial} \cdot w_{i(k)}^{shape} \cdot f_k$, where $w_{i(k)}^{spatial}$ refers to the spatial location weight and $w_{i(k)}^{shape}$ refers to the shape similarity weight, defined in the following:

$$w_{i(k)}^{spatial} = 1.0 - \frac{\exp(d_k)}{\sum_{j=1}^{i-1} \exp(d_j)}, \qquad w_{i(k)}^{shape} = \frac{\exp(f_k^T f_i)}{\sum_{j=1}^{i-1} \exp(f_j^T f_i)}. \tag{3}$$

d_k measures the distance between centers of the k-th and the i-th proposals, while the shape similarity weight measures the feature similarity. Lastly, we combine the correlation features r_i^{up} and r_i^{down} with the proposal feature f_i to derive the corresponding label of vertebra i by several Fully-Connected (FC) layers.

Continuous Label Prediction. The straightforward classification approach with sparse, discrete labels for different proposals receives insufficient supervision and usually leads to inaccurate results. To tackle this problem, we introduce a continuous label map, a novel representation of the vertebra labels, to enhance the network to learn more reliable, fine-grained features. Specifically, as shown in Fig. 3, we first fit a B-spline curve to the 3D coordinates of vertebra centers. Then, we sample a dense set of points on the curve and assign a floating label by linear interpolation of the nearest upward and downward vertebra center labels. Finally, the label values on the curve is mapped to the whole 3D image space based on the nearest neighbor searching. Finally, we add an extra decoder branch to regress the voxel-wise continuous labels in each vertebra patch.

Overall, the training loss $\mathcal{L}_{id}$ for identification is defined as $\mathcal{L}_{id} = \mathcal{L}_{cls} + \eta \mathcal{L}_{reg}$, where $\mathcal{L}_{cls}$ is the cross-entropy loss for classifying the vertebra labels, and $\mathcal{L}_{reg}$ is the smooth L1 loss for regressing the continuous label map. The hyper-parameter η is empirically chosen as 0.5 to balance the loss terms.

Post-label Voting. Although the identification network already achieves excellent performance, we still observed a few incorrect predictions of the vertebra labels due to insufficient image quality such as motion effects. To correct the prediction error, we employ a voting-based post-processing based on two facts that: 1) vertebra labels always monotonically increase or decrease, and 2) most of our labels are correct. Because every vertebra also implicitly carries the labels of others based on their own labels and adjacency, we finally use the voted label as the resulting label for each vertebra.

2.3   Implementation Details

We employed V-Net [11] as the network backbone of our two-stage framework. In the localization network, all CT scans were randomly cropped into the same input size of $256 \times 256 \times 256$, and the cropped patch size of the identification network was set as $96 \times 96 \times 96$ to ensure that the target vertebra can be entirely enclosed. We used the Adam optimizer with an initial learning rate of 0.01 divided by 10 every 5000 iterations. Both networks were trained 20K iterations. It took about 10 h to train the localization network and 12 h for the identification network on a single Tesla M40 GPU.

3   Experiments

3.1   Dataset and Evaluation Metric

We have extensively evaluated our framework on 1000 chest CT images collected from real-world clinics which contain many cases with severe pathological spine problems. These scans have been pre-processed with isotropic resolution of $1.0 \times 1.0 \times 1.0$ mm^3. All vertebra centers and labels are annotated by experienced radiologists. We randomly split the 1000 scans into 600 for training, 100 scans for validation, and the remaining 300 scans for testing.

To quantitatively evaluate the performance of our framework, we first measure the localization error (mm) as the distance of each predicted vertebra center to its nearest ground-truth vertebra center. As for the vertebra labeling task, we compute the identification accuracy (%) at the vertebra and patient levels (denoted as Id_{acc}-V and Id_{acc}-P, respectively). The former measures the percentage of the correctly identified vertebrae per-patient, while the later measures the percentage of the patients whose vertebrae are all correctly identified. For both networks, we also reported the specific metrics of cervical vertebrae (Cer.), thoracic vertebrae (Tho.), lumbar vertebrae (Lum.), as shown in Tables 1 and 2.

Table 1. Quantitative vertebra localization results of alternative networks. The identification task is performed using the baseline network bNet$_{id}$.

Method	Localization [mm] $\downarrow$				Identification [%] $\uparrow$	
	Cer.	Tho.	Lum.	All	Id$_{acc}$-V	Id$_{acc}$-P
bNet$_{vl}$	3.1 ± 3.1	3.8 ± 1.9	3.3 ± 2.5	3.3 ± 2.3	82.1	51.3
bNet$_{vl}$-W	1.7 ± 1.1	2.4 ± 1.5	2.1 ± 1.6	2.0 ± 1.3	86.4	57.3

Table 2. Quantitative vertebra identification results of alternative networks. The higher the percentage value (%), the better the identification accuracy ($\uparrow$).

Method	Cer.	Tho.	Lum.	Id$_{acc}$-V	Id$_{acc}$-P
bNet$_{id}$	87.2	86.0	86.9	86.4	57.3
bNet$_{id}$-R	91.3	89.0	90.2	90.5	82.3
bNet$_{id}$-R-C	96.8	94.7	96.1	96.1	92.3
FullNet	**99.1**	**98.6**	**98.7**	**98.9**	**98.7**

3.2 Ablation Study of Key Components

We conduct ablative experiments to demonstrate the effectiveness of the proposed components. We first build the baseline networks for the vertebra localization and identification tasks, denoted as bNet$_{vl}$ and bNet$_{id}$, respectively. bNet$_{vl}$ directly detects vertebrae using a 3D heatmap, while bNet$_{id}$ simply builds upon a multi-label classification network without the bidirectional relation module and continuous label prediction in our identification task. All alternative networks are derived from the baseline networks by augmenting different components.

Benefits of Weighted-Voting Scheme. Unlike the direct heatmap regression, we utilize the weighted-voting scheme by combining a 3D offset map with the 3D heatmap with the Chamfer loss as an additional supervision. For validation, we augment the baseline bNet$_{vl}$ with an extra output branch to generate 3D offsets followed by the post-clustering, denoted as bNet$_{vl}$-W. The quantitative results are shown in Table 1. Compared to bNet$_{vl}$, bNet$_{vl}$-W consistently improves the localization and identification accuracy. The mean and variance of the localization error are decreased, leading to more robust results. For cervical vertebrae smaller in size and tightly packed, benefit of the weighted-voting scheme is much clearer (reduced by 1.4 mm). This results in an increase of 4.3% (and 6.0%) in Id$_{acc}$-V (and Id$_{acc}$-P) with bNet$_{id}$.

Benefits of Bidirectional Relation Module. We utilize the bNet$_{vl}$-W as the localization network and augment the identification baseline network (bNet$_{id}$) with the bidirectional relation module, denoted as bNet$_{id}$-R. Table 2 shows that the bNet$_{id}$-R consistently improves the identification accuracy, 25.0% in Id$_{acc}$-P, illustrating high efficacy of this module for the sequential prediction task.

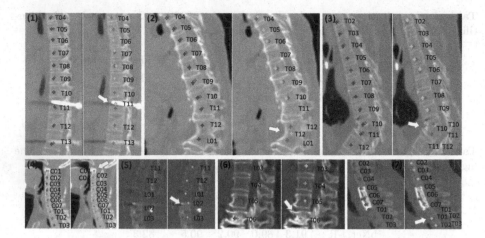

Fig. 4. Comparison between our results (blue) and those by Deep-HMM (yellow) against the ground-truth (GT) (red). Seven typical examples are presented: metal artifacts (1, 7), pathological spines (2, 3), and limited field of view (4, 5, 6, 7). The GT label is annotated if incorrect prediction occurs. (Color figure online)

Table 3. Quantitative comparison with state-of-the-art methods.

Method	Localization	Id_{acc}-V	Id_{acc}-P
J-CNN [1]	7.6 ± 12.4	86.7	60.7
ML-LSTM [13]	2.7 ± 2.8	90.0	79.3
Deep-HMM [2]	2.5 ± 2.3	95.2	89.3
Ours	$\mathbf{2.0 \pm 1.3}$	**98.9**	**98.7**

Benefits of Continuous Label Prediction. We augment $bNet_{id}$-R with this extra task, denoted as $bNet_{id}$-R-C. As shown in Table 2, compared to the $bNet_{id}$-R, Id_{acc}-V improves significantly from 90.5% to 96.1%. This shows the task of continuous label prediction can effectively facilitate the network to learn more discriminative features for fine-grained classification.

Benefits of Post-label Voting. In our FullNet (*VertNet*), a voting-based post-processing step is added on top of $bNet_{id}$-R-C, to generate consistent and correct labels. As presented in Table 2, FullNet obtains the best performance. Notably, the Id_{acc}-P is significantly boosted to 98.7%, which suggests the potential applicability of our framework in real-world clinical scenarios. A set of typical visual results are shown in Fig. 4 with centers projected to the specific CT slice.

3.3   Comparison with State-of-the-Art Methods

We implemented three state-of-the-art vertebra localization and identification methods for comparison, including a one-stage method (J-CNN [1]) and two-stage methods (ML-LSTM [13] and Deep-HMM [2]). Instead of using the CSI

2014 dataset, we train and test all the networks on our newly collected dataset which has a larger sample size and contains more challenging cases. Table 3 shows ML-LSTM and Deep-HMM outperform J-CNN by a large margin. Compared with ML-LSTM, our network with the bidirectional relation module and continuous label prediction generates more reliable vertebra labels. Compared to our bidirectional module for modeling the global relationship among vertebrae, Deep-HMM employs the Markov modeling of vertebra labels which is limited to short-range relationships. As a result, our method significantly outruns DeepHMM in the identification accuracy (3.7% and 9.4% increases on Id_{acc}-V and Id_{acc}-P, respectively). Figure 4 shows qualitative results between ours (blue) and those by DeepHMM (yellow) against GT labels (red) on challenging cases with metal artifacts, pathological spines, or limited field of views, further supporting our design choices.

4   Conclusion

We investigated the problem of vertebra localization and labeling from CT images. The proposed two-stage framework accurately detects all the vertebra centers and successfully identifies all the vertebra labels with satisfactory high accuracy. We qualitatively and quantitatively evaluated our method on our representative clinical dataset and compared against the state-of-the-art approaches. The superior performance suggests the potential applicability of our framework in real-world clinical scenarios.

References

1. Chen, H., et al.: Automatic localization and identification of vertebrae in spine CT via a joint learning model with deep neural networks. In: Navab, N., Hornegger, J., Wells, W.M., Frangi, A.F. (eds.) MICCAI 2015. LNCS, vol. 9349, pp. 515–522. Springer, Cham (2015). https://doi.org/10.1007/978-3-319-24553-9_63
2. Chen, Y., Gao, Y., Li, K., Zhao, L., Zhao, J.: vertebrae identification and localization utilizing fully convolutional networks and a hidden Markov model. IEEE Trans. Med. Imaging 39(2), 387–399 (2019)
3. Cui, Z., et al.: TSegNet: an efficient and accurate tooth segmentation network on 3D dental model. Med. Image Anal. 69, 101949 (2021)
4. Cui, Z., Li, C., Wang, W.: Toothnet: automatic tooth instance segmentation and identification from cone beam ct images. In: Proceedings of the IEEE/CVF Conference on Computer Vision and Pattern Recognition. pp. 6368–6377 (2019)
5. Haldeman, S., et al.: The global spine care initiative: classification system for spine-related concerns. Eur. Spine J. 27(6), 889–900 (2018)
6. Klinder, T., Ostermann, J., Ehm, M., Franz, A., Kneser, R., Lorenz, C.: Automated model-based vertebra detection, identification, and segmentation in CT images. Med. Image Anal. 13(3), 471–482 (2009)
7. Knez, D., Likar, B., Pernuš, F., Vrtovec, T.: Computer-assisted screw size and insertion trajectory planning for pedicle screw placement surgery. IEEE Trans. Med. Imaging 35(6), 1420–1430 (2016)

8. Kumar, R.: Robotic assistance and intervention in spine surgery. In: Li, S., Yao, J. (eds.) Spinal Imaging and Image Analysis. LNCVB, vol. 18, pp. 495–506. Springer, Cham (2015). https://doi.org/10.1007/978-3-319-12508-4_16
9. Létourneau, D., et al.: Semiautomatic vertebrae visualization, detection, and identification for online palliative radiotherapy of bone metastases of the spine. Med. Phys. **35**(1), 367–376 (2008)
10. Liao, H., Mesfin, A., Luo, J.: Joint vertebrae identification and localization in spinal CT images by combining short-and long-range contextual information. IEEE Trans. Med. Imaging **37**(5), 1266–1275 (2018)
11. Milletari, F., Navab, N., Ahmadi, S.A.: V-Net: fully convolutional neural networks for volumetric medical image segmentation. In: 2016 Fourth International Conference on 3D Vision (3DV), pp. 565–571. IEEE (2016)
12. Qi, C.R., Su, H., Mo, K., Guibas, L.J.: PointNet: deep learning on point sets for 3D classification and segmentation. In: Proceedings of the IEEE Conference on Computer Vision and Pattern Recognition (CVPR), July 2017
13. Qin, C., Yao, D., Zhuang, H., Wang, H., Shi, Y., Song, Z.: Residual block-based multi-label classification and localization network with integral regression for vertebrae labeling. arXiv preprint arXiv:2001.00170 (2020)
14. Rodriguez, A., Laio, A.: Clustering by fast search and find of density peaks. Science **344**(6191), 1492–1496 (2014)
15. Schmidt, S., et al.: Spine detection and labeling using a parts-based graphical model. In: Karssemeijer, N., Lelieveldt, B. (eds.) IPMI 2007. LNCS, vol. 4584, pp. 122–133. Springer, Heidelberg (2007). https://doi.org/10.1007/978-3-540-73273-0_11
16. Sekuboyina, A., Rempfler, M., Valentinitsch, A., Kirschke, J.S., Menze, B.H.: Adversarially learning a local anatomical prior: vertebrae labelling with 2D reformations. arXiv preprint arXiv:1902.02205 (2019)
17. Wang, F., Zheng, K., Lu, L., Xiao, J., Wu, M., Miao, S.: Automatic vertebra localization and identification in CT by spine rectification and anatomically-constrained optimization. arXiv preprint arXiv:2012.07947 (2020)
18. Yang, D., et al.: Deep image-to-image recurrent network with shape basis learning for automatic vertebra labeling in large-scale 3D CT volumes. In: Descoteaux, M., Maier-Hein, L., Franz, A., Jannin, P., Collins, D.L., Duchesne, S. (eds.) MICCAI 2017. LNCS, vol. 10435, pp. 498–506. Springer, Cham (2017). https://doi.org/10.1007/978-3-319-66179-7_57
19. Zhan, Y., Maneesh, D., Harder, M., Zhou, X.S.: Robust MR spine detection using hierarchical learning and local articulated model. In: Ayache, N., Delingette, H., Golland, P., Mori, K. (eds.) MICCAI 2012. LNCS, vol. 7510, pp. 141–148. Springer, Heidelberg (2012). https://doi.org/10.1007/978-3-642-33415-3_18

VinDr-SpineXR: A Deep Learning Framework for Spinal Lesions Detection and Classification from Radiographs

Hieu T. Nguyen[1,2], Hieu H. Pham[1,3]([✉]), Nghia T. Nguyen[1], Ha Q. Nguyen[1,3], Thang Q. Huynh[2], Minh Dao[1], and Van Vu[1,4]

[1] Medical Imaging Center, Vingroup Big Data Institute, Hanoi, Vietnam
v.hieuph4@vinbigdata.org
[2] School of Information and Communication Technology,
Hanoi University of Science and Technology, Hanoi, Vietnam
[3] College of Engineering and Computer Science, VinUniversity, Hanoi, Vietnam
[4] Department of Mathematics, Yale University, New Heaven, USA

Abstract. Radiographs are used as the most important imaging tool for identifying spine anomalies in clinical practice. The evaluation of spinal bone lesions, however, is a challenging task for radiologists. This work aims at developing and evaluating a deep learning-based framework, named VinDr-SpineXR, for the classification and localization of abnormalities from spine X-rays. First, we build a large dataset, comprising 10,468 spine X-ray images from 5,000 studies, each of which is manually annotated by an experienced radiologist with bounding boxes around abnormal findings in 13 categories. Using this dataset, we then train a deep learning classifier to determine whether a spine scan is abnormal and a detector to localize 7 crucial findings amongst the total 13. The VinDr-SpineXR is evaluated on a test set of 2,078 images from 1,000 studies, which is kept separate from the training set. It demonstrates an area under the receiver operating characteristic curve (AUROC) of 88.61% (95% CI 87.19%, 90.02%) for the image-level classification task and a mean average precision (mAP@0.5) of 33.56% for the lesion-level localization task. These results serve as a proof of concept and set a baseline for future research in this direction. To encourage advances, the dataset, codes, and trained deep learning models are made publicly available.

Keywords: Spine X-rays · Classification · Detection · Deep learning

Electronic supplementary material The online version of this chapter (https://doi.org/10.1007/978-3-030-87240-3_28) contains supplementary material, which is available to authorized users.

© Springer Nature Switzerland AG 2021
M. de Bruijne et al. (Eds.): MICCAI 2021, LNCS 12905, pp. 291–301, 2021.
https://doi.org/10.1007/978-3-030-87240-3_28

1 Introduction

1.1 Spine X-Ray Interpretation

Conventional radiography or X-ray has the ability to offer valuable information than many other imaging modalities in the assessment of spinal lesions [20, 29]. It has been the primary tool widely used to identify and monitor various abnormalities of the spine. A wide range of spine conditions can be observed from the X-rays like fractures, osteophytes, thinning of the bones, vertebral collapse, or tumors [2,24]. In clinical practice, radiologists usually interpret and evaluate the spine on X-ray scans stored in Digital Imaging and Communications in Medicine (DICOM) standard. Abnormal findings could be identified based on differences in density, intensity, and geometry of the lesions in comparison with normal areas. In many cases, those differences might be subtle and it requires an in-depth understanding of diagnostic radiography to spot them out. Large variability in the number, size, and general appearance of spine lesions makes the interpretation of spinal X-rays a complex and time-consuming task. These factors could lead to the risk of missing significant findings [19], resulting in serious consequences for patients and clinicians.

The rapid advances in machine learning, especially deep neural networks, have demonstrated great potential in identifying diseases from medical imaging data [26]. The integration of such systems in daily clinical routines could lead to a much more efficient, accurate diagnosis and treatment [12]. In this study, we aim to develop and validate a deep learning-based computer-aided diagnosis (CAD) framework called VinDr-SpineXR that is able to classify and localize abnormal findings from spine X-rays. Both the training and validation of our proposed system are performed on our own dataset where radiologists' annotations serve as a strong ground truth.

1.2 Related Work

Developing CAD tools with a high clinical value to support radiologists in interpreting musculoskeletal (MSK) X-ray has been intensively studied in recent years [6]. Early approaches to the analysis of spine X-rays focus on using radiographic textures for the detection of several specific pathologies such as vertebral fractures [10], osteoporosis [9], or osteolysis [31]. Currently, deep convolutional networks [14] (CNNs) have shown their significant improvements for the MSK analysis from X-rays [11,16,17,21,25,30,33]. Most of these studies focus on automated fracture detection and localization [11,16,21,30,33]. To the best of our knowledge, no existing studies devoted to the development and evaluation of a comprehensive system for classifying and localizing multiple spine lesions from X-ray scans. The lack of large datasets with high-quality images and human experts' annotations is the key obstacle. To fill this gap, this work focuses on creating a significant benchmark dataset of spine X-rays that are manually annotated at the lesion level by experienced radiologists. We also propose to develop and evaluate, based on our dataset, a deep learning-based framework that includes a normal versus abnormal classifier followed by a lesion detector that localizes multiple categories of findings with bounding boxes. Both of these tasks can be beneficial in clinical practice: the normal versus abnormal classifier helps triage patients, while the lesion detector helps speed up the reading procedure and complements radiologist's observations.

1.3 Contribution

Our contributions in this paper are two folds. First, we present a new large-scale dataset of 10,469 spine X-ray images from 5,000 studies that are manually annotated with 13 types of abnormalities by radiologists. This is the largest dataset to date that provides radiologist's bounding-box annotations for developing supervised-learning *object detection* algorithms. Table 1 provides a summary of publicly available MSK datasets, in which two previous spine datasets [5, 11] are significantly smaller than ours in size. Furthermore, human expert's annotations of spinal abnormalities are not available in those datasets. Second, we develop and evaluate VinDr-SpineXR – a deep learning framework that is able to classify and localize multiple spine lesions. Our main goal is to provide a nontrivial baseline performance of state-of-the-art deep learning approaches on the released dataset, which could be useful for further model development and comparison of novel CAD tools. To facilitate new advances, we have made the dataset available for public access on our project's webpage[1]. The codes used in the experiments are available at Github[2].

Table 1. Overview of publicly available MSK image datasets.

Dataset	Year	Study type	Label	# Im.
Digital hand Atlas [5]	2007	Left hand	Bone age	1,300
Osteoarthritis initiative [1]	2013	Knee	K & L Grade	8,892
MURA [21]	2017	Upper body	Abnormalities	40,561
RSNA pediatric bone age [7]	2019	Hand	Bone age	14,236
Kuok *et al.* [13]	2018	Spine	Lumbar vertebrae mask	60
Kim *et al.* [11]	2020	Spine	Spine position	797
Ours	**2021**	**Spine**	**Multiple abnormalities**	**10,469**

2 Proposed Method

2.1 Overall Framework

The proposed deep learning framework (see Fig. 1) includes two main parts: (1) a classification network, which accepts a spine X-ray as input and predicts if

---

[1] https://vindr.ai/datasets/spinexr.
[2] https://github.com/vinbigdata-medical/vindr-spinexr.

it could be a normal or abnormal scan; (2) a detection network receives the same input and predicts the location of abnormal findings. To maximize the framework's detection performance, we propose a decision rule to combine the outputs of the two networks.

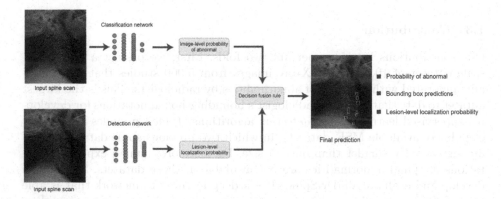

Fig. 1. Overview of the VinDr-SpineXR for spine abnormalities classification and localization. A binary classifier takes as input one spine scan and predicts its probability of abnormality. A detector takes the same scan as input and provides bounding boxes along with probabilities of abnormal findings at lesion-level. A decision rule is then proposed to combine the two outputs and maximize the detection performance.

2.2   Dataset

Data Collection. Spine X-rays with lesion-level annotations are needed to develop automated lesion detection systems. In this work, more than 50,000 raw spine images in DICOM format were retrospectively collected from the Picture Archive and Communication System (PACS) of different primary hospitals between 2010–2020. The data collection process was conducted under our cooperation with participating hospitals. Since this research did not impact clinical care, patient consent was waived. To keep patient's Protected Health Information (PHI) secure, all patient-identifiable information associated with the images has been removed. Several DICOM attributes that are important for evaluating the spine conditions like patient's age and sex were retained.

Data Annotation. To annotate data, we developed an in-house labeling framework called VinDr Lab [18] which was built on top of a PACS. This is a web-based framework that allows multiple radiologists to work remotely at the same time. Additionally, it also provides a comprehensive set of annotation tools that help maximize the performance of human annotators. A set of 5,000 spine studies were randomly selected from the raw data after removing outliers (*e.g.* scans of other body parts or low-quality). All these scans were then uploaded to the labeling framework and assigned to 3 participating radiologists, who have at least 10 years of experience, such that each scan is annotated by exactly one radiologist. In this step, the radiologists decide whether a scan was abnormal or normal as well as marked the exact location of each abnormality on the scan. The labels and annotations provided by the radiologists were served as ground truth for model development and validation later. During this process, the radiologists were blinded to relevant clinical information except patient's age and sex. Finally, a total of 10,468 spine images from 5,000 studies were annotated for the presence of 13 abnormal findings: ankylosis, disc space narrowing, enthesophytes, foraminal stenosis, fracture, osteophytes, sclerotic lesion, spondylolysthesis, subchondral sclerosis, surgical implant, vertebral collapse, foreign body, and other lesions. The "no finding" label was intended to represent the absence of all abnormalities. For the development of the deep learning algorithms, we randomly stratified the labeled dataset, at study-level, into a development set of 4,000 studies and a test set of 1,000 studies. Table 2 summarizes the data characteristics, including patient demographic and the prevalence of each label via

Table 2. Characteristics of patients in the training and test datasets.

	Characteristic	Training set	Test set	Total
Statistics	Years	2011 to 2020	2015 to 2020	2011 to 2020
	Number of studies	4000	1000	5000
	Number of images	8390	2078	10468
	Number of normal images	4257	1068	5325
	Number of abnormal images	4133	1010	5143
	Image size (pixel×pixel, mean)	2455 × 1606	2439 × 1587	2542 × 1602
	Age (mean, years [range])*	49 [6–94]	49 [13–98]	49 [6–98]
	Male (%)*	37.91	37.53	37.83
	Female (%)*	62.08	62.47	62.17
	Data size (GiB)	29.04	7.17	36.21
Lesion type	1. Ankylosis	6	1	7
	2. Disc space narrowing	924	231	1155
	3. Enthesophytes	54	13	67
	4. Foraminal stenosis	387	95	482
	5. Fracture	10	2	12
	6. Osteophytes	12562	3000	15562
	7. Sclerotic lesion	24	5	29
	8. Spondylolysthesis	280	69	349
	9. Subchondral sclerosis	87	23	110
	10. Surgical implant	429	107	536
	11. Vertebral collapse	268	69	337
	12. Foreign body	17	4	21
	13. Other lesions	248	63	311

(*) These calculations were performed on studies where sex and age were available.

the number of bounding boxes. Figure 2 shows several representative samples with and without abnormal findings from our dataset.

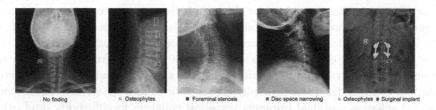

No finding   Osteophytes   ■ Foraminal stenosis   ■ Disc space narrowing   Osteophytes ■ Surgical implant

Fig. 2. Examples of spine X-rays with radiologist's annotations, in which abnormal findings are marked by rectangular bounding boxes.

2.3   Model Development

Network Architecture and Training Methodology To classify a spine X-ray image as either normal or abnormal, three CNNs (DenseNet-121, DenseNet-169, DenseNet-201) have been deployed. These networks [8] are well-known to be effective for X-ray interpretation [21,22]. Each network accepts an image of the spine as input and outputs a binary label corresponding to normal or abnormal scan. A total of 4,257 spine images with normal findings and 4,133 spine images with abnormal findings (reflecting any of the lesions) from the training set was used to optimize networks' weights. During this stage, we optimized cross-entropy loss between the image-level labels and network's outputs using SGD optimizer. The average ensemble of three models serves as the final classifier. An image is considered abnormal in the inference stage if its probability of abnormality is greater than an optimal threshold. In particular, we determine the optimal threshold c^* for the classifier by maximizing Youden's index [32], $J(c) = q(c) + r(c) - 1$, where the sensitivity q and the specificity r are functions of the cutoff value c. For the detection task, we aim to localize 7 important lesions: osteophytes, disc space narrowing, surgical implant, foraminal stenosis, spondylolysthesis, vertebral collapse, and other lesions. Due to the limited number of positive samples, the rest lesions were considered "other lesions". State-of-the-art detectors, namely Faster R-CNN [23], RetinaNet [15], EfficientDet [28], and Sparse R-CNN [27], have been deployed for this task. Faster R-CNN [23] was chosen as a representative for anchor-based two-stage detectors, which firstly propose a set of candidate regions then categorize and refine their locations. RetinaNet [15] and EfficientDet [28] are both one-state detectors that directly localize and classify the densely proposed anchor boxes. Different from previous detectors, Sparse R-CNN [27] starts from a set of initial proposals, then repeatedly refines and classifies these boxes using attention mechanism. All the networks were trained to localize spine lesions using the stochastic gradient descent (SGD) optimizer. During the learning phase, the bounding box regression loss and region-level classification loss were jointly minimized. To improve the generalization performance of the detectors, learned data augmentation strategies for object detection [34] were incorporated in the training procedure.

Decision Fusion Rule. Given an input image $\mathbf{x}$, we denote $\hat{p}(\texttt{abnormal}|\mathbf{x})$ as the classifier's output that reflects the probability of the image being abnormal. To maximize the performance of the lesion detector, we propose a heuristic fusion rule as follows. For any $\mathbf{x}$ with the prediction $\hat{p}(\texttt{abnormal}|\mathbf{x}) \geq c^*$, all lesion detection results are retained. For the case $\hat{p}(\texttt{abnormal}|\mathbf{x}) < c^*$, only predicted bounding boxes with confidence higher than 0.5 are kept.

3 Experiments and Results

3.1 Experimental Setup and Implementation Details

Several experiments were conducted to evaluate the performance of the VinDr-SpineXR. First, we evaluated the classification and detection networks independently on the test set. We then investigated the effect of the fusion rule on the detection performance of the whole framework. All networks were implemented and trained using PyTorch framework (version 1.7.1) on a server with one NVIDIA V100 32 GiB GPU. For the classification task, all training images were rescaled to 224×224 pixels and normalized by the mean and standard deviation of images from the ImageNet dataset. The networks were initialized with pre-trained weights on ImageNet. Each network was trained end-to-end using SGD for 10,000 iterations, approximately 2 h. We used mini-batches of size 32 and set the learning rate to 0.001. An ensemble of the three best models served as the final classifier. For the detection task, all training images were randomly downsampled such that the shorter edge ranging between 640 and 800 pixels, then random augmentation transforms were applied before grouping into mini-batches of 16 samples. The detectors were initialized with weights pre-trained on COCO dataset then trained for 50,000 iterations (about 24 h) using SGD with the learning rate reduced by $10\times$ at the $30,000\text{-}th$ iteration.

3.2 Evaluation Metrics

We report the classification performance using AUROC, sensitivity, specificity, and F1 score. We also estimate the 95% confidence interval (CI) by bootstrapping with 10,000 bootstraps for each measure. In each bootstrap, a new dataset is constructed by sampling with replacement [3]. For the detection task, we followed PASCAL VOC metric, mean average precision (mAP@0.5) [4]. A predicted finding is a true positive if it has an intersection over union (IoU) of at least 0.5 with a ground truth lesion of the same class. For each class, average precision (AP) is the mean of 101 precision values, corresponding to recall values ranging from 0 to 1 with a step size of 0.01. The final metric, mAP@0.5, is the mean of AP over all lesion categories.

3.3 Experimental Results

Classification Performance. On the test set of 2,078 images, our ensemble model reported an AUROC of 88.61% (95% CI 87.19%, 90.02%). At the optimal operating point of 0.2808 (see Supplementary Material), the network achieved a sensitivity of 83.07% (95% CI 80.64%, 85.34%), a specificity of 79.32% (95% CI 77.84%, 81.71%), and an F1 score of 81.06% (95% CI 79.20%, 82.85%), respectively. Table 3 shows quantitative results over all the classification networks.

Table 3. Classification performance on the test set (in percent with 95% CI).

Classifier	AUROC	F1 score	Sensitivity	Specificity
DenseNet121	86.93 (85.4,88.4)	79.55 (77.6,81.4)	80.39 (77.9,82.8)	79.32 (76.9,81.7)
DenseNet169	87.29 (85.8,88.8)	80.25 (78.3,82.1)	81.74 (79.3,84.1)	79.02 (76.6,81.4)
DenseNet201	87.14 (85.6,88.6)	79.03 (77.1,80.9)	77.97 (75.4,80.5)	81.46 (79.1,83.8)
Ensemble	88.61 (87.2,90.0)	81.06 (79.2,82.9)	83.07 (80.6,85.3)	79.32 (77.8,81.7)

Detection Performance. The AP of each abnormal finding detected by 4 detectors is shown in Table 4. Sparse R-CNN [27] showed it as the best performing detector for this task with a mAP@0.5 of 33.15% (see Supplementary Material). Meanwhile, RetinaNet [15] showed the worst performance with a mAP@0.5 of 28.09%. We observed that the reported performances varied over the target lesions, e.g. all the detectors performed best on the LT10 label (surgical implant) and worst on the LT13 (other lesions).

Table 4. Spine X-ray detection performance on the test set.

Detector	LT2(*)	LT4	LT6	LT8	LT10	LT11	LT13	mAP@0.5
Faster R-CNN [23]	22.66	35.99	49.24	31.68	65.22	51.68	2.16	31.83
RetinaNet [15]	14.53	25.35	41.67	32.14	65.49	51.85	5.30	28.09
EfficientDet [28]	17.05	24.19	42.69	35.18	61.85	52.53	2.45	28.73
Sparse R-CNN [27]	20.09	32.67	48.16	45.32	72.20	49.30	5.41	33.15

(*) LT2, LT4, LT6, LT8, LT10, LT11, LT13 denotes for disc space narrowing, foraminal stenosis, osteophytes, spondylolysthesis, surgical implant, vertebral collapse and other lesions, respectively, following the same indexing in Table 2.

Effect of the Decision Fusion Rule. Table 5 provides a comparison of the detection performance between the detector network and the whole framework. We observed that by combining the detection and classification networks via the proposed decision rule, the whole framework leads to a higher performance over all lesions, except LT4 and LT8. Visualization of predicted lesions by the VinDr-SpineXR is provided in the Supplementary Material.

Table 5. Spine X-ray detection performance of the whole framework on the test set.

Method	LT2	LT4	LT6	LT8	LT10	LT11	LT13	mAP@0.5
Detector only	18.06	28.89	34.23	41.52	62.45	42.85	4.03	33.15
Whole framework	21.43	27.36	34.78	41.29	62.53	43.39	4.16	33.56

4 Discussion

The proposed decision fusion uses the result of the classifier to influence the detector. As shown in Table 5, this rule, although very simple, helps improve the mAP of the detector by **0.41%**. We have also experimented with a counterpart fusion rule to use the result of the detector to influence the classifier. Specifically, we averaged classifier's output with the highest box score from detector's outputs, boosting the AUROC and F1 score of the classifier by **1.58%** and **0.46%**, respectively. These experiments have highlighted the effectiveness of the proposed mutual ensemble mechanism.

5 Conclusions

In this work, we developed and validated VinDr-SpineXR – a deep learning-based framework to detect and localize abnormalities in radiographs of spine. We contributed a new labeled dataset and models for spine X-ray analysis. To the best of our knowledge, this is the first effort to address the problem of multiple lesions detection and localization in spine X-rays. Our experiments on the dataset demonstrated the effectiveness of the proposed method. For future work, we expect to extend our dataset and consider more lesion labels for further experiments, including rare labels. We also plan to conduct more experiments and evaluate the impact of the proposed framework in real-world clinical scenarios.

References

1. Osteoarthritis initiative: a multi-center observational study of men and women. https://oai.epi-ucsf.org/datarelease/. Accessed 22 Feb 2021
2. Baert, A.: Spinal Imaging: Diagnostic Imaging of the Spine and Spinal Cord. Springer, Heidelberg (2007). https://doi.org/10.1007/978-3-540-68483-1
3. Efron, B., Tibshirani, R.J.: An Introduction to the Bootstrap. Monographs on Statistics and Applied Probability, vol. 57. Chapman & Hall/CRC (1993)
4. Everingham, M., Van Gool, L., Williams, C.K., Winn, J., Zisserman, A.: The Pascal visual object classes (VOC) challenge. Int. J. Comput. Vis. 88(2), 303–338 (2010)
5. Gertych, A., Zhang, A., Sayre, J., Pospiech-Kurkowska, S., Huang, H.: Bone age assessment of children using a digital hand atlas. Comput. Med. Imaging Graph. 31(4–5), 322–331 (2007)

6. Gundry, M., Knapp, K., Meertens, R., Meakin, J.: Computer-aided detection in musculoskeletal projection radiography: a systematic review. Radiography **24**(2), 165–174 (2018)
7. Halabi, S.S., et al.: The RSNA pediatric bone age machine learning challenge. Radiology **290**(2), 498–503 (2019)
8. Huang, G., Liu, Z., Van Der Maaten, L., Weinberger, K.Q.: Densely connected convolutional networks. In: IEEE Conference on Computer Vision and Pattern Recognition (CVPR), pp. 4700–4708 (2017)
9. Kasai, S., Li, F., Shiraishi, J., Li, Q., Doi, K.: Computerized detection of vertebral compression fractures on lateral chest radiographs: preliminary results with a tool for early detection of osteoporosis. Med. Phys. **33**(12), 4664–4674 (2006)
10. Kasai, S., Li, F., Shiraishi, J., Li, Q., Nie, Y., Doi, K.: Development of computerized method for detection of vertebral fractures on lateral chest radiographs. In: Medical Imaging: Image Processing, vol. 6144, p. 61445D (2006)
11. Kim, K.C., Cho, H.C., Jang, T.J., Choi, J.M., Seo, J.K.: Automatic detection and segmentation of lumbar vertebrae from x-ray images for compression fracture evaluation. Comput. Meth. Programs Biomed. **200**, 105833 (2020)
12. Krupinski, E.A., Berbaum, K.S., Caldwell, R.T., Schartz, K.M., Kim, J.: Long radiology workdays reduce detection and accommodation accuracy. J. Am. Coll. Radiol. **7**(9), 698–704 (2010)
13. Kuok, C.P., Fu, M.J., Lin, C.J., Horng, M.H., Sun, Y.N.: Vertebrae segmentation from X-ray images using convolutional neural network. In: International Conference on Information Hiding and Image Processing (IHIP), pp. 57–61 (2018)
14. LeCun, Y., Bengio, Y., Hinton, G.: Deep learning. Nature **521**(7553), 436–444 (2015)
15. Lin, T.Y., Goyal, P., Girshick, R., He, K., Dollár, P.: Focal loss for dense object detection. In: IEEE International Conference on Computer Vision (ICCV), pp. 2980–2988 (2017)
16. Lindsey, R., et al.: Deep neural network improves fracture detection by clinicians. Proc. Natl. Acad. Sci. **115**(45), 11591–11596 (2018)
17. Mandal, I.: Developing new machine learning ensembles for quality spine diagnosis. Knowl. Based Syst. **73**, 298–310 (2015)
18. Nguyen, N.T., et al.: VinDr Lab: a data platform for medical AI (2021). https://github.com/vinbigdata-medical/vindr-lab
19. Pinto, A., et al.: Traumatic fractures in adults: missed diagnosis on plain radiographs in the emergency department. Acta Bio Medica Atenei Parmensis **89**(1), 111 (2018)
20. Priolo, F., Cerase, A.: The current role of radiography in the assessment of skeletal tumors and tumor-like lesions. Eur. J. Radiol. **27**, S77–S85 (1998)
21. Rajpurkar, P., et al.: MURA: large dataset for abnormality detection in musculoskeletal radiographs. arXiv preprint arXiv:1712.06957 (2017)
22. Rajpurkar, P., et al.: ChexNet: radiologist-level pneumonia detection on chest X-rays with deep learning. arXiv preprint arXiv:1711.05225 (2017)
23. Ren, S., He, K., Girshick, R., Sun, J.: Faster R-CNN: towards real-time object detection with region proposal networks. In: Cortes, C., Lawrence, N., Lee, D., Sugiyama, M., Garnett, R. (eds.) Advances in Neural Information Processing Systems, vol. 28. Curran Associates, Inc. (2015). https://proceedings.neurips.cc/paper/2015/file/14bfa6bb14875e45bba028a21ed38046-Paper.pdf
24. Rodallec, M.H., et al.: Diagnostic imaging of solitary tumors of the spine: what to do and say. Radiographics **28**(4), 1019–1041 (2008)

25. Sa, R., et al.: Intervertebral disc detection in X-ray images using Faster R-CNN. In: Annual International Conference of the IEEE Engineering in Medicine and Biology Society (EMBC), pp. 564–567 (2017)
26. Shen, D., Wu, G., Suk, H.I.: Deep learning in medical image analysis. Annu. Rev. Biomed. Eng. **19**, 221–248 (2017)
27. Sun, P., et al.: Sparse R-CNN: end-to-end object detection with learnable proposals. arXiv preprint arXiv:2011.12450 (2020)
28. Tan, M., Pang, R., Le, Q.V.: EfficientDet: scalable and efficient object detection. In: IEEE Conference on Computer Vision and Pattern Recognition (ECCV), pp. 10781–10790 (2020)
29. Tang, C., Aggarwal, R.: Imaging for musculoskeletal problems. InnovAiT **6**(11), 735–738 (2013)
30. Thian, Y.L., Li, Y., Jagmohan, P., Sia, D., Chan, V.E.Y., Tan, R.T.: Convolutional neural networks for automated fracture detection and localization on wrist radiographs. Radiol. Artif. Intell. **1**(1), e180001 (2019)
31. Wilkie, J.R., Giger, M.L., Pesce, L.L., Engh Sr., C.A., Hopper Jr., R.H., Martell, J.M.: Imputation methods for temporal radiographic texture analysis in the detection of periprosthetic osteolysis. In: Medical Imaging: Computer-Aided Diagnosis, vol. 6514, p. 65141L (2007)
32. Youden, W.J.: Index for rating diagnostic tests. Cancer **3**(1), 32–35 (1950)
33. Zhang, X., et al.: A new window loss function for bone fracture detection and localization in X-ray images with point-based annotation. arXiv preprint arXiv:2012.04066 (2020)
34. Zoph, B., Cubuk, E.D., Ghiasi, G., Lin, T.Y., Shlens, J., Le, Q.V.: Learning data augmentation strategies for object detection. In: IEEE European Conference on Computer Vision (ECCV), pp. 566–583 (2020)

Multi-frame Collaboration for Effective Endoscopic Video Polyp Detection via Spatial-Temporal Feature Transformation

Lingyun Wu[1], Zhiqiang Hu[1], Yuanfeng Ji[2], Ping Luo[2], and Shaoting Zhang[1(✉)]

[1] SenseTime Research, Hong Kong, China
zhangshaoting@sensetime.com
[2] The University of Hong Kong, Hong Kong, China

Abstract. Precise localization of polyp is crucial for early cancer screening in gastrointestinal endoscopy. Videos given by endoscopy bring both richer contextual information as well as more challenges than still images. The camera-moving situation, instead of the common camera-fixed-object-moving one, leads to significant background variation between frames. Severe internal artifacts (e.g. water flow in the human body, specular reflection by tissues) can make the quality of adjacent frames vary considerably. These factors hinder a video-based model to effectively aggregate features from neighborhood frames and give better predictions. In this paper, we present Spatial-Temporal Feature Transformation (STFT), a multi-frame collaborative framework to address these issues. Spatially, STFT mitigates inter-frame variations in the camera-moving situation with feature alignment by proposal-guided deformable convolutions. Temporally, STFT proposes a channel-aware attention module to simultaneously estimate the quality and correlation of adjacent frames for adaptive feature aggregation. Empirical studies and superior results demonstrate the effectiveness and stability of our method. For example, STFT improves the still image baseline FCOS by 10.6% and 20.6% on the comprehensive F1-score of the polyp localization task in CVC-Clinic and ASUMayo datasets, respectively, and outperforms the state-of-the-art video-based method by 3.6% and 8.0%, respectively. Code is available at https://github.com/lingyunwu14/STFT.

1 Introduction

Gastrointestinal endoscopy is widely used for early gastric and colorectal cancer screening, during which a flexible tube with a tiny camera is inserted and guided through the digestive tract to detect precancerous lesions [2]. Identifying and removing adenomatous polyp are routine practice in reducing gastrointestinal cancer-based mortality [14]. However, the miss rate of polyp is as high as 27% due to subjective operation and endoscopist fatigue after long duty [1]. An automatic polyp detection framework is thus desired to aid in endoscopists and reduce the risk of misdiagnosis.

For accurate and robust polyp detection, it is necessary to explore the correlation and complementarity of adjacent frames, to compensate for the possible image corruption or model errors in single images [16]. Nevertheless, there have been two long-standing and serious challenges in endoscopic video polyp detection:

© Springer Nature Switzerland AG 2021
M. de Bruijne et al. (Eds.): MICCAI 2021, LNCS 12905, pp. 302–312, 2021.
https://doi.org/10.1007/978-3-030-87240-3_29

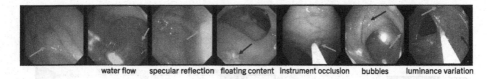

water flow   specular reflection   floating content   instrument occlusion   bubbles   luminance variation

Fig. 1. The easier frame (the first) and typical challenging frames (last six) associated with polyp detection in endoscopic video. Green arrows point to the polyp and blue arrows point to the internal artifacts that often cause false positives. (Color figure online)

How to align object features across frames given the complex motion of the endoscopic camera? One key difference of endoscopic videos from common videos is the camera-moving instead of the common camera-fixed-object-moving situation. The complex motion of the endoscopic camera leads to significant background variation between frames. As a result, the mainstream video methods [26,27] based on optical flow alignment are not suitable, since you do not have a reference, which leads to poor performance of optical flow evaluation [25]. Intuitively, attempting a global alignment is both difficult and unnecessary, since the variable background distracts the focus of the network and overwhelms the foreground modeling for feature alignments. We thus argue that an object-centered and proposal-guided feature alignment is required to mask out background trifles and focus on the concerned foreground variation.

How to assemble features of neighborhood frames given the varied image quality resulting from water flow, reflection, bubbles, etc.? As shown in Fig. 1, frames in endoscopic videos are always and inevitably encountered with image corruptions such as water flow, specular reflection, instrument occlusion, bubbles, etc. These internal artifacts can make the quality of adjacent frames vary considerably. The *quality*, as a result, should be given equal consideration as the *correlation* between frames, in the stage of adjacent feature aggregation. We further notice that different internal artifacts are handled by different kernels in convolutional neural networks, and results in varied activation patterns in different channels. The combination of channel-by-channel selection and position-wise similarity [4,8] are thus believed to be necessary for the simultaneous assessment of foreground correlation and feature quality.

We aim to tackle the two challenges with carefully designed spatial alignment and temporal aggregation, and propose the multi-frame collaborative framework named Spatial-Temporal Feature Transformation (STFT). Spatially, we choose deformable convolution [7] as building blocks for feature alignments, for its adaptability in modeling large variations. We further enhance its object-centered awareness and avoid background distraction by conditioning the offset prediction of the deformable convolution on the object proposals extracted by the image-based detector. Temporally, we design a channel-aware attention module that combines both the cosine similarity to model the foreground correlation between frames and the learned per-channel reweighting to estimate the inter-frame quality variation. The modules achieve a balance of expressiveness and efficiency without much additional computational complexity. Note that the two components are mutually beneficial in that spatial alignment acts as the prerequisite and temporal aggregation looks for more advantage, which is also demonstrated in experimental results.

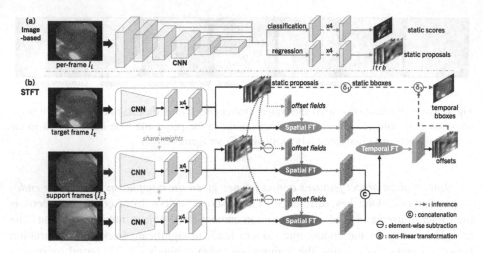

Fig. 2. Illustration of polyp detection using an image-based baseline (a) and our proposed Spatial-Temporal Feature Transformation (b). See Sect. 2 for more details. (Color figure online)

The contribution of this work can be summarized as three folds. Firstly, we present a proposal-guided spatial transformation to enhance the object-centered awareness of feature alignment and mitigate the feature inconsistency between adjacent frames in the camera-moving situation of endoscopic videos. Secondly, we design a novel channel-aware attention module for feature aggregation that achieves a balance of expressiveness and efficiency and shows superiority over other counterparts in experimental results. Lastly, we propose an effective multi-frame collaborative framework STFT on top of the two components. STFT sets new state-of-the-arts on both two challenging endoscopic video datasets and two polyp tasks. Noticeably, STFT shows a far more significant improvement over still image baselines than other video-based counterparts (for example, 10.6% and 20.6% localization F1-score improvements on the CVC-Clinic and ASUMayo datasets, respectively).

2   Method

An Image-Based Baseline. Given the endoscopic video frames $\{I_i\}, i = 1, \ldots, \infty$, a baseline approach for polyp detection is to apply an image-based detector to each frame individually. We adopt a simple one-stage detector FCOS [21] as our baseline. As Fig. 2(a), it firstly generates a set of multi-level feature maps with FPN [15] over the input image I_i. Then, it outputs static classification scores and regression proposals by classification branch and regression branch respectively. Each branch is implemented by four convolutional layers, sharing weights between different feature levels. For level l, Let $F_i^l \in \mathbb{R}^{C \times H \times W}$ be the final feature map output by the regression branch. $G_i = (x_0, y_0, x_1, y_1)$ is the associated ground-truth bounding box, where (x_0, y_0) and (x_1, y_1) denote the coordinates of the left-top and right-bottom corners. Static proposals are $\{p_i\} \in \mathbb{R}^{4 \times H \times W}$, and their regression targets $\{g_i\} \in \mathbb{R}^{4 \times H \times W}$ are offsets between

G_i and all spatial locations on F_i^l. For each spatial location (x, y), $g_i = (l_i, t_i, r_i, b_i)$ is a 4D real vector, which represents distances from (x, y) to four boundaries of G_i. It can be formulated as

$$l_i = x - x_0, t_i = y - y_0, r_i = x_1 - x, b_i = y_1 - y. \tag{1}$$

Given the challenging frame with water flow in Fig. 1, the baseline showed low confidence on the ground-truth and missed polyp detection (see results in Fig. 4).

STFT Architecture. Given the target frame I_t and its adjacent support frames $\{I_s\}, s = 1, \ldots, N$, our aim is to accurately detect polyp in I_t by using the features from $\{I_s\}$. Firstly, we generate multi-level feature maps and static prediction results for all frames via the same architecture as the image-based baseline (only the regression branch is shown in Fig. 2(b) for ease of explanation). At l-th feature level, we use predicted static proposals $\{p_t\}$ of I_t to guide the spatial transformation of target feature F_t^l (as shown in green in Fig. 2(b)). Meanwhile, we leverage the difference between $\{p_t\}$ and static proposals $\{p_s\}$ of each I_s to guide the spatial transformation of each support feature F_s^l to align it with the target (blue operations in Fig. 2(b)). Then, we model channel-aware relations of all spatially aligned features via a temporal feature transformation module (orange in Fig. 2(b)). For each level, the temporal transformed features of the classification branch and the regression branch predict offsets for static scores and static proposals, respectively. The ultimate temporal bounding box are computed in non-linear transformations between static proposals and proposal offsets (red dashed line in Fig. 2(b)), while the ultimate classification scores are obtained by multiplying the static scores and score offsets. Finally, the predictions from all levels are combined using non-maximum suppression just like FCOS.

Proposal-Guided Spatial Feature Transformation. Ideally, the feature for a large proposal should encode the content over a large region, while those for small proposals should have smaller scopes accordingly [23]. Following this intuition, we spatially transform F_t^l based on proposals $\{p_t\}$ to make the feature sensitive to the object. In practice, the range of each p_t is image-level. In order to generate feature-level offset fields required for each spatial location deformation, we first calculate normalized proposals $p_t^* = (l_t^*, t_t^*, r_t^*, b_t^*)$ with

$$l_t^* = \frac{-l_t}{s_l}, t_t^* = \frac{-t_t}{s_l}, r_t^* = \frac{r_t}{s_l}, b_t^* = \frac{b_t}{s_l}, \tag{2}$$

where s_l is the FPN stride until the l-level layer. We devise a 1×1 convolutional layer $\mathcal{N}_o$ on $\{p_t^*\}$ to generate the proposal-guided offset fields and a 3×3 deformable convolutional layer $\mathcal{N}_{dcn}$ to implement spatial feature transformation, as follows:

$$F_t^{ll} = \mathcal{N}_{dcn}(F_t^l, \mathcal{N}_o(\{p_t^*\})), \tag{3}$$

where F_t^{ll} is spatial transformed features. We also perform this spatial transformation scheme on support features $\{F_s^l\}$. Specially, we propagate predicted proposals from I_t to each I_s and leverage the difference between them to generate offset fields for deformation of each F_s^l. In other words, we make each spatial transformed feature $F_s^{l\prime}$ sensitive to both the object and the difference. This step plays a key role in improving the recall rate (see Table 2).

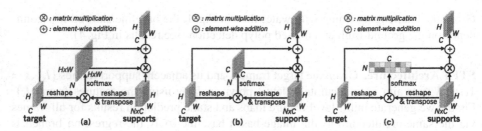

Fig. 3. Comparisons of different attention mechanisms. (a) Point-wise. (b) Channel-wise. (c) proposed Channel-aware for Temporal Feature Transformation.

Channel-Aware Temporal Feature Transformation. Because I_t and $\{I_s\}$ share weights of all layers in our STFT, it can be considered that features of all frames on each channel have been spatially aligned after spatial feature transformation. On the other hand, based on the principle of deep convolutional network learning, the features of certain channels are bound to be sparse, and their activations are close to zero. Naturally, we propose channel-aware temporal feature transformation, aiming to mine the most representative channel features in the neighborhood for feature aggregation. It is implemented by the attention mechanism. We calculate the channel-aware attention map A_{ts}^l from spatial aligned features $F_t^{l\prime}$ and $\{F_s^{l\prime}\}$ by

$$A_{ts}^l = softmax(\frac{\mathscr{R}(F_t^{l\prime})\mathscr{R}(\{F_s^{l\prime}\})^T}{\sqrt{d_f}})\mathscr{R}(\{F_s^{l\prime}\}), \tag{4}$$

where $\mathscr{R}$ is the reshape and T is the transpose for matrix multiplication, more details shown in Fig. 3(c). d_f is a scaling factor [22]. In our algorithm, d_f is equal to $H \times W$ that represents the dimension of each channel feature.

Target Assignment. According to Eq. 1, for each location (x,y), we can obtain the predicted static bounding box $y_t = (x - l_t, y - t_t, x + r_t, y + b_t)$ from static proposals p_t. In our STFT, if the intersection-over-union between y_t and G_t is larger than a threshold (0.3 by default), the temporal classification label of (x,y) is assigned to positive and p_t is considered a significative proposal guide. Then, the temporal regression target δ_t for (x,y) is offsets between y_t and G_t. $\delta_t = (\delta_{x_0}, \delta_{y_0}, \delta_{x_1}, \delta_{y_1})$ are computed by

$$\delta_{x_0} = \frac{x_0 - (x - l_t)}{w * \sigma}, \delta_{y_0} = \frac{y_0 - (y - t_t)}{h * \sigma}, \delta_{x_1} = \frac{x_1 - (x + r_t)}{w * \sigma}, \delta_{y_1} = \frac{y_1 - (y + b_t)}{h * \sigma}, \tag{5}$$

where w, h are the width and height of y_t, and $\sigma = 0.5$ is the variance to improve the effectiveness of offsets learning.

Table 1. Quantitative comparison with SOTA Image-based and Video-based methods on CVC-Clinic and ASUMayo video datasets. The subscript list the relative gains compared to the corresponding Image-based baseline. 'N/A' denotes not available.

		#	Methods	Polyp detection			Polyp localization		
				Precision	Recall	F1-score	Precision	Recall	F1-score
CVC-Clinic	Image-based	1	UNet [18] (*MICCAI'*15)	89.7	75.9	82.2	81.7	72.0	76.5
		2	Faster R-CNN [12] (*ICCV'*15)	84.6	98.2	90.9	78.5	87.9	82.9
		3	R-FCN [6] (*NIPS'*16)	91.7	87.1	89.3	81.4	83.2	82.3
		4	RetinaNet [19] (*CVPR'*17)	93.7	86.2	89.8	87.8	83.1	85.4
		5	Yolov3 [17] (*arXiv'*18)	N/A	N/A	N/A	98.3	70.5	82.1
		6	FCOS [21] (*ICCV'*19)	92.1	74.1	82.1	94.7	70.4	80.8
		7	PraNet [10] (*MICCAI'*20)	94.8	82.2	88.1	96.7	82.1	88.8
	Video-based	3*	FGFA [26] (*ICCV'*17)	**94.5**$_{\uparrow2.8}$	89.2$_{\uparrow2.1}$	91.7$_{\uparrow2.4}$	88.7$_{\uparrow7.3}$	86.4$_{\downarrow3.2}$	87.6$_{\uparrow5.3}$
		2*	RDN [8] (*ICCV'*19)	91.2$_{\uparrow6.6}$	91.3$_{\downarrow6.9}$	91.2$_{\uparrow0.3}$	88.7$_{\uparrow10.2}$	85.9$_{\downarrow2.0}$	87.3$_{\uparrow4.4}$
		1*	OptCNN [25] (*ISBI'*19)	84.6$_{\downarrow5.1}$	**97.3**$_{\uparrow21.4}$	90.5$_{\uparrow8.3}$	74.3$_{\uparrow7.4}$	**96.4**$_{\uparrow24.4}$	83.9$_{\uparrow7.4}$
		5*	AIPDT [24] (*MICCAI'*20)	N/A	N/A	N/A	90.6$_{\uparrow7.7}$	84.5$_{\uparrow14.0}$	87.5$_{\uparrow5.4}$
		2**	MEGA [5] (*CVPR'*20)	91.6$_{\uparrow7.0}$	87.7$_{\downarrow10.5}$	89.6$_{\downarrow1.3}$	91.8$_{\uparrow13.3}$	84.2$_{\downarrow3.7}$	87.8$_{\uparrow4.9}$
		6*	**STFT** (*Ours*)	91.9$_{\downarrow0.2}$	92.0$_{\uparrow17.9}$	**92.0**$_{\uparrow9.9}$	**95.0**$_{\downarrow0.3}$	88.0$_{\uparrow17.6}$	**91.4**$_{\uparrow10.6}$
ASUMayo	Image-based	2	Faster R-CNN [12] (*ICCV'*15)	95.8	98.8	97.2	78.4	98.4	87.3
		3	R-FCN [6] (*NIPS'*16)	96.1	96.4	96.3	80.1	96.2	87.4
		4	RetinaNet [19] (*CVPR'*17)	98.8	84.0	90.8	91.8	83.8	87.6
		6	FCOS [21] (*ICCV'*19)	99.5	68.0	80.8	95.7	65.4	77.7
		7	PraNet [10] (*MICCAI'*20)	98.7	82.3	89.8	94.8	82.1	87.9
	Video-based	3*	FGFA [26] (*ICCV'*17)	98.3$_{\uparrow2.2}$	91.4$_{\downarrow5.0}$	94.8$_{\downarrow1.5}$	88.2$_{\uparrow8.1}$	91.1$_{\downarrow5.1}$	89.6$_{\uparrow2.2}$
		2*	RDN [8] (*ICCV'*19)	97.9$_{\uparrow2.1}$	94.1$_{\downarrow4.7}$	95.9$_{\downarrow1.3}$	87.1$_{\uparrow8.7}$	93.7$_{\downarrow4.7}$	90.3$_{\uparrow3.0}$
		2**	MEGA [5] (*CVPR'*20)	96.8$_{\uparrow1.0}$	95.9$_{\downarrow2.9}$	96.3$_{\downarrow0.9}$	82.6$_{\uparrow4.2}$	94.3$_{\downarrow4.1}$	88.1$_{\downarrow0.8}$
		6*	**STFT** (*Ours*)	**98.9**$_{\downarrow0.6}$	**97.8**$_{\uparrow29.8}$	**98.3**$_{\uparrow17.5}$	**98.3**$_{\downarrow3.5}$	**99.2**$_{\uparrow32.0}$	**98.3**$_{\uparrow20.6}$

Loss Function. Objects with different sizes are assigned to different feature levels. Combining outputs from each level, our STFT is easy to optimize in an end-to-end way using a multi-task loss function as follows:

$$\mathscr{L} = \mathscr{L}_{cls} + \mathscr{L}_{reg} + \frac{1}{N_{pos}} \sum_{x,y} \mathscr{L}_{cls}^{st}(\mathscr{C}^{st}, \mathscr{C}^*) + \mathscr{L}_{reg}^{st} \mathbb{1}_{\{\mathscr{C}^*>0\}}(\Delta^{st}, \Delta^*), \qquad (6)$$

where $\mathscr{L}_{cls}$ and $\mathscr{L}_{reg}$ are the static classification and regression loss respectively [21]. $\mathscr{L}_{cls}^{st}$ is the temporal classification loss implemented by focal loss and $\mathscr{L}_{reg}^{st}$ is the temporal regression loss implemented by $\mathscr{L}_1$ loss. $\mathscr{C}^{st}$ and Δ^{st} are predicted offsets for scores and proposals by STFT. $\mathscr{C}^*$ and Δ^* are assigned classification label and regression target. $\mathbb{1}_{\{\mathscr{C}^*>0\}}$ is the indicator function, being 1 if $\mathscr{C}^* > 0$ and 0 otherwise.

3 Experiments

3.1 Datasets and Settings

We evaluate the proposed STFT on two public video format polyp detection benchmarks. (1) CVC-VideoClinicDB [3]: 18 video sequences were split into test sets (4 videos, number of #2, 5, 10, 18; 2484 images) and training sets (the rest 14 videos;

9470 images) following [25]; (2) ASU-Mayo Clinic Colonoscopy Video [20]: 10 annotated videos containing polyps were split into test sets (4 videos, number of #4, 24, 68, 70; 2098 images) and training sets (the rest 6 videos; 3304 images). All methods in our experiments follow the same data partitioning strategy.

We use ResNet-50 [13] as our backbone and FCOS [21] as our baseline for all experiments. STFT is trained on 4 Tesla V100 GPUs by synchronized SGD, with one target frame and N support frames holding in each GPU. N is limited by GPU memory. We adopt a temporal dropout [26], that is, randomly discard support frames in the neighborhoods $[-9, 9]$ around the target frame. We set $N = 10$ in inference but 2 in training by default. The model will be deployed on SenseCare [9]. For more training details and external experiments, please refer to https://github.com/lingyunwu14/STFT.

Fig. 4. Qualitative comparison of polyp localization. The green and yellow boxes denote correct and incorrect detections, respectively. (Color figure online)

Table 2. The effects of each module in our STFT design.

Methods	Temporal aggregation?	Channel-aware?	Spatial adaptation?	Proposal-guided?	Precision	Recall	F1
(a)					94.7	70.4	80.8
(b)	✓				$64.4_{\downarrow 30.3}$	$77.9_{\uparrow 7.5}$	$70.5_{\downarrow 10.3}$
(c)	✓	✓			$94.6_{\downarrow 0.1}$	$82.5_{\uparrow 12.1}$	$88.2_{\uparrow 7.4}$
(d)	✓	✓	✓		$94.3_{\downarrow 0.4}$	$83.7_{\uparrow 13.3}$	$88.7_{\uparrow 7.9}$
(e)	✓	✓	✓	✓	$\mathbf{95.0}_{\downarrow 0.3}$	$\mathbf{88.0}_{\uparrow 17.6}$	$\mathbf{91.4}_{\uparrow 10.6}$

3.2   Quantitative and Qualitative Comparison

Table 1 shows performance comparisons between state-of-the-art methods without any post-processing on two polyp datasets. Following [25], precision, recall, and F1-score (the harmonic mean of precision and recall) are evaluated on two different tasks: polyp detection and polyp localization. All compared methods are divided into two groups, image-based and video-based. The number in front of the method represents the correspondence between video-based methods and image-based methods, such as the static baseline of #1* is #1, #2 is the baseline of #2* and #2**, STFT's baseline is #6, etc.

Overall, STFT outperforms all SOTAs across both two datasets and two tasks, in the comprehensive metric F1-score. On the ASUMayo, STFT performs best in all metrics. On the CVC-Clinic, #1* achieves higher recall, but its lower precision means a higher false-positive rate, which is not acceptable in clinical practice. Second, STFT achieves a larger overall performance gain (F1-score: about $\geq 10\%$) relative to its image-level baseline than other video-based methods. Our baseline #6 shows the lowest recall on two tasks; STFT improves it to a comparable level to the SOTAs and outperforms all methods on the ASUMayo. This suggests that STFT has a strong learning ability to effectively detect polyps. Moreover, Fig. 4 provides qualitative comparisons of STFT with the baseline #6 and the flow-based method #3*. STFT can precisely locate polyps in various challenging cases, such as water flow, floating content, and bubbles. ·

3.3 Ablation Study

STFT Module Design. Table 2 compares our STFT (e) and its variants with the image-based baseline (a). Metrics are evaluated on the polyp localization task with the CVC-Clinic dataset. Method (b) is a naive temporal aggregation approach that directly adds adjacent features together. The F1-score decreases to 70.5%. Method (c) adds our channel-aware transformation into (b) for adaptive weighting. It obtains an F1-score of 88.2%, 17.7% higher than that of (b). This indicates that it is critical to consider the quality weight of adjacent features. Method (d) is a degenerated variant of (e). It uses the original deformable convolution [7] to achieve spatial adaptation without our proposal-guided. It has almost no improvement compared to (c). (e) is the proposed STFT, which adds the proposal-guided spatial transformation module to (d). It increases the F1-score by 10.6% to 91.4%. The improvement for the recall is more significant (70.4% to 88.0%). This proves that our proposal-guide transformation plays a key role, and STFT effectively mines useful feature representations in the neighborhood.

Table 3. Performance and complexity comparisons in different weighting manners.

Methods	Params	Complexity	Precision	Recall	F1
STFT-*CosineSimilarity* [4,26]	–	$\mathcal{O}(NHW)$	$94.2_{\downarrow 0.5}$	$79.4_{\uparrow 7.0}$	$86.2_{\uparrow 5.4}$
STFT-*PointWise* [11]	α	$\mathcal{O}(NH^2W^2)$	$93.2_{\downarrow 1.5}$	$81.7_{\uparrow 11.3}$	$87.1_{\uparrow 6.3}$
STFT-*ChannelWise* [11]	β	$\mathcal{O}(NC^2)$	$\mathbf{95.2}_{\downarrow 0.5}$	$80.5_{\uparrow 10.1}$	$87.3_{\uparrow 6.5}$
STFT-*ChannelAware*	–	$\mathcal{O}(NC)$	$95.0_{\uparrow 0.3}$	$\mathbf{88.0}_{\uparrow 17.6}$	$\mathbf{91.4}_{\uparrow 10.6}$

Table 4. Results of using different number of support frames. * indicates default setting.

# Training frames	2*					6				
# Inference frames	2	6	10*	14	18	2	6	10	14	18
CVC-Clinic	91.1	91.3	91.4	**91.5**	91.4	90.3	90.5	90.7	90.6	90.7
ASUMayo	98.2	98.2	**98.3**	98.3	98.3	95.6	95.7	95.8	95.8	95.8

Effectiveness of Channel-Aware. As noted in Table 3, we use various adaptive weighting manners to replace the proposed channel-aware in temporal transformation for comparison. In existing video detection works [4,26], calculating cosine similarity is a common weighting method. Point-wise (in Fig. 3(a)) and Channel-wise (in Fig. 3(b)) are the mainstream attention mechanisms [11]. Compared with them, Channel-aware has the lowest computational complexity without any hyperparameters. In addition, Channel-aware achieves the largest gain over the baseline (a) in Table 2.

Impact of Support Frame Numbers. We investigated the impact of different support frame numbers on STFT in Table 4. Under the localization F1-score metric on two datasets, training with 2 frames achieves better accuracy (6 frames reach the memory cap). For inference, as expected, performance improves slowly as more frames are used and stabilizes. Combining Table 1, STFT always achieves the highest localization F1-score and is insensitive to support frame numbers.

Learning Under Sparse Annotation. It is worth noting that we only use the ground-truth of target frames to optimize all losses of STFT. Considering that clinical annotation is very expensive, target frames in training set are uniformly sampled to verify the learning capacity of STFT in the case of sparse labeling. Combining Table 5 and 1, STFT shows stable comprehensive performance in both detection and localization tasks.

Table 5. Results of using different ratios of annotation frames in training.

# Ratios	1	1/2	1/4	1/6	1/8	1/10	1/12	1/14	1/16	1/18	1/20
Detection F1-score	92.0	92.1	92.6	91.9	91.4	**92.9**	91.2	91.4	90.2	91.2	90.9
Localization F1-score	**91.4**	91.3	91.4	91.1	90.8	91.0	90.9	90.5	90.3	90.4	90.3

4   Conclusion

We propose Spatial-Temporal Feature Transformation (STFT), an end-to-end multi-frame collaborative framework for automatically detect and localize polyp in endoscopy video. Our method enhances adaptive spatial alignment and effective temporal aggregation of adjacent features via proposal-guided deformation and channel-aware attention. Extensive experiments demonstrate the strong learning capacity and stability of STFT. Without any post-processing, it outperforms all state-of-the-art methods by a large margin across both two datasets and two tasks, in the comprehensive metric F1-score.

Acknowledgments. This work is partially supported by the funding of Science and Technology Commission Shanghai Municipality No. 19511121400, the General Research Fund of Hong Kong No. 27208720, and the Research Donation from SenseTime Group Limited.

References

1. Ahn, S.B., Han, D.S., Bae, J.H., Byun, T.J., et al.: The miss rate for colorectal adenoma determined by quality-adjusted, back-to-back colonoscopies. Gut Liver **6**(1), 64 (2012)
2. Ali, S., et al.: Deep learning for detection and segmentation of artefact and disease instances in gastrointestinal endoscopy. Med. Image Anal. **70**, 102002 (2021)
3. Bernal, J., et al.: Polyp detection benchmark in colonoscopy videos using GTcreator: a novel fully configurable tool for easy and fast annotation of image databases. In: Proceedings of 32nd CARS conference (2018)
4. Bertasius, G., Torresani, L., Shi, J.: Object detection in video with spatiotemporal sampling networks. In: Ferrari, V., Hebert, M., Sminchisescu, C., Weiss, Y. (eds.) ECCV 2018. LNCS, vol. 11216, pp. 342–357. Springer, Cham (2018). https://doi.org/10.1007/978-3-030-01258-8_21
5. Chen, Y., Cao, Y., Hu, H., Wang, L.: Memory enhanced global-local aggregation for video object detection. In: Proceedings of the IEEE/CVF Conference on Computer Vision and Pattern Recognition, pp. 10337–10346 (2020)
6. Dai, J., Li, Y., He, K., Sun, J.: R-FCN: object detection via region-based fully convolutional networks. In: Advances in Neural Information Processing Systems, pp. 379–387 (2016)
7. Dai, J., Qi, H., Xiong, Y., Li, Y., Zhang, G., et al.: Deformable convolutional networks. In: Proceedings of the IEEE International Conference on Computer Vision, pp. 764–773 (2017)
8. Deng, J., Pan, Y., Yao, T., Zhou, W., Li, H., Mei, T.: Relation distillation networks for video object detection. In: Proceedings of the IEEE/CVF International Conference on Computer Vision, pp. 7023–7032 (2019)
9. Duan, Q., et al.: SenseCare: a research platform for medical image informatics and interactive 3D visualization. arXiv preprint arXiv:2004.07031 (2020)
10. Fan, D.P., et al.: PraNet: parallel reverse attention network for polyp segmentation. In: Martel, A.L., et al. (eds.) MICCAI 2020. LNCS, vol. 12266, pp. 263–273. Springer, Cham (2020). https://doi.org/10.1007/978-3-030-59725-2_26
11. Fu, J., et al.: Dual attention network for scene segmentation. In: Proceedings of the IEEE/CVF Conference on Computer Vision and Pattern Recognition, pp. 3146–3154 (2019)
12. Girshick, R.: Fast R-CNN. In: Proceedings of the IEEE International Conference on Computer Vision, pp. 1440–1448 (2015)
13. He, K., Zhang, X., Ren, S., Sun, J.: Deep residual learning for image recognition. In: Proceedings of the IEEE Conference on Computer Vision and Pattern Recognition, pp. 770–778 (2016)
14. Jemal, A., et al.: Cancer statistics, 2008. CA Cancer J. Clin. **58**(2), 71–96 (2008)
15. Lin, T.Y., Dollár, P., Girshick, R., He, K., Hariharan, B., Belongie, S.: Feature pyramid networks for object detection. In: Proceedings of the IEEE Conference on Computer Vision and Pattern Recognition, pp. 2117–2125 (2017)
16. Qadir, H.A., Balasingham, I., Solhusvik, J., Bergsland, J., Aabakken, L., Shin, Y.: Improving automatic polyp detection using CNN by exploiting temporal dependency in colonoscopy video. IEEE J. Biomed. Health Inform. **24**(1), 180–193 (2019)
17. Redmon, J., Farhadi, A.: YOLOv3: an incremental improvement. arXiv preprint arXiv:1804.02767 (2018)
18. Ronneberger, O., Fischer, P., Brox, T.: U-Net: convolutional networks for biomedical image segmentation. In: Navab, N., Hornegger, J., Wells, W.M., Frangi, A.F. (eds.) MICCAI 2015. LNCS, vol. 9351, pp. 234–241. Springer, Cham (2015). https://doi.org/10.1007/978-3-319-24574-4_28
19. Ross, T.Y., Dollár, G.: Focal loss for dense object detection. In: Proceedings of the IEEE Conference on Computer Vision and Pattern Recognition, pp. 2980–2988 (2017)

20. Tajbakhsh, N., Gurudu, S.R., Liang, J.: Automated polyp detection in colonoscopy videos using shape and context information. IEEE Trans. Med. Imaging 35(2), 630–644 (2015)
21. Tian, Z., Shen, C., Chen, H., He, T.: FCOS: fully convolutional one-stage object detection. In: Proceedings of the IEEE/CVF International Conference on Computer Vision, pp. 9627–9636 (2019)
22. Vaswani, A., et al.: Attention is all you need. arXiv preprint arXiv:1706.03762 (2017)
23. Wang, J., Chen, K., Yang, S., Loy, C.C., Lin, D.: Region proposal by guided anchoring. In: Proceedings of the IEEE/CVF Conference on Computer Vision and Pattern Recognition, pp. 2965–2974 (2019)
24. Zhang, Z., et al.: Asynchronous in parallel detection and tracking (AIPDT): real-time robust polyp detection. In: Martel, A.L., et al. (eds.) MICCAI 2020. LNCS, vol. 12263, pp. 722–731. Springer, Cham (2020). https://doi.org/10.1007/978-3-030-59716-0_69
25. Zheng, H., Chen, H., Huang, J., Li, X., Han, X., Yao, J.: Polyp tracking in video colonoscopy using optical flow with an on-the-fly trained CNN. In: 2019 IEEE 16th International Symposium on Biomedical Imaging, ISBI 2019, pp. 79–82. IEEE (2019)
26. Zhu, X., Wang, Y., Dai, J., Yuan, L., Wei, Y.: Flow-guided feature aggregation for video object detection. In: Proceedings of the IEEE International Conference on Computer Vision, pp. 408–417 (2017)
27. Zhu, X., Xiong, Y., Dai, J., Yuan, L., Wei, Y.: Deep feature flow for video recognition. In: Proceedings of the IEEE Conference on Computer Vision and Pattern Recognition, pp. 2349–2358 (2017)

MBFF-Net: Multi-Branch Feature Fusion Network for Carotid Plaque Segmentation in Ultrasound

Shiyu Mi[1], Qiqi Bao[1], Zhanghong Wei[2], Fan Xu[3], and Wenming Yang[1,3](✉)

[1] Department of Electronic Engineering, Shenzhen International Graduate School, Tsinghua University, Shenzhen, Guangdong, China
yang.wenming@sz.tsinghua.edu.cn
[2] Department of ultrasound, Shenzhen People's Hospital, Shenzhen, Guangdong, China
[3] Peng Cheng Laboratory, Shenzhen, Guangdong, China

Abstract. Stroke is one of the leading causes of death around the world. Segmenting atherosclerotic plaques in carotid arteries from ultrasound images is of great value for preventing and treating ischemic stroke, yet still challenging due to the ambiguous boundary of plaque and intense noise in ultrasound. In this paper, we introduce a new approach for carotid plaque segmentation, namely Multi-Branch Feature Fusion Network (MBFF-Net). Inspired by the prior knowledge that carotid plaques generally grow in carotid artery walls (CAWs), we design a Multi-Branch Feature Fusion (MBFF) module with three branches. Specifically, the first two branches are well-designed to extract plaque features of multiple scales and different contexts, and the other branch is to exploit the prior information of CAWs. In addition, a boundary preserving structure is applied to alleviate the ambiguity of plaque boundary. With the proposed MBFF and the novel structure, our model is capable of extracting discriminative features of plaques and integrating the location information of CAWs for better segmentation. Experiments on the clinical dataset demonstrate that our model outperforms state-of-the-art methods. Code is available at https://github.com/mishiyu/MBFF.

Keywords: Carotid plaque · Segmentation · Ultrasound image · Deep learning · Stroke

1 Introduction

Stroke is the second leading cause of death in adults worldwide [1,2]. And atherosclerotic plaque in the carotid artery is an important cause of ischemic stroke. Currently, B-mode ultrasound is the routine imaging modality to examine carotid atherosclerotic plaques. Evaluating the condition of plaques requires

---

S. Mi and Q. Bao—These authors contributed equally to this work.

© Springer Nature Switzerland AG 2021
M. de Bruijne et al. (Eds.): MICCAI 2021, LNCS 12905, pp. 313–322, 2021.
https://doi.org/10.1007/978-3-030-87240-3_30

manually delineating lesion regions by radiologists. However, such subjective examination is prone to a high diagnostic error rate. Thus, an automatic and reliable method needs to be designed to segment plaques, improving clinical diagnosis efficiency.

The problem of carotid plaque segmentation in carotid longitudinal section (CLS) ultrasound images and transverse section ultrasound images has been extensively explored in the literature. Since carotid plaques generally grow in CAWs, there are many studies on the detection of CAW. Methods proposed in [6–8] were dedicated to solving the problem of arterial wall segmentation in images of transverse sections. For CLS images, Sifakis et al. [4] exploited basic statistics along with anatomical knowledge to recognize the carotid artery. The predicted result can further facilitate arterial segmentation in CLS ultrasound images. Golemati et al. [5] investigated the possibility of applying the Hough transform to segment the CAW from ultrasound image sequences. Azzopardi et al. [20] explored the boundary segmentation of the carotid artery in ultrasound images with convolutional neural networks. However, these methods only detect CAW and do not further segment plaques.

Many methods explicitly detect plaques in ultrasound images. Some of them [11–13] tried to segment plaque in images of transverse sections. For CLS images, Destrempes et al. [9] utilized motion estimation and Bayesian model to segment the plaque in CLS images. However, since it needs manual annotation of the first frame, the operator's subjectivity may affect the segmentation results. Yoneyama et al. [10] compared a semi-automatic method called morphology-enhanced probabilistic with the human reading method in plaque segmentation, showing a good correlation between these two methods. Nevertheless, these traditional methods need to extract features manually, which is not flexible and universal. With the rapid development of deep learning, many methods [14–19] have achieved great success in the image segmentation task. Recently, deep learning methods have also been applied to the segmentation of carotid plaques. Meshram et al. [21] presented a dilated U-Net architecture to segment plaques, finding the performance of the semi-automatic method with bounding box are significantly better than those of the fully automatic method. In general, the current plaque segmentation methods don't make good use of the prior information of CAW. Therefore, how to effectively utilize attributes of CLS ultrasound images for plaque segmentation remains a challenge.

In this paper, we propose the MBFF-Net for automatic carotid plaques segmentation in CLS ultrasound images. It is known that CAW is divided into intima, media, and adventitia. Since plaques usually appear between the intima and the media, detecting the carotid intima-media (IM) region can be of vital importance for detecting plaques [3]. Our model is well-designed based on the above-mentioned pathological knowledge of plaques and achieves discriminative feature extraction and segmentation performance. The main contributions of our work are outlined as follows.

Firstly, as plaques usually exist between the IM regions of CAWs, we refer our previous work of IM detection and use it as prior information to assist plaque segmentation.

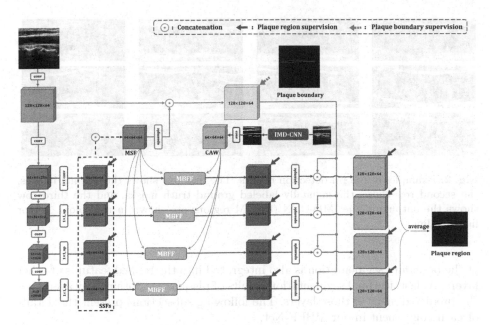

Fig. 1. The overview framework of the proposed MBFF-Net, in which the orange lines show the boundary preserving structure, the red lines show the extraction process of CAWs, and the blue lines show the extraction process of plaques. (Color figure online)

Secondly, we design a new MBFF module to use the prior information, fusing the features of multi-scales and multi-contexts.

Thirdly, we present a boundary preserving structure to enhance the plaque boundary, which diminishes the influence of blurred boundaries of plaque in ultrasound images.

Finally, through comparative experiments, we show that MBFF-Net is superior to existing general segmentation networks in plaque segmentation of CLS ultrasound images and achieves state-of-the-art.

2   Method

The overview of the proposed MBFF-Net is illustrated in Fig. 1. It takes carotid ultrasound images as inputs and outputs the segmentation results. Specifically, the original image is passed through the IMD-CNN and a CNN (ResNeXt101 [22] is used here) to detect the CAW and generate a set of feature maps with different resolutions, respectively. Then, the set of feature maps (except the feature maps at the shallowest layer) is used to generate Multi-Scale Feature (MSF). Afterwards, the MBFF module integrates MSF, CAW, and Single-Scale Feature (SSF) of each layer to produce the refined feature maps.

Furthermore, the feature maps at the shallowest layer are used for segmenting the plaque boundary after combining with the MSF. Subsequently, the feature

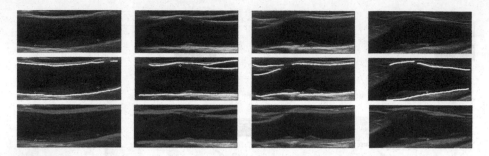

Fig. 2. Examples of the IM detection results. The first row shows the original images, the second row shows the manually labeled ground truth masks, and the third row shows the outputs of the IMD-CNN (the IM regions are shown in red color). (Color figure online)

of the boundary segmentation is also integrated into the refined features of other layers. At last, the final segmentation results of plaques are obtained by averaging the predicted maps of those layers. The following subsections present the details of each component in our MBFF-Net.

2.1   CAW Detecting Module

To embed the prior information of CAWs in the plaque segmentation model, a CAW detecting module is included in the MBFF-Net. This module is from our previous published work denoted as IMD-CNN [23], which uses the patch-wise training method to segment the IM regions in the CAW. Since collecting training data of medical images is a time-consuming, cumbersome and costly task, IMD-CNN is trained in a patch-wise manner instead of an image-wise manner. More concretely, each pixel is taken as the center to crop the patch. Then, IMD-CNN takes these patches as inputs and outputs pixel-wise labels indicating whether a pixel is in the IM regions. The IM detection results are shown in Fig. 2.

In this study, we collect a new CLS ultrasound images dataset different from the previous dataset. The IMD-CNN trained in a patch-wise manner is not affected by the change of image size. Therefore, in this new dataset, CAWs are obtained by adopting the original network structure and parameters directly.

2.2   MBFF Module

The structure of the MBFF module is shown in Fig. 3. Given the SSF of the m-th layer, denoted as Φ_m, it is sent into three branches, i.e., the multi-scale branch, the multi-context branch and the CAW-guided branch, boosting the performance of plaque segmentation. Each branch is described in detail below.

The first branch aims to fuse multi-scale features, integrating feature maps at shallow and deep layers to contain detailed information as well as capturing highly semantic information. In this branch, the Φ_m and MSF are concatenated

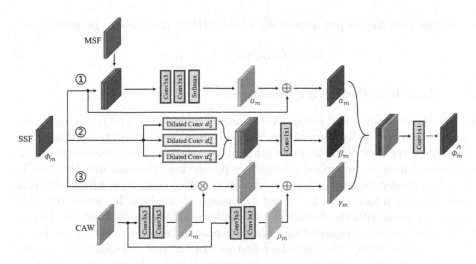

Fig. 3. The details of the MBFF Module at the m-th layer.

and sent into two convolutional layers followed by a Softmax operation to generate the intermediate feature a_m, and then the new feature α_m can be generated as

$$\alpha_m = \Phi_m + a_m \tag{1}$$

The second branch is designed to extract features of plaques considering multi-scale contextual information. By employing the dilated convolution, we extract features with a numerous range of receptive fields in this branch. Afterwards, these features with different receptive fields are concatenated across channel-wise. The generated feature β_m can be defined as

$$\beta_m = f_{conv}(d_2^3(\Phi_m) \odot d_4^3(\Phi_m) \odot d_6^3(\Phi_m)) \tag{2}$$

where $f_{conv}(\cdot)$ denotes the 1×1 convolution function, $d_q^p(\cdot)$ denotes dilated convolution function with a dilation rate q and filter size of $p \times p$, and $\odot$ denotes concatenation.

The third branch combines the location priors of CAWs, which are acquired from the feature around plaques. In this branch, spatial feature modulation is constructed, where affine transformation parameters are formulated with spatial feature transformation [24]. With the priors of the CAW, this branch learns a mapping function T to produce a modulation parameter pair (λ_m, ρ_m), which is written as:

$$(\lambda_m, \rho_m) = T(\Phi_m) \tag{3}$$

Afterwards, intermediate feature maps are adjusted by scaling and shifting according to the modulation parameters. The modified feature γ_m is defined as:

$$\gamma_m = \Phi_m \otimes \lambda_m + \rho_m \tag{4}$$

where $\otimes$ indicates element-wise multiplication.

In the end, the output feature $\hat{\Phi}_m$ of the MBFF module can be written as:

$$\hat{\Phi}_m = f_{conv}(\alpha_m \odot \beta_m \odot \gamma_m) \tag{5}$$

2.3 Boundary Preserving Structure

Due to the high noise and low quality of ultrasound images, the boundary region of the lesion is usually blurred, leading to difficulty in accurate segmentation. Therefore, to preserve the accurate information of the plaques' boundary in CLS ultrasound images, we embed a boundary preserving structure into the MBFF-Net. Specifically, we use the feature maps at the shallowest layer which retain the most detailed information to extract the plaque boundary. In detail, the MSF is concatenated with the feature of the shallowest layer to diminish the impact of missing semantic information. Then, a plaque boundary loss is used as a supervision to extract the boundary feature. In the end, the feature extracted from the boundary segmentation is integrated into the feature of each refined SSF to enhance the structure boundary of plaques.

3 Experiments

Dataset. Our experiments are conducted on the CLS ultrasound images of the carotid artery collected by Philips IU22 with the L9-3 probe and GE logiq E9 with the 9L probe. The two devices are of the same center frequency with 9MHz. A total of 430 images of different patients are included in this dataset with the size of 768×576. Since the ultrasound acquisition device generates ultrasound images surrounded by irrelevant parameter information, we crop the ultrasound images from the original images and then rescale them into the size of 256×256 to adjust the network's input size. In addition, randomly selected 330 images are used as the training dataset, and the remaining 100 images are used as the testing dataset.

Loss Function. Binary cross-entropy with logits loss is used as the loss function to optimize the whole network. The total loss (denoted as L_t) is defined as:

$$L_t = \sum_{i=1}^{n} \omega_i L_{1i} + \sum_{j=1}^{n} \omega_j L_{2j} + \sum_{k=1}^{n} \omega_k L_{3k} + \omega_b L_b \tag{6}$$

where L_{1i} denotes the loss of the feature before the MBFF module at the i-th layer, L_{2j} denotes the loss of the refined feature extracted by the MBFF module at the j-th layer, L_{3k} denotes the loss of the refined feature integrated with the boundary feature at the k-th layer, and L_b denotes the loss on plaque boundary. These supervised features are specifically identified by arrows in Fig. 1. Meanwhile, n denotes the number of layers, and ω denotes weight. All weights here are set as 1 empirically.

Evaluation Metrics. Widely-used evaluation indicators are employed to measure the performance of the proposed model, including Dice score, Intersection over Union (IoU), Precision, and Recall.

Implementation. Our model is implemented by PyTorch on the NVIDIA 2080Ti GPU. The SGD optimizer with the learning rate being 0.005 and the momentum being 0.9 is adopted to train the MBFF-Net. The maximum number of epochs is set to 100. Each epoch takes about forty seconds.

Ablation Experiment. To evaluate the efficiency of proposed components, we conduct ablation experiments and show the quantitative results in Table 1. It can be clearly observed that each component has a certain positive effect. We remove the MBFF module and the boundary preserving structure from MBFF-Net as the baseline. Results of the baseline are represented in the first row. Benefiting from the dilated convolution branch (denoted as DC), the second row indicates that the model is able to utilize contextual information for better segmentation. In the third row, it shows the results with the CAW-guided branch (denoted as CAW) added to the baseline. And the improved visualization results illustrate the effectiveness of the integrated prior knowledge. In the fourth row, the multi-scale feature fusion branch (denoted as MSF) is evaluated. Since the model is equipped with the ability of multi-scale feature extraction, it can be seen that the segmentation effect is improved. Then we combine these three branches together and obtain a further improvement, demonstrating the power of the proposed MBFF module. Finally, the boundary preserving structure (denoted as BPS) is included, and our proposed MBFF-Net achieves the best performance.

Table 1. Ablation results of different components.

DC	CAW	MSF	BPS	Dice	IoU	Precision	Recall
				0.666	0.577	0.900	0.627
✓				0.683	0.597	0.852	0.685
	✓			0.688	0.611	0.840	0.707
		✓		0.710	0.625	0.910	0.671
✓	✓	✓		0.736	0.647	0.896	0.710
✓	✓	✓	✓	0.780	0.702	0.849	0.797

Comparative Experiment. We quantitatively compare the results of our MBFF-Net with some state-of-the-art methods, including PSPNet [14], ResUNet [15], U-Net [16], SegNet [17] and DenseUNet [18]. The results are reported in Table 2. For fair comparison, we retrain these models with common implementations and adjust training parameters to obtain the best segmentation results. It can be observed in Table 2 that our method is superior to other methods in

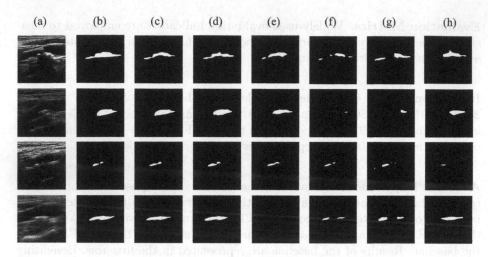

Fig. 4. Examples of plaque segmentation results produced by different methods. (a): CLS ultrasound images; (b): ground truth masks; (c)–(h): corresponding segmentations from our method, DenseUNet [18], SegNet [17], U-Net [16], ResUNet [15] and PSPNet [14], respectively.

almost all indicators. Figure 4 shows some of the segmentation results. Obviously, the plaque regions obtained by our method are more convincing.

Table 2. Metric results of diffrent methods.

Method	Dice	IoU	Precision	Recall
PSPNet[14]	0.585	0.472	0.760	0.586
ResUNet[15]	0.621	0.521	0.755	0.635
U-Net[16]	0.630	0.552	0.817	0.658
SegNet[17]	0.687	0.593	0.856	0.676
DenseUNet[18]	0.778	0.690	**0.872**	0.763
Ours	**0.780**	**0.702**	0.849	**0.797**

4 Conclusion

In this paper, we propose a novel framework for carotid plaque segmentation in CLS ultrasound images. Complementary information is extracted from the features of multiple scales and different contexts. Besides, inspired by the physiological characteristics of plaque, we use the location priors of CAWs to guide plaque segmentation. Furthermore, the boundary preserving structure is designed to handle the fuzzy boundary of lesions and the noise in ultrasound. Experiments on the challenging CLS images demonstrate the superior performance of the MBFF-Net compared to the most advanced methods.

Acknowledgments. This work was supported by the Natural Science Foundation of Guangdong Province (No. 2020A1515010711) and the Special Foundation for the Development of Strategic Emerging Industries of Shenzhen (Nos. JCYJ20170818161845824, JCYJ20200109143010272 and JCYJ20200109143035495).

References

1. Feigin, V.L., Nguyen, G., Cercy, K., et al.: Global, regional, and country-specific lifetime risks of stroke, 1990 and 2016. N. Engl. J. Med. **379**(25), 2429–2437 (2018)
2. Johnson, C.O., Nguyen, M., Roth, G.A.: Global, regional, and national burden of stroke, 1990–2016: a systematic analysis for the Global Burden of Disease Study 2016. Lancet Neurol. **18**(5), 439–458 (2019)
3. Touboul, P.J., Hennerici, M.G., Meairs, S., et al.: Mannheim carotid intima-media thickness and plaque consensus (2004-2006-2011). Cerebrovasc. Dis. **34**(4), 290–296 (2012)
4. Sifakis, E.G., Golemati, S.: Robust carotid artery recognition in longitudinal b-mode ultrasound images. IEEE Trans. Image Process. **23**(9), 3762–3772 (2014)
5. Golemati, S., Stoitsis, J., Sifakis, E.G., et al.: Using the Hough transform to segment ultrasound images of longitudinal and transverse sections of the carotid artery. Ultrasound Med. Biol. **33**(12), 1918–1932 (2008)
6. Zhang, J., Teng, Z., Guan, Q., et al.: Automatic segmentation of MR depicted carotid arterial boundary based on local priors and constrained global optimisation. IET Image Proc. **13**(3), 506–514 (2019)
7. China, D., Nag, M.K., Mandana, K.M., et al.: Automated in vivo delineation of lumen wall using intravascular ultrasound imaging. In: 38th Annual International Conference of the IEEE Engineering in Medicine and Biology Society, pp. 4125–4128 (2016)
8. Samber, D.D., Ramachandran, S., Sahota, A., et al.: Segmentation of carotid arterial walls using neural networks. World J. Radiol. **12**, 1–9 (2020)
9. Destrempes, F., Soulez G., Giroux, M.F., et al.: Segmentation of plaques in sequences of ultrasonic B-mode images of carotid arteries based on motion estimation and Nakagami distributions. In: IEEE International Ultrasonics Symposium, pp. 2480–2483 (2010)
10. Yoneyama, T., et al.: In vivo semi-automatic segmentation of multicontrast cardiovascular magnetic resonance for prospective cohort studies on plaque tissue composition: initial experience. Int. J. Cardiovasc. Imaging **32**(1), 73–81 (2015). https://doi.org/10.1007/s10554-015-0704-0
11. Wei, M., Ran, Z., Yuan, Z., et al.: Plaque recognition of carotid ultrasound images based on deep residual network. In: IEEE 8th Joint International Information Technology and Artificial Intelligence Conference, pp. 931–934 (2019)
12. Van't, K.R., Naggara, O., Marsico, R., et al.: Automated versus manual in vivo segmentation of carotid plaque MRI. Am. J. Neuroradiol. **33**(8), 1621–1627 (2012)
13. Bonanno, L., Sottile, F., Ciurleo, R., et al.: Automatic algorithm for segmentation of atherosclerotic carotid plaque. J. Stroke Cerebrovasc. Dis. **26**(2), 411–416 (2017)
14. Zhao, H., Shi, J., Qi, X., et al.: Pyramid scene parsing network. In: IEEE Conference on Computer Vision and Pattern Recognition, pp. 2881–2890 (2017)
15. Zhang, Z., Liu, Q., Wang, Y.: Road extraction by deep residual U-Net. IEEE Geosci. Remote Sens. Lett. **15**(5), 749–753 (2018)

16. Ronneberger, O., Fischer, P., Brox, T.: U-Net: convolutional networks for biomedical image segmentation. In: International Conference on Medical Image Computing & Computer Assisted Intervention, pp. 234–241 (2015)
17. Badrinarayanan, V., Kendall, A., Cipolla, R.: SegNet: a deep convolutional encoder-decoder architecture for image segmentation. IEEE Trans. Pattern Anal. Mach. Intell. **39**(12), 2481–2495 (2019)
18. Li, X., Chen, H., Qi, X., et al.: H-DenseUNet: hybrid densely connected UNet for liver and tumor segmentation from CT volumes. IEEE Trans. Med. Imaging **37**(12), 2663–2674 (2018)
19. Wang, Y., Deng, Z., Hu, X., et al.: Deep attentional features for prostate segmentation in ultrasound. In: International Conference on Medical Image Computing & Computer Assisted Intervention (2018)
20. Azzopardi, C., Hicks, Y.A., Camilleri, K.P.: Automatic carotid ultrasound segmentation using deep convolutional Neural Networks and phase congruency maps. In: IEEE 14th International Symposium on Biomedical Imaging (2017)
21. Meshram, N.H., Mitchell, C.C., Wilbrand, S., et al.: Deep learning for carotid plaque segmentation using a dilated U-Net architecture. Ultrason. Imaging **42**(4–5), 221–230 (2020)
22. Xie, S., Girshick, R., Dollár, P., et al.: Aggregated residual transformations for deep neural networks. In: IEEE Conference on Computer Vision and Pattern Recognition, pp. 5987–5995 (2017)
23. Mi, S., Wei, Z., Xu, J., et al.: Detecting carotid intima-media from small-sample ultrasound images. In: Annual International Conference of the IEEE Engineering in Medicine & Biology Society, pp. 2129–2132 (2020)
24. Wang, X., Yu, K., Dong, C., et al.: Recovering realistic texture in image super-resolution by deep spatial feature transform. In: IEEE/CVF Conference on Computer Vision and Pattern Recognition, pp. 606–615 (2018)

Balanced-MixUp for Highly Imbalanced Medical Image Classification

Adrian Galdran[1]([✉]), Gustavo Carneiro[2], and Miguel A. González Ballester[3,4]

[1] Bournemouth University, Poole, UK
agaldran@bournemouth.ac.uk
[2] University of Adelaide, Adelaide, Australia
gustavo.carneiro@adelaide.edu
[3] BCN Medtech, Department of Information and Communication Technologies,
Universitat Pompeu Fabra, Barcelona, Spain
ma.gonzalez@upf.edu
[4] Catalan Institution for Research and Advanced Studies (ICREA), Barcelona, Spain

Abstract. Highly imbalanced datasets are ubiquitous in medical image classification problems. In such problems, it is often the case that rare classes associated to less prevalent diseases are severely under-represented in labeled databases, typically resulting in poor performance of machine learning algorithms due to overfitting in the learning process. In this paper, we propose a novel mechanism for sampling training data based on the popular MixUp regularization technique, which we refer to as Balanced-MixUp. In short, Balanced-MixUp simultaneously performs regular (*i.e.*, instance-based) and balanced (*i.e.*, class-based) sampling of the training data. The resulting two sets of samples are then mixed-up to create a more balanced training distribution from which a neural network can effectively learn without incurring in heavily under-fitting the minority classes. We experiment with a highly imbalanced dataset of retinal images (55K samples, 5 classes) and a long-tail dataset of gastro-intestinal video frames (10K images, 23 classes), using two CNNs of varying representation capabilities. Experimental results demonstrate that applying Balanced-MixUp outperforms other conventional sampling schemes and loss functions specifically designed to deal with imbalanced data. Code is released at https://github.com/agaldran/balanced_mixup

Keywords: Imbalanced learning · Long-tail image classification

1 Introduction

Backed by the emergence of increasingly powerful convolutional neural networks, medical image classification has made remarkable advances over the last years, reaching unprecedented levels of accuracy [20]. However, due to the inherent difficulty in collecting labeled examples of rare diseases or other unusual instances in a medical context, these models are often trained and tested on large datasets

© Springer Nature Switzerland AG 2021
M. de Bruijne et al. (Eds.): MICCAI 2021, LNCS 12905, pp. 323–333, 2021.
https://doi.org/10.1007/978-3-030-87240-3_31

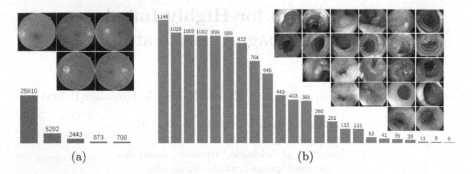

Fig. 1. The two different data imbalance scenarios considered in this paper. Left: Highly imbalanced problem (DR grading from retinal images, $K = 5$ [17]). Right: Long-tailed data (Gastro-intestinal image classification, $K = 23$ [2]).

containing a more balanced distribution with respect to image classes than the one found in real-life clinical scenarios, which typically has a long-tailed distribution. Under such severe data imbalance, over-represented (i.e., majority) classes tend to dominate the training process, resulting in a decrease of performance in under-represented (i.e., minority) classes [23,31]. Therefore, developing training methods that are adapted to strong data imbalance is essential for the advance of medical image classification.

Common solutions to address data imbalance involve data re-sampling to achieve a balanced class distribution [4], curriculum learning [13], adapted loss functions, *e.g.* cost-sensitive classification [9,30], or weighting the contribution of the different samples [7,19]. Another approach uses synthetic manipulation of data and/or labels to drive the learning process towards a more suitable solution, like label smoothing [8] or SMOTE [5]. Our proposed technique is a combination of modified training data sampling strategies with synthetic data manipulation via the well-known MixUp regularization method [28]. Therefore it is deeply connected with MixUp, with the fundamental difference that, while MixUp randomly combines training samples without taking into account their classes, we carefully mix examples from minority categories with examples from the other classes in order to create a more suitable training data distribution.

Some approaches have been recently explored involving MixUp and data imbalance. ReMix [6] mixes up by keeping the minority class label, instead of mixing up the labels. Similarly, MixBoost [14] attempts to combine active learning with MixUp to select which training samples to mix from each category, adding an extra complexity layer to the sampling process. Another popular technique that is related to our approach is SMOTE [5]. However, SMOTE generates convex combinations of input samples only between nearest neighbors of the same class. While this creates extra examples in regions of the space where there is less data, it does not provide the same regularizing benefits as MixUp.

In this paper, we propose Balanced-MixUp, a new imbalanced-robust training method that mixes up imbalanced (instance-based) and balanced (class-based) sampling of the data. Experiments on two different medical image classification tasks with highly imbalanced and long-tailed data (as shown in Fig. 1), using neural networks of different complexities, show that Balanced-MixUp is effective in creating a more evenly distributed training data and also regularizing over-parametrized models, resulting in better performance than other conventional and recent approaches to handle data imbalance in all the considered cases.

2 Methodology

In this paper we intend to combine the MixUp regularization technique with a modified sampling strategy for imbalanced datasets. This section introduces each of these concepts and then our proposed combined technique.

2.1 MixUp Regularization

The MixUp technique was initially introduced in [28] as a simple regularization technique to decrease overfitting in deep neural networks. If we denote (x_i, y_i) a training example composed of an image x_i and its associated one-hot encoded label y_i, with f_θ being a neural network to be trained to approximate the mapping $f(x_i) = y_i \ \forall i$, then MixUp creates synthetic examples and labels as follows:

$$\hat{x} = \lambda x_i + (1 - \lambda)x_j, \quad \hat{y} = \lambda y_i + (1 - \lambda)y_j \tag{1}$$

where $\lambda \sim \text{Beta}(\alpha, \alpha)$, with $\alpha > 0$. Since $\lambda \in [0, 1]$, $\hat{x}$ and $\hat{y}$ are random convex combinations of data and label inputs. Typical values of α vary in $[0.1, 0.4]$, which means that in practice $\hat{x}$ will likely be relatively close to either x_i or x_j. Despite its simplicity, it has been shown that optimizing f_θ on mixed-up data leads to better generalization and improves model calibration [26].

2.2 Training Data Sampling Strategies

When dealing with extremely imbalanced data, the training process is typically impacted by different per-class learning dynamics that result in underfitting of minority classes, which are rarely presented to the model and may end up being entirely ignored [15]. Modified sampling strategies can be utilized to mitigate this effect, like oversampling under-represented categories, although merely doing so often leads to counter-productive outcomes, e.g. repeatedly showing to the model the same training examples may lead to the overfitting of minority classes [5].

Before detailing how to combine oversampling of minority classes with MixUp regularization, we introduce some further notation to describe sampling strategies. Given a training set $\mathcal{D} = \{(x_i, y_i), \ i = 1, ..., N\}$ for a multi-class problem with K classes, if each class k contains n_k examples, we have that $\sum_{k=1}^{K} n_k = N$. We can then describe common data sampling strategies as follows:

$$p_j = \frac{n_j^q}{\sum_{k=1}^{K} n_k^q}, \tag{2}$$

being p_j the probability of sampling from class j during training. With this, selecting $q = 1$ amounts to picking examples with a probability equal to the frequency of their class in the training set (*instance-based sampling*), whereas choosing $q = 0$ leads to a uniform probability $p_j = 1/K$ of sampling from each class, this is, *class-based sampling* or oversampling of minority classes. Another popular choice is *square-root sampling*, which stems from selecting $q = 1/2$.

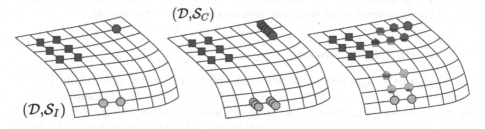

Fig. 2. Schematic illustration of Balanced-MixUp (right) filling underpopulated areas in the data space - linked to infrequent categories - with synthetic examples.

2.3 Balanced-MixUp

We propose to combine the two training enhancements described above into a single technique that is suitable for learning under strong data imbalance. We introduce an adaptation to this scenario of MixUp, referred to as Balanced-MixUp, in which two data points are not randomly sampled without considering their category, but rather sampling one following an instance-based strategy and the other following a class-based sampling.

Let us refer to a training dataset together with a sampling strategy as $(\mathcal{D}, \mathcal{S})$, with $\mathcal{S}_I$ and $\mathcal{S}_C$ referring to instance-based and class-based sampling respectively. With this notation, Balanced-MixUp can be described as follows:

$$\hat{x} = \lambda x_I + (1 - \lambda)x_C, \quad \hat{y} = \lambda y_I + (1 - \lambda)y_C, \tag{3}$$

for $(x_I, y_I) \in (\mathcal{D}, \mathcal{S}_I), (x_C, y_C) \in (\mathcal{D}, \mathcal{S}_C)$. This induces a more balanced distribution of training examples by creating synthetic data points around regions of the space where minority classes provide less data density, as illustrated in Fig. 2. At the same time, adding noise to the labels helps regularizing the learning process.

In our case, departing from the original MixUp formulation, we obtain the mixing coefficient as $\lambda \sim \text{Beta}(\alpha, 1)$. This results in an exponential-like distri bution as shown in Fig. 3, which leads to convex combinations in which examples from $(\mathcal{D}, \mathcal{S}_I)$ receive more weight, preventing overfitting in minority classes. In addition, it allows us to formulate our technique as depending on a single hyperparameter with an intuitive behavior: as α grows,

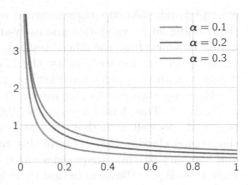

Fig. 3. $\text{Beta}(\alpha, 1)$ distribution for varying α

examples from minority classes are mixed up with greater weights, increasing class balance at the expense of risking some overfitting.

2.4 Training Details

All models in this paper are CNNs implemented in Pytorch [22] and optimized to reduce the Cross-Entropy loss using Stochastic Gradient Descent with a batch size of 8 and a learning rate of 0.01. During training, standard data augmentation operations are applied, and the learning rate is cyclically annealed towards 0. The metric of interest in each task (see next section) is used to monitor model's performance on the validation set and the best one is kept. After training, we generate predictions by averaging the result of the input image and its horizontally flipped version, which is common practice to regularize predictions [25].

3 Experimental Analysis

In this section we detail our experimental setup: considered datasets with relevant statistics, baseline techniques for comparison, evaluation metrics, and a discussion on the numerical differences in performance between methods.

3.1 Experimental Details

We consider two different image classification tasks: Diabetic Retinopathy (DR) grading from retinal fundus images and gastrointestinal image (GI) classification.

- For DR grading, we use the Eyepacs database[1] the largest publicly available dataset for this task. This database has images labeled as belonging to one out of five possible DR grades, ranging from DR0 (no DR) to DR4 (proliferative DR). Due to the nature of this disease, and the very subtle signs associated to the DR1 class (just a few microaneurysms present), the class distribution is highly imbalanced, as shown in Fig. 1. The database contains ~35,000 training

---

[1] https://www.kaggle.com/c/diabetic-retinopathy-detection.

examples and ~55,000 images assigned to a separate test set. We use 10% of the training set for validation and early-stopping purposes, and derive performance measures from the official test set. As an external database to examine model generalization we also use the 1748 retinal images in the Messidor-2 dataset[2], with the grades provided[3] in [17].

- For GI image classification, we use the Hyper-Kvasir dataset[4], recently made available [2]. This database contains 10,662 images labeled with anatomical landmarks as well as pathological and normal findings, and it represents a challenging classification problem due to the larger amount of classes (23) and the varying class frequencies, with minority classes very rarely represented, see Fig. 1. Please note that there is also an official test set [12], but we found this to contain only 11 different classes. Since our focus is on assessing model performance under a long-tail distribution, we prefer to discard this set, running all our experiments and computing performance for all models in a stratified 5-fold cross-validation manner.

For performance assessment in the DR case, we adopt as the main evaluation a measure based on the quadratic-weighted kappa score (quad-κ), which is the standard performance measure in grading problems, as it considers the correct ordering of predictions. We also report Matthews Correlation Coefficient (MCC), which has been shown to be meaningful under strong class imbalance [3], and the Kendall-τ coefficient. For the GI image classification task, we adopt as the main assessment measure MCC, as it was used for performance ranking in a recent challenge associated to this database [12]. Since we are dealing with a long-tail data distribution, we also report Balanced Accuracy (the mean of the diagonal in a normalized confusion matrix) and the macro-F1 score, which capture well performance in minority classes.

For comparison purposes, in both tasks we contrast Balanced-MixUp with other conventional sampling approaches: 1) Class-based sampling (oversampling of minority classes, as recommended in [4]), 2) Instance-based sampling, and 3) sampling based on the square root of class-frequencies [21]. We also include training with loss functions that are meant to perform better under severe class imbalance, namely 4) the popular Focal Loss [18], which modulates the contribution of each example according to the amount of loss it generates, and 5) a Class-Balanced loss recently introduced that weighs the loss of each class during training based on the *effective* number of samples, see [7].

In all cases, in order to observe the behavior of the different techniques with models of varying sizes, we train two different CNN baseline architectures: a MobileNet V2 [24] and more powerful ResNext50, modified as in [16]. For sensitivity analysis purposes, we measure the impact of varying the only hyperparameter α in Balanced-MixUp, with $\alpha \in \{0.1, 0.2, 0.3\}$ and report the corresponding results.

---

[2] https://www.adcis.net/en/third-party/messidor2/.
[3] https://www.kaggle.com/google-brain/messidor2-dr-grades.
[4] https://endotect.com/.

3.2 Discussion on the Numerical Results on DR Grading

Assessment measures for the task of DR grading from retinal images are displayed in Table 1. The leftmost three columns show performance of a MobileNet V2 CNN in terms of quad-κ, MCC and Kendall-τ. In this case, it can be seen that Balanced-MixUp with $\alpha = 0.1$ outperforms all other approaches in terms of the three considered figures of merit. Note that setting $\alpha = 0.2$ still outperforms all other techniques. It is also interesting to observe the differences when we switch to a larger model, ResNext50. In particular, we can see that: 1) the performance of almost all methods is increased, indicating that ResNext50 is a more suitable choice for this problem; and 2) Balanced-MixUp with $\alpha = 0.2$ is now the dominant approach, reaching the highest scores in all considered measures. This is consistent with the intuition that Balanced-MixUp can regularize the learning process, and that increasing its hyperparameter α leads to a stronger regularization Indeed, for this architecture the second best method is Balanced-MixUp with $\alpha = 0.3$. It is also worth stressing that, again, the three values of α translate in better results than any of the other considered alternatives.

Table 1. Performance comparison with two different CNN architectures for DR grading on the Eyepacs test set. Best results are marked bold.

	MobileNet V2			ResNeXt-50		
	quad-κ	MCC	Kendall-τ	quad-κ	MCC	Kendall-τ
Class-sampling	74.58	49.34	63.30	74.45	51.12	64.14
Instance-sampling	78.75	61.79	73.39	80.17	62.78	74.32
Sqrt-sampling [21]	79.32	59.66	72.22	79.38	58.77	72.22
Focal Loss [18]	78.59	60.74	72.70	79.73	62.66	73.37
CB Loss [7]	77.84	61.08	72.85	79.19	61.83	74.32
Bal-Mxp: $\alpha = 0.1$	**79.61**	**62.41**	**74.05**	80.35	63.31	74.74
Bal-Mxp: $\alpha = 0.2$	79.43	62.22	73.69	**80.78**	**63.78**	**75.05**
Bal-Mxp: $\alpha = 0.3$	78.34	61.65	73.25	80.62	63.51	74.89

It is important to stress that the results shown in Table 1 do not represent the state-of-the-art in this dataset. Recent methods are capable of reaching greater quad-κ values by means of specialized architectures with attention mechanisms [11], or semi-supervised learning with segmentation branches [29]. Our results are however aligned with other recently published techniques, as shown in Table 2, which also contains test results in the Messidor-2 dataset (without any re-training), demonstrating their competitiveness and generalization ability. Moreover we consider Balanced-MixUp to be a supplementary regularizing technique of the learning process from which other methods could also benefit.

3.3   Discussion on the Numerical Results on GI Image Classification

The second task we consider is Gastro-Intestinal image classification. Here the class imbalance is greater than in the DR problem, in addition to the presence of 23 categories, which turn this into a more challenging dataset. Table 3 shows the median performance in a stratified 5-fold cross-validation process, when considering exactly the same techniques and architectures as in the previous section.

Although in this task the performance measures appear to be more mixed than in the DR problem, we can still see how Balanced-MixUp appears to bring improvements over the other competing approaches. Again, the ResNext50 architecture delivers better performance in terms of MCC, although differences are not as high as before, and some methods show a degraded performance when applied with this larger architecture (compared to using MobileNet).

Table 2. Performance comparison in terms of quad-κ for methods trained on the Eyepacs dataset and tested (no retraining) on Eyepacs and Messidor-2.

| | QWKL | DR\|graduate | Cost-sensitive | Iter. Aug. | Bal-Mxp |
	PRL'18 [27]	MedIA'20[1]	MICCAI'20 [9]	TMI'20 [10]	Rx50, $\alpha = 0.2$
Eyepacs	74.00	74.00	78.71	80.00	**80.78**
Messidor-2	–	71.00	79.79	–	**85.14**

Table 3. Stratified 5-Fold cross-validation (median) results for GI image classification on the Hyper-Kvasir dataset. Best results are marked bold.

| | MobileNet V2 | | | ResNeXt-50 | | |
	MCC	B-ACC	Macro-F1	MCC	B-ACC	Macro-F1
Class-sampling	89.74	61.84	61.84	88.97	59.57	58.40
Instance-sampling	90.49	62.13	62.16	90.74	61.41	61.69
Sqrt-sampling [21]	90.11	63.01	62.66	90.29	62.84	62.85
Focal Loss [18]	90.30	62.09	61.78	90.23	62.12	62.36
CB Loss [7]	85.84	54.93	54.87	89.67	**63.84**	63.71
Bal-Mxp: $\alpha = 0.1$	**90.90**	63.49	62.77	91.05	62.55	62.92
Bal-Mxp: $\alpha = 0.2$	90.54	63.44	63.85	**91.15**	62.80	**64.00**
Bal-Mxp: $\alpha = 0.3$	90.39	**64.76**	**64.07**	90.84	62.34	62.35

Focusing first on the MCC score, we can appreciate again the regularizing effect of Balanced-MixUp: for both architectures, applying it leads to improved performance. Moreover, we can again see the same trend as above: the MobileNet model performs better with less regularization ($\alpha = 0.1$), and increasing α slowly deteriorates the MCC. As expected, when using the ResNeXt architecture, a greater $\alpha = 0.2$ delivers better performance. It is worth noting that the MCC reached by the MobileNet model with $\alpha = 0.1$ is greater than in any of the considered methods and architectures, excluding models regularized with Balanced-MixUp and varying α.

Finally, it is relevant to observe that when we consider balanced accuracy and macro-F1, which can capture better performance in minority classes, for almost all methods it appears to be beneficial to use a smaller architecture. This appears to reduce to some extent the minority class overfitting due to oversampling, observed in the low performance of class-based sampling. Interestingly, even with this small architecture, Balanced MixUp increases the performance for all values of α, which further shows that our approach not only helps regularizing large models but also contributes to an improved learning on minority classes.

4 Conclusion

Imbalanced data is present in almost any medical image analysis task. Therefore, designing techniques for learning in such regimes is a challenge of great clinical significance. This paper introduces Balanced-MixUp to deal with heavy imbalanced data distributions by means of the combination of the popular MixUp regularization technique and modified training data sampling strategies. Balanced-MixUp is easy to implement and delivers consistent performance improvements when compared with other popular techniques. Its extension to other tasks beyond image classification represents a promising future research direction.

Acknowledgments. This work was partially supported by a Marie Skłodowska-Curie Global Fellowship (No. 892297) and by Australian Research Council grants (DP180103232 and FT190100525).

References

1. Araujo, T., et al.: DR|GRADUATE: uncertainty-aware deep learning-based diabetic retinopathy grading in eye fundus images. Med. Image Anal. **63**, 101715 (2020)
2. Borgli, H., et al.: HyperKvasir, a comprehensive multi-class image and video dataset for gastrointestinal endoscopy. Sci. Data **7**(1), 283 (2020)
3. Boughorbel, S., Jarray, F., El-Anbari, M.: Optimal classifier for imbalanced data using Matthews Correlation Coefficient metric. PLOS ONE **12**(6), 0177678 (2017)
4. Buda, M., Maki, A., Mazurowski, M.A.: A systematic study of the class imbalance problem in convolutional neural networks. Neural Netw. **106**, 249–259 (2018)
5. Chawla, N.V., Bowyer, K.W., Hall, L.O., Kegelmeyer, W.P.: SMOTE: synthetic minority over-sampling technique. J. Artif. Intell. Res. **16**(1), 321–357 (2002)
6. Chou, H.-P., Chang, S.-C., Pan, J.-Y., Wei, W., Juan, D.-C.: Remix: rebalanced mixup. In: Bartoli, A., Fusiello, A. (eds.) ECCV 2020. LNCS, vol. 12540, pp. 95–110. Springer, Cham (2020). https://doi.org/10.1007/978-3-030-65414-6_9
7. Cui, Y., Jia, M., Lin, T.Y., Song, Y., Belongie, S.: Class-balanced loss based on effective number of samples. In: Proceedings of the IEEE/CVF Conference on Computer Vision and Pattern Recognition (CVPR) (June 2019)

8. Galdran, A., et al.: Non-uniform label smoothing for diabetic retinopathy grading from retinal fundus images with deep neural networks. Trans. Vis. Sci. Technol. 9(2), 34–34 (2020)
9. Galdran, A., Dolz, J., Chakor, H., Lombaert, H., Ben Ayed, I.: Cost-sensitive regularization for diabetic retinopathy grading from eye fundus images. In: Martel, A.L., et al. (eds.) MICCAI 2020. LNCS, vol. 12265, pp. 665–674. Springer, Cham (2020). https://doi.org/10.1007/978-3-030-59722-1_64
10. González-Gonzalo, C., Liefers, B., Ginneken, B., Sánchez, C.I.: Iterative augmentation of visual evidence for weakly-supervised lesion localization in deep interpretability frameworks: application to color fundus images. IEEE Trans. Med. Imaging 39(11), 3499–3511 (2020)
11. He, A., Li, T., Li, N., Wang, K., Fu, H.: CABNet: category attention block for imbalanced diabetic retinopathy grading. IEEE Trans. Med. Imaging 40(1), 143–153 (2021)
12. Hicks, S., Jha, D., Thambawita, V., Halvorsen, P., Hammer, H.L., Riegler, M.: The EndoTect 2020 challenge: evaluation and comparison of classification, segmentation and inference time for endoscopy. In: 25th International Conference on Pattern Recognition (ICPR) (2020)
13. Jiménez-Sánchez, A., et al.: Medical-based deep curriculum learning for improved fracture classification. In: Shen, D., et al. (eds.) MICCAI 2019. LNCS, vol. 11769, pp. 694–702. Springer, Cham (2019). https://doi.org/10.1007/978-3-030-32226-7_77
14. Kabra, A., et al.: MixBoost: synthetic oversampling with boosted mixup for handling extreme imbalance. arXiv arXiv: 2009.01571 (September 2020)
15. Kang, B., et al.: Decoupling representation and classifier for long-tailed recognition. In: ICLR (2020)
16. Kolesnikov, A., et al.: Big Transfer (BiT): general visual representation learning. In: Vedaldi, A., Bischof, H., Brox, T., Frahm, J.-M. (eds.) ECCV 2020. LNCS, vol. 12350, pp. 491–507. Springer, Cham (2020). https://doi.org/10.1007/978-3-030-58558-7_29
17. Krause, J., et al.: Grader variability and the importance of reference standards for evaluating machine learning models for diabetic retinopathy. Ophthalmology 125(8), 1264–1272 (2018)
18. Lin, T.Y., Goyal, P., Girshick, R., He, K., Dollar, P.: Focal loss for dense object detection, pp. 2980–2988 (2017)
19. Lin, T.Y., Goyal, P., Girshick, R., He, K., Dollár, P.: Focal loss for dense object detection. IEEE Trans. Pattern Anal. Mach. Intell. 42(2), 318–327 (2020)
20. Litjens, G., et al.: A survey on deep learning in medical image analysis. Med. Image Anal. 42, 60–88 (2017)
21. Mahajan, D., et al.: Exploring the limits of weakly supervised pretraining. In: Ferrari, V., Hebert, M., Sminchisescu, C., Weiss, Y. (eds.) ECCV 2018. LNCS, vol. 11206, pp. 185–201. Springer, Cham (2018). https://doi.org/10.1007/978-3-030-01216-8_12
22. Paszke, A., et al.: PyTorch: an imperative style, high-performance deep learning library. In: NEURIPS 2019, pp. 8024–8035 (2019)
23. Quellec, G., Lamard, M., Conze, P.H., Massin, P., Cochener, B.: Automatic detection of rare pathologies in fundus photographs using few-shot learning. Med. Image Anal. 61, 101660 (2020)
24. Sandler, M., Howard, A., Zhu, M., Zhmoginov, A., Chen, L.C.: MobileNetV2: inverted residuals and linear bottlenecks. In: 2018 IEEE/CVF Conference on Computer Vision and Pattern Recognition, pp. 4510–4520 (June 2018)

25. Shanmugam, D., Blalock, D., Balakrishnan, G., Guttag, J.: When and why test-time augmentation works. arXiv arXiv:2011.11156 (November 2020)
26. Thulasidasan, S., Chennupati, G., Bilmes, J.A., Bhattacharya, T., Michalak, S.: On mixup training: improved calibration and predictive uncertainty for deep neural networks. In: Advances in Neural Information Processing Systems (2019)
27. de la Torre, J., Puig, D., Valls, A.: Weighted kappa loss function for multi-class classification of ordinal data in deep learning. Pattern Recogn. Lett. **105**, 144–154 (2018)
28. Zhang, H., Cisse, M., Dauphin, Y.N., Lopez-Paz, D.: mixup: beyond empirical risk minimization. In: International Conference on Learning Representations (2018)
29. Zhou, Y., et al.: Collaborative learning of semi-supervised segmentation and classification for medical images. In: Conference on Computer Vision and Pattern Recognition (June 2019)
30. Zhou, Z.H., Liu, X.Y.: Training cost-sensitive neural networks with methods addressing the class imbalance problem. IEEE Trans. Knowl. Data Eng. **18**(1), 63–77 (2006)
31. Zhuang, J., Cai, J., Wang, R., Zhang, J., Zheng, W.: CARE: class attention to regions of lesion for classification on imbalanced data. In: International Conference on Medical Imaging with Deep Learning, pp. 588–597. PMLR (May 2019)

Transfer Learning of Deep Spatiotemporal Networks to Model Arbitrarily Long Videos of Seizures

Fernando Pérez-García[1,2,3](✉) ⓘ, Catherine Scott[4,5], Rachel Sparks[3] ⓘ,
Beate Diehl[4,5] ⓘ, and Sébastien Ourselin[3] ⓘ

[1] Department of Medical Physics and Biomedical Engineering,
University College London, London, UK
fernando.perezgarcia.17@ucl.ac.uk
[2] Wellcome/EPSRC Centre for Interventional and Surgical Sciences (WEISS),
University College London, London, UK
[3] School of Biomedical Engineering and Imaging Sciences (BMEIS),
King's College London, London, UK
[4] Department of Clinical and Experimental Epilepsy, UCL Queen Square Institute
of Neurology, London, UK
[5] Department of Clinical Neurophysiology, National Hospital for Neurology
and Neurosurgery, London, UK

Abstract. Detailed analysis of seizure semiology, the symptoms and
signs which occur during a seizure, is critical for management of epilepsy
patients. Inter-rater reliability using qualitative visual analysis is often
poor for semiological features. Therefore, automatic and quantitative analysis of video-recorded seizures is needed for objective assessment. We
present GESTURES, a novel architecture combining convolutional neural networks (CNNs) and recurrent neural networks (RNNs) to learn deep
representations of arbitrarily long videos of epileptic seizures. We use a spatiotemporal CNN (STCNN) pre-trained on large human action recognition
(HAR) datasets to extract features from short snippets (≈ 0.5 s) sampled
from seizure videos. We then train an RNN to learn seizure-level representations from the sequence of features. We curated a dataset of seizure
videos from 68 patients and evaluated GESTURES on its ability to classify
seizures into focal onset seizures (FOSs) ($N = 106$) vs. focal to bilateral
tonic-clonic seizures (TCSs) ($N = 77$), obtaining an accuracy of 98.9%
using bidirectional long short-term memory (BLSTM) units. We demonstrate that an STCNN trained on a HAR dataset can be used in combination with an RNN to accurately represent arbitrarily long videos of
seizures. GESTURES can provide accurate seizure classification by modeling sequences of semiologies. The code, models and features dataset are
available at https://github.com/fepegar/gestures-miccai-2021.

Electronic supplementary material The online version of this chapter (https://
doi.org/10.1007/978-3-030-87240-3_32) contains supplementary material, which is
available to authorized users.

The original version of this chapter was revised: Equation (1) and the first sentence of
Sect. 2.2 were corrected. The correction to this chapter is available at
https://doi.org/10.1007/978-3-030-87240-3_80

Keywords: Epilepsy video-telemetry · Temporal segment networks · Transfer learning

1   Introduction

Epilepsy is a neurological condition characterized by abnormal brain activity that gives rise to seizures, affecting about 50 million people worldwide [8]. Seizure semiology, "the historical elicitation or observation of certain symptoms and signs" during seizures, provides context to infer epilepsy type [9]. Focal onset seizures (FOSs) start in a region of one hemisphere. If they spread to both hemispheres, they are said to *generalize*, becoming focal to bilateral tonic-clonic seizures (TCSs) [9]. In TCSs, the patient first presents semiologies associated with a FOS, such as head turning or mouth and hand automatisms. This is followed by a series of phases, in which muscles stiffen (tonic phase) and limbs jerk rapidly and rhythmically (clonic phase). TCSs put patients at risk of injury and, if the seizure does not self-terminate rapidly, can result in a medical emergency. SUDEP is the sudden and unexpected death of a patient with epilepsy, without evidence of typical causes of death. Risk of SUDEP depends on epilepsy and seizure characteristics as well as living conditions. TCSs in particular increase SUDEP risk substantially [17]. In a small number of SUDEP cases occurring in epilepsy monitoring units (EMUs), death was preceded by a TCS followed by cardiorespiratory dysfunction minutes after seizure offset [20]. Identifying semiologies related to increased risk of SUDEP to appropriately target treatment is an open research question. One limitation determining SUDEP risk factors is that inter-rater reliability based on qualitative visual analysis is poor for most semiological features (e.g., limb movement, head pose or eye gaze), especially between observers from different epilepsy centers [22]. Therefore, automatic and quantitative analysis of video-recorded seizures is needed to standardize assessment of seizure semiology across multicenter studies [3].

Early quantitative analysis studies of epileptic seizures evaluated patient motion by attaching infrared reflective markers to key points on the body or using cameras with color and depth streams [6,7,14,18]. These methods are not robust to occlusion by bed linens or clinical staff, differences in illumination and pose, or poor video quality caused by compression artifacts or details out of focus.

Neural networks can overcome these challenges by automatically learning features from the training data that are more robust to variations in the data distribution. Most related works using neural networks focus on classifying the *epilepsy type* by predicting the location of the epileptogenic zone (EZ), e.g., "temporal lobe epilepsy" vs. "extratemporal lobe epilepsy", from short (≤ 2 s) snippets extracted from videos of one or more seizures [1,2,4,13,16]. Typically, this is done as follows. First, the bed is detected in the first frame and the entire video is cropped so the field of view (FOV) is centered on the bed. During training, a convolutional neural network (CNN) is used to extract features for each frame in a sampled snippet. Then, a recurrent neural network (RNN) aggregates the features into a *snippet-level* representation and a fully-connected layer predicts the epilepsy type. Finally, a *subject-level* prediction is obtained by averaging all snippet-level predictions. This approach has several disadvantages. First, it is not robust to incorrect bed detection or changes in the FOV due to zooming or panning. Second, the order of semiologies is ignored, as the epilepsy type is predicted from short snippets independently of their occurrence during a seizure. Moreover, patients with the same epilepsy type may present different seizure types. Finally, training neural networks from small datasets, as is often the case in clinical settings, leads to limited results.

The goal of this work is to compute *seizure-level* representations of arbitrarily long videos when a small dataset is available, which is typically the case in EMUs.

To overcome the challenge of training with small datasets, transfer learning from spatiotemporal CNNs (STCNNs) trained for human action recognition (HAR) can be used [13]. Although seizures are, strictly speaking, not actions, HAR models are expected to encode strong representations of human motion that may be relevant for seizure characterization. These methods are typically designed to classify human actions by aggregating predictions for snippets sampled from short clips (≈ 10 s). Epileptic seizures, however, can last from seconds to tens of minutes [12]. A common aggregation method is to average predictions from randomly sampled snippets [5,10,21]. Averaging predictions typically works because most video datasets considered are trimmed, i.e., the same action occurs along most of the video duration. In our dataset, due to the nature of TCSs, more than half the frames are labeled as non-generalizing in $49/79$ (62%) of the TCS videos. Therefore, simply averaging snippet-level predictions would result in a large number of seizures being misclassified as FOSs. Temporal segment networks (TSNs) [23] split videos of any duration into n non-overlapping segments and a consensus function aggregates features extracted from each segment. Therefore, we propose the use of TSNs to capture semiological features across the entirety of the seizure. We use an RNN as a consensus function to model the sequence of feature vectors extracted from the segments.

We present a novel neural network architecture combining TSNs and RNNs, which we denote Generalized Epileptic Seizure classification from video–Telemetry Using REcurrent convolutional neural networkS (GESTURES), that provides full representations of arbitrarily long seizure videos. These representations could be used for tasks such as classification of seizure types, seizure description using natural language, or triage. To model the relevant patient motion during seizure without the need for object detection, we use a STCNN trained on largescale HAR datasets (over 65 million videos from Instagram and 250,000 from YouTube) [10] to extract features from short snippets. Then, an RNN is used to learn a representation for the full duration of the seizure.

We chose as a proof of concept to distinguish between FOSs and TCSs, because the key distinction, if the discharge spreads across hemispheres, is only observed later in the seizure. This task demonstrates that we can train a model to take into account features across the entirety of the seizure. The main challenge, apart from the typical challenges in video-telemetry data described above, is distinguishing between TCSs and hyperkinetic FOSs, which are characterized by intense motor activity involving the extremities and trunk.

2 Materials and Methods

2.1 Video Acquisition

Patients were recorded using two full high-definition (1920 × 1080 pixels, 30 frames per second (FPS)) cameras installed in the EMU as part of standard clinical practice. Infrared is used for acquisition in scenes with low light intensity, such as during nighttime. The acquisition software (Micromed, Treviso, Italy) automatically resizes one of the video streams (800 × 450), superimposes it onto the top-left corner of the other stream and stores the montage using MPEG-2. See the supplementary materials for six examples of videos in our dataset.

2.2 Dataset Description and Ground-Truth Definitions

A neurophysiologist (C.S.) annotated for each seizure the following times: clinical seizure onset t_0, onset of the clonic phase t_G (TCSs only) and clinical seizure offset t_1. The annotations were confirmed using electroencephalography (EEG).

We curated a dataset comprising 141 FOSs and 77 TCSs videos from 68 epileptic patients undergoing presurgical evaluation at the National Hospital for Neurology and Neurosurgery, London, United Kingdom. To reduce the seizure class imbalance, we discarded seizures where $t_1 - t_0 < 15$ s, as this threshold is well under the shortest reported time for TCSs [12]. After discarding short videos, there were 106 FOSs. The 'median (min, max)' number of seizures per patient is 2 (1, 16). The duration of FOS and TCS is 53 (16, 701) s and 93 (51, 1098) s, respectively. The total duration of the dataset is 298 min, 20% of which correspond to TCS phase (i.e., the time interval $[t_G, t_1]$). Two patients had only FOS, 32 patients had only TCS, and 34 had seizures of both types. The 'mean (standard deviation)' of the percentage of the seizure duration before the appearance of generalizing semiology, i.e., $r = (t_G - t_0)/(t_1 - t_0)$, is 0.56 (0.18), indicating that patients typically present generalizing semiological features in the second half of the seizure.

Let a seizure video be a sequence of K frames starting at t_0. Let the time of frame $k \in \{0, \ldots, K - 1\}$ be $t_k = t_0 + \frac{k}{f}$, where f is the video frame rate. We use 0 and 1 to represent FOS and TCS labels, respectively. The ground-truth label $y_k \in \{0, 1\}$ for frame k is defined as $y_k := 0$ if $t_k < t_G$ and 1 otherwise, where $t_G \to \infty$ for FOSs.

Let $\mathbf{x} \in \mathbb{R}^{3 \times l \times h \times w}$ be a stack of frames or *snippet*, where 3 denotes the RGB channels, l is the number of frames, and h and w are the number of rows and columns in a frame, respectively. The label for a snippet starting at frame k is

$$Y_k := \begin{cases} 0 \text{ if } \frac{t_k + t_{k+l}}{2} < t_G \\ 1 \text{ otherwise} \end{cases} \tag{1}$$

2.3 Snippet-Level Classification

The probability $\hat{Y}_k$ that a patient presents generalizing features within snippet $\mathbf{x}_k$ starting at frame k is computed as

$$\hat{Y}_k = \Pr(Y_k = 1 \mid \mathbf{x}_k) = \mathcal{F}_{\theta_{\mathbf{z}, \mathbf{x}}}(\mathcal{C}_{\theta_{\mathbf{x}}}(\mathbf{x}_k)) = \mathcal{F}_{\theta_{\mathbf{z}, \mathbf{x}}}(\mathbf{z}_k) \tag{2}$$

where $\mathcal{C}_{\theta_\mathbf{x}}$ is an STCNN parameterized by $\theta_\mathbf{x}$ that extracts features, $\mathbf{z}_k \in \mathbb{R}^m$ is a vector of m features representing $\mathbf{x}_k$ in a latent space, and $\mathcal{F}_{\theta_{\mathbf{z},\mathbf{x}}}$ is a fully-connected layer parameterized by $\theta_{\mathbf{z},\mathbf{x}}$ followed by a sigmoid function that maps logits to probabilities. In this work, we do not update $\theta_\mathbf{x}$ during training.

2.4   Seizure-Level Classification

Temporal Segment Network. Let $V = \{\mathbf{x}_k\}_{k=1}^{K-l}$ be the set of all possible snippets sampled from a seizure video. We define a sampling function $f : (V, n, \gamma) \mapsto S$ that extracts a sequence S of n snippets by splitting V into n non-overlapping segments and randomly sampling one snippet per segment. There are two design choices: the number of segments n and the probability distribution used for sampling within a segment. If a uniform distribution is used, information from two adjacent segments might be redundant. Using the middle snippet of a segment minimizes redundancy, but reduces the proportion of data leveraged during training. We propose using a symmetric beta distribution $(\mathrm{Beta}(\gamma, \gamma))$ to model the sampling function, where γ controls the dispersion of the probability distribution (Fig. 1(b)). The set of latent snippet representations is $Z = \{\mathcal{C}_{\theta_\mathbf{x}}(\mathbf{x}_i)\}_{i=1}^n$.

Recurrent Neural Network. To perform a seizure-level prediction $\hat{\mathbf{Y}}$, Z is aggregated as follows:

$$\hat{\mathbf{Y}} = \mathrm{Pr}(\mathbf{Y} = 1 \mid S) = \mathcal{F}_{\theta_{\mathbf{z},\mathbf{s}}}(\mathcal{R}_{\theta_\mathbf{s}}(Z)) = \mathcal{F}_{\theta_{\mathbf{z},\mathbf{s}}}(\mathbf{z}) \qquad (3)$$

where $\mathcal{R}_{\theta_\mathbf{s}}$ is an RNN parameterized by $\theta_\mathbf{s}$, $\mathcal{F}_{\theta_{\mathbf{z},\mathbf{s}}}$ is a fully-connected layer parameterized by $\theta_{\mathbf{z},\mathbf{s}}$ which uses a softmax function to output probabilities, and $\mathbf{z}$ is a feature-vector representation of the entire seizure video, corresponding to the last hidden state of $\mathcal{R}_{\theta_\mathbf{s}}$.

3   Experiments and Results

All videos were preprocessed by separating the two streams into different files (replacing the small embedded view with black pixels), resampling to 15 FPS and 320×180 pixels, and reencoding using High Efficiency Video Coding (HEVC). To avoid geometric distortions while maximizing the FOV and resolution, videos were cropped horizontally by removing 5% of the columns from each side, and padded vertically so frames were square. Snippets were resized to 224×224 or 112×112, as imposed by each architecture. For realism, six video streams in which the patient was completely outside of the FOV were discarded for training but used for evaluation.

Experiments were implemented in PyTorch 1.7.0. We used a stratified 10-fold cross-validation, generated to ensure the total duration of the videos and ratio of FOSs to TCSs were similar across folds. Both views from the same video were assigned to the same fold, but videos from the same patient were not. This is

Table 1. Performance of the feature extractors. The number of parameters is shown in millions. AUC is the area under the precision-recall curve. Accuracy is computed for TCSs and FOSs, while F_1-score and AUC only for TCSs, represented by an asterisk (*). Metrics are expressed as 'median (interquartile range)'.

Model (frames)	Parameters	Features	Accuracy	F_1-score*	AUC*
Wide R2D-50-2 (1)	66.8 M	2048	80.3 (33.2)	67.4 (30.2)	75.7 (38.4)
R2D-34 (1)	21.2 M	512	89.7 (27.7)	73.9 (23.6)	84.3 (28.7)
R(2+1)D-34 (8)	63.5 M	512	93.9 (18.3)	81.6 (16.9)	93.7 (13.4)
R(2+1)D-34 (32)	63.5 M	512	96.9 (12.9)	84.7 (13.4)	94.7 (11.9)

because individual patients can present with both FOSs or TCSs, so data leakage at the patient level is not a concern. We minimized the weighted binary cross-entropy loss to overcome dataset imbalance, using the AdamW optimizer [15]. The code is available at https://github.com/fepegar/gestures-miccai-2021.

For each fold, evaluation is performed using the model from the epoch with the lowest validation loss. At inference time, the network predicts probabilities for both video streams of a seizure, and these predictions are averaged. The final binary prediction is the consensus probability thresholded at 0.5. We analyzed differences in model performance using a one-tailed Mann-Whitney U test (as metrics were not normally distributed) with a significance threshold of $\alpha = 0.05$, and Bonferroni correction for each set of c experiments: $\alpha_{\text{Bonf}} = \frac{\alpha}{e(e-1)}$.

3.1 Evaluation of Feature Extractors for Snippets

Despite recent advances in STCNNs for HAR, these architectures do not always outperform single-frame CNNs (SFCNNs) pre-trained on large generic datasets [11]. We assessed the ability of different feature extractors to model semiologies by training a classifier for snippet-level classification (Sect. 2.3).

We used two pre-trained versions of the STCNN R(2+1)D-34 [10] that take as inputs 8 frames (≈ 0.5 s) or 32 frames (≈ 2.1 s). Models were trained using weakly supervised learning on over 65 million Instagram videos and fully supervised learning on over 250,000 YouTube videos of human actions. We selected two pre-trained SFCNNs with 34 (R2D-34) and 50 (Wide R2D-50-2) layers, trained on ImageNet [24]. The SFCNNs were chosen so the numbers of layers (34) and parameters (≈ 65 million) were similar to the STCNNs.

To ensure that all features datasets have the same number of training instances, we divided each video into segments of 32 frames. Then, we use the models to extract features from snippets of the required length (8, 32, or 1) such that all snippets are centered in the segments. The datasets of extracted feature vectors are publicly available [19]. We trained a fully-connected layer for 400 epochs on each feature set, treating views from the same video independently. We used an initial learning rate 10^{-3} and mini-batches of 1024 feature vectors. We minimized a weighted binary cross-entropy loss, where the weight for TCSs was computed as the ratio of FOS frames to TCS frames.

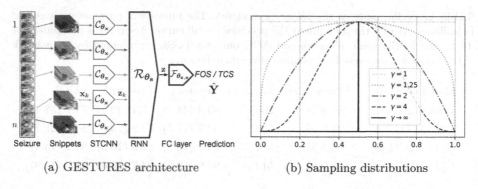

(a) GESTURES architecture   (b) Sampling distributions

Fig. 1. Left: The GESTURES architecture. We train only the models with thick red borders. Right: Probability distributions used to sample snippets from video segments.

For evaluation, a sliding window was used to infer probabilities for all snippets. STCNNs performance was significantly better than SFCNNs ($p < 10^{-7}$) (Table 1). The difference between SFCNNs was not significant ($p = 0.012$).

3.2 Aggregation for Seizure-Level Classification

In this experiment, we compared the performance of three aggregation methods to perform seizure-level classification, using 1) the mean, 2) an RNN with 64 long short-term memory (LSTM) units and 3) an RNN with 64 bidirectional LSTM (BLSTM) units to aggregate the n feature vectors sampled from the video segments. We used the dataset of feature vectors generated by R(2+1)D-34 (8) (Sect. 3.1). For the task of classifying FOS and TCS, the number of segments should be selected to ensure snippets after t_G, when generalizing semiologies begin, are sampled. The theoretical minimum number of segments needed to sample snippets after t_G is $n_{min} = \lceil 1/(1 - r_{max})\rceil$, where r_{max} is the largest possible ratio of non-generalizing to generalizing seizure durations (Sect. 2.2). We can estimate r_{max} from our dataset: $r_{max} = \max(r_1, \ldots, r_{n_{TCS}}) = 0.93$, where n_{TCS} is the number of TCSs, which yields $n_{min} = 15$ segments. We evaluated model performance using $n \in \{2, 4, 8, 16\}$ segments per video and a sampling distribution using $\gamma \in \{1, 1.25, 2, 4, \infty\}$, corresponding to uniform, near semi-elliptic, parabolic, near Gaussian and Dirac's delta distributions, respectively. For evaluation, we used $\gamma \to \infty$, i.e., only the central snippet of each segment. We trained using mini-batches with sequences sampled from 64 videos, and an initial learning rate of 10^{-2}. We used a weighted binary cross-entropy loss for training, where the weight for TCSs was the ratio of FOSs to TCSs.

The highest accuracies were obtained using $n = 16$ segments, $\gamma \in \{2, 4\}$ and the BLSTM aggregator (Fig. 2). The model with the highest accuracy (98.9%) and F_1-score (98.7%) yielded 77 true positives, 104 true negatives, 2 false positives and 0 false negatives, where TCS is the positive class (Sect. 2.2). See the supplementary materials for examples of videos classified correctly and incorrectly, with different levels of confidence.

4   Discussion and Conclusion

Objective assessment of seizure semiology from videos is important to determine appropriate treatment for the diagnosed epilepsy type and help reduce SUDEP risk. Related works focus on EZ localization by averaging classifications of short snippets from multiple seizures, ignoring order of semiologies, and are not robust to variations seen in real world datasets such as changes in the FOV. Moreover, their performance is limited by the size of the training datasets, which are small due to the expense of curating datasets. Methods that take into account the sequential nature of semiologies and represent the entirety of seizures are needed.

We presented GESTURES, a method combining TSNs and RNNs to model long-range sequences of seizure semiologies. GESTURES can classify seizures into FOSs and TCSs with high accuracy. To overcome the challenge of training on limited data, we used a network pre-trained on large HAR datasets to extract relevant features from seizure videos, highlighting the importance of transfer learning in medical applications. GESTURES can take videos from multiple cameras, which makes it robust to patients being out of the FOV.

In Sect. 3.1 we compared STCNNs to SFCNNs for snippet-level classification. To make comparisons fair, we selected models with a similar number of layers (R2D-34) or parameters (Wide R2D-50-2). We found the larger SFCNN had worse performance, due to overfitting to the training dataset. Classification accuracy was proportional to snippet duration (Table 1), meaning that both STCNNs outperformed SFCNNs. We selected R(2+1)D-34 (8) for the aggregation experiment (Sect. 3.2), as performance between the two STCNNs was similar and this model is less computationally expensive.

Using LSTM or BLSTM units to aggregate features from snippets improved accuracy compared to averaging (Fig. 2), confirming that modeling the order of semiologies is important for accurate seizure representation. Model performance was proportional to the number of temporal segments, with more segments providing a denser sampling of seizure semiologies. Ensuring some dispersion in the

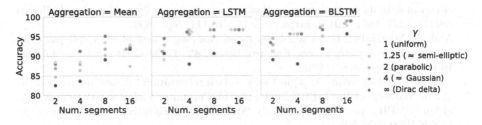

Fig. 2. Quantitative results for seizure-level classification. Marker brightness is proportional to the dispersion associated with the probability distribution used to sample snippets from the video segments (see Fig. 1b).

probability distributions used to sample snippets improved classification. One of the two false positives was caused by the patient being out of the FOV in one of the video streams. We did not observe overfitting to unrelated events in the videos, such as nurses in the room, to predict TCS, and models correctly discriminated between TCSs and hyperkinetic FOSs.

We demonstrated that methods designed for HAR can be adapted to learn deep representations of epileptic seizures. This enables a fast, automated and quantitative assessment of seizures. GESTURES takes arbitrarily long videos and is robust to occlusions, changes in FOV and multiple people in the room. In the future, we will investigate the potential of GESTURES to classify different types of TCSs and to localize the EZ, using datasets from multiple EMUs.

Acknowledgments. This work is supported by the Engineering and Physical Sciences Research Council (EPSRC) [EP/R512400/1]. This work is additionally supported by the EPSRC-funded UCL Centre for Doctoral Training in Intelligent, Integrated Imaging in Healthcare (i4health) [EP/S021930/1] and the Wellcome/EPSRC Centre for Interventional and Surgical Sciences (WEISS, UCL) [203145Z/16/Z]. The data acquisition was supported by the National Institute of Neurological Disorders and Stroke [U01-NS090407].

This publication represents, in part, independent research commissioned by the Wellcome Innovator Award [218380/Z/19/Z/]. The views expressed in this publication are those of the authors and not necessarily those of the Wellcome Trust.

The weights for the 2D and 3D models were downloaded from TorchVision and https://github.com/moabitcoin/ig65m-pytorch, respectively.

References

1. Ahmedt-Aristizabal, D., Nguyen, K., Denman, S., Sridharan, S., Dionisio, S., Fookes, C.: Deep motion analysis for epileptic seizure classification. In: 2018 40th Annual International Conference of the IEEE Engineering in Medicine and Biology Society (EMBC), pp. 3578–3581 (July 2018). ISSN 1558-4615. https://doi.org/10.1109/EMBC.2018.8513031
2. Ahmedt-Aristizabal, D., et al.: A hierarchical multimodal system for motion analysis in patients with epilepsy. Epilepsy Behav. **87**, 46–58 (2018). https://doi.org/10.1016/j.yebeh.2018.07.028
3. Ahmedt-Aristizabal, D., et al.: Automated analysis of seizure semiology and brain electrical activity in presurgery evaluation of epilepsy: a focused survey. Epilepsia **58**(11), 1817–1831 (2017). https://doi.org/10.1111/epi.13907
4. Ahmedt-Aristizabal, D., et al.: Deep facial analysis: a new phase I epilepsy evaluation using computer vision. Epilepsy Behav. **82**, 17–24 (2018). https://doi.org/10.1016/j.yebeh.2018.02.010
5. Carreira, J., Zisserman, A.: Quo Vadis, action recognition? A new model and the kinetics dataset. In: 2017 IEEE Conference on Computer Vision and Pattern Recognition (CVPR), pp. 4724–4733 (July 2017). ISSN 1063-6919. https://doi.org/10.1109/CVPR.2017.502

6. Cunha, J.P.S., Vollmar, C., Li, Z., Fernandes, J., Feddersen, B., Noachtar, S.: Movement quantification during epileptic seizures: a new technical contribution to the evaluation of seizure semiology. In: Proceedings of the 25th Annual International Conference of the IEEE Engineering in Medicine and Biology Society (IEEE Cat. No.03CH37439), vol. 1, pp. 671–673 (September 2003). ISSN 1094-687X. https://doi.org/10.1109/IEMBS.2003.1279851

7. Cunha, J.P.S., et al.: NeuroKinect: a novel low-cost 3Dvideo-EEG system for epileptic seizure motion quantification. PLOS ONE 11(1), e0145669 (2016). https://doi.org/10.1371/journal.pone.0145669

8. Fiest, K.M., et al.: Prevalence and incidence of epilepsy. Neurology 88(3), 296–303 (2017). https://doi.org/10.1212/WNL.0000000000003509

9. Fisher, R.S., et al.: Operational classification of seizure types by the international league against Epilepsy: Position Paper of the ILAE commission for classification and terminology. Epilepsia 58(4), 522–530 (2017). https://doi.org/10.1111/epi.13670

10. Ghadiyaram, D., Feiszli, M., Tran, D., Yan, X., Wang, H., Mahajan, D.: Large-scale weakly-supervised pre-training for video action recognition. arXiv arXiv:1905.00561 [cs] (May 2019)

11. Hutchinson, M., et al.: Accuracy and performance comparison of video action recognition approaches. In: 2020 IEEE High Performance Extreme Computing Conference (HPEC), pp. 1–8 (September 2020). ISSN 2643-1971. https://doi.org/10.1109/HPEC43674.2020.9286249

12. Jenssen, S., Gracely, E.J., Sperling, M.R.: How long do most seizures last? A systematic comparison of seizures recorded in the epilepsy monitoring unit. Epilepsia 47(9), 1499–1503 (2006). https://doi.org/10.1111/j.1528-1167.2006.00622.x

13. Karácsony, T., Loesch-Biffar, A.M., Vollmar, C., Noachtar, S., Cunha, J.P.S.: A deep learning architecture for epileptic seizure classification based on object and action recognition. In: 2020 IEEE International Conference on Acoustics, Speech and Signal Processing (ICASSP), ICASSP 2020, pp. 4117–4121 (May 2020). ISSN 2379-190X. https://doi.org/10.1109/ICASSP40776.2020.9054649

14. Li, Z., Silva, A.M., Cunha, J.P.S.: Movement quantification in epileptic seizures: a new approach to video-EEG analysis. IEEE Trans. Biomed. Eng. 49(6), 565–573 (2002). Conference Name: IEEE Transactions on Biomedical Engineering. https://doi.org/10.1109/TBME.2002.1001971

15. Loshchilov, I., Hutter, F.: Decoupled weight decay regularization. arXiv arXiv:1711.05101 [cs, math] (January 2019)

16. Maia, P., Hartl, E., Vollmar, C., Noachtar, S., Cunha, J.P.S.: Epileptic seizure classification using the NeuroMov database. In: 2019 IEEE 6th Portuguese Meeting on Bioengineering (ENBENG), pp. 1–4 (February 2019). https://doi.org/10.1109/ENBENG.2019.8692465

17. Nashef, L., So, E.L., Ryvlin, P., Tomson, T.: Unifying the definitions of sudden unexpected death in epilepsy. Epilepsia 53(2), 227–233 (2012). https://doi.org/10.1111/j.1528-1167.2011.03358.x

18. O'Dwyer, R., et al.: Lateralizing significance of quantitative analysis of head movements before secondary generalization of seizures of patients with temporal lobe epilepsy. Epilepsia 48(3), 524–530 (2007). https://doi.org/10.1111/j.1528-1167.2006.00967.x

19. Pérez-García, F., Scott, C., Sparks, R., Diehl, B., Ourselin, S.: Data to support the paper "transfer learning of deep spatiotemporal networks to model arbitrarily long videos of seizures" (July 2021). Publisher: University College London Type: dataset. https://doi.org/10.5522/04/14781771.v1

20. Ryvlin, P., et al.: Incidence and mechanisms of cardiorespiratory arrests in epilepsy monitoring units (MORTEMUS): a retrospective study. Lancet Neurol. **12**(10), 966–977 (2013). https://doi.org/10.1016/S1474-4422(13)70214-X

21. Simonyan, K., Zisserman, A.: Two-stream convolutional networks for action recognition in videos. In: Proceedings of the 27th International Conference on Neural Information Processing Systems, NIPS 2014, vol. 1, pp. 568–576. MIT Press, Cambridge (December 2014)

22. Tufenkjian, K., Lüders, H.O.: Seizure semiology: its value and limitations in localizing the epileptogenic zone. J. Clin. Neurol. (Seoul, Korea) **8**(4), 243–250 (2012). https://doi.org/10.3988/jcn.2012.8.4.243

23. Wang, L., et al.: Temporal segment networks for action recognition in videos. IEEE Trans. Pattern Anal. Mach. Intell. **41**(11), 2740–2755 (2019). Conference Name: IEEE Transactions on Pattern Analysis and Machine Intelligence. https://doi.org/10.1109/TPAMI.2018.2868668

24. Zagoruyko, S., Komodakis, N.: Wide residual networks. arXiv arXiv:1605.07146 [cs] (June 2017)

Retina-Match: Ipsilateral Mammography Lesion Matching in a Single Shot Detection Pipeline

Yinhao Ren[1]([✉]), Jiafeng Lu[3], Zisheng Liang[1], Lars J. Grimm[2], Connie Kim[2], Michael Taylor-Cho[2], Sora Yoon[2], Jeffrey R. Marks[4], and Joseph Y. Lo[1,2,3]

[1] Department of Biomedical Engineering, Duke University, Durham, USA
{yinhao.ren,ziheng.liang}@duke.edu
[2] Department of Radiology, Duke University School of Medicine, Durham, USA
{lars.grimm,connie.kim,michael.taylorcho,sora.yoon,joseph.lo}@duke.edu
[3] Department of Electrical Engineering, Duke University, Durham, USA
jiafeng.lu@duke.edu
[4] Department of Surgery, Duke University School of Medicine, Durham, USA
jeffrey.marks@duke.edu

Abstract. In mammography and tomosynthesis, radiologists use the geometric relationship of the four standard screening views to detect breast abnormalities. To date, computer aided detection methods focus on formulations based only on a single view. Recent multi-view methods are either black box approaches using methods such as relation blocks, or perform extensive, case-level feature aggregation requiring large data redundancy. In this study, we propose Retina-Match, an end-to-end trainable pipeline for detection, matching, and refinement that can effectively perform ipsilateral lesion matching in paired screening mammography images. We demonstrate effectiveness on a private, digital mammography data set with 1,016 biopsied lesions and 2,000 negative cases.

Keywords: Computer Aided Diagnoisis · Breast cancer screening · Mammogram · Object detection · Object re-identification

1 Introduction

Computer Aided Diagnosis (CAD) for breast cancer detection has converged from heavily handcrafted features and cascade of rule-based models [15,23] to fully deep-learning-based, end-to-end detection frameworks [3,5,8,13,24]. Existing object detection frameworks can be applied to the single-view lesion detection task because most of the relevant information is near the lesion and can be extracted by CNNs. However, radiologists are trained to localize lesions by referencing all four standard views in order to reduce false positives and increase sensitivity. There exists a significant challenge to bridge the gap between current, single-view algorithm performance and the radiologist's performance in detecting both mass and calcification lesions.

© Springer Nature Switzerland AG 2021
M. de Bruijne et al. (Eds.): MICCAI 2021, LNCS 12905, pp. 345–354, 2021.
https://doi.org/10.1007/978-3-030-87240-3_33

Multi-view lesion detection includes three approaches. First, *ipsilateral matching* finds the correspondence between lesions on craniocaudal (CC) and mediolateral oblique (MLO) views of the same breast. Such matching is difficult due to the positioning and compression of the soft breast tissue with almost no anatomical landmarks. Techniques such as relation blocks [22], deep aggregation, and bipartile matching [15,23] take advantage of the two views' correspondence. Second, *contralateral matching* compares lesion candidates on the same view (e.g., CC) across left versus right breasts, using registration to distinguish high density symmetric tissues versus suspicious lesions [6,13]. Third, *temporal matching* matches lesions on the same view of the same breast between current versus prior exams, again using registration methods similar to contralateral matching.

Recent multi-view-based mass detection frameworks [12,24] utilize the geometric and similarity relationships of lesion candidates in paired mammography images. McKinney [13] used spatial aggregation of latent features from the same breast and contralateral aggregation of lesion candidates from coarse registration. Similarly, Geras [5] employed latent feature aggregation, but the spatially concatenated features from a soft-tissue organ are not well suited for such matching scenarios. It also requires great data redundancy to properly train the model. On the other hand, CVR-RCNN [12] and MommiNet [24] use cross-view relation blocks to aggregate latent features of generated lesion candidates from different views in an end-to-end approach. Nevertheless, this design assumes the latent features of each lesion candidate can affect all other candidates, breaking the clinical assumption of one-to-one matching. Furthermore, since the relation block is not directly supervised by lesion matching, it is difficult to interpret the model output as well as to fine-tune performance. Therefore we need an interpretable, multi-view lesion detection model that is lightweight and can be trained with reasonable sized data such as from a single institution.

In this study, we propose Retina-Match, an end-to-end trainable RetinaNet-based model that incorporates multi-view lesion detection for mammography. The model is inspired by RetinaTrack [11] that introduced an object re-identification task into the single-shot detection pipeline. RetinaTrack is trained only using triplet sampling and performs object tracking during testing as post-processing. In contrast, we use the lesion matching result to refine lesion detection scores to directly supervise the model training. We summarize our contributions as follows:

1. We propose the Retina-Match framework that improves cancer detection using ipsilateral matching of lesion candidates from multiple mammography views.
2. We curated a large screening mammography data set and used it to develop the proposed framework.
3. We show our proposed framework is superior to relation-block-based approaches in lesion detection in screening mammography.

2 Related Work

RetinaTrack [11] was the first modern, single-shot detection framework for tracking dense objects. The task of car tracking in videos is realized by training the detection task and object appearance embedding concurrently using randomly mined quadruplets [4,19]. The appearance embedding network minimizes the euclidean distance of embedding vectors from the same objects while maximizing distances from different objects. Note the object detection score in tracking tasks is not designed to be affected by its correspondence in the consecutive frames. However, in mammography lesion detection, the co-occurrence of similar lesions in CC and MLO views provides confidence for a true positive detection.

CVR-RCNN [12] and MommiNet [24] use relation block to learn the latent relationship of lesion candidates for better detection performance. Relation block is a variant of non-local block that computes the dot product similarity between each lesion candidate's feature embedding and modify the latent features through addition. This approach is effective in learning complex relationships among objects [7,10,21], but has many drawbacks in multi-view lesion detection application. First, a relation block is non-intepretable. The input and output of a relational block are usually latent features, making it impossible to understand the lesion matching relationship. Second, the computational complexity is quadratic to embedded feature length, making it easy to over-fit on relatively small data sets. Lastly, GCNet [2] showed that general non-local blocks often degrade into a global feature extractor. Rather than learning the lesion-level relationships of each detected candidate, all candidates are instead modified by a global summary extracted from the paired view. This is concerning since the imaging features that distinguish different physical lesions can be subtle, resulting in similar

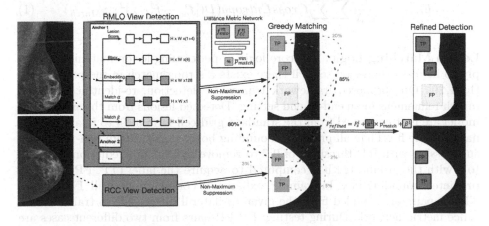

Fig. 1. Proposed Retina-Match architecture with an illustrative example. All trainable weights are front-loaded in the pipeline by design so that greedy matching operation can be treated as part of the detection refinement loss function. In this example, one TP-TP pair and one FP-FP survived the matching process. Greedy matching result is available as model output

feature embedding. Moreover, radiologists consider the ranking of lesion similarities when referencing two views, however, this cannot be easily implemented in the relation block formulation.

3 Method

Our RetinaMatch multi-view, lesion-detection pipeline has three subcomponents.

Single-View Lesion Detection. We implemented our single-view detector as in RetinaTrack [11]. We split our bounding boxes into 2 anchors by size. Each anchor contains independent regressors for bounding box predictions and lesion appearance embedding. The shared feature extractors are split early in the network to allow effective individual lesions appearance embedding. In order to have a unified logit P_s to apply the multi-view refinement score, we adopted a modified Yolo V2 [16,17] object detection loss formulation that treats object detection and classification separately.

Lesion Re-identification. Given two sets of true-positive (TP) and false-positive (FP) from CC and MLO view's single-view detection pipeline, TP-TP pairs and TP-FP pairs are generated on the fly. The kth pair label y_{Match}^k is 1 if the two lesion ID matches, otherwise 0. A learned distance metric network $D(x)$ optimized together with the overall framework is applied to classify the ith and jth embedding features f_{CC}^i and f_{MLO}^j into a positive or negative pair. The distance network is trained using sigmoid cross entropy loss as in Eq. 1, where M and N is the number of pairs generated in each view.

$$L_{reID} = \frac{1}{MN} \sum_i^M \sum_j^N CrossEntropy(D([f_{cc}^i - f_{MLO}^j]^2), y_{Match}^k) \quad (1)$$

Lesion Matching Logic. The core logic of performing lesion matching is simply to increase the lesion score if there exists a strong match in the paired view [1,14,20]. Our proposed framework trains the detection, re-identification and match refinement in an end-to-end setting. Detected lesions from the single-view model are first converted from the standard grid-cell format [9] to image coordinates on the fly. Than all predicted bounding boxes pass through non-maximum suppression with IoU threshold of 0.7 to remove redundant predictions. Finally IoU with the ground truth is computed to acquire the label (TP, FP) for each predicted bounding box. In training, exhaustive combination of all TP-TP and TP-FP pairs are sampled from the given ipsilateral image pair to train the distance metric network. During testing, FP-FP pairs from two different cases are also included to enrich the non-pair sample pool. The resulting exhaustive pairs are ranked and greedily matched so that each lesion can only be matched to one candidate in the paired view. This significantly reduces the total number of possible pairs. Surviving lesion candidates are than refined using the matching results as follows.

Given the ith single-view lesion score P_s^i from the single-view regressor, the matching-refined lesions score $P_{refined}^i$ is computed as:

$$P_{refined}^i = P_s^i + \alpha_i \times p_{match}^i + \beta_i \tag{2}$$

where α_i and β_i are the per lesion scaling and bias factor for performing the matching. The value of α_i and β_i are predicted per anchor box so that no hand-crafting of multi-view matching rule is needed. The training of α and β value is only possible in an end-to-end setting, where $P_{refined}$ is included as part of the loss function. The proposed lesion refinement loss is then computed as weighted MSE for TP and FP lesions respectively as in Eq. 3. In contrast, RetinaTrack [11] only trains the detection and re-identification components and applies the object-tracking logic as post-processing.

$$L_{match\_refine} = \frac{1}{N}(\sum_i^N P_{refined}^i - y^i)^2 + \frac{1}{M}(\sum_j^M P_{refined}^j - y^j)^2 \tag{3}$$

We emphasize the prediction of per-anchor α and β is the key difference of our methods to existing, matching-based detection algorithms. Without α, handcrafting is needed to determine how much a matching result influences the final lesion score predictions. Without β, lesions visible only on one view such as asymmetry and architectural distortion are inherently de-prioritized, causing unnecessary loss in sensitivity. Training the model end-to-end with a lesion score bias term allows the model to compensate for such single-view lesions.

4 Experiments

4.1 Dataset

The data set used for this project is a large set of General Electric (GE) screening full-field digital mammography images collected from our institution between 2010 and 2018. In total there are 1016 cases with biopsy-proven benign or malignant soft-tissue lesions. The detailed train/test split and benign/malignant distribution are shown in Table 1. Annotations are provided by 4 radiologists with access to the screening radiology reports. Tight bounding boxes are drawn for lesions that triggered either a biopsy or surgery within 3 months of the screening exam. Case-level lesion IDs are assigned to establish multi-view, lesion-pair relationships. For testing, we randomly sampled 2,000 negative cases (from over 10,000 available with BI-RADS rating of 1 or 2) and report our FROC curves based on the average false positives detection on only the negative images.

An in-house pectoral muscle segmentation model is applied on each MLO view to determine the chestwall-to-nipple datum line. A 2D breast depth encoding is than constructed by calculating the estimated nipple distance of each pixel within the breast along the datum line.

We normalize all images using the DICOM window-level setting and flip them such that the breast is always facing right. Data augmentation (random cropping, flipping, scaling) are implemented on the fly in training. Ipsilaterally paired images are always augmented in sync so that the lesion appearance and lesion depth relationships are preserved.

Table 1. Mass lesion distribution.

	Train			Test		
	Case	View	Annotation	Case	View	Annotation
Benign	529	1099	1205	209	447	503
Malignant	195	409	437	83	166	195
All Pos	724	1508	1642	292	613	698
Negative	/	/	/	2000	8000	/

4.2   Network Architecture

The network architecture is illustrated in Fig. 1. We implemented our baseline single-view architecture mainly as in RetinaTrack [11]. We use pre-trained MobileNetV2 as the main feature extractor. A single-channel depth encoding is resized and concatenated to extracted latent features as input to each anchor regressor. Each anchor regressor contains one multi-head self-attention block [22] followed by four almost identical regressors for each auxiliary task. Each regressor consists of four repetitions of 3×3 followed by 1×1 convolution layers with Leaky Relu activation function. The number of features output at each regressor's last convolution layer is determined by the auxiliary task.

Specifically, (1) **classification regressor** predicts the probability a grid cell contains a suspicious lesion as well as abnormality type of the lesion; (2) **bounding box regressor** predicts the x and y offset of the box as well as the width and height; (3) **embedding regressor** produces a vector length of 128 to describe the appearance of the grid cell; and (4) **two match modifier regressors** produces the α and β value for each grid cell.

The additional distance metric network contains 3 fully connected layers that have 128, 64, 1 outputs respectively. The last layer's output is constrained by sigmoid activation function.

4.3   Training

The proposed model is implemented in Tensorflow 2.1, trained using two 2080ti graphic cards. The parameter of the shared feature extractor is initialized by ImageNet pretrained MobileNetV2 weights. Adam optimizer is used with a learning rate of 0.0001 with default settings. The single-view detection components are all trainable and fine-tuned for the first 25 epochs with a mini-batch size of 2 ipsilateral pairs. Than we freeze the single-view component and fine-tune the lesion re-identification and lesion matching components for another 25 epochs.

4.4 Choice of Single-View Detection Model

In order to evaluate weather a single-view detection models is sufficient for this application in the early experiments, we compare the RetinaNet with Yolo v2 [17] and Faster-RCNN [18]. We found in our application, the focal-loss, feature-pyramid network and self-attention blocks significantly improve detection performance from Yolo V2 to RetinaNet. There is also no noticeable gain in the single-view detection performance using an off-the-shelf Faster-RCNN implementation. Due to the much faster inference speed, we build our proposed model upon the RetinaNet design for future scalability into tomosynthesis applications.

4.5 Ablation Study

We evaluated the effectiveness of our proposed methods in a step-by-step ablation study with lesion-level FROC. Detections with IoU w.r.t ground truth larger than 0.25 are marked as TP. The results are shown in Fig. 2.

We first evaluated the patch variant of Retina-Match trained on only the single-view object detection and lesion re-identification as in Eq. 4. The lesion score refinement is conducted offline using hand-crafted constant value of $\alpha = 0.35$. This configuration shows the least gain in FROC compare to other configurations but is still better than the single-view baseline.

$$P^i_{refined\_patch} = P^i_s + 0.35 \times p^i_{match} \qquad (4)$$

Then we train the lesion score refinement module but without the β bias term as in Eq. 5. This forces the model to sacrifice single-view-only findings such as lymph nodes and focal asymmetries to favor lesions with strong ipsilateral matching, thus reducing sensitivity.

$$P^i_{refined} = P^i_s + \alpha_i \times p^i_{match} \qquad (5)$$

Finally we run the full model with α and β both available per anchor prediction as in Eq. 2. This configuration gives the most gain. Additionally, ipsilateral

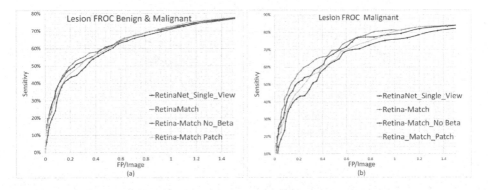

Fig. 2. (a) FROC of benign and cancer lesions. (b) FROC for cancer lesions

matching produced greater gain in detection of cancer cases compared to all biopsied cases. We believe this is due to cancer lesions have more distinctive structures that can be used as matching features.

To verify our performance gain is not a result of model fine-tuning, we also experimented with the trivial configuration of single-view detection ($\alpha = 0$) but with the β term trainable, and confirmed that the model performs identically to our single-view baseline after fine-tuning.

4.6   Comparison with the Relation Block Approach

We also compared our method against a re-implementation of [12], which uses only relation blocks for multi-view information fusion. The implementation detail follows CVR-RCNN [12]) with 2 relation blocks. Only the lesion score logits prediction are refined by the relation block. All other model designs are kept the same as our method for controlled comparison. We feed the 128-length embedding vector to the relation block to represent visual features from each predicted ROI. The embedding layer is randomly initialized and only supervised by the final lesion-refinement loss. This only provides slight gains in the high-specificity region of the FROC when compared against our method. The comparison results are shown in Fig. 3.

We believe the reasons are three fold. First, the general form of the non-local block [22] uses softmax operation to represent attentions, however, the relation block removed that mechanism to modal object-instance similarity. This allows ambiguous one-to-many relations in two sets of predicted lesion candidates, which isn't clinically common. Additionally, relation blocks do not consider the ranking of lesion similarities. Consequently, it cannot efficiently approximate matching algorithms such as greedy or bipartile matching, which is critical when differences between candidates are subtle. Finally, our task is inherently more difficult because both benign and malignant lesions are treated as region of interests. There also exist many benign-looking lesions that didn't trigger a biopsy and were thus not annotated. Overall, the net result is more false posi-

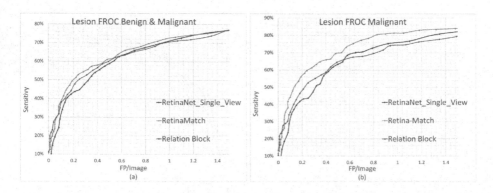

Fig. 3. (a) FROC of benign and cancer lesions. (b) FROC for cancer lesions

tive detection per image that further degrades the effectiveness of the relation block. On the other hand, Retina-Match performs better in the same problem setting with less parameters and faster inference time. The feature embedding for lesion matching are directly supervised by lesion ID labels, providing better update gradients than using only the refined detection result. Moreover, greedy matching is part of the end-to-end training pipeline, allowing the model to rank lesion pairs to reduce false positive matches.

5 Conclusion

We demonstrate a multi-view mammography lesion detection pipeline designed for end-to-end training. Our design not only obviates traditional, hand-crafted lesion-matching rules, but also preserves detection sensitivity by recovering some of the obvious single-view-only lesions. In future studies, we will continue to improve our multi-view lesion detection pipeline. We will also take advantage of our modular, computationally lightweight design to extend this application beyond 2D mammography to tomosynthesis.

References

1. Arasu, V.A., et al.: Benefit of semiannual ipsilateral mammographic surveillance following breast conservation therapy. Radiology **264**(2), 371–377 (2012)
2. Cao, Y., Xu, J., Lin, S., Wei, F., Hu, H.: GCNet: non-local networks meet squeeze-excitation networks and beyond. In: 2019 IEEE/CVF International Conference on Computer Vision Workshop (ICCVW), pp. 1971–1980 (2019). https://doi.org/10.1109/ICCVW.2019.00246
3. Sainz de Cea, M.V., Diedrich, K., Bakalo, R., Ness, L., Richmond, D.: Multi-task learning for detection and classification of cancer in screening mammography. In: Martel, A.L., et al. (eds.) MICCAI 2020. LNCS, vol. 12266, pp. 241–250. Springer, Cham (2020). https://doi.org/10.1007/978-3-030-59725-2_24
4. Chen, W., Chen, X., Zhang, J., Huang, K.: Beyond triplet loss: a deep quadruplet network for person re-identification (July 2017). https://doi.org/10.1109/CVPR.2017.145
5. Geras, K., Wolfson, S., Kim, S., Moy, L., Cho, K.: High-resolution breast cancer screening with multi-view deep convolutional neural networks (March 2017)
6. Hagos, Y., Gubern-Mérida, A., Teuwen, J.: Improving breast cancer detection using symmetry information with deep learning. In: 3rd International Workshop, RAMBO 2018, 4th International Workshop, BIA 2018, and 1st International Workshop, TIA 2018, Held in Conjunction with MICCAI 2018, Granada, Spain, September 16 and 20, 2018, Proceedings, pp. 90–97 (September 2018)
7. Hu, H., Gu, J., Zhang, Z., Dai, J., Wei, Y.: Relation networks for object detection. In: 2018 IEEE/CVF Conference on Computer Vision and Pattern Recognition, pp. 3588–3597 (2018)
8. Kooi, T., Litjens, G., et al.: Large scale deep learning for computer aided detection of mammographic lesions. Med. Image Anal. **35**, 303–312 (2017)
9. Lin, T.Y., Goyal, P., Girshick, R.B., He, K., Dollár, P.: Focal loss for dense object detection. In: 2017 IEEE International Conference on Computer Vision (ICCV) pp. 2999–3007 (2017)

10. Liu, L., et al.: Deep learning for generic object detection: a survey. Int. J. Comput. Vis. **128**(2), 261–318 (2020)
11. Lu, Z., Rathod, V., Votel, R., Huang, J.: RetinaTrack: online single stage joint detection and tracking. In: 2020 IEEE/CVF Conference on Computer Vision and Pattern Recognition (CVPR), pp. 14656–14666 (2020)
12. Ma, J., Liang, S., Li, X., Li, H., Menze, B., Zhang, R., Zheng, W.: Cross-view relation networks for mammogram mass detection. arXiv arxiv:abs/1907.00528 (2019)
13. McKinney, S.M., et al.: International evaluation of an AI system for breast cancer screening. Nature **577**(7788), 89–94 (2020)
14. Padayachee, J., Alport, M., Rae, W.: Mammographic CAD: correlation of regions in ipsilateral views - a pilot study. S. Afr. J. Radiol. **13** (2009). https://doi.org/10.4102/sajr.v13i3.497
15. Qian, W., Song, D., Lei, M., Sankar, R., Eikman, E.: Computer-aided mass detection based on ipsilateral multiview mammograms. Acad. Radiol. **14**(5), 530–538 (2007)
16. Redmon, J., Divvala, S., Girshick, R., Farhadi, A.: You only look once: unified, real-time object detection. In: 2016 IEEE Conference on Computer Vision and Pattern Recognition (CVPR), pp. 779–788 (2016). https://doi.org/10.1109/CVPR.2016.91
17. Redmon, J., Farhadi, A.: Yolo9000: Better, faster, stronger (2016)
18. Ren, S., He, K., Girshick, R., Sun, J.: Faster R-CNN: towards real-time object detection with region proposal networks. In: Cortes, C., Lawrence, N., Lee, D., Sugiyama, M., Garnett, R. (eds.) Advances in Neural Information Processing Systems, vol. 28. Curran Associates, Inc. (2015). https://proceedings.neurips.cc/paper/2015/file/14bfa6bb14875e45bba028a21ed38046-Paper.pdf
19. Schroff, F., Kalenichenko, D., Philbin, J.: FaceNet: a unified embedding for face recognition and clustering. In: Proceedings of the IEEE Conference on Computer Vision and Pattern Recognition (CVPR) (June 2015)
20. Sun, X., Qian, W., Song, D.: Ipsilateral-mammogram computer-aided detection of breast cancer. Comput. Med. Imaging Graph. **28**(3), 151–158 (2004)
21. Wang, J., et al.: Deep high-resolution representation learning for visual recognition. IEEE Trans. Pattern Anal. Mach. Intell. **43**, 3349–3364 (2020). https://doi.org/10.1109/TPAMI.2020.2983686
22. Wang, X., Girshick, R., Gupta, A., He, K.: Non-local neural networks. In: 2018 IEEE/CVF Conference on Computer Vision and Pattern Recognition, pp. 7794–7803 (2018). https://doi.org/10.1109/CVPR.2018.00813
23. Wei, J., et al.: Computer-aided detection of breast masses on mammograms: dual system approach with two-view analysis. Med. Phys. **36**(10), 4451–4460 (2009)
24. Yang, Z., et al.: MommiNet: mammographic multi-view mass identification networks. In: Martel, A.L., et al. (eds.) MICCAI 2020. LNCS, vol. 12266, pp. 200–210. Springer, Cham (2020). https://doi.org/10.1007/978-3-030-59725-2_20

Towards Robust Dual-View Transformation via Densifying Sparse Supervision for Mammography Lesion Matching

Junlin Xian[1], Zhiwei Wang[2,3], Kwang-Ting Cheng[4], and Xin Yang[1,5(✉)]

[1] School of Electronic Information and Communication,
Huazhong University of Science and Technology, Wuhan, China
[2] Britton Chance Center for Biomedical Photonics, Wuhan National Laboratory
for Optoelectronics, Huazhong University of Science and Technology, Wuhan, China
[3] MoE Key Laboratory for Biomedical Photonics, Collaborative Innovation Center
for Biomedical Engineering, School of Engineering Sciences,
Huazhong University of Science and Technology, Wuhan, China
[4] Hong Kong University of Science and Technology, Hong Kong, China
[5] Wuhan National Laboratory for Optoelectronics, Huazhong University of Science
and Technology, Wuhan, China
xinyang2014@hust.edu.cn

Abstract. A holistic understanding of dual-view transformation (DVT) is an enabling technique for computer-aided diagnosis (CAD) of breast lesion in mammogram, e.g., micro-calcification (μC) or mass matching, dual-view feature extraction etc. Learning a complete DVT usually relies on a dense supervision which indicates a corresponding tissue in one view for each tissue in another. Since such dense supervision is infeasible to obtain in practical, a sparse supervision of some traceable lesion tissues across two views is thus an alternative but will lead to a defective DVT, limiting the performance of existing CAD systems dramatically. To address this problem, our solution is simple but very effective, i.e., densifying the existing sparse supervision by synthesizing lesions across two views. Specifically, a Gaussian model is first employed for capturing the spatial relationship of real lesions across two views, guiding a following proposed LT-GAN where to synthesize fake lesions. The proposed novel LT-GAN can not only synthesize visually realistic lesions, but also guarantee appearance consistency across views. At last, a denser supervision can be composed based on both real and synthetic lesions, enabling a robust DVT learning. Experimental results show that a DVT can be learned via our densified supervision, and thus result in a superior performance of cross-view μC matching on INbreast and CBIS-DDSM dataset to the state-of-the-art methods.

Keywords: Dual-view transformation · Supervision-densifying · Lesion matching

---

J. Xian and Z. Wang are the co-first authors.

© Springer Nature Switzerland AG 2021
M. de Bruijne et al. (Eds.): MICCAI 2021, LNCS 12905, pp. 355–365, 2021.
https://doi.org/10.1007/978-3-030-87240-3_34

1   Introduction

Mammography consists of X-ray images usually from two views of breast, i.e., craniocaudal (CC) and mediolateral-oblique (MLO), which can provide important biomarkers for early breast cancer diagnosis. Holistically understanding dual-view transformation (DVT), which reflects how spatial location and appearance of each tissue changes across CC and MLO views, can act as a significant guidance for both manual mammography screening and computer-aided diagnosis (CAD) of breast cancer. For instance, radiologists often identify different lesions and false positives in one view by cross-checking with those in another.

However, learning a complete DVT must rely on a dense supervision which indicates every paired region in CC and MLO images projected from a same breast tissue. Such dense supervision is practically infeasible to obtain for information loss of the 3D construction during the mammogram [2]. Fortunately, a sparse supervision obtained by some visually traceable lesions, e.g., microcalcification (μC) or mass, has been demonstrated to be an alternative for learning DVT in recent works [12,17]. Shaked *et al.* [12] generated patch pairs for each mass lesion and used a CNN approach to learn pairwise feature representations. Yan *et al.* [17] further extended via an auxiliary classification task to tell if patches contain lesions or not.

The key of their success is Siamese network [5,7,15] whose mechanism of network sharing shows effectiveness in dual-view related tasks, e.g., cross-view matching. Depending on whether there is a metric network or not, the Siamese network-based approaches can be categorized into two classes. The former first utilizes a network to extract features from two regions and then predicts a probability that the two regions belong to the same object or not. For instance, MatchNet [5] used two weight-shared networks as a feature extractor and three fully-connected layers as a metric network. Zagoruyko *et al.* [20] concatenated two patches into a 2-channel meta-patch and fed it into a single-branch CNN as both feature extractor and metric network. SCFM [13] further exploited for combination in spatial-dimension. In contrast, the latter aims at reshaping the distribution of matched and unmatched samples, i.e., minimizing feature distances of matched samples and maximizing those of unmatched ones. Melekhov *et al.* [10] used Euclidean distance for pairwise distance measuring, while PN-Net [1] employed a triplet-loss.

Based on those Siamese network-based approaches, transformation between two views can be easily learned by exhaustively searching every matched region across different views. However, without a dense supervision, only a defective DVT can be learned via the Siamese network. For instance, previous works [9,12,17,19] can only learn an inadequate DVT around mass regions rather than a whole breast. To this end, we aim at synthesizing some trackable lesion tissues like μC or mass across two views. These synthesized cross-view lesions are guaranteed that 1) the correlation of spatial locations between two views is correct, and 2) changes of appearance across two views are consistent with those of real lesions under breast compression and view transferring. Both synthesized and real trackable lesion tissues make the original sparse supervision denser,

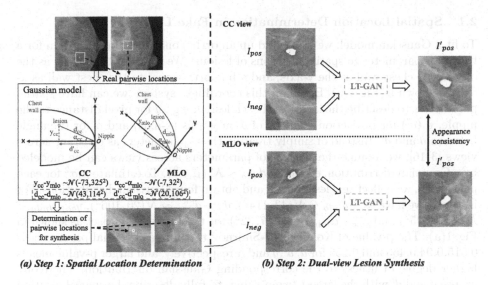

(a) Step 1: Spatial Location Determination   *(b) Step 2: Dual-view Lesion Synthesis*

Fig. 1. An overall framework of our proposed supervision-densifying method. (a) illustrates the corresponding four Gaussian distributions in INbreast.

which thus leads to a more robust DVT learning. Specifically, our proposed framework for densifying sparse supervision is shown Fig. 1, which consists of two key steps. In the first step, a Gaussian-model is employed to analysis cross-view spatial correlation of real lesions, and further guide determination of candidate locations for synthesizing. Next, we propose a novel Lesion-Translation GAN (LT-GAN) based on Cycle-GAN [21] which can synthesize lesions by inheriting real pairwise lesion information so as to make cross-view changes of appearance realistic.

To summarize, major contributions of this work are as follows: (1) We introduce a novel supervision-densifying approach, which helps capture a robust DVT in mammogram. (2) We present a novel LT-GAN for realistic trackable lesion tissue synthesis, which prevents limitations of regular GANs and remains appearance consistency. (3) We conduct several experiments of cross-view μC matching on both INbreast [8] and CBIS-DDSM [11] dataset, demonstrating that our proposed supervision-densifying approach superior on various general architectures, and greatly surpasses state-of-the-art methods of mammography matching.

2   Method

Figure 1 outlines the overall framework of our proposed supervision-densifying approach, which consists of two key steps, i.e., a Gaussian model towards pairwise location determination, and a LT-GAN to synthesize realistic μC lesions across views. We will detail the two sequential steps in the following.

2.1   Spatial Location Determination in Fake Lesions

To fit a Gaussian model, we first build up an orthogonal coordinate system for a breast to parameterize spatial locations of lesions. We consider the nipple as the origin, the chest wall as the y-axis, and a line orthogonal to the chest wall as x-axis as shown in Fig. 1(a). Based on this coordinate system, we can have spatial parameters to describe the location of each lesion, e.g., d for pixel distance to the nipple, y (d') for projection distance of d on y-axis (x-axis) and α for the angle between d and d'. Instead of simply treating these parameters as constant across views [3,16], we assume the changes of parameters across views can be modeled by a Gaussian distribution, e.g., $d_{cc}-d_{mlo} \sim \mathcal{N}(\mu, \sigma^2)$. To estimate μ, σ^2 for each parameter, we collect real lesion pairs and obtain four Gaussian distributions. For instance, we have $d_{cc}-d_{mlo} \sim \mathcal{N}(43, 114^2)$, $d'_{cc}-d'_{mlo} \sim \mathcal{N}(20, 106^2)$, $y_{cc}-y_{mlo} \sim \mathcal{N}(-73, 325^2)$ and $\alpha_{cc}-\alpha_{mlo} \sim \mathcal{N}(-7, 32^2)$ for the INbreast dataset as shown in Fig. 1(a). The p-value of Kolmogorov-Smirnov test for each spatial parameter is $0.515, 0.94, 0.463$ and 0.255 for d, d', y and α respectively, and larger p-value means higher degree of fitness to the corresponding Gaussian distribution. Therefore, we use d and d' with the largest two p-values to fully describe the spatial location of a lesion in INbreast. In the inference, we first randomly select a point (d_{mlo}, d'_{mlo}) in MLO view based on the orthogonal coordinate, and then generate the changes from MLO to CC view $(d_{cc} - d_{mlo}, d'_{cc} - d'_{mlo})$ using corresponding Gaussian models. By adding (d_{mlo}, d'_{mlo}) and $(d_{cc} - d_{mlo}, d'_{cc} - d'_{mlo})$ together, we can get a counterpart point in CC view (d_{cc}, d'_{cc}) of the original point in MLO view (d_{mlo}, d'_{mlo}). These paired locations can be used for guiding where to synthesize fake lesions.

2.2   LT-GAN for Cross-View Lesion Synthesis

As shown in Fig. 2, our proposed LT-GAN consists of two different domain-translation modules: a lesion-depositing translator (i.e., De-lesion) and a conditional generative translator (i.e., Conditional-GAN), denoted as T_{p2n} and T_{n2p} respectively. The two translators share identical backbone of U-net [14] but not trainable weights. Inspired by [18], T_{p2n} takes a patch cropped around the center of a lesion I_{pos} as input, and separate it as two parts, i.e., Z a 256-dimentional lesion-relative latent vectors and I'_{neg} a pure normal tissue of I_{pos}. Conversely, T_{n2p} recovers an encoded 256-d latent vector based on a pure background patch I_{neg} so as to generate a synthesized patch I'_{pos} with lesion in I_{pos} and normal tissue in I_{neg} combined together. As can be seen in Fig. 2, LT-GAN is trained in a weakly-supervised and cycling fashion, which only requires image-level labels indicating containing lesions or not instead of pixel-level annotations of each lesion. Loss functions designed for training LT-GAN are detailed below.

Adversarial Loss: For stable training based on unpaired I_{pos} and I_{neg}, we utilize two discriminators $(D_{pos}$ and $D_{neg})$ and follow WGAN-GP [4] to measure W-distances between the fake and real samples in both domains:

$$W_{pos}(T_{n2p}, D_{pos}) = \sum_{x \in I_{pos}} [D_{pos}(x)] - \sum_{x' \in I'_{pos}} [D_{pos}(x')] - \mu_{pos} R_{pos} \quad (1)$$

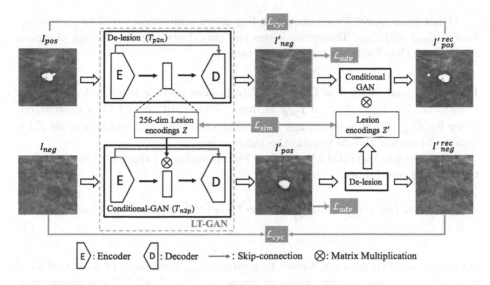

Fig. 2. An overview of our proposed LT-GAN training in a cycling fashion.

$$W_{neg}\left(T_{p2n}, D_{neg}\right) = \sum_{y \in I_{neg}} \left[D_{neg}(\mathbf{y})\right] - \sum_{y' \in I'_{neg}} \left[D_{neg}\left(\mathbf{y}'\right)\right] - \mu_{neg}R_{neg} \quad (2)$$

where T_{n2p} (T_{p2n}) aims to minimize the above objective against an adversary D_{pos} (D_{neg}) that tries to maximize it, $\mu_{pos}R_{pos}$ $(\mu_{neg}R_{neg})$ is the gradient penalty term. Accordingly, the loss function of adversarial training translators is:

$$\mathcal{L}_{adv} = W_{pos} + W_{neg} \quad (3)$$

Cycle-Consistency Loss: Same as Cycle-GAN, we train our LT-GAN in a cycling-fashion so as to well preserve background consistency between fake and real images. Reconstruction patches of two domains can be denoted as:

$$I'^{\,rec}_{pos} = T_{n2p}\left(T_{p2n}\left(I_{pos}\right), Z'\right), I'^{\,rec}_{neg} = T_{p2n}\left(T_{n2p}\left(I_{neg}, Z\right)\right) \quad (4)$$

The overall cycle-consistency loss can be expressed as:

$$\mathcal{L}_{cyc}\left(T_{n2p}, T_{p2n}\right) = \left\|I'^{\,rec}_{pos} - I_{pos}\right\|_1 + \left\|I'^{\,rec}_{neg} - I_{neg}\right\|_1 \quad (5)$$

Lesion Similarity Loss: It is desirable that lesion-relative latent vectors Z and Z' should share same information, while existing noises during training usually bring inconsistency and oppositely confuse the synthesis process. To alleviate noise of lesion encodings extracted from T_{p2n}, we build up a L1-based similarity loss $\mathcal{L}_{sim}$ between lesion encodings Z and Z':

$$\mathcal{L}_{sim} = \|Z - Z'\|_1 \quad (6)$$

Other than above losses shown in Fig. 2, we additionally add two auxiliary losses: $\mathcal{L}_{app}$ and $\mathcal{L}_{idt}$. These two losses can effectively alleviate wrong mapping results and thus facilitate our training process.

Identity Loss: We follow Cycle-GAN to impose specific identity loss for each translator. For T_{p2n}, if given I_{neg} as input, its output should keep unchanged. Then for T_{n2p}, if lesion encodings $\hat{Z}$ are extracted from I_{neg} or there is no Z for fusion, its output should remain the same as input.

We formulate the total identity loss by measuring L1 distance between identity patches and inputs:

$$\mathcal{L}_{idt} = \left\| T_{n2p}\left(I_{neg}, \hat{Z}\right) - I_{neg} \right\|_1 + \left\| T_{n2p}\left(I_{neg}\right) - I_{neg} \right\|_1 + \left\| T_{p2n}\left(I_{neg}\right) - I_{neg} \right\|_1 \tag{7}$$

Appearance Translation Loss: To improve T_{n2p}'s tolerance of noise and force it to effectively fuse specific lesion information from Z, a L1-based appearance translation loss is applied for consistency between $I'_{pos} - I_{neg}$ and $I_{pos} - I'_{neg}$, where these two difference terms represent lesion apperance information from I'_{pos} and I_{pos} respectively:

$$\mathcal{L}_{app} = \left\| \left(I'_{pos} - I_{neg}\right) - \left(I_{pos} - I'_{neg}\right) \right\|_1 \tag{8}$$

Our goal of training is to minimize sum of above objectives:

$$\mathcal{L} = \lambda_1 \mathcal{L}_{adv} + \lambda_2 \mathcal{L}_{cyc} + \lambda_3 \mathcal{L}_{idt} + \lambda_4 \mathcal{L}_{app} + \lambda_5 \mathcal{L}_{sim} \tag{9}$$

where λ_i (i=1,2,. . . ,5) are weights to control importance of different losses. We experimentally set λ_1=1, λ_2=1, λ_3=0.5, λ_4=40 and λ_5=1.

3   Experiment

3.1   Dataset and Implementation Details

To validate effectiveness of our proposed method on DVT learning, we conduct several experiments of cross-view μC matching and evaluate on two public mammography dataset, i.e., the INbreast (115 patients and 410 images in total) and the CBIS-DDSM (3103 images in total). We selected 38 (44) cases with a single lesion, and 9 (16) cases with multiple lesions in INbreast (CBIS-DDSM). Each case has paired CC and MLO view images. We preprocess the dataset by operating contrast enhancement and brightness normalization to reduce brightness variance across views.

For each ground-truth (GT) point of μC, we define a 200 * 200 square with the μC in the center as the GT mask. Candidate lesion patches are then generated by considering their Dice-Sørensen Coefficient (DSC) with the GT mask, i.e., true lesion patches if DSC > 0.25, and false lesion patches otherwise. For each

μC, we generate 7 true lesion patches and 3 false ones in each view. We pair these patches of two views, and obtain a positive pair which consists of true lesion patches of the same μC from two views, and a negative pair which is any other combination except two false lesion patches of two views.

Evaluations are based on 5-fold cross-validation and metrics are balanced-accuracy (i.e. b-Acc) and AUC for classifying positive and negative pairs. Each candidate patch is resized into 64 * 64 pixels. All matching networks are trained with the cross-entropy loss and optimized using Adam solver under a batch size of 64. The learning rate starts at 5e−5 and gradually decays by 0.9.

3.2   Ablation Study

First, we conduct an ablation study on INbreast to analysis impacts of two requirements of synthesis lesions, i.e., 1) the correlation of spatial locations between two views is correct, and 2) changes of appearance across two views are consistent with those of real lesions. We implement several methods on MatchNet [5] to compare including densifying supervision by randomly locations to synthesize (*Location inconsistent*), by Cycle-GAN to synthesize (*Appearance inconsistent*). To validate the visually realistic appearance of synthesized lesions, we also implement a supervision-densifying methods by simply placing calcification-like white dots across two views (*Unrealistic lesion*).

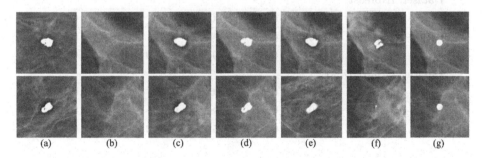

(a)   (b)   (c)   (d)   (e)   (f)   (g)

Fig. 3. Visualization of lesion generation using different methods. (a) and (b) are pairwise lesion and background GTs. (c) and (d) are respectively generated using our proposed approach and lesion-pasting, while (e)–(g) are respectively location inconsistent, appearance inconsistent and unrealistic lesion samples.

Visual examples of these four methods are shown in Fig. 3. Note that the locations of all examples except Fig. 3(e) are selected by our Gaussian model, which guarantees a correct spatial correlation between two views. Table 1 gives comparison results of the four methods, and our method achieves the best results for both b-Acc and AUC. By comparing b-ACC value of our method with the 1st and 2nd rows, we can see that 1) a correct spatial location can result in 1.971% improvement, and 2) appearance consistency can result in 2.948% improvement. By comparing the 2nd and 3rd rows, we surprisingly find that Cycle-GAN is even worse than placing calcification-like dots by 2.094% of b-ACC, which implies that the inconsistency of appearance dramatically confused the DVT learning, resulting in unreasonable performance of cross-view matching.

Table 1. An ablation study of our proposed supervision-densifying method on INbreast. We report average results of balanced-Acc (%) and AUC.

Method	b-Acc (%)	AUC
Location inconsistent	79.631	0.888
Lesion inconsistent	78.654	0.888
Unrealistic lesion	80.748	0.892
Ours	**81.602**	**0.894**

Table 2. Quantitative comparison of different DVT-learning methods on both INbreast and CBIS-DDSM dataset. We evaluate on three general architectures and use ResNet-101 [6] as their unified feature network.

Method	MatchNet [5]		2-channel [20]		SCFM [13]	
	b-Acc(%)	AUC	b-Acc(%)	AUC	b-Acc(%)	AUC
Dataset: INbreast						
Sparse supervision	77.651	0.853	77.900	0.838	77.119	0.851
Lesion-pasting	80.200	0.890	77.840	0.848	80.103	0.881
Ours	**81.602**	**0.894**	**79.085**	**0.859**	**81.179**	**0.891**
Dataset: CBIS-DDSM						
Sparse supervision	72.904	0.796	70.664	0.776	71.686	0.788
Lesion-pasting	73.805	0.823	68.488	0.749	72.401	0.789
Ours	**76.092**	**0.839**	**73.529**	**0.812**	**76.294**	**0.824**

3.3 Comparison with Other DVT Learning Methods

Table 2 shows final results of different DVT learning methods, including training under sparse supervision, lesion-pasting and our proposed denser supervision. Lesion pasting directly crops areas of real lesions using masks and paste on pure backgrounds. Note that lesion-pasting requires a GT mask of each lesion while our method is trained in a weakly-supervised fashion and only need the location of each lesion. To eliminate the negative impact of less training sample by sparse supervision, we partially select true and false lesion patches in each view for lesion-pasting and ours to make the number of pairs for training equal. As shown in Table 2, we implement three state-of-the-art methods [5,13,20] for comparison, each of which is trained three times based on sparse supervision, lesion-pasting and our method respectively.

Two observations can be made from the results. First, sparse supervision achieves the worse performance, indicating the effectiveness of densified supervision for DVT learning. More specifically, our method greatly surpasses 3.951% (3.188%) of b-ACC and 0.041 (0.034) of AUC (1st and 2nd columns) to the state-of-the-art methods of mammogram matching adopting MatchNet [1] under sparse supervision on INbreast (CBIS-DDSM). Second, comparing with lesion pasting, our proposed approach improves even greater. Despite visually similar of both methods as shown in Fig. 3(c) and (d), without employing adversarial techniques, lesion pasting fails to consider lesion-surrounding tissue variation (e.g., shade) which possibly acts as an important clue of DVT. Moreover, comparing with lesion pasting utilizing pixel-wise masks, our supervision-densifying method only exploits weaker supervision of lesion locations, which shows more convenience in clinical usage and greater universality on various tasks.

We have calculated statistical significance of our proposed method compared to sparse supervision and lesion pasting. On INbreast, improvements by our method are mostly significant on MatchNet and SCDM (p<0.05 except for AUC vs. lesion pasting), while on CBIS-DDSM, our method is significant on all baselines (p<0.05 except for b-Acc vs. lesion pasting on 2-channel network).

4   Conclusions

In this paper, we propose a novel supervision-densifying approach towards a robust DVT learning. Those supervision-densified synthetic lesions remain both correct correlation of spatial locations by a Gaussian model and appearance consistency across two views by our proposed LT-GAN. Experiment results demonstrate that our densified supervision strongly contributes to DVT learning and performs superiority in cross-view μC matching. Future works will explore more advanced location correspondence measuring techniques and pairwise lesion synthesis architectures.

Acknowledgement. This work was supported by the National Natural Science Foundation of China (6187241762061160490), the project of Wuhan Science and Technology Bureau (2020010601012167), the Open Project of Wuhan National Laboratory for Optoelectronics (2018WNLOKF025), the Fundamental Research Funds for the Central Universities (2021XXJS033).

References

1. Balntas, V., Johns, E., Tang, L., Mikolajczyk, K.: Pn-net: Conjoined triple deep network for learning local image descriptors. arXiv preprint arXiv:1601.05030 (2016)
2. Beatty, J.: The radon transform and the mathematics of medical imaging (2012)
3. Engeland, S.V., Timp, S., Karssemeijer, N.: Finding corresponding regions of interest in mediolateral oblique and craniocaudal mammographic views. Med. Phys. **33**(9), 3203–3212 (2006)

4. Gulrajani, I., Ahmed, F., Arjovsky, M., Dumoulin, V., Courville, A.: Improved training of wasserstein gans. arXiv preprint arXiv:1704.00028 (2017)
5. Han, X., Leung, T., Jia, Y., Sukthankar, R., Berg, A.C.: Matchnet: unifying feature and metric learning for patch-based matching. In: Proceedings of the IEEE Conference on Computer Vision and Pattern Recognition, pp. 3279–3286 (2015)
6. He, K., Zhang, X., Ren, S., Sun, J.: Deep residual learning for image recognition. In: Proceedings of the IEEE Conference on Computer Vision and Pattern Recognition, pp. 770–778 (2016)
7. Koch, G., Zemel, R., Salakhutdinov, R.: Siamese neural networks for one-shot image recognition. In: ICML Deep Learning Workshop, vol. 2. Lille (2015)
8. Lee, R.S., Gimenez, F., Hoogi, A., Miyake, K.K., Gorovoy, M., Rubin, D.L.: A curated mammography data set for use in computer-aided detection and diagnosis research. Sci. Data 4(1), 1–9 (2017)
9. Ma, J., Li, X., Li, H., Wang, R., Menze, B., Zheng, W.S.: Cross-view relation networks for mammogram mass detection. In: 2020 25th International Conference on Pattern Recognition (ICPR), pp. 8632–8638. IEEE (2021)
10. Melekhov, I., Kannala, J., Rahtu, E.: Siamese network features for image matching. In: 2016 23rd International Conference on Pattern Recognition (ICPR), pp. 378–383. IEEE (2016)
11. Moreira, I.C., Amaral, I., Domingues, I., Cardoso, A., Cardoso, M.J., Cardoso, J.S.: Inbreast: toward a full-field digital mammographic database. Acad. Radiol. 19(2), 236–248 (2012)
12. Perek, S., Hazan, A., Barkan, E., Akselrod-Ballin, A.: Siamese network for dual-view mammography mass matching. In: Stoyanov, D., et al. (eds.) RAMBO/BIA/TIA -2018. LNCS, vol. 11040, pp. 55–63. Springer, Cham (2018). https://doi.org/10.1007/978-3-030-00946-5_6
13. Quan, D., Fang, S., Liang, X., Wang, S., Jiao, L.: Cross-spectral image patch matching by learning features of the spatially connected patches in a shared space. In: Jawahar, C.V., Li, H., Mori, G., Schindler, K. (eds.) ACCV 2018. LNCS, vol. 11362, pp. 115–130. Springer, Cham (2019). https://doi.org/10.1007/978-3-030-20890-5_8
14. Ronneberger, O., Fischer, P., Brox, T.: U-Net: convolutional networks for biomedical image segmentation. In: Navab, N., Hornegger, J., Wells, W.M., Frangi, A.F. (eds.) MICCAI 2015. LNCS, vol. 9351, pp. 234–241. Springer, Cham (2015). https://doi.org/10.1007/978-3-319-24574-4_28
15. Simo-Serra, E., Trulls, E., Ferraz, L., Kokkinos, I., Fua, P., Moreno-Noguer, F.: Discriminative learning of deep convolutional feature point descriptors. In: Proceedings of the IEEE International Conference on Computer Vision, pp. 118–126 (2015)
16. Wilms, M., Krüger, J., Marx, M., Ehrhardt, J., Bischof, A., Handels, H.: Estimation of corresponding locations in ipsilateral mammograms: a comparison of different methods. In: Medical Imaging 2015: Computer-Aided Diagnosis, vol. 9414, p. 94142B. International Society for Optics and Photonics (2015)
17. Yan, Y., Conze, P.H., Lamard, M., Quellec, G., Cochener, B., Coatrieux, G.: Multi-tasking siamese networks for breast mass detection using dual-view mammogram matching. In: International Workshop on Machine Learning in Medical Imaging, pp. 312–321. Springer (2020)
18. Yang, H., Ciftci, U., Yin, L.: Facial expression recognition by de-expression residue learning. In: Proceedings of the IEEE Conference on Computer Vision and Pattern Recognition, pp. 2168–2177 (2018)

19. Yang, Z., et al.: MommiNet: mammographic multi-view mass identification networks. In: Martel, A.L., et al. (eds.) MICCAI 2020. LNCS, vol. 12266, pp. 200–210. Springer, Cham (2020). https://doi.org/10.1007/978-3-030-59725-2_20
20. Zagoruyko, S., Komodakis, N.: Learning to compare image patches via convolutional neural networks. In: Proceedings of the IEEE Conference on Computer Vision and Pattern Recognition, pp. 4353–4361 (2015)
21. Zhu, J.Y., Park, T., Isola, P., Efros, A.A.: Unpaired image-to-image translation using cycle-consistent adversarial networks. In: Proceedings of the IEEE International Conference on Computer Vision, pp. 2223–2232 (2017)

DeepOPG: Improving Orthopantomogram Finding Summarization with Weak Supervision

Tzu-Ming Harry Hsu[1(✉)] and Yin-Chih Chelsea Wang[2]

[1] MIT CSAIL, Cambridge, USA
stmharry@mit.edu
[2] Chung Shan Medical University, Taichung, Taiwan

Abstract. Clinical finding summaries from an orthopantomogram, or a dental panoramic radiograph, have significant potential to improve patient communication and speed up clinical judgments. While orthopantomogram is a first-line tool for dental examinations, no existing work has explored the summarization of findings from it. A finding summary has to find teeth in the imaging study and label the teeth with several types of past treatments. To tackle the problem, we develop DeepOPG that breaks the summarization process into functional segmentation and tooth localization, the latter of which is further refined by a novel dental coherence module. We also leverage weak supervision labels to improve detection results in a reinforcement learning scenario. Experiments show high efficacy of DeepOPG on finding summarization, achieving an overall AUC of 88.2% in detecting six types of findings. The proposed dental coherence and weak supervision are shown to improve DeepOPG by adding 5.9% and 0.4% to AP@IoU = 0.5. The dataset and code are made available online.

Keywords: Orthopantomogram · Dental panoramic radiograph · Reinforcement learning · Weak supervision

1 Introduction

An orthopantomogram (OPG), or a dental panoramic radiograph, is a half-circle X-ray scanning of the oral region that compresses the complicated 3D structures to a 2D representation as shown in Fig. 1. OPG has many advantages including short acquisition time and convenience of examination. Moreover, its capability to deliver rich information about the oral and maxillofacial regions makes it a first-line dental screening tool [1]. With that said, it is this structural complexity that unavoidably limits the interpretation of OPG to only dental experts [2].

Electronic supplementary material The online version of this chapter (https://doi.org/10.1007/978-3-030-87240-3_35) contains supplementary material, which is available to authorized users.

Even for these dental experts, interpretation of findings can suffer from insufficient inter-rater agreement [3] and low time efficiency in clinical practices [4,5]. As such, an automatic system to provide finding summaries on the fly can be beneficial in terms of both patient communication and clinical assistance. The systematically collected summaries can further provide an invaluable source for subsequent dental research and statistical analysis, which the current clinical workflow cannot offer.

There have been attempts to provide information about teeth in radiographs with convolutional neural networks (CNNs). In [6], they offer pixel-wise segmentation maps that label seven different parts of teeth. [7] classifies teeth images into eight categories but requires that the bounding boxes be manually annotated first. [8] identifies silhouettes for natural teeth in OPG with semantic segmentation but treats all teeth as a single connected region. [9,10] use a novel OPG dataset and object detection to treat teeth as individual instances for object detection, yet they both do not number the teeth. [11] addresses both detection and numbering, but fails to include dental implants. [12] provides detection of teeth, implant, and crowns but does not associate them with findings. Moreover, the vast majority of past research relies on annotations on dense attribute maps which is resource-intensive, and the use of weaker (and faster to collect) supervision has not been explored.

In this work, we aim to provide a summary of findings in an OPG image, including all teeth found in the image, their FDI notations, and all the clinical findings on each. We propose DeepOPG, which breaks the finding summarization process into two sub-tasks: functional segmentation and tooth localization, the latter of which is further refined at inference-time by maximizing the novel *Dental Coherence Reward (DCR)*. DCR can also be used by reinforcement learning (RL) for training-time optimization, leveraging the *missing teeth annotation* that are quick for dentists to label as weak supervision. We curate a set of annotations on OPG including semantic segmentation, instance segmentation, and finding summaries for 298 studies on a dataset in the public domain. Our experiments show that DeepOPG achieves an overall AUC of 88.2% on finding detection, which is 1.6% higher than without weak supervision. The tooth/implant localization yields an average precision at zero IoU of 98.6%, which is 5.6% higher than without injecting dental domain knowledge and 0.9% higher than without feeding in segmentation maps. The numbers demonstrate the effectiveness of each component of DeepOPG. To our knowledge, this is the first work to explore the summarization of findings in OPG images and to use weak supervision to improve finding summarization. To enable reproducible research and encourage subsequent works, we are making our dataset and code available.[1]

2   Methods

Our ultimate goal for the DeepOPG system is to generate a finding summary entailing six different types of findings on each tooth in a OPG. The resulting

---

[1] https://github.com/stmharry/deepopg.

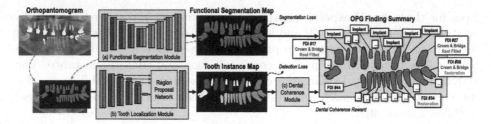

Fig. 1. System Overview of DeepOPG. There are three modules in DeepOPG working together to obtain orthopantomogram finding summary on each tooth. (a) Functional Segmentation Module performs per-pixel classification, (b) Tooth localization Module localizes each tooth, and (c) Dental Coherence Module ensures that output is clinically reasonable. Ultimately, we combine the segmentation maps and teeth identity maps to produce findings with explainable predictive values.

findings are formulated as binary attribute labels on the teeth found in the OPG. We decompose the problem into two main tasks: localizing the objects of interest (in this case, teeth and implants) and determining the visual features that result in the findings. Illustrated in Fig. 1, there are three modules in DeepOPG, and they operate at original resolutions of the images. This is essential since some findings (e.g., fillings found in the root canal) are visually tiny, and any down-sampling would result in a loss of information. We combine the results from both tasks of localization and function determination to output predictive values for each of the finding types on individual teeth.

2.1   Model Architecture

First of all, the *functional segmentation* module as shown in Fig. 1a consumes a radiographic image as the input and generates a map that shows the dental functionality of each pixel. The functional segmentation map and the original image is then concatenated to go into the *tooth localization* module in Fig. 1b that picks out the individual dental object of interest including teeth and implants. The resulting detection outcomes are further refined by the *dental coherence* module where clinical heuristics are applied to ensure coherence with dental knowledge.

Functional Segmentation via Semantic Segmentation. Given the input gray-scale image $I \in \mathbb{R}^{H \times W \times 1}$, where H and W are the height and width of the image, we employ a network with an U-Net-like [6] structure to predict $S \in \mathbb{R}^{H \times W \times C_{\text{seg}}}$, a per-class probability for each pixel that determines its functional class $c \in \{1, 2, \ldots, C_{\text{seg}}\}$, where C_{seg} is the number of classes. In our experiments, $C_{\text{seg}} = 7$ and includes the following classes for finding summarization: (1) background, (2) normal (non-impacted) teeth, (3) impaction, (4) crown & bridge, (5) restoration, (6) root filling material, and (7) implant. Note that as these classes are mutually exclusive, a ground truth segmentation map

$S^{\mathrm{gt}} \in \mathbb{R}^{H \times W \times C_{\mathrm{seg}}}$ is one-hot encoded and the activation function of network output is thus softmax. Specifically, we choose ResNet-50 [13] to be the encoder and ResNet-18 with transposed convolutions to be the decoder.

Tooth Localization via Object Detection. The tooth localization module takes the concatenated image $[I, S] \in \mathbb{R}^{H \times W \times (1+C_{\mathrm{seg}})}$ and produces N *detections*, each including a class probability vector $\mathbf{p}_n \in \mathbb{R}^{C_{\mathrm{det}}}$, a region of interest (ROI) $\mathbf{b}_n \in \mathbb{R}^4$, and the class-wise masks $M_n \in \mathbb{R}^{H \times W \times C_{\mathrm{det}}}$, where C_{det} is the number of classes in detection. Concretely, we adopt Mask-RCNN [14] that proposes a pool of candidate ROIs with a region proposal network (RPN) before using a small sub-network to derive the aforementioned detection properties.

As we are interested in not only natural teeth but also dental implants, in total there are $C_{\mathrm{det}} = 34$ classes representing the background, 32 different teeth in permanent dentition, and the implant. Hereafter the natural teeth are annotated using the FDI World Dental Federation notation as shown in Fig. 2a.

Inference-Time Dental Coherence Decoding. One major downside of directly using off-the-shelf detection algorithms is that they mostly consider the detection efficacy of individual objects rather than the conglomerate of several objects. As a result, in pilot experiments, we often observe the detection module to output several objects of interest with the same FDI tooth number, which is highly unlikely in practice. Equally frequently, there are cases where an image patch can be detected as multiple different classes at the same time, with largely overlapping masks. Even with existing techniques such as non-maximum suppression (NMS) that filters out overlapping objects with lower confidence scores, we are only able to partially resolve the latter problem.

To this end, we propose to look at this problem from an optimization perspective and *decode* an assignment $\mathbf{E}$ of teeth number to the detected objects by maximizing the Dental Coherence Reward (DCR) defined as

$$r_{\mathrm{DCR}}(\mathbf{P}, \mathbf{E}, \mathbf{M}) \equiv \sum_{n,c} p_{nc} \cdot e_{nc} - \sum_{n,c,m,d} q_{ncmd} \cdot e_{nc} \cdot e_{md}, \qquad (1)$$

subject to $\sum_n e_{nc} \leq 1 \; \forall c \in \{1, 2, \ldots, C\}$, and $e_{nc} \in \{0, 1\} \; \forall c \in \{1, 2, \ldots, C\} \, \forall n \in \{1, 2, \ldots, N\}$, where $p_{nc} = (\mathbf{p}_n)_c$ is the probability of object n belonging to class c, $q_{ncmd} = \frac{(M_n)_c \cap (M_m)_d}{(M_n)_c \cup (M_m)_d}$ is the intersection-over-union between masks $(M_n)_c$ (the class-c mask of object n) and $(M_m)_d$, and e_{nc} is an indicator whether we *assign* tooth c to object n. Note that an object n can be *suppressed* (*i.e.*, discarded) if $\sum_c e_{nc} = 0$. This formulation happens to be the *Generalized Quadratic Assignment Problem* (GQAP) [15] which is extensively studied in optimization theory and has solvers widely available. Implants are not modified in this module, and hence $C = C_{\mathrm{det}} - 1$ with implants excluded for optimization.

The idea to maximize DCR closely resembles how our dental experts parse an OPG, where they explain they would (1) identify all minimally overlapping objects and mentally assign each a number, followed by (2) ensuring that across a single image, no teeth share the same FDI number (obviously, multiple dental implants can still present simultaneously). While it is certainly possible in the clinics to observe the extremely rare cases where two natural teeth overlap on the OPG, oftentimes highly overlapping masks simply indicate that a tooth is independently recognized by two RPN proposals.

Explainable OPG Finding Summary. We assemble the information from the semantic segmentation and the detection outputs to derive the finding summary. For each of the teeth or implants, we use its mask M to select the corresponding regions in the segmentation map S and calculate the percentage of pixel counts for each functional class c in that area as $f_c = \sum_{i \in M} \frac{\mathbb{1}[S_i = c]}{|M|}$, where $\mathbb{1}[\cdot]$ is the indicator function. The percentage area f_c is then used as the predictive value for finding type c on that tooth. Doing so not only allows us to provide an explainable finding output that dentists can easily reason, but we also can adjust the threshold on f_c based on our sensitivity/specificity requirements.

2.2 Improved Tooth Localization with Weakly Supervised Reinforcement Learning

The annotation for tooth localization usually requires that the dental experts carefully outline the silhouettes of each tooth and provide an FDI number for it. This type of annotation is labor-intensive and is usually not available at most data registries. What is more likely to be available is a description of whether a tooth is missing or not in a text report (*i.e.*, $\sum_n e_{nc} = 1$ if the tooth c is present and 0 otherwise). We hereby are interested to find if weak supervision in the form of tooth missingness is helpful to train the tooth localization module in a reinforcement learning (RL) scenario.

We utilize the REINFORCE [16] algorithm where as long as a probability and a reward are defined for output, the network can learn to maximize the reward function. At training time, instead of decoding the GQAP problem, we sample an one-hot vector $\hat{\mathbf{e}}_n = [\hat{e}_{n1}, \hat{e}_{n2}, \ldots, \hat{e}_{nC}] \sim \mathbf{p}_n$ from the class distribution $\mathbf{p}_n$ for each object n independently. As the random samples might violate the constraint that each FDI number cannot be taken by multiple teeth (*i.e.*, $\sum_n e_{nc} > 1$), we penalize this situation by setting the reward for extra teeth to be negative

$$\hat{p}_{nc} = \begin{cases} +p_{nc} & \text{if tooth } c \text{ is present and } p_{nc} \text{ is the largest probability for it} \\ -p_{nc} & \text{otherwise (for extra teeth)}, \end{cases} \quad (2)$$

and calculate $r_{\text{DCR}}\left(\hat{\mathbf{P}}, \hat{\mathbf{E}}, \mathbf{M}\right)$ on the samples. The loss as given by REINFORCE is thus

$$\nabla_\theta \mathcal{L}_{\text{DCR}} = -\mathbb{E}_{\hat{\mathbf{E}} \sim p_\theta(\hat{\mathbf{E}})} \left[r_{\text{DCR}}\left(\hat{\mathbf{P}}, \hat{\mathbf{E}}, \mathbf{M}\right) \nabla_\theta \sum_{n,c} \hat{e}_{nc} \log p_{nc} \right], \quad (3)$$

where $p_\theta(\cdot)$ is the distribution characterized by the network. We can approximate the above gradient with Monte-Carlo samples and average gradients across training examples in the batch. Different from the aforementioned inference-time decoding, we can explicitly optimize the network for DCR with reinforcement learning here.

To learn DeepOPG, we employ a multi-stage learning procedure since the RPN in Mask-RCNN is non-differentiable. First, we train the functional segmentation module, optimizing the segmentation cross-entropy loss $\mathcal{L}_{seg}$. Following this, the tooth localization module learns using the inference-time predicted segmentation maps and minimizes a loss $\mathcal{L}_{det}$ as detailed in [14]. Finally, we fine-tune the tooth localization module with DCR weak supervision, minimizing the joint loss $\mathcal{L} = \mathcal{L}_{det} + \mathcal{L}_{DCR}$, freezing all network layers except for the last. For implementation details, please refer to the supplementary material.

3 Experiments

In this section, we provide validation of individual modules as well as DeepOPG as a whole. First of all, we present a dataset with novel annotations on segmentation, detection, and finding summary. We then offer an overview of the finding summarization efficacy for each of the finding types. Following this, we provide an ablation study on the tooth localization module including our proposed DCR decoding and reinforcement learning. Finally, we compare our DeepOPG with existing works under comparable settings. For brevity, the performance of the functional segmentation module is provided in the supplementary material.

3.1 Dataset

In this work, we use the UFBA-UESC Dental Images Deep dataset [9] where there are 1,500 OPG images in total, 267 out of which are annotated for tooth localization (implant annotations are not provided in the original data). The OPG images can be split into four major categories: (1) studies with all permanent dentition present and no implants, (2) studies with missing teeth and no implants, (3) studies with implants, and (4) studies with mixed dentition.

We exclude all studies with mixed dentition and supernumerary teeth as they are outside the scope of this work.

To enrich the dataset for learning DeepOPG, we ask 3 board-certified dentists to provide additional annotations including (1) functional segmentation maps on 68 studies, (2) tooth/implant localization maps on 39 studies, (3) tooth/implant missingness summary (weak supervision, in the form of 32 binary labels per study) on 144 studies, and (3) finding summary (in the form of 32×6 binary labels per study) on 47 studies.

To avoid overfitting the data, no study is annotated for two or more annotation types. It is important to note that segmentation/localization maps take, on average, 30 min to annotate per study, while the teeth missingness information only takes 30 s each. In each stage of DeepOPG learning, data is split into 70/30 training/test randomly, and the finding summary is exclusively used as test data.

Table 1. AUC Comparisons. We compare the AUCROC for six OPG finding types for two settings of DeepOPG. See Sect. 3.3 for method descriptions.

Method	AUCROC (%)						
	Missing Teeth	Impacted Teeth	w/Crown & Bridge	w/Restoration	Root Filled	Implants	Macro Avg.
DeepOPG (full)	**90.6**	**96.9**	86.5	**89.3**	**88.2**	77.6	**88.2**
w/o RL	87.6	96.5	**88.2**	86.4	82.9	**78.1**	86.6
Area Threshold @ max F1	24.2%	34.5%	25.9%	2.70%	0.33%	–	–

3.2   Overall Evaluation of DeepOPG for Findings Summarization

As mentioned before, DeepOPG combines the functional segmentation map and the tooth localization results by calculating the percentage area of each functional class for each tooth. Using the percentage area as the predictive value for the binary finding labels, we are able to evaluate the overall performance of DeepOPG by calculating the receiver operating characteristic (ROC) curve where we plot the true positive rate TPR $= \frac{TP}{TP+FN}$ against the true negative rate TNR $= \frac{TN}{TN+FP}$. Note that for a finding prediction to be TP, it not only has to have enough pixels of that finding in the tooth, but the tooth number itself has to be correctly detected.

The ROC curves for the six types of findings are shown in detail in the supplementary material. We can calculate the area under curve (AUC) for each of the findings as summarized in Table 1. In the table, we also compare a setting where the RL with DCR is disabled. It is clear that the weak supervisions with RL can improve the finding summarization. Of the six findings, impacted teeth with an AUC of 96.9% is the easiest task, possibly because it is a large object and that is often found at fixed locations such as the wisdom teeth. We also show, on the last row, the threshold on the percentage area at the operating point with the largest F1 $= \frac{2 \times TP}{2 \times TP+FN+FP}$. It is interesting to see that root-filled teeth only require 0.33% of the area to be finding-positive while impacted teeth require 34.5% of the tooth to be labeled impacted.

To highlight the usefulness of weak supervision, the "w/o RL" model (86.6% AUC) trains with 273 per-pixel annotations which take 136 expert hours to prepare. The "DeepOPG (full)" model adds 100 weak supervision annotations which only take an additional 0.8 expert hours, but a gain of 1.6% overall AUC. This demonstrates weak supervision is effective in boosting AUC while requiring substantially less expert effort (<1% extra time) than per-pixel annotations.

3.3   Tooth Localization with Dental Coherence

To verify the efficacy of the proposed modifications to the off-the-shelf object detection networks, we perform several ablation studies to inspect the contribution of these modifications. In particular, we assess

Table 2. Comparisons of Detection Metrics. We show detection metrics for various settings of DeepOPG. We report the metric values and their standard errors. AP_x denotes AP@IoU $= x$. See Sect. 3.3 for method descriptions.

Method	Per-Object				Per-Image
	$AP_{0.0}$ (%)	$AP_{0.5}$ (%)	DA (%)	FA (%)	IoU (%)
DeepOPG (full)	$\mathbf{98.6}_{0.1}$	$\mathbf{97.6}_{0.3}$	$\mathbf{98.7}_{0.4}$	$\mathbf{97.5}_{0.6}$	$\mathbf{80.5}_{1.5}$
w/o RL	$98.4_{0.1}$	$97.2_{0.4}$	$\mathbf{98.7}_{0.4}$	$\mathbf{97.5}_{0.6}$	$80.1_{1.5}$
w/o RL and dental coherence	$93.0_{0.1}$	$91.3_{0.4}$	$93.7_{0.9}$	$87.4_{1.2}$	$79.7_{1.6}$
w/o segmentation	$97.7_{0.1}$	$96.2_{0.3}$	$97.9_{0.5}$	$95.7_{0.8}$	$80.2_{1.5}$

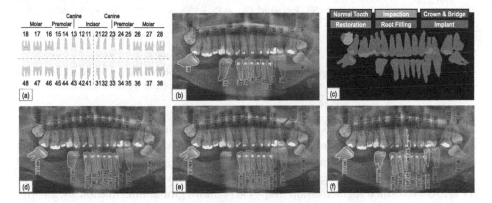

Fig. 2. Illustrations of Tooth localization Results. (a) The FDI tooth numbering system, (b) input OPG with ground truth localization, (c) functional segmentation map, (d) localization for DeepOPG (full), (e) localization w/o reinforcement learning, and (f) localization w/o segmentation input.

1. **DeepOPG (full):** We enable all model features, including feeding segmentation maps as the input for the tooth localization, using dental coherence module at inference, and training the model with reinforcement learning.
2. **w/o RL:** All model features, except training with RL.
3. **w/o RL and dental coherence:** We remove both the dental coherence module and the reinforcement learning components.
4. **w/o segmentation:** Segmentation maps are not fed into the tooth localization module in this case.

Metrics. The performance of different models are compared using various metrics, including the commonly used average precision (AP) defined in PASCAL VOC [17] for detection tasks, the detection accuracy DA $\equiv \frac{TP+FN}{TP+FN+FP}$ and identification accuracy FA $\equiv \frac{TP}{TP+FN+FP}$ [18]. On a per-image level, we evaluate the intersection over union as IoU $\equiv \frac{\sum_n M_n \cap M_n^{gt}}{\sum_n M_n \cup M_n^{gt}}$.

374   T.-M. H. Hsu and Y.-C. C. Wang

Table 3. Comparison to Prior Works. We compare our full model DeepOPG under similar conditions to prior works on various tasks. Note the evaluations are done on different dataset in each work. We report the metric values and their standard errors. AP_x denotes $AP@IoU = x$.

Method	Tooth Segmentation		
	Precision (%)	Recall (%)	F1 (%)
Wirtz et al. [19]	79.0	82.7	80.3
Jader et al. [10]	94_6	84_7	88_5
DeepOPG (Ours)	$90.7_{2.7}$	$90.6_{2.0}$	$90.6_{1.8}$

Method	Natural Tooth Detection				Implant Detection	
	Sensitivity[a] (%)	Precision[a] (%)	$AP_{0.5}$ (%)	$AP_{0.7}$ (%)	$AP_{0.5}$ (%)	$AP_{0.7}$ (%)
Tuzo et al. [11]	99.4	99.4	–	–	–	–
Kim et al. [12]	–	–	96.7	75.4	45.1	26.6
DeepOPG (Ours)	100.0	99.8	$97.6_{0.3}$	$89.4_{1.0}$	$75.0_{0.1}$	$75.0_{0.1}$

[a]They considered detection of teeth as 32 one-vs-all sub-problems. Even when a tooth is mis-labelled, it still is correct on 30 problems, and hence the high metrics.

In Table 2, we observe consistent gains in performance across all metrics when we incorporate different proposed features. Most notably, the dental coherence module constitutes most of the gains, providing +5.6% in AP@IoU=0.0 and +5.0% in DA. Using segmentation maps also provides +0.9% gain in AP@IoU=0.0 since segmentation maps carry more global information by nature. The weak supervision, while seemingly provides less compelling improvements, is, in fact, remarkable as annotating the teeth missingness summary is faster than annotating the localization maps by orders of magnitudes.

Figure 2 showcases localization results on a test image with three different configurations. This study contains a maloccluded tooth, on which all three configurations predict incorrectly. It is also worth noting that by removing the segmentation input, the localization depends totally on the input OPG and can be over-sensitive as indicated by red arrows in Fig. 2f.

3.4 Comparing Existing Works

Comparison of model performances across works suffers from not only dataset difference but also clinical task difference. While we are unable to obtain proprietary datasets from previous works for evaluation, we can set up DeepOPG to similar settings to allow fairer comparisons. For example, in Table 3, [19] and [10] tackled teeth-only segmentation, and hence we ignore error resulting from classes other than the teeth and the background in our segmentation module for a fair comparison. [11] and [12] addressed detection of natural teeth and implants, and thus we compare only detection results. Across all tasks except for precision in tooth segmentation, we are able to show superior performance.

Finally, for the *missing teeth* finding summary, [12] reached a sensitivity of 75.5% and a precision of 84.5% at a specificity of 80.4%. Under the same specificity, we have a sensitivity of 94.3% and a precision of 96.4%.

4   Conclusion

In this work, we provide an initial study, showing the possibilities to summarize findings for individual teeth from an orthopantomogram. By dividing the summarization process into two tasks: semantic segmentation and object detection, we can leverage weaker but faster-to-collect annotations to improve the detection model with reinforcement learning. The experiments demonstrate the efficacy of each module in the DeepOPG system, and, we hope to point the way for future works in this line and encourage dental imaging research.

References

1. Perschbacher, S.: Interpretation of panoramic radiographs. Aust. Dent. J. **57**, 40–45 (2012)
2. Henzler, P., Rasche, V., Ropinski, T., Ritschel, T.: Single-image tomography: 3d volumes from 2d cranial x-rays. In: Computer Graphics Forum, vol. 37, pp. 377–388. Wiley Online Library (2018)
3. Hye-In Kweon, H., Lee, J.-H., Youk, T., Lee, B.-A., Kim, Y.-T.: Panoramic radiography can be an effective diagnostic tool adjunctive to oral examinations in the national health checkup program. J. Periodontal Implant Sci. **48**(5), 317–325 (2018)
4. Plessas, A., Nasser, M., Hanoch, Y., O'Brien, T., Delgado, M.B., Moles, D.: Impact of time pressure on dentists' diagnostic performance. J. Dentistry 82, 38–44 (2019)
5. Rozylo-Kalinowska, I.: Artificial intelligence in dentomaxillofacial radiology: hype or future? J. Oral Maxillofacial Radiol. **6**(1), 1–1 (2018)
6. Ronneberger, O., Fischer, P., Brox, T.: Dental x-ray image segmentation using a u-shaped deep convolutional network. In: International Symposium on Biomedical Imaging (2015)
7. Miki, Y., Muramatsu, C., Hayashi, T., Zhou, X., Hara, T., Katsumata, A., Fujita, H.: Classification of teeth in cone-beam ct using deep convolutional neural network. Comput. Biol. Med. **80**, 24–29 (2017)
8. Koch, T.L., Perslev, M., Igel, C., Brandt, S.S.: Accurate segmentation of dental panoramic radiographs with u-nets. In: 2019 IEEE 16th International Symposium on Biomedical Imaging (ISBI 2019), pp. 15–19. IEEE (2019)
9. Silva, G., Oliveira, L., Pithon, M.: Automatic segmenting teeth in x-ray images: trends, a novel data set, benchmarking and future perspectives. Expert Syst. Appl. **107**, 15–31 (2018)
10. Jader, G., Fontineli, J., Ruiz, M., Abdalla, K., Pithon, M., Oliveira, L.: Deep instance segmentation of teeth in panoramic x-ray images. In: 2018 31st SIBGRAPI Conference on Graphics, Patterns and Images (SIBGRAPI), pp. 400–407. IEEE (2018)
11. Tuzoff, D.V., et al.: Tooth detection and numbering in panoramic radiographs using convolutional neural networks. Dentomaxillofacial Radiology **48**(4), 20180051 (2019)

12. Kim, C., Kim, D., Jeong, H.G., Yoon, S.-J., Youm, S.: Automatic tooth detection and numbering using a combination of a CNN and heuristic algorithm. Appl. Sci. **10**(16), 5624 (2020)
13. He, K., Zhang, X., Ren, S., Sun, J.: Deep residual learning for image recognition. In: Proceedings of the IEEE Conference on Computer Vision and Pattern Recognition, pp. 770–778 (2016)
14. He, K., Gkioxari, G., Dollár, P., Girshick, R.: Mask r-cnn. In: Proceedings of the IEEE International Conference on Computer Vision, pp. 2961–2969 (2017)
15. Lee, C.-G., Ma, Z.: The generalized quadratic assignment problem. Research Rep., Dept., Mechanical Industrial Eng., Univ. Toronto, Canada, p. M5S (2004)
16. Williams, R.J.: Simple statistical gradient-following algorithms for connectionist reinforcement learning. Mach. Learn. **8**(3–4), 229–256 (1992)
17. Everingham, M., Van Gool, L., Williams, C.K.I., Winn, J., Zisserman, A.: The pascal visual object classes (voc) challenge. Int. J. Comput. Vision **88**(2), 303–338 (2010)
18. Cui, Z., Li, C., Wang, W.: Toothnet: automatic tooth instance segmentation and identification from cone beam ct images. In: Proceedings of the IEEE/CVF Conference on Computer Vision and Pattern Recognition, pp. 6368–6377 (2019)
19. Wirtz, A., Mirashi, S.G., Wesarg, S.: Automatic teeth segmentation in panoramic X-Ray images using a coupled shape model in combination with a neural network. In: Frangi, A.F., Schnabel, J.A., Davatzikos, C., Alberola-López, C., Fichtinger, G. (eds.) MICCAI 2018. LNCS, vol. 11073, pp. 712–719. Springer, Cham (2018). https://doi.org/10.1007/978-3-030-00937-3_81

Joint Spinal Centerline Extraction and Curvature Estimation with Row-Wise Classification and Curve Graph Network

Long Huo[1,2], Bin Cai[1,2], Pengpeng Liang[3], Zhiyong Sun[1], Chi Xiong[1,2,5], Chaoshi Niu[1,2,5], Bo Song[1], and Erkang Cheng[1,4(✉)]

[1] Institute of Intelligent Machines, HFIPS, Chinese Academy of Sciences, Hefei, China
ekcheng@iim.ac.cn
[2] University of Science and Technology of China, Hefei, China
[3] School of Information Engineering, Zhengzhou University, Zhengzhou, China
[4] Nullmax, Shanghai, China
[5] The First Affiliated Hospital of USTC, Division of Life Sciences and Medicine, University of Science and Technology of China, Hefei, China

Abstract. Spinal curvature estimation plays an important role in adolescent idiopathic scoliosis (AIS) evaluation and treatment. The Cobb angle is a well-established standard for spinal curvature estimation. Recent studies of Cobb angle estimation usually rely on detection of vertebra landmarks which requires complex post-processing of curvature calculation. Approaches directly regressing the Cobb angles apply entire image or centerline segmentation results as the network input, which limits exploring the specific curve structure of the spine. In this paper, we propose a deep learning-based approach to simultaneously estimate spine centerline and spinal curvature with shared convolutional backbone. The spine centerline extraction is formulated as a row-wise classification task. To directly regress Cobb angles, we adopt curve graph convolution to exploit curve structure of the spine centerline. In addition, given a spine centerline, a Curve Feature Pooling (CFP) module is designed to aggregate features used as the input of Curve Graph Network (CGN) to regress the Cobb angles. We evaluate our method on the accurate automated spinal curvature estimation (AASCE) challenge 2019, and the proposed approach achieves promising results on both spine centerline extraction and Cobb angles estimation tasks.

Keywords: Curve graph network · Curve feature pooling · Row-wise classification

1 Introduction

Analysis of spine structure information is an essential step for spinal screening and examination that assist pathological progression evaluation, therapies, and

M. de Bruijne et al. (Eds.): MICCAI 2021, LNCS 12905, pp. 377–386, 2021.
https://doi.org/10.1007/978-3-030-87240-3_36

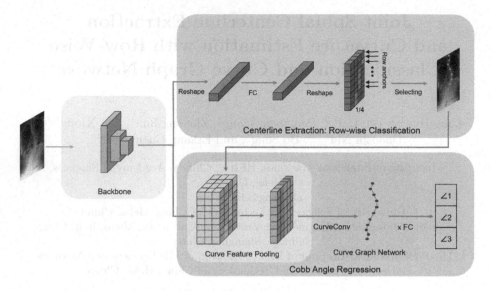

Fig. 1. Overview of the proposed network. The network has two branches, spine centerline extraction branch and Cobb angle estimation branch. The spine centerline estimation is formulated to a row-wise classification problem. Cobb angle estimation is obtained by a Curve Graph Network (CGN) with Curve Feature Pooling (CFP) module.

disease diagnosis. Patients with serious adolescent idiopathic scoliosis (AIS) have chances of being disabled if they are not treated in time. AIS is also a threat mostly in children and teenagers. In clinical studies, scoliosis is often defined as a curvature of the spine and the Cobb angle is the current clinical standard for AIS assessment. Therefore, the accuracy of Cobb angle is especially important for AIS treatment. Automated Cobb angle estimation is challenging due to complex structure of the spine from patient to patient and the variations of imaging quality.

Benefiting from the progress of deep learning techniques, automated Cobb angle estimation and spine centerline extraction have advanced in recent years. We categorize deep learning-based Cobb angle estimation approaches into two categories: (1) Curvature-based Cobb angle estimation that depends on vertebra landmarks [1,4,12,17] or spine centerline [2,16]. These approaches usually require complex post-processing of curvature calculation which is difficult to generalize. (2) Directly estimating the Cobb angles by a neural network [5,13,14]. Several works also jointly output spine centerline and Cobb angle estimation. For example, spine centerline is the result of a separate branch [13] or the intermediate result of the network [5]. These regression networks take entire image or centerline segmentation masks as input, which limits exploring the specific curve structure of the spine.

In this paper, we propose to locate the spine centerline and estimate the Cobb angle simultaneously with shared network backbone. First, to enable the network to have larger receptive field by using global features and significantly reduces the complexity of the network, inspired by recent works on lane detection [9,11] for autonomous driving, we adopt the row-wise classification strategy for spine centerline extraction. Instead of segmenting every pixel of spine centerline in the image, our method finds locations of spine centerline with a set of pre-defined rows. Then, to explore curve structure characteristics of the spine centerline, we present a curve graph sub-network to regress Cobb angles. A curve convolution operator is designed to capture curve structure context. Rather than use spine centerline mask or original image as input to regress Cobb angles [5,13,14], the curve graph network takes features aggregated by curve feature pooling (CFP) module. Given a spine centerline, the CFP extracts features from the spine centerline locations in feature layers of the network. By using CFP module and curve graph convolution, the regression network can largely exploring the anatomic curve structure information of spinal centerline which defines the Cobb angles. Figure 1 demonstrates the overview of our method. We evaluate our method on the accurate automated spinal curvature estimation (AASCE) challenge 2019, and it achieves promising results on both spine centerline extraction and Cobb angels estimation tasks. The effectiveness of each proposed component is validated via ablation study.

To summarize, the contributions of this study are three folds:

- **Joint Network for Spinal Centerline Extraction and Cobb Angle Estimation:** We introduce a joint network that can directly estimate both spinal centerline and Cobb angles without the need of post-processing steps.
- **Spinal Angle Regression by Curve Graph Network:** In order to explore curve structure of the spine, we propose a curve graph network for Cobb angle estimation. The curve graph network consists of curve graph convolution and Curve Feature Pooling modules.
- **Spine Centerline Extraction by Row-wise Classification:** We formulate the spine centerline extraction task to an efficient row-wise classification problem, which has benefit of using global context and brings more accurate results.

2   Method

As shown in Fig. 1, the proposed network consists of two sub networks, a spine centerline extraction network and a curve graph network for Cobb angle estimation.

2.1   Spine Centerline Extraction by Row-Wise Classification

In computer vision and autonomous driving domain, deep segmentation methods dominate the lane detection fields. For example, a neural network is designed to capture long-range context information of lane lines [7]. Recently, instead of dense segmentation, researchers treat lane detection task as a row-wise classification problem for lane detection [9,11]. The row-wise classification network

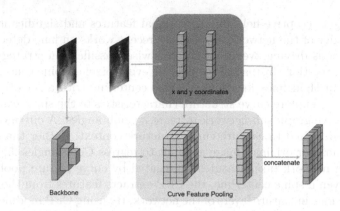

Fig. 2. Curve Feature Pooling (CFP). Given a spine centerline curve (a set of points), features from backbone are firstly pooled from these points of interest (PoI). Spatial coodinates of PoI are then concatenated to form the final output.

is computation efficient and brings global context feature to curve structure extraction. By observing that spine centerline has inherent curve structure, we also formulate spine centerline extraction as a row-wise classification task.

Throughout, we follow the definition of the terms of UltraFast [9]. Spine centerline is described as a set of horizontal locations at predefined rows, i.e., row anchors, and the location of each row is divided into many cells. The detection of spine is to find the corresponding cells over predefined row anchors. Mathematically, a feature map is defined as $X \in \mathbb{R}^{\{C \times h \times w\}}$ with height h and width w . The prediction of centerline can be written as:

$$p_{i,:} = f_i(X), \ s.t. \ i \in [1, h], \tag{1}$$

where $p_{i,:}$ is predicted probability of $(w + 1)$ gridding cells for the i-th row, and f_i is a centerline point location classifier. We set the gridding cells number to (w+1) instead of w by adding an extra dimension to indicate the absence of the centerline. Given the one-hot label of centerline on the i-th row anchor $t_{i,:}$, cross entropy loss L_{CE} is used for row-wise classification optimization:

$$L_{cls} = \sum_{i=1}^{h} L_{CE}(p_{i,:}, t_{i,:}) \tag{2}$$

Because of the continuity of spine centerline, centerline points should be close to each other in adjacent row. Therefore, the similarity loss L_{sim} [9] is also applied.

$$L_{sim} = \sum_{i=1}^{h-1} \|P_{i,:} - P_{i+1,:}\| \tag{3}$$

In this way, the total loss of row-wise centerline prediction can be denoted as:

$$L_{centerline} = L_{cls} + \lambda_1 L_{sim} \tag{4}$$

In this formulation, our method predicts the probability distribution of spine centerline on each row with global features, then the correct location of spine centerline can be computed based on the probability output. In addition, compared to segmentation task, our formulation is efficient due to smaller size of the output by gridding. Also, as in [9], row-wise classification directly ouputs centerline locations using expectation value of the prediction probability without the need of complex post-processing steps. Spine centerline locations of the input resolution are obtained by simple interpolation from adjacent rows.

2.2 Curve Feature Pooling

RoI pooling [3] is applied to extract feature in a rectangle region of interest for object detection task. In order to provide a curve structure representation of the spine centerline, we introduce a Curve Feature Pooling (CFP) module. Similar to RoI pooling, given a spine centerline, CFP simply pools features by projection on each point of interest (PoI). We find that spatial location information is important for Cobb angle estimation. Inspired by **CoordConv** operator [6,15], we directly use normalized pixel coordinates as additional features of PoI.

Figure 2 illustrates the CFP operator. A backbone generates feature maps $F \in \mathbb{R}^{C \times h \times w}$ from an input image. Then, each spine centerline point is projected onto the feature maps to pool features. Typically, a spine centerline is defined as $S = \{p_n\}_{n=1}^{N}$. Thus, a centerline point p_n has its corresponding feature vector $F'(p_n) \in \mathbb{R}^{C}$ by pooling from F that carries local feature information. Combing the coordinate information as **CoordConv** operator, the final feature has dimension of $C + 2$. Finally, a feature map $F_{curve} \in \mathbb{R}^{(C+2) \times N}$ is caculated by the proposed CFP module.

2.3 Curve Graph Network (CGN)

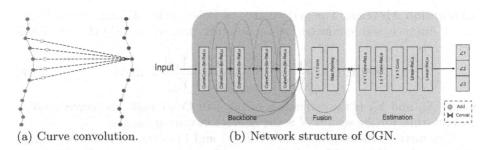

(a) Curve convolution.   (b) Network structure of CGN.

Fig. 3. Curve Graph Network. (a) Curve convolution is a standard 1d conv for a curve-like graph. (b) Curve graph network is built with three components: a backbone, a fusion module and a angle estimation head.

Having representation of spline centerline by the CFP module, we then present to use graph network for the Cobb angle regression. In order to learn the structure feature of object contour, DeepSake [8] applies circular convolution which better exploits the cycle-graph structure of a contour compared against generic graph convolution. Since spine centerline is a open-end curve, we introduce curve graph netwok (Fig. 3) to explore its spline-like structure.

Curve convolution is a standard 1d conv operator for a curve-like graph, as illustrated in Fig. 3. The bright green nodes are the input features defined on a curve by CFP, the yellow nodes represent the kernel function, and the dark green nodes are the output features which has the same dimension of the input. Detailed curve graph network is listed in Fig. 3. Similar to [8], the curve graph network has three components: a backbone, a fusion block, and a angle estimation head. The backbone is built of several "CurveConv-Bn-ReLU" layers with residual skip connections, where "CurveConv" is the curve convolution. The fusion module is applied to fuse the context across all curve points at multiple scales by a 1×1 convolution layer and a max pooling. The angle estimation head uses 1×1 convolution layers followed by FC layers and outputs the Cobb angles. The smooth $\ell1$ loss is adopted as the regression loss L_{angle}.

3   Experiments

3.1   Dataset and Evaluation

Dataset. The public dataset of 609 spinal anterior-posterior X-ray images, which is used as the training set in the Accurate Automated Spinal Curvature Estimation (AASCE) challenge 2019, is used for the evaluation of our method. All of the 609 images are split as the training set and testing set with 481 and 128 images respectively by the organizers. In addition, there are 98 blind testing images in the challenge. There are 68 ground truth landmarks corresponding to 4 corners of 17 vertebrae and 3 Cobb Angles in each image. The centerline ground truth is calculated by connecting the midpoints of vertebrae landmarks.

Evaluation Metrics. The performance of Cobb angle estimation is evaluated by $\ell1$ distance and symmetric mean absolute percentage (SMAPE) which is defined as:

$$\text{SMAPE} = \frac{200}{N} \sum_{i=1}^{N} \frac{\sum_{j=1}^{3} \|X_{ij} - Y_{ij}\|}{\sum_{j=1}^{3} (X_{ij} + Y_{ij})} \times 100\%, \tag{5}$$

where X_{ij} and Y_{ij} is the estimation of the j-th Cobb angle and corresponding ground truth for the scan i. N is the number of testing images.

Two metrics, Identificaiton Rate (id_{rate}) and Localization Distance (d_{mean}) [10], are adopted to evaluate the accuracy of spine centerline prediction. d_{mean} can be computed as:

$$d_{\text{mean}} = \frac{1}{M} \sum_{i=1}^{M} \|\hat{P}_i - P_i\|_2, \tag{6}$$

where $\hat{P}_i$ and P_i represent the i-th predicted point and the groundtruth point, respectively. M is the number of the predicted points. id_{rate} is the percentage points whose Euclidean distance between prediction and ground truth is less than a threshold τ, which can be written as:

$$\text{id}_{\text{rate}} = \frac{\sum \mathbb{1}\{\|\hat{P}_i - P_i\|_2 \leq \tau\}}{M} \times 100\% \tag{7}$$

3.2 Implementation Details

We implement our method in Pytorch on a NVIDIA RTX 2080TI GPU. The backbone of the network is Resnet-34. Only random rotation and translation is adopted as the augmentation strategy. The network is optimized by Adam with an initial learning rate 1e−4, and weight decay of 1e−4. We resize the images to a fixed size of 1024 × 512 and train the network for 500 epochs. The output of spine centerline network is of size $h = 256$ and $w = 129$, which gives 256 row anchors. Then we uniformly sample 128 points in CFP module to construct input for the curve graph network. $\lambda_1 = 1$ is used in the experiments. The kernel size of curve convolution is set to 9. In addition, we firstly train the centerline sub-network and then only tune parameters of the network related to the angle regression task. Spine centerline prediction results of the centerline branch are reported as the joint network by using the parameters saved in the first step.

3.3 Result and Analysis

The results of Cobb angles estimation in the AASCE 2019 challenge are reported in Table 1. The top ranking SAMPE results using 98 blind testing cases are included. Although not using the same testing images in SMAPE evaluation, our method achieves comparable SMAPE score with top ranking approaches, 1.2 lower than the winner method. By evaluating each Cobb angle on the same 128 testing images, our method outperforms the winner method [5].

Table 1. Cobb angle performance on AASCE 2019 Challenge.

Ranking	Team	SMAPE		Angles			
		#test data	value	#test data	Angle1	Angle2	Angle3
1	X [5]	98	21.71	128	4.72	5.80	5.66
2	iFLYTEK [1]	98	22.17	–	–	–	–
3	Erasmus MC [2]	98	22.96	–	–	–	–
4	vipsl [17]	98	24.80	–	–	–	–
5	JLD [12]	98	25.48	–	–	–	–
N/A	Bidur Khanal [4]	98	25.69	–	–	–	–
N/A	Top-Down [16]	98	26.05	–	–	–	–
N/A	Ours	128	**20.51**	128	**4.26**	**4.87**	**5.18**

We also perform experimental comparison of different input type for the Cobb angle regression sub-network. Experiment results are listed in Table 2. Our proposed CFP module gives the best prediction results on all the evaluation metrics.

Simply using coordinate as input provides worst results due to no image context is applied in the regression network. Compare to taking coordinate as input, using image feature F learned by the backbone yileds better angle estimation results on Angle1 and SMAPE metrics. Our CFP which embeds spatial location information achieves more accurate results than the one without coordinate information. The results validate that spatial information of spine centerline points is critical for Cobb angle estimation.

Table 2. Ablation study on input types for Cobb angle regression. F is the feature map generated by the network backbone.

Input Type	Angle1	Angle2	Angle3	SMAPE
F	4.80	7.17	5.87	26.28
Coord	6.25	5.20	5.72	25.19
CFP w/o Coord	5.13	7.39	7.07	29.74
CFP	**4.26**	**4.87**	**5.18**	**20.51**

Table 3. Segmentation vs. Row-wise Classification. Evaluation of centerline extraction. Seg: segmentation; Clf: row-wise classification; Clf+Seg: Row-wise classification with an auxiliary segmentation output.

Method	d_{mean}	id_{rate}						GFLOPs	Params
		$\tau = 6$	$\tau = 8$	$\tau=10$	$\tau = 12$	$\tau = 15$	$\tau = 20$		
Seg	7.82	18.84	27.13	29.59	33.81	37.93	41.49	93.22	**24.68**
Clf	6.22	57.73	70.82	79.77	86.00	91.40	95.14	**38.48**	97.35
Clf+Seg	**6.03**	**59.28**	**72.15**	**81.07**	**87.13**	**92.25**	**95.82**	93.30	100.74

In addition, we study the effectiveness of the row-wise classification for spine centerline detection. As shown in Table 3, by reformulating the spine centerline extraction as a row-wise classification task, the row-wise based method provides superior prediction accuracy than the segmentation based method and significantly reduces the complexity of the network. The segmentation network is the same as the one in [9]. Segmentation results are recovered back to original image resolution using bilinear interpolation and argmax operation is applied to find spine centerline. The centerline performance of our joint network can be recognized as the Clf result since we freeze the classification sub-network and then update parameters related to the regression network. As in [9], we also construct a row-wise classification network with an auxiliary segmentation task. The Clf+Seg combination gives more accurate results than each single task.

4 Conclusion

In this paper, we propose a joint neural network which automatically estimates both spine centerline and the Cobb angle. The proposed curve graph sub-network and curve feature pooling module give superior results on Cobb angle estimation by exploring spine structure information. The row-wise based classification sub-network provides a efficient spine centerline detector. Experiments and analysis on a public dataset shows the effectiveness of our method.

Acknowledgments. This work is supported in part by the grant of NSFC (61804100, 61973294, 61806181), KRDP of Anhui Province (201904a05020086) and CAS (GJTD-2018-15).

References

1. Chen, K., Peng, C., Li, Y., Cheng, D., Wei, S.: Accurate automated keypoint detections for spinal curvature estimation. In: Cai, Y., Wang, L., Audette, M., Zheng, G., Li, S. (eds.) CSI 2019. LNCS, vol. 11963, pp. 63–68. Springer, Cham (2020). https://doi.org/10.1007/978-3-030-39752-4_6
2. Dubost, F., et al.: Automated estimation of the spinal curvature via spine centerline extraction with ensembles of cascaded neural networks. In: Cai, Y., Wang, L., Audette, M., Zheng, G., Li, S. (eds.) CSI 2019. LNCS, vol. 11963, pp. 88–94. Springer, Cham (2020). https://doi.org/10.1007/978-3-030-39752-4_10
3. Girshick, R.: Fast R-CNN. In: Proceedings of the IEEE International Conference on Computer Vision, pp. 1440–1448 (2015)
4. Khanal, B., Dahal, L., Adhikari, P., Khanal, B.: Automatic cobb angle detection using vertebra detector and vertebra corners regression. In: Cai, Y., Wang, L., Audette, M., Zheng, G., Li, S. (eds.) CSI 2019. LNCS, vol. 11963, pp. 81–87. Springer, Cham (2020). https://doi.org/10.1007/978-3-030-39752-4_9
5. Lin, Y., Zhou, H.-Y., Ma, K., Yang, X., Zheng, Y.: Seg4Reg networks for automated spinal curvature estimation. In: Cai, Y., Wang, L., Audette, M., Zheng, G., Li, S. (eds.) CSI 2019. LNCS, vol. 11963, pp. 69–74. Springer, Cham (2020). https://doi.org/10.1007/978-3-030-39752-4_7
6. Liu, R., et al.: An intriguing failing of convolutional neural networks and the coord-conv solution. arXiv preprint arXiv:1807.03247 (2018)
7. Pan, X., Shi, J., Luo, P., Wang, X., Tang, X.: Spatial as deep: spatial CNN for traffic scene understanding. In: Proceedings of the AAAI Conference on Artificial Intelligence, vol. 32 (2018)
8. Peng, S., Jiang, W., Pi, H., Li, X., Bao, H., Zhou, X.: Deep snake for real-time instance segmentation. In: Proceedings of the IEEE/CVF Conference on Computer Vision and Pattern Recognition, pp. 8533–8542 (2020)
9. Qin, Z., Wang, H., Li, X.: Ultra fast structure-aware deep lane detection. arXiv preprint arXiv:2004.11757 (2020)
10. Sekuboyina, A., et al.: Verse: A vertebrae labelling and segmentation benchmark for multi-detector CT images (2020)
11. Tabelini, L., Berriel, R., Paixão, T.M., Badue, C., De Souza, A.F., Olivera-Santos, T.: Keep your eyes on the lane: attention-guided lane detection. arXiv preprint arXiv:2010.12035 (2020)
12. Tao, R., Xu, S., Wu, H., Zhang, C., Lv, C.: Automated spinal curvature assessment from X-ray images using landmarks estimation network via rotation proposals. In: Cai, Y., Wang, L., Audette, M., Zheng, G., Li, S. (eds.) CSI 2019. LNCS, vol. 11963, pp. 95–100. Springer, Cham (2020). https://doi.org/10.1007/978-3-030-39752-4_11

13. Wang, J., Wang, L., Liu, C.: A multi-task learning method for direct estimation of spinal curvature. In: Cai, Y., Wang, L., Audette, M., Zheng, G., Li, S. (eds.) CSI 2019. LNCS, vol. 11963, pp. 113–118. Springer, Cham (2020). https://doi.org/10.1007/978-3-030-39752-4_14

14. Wang, S., Huang, S., Wang, L.: Spinal curve guide network (SCG-Net) for accurate automated spinal curvature estimation. In: Cai, Y., Wang, L., Audette, M., Zheng, G., Li, S. (eds.) CSI 2019. LNCS, vol. 11963, pp. 107–112. Springer, Cham (2020). https://doi.org/10.1007/978-3-030-39752-4_13

15. Wang, X., Kong, T., Shen, C., Jiang, Y., Li, L.: SOLO: segmenting objects by locations. In: Vedaldi, A., Bischof, H., Brox, T., Frahm, J.-M. (eds.) ECCV 2020. LNCS, vol. 12363, pp. 649–665. Springer, Cham (2020). https://doi.org/10.1007/978-3-030-58523-5_38

16. Zhao, S., Wang, B., Yang, K., Li, Y.: Automatic spine curvature estimation by a top-down approach. In: Cai, Y., Wang, L., Audette, M., Zheng, G., Li, S. (eds.) CSI 2019. LNCS, vol. 11963, pp. 75–80. Springer, Cham (2020). https://doi.org/10.1007/978-3-030-39752-4_8

17. Zhong, Z., Li, J., Zhang, Z., Jiao, Z., Gao, X.: A coarse-to-fine deep heatmap regression method for adolescent idiopathic scoliosis assessment. In: Cai, Y., Wang, L., Audette, M., Zheng, G., Li, S. (eds.) CSI 2019. LNCS, vol. 11963, pp. 101–106. Springer, Cham (2020). https://doi.org/10.1007/978-3-030-39752-4_12

LDPolypVideo Benchmark: A Large-Scale Colonoscopy Video Dataset of Diverse Polyps

Yiting Ma[1], Xuejin Chen[1,2(✉)], Kai Cheng[1], Yang Li[3], and Bin Sun[3]

[1] National Engineering Laboratory for Brain-inspired Intelligence Technology and Application, University of Science and Technology of China, Hefei, China
xjchen99@ustc.edu.cn
[2] Institute of Artificial Intelligence, Hefei, China
[3] The First Affiliated Hospital of Anhui Medical University, Hefei, China

Abstract. Computer-Aided Diagnosis (CAD) systems for polyp detection provide essential support for colorectal cancer screening and prevention. Recently, deep learning technology has made breakthrough progress in medical image computation and computer-aided diagnosis. However, the deficiency of training data seriously impedes the development of polyp detection techniques. Existing fully-annotated databases, including CVC-ClinicDB, ETIS-Larib, CVC-Colon dataset, Kvasir-Seg, and CVC-ClinicVideoDB, are very limited in polyp size and shape diversity, which is far from the significant complexity in the actual clinical situation. In this paper, we propose LDPolypVideo, a large-scale colonoscopy video database that contains a variety of polyps and more complex bowel environments. Our database contains 160 colonoscopy videos and 40,266 frames in total with polyp annotations, which are four times the size of the largest existing colonoscopy video database CVC-ClinicVideoDB. In order to improve the efficiency of polyp annotation, we design an intelligent annotation tool based on object tracking. Extensive experiments have been conducted to evaluate state-of-the-art object detection approaches on our LDPolypVideo dataset. The average drops of Recall and Precision of four SOTA approaches on this dataset are 26% and 15%, respectively. The great performance drop demonstrates the significant challenges but also the great value of our large-scale and diverse polyp video dataset to facilitate the research on polyp detection. Our dataset is available at https://github.com/dashishi/LDPolypVideo-Benchmark.

Keywords: Colonoscopy video · Large-scale dataset · Polyp detection · Tracking-assisted annotation

1 Introduction

Colorectal cancer (CRC) is one of the leading causes of death worldwide [17]. Polyps are considered as precursors to colorectal cancer. Early detection of

Electronic supplementary material The online version of this chapter (https://doi.org/10.1007/978-3-030-87240-3_37) contains supplementary material, which is available to authorized users.

© Springer Nature Switzerland AG 2021
M. de Bruijne et al. (Eds.): MICCAI 2021, LNCS 12905, pp. 387–396, 2021.
https://doi.org/10.1007/978-3-030-87240-3_37

polyps can efficiently reduce the cancer rate. Video colonoscopy is now commonly used for polyp detection. However, the detection rate of polyps highly depends on the doctor's skill and experience on colonoscopy screening. Polyps are easily missed due to complex intestinal environments, specular reflections, and limited field of view.

In order to improve the detection rate of polyps, the computer-aided diagnosis has attracted great attention from researchers. Many approaches have been proposed to apply computer vision algorithms to the automatic detection of polyps in colonoscopy images or videos. Earlier studies explore hand-crafted features including shape, color, and texture for polyp detection and diagnosis [1,20,23,24]. Recently, convolutional neural networks (CNN) have gained much more attention in medical image computation. Utilizing common object detection networks with colonoscopy imagery, many polyp detection approaches have been developed [5,19,21,22]. However, learning-based approaches significantly rely on training data. Acquiring large-scale colonoscopy images with accurate polyp annotation is arduous due to the large diversity of polyps and the requirement of expertise. A few small-scale datasets that contain high-quality colonoscopy images have been built for polyp detection in single colonoscopy images [2,4,18]. Hyper-Kvasir [6] is the largest dataset for gastrointestinal disease detection. It contains 73 videos that are classified as polyp but no localization annotations of the polyps. Thus it does not support supervised methods for polyp localization. Among it, a subset of 1000 noncontinuous polyp images, called as Kvasir-Seg, contain polyp masks and can be used for polyp localization.

Real clinical colonoscopy videos, however, have more diverse appearances and lower image quality with motion blurs, which pose significant challenges for these CNN-based models on accurate and continuous polyp detection. In order to improve the detection rate of polyps in colonoscopy videos, some studies exploit temporal correlations to guide continuous polyp detection [13,26,27]. These approaches have achieved promising results on existing colonoscopy video datasets that are not large and diverse enough for both training and comprehensive evaluation. We make an extensive survey on existing colonoscopy datasets for polyp detection and list the parameters in Table 1. The widely used video dataset, CVC-ClinicVideoDB, only contains 18 videos with annotations and 18

Table 1. Summary of public annotated colonoscopy datasets. Our LDPolypVideo dataset is the largest one up to now.

Dataset	Format	Resolution	N_{images}	N_{videos}	N_{polyps}
CVC-ClinicDB [2]	Image	384×288	612	29	29
CVC-ColonDB [4]	Image	574×500	380	15	15
ETIS-Larib [18]	Image	1225×966	194	34	44
Kvasir-Seg [6]	Image	Various	1000	N/A	N/A
CVC-ClinicVideoDB [1,3]	Video	384×288	11945	18	18
LDPolypVideo	Video	560×480	40266	160	200

videos without polyp annotations. Besides, there is at most one polyp in each frame of the CVC-ClinicVideoDB dataset. In contrast, it is a common case that multiple polyps occur at the same time in clinical examination.

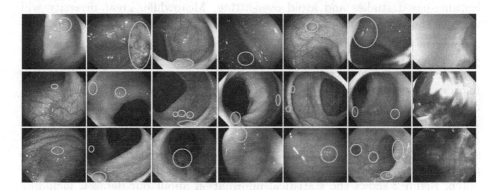

Fig. 1. Examples of images and polyp annotations in our LDPolypVideo dataset. The last column shows a few challenging frames with motion blur.

In order to increase the size and diversity of colonoscopy data for training and evaluation of polyp detection approaches, we present a large-scale and diverse colonoscopy video dataset named LDPolypVideo. It consists of 160 videos with 40,266 frames, nearly four times the size of the largest existing fully-annotated dataset. There are 33,884 frames that contain at least one polyp and in total 200 labeled polyps, which are more than 11 times the polyps in CVC-ClinicVideoDB. The polyps present more diverse morphologies. Figure 1 shows a set of example images and the annotated masks for polyps. Besides, we also provide 103 videos, including 861,400 frames without full annotations. Each video has a label indicating whether it contains polyps. These videos enrich the data diversity and will support unsupervised and semi-supervised methods. Based on our LDPolypVideo dataset, we evaluate a number of state-of-the-art approaches for polyp detection to analyze their strengths and weaknesses, demonstrating the challenges of colonoscopy polyp detection in clinical examination.

In summary, our contributions mainly are two-folds:

1. We propose a large-scale colonoscopy video dataset, LDPolypVideo, about four times the size of the existing largest database. We provide a better benchmark for polyp detection.
2. In order to improve the efficiency of data labeling, we develop an intelligent labeling tool based on a video tracking algorithm, which greatly reduces the labeling effort and can be used for other datasets.
3. Based on our LDPolypVideo dataset, we conduct extensive experiments to analyze the challenges of polyp detection in practical clinical data.

2   Our LDPolypVideo Dataset

Our goal to establish a colonoscopy polyp dataset consists of two key points, **large scale** and **diversity**. A large-scale dataset can meet the demand of deep learning-based studies and avoid over-fitting. Meanwhile, great diversity will improve the practical significance of trained models in the clinic.

2.1   Data Acquisition

In order to get a large-scale dataset, we collected clinical colonoscopy videos from routine clinical examinations at a state hospital. All metadata about the patients were removed. An experienced clinician selects video clips that contain polyps. After cutting off the black borders, we resize the video frames to 560 × 480. We sampled 160 videos with 40,266 frames while ensuring that there is at least one polyp in each video. There are in total 33,884 frames that contain at least one polyp. Figure 2 shows the statistical information about our dataset, including the number of frames (blue bars), the number of positive frames where there is at least a polyp (orange bars), and the total number of polyps in each video.

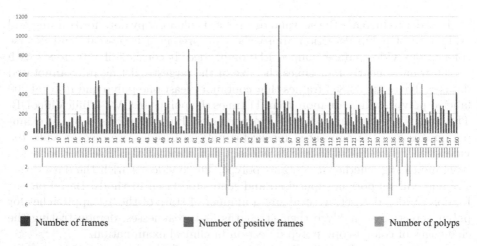

Fig. 2. Summary of our LDPolypVideo dataset.

We increase the diversity of colonoscopy images in two aspects. First, motion blur caused by camera movement, colon folds, and intestinal peristalsis frequently occur in clinical colonoscopy. However, previous colonoscopy image datasets only contain high-quality images and distinct polyps. As Fig. 3 shows, motion blurs and specular reflections bring great disturbance for polyp detection. In the second row, the polyp appears in continuous frames of a video but is difficult to detect at the beginning due to colon folds. In this case, temporal context plays a critical role for accurate detection. Second, there might be multiple polyps in a single frame. In our LDPolypVideo dataset, 22 videos contain multiple polyps, as shown in Fig. 1 and Fig. 2. Particularly, three videos in our dataset contain five polyps, which are very challenging to completely detect, even for an experienced clinician. In this situation, a robust computer-aided polyp detection tool is valuable to improve the detection rate in polyp screening.

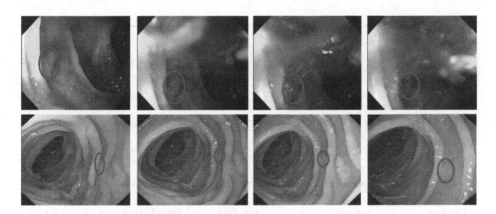

Fig. 3. Video frames with motion blur and reflections (top) and a polyp example that is difficult to detect due to colon folds.

2.2 Tracking-Assisted Annotation Tool

The most labor-intensive but important work in establishing a database is polyp annotation in all frames. Instead of asking experts to manually label the polyps frame by frame, we develop an intelligent annotation tool by combining manual labeling with object tracking algorithms. Considering efficiency and accuracy, we choose the Kernelized correlation filtering method [11] as the tracker. It trains a correlation filter from two adjacent frames and calculates the correlation of a frame with the previous labeled frame to predict the presence probability of the labeled object. We select the point with the highest probability as the candidate position in the new frame. Utilizing this tracker, for each video, we ask an expert to label a polyp when it first appears and the tracker automatically predicts its position in the next frame. The expert only needs to check whether the tracking result is acceptable. If the tracking result is not satisfied, manual annotation is performed on this new frame and tracking continues. Figure 4 illustrates our annotation process. After a round of labeling for the entire video, we ask another expert to verify the annotations and make modifications if necessary. This tracking-assisted annotation pipeline is very effective and efficient. We tested our tool on two videos in HyperKvasir. For the 894 frames in these two videos, the polyps in 681 frames can be automatically annotated and only 80 frames require manual re-annotation, which shows the usefulness of our tool on other datasets.

3 Experiments on LDPolypVideo Dataset

The large size and diversity of our LDPolypVideo dataset pose challenges for polyp detection. We conduct a series of experiments to comprehensively evaluate state-of-the-art (SOTA) detection approaches and demonstrate the challenges of clinical colonoscopy polyp screening.

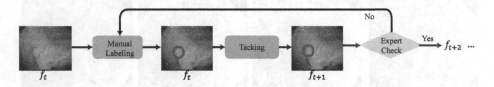

Fig. 4. The process of tracking-assisted polyp annotation in colonoscopy videos.

Table 2. Performance of recent polyp detection approaches on existing datasets.

Dataset	Method	Baseline	Recall	Precision	F1	F2
ETIS-Larib	Shin et al. [16]	Faster RCNN	0.80	0.73	0.78	0.78
	Wittenberg et al. [25]	Mask RCNN	0.80	**0.87**	**0.82**	**0.82**
	Brandao et al. [7]	FCN	**0.83**	0.74	0.81	0.81
	Sornapudi et al. [19]	FPN+RCNN	0.74	0.77	0.74	0.74
CVC-ColonDB	Sornapudi et al. [19]	FPN+RCNN	**0.92**	**0.90**	**0.97**	**0.91**
	Brandao et al. [7]	FCN	0.90	0.80	0.85	0.88
CVC-ClinicDB	Wang et al. [22]		**0.88**	**0.93**	**0.91**	**0.89**
	Wittenberg et al. [25]	Mask RCNN	0.86	0.80	0.82	0.85
ASU-Mayo	Yu et al. [26]	3D-FCN	0.71	0.88	0.78	0.73
	Zhang et al. [27]	Yolo	0.72	**0.89**	**0.9**	0.74
	Shin et al. [16]	Faster RCNN	**0.84**	0.83	0.83	**0.84**
CVC-ClinicVideo	Shin et al. [16]	Faster RCNN	**0.84**	**0.89**	**0.87**	**0.85**
	Qadir et al. [13]	Faster RCNN	0.81	0.87	0.84	0.83

We first summarize recent approaches for colonoscopy polyp detection on existing datasets in Table 2. Generally, the higher performance is achieved for polyp detection on the colonoscopy image datasets CVC-ColonDB [4] and CVC-ClinicDB [2] compared to the results on video polyp detection. On the video dataset CVC-ClinicVideo [1,3], Both Shin et al. [16] and Qadir et al. [13] use Faster R-CNN [15] as the backbone network. The former adopts post learning to fine-tune the detector with detected positives in the test video and the latter exploits temporal dependency between frames to reduce the false positives.

Following these methods, we evaluate four commonly used object detection networks, including Faster R-CNN [15], Yolo3 [14], RetinaNet [12], and Center-Net [9], on our LDPolypVideo dataset. We choose ResNet-50 [10] as the backbone for Faster R-CNN, RetinaNet, and CenterNet, while Yolo3 adopts DarkNet [14] as its backbone. We do not implement any post-training on specific videos for a fair comparison. We train all detection networks on CVC-ClinicDB, which consists of 612 colonoscopy images with single polyps. To prevent overfitting, data augmentation including image rotation, flipping, scaling, shearing, and motion blur is performed. All the models are pre-trained on ImageNet [8] and then fine-tuned on the augmented dataset. We test these four trained detection models on both CVC-ClinicVideo and our LDPolypVideo dataset. The results are shown in Table 3. Overall, RetinaNet performs the best on the two datasets. Compared to CVC-ClinicVideo which only contains 18 videos, the performance of these approaches significantly degrades on our LDPolypVideo dataset. The average drops of Recall and Precision are 26% and 15%, respectively.

Table 3. Performance comparison on CVC-ClinicVideo and LDPolypVideo datasets.

Methods	CVC-ClinicVideo				LDPolypVideo			
	Recall	Precision	F1	F2	Recall	Precision	F1	F2
Faster RCNN [15]	0.61	0.83	0.70	0.65	0.30	0.68	0.42	0.34
Yolo3 [14]	0.45	0.82	0.58	0.49	0.29	0.67	0.41	0.33
RetinaNet [12]	0.64	0.85	0.73	0.67	0.47	0.65	0.55	0.50
CenterNet [9]	0.57	0.98	0.72	0.62	0.17	0.88	0.28	0.20

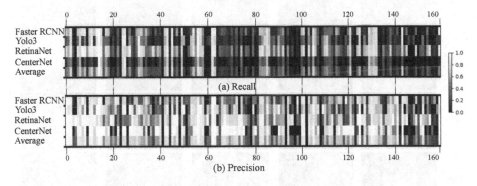

Fig. 5. Recall and precision of four detection networks on 160 videos in our dataset. Darker colors represent poorer performances.

In order to analyze the challenges of our LDPolypVideo dataset for SOTA models trained on colonoscopy image datasets, we analyze their performance on each individual video and visualize the precision and recall in Fig. 5. It shows that the performance varies in a wide range for the 160 videos. We select several typical challenging cases and analyze their common characteristics. Figure 6 shows nine representative cases. For the first two cases (a)(b), the small polyps are very difficult to detect. None of the four detectors succeed in detecting them at any frame. For the cases (c) to (f), the four detectors fail to detect the polyps in successive frames because of specular reflections, small size, motion blur, colon folds, etc. The polyp in (c) is not detected when it is partially visible with indistinguishable appearances with the background. In case (d), a colorectal ileocecal valve is detected by Yolo3 by mistake. In cases (e) and (g), there are multiple polyps in the same video. They are very challenging for the detectors to completely localize them, especially for small polyps. We also show typical false positives in Fig. 6 (h)(i). All detectors are easily deceived by the colorectal ileocecal valves and colon folds which look similar to polyps.

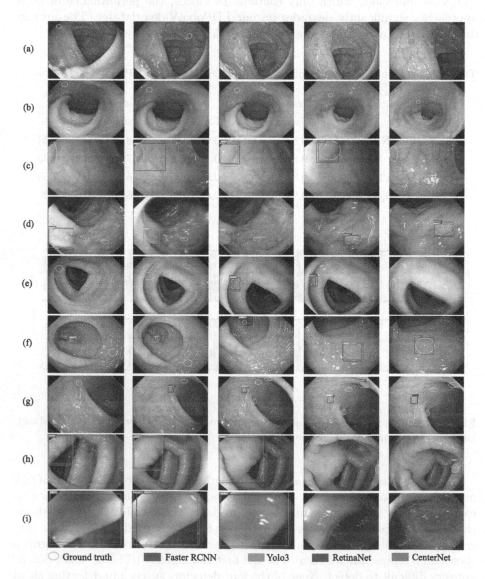

Fig. 6. Representative cases in our dataset and the detection results of SOTA methods.

4 Conclusion

In this paper, we present a large-scale colonoscopy video dataset, named LDPolypVideo, with polyp annotations in all frames. This new dataset contains 160 colonoscopy videos that are decomposed into 40,266 frames with 200 polyps. Among them, there are 33,884 frames that contains at least one polyp. We developed an efficient labeling tool based on object tracking to facilitate the labeling process. The large dataset size, great diversity in polyp morphology, multiple polyps in single frames, motion blur, and specular reflections in the colonoscopy videos pose significant challenges for polyp detection. We conducted extensive experiments to evaluate a number of state-of-the-art detection approaches and comprehensively analyze the characteristics of our dataset. We believe that our LDPolypVideo dataset provides a new benchmark and will facilitate researches on computer-aided polyp detection and diagnosis.

Acknowledgment. This work was supported by the National Natural Science Foundation of China under Grants 61976007 and 62076230.

References

1. Angermann, Q., et al.: Towards real-time polyp detection in colonoscopy videos: Adapting still frame-based methodologies for video sequences analysis. In: Proceedings of the International Workshop Comput. Assisted Robot. Endoscopy Clinical Image-Based Procedures, pp. 29–41 (2017)
2. Bernal, J., et al.: Wm-dova maps for accurate polyp highlighting in colonoscopy: validation vs. saliency maps from physicians. Comput. Med. Imaging Graph **43**, 99–111 (2015)
3. Bernal, J., et al.: Polyp detection benchmark in colonoscopy videos using gtcreator: a novel fully configurable tool for easy and fast annotation of image databases. In: Proceedings of 32nd CARS Conference (2018)
4. Bernal, J., Snchez, J., et al.: Towards automatic polyp detection with a polyp appearance model. Pattern Recognition **45**(9), 3166–3182 (2012)
5. Billah, M., Waheed, S., et al.: An automatic gastrointestinal polyp detection system in video endoscopy using fusion of color wavelet and convolutional neural network features. Int. J. Biomed. Imaging, 1–9 (2017)
6. Borgli, H., et al.: Hyper-kvasir: a comprehensive multi-class image and video dataset for gastrointestinal endoscopy. Sci. Data **7**(283), 1–14 (2020)
7. Brandao, P., Zisimopoulos, O., et al.: Towards a computed-aided diagnosis system in colonoscopy: automatic polyp segmentation using convolution neural networks. Journal of Medical Robotics Research **3**(02), 1840002 (2018)
8. Deng, J., Dong, W., other: Imagenet: A large-scale hierarchical image database. In: 2009 IEEE conference on computer vision and pattern recognition. pp. 248–255. Ieee (2009)
9. Duan, K., Bai, S., et al.: Centernet: keypoint triplets for object detection. In: Proceedings of the IEEE/CVF International Conference on Computer Vision, pp. 6569–6578 (2019)
10. He, K., Zhang, X., et al.: Deep residual learning for image recognition. In: Proceedings of the IEEE Conference on Computer Vision and Pattern Recognition, pp. 770–778 (2016)

11. Henriques, J., et al.: High-speed tracking with kernelized correlation filters. IEEE Trans. Pattern Anal. Mach. Intell. **37**, 583–596 (2015)
12. Lin, T., Goyal, P., et al.: Focal loss for dense object detection. In: Proceedings of the IEEE International Conference on Computer Vision, pp. 2980–2988 (2017)
13. Qadir, H., et al.: Improving automatic polyp detection using cnn by exploiting temporal dependency in colonoscopy video. IEEE J. Biomed. Health Inf. (2019)
14. Redmon, J., Farhadi, A.: Yolov3: An incremental improvement. arXiv preprint arXiv:1804.02767 (2018)
15. Ren, S., He, K., et al.: Faster r-cnn: towards real-time object detection with region proposal networks. IEEE Trans. Pattern Anal. Mach. Intell. **39**(6), 1137–1149 (2016)
16. Shin, Y., et al.: Automatic colon polyp detection using region-based deep cnn and posting learning approached. IEEE Access **6**, 40950–40962 (2018)
17. Siegel, R., Miller, K., Jemal, A.: Cancer statistics. CA A Cancer J. Clinicians **69**(1), 7–34 (2019)
18. Silva, J., Histace, A., et al.: Toward embedded detection of polyps in wce images for early diagnosis of colorectal cancer. Int. J. Comput. Assisted Radiol. Surg. **9**(2), 283–293 (2013)
19. Sornapudi, S., Meng, F., Yi, S.: Region-based automated localization of colonoscopy and wireless capsule endoscopy polyps. Appl. Sci. **9**(12) (2019)
20. Tajbakhsh, N., et al.: Automated polyp detection in colonoscopy videos using shape and context information. IEEE Trans. Med. Imaging **35**(2), 630–644 (2016)
21. Tajbakhsh, N., et al.: Convolutional neural networks for medical image analysis: full training or fine tuning? IEEE Trans. Med. Imaging **35**(5), 1299–1312 (2016)
22. Wang, P., Xiao, X., et al.: Development and validation of a deep-learning algorithm for the detection of polyps during colonoscopy. Nature Biomed. Eng. **2**(10), 741–748 (2018)
23. Wang, S., Tao, J., et al.: Improved classifier for computer-aided polyp detection in ct colonography by nonlinear dimensionality reduction. Med. Phys. **35**(4), 1377–1386 (2008)
24. Wang, Z., et al.: Computer aided detection and diagnosis of colon polyps with morphological and texture features. In: Proceedings of SPIE - The International Society for Optical Engineering, vol. 5370, pp. 972–980 (2004)
25. Wittenberg, T., Zobel, P., et al.: Computer aided detection of polyps in whitelight-colonoscopy images using deep neural networks. Current Directions Biomed. Eng. **5**(1), 231–234 (2019)
26. Yu, Y., Chen, H., et al.: Integrating online and offline three-dimensional deep learning for automated polyp detection in colonoscopy videos. IEEE J. Biomed. Health Inf. **21**(1), 65–75 (2017)
27. Zhang, R., Zheng, Y., et al.: Polyp detection during colonoscopy using a regression-based convolutional neural network with a tracker. Pattern Recogn. **83**, 209–219 (2018)

Continual Learning with Bayesian Model Based on a Fixed Pre-trained Feature Extractor

Yang Yang[1,2], Zhiying Cui[1,2], Junjie Xu[1,2], Changhong Zhong[1,2], Ruixuan Wang[1,2(✉)], and Wei-Shi Zheng[1,2,3]

[1] School of Computer Science and Engineering, Sun Yat-sen University, Guangzhou, China
[2] Key Laboratory of Machine Intelligence and Advanced Computing, MOE, China
wangruix5@mail.sysu.edu.cn
[3] Pazhou Lab, Guangzhou, China

Abstract. Current deep learning models are characterised by catastrophic forgetting of old knowledge when learning new classes. This poses a challenge in intelligent diagnosis systems where initially only training data of a limited number of diseases are available. In this case, updating the intelligent system with data of new diseases would inevitably downgrade its performance on previously learned diseases. Inspired by the process of learning new knowledge in human brains, we propose a Bayesian generative model for continual learning built on a fixed pre-trained feature extractor. In this model, knowledge of each old class can be compactly represented by a collection of statistical distributions, e.g. with Gaussian mixture models, and naturally kept from forgetting in continual learning. Experiments on two skin image sets showed that the proposed approach outperforms state-of-the-art approaches which even keep some images of old classes during continual learning of new classes.

Keywords: Continual learning · Bayesian model · Generative approach

1 Introduction

Deep learning models, particularly convolutional neural networks (CNNs), have shown human-level performance in diagnosis of various diseases [1,6,12]. However, most intelligent diagnosis systems are limited to diagnosis of only one or a few diseases and cannot be easily extended once deployed, therefore cannot diagnose all diseases of certain tissue or organ (e.g., skin or lung) as medical specialists do. Since collecting data of all (e.g., skin or lung) diseases is challenging due to various reasons (e.g., privacy and limited data sharing), it is impractical to train an intelligent system diagnosing all diseases all at once. One possible solution is to make the intelligent system have the continual or lifelong learning ability, such that it can continually learn to diagnose more and more diseases

© Springer Nature Switzerland AG 2021
M. de Bruijne et al. (Eds.): MICCAI 2021, LNCS 12905, pp. 397–406, 2021.
https://doi.org/10.1007/978-3-030-87240-3_38

without resourcing (or resourcing few) original data of previously learned diseases [3]. However, current intelligent models are characterised by catastrophic forgetting of old knowledge when learning new classes.

Researchers have recently tried to reduce the catastrophic forgetting issue mainly in the natural image domain. One approach is to find model components (e.g., kernels in CNNs) crucial for old knowledge, and then try to change them as little as possible when learning new knowledge [9]. However, it would become increasingly more difficult to continually learn new knowledge because more and more kernels in CNNs become crucial for increasingly old knowledge. To make models more flexibly learn new knowledge, another approach tries to modify model structures by add new layers or kernels when learning new knowledge [16]. Knowledge distillation has also been widely used during learning new classes [4,7, 8,10,11,14], where the old knowledge is implicitly represented by soft outputs of the old classifier with stored small old images or new classes of images as inputs. An alternative approach is to train a generative model to produce enough number of synthetic training data for each old class when learning new classes [19,20].

However, almost all existing approaches modify the feature extraction part of the classifiers either in parameter values or in structures during continual learning of new classes. In contrast, humans seem to learn new knowledge by adding memory of the learned new information without modifying the (e.g., visual) perceptual pathway. Therefore, one possible cause to catastrophic forgetting in existing models is the change in the feature extraction part (corresponding to the perceptual pathway in human brains) when learning new knowledge. With this consideration, we propose a generative model for continual learning built on a fixed pre-trained feature extractor, which is different from all existing (discriminative) models. The generative model can naturally keep knowledge of each old class from forgetting, without storing original images of old classes or regenerating synthetic images during continual learning. Experiments on two skin disease classification tasks showed the proposed approach outperforms state-of-the-art approaches which even keep some images of old classes during continual learning.

2   A Generative Model for Continual Learning

The proposed method is inspired by two interesting findings in neuroscience. One finding is that most infants cannot form episodic memory before 3 years old [2,17], and the other finding is that humans continually form memory from infants to elderly people [13]. One hypothetical explanation is that the visual pathway in younger infant's brain might be rapidly changing with daily visual stimuli from surroundings and then become firm with little change since 3 years old or so. Humans can continually learn new visual knowledge through their whole lives probably because they formed new memories about the new knowledge, but without changing the visual pathway which works as a visual feature extractor. This could explain why current deep learning models are characterised by catastrophic forgetting of old knowledge, i.e., model parameters or model structures from the the feature extractor part are always changed to some extent in almost all continual learning approaches. With this consideration, we propose a human-like continual learning framework, i.e., first pre-training a feature extractor, then fixing the feature extractor and forming new memory for each new knowledge. In the following, we will introduce one general way to pre-train the feature extractor, one statistical method to represent the memory, and one Bayesian model to predict class of any new (test) data after continual learning each time.

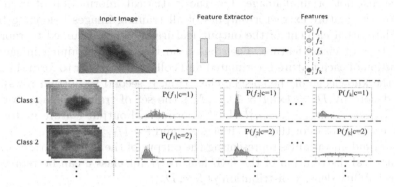

Fig. 1. Fixed pre-trained feature extractor (top) and memory formation (middle to bottom). Feature extractor is pre-trained and fixed during continual learning. Memory of each class is represented by a set of statistical distributions over features.

2.1   Fixed Pre-trained Feature Extractor

An ideal feature extractor should output two different feature vectors if two input data were visually different, meanwhile visually more similar inputs should result in more similar feature vectors from the feature extractor. The visual feature extractor (i.e., visual pathway) in younger infants are probably taught in certain self-supervised way, although the mechanism of self-supervision in infant brain has not been explicitly understood [15]. While it is worth exploring various self-supervised learning approaches (e.g., auto-encoder) to train a feature extractor, here we leave the self-supervision exploration for future work, and adopt a simpler but widely used approach, i.e., pre-training a CNN classifier with relatively large number images whose classes or domains are relevant but different from those in the task of interest, and then using the pre-trained CNN feature extractor (often consisting of all the convolutoinal layers) for the continual learning classification task of interest (Fig. 1, top row). It is expected that the pre-trained feature extractor would probably be powerful enough to discriminate different input images from the task of interest. Experiments showed that even such a simple approach to a fixed pre-trained feature extractor can already significantly help reduce catastrophic forgetting of old knowledge with the proposed generative approach. It is worth noting that, during continual learning of new classes in the classification task of interest, the pre-trained feature extractor is fixed and not updated. The knowledge of each learned new class is represented and stored as described in the following subsection.

2.2 Memory Formation

Different from the state-of-the-art continual learning approaches which often store a small number of original images for each old class, the proposed approach stores not original images but the statistical information of each class based on the feature extractor outputs of all training images belonging to the class. Here, each element of the output feature vector is assumed to represent certain type of visual feature. Then based on the class of training images, the distribution of each feature is estimated and collected together to form the memory of the knowledge of the specific class (Fig. 1, second to bottom rows). Formally, denote by $D_c = \{\mathbf{x}_i, i = 1, \ldots, N_c\}$ the set of training images for class c, $\mathbf{z}_i = [z_{i1}, z_{i2}, \ldots, z_{ik}, \ldots, z_{iK}]^\mathsf{T}$ the L_2-normalized output feature vector from the feature extractor for the input image $\mathbf{x}_i$, and $\mathbf{f} = [f_1, f_2, \ldots, f_k, \ldots, f_K]^\mathsf{T}$ the vector of random variables representing the output of the feature extractor, then the statistical distribution of the k-th feature f_k for class c can be represented by a probability density distribution $p(f_k|c, D_c)$,

$$p(f_k|c, D_c) = g(\{z_{ik}, i = 1, \ldots, N_c\}), \quad \forall k \in \{1, \ldots, K\} \tag{1}$$

where $g(\cdot)$ could be any appropriate distribution estimator. Here a Gaussian mixture model (GMM) with a small number of S components is adopted to represent $g(\cdot)$ for its simplicity. Since each Gaussian component can be compactly represented by its mean and standard deviation, totally only $2 \cdot S \cdot K$ numbers are stored in the memory to represent the knowledge of each class. D_c would be omitted from $p(f_k|c, D_c)$ in the following for simplicity.

2.3 Bayesian Model for Prediction

Based on the statistical distributions of visual features for each class, we propose a generative classification model based on the Bayesian rule for prediction. Given a test image $\mathbf{x}_j$, denote by $\mathbf{z}_j = [z_{j1}, z_{j2}, \ldots, z_{jk}, \ldots, z_{jK}]^\mathsf{T}$ the corresponding output from the feature extractor, and $p(c|\mathbf{z}_j)$ the probability of the test image belonging to class c. Then based on the Bayes rule, we can get

$$p(c|\mathbf{z}_j) = \frac{p(\mathbf{z}_j|c) \cdot p(c)}{\sum_{m=1}^{M} p(\mathbf{z}_j|m) \cdot p(m)}, \tag{2}$$

where M is the number of classes learned so far. Considering that potential correlations between certain feature components are probably caused by co-occurred visual parts of a specific class of objects, it can be assumed that different feature components f_k's are conditionally independent given specific class c. Then, the logarithm of Eq. 2 gives

$$\log p(c|\mathbf{z}_j) = \sum_k \log p(f_k = z_{jk}|c) + \log p(c) - \alpha, \tag{3}$$

where $\alpha = \log \sum_m p(\mathbf{z}_j|m)p(m)$ can be considered a constant for different classes. In Eq. (3), the likelihood function value $p(f_k = z_{jk}|c)$ for each feature element k can be directly obtained based on the previously stored knowledge $p(f_k|c)$ (Eq. 1) in the memory. The prior $p(c)$ for class c can be simply estimated based on the ratio of the number of training images for this class over the total number of training images of all learned classes so far, i.e., $p(c) = N_c / \sum_m N_m$. Note that in this case, the number of training images for each class needs to be stored in the memory such that $p(c)$ can be easily updated when new classes' knowledge is learned as above (Eq. 1). Based on Eq. (3), the class of the test image $\mathbf{x}_j$ would be directly predicted as the one with the highest value of $\log p(c|\mathbf{z}_j)$ over all classes learned so far.

The advantages of the the proposed approach over existing continual learning approaches are obvious. First, the knowledge of each old class is statistically represented by the set of likelihood functions (Eq. 1) and compactly stored in the memory. Therefore, old knowledge will not be forgotten over continual learning of new classes. In comparison, old knowledge will be inevitably and gradually forgotten over multiple rounds of continual learning in existing approaches, either due to the changes in feature extractor or due to the reduced number of original images to be stored in the limited memory. Second, the final performance of the proposed approach over multiple rounds of continual learning is not affected by the number of learning rounds and the number of new classes added in each round. In contrast in existing approaches, more rounds of continual learning with smaller number of new classes added each time would often lead to worse classification performance. Therefore, the proposed approach is more robust to various learning conditions with little forgetting of old knowledge.

3 Experimental Evaluations

3.1 Experimental Setup

The proposed approach was extensively evaluated on two medical skin image datasets. Skin7 [5] is a skin lesion dataset from the challenge of dermoscopic image classification held by the International Skin Imaging Collaboration (ISIC) in 2018. It consists of 7 disease categories, and each image is of size 600×450 pixels. This dataset presents severe class imbalance, with the largest class 60 times larger than the smallest one. Skin40 is a subset of 193 classes of skin disease images collected from the internet [18]. Skin40 contains two types of images, dermoscopic images which have relatively consistent imaging conditions (e.g., similar illumination) and therefore low levels of imaging noise, and clinical images captured mostly with digital cameras or mobile phones. The 40 classes with relatively more number of images (60 images per class) were chosen from the 193 classes to form the Skin40 dataset, while the remaining 153 classes (10 to 40 image per class) were used to train a CNN classifier whose final classification layer was then removed to form the fixed feature extractor for our approach or to be used as tunable feature extractor for baseline methods in most experiments. It is worth noting that there is no overlap between the 153 classes (for training the feature extractor in advance) and the classes in Skin7 and Skin40 (for continual learning evaluation).

During feature extractor training based on the 153 skin image classes, each image was randomly cropped within the scale range [0.8, 1.0] and then resized to 224 × 224 pixels, followed by random horizontal and vertical flipping. The mini-batch stochastic gradient descent (batch size 32) was used, with initial learning rate 0.01 and then divided by 10 at the 35th, 70th, and 105th epoch respectively. Weight decay (0.0005) and momentum (0.9) were also applied. The feature extractor was trained for 120 epochs with observed convergence.

In each experiment, multiple rounds of continual learning were performed, with a few (e.g., 2, 5) new classes learned each time. After each round of continual learning, the mean class recall (MCR) over all classes learned so far was calculated. The mean and standard deviation of MCR over five runs were reported, where the five orders of classes to be continually learned were fixed and used in all methods. Unless otherwise mentioned, ResNet-101 was used as the backbone for the feature extractor, and the dimension of feature vector K was 2048 and the number of Gaussian components in each GMM model was empirically set to 2 based on a small validation set for each dataset.

3.2   Effectiveness of the Generative Model

This section evaluates the effectiveness of the proposed approach by comparing with state-of-the-art strong baselines, including iCaRL [14], End-to-End Incremental Learning (End2End) [4], learning a Unified Classifier Incrementally via Rebalancing (UCIR) [8], Distillation and Retrospection (DR) [7], and Learning without Forgetting (LwF) [10]. The suggested hyper-parameter settings in the original work were adopted. In each round of continual learning, for the iCaRL, End2End, DR, and UCIR which need certain number of old data, the number of images stored (i.e., memory size) for all old classes is respectively 50 for Skin7 and 100 for Skin40. The memory size was chosen such that stored number of images for each class was only a small portion of the original training images at the last round of learning. An upper-bound result was also reported (Fig. 2, green star) by training a non-continual classifier with all training data.

Figure 2 shows that, with certain number of new classes to be continually learned at each round, the proposed approach always performs better than all the strong baselines particularly at later round of continual learning, although the same pre-trained feature extractor was used to fine-tune the CNN classifier for each baseline method. Even with more images of old classes stored for the representative strong baseline iCaRL, the proposed approach still performs better (Fig. 2, second row, last). The results also tell us the final-round performance of the proposed approach is not affected by the number of new classes to be learned each time. In comparison, the final performance of each baseline becomes worse with smaller number of new classes to be learned each time. These results clearly support that the proposed approach is more effective in keeping old knowledge from forgetting.

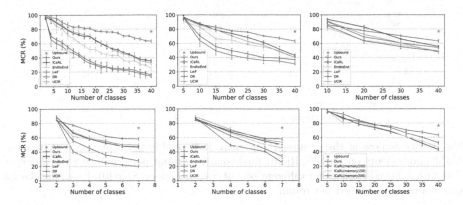

Fig. 2. Performance comparison on Skin40 and Skin7. First row (from left): learning 2, 5, 10 classes each time on Skin40. Second row (from left): learning 1 and 2 classes each time on Skin7, and comparison with iCaRL with varying memory size on Skin40. X-axis in each sub-figure represents the accumulated number of learned classes in the corresponding continual learning task.

Table 1. Performance on various feature extractor backbones. Results after last round of continual learning were reported, with 1 (Skin7) or 5 (Skin40) new classes per round.

Dataset	VGG19				ResNet18				ResNet34				ResNet101			
	LwF	iCaRL	IR	Ours	LwF	iCaRL	IR	Ours	LwF	iCaRL	IR	Ours	LwF	iCaRL	IR	Ours
Skin7	18.9	39.7	38.3	**46.5**	19.8	44.3	46.2	**55.6**	20.1	46.9	48.3	**56.8**	21.0	48.5	50.0	**58.4**
Skin40	27.4	33.6	32.5	**52.8**	30.4	41.8	37.1	**61.9**	31.1	42.3	39.5	**62.8**	31.2	43.1	40.2	**63.1**

3.3   Generalizability and Robustness of the Generative Model

The proposed approach is a general framework and therefore can employ different feature extractor backbones. As Table 1 shows, the proposed approach performs consistently better than strong baselines with different feature extractor backbones, supporting that the proposed approach is not limited to specific feature extractor structures. To evaluate the robustness of the generative model, the GMM with varying number of Gaussian components and different orders of classes to be continually learned were tried during continual learning. Figure 3 clearly shows that the generative model works stably well with different number of Gaussian components in GMM and with different class orders, with little change in performance. In comparison, the performance of the representative iCaRL varied a lot with different class orders. This is because knowledge of each old class is compactly stored and not changed throughout the whole process of continual learning by our approach. In comparison, all the strong baselines modify the feature extractor during continual learning, which would then change the representation of each stored old data and further change the representation of old knowledge, differently with different orders of classes to be learned.

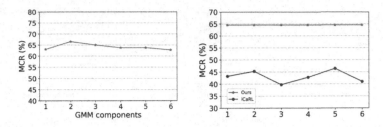

Fig. 3. Robustness of the proposed approach. Left: stable performance with varying GMM components on Skin40. Right: final-round learning performance with different class orders (x-axis indices for different sets of class orders on Skin40). Five new classes were continually learned per round.

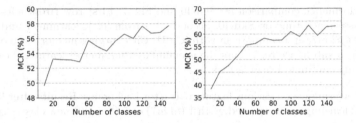

Fig. 4. Effect of feature extractor on continual learning. More classes (x-axis) used to train feature extractor result in better performance on Skin7 (left) and Skin40 (right).

3.4   Effect of Feature Extractor

The proposed approach is based on a fixed pre-trained feature extractor. To confirm that better feature extractors would help the generative model perform better in continual learning, the original 153 classes of skin images used for training the feature extractor were reduced gradually to only 10 classes, each time using such reduced number of classes to train the feature extractor and then the performance of the proposed approach at last round of continual learning on both the Skin7 and Skin40 datasets was calculated. Figure 4 does show that more classes used for training the feature extractor would generally result in better performance of the proposed approach. The feature extractor trained by more classes of data would probably have learned to extract more types of features and therefore could be more generalizable to a new but relevant domain. Consistent with the observation and explanation, when the feature extractor is fixed by random parameter weights (i.e., without any training), the classifier in continual learning showed the worst performance (MCR is 21% on Skin7, 6% on Skin40; not shown in Fig. 4). These results strongly suggest that exploring better ways to obtain a better feature extractor would further improve performance of the generative model in continual learning.

4 Conclusion

In this study, we propose a Bayesian generative model for continual learning of new classes. The model does not update the feature extractor but generates statistical information to represent knowledge of each class. Without storing any original data, the generative model can keep knowledge of each old class from forgetting and outperforms existing state-of-the-art approaches which often store small number of old data. The model is not limited to any specific feature extractor, and the final-round performance is not affected by the process of continual learning such as the number of new classes to be learned each time or the number of rounds of continual learning. Better pre-trained feature extractor could be explored to further improve the performance of the generative approach.

Acknowledgement. This work is supported in part by the National Natural Science Foundation of China (grant No. 62071502, U1811461), the Guangdong Key Research and Development Program (grant No. 2020B1111190001, 2019B020228001), and the Meizhou Science and Technology Program (grant No. 2019A0102005).

References

1. Ardila, D., et al.: End-to-End lung cancer screening with three-dimensional deep learning on low-dose chest computed tomography. Nature Medicine (2019)
2. Bauer, P.J.: A complementary processes account of the development of childhood amnesia and a personal past. Psychological Review (2015)
3. Baweja, C., Glocker, B., Kamnitsas, K.: Towards continual learning in medical imaging. In: NIPS Workshop (2018)
4. Castro, F.M., Marín-Jiménez, M.J., Guil, N., Schmid, C., Alahari, K.: End-to-End incremental learning. In: ECCV (2018)
5. Codella, N.C.F., et al.: Skin lesion analysis toward melanoma detection 2018: A challenge hosted by the international skin imaging collaboration (ISIC). CoRR abs/1902.03368 (2019). http://arxiv.org/abs/1902.03368
6. De Fauw, J., et al.: Clinically applicable deep learning for diagnosis and referral in retinal disease. Nature Medicine (2018)
7. Hou, S., Pan, X., Change Loy, C., Wang, Z., Lin, D.: Lifelong learning via progressive distillation and retrospection. In: ECCV (2018)
8. Hou, S., Pan, X., Loy, C.C., Wang, Z., Lin, D.: Learning a unified classifier incrementally via rebalancing. In: CVPR (2019)
9. Kirkpatrick, J., et al.: Overcoming catastrophic forgetting in neural networks. Proceedings of the National Academy of Sciences (2017)
10. Li, Z., Hoiem, D.: Learning without forgetting. IEEE Trans. Pattern Anal. Mach. Intell. **40**, 2935–2947 (2017)
11. Li, Z., Zhong, C., Wang, R., Zheng, W.-S.: Continual learning of new diseases with dual distillation and ensemble strategy. In: Martel, A.L., Abolmaesumi, P., Stoyanov, D., Mateus, D., Zuluaga, M.A., Zhou, S.K., Racoceanu, D., Joskowicz, L. (eds.) MICCAI 2020. LNCS, vol. 12261, pp. 169–178. Springer, Cham (2020). https://doi.org/10.1007/978-3-030-59710-8_17

12. McKinney, S.M., et al.: International evaluation of an ai system for breast cancer screening. Nature (2020)
13. Nadel, L., Hupbach, A., Gomez, R., Newman-Smith, K.: Memory formation, consolidation and transformation. Neuroscience & Biobehavioral Reviews (2012)
14. Rebuffi, S.A., Kolesnikov, A., Sperl, G., Lampert, C.H.: iCaRL: Incremental classifier and representation learning. In: CVPR (2017)
15. Ribordy, F., Jabès, A., Lavenex, P.B., Lavenex, P.: Development of allocentric spatial memory abilities in children from 18 months to 5 years of age. Cognitive Psychology (2013)
16. Rusu, A.A., et al.: Progressive neural networks. arXiv preprint arXiv:1606.04671 (2016)
17. Scarf, D., Gross, J., Colombo, M., Hayne, H.: To have and to hold: Episodic memory in 3-and 4-year-old children. Developmental Psychobiology (2013)
18. Sun, X., Yang, J., Sun, M., Wang, K.: A benchmark for automatic visual classification of clinical skin disease images. In: ECCV (2016)
19. Xiang, Y., Fu, Y., Ji, P., Huang, H.: Incremental learning using conditional adversarial networks. In: ICCV (2019)
20. Zhai, M., Chen, L., Tung, F., He, J., Nawhal, M., Mori, G.: Lifelong GAN: Continual learning for conditional image generation. In: ICCV (2019)

Alleviating Data Imbalance Issue with Perturbed Input During Inference

Kanghao Chen[1], Yifan Mao[1], Huijuan Lu[1], Chenghua Zeng[1],
Ruixuan Wang[1,2(✉)], and Wei-Shi Zheng[1,2]

[1] School of Computer Science and Engineering, Sun Yat-sen University,
Guangzhou, China
wangruix5@mail.sysu.edu.cn
[2] Key Laboratory of Machine Intelligence and Advanced Computing, MOE,
Guangzhou, China

Abstract. Intelligent diagnosis is often biased toward common diseases due to data imbalance between common and rare diseases. Such bias may still exist even after applying re-balancing strategies during model training. To further alleviate the bias, we propose a novel method which works not in the training but in the inference phase. For any test input data, based on the difference between the temperature-tuned classifier output and a target probability distribution derived from the inverse frequency of different diseases, the input data can be slightly perturbed in a way similar to adversarial learning. The classifier prediction for the perturbed input would become less biased toward common diseases compared to that for the original one. The proposed inference-phase method can be naturally combined with any training-phase re-balancing strategies. Extensive evaluations on three different medical image classification tasks and three classifier backbones support that our method consistently improves the performance of the classifier which even has been trained by any re-balancing strategy. The performance improvement is substantial particularly on minority classes, confirming the effectiveness of the proposed method in alleviating the classifier bias toward dominant classes.

Keywords: Data imbalance · Perturbed input · Prediction bias

1 Introduction

Deep learning has been widely applied to intelligent diagnosis of various diseases from medical images [6,7,17]. The success of intelligent diagnosis often depends on large annotated data for model training. However, while it is relatively easy to collect and annotate large amount of data for commonly encountered diseases, it is very challenging (if not impossible) to collect enough data for various rare diseases. Such data imbalance across diseases in nature often causes diagnostic bias toward common diseases by the intelligent system [1,11]. To improve

---

K. Chen and Y. Mao—The authors contribute equally to this paper.

the diagnostic performance of the intelligent system especially for those rare diseases, it is crucial to investigate effective learning strategies which can help the intelligent system successfully learn the features of both common and rare diseases from the imbalanced disease dataset.

Multiple re-balancing approaches have been developed to alleviate the data imbalance issue. Among them, data re-balancing and cost-sensitive re-weighting have been well explored and commonly adopted. The basic idea of data re-balancing is to use similar amount of data for each class to train the intelligent system, either by over-sampling the limited data for the small-sample (minority) classes [3, 9] or under-sampling the data for larger-sample (dominant) classes [15]. One special over-sampling strategy especially for training deep learning models is data augmentation [23] which can generate almost unlimited transformed data for minority (and dominant) classes. Different from the data re-balancing strategies on the model input side, cost-sensitive re-weighting strategies adjust the importance of loss terms in the loss function during model training, either at the class level or at the instance (individual data) level. At the class level, setting larger cost weight for minority classes has been widely adopted, where the weight is inversely proportional to the class frequency [12, 25]. At the instance level, the weight for each individual training data can be adjusted based on the difficulty of being correctly classified, with well-known techniques like boosting [3] or focal loss [16]. Besides data re-balancing and cost-sensitive re-weighting strategies, another set of strategies focus on the intelligent model itself, including transfer learning and model ensembling which have become routine to improve classification performance [25, 26]. However, all these strategies can just alleviate the data imbalance issue to some extent, in the sense that the well-trained model is still more or less biased toward dominant classes during inference [14, 28]. Recent studies found that widely used strategies to handle data imbalance often downgrade feature representation ability in the deep learning model, while the deep learning model without adopting any re-balancing strategy has a more generalizable feature extractor [28]. With this observation, it is proposed to first learn a generalizable feature extractor regardless of data imbalance, and then the model head for classification is re-trained with certain re-balancing strategy [2, 14, 28].

Different from the aforementioned approaches which alleviate the imbalance issue in the training phase, this paper proposes a simple yet effective approach which works not in the training phase but in the testing phase, aiming to further alleviate the model's prediction bias toward dominant classes if existing. This is achieved by slightly perturbing the test data before fed to the model based on a special single-data loss function. Different from adversarial attack methods [4, 8, 19, 21] which try to make models make wrong predictions, the proposed approach here aims to alleviate the model bias toward dominant classes. Extensive evaluations on multiple medical image datasets and model backbones support that the proposed approach, built on models trained with various re-balancing strategies, is effective in further improving the classification performance particularly on minority classes.

2   Methodology

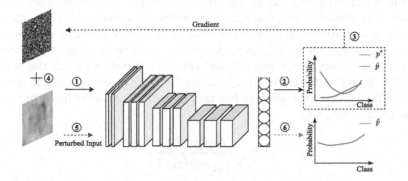

Fig. 1. Demonstrative procedure of our method. Circled number indicates the order of processing or corresponding signal flow. During inference, perturbation is computed based on gradient of a difference measure function over input pixels, and then the perturbed input is fed to the CNN model to obtain the final prediction.

The goal of this study is to improve the performance of any classifier trained on an imbalanced dataset, such that the classifier is less biased toward dominant classes in prediction. Different from most existing methods which focus on the classifier training phase, our method focuses on the testing (i.e., inference) phase. Given any classifier already well-trained on the imbalanced dataset and one test data to be classified, the intuitive idea of our method is to perturb the test data such that the classifier would be slightly inclined toward minority classes during inference. While such perturbation could downgrade the classification performance on dominant classes, it largely improves the performance on minority classes and the overall classification performance. The classification improvement on minority classes is crucial especially when missing diagnosis of rare diseases would cause serious consequence for patients.

The proposed approach is demonstrated in Fig. 1. Consider a convolutional neural network (CNN) classifier well trained based on an imbalanced dataset, where the number of training data for the c-th class is denoted by n_c, $c \in \{1, 2, \ldots, C\}$. Assume the classifier predictions over multiple test data are statistically biased toward dominant classes due to imperfect model training with the imbalanced training dataset. Then, for any test data $\mathbf{x}$ and the correspondingly original probability distribution output $\mathbf{p}$ of the classifier, the higher probability prediction in $\mathbf{p}$ would be likely biased toward the dominant classes. To alleviate such prediction bias, one naive way is to manually decrease the probability predictions by certain amount for dominant classes and increase the probability predictions by certain amount for minority classes. However, it would be very challenging and ad-hoc to manually determine the amount of prediction adjustment for each class. In this study, inspired by the strategy of generating adversarial examples, we propose a strategy to automatically perturb the input data $\mathbf{x}$ such that the classifier output $\tilde{\mathbf{p}}$ for the perturbed input $\tilde{\mathbf{x}}$ is slightly biased toward minority classes compared to the original output $\mathbf{p}$.

As in adversarial learning for adversarial example generation, a specific loss function with classifier input as variables needs to be designed. Here, the temperature scaling for the softmax operation of the CNN classifier and the prior frequency distribution $\{n_c\}_{c=1}^C$ over classes are employed to help design the loss function and subsequently perturb the classifier input. Suppose the softmax input (i.e., logit) vector is $\mathbf{z} = [z_1, z_2, \ldots, z_C]^T$ for the classifier input $\mathbf{x}$. Temperature scaling modifies the softmax function by including the temperature scaling parameter $T \in \mathbb{R}^+$, i.e.,

$$\hat{p}_c = \frac{\exp(z_c/T)}{\sum_{k=1}^C \exp(z_k/T)} , \tag{1}$$

where $\hat{p}_c$ is the c-th element of the temperature-tuned classifier output $\hat{\mathbf{p}} = [\hat{p}_1, \hat{p}_2, \ldots, \hat{p}_C]^T$. By setting a large temperature value (e.g., $T = 1000$), $\hat{p}_c$'s will become approximately equivalent to each other (e.g., $\hat{p}_c \approx 1/C$), but note that each $\hat{p}_c$ is a function of the classifier input $\mathbf{x}$ no matter which value T is set. With the almost-known output $\hat{\mathbf{p}}$ thanks to a large temperature value, the difference between $\hat{\mathbf{p}}$ and any specific target vector $\mathbf{p}^*$ would always exist if $\mathbf{p}^* \neq \hat{\mathbf{p}}$. With appropriate target $\mathbf{p}^*$, the difference between $\hat{\mathbf{p}}$ and $\mathbf{p}^*$ can be used to help perturb the classifier input in analogy to the well-known adversarial learning strategy. Considering that the objective of input perturbation is to bias the classifier output slightly toward minority classes, the target vector $\mathbf{p}^* = [p_1^*, p_2^*, \ldots, p_C^*]^T$ is designed as

$$p_c^* = \frac{g(n_c)}{\sum_{k=1}^C g(n_k)} , \tag{2}$$

where $g(n_c)$ is a scalar function of the inverse frequency n_c. In this study, $g(n_c) = \log(M/n_c)$, with M being a relatively larger constant such that $g(n_c)$ is non-negative for all classes (M was set to the number of training data from the largest class in experiments). The logarithmic function was adopted here such that the discrete probability distribution $\mathbf{p}^*$ is smoother across classes, which in turn would help cause smaller bias toward the minority classes. It can be seen that p_c^* is relatively larger ($p_c^* > 1/C$) for minority classes and smaller ($p_c^* < 1/C$) for dominant classes. The difference between the temperature-tuned output $\hat{p}_c$ ($\approx 1/C$) and the target output p_c^* is limited to a relatively smaller range $(-1/C, 0)$ for dominant classes and a larger range $(0, 1-1/C)$ for minority classes. Therefore, the overall difference between $\hat{\mathbf{p}}$ and $\mathbf{p}^*$ is dominated by the minority classes. This indicates that perturbing the classifier input based on the overall difference between $\hat{\mathbf{p}}$ and $\mathbf{p}^*$ would change the pre-softmax logits more largely for minority classes (i.e., larger increasing in logits) than for dominant classes (i.e., smaller decreasing in logits). As a result, slightly drawing the classifier output $\hat{\mathbf{p}}$ closer to the target $\mathbf{p}^*$ by perturbing the classifier input would bias the classifier prediction slightly toward minority classes compared to the original prediction. With an appropriate difference measure $\ell(\hat{\mathbf{p}}, \mathbf{p}^*)$ (e.g., cross entropy) which is essentially a function of the classifier input, the perturbed classifier input $\tilde{\mathbf{x}}$ can be obtained by the signed gradient of $\ell(\hat{\mathbf{p}}, \mathbf{p}^*)$ over input $\mathbf{x}$ [8], i.e.,

$$\tilde{\mathbf{x}} = \mathbf{x} - \varepsilon \cdot \text{sign}(\nabla \ell(\hat{\mathbf{p}}, \mathbf{p}^*)), \tag{3}$$

where ε is a scalar constant controlling the maximum perturbation on each data element (e.g., image pixel), $\nabla \ell(\hat{\mathbf{p}}, \mathbf{p}^*)$ is the gradient of difference measure (i.e., loss) function over the classifier input, and $\text{sign}(\cdot)$ is the pixel-wise sign function. Once the perturbed input $\tilde{\mathbf{x}}$ is obtained, it can be fed to the classifier to get the final output $\tilde{\mathbf{p}}$, in which the class with the maximum output is considered as the final prediction for the original input $\mathbf{x}$.

Comparison with Relevant Studies: Our method can be considered as one type of post-hoc logit adjustment during inference [18]. In contrast to the post-hoc logit adjustment at the output side of the classifier model [18,22], our method adjusts the logit by perturbing the model input. In addition, since our method is applied during inference, it can be naturally combined with existing methods which focus on classifier training, and the combinations would often improve the classification performance compared to those original methods.

3 Experiments

3.1 Experimental Settings

The proposed method was extensively evaluated on three imbalanced medical image datasets, Skin7 [5], OCTMNIST [27], and X-ray6 (Table 1). Specially, X-ray6 contains six diseases of X-ray images (Atelectasis, Cardiomegaly, Emphysema, Hernia, Mass, Effusion), where the six classes were selected from the original 14-class dataset ChestX-ray14 [24] by removing those classes of images which may contain multiple or ambiguous diseases in single images. Although dataset scale varying a lot, all three datasets present clear data imbalance (Table 1, last column). For OCTMNIST, all the images were used for model training, and an additional set of images (250 per class) officially provided were used for testing. For Skin7 and X-ray6, a five-fold cross-validation scheme was adopted, with four folds for training and another fold for testing each time.

Table 1. Dataset statistics. Last column: imbalance ratio = image number in the largest classes divided by image number in the smallest class.

Dataset	ImageType	#Class	ImageSize	#SmallestClass	#LargestClass	Imbalance
Skin7	Dermoscopy	7	600 * 450	115	6705	58.3
OCTMNIST	OCT	4	28 * 28	7754	46026	5.9
X-ray6	X-ray	6	1024 * 1024	88	3368	38.3

Since our method was applied to a well-trained classifier model, an convolutional neural network (CNN) classifier needs to be trained in advance, either using certain re-balancing strategy or not. In experiments, three CNN backbones pre-trained on ImageNet were used, including ResNet50 [10], MobilenetV2 [20], and DenseNet169 [13]. All experiments are conducted on a single 2080Ti GPU. For model training, Skin7 images were resized to 300×300 and randomly cropped to 224×224 pixels, followed by a random horizontal flip, while X-ray6 images were resized to 224×224 pixels and OCTMNIST images were resized to 32×32 pixels followed by random horizontal flip. For testing, only similar resizing operation was performed for each test image. During model training, the stochastic gradient descent with momentum (0.9) and weight decay (0.0005) were adopted. The batch size was set 32 on Skin7 and X-ray6, and 128 on OCTMNIST. The learning rate was set 0.01 for MobilenetV2 and 0.001 for ResNet50 and DenseNet169, which was then decayed by 0.1 after every 50 epochs. Linear warm-up of learning rate was used in the first epoch. All models were trained for 200 epochs with clear convergence. During testing, unless otherwise mentioned, the difference measure $\ell(\hat{\mathbf{p}}, \mathbf{p}^*)$ was based on the cross-entropy loss. The temperature T was set 1000, and the constant ε was set 0.001 on Skin7 and OCTMNIST and 0.0001 on X-ray6, based on an extra small validation set for each dataset on the ResNet50 backbone. Because of the imbalance property in testing for Skin7 and X-ray6, the mean class recall over all classes (MCR) and the recall on the smallest class (SCR) were used for evaluation. The standard deviation of MCR and SCR over the five folds (with five-fold cross validation) were also reported when evaluated on the Skin7 and X-ray6 datasets. Note that the proposed method is only slightly slower than corresponding baseline during inference, e.g., with the average inference time 0.283 s per image by the proposed method versus 0.107 s by the corresponding baseline.

3.2   Effectiveness and Generalizability Evaluation

The effectiveness of our method was extensively evaluated by comparing with the widely used strategies to handle data imbalance, including the data re-sampling (RS) for class-balanced mini-batch, the class-level re-weighting (RW), the instance-level re-weighting with focal loss (FL) [16], and the recently proposed state-of-the-art methods, including the two-stage deferred re-sampling (DRS) [28] and the margin-based method LDAM with deferred re-weighting [2]. The model trained with conventional cross-entropy loss (CE) (i.e., without using any re-balancing strategy) was also used for comparison.

From Table 2, it can be observed that, although the performance varies across baselines on each of the three datasets, our method (built on the model trained by the baseline method) always performs better than the corresponding baseline when measured by MCR for all classes. Importantly, the performance boosting on the smallest class is much more significant than for all classes, as seen in the SCR columns. Detail performance on each Skin7 class from Fig. 2 also shows that our method obtains substantial improvement on small classes, although with certain

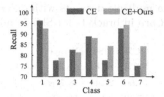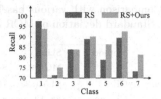

Fig. 2. Performance comparison with corresponding baseline (CE and RS) on each Skin7 class with ResNet50 backbone. X-axis: 1 for the largest class and decreasingly 7 for the smallest class.

decreased performance on dominant classes as often observed from state-of-the-art methods (e.g., LDAM [2]). These results clearly support that our method is effective in alleviating the model's bias toward dominant classes when the model was trained on imbalanced dataset with or without any re-balancing training strategy. Note that the larger variance of SCR than that of MCR on the X-ray6 and Skin7 datasets is probably due to the relatively smaller testing images on the smallest classes (only 22 and 23 images in each fold for the smallest class). Interestingly, some re-balancing baselines (e.g., RS, FL) performed worse than the plane cross-entropy baseline (CE). This may be due to the heavy imbalance in the datasets which cannot be well addressed with those re-balancing strategies. However, the inclusion of our method during inference consistently improves the performance of all the models trained with different strategies.

Table 2 also suggests that our method has desired generalizability. Our method consistently improves the performance particularly on minority classes for multiple classification tasks (Skin7, OCTMNIST, X-ray6), in combination with various re-balancing strategies (RW, DRS, LDAM, etc.), and with different classifier backbones (ResNet50, MobileNetV2, and DenseNet169). Because of its simple and independent operation on the inference phase, our method is expected to work well for more types of tasks and on various model architectures.

3.3   Robustness to Hyper-parameters

Our method is robust to the choice of perturbation magnitude ε. As shown in Fig. 3 (left and middle), when ε is smaller enough (e.g., in the range $(0, 0.001]$), our method performs consistently better than the corresponding baseline (with $\varepsilon = 0$ on the curve), no matter which baseline and CNN backbone is used. From this figure, we can also see that the best choice of ε varies when our method is combined with different baselines. This also indicates that the previously reported performance (Table 2) of our method on the Skin7 dataset is indeed conservative, where $\varepsilon = 0.001$ (not the best choice in most cases) was adopted in all comparisons. Actually, from Fig. 3 (middle), it can be expected that consistently better performance than reported in Table 2 would be obtained if setting ε smaller (e.g., 0.0005) when combining our method with most baselines on the MobileNetV2 backbone.

Table 2. Comparison with various baselines on multiple datasets with different CNN backbones. Standard deviation of MCR & SCR are in brackets for Skin7 and X-ray6.

Model	Method	Skin7		OCTMNIST		X-ray6	
		MCR	SCR	MCR	SCR	MCR	SCR
ResNet50	CE	84.54(0.86)	75.66(6.63)	75.60	26.40	58.77(2.63)	40.00(15.56)
	CE+ours	**86.42**(1.29)	**84.36**(5.00)	**77.20**	**43.60**	**59.81**(2.39)	**43.64**(14.59)
	RS	83.23(1.36)	73.04(9.95)	78.70	37.60	57.90(2.06)	34.56(13.11)
	RS+ours	**86.34**(1.11)	**81.76**(7.14)	**78.90**	**52.40**	**58.82**(1.98)	**37.28**(14.18)
	RW	85.03(0.97)	74.78(8.43)	76.00	28.40	62.17(0.76)	53.64 6.68
	RW+ours	**87.82**(1.10)	**87.83**(5.07)	**78.10**	**43.20**	**62.89**(1.43)	**56.36**(7.39)
	FL	83.10(0.94)	73.92(6.88)	74.80	23.20	57.69(2.42)	33.64(9.43)
	FL+ours	**85.90**(1.50)	**86.12**(4.75)	**77.60**	**38.80**	**58.23**(2.38)	**34.56**(9.43)
	DRS	84.12(1.46)	75.66(9.00)	79.20	39.20	54.85(2.20)	30.00(9.43)
	DRS+ours	**86.37**(0.81)	**84.36**(5.00)	**80.60**	**61.20**	**56.42**(2.54)	**32.72**(11.31)
	LDAM	83.48(1.47)	73.94(9.73)	79.60	40.00	59.44(2.69)	41.82(13.03)
	LDAM+ours	**84.81**(0.92)	**79.12**(9.92)	**81.60**	**54.40**	**60.04**(2.69)	**51.82**(10.44)
MobileNetV2	CE	84.46(2.05)	77.40(8.37)	76.40	25.20	56.44(1.20)	26.36(8.77)
	CE+ours	**85.56**(1.54)	**88.70**(3.87)	**77.60**	**45.20**	**57.22**(1.28)	**29.10**(11.43)
	RS	82.42(1.42)	66.98(11.36)	76.20	27.60	56.58(1.64)	30.92(5.92)
	RS+ours	**86.07**(0.78)	**84.36**(7.87)	**78.40**	**45.60**	**59.30**(2.20)	**41.82**(8.72)
	RW	85.75(0.91)	80.00(4.43)	77.70	32.40	60.70(1.82)	50.00(11.85)
	RW+ours	**85.96**(1.33)	**90.43**(3.25)	**78.20**	**52.40**	**61.74**(1.76)	**53.64**(10.52)
	FL	84.23(1.40)	76.54(10.05)	77.10	35.20	57.24(1.96)	35.46(9.90)
	FL+ours	**84.81**(0.99)	**89.58**(5.00)	**79.20**	**51.60**	**58.10**(2.32)	**38.18**(10.95)
	DRS	84.80(1.52)	78.26(8.72)	77.80	32.80	54.41(1.44)	25.46(10.49)
	DRS+ours	**85.73**(1.57)	**88.70**(3.87)	**78.40**	**56.00**	**56.07**(1.00)	**29.10**(8.25)
	LDAM	83.63(1.05)	76.54(10.01)	80.90	48.00	60.25(3.65)	17.26(16.84)
	LDAM+ours	**84.42**(0.61)	**82.64**(7.52)	**82.70**	**64.00**	**60.70**(3.59)	**36.36**(23.40)
DenseNet169	CE	84.92(1.10)	76.56(9.54)	73.90	21.60	60.74(2.18)	40.92(11.58)
	CE+ours	**86.20**(1.72)	**86.98**(6.86)	**76.10**	**36.00**	**61.44**(2.43)	**41.84**(11.76)
	RS	82.43(1.48)	69.56(9.75)	75.20	23.60	58.86(2.27)	36.38(14.35)
	RS+ours	**85.33**(1.67)	**79.14**(7.14)	**78.00**	**38.40**	**60.17**(1.86)	**40.90**(11.57)
	RW	84.50(0.68)	76.52(7.58)	75.50	20.40	63.68(1.88)	53.64(11.28)
	RW+ours	**86.48**(0.85)	**84.35**(8.95)	**77.40**	**38.00**	**64.44**(2.04)	**56.36**(10.98)
	FL	84.00(1.73)	80.00(3.87)	75.10	20.00	59.46(2.75)	40.00(12.21)
	FL+ours	**85.64**(1.23)	**87.84**(3.61)	**78.80**	**34.80**	**60.53**(2.14)	**43.64**(8.86)
	DRS	83.53(1.82)	71.32(10.00)	73.70	23.20	57.95(2.39)	35.46(12.61)
	DRS+ours	**85.81**(1.22)	**82.64**(7.55)	**75.00**	**40.40**	**59.26**(2.29)	**39.10**(13.08)
	LDAM	83.71(1.46)	72.18(12.54)	81.60	46.40	62.14(4.76)	44.54(17.83)
	LDAM+ours	**85.77**(1.00)	**82.62**(8.15)	**83.40**	**63.20**	**62.61**(4.54)	**46.36**(17.58)

The robustness of our method to the temperature scaling T is demonstrated in Fig. 3 (right). It shows that our method would work stably better than corresponding baselines when T is larger than 100. This also confirms the effectiveness of the temperature scaling at a larger value. In addition, besides the cross-entropy loss for the difference measure $\ell(\hat{\mathbf{p}}, \mathbf{p}^*)$, some other choices including the mean squared error and focal loss were also tried for $\ell(\hat{\mathbf{p}}, \mathbf{p}^*)$, resulting in equivalent performance compared to that from the cross-entropy loss.

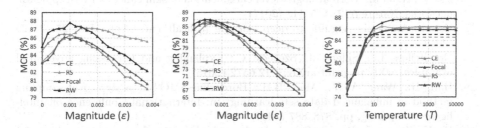

Fig. 3. Robustness to hyper-parameters. Left & middle: performance of our method with varying perturbation ε respectively on ResNet50 & MobileNetV2. Right: with varying temperature T on ResNet50 (dashed curves for corresponding baselines). Different curves for combinations of ours with different baselines. Skin7 was used here.

4 Conclusion

Here we propose a simple yet effective method to alleviate the data imbalance issue not during model training but during inference. The natural combination of our method with existing methods further alleviates the classifier's bias toward dominant classes, as supported by extensive evaluations on three medical datasets with different data-imbalance methods and model backbones. The applications of our method to more tasks like lesion detection will be explored.

Acknowledgments. This work is supported by the National Natural Science Foundation of China (No. 62071502, U1811461), the Guangdong Key Research and Development Program (No. 2020B1111190001, 2019B020228001), and the Meizhou Science and Technology Program (No. 2019A0102005).

References

1. Buda, M., Maki, A., Mazurowski, M.: A systematic study of the class imbalance problem in convolutional neural networks. Neural Netw. **106**, 249–259 (2018)
2. Cao, K., Wei, C., Gaidon, A., Arechiga, N., Ma, T.: Learning imbalanced datasets with label-distribution-aware margin loss. In: Advances in Neural Information Processing Systems, vol. 32 (2019)
3. Chawla, N.V., Bowyer, K.W., Hall, L.O., Kegelmeyer, W.P.: Smote: synthetic minority over-sampling technique. J. Artif. Intell. Res. **16**, 321–357 (2002)
4. Chen, P., Sharma, Y., Zhang, H., Yi, J., Hsieh, C.J.: EAD: elastic-net attacks to deep neural networks via adversarial examples. In: AAAI (2018)

5. Codella, N.C.F., et al.: Skin lesion analysis toward melanoma detection: a challenge at the 2017 international symposium on biomedical imaging (ISBI), hosted by the international skin imaging collaboration (ISIC). In: IEEE International Symposium on Biomedical Imaging, pp. 168–172 (2018)

6. Esteva, A., et al.: Dermatologist-level classification of skin cancer with deep neural networks. Nature **542**, 115–118 (2017)

7. Esteva, A., et al.: A guide to deep learning in healthcare. Nat. Med. **25**, 24–29 (2019)

8. Goodfellow, I.J., Shlens, J., Szegedy, C.: Explaining and harnessing adversarial examples. arXiv preprint arXiv:1412.6572 (2015)

9. Han, H., Wang, W.Y., Mao, B.H.: Borderline-SMOTE: a new over-sampling method in imbalanced data sets learning. In: International Conference on Intelligent Computing, pp. 878–887 (2005)

10. He, K., Zhang, X., Ren, S., Sun, J.: Deep residual learning for image recognition. In: Proceedings of the IEEE Conference on Computer Vision and Pattern Recognition, pp. 770–778 (2016)

11. Horn, G.V., Perona, P.: The devil is in the tails: fine-grained classification in the wild. arXiv preprint arXiv:1709.01450 (2017)

12. Huang, C., Li, Y., Loy, C.C., Tang, X.: Learning deep representation for imbalanced classification. In: Proceedings of the IEEE Conference on Computer Vision and Pattern Recognition, pp. 5375–5384 (2016)

13. Huang, G., Liu, Z., Weinberger, K.Q.: Densely connected convolutional networks. In: Proceedings of the IEEE Conference on Computer Vision and Pattern Recognition, pp. 2261–2269 (2017)

14. Kang, B., et al: Decoupling representation and classifier for long-tailed recognition. In: Proceedings of the International Conference on Learning Representations (2020)

15. Kubat, M., Matwin, S.: Addressing the curse of imbalanced training sets: one-sided selection. In: International Conference on Machine Learning, vol. 97, pp. 179–186 (1997)

16. Lin, T.Y., Goyal, P., Girshick, R., He, K., Dollár, P.: Focal loss for dense object detection. In: Proceedings of the IEEE International Conference on Computer Vision, pp. 2980–2988 (2017)

17. Litjens, G., et al.: A survey on deep learning in medical image analysis. Med. Image Anal. **42**, 60–88 (2017)

18. Menon, A., Jayasumana, S., Rawat, A., Jain, H., Veit, A., Kumar, S.: Long-tail learning via logit adjustment. arXiv preprint arXiv:2007.07314 (2020)

19. Moosavi-Dezfooli, S.M., Fawzi, A., Frossard, P.: DeepFool: a simple and accurate method to fool deep neural networks. In: 2016 IEEE Conference on Computer Vision and Pattern Recognition (CVPR), pp. 2574–2582 (2016)

20. Sandler, M., Howard, A., Zhu, M., Zhmoginov, A., Chen, L.C.: MobilenetV2: inverted residuals and linear bottlenecks. In: Proceedings of the IEEE Conference on Computer Vision and Pattern Recognition, pp. 4510–4520 (2018)

21. Su, J., Vargas, D.V., Sakurai, K.: One pixel attack for fooling deep neural networks. IEEE Trans. Evol. Comput. **23**, 828–841 (2019)

22. Tang, K., Huang, J., Zhang, H.: Long-tailed classification by keeping the good and removing the bad momentum causal effect. Adv. Neural. Inf. Process. Syst. **33**, 1513–1524 (2020)

23. Wang, J., Perez, L.: The effectiveness of data augmentation in image classification using deep learning. Stanford University Research Report (2017)

24. Wang, X., Peng, Y., Lu, L., Lu, Z., Bagheri, M., Summers, R.M.: ChestX-ray8: hospital-scale chest X-ray database and benchmarks on weakly-supervised classification and localization of common thorax diseases. In: Proceedings of the IEEE Conference on Computer Vision and Pattern Recognition, pp. 3462–3471 (2017)
25. Wang, Y.X., Ramanan, D., Hebert, M.: Learning to model the tail. Adv. Neural. Inf. Process. Syst. **30**, 7032–7042 (2017)
26. Xiang, L., Ding, G., Han, J.: Learning from multiple experts: self-paced knowledge distillation for long-tailed classification. In: Vedaldi, A., Bischof, H., Brox, T., Frahm, J.-M. (eds.) ECCV 2020. LNCS, vol. 12350, pp. 247–263. Springer, Cham (2020). https://doi.org/10.1007/978-3-030-58558-7_15
27. Yang, J., Shi, R., Ni, B.: MedMNIST classification decathlon: a lightweight AutoML benchmark for medical image analysis. arXiv preprint arXiv:2010.14925 (2020)
28. Zhou, B., Cui, Q., Wei, X.S., Chen, Z.: BBN: bilateral-branch network with cumulative learning for long-tailed visual recognition. In: Proceedings of the IEEE Conference on Computer Vision and Pattern Recognition, pp. 9716–9725 (2020)

A Deep Reinforced Tree-Traversal Agent for Coronary Artery Centerline Extraction

Zhuowei Li, Qing Xia[✉], Zhiqiang Hu, Wenji Wang, Lijian Xu,
and Shaoting Zhang

SenseTime Research, Beijing, China
{lizhuowei,xiaqing,huzhiqiang,wangwenji,xulijian,
zhangshaoting}@sensetime.com

Abstract. Vessel centerline extraction is fundamental for plentiful medical applications. Majority of current methods require pre-segmentations, distance maps or similar sorts of scanning whole volume action and followed by minimal-path or skeletonization algorithms. In this paper, we demonstrate a deep reinforced tree-traversal agent that automatically traces tree-structure centerlines assuming no post-prune or post-merging. It takes raw images as input and generates tree-structure centerlines naturally. To this end, road mark and dynamic reward mechanisms are proposed to make tree-structure vessels learnable and impart the agent how to learn correspondingly. Besides, a multi-task discriminator is raised to simultaneously detect bifurcations and decide terminations. We experimentally show that traced centerlines have an overlap of more than 90% and a distance less than 0.25 mm with annotated reference centerlines on coronary arteries. Beyond the promising accuracy, the proposed method also surpasses other existing methods by a large margin in terms of the time and memory efficiency. And a flexible trade-off between accuracy and time efficiency is exhibited at the inference. Codes are available at https://github.com/LzVv123456/Deep-Reinforced-Tree-Traversal.

Keywords: Coronary artery · Centerline extraction · Reinforcement learning · Deep learning

1 Introduction

Vessel centerline extraction is fundamental for plentiful medical applications. It provides, beyond what segmentation and detection provide as *what* and *where*, a more semantic representation of topology and geometry. And as a result, it can facilitate clinical diagnosis and treatment planning.

The vessel centerline extraction problem has been studied for decades. This line of research falls into two general categories: Two-stage methodologies and tracing-based techniques. Majority of current approaches require segmentations, distance maps or similar sorts of scanning whole volume action and followed by

© Springer Nature Switzerland AG 2021
M. de Bruijne et al. (Eds.): MICCAI 2021, LNCS 12905, pp. 418–428, 2021.
https://doi.org/10.1007/978-3-030-87240-3_40

minimal-path or skeletonization algorithms [3–7,10,11,14,16,17,20,23]. On the contrary, tracing based methods explore local features sequentially [1,2,4,8,18, 22,24]. In spite of speed, memory and data efficiency of tracing-based methods, previous tracing methods lack the generalization ability and facing difficulties in handling intricate tree structures. Most recently, Zhang et al. [22] proposed a deep reinforcement learning (DRL) pipeline for aorta centerline tracing. Despite the heuristic functionality of the work, it only deals with single-tubular structures with few orientational variances. Wolterink et al. [18] and Yang et al. [21]. On the other hand, train a local navigator with supervised learning. However, it still needs pre-disposed seed points for tree-structure extraction and suffers fussy centerline post-prune and post-merging.

Inspired by the sequential nature of both DRL and tree-traversal process, we here present a **Deep Reinforced Tree-traversal (DRT)** agent that infers tree-structure centerlines from a given initial point. This framework takes raw local patches and generates tree-structured centerlines sequentially. In order to achieve this functionality, three main challenges are addressed as listed:

- *Where is the bifurcation, and when the vessel terminate?* We propose a regression-based **multi-task discriminator** to detect bifurcations and terminations simultaneously. The discriminator essentially models distances between the point of interest and its nearest bifurcation and reference point.
- *In what order should the agent tracing at bifurcations?* We leverage the agent to decide for itself. However, during the training phase, we propose the **dynamic reward** mechanism to best support and supervise the agent's choice. Specifically, we observe the agent for several steps once meeting a bifurcation, and then we match the path that the agent walks against all the candidate branches. Finally, the best-matched branch wins for the subsequent reward supervision.
- *How to ensure the agent backtrack different branches?* This problem is critical to keep the agent away from trapping in the endless loop. To this end, we come up the **road mark** mechanism to remind the agent of trajectories already passed and serving as a navigation map implicitly.

To evaluate the proposed framework, a dataset contains 280 cardiac CTA images from multiple clinical institutes is collected. Masks of coronary arteries are annotated by experts and centerlines are calculated according to masks followed by manual refinements. In general, our framework surpasses three reproduced baselines (detailed in experiments) by achieving more than 90% overlap rate with less than 0.25 mm average distances referencing ground-truth centerlines. Beyond the promising accuracy, our framework also outperforms all baselines by a large margin in terms of time and memory efficiency. Extensive ablation studies are further conducted to substantiate proposed innovations.

2   Methods

The centerline tracing process can be viewed as a sequential decision-making process that satisfies the finite Markov Decision Process (MDP) and thus can be

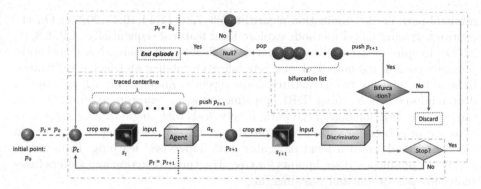

Fig. 1. The general workflow for tree-structured centerline traversal. The red dashed block contains sequential tracing process within a branch and the green dashed block illustrates backtrack routine among separate branches. (Color figure online)

fitted into a standard RL pattern. Then the best policy $q_*(s,a)$ can be approximated by training parameterized network $q(s,a,\theta)$ to minimize:

$$\mathcal{L} = (r_{t+1} + \gamma \max_{a'} q(s_{t+1},a',\theta^-) - q(s_t,a,\theta))^2 \tag{1}$$

where θ and θ^- parameterize *target network* and *policy network* correspondingly [12]. In our case, *state s* is a 3D local patch cropped at a given point p, *action a* is sampled from the action space A which is N orientations uniformly distributed on a sphere. At each time step, the agent move from current point p_t to p_{t+1} according to action a_t with pre-defined step-size. Once the current trajectory finished, collected points $\{p_0, p_1, \ldots, p_t\}$ will be deemed as the centerline. In the view of tree structures, we perform tree traversal process as illustrated in Fig. 1. The general process is analog to a depth-first traversal. Despite this intuitive formation, multiple obstacles need to be addressed, and we will plumb these challenges in the following parts of this section.

Reward Design. The goal, *trace centerline*, here can be further decomposed into two sub-goals: (1) trace along the correct direction; (2) trace as close as possible along reference centerlines; Here we propose a target that merges two subgoals into one by directly pulling current proposed centerline point p_t to next target reference point g_{t+k}. Figure 2 provides a straight forward view of how reward mechanism works (k in Fig. 2 is set to 1 for the sake of interpretation). Given a current proposed point p_t, a corresponding reference point g_t is paired by finding closest point in Euclidean space from the reference centerline. Then the target reference point g_{t+k} is selected from reference centerline sequence with index $t + k$. Then a scalar *trend T* defined as:

$$T = \|v_1\|_2 - \|v_2\|_2 \tag{2}$$

is used to evaluate the pulling action. If $T > 0$, then p_{t+1} is getting closer to g_{t+k} comparing with p_t. Otherwise, it's being pushed away. This scalar T implicitly

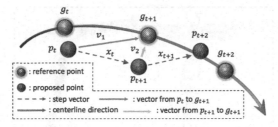

Fig. 2. A straight forward view of reward design. The goal of reward is to encourage the pulling action that pulls current point p_t to objective reference point g_{t+1}.

convey the goodness of the sampled orientation. Depending on the *trend*, the *reward* is designed as:

$$R = \begin{cases} \frac{T}{2\|x_t\|_2} + A & \text{if } T > 0. \\ 0 & \text{otherwise.} \end{cases} \tag{3}$$

A is an auxiliary signal employed to encourage proposed point to stay around the reference centerline. Explicit expression for A is: $A = \frac{1}{1+e^{-x}}$ where x is point to curve distance defined as L_2 distance from p_{t+1} to it's nearest reference point.

Road Mark. Considering the single vessel situation where only a single tubular structure exists. The main challenge emerges as ambiguous orientation information. The vessel centerline can be viewed as a directed curve from the proximal point to the distal point. Under a tracing framework, global information is commonly not available, and the local patch can have a similar appearance towards both proximal and distal directions. This will cause conflicts during training and potentially cause 'look back' or even 'stay around' during inference. Regarding tree structures at the inference process, the agent will trace the identical branch due to its deterministic essence unless manual interventions are imposed. Here, we propose a simple yet effective solution, namely **road mark**, for both scenarios. As indicated by the name and showed in Fig. 3, a distinctive mark (a cube with size n and value c in our setting) is left to the raw environment where the agent has passed. By doing so, we transfer the directional and ordinal information into visual features that can be directly encoded into the neural network. The road mark prevents the agent from hesitation as well as indicating tracing order at bifurcations.

Dynamic Reward. With the assistance of the road mark, the topological information can be reconstructed by purely learning local patches. Nevertheless, it is still an open question on how to learn at bifurcations. Specifically, in what order should the agent trace at bifurcations. We speculate that any hand-crafted rules can lead to sub-optimum. Based on this conjecture, we propose the **dynamic reward** mechanism to leverage the agent itself to decide the tracing order during training. A buffer zone equaling to Z time steps is created when reaching a bifurcation. Then accumulated rewards $R_n = \sum_{i=1}^{Z} r_i$ within Z continuous time steps according to all N potential candidates are collected. And the candidate

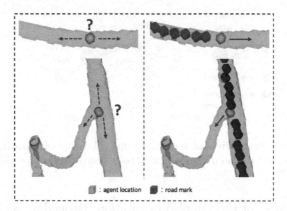

Fig. 3. Left side and right side of the figure displays a real environment with and without the road mark correspondingly. Road marks in the vessel served as the orientation indicator.

with index equal to $argmax(R_0, R_1, \ldots, R_{N-1})$ is set as the reference centerline. Once a reference centerline is decided, other candidates will be ignored and get back to single vessel mode. Within the buffer zone range, the agent is always awarded the largest reward regardless of which candidates it relies on. When the agent backtracks to the same bifurcation, the reference centerline already used will be disregarded from the candidate pool, then the same strategy is executed again. This mechanism provides a dynamic control over rewards and reference centerline at bifurcation during the training.

Multi-task Discriminator. Herein, a *multi-task discriminator* is trained separately to regress distances from a point to it's nearest bifurcation and reference point simultaneously. The training data is generated by random sampling points from reference centerlines with jittering (a 3D Gaussian distribution parameterized by $\mu = 0, \sigma = 10$ is used for jittering). Two ground truth scalars for each training point are decided using an identical formula:

$$G = \begin{cases} e^{a(1-\frac{D(x)}{d_M})} - 1 & \text{if } D(x) < d_M. \\ 0 & \text{otherwise.} \end{cases} \quad (4)$$

proposed by Sironi et al. [15]. $D(x)$ is the $\mathcal{L}_2$ distance between the sampled point x and it's target point. a is set to 6 as explored in the original paper and d_M is set to half of the patch size. Then a weighed MSE loss defined as: $\mathcal{L} = \frac{\mathcal{L}_1}{2} + \frac{\mathcal{L}_2}{2}$ is minimized. During the inference, the discriminator slides together with the agent along the vessel and a window of passed proximity values are maintained. Then the local maximums are extracted immediately as bifurcations and the termination will be triggered if the average value of the corresponding proximity map exceeds certain threshold $\mathcal{T}$ which means the current location is no longer close to any reasonable reference centerline.

Train and Inference. The tracing agent is trained episodically. Within an episode, the 3D volumetric image and centerlines formed as directed graphs are

collected. The agent is initialized around the root of the tree structure and training by conducting tracing. Once the agent successfully reaches a bifurcation, it
will act according to the dynamic reward mechanism and activate this bifurcation. The current trajectory terminates if one of the following criteria satisfied:
(1) reach a distal point; (2) go out of the safe range; (3) reach predefined maximum steps; Once a trajectory is ended, the agent will backtrack from an activated
bifurcation and deactivate this bifurcation once there is no trainable candidate
branch. The whole episode ends until there is no existing activated bifurcation.
During inference, given a root point, the agent has already equipped with the
knowledge on how to trace and in what order to trace. Noticing that the discriminator is trained independently and no road mark mechanism is applied for
the discriminator. So we also maintain a clear environment for the discriminator
during inference.

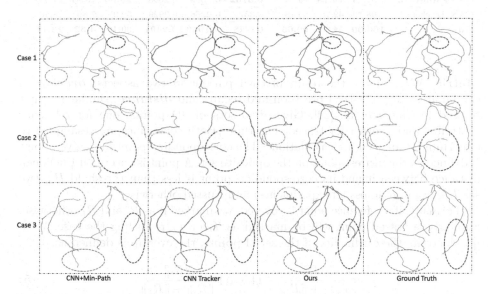

Fig. 4. Centerlines are displayed using scatters and all results remain their original
appearance without post-processing. Tiny green points in our methods are bifurcations detected and two larger purple points are ostium location. According to this
visualization result, our method achieved better completeness and acting less prone to
over-tracing. (Color figure online)

3   Experiments and Results

Dataset. We collect a cardiac CTA dataset that contains 280 patients from 4
clinical institutions. Most of these patients contain a certain degree of stenosis
and plaques. And masks of all feasible coronary arteries are deliberately annotated by experts, and centerlines are extracted using TEASAR [13]. Furthermore, centerlines extracted are scrutinized and manually refined. All centerlines

are formed as directed acyclic graph (DAG) for the final input. In experiments, images are resampled isotropically to spacing 0.5 mm, and intensities are normalized between 0 and 1. The whole dataset is randomly split into 140 training and 140 validation.

Table 1. Comparison of proposed method against three baselines with respect to metrics of both accuracy and efficiency. Our method surpasses three baselines in terms of both accuracy and efficiency.

Method	Overlap (%)			Distance (mm)			Time cost (s)		
	Max	Min	Average	Max	Min	Average	Max	Min	Average
MSCAR-DBT+ skeletonization	96.41	47.68	81.66	0.4193	0.2865	0.3437	26.32	14.01	20.12
CNN+Min-Path	97.46	**74.78**	89.29	**0.3102**	0.2055	0.2608	**25.50**	18.84	21.93
CNN Tracker	98.79	42.78	89.18	0.357	0.2266	0.2929	59.9	7.89	28.40
Ours	**100.00**	54.93	**90.05**	0.3478	**0.1966**	**0.2491**	30.48	**4.32**	**13.11**

Metric. We evaluate the result of each patient in two aspects: ***overlap*** (a combination of point-level precision and recall) and ***distance***, similar to what is used in Zhang et al. [22]. Given two sets of 3D points, one for reference centerline and another for traced centerline (Both centerlines are resampled to 0.025 mm for an accurate calculation). The corresponding point for a given point is defined as the nearest point on the opposite set. A point is covered if Euclidean distance between it and its corresponding point is less than threshold H_m (we set $H_m = 1$ mm across all experiments) and missed otherwise.

Formally, if a reference point is covered, we marked it as R_t and R_f otherwise. Similarly, a traced point will be marked as T_t or T_f regarding the case. With $\|.\|$ denoting the cardinality of the set of points, the overlap is defined as:

$$\alpha \frac{\|T_t\|}{\|T_t\| + \|T_f\|} + (1 - \alpha) \frac{\|R_t\|}{\|R_t\| + \|R_f\|} \tag{5}$$

As for distance, D_r is defined as average Euclidean distance between matched points in reference and their corresponding traced points. And D_t is defined analogously with reversed direction. Then the distance is defined as: $\alpha D_r + (1 - \alpha)D_t$. α is set to 0.5 for both metrics.

Coronary Artery Centerline Extraction. Herein, three methods range from pure traditional methods to solo deep learning methods are reproduced as baselines. We first reproduced SOTA tracing method **CNN Tracker** [18]. Since we do not assume the acquisition of vessel radius, all radius-related parameters are set to 1mm. Other parameters remained the same as the original work. The second baseline is a combination of CNN segmentation with minimal path extraction (**CNN+Min-Path**). Two segmentation models from coarse to fine are trained according to Xia et al. [19] with our annotated coronary artery masks. Then centerlines are extracted from a fine-grained segmentation mask using a TEASAR [13] algorithm. At last, we also reproduced a traditional centerline extraction method (*MSCAR-DBT+skeletonization*), coronary artery masks are extracted according to MSCAR-DBT [24] and centerlines are acquired through skeletonization operation [9]. The architecture used in our work is the same as what was proposed in the CNN Tracker to perform an impartial comparison. Ostium locations are provided for both CNN Tracker and our method for the sake of justice. As displayed in Table 1, Our methods surpass all baselines in terms of both accuracy and efficiency. CNN Tracker failed to track several sub-branches, and it is prone to over-tracing. CNN+Min-Path suffers from segmentation errors and shortcuts caused by the minimal path. And traditional method falls behind in general due to its weak generalization ability. Figure 4 visualizes three cases. Step-size at inference can be set arbitrarily within a reasonable range regardless of training step-size. Here we measure step-size from 0.5 to 2.0. As shown in Table 2, there is a trade-off between speed and quality. According to experiments, step-size equaling to 1.0 generates the highest cost performance. It only takes around 5 s for each patient with little damage in accuracy.

Table 2. Detail results for different step-size.

Step-size	Average overlap (%)	Average distance (mm)	Average time cost (s)
0.5	90.05	0.2491	13.11
0.6	89.64	0.2511	10.52
0.7	89.44	0.2549	9.76
0.8	89.58	0.2574	7.11
0.9	89.19	0.2638	6.06
1.0	89.36	0.2673	5.68
1.1	89.38	0.2758	5.57
1.2	89.39	0.2830	5.16
1.3	88.49	0.2884	4.55
1.4	87.63	0.2947	4.16
1.5	88.03	0.3021	4.05
1.6	86.31	0.3092	3.57
1.7	86.57	0.3175	3.39
1.8	85.82	0.3246	3.01
1.9	84.48	0.3328	2.72
2.0	83.54	0.3433	2.49

Table 3. Road mark combined with dynamic reward outperform all other combinations.

Variations	Average overlap (%)	Average distance (mm)
No-mark+random	60.59	0.3443
No-mark+order	62.67	0.3135
No-mark+angle	63.54	0.3102
No-mark+dynamic	61.07	0.3029
Mark+random	76.41	0.3160
Mark+order	89.05	0.2632
Mark+angle	92.86	0.2482
Mark+dynamic	**93.74**	**0.2447**

Ablation Studies. We demonstrate the effectiveness of the road mark and dynamic reward mechanism in this section. Ground truth bifurcations and terminations are provided across ablation studies to avoid perturbation caused by the discriminator. 8 variations are implemented to interpret road mark and dynamic reward, **mark** and **no-mark** represents whether utilize road mark or not. Four different training modes at bifurcations are disposed: (1) randomly select a sub-branch at bifurcations (**random**); (2) always choose the sub-branch with least accumulated x coordinate values (**order**); (3) always choose the sub-branch with the least angle referencing it's father-branch (**angle**); (4) our dynamic reward mode (**dynamic**). As showed in Table 3, road mark enabled tree structure tracing. According to our off-line visualization results, the agent will always trace the same path and sometimes traces loops without a road mark. Dynamic reward surpassed all three other manually designed training modes. The most allied results came from the angle mode, which also satisfies the common intuition.

4   Conclusion and Future Work

This work presents a novel framework for tree-structure vessel centerline tracing and demonstrates promising results on coronary artery centerline extraction. Unlike other existing learning-based methods. This framework is designed and boosted to consume DAG and infer the tree-structure naturally. However, due to the sequential tracing nature, its advantage in efficiency will degrade confronting non-sparse environments. For future improvement, it will be more elegant and efficient to further merge the agent with the discriminator.

Acknowledgements. This work was supported by the Beijing Postdoctoral Research Foundation, the Beijing Nova Program (Z201100006820064), the Shanghai Xuhui District HealthCare AI Cooperation Project (2020-011) and the National Key Research and Development Project of China (2020YFC2004800).

References

1. Aylward, S.R., Bullitt, E.: Initialization, noise, singularities, and scale in height ridge traversal for tubular object centerline extraction. IEEE Trans. Med. Imaging **21**(2), 61–75 (2002)
2. Cetin, S., Unal, G.: A higher-order tensor vessel tractography for segmentation of vascular structures. IEEE Trans. Med. Imaging **34**(10), 2172–2185 (2015)
3. Cui, H., Xia, Y.: Automatic coronary centerline extraction using gradient vector flow field and fast marching method from CT images. IEEE Access **6**, 41816–41826 (2018)
4. Friman, O., Kuehnel, C., Peitgen, H.: Coronary centerline extraction using multiple hypothesis tracking and minimal paths (July 2008)
5. Gülsün, M.A., Funka-Lea, G., Sharma, P., Rapaka, S., Zheng, Y.: Coronary centerline extraction via optimal flow paths and CNN path pruning. In: Ourselin, S., Joskowicz, L., Sabuncu, M.R., Unal, G., Wells, W. (eds.) MICCAI 2016. LNCS, vol. 9902, pp. 317–325. Springer, Cham (2016). https://doi.org/10.1007/978-3-319-46726-9_37
6. Guo, Z., et al.: DeepCenterline: a multi-task fully convolutional network for centerline extraction. arXiv arXiv:abs/1903.10481 (2019)

7. Jin, D., Iyer, K.S., Chen, C., Hoffman, E.A., Saha, P.K.: A robust and efficient curve skeletonization algorithm for tree-like objects using minimum cost paths. Pattern Recogn. Lett. **76**(C), 32–40 (2016)

8. Lesage, D., Angelini, E.D., Funka-Lea, G., Bloch, I.: Adaptive particle filtering for coronary artery segmentation from 3D CT angiograms. Comput. Vis. Image Underst. **151**, 29–46 (2016). Probabilistic Models for Biomedical Image Analysis

9. Maragos, P., Schafer, R.: Morphological skeleton representation and coding of binary images. IEEE Trans. Acoust. Speech Sig. Process. **34**, 1228–1244 (1986)

10. Metz, C.T., Schaap, M., Weustink, A.C., Mollet, N.R., van Walsum, T., Niessen, W.J.: Coronary centerline extraction from CT coronary angiography images using a minimum cost path approach. Med. Phys. **36**(12), 5568–5579 (2009)

11. Mirikharaji, Z., Zhao, M., Hamarneh, G.: Globally-optimal anatomical tree extraction from 3d medical images using pictorial structures and minimal paths. In: Descoteaux, M., Maier-Hein, L., Franz, A., Jannin, P., Collins, D.L., Duchesne, S. (eds.) MICCAI 2017. LNCS, vol. 10434, pp. 242–250. Springer, Cham (2017). https://doi.org/10.1007/978-3-319-66185-8_28

12. Mnih, V., et al.: Playing atari with deep reinforcement learning. arXiv preprint arXiv:1312.5602 (2013). https://arxiv.org/pdf/1312.5602.pdf

13. Sato, M., Bitter, I., Bender, M.A., Kaufman, A.E., Nakajima, M.: TEASAR: tree-structure extraction algorithm for accurate and robust skeletons. In: Proceedings the 8th Pacific Conference on Computer Graphics and Applications, pp. 281–449 (October 2000). https://doi.org/10.1109/PCCGA.2000.883951

14. Schaap, M., et al.: Robust shape regression for supervised vessel segmentation and its application to coronary segmentation in CTA. IEEE Trans. Med. Imaging **30**(11), 1974–1986 (2011)

15. Sironi, A., Lepetit, V., Fua, P.: Multiscale centerline detection by learning a scale-space distance transform. In: 2014 IEEE Conference on Computer Vision and Pattern Recognition, pp. 2697–2704 (June 2014)

16. Stefancik, R., Sonka, M.: Highly automated segmentation of arterial and venous trees from three-dimensional magnetic resonance angiography (MRA). Int. J. Cardiovasc. Imaging **17**, 37–47 (2001)

17. Wink, O., Frangi, A.F., Verdonck, B., Viergever, M.A., Niessen, W.J.: 3D MRA coronary axis determination using a minimum cost path approach. Magn. Reson. Med. **47**(6), 1169–1175 (2002)

18. Wolterink, J.M., van Hamersvelt, R.W., Viergever, M.A., Leiner, T., Išgum, I.: Coronary artery centerline extraction in cardiac CT angiography using a CNN-based orientation classifier. Med. Image Anal. **51**, 46–60 (2019)

19. Xia, Qing, Yao, Yuxin, Hu, Zhiqiang, Hao, Aimin: Automatic 3D atrial segmentation from GE-MRIs using volumetric fully convolutional networks. In: Pop, M., et al. (eds.) STACOM 2018. LNCS, vol. 11395, pp. 211–220. Springer, Cham (2019). https://doi.org/10.1007/978-3-030-12029-0_23

20. Yang, G., et al.: Automatic centerline extraction of coronary arteries in coronary computed tomographic angiography. Int. J. Cardiovasc. Imaging **28**(4), 921–933 (2012)

21. Yang, H., Chen, J., Chi, Y., Xie, X., Hua, X.: Discriminative coronary artery tracking via 3D CNN in cardiac CT angiography. In: Shen, D., et al. (eds.) MICCAI 2019. LNCS, vol. 11765, pp. 468–476. Springer, Cham (2019). https://doi.org/10.1007/978-3-030-32245-8_52

22. Zhang, P., Wang, F., Zheng, Y.: Deep reinforcement learning for vessel centerline tracing in multi-modality 3D volumes. In: Frangi, A.F., Schnabel, J.A., Davatzikos, C., Alberola-López, C., Fichtinger, G. (eds.) MICCAI 2018. LNCS, vol. 11073, pp. 755–763. Springer, Cham (2018). https://doi.org/10.1007/978-3-030-00937-3_86
23. Zheng, Y., Tek, H., Funka-Lea, G.: Robust and accurate coronary artery centerline extraction in CTA by combining model-driven and data-driven approaches. In: Mori, K., Sakuma, I., Sato, Y., Barillot, C., Navab, N. (eds.) MICCAI 2013. LNCS, vol. 8151, pp. 74–81. Springer, Heidelberg (2013). https://doi.org/10.1007/978-3-642-40760-4_10
24. Zhou, C., et al.: Automated coronary artery tree extraction in coronary CT angiography using a multiscale enhancement and dynamic balloon tracking (MSCAR-DBT) method. Comput. Med. Imaging Graph. 36(1), 1–10 (2012)

Sequential Gaussian Process Regression for Simultaneous Pathology Detection and Shape Reconstruction

Dana Rahbani[✉], Andreas Morel-Forster, Dennis Madsen, Jonathan Aellen, and Thomas Vetter

Department of Mathematics and Computer Science, University of Basel, Basel, Switzerland
dana.rahbani@unibas.ch

Abstract. In this paper, we view pathology segmentation as an outlier detection task. Hence no prior on pathology characteristics is needed, and we can rely solely on a statistical prior on healthy data. Our method is based on the predictive posterior distribution obtained through standard Gaussian process regression. We propose a region-growing strategy, where we incrementally condition a Gaussian Process Morphable Model on the part considered healthy, as well as a dynamic threshold, which we infer from the uncertainty remaining in the resulting predictive posterior distribution. The threshold is used to extend the region considered healthy, which in turn is used to improve the regression results. Our method can be used for detecting missing parts or pathological growth like tumors on a target shape. We show segmentation results on a range of target surfaces: mandible, cranium and kidneys. The algorithm itself is theoretically sound, straight-forward to implement and extendable to other domains such as intensity-based pathologies. Our implementation is made open source with the publication.

Keywords: Gaussian process regression · Statistical shape models · Anomaly detection · Sequential learning

1 Introduction

Pathology detection aims to segment a given target into healthy and unhealthy regions. Common organ segmentation approaches rely on labeled data for training discriminative classification algorithms, such as neural networks, or on an underlying distribution that characterizes the organ's shape or intensity, such as with Bayesian classification approaches [6,23]. Both approaches assume that the provided data is sufficient to model the true expected variations of the target. However, when the target is pathological, this requirement is not met.

Code available at https://github.com/unibas-gravis/sequential-gpmm.

M. de Bruijne et al. (Eds.): MICCAI 2021, LNCS 12905, pp. 429–438, 2021.
https://doi.org/10.1007/978-3-030-87240-3_41

Due to the high variability of pathologies, it is often impossible to construct an underlying pathology distribution from a given dataset. Some specific pathologies are only rarely observed, while others demand high costs and risks to be imaged. Further, unlike healthy segmentation targets, pathologies are not limited to specific shapes, sizes, or intensities. As a result, available pathological data is not a representative sample from the full underlying class distribution. Therefore, algorithms that solely rely on data have low generalization, which motivates the use of more general characterizations of pathologies [11,18]. This problem has been encountered in signal processing, where it is known as anomaly or outlier detection. Outlier detection aims to find novel data that are different from previously evaluated datapoints. The core idea behind anomaly detection is one-class classification. From this viewpoint, outlier detection is not based on an outlier model. Instead, it is achieved by learning a threshold on the inlier model [18]. The main challenge in this approach is determining the threshold when the inliers are not labeled in advance, which is the case in practice.

In this paper, we view pathology segmentation as an outlier detection problem. For a target surface, e.g. extracted from a medical image volume, we perform classification of its vertices; healthy points on the surface are inliers, while pathological ones are outliers. We utilize Gaussian process regression to infer the underlying inlier distribution and show how to learn a dynamic threshold on the regressed posterior distribution which adapts to the regression uncertainty. In the experiments, we apply our method to missing data, where outliers are points that exist on a reference shape but not on the target, as well as deformed data and cancerous growth, where outliers are points that exist on the target but not on the reference shape. This general approach detects shape deviations significantly different from a shape prior learned solely from healthy data. We show results of successful pathology detection on two MICCAI challenges. Further work is needed to test the validity of the approach on other pathology types or data modalities.

1.1 Related Work

One-class classification methods are not new to outlier detection in images and surfaces. In this section, we group some of the major previous works into two categories: reconstruction-based and probabilistic [18].

Reconstruction-based approaches regress a healthy data model on a novel target. Points in the reconstruction with high errors are labeled as pathologies. Recent approaches utilize generative adversarial networks (GANs) or variational autoencoders (VAEs) trained on healthy datasets [22,24]. The threshold in this case is the pixel-wise average intensity of healthy data. Other strategies designed specifically for mesh reconstruction are robust variations of the Iterative Closest Point (ICP) algorithm, where the sorted reconstruction errors are trimmed to classify outliers [4] or used to assign weights that filter outlier points out [1,11]. The threshold in this case is often based on the sorted vertex reconstruction errors. There are two main disadvantages behind these approaches. The first comes from the use of the full healthy model to perform the reconstruction.

The success of the method is heavily dependent on the balance between generalization and specificity of the model. It is not possible to ensure that reconstruction errors in inlier regions will be low, since targets with pathologies can result in a poor reconstruction from the full model. The second disadvantage is the predetermined threshold which fails to take into account the reconstruction quality. We tackle these problems with our region-growing strategy that fits the target locally, before attempting a full reconstruction with a varying threshold. Previous reconstruction-based approaches with region-growing also exist. Perhaps the most famous of these strategies is RANSAC [7], where a random subset of the target points is first regressed. If the reconstruction quality across the full target has high errors, then the initial subset is considered to include outliers. The process is repeated until the reconstruction errors are below an acceptance threshold. The major disadvantage of RANSAC is that it assumes that correct correspondences are available. Another example more specific to mesh reconstruction is the robust moving least squares (RMLS) fitting algorithm, in which locally smooth reconstructions are first obtained and used to classify neighboring vertices in a forward-search algorithm [8]. We improve on these strategies by taking informed region-growing steps instead of sparse samples, giving a more accurate outlier map without requiring correspondences as input.

In contrast to reconstruction based methods, probabilistic approaches make use of the healthy model distribution to learn a classification threshold. The underlying distribution can be obtained by mapping the data to a space where it is Gaussian distributed. This enables the use of confidence intervals as classification thresholds [12, 21]. Early approaches use statistical shape models (SSMs) for the healthy model. Reconstruction residual errors are assumed to be Gaussian i.i.d., based on which points are labeled as pathologies [5] when their residual errors are some standard deviations away from the mean. These approaches impose a fixed threshold on the reconstruction errors. This causes problems in scenarios with low signal-to-noise ratios or with residual errors that vary along the target. For example, the i.i.d. assumption used to obtain the threshold does not hold since reconstruction errors have different distributions in the inlier and outlier regions. We provide a solution that uses the reconstruction confidence itself as quantified by the predictive posterior distribution from Gaussian process regression to update the expected distribution of each residual and the threshold.

1.2   Main Contributions

We propose a pathology segmentation algorithm inspired from general anomaly detection algorithms that uses the predictive posterior distribution (PPD) obtained by regressing the Gaussian process (GP) on inlier points as in GP one-class classification [12]. Our key contributions are:

- A pathology detection algorithm, independent of the target pathology.
- Sequential learning integration into the traditional GPMM and SSM pipelines.

– A dynamic threshold for pathology detection based on the regressed PPD to account for the number, locations, and fitting quality of the inliers.

2   Background

We provide the necessary background on GP regression for calculating the PPD using the 3D discrete case as used in shape analysis. A detailed explanation for the 1D case over a continuous domain can be found in Sect. 6.4 of [3].

Gaussian process morphable models (GPMMs) [14] model a shape s as a deformation $u : \Omega \to \mathbb{R}^3$ from a reference shape $\Gamma_R \subset \mathbb{R}^3$

$$s = \{x + u(x) | x \in \Gamma_R\}. \tag{1}$$

The deformation is modeled as $u \sim GP(\mu, k)$. We refer to common statistical shape models (SSMs) represented as a GP by writing μ_{SSM} and k_{SSM} for the mean and covariance, both estimated from a set of training shapes in correspondence. Given deformations that map a subset $\{x_1, x_2, \ldots, x_n\}$ of the m reference vertices to those on a target shape, the PPD on the remaining vertices $\{x_{n+1}, x_{n+2}, \ldots, x_m\}$ is available in closed-form such that

$$\mu_p(x) = \mu(x) + K_X(x)^T (K_{XX} + \sigma^2 I)^{-1} \hat{U} \tag{2}$$

$$k_p(x, x') = k(x, x') - K_X(x)^T (K_{XX} + \sigma^2 I)^{-1} K_X(x'). \tag{3}$$

The closed form solutions are used to infer the parameters of the marginal distributions in the conditioned GP at each of the remaining vertices. For a vertex x, the mean depends on the mean $\mu(x)$ of the initial GP, the covariance between x and the n observed points $K_X(x) = (k(x, x_i))_{i=1}^{n} \in \mathbb{R}^{3n \times 3}$, the variance of the training points $K_{XX} = (k(x_i, x_j))_{i,j=1}^{n} \in \mathbb{R}^{3n \times 3n}$ and the mean free observed deformation values $\hat{U} = ((u(x_1) - \mu(x_1))^T, \ldots, (u(x_n) - \mu(x_n))^T)^T \in \mathbb{R}^{3n}$. On the other hand, the remaining covariance is independent of the observed deformation values and only requires the evaluation of the kernel between the novel point x and the location of the training points on the domain [20].

The PPD has multiple known characteristics [3] that our method can take advantage of. First, given the smooth Gaussian kernel and low-rank approximation, the inferred variance is smaller for points on the domain which are closer to the observed points $\{x_1, \ldots, x_n\}$. This implies that, under the PPD, the model is more certain about deformations at the mean at x_{n+1} if it lies close to any of the observation locations $\{x_1, \ldots, x_n\}$ and less certain about faraway regions. This motivates our decision to only look at direct neighboring points of the inliers. Second, under the same conditions, the variance remaining at any point under the PPD decreases as the number of observations increases. We use the inverse of the PPD variance to quantify model confidence about unobserved regions. To counter the variance drop [12], we introduce a method to update the threshold as more observations are included in the PPD. Finally, the order of introducing the observations does not influence the final predictive posterior distribution;

conditioning on x_1 then on x_2 gives the same predictive posterior distribution for x_{n+1} that would be obtained by directly conditioning on x_1, x_2. Similar to sequential Bayesian regression, which performs online learning by evaluating the observations one after the other [3], we introduce observations gradually using a region-growing approach on the model vertices.

3   Method: Sequential GPMM Regression

Our algorithm uses sequential GP regression by iterating the following steps: 1. Calculate the PPD based on the current inliers. 2. Update the dynamic threshold based on the remaining uncertainty and residuals of the regressed inliers. 3. Update the set of inliers by applying the threshold on the remaining residuals and PPD uncertainty. The detailed algorithm for our sequential GPMM regression is shown in Algorithm 1:

Algorithm 1: Sequential GPMM regression

> **input**   : prior model GP^0, target Γ_T, inlier pairs $L^{t=0}$, confidence interval: CI
> **Output:** label map $\mathbf{z}^t$, PPD GP^t
> **repeat**
> > 1. Update PPD GP^t using GP regression with L^{t-1} as observations
> > 2. Compute μ^t and σ^2 from logged Mahalanobis distances $\boldsymbol{d}^t := \log(d_M^t(\hat{x}_i^t))$
> > $\forall i : z_i^{t-1} = 1$
> > 3. Update the threshold using the percentile value:
> > $\tau_{upper}^t := P_{0.5 + \frac{CI}{2.0}}(\mathcal{N}(\mu^t, \sigma^t))$
> > 4. Update labels: $z_j^t := 1$ if $\log(d_M^t(\hat{x}_j)) < \tau_{upper}^t \ \forall j \in N(x_i)$ and $z_i = 1$
> > 5. Update L^t using $x_i^t \ \forall i : z_i^t = 1$
> **until** $\mathbf{z}^t = \mathbf{z}^{t-1}$;

Initialization. Prior to the algorithm, the target is rigidly aligned to the model based on a set of initial inlier pairs $L^{t=0} := (x_i, \hat{x}_i)$ where $\hat{x}_i$ is the landmark on the target and x_i is its corresponding landmark on the reference shape. An initial label map is created with $z_i^{t=0} := 1$ if x_i is one of the landmarks else $z_i^{t=0} := 0$.

Dynamic Threshold and CI. A fixed threshold cannot be used, because the uncertainty of the model decreases as more points are used in the regression and hence it would start to also exclude inliers. In [2], the authors proposed to scale the errors with a manually adjusted factor, which is difficult to tune. Instead, we update the threshold automatically based on the current inlier distribution. The inlier residuals during regression are not Gaussian but right-skewed, shown in the Mahalanobis distance histograms in Fig. 1. By taking the logarithm of the residuals, we map them to a Gaussian distribution where a confidence interval (CI) can be computed. The single-sided CI is computed using the upper percentile $0.5 + \frac{CI}{2.0}$. The CI represents a trade-off between safety and speed. We opt

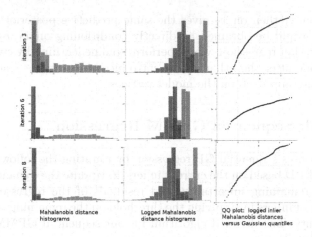

iteration 3

iteration 6

iteration 8

Mahalanobis distance
histograms

Logged Mahalanobis
distance histograms

QQ plot: logged inlier
Mahalanobis distances
versus Gaussian quantiles

Fig. 1. Mahalanobis distance histograms of the current inlier (blue) and outlier (red) obtained during sequential GPMM regression of a target with true inlier ratio = 0.66. First two rows (iteration steps 3 and 6): in red are domain points not yet evaluated. Third row (final iteration): in red are outliers detected as pathologies. Neighboring points of inliers with their distances below the threshold (vertical black line) are updated to inliers. The distribution of logged Mahalanobis distances in the second column is Gaussian distributed, confirmed by the QQ-plots in the third column. (Color figure online)

for a 0.3 CI, meaning a threshold at the 0.65 percentile which takes safe growing steps at the cost of more iterations for synthetic experiments, and increase it to 0.5 CI for real experiments since the residuals are larger.

Classification. Only outlier vertices with $z_j^{t-1} = 0$ which have an inlier vertex as neighbor are evaluated. If the Mahalanobis distance of $\hat{x}_j$ to x_j is smaller than the dynamic threshold, then its label is updated to $z_j^t := 1$.

4   Evaluation

To build a GPMM, we first extract meshes from volumetric CT images using thresholding and align them afterwards. We then bring the aligned meshes into correspondence and build the GPMM using the pipeline of [9].

We compare our sequential GPMM results to algorithms introduced earlier: nonrigid ICP (NICP) for reconstruction-based over the full model, RANSAC for reconstruction-based with region growing, and CPD as probabilistic approach. *NICP* is a robust extension of ICP with optimal update steps [1]. We initialize the regularization hyperparameter for template stiffness to $\alpha = 512$ and half it every 2 iterations until it reaches 0.5. We use the vertex weights determined by the algorithm to obtain the pathology map [1].

RANSAC [7] starts with a random subset of model vertices (set at 30%) labeled as initial inliers in **z**. The predictive posterior distribution is then used to compute the consensus set: points in the reconstruction with Mahalanobis distance less than a threshold (set at 3.0). If the consensus set is big enough (set at 60% of the reference points), the labels are updated and reconstruction continues with the new **z**. Fitting is repeated, each time with a different random sample of inlier vertices. Finally, the subset which has the lowest errors in the reconstruction is chosen. Our implementation is inspired by the illumination parameter estimation strategy of [6].

CPD performs robust probabilistic registration of two point sets as Gaussian mixture model density estimation. We set the weight vertices w to the true pathology ratio and obtain the label map from the sparse matrix after convergence [16].

The same post-processing cleanup step is applied to all results: all outliers in the final label map without any neighboring outlier vertices are labeled as healthy.

Synthetic Pathologies. We use a shape model of a mandible built from 9 outlier-free shapes extracted from CT scans. We generate an instance from the model, then introduce deformations that influence $\frac{1}{20}$, $\frac{1}{10}$, $\frac{1}{5}$ or $\frac{1}{3}$. These deformations represent synthetic pathologies that should be detectable and can be seen in Fig. 3. In this way, we have ground truth outlier labels as well as reconstructions. We use the F1 score (best at 1.0) to evaluate the pathology segmentation on the reference and the average and Hausdorff distances to evaluate the reconstruction in the inlier region (best at 0.0 mm). Results in Fig. 2 show stable performance from our sequential GPMM regression as the pathology size increases, outperforming the other three approaches. In comparison, the performance of the three

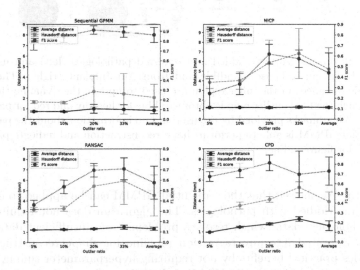

Fig. 2. The comparison of average and Hausdorff distance (AD and HD) and F1 score of 40 synthetic test cases shows the superiority of our method.

other approaches deteriorates as the pathology size increases. Although CPD performs similarly in detection, the reconstruction scores are poor because of the missing shape prior. On the other hand, RANSAC succeeds in the reconstruction but fails in the detection task because the sparse random samples do not result in a smooth label map.

Real Data. We use data from the AutoImplant [13] challenge for cranium reconstruction and the KiTS19 [10] challenge for kidney tumor detection. For model building, the pathological training data is never used. Instead, we use an in-house cranium GPMM and build a kidney GPMM from the healthy sides in the provided training set. This is possible because the method is generalizable: the SSM and novel target should represent the same shape family but do not have to come from the same dataset. We obtain an average dice score of 0.68 compared to 0.942 on the AutoImplant challenge leaderboard. Our solution is most similar to that from [19], but using our sequential GPMM regression for pathology detection removes the need to train an additional GAN. Our mean of the kidney and tumor dice scores is 0.74 compared to 0.9168 on the KiTS19 leaderboard. Note, we only use the information from shape while others also take intensities into account. Examples in Fig. 3 show reconstruction and detection results.

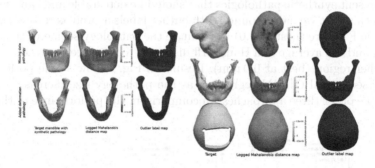

Fig. 3. Results using synthetic added and removed pathologies (left) and using real data from KiTS, an in-house mandible and from AutoImplant (right). The logged Mahalanobis distances in the inlier region are used to update the dynamic threshold, based on which the final outlier map is obtained: blue for healthy and red for pathology. Using up to 3 landmarks provided by the user and no information about the pathology, the sequential GPMM is able to perform shape reconstruction and indicate pathology regions. (Color figure online)

Discussion. We showed that the sequential GPMM can be applied to different target organs riddled with pathologies. The algorithm does not require training on pathological data, which resolves problems from insufficient datasets. It can be incorporated into any regression pipeline for shape reconstruction. It also provides practical benefits by not requiring hyperparameter tuning such as setting the fraction of inliers (RANSAC and CPD) or the regularization parameters (NICP). A faster growth would be possible with probabilistic CPD-like or surface feature-based predicted correspondences [1]. Further, the provided landmarks can be replaced by automatic landmarking tools to increase the number of the initial landmarks [17]. As these tools come with the risk of low accuracy because they are not trained on pathological data, we could add translation and rotation to the estimated shape parameters [15].

5   Conclusion

We presented the sequential GPMM algorithm as a one-class classification solution which is able to detect diverse pathologies. One key contribution of this work is the dynamic threshold necessary to obtain an accurate PPD. This threshold is estimated based on the reconstruction quality and regression uncertainties of the inliers, and is applied on the marginalized outlier PPD for classification. Unlike previous strategies, the sequential GPMM does not have to be tuned to account for different pathologies. Only the confidence interval that indicates how similar to healthy data new inliers can be is set by the user. The sequential GPMM is promising because it does not rely on any pathological data for training, avoids organ-specific tuning and performs the detection task regardless of pathology type or size. Future directions include extending the algorithm to the image domain through a voxel intensity PPD.

Acknowledgments. We thank the Zurich Institute of Forensic Medicine for providing the mandible with missing teeth as an example application.

References

1. Amberg, B., Romdhani, S., Vetter, T.: Optimal step nonrigid ICP algorithms for surface registration. In: 2007 IEEE Conference on Computer Vision and Pattern Recognition, pp. 1–8. IEEE (2007)
2. Babin, P., Giguere, P., Pomerleau, F.: Analysis of robust functions for registration algorithms. In: 2019 International Conference on Robotics and Automation (ICRA), pp. 1451–1457. IEEE (2019)
3. Bishop, C.M.: Pattern Recognition and Machine Learning. Springer, New York (2006)
4. Chetverikov, D., Svirko, D., Stepanov, D., Krsek, P.: The trimmed iterative closest point algorithm. In: Object Recognition Supported by User Interaction for Service Robots, vol. 3, pp. 545–548 (2002). https://doi.org/10.1109/ICPR.2002.1047997
5. Dufour, P.A., Abdillahi, H., Ceklic, L., Wolf-Schnurrbusch, U., Kowal, J.: Pathology hinting as the combination of automatic segmentation with a statistical shape model. In: Ayache, N., Delingette, H., Golland, P., Mori, K. (eds.) MICCAI 2012. LNCS, vol. 7512, pp. 599–606. Springer, Heidelberg (2012). https://doi.org/10.1007/978-3-642-33454-2_74
6. Egger, B., Schneider, A., Blumer, C., Forster, A., Schönborn, S., Vetter, T.: Occlusion-aware 3D morphable face models. In: BMVC, vol. 2, p. 4 (2016)
7. Fischler, M.A., Bolles, R.C.: Random sample consensus: a paradigm for model fitting with applications to image analysis and automated cartography. Commun. ACM **24**(6), 381–395 (1981)

8. Fleishman, S., Cohen-Or, D., Silva, C.T.: Robust moving least-squares fitting with sharp features. ACM Trans. Graph. (TOG) **24**(3), 544–552 (2005)
9. Gerig, T., et al.: Morphable face models-an open framework. In: 2018 13th IEEE International Conference on Automatic Face & Gesture Recognition, FG 2018, pp. 75–82. IEEE (2018)
10. Heller, N., et al.: The KiTS19 challenge data: 300 kidney tumor cases with clinical context, CT semantic segmentations, and surgical outcomes. arXiv preprint arXiv:1904.00445 (2019)
11. Hontani, H., Matsuno, T., Sawada, Y.: Robust nonrigid ICP using outlier-sparsity regularization. In: 2012 IEEE Conference on Computer Vision and Pattern Recognition, pp. 174–181. IEEE (2012)
12. Kemmler, M., Rodner, E., Wacker, E.S., Denzler, J.: One-class classification with Gaussian processes. Pattern Recogn. **46**(12), 3507–3518 (2013)
13. Li, J., Egger, J. (eds.): AutoImplant 2020. LNCS, vol. 12439. Springer, Cham (2020). https://doi.org/10.1007/978-3-030-64327-0
14. Lüthi, M., Gerig, T., Jud, C., Vetter, T.: Gaussian process morphable models. IEEE Trans. Pattern Anal. Mach. Intell. **40**(8), 1860–1873 (2017)
15. Morel-Forster, A., Gerig, T., Lüthi, M., Vetter, T.: Probabilistic fitting of active shape models. In: Reuter, M., Wachinger, C., Lombaert, H., Paniagua, B., Lüthi, M., Egger, B. (eds.) ShapeMI 2018. LNCS, vol. 11167, pp. 137–146. Springer, Cham (2018). https://doi.org/10.1007/978-3-030-04747-4_13
16. Myronenko, A., Song, X.: Point set registration: coherent point drift. IEEE Trans. Pattern Anal. Mach. Intell. **32**(12), 2262–2275 (2010)
17. Payer, C., Štern, D., Bischof, H., Urschler, M.: Integrating spatial configuration into heatmap regression based CNNs for landmark localization. Med. Image Anal. **54**, 207–219 (2019)
18. Pimentel, M.A., Clifton, D.A., Clifton, L., Tarassenko, L.: A review of novelty detection. Sig. Process. **99**, 215–249 (2014)
19. Pimentel, P., et al.: Automated virtual reconstruction of large skull defects using statistical shape models and generative adversarial networks. In: Li, J., Egger, J. (eds.) AutoImplant 2020. LNCS, vol. 12439, pp. 16–27. Springer, Cham (2020). https://doi.org/10.1007/978-3-030-64327-0_3
20. Rasmussen, C.E., Williams, C.K.I.: Gaussian Processes for Machine Learning. Adaptive Computation and Machine Learning. MIT Press, Cambridge, Mass (2006)
21. Roth, V.: Kernel fisher discriminants for outlier detection. Neural Comput. **18**(4), 942–960 (2006)
22. Schlegl, T., Seeböck, P., Waldstein, S.M., Schmidt-Erfurth, U., Langs, G.: Unsupervised anomaly detection with generative adversarial networks to guide marker discovery. In: Niethammer, M., et al. (eds.) IPMI 2017. LNCS, vol. 10265, pp. 146–157. Springer, Cham (2017). https://doi.org/10.1007/978-3-319-59050-9_12
23. Schnider, E., Horváth, A., Rauter, G., Zam, A., Müller-Gerbl, M., Cattin, P.C.: 3D segmentation networks for excessive numbers of classes: distinct bone segmentation in upper bodies. In: Liu, M., Yan, P., Lian, C., Cao, X. (eds.) MLMI 2020. LNCS, vol. 12436, pp. 40–49. Springer, Cham (2020). https://doi.org/10.1007/978-3-030-59861-7_5
24. You, S., Tezcan, K., Chen, X., Konukoglu, E.: Unsupervised lesion detection via image restoration with a normative prior. In: International Conference on Medical Imaging with Deep Learning - Full Paper Track, London, United Kingdom, 08–10 July 2019 (2019). https://openreview.net/forum?id=S1xg4W-leV

Predicting Symptoms from Multiphasic MRI via Multi-instance Attention Learning for Hepatocellular Carcinoma Grading

Zelin Qiu[1,2,3], Yongsheng Pan[1,2,3], Jie Wei[1,2,3], Dijia Wu[4(✉)], Yong Xia[1,2(✉)], and Dinggang Shen[3,4(✉)]

[1] National Engineering Laboratory for Integrated Aero-Space-Ground-Ocean Big Data Application Technology, School of Computer Science and Engineering, Northwestern Polytechnical University, Xi'an 710072, Shaanxi, China
[2] Research and Development Institute of Northwestern Polytechnical University in Shenzhen, Shenzhen 518057, China
yxia@nwpu.edu.cn
[3] School of Biomedical Engineering, ShanghaiTech University, Shanghai, China
dgshen@shanghaitech.edu.cn
[4] Shanghai United Imaging Intelligence Co., Ltd., Shanghai, China
dijia.wu@united-imaging.com

Abstract. Liver cancer is the third leading cause of cancer death in the world, where the hepatocellular carcinoma (HCC) is the most common case in primary liver cancer. In general diagnosis, accurate prediction of HCC grades is of great help to the subsequent treatment to improve the survival rate. Rather than to straightly predict HCC grades from images, it will be more interpretable in clinic to first predict the symptoms and then obtain the HCC grades from the Liver Imaging Reporting and Data System (LI-RADS). Accordingly, we propose a two-stage method for automatically predicting HCC grades according to multiphasic magnetic resonance imaging (MRI). The first stage uses multi-instance learning (MIL) to classify the LI-RADS symptoms while the second stage resorts LI-RADS to grade from the predicted symptoms. Since our method provides more diagnostic basis besides the grading results, it is more interpretable and closer to the clinical process. Experimental results on a dataset with 439 patients indicate that our two-stage method is more accurate than the straight HCC grading approach.

Keywords: Hepatocellular carcinoma · Multiphasic MRI · LI-RADS · Multi-instance learning

1 Introduction

Liver cancer is the third common leading cause among all the cancer-related death in the world, where hepatocellular carcinoma (HCC) accounts for nearly

This work is mainly completed under the collaboration of Z. Qiu and Y. Pan. Z. Qiu and Y. Pan—Contribute equally.

© Springer Nature Switzerland AG 2021
M. de Bruijne et al. (Eds.): MICCAI 2021, LNCS 12905, pp. 439–448, 2021.
https://doi.org/10.1007/978-3-030-87240-3_42

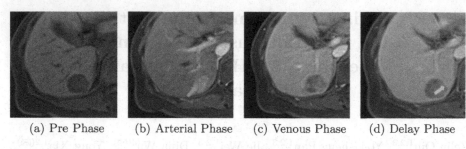

(a) Pre Phase   (b) Arterial Phase   (c) Venous Phase   (d) Delay Phase

Fig. 1. Illustration of phases and symptoms in LI-RADS. (a) Pre-contrast phase (Pre): before the contrast injection; (b) Hepatic arterial phase (Arterial): hepatic artery and branches are fully enhanced; (c) Portal venous phase (Venous): portal veins are fully enhanced; (d) Delayed phase (Delay): time after Venous phase where the enhancement is less. In addition, the LI-RADS symptoms are pointed out with arrows in the figures, such as norim APHE (APHE): norim-like enhancement in Arterial phase (orange arrow); nonperipheral washout (Washout): reduction of enhancement between Arterial and Venous phases (blue arrow); enhancing "capsule" (Capsule): enhanced border around the tumor in Venous and Delay phases (green arrow). (Color figure online)

85% cases in primary liver cancer. HCC generally has a poor prognosis and a high recurrent rate, with 5-year survival rate less than 10%, and even less than 5% for advanced patients [1,2]. Since its incidence and mortality increase in many countries around the world, HCC is very likely to become a public health issue in the coming future [3]. This will lead to a great clinical demand of a rapid and accurate diagnosis approach for HCC, which can help patients to gain treatment in the early stage.

Since multiphasic magnetic resonance imaging (MRI) can provide great convenience in clinical diagnosis without invasive diagnostic biopsies, it is widely used in HCC diagnosis [4,5]. Specially, each multiphasic MRI is a dynamic 3D gradient-echo sequence obtained before and after the injection of contrast medium. A typical sequence of multiphasic MRI is shown in Fig. 1. These multiphasic images contain the information of both contrast change and time sequence, and thus can help doctors to identify the histologic subtypes of HCC and predict recurrences [4,5].

Meanwhile, there is a widely accepted standard in the diagnoses of HCC, *i.e.*, the Liver Imaging Reporting and Data System (LI-RADS). This system can provide comprehensive criteria for multiphasic MRI and CT, as well as screening, diagnosis and treatment for HCC [6,7]. It categorizes liver tumors into 5 main grades, denoted by LR-1 to LR-5, among which LR-3 (probably malignancy), LR-4 (probably HCC), and LR-5 (definitely HCC) are more important in practical diagnosis [8].

Recently, deep-learning-based methods have made progress in liver cancer diagnosis with small reliance to subjective experience and high possibility of high-volume practice [8–12]. For example, some studies use adversarial networks [13] or 3D residual networks [14–16] for HCC grading from MR or CT

Table 1. Diagnostic table from LI-RADS v2018 [6].

Arterial phase hyperenhancement (APHE)		No APHE		Nonrim APHE		
Observation size (mm)		<20	≥20	<10	10–19	≥20
Count additional major features: • Capsule • Washout • Threshold growth^a	None	LR-3	LR-3	LR-3	LR-3	LR-4
	One	LR-3	LR-4	LR-4	LR-4 ^{b} / LR-5	LR-5
	≥Two	LR-4	LR-4	LR-4	LR-5	LR-5

^{a} Considering clinical practice, threshold growth (size of tumor increase of a mass ≥50% within six months) is neglected in this paper.
^{b} Choose LR-4 if "Capsule" or LR-5 if "Washout" or threshold growth.

images. As acquiring some clinical data [17], such as laboratory test result, is very time-consuming and costly, previous approaches mostly focus on how to determine HCC grade directly from images. However, deep neural networks, which map straightforward from high dimensional images into a single value (*i.e.*, the HCC grade), may easily sink into over-fitting due to small amount of training data. Meanwhile, these methods also lack interpretability in clinical practice.

It is noticeable that LI-RADS v2018 [6] provides a diagnostic table with grading from LR-3 to LR-5 for CT/MRI using observation size of tumor and four major features (to avoid ambiguity, in this paper we use "symptoms" instead) from the images (see Fig. 1 and Table 1). In order to avoid direct grading, as well as to keep compatible with LI-RADS v2018, we propose a method to classify the symptoms and utilize Table 1 for HCC grading. We separate the image-to-grade pipeline into two stages: "images to symptoms" and "symptoms to grade".

Accordingly, we propose a two-stage method for automated prediction of HCC grade using multiphasic MRI. In the first stage, multi-instance learning (MIL) [18] with a 2D ResNet backbone is used to classify LI-RADS symptoms. In the second stage, LI-RADS is adopted to grade HCC based on the classification of symptoms. Experimental results on a dataset with 439 subjects indicate that our two-stage HCC grading method is more accurate than the existing one-stage approaches. Meanwhile, since our method provides the classification of symptoms besides the grading results, it is more interpretable and closer to the clinical process.

2 Method

2.1 Model Overview

Figure 2 illustrates the framework of our proposed method, which contains four image encoders (*i.e.*, E_{pre}, E_{art}, E_{ven}, and E_{del}) to extract image features from four phasic images and also four symptom encoders (*i.e.*, E_a, E_w, E_{cv}, and E_{cd}) to extract semantic features associated with symptoms from image features. The

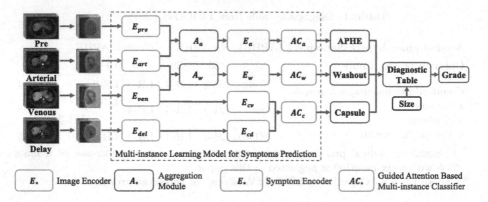

Fig. 2. Overview of the proposed approach. Symptoms are predicted by end-to-end multi-instance learning, while HCC grade is obtained by symptom predictions and tumor size according to the diagnostic table.

inputs of these image encoders are image patches with tumor cropped from the phasic images.

Since each symptom is associated with different phasic images, these encoders are combined in different ways to classify three symptoms. For symptoms "APHE" and "Washout", since they need to compare different phases, we first introduce the aggregation module to combine the features extracted from the corresponding phases, and then input the aggregated features to an encoder to extract semantic information. The features extracted by E_{pre} and E_{art} are aggregated and input to E_a, while the features extracted by E_{art} and E_{ven} are aggregated and input to E_w. Subsequently, the symptom features obtained by E_a and E_w are fed into classifiers AC_a and AC_w, respectively, to classify the "APHE" and "Washout" symptoms. For the symptom "Capsule", since it does not need comparing different phases but directly relies on either the Venous or the Delay phases, we directly input the features extracted by E_{ven} and E_{del} to E_{cv}, and E_{cd}, respectively. The symptom features generated by E_{cv} and E_{cd} are then concatenated and fed into classifier AC_c to classify the "Capsule" symptom. Finally, the HCC grade is obtained from LI-RADS diagnostic table based on the symptom predictions as well as the tumor size.

Considering the fact that these symptoms are mostly defined in 2D space and the structural information between slices may not be taken into account in clinical practice, we utilize 2D convolutional neural networks to build our model. Besides, we use multi-instance learning with guided attention module for classifying symptoms.

2.2 Modules

Encoder. There are two types of encoders in our framework, i.e., image encoders and symptom encoders. Notice that the same-type encoders share the same structure but independent in weights. We follow the structure of ResNet18 [19]

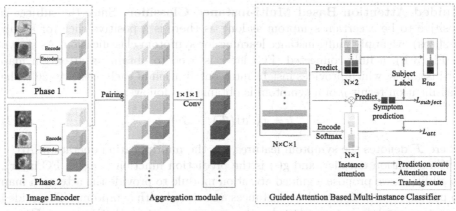

(a) Feature extraction and aggregation   (b) Symptom classification

Fig. 3. Detailed aggregation module and symptom prediction module. Each phasic image is encoded into features and then paired with neighbouring features from another phasic image. Next, the paired features are aggregated by a $1 \times 1 \times 1$ convolution (Conv). The symptom features are considered as instances to result a multi-instance learning for classification. Specially, an instance attention is generated from the feature and multiplied to the feature for symptom prediction.

to construct our model since it has good performance in many classification tasks. ResNet18 could be divided into four parts according to the size of feature maps, where we regard the first 2 parts as an image encoder and the rest 2 parts as a symptom encoder. Each image encoder extracts contrast information from a 3D phasic image by inputting it as serial 2D slice-level images. In this case, the output features keep same orders as original slices. Similarly, each symptom encoder accepts the slice-by-slice image features obtained by image encoders and outputs the slice-by-slice symptom features with same input orders.

Aggregation Module. Aggregation modules are designed to pair and aggregate features from different phases for symptoms "APHE" and "Washout". Considering the size of tumor may vary between far-apart slices, we only pair features with neighbouring features associated to another phase, e.g., the light green features in Fig. 3 will only be paired with light orange features and pale orange features. After that, the paired features are concatenated and fused by a $1 \times 1 \times 1$ 3D convolution. This procedure can be formulated as

$$\mathfrak{F}_n = \mathrm{Agg}(\boldsymbol{\mathcal{X}}_{1_i}, \ \boldsymbol{\mathcal{X}}_{2_j}), \ |i - j| \leq 1 \tag{1}$$

where $\mathfrak{F}_n$ denotes the n-th feature map after aggregation, $\boldsymbol{\mathcal{X}}_{m_i}$ is the i-th feature from the m-th phase, and $\mathrm{Agg}(\cdot)$ is the aggregation function (i.e., the $1 \times 1 \times 1$ convolution).

Guided Attention Based Multi-instance Classifier. Since the subject is positive to be a certain symptom as long as there is a positive slice (or a pair of slices), we apply multi-instance learning for symptom classification. As shown in Fig. 3(b), a fully-connected (FC) layer is used to obtain an attention map for instances, which provides insight into contribution of each instance [20]. The classification result $\hat{y}$ of a symptom is obtained as follows:

$$\hat{y} = g\left(\text{Att}(\mathcal{F})^{\text{T}} \cdot \mathcal{F}\right) \tag{2}$$

where $\mathcal{F}$ denotes the symptom features, $\cdot$ is the multiplication operation, $\text{Att}(\cdot)$ is the attention encoder, and $g(\cdot)$ is the prediction function (*i.e.*, an FC layer). In addition, we propose a guided attention module to provide the attention map to pay more attention to those instances associated with symptoms. The guided attention module is trained under the supervision of prediction error $E_{\text{ins}} = \text{CE}(g(\mathcal{F}),\ y)$ of each instance. Thus the total loss $\mathcal{L}_{\text{total}}$ can be formulated as:

$$\mathcal{L}_{\text{total}} = \text{FL}(\hat{y},\ y) + \alpha\text{MSE}\left(\text{Att}(\mathcal{F}),\ E_{\text{ins}}\right) \tag{3}$$

where $\text{FL}(\cdot)$, $\text{CE}(\cdot)$, and $\text{MSE}(\cdot)$ represent focal loss [21], cross-entropy loss, and mean-square error (MSE) loss, respectively, and α is the hyper-parameter.

2.3 Implementation Details

Medical images of small quantity are usually not sufficient for a deep model to train from scratch, hence we use the ResNet18 [22] pre-trained on ImageNet [23] to initialize encoders. To focus on tumor regions, we only extract features from the cropped patches with tumor contained. We use a segmentation task performed by a 3D-UNet as an upstream task to provide bounding box for each tumor. Herein, the tumor region is extended by $\lfloor -a \times Size_{\text{tumor}} + b \rfloor$ pixels along the slice plane to keep more boundary information for small tumors, where a and b are hyper-parameters. We consider every feature for symptom as an instance and the subject as a bag that contains all instances. Every instance accounts for labeling the bag, *i.e.*, classifying the symptoms. Notice that E_{ins} relies on the prediction of each instance, hence we train this part in two steps: First we train prediction layer without $\mathcal{L}_{\text{att}}$, then train attention encoder with guidance.

3 Experiment

3.1 Dataset and Experimental Setup

We evaluate our method on a dataset of 439 subjects, including 93, 98, and 248 HCC cases of grading LR-3, LR-4, and LR-5, respectively. All data were obtained using either a 1.5 T (Magnetom Aera, Siemens Healthcare, Erlangen, Germany) or a 3.0 T (uMR 770, United Imaging Healthcare, Shanghai, China) MR scanner with Gd-EOB-DTPA as contrast agent in same hospital and labeled by three radiologists with experience over 5 years. Each subject has four phasic images taken in Pre, Arterial, Venous, and Delay phases chronologically with the same spatial resolution of $1.0 \times 1.0 \times 3.0$ mm^3. We apply zero-mean normalization for each subject with its mean and standard deviation. The grade label, symptoms, and tumor region with observation size required in our experiments are provided in this dataset. The size of tumor in this dataset varies from 5 mm to 128 mm, hence we set the extending hyper-parameters $a = 0.05$, $b = 5$. In addition, we set α in Eq. 3 to 0.5.

Our experiments were conducted by MindSpore framework. All the competing methods are trained for 150 epochs using Adam optimizer with the initial learning rate of 0.025 and cosine annealing decay [24]. For data augmentation, each image is randomly shifted 0–3 pixel along the slice dimension before being cropped into patch. Each cropped patch is then resized to $28 \times 28 \times 28$ with trilinear interpolation.

3.2   Results

We compare the proposed method with four approaches, including two variants of the baseline method [16], and two variants of our proposed multi-instance learning. The baseline method (i) consists of two parts, with the first part using four individual 3D-ResNet to extract features from four phasic images and the second part using an FC layer to predict the HCC grade from the concatenation of previous features directly. The second method (ii) switches the target of FC layer in the baseline method from predicting HCC grade to first classify the LI-RADS symptoms and then obtain the HCC grade using diagnostic table. The third method (iii) replaces the backbone with 2D ResNet and applies multi-instance learning (MIL) for LI-RADS symptom classification. The fourth method (iv) is our method without guidance (G) of attention (A). The fifth method (v) is our method that uses 2D-ResNet backbone and multi-instance learning (MIL) for LI-RADS symptoms classification with attention guidance. For fair comparison, all these methods are trained and validated with the same dataset and same experimental settings.

The results of four metrics (accuracy, precision, recall, F1 Score) for HCC grading are given in Table 2. From Table 2, the following observations could be found. (1) Four advanced methods (ii, iii, iv, and v) consistently achieve significant improvements on all metrics compared to the baseline method, which demonstrates that the classification of LI-RADS symptoms largely contributes to HCC grading. (2) The deviation of three methods (iii, iv, and v) are obviously lower than the baseline method and its variant (i and ii). This indicates that using 2D ResNet and multi-instance learning can improve the robustness of these methods on such a small quantity of data. (3) The results of method iv drops slightly comparing to method iii while the deviation increases a little. This may be due to the fact that the attention module increases the model complexity and thus is easier to sink into over-fitting. (4) Our method (v) with attention guidance outperforms its two variants (iii and iv), which indicates that guiding attention map to focus on certain instances via prediction error (G) can further

Table 2. Grading results (%) of our experiment (Mean ± Standard deviation). L, A, G denote using LI-RADS, attention module, and guidance for attention, respectively.

	Methods	Accuracy	Precision	Recall	F1 score
i	Baseline	56.2 ± 3.4	60.2 ± 8.7	56.2 ± 3.4	57.8 ± 5.3
ii	Baseline + L	70.4 ± 5.9	70.8 ± 6.1	70.4 ± 5.9	70.2 ± 6.0
iii	MIL + L	70.2 ± 1.8	70.2 ± 2.2	70.2 ± 1.8	70.0 ± 2.1
iv	MIL + L + A	69.6 ± 3.0	70.2 ± 2.8	69.6 ± 3.0	69.8 ± 2.8
v	MIL + L + A + G (ours)	**72.0 ± 2.9**	**72.6 ± 1.9**	**72.0 ± 2.9**	**72.0 ± 2.1**

Table 3. AUC for classification of each symptom. L, A, G denote using LI-RADS, attention module, and guidance for attention, respectively.

	Method	APHE	Washout	Capsule	Average
ii	Baseline + L	0.91	0.79	0.64	0.78
iii	MIL + L	0.92	0.81	0.63	0.79
iv	MIL + L + A	0.89	0.82	0.61	0.77
v	MIL + L + A + G	**0.92**	**0.87**	**0.64**	**0.81**

improve the grading performance. In general, our method (**v**) gets performance gains from all four components, and is more potential to be used in clinic.

For quantitative comparison, we further report the area under the receiver operating characteristic curve (AUC) for classification of each single symptom in Table 3. It can be seen from Table 3 that our method achieves the highest AUC on all symptoms. Comparing to the baseline variant (**ii**), our method (**v**) increases differently on different symptoms. The most significant improvement is achieved on the "Washout" symptom, which sparsely appears in these MRI slices. This indicates that our guided attention module can focus on the most representative slices. In particular, the classification of "Capsule" is the most difficult symptom for all methods to classify. The underlying reason may lie in that the enhancement on the surface of liver organ is very likely to the enhancement of "Capsule", thus making it too hard for these methods to learn sufficient hidden information in small data. This also verifies the strong robustness of LI-RADS to misclassifed symptoms, considering that our method still yields good results in grading of HCC. This further suggests that our two-stage approach is reasonable, and more accurate than the straight HCC grading approach.

4   Conclusion

In this paper, we proposed a two-stage method for automatically hepatocellular carcinoma (HCC) grading. The first stage used the multi-instance learning (MIL) with 2D ResNet backbones to classify the LI-RADS symptoms, while the second stage resorted LI-RADS to grade according to the grading table from the predicted symptoms. This method is more interpretable and closer to the clinical process as these symptoms have clinical implications. Meanwhile, experimental results on a dataset with 439 patients indicate that our two-stage approach is also more accurate than the straight HCC grading approach.

Acknowledgments. This work was supported in part by the National Natural Science Foundation of China under Grants 61771397, in part by the CAAI-Huawei Mind-Spore Open Fund under Grants CAAIXSJLJJ-2020-005B, and in part by the China Postdoctoral Science Foundation under Grants BX2021333.

References

1. Sung, H., et al.: Global cancer statistics 2020: GLOBOCAN 2020 estimates of incidence and mortality worldwide for 36 cancers in 185 countries. CA Cancer J. Clin. **71**, 209–249 (2021)
2. El-Serag, H.B., Rudolph, K.L.: Hepatocellular carcinoma: epidemiology and molecular carcinogenesis. Gastroenterology **132**(7), 2557–2576 (2007)
3. Liu, Z., et al.: The trends in incidence of primary liver cancer caused by specific etiologies: results from the global burden of disease study 2016 and implications for liver cancer prevention. J. Hepatol. **70**(4), 674–683 (2019)
4. Mulé, S., et al.: Multiphase liver MRI for identifying the macrotrabecular-massive subtype of hepatocellular carcinoma. Radiology **295**(3), 562–571 (2020)
5. Block, K.T., Uecker, M., Frahm, J.: Undersampled radial MRI with multiple coils. Iterative image reconstruction using a total variation constraint. Magn. Reson. Med. **57**(6), 1086–1098 (2007)
6. American College of Radiology: Liver imaging reporting and data system version 2018. https://www.acr.org/Clinical-Resources/Reporting-and-Data-Systems/LI-RADS/
7. Chernyak, V., et al.: Liver imaging reporting and data system (LI-RADS) version 2018: imaging of hepatocellular carcinoma in at-risk patients. Radiology **289**(3), 816–830 (2018)
8. Wu, Y., et al.: Deep learning LI-RADS grading system based on contrast enhanced multiphase MRI for differentiation between LR-3 and LR-4/LR-5 liver tumors. Ann. Trans. Med. **8**(11), 701 (2020)
9. Li, X., Chen, H., Qi, X., Dou, Q., Fu, C.W., Heng, P.A.: H-DenseUNet: hybrid densely connected UNet for liver and tumor segmentation from CT volumes. IEEE Trans. Med. Imaging **37**(12), 2663–2674 (2018)
10. Kawka, M., Dawidziuk, A., Jiao, L.R., Gall, T.M.H.: Artificial intelligence in the detection, characterisation and prediction of hepatocellular carcinoma: a narrative review. Transl. Gastroenterol. Hepatol. (2020)
11. Hamm, C.A., et al.: Deep learning for liver tumor diagnosis part i: development of a convolutional neural network classifier for multi-phasic MRI. Eur. Radiol. **29**(7), 3338–3347 (2019)
12. Yamashita, R., et al.: Deep convolutional neural network applied to the liver imaging reporting and data system (LI-RADS) version 2014 category classification: a pilot study. Abdom. Radiol. **45**(1), 24–35 (2020)
13. Frid-Adar, M., Diamant, I., Klang, E., Amitai, M., Goldberger, J., Greenspan, H.: Gan-based synthetic medical image augmentation for increased CNN performance in liver lesion classification. Neurocomputing **321**, 321–331 (2018)

14. Liang, D., et al.: Residual convolutional neural networks with global and local pathways for classification of focal liver lesions. In: Geng, X., Kang, B.-H. (eds.) PRICAI 2018. LNCS (LNAI), vol. 11012, pp. 617–628. Springer, Cham (2018). https://doi.org/10.1007/978-3-319-97304-3_47

15. Yasaka, K., Akai, H., Abe, O., Kiryu, S.: Deep learning with convolutional neural network for differentiation of liver masses at dynamic contrast-enhanced CT: a preliminary study. Radiology **286**(3), 887–896 (2018)

16. Trivizakis, E., et al.: Extending 2-D convolutional neural networks to 3-D for advancing deep learning cancer classification with application to MRI liver tumor differentiation. IEEE J. Biomed. Health Inform. **23**(3), 923–930 (2018)

17. Zhen, S., et al.: Deep learning for accurate diagnosis of liver tumor based on magnetic resonance imaging and clinical data. Front. Oncol. **10**, 680 (2020)

18. Wu, J., Yu, Y., Huang, C., Yu, K.: Deep multiple instance learning for image classification and auto-annotation. In: Proceedings of the IEEE Conference on Computer Vision and Pattern Recognition, pp. 3460–3469 (2015)

19. He, K., Zhang, X., Ren, S., Sun, J.: Deep residual learning for image recognition. In: Proceedings of the IEEE Conference on Computer Vision and Pattern Recognition, pp. 770–778 (2016)

20. Ilse, M., Tomczak, J., Welling, M.: Attention-based deep multiple instance learning. In: International Conference on Machine Learning, pp. 2127–2136. PMLR (2018)

21. Lin, T.Y., Goyal, P., Girshick, R., He, K., Dollár, P.: Focal loss for dense object detection. In: Proceedings of the IEEE International Conference on Computer Vision, pp. 2980–2988 (2017)

22. Kermany, D.S., et al.: Identifying medical diagnoses and treatable diseases by image-based deep learning. Cell **172**(5), 1122–1131 (2018)

23. Deng, J., Dong, W., Socher, R., Li, L.J., Li, K., Fei-Fei, L.: ImageNet: a large-scale hierarchical image database. In: IEEE Conference on Computer Vision and Pattern Recognition 2009, pp. 248–255. IEEE (2009)

24. Loshchilov, I., Hutter, F.: SGDR: stochastic gradient descent with warm restarts. arXiv preprint arXiv:1608.03983 (2016)

Triplet-Branch Network with Prior-Knowledge Embedding for Fatigue Fracture Grading

Yuexiang Li[1(✉)], Yanping Wang[2], Guang Lin[2], Yi Lin[1], Dong Wei[1],
Qirui Zhang[2], Kai Ma[1], Guangming Lu[2], Zhiqiang Zhang[2], and Yefeng Zheng[1]

[1] Tencent Jarvis Lab, Shenzhen, China
vicyxli@tencent.com
[2] Department of Diagnostic Radiology, Jinling Hospital, Medical School
of Nanjing University, Nanjing, China

Abstract. In recent years, there has been increasing awareness of the occurrence of fatigue fractures. Athletes and soldiers, who engaged in unaccustomed, repetitive or vigorous activities, are potential victims of such a fracture. Due to the slow-growing process of fatigue fracture, the early detection can effectively protect athletes and soldiers from the material bone breakage, which may result in the catastrophe of career retirement. In this paper, we propose a triplet-branch network (TBN) for the accurate fatigue fracture grading, which enables physicians to promptly take appropriate treatments. Particularly, the proposed TBN consists of three branches for representation learning, classifier learning and grade-related prior-knowledge learning, respectively. The former two branches are responsible to tackle the problem of class-imbalanced training data, while the latter one is implemented to embed grade-related prior-knowledge into the framework via an auxiliary ranking task. Extensive experiments have been conducted on our fatigue fracture X-ray image dataset. The experimental results show that our TBN can effectively address the problem of class-imbalanced training samples and achieve a satisfactory accuracy for fatigue fracture grading.

Keywords: Fatigue fracture · Disease grading · X-ray image

1 Introduction

Fatigue/stress fracture, which occurs in bone subjected to repetitive stress, has been recognized in military recruits for many years. In recent years, there has been increasing awareness of the occurrence of fatigue fractures in the civilian population. Apart from soldiers, athletes engaged in unaccustomed, repetitive or vigorous activities are also the potential victim of such a fracture, *e.g.,* Yao Ming, a famous Chinese basketball player, retired early from his athletic career due to a

---

Y. Li and Y. Wang—Equal contribution.

© Springer Nature Switzerland AG 2021
M. de Bruijne et al. (Eds.): MICCAI 2021, LNCS 12905, pp. 449–458, 2021.
https://doi.org/10.1007/978-3-030-87240-3_43

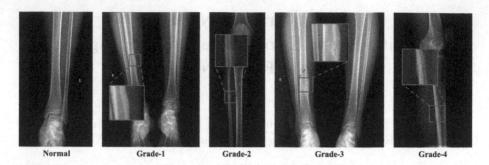

Normal   Grade-1   Grade-2   Grade-3   Grade-4

Fig. 1. Exemplars of fatigue fractures on different grades. The fractures in red boxes are zoomed for easy observations. (Color figure online)

fatigue fracture of navicular bone. Nevertheless, the cyclic stress loading to bone may repeat millions of times before the fracture is significant enough to cause material failure. Therefore, the early detection of fatigue fracture is extremely crucial to protect athletes and soldiers from material bone breakage. According to clinical guidelines, the fatigue fracture can be classified to four grades, which require different treatments—the low-grade (1 and 2) patients will recover after adequate rest, while the high-grade (3 and 4) ones may need surgery.

Witness the success of deep learning in computer vision, researchers [3,4,7] began to apply the convolutional neural networks to perform the automated fracture detection due to the potential to improve the speed and accuracy of diagnostics and offloading radiologists from time-intensive tasks. A thorough review of deep learning based applications on fracture detection can be found in [8]. However, the fatigue fracture grading is still an untouched problem. One of the major challenges for developing an automated grading system is the class-imbalance problem occurred in the training data, *i.e.*, the number of grade-2 samples is much higher than the other fracture grades. The underlying reason is that the grade-2 fatigue fracture begins to cause obvious pains to patients; thereby, they would go to hospital for screening and treatment.

Recently, various approaches have been proposed to alleviate the influence caused by class-imbalance problem. For examples, Cao *et al.* [1] proposed a label-distribution-aware margin (LDAM) loss to encourage larger margins for minority classes, which therefore improves the generalization error of minority classes without sacrificing the model's ability to fit the frequent ones. Zhou *et al.* [15] proposed a bilateral-branch network (BBN) to deal with the class-imbalance problem. The proposed BBN consists of two branches for representation learning and classifier learning, respectively, and is further equipped with a cumulative learning strategy to balance the learning phases of the two branches. However, the existing approaches share a common drawback—they do not explicitly include 'ranking' of visual similarity within the model training, which is obvious in fatigue fractures. As shown in Fig. 1, the grade-3 fatigue fracture is visually similar to the adjacent grades (2 and 4), rather than the grade-1 and normal.

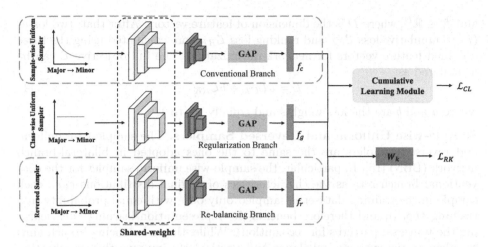

Fig. 2. The pipeline of the proposed triplet-branch network (TBN), which consists of three branches (*i.e.,* conventional, re-balancing and regularization) for feature representation learning, classifier learning and grade-related prior-knowledge learning, respectively. The features generated by the three branches (*i.e.,* f_c, f_r and f_g) are fed into the cumulative learning module and a fully-connected layer $\boldsymbol{W}_k$ for loss calculation.

Incorporating such class dependency prior-knowledge is prone to boost the disease grading accuracy of models by recent studies [6,11].

In this paper, we propose a triplet-branch network (TBN) to simultaneously deal with class-imbalance problem and class dependency embedding within a unified framework. Inspired by BBN [15], our TBN also consists of conventional and re-balancing branches to extract universal patterns and minority-class-specific features, respectively, which alleviate the influence caused by class-imbalanced training data. Apart from them, we integrate an additional regularization branch to enforce the model to learn grading-related prior-knowledge via an auxiliary ranking task. Extensive experiments have been conducted on our fatigue fracture X-ray image dataset. The experimental results show that our TBN can effectively address the problem of class-imbalanced training samples and achieve a satisfactory accuracy for fatigue fracture grading.

2 Method

As shown in Fig. 2, the proposed triplet-branch network (TBN) consists of three branches (termed conventional branch, re-balancing branch and regularization branch) for feature representation, classifier and grade-related prior-knowledge learning, respectively. Denote x as a training sample and $y \in \{1, 2, \cdots, C\}$ as its corresponding label, where C is the number of grades. For triplet branches, triplet-wise samples (x_c, x_r and x_g) are obtained using different samplers (*i.e.,* sample-wise uniform, reversed and class-wise uniform samplers) and then fed into their own corresponding branch to yield the feature vectors ($f_c \in \mathbb{R}^D$, $f_r \in \mathbb{R}^D$

and $f_g \in \mathbb{R}^D$, where D is the dimension of feature vector). After that, two losses (*i.e.*, cumulative loss $\mathcal{L}_{CL}$ and ranking loss $\mathcal{L}_{RK}$) are calculated using the three obtained feature vectors for network optimization. The full objective $\mathcal{L}$ can be formulated as:

$$\mathcal{L} = a\mathcal{L}_{CL} + b\mathcal{L}_{RK} \tag{1}$$

where a and b are the loss weights and equally set to 1.

Sample-wise Uniform and Reversed Samplers. The sample-wise uniform and reversed samplers are the same to the ones adopted in bilateral-branch network (BBN) [15]. In particular, the sample-wise uniform sampler for the conventional branch retains the characteristics of original distributions (*i.e.*, each sample in the training dataset is sampled only once with equal probability in a training epoch), and therefore benefits the representation learning (*i.e.*, learning the universal patterns for recognition). While, the re-balancing branch aims to alleviate the extreme imbalance and particularly improve the classification accuracy on minority classes, whose input data comes from a reversed sampler. For the reversed sampler, the sampling probability of each class is proportional to the reciprocal of its sample size, *i.e.*, the more samples in a class, the smaller sampling probability that class has.

Class-wise Uniform Sampler. Here, we reveal the drawback of BBN—the previous two data samplers are all based on the statistical probabilities, which may easily suffer from the overfitting problem, *i.e.*, the trained model is biased to the classes with higher sampling probabilities. To address the problem, we implement a random-probability-based sampler, termed class-wise uniform sampler, to the framework, which assigns equal sampling probabilities ($\frac{1}{C}$) to all classes and accordingly increases the variety of the selected triplet-wise samples.

2.1   Cumulative Learning Module

Our TBN consists of a cumulative learning module to integrate the knowledge learned by different branches. Concretely, a set of adaptive trade-off parameters (α and β) is utilized to control the weights of f_c, f_r and f_g. The weighted features αf_c, $(1-\alpha)f_r$ and βf_g are sent into the classifiers $\boldsymbol{W}_c \in \mathbb{R}^{D \times C}$, $\boldsymbol{W}_r \in \mathbb{R}^{D \times C}$ and $\boldsymbol{W}_g \in \mathbb{R}^{D \times C}$, respectively, for the calculation of classification error. Denoting $\mathbb{E}(.)$ as the cross-entropy loss function, the cumulative learning loss ($\mathcal{L}_{CL}$) can be formulated as:

$$\mathcal{L}_{CL} = \alpha\mathbb{E}(\boldsymbol{W}_c^\top f_c, y_c) + (1 - \alpha)\mathbb{E}(\boldsymbol{W}_r^\top f_r, y_r) + \beta\mathbb{E}(\boldsymbol{W}_g^\top f_g, y_g) \tag{2}$$

where y_c, y_r and y_g are the ground truth for the input samples of each branch.

As depicted in Eq. (2), the learning focus among the triplet branches can be shifted by controlling the set of weight parameters (α and β). Since the discriminative feature representation is the foundation for learning a robust classifier, the learning focus of our TBN gradually changes from the feature representation learning (conventional branch) to the classifier learning (re-balancing branch) to gradually improve the accuracy with class-imbalanced training data. Hence, the trade-off parameter α between the two branches is defined as [15]:

$$\alpha = 1 - \left(\frac{N}{N_{max}}\right)^2 \tag{3}$$

where N and N_{max} are the current epoch and the number of total training epochs, respectively.

Since the number of samples from majority classes is usually extremely larger than that of minority classes, the whole framework may overfit to the majority classes during the former feature learning stage, which may accordingly influence the subsequent feature extraction from the minority classes. Hence, the regularization branch (*i.e.,* equal sampling possibilities for all classes), serving as a regularizer, is assigned with a larger weight β to alleviate the overfitting problem in the early stage, which gradually decreases for sophisticated tuning of minority classes. Therefore, we set $\beta = \lambda\alpha$ (*i.e.,* $\lambda = 0.5$ in our experiments).

2.2   Auxiliary Ranking Task

As aforementioned, the conventional approaches dealing with class-imbalance problem are lack of the consideration of class dependency prior-knowledge. In this regard, we embed such grading-related prior-knowledge into the framework via an auxiliary ranking task, which simultaneously takes the three obtained feature vectors (f_c, f_r and f_g) as input, and requires the framework to rank the features in descending order of their fracture grades. We formulate the auxiliary task as listwise ranking [2,13,14], and the ranking loss proposed by Cao *et al.* [2] is adopted for optimization. In particular, the obtained feature vectors (f_c, f_r and f_g) are fed into a fully connected layer $\boldsymbol{W}_k \in \mathbb{R}^{KD \times K}$, where K is the number of input features ($K = 3$ in our framework). Therefore, the ranking scores z for the input features can be obtained by

$$z = Softmax(\boldsymbol{W}_k^\top Cat(f_c, f_r, f_g)) \tag{4}$$

where $Softmax(.)$ and $Cat(.)$ are the softmax activation and concatenation functions, respectively. Denote the ground truth for the auxiliary task as y_{aux}. The ranking loss $\mathcal{L}_{RK}$ can be written as:

$$\mathcal{L}_{RK} = D_{KL}(y_{aux}, z) \tag{5}$$

where $D_{KL}(.)$ is the KL-divergence, and $y_{aux} = (\frac{y_c}{y_c+y_r+y_g}, \frac{y_r}{y_c+y_r+y_g}, \frac{y_g}{y_c+y_r+y_g})$. Such a ranking task enables the framework to learn the similarity between grades and exploit the class dependancy prior-knowledge for fatigue fracture grading.

3   Experiments

In this section, we validate the effectiveness of the proposed triplet-branch network (TBN) on our fatigue fracture X-ray image dataset and present the experimental results.

Fatigue Fracture X-ray Image Dataset. We collected 2,725 X-ray images from the collaborating hospital, which consists of 1,785 normal, 190 grade-1, 452 grade-2, 196 grade-3 and 102 grade-4 cases, respectively. The patients may have fatigue fracture on different positions, *i.e.*, metatarsal bone, navicular bone, tibia and fibula. We invite three physicians to annotate the grade of fatigue fracture and decide the final grade via majority-voting. The resolution of X-ray images is around $1,000 \times 1,500$ pixels, which is resized to a uniform size (352×352) for network processing.

The dataset is split to training, validation and test sets according to the ratio of 70:10:20. However, it can be observed from the data distribution that there are two class-imbalance problems occurred in our fatigue fracture dataset—normal *vs.* fatigue fracture and grade-2 *vs.* other fracture grades. Hence, a prediction biased to normal and grade-2 can still achieve a relatively high test accuracy. In this regard, we randomly sample an equal number (20) of test images from each category to form a new test set, termed uniform test set, to validate the effectiveness of different approaches dealing with the problem of imbalanced training data.

Baselines and Evaluation Criterion. Apart from the vanilla network trained with a single cross-entropy loss, the state-of-the-art approaches proposed to tackle the problem of imbalanced training data are also involved as baselines for performance comparison, *i.e.*, label-distribution-aware margin (LDAM) loss [1] and bilateral branch network (BBN) [15]. The average classification accuracy (ACC) and F1 score are adopted as metric for the performance evaluation.

Implementation Details. The proposed triplet-branch network (TBN) is implemented using PyTorch. As illustrated in Fig. 2, all branches share the same backbone, *i.e.*, ResNet-50 [5] in our experiments, which facilitate the information flow between different branches and reduce the computational complexity in the inference phase [15].[1] The proposed TBN is trained with a mini-batch size of 16. The initial learning rate is set to 0.001. The Adam solver [9] is used for network optimization. The whole framework converges after 100 epochs of training.

3.1 Ablation Study

We conduct an ablation study to evaluate the importance of different components of our TBN and present the results in Table 1. It can be observed that the frameworks using single sampler achieve similar ACC, *i.e.*, 36–40%. The reversed-sampler-only framework achieves a slightly higher ACC than the other two samplers, as it pays more attentions to the minority classes. The combination of two samplers consistently boosts the ACC to 45%–48%, which demonstrates the diversity of information learned by different branches. Hence, by using all the three samplers, the accuracy can be further improved to 51.0%. Furthermore, the results listed in Table 1 also demonstrate the effectiveness of our auxiliary

---

[1] The inference phase of our TBN is the same to [15], *i.e.*, aggregating the predictions yielded by the three branches by element-wise addition.

Table 1. Comparison of grading accuracy (%) yielded by frameworks with different combinations of components. (U. S.–Uniform Sampler)

Sample-wise U. S.	Reversed S.	Class-wise U. S.	Ranking Task	ACC	F1
✓				36.0	29.2
	✓			40.0	36.4
		✓		38.0	34.9
✓	✓			48.0	45.6
✓		✓		47.0	46.1
	✓	✓		45.0	44.3
✓	✓	✓		51.0	50.0
✓	✓	✓	✓	**55.0**	**54.2**

ranking task. The TBN framework taking class dependency prior-knowledge into consideration achieves the best ACC of 55.0% on the uniform test set.

Analysis of Loss Weight β**.** As aforementioned, the loss weight $\beta = \lambda\alpha$. Here, the λ is a hyper-parameter. We therefore conduct an experiment to evaluate the influence caused by different values of λ to the grading accuracy. The evaluation results of λ with different values are presented in Table 2.

Table 2. Variation trend of grading performance (%) with different values of λ. The grading accuracy (ACC) of our TBN reaches the best (55.0%) as $\lambda = 0.5$.

λ	0	0.25	0.5	0.75	1
ACC	49.0	52.0	**55.0**	49.0	45.0
F1	46.3	51.1	**54.2**	44.1	41.8

In summary, the grading accuracy (ACC) of our TBN gradually increases and reaches the best (55.0%) as λ changes from 0 to 0.5. The accuracy is observed to degrade while λ further increases. The underlying reason may be that the class-wise sampler works as a regularizer at the beginning of network training. A large weight β enlarges the penalty, which may accordingly influence the representation learning supervised by the sample-wise uniform sampler.

3.2   Comparison with State-of-the-art

To further demonstrate the effectiveness of our TBN, we evaluate the performance of state-of-the-art approaches dealing with the class-imbalance problem on the uniform test set. The confusion matrix yielded by different approaches is presented in Fig. 3.

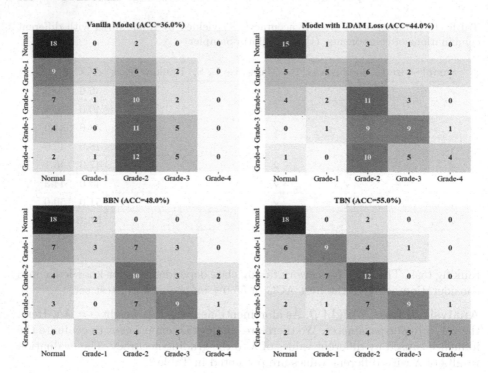

Fig. 3. The confusion matrix of our TBN and the state-of-the-art approaches on the uniform test set.

As shown in Fig. 3, our TBN achieves better accuracies for minority classes, especially for grade-1, without sacrificing the ability for majority class identification (grade-2 and normal), compared to the listed benchmarking algorithms. Additionally, due to the use of class dependency prior-knowledge, most of the mistakes made by our TBN are wrongly identifying the patient to the adjacent grade of ground truth, which reduces risks of under and over treatments in the clinical application. It is worthwhile to mention that the state-of-the-art approach for fracture grading (*i.e.,* grading loss [6]) is also evaluated. However, it suffers from the class-imbalance problem and only achieves an ACC of 42.0%.

3.3   Generalization Evaluation

To validate the generalization of our TBN, we also conduct an experiment on the publicly available APTOS 2019 blindness detection dataset.² The dataset consists of 3,662 fundus images, which can be categorized to four grades according to the severity of diabetic retinopathy (DR), *i.e.,* 1,805 normal, 370 mild DR, 999 moderate DR, 193 severe DR and 295 proliferative DR cases. Consistent to [11], we split the dataset into five folds for cross-validation for a fair comparison.

---

² https://www.kaggle.com/c/aptos2019-blindness-detection.

Table 3 shows the comparison results. With the same backbone network, the proposed TBN performs favorably against the state-of-the-art approaches, *i.e.*, an ACC of 85.0% is achieved, which is +0.6% higher than the runner-up.

Table 3. Five fold cross-validation result (%) of our TBN and the state-of-the-art approaches on the APTOS2019 dataset. All the approaches use ResNet-50 as backbone.

	DLI [12]	CANet [10]	GREEN [11]	TBN (ours)
ACC	82.5	83.2	84.4	**85.0**
F1	80.3	81.3	83.6	**84.3**

4 Conclusion

In this paper, we uncovered a practically relevant yet previously untouched problem, *i.e.*, fatigue fracture grading. A triplet-branch network (TBN) was proposed for the accurate fatigue fracture grading, which enabled physicians to promptly take appropriate treatments for the potential fatigue fracture. The proposed TBN has been evaluated on our fatigue fracture X-ray image dataset. The experimental results demonstrated the effectiveness of our TBN dealing with the problem of class-imbalanced training samples.

Acknowledgements. This work was funded by Key-Area Research and Development Program of Guangdong Province, China (No. 2018B010111001), and the Scientific and Technical Innovation 2030–"New Generation Artificial Intelligence" Project (No. 2020AAA0104100).

References

1. Cao, K., Wei, C., Gaidon, A., Arechiga, N., Ma, T.: Learning imbalanced datasets with label-distribution-aware margin loss. In: Advances in Neural Information Processing Systems, pp. 1567–1578 (2019)
2. Cao, Z., Qin, T., Liu, T.Y., Tsai, M.F., Li, H.: Learning to rank: from pairwise approach to listwise approach. In: International Conference on Machine Learning, pp. 129–136 (2007)
3. Chen, H., et al.: Anatomy-aware Siamese network: Exploiting semantic asymmetry for accurate pelvic fracture detection in X-ray images. In: European Conference on Computer Vision, pp. 239–255 (2020)
4. Guan, B., Zhang, G., Yao, J., Wang, X., Wang, M.: Arm fracture detection in X-rays based on improved deep convolutional neural network. Comput. Electr. Eng. **81**, 106530 (2020)
5. He, K., Zhang, X., Ren, S., Sun, J.: Deep residual learning for image recognition. In: IEEE Conference on Computer Vision and Pattern Recognition, pp. 770–778 (2016)

6. Husseini, M., Sekuboyina, A., Loeffler, M., Navarro, F., Menze, B.H., Kirschke, J.S.: Grading loss: a fracture grade-based metric loss for vertebral fracture detection. In: International Conference on Medical Image Computing and Computer Assisted Intervention, pp. 733–742 (2020)

7. Iyer, S., Sowmya, A., Blair, A., White, C., Dawes, L., Moses, D.: A novel approach to vertebral compression fracture detection using imitation learning and patch based convolutional neural network. In: IEEE International Symposium on Biomedical Imaging, pp. 726–730 (2020)

8. Kalmet, P.H.S., et al.: Deep learning in fracture detection: a narrative review. Acta Orthopaedica **91**(2), 215–220 (2020)

9. Kingma, D.P., Ba, J.: Adam: A method for stochastic optimization. arXiv preprint arXiv:1412.6980 (2014)

10. Li, X., Hu, X., Yu, L., Zhu, L., Fu, C.W., Heng, P.A.: CANet: cross-disease attention network for joint diabetic retinopathy and diabetic macular edema grading. IEEE Trans. Med. Imaging **39**(5), 1483–1493 (2020)

11. Liu, S., Gong, L., Ma, K., Zheng, Y.: GREEN: a graph residual re-ranking network for grading diabetic retinopathy. In: International Conference on Medical Image Computing and Computer Assisted Intervention, pp. 585–594 (2020)

12. Rakhlin, A.: Diabetic retinopathy detection through integration of deep learning classification framework. bioRxiv preprint (2018)

13. Xia, F., Liu, T.Y., Wang, J., Zhang, W., Li, H.: Listwise approach to learning to rank: theory and algorithm. In: International Conference on Machine Learning, pp. 1192–1199 (2008)

14. Yu, H.T., Jatowt, A., Joho, H., Jose, J.M., Yang, X., Chen, L.: WassRank: listwise document ranking using optimal transport theory. In: ACM International Conference on Web Search and Data Mining, pp. 24–32 (2019)

15. Zhou, B., Cui, Q., Wei, X.S., Chen, Z.M.: BBN: bilateral-branch network with cumulative learning for long-tailed visual recognition. In: IEEE Conference on Computer Vision and Pattern Recognition (2020)

DeepMitral: Fully Automatic 3D Echocardiography Segmentation for Patient Specific Mitral Valve Modelling

Patrick Carnahan[1,2]([✉]), John Moore[1], Daniel Bainbridge[3], Mehdi Eskandari[4], Elvis C. S. Chen[1,2,5][iD], and Terry M. Peters[1,2,5]

[1] Imaging Laboratories, Robarts Research Institute, London, Canada
[2] School of Biomedical Engineering, Western University, London, Canada
pcarnah@uwo.ca
[3] Department of Anesthesiology, London Health Sciences Centre, Western University, London, Canada
[4] King's College Hospital, Denmark Hill, London, UK
[5] Department of Medical Biophysics, Western University, London, Canada

Abstract. Recently, developments have been made towards modelling patient-specific deformable mitral valves from transesophageal echocardiography (TEE). Thus far, a major limitation in the workflow has been the manual process of segmentation and model profile definition. Completing a manual segmentation from 3D TEE can take upwards of two hours, and existing automated segmentation approaches have limitations in both computation time and accuracy. Streamlining the process of segmenting the valve and generating a surface mold is important for the scalability and accuracy of patient-specific mitral valve modelling. We present DeepMitral, a fully automatic, deep learning based mitral valve segmentation approach that can quickly and accurately extract the geometry of the mitral valve directly from TEE volumes. We developed and tested our model on a data set comprising 48 diagnostic TEE volumes with corresponding segmentations from mitral valve intervention patients. Our proposed pipeline is based on the Residual UNet architecture with five layers. Evaluation of our proposed pipeline was assessed using manual segmentations performed by two clinicians as a gold-standard. The comparisons are made using the mean absolute surface distance (MASD) between the boundaries of the complete segmentations, as well as the 95% Hausdorff distances. DeepMitral achieves a MASD of 0.59 ± 0.23mm and average 95% Hausdorff distance of 1.99 ± 1.14mm. Additionally, we report a Dice score of 0.81. The resulting segmentations from our approach successfully replicate gold-standard segmentations with improved performance over existing state-of-the-art methods. DeepMitral improves the workflow of the mitral valve modelling process by reducing the time required for completing an accurate mitral valve segmentation, and providing more consistent results by removing user variability from the segmentation process.

Keywords: Mitral valve · 3D Echocardiography · Segmentation · Patient-specific modelling · Deep learning · Ultrasound

© Springer Nature Switzerland AG 2021
M. de Bruijne et al. (Eds.): MICCAI 2021, LNCS 12905, pp. 459–468, 2021.
https://doi.org/10.1007/978-3-030-87240-3_44

1 Introduction

Mitral valve (MV) disease is a common pathologic problem occurring in approximately 2% of the general population but climbing to 10% in those over the age of 75 [3]. Of this group, approximately 20% have a sufficiently severe form of the disease that may require surgical intervention to restore normal valve function and prevent early mortality [21]. The preferred intervention for mitral regurgitation is valve repair, due to superior patient outcomes compared to those following valve replacement [1,17]. However, the repair must be tailored to the patient-specific anatomy and pathology, which requires expert training and experience. Consequently, there is a need for patient-specific models that can permit the training and procedure-planning of patient-specific repairs to minimize its learning curve and preventable errors [7,9]. Heart simulator technology has been adopted widely by both industry for evaluation of technologies for imaging heart valves [14], and academia for the assessment of modelled heart valves [16]. Recently, developments have been made on a workflow to create 3D, patient-specific valve models directly from trans-esophageal echocardiography (TEE) images. When viewed dynamically using TEE within a pulse duplicator simulator, it has been demonstrated that these models result in pathology-specific TEE images similar to those acquired from the patient's valves in-vivo [8].

A key step in patient-specific modelling workflows is delineating the mitral valve leaflets in patient ultrasound image data, a necessary operation to extract the patient-specific leaflet geometry, which will be used to form the basis of the model. Performing manual segmentations is very time consuming, taking upwards of 2 hours, which is a serious bottleneck in modelling workflows. Several mitral leaflet segmentation methods have been proposed, targeting a number of different applications. These methods focus on varying goals between deriving quantitative valve measurements and extracting annular and leaflet geometry from 3D TEE images. These methods can be divided into two categories: semi-automatic and fully automatic approaches. The semi-automatic approaches all require some level of user intervention during the segmentation process, while fully automatic methods do not. Scheinder *et.al.* proposed a semi-automatic method for segmenting the mitral leaflets in 3D TEE over all phases of the cardiac cycle [23]. This method utilizes geometric priors and assumptions about the mechanical properties of the valve to model the leaflets through coaptation with a reported surface error of 0.84mm. However, their method only represents the mitral leaflets as a single medial surface, rather than structures with thickness. Burlina *et.al.* [4] proposed a semi-automatic segmentation method based on active contours and thin tissue detection for the purpose of computational modelling, reporting errors in the range of 4.00 mm to 5.00 mm . An additional semi-automatic approach designed for patient specific valve modelling reported a surface error of 1.4mm overall, and an surface error of 1.01mm for the atrial surface critical in mitral valve model creation [5]. Several fully automatic methods have been proposed that are based on population average atlases. Ionasec *et.al.* [10] describe a technique which uses a large database of manually labelled images and machine learning algorithms to locate and track valve landmarks, reporting a surface error of 1.54mm. While this method is fully automatic, the use of sparse landmarks potentially limits the amount of patient-specific detail that can be extracted. Pouch *et.al.* [20] also describe a fully automatic method that employs a set of atlases to generate a deformable template, which is then guided to the leaflet geometry using joint label fusion. The surface error of this method is reported as 0.7mm, however this is only achieved on healthy valves and performance is reduced when segmenting diseased valves. Atlas based methods relying on sparse deformation could be biased towards the atlas geometry, limiting the potential for capturing patient specific detail, which is not ideal for valve modelling applications.

Automatic 3D segmentation methods offer significant implications for the feasibility of patient-specific modelling in clinical use. While existing methods have demonstrated the ability to accurately segment the mitral valve structure,they remain highly time-intensive. Furthermore, some of these published methods show decreased performance when applied to highly diseased valves, demonstrating limitations in patient-specificity. Convolutional neural networks (CNNs) have been widely demonstrated to be effective for segmentation tasks. However, to our knowledge no CNN segmentation approaches have been reported for mitral valve segmentation in 3D TEE imaging, although 3D Unet based approaches have been used in other cardiac ultrasound applications such as automatic annulus detection [2]. Working in 2D, UNet has been used for mitral leaflet segmentation [6]. In this paper, we present DeepMitral, a 3D segmentation pipeline for mitral valve segmentation based on the 3D Residual UNet architecture [12]. We demonstrate the feasibility of CNN based segmentation for 3D TEE images, and establish a baseline of performance for future methods. DeepMitral will have applications in patient specific valve modelling, enabling improvements in the workflow. DeepMitral has been made open source including our trained model and is freely available on GitHub[1].

2   Methods

2.1   Data Acquisition

Patients with mitral valve regurgitation undergoing clinical interventions were imaged preoperatively as per clinical protocol with appropriate ethics approval using a Philips Epiq system with an X8-2T transducer,. The 3D TEE images were exported into Cartesian format, and the SlicerHeart module was used to import the Cartesian DICOM files into 3D Slicer[2] [22]. Images at end-diastole were selected for image analysis. The exported Cartesian format images have an axial resolution of approximately 0.5mm. We collected a total of 48 volumes, which were divided into training, validation and testing partitions with 36, 4, and 8 volumes respectively. Annotations for the training and validation sets were performed in 3D Slicer by multiple different trained users. These segmentations were performed via manual refinement of the output of a semi-automatic

---

[1] https://github.com/pcarnah/DeepMitral.

[2] https://slicer.org.

segmentation tool [5], and were then reviewed and modified as necessary by a single experienced user to ensure consistency. The test set was annotated entirely manually by two cardiac imaging clinicians using 3D Slicer.

2.2 Model Selection

Our training and validation sets were used to perform model selection and hyper-parameter tuning on a variety of network architectures including Residual UNet, VNet, AHNet and SegResNetVAE [12,15,18,19]. We trained each network with a selection of hyper-parameters and computed mean Dice coefficient scores, mean surface error scores, and mean 95% Hausdorff distance on the validation set. The best performing version of each network architecture is shown in Table 1. Of the models, the Residual UNet architecture achieved the best performance with respect to all validation scores, so it was chosen as our final network for use in the DeepMitral pipeline.

Table 1. Validation metrics for the tested network architectures.

Network	MASD (mm)	95% Hausdorff (mm)	Dice Coefficient
Residual UNet	0.50	3.41	0.83
VNet	0.52	4.13	0.80
AHNet	0.66	5.15	0.76
SegResNetVAE	0.74	4.17	0.79

2.3 DeepMitral Pipeline

Our 3D TEE volume segmentation platform was built using the MONAI[3] framework, that provides domain-optimized foundational capabilities for developing healthcare imaging training workflows. This platform includes the implementation of many common network architectures for both 2D and 3D data, as well as a number of medical imaging focused pre-processing methods.

Our workflow begins with a sequence of pre-processing operations from the MONAI framework. First, we load the images, and add a channel along the first dimension to transform them into channel-first representation. Next, we isotropically re-sample the volumes to 0.3mm spacing, using bilinear re-sampling for the image data and nearest neighbour re-sampling for the label. Following re-sampling, we re-scale the image intensities to the range of 0.00 to 1.00, then crop the images to the foreground using the smallest possible bounding box that includes all non-zero voxels. Finally, random sampling is performed on the volumes, taking 4 samples of size 96 × 96 × 96, centered on voxels labelled as leaflet. The final random sampling step is recomputed at every epoch during network

---

[3] https://monai.io/.

training. No data-augmentation was performed as adding rigid or deformable spatial transformations resulted in no improvements in validation metrics, and a reduction of training speed.

Our network uses a Residual UNet architecture [12], implemented by the MONAI framework, with 5 layers of 16, 32, 64, 128 and 256 channels respectively. Each of these layers is created using a residual unit with 2 convolutions and a residual connection. Convolutions are performed with stride 2 at every residual unit for up-sampling and down-sampling.

We trained our model using batch sizes of 32, composed of 8 different volumes, with 4 random samples being taken from each volume. Training was performed for 2000 epochs using an Nvidia GTX 1080 graphics card with 8GB of ram, and took roughly 5 hours. We employed the Adam optimizer, with an initial learning rate of 1.00×10^{-3}, which is reduced to 1.00×10^{-3} after 1000 epochs [13]. Batch normalization is used to help prevent over-fitting of the model.

2.4   Evaluation

Final evaluation of our pipeline is performed using a separate test set consisting of eight volumes with ground truth annotations that were performed manually by two cardiac clinicians. Prior to evaluating our model, we retrained the network using combined training and validation sets. The primary comparison metrics are the mean absolute surface distance (MASD) between the boundaries of the complete segmentations, as well as the 95% Hausdorff distances. We also report the Dice coefficient scores.

3   Results

DeepMitral achieves a MASD of 0.59 ± 0.23mm, average 95% Hausdorff distance of 1.99 ± 1.14mm, and a Dice score of 0.81. In all 8 volumes used for testing the mitral leaflets were successfully segmented, with no cases of complete failure to identify the leaflets. Overall, the scores are consistent among 6 of the 8 examples, with 2 examples exhibiting lower performance. In one instance, case P3, the leaflets are under-segmented near the leaflet tips, and in another instance, P8, the chordae tendineae are mis-labelled as leaflet, as seen in Fig. 1. We see the corresponding metrics for these cases in Fig. 2, which show higher surface distance errors and worse dice scores than the other cases.

3.1   Inference Runtime Performance

Deep learning segmentation methods enable predictions to be performed for low computational cost. We evaluated DeepMitral's inference speed on our test set using both CPU only (Intel i7-6700K) and GPU acceleration (Nvidia GeForce GTX 1080). The size of the volumes range from approximately voxels. Using only the CPU for inference, DeepMitral takes on average $9.5 \pm 2.26s$ to perform the inference itself, and $11.40 \pm 2.28s$ for overall runtime including startup overhead. When using GPU acceleration, these times are reduced on average to $3.5 \pm 0.53s$ for inference and $5.69 \pm 0.66s$ for overall runtime. DeepMitral achieves fast inference times on both CPU and GPU, with GPU acceleration reducing runtime by a factor of two on average. Additionally, startup overhead is consistently around 2s for both CPU and GPU. This overhead would only occur once in the case of performing inference on multiple volumes in a single run.

Table 2. MASD, 95% hausdorff distance and dice coefficient scores for each of the eight volumes in the test set

Test ID	MASD (mm)	95% Hausdorff (mm)	Dice coefficient
P1	0.43	1.18	0.84
P2	0.37	1.10	0.87
P3	1.04	3.53	0.65
P4	0.53	1.49	0.83
P5	0.49	1.25	0.83
P6	0.42	1.37	0.85
P7	0.71	1.97	0.76
P8	0.76	4.02	0.82
Average	0.59 ± 0.23	1.99 ± 1.14	0.81 ± 0.07

4   Discussion

These results demonstrate the feasibility of CNN based techniques for mitral valves segmentations in 3D TEE volumes. Trained with a relatively small dataset, DeepMitral achieves an improvement in accuracy over the existing state of the art approaches. Average surface error is reduced to 0.6mm on average, where the best performing existing methods report an error of 0.7mm. Additionally, our reported error is almost equal to typical inter-user variability, which was previously reported as $0.6 \pm 0.17mm$ [11]. Our reported MASD is approaching the axial resolution of the ultrasound volumes, which is approximately 0.5mm on average. These results indicate that while DeepMitral is accurately labelling the valve leaflets overall, we note the 95% Hausdorff distances are typically larger, in the range of 1.00mm to 4.00mm. We can see in Fig. 2 that there are small regions of the leaflets where accuracy is worse, contributing to these larger 95% Hausdorff errors, while the majority of the leaflet surface maintains sub-millimetre error. For P8, we see that the leaflets themselves are well identified, however the large protrusion where the chordae were mis-identified contributes to the poor error metrics in the case.

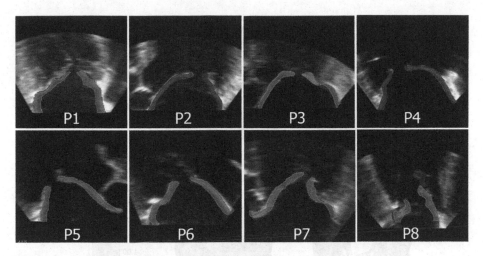

Fig. 1. Cross sectional views of 3D TEE images and segmentations for each volume in our test set. Ground-truth label is shown in green, and predicted label is shown in blue. (Color figure online)

Our results suggest deep learning based approaches perform better than atlas based approaches for capturing unique valve features. Although both methods use a collection of prior data to inform the segmentation process, CNNs are much more flexible at applying this prior knowledge to new problems. Atlas based methods, however, can transfer particular geometric patterns to a new segmentation, and thus tend to perform best only on healthy valves [20]. We tested DeepMitral on exclusively diseased valves, but as demonstrated by our test data, there is a range of distinct valve geometries, all of which are accurately identified by our model. The areas of poor performance are not due to systematic geometrical bias, but are instead caused by poor image data and mis-identified structures.

CNN based approaches are particularly beneficial for use in valve modelling applications since they eliminate the computational time that prior methods have reported, ranging from 15 minutes to 3 hours for a single segmentation. Deep learning methods instead can perform a segmentation in seconds, which removes a large bottleneck in valve modelling workflows. Additionally, since these methods are fully automatic, the resulting segmentations will be more consistent than semi-automatic or manual approaches, where individual users can vary greatly on how much of the atrial wall they label as leaflet. DeepMitral produces accurate segmentations in most cases, however in instances where the segmentation is sub-optimal manual editing of the result is still possible. DeepMitral can be easily integrated into 3D Slicer, which would allow for an initial segmentation to be created very quickly, and then be verified and edited if necessary before being used in any downstream applications.

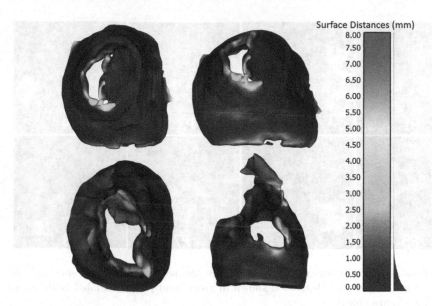

Fig. 2. Distance comparison heatmaps for P4 (top) and P8 (bottom), showing the distribution of error across the leaflet segmentation.

4.1   Limitations and Future Work

DeepMitral fails to differentiate between chordae and leaflet in some images where the chordae are very clear, as seen in case P8 and to a lesser extent case P6 in Fig. 1. Currently we have a lack of training data where chordae are strongly delineated in the image, as this rarely occurs in TEE imaging. As a result, our model tends to classify the chordae as leaflet, as it presents as a similar image feature when visible. This will be addressed in future work by including an additional label for the chordae in our training set. This will allow our model to learn how to differentiate between leaflet and chordae. Additionally, sub-optimal image quality can cause the segmentation to perform poorly. This is a fundamental limitation when working with cardiac ultrasound, as it is possible for acquisitions to be very noisy, or lack detail of the mitral leaflets due to signal dropout. Expanding our data-set to include wider variations in image quality will allow us to better evaluate the conditions in which this approach can be successful. We plan on expanding our results beyond single frames to a 4D segmentation problem, that will allow us to incorporate the cyclical nature of cardiac motion into a segmentation workflow. Since different structures are better imaged at different phases of the cardiac cycle, this technique has the potential to further improve our results and overcome limitations due to image quality. Our methods could also be extended to adult tricuspid valves through transfer learning, enabling improved results as data availability for the tricuspid valve is more limited and image quality is generally poor.

5   Conclusions

The results from DeepMitral successfully replicate the gold standard segmentations with improved performance over existing state-of-the-art methods. Submillimetre average surface error in the segmentation stage are sufficient for use in patient specific valve modelling without manual intervention. We demonstrate the effectiveness of CNN based segmentation approaches for mitral valves from 3D TEE volumes. Improved mitral valve segmentation methods have wide applications including basic valve research and improved patient diagnostics. DeepMitral improves the workflow of the mitral valve modelling process by reducing the time required for completing an accurate mitral valve segmentation and providing more consistent and accurate results. Improvements in the mitral valve modelling workflow will lead to easier clinical translation, and will have implications in both surgical planning and training.

Conflict of Interest

The authors declare that they have no conflict of interest.

Acknowledgements. We would like to acknowledge the following sources of funding: Canadian Institutes for Health Research, Natural Sciences and Engineering Research Council of Canada; Canadian Foundation for Innovation.

References

1. Ailawadi, G., et al.: Is mitral valve repair superior to replacement in elderly patients? Ann. Thorac. Surg. **86**(1), 77–86 (2008)
2. Andreassen, B.S., Veronesi, F., Gerard, O., Solberg, A.H.S., Samset, E.: Mitral annulus segmentation using deep learning in 3-D transesophageal echocardiography. IEEE J. Biomed. Health Inf. **24**(4), 994–1003 (2020)
3. Benjamin, E.J., et al.: Heart disease and stroke statistics—2018 update: a report from the American heart association. Circulation **137**(12), E67–E492 (2018)
4. Burlina, P., et al.: Patient-specific modeling and analysis of the mitral valve using 3D-TEE. In: Navab, N., Jannin, P. (eds.) IPCAI 2010. LNCS, vol. 6135, pp. 135–146. Springer, Heidelberg (2010). https://doi.org/10.1007/978-3-642-13711-2_13
5. Carnahan, P., et al.: Interactive-automatic segmentation and modelling of the mitral valve. In: Coudière, Y., Ozenne, V., Vigmond, E., Zemzemi, N. (eds.) FIMH 2019. LNCS, vol. 11504, pp. 397–404. Springer, Cham (2019). https://doi.org/10.1007/978-3-030-21949-9_43
6. Costa, E., et al.: Mitral valve leaflets segmentation in echocardiography using convolutional neural networks. In: 2019 IEEE 6th Portuguese Meeting on Bioengineering (ENBENG), IEEE (February 2019)
7. Eleid, M.F., et al.: The learning curve for transcatheter mitral valve repair with MitraClip. J. Interv. Cardiol. **29**(5), 539–545 (2016)
8. Ginty, O.K., et al.: Dynamic, patient-specific mitral valve modelling for planning transcatheter repairs. Int. J. Comput. Assist. Radiol. Surg. **14**(7), 1227–1235 (2019)
9. Holzhey, D.M., Seeburger, J., Misfeld, M., Borger, M.A., Mohr, F.W.: Learning minimally invasive mitral valve surgery. Circulation **128**(5), 483–491 (2013)

10. Ionasec, R.I., et al.: Patient-specific modeling and quantification of the aortic and mitral valves from 4-D cardiac CT and TEE. IEEE Trans. Med. Imaging 29(9), 1636–1651 (2010)
11. Jassar, A.S., et al.: Quantitative mitral valve modeling using real-time three-dimensional echocardiography: technique and repeatability. Ann. Thorac. Surg. 91(1), 165–171 (2011)
12. Kerfoot, E., Clough, J., Oksuz, I., Lee, J., King, A.P., Schnabel, J.A.: Left-Ventricle Quantification Using Residual U-Net. In: Pop, M., et al. (eds.) STACOM 2018. LNCS, vol. 11395, pp. 371–380. Springer, Cham (2019). https://doi.org/10.1007/978-3-030-12029-0_40
13. Kingma, D.P., Ba, J.L.: Adam: A Method for Stochastic Optimization. In: CoRR, vol. 1412.6980 (2014)
14. Kozlowski, P., Bandaru, R.S., D'hooge, J., Samset, E.: Real-time catheter localization and visualization using three-dimensional echocardiography. In: Webster, R.J., Fei, B. (eds.) Medical Imaging 2017: Image-Guided Procedures, Robotic Interventions, and Modeling. SPIE (Mar 2017)
15. Liu, S., et al.: 3D anisotropic hybrid network: transferring convolutional features from 2D images to 3D anisotropic volumes. In: Frangi, A.F., Schnabel, J.A., Davatzikos, C., Alberola-López, C., Fichtinger, G. (eds.) MICCAI 2018. LNCS, vol. 11071, pp. 851–858. Springer, Cham (2018). https://doi.org/10.1007/978-3-030-00934-2_94
16. Mashari, A., et al.: Hemodynamic testing of patient-specific mitral valves using a pulse duplicator: a clinical application of three-dimensional printing. J. Cardiothorac. Vasc. Anesth. 30(5), 1278–1285 (2016)
17. McNeely, C.A., Vassileva, C.M.: Long-term outcomes of mitral valve repair versus replacement for degenerative disease: a systematic review. Curr. Cardiol. Rev. 11(2), 157–62 (2015). http://www.ncbi.nlm.nih.gov/pubmed/25158683
18. Milletari, F., Navab, N., Ahmadi, S.A.: V-Net: fully convolutional Neural Networks for Volumetric Medical Image Segmentation. In: 2016 Fourth International Conference on 3D Vision (3DV), pp. 565–571. IEEE (October 2016)
19. Myronenko, A.: 3D MRI brain tumor segmentation using autoencoder regularization. In: Crimi, A., Bakas, S., Kuijf, H., Keyvan, F., Reyes, M., van Walsum, T. (eds.) BrainLes 2018. LNCS, vol. 11384, pp. 311–320. Springer, Cham (2019). https://doi.org/10.1007/978-3-030-11726-9_28
20. Pouch, A.M., et al.: Fully automatic segmentation of the mitral leaflets in 3D transesophageal echocardiographic images using multi-atlas joint label fusion and deformable medial modeling. Med. Image Anal. 18(1), 118–129 (2014)
21. Ray, S.: Changing epidemiology and natural history of valvular heart disease. Clin. Med. 10(2), 168–171 (2010)
22. Scanlan, A.B., et al.: Comparison of 3D echocardiogram-derived 3D printed valve models to molded models for simulated repair of pediatric atrioventricular valves. Pediatr. Cardiol. 39(3), 538–547 (2017)
23. Schneider, R.J., Tenenholtz, N.A., Perrin, D.P., Marx, G.R., del Nido, P.J., Howe, R.D.: Patient-specific mitral leaflet segmentation from 4D ultrasound. In: Fichtinger, G., Martel, A., Peters, T. (eds.) MICCAI 2011. LNCS, vol. 6893, pp. 520–527. Springer, Heidelberg (2011). https://doi.org/10.1007/978-3-642-23626-6_64

Data Augmentation in Logit Space for Medical Image Classification with Limited Training Data

Yangwen Hu[1], Zhehao Zhong[1], Ruixuan Wang[1,2(✉)], Hongmei Liu[1], Zhijun Tan[1], and Wei-Shi Zheng[1,2,3]

[1] School of Computer Science and Engineering, Sun Yat-sen University, Guangzhou, China
wangruix5@mail.sysu.edu.cn
[2] Key Laboratory of Machine Intelligence and Advanced Computing, MOE, Guangzhou, China
[3] Pazhou Lab, Guangzhou, China

Abstract. Successful application of deep learning often depends on large amount of training data. However in practical medical image analysis, available training data are often limited, often causing over-fitting during model training. In this paper, a novel data augmentation method is proposed to effectively alleviate the over-fitting issue, not in the input space but in the logit space. This is achieved by perturbing the logit vector of each training data within the neighborhood of the logit vector in the logit space, where the size of neighborhood can be automatically and adaptively estimated for each training data over training stages. The augmentations in the logit space may implicitly represent various transformations or augmentations in the input space, and therefore can help train a more generalizable classifier. Extensive evaluations on three small medical image datasets and multiple classifier backbones consistently support the effectiveness of the proposed method.

Keywords: Data augmentation · Logit space · Limited data

1 Introduction

Deep learning techniques have been successfully applied to intelligent diagnosis of various diseases [6,7,13]. In general, expert-level diagnosis from the intelligent systems are often based on large set of annotated training data for each disease. However, due to the existence of lots of rare diseases, costly and little time resource from clinicians, privacy concerns, and difficulty in data sharing across medical centres etc., annotated and publicly available large datasets for disease diagnosis are very limited. As a result, practical investigations of intelligent diagnosis often face the challenge of limited training data for model training.

Y. Hu and Z. Zhong—The authors contribute equally to this work.

© Springer Nature Switzerland AG 2021
M. de Bruijne et al. (Eds.): MICCAI 2021, LNCS 12905, pp. 469–479, 2021.
https://doi.org/10.1007/978-3-030-87240-3_45

To alleviate the over-fitting issue due to limited training data for the current task of interest, various approaches have been developed particularly for deep learning models. One group of approaches are based on transferring knowledge from a relatively large auxiliary dataset which contains different classes but in content is often similar to the dataset of the current task. The auxiliary dataset can be used to train and then fix a feature extractor for the current task, as in the matching network [22], prototypical network [16], and relation network [18], or to train a feature extractor which is then fine-tuned by the dataset of the current task, as in the meta-learning methods MAML [8] and LEO [14], or to jointly train a feature extractor with the dataset of the current task [24]. Such transfer learning techniques assume that the auxiliary dataset is annotated and similar to the dataset of the current task. However, large annotated auxiliary medical image dataset is generally not available in the scenario of intelligent diagnosis. Another group of approaches are based on various data augmentation techniques to increase the amount of the original training data. Besides the conventional data augmentations like random cropping, scaling, rotating, flipping, and color jittering of each training image, more advanced augmentation techniques have been recently developed, including Mixup [27], Cutout [4], Cutmix [26], AutoAugment [2], and RandAugment [3]. All these augmentations are performed directly on images and the types of basic augmentations (transformations) on images need to be manually designed. Besides data augmentation in the input space, augmentation in the semantic feature space has also been proposed [25], where various semantic transformations on images may be implicitly realized by perturbing each feature vector along certain feature dimensions.

This study follows the direction of data augmentation for over-fitting alleviation. Different from all the existing augmentations either in the input space or in the feature space, the proposed novel augmentation is in the (classifier's pre-softmax) logit space. Perturbations of each data in the logit space can implicitly represent various transformations in the input or feature space, and the augmented data in the logit space can help train the classifier to directly satisfy the desired property of generalizability, i.e., similar data in the logit space should come from the same class. Innovatively, the magnitude of perturbation can be adaptively estimated over the training process based on uncertainty for each logit element, where the logit uncertainty is part of the classifier model output. Experimental evaluations on multiple datasets with various classifier backbones prove the effectiveness and generalizability of the proposed method.

2 Methodology

The objective is to train a generalizable classifier with limited available training data. Assume a classifier is represented by a convolutional neural network (CNN), consisting of multiple convolutional layers (i.e., feature extractor) and one last fully connected layer (i.e., classifier head). One desired property of any generalizable classifier is that, two images should come from the same class if their feature representations (i.e., feature vectors in the feature space) from the

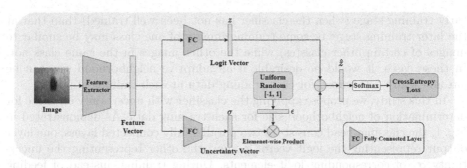

Fig. 1. Classifier training with augmentation in the logit space. Feature vector is extracted from the input image and then forwarded to two parallel fully connected layers to obtain the original logit vector and the uncertainty vector. Multiples samples of logit vectors are then generated based on the combination of the original logit vector and the uncertainty vector, with each sample fed to the softmax operator finally.

feature extractor output are similar enough. This property can be extended to the output space (i.e., from the linear transformation of feature vector, also called logit space) of the pre-softmax in the last fully connected layer. Most existing data augmentation methods try to help classifiers satisfy this property during classifier training indirectly by generating multiple transformed versions of the same image and expecting the classifier to generate correspondingly similar feature vectors (and logit vectors). However, such data augmentations cannot assure that similar feature or logit vectors would always come from the same class. With this observation, we propose a simple yet effective augmentation method in the logit space, directly helping classifier satisfy the property that similar logit feature vectors should come from the same class. Particularly, we propose training a CNN classifier which can generate not only the conventional logit vector but also the uncertainty for each logit element. Based on the logit and its uncertainty, multiple logit vector samples for each single input image can be obtained.

2.1   Classifier with Logit Uncertainty for Data Augmentation

For the logit vector $\mathbf{z}_i$ of any input image $\mathbf{x}_i$, suppose those logit vectors within its neighborhood in the logit space correspond to similar input images in content. Then sampling from the neighborhood would naturally generate multiple logit vectors associating with various input data instances from the same class. By enforcing the classifier to have same prediction for these sampled logit vectors during training, the classifier would satisfy the desire property of generalizability.

The challenge for data augmentation in the logit space is to determine the size of the neighborhood in the logit space for each input image. Manually setting a fixed neighborhood size often results in undesirable augmentation effect without considering the particular training data, training stage, and the employed CNN classifier architecture. For example, the distribution of logit vectors for each class of data may be spread in a much larger region in the logit space at the early training stage (when the classifier has not been well trained) than that at the later training stage, or some training images of one class may be similar to images of certain other class(es) while the other images in the same class not. In these cases, it would be desirable if an adaptive neighborhood size can be automatically determined for each training data at each training stage.

In this study, we propose applying the classifier with uncertainty estimate for determination of neighborhood size for each training data. As demonstrated in Fig. 1, the classifier head consists of two parallel fully connected layers, one layer output representing the logit vector $\mathbf{z}_i$, and the other representing the uncertainty $\boldsymbol{\sigma}_i$ of corresponding logit elements. During training, instead of feeding the logit vector $\mathbf{z}_i$ to the softmax operator as in conventional classifier, here the classifier feeds multiple samples of logit vectors $\{\hat{\mathbf{z}}_{i,k}, k = 1, \ldots, K\}$ around the original $\mathbf{z}_i$, each generated by

$$\hat{\mathbf{z}}_{i,k} = \mathbf{z}_i + \boldsymbol{\epsilon}_k \odot \boldsymbol{\sigma}_i. \tag{1}$$

$\boldsymbol{\epsilon}_k$ is a vector whose element is randomly sampled from the uniform distribution within the range $[-1, 1]$, and $\odot$ is the Hadamard (i.e., element-wise) product. Note that the element in the uncertainty vector $\boldsymbol{\sigma}_i$ is not constrained to be non-negative (although it should be in principle) considering that the negative sign of any element in $\boldsymbol{\sigma}_i$ can be absorbed into the corresponding element in the random vector $\boldsymbol{\epsilon}_k$. For each sampled logit vector $\hat{\mathbf{z}}_{i,k}$, denote the corresponding classifier output by $\hat{\mathbf{y}}_{i,k}$ which is the softmax output with $\hat{\mathbf{z}}_{i,k}$ as input, and the ground-truth one-hot vector by $\mathbf{y}_i$ for the original input image $\mathbf{x}_i$. Then, the classifier can be trained with the training set $\{(\mathbf{x}_i, \mathbf{y}_i), i = 1, \ldots, N\}$ by minimizing the cross-entropy loss L,

$$L = \frac{1}{NK} \sum_{i=1}^{N} \sum_{k=1}^{K} l(\mathbf{y}_i, \hat{\mathbf{y}}_{i,k}) = -\frac{1}{NK} \sum_{i=1}^{N} \sum_{k=1}^{K} \left(\hat{z}_{i,k,c} - \log \sum_{c'=1}^{C} \exp(\hat{z}_{i,k,c'}) \right), \tag{2}$$

where $l(\mathbf{y}_i, \hat{\mathbf{y}}_{i,k})$ is the well-known cross-entropy function to measure the difference between $\mathbf{y}_i$ and $\hat{\mathbf{y}}_{i,k}$, which can be transformed to the format on the right side of Eq. (2). $\hat{z}_{i,k,c}$ is the c-th element in $\hat{\mathbf{z}}_{i,k}$ and c is the index associated with the ground-truth class for the input $\mathbf{x}_i$, and C is the number of logit or classifier output elements. At the early stage of classifier training, the classifier has not been well trained and therefore the logit vector $\mathbf{z}_i$ would often correspond to undesired classifier output. In this case, the classifier would not be prone to generate smaller uncertainty values in $\boldsymbol{\sigma}_i$, because smaller uncertainty would cause the multiple sampled logit vectors $\{\hat{\mathbf{z}}_{i,k}, k = 1, \ldots, K\}$ often having similar undesired classifier output as from $\mathbf{z}_i$. In contrast, at later training stage when the classifier has been well trained, the classifier may correctly predict the class of the input $\mathbf{x}_i$ based on the corresponding logit vector $\mathbf{z}_i$. This generally would be associated with smaller uncertainty estimate, because larger uncertainty would cause more varied logit vector samples $\{\hat{\mathbf{z}}_{i,k}\}$ and part of the samples which are more different from $\mathbf{z}_i$ could cause undesired classifier output. Therefore, the logit uncertainty estimate can be automatically adapted during model training. The trend of uncertainty estimate over training process has been confirmed in experiments (Sect. 3.3). Similar analysis can be performed for individual training images x_i's, with harder images (i.e., prone to be incorrectly classified) often associated with larger logit uncertainty. As a result, the uncertainty logit estimate σ_i from the classifier can be naturally used to determine the size of neighborhood around z_i in the logit space, based on which multiple samples in the neighborhood can be generated for each input image and used for data augmentation during model training.

2.2 Comparison with Relevant Techniques

The multiple sampled logit vectors $\{\hat{z}_{i,k}\}$ for each input image x_i can be considered as the classifier logit outputs of various input images similar to the original input x_i, whether they correspond to the transformed versions of x_i or different instances (e.g., more lesion spots in images) from the same class. Therefore, the data augmentation in the logit space is a generalization of existing augmentation strategies in the input data space [4,26–28], but without requiring manual choice of augmentation types (e.g., various spatial transformations) as usually used in the input data space. Our method can also be considered as an extension of the data augmentation in the feature space [21,25]. However, augmentation in feature space does not assure that similar logit vectors would come from the same class. In addition, the general decreasing trend of uncertainty values over training stages from our method reminds people of the traditional simulated annealing technique for optimization [20]. From this aspect, the uncertainty estimate in our method can be considered as the temperature parameters, tuning the classifier training process such that the optimization is less likely trapped to a poor local solution. Note that the classifier with uncertainty estimate has been proposed previously [12] for uncertainty estimate of pixel classification. We used the uncertainty estimate novelly for data augmentation to improve the performance of classification with limited training data. Different from the method [12] whose loss function is based on the difference between the ground-truth vector and the mean vector of multiple output vectors, the loss function in our method is based on the average of the differences between the ground-truth vector and each output vector. Another difference is that our method adopts the uniform sampling rather than Gaussian normal random sampling for sample generation. Such differences cause significant performance improvement from our method.

3 Experiment

3.1 Experimental Setting

The proposed method was extensively evaluated on three medical image classification datasets, Skin40, Skin8, and Xray6. Skin40 is a subset of the 198-disease skin image dataset [17]. It contains 40 skin diseases, with 60 images for each disease. Skin8 is from the originally class-imbalanced ISIC2019 challenge dataset [1]. Based on the number (i.e., 239) of images from the smallest class, 239 images were randomly sampled from every other class, resulting in the Skin8 dataset. Xray6 is a subset of ChestXray14 dataset [23], containing six diseases of X-ray images (Atelectasis, Cardiomegaly, Emphysema, Hernia, Mass, Effusion). Based on the smallest class (i.e., Hernia) which has only 110 images, the same number of images were randomly sampled from every other class, forming the small-sample Xray6 dataset. Due to limited images for each dataset, a five-fold cross-validation scheme was adopted, with four folds for training and another fold for testing each time.

For model training on each dataset, each image was resized to 224×224 pixels after random cropping, scaling, and horizontal flipping. For testing, only the same resizing was adopted. During model training, the stochastic gradient descent (batch size 64) with momentum (0.9) and weight decay(0.0001) were adopted. The initial learning rate 0.001 was decayed by 0.1 respectively on epoch 80, 100, 110. All models were trained for 120 epochs with observed convergence. As widely adopted, the initial model parameters were from the pre-trained model based on the natural image dataset Imagenet. Unless otherwise mentioned, the number of samples K was set 5 (different numbers generally lead to similar performance from the proposed method). The average and standard deviation of classification accuracy over the five folds based on the five-fold cross-validation scheme were used to evaluate the performance of each method.

3.2 Effectiveness Evaluation

The effectiveness of our method was evaluated by comparing with various data augmentation methods, including the basic augmentation (random cropping + scaling + flipping, 'Basic' in Table 1), Mixup [27], Manifold Mixup (MM) [21], Dropblock (DB) [9], Cutmix [26], Cutout [4], and RandomErase (RE) [28]. The originally proposed training loss for uncertainty estimate in the related study [12] was also used as a baseline ('UC' in Table 1). The suggested hyper-parameter settings in the original studies were adopted here. Note that the basic data augmentation was used in all the baselines and our method by default. As Table 1 shows, our method outperforms all the data augmentation methods by a clear margin (1%–4%) on all the three medical image classifications tasks, supporting the better effectiveness of our method in alleviating the over-fitting issue. The relatively smaller average accuracy and larger standard deviation on the Xray6 dataset might be due to the smaller dataset (totally 660 images) and the highly inter-class similarity. The little performance improvement from the UC method compared to the Basic method also confirms that the different formulation of the loss function (based on individual augmented logit vector rather than mean logit vector) in our method is crucial. Unexpectedly, some advanced augmentation techniques like Mixup, MM, and RE did not perform better than the Basic method. The fine-grained difference between different diseases in the medical classification tasks might cause the failure of these augmentation techniques.

The effectiveness of our method is further confirmed with various model backbones, including VGG16 [15], ResNet18 [10], ResNet50, SE-ResNet50 [11],

Table 1. Comparison with existing data augmentation methods on three medical image datasets. The model backbone is ResNet50.

Dataset	Basic	Mixup	MM	DB	CutMix	Cutout	RE	UC	Ours
Skin40	73.54	73.67	73.38	73.37	74.29	74.33	73.58	73.38	**76.13**
	(1.21)	(0.75)	(1.91)	(1.05)	(1.78)	(1.29)	(1.25)	(1.65)	(1.98)
Skin8	68.52	68.95	68.47	69.77	68.42	69.00	68.62	70.04	**72.03**
	(2.14)	(3.14)	(1.27)	(3.10)	(1.81)	(2.05)	(1.31)	(2.87)	(1.75)
Xray6	51.22	48.18	50.73	51.73	51.21	51.36	50.61	51.82	**52.61**
	(5.81)	(4.85)	(3.74)	(3.98)	(3.41)	(5.08)	(4.61)	(6.83)	(4.57)

Table 2. Performance comparison on different model backbones.

Backbone	Skin40			Skin8			Xray6		
	Basic	Cutout	Ours	Basic	Cutout	Ours	Basic	Cutout	Ours
ResNet18	71.71	69.58	**73.67**	65.75	67.05	**68.74**	46.52	46.21	**49.37**
	(1.77)	(1.06)	(2.10)	(2.43)	(0.67)	(2.88)	(2.42)	(3.12)	(2.38)
ResNet50	73.54	74.33	**76.13**	68.52	69.00	**72.03**	51.22	51.36	**52.61**
	(1.21)	(1.29)	(1.98)	(2.14)	(2.05)	(1.75)	(5.81)	(5.08)	(4.57)
VGG16	72.63	72.71	**74.92**	68.42	70.62	**72.29**	53.36	48.18	**54.34**
	(1.74)	(1.71)	(1.68)	(2.70)	(3.09)	(1.66)	(4.85)	(4.80)	(6.37)
SE-ResNet50	72.33	74.33	**75.88**	68.27	69.32	**70.36**	48.77	49.70	**52.34**
	(1.30)	(1.23)	(1.88)	(4.32)	(1.18)	(2.81)	(5.39)	(4.63)	(7.31)
EfficientNet-B3	69.67	69.54	**75.42**	68.11	67.28	**70.98**	50.96	47.42	**54.08**
	(1.36)	(2.27)	(1.97)	(1.80)	(2.13)	(1.90)	(4.63)	(4.36)	(6.21)
ViT-B_224	74.71	75.38	**77.38**	67.16	67.38	**69.73**	43.18	44.21	**45.30**
	(0.60)	(0.84)	(1.89)	(1.46)	(1.67)	(2.52)	(6.81)	(5.96)	(2.16)

EfficientNet-B3 [19], and the Transformer architecture ViT [5]. As shown in Table 2, although the classification performance varies across model backbones, our method consistently performs better than the representative baselines Basic and Cutout on each model backbone. This also suggests that our method is not limited to any specific model structure and can be applied to the training of models with various architectures. Note that more advanced backbones (e.g., EfficientNet) did not always perform better than more basic backbones (e.g., VGG and ResNet50). This again might be caused by various factors such as the fine-grained difference between different diseases, and the limited transferability of backbones from natural image dataset to small medical dataset.

3.3 Model Component Choice and Effect of Hyper-parameters

Effect of Random Sampler: During model training, uniform random sampler was adopted to generate multiple samples in the logit space based on each training data. Compared to the samples from Gaussian normal random sampler, samples from uniform random sampler are more varied within the neighborhood of each original logit vector, and therefore may correspond to more different augmentations in the input space. Since more augmentations can represent more different training data, it is expected that uniform random sampler would perform better than Gaussian normal random sampler in improving model performance. This is confirmed with different model backbones on Skin40 (Fig. 2).

Although samples from Gausian normal random sampler already increase the model performance compared to basic data augmentation in the input space, uniform random sampler helps further improve the performance consistently.

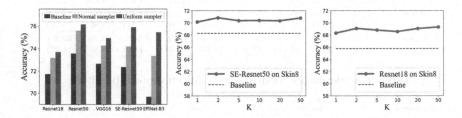

Fig. 2. Effect of random sampler (left) and sampling number K (middle and right).

Effect of Sampling Number K: While K was set 5 for all above experiments, we expect different K would result in similar classification performance because multiple iterations of training would equivalently generate various samples in the logit space. As Fig. 2 (middle and right) shows, with the two backbones SE-Resnet50 (middle) and Resnet18 (right) respectively, the performance of the classifier is relatively stable with respect to the number of samples K and all are clearly better than that from the Basic augmentation ('Baseline' in Fig. 2).

Effect of Uncertainty Estimate: Our method can directly estimate the uncertainty of logit elements for each training data at each training stage, and we expect the uncertainty decreases over training stages as analyzed in Sect. 2.1. This is confirmed in Fig. 3, which shows that the average (absolute) uncertainty over all training images generally decrease over training epochs. The adaptive uncertainty estimate can help adjust the size of neighborhood for sampling over training stages and training images. From Table 3, it is clear that automatic and adaptive uncertainty estimate works consistently well, while fixed uncertainty value even within the range of automatically estimated uncertainty (e.g., $\sigma = 0.05, 0.15, 0.35$) works well only on specific dataset with specific model backbone, supporting the necessity and effectiveness of adaptive uncertainty estimate for data augmentation in the logit space.

Table 3. Effect of adaptive uncertainty estimate on classification performance.

Uncertainty σ	Skin40			Skin8		
	ResNet18	ResNet50	SE-ResNet50	ResNet18	ResNet50	SE-ResNet50
Adaptive σ	73.67	76.13	75.88	68.74	72.03	70.36
Fixed $\sigma = 0.05$	73.79	74.08	75.54	69.01	71.35	70.35
Fixed $\sigma = 0.15$	73.45	76.20	74.92	68.36	70.99	69.47
Fixed $\sigma = 0.35$	73.38	75.67	75.87	68.31	72.08	69.62

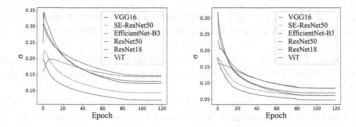

Fig. 3. Uncertainty estimate over training stages on Skin40 (left) and Skin8 (right).

4 Conclusion

To alleviate the over-fitting due to limited training data, a novel data augmentation method is proposed, not in the input or feature space, but in the logit space. Experimental evaluations on multiple datasets and model backbones confirm the effectiveness of the proposed method for improving classification performance. In future work, the proposed method will be applied to more medical image analysis tasks including imbalanced classification and classification on large datasets.

Acknowledgement. This work is supported by the National Natural Science Foundation of China (No. 62071502, U1811461), the Guangdong Key Research and Development Program (No. 2020B1111190001, 2019B020228001), and the Meizhou Science and Technology Program (No. 2019A0102005).

References

1. Combalia, M., et al.: Bcn20000: Dermoscopic lesions in the wild. arXiv preprint arXiv:1908.02288 (2019)
2. Cubuk, E.D., Zoph, B., Mane, D., Vasudevan, V., Le, Q.V.: Autoaugment: learning augmentation strategies from data. In: Proceedings of the IEEE Conference on Computer Vision and Pattern Recognition (2019)
3. Cubuk, E.D., Zoph, B., Shlens, J., Le, Q.: Randaugment: practical automated data augmentation with a reduced search space. Adv. Neural. Inf. Process. Syst. **33**, 18613–18624 (2020)
4. DeVries, T., Taylor, G.W.: Improved regularization of convolutional neural networks with cutout. arXiv preprint arXiv:1708.04552 (2017)
5. Dosovitskiy, A., et al.: An image is worth 16x16 words: Transformers for image recognition at scale. arXiv preprint arXiv:2010.11929 (2020)
6. Esteva, A., Kuprel, B., Novoa, R.A., Ko, J., Swetter, S.M., Blau, H.M., Thrun, S.: Dermatologist-level classification of skin cancer with deep neural networks. Nature **542**, 115–118 (2017)
7. Esteva, A., Robicquet, A., Ramsundar, B., Kuleshov, V., DePristo, M., Chou, K., Cui, C., Corrado, G., Thrun, S., Dean, J.: A guide to deep learning in healthcare. Nat. Med. **25**, 24–29 (2019)
8. Finn, C., Abbeel, P., Levine, S.: Model-agnostic meta-learning for fast adaptation of deep networks. In: Proceedings of the 34th International Conference on Machine Learning, vol. 70, pp. 1126–1135 (2017)

9. Ghiasi, G., Lin, T.Y., Le, Q.V.: Dropblock: a regularization method for convolutional networks. In: Advances in Neural Information Processing Systems, vol. 31 (2018)
10. He, K., Zhang, X., Ren, S., Sun, J.: Deep residual learning for image recognition. In: Proceedings of the IEEE Conference on Computer Vision and Pattern Recognition, pp. 770–778 (2016)
11. Hu, J., Shen, L., Sun, G.: Squeeze-and-excitation networks. In: Proceedings of the IEEE Conference on Computer Vision and Pattern Recognition, pp. 7132–7141 (2018)
12. Kendall, A., Gal, Y.: What uncertainties do we need in bayesian deep learning for computer vision? In: Advances in Neural Information Processing Systems, vol. 30 (2017)
13. Litjens, G., Kooi, T., Bejnordi, B.E., Setio, A.A.A., Ciompi, F., Ghafoorian, M., Van Der Laak, J.A., Van Ginneken, B., Sánchez, C.I.: A survey on deep learning in medical image analysis. Med. Image Anal. **42**, 60–88 (2017)
14. Rusu, A.A., Rao, D., Sygnowski, J., Vinyals, O., Pascanu, R., Osindero, S., Hadsell, R.: Meta-learning with latent embedding optimization. In: 7th International Conference on Learning Representations (2019)
15. Simonyan, K., Zisserman, A.: Very deep convolutional networks for large-scale image recognition. In: 3rd International Conference on Learning Representations (2015)
16. Snell, J., Swersky, K., Zemel, R.: Prototypical networks for few-shot learning. In: Advances in Neural Information Processing Systems, vol. 30 (2017)
17. Sun, X., Yang, J., Sun, M., Wang, K.: A benchmark for automatic visual classification of clinical skin disease images. In: Leibe, B., Matas, J., Sebe, N., Welling, M. (eds.) ECCV 2016. LNCS, vol. 9910, pp. 206–222. Springer, Cham (2016). https://doi.org/10.1007/978-3-319-46466-4_13
18. Sung, F., Yang, Y., Zhang, L., Xiang, T., Torr, P.H.S., Hospedales, T.M.: Learning to compare: relation network for few-shot learning. In: Proceedings of the IEEE Conference on Computer Vision and Pattern Recognition, pp. 1199–1208 (2018)
19. Tan, M., Le, Q.V.: Efficientnet: rethinking model scaling for convolutional neural networks. In: Proceedings of the 36th International Conference on Machine Learning, vol. 97, pp. 6105–6114 (2019)
20. Van Laarhoven, P.J., Aarts, E.H.: Simulated annealing. In: Simulated Annealing: Theory and Applications, pp. 7–15 (1987)
21. Verma, V., Lamb, A., Beckham, C., Najafi, A., Mitliagkas, I., Lopez-Paz, D., Bengio, Y.: Manifold mixup: Better representations by interpolating hidden states. In: Proceedings of the 36th International Conference on Machine Learning. vol. 97, pp. 6438–6447 (2019)
22. Vinyals, O., Blundell, C., Lillicrap, T., kavukcuoglu, k., Wierstra, D.: Matching networks for one shot learning. In: Advances in Neural Information Processing Systems, vol. 29 (2016)
23. Wang, X., Peng, Y., Lu, L., Lu, Z., Bagheri, M., Summers, R.M.: ChestX-ray8: Hospital-scale chest X-ray database and benchmarks on weakly-supervised classification and localization of common thorax diseases. In: Proceedings of the IEEE Conference on Computer Vision and Pattern Recognition. pp. 3462–3471 (2017)
24. Wang, Y., Yao, Q., Kwok, J.T., Ni, L.M.: Generalizing from a few examples: a survey on few-shot learning. ACM Comput. Surv. **53**(3), 63:1–63:34 (2020)
25. Wang, Y., Huang, G., Song, S., Pan, X., Xia, Y., Wu, C.: Regularizing deep networks with semantic data augmentation. IEEE Transactions on Pattern Analysis and Machine Intelligence (2021)

26. Yun, S., Han, D., Chun, S., Oh, S.J., Yoo, Y., Choe, J.: Cutmix: regularization strategy to train strong classifiers with localizable features. In: Proceedings of the IEEE International Conference on Computer Vision, pp. 6022–6031 (2019)
27. Zhang, H., Cissé, M., Dauphin, Y.N., Lopez-Paz, D.: mixup: beyond empirical risk minimization. In: 6th International Conference on Learning Representations (2018)
28. Zhong, Z., Zheng, L., Kang, G., Li, S., Yang, Y.: Random erasing data augmentation. In: The Thirty-Fourth AAAI Conference on Artificial Intelligence, pp. 13001–13008 (2020)

Collaborative Image Synthesis and Disease Diagnosis for Classification of Neurodegenerative Disorders with Incomplete Multi-modal Neuroimages

Yongsheng Pan[1,2,3](✉), Yuanyuan Chen[1,2], Dinggang Shen[3](✉),
and Yong Xia[1,2](✉)

[1] National Engineering Laboratory for Integrated Aero-Space-Ground-Ocean Big
Data Application Technology, School of Computer Science and Engineering,
Northwestern Polytechnical University, Xi'an 710072, Shaanxi, China
yspan@mail.nwpu.edu.cn

[2] Research and Development Institute of Northwestern Polytechnical University
in Shenzhen, Shenzhen 518057, China
yxia@nwpu.edu.cn

[3] School of Biomedical and Engineering,
ShanghaiTech University, Shanghai 201210, China
dgshen@shanghaitech.edu.cn

Abstract. The missing data issue is a common problem in multi-modal
neuroimage (e.g., MRI and PET) based diagnosis of neurodegenerative disorders. Although various generative adversarial networks (GANs)
have been developed to impute the missing data, most current solutions
treat the image imputation and disease diagnosis as two standalone tasks
without considering the impact of diagnosis on image synthesis, leading
to less competent synthetic images to the diagnosis task. In this paper,
we propose the *collaborative diagnosis-synthesis framework* (**CDSF**) for
joint missing neuroimage imputation and multi-modal diagnosis of neurodegenerative disorders. Under the CDSF framework, there is an image
synthesis module (ISM) and a multi-modal diagnosis module (MDM),
which are trained in a collaborative manner. Specifically, ISM is trained
under the supervision of MDM, which poses the feature-consistent constraint to the cross-modality image synthesis, while MDM learns the
disease-related multi-modal information from both real and synthetic
multi-modal neuroimages. We evaluated our CDSF model against five
image synthesis methods and three multi-modal diagnosis models on an
ADNI datasets with 1464 subjects. Our results suggest that the proposed CDSF model *not only* generates neuroimages with higher quality,
but also achieves the state-of-the-art performance in AD identification
and MCI-to-AD conversion prediction.

1 Introduction

Multi-modal neuroimages provide complementary functional (PET) and structural (MRI) information, and hence have unique advantages over single-modal

© Springer Nature Switzerland AG 2021
M. de Bruijne et al. (Eds.): MICCAI 2021, LNCS 12905, pp. 480–489, 2021.
https://doi.org/10.1007/978-3-030-87240-3_46

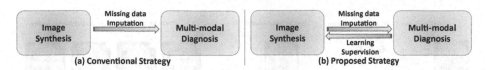

Fig. 1. Strategies for diagnosis with incomplete multi-modal data: (a) conventional strategy treats image synthesis and multi-modal diagnosis as two standalone tasks, and (b) proposed strategy treats them as two collaborative tasks.

images in the diagnosis of neurodegenerative disorders such as the Alzheimer's disease (AD) and mild cognitive impairment (MCI) [1,2]. Multi-modal diagnosis, however, often suffers from the missing data problem, *i.e.*, some subjects may lack the data of a specific modality (mostly PET) due to patient dropouts or systematic errors. Rather than discarding modality-incomplete subjects [1], imputing the missing data makes it possible to use all available subjects for model training. The enlarged training dataset is usually able to boost the performance of diagnosis model [2,3].

Generative adversarial networks (GANs) have achieved remarkable success in image-to-image translation [4–7]. Such success has prompted the application of various GANs to imputing the missing modality of neuroimages [2,8]. Pan et al. [2,8] applied the cycle-consistent GAN (cycGAN) and voxel-wise-consistent GAN (voxGAN) to synthesize the missing PET in incomplete multi-modal neuroimages. Choi et al. [5] proposed the star GAN to synthesize multiple modalities from only one available modality, while Lee et al. [6] designed the collaborative GAN to synthesize the missing modality from multiple available modalities. In these solutions, synthesizing the missing image is treated as a standalone task, without considering the subsequent disease diagnosis. The information flow is one-way from image synthesis to disease diagnosis (see Fig. 1(a)). As a result, the diagnosis-specific information may be lost during the image synthesis process, and the synthetic image may be less competent to the diagnosis task. It has been reported that, in an even worse scenario, the diagnosis model can hardly learn any useful information from the synthetic images produced by GANs [9].

To bridge the gap between image synthesis and disease diagnosis, a disease-image specific deep learning (DSDL) framework [10] was proposed to complete both tasks jointly. Under the DSDL framework, the missing neuroimages are imputed in a diagnosis-oriented manner, resulting in a lot of discriminative information being preserved in synthetic images. DSDL can also synthesize PET images to improve the diagnosis of subjective cognitive decline without using any real PET scans [11]. However, DSDL has two major drawbacks. Fist, it still treats image synthesis and multi-modal diagnosis as two sequential learning processes. Second, although two single-modal diagnosis models are used to guide the image synthesis process, they cannot provide the complementary multi-modal information to the synthesis process. These two blemishes lead to the degradation of diagnostic performance. Therefore, a collaborative learning process for image synthesis and multi-modal diagnosis is intuitively desirable.

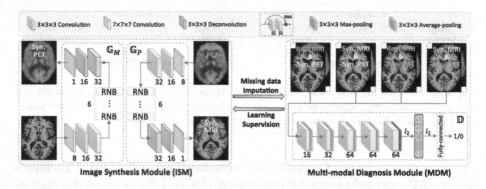

Fig. 2. Diagram of the proposed collaborative diagnosis-synthesis framework (CDSF), including (left) an image synthesis module (ISM) and (right) a multi-modal diagnosis module (MDM). RNB: Residual Network Block.

In this paper, we propose a *collaborative diagnosis-synthesis framework* (**CDSF**) for the computer-aided diagnosis of AD and MCI using incomplete multi-modal PET and MRI scans. As shown in Fig. 2, CDSF contains an image synthesis module (ISM) and a multi-modal diagnosis module (MDM), which are trained collaboratively with a two-way information flow (see Fig. 1(b)). Specifically, ISM synthesizes cross-modality images under the supervision of MDM, which poses the feature-consistent constraint to the synthesis process, while MDM learns the disease-related information and multi-modal diagnosis on both real and ISM-synthesized multi-modal neuroimages.

The proposed CDSF has three advantages over DSDL and other models: (1) both ISM and MDM are learned collaboratively and thus can cooperate well, (2) MDM can directly provide the multi-modal feature consistency for ISM, avoiding the use of two additional single-modal classifiers, and (3) our CDSF achieves substantially improved performance in both neuroimage synthesis and neurodegenerative disorder diagnosis.

2   Method

2.1   Problem Formulation

Let a multi-modal neuroimage dataset contain N_c modality-complete subjects (*i.e.*, having paired MRI and PET scans) and $(N - N_c)$ subjects with only one imaging modality (e.g., MRI). A multi-modal diagnosis system requires a pair of MRI (denoted by $\mathbf{M}_i$) and PET (denoted by $\mathbf{P}_i$) scans as its input. On these N_c modality-complete subjects, this system (denoted as $\mathbb{D}$) can be formulated as

$$\hat{\mathbf{y}}_i = \mathbb{D}\left(\mathbf{M}_i, \mathbf{P}_i\right), \tag{1}$$

where $\hat{\mathbf{y}}_i$ is the estimated label of the ith subject whose ground truth label is $\mathbf{y}_i$.

To train the diagnosis system $\mathbb{D}$ with incomplete data, an image synthesis model should be constructed to impute the missing images based on the relevance between two imaging modalities. The image synthesis model learns a mapping from MRI scans to PET scans (denoted by $\mathbb{G}_M : \mathbf{M}_i \rightarrow \mathbf{P}_i$) [2,10], which can be used to estimate a virtual PET scan $\hat{\mathbf{P}}_i$ for the missing one based on the available MRI scan $\mathbf{M}_i$. Then, the diagnosis model can be trained with both real and synthetic modality-complete data, shown as follows

$$\hat{\mathbb{D}} = \underset{\mathbb{D}}{\arg\min} \left\{ \sum_{i=1}^{N_c} \|\mathbb{D}(\mathbf{M}_i, \mathbf{P}_i) - \mathbf{y}_i\| + \sum_{i=N_c+1}^{N} \|\mathbb{D}(\mathbf{M}_i, \mathbb{G}_M(\mathbf{M}_i)) - \mathbf{y}_i\| \right\}. \quad (2)$$

2.2 Model Overview

The proposed CDSF consists of an image synthesis module (ISM) and a multimodal diagnosis module (MDM). ISM contains two synthesis branches with opposite directions, and MDM contains only one diagnosis network which takes a pair of MRI and PET scans as its two-channel input. Since the AD-associated regions in MRI scans are usually different from those in PET scans, we process MRI images and PET images using two branches in ISM, rather than a single branch with multi-channel input or output [5,6]. In MDM, we follow the early-fusion strategy that utilizes a single branch to extract closely relevant multimodal features, and abandon the late-fusion strategy [2,10] that utilizes two branches to extract the features of two modalities, respectively. The diagram of our CDSF is shown in Fig. 2. We now delve into the details of each part.

2.3 Image Synthesis

As shown in the left half of Fig. 2, the proposed ISM has two synthesis branches (i.e., $\mathbb{G}_M$ and $\mathbb{G}_P$) with the opposite directions. Each branch consists of an encoding part, a transferring part, and a decoding part. The encoding part contains three convolutional (Conv) layers with 8, 16, and 32 channels, respectively. The transferring part contains six residual network blocks (RNBs) [12]. The decoding part contains two deconvolutional (Deconv) layers with 32 and 16 channels, respectively, and one Conv layer with 1 channel. Except for the last Conv layer that uses the "tanh" activation, all Conv/Deconv layers use the "relu" activation. These two Deconv layers, the second layer, and the third Conv layer are with a stride of 2 to construct a pyramid structure. Meanwhile, instance normalization is performed for all layers. During the training phase, both the feature-consistent constraint [10] and voxel-wise-consistent constraint [2] are used to guide the learning of ISM (see Eqs. (3) (4)). With these two constraints, each pair of real and synthetic images are encouraged to keep consistent in both visual and diagnostic aspects. It should be noted that we do not use any discriminators since discriminators can hardly benefit the disease diagnosis [9].

2.4 Multi-modal Diagnosis

As shown in the right half of Fig. 2, the proposed MDM contains only one diagnosis network (*i.e.*, $\mathbb{D}$) with a feature extraction part and a spatial representation part. The feature extraction part has five Conv layers (followed by instance normalization and "relu" activation) with 16, 32, 64, 64, and 64 channels, respectively. The first four Conv layers and the last Conv layer are followed, respectively, by $3 \times 3 \times 3$ max-pooling and average-pooling layers with a stride of 2. To deal with multi-modal inputs, each pair of neuroimages are first concatenated into a two-channel image and then fed into the feature extraction part. The feature extraction part generates feature maps at each Conv layer, where the feature map at the jth layer are denoted by $\mathbb{D}_j$ ($j = 1, \cdots, 5$). The output of the feature extraction part is first l_2-normalized along the channel dimension and then reshaped to spatial representation. Then, the spatial representation is l_2-normalized and fed to a fully-connected layer with the "softmax" activation to compute the probabilities of a subject belonging to the positive class and negative class. This twice l_2-normalization is a key to boost the diagnosis performance, which is same as the spatial cosine kernel in [10]. During the training phase, the cross-entropy loss is used to guide the learning of MDM.

2.5 Collaborative Learning

Training Phase: Different from [2], the proposed ISM and MDM are trained collaboratively with both complete subjects and incomplete subjects. Given a modal-complete input with both $\mathbf{M}_i$ and $\mathbf{P}_i$, two synthesis branches $\mathbb{G}_M$ and $\mathbb{G}_P$ in ISM are trained as follows

$$\hat{\mathbb{G}}_M = \underset{\mathbb{G}_M}{\arg\min}\{\sum_{j=1}^{5} \|\mathbb{D}_j(\mathbf{M}_i, \mathbb{G}_M(\mathbf{M}_i)) - \mathbb{D}_j(\mathbf{M}_i, \mathbf{P}_i)\|_{l_1} + \|\mathbb{G}_M(\mathbf{M}_i) - \mathbf{P}_i\|_{l_1}\}, \quad (3)$$

$$\hat{\mathbb{G}}_P = \underset{\mathbb{G}_P}{\arg\min}\{\sum_{j=1}^{5} \|\mathbb{D}_j(\mathbb{G}_P(\mathbf{P}_i), \mathbf{P}_i) - \mathbb{D}_j(\mathbf{M}_i, \mathbf{P}_i)\|_{l_1} + \|\mathbb{G}_P(\mathbf{P}_i) - \mathbf{M}_i\|_{l_1}\}, \quad (4)$$

where $\|*\|_{l_1}$ represents the mean absolute error (MAE) loss. The former and later items in Eqs. 3 and 4 are, respectively, the feature-consistent constraint [10] and the voxel-wise-consistent constraint [2]. Meanwhile, the diagnosis network $\mathbb{D}$ in MDM is trained as follows

$$\hat{\mathbb{D}} = \underset{\mathbb{D}}{\arg\min}\{ \|\mathbb{D}(\mathbf{M}_i, \mathbf{P}_i), \mathbf{y}_i\|_{ce} + \|\mathbb{D}(\mathbb{G}_P(\mathbf{P}_i), \mathbb{G}_M(\mathbf{M}_i)), \mathbf{y}_i\|_{ce}$$
$$+ \|\mathbb{D}(\mathbb{G}_P(\mathbf{P}_i), \mathbf{P}_i), \mathbf{y}_i\|_{ce} + \|\mathbb{D}(\mathbf{M}_i, \mathbb{G}_P(\mathbf{M}_i)), \mathbf{y}_i\|_{ce}\}, \quad (5)$$

where $\|*, *\|_{ce}$ represents the cross entropy loss. Equation 5 means that MDM learns from four combinations, *i.e.*, real MRI and real PET, real MRI and synthetic PET, synthetic MRI and real PET, synthetic MRI and synthetic PET.

If the input is a modal-incomplete data with only $\mathbf{M}_i$ (or $\mathbf{P}_i$), we first resort to ISM to generate a synthetic version of the missing data $\hat{\mathbf{P}}_i = \mathbb{G}_M(\mathbf{M}_i)$ (or $\hat{\mathbf{M}}_i = \mathbb{G}_P(\mathbf{P}_i)$), and then train ISM and MDM in the same way.

Test Phase: For each input subject, if it is a complete subject with both $\mathbf{M}_i$ and $\mathbf{P}_i$, we predict its diagnostic label as $\hat{\mathbf{y}}_i = \mathbb{D}(\mathbf{M}_i, \mathbf{P}_i)$. If it is an incomplete subject with only $\mathbf{M}_i$, we first impute it by $\hat{\mathbf{P}}_i = \mathbb{G}_M(\mathbf{M}_i)$, and then predict its diagnostic label as $\hat{\mathbf{y}}_i = \mathbb{D}\left(\mathbf{M}_i, \hat{\mathbf{P}}_i\right)$.

3 Materials

Two subsets of the Alzheimer's Disease Neuroimaging Initiative (ADNI) studies [13], including ADNI-1 phase and ADNI-2 phase, were used for this study. All collected subjects in ADNI-1/-2 have baseline MRI data, while only part of them have PET images. We follow the same steps in [2,10] to process the collected data. The subjects in ADNI-1/2 were divided into three categories: AD, cognitively normal (CN), and MCI. MCI could be further divided into progressive MCI (pMCI) and static MCI (sMCI) that would or would not progress to AD within 36 months after the baseline. Totally, we also have 205 AD, 231 CN, 165 pMCI, and 147 sMCI subjects in ADNI-1, and 162 AD, 209 CN, 89 pMCI, and 256 sMCI subjects in ADNI-2.

4 Experiments and Results

4.1 Experimental Settings

Subjects from ADNI-1 and ADNI-2 were respectively used as the training set and the independent test set. Our models were iteratively trained 100 epochs via the Adam solver and stochastic gradient descent (SGD) solver [14] for ISM and MDM, respectively, with a batch size of 1 and a learning rate of 1×10^{-3} and 1×10^{-2}.

The proposed CDSF was evaluated in terms of image synthesis and disease diagnosis. The quality of synthetic images was measured by the mean absolute error (MAE), mean square error (MSE), peak signal-to-noise ratio (PSNR), and structural similarity index measure (SSIM) [15]. The performance on AD identification (AD vs. CN classification) and MCI conversion prediction (pMCI vs. sMCI classification) was measured by six metrics, including the area under ROC curve (AUC), accuracy (ACC), sensitivity (SEN), specificity (SPE), F1-Score (F1S), and Matthews correlation coefficient (MCC) [16].

4.2 Results of Neuroimage Synthesis

To evaluate the performance of our CDSF in image synthesis, we compared it to five generative models, including (1) a conventional GAN with only the adversarial loss, (2) cycle-consistent GAN (CGAN) [8], (3) voxel-wise-consistent GAN (VGAN) that is a 3D version of pix2pixGAN [4], (4) (L1GAN) [9] with only the MAE loss, and (5) feature-consistent GAN (FGAN) [10]. We trained these models (*i.e.*, GAN, CGAN, VGAN, L1GAN and FGAN) using the complete

Table 1. Performance of six models trained on ADNI-1 in synthesizing MRI scans or PET scans for subjects in ADNI-2.

Method	Synthetic MRI				Synthetic PET			
	MAE (%)	MSE (%)	SSIM (%)	PSNR	MAE (%)	MSE (%)	SSIM (%)	PSNR
GAN	10.66	3.68	59.34	26.43	10.79	3.05	57.41	27.27
CycGAN	10.16	3.33	60.55	26.86	10.36	2.53	57.63	27.56
VoxGAN	10.12	3.14	61.35	27.20	9.47	2.51	64.51	28.85
FGAN	8.64	2.27	68.20	28.56	8.24	1.80	67.34	29.70
L1GAN	**7.30**	**1.85**	**70.54**	**29.56**	8.03	2.05	68.77	30.04
CDSF (Ours)	8.24	2.26	67.14	28.66	8.04	**1.98**	**69.52**	**30.17**

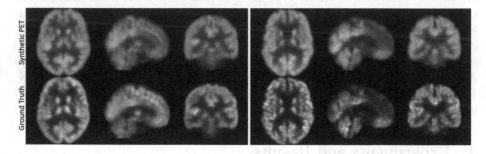

Fig. 3. PET scans synthesized by CDSF for two typical subjects (Roster ID: 4757, 4784) in ADNI-2, along with their corresponding ground-truth images.

subjects (with both MRI and PET scans) in ADNI-1, and tested the trained models on the complete subjects in ADNI-2. Performance of these models in image synthesis was reported in Table 1.

Based on the results listed in Table 1, three conclusions can be drawn. First, the conventional GAN performs worse than other five synthesis models, which use additional constraints other than the adversarial loss. It suggests that the adversarial loss is a weak constraint to learn pairwise patterns. Second, the synthesis models without using the adversarial loss (L1GAN and CDSF) consistently outperform those using the adversarial loss (GAN, CGAN, VGAN, FGAN). It verifies that the adversarial loss causes too much "hallucination" [9] and may not be helpful to the corss-modal image generation task. Third, our CDSF achieves similar performance to L1GAN in synthesizing PET scans in terms of all four metrics. It reveals that the feature-consistent loss can co-exist well with the MAE loss so as to keep the image quality for the synthesis model.

The synthetic PET images generated by our CDSF for two typical subjects in ADNI-2, together with the corresponding ground-truth images, were visualized in Fig. 3. It reveals that these synthetic PET images are visually similar to real ones, though there is still some blur in synthetic images.

4.3   Results of Disease Diagnosis

To evaluate the performance of the proposed CDSF in disease diagnosis, we compared it to three state-of-the-art deep learning models, including the landmark-based deep multi-instance learning (LDMIL) [8], spatially-constrained Fisher representation (SCFR) [2], and the DSDL model [10]. Since each competing model has a corresponding imputation module, it can be trained with both complete and incomplete subjects. We also compared our CDSF to two of its variants denoted by CDSF-V and CDSF-F. CDSF-V uses only the voxel-wise-consistent constraint, and CDSF-F uses only the feature-consistent constraint. In additional, we reported the performance of baseline model, denoted by CDSF-M, which uses only MRI scans for training and testing.

The performance of these models in AD-CN classification and pMCI-sMCI classification was reported in Table 2. It shows that our CDSF is superior to other competing methods in terms of five out of six performance metrics in both classification tasks. For pMCI-sMCI classification, our CDSF achieves the best AUC value (86.28%), and its variant CDSF-V achieves the second best AUC (84.96%). It implies that our CDSF is more effective in predicting the progression of MCI patients and the collaborative strategy is promising for the multi-modal diagnosis with incomplete subjects. Besides, on both classification tasks, five multi-modal diagnosis models generally outperform the baseline model, showing the advantage of multi-modal diagnosis over single-modal diagnosis. Note that LDMIL does not achieve good performance in MCI conversion prediction, which may be attributed to the fact that it is not easy to adapt the manually defined landmarks to the prediction task. Furthermore, CDSF and CDSF-F perform better than CDSF-P, though they use the same integrating strategy. It suggests that preserving the disease-specific information in synthetic images is a necessity for improving the diagnosis performance. In addition, CDSF outperforms CDSF-V and slightly outperforms CDSF-F, suggesting that both feature-consistent constraint and voxel-wise-consistent constraint contribute to promoting the performance of neurodegenerative disease diagnosis.

Table 2. Performance (%) of seven models in diagnosis of neurodegenerative disorders. CDSF-M is the baseline method that uses only MRI scans.

Method	AD-CN Classification						pMCI-sMCI Classification					
	AUC	ACC	SEN	SPE	F1S	MCC	AUC	ACC	SEN	SPE	F1S	MCC
CDSF-M	95.68	90.11	89.70	90.43	88.89	79.99	82.36	75.65	73.03	76.56	60.75	45.14
LDMIL	95.89	92.50	89.94	94.53	91.37	84.78	75.84	79.06	55.26	82.85	40.86	30.13
SCFR	96.95	**93.58**	91.52	**95.22**	92.64	**86.97**	83.87	76.81	75.28	77.34	62.62	47.90
DSDL	97.02	92.25	90.91	93.30	91.18	84.27	83.94	80.00	70.79	83.20	64.61	51.20
CDSF-V	96.01	92.78	89.70	**95.22**	92.64	85.35	84.48	78.84	70.79	81.64	74.22	49.22
CDSF-F	**97.13**	92.78	90.91	94.25	92.66	85.34	84.96	79.13	**77.53**	79.69	75.36	52.46
CDSF (Ours)	97.09	**93.58**	**92.12**	94.74	**93.48**	**86.97**	**86.28**	81.45	73.03	**84.38**	**77.05**	**54.59**

5   Conclusion

We proposed the *collaborative diagnosis-synthesis framework* (CDSF) to address the missing data problem in multi-modal diagnosis of neurodegenerative disorders. Under this framework, the image synthesis module and multi-modal diagnosis module are learned collaboratively with a two-way information flow. The cross-modality images synthesis module is trained under the supervision of the diagnosis via the feature-consistent constraint, whereas the multi-model diagnosis module is trained with both real and synthetic neuroimages. Comparative experiments on ADNI demonstrate that our CDSF outperforms both neuroimage synthesis methods and multi-modal diagnosis methods, showing great potential for clinical use.

Acknowledgments. This work was supported in part by the National Natural Science Foundation of China under Grants 61771397, in part by the CAAI-Huawei Mind-Spore Open Fund under Grants CAAIXSJLJJ-2020-005B, and in part by the China Postdoctoral Science Foundation under Grants BX2021333.

References

1. Zhou, J., Yuan, L., Liu, J., Ye, J.: A multi-task learning formulation for predicting disease progression. In: ACM SIGKDD, pp. 814–822. ACM (2011)
2. Pan, Y., Liu, M., Lian, C., Xia, Y., Shen, D.: Spatially-constrained fisher representation for brain disease identification with incomplete multi-modal neuroimages. IEEE Trans. Med. Imag. **39**(9), 2965–2975 (2020)
3. Baraldi, A.N., Enders, C.K.: An introduction to modern missing data analyses. J. Sch. Psychol. **48**(1), 5–37 (2010)
4. Isola, P., Zhu, J.Y., Zhou, T., Efros, A.A.: Image-to-image translation with conditional adversarial networks. In: Proceedings of the IEEE Conference Computer Vision Pattern Recognition, pp. 1125–1134 (2017)
5. Choi, Y., Choi, M., Kim, M., Ha, J., Kim, S., Choo, J.: Stargan: Unified generative adversarial networks for multi-domain image-to-image translation. In: Proceedings of the IEEE/CVF Conference Computer Vision Pattern Recognition, pp. 8789–8797 (2018)
6. Lee, D., Kim, J., Moon, W., Ye, J.C.: Collagan: Collaborative gan for missing image data imputation. In: Proceedings of the IEEE/CVF Conference Computer Vision Pattern Recognition, pp. 2482–2491 (2019)
7. Huang, X., Liu, M.Y., Belongie, S., Kautz, J.: Multimodal unsupervised image-to-image translation. In: Proceedings of the European Conference Computer Vision, pp. 179–196 (2018)
8. Pan, Y., Liu, M., Lian, C., Zhou, T., Xia, Y., Shen, D.: Synthesizing missing PET from MRI with cycle-consistent generative adversarial networks for Alzheimer's disease diagnosis. In: Proceedings of the International Conference Medical Image Comput. Computer Assisted Intervention, pp. 455–463 (2018)
9. Cohen, J.P., Luck, M., Honari, S.: Distribution matching losses can hallucinate features in medical image translation. In: Frangi, A.F., Schnabel, J.A., Davatzikos, C., Alberola-López, C., Fichtinger, G. (eds.) MICCAI 2018. LNCS, vol. 11070, pp. 529–536. Springer, Cham (2018). https://doi.org/10.1007/978-3-030-00928-1_60

10. Pan, Y., Liu, M., Lian, C., Xia, Y., Shen, D.: Disease-image specific generative adversarial network for brain disease diagnosis with incomplete multi-modal neuroimages. In: Shen, D., Liu, T., Peters, T.M., Staib, L.H., Essert, C., Zhou, S., Yap, P.-T., Khan, A. (eds.) MICCAI 2019. LNCS, vol. 11766, pp. 137–145. Springer, Cham (2019). https://doi.org/10.1007/978-3-030-32248-9_16

11. Liu, Y., Pan, Y., Yang, W., Ning, Z., Yue, L., Liu, M.: Joint neuroimage synthesis and representation learning for conversion prediction of subjective cognitive decline. In: Proceedings of the International Conference Medical Image Comput. Computer Assisted Intervention, pp. 583–592, September 2020

12. He, K., Zhang, X., Ren, S., Sun, J.: Deep residual learning for image recognition. In: Proceedings of the IEEE/CVF Conference Computer Vision Pattern Recognition, pp. 770–778 (2016)

13. Jack, C., Bernstein, M., Fox, N., et al.: The Alzheimer's disease neuroimaging initiative (ADNI): MRI methods. J. Magn. Reson. Imaging **27**(4), 685–691 (2008)

14. Kingma, D.P., Ba, J.: Adam: A method for stochastic optimization. arXiv preprint arXiv:1412.6980 (2014)

15. Hore, A., Ziou, D.: Image quality metrics: PSNR vs. SSIM. In: Proceedings of the International Conference Pattern Recognition, pp. 2366–2369 (2010)

16. Koyejo, O.O., Natarajan, N., Ravikumar, P.K., Dhillon, I.S.: Consistent binary classification with generalized performance metrics. In: Conference Neural Information Processing System, pp. 2744–2752 (2014)

Seg4Reg+: Consistency Learning Between Spine Segmentation and Cobb Angle Regression

Yi Lin[✉], Luyan Liu, Kai Ma, and Yefeng Zheng

Tencent Jarvis Lab, Shenzhen, China

Abstract. Automated methods for Cobb angle estimation are of high demand for scoliosis assessment. Existing methods typically calculate the Cobb angle from landmark estimation, or simply combine the low-level task (e.g., landmark detection and spine segmentation) with the Cobb angle regression task, without fully exploring the benefits from each other. In this study, we propose a novel multi-task framework, named Seg4Reg+, which jointly optimizes the segmentation and regression networks. We thoroughly investigate both local and global consistency and knowledge transfer between each other. Specifically, we propose an attention regularization module leveraging class activation maps (CAMs) from image-segmentation pairs to discover additional supervision in the regression network, and the CAMs can serve as a region-of-interest enhancement gate to facilitate the segmentation task in turn. Meanwhile, we design a novel triangle consistency learning to train the two networks jointly for global optimization. The evaluations performed on the public AASCE Challenge dataset demonstrate the effectiveness of each module and superior performance of our model to the state-of-the-art methods.

1 Introduction

Adolescent idiopathic scoliosis (AIS) causes a structural, lateral, rotated curvature of the spine that arises in children at or around puberty [12]. The Cobb angle, derived from a anterior-posterior X-ray and measured by selecting the most tilted vertebra, is the primary means for clinical diagnosing of AIS. However, manual measurement of the Cobb angle requires professional radiologists to carefully identify each vertebra and measure angle, which is time-consuming and could suffer from a large inter-/intra-observer variety. Hence, it is needed to provide an accurate and robust method for quantitative measurement of Cobb angle automatically.

Numerous computer-aided methods have been developed for automated Cobb angle estimation. Conventional methods utilized active contour model [2], customized filter [1] and charged-particle models [8] for spine segmentation to calculate the Cobb angle, which are computationally expensive and the unclear spine boundary will result in inaccurate estimations. Recently, deep learning based

M. de Bruijne et al. (Eds.): MICCAI 2021, LNCS 12905, pp. 490–499, 2021.
https://doi.org/10.1007/978-3-030-87240-3_47

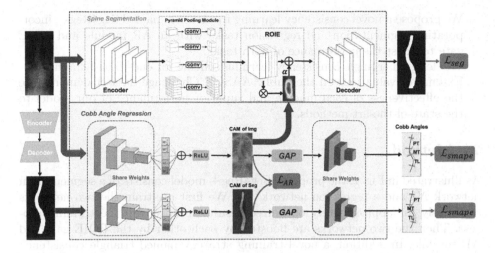

Fig. 1. An overview of the proposed framework.

methods [3,6,9,13,15,17] have been proposed to consolidate the tasks of verte-bral landmark detection with Cobb angle estimation to improve the robustness of spinal curvature assessment. And Seg4Reg [7], which won the 1st place in the AASCE[1] challenge, regarded the segmentation results of spine as the input of the regression network for Cobb angle estimation. Its superior performance owns to the segmentation mask of spine, which retains the shape information and filters the distractions (e.g., artifacts and local contrast variation). However, the performance of this method depends heavily on the segmentation results.

Although the above methods have achieved great success, their applications to Cobb angle estimation suffer from three limitations: 1) The methods [3,6,9, 13,15,17] relying on the landmark coordinates are susceptible since a small error in landmark coordinates may cause a huge mistake in angle estimation; 2) the two-stage frameworks [3,6,7,17] often suffers from the error accumulation; and 3) the cascaded networks [3,6,7] cannot guarantee a global optimum.

In this paper, we propose a novel consistency learning framework, named Seg4Reg+, which incorporates segmentation into the regression task, as shown in Fig. 1. The segmentation task extracts representative features for the regression task by an **attention regularization** (AR) module with auxiliary constraint on the class activation map (CAM). And the regression task is able to provide specific hints for the segmentation task by a **region-of-interest enhancement** (ROIE) gate to force the segmentation network to pay more attention to the important area. To reach the global optimum, we further design a novel **triangle consistency learning** scheme for end-to-end training. In summary, our main contributions are as follows:

[1] https://aasce19.grand-challenge.org.

- We propose a novel consistency learning framework, named Seg4Reg+, incorporating segmentation and regression tasks with an AR module and ROIE gate to boost the performance of both tasks.
- We design a triangle consistency learning scheme for end-to-end training.
- Extensive evaluations on the public AASCE Challenge dataset demonstrate the effectiveness of each module and superior performance of our model to the state-of-the-art methods.

2   Method

As illustrated in Fig. 1, the proposed Seg4Reg+ model consists of a segmentation network N_S and a regression network N_R. We first pre-train the two networks separately for approximately optimized results to speed up the training process. Then the two networks are boosted by each other by the ROIE gate and AR module. In addition, a novel training strategy named triangle consistency learning scheme is designed for end-to-end training. In the following, we first introduce the proposed ROIE gate and AR module, then illustrate the triangle consistency learning scheme.

ROIE Gate. We first train N_S to roughly segment the spine region. We adopt the same network as Seg4Reg [7] for a fair comparison. Specifically, we modify the PSPNet [18] by replacing the pooling layer with the dilated convolution in the pyramid pooling module and take ResNet-50 [4] as the backbone. The objective function is the weighted Dice loss and cross-entropy loss:

$$\mathcal{L}_{\text{seg}}(s(x), y) = \sum_i \left(1 - \frac{2 \times s(x_i)y_i}{s(x_i) + y_i}\right) + \lambda \sum_i (-y_i \log(s(x_i))), \quad (1)$$

where x denotes the input image, and $s(x_i)$ and y_i denote the label of pixel x_i of prediction and ground truth, respectively, and λ denotes the hyperparameter weighting the two losses.

To boost the performance of N_S with N_R, the ROIE gate is designed as an attention mechanism to transfer the specific hints from N_R to N_S. The proposed ROIE gate is inspired by CAM, which is the most common technique in weakly supervised segmentation methods [19]. We expect that the CAM of N_R can incorporate the refined prior information about the spine area into the segmentation process. In addition, the value of each pixel on the CAM represents its significance to the regression output, which in turn guides N_S to pay more attention to the important areas (i.e., the most tilted vertebra endplates). Specifically, we treat CAM of N_R as attention map, and perform a matrix multiplication between CAM and the feature map $f_m(x)$ from the middle layer of N_S. Then, we multiply the result by a scalar parameter α and perform an element-wise sum operation with the feature $f_m(x)$ to obtain the final output $f'_m(x)$ as follows:

$$f'_m(x) = \alpha(C(r(x)) \circ f_m(x)) + f_m(x), \quad (2)$$

where $C(\cdot)$ is a CAM that indicates the discriminative part with respect to the regression results, α is a learnable parameter which is initialized as 0, $\circ$ denotes

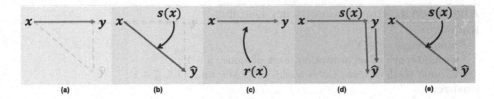

Fig. 2. The training strategy of triangle consistency learning: (a) training N_S, (b) training N_R with AR, (c) fine-tuning N_S with ROIE, (d) fine-tuning N_S by the SMAPE loss, and (e) fine-tuning N_R with refined segmentation.

multiplication function. It can be inferred from Eq. (2) that the resulting feature $f'_m(x)$ combines global contextual view and selectively aggregates contexts according to the CAM, thus improving intra-class compact and semantic consistency.

AR Module. For N_r, we modify the state-of-the-art classification network by replacing the last convolutional layer with the output channel corresponding to three clinically relevant Cobb angles: proximal thoracic (PT), main thoracic (MT) and thoracolumbar/lumbar (TL). And the activation function in the last layer is set to the sigmoid function. Here, we design a novel objective function based on symmetric mean absolute percentage error, named SMAPE loss:

$$\mathcal{L}_{\text{SMAPE}}(r(x), \hat{y}) = \frac{\sum_{i=1}^{n} |\hat{y}_i - r(x_i)|}{\sum_{i=1}^{n} |\hat{y}_i + r(x_i) + \epsilon|}, \quad (3)$$

where $\hat{y}_i$ and $r(x_i)$ denote the ground truth and prediction of ith angle of the total $n = 3$ Cobb angles (i.e., PT, MT and TL), and ϵ is a smooth factor.

To boost the performance of N_R with N_S, the AR module is designed to explore the hidden state representation of the N_R via classification activation mapping (CAM) to force it to focus on the spine area. Specifically, to integrate regularization on N_R, we expand the N_R into a shared-weight Siamese structure. One branch takes the concatenation of the raw image and its corresponding segmentation mask as input, and the other directly takes the segmentation as input. The output activation maps from two branches are regularized by mean absolute error to guarantee the consistency of CAMs, and the regression network in consequence is forced to focus on the spine area. The objective function is:

$$\mathcal{L}_{\text{AR}} = \|C(x, s(x)) - C(s(x))\|_1, \quad (4)$$

Triangle Consistency Learning. Inspired by the inference-path invariance theory [16] which declares that inference paths with the same endpoints, but different intermediate domains, yield similar results. The segmentation process is essentially an auxiliary task for the regression task, thus the regression network has the potential to optimize the segmentation network. Based on this assumption, we design a novel training strategy named as triangle consistency learning for end-to-end training, are shown in Fig. 2. The details of the proposed

Algorithm 1. Triangle Consistency Learning.

Require:
 Input image, $x \in X$;
 Ground truth of segmentation mask of spine, $y \in Y$;
 Ground truth of three Cobb angles (PT, MT and TL), $\hat{y} \in \hat{Y}$;

Ensure:
 $s(\cdot)$: Segmentation task of network N_S with parameter θ_1;
 $r(\cdot)$: Regression task of network N_R with parameter θ_2;
 $C(\cdot)$: Class activation map (CAM) generated by N_R;
 Training basic N_S.

1: **while** stopping criterion not met **do**
2: Compute the segmentation loss $\mathcal{L}_{\text{seg}}(s(x), y)$ with Equation (1);
3: Update parameters θ_1 of N_S by backpropagation;
4: **end while**
 Training N_R with AR.
5: **while** stopping criterion not met **do**
6: Compute the SMAPE loss $\mathcal{L}_{\text{SMAPE}}(r(x, s(x)), \hat{y})$ with Equation (3);
  $(x, s(x))$ means concatenation of raw image and its segmentation mask.
7: Compute the SMAPE loss $\mathcal{L}_{\text{SMAPE}}(r(s(x)), \hat{y})$ with Equation (3);
8: Compute the attention regularization loss $\mathcal{L}_{\text{AR}} = \|C(x, s(x)) - C(s(x))\|_1$;
9: Update the regression network parameters θ_2;
10: **end while**
 Fine-tuning N_S with ROIE.
11: **while** stopping criterion not met **do**
12: Add local consistency constraints to N_S with Equation (2);
13: Update the parameters θ_1 of N_S;
14: **end while**
 Fine-tuning N_S by SMAPE loss.
15: **while** stopping criterion not met **do**
16: Compute the regression loss $\mathcal{L}_{\text{SMAPE}}(r(s), r(s(x, C(x))))$;
17: Freeze the θ_2 and update the θ_1;
18: **end while**
 Fine-tuning N_R with refined segmentation.
19: Repeat steps 5-10
20: **return** θ_1 and θ_2

---

training strategy are shown in Algorithm 1, which can be divided into five processes: 1) training basic N_S (steps 1–4); 2) training N_R with AR (steps 5–10); 3) fine-tuning N_S with ROIE (steps 11–14); 4) fine-tuning N_S with SMAPE loss on the regression output of $s(x)$ and y (steps 15–18); and 5) fine-tuning N_R with refined segmentation (step 19). In the fourth process, we compute the SMAPE loss for the regression outputs of segmentation results and the corresponding ground truth, denoted $r(s(x))$ and $r(y)$, respectively. Then we freeze the parameters of N_R and optimize N_S by the backpropagation. In this way, we can generate more suitable segmentation mask for N_R, superior to traditional segmentation.

Table 1. The ablation study for each part of Seg4Reg+. AR: attention regularization module, ROIE: region-of-interest enhancement, TCL: triangle consistency learning, Img: raw image as input, and Seg: segmentation mask.

Baseline	AR	ROIE	TCL	Img	Seg	MAE	SMAPE (%)
✓				✓		6.34, 7.77, 8.01	12.32
✓	✓			✓		4.55, 5.75, 5.92	9.39
✓	✓	✓		✓		4.21, 5.32, **5.22**	9.32
✓	✓	✓	✓	✓		**4.01, 5.16**, 5.51	**9.17**
✓					✓	6.51, 6.22, 7.17	10.95
✓	✓				✓	4.03, 6.38, 5.80	9.52
✓	✓	✓			✓	**3.61**, 4.90, 5.53	9.01
✓	✓	✓	✓		✓	5.13, **4.73**, 5.24	**8.92**
✓	✓	✓	✓	✓	✓	**3.88, 4.62, 4.99**	8.47

3   Experiments

Data and Implementation Details. We use the public dataset of MICCAI 2019 AASCE Challenge [13], which consists of 609 spinal anterior-posterior X-ray images to evaluate our method. The dataset is divided by the provider into 481 images for training and 128 images for testing. We evaluate the proposed method using two metrics, symmetric mean absolute percent error (SMAPE) and mean absolute error (MAE). And we evaluate the segmentation results using five performance metrics including the Jaccard index (JA), Dice coefficient (Dice), pixel-wise accuracy (pixel-AC), pixel-wise sensitivity (pixel-SE), and pixel-wise specificity (pixel-SP).

We pre-process the data before inputting it into our network. First, we resize the image to [512, 256]. Then, we linearly transform the Cobb angles into [0, 1], and augment the dataset by randomly flipping, rotating ($-45°$, $45°$), and rescaling with the factor between (0.85, 1.25). We train the N_S for 90 epochs using ADAM optimization with learning rate 1×10^{-4} and weight decay 1×10^{-5}. And for regression, we train the network for 200 epochs in total with learning rate 1×10^{-3} and weight decay 1×10^{-5}.

Ablation Studies. To verify the effectiveness of each module in the proposd Seg4Reg+ approach, we conduct a set of experiments for ablation study, which are shown in Table 1. We first test with the raw image as input, the AR module has a 2.93% improvement in SMAPE compared to baseline, which validates that the AR module can gain from segmentation and benefit for the regression task. By combining the ROIE gate, the performance can be further improved by 0.07% (the improvement with the segmentation mask as input is more significant with 0.51% boost in SMAPE), which demonstrates N_R can boost N_S in turn. And applying triangle consistency learning could improve the SMAPE by 0.15%. Similar trend can also be observed when we test with segmentation results as input. Finally, when we concatenate the raw image and segmentation mask together as input, the performance can by further improved up to 8.47%.

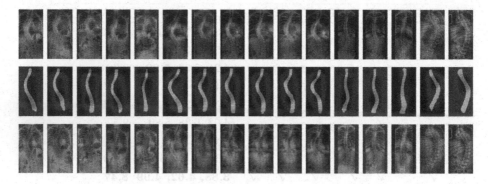

Fig. 3. The visualization of CAMs with and without the AR module. The top and middle rows show the result of baseline method (without AR) using raw image and segmentation mask as input, respectively. And the bottom row shows the CAMs generated from the baseline method with AR.

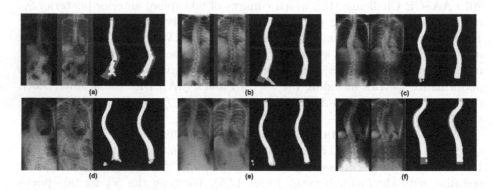

Fig. 4. Examples of segmentation results. Each group shows the original image, CAM, and segmentations without and with ROIE, respectively. The yellow mask is true positive, red mask is false negative, and green mask is false positive.

Effectiveness of AR Module. Figure 3 shows the results of the baseline method and the AR module. From the first row in Fig. 3, we can see several drawbacks of the CAM generated by the baseline method: 1) focusing only on the local region (e.g., columns 1–11), 2) the relative inferior ability of feature extraction (e.g., columns 12–14), and 3) the vulnerability to the blurred images (e.g., columns 15–16). And the bottom row in Fig. 3 shows that with the AR module, N_R can make more precise prediction with more reasonable perspective view. Particularly, in column 14, our AR module predicts more precise attention area while the baseline method generates very weak attention maps.

Table 2. Comparison of segmentation performance (%) of ROIE with other methods.

Methods	JA	Dice	pixel-AC	pixel-SE	pixel-SP
W/o CAMs	75.47	86.02	94.83	86.27	98.21
MDC [11]	75.91	86.31	**95.73**	**89.08**	98.12
D-Net [5]	76.99	87.00	95.02	86.61	98.46
MB-DCNN [14]	76.17	86.47	95.12	87.18	98.19
Ours	**77.86**	**87.55**	95.49	88.06	**98.42**

Table 3. Comparison of regression performance with the state-of-the-art methods.

Methods	A-Net [3]	L-Net [3]	B-Net [13]	PFA [10]	VF [6]		Seg4Reg [7]		Ours	
Backbone	–	–	–	res50	UNet	MNet	res18	eff_b1	res18	eff_b1
MAE	8.58	10.46	9.31	6.69	3.90	**3.51**	6.63	3.96	4.50	3.73
SMAPE (%)	20.35	26.94	23.44	12.97	8.79	7.84	10.95	7.64	8.47	**7.32**

Effectiveness of ROIE Gate. We evaluate both qualitative and quantitative segmentation results of our ROIE module. Table 2 shows that applying the ROIE gate to the origin segmentation model promotes JA from 75.47% to 77.86%, and the visualization results in Fig. 4 show that the ROIE gate is helpful for false positive reduction. Furthermore, we compare different ways of transferring CAM to the segmentation network, i.e., MDC [11], D-Net [5] and MB-DCNN [14]. However, simply delivering CAMs to N_S, these methods would be affected by inaccurate CAMs. And our ROIE gate alleviates this problem by fusing the CAM adaptively with a learnable parameter. For a fair comparison, all of these methods use the same segmentation architecture. It shows that our ROIE gate improves JA by 0.9% and 1.7% over D-Net and MB-DCNN, respectively.

Comparison with State-of-the-Art. We compare the proposed method with state-of-the-art methods including A-Net [3], L-Net [3], BoostNet [13], PFA [10], VF [6], and Seg4Reg [7], which won the 1st place in AASCE challenge. We follow the same experiment setting with VF [6], and the performance of all competing methods is adopted from the original publications for a fair comparison.

As depicted in Table 3, promising results are observed in predicting Cobb angle using the proposed framework. The SMAPE is 7.32% with our Seg4Reg+ framework, whereas PFA, VF-MNet and Seg4Reg achieve 12.97%, 7.84% and 7.64%, respectively. The MAE of our method is 3.73, slightly inferior to VF-MNet. We believe the superior performance of our model in SMAPE is more convincing as SMAPE is the only evaluation metric in the AASCE challenge.

4   Conclusion

In this paper, we proposed a novel Seg4Reg+ model for automated Cobb angle estimation. Our Seg4Reg+ model incorporates segmentation into regression task via an attention regularization module and a region-of-interest enhancement gate to boost the performance of both tasks. Further, the two networks are integrated with global consistency learning for global optimization. Experimental results demonstrated the effectiveness of each module and the superior performance of our model to the state-of-the-art methods.

Acknowledgments. This work was funded by Key-Area Research and Development Program of Guangdong Province, China (No. 2018B010111001), and the Scientific and Technical Innovation 2030-"New Generation Artificial Intelligence" Project (No. 2020AAA0104100).

References

1. Anitha, H., Karunakar, A.K., Dinesh, K.V.N.: Automatic extraction of vertebral endplates from scoliotic radiographs using customized filter. Biomed. Eng. Lett. **4**(2), 158–165 (2014). https://doi.org/10.1007/s13534-014-0129-z
2. Anitha, H., Prabhu, G.: Automatic quantification of spinal curvature in scoliotic radiograph using image processing. J. Med. Syst. **36**(3), 1943–1951 (2012)
3. Chen, B., Xu, Q., Wang, L., Leung, S., Chung, J., Li, S.: An automated and accurate spine curve analysis system. IEEE Access **7**, 124596–124605 (2019)
4. He, K., Zhang, X., Ren, S., Sun, J.: Deep residual learning for image recognition. In: Proceedings of the IEEE/CVF Conference on Computer Vision and Pattern Recognition, pp. 770–778 (2016)
5. Hong, S., Noh, H., Han, B.: Decoupled deep neural network for semi-supervised semantic segmentation. arXiv preprint arXiv:1506.04924 (2015)
6. Kim, K.C., Yun, H.S., Kim, S., Seo, J.K.: Automation of spine curve assessment in frontal radiographs using deep learning of vertebral-tilt vector. IEEE Access **8**, 84618–84630 (2020)
7. Lin, Y., Zhou, H.-Y., Ma, K., Yang, X., Zheng, Y.: Seg4Reg networks for automated spinal curvature estimation. In: Cai, Y., Wang, L., Audette, M., Zheng, G., Li, S. (eds.) CSI 2019. LNCS, vol. 11963, pp. 69–74. Springer, Cham (2020). https://doi.org/10.1007/978-3-030-39752-4_7
8. Sardjono, T.A., Wilkinson, M.H., Veldhuizen, A.G., van Ooijen, P.M., Purnama, K.E., Verkerke, G.J.: Automatic Cobb angle determination from radiographic images. Spine **38**(20), E1256–E1262 (2013)
9. Sun, H., Zhen, X., Bailey, C., Rasoulinejad, P., Yin, Y., Li, S.: Direct estimation of spinal cobb angles by structured multi-output regression. In: Niethammer, M., Styner, M., Aylward, S., Zhu, H., Oguz, I., Yap, P.-T., Shen, D. (eds.) IPMI 2017. LNCS, vol. 10265, pp. 529–540. Springer, Cham (2017). https://doi.org/10.1007/978-3-319-59050-9_42
10. Wang, J., Wang, L., Liu, C.: A multi-task learning method for direct estimation of spinal curvature. In: Cai, Y., Wang, L., Audette, M., Zheng, G., Li, S. (eds.) CSI 2019. LNCS, vol. 11963, pp. 113–118. Springer, Cham (2020). https://doi.org/10.1007/978-3-030-39752-4_14
11. Wei, Y., Xiao, H., Shi, H., Jie, Z., Feng, J., Huang, T.S.: Revisiting dilated convolution: A simple approach for weakly- and semi-supervised semantic segmentation. In: Proceedings of the IEEE/CVF Conference on Computer Vision and Pattern Recognition, pp. 7268–7277 (2018)

12. Seifert, J., Thielemann, F., Bernstein, P.: Der Orthopäde **45**(6), 509–517 (2016). https://doi.org/10.1007/s00132-016-3274-5
13. Wu, H., Bailey, C., Rasoulinejad, P., Li, S.: Automatic landmark estimation for adolescent idiopathic scoliosis assessment using BoostNet. In: Descoteaux, M., Maier-Hein, L., Franz, A., Jannin, P., Collins, D.L., Duchesne, S. (eds.) MICCAI 2017. LNCS, vol. 10433, pp. 127–135. Springer, Cham (2017). https://doi.org/10.1007/978-3-319-66182-7_15
14. Xie, Y., Zhang, J., Xia, Y., Shen, C.: A mutual bootstrapping model for automated skin lesion segmentation and classification. IEEE Trans. Med. Imaging **39**(7), 2482–2493 (2020)
15. Yi, J., Wu, P., Huang, Q., Qu, H., Metaxas, D.N.: Vertebra-focused landmark detection for scoliosis assessment. In: IEEE 17th International Symposium on Biomedical Imaging, pp. 736–740. IEEE (2020)
16. Zamir, A.R., et al.: Robust learning through cross-task consistency. In: Proceedings of the IEEE/CVF Conference on Computer Vision and Pattern Recognition, pp. 11197–11206 (2020)
17. Zhang, C., Wang, J., He, J., Gao, P., Xie, G.: Automated vertebral landmarks and spinal curvature estimation using non-directional part affinity fields. Neurocomputing (2021)
18. Zhao, H., Shi, J., Qi, X., Wang, X., Jia, J.: Pyramid scene parsing network. In: Proceedings of the IEEE/CVF Conference on Computer Vision and Pattern Recognition, pp. 2881–2890 (2017)
19. Zhou, B., Khosla, A., Lapedriza, A., Oliva, A., Torralba, A.: Learning deep features for discriminative localization. In: Proceedings of the IEEE/CVF Conference on Computer Vision and Pattern Recognition. pp. 2921–2929 (2016)

Meta-modulation Network for Domain Generalization in Multi-site fMRI Classification

Jaein Lee[1], Eunsong Kang[1], Eunjin Jeon[1], and Heung-Il Suk[1,2(✉)]

[1] Department of Brain and Cognitive Engineering, Korea University,
Seoul, Republic of Korea
{wodls9212,eunsong1210,eunjinjeon,hisuk}@korea.ac.kr
[2] Department of Artificial Intelligence, Korea University, Seoul, Republic of Korea

Abstract. In general, it is expected that large amounts of functional magnetic resonance imaging (fMRI) would be helpful to deduce statistically meaningful biomarkers or to build generalized predictive models for brain disease diagnosis. However, the site-variation inherent in rs-fMRI hampers the researchers to use the entire samples collected from multiple sites because it involves the unfavorable heterogeneity in data distribution, thus negatively impact on identifying biomarkers and making a diagnostic decision. To alleviate this challenging multi-site problem, we propose a novel framework that adaptively calibrates the site-specific features into site-invariant features via a novel modulation mechanism. Specifically, we take a learning-to-learn strategy and devise a novel meta-learning model for domain generalization, *i.e.*, applicable to samples from unseen sites without retraining or fine-tuning. In our experiments over the ABIDE dataset, we validated the generalization ability of the proposed network by showing improved diagnostic accuracy in both seen and unseen multi-site samples.

Keywords: Multi-site · Domain generalization · Modulation network · Meta-learning · Resting-state functional magnetic resonance imaging · Autism spectrum disorder

1 Introduction

Resting state fMRI (rs-fMRI) that measures spontaneous brain activity patterns of task-negative state has shown its potential to identify functional biomarkers. Especially, it has been used for individuals who are unable to perform tasks of cognitive functions [10], including patients such as young (attention deficit hyperactivity disorder) [4], sedated (Schizophrenia) [19] or cognitively impaired (Alzheimer's disease) [2].

Electronic supplementary material The online version of this chapter (https://doi.org/10.1007/978-3-030-87240-3_48) contains supplementary material, which is available to authorized users.

© Springer Nature Switzerland AG 2021
M. de Bruijne et al. (Eds.): MICCAI 2021, LNCS 12905, pp. 500–509, 2021.
https://doi.org/10.1007/978-3-030-87240-3_48

In recent studies of identifying functional diagnostic biomarkers, machine learning or deep learning methods have been widely applied to rs-fMRI, attributed to the superiority of extracting meaningful features from complex data. For machine/deep learning models, a large amount of data is a prerequisite for accurate models that show high generalizability and predictability for brain disease diagnosis. Therefore, in order to achieve productive researches, obtaining and sharing independent rs-fMRI data from multiple institutions or sites has been taken recently [17]. However, it raises new challenges, so-called multi-site problems, that occur due to 'domain shift' caused by inter-site variation in acquisition protocols or scanners among sites [1,9,23]. Since such heterogeneity can result in distorted afterward analysis, many studies still analyze functional biomarkers with a single site rs-fMRI, at the expense of sample size.

To handle the multi-site heterogeneity, the common approach is to aggregate data from all trainable sites, which is based on the assumption that the model can implicitly capture site-invariant features robust to domain shift with the increasing diversity of domains [1,11,18]. However, a simple aggregation method without considering domain shifts can potentially cause conflicting results in analysis, especially in resting-state fMRI studies. Another approach is a Combat Harmonization method [8,12,23], which treats site-differences of the scanner and protocol as covariates and eliminates them in an empirical Bayes manner. It has been widely applied to various data from gene to neuroimaging. However, because it needs re-parameterization of covariates each time a new site is added, it has a limitation of taking additional efforts to deal with unseen sites.

Meanwhile, from a machine learning perspective, there are domain adaptation and domain generalization approaches to directly address the domain shift problem in latent feature learning. Given rs-fMRI of two sites, 'domain adaptation' methods aim to tackle domain shift by transferring the information from a source site to a target site during representation learning [16,21]. However, it has its own limitation of building a new model for each unseen site because an adapted model is specifically trained to a given target site. On the other hand, 'domain generalization' aims to generalize the model by training with multi-domain source data so that it can even work to a new target domain without retraining the model [14].

In this paper, by taking the concept of domain generalization, we focus on building a site-invariant model for diagnostic classification, which can also work on a new unseen data. To become a generalized model, the model should be able to address diversity or heterogeneity among sites with a limited amount of samples from multiple sites. To this end, we propose a modulation network that learns to adapt feature representations within a meta-learning framework via an episodic-learning strategy [7,14,15]. Given episodes of the target task to simulate differences between sites, the modulation network learns diverse patterns that can cope with various domain shifts. To the best of our knowledge, our proposed method is the first meta-learning based domain generalization model for multi-site fMRI dataset analysis.

The main contributions of our proposed meta-learning framework are three folds:

1. We propose a site-invariant model for diagnostic classification of multi-site fMRIs.
2. We devise a modulation network that enables to handle site-specific properties, explicitly.
3. We significantly improved a diagnostic classification performance in unseen sites as well as seen sites.

We validated our proposed network over the public Autism Brain Imaging Data Exchange (ABIDE) dataset comprised of data from multiple sites.

2   Materials and Image Preprocessing

We used rs-fMRI datasets collected from 16 sites[1] in Autism Imaging Data Exchange (ABIDE I) repository[2] Data were preprocessed with Data Processing Assistant for Resting-State fMRI pipeline [5,6]. Specifically, the preprocessing step included slice timing correction, motion realignment, and intensity normalization. Nuisance signal removal was performed with 24 parameters and low-frequency drifts. The data was band-pass filtered within $0.01 \sim 0.1$ Hz and then registered to MNI 152 space. After spatial normalization, the brain was parcellated using the AAL atlas [20] and the mean time signal was extracted for each of the 116 regions of interest (ROIs) per subject. We used $1,032$ samples from a total of 16 sites[3], in which 496 subjects are with autism spectrum disorders (ASD) and 536 subjects are with typical developments (TD).

3   Proposed Method

3.1   Problem Setup

Assuming that we are given $\{\mathcal{D}_i\}_{i=1}^{p}$ datasets from p sites, where $\mathcal{D}_i = \{(\mathbf{x}_j^i, y_j^i)\}_{j=1}^{N_i}$ represents the dataset with N_i samples from the ith site, each site dataset consists of an input space $\mathcal{X}$, e.g., rs-fMRI BOLD signals or functional connectivity, and its label space $\mathcal{Y}$, i.e., TD or ASD, for a diagnosis task. Since each site has different scanner and imaging protocols, the distribution of instances $p(\mathcal{D}_i)$ is diverse over sites. On the other hand, the label space $\mathcal{Y}$ is shared across all sites because it aims for the same diagnostic task. Thus, the distribution of diagnostic information in multi-site exists on various joint spaces of $\mathcal{X} \times \mathcal{Y}$. Therefore, even if the model was trained with data from several sites, it shows poor diagnostic performance for new sites since the distribution of information on new sites differs from the distribution learned.

---

[1] CALTECH, KKI, LEUVEN, MAX_MUN, NYU, OHSU, OLIN, PITT, SBL, SDSU, STANFORD, TRINITY, UCLA, UM, USM, YALE.

[2] Scan procedure and parameters can be found at http://fcon_1000. projects.nitrc.org/indi/abide/.

[3] Note that we excluded CMU site in ABIDE because only ASD subjects left after removing subjects with pre-processing problem.

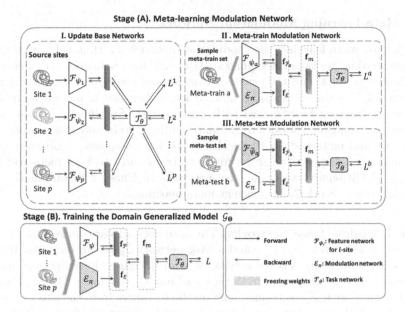

Fig. 1. Overview of our proposed framework. Our proposed framework is divided into two stages (meta-learning the modulation network and training the domain generalized model) with three networks (feature network, task network, and modulation network).

3.2 Overview Method

In multi-site fMRI-based diagnostic classification, the main goal is to learn complex fMRI patterns for more accurate diagnosis without hindrance of domain shifts. To this end, we propose a novel framework, as shown in Fig. 1, that is mainly divided into two stages: (A) *meta-learning the modulation network* and (B) *training the domain generalized model*. Our domain generalized model in stage (B) for multi-site diagnostic classification is composed of a feature network $\mathcal{F}$, a modulation network $\mathcal{E}$, and a task network $\mathcal{T}$, having their respective parameters as $\Theta = \{\psi, \pi, \theta\}$. As typical architectures of neural networks for a diagnostic task, $\mathcal{F}$ extracts suitable latent representations from an instance $\mathbf{x}$ and $\mathcal{T}$ predicts whether the instance $\mathbf{x}$ is ASD or TD with latent representation. To handle with diagnostic information in multi-site data, we devise a modulation network $\mathcal{E}_\pi$ that allows for site-specific features to adapt the general task network $\mathcal{T}_\theta$. With the meta-learning based training strategy of adapting new sites in stage (A), $\mathcal{E}_\pi$ mitigates distributional heterogeneity among multiple sites, rather than being optimized for a single site. With the trained $\mathcal{E}_\pi$ and $\mathcal{T}_\theta$ in stage (A), a site-invariant feature extractor $\mathcal{F}_\psi$ in stage (B) extracts latent representation of any site. For inference, we predict the final label by using $\mathcal{F}_\psi$, $\mathcal{E}_\pi$, and $\mathcal{T}_\theta$ together.

3.3   Meta-Learning Modulation Network

In order for meta-learning the modulation network $\mathcal{E}_\pi$ that calibrates multi-site information, we take three phases: (I) *update base networks*, (II) *meta-train modulation network*, and (III) *meta-test modulation network*. Meta-learning [7] is commonly known as 'learning how to learn' because it has an adaptation process for learning how to deal with unseen data. Specifically, in our study, we construct multiple episodes to simulate multiple train and test sessions (*i.e.*, meta-train and meta-test) in Phase II and III. Let an episode is organized with two randomly selected sites, site a for meta-train and site b for meta-test from $\{\mathcal{D}_i\}_{i=1}^p$. Additionally, we train base networks in Phase I to obtain the site-invariant task network $\mathcal{T}_\theta$ for efficient meta-learning.

Phase I. Update Base Networks. The base networks are devised for training of p feature networks $\{\mathcal{F}_{\psi_i}\}_{i=1}^p$ and a task network $\mathcal{T}_\theta$ for all datasets. If there are already trained models for p sites, we can initialize the weights of feature networks $\{\mathcal{F}_{\psi_i}\}_{i=1}^p$ with them in the sense of model reusability. Based on supervised learning, each $\mathcal{F}_{\psi_i}$ is trained only on the instances from the site i to obtain site-specific features, whereas $\mathcal{T}_\theta$ is updated with all samples from p sites.

For a classification loss L, we used the cosine loss [3] and the cross-entropy loss harmoniously

$$L(\cdot, y) = 1 - \left\langle \varphi(y), \frac{\mathcal{T}_\theta(\cdot)}{\|\mathcal{T}_\theta(\cdot)\|_2} \right\rangle - \lambda \left\langle \varphi(y), \log\left(\frac{\exp(\mathcal{T}_\theta(\cdot))}{\|\exp(\mathcal{T}_\theta(\cdot))\|_1} \right) \right\rangle \quad (1)$$

where $\cdot$ is the features to predict, φ is a function for one-hot vector transformation, $\langle\rangle$ is dot product, and λ is a scaling factor of the cross-entropy loss. A cosine loss helps prevent overfitting when learning with a small number of data [3]. Thereby, in Phase I, we obtain the site-specific feature network $\{\mathcal{F}_{\psi_i}\}_{i=1}^p$ and the site-invariant task network $\mathcal{T}_\theta$ for meta-learning by updating Eq. (2) and Eq. (3) k times with a learning rate α, as follows:

$$\psi_i \leftarrow \psi_i - \alpha\nabla L(\mathbf{x}^i, y^i; \mathcal{F}_{\psi_i}, \mathcal{T}_\theta) \quad (2)$$

$$\theta \leftarrow \theta - \alpha\nabla L(\mathbf{x}^i, y^i; \mathcal{F}_{\psi_i}, \mathcal{T}_\theta). \quad (3)$$

Phase II. Meta-Train Modulation Network. In Phase II and III, we train the model, which consists of a feature network $\mathcal{F}_{\psi_a}$, a task network $\mathcal{T}_\theta$, and a modulation network $\mathcal{E}_\pi$. As for the task network $\mathcal{T}_\theta$, the task network trained in Phase I is used. In the meantime, the feature network $\mathcal{F}_{\psi_a}$ is initialized with ψ_a corresponding to the meta-train site a in the episode, which is trained in Phase I. For meta-training, we optimize $\mathcal{F}_{\psi_a}$ to learn feature representations of the samples of the seen site a, while freezing the networks of $\mathcal{E}_\pi$ and $\mathcal{T}_\theta$. However, note that the features of $\mathcal{F}_{\psi_a}$ is further calibrated by those of $\mathcal{E}_\pi$ before feeding into the task network. Influenced by modulated feature $\mathbf{f}_m$ resulting from an element-wise product with features of $\mathbf{f}_{\mathcal{F}_a}$ from $\mathcal{F}_{\psi_a}$ and $\mathbf{f}_{\mathcal{E}}$ from $\mathcal{E}_\pi$, we update the ψ_a to $\widehat{\psi}_a$ l times with Eq. (4). The same form of the classification loss L in Eq. (1) is used

$$\widehat{\psi}_a \leftarrow \psi_a - \alpha \nabla L(\mathbf{x}^a, y^a; \mathcal{F}_{\psi_a}, \mathcal{E}_\pi, \mathcal{T}_\theta). \tag{4}$$

Phase III. Meta-Test Modulation Network. After meta-training for site a, we conduct a meta-test for site b with the trained networks $\mathcal{F}_{\widehat{\psi}_a}$ from Phase II. With the same loss of Eq. (1), only the modulation network $\mathcal{E}_\pi$ is updated as Eq. (5), while the feature network $\mathcal{F}_{\widehat{\psi}_a}$ and the task network $\mathcal{T}_\theta$ are fixed.

$$\pi \leftarrow \pi - \alpha \nabla L(\mathbf{x}^b, y^b; \mathcal{F}_{\widehat{\psi}_a}, \mathcal{E}_\pi, \mathcal{T}_\theta). \tag{5}$$

Since the meta-test phase is trained with site b, the $\mathcal{E}_\pi$ should extract the feature $\mathbf{f}_\mathcal{E}$ that calibrates site differences between the seen site a and the unseen site b. Consequently, the modulated features $\mathbf{f}_m$ integrated with $\mathbf{f}_{\mathcal{F}_b}$ and $\mathbf{f}_\mathcal{E}$ contain the combined patterns of features for the site b extracted from the perspective of site a-specified $\mathcal{F}_{\widehat{\psi}_a}$ and features of calibrating information for the site b. Going through multiple episodes by updating networks via Eq. (2), Eq. (3), Eq. (4), and Eq. (5) repeatedly, $\mathcal{E}_\pi$ could finally grasp the overall site differences given any site.

3.4   Training the Domain Generalized Model

The domain generalized model (GenM), which will be used for test finally, consists of the multi-site feature network $\mathcal{F}_\psi$, the modulation network $\mathcal{E}_\pi$, and the task network $\mathcal{T}_\theta$ as follows:

$$\mathcal{G}_\Theta(\mathbf{x}) = \mathcal{T}_\theta(\mathcal{F}_\psi(\mathbf{x}) \circ \mathcal{E}_\pi(\mathbf{x})), \tag{6}$$

where $\circ$ denotes an element-wise product. The modulation network $\mathcal{E}_\pi$ and the task network $\mathcal{T}_\theta$, trained in stage (A) are used for initialization. With $\mathcal{E}_\pi$ and $\mathcal{T}_\theta$ fixed, the site-invariant feature network $\mathcal{F}_\psi$ is trained on all p seen sites by minimizing the loss given by Eq. (1), as follows:

$$\psi \leftarrow \psi - \alpha \nabla L(\mathbf{x}, y; \mathcal{F}_\psi, \mathcal{E}_\pi, \mathcal{T}_\theta). \tag{7}$$

If the $\mathcal{E}_\pi$ is trained well, the $\mathbf{f}_\mathcal{F}$ of multi-sites will be revised to site-invariant feature $\mathbf{f}_m$ that results in high classification performance. It should be distinguished that the feature network $\mathcal{F}_{\psi_a}$ in stage (A) aims for training the modulation network $\mathcal{E}_\pi$, whereas $\mathcal{F}_\psi$ in stage (B) is to extract multi-site representations. The overall methodological flow is summarized in the algorithm of supplementary Sect. 1.

4   Experiments and Analysis

To validate the effectiveness of our method, we compared our method with three existing methods: (1) Aggregation learning method (Agg) that uses all instance

Table 1. Intra-site classification results. Average of 5-fold results from 4 seen sites (NYU, UM, USM, UCLA) $(* : p < 0.05)$

Models	AUC	ACC (%)	SEN (%)	SPEC (%)
Agg (baseline)	$0.6517 \pm 0.062^*$	$64.31 \pm 6.05^*$	65.44 ± 7.97	66.40 ± 12.08
Combat	$0.6701 \pm 0.0582^*$	$65.50 \pm 5.59^*$	64.58 ± 8.69	67.66 ± 10.23
MLDG	$0.6916 \pm 0.0468^*$	$67.69 \pm 3.94^*$	65.04 ± 8.96	69.63 ± 9.35
GenM (proposed)	$\mathbf{0.7117 \pm 0.0295}$	$\mathbf{70.80 \pm 3.16}$	$\mathbf{67.41 \pm 8.45}$	$\mathbf{73.25 \pm 9.90}$

Table 2. Inter-site classification results. Average of 5-fold results from 12 unseen sites $(* : p < 0.05)$

Models	AUC	ACC (%)	SEN (%)	SPEC (%)
Agg (baseline)	$0.6320 \pm 0.0619^*$	$62.13 \pm 5.89^*$	59.55 ± 7.70	66.15 ± 9.08
Combat	$0.6464 \pm 0.0462^*$	$63.26 \pm 4.95^*$	60.79 ± 7.20	66.06 ± 8.95
MLDG	$0.6406 \pm 0.0526^*$	$63.11 \pm 5.60^*$	60.55 ± 8.43	66.01 ± 10.18
GenM (proposed)	$\mathbf{0.6657 \pm 0.0387}$	$\mathbf{65.57 \pm 3.71}$	$\mathbf{64.23 \pm 5.03}$	$\mathbf{68.08 \pm 8.99}$

of entire sites for training, (2) Combat Harmonization method (Combat) [8, 12,23] that removes multi-site effects using an empirical Bayes methods and then train the aggregation model with those post-processed data, (3) Meta-Learning for Domain Generalization (MLDG) [14] that applies a gradient-based meta-learning to domain generalization task. Given the relatively large sized datasets of NYU, UM, USM, and UCLA site $(p = 4)$ for training among the 16 existing sites in ABIDE, we tested all these comparative methods with $\mathcal{F}$ and $\mathcal{T}$ of the same configuration of those in GenM. It should be noted that a train session includes both meta-learning modulation network and training the final generalized model stages. For the test, we tested our method in two ways: inter-site setting that tests unseen sites, $i.e.$, 12 sites, and intra-setting that tests seen sites, $i.e.$, 4 sites. We conducted 5-fold cross validation for all the experiments.

4.1   Experimental Settings

First, we took the upper triangular part of Fisher z-transformed functional connectivity maps $M \in \mathbb{R}^{R \times (R-1)/2}$ as an input. Both modulation network $\mathcal{E}$ and feature networks $\mathcal{F}$ in stage (A) and stage (B) consist of a fully-connected layer (FCL) with $\{8\}$ units, while the task network $\mathcal{T}$ is composed of that with $\{2\}$ units. Here, R denotes 116 number of ROIs. We used Adam optimizer [13] with a learning rate of $1e^{-4}$. In (A) stage, a batch size was 8 in Phase I and 5 in Phase II and III. For GenM, the batch size was 8. In order for the modulation network training, we set two stopping criteria: (1) when the difference between meta-train and meta-test loss is less than a predefined threshold, $1e^{-8}$ or (2) meta-test loss is not reduced for 5000 consecutive iterations. Meanwhile, in order to avoid an

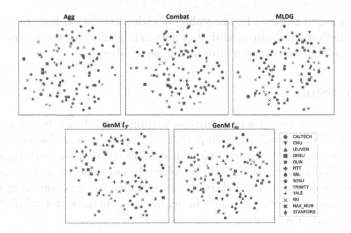

Fig. 2. t-SNE visualization of features from Agg, Combat, MLDG, and before and after being modulated ($\mathbf{f}_{\mathcal{F}}$ and $\mathbf{f}_m$) of GenM. Shapes represent sites. Red and blue represent ASD and TD, respectively. (Color figure online)

overfitting problem, possibly caused due to small samples and class-imbalances, we combined a cosine loss and a cross-entropy loss, and λ is 0.1 to scale a cross-entropy loss. The number of samples of each site used in the experiments, data partition criterion, and the number of training epochs in each phase are described in the supplementary Sect. 1 and 2, respectively.

4.2 Results and Discussion

We conducted ASD classification in two settings: intra-site and inter-site setting. The performance is the average of the 5-fold results from test sites in terms of four metrics: Area Under the receiver operating characteristic Curve (AUC), accuracy (ACC), sensitivity (SEN), and specificity (SPEC).

Table 1 shows the averaged performance of the comparative models on intra-site setting where we tested the final generalized model (GenM) with NYU, UM, USM, and UCLA that we used for training. The proposed method achieved the best AUC, ACC, SEN, and SPEC in all intra-sites compared to the competing methods. As an inter-site setting, we validated our proposed model with 12 unseen sites except for four sites used in training, which is summarized in Table 2. The results show the highest AUC, ACC, SEN, and SPEC with our proposed method. Moreover, we observed statistically significant differences between our proposed method and each of the comparative methods in both settings by means of a two-tailed Wilcoxon's signed rank test ($p < 0.05$) [22]. The test results of each site in inter-site and intra-site settings are presented in the supplementary Sect. 3 and 4, respectively.

Furthermore, we demonstrated the importance of modulation network in our multi-site experiment by t-SNE visualization in Fig. 2. The figure shows that the modulated features $\mathbf{f}_m$ are more suitable for diagnostic classification compared to the distribution of features $f_{\mathcal{F}}$ uncombined with modulation feature. From these results, we believe that our modulation network is effective to the site-invariant diagnostic model both on seen sites and unseen sites.

5   Conclusion

In this work, we proposed a novel meta-learning based framework for multi-site resting-state fMRI classification. In particular, we designed a domain generalized model with a concept of feature modulation and devised its learning algorithm to handle a challenging domain-shift problem. In our experiments on the ABIDE dataset, the proposed method showed a superiority to the competing multi-site methods on various metrics, which indicates its generalization ability on both various seen and unseen sites.

Acknowledgement. This work was supported by National Research Foundation of Korea (NRF) grant funded by the Korea government (MSIT) (No. 2019R1A2C1006543) and partially by Institute of Information & communications Technology Planning & Evaluation (IITP) grant funded by the Korea government (MSIT) (No. 2019-0-00079, Artificial Intelligence Graduate School Program (Korea University)).

References

1. Abraham, A., et al.: Deriving reproducible biomarkers from multi-site resting-state data: an autism-based example. Neuroimage **147**, 736–745 (2017)
2. Badhwar, A., Tam, A., Dansereau, C., Orban, P., Hoffstaedter, F., Bellec, P.: Resting-state network dysfunction in Alzheimer's disease: a systematic review and meta-analysis. Alzheimer's Dementia Diagnosis Assessment Disease Monitoring **8**, 73–85 (2017)
3. Barz, B., Denzler, J.: Deep learning on small datasets without pre-training using cosine loss. In: Proceedings of the IEEE/CVF Winter Conference on Applications of Computer Vision, pp. 1371–1380 (2020)
4. Castellanos, F.X., Aoki, Y.: Intrinsic functional connectivity in attention-deficit/hyperactivity disorder: a science in development. Biological Psychiatry Cognitive Neuroscience Neuroimaging **1**(3), 253–261 (2016)
5. Chao-Gan, Y., Yu-Feng, Z.: DPARSF: a MATLAB toolbox for "pipeline" data analysis of resting-state fMRI. Front. Syst. Neurosci. **4** (2010)
6. Craddock, C., et al.: The neuro bureau preprocessing initiative: open sharing of pre-processed neuroimaging data and derivatives. Frontiers Neuroinformatics **7** (2013)
7. Finn, C., Abbeel, P., Levine, S.: Model-agnostic meta-learning for fast adaptation of deep networks. In: Proceedings of the 34th International Conference on Machine Learning-Volume 70, pp. 1126–1135. JMLR. org (2017)
8. Fortin, J.P., Cullen, N., Sheline, Y.I., Taylor, W.D., Aselcioglu, I., Cook, P.A., Adams, P., Cooper, C., Fava, M., McGrath, P.J., et al.: Harmonization of cortical thickness measurements across scanners and sites. Neuroimage **167**, 104–120 (2018)
9. Friedman, J., Hastie, T., Tibshirani, R.: Sparse inverse covariance estimation with the graphical lasso. Biostatistics **9**(3), 432–441 (2008)
10. Greicius, M.: Resting-state functional connectivity in neuropsychiatric disorders. Curr. Opin. Neurol. **21**(4), 424–430 (2008)

11. Heinsfeld, A.S., Franco, A.R., Craddock, R.C., Buchweitz, A., Meneguzzi, F.: Identification of autism spectrum disorder using deep learning and the abide dataset. NeuroImage: Clin. **17**, 16–23 (2018)
12. Johnson, W.E., Li, C., Rabinovic, A.: Adjusting batch effects in microarray expression data using empirical bayes methods. Biostatistics **8**(1), 118–127 (2007)
13. Kingma, D.P., Ba, J.: Adam: a method for stochastic optimization. arXiv preprint arXiv:1412.6980 (2014)
14. Li, D., Yang, Y., Song, Y.Z., Hospedales, T.M.: Learning to generalize: meta-learning for domain generalization. In: Thirty-Second AAAI Conference on Artificial Intelligence (2018)
15. Li, D., Zhang, J., Yang, Y., Liu, C., Song, Y.Z., Hospedales, T.M.: Episodic training for domain generalization. In: Proceedings of the IEEE/CVF International Conference on Computer Vision, pp. 1446–1455 (2019)
16. Li, X., Gu, Y., Dvornek, N., Staib, L.H., Ventola, P., Duncan, J.S.: Multi-site fMRI analysis using privacy-preserving federated learning and domain adaptation: ABIDE results. Med. Image Anal. **65**, 101765 (2020)
17. Lombardo, M.V., Lai, M.C., Baron-Cohen, S.: Big data approaches to decomposing heterogeneity across the autism spectrum. Mol. Psychiatry **24**(10), 1435–1450 (2019)
18. Saeed, F., Eslami, T., Mirjalili, V., Fong, A., Laird, A.: ASD-DiagNet: a hybrid learning approach for detection of autism spectrum disorder using fMRI data. Front. Neuroinform. **13**, 70 (2019)
19. Sheffield, J.M., Barch, D.M.: Cognition and resting-state functional connectivity in schizophrenia. Neurosci. Biobehav. Rev. **61**, 108–120 (2016)
20. Tzourio-Mazoyer, N., Landeau, B., Papathanassiou, D., Crivello, F., Etard, O., Delcroix, N., Mazoyer, B., Joliot, M.: Automated anatomical labeling of activations in SPM using a macroscopic anatomical parcellation of the MNI MRI single-subject brain. Neuroimage **15**(1), 273–289 (2002)
21. Wang, M., Zhang, D., Huang, J., Yap, P.T., Shen, D., Liu, M.: Identifying autism spectrum disorder with multi-site fmri via low-rank domain adaptation. IEEE Trans. Med. Imaging **39**(3), 644–655 (2019)
22. Wilcoxon, F.: Individual comparisons by ranking methods. In: Breakthroughs in Statistics, pp. 196–202. Springer, New York (1992). https://doi.org/10.1007/978-1-4612-4380-9_16
23. Yu, M., Linn, K.A., Cook, P.A., Phillips, M.L., McInnis, M., Fava, M., Trivedi, M.H., Weissman, M.M., Shinohara, R.T., Sheline, Y.I.: Statistical harmonization corrects site effects in functional connectivity measurements from multi-site fmri data. Hum. Brain Mapp. **39**(11), 4213–4227 (2018)

3D Brain Midline Delineation for Hematoma Patients

Chenchen Qin[1], Haoming Li[1], Yixun Liu[1], Hong Shang[1], Hanqi Pei[1],
Xiaoning Wang[1], Yihao Chen[2], Jianbo Chang[2], Ming Feng[2], Renzhi Wang[2],
and Jianhua Yao[1(✉)]

[1] Tencent AI Lab, Shenzhen, China
jianhuayao@tencent.com
[2] Peking Union Medical College Hospital, Beijing, China

Abstract. Brain midline delineates the boundary between the two cerebral hemispheres of the human brain, which plays a significant role in guiding intracranial hemorrhage surgery. Large midline shift caused by hematomas remains an inherent challenge for delineation. However, most previous methods only handle normal brains and delineate the brain midline on 2D CT images. In this study, we propose a novel hemisphere-segmented framework (HSF) for generating smooth 3D brain midline especially when large hematoma shifts the midline. Our work has four highlights. First, we propose to formulate the brain midline delineation as a 3D hemisphere segmentation task, which recognizes the midline location via enriched anatomical features. Second, we employ a distance-weighted map for midline aware loss. Third, we introduce rectificative learning for the model to handle various head poses. Finally, considering the complexity of hematomas distribution in human brain, we build a classification model to automatically identify the situation when hematoma breaks into brain ventricles and formulate a midline correction strategy to locally adjust the midline according to the location and boundary of hematomas. To our best knowledge, it is the first study focusing on delineating the brain midline on 3D CT images of hematoma patients and handling the situation of ventricle break-in. Through extensive validations on a large in-house dataset, our method outperforms state-of-the-art methods in various evaluation metrics.

Keywords: 3D midline delineation · Intracranial hemorrhage ·
Hemisphere-segmentation Framework

1 Introduction

Intracranial hemorrhage (ICH) refers to serious bleeding within the intracranial vault, including the brain parenchyma and surrounding meningeal spaces, which is generally considered as a devastating disease [1]. In the last decade,

---

C. Qin and H. Li—The two authors contribute equally to this work.

© Springer Nature Switzerland AG 2021
M. de Bruijne et al. (Eds.): MICCAI 2021, LNCS 12905, pp. 510–518, 2021.
https://doi.org/10.1007/978-3-030-87240-3_49

minimally invasive surgery, a stereotactic cranio-puncture technique, has been widely adopted to treat ICH [10].

During the surgery, the surgeon needs to manipulate the needle, penetrating the hematoma center without breaking the midline boundary between the two hemispheres (Fig. 1(d) (e)). Ideally, the midline should be a relatively straight line, with clear anatomical texture (Fig. 1(a)). However, high intracranial pressure could shift the midline and break the anatomical symmetry around the midline (Fig. 1(b)). Particularly, when the hemorrhage breaking into ventricles, the part of brain is covered by large hemorrhage regions (Fig. 1(c)), which adds the difficulty for surgeon to identify the midline boundary. In such cases, the risk of secondary brain injury (SBI) caused by needle puncture is significantly increased. To address this issue, it is highly desirable to develop an efficient algorithm for automatically delineating the midline of hematoma cases for the surgery guidance.

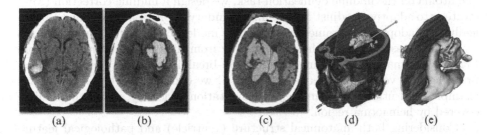

(a)   (b)   (c)   (d)   (e)

Fig. 1. (a)(b)(c) denote the axial slice of non-interaction, non-break-in and break-in cases respectively. (d) denotes the 3D brain CT volume of (c). (e) represents the 3D segmentation of brain midline and hematoma of (d). Green line in (a) denotes a normal brain midline. Yellow line in (d) represents the puncture needle in ICH surgery. Red and blue colors denote the deformed midlines and hematoma respectively. (Color figure online)

Previous methods mainly focus on delineating midlines on 2D images using anatomical markers, symmetrical model and segmentation model [4–6,11]. Liao et al. [4] proposed a method that decomposes the midline into three isolated parts and then generates these segments based on bilateral symmetry of intensity. Liu et al. [5] proposed to delineate the deformed brain midline by automatically detecting the anatomical markers. Wei et al. [11] presented a regression-based method to generate refined midline segmentation via fusing multi-scale outputs. Wang et al. Pisov et al. [6] proposed to segment the midline based on U-Net and adopted a limited head as a constraint. Although effective, these methods utilize the anatomical feature or brain symmetry to delineate the midline boundary, but still have the following limitations: (1) they can not generate a smooth 3D midline due to ignoring the continuity of midlines in three-dimensional space. (2) they resort to using aligned images to train the network, which may cause the failure in some extreme poses of the brain. Wang et al. [9] proposed a rectification network to align images to a standard pose before feeding into the delineation network. This method cannot to be trained end-to-end, which is sub-optimal.

To address the aforementioned issues, we propose a hemisphere-segmented framework (HSF) to automatically delineate the 3D brain midline. Instead of segmenting the midline directly, we define the brain midline delineation as a 3D hemisphere segmentation task. The workflow includes three parts. First, a segmentation network equipped with atrous spatial pyramid pooling (ASPP) [2] module is proposed to yield the probability maps (including hemispheres and hematoma) and acquire the initial brain midline via gradient detection. To further distill discriminate features for better recognition of midline shape, we design a distance-weighted map (DWM) to assist the model paying more attention to the anatomical structure around the midline. Second, a rectificative learning (RL) module is integrated into our framework for capturing the transformation between the input image and the standard-pose image, which enhances the model adaptability for various brain poses. Third, to decrease the effect of hematoma on the midline delineation task, we design a midline correction (MC) strategy to adaptively adjust the midline boundary in different situations. Specifically, MC adopts a pretrained classification model to classify the input images into three classes, including hematoma away from the midline (non-interaction, Fig. 1(a)), hematoma squeezing midline (non-break-in, Fig. 1(b)) and hematoma breaking into ventricles (break-in, Fig. 1(c)). we consider the break-in as hard examples and utilize the hematoma segmentation to locally adjust the midline covered by hemaotma region.

Considering both anatomical structure (ventricle) and pathological feature (hematomas) in the 3D images, our method can extract abundant texture features, which is significant for delineating the brain midine. As a proof-of-concept, we perform extensive experiments on a challenging in-house dataset containing 300 3D brain CT images with hematoma. Results show that our method outperforms the state-of-the-art methods in various metrics. Moreover, to our best knowledge, it is the first work adopting fully-convolutional method for 3D brain midline delineation for hematoma patients and the first work handling break-in situation.

2   Methodology

Figure 2 illustrates the pipeline of the proposed HSF for 3D brain midline delineation. As mentioned above, The proposed framework consists of three main components: (1) a segmentation network with encoder-decoder architecture to generate the hemispheres and hematomas segmentation. (2) rectificative learning to enhance the model generalization for various head poses. (3) midline correction to locally adjust the degree of midline shift caused by hematoma breaking in. The correction strategy is performed according to the classification segmentation results.

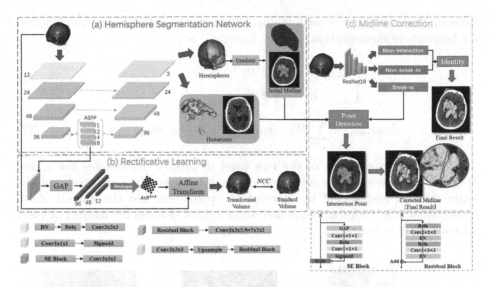

Fig. 2. The overflow of our proposed hemisphere-segmentation based 3D brain midline delineation framework for hematoma patients. (Color figure online)

2.1 Hemisphere Segmentation Network

As presented in Fig. 2(a), we build a strong segmentation network based on classic 3D-Unet architecture [7]. The encoder contains 4 blocks. In the last three blocks (blue cubes), each consists of a residual block and one convolution block with stride 2 for down-sampling. At the bottom of the encoder, we introduce the ASPP module, which can effectively capture multi-scale information and enlarge the receptive field for localizing the brain midline. The high-level feature from the ASPP is input to the rectificative learning module. Correspondingly, in the decoder, the first three blocks (green cubes) contain interpolation operation for up-sampling. The segmentation outputs the probability map with three channels, representing the semantic classes of left hemisphere, right hemisphere and hematomas respectively. Then, we adopt the gradient detection to yield the initial midline between the two hemisphere masks. Thus, the accuracy of hemisphere boundary is significant for midline delineation.

2.2 Distance-Weighted Map

However, simply adopting the segmentation loss (dice loss or cross-entropy) is hard for optimizing the model to generate a fine-grained boundary. To address the problem, we introduce the DWM in the training stage. As shown in Fig. 3, the DWM is generated by a soft attention mechanism based on brain and midline annotations, which smoothly decreases the weights in regions away from the midline while maintaining the high weights in midline-surrounding regions. In detail, the DWM weight for each voxel is determined by the minimum distance between the voxel and the brain midline. Equipped with DWM, the loss function for hemisphere segmentation, namely $Loss_{hemi}$, is:

$$W_i = \begin{cases} exp(\dfrac{\alpha - d_i}{\alpha}) & i \in R_b \\ 1 & i \notin R_b \end{cases} \tag{1}$$

$$Loss_{hemi} = \sum_{i=1}^{N} W_i * [y_i log(\hat{y}_i) + (1 - y_i)log(1 - \hat{y}_i)] \tag{2}$$

W_i denotes the weight of the i-th voxel in images. R_b denotes the region of brain. d_i represents the minimum distance between the i-th voxel and the brain midline, and $\hat{y}_i$ and y_i represent the prediction and the ground truth, respectively. α is a constant term and set as 200. N denotes the number of voxel for each input volume.

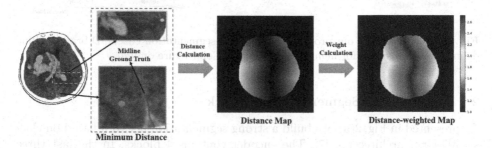

Fig. 3. The details of distance-weighted map. The yellow dot denotes a voxel away from the midline, the green one denotes a voxel close to the midline. (Color figure online)

2.3 Rectificative Learning

Considering the midline delineation is sensitive to the head pose in the image, we introduce the rectificative learning (RL) module to assist the model capturing the transformation between the input volume and the standard-pose volume. As shown in Fig. 2(b), the proposed RL module includes the Squeeze-and-Excitation block (SE block) [8], global average pooling (GAP) and fully connection layers. Taking the ASPP feature as input, the RL generates an affine transform matrix $A \in R^{3 \times 4}$. Then, using this matrix, we transform the input volume to the transformed volume. It should be noticed that the introduction of the SE block can effectively boost the model generalization for various poses. We adopt normalization cross correlation (NCC) to measure the divergence between the transformed volume and the standard volume. The loss function of our RL module is defined as:

$$Loss_{ncc} = \frac{\sum V_t \cdot V_s}{\sqrt{(\sum V_t \cdot V_t) \times (\sum V_s \cdot V_s)}} \tag{3}$$

where V_t and V_s denote the transformed volume and the standard volume respectively. The overall loss function of our framework is:

$$Loss_{total} = \lambda_1 Loss_{hemi} + \lambda_2 Loss_{hema} + \lambda_3 Loss_{ncc} \qquad (4)$$

where $Loss_{hema}$ denotes the Dice loss on the hematoma prediction of decoder. $\lambda_1 = 1$, $\lambda_2 = 0.4$ and $\lambda_3 = 0.6$ denote the weight for $Loss_{hemi}$, $Loss_{hema}$ and $Loss_{ncc}$ respectively.

2.4 Midline Correction

In a normal brain, the midline is a continuous line and surrounded by clear brain textures. Hematoma around the midline may cause the midline to shift due to high pressure. However, when hematoma breaks the midline boundary and floods into the ventricle, the pressure decreases which leads to the midline shift reverted. In this case, the direct output of the network cannot guarantee the accuracy of midline delineation. To address such issues, To address such issues, we formulate a midline correction strategy locally adjusting the midline boundary covered by hematoma regions.

The core of the correction strategy is to decrease the shifting degree predicted by the segmentation network. As shown in Fig. 2(c), we build a classification model based on ResNet architecture (ResNet18) [3] to estimate three situations: 1) hematoma has no contact with midline (non-interaction), 2) hematoma squeezes the midline without breaking into the ventricle (non-break-in), or 3) hematoma breaks into the ventricle (break-in). If the estimation is break-in (hard example), the correction strategy is performed. Specifically, we capture the two intersection points between the initial midline and hematoma boundary predictions and adopt the straight line between the two points as the partial correction result. If the estimation is non-interaction or non-break-in, the output from the segmentation network will be the final result.

3 Experiments

3.1 Dataset and Implementation Details

To comprehensively present the performance of our HSF framework on 3D brain delineation, the experiments were implemented on a dataset consisting of 300 CT volumes collected from our collaborating hospitals with approval from the local research ethics committee. The dataset contains 170 non-interaction, 90 non-break-in and 40 break-in examples. All the ground truths were manually annotated by clinicians. Annotation results were double-checked under strict quality control of an experienced expert. The labeled dataset was randomly divided into a training, validation and testing groups at a ratio of $3 : 1 : 1$. Volume size in our dataset varies from subject to subject. Specifically, the maximum original size in the dataset is $512 \times 512 \times 172$. Voxel spacing is resampled as $1.0 \times 1.0 \times 1.0\,\mathrm{mm}^3$. Limited by GPU memory, we randomly cropped the volume size to $192 \times 192 \times 160$. In this study, all the methods were implemented in PyTorch, in which our models ran 150 epochs in a single P40 GPU with 24 GB RAM. Adam with a batch size of 2 and an initial learning rate of 0.0001 was adopted to optimize the models.

Ventricle Break-In Classification. In the midline correction strategy, we pretrained the classification model with an extra dataset consisting of 379 volumes annotated as break-in, 985 volumes annotated as non-break-in and 492 volumes annotated as non-interaction. On the testing dataset, the classification accuracy is about 91.4%.

3.2 Quantitative and Qualitative Evaluation

In this paper, we adopt two evaluation metrics to evaluate the midline delineated by different methods, including Hausdorff Distance (HD-mm) and Average surface Distance (ASD-mm). ASD and HD represent the average and longest distance over the shortest voxel distances between surfaces of prediction and ground truth.

Table 1 shows the quantitative comparisons of midline delineation among our HSF and three competitive methods, including RLDN [11], CAR-Net [9] and Pisov et al. [6]. We re-implemented those three methods and adjusted the training parameters to obtain the best performance on the same dataset. The baseline is a U-Net structure followed by the gradient detection. It can be observed that the our method achieves the superior performance in terms of HD and ASD. Specifically, HSF improves the best HD to 11.83 mm and ASD to 1.20 mm, which significantly outperforms the other competitors.

Table 1. Quantitative comparisons of different methods on midline delineation performance.

Method	HD	ASD
U-Net [7]	19.08 ±12.1	1.24 ±1.5
RLDN [11]	17.88 ±9.8	2.08 ±1.6
CAR-Net [9]	14.62 ±14.5	1.52 ±1.8
Pisov et al. [6]	12.75 ±11.0	1.43 ±1.3
HSF(Ours)	**11.83** ±10.6	**1.20** ±2.5

Table 2. Quantitative comparisons of ablation experiments (superscript for standard deviation).

Method	Modules			Midline	
	RL	DWM	MC	HD	ASD
U-Net [7]	×	×	×	$19.08^{12.1}$	$1.24^{1.5}$
HSF (Ours)	✓	×	×	$17.94^{12.8}$	$1.35^{2.2}$
HSF (Ours)	✓	✓	×	$11.91^{10.6}$	$1.25^{2.5}$
HSF (Ours)	✓	✓	✓	$11.83^{10.6}$	$1.20^{2.5}$

We further conducted ablation study to verify the contribution of RL, DWM and MC strategies. As illustrated in Table 2, only with RL module, the improvement on midline delineation is relatively minor (1.14 mm HD). After introducing the DWM, the midline delineation performance is significantly improved,

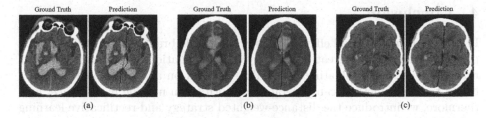

Fig. 4. Visualization results of midline correction. (a) (b) (c) represent three different cases respectively. Yellow, red and green line denote the ground truth, initial prediction and corrected part respectively. (Color figure online)

especially in terms of HD (6.03 mm). When the midline correction strategy is employed, the mean HD and ASD decrease by 0.08 mm and 0.05 mm respectively.

Figure 4 displays the corrected results. It can be observed that our method can generate more accurate midline boundary after performing the midline correction strategy. Figure 5 visualizes the delineation difference among the RLDN, CAR-Net, Pisov and our method. It is obvious that the proposed method yields more accurate results in the cases of hemorrhage break-in. Furthermore, the 3D visualization indicates the our HSF presents the smoother results compared with other methods and even the ground truth.

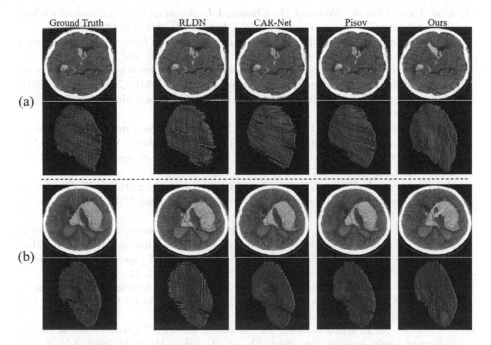

Fig. 5. Visualization comparisons of different methods. (a) (b) represent two different examples respectively. Yellow arrow indicates the right midline correction. (Color figure online)

4   Conclusion

In this study, a novel and efficient framework for 3D brain midline delineation is presented, which holds great potentials for the navigation of intracranial hemorrhage surgery. By formulating the midline delineation as a hemisphere segmentation task, our method can generate a smooth brain midline in 3D space. Furthermore, we introduce the distance-weighted strategy and rectificative learning to help improve the delineation performance and enhance the model adaptability in various head poses. Finally, considering the effect of hematoma on the midline shift, we design the midline correction strategy to adaptively adjust the midline boundary. Such strategy effectively improves the performance of our method on some extreme cases. Both quantitative and qualitative evaluations demonstrate the superiority of the proposed method in term of various metrics.

References

1. Caceres, J.A., Goldstein, J.N.: Intracranial hemorrhage. Emerg. Med. Clin. North Am. **30**(3), 771 (2012)
2. Chen, L.C., Papandreou, G., Schroff, F., Adam, H.: Rethinking atrous convolution for semantic image segmentation. arXiv preprint arXiv:1706.05587 (2017)
3. He, K., Zhang, X., Ren, S., Sun, J.: Deep residual learning for image recognition. In: Proceedings of the IEEE Conference on Computer Vision and Pattern Recognition, pp. 770–778 (2016)
4. Liao, C.C., Xiao, F., Wong, J.M., Chiang, I.J.: Automatic recognition of midline shift on brain CT images. Comput. Biol. Med. **40**(3), 331–339 (2010)
5. Liu, R., et al.: Automatic detection and quantification of brain midline shift using anatomical marker model. Comput. Med. Imaging Graph. **38**(1), 1–14 (2014)
6. Pisov, M., et al.: Incorporating task-specific structural knowledge into CNNs for brain midline shift detection. In: Suzuki, K., et al. (eds.) ML-CDS/IMIMIC -2019. LNCS, vol. 11797, pp. 30–38. Springer, Cham (2019). https://doi.org/10.1007/978-3-030-33850-3_4
7. Ronneberger, O., Fischer, P., Brox, T.: U-Net: convolutional networks for biomedical image segmentation. In: Navab, N., Hornegger, J., Wells, W.M., Frangi, A.F. (eds.) MICCAI 2015. LNCS, vol. 9351, pp. 234–241. Springer, Cham (2015). https://doi.org/10.1007/978-3-319-24574-4_28
8. Rundo, L., et al.: USE-NET: incorporating squeeze-and-excitation blocks into U-Net for prostate zonal segmentation of multi-institutional MRI datasets. Neurocomputing **365**, 31–43 (2019)
9. Wang, S., Liang, K., Li, Y., Yu, Y., Wang, Y.: Context-aware refinement network incorporating structural connectivity prior for brain midline delineation. In: Martel, A.L., et al. (eds.) MICCAI 2020. LNCS, vol. 12267, pp. 208–217. Springer, Cham (2020). https://doi.org/10.1007/978-3-030-59728-3_21
10. Wang, W.Z., et al.: Minimally invasive craniopuncture therapy vs. conservative treatment for spontaneous intracerebral hemorrhage: results from a randomized clinical trial in china. Int. J. Stroke **4**(1), 11–16 (2009)
11. Wei, H., et al.: Regression-based line detection network for delineation of largely deformed brain midline. In: Shen, D., et al. (eds.) MICCAI 2019. LNCS, vol. 11766, pp. 839–847. Springer, Cham (2019). https://doi.org/10.1007/978-3-030-32248-9_93

Unsupervised Representation Learning Meets Pseudo-Label Supervised Self-Distillation: A New Approach to Rare Disease Classification

Jinghan Sun[1,2], Dong Wei[2], Kai Ma[2], Liansheng Wang[1(✉)], and Yefeng Zheng[2]

[1] Xiamen University, Xiamen, China
jhsun@stu.xmu.edu.cn, lswang@xmu.edu.cn
[2] Tencent Jarvis Lab, Shenzhen, China
{donwei,kylekma,yefengzheng}@tencent.com

Abstract. Rare diseases are characterized by low prevalence and are often chronically debilitating or life-threatening. Imaging-based classification of rare diseases is challenging due to the severe shortage in training examples. Few-shot learning (FSL) methods tackle this challenge by extracting generalizable prior knowledge from a large base dataset of common diseases and normal controls, and transferring the knowledge to rare diseases. Yet, most existing methods require the base dataset to be labeled and do not make full use of the precious examples of the rare diseases. To this end, we propose in this work a novel hybrid approach to rare disease classification, featuring two key novelties targeted at the above drawbacks. First, we adopt the unsupervised representation learning (URL) based on self-supervising contrastive loss, whereby to eliminate the overhead in labeling the base dataset. Second, we integrate the URL with pseudo-label supervised classification for effective self-distillation of the knowledge about the rare diseases, composing a hybrid approach taking advantages of both unsupervised and (pseudo-) supervised learning on the base dataset. Experimental results on classification of rare skin lesions show that our hybrid approach substantially outperforms existing FSL methods (including those using fully supervised base dataset) for rare disease classification via effective integration of the URL and pseudo-label driven self-distillation, thus establishing a new state of the art.

Keywords: Rare disease classification · Unsupervised representation learning · Pseudo-label supervised self-distillation

J. Sun and D. Wei—Contributed equally; J. Sun contributed to this work during an internship at Tencent.

Electronic supplementary material The online version of this chapter (https://doi.org/10.1007/978-3-030-87240-3_50) contains supplementary material, which is available to authorized users.

© Springer Nature Switzerland AG 2021
M. de Bruijne et al. (Eds.): MICCAI 2021, LNCS 12905, pp. 519–529, 2021.
https://doi.org/10.1007/978-3-030-87240-3_50

1   Introduction

Rare diseases are a significant public health issue and a challenge to healthcare. On aggregate, the number of people suffering from rare diseases worldwide is estimated over 400 million, and there are about 5000–7000 rare diseases—with 250 new ones appearing each year [27]. Patients with rare diseases face delayed diagnosis: 10% of patients spent 5–30 years to reach a final diagnosis. Besides, many rare diseases can be misdiagnosed. Therefore, image-based accurate and timely diagnosis of rare diseases can be of great clinical value. In recent years, deep learning (DL) methods have developed into the state of the art (SOTA) for image-based computer-aided diagnosis (CAD) of many diseases [13,19,24]. However, due to the limited number of patients for a specific rare disease, collecting sufficient data for well training of generic DL classification models can be practically difficult or even infeasible for rare diseases.

To cope with the scarcity of training samples, a machine learning paradigm called few-shot learning (FSL) has been proposed [17] and achieved remarkable advances in the natural image domain [5,9,14,25,31]. In FSL, generalizable prior knowledge is learned on a large dataset of base classes, and subsequently utilized to boost learning of previously unseen novel classes given limited samples (the target task). Earlier approaches [5,9,14,25,31] to FSL mostly resorted to the concept of meta-learning and involved complicated framework design and task construction. Recently, Tian *et al.* [29] showed that superior FSL performance could be achieved by simply learning a good representation on the base dataset using basic frameworks, followed by fitting a simple classifier to few examples of the novel classes. Additional performance boosts were achieved through self-distillation [6,8]. How to implement a similar self-distilling strategy on an unsupervised base dataset, though, is not obvious. On the other hand, we are aware of only few methods [12,18,21,34] for FSL of medical image classification, and to the best of our knowledge, all of them relied on heavy labeling of the base dataset, causing a great burden for practical applications. Lastly, the meta-learning process and the target task are often isolated in most existing FSL approaches, and the meta-learner has little knowledge about its end task. For natural images, this setting is consistent with the general purpose of pretraining a classifier that can be quickly adapted for diverse tasks. For the scenario we consider, however, known types of rare diseases are mostly fixed, and their recognition constitutes a definite task. We hypothesize that, by bridging the base dataset and the definite task, the performance can be boosted for the rare disease classification. In this regard, we consider this work a specific and practically meaningful application of the general FSL.

In this work, we propose a novel hybrid approach to rare disease classification, which combines unsupervised representation learning (URL) and pseudo-label supervised self-distillation [6,8]. Motivated by the recent surge of representation learning in FSL [2,29], we first build a simple yet effective baseline model based on URL, where a good representation is learned on a large unlabeled base dataset consisting of common diseases and normal controls (CDNC) using contrastive learning [7], and applied to rare disease classification. So far as we are

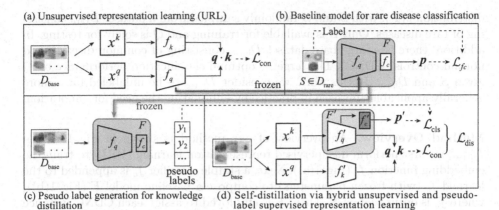

Fig. 1. Overview of the proposed approach. Solid line: information flow; dashed line: loss computation. Note that $\mathcal{L}_{f_c}$ in (b) can be any loss suitable for the classifier f_c.

aware of, this is the first study that explores few-shot medical image classification using an unsupervised base dataset. Then, we further propose to inject knowledge about the rare diseases into representation learning, to exploit the CDNC data for a more targeted learning of the rare diseases. Specifically, we use the baseline model as a teacher model to generate pseudo labels for instances of CDNC *belonging to the rare diseases*, to supervise knowledge distillation to the student model. Our rationale is that CDNC and rare diseases often share common characteristics, thus we can steer the representation learning on the former towards characteristics that can better distinguish the latter via supervision by the pseudo labels. In addition, we empirically explore design options for the distillation, and find that a hybrid self-distillation integrating URL and pseudo-label supervised classification yields the best performance. Therefore, our main contributions are two folds: the novel application of URL to rare disease classification, and hybrid distillation combining contrastive and pseudo-label learning. Experimental results on the ISIC 2018 skin lesion classification dataset show that the URL-based baseline model already outperforms previous SOTA FSL methods (including those using fully supervised base dataset), and that further boosts in performance are achieved by the proposed hybrid approach.

2   Methods

Problem Setting. Similar to Li *et al.* [18], we formulate the task of rare disease classification as a few-shot learning (FSL) problem. Specifically, we model a specific task as $\mathcal{T} = \{S, Q\}$ consisting of a support set $S = \{(x, y)\}$ and a query set $Q = \{(x, y)\}$, where x is an image and y is its label. An N-way K-shot task includes N rare diseases, each with K instances in S, where K is small. Thus $y \in [1, \ldots, N]$ and $|S| = N \times K$. The task instance $\mathcal{T}$ is randomly drawn from a distribution $p(\mathcal{T})$, with S and Q randomly sampled from a dataset D_{rare} consisting of rare diseases. Only S is available for training and Q is solely for testing. In addition, there is a large base dataset D_{base} consisting of common diseases and normal controls (CDNC). The target is optimal classification performance on Q given S and D_{base}. In this work, we consider D_{base} to be unlabeled for a more generally applicable approach in practice by eliminating the need for annotation.

Method Overview. An overview of our method is shown in Fig. 1. Given D_{base}, we first perform unsupervised representation learning (URL) to train the embedding function f_q (Fig. 1(a)). Next, a simple classifier f_c is appended to the learned f_q (with frozen parameters) to compose a baseline model F (Fig. 1(b)), where f_c is optimized on S. Then, F is employed to assign each CDNC instance in D_{base} a pseudo label of the rare diseases (Fig. 1(c)). Lastly, a self-distillation via hybrid unsupervised and pseudo-label supervised representation learning is performed on D_{base} to produce the final (student) model F' (Fig. 1(d)).

URL on CDNC for Rare Disease Classification. Inspired by the recent success of representation learning in FSL [2,29] and based on the recent advances in URL [1,7], we propose to perform URL on the big yet unlabeled CDNC dataset for rare disease classification. Specifically, we adopt instance discrimination with the contrastive loss [7,20] as our self-supervising task. We employ MoCo_v1 [7], where each image x_i in D_{base} is augmented twice to obtain x_i^q and x_i^k, whose embedded representations are subsequently obtained by $q_i = f_q(x_i^q; \theta_q)$ and $k_i = f_k(x_i^k; \theta_k)$, where f_q and f_k are the query and key encoders parameterized by θ_q and θ_k, respectively. The contrastive loss $\mathcal{L}_{\text{con}}$ is defined as [20]:

$$\mathcal{L}_{\text{con}}(x_i) = -\log\left[\exp\left(q_i \cdot k_i / \tau\right) / \left(\exp\left(q_i \cdot k_i / \tau\right) + \sum_{j=1}^{L} \exp\left(q_i \cdot k_j / \tau\right)\right)\right],$$
(1)

where L is the number of keys stored in the dynamic dictionary implemented as a queue, τ is a temperature hyperparameter. Intuitively, this loss is the log loss of an $(L+1)$-way softmax-based classifier trained to discriminate the (augmented) instance x_i from other images stored in the dictionary queue (represented by their embeddings). Then, θ_q is updated by back-propagation whereas θ_k is updated with a momentum m: $\theta_k \leftarrow m\theta_k + (1-m)\theta_q$, where $m \in [0,1)$. Another notable difference of our work from the prevailing meta-learning based approaches is that we randomly sample mini-batches of training images from D_{base}, instead of iteratively constructing episodic training tasks as in most previous works [5,9,14,18,25,28]. This back-to-basic training scheme has proven effective despite being simpler [29], especially for an unsupervised D_{base} where category information is missing for episode construction.

After the URL, we freeze θ_q and append a simple classifier f_c to f_q to form a baseline model $F = f_c(f_q)$ for rare disease classification. Like [29], we use logistic regression for f_c, whose parameters θ_c are optimized on the support set S.

Self-Distillation of Rare Disease Knowledge. Despite its decent performance, the baseline model completely ignores the precious knowledge about target rare diseases contained in the support set S during representation learning. We hypothesize that a better representation learning for classification of the target rare diseases can be achieved by fully exploiting this knowledge while at the same time utilizing the big unlabeled data in D_{base}. To do so, we propose to inject target task knowledge extracted from S into the representation learning process via knowledge distillation [8], which can transfer knowledge embedded in a teacher model to a student model. In addition, we adopt the born-again strategy where the teacher and student models have an identical architecture, for its superior performance demonstrated by Furlanello et al. [6].

The key idea behind our knowledge distillation scheme is that, although D_{base} and D_{rare} comprise disjoint classes, it is common that certain imaging characteristics (e.g., color, shape and/or texture) of the CDNC data are shared by the rare diseases. Therefore, it is feasible to learn rare-disease-distinctive representations and classifiers by *training the networks to classify CDNC instances as rare diseases of similar characteristics*. Mathematically, we use the baseline model F as the teacher model to predict the probabilities of each image x in D_{base} belonging to the rare diseases in D_{rare}: $\boldsymbol{p} = F(x) = [p_1, \ldots, p_N]^T$, where $\sum_{n=1}^{N} p_n = 1$. Next, we define the pseudo label $\boldsymbol{y} = [y_1, \ldots, y_N]^T$ based on $\boldsymbol{p}$ with two alternative strategies: (i) hard labeling where $y_n = 1$ if $n = \text{argmax}_n p_n$ and 0 otherwise, and (2) soft labeling where $y_n = p_n$. In effect, the first strategy indicates the rare disease that x resembles most, whereas the second reflects the extents of resemblance between x and all the rare diseases in D_{rare}. In addition, we propose a hybrid distilling loss integrating pseudo-label supervised classification and contrastive instance discrimination. As we will show, the hybrid distillation scheme is important for preventing overfitting to noise and bias in the small support set. Then, adopting the born-again [6] strategy, a randomly initialized student model $F' = f'_c(f'_q)$ parameterized by θ'_c and θ'_q is trained to minimize the hybrid loss $\mathcal{L}_{dis}$:

$$\mathcal{L}_{dis} = \mathcal{L}_{con}(x; \theta'_q, \theta'_k) + \mathcal{L}_{cls}(\boldsymbol{y}, F'(x; \theta'_q, \theta'_c)), \tag{2}$$

where θ'_k are parameters of the key encoder f'_k (for computation of $\mathcal{L}_{con}$) and updated along with θ'_q, and $\mathcal{L}_{cls}$ is the pseudo-label supervised classification loss. For $\mathcal{L}_{cls}$, the cross-entropy and Kullback-Leibler divergence losses are used for the hard and soft labels, respectively. Lastly, to allow for an end-to-end training, a fully connected layer (followed by softmax) is used for f'_c.

After distillation, the student model $F' = f'_c(f'_q)$ can be directly used for rare disease classification. One might argue for an alternative way of usage: discarding f'_c but appending a logistic regression classifier fit to the support set to f'_q, just like in the baseline model. However, as confirmed by our comparative experiment (supplement Table S1), direct use of F' performs much better. This is because, the knowledge about the rare diseases is distilled into not only f'_q but also f'_c, thus discarding the latter results in performance degradation. Lastly, through preliminary experiments we find that distilling more than once does not bring further improvement. Therefore, we perform the self-distillation only once.

Adaptive Pseudo Labels. In practice, the pseudo labels defined by F may not be entirely trustworthy, given the tiny size and potential noise and bias of the support set. This may adversely affect performance of the student model. To alleviate the adverse effect, we further propose adaptive pseudo labels based on the self-adaptive training [11] strategy. Concretely, given the prediction p' by the student model and pseudo labels y defined above, we combine them as our new training target: $y^{\text{adpt}} = (1 - \alpha) \times y + \alpha \times p'$, where α is a confidence parameter controlling how much we trust the teacher's knowledge. y^{adpt} is termed adaptive hard/soft labels depending on y being hard or soft pseudo labels. Many previous works used a constant α [11,32]. In the first few epochs, however, the student model lacks reliability—it gradually develops a better ability of rare disease classification as the training goes on. Therefore, we adopt a linear growth rate [15] for α at the tth epoch: $\alpha_t = \alpha_T \times (t/T)$, where α_T is the last-epoch value and set to 0.7 as in [15], and T is the total number of epochs. As suggested by the comparative experiments (supplement Table S2), the adaptive hard labels work the best, thus are used in our full model for comparison with other methods.

3 Experiments

Dataset and Evaluation Protocol. The ISIC 2018 skin lesion classification dataset [4,30][1] includes 10,015 dermoscopic images from seven disease categories: melanocytic nevus (6,705), benign keratosis (1,099), melanoma (1,113), basal cell carcinoma (514), actinic keratosis (327), dermatofibroma (115), and vascular lesion (142). From that, we simulate the task of rare disease classification as below. Following Li *et al.* [18], we use the four classes with the most cases as the CDNC dataset D_{base}, and the other three as the rare disease dataset D_{rare}. K images are sampled for each class in D_{rare} to compose the support set S for a 3-way K-shot task. Compared to the binary classification tasks [18], the multi-way evaluation protocol more genuinely reflects the practical clinical need where more than two rare diseases are present, albeit more challenging. As to K, we experiment with 1, 3, and 5 shots in this work. All remaining images in D_{rare} compose the query set Q for performance evaluation. Again, this task construction more genuinely reflects the intended scenario of rare disease classification—where only few examples are available for training a classifier to be applied to all future test cases—than the repeated construction of small Q's [18]. We sample three random tasks $\mathcal{T} \sim p(\mathcal{T})$ and report the mean and standard deviation of these tasks. Besides accuracy, we additionally employ the F1 score as the evaluation metric considering the data imbalance between the rare disease classes.

---

[1] https://challenge2018.isic-archive.com/task3/.

Implementation. The PyTorch [26] framework (1.4.0) is used for experiments. We use the ResNet-12 [16, 22, 29] architecture as backbone network for f_q and f_k, for its superior performance in the comparative experiments (supplement Table S3 [10, 23, 33]). We train the networks for 200 epochs with a mini-batch size of 16 images on four NVIDIA Tesla V100 GPUs. We adopt the stochastic gradient descent optimizer with a momentum of 0.9 and a weight decay of 0.0001. The learning rate is initialized to 0.03 and decays at 120 and 160 epochs by multiplying by 0.1. The feature dimension of the encoded representation is 128, and the number of negatives in the memory bank [7] is 1280. The temperature τ in Eq. (1) is set to 0.07 as in [7]. All images are resized to 224×224 pixels. Online data augmentation including random cropping, flipping, color jittering, and blurring [1] is performed. The source code is available at: https://github.com/SunjHan/Hybrid-Representation-Learning-Approach-for-Rare-Disease-Classification.

Comparison to SOTA Methods. According to the labeling status of the base dataset and genre of the methodologies, all the compared methods are grouped into four quadrants: (i) supervised meta-learning (SML) including MAML [5], Relation Networks [28], Prototypical Networks [25], and DAML [18], (ii) unsupervised meta-learning (UML) including UMTRA [14] and CACTUs [9], (iii) supervised representation learning (SRL; with or without self-distillation) [29], and (iv) URL including SimCLR [1], MoCo_v2 [3], MoCo_v1 [7] (composing the baseline model in our work), and our proposed method. These methods cover a wide range of the latest advances in FSL for image classification. For reference purpose, we also show the results of training a classifier from scratch solely on S. Besides the ResNet-12 backbone, we additionally show the results using the 4 conv blocks [31] as the backbone network considering its prevalent usage in the FSL literature [5, 9, 14, 25, 28]. Note that for all compared methods, we optimize their performance via empirical parameter tuning.

The results are shown in Table 1 (ResNet-12) and supplement Table S4 (4 conv blocks), on which we make the following observations. First, the representation learning based methods generally achieve better performance than the meta-learning based irrespective of the labeling status of D_{base}, which is consistent with the findings in the natural image domain [29]. Second, the URL based methods surprisingly outperform the SRL based in most circumstances. A possible explanation may be that the few classes in D_{base} present limited variations and make the representations overfit to their differentiation, whereas the task of instance discrimination in URL forces the networks to learn more diverse representations that are more generalizable on novel classes. Especially, the baseline model presented in this work (URL with MoCo_v1 [7]) wins over the SRL plus self-distillation [29] by large margins. Third, our proposed hybrid approach brings further improvements upon the baseline model in both accuracy ($\sim$1–2% with ResNet-12) and F1 score ($\sim$3–5% with ResNet-12). Notably, it achieves an accuracy of 81.16% in the 5-shot setting without any label of the base dataset. These results strongly support our hypothesis that, by extracting the knowledge about the rare diseases from the small support set and injecting it into the rep-

Table 1. Evaluation results and comparison with SOTA FSL methods with ResNet-12 as backbone network. Data format: mean (standard deviation).

Method	$(N, K) = (3, 1)$		$(N, K) = (3, 3)$		$(N, K) = (3, 5)$	
	Accuracy (%)	F1 score (%)	Accuracy (%)	F1 score (%)	Accuracy (%)	F1 score (%)
Training from scratch	37.74 (1.07)	29.90 (3.65)	39.76 (0.88)	35.60 (1.72)	45.36 (3.76)	38.41 (4.06)
▷ SML MAML [5]	47.49 (5.38)	42.33 (6.16)	55.55 (3.12)	49.19 (4.20)	58.94 (2.59)	53.51 (2.46)
RelationNet [28]	46.10 (4.80)	39.98 (6.73)	47.29 (2.77)	43.37 (3.65)	55.71 (3.30)	49.34 (3.57)
ProtoNets [25]	35.18 (3.12)	30.81 (3.09)	38.59 (1.91)	33.11 (2.08)	42.45 (2.45)	34.92 (3.70)
DAML [18]	50.05 (5.18)	41.65 (3.98)	55.57 (3.55)	49.01 (6.62)	59.44 (3.17)	54.66 (2.43)
▷ UML UMTRA [14]	45.88 (3.63)	41.44 (4.37)	51.29 (3.54)	45.91 (3.96)	57.33 (1.76)	53.06 (0.89)
CACTUs-MAML [9]	42.98 (2.91)	35.38 (3.08)	44.44 (3.35)	39.94 (3.65)	48.11 (4.20)	44.32 (3.65)
CACTUs-ProtoNets [9]	42.67 (2.43)	39.24 (2.72)	45.00 (3.26)	39.69 (2.66)	47.95 (3.52)	44.08 (2.63)
▷ SRL SRL-simple [29]	54.45 (5.82)	51.02 (6.93)	61.31 (6.31)	57.65 (3.46)	70.53 (2.17)	65.58 (3.72)
SRL-distil [29]	55.43 (7.36)	51.18 (5.50)	64.92 (6.00)	59.88 (4.87)	72.78 (1.67)	65.89 (2.54)
▷ URL SimCLR [1]	52.43 (5.01)	44.70 (8.24)	63.82 (3.70)	57.55 (3.67)	70.18 (1.76)	63.73 (1.78)
MoCo_v2 [3]	59.95 (4.73)	55.98 (3.81)	70.84 (2.91)	64.77 (3.69)	75.80 (1.85)	70.69 (2.13)
MoCo_v1 (baseline) [7]	61.90 (2.92)	56.30 (1.48)	74.92 (2.96)	69.50 (5.72)	79.01 (2.00)	74.47 (3.03)
Hybrid distil (ours)	**64.15** (2.86)	**61.01** (1.30)	**75.82** (2.47)	**73.34** (2.30)	**81.16** (2.60)	**77.35** (4.21)

resentation learning process via pseudo-label supervision, we can fully exploit the large CDNC dataset to learn representations and classifiers that can better distinguish the rare diseases. To further investigate whether the improvements sustain higher K values, we conduct extra experiments (with ResNet-12) for the baseline and hybrid models with $K = 10$ and 20. The results confirm that our proposed hybrid approach still wins over the baseline model by ~1% absolute differences in both metrics and settings, yielding the accuracies and F1 scores of 83.65% and 79.98% when $K = 10$, and of 86.91% and 83.40% when $K = 20$. Lastly, the results using ResNet-12 as backbone are generally better than those using the 4 conv blocks, as expected.

Ablation Study. We probe the effect of the proposed hybrid distillation by comparing performance of distilling with only $\mathcal{L}_{cls}$ and the full model. In addition, we experiment with a variant of $\mathcal{L}_{dis}$ where $\mathcal{L}_{cls}$ is replaced by an L1 loss $\mathcal{L}_{reg}$ to directly regress the output of f_q, to evaluate the effect of injecting knowledge about the rare diseases. Results (Table 2) show that distilling with $\mathcal{L}_{cls}$ alone brings moderate improvement upon the baseline model, suggesting the efficacy of injecting rare disease knowledge into representation learning. Yet, distilling with the proposed hybrid loss achieves further appreciable improvement. We conjecture this is because $\mathcal{L}_{con}$ helps avoid overfitting to the support set of the rare diseases, which may be affected by noise and bias due to its small size. On the other hand, distilling with $\mathcal{L}_{reg}$ (plus $\mathcal{L}_{con}$) gives performance similar to the baseline model, implying that it is the distilled knowledge about the rare diseases that matters, rather than the distillation procedure. These results confirm the efficacy of the hybrid distillation.

Table 2. Ablation study on the hybrid distillation (with ResNet-12 backbone and adaptive hard labels). Data format: mean (standard deviation).

$\mathcal{L}_{dis}$	$(N, K) = (3, 1)$		$(N, K) = (3, 3)$		$(N, K) = (3, 5)$	
	Accuracy (%)	F1 score (%)	Accuracy (%)	F1 score (%)	Accuracy (%)	F1 score (%)
N.A.	61.90 (2.92)	56.30 (1.48)	74.92 (2.96)	69.50 (5.72)	79.01 (2.00)	74.47 (3.03)
$\mathcal{L}_{cls}$	63.70 (3.39)	57.31 (7.73)	74.92 (2.10)	70.28 (3.97)	80.24 (1.61)	77.29 (2.91)
$\mathcal{L}_{con} + \mathcal{L}_{cls}$	**64.15** (2.86)	**61.01** (1.30)	**75.82** (2.47)	**73.34** (2.30)	**81.16** (2.60)	**77.35** (4.21)
$\mathcal{L}_{con} + \mathcal{L}_{reg}$	62.20 (5.18)	56.19 (4.28)	74.43 (2.88)	69.74 (4.13)	79.14 (2.09)	74.41 (2.65)

4   Conclusion

In this work, we proposed a novel approach to rare disease classification in two steps. First, we built a baseline model on unsupervised representation learning for a simple and label-free (w.r.t. the base dataset of common diseases and normal controls) framework, which achieved superior performance to existing FSL methods on skin lesion classification. Second, we further proposed to utilize the baseline model as the teacher model for a hybrid self-distillation integrating unsupervised contrastive learning and pseudo-label supervised classification. Experimental results suggested that the hybrid approach could effectively inject knowledge about the rare diseases into the representation learning process through the pseudo-labels, and meanwhile resist overfitting to noise and bias in the small support set thanks to the contrastive learning, and that it had set a new state of the art for rare disease classification.

Acknowledgments. This work was supported by the Fundamental Research Funds for the Central Universities (Grant No. 20720190012), Key-Area Research and Development Program of Guangdong Province, China (No. 2018B010111001), and Scientific and Technical Innovation 2030 - "New Generation Artificial Intelligence" Project (No. 2020AAA0104100).

References

1. Chen, T., Kornblith, S., Norouzi, M., Hinton, G.: A simple framework for contrastive learning of visual representations. In: International Conference on Machine Learning, pp. 1597–1607. PMLR (2020)
2. Chen, W.Y., Liu, Y.C., Kira, Z., Wang, Y.C.F., Huang, J.B.: A closer look at few-shot classification. In: International Conference on Learning Representations (2019)
3. Chen, X., Fan, H., Girshick, R., He, K.: Improved baselines with momentum contrastive learning. arXiv preprint arXiv:2003.04297 (2020)
4. Codella, N., et al.: Skin Lesion Analysis Toward Melanoma Detection 2018: A Challenge hosted by the International Skin Imaging Collaboration (ISIC). arXiv preprint arXiv:1902.03368 (2019)
5. Finn, C., Abbeel, P., Levine, S.: Model-agnostic meta-learning for fast adaptation of deep networks. In: International Conference on Machine Learning, pp. 1126–1135. PMLR (2017)

6. Furlanello, T., Lipton, Z., Tschannen, M., Itti, L., Anandkumar, A.: Born again neural networks. In: International Conference on Machine Learning, pp. 1607–1616. PMLR (2018)
7. He, K., Fan, H., Wu, Y., Xie, S., Girshick, R.: Momentum contrast for unsupervised visual representation learning. In: Proceedings of the IEEE/CVF Conference on Computer Vision and Pattern Recognition, pp. 9729–9738 (2020)
8. Hinton, G., Vinyals, O., Dean, J.: Distilling the knowledge in a neural network. In: NIPS Deep Learning and Representation Learning Workshop (2015)
9. Hsu, K., Levine, S., Finn, C.: Unsupervised learning via meta-learning. In: International Conference on Learning Representations (2018)
10. Huang, G., Liu, Z., van der Maaten, L., Weinberger, K.Q.: Densely connected convolutional networks. In: Proceedings of the IEEE Conference on Computer Vision and Pattern Recognition (2017)
11. Huang, L., Zhang, C., Zhang, H.: Self-adaptive training: Beyond empirical risk minimization. arXiv preprint arXiv:2002.10319 (2020)
12. Jiang, X., Ding, L., Havaei, M., Jesson, A., Matwin, S.: Task adaptive metric space for medium-shot medical image classification. In: Shen, D., et al. (eds.) MICCAI 2019. LNCS, vol. 11764, pp. 147–155. Springer, Cham (2019). https://doi.org/10.1007/978-3-030-32239-7_17
13. Ker, J., Wang, L., Rao, J., Lim, T.: Deep learning applications in medical image analysis. IEEE Access 6, 9375–9389 (2018)
14. Khodadadeh, S., Boloni, L., Shah, M.: Unsupervised meta-learning for few-shot image classification. In: Advances in Neural Information Processing Systems, vol. 32. Curran Associates, Inc. (2019)
15. Kim, K., Ji, B., Yoon, D., Hwang, S.: Self-knowledge distillation: A simple way for better generalization. arXiv preprint arXiv:2006.12000 (2020)
16. Lee, K., Maji, S., Ravichandran, A., Soatto, S.: Meta-learning with differentiable convex optimization. In: Proceedings of the IEEE/CVF Conference on Computer Vision and Pattern Recognition, pp. 10657–10665 (2019)
17. Li, F.F., Fergus, R., Perona, P.: One-shot learning of object categories. IEEE Trans. Pattern Anal. Mach. Intell. 28(4), 594–611 (2006)
18. Li, X., Yu, L., Jin, Y., Fu, C.-W., Xing, L., Heng, P.-A.: Difficulty-Aware Meta-learning for Rare Disease Diagnosis. In: Martel, A.L., et al. (eds.) MICCAI 2020. LNCS, vol. 12261, pp. 357–366. Springer, Cham (2020). https://doi.org/10.1007/978-3-030-59710-8_35
19. Litjens, G., et al.: A survey on deep learning in medical image analysis. Med. Image Anal. 42, 60–88 (2017)
20. Oord, A.v.d., Li, Y., Vinyals, O.: Representation learning with contrastive predictive coding. arXiv preprint arXiv:1807.03748 (2018)
21. Paul, A., Tang, Y.X., Shen, T.C., Summers, R.M.: Discriminative ensemble learning for few-shot chest X-ray diagnosis. Med. Image Anal. 68, 101911 (2021)
22. Ravichandran, A., Bhotika, R., Soatto, S.: Few-shot learning with embedded class models and shot-free meta training. In: Proceedings of the IEEE/CVF International Conference on Computer Vision, pp. 331–339 (2019)
23. Sandler, M., Howard, A., Zhu, M., Zhmoginov, A., Chen, L.C.: MobileNetV2: Inverted residuals and linear bottlenecks. In: Proceedings of the IEEE Conference on Computer Vision and Pattern Recognition (2018)
24. Shen, D., Wu, G., Suk, H.I.: Deep learning in medical image analysis. Annu. Rev. Biomed. Eng. 19(1), 221–248 (2017)
25. Snell, J., Swersky, K., Zemel, R.: Prototypical networks for few-shot learning. In: Advances in Neural Information Processing Systems, pp. 4080–4090 (2017)

26. Steiner, B., et al.: PyTorch: an imperative style, high-performance deep learning library. Adv. Neural Inf. Process. Syst. **32**, 8026–8037 (2019)
27. Stolk, P., Willemen, M.J., Leufkens, H.G.: Rare essentials: drugs for rare diseases as essential medicines. Bull. World Health Organ. **84**, 745–751 (2006)
28. Sung, F., Yang, Y., Zhang, L., Xiang, T., Torr, P.H., Hospedales, T.M.: Learning to compare: relation network for few-shot learning. In: Proceedings of the IEEE Conference on Computer Vision and Pattern Recognition (June 2018)
29. Tian, Y., Wang, Y., Krishnan, D., Tenenbaum, J.B., Isola, P.: Rethinking few-shot image classification: a good embedding is all you need? In: Proceedings of the European Conference on Computer Vision (2020)
30. Tschandl, P., Rosendahl, C., Kittler, H.: The HAM10000 dataset, a large collection of multi-source dermatoscopic images of common pigmented skin lesions. Sci. Data **5**(1), 1–9 (2018)
31. Vinyals, O., Blundell, C., Lillicrap, T., Kavukcuoglu, K., Wierstra, D.: Matching networks for one shot learning. In: Advances in Neural Information Processing Systems, pp. 3637–3645 (2016)
32. Zhang, L., Song, J., Gao, A., Chen, J., Bao, C., Ma, K.: Be your own teacher: improve the performance of convolutional neural networks via self distillation. In: Proceedings of the IEEE/CVF International Conference on Computer Vision, pp. 3713–3722 (2019)
33. Zhang, X., Zhou, X., Lin, M., Sun, J.: ShuffleNet: an extremely efficient convolutional neural network for mobile devices. In: Proceedings of the IEEE Conference on Computer Vision and Pattern Recognition (2018)
34. Zhu, W., Liao, H., Li, W., Li, W., Luo, J.: Alleviating the incompatibility between cross entropy loss and episode training for few-shot skin disease classification. In: Martel, A.L., et al. (eds.) MICCAI 2020. LNCS, vol. 12266, pp. 330–339. Springer, Cham (2020). https://doi.org/10.1007/978-3-030-59725-2_32

nnDetection: A Self-configuring Method for Medical Object Detection

Michael Baumgartner[1(✉)], Paul F. Jäger[2], Fabian Isensee[1,3], and Klaus H. Maier-Hein[1,4]

[1] Division of Medical Image Computing, German Cancer Research Center, Heidelberg, Germany
m.baumgartner@dkfz.de
[2] Interactive Machine Learning Group, German Cancer Research Center, Heidelberg, Germany
[3] HIP Applied Computer Vision Lab, German Cancer Research Center, Heidelberg, Germany
[4] Pattern Analysis and Learning Group, Heidelberg University Hospital, Heidelberg, Germany

Abstract. Simultaneous localisation and categorization of objects in medical images, also referred to as medical object detection, is of high clinical relevance because diagnostic decisions often depend on rating of objects rather than e.g. pixels. For this task, the cumbersome and iterative process of method configuration constitutes a major research bottleneck. Recently, nnU-Net has tackled this challenge for the task of image segmentation with great success. Following nnU-Net's agenda, in this work we systematize and automate the configuration process for medical object detection. The resulting self-configuring method, nnDetection, adapts itself without any manual intervention to arbitrary medical detection problems while achieving results en par with or superior to the state-of-the-art. We demonstrate the effectiveness of nnDetection on two public benchmarks, ADAM and LUNA16, and propose 11 further medical object detection tasks on public data sets for comprehensive method evaluation. Code is at https://github.com/MIC-DKFZ/nnDetection.

1 Introduction

Image-based diagnostic decision-making is often based on rating objects and rarely on rating individual pixels. This process is well reflected in the task of medical object detection, where entire objects are localised and rated. Nevertheless, semantic segmentation, i.e. the categorization of individual pixels, remains the predominant approach in medical image analysis with 70% of biomedical

---

M. Baumgartner and P. F. Jäger—Equal contribution.

Electronic supplementary material The online version of this chapter (https://doi.org/10.1007/978-3-030-87240-3_51) contains supplementary material, which is available to authorized users.

© Springer Nature Switzerland AG 2021
M. de Bruijne et al. (Eds.): MICCAI 2021, LNCS 12905, pp. 530–539, 2021.
https://doi.org/10.1007/978-3-030-87240-3_51

challenges revolving around segmentation [14]. To be of diagnostic relevance, however, in many use-cases segmentation methods require ad-hoc postprocessing that aggregates pixel predictions to object scores. This can negatively affect performance compared to object detection methods that already solve these steps within their learning procedure [9].

Compared to a basic segmentation architecture like the U-Net, the set of hyper-parameters in a typical object detection architecture is extended by an additional detection head with multiple loss functions including smart sampling strategies ("hard negative mining"), definition of size, density and location of prior boxes ("anchors"), or the consolidation of overlapping box predictions at test time ("weighted box clustering"). This added complexity might be an important reason for segmentation methods being favoured in many use-cases. It further aggravates the already cumbersome and iterative process of method configuration, which currently requires expert knowledge, extensive compute resources, sufficient validation data, and needs to be repeated on every new tasks due to varying data set properties in the medical domain [8].

Recently, nnU-net achieved automation of method configuration for the task of biomedical image segmentation by employing a set of fixed, rule-based, and empirical parameters to enable fast, data-efficient, and holistic adaptation to new data sets [8]. In this work, we follow the recipe of nnU-Net to systematize and automate method configuration for medical object detection. Specifically, we identified a novel set of fixed, rule-based, and empirical design choices on a diverse development pool comprising 10 data sets. We further follow nnU-Net in deploying a clean and simple base-architecture: Retina U-Net [9]. The resulting method, which we call nnDetection, can now be fully automatically deployed on arbitrary medical detection problems without requiring compute resources beyond standard network training.

Without manual intervention, nnDetection sets a new state of the art on the nodule-candidate-detection task of the well-known LUNA16 benchmark and achieves competitive results on the ADAM leaderboard. To address the current lack of public data sets compared to e.g. medical segmentation, we propose a new large-scale benchmark totaling 13 data sets enabling sufficiently diverse evaluation of medical object detection methods. To this end, we identified object detection tasks in data sets of existing segmentation challenges and compare nnDetection against nnU-Net (with additional postprocessing for object scoring) as a standardized baseline.

With the hope to foster increasing research interest in medical object detection, we make nnDetection publicly available (including pre-trained models and object annotations for all newly proposed benchmarks) as an out-of-the-box method for state-of-the-art object detection on medical images, a framework for novel methodological work, as well as a standardized baseline to compare against without manual effort.

2   Methods

Figure 1 shows how nnDetection systematically addresses the configuration of entire object detection pipelines and provides a comprehensive list of design choices.

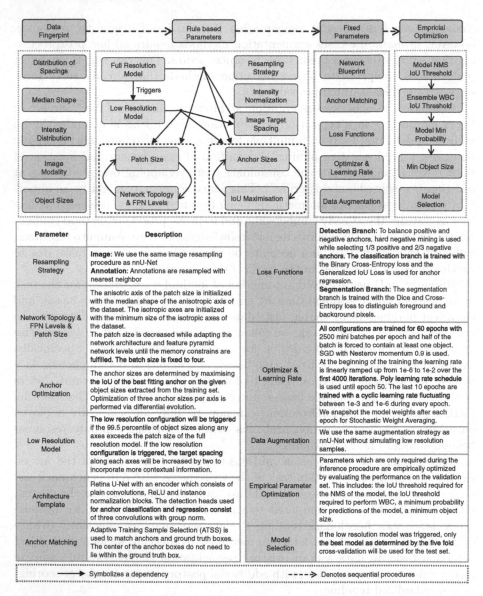

Fig. 1. Overview of the high level design choices and mechanisms of nnDetection (For details and reasonings about all design decisions we refer to our code repository at https://github.com/MIC-DKFZ/nnDetection). Due to the high number of dependencies between parameters, only the most important ones are visualized as arrows. Given a new medical object detection task, a fingerprint covering relevant data set properties is extracted (purple). Based on this information, a set of heuristic rules is executed to determine the rule-base parameters of the pipeline (green). These rules act in tandem with a set of fixed parameters which do not require adaptation between data sets (blue). After training, empirical parameters are optimized on the training set (orange). All design choices were developed and extensively evaluated upfront on our development pool, thus ensuring robustness and enabling rapid application of nnDetection to new data sets without requiring extensive additional compute resources.

nnDetection Development. To achieve automated method configuration in medical object detection, we roughly follow the recipe outlined in nnU-Net, where domain knowledge is distilled in the form of fixed, rule-based, and empirical parameters. Development was performed on a pool of 10 data sets (see supplementary material).

Fixed Parameters: (For a comprehensive list see Fig. 1). First, we identified design choices that do not require adaptation between data sets and optimized a joint configuration for robust generalization on our 10 development data sets. We opt for Retina U-Net as our architecture template, which builds on the simple RetinaNet to enable leveraging of pixel-level annotations [9], and leave the exact topology (e.g. kernel sizes, pooling strides, number of pooling operations) to be adapted via rule-based parameters. To account for varying network configurations and object sizes across data sets we employ adaptive training sample selection [24] for anchor matching. However, we discarded the requirement as to which the center point of selected anchors needs to lie inside the ground truth box because, as we found it often resulted in the removal of all positive anchors for small objects. Furthermore, we increased the number of anchors per position from one to 27, which we found improves results especially on data sets with few objects.

Rule-based Parameters: (For a comprehensive list see Fig. 1). Second, for as many of the remaining decisions as possible, we formulate explicit dependencies between the Data Fingerprint and design choices in the form of interdependent heuristic rules. Compared to nnU-Net our Data Fingerprint additionally extracts information about object sizes (see Fig. 1). We use the same iterative optimization process as nnU-Net to determine network topology parameters such as kernel sizes, pooling strides, and the number of pooling operations, but fixed the batch size at 4 as we found this to improve training stability. Similar to nnU-Net, an additional low-resolution model is triggered to account for otherwise missing context in data sets with very large objects or high resolution images. Finding an appropriate anchor configuration is one of the most important design choices in medical object detection [15,26]. Following Zlocha et al. [26], we iteratively maximize the intersection over union (IoU) between anchors and ground-truth boxes. In contrast to their approach, we found performing this optimization on the training split instead of the validation split led to more robust anchor configurations due to a higher number of samples. Also, we fit three anchor sizes per axis and use the euclidean product to produce the final set of anchors for the highest resolution pyramid level the detection head operates on.

Empirical Parameters: (For a comprehensive list see Fig. 1). Postprocessing in object detection models mostly deals with clustering overlapping bounding box predictions. There are different sources for why predictions might overlap. The inherent overlap of predictions from dense anchors is typically accounted for by Non-maximum Suppression (NMS). Due to limited GPU memory, nnDetection uses sliding window inference with overlapping patches. Overlaps across neighboring patches are clustered via NMS while weighing predictions near the center of a patch higher than predictions at the border. To cluster predictions from multiple models or different test time augmentations Weighted Box Clustering [9] is used. Empirical Parameters which are only used at test time (see a full list in the Table in Fig. 1) are optimized empirically on the validation set. Due to their interdependencies, nnDetection uses a pre-defined initialization of the parameters and sequentially optimizes them by following the order outlined in Fig. 1. If the low resolution model has been triggered, the best model will be selected empirically for testing based on the validation results.

nnDetection Application. Given a new data set, nnDetection runs automatic configuration without manual intervention. Thus, no additional computational cost beyond a standard network training procedure is required apart from the few required empirical choices. First, nnDetection extracts the Data Fingerprint and executes the heuristic rules to determine the rule-based parameters. Subsequently, the full-resolution and, if triggered, the low-resolution model will be trained via five-fold cross-validation. After training, empirical parameters are determined and the final prediction is composed by ensembling the predictions of the five models obtained from cross-validation of the empirically selected configuration. We evaluate the generalization ability of nnDetection's automated configuration on 3 additional data sets (see supplementary material).

nnU-Net as an Object Detection Baseline. Our first nnU-Net baseline, called *nnU-Net Basic*, reflects the common approach to aggregating pixel predictions: Argmax is applied over softmax predictions, followed by connected component analysis per foreground class, and finally an object score per component is obtained as the maximum pixel softmax score of the assigned category. *nnU-Net Plus:* To ensure the fairest possible comparison, we enhance the baseline by empirically choosing the following postprocessing parameters based on the training data for each individual task: Replacement of argmax by a minimum threshold on the softmax scores to be assigned to a component, a threshold on the minimum number of pixels per object, and the choice of the aggregation method (max, mean, median, 95% percentile). During our experiments on LIDC [1] we observed convergence issues of nnU-Net. Thus, we identified an issue with numerical constants inside the Dice loss and were able to improve results significantly by removing those.

3   Experiments and Results

Proposed Benchmark for Medical Object Detection. Recently, strong evidence has been provided for the importance of evaluating segmentation methods on a large and diverse data set pool [8]. This requirement arises from volatility of evaluation metrics caused by limited data set size as well as considerable label noise in the medical domain. Furthermore, covering data set diversity prevents

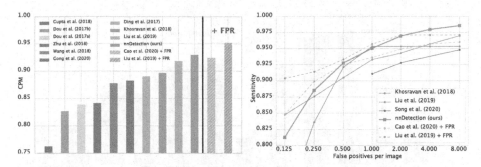

Fig. 2. *Left:* nnDetection outperforms all competing approaches on the nodule-candidate-detection task of LUNA16 and is only beaten by Liu et al. [13] in the general task, where additional False Positive Reduction (FPR) models are employed (we consider such task-specific intervention to be out of scope for this work). *Right:* FROC curves for the top 7 methods. Starting from 1/4 False Positives per Scan, nnDetection outperforms Liu et al. [13] without FPR.

general methodological claims from being overfit to specific tasks. We argue these aspects directly translate to medical object detection and thus propose a new benchmark based a diverse pool of 13 existing data sets. Since public benchmarks are less abundant compared to segmentation tasks, we identified object detection tasks in 5 data sets of existing segmentation challenges (where we focus on detecting tumors and consider organs as background, see supplementary material for details). To generate object annotations from pixel-level label maps, we performed connected component analysis and discarded all objects with a diameter less than $3mm$. Further, object annotations originating from obvious segmentation errors were manually removed (see supplementary material). Reflecting clinical relevance regarding coarse localisation on medical images and the absence of overlapping objects in 3D images, we report mean Average Precision (mAP) at an IoU threshold of 0.1 [9].

Data Sets. An overview of all data sets and their properties can be found in the supplementary material. Out of the 13 data sets, we used 10 for development and validation of nnDetection. These are further divided into a training pool (4 data sets: CADA [21], LIDC-IDRI [1,9], RibFrac [10] and Kits19 [6].) and validation pool (6 data sets: ProstateX [3,12], ADAM [22], Medical Segmentation Decathlon Liver, Pancreas, Hepatic Vessel and Colon [19]). While in the training pool we used all data for development and report 5-fold cross-validation results, in the validation pool roughly 40% of each data set was split off as held-out test set before development. These test splits were only processed upon final evaluation (for ADAM we used the public leaderboard as hold-out test split). The test pool consists of 3 additional data sets (LUNA16 [18], and TCIA Lymph-Node [16,17]) that were entirely held-out from method development to evaluate the generalization ability of our automated method configuration.

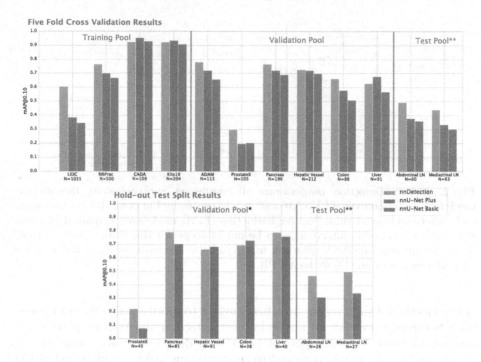

Fig. 3. Large-scale benchmark against nnU-Net on 12 data sets (cross-validation results on the top and test split results on the bottom panel). *The test split result of ADAM is represented by our submission to the live leaderboard and can be found in the supplementary material). **LUNA16 results are visualized in Fig. 2. Numerical values for all experiments can be found at https://github.com/MIC-DKFZ/nnDetection.

Compared Methods. While there exist reference scores in literature for the well-known LUNA16 benchmark and ADAM provides an open leaderboard, there exists no standardized evaluation protocol for object detection methods on the remaining 11 data sets. Thus, we initiate a new benchmark by comparing nnDetection against nnU-Net that we modified to serve as a standardized baseline for object detection (see Sect. 2). This comparison is relevant for three reasons: 1) Segmentation methods are often modified to be deployed on detection tasks in the medical domain [22]. 2) nnU-Net is currently the only available method that can be readily deployed on a large number of data sets without manual adaptation. 3) The need for tackling medical object detection task with dedicated detection methods rather than segmentation-based substitutes has only been studied on two medical data sets before [9], thus providing large-scale evidence for this comparison is scientifically relevant.

Public Leaderboard Results. LUNA16 [18] is a long standing benchmark for object detection methods [2,4,5,11,13,20,23,25] which consists of 888 CT scans with lung nodule annotations. While LUNA16 images represent a subset of LIDC-IDRI, the task is different since LUNA16 does not differentiate between benign and malignant classes and the annotations differ in that they were reduced to a center point plus radius (for training we generated segmentation labels in the form of circles from this information). As LUNA16 is part of our test pool, nnDetection was applied by simply executing automated method configuration once and without any manual intervention. Our method achieves a Competition Performance Metric (CPM) of 0.930 outperforming all previous methods on the nodule-candidate-detection task (see Fig. 2 and supplementary material for details). Our submission to the public leaderboard of the Aneurysm Detection And segMentation (ADAM) [22] challenge currently ranks third with a sensitivity of 0.64 at a false positive count of 0.3 (see supplementary material for more details). One of the two higher ranking submissions is a previous version of nnDetection, which hints upon a natural performance spread on limited test sets in object detection tasks (the previous version represented our original submission to the 2020 MICCAI event, there were only these two submissions to ADAM from our side in total).

Large-scale Comparison Against nnU-Net. nnDetection outperforms the enhanced baseline *nnU-Net Plus* on 9 out of 12 data sets in the cross-validation protocol (Fig. 3 top panel). Thereby, substantial margins ($> 5\%$) are observed in 7 data sets and substantially lower performance only on the liver data set [19]. The baseline with fixed postprocessing strategy (*nnU-Net Basic*) shows worse results than nnDetection on 11 out of 12 data sets. On the hold-out test splits (Fig. 3 bottom panel), nnDetection outperforms *nnU-Net Plus* on 5 out of 7 data sets with substantial margins in 4 of them and only substantially lower performance on the colon data set [19]. Notably, 4 of the 7 data sets were part of nnU-Net's development pool and thus not true hold-out splits for the baseline algorithm [8]. High volatility of evaluation metrics between cross-validation and test results is observed, especially on the liver and colon data sets, hinting upon the importance of evaluation across many diverse data sets.

4   Discussion

nnDetection opens a new perspective on method development in medical object detection. All design choices have been optimized on a data set-agnostic meta-level, which allows for out-of-the-box adaptation to specific data sets upon application and removes the burden of manual and iterative method configuration. Despite this generic functionality, nnDetection shows performance superior to or on par with the state-of-the-art on two public leaderboards and 11 benchmarks that were newly proposed for object detection. Our method can be considered as a starting point for further manual task-specific optimization. As seen on LUNA16, an additional false-positive-reduction component can further improve results. Also, data-driven optimization along the lines of AutoML [7] could be computationally feasible for specific components of the object detection pipeline and thus improve results even further.

In making nnDetection available including models and object annotations for all newly proposed benchmarks we hope to contribute to the rising interest in object detection on medical images by providing a tool for out-of-the-box object predictions, a framework for method development, a standardized baseline to compare against, as well as a benchmark for large-scale method evaluation.

Acknowledgements. Part of this work was funded by the Deutsche Forschungsgemeinschaft (DFG, German Research Foundation) – 410981386 and the Helmholtz Imaging Platform (HIP), a platform of the Helmholtz Incubator on Information and Data Science.

References

1. Armato, S.G., III., et al.: The lung image database consortium (lidc) and image database resource initiative (idri): a completed reference database of lung nodules on ct scans. Medical physics **38**(2), 915–931 (2011)
2. Cao, H., et al.: A two-stage convolutional neural networks for lung nodule detection. IEEE J. Biomed. Health Inf. **24**(7), 2006–2015 (2020)
3. Cuocolo, R. et al.: Deep learning whole-gland and zonal prostate segmentation on a public MRI dataset. J. Magn. Reson. Imaging **54**(2), 452–459 (2021). https://doi.org/10.1002/jmri.27585. PMID: 33634932
4. Ding, J., Li, A., Hu, Z., Wang, L.: Accurate pulmonary nodule detection in computed tomography images using deep convolutional neural networks. In: Descoteaux, M., Maier-Hein, L., Franz, A., Jannin, P., Collins, D.L., Duchesne, S. (eds.) MICCAI 2017. LNCS, vol. 10435, pp. 559–567. Springer, Cham (2017). https://doi.org/10.1007/978-3-319-66179-7_64
5. Dou, Q., Chen, H., Jin, Y., Lin, H., Qin, J., Heng, P.-A.: Automated pulmonary nodule detection via 3D ConvNets with online sample filtering and hybrid-loss residual learning. In: Descoteaux, M., Maier-Hein, L., Franz, A., Jannin, P., Collins, D.L., Duchesne, S. (eds.) MICCAI 2017. LNCS, vol. 10435, pp. 630–638. Springer, Cham (2017). https://doi.org/10.1007/978-3-319-66179-7_72
6. Heller, N., et al.: The kits19 challenge data: 300 kidney tumor cases with clinical context, ct semantic segmentations, and surgical outcomes. arXiv preprint arXiv:1904.00445 (2019)
7. Hutter, F., Kotthoff, L., Vanschoren, J.: Automated Machine Learning: Methods, Systems, Challenges. Springer Nature, Cham (2019) https://doi.org/10.1007/978-3-030-05318-5
8. Isensee, F., Jaeger, P.F., Kohl, S.A., Petersen, J., Maier-Hein, K.H.: nnU-net: a self-configuring method for deep learning-based biomedical image segmentation. Nat. Methods **18**(2), 203–211 (2021)
9. Jaeger, P.F. et al.: Retina u-net: Embarrassingly simple exploitation of segmentation supervision for medical object detection. In: ML4H, pp. 171–183. PMLR (2020)
10. Jin, J., et al.: Deep-learning-assisted detection and segmentation of rib fractures from CT scans: Development and validation of FracNet. 62. Publisher: Elsevier (2020)
11. Khosravan, N., Bagci, U.: *S4ND*: single-shot single-scale lung nodule detection. In: Frangi, A.F., Schnabel, J.A., Davatzikos, C., Alberola-López, C., Fichtinger, G. (eds.) MICCAI 2018. LNCS, vol. 11071, pp. 794–802. Springer, Cham (2018). https://doi.org/10.1007/978-3-030-00934-2_88

12. Litjens, G., Debats, O., Barentsz, J., Karssemeijer, N., Huisman, H.: Computer-aided detection of prostate cancer in mri. IEEE TMI **33**(5), 1083–1092 (2014)
13. Liu, J., Cao, L., Akin, O., Tian, Y.: 3DFPN-HS^2: 3D feature pyramid network based high sensitivity and specificity pulmonary nodule detection. In: Shen, D., et al. (eds.) MICCAI 2019. LNCS, vol. 11769, pp. 513–521. Springer, Cham (2019). https://doi.org/10.1007/978-3-030-32226-7_57
14. Maier-Hein, L.: Why rankings of biomedical image analysis competitions should be interpreted with care. Nat. Commun. **9**(1), 5217 (2018). Number: 1 Publisher: Nature Publishing Group
15. Redmon, J., Farhadi, A.: Yolo9000: better, faster, stronger. In: CVPR, pp. 7263–7271 (2017)
16. Roth, H.R., et al.: A new 2.5D representation for lymph node detection using random sets of deep convolutional neural network observations. In: Golland, P., Hata, N., Barillot, C., Hornegger, J., Howe, R. (eds.) MICCAI 2014. LNCS, vol. 8673, pp. 520–527. Springer, Cham (2014). https://doi.org/10.1007/978-3-319-10404-1_65
17. Seff, A., Lu, L., Barbu, A., Roth, H., Shin, H.-C., Summers, R.M.: Leveraging mid-level semantic boundary cues for automated lymph node detection. In: Navab, N., Hornegger, J., Wells, W.M., Frangi, A.F. (eds.) MICCAI 2015. LNCS, vol. 9350, pp. 53–61. Springer, Cham (2015). https://doi.org/10.1007/978-3-319-24571-3_7
18. Setio, A.A.A., et al.: Validation, comparison, and combination of algorithms for automatic detection of pulmonary nodules in computed tomography images: the luna16 challenge. MedIA **42**, 1–13 (2017)
19. Simpson, A.L., et al.: A large annotated medical image dataset for the development and evaluation of segmentation algorithms. arXiv preprint arXiv:1902.09063 (2019)
20. Song, T., et al.: CPM-Net: a 3D center-points matching network for pulmonary nodule detection in CT scans. In: Martel, A.L., et al. (eds.) MICCAI 2020. LNCS, vol. 12266, pp. 550–559. Springer, Cham (2020). https://doi.org/10.1007/978-3-030-59725-2_53
21. Tabea Kossen, C., et al.: Cerebral aneurysm detection and analysis (March 2020)
22. Timmins, K., Bennink, E., van der Schaaf, I., Velthuis, B., Ruigrok, Y., Kuijf, H..: Intracranial Aneurysm Detection and Segmentation Challenge (2020)
23. Wang, B., Qi, G., Tang, S., Zhang, L., Deng, L., Zhang, Y.: Automated pulmonary nodule detection: high sensitivity with few candidates. In: Frangi, A.F., Schnabel, J.A., Davatzikos, C., Alberola-López, C., Fichtinger, G. (eds.) MICCAI 2018. LNCS, vol. 11071, pp. 759–767. Springer, Cham (2018). https://doi.org/10.1007/978-3-030-00934-2_84
24. Zhang, S., Chi, C., Yao, Y., Lei, Z., Li, S.Z.: Bridging the gap between anchor-based and anchor-free detection via adaptive training sample selection. In: CVPR, pp. 9759–9768 (2020)
25. Zhu, W., Liu, C., Fan, W., Xie, X.: Deeplung: deep 3d dual path nets for automated pulmonary nodule detection and classification. In: WACV, pp. 673–681. IEEE (2018)
26. Zlocha, M., Dou, Q., Glocker, B.: Improving RetinaNet for CT lesion detection with dense masks from weak RECIST labels. In: Shen, D., et al. (eds.) MICCAI 2019. LNCS, vol. 11769, pp. 402–410. Springer, Cham (2019). https://doi.org/10.1007/978-3-030-32226-7_45

Automating Embryo Development Stage Detection in Time-Lapse Imaging with Synergic Loss and Temporal Learning

Lisette Lockhart[1(✉)], Parvaneh Saeedi[1], Jason Au[2], and Jon Havelock[2]

[1] Simon Fraser University, Burnaby, Canada
{llockhar,psaeedi}@sfu.ca
[2] Pacific Centre for Reproductive Medicine, Burnaby, Canada
{jau,jhavelock}@pacificfertility.ca

Abstract. In Vitro Fertilization (IVF) treatment is increasingly chosen by couples suffering from infertility as a means to conceive. Time-lapse imaging technology has enabled continuous monitoring of embryos in vitro and time-based development metrics for assessing embryo quality prior to transfer. Timing at which embryos reach certain development stages provides valuable information about their potential to become a positive pregnancy. Automating development stage detection of day 4–5 embryos remains difficult due to small variation between stages. In this paper, a classifier is trained to detect embryo development stage with learning strategies added to explicitly address challenges of this task. Synergic loss encourages the network to recognize and utilize stage similarities between different embryos. Short-range temporal learning incorporates chronological order to embryo sequence predictions. Image and sequence augmentations complement both approaches to increase generalization to unseen sequences. The proposed approach was applied to human embryo sequences with labeled morula and blastocyst stage onsets. Morula and blastocyst stage classification was improved by 5.71% and 1.11%, respectively, while morula and blastocyst stage mean absolute onset error was reduced by 19.1% and 8.7%, respectively. Code is available: https://github.com/llockhar/Embryo-Stage-Onset-Detection.

Keywords: Image classification · Embryology · Time-lapse imaging

1 Introduction

With infertility rates rising, more than 100,000 embryo transfers for IVF treatment are performed annually in the US [1]. A crucial component is selecting the highest quality embryo from each batch for implantation. Onset of certain embryo development stages has shown to be correlated with blastocyst quality and likelihood of implantation, including morula [2–4] and blastocyst [4,5] stages. Onset and duration of embryonic 2–5 cell stages have also shown to provide valuable insight into embryo quality and implantation potential [5–8].

© Springer Nature Switzerland AG 2021
M. de Bruijne et al. (Eds.): MICCAI 2021, LNCS 12905, pp. 540–549, 2021.
https://doi.org/10.1007/978-3-030-87240-3_52

Predicting onset of morula stage is challenging because cell adhesion of an embryo throughout morula stage often varies, reverting in appearance to cleavage stage. Similarly, blastocyst stage embryos can contract such that they appear more like a cleavage or morula stage embryos. Difference in appearance between frames is small at morula and blastocyst stage onsets, as seen in Fig. 1. This work uses deep convolutional neural networks (CNNs) to classify embryo frames as cleavage, morula, or blastocyst stage and compute the morula and blastocyst stage onset in embryo sequences. Automating these tasks can assist embryologists to select the highest quality embryo(s) for implantation by removing the need for manual stage onset detection.

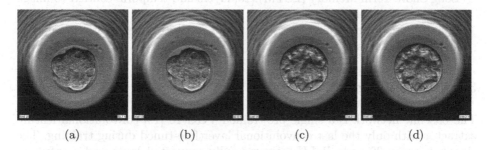

(a)   (b)   (c)   (d)

Fig. 1. Frames directly before (a) and after (b) morula stage onset, and before (c) and after (d) blastocyst stage onset. Differences are small between consecutive frames at stage onsets.

2   Related Work

Embryo staging of embryonic cleavage times and morula through expanded blastocyst stage onset was performed using image processing techniques [9]. A fine-tuned CNN classifier took as input multiple focal planes of each time-lapse imaging frame to classify embryos as 1–5 cell, morula, or blastocyst stage [10].

Embryonic cell stage was first classified using a CNN in dark-field microscopy image sequences [11]. More recently, embryonic cell stage has been classified in HMC microscopy time-lapse sequences. Temporal fusion combined extracted features of multiple frames for single-frame cell stage classification [12]. Multiple frames were used as input for multi-frame cell stage classification with temporal ensembling [13]. Confidence-based majority voting between two single-frame CNN cell stage classifiers was used in [14]. Embryos were localized with hand-crafted feature classification then used as input for cell stage classification [15]. A two-stage network was trained for simultaneous embryo localization and embryonic cell stage classification [16]. Most recently, a global optimization method was used to restructure embryo cell stage predictions [17].

CNN-based embryo stage classification methods analyze image frames individually then add dynamic decoding to incorporate temporal information when restructuring sequence predictions [10,12,13,16,17].

While these works show promise for automating embryonic cell stage classification with deep neural networks, the subtle differences between morula and blastocyst embryos makes predicting these later stage onsets difficult. In this work, we propose a method to improve embryo stage classification and morula and blastocyst stage onset prediction using the following techniques:

- Synergic loss provides additional feedback to learning of embryo-independent stage similarities,
- Long Short-Term Memory (LSTM) [18] layers add temporal context to image stage classification, and
- Image and sequence training augmentations promote generalization to unseen embryo sequences.

3 Methodology

The baseline network is a single pre-trained VGG-16 [19] convolutional feature extractor with only the last convolutional layer fine-tuned during training. The classifier was a 32-node ReLU-activated fully-connected layer and a softmax-activated output layer. This network architecture was chosen to reduce overfitting to the small training dataset with minor inter-class variation.

3.1 Synergic Network

Embryos vary considerably throughout each stage but vary little near stage onsets. Encoded convolutional features from frames throughout a stage or from different embryo sequences could differ significantly. Synergic loss was incorporated to explicitly encourage the network to learn embryo-independent features similar between frames within each stage. This pairwise-learning was implemented using two identical baseline networks with unshared weights [20].

An additional classifier concatenated nodes from the 2nd last (fully-connected) layer of each branch. This tensor was fed through a 32-node fully-connected layer followed by a single-node sigmoid-activated output layer. Stage similarity was encoded as 1 for input pairs of frames at the same embryo stage and 0 otherwise. Synergic loss was implemented as binary cross-entropy between actual (y_s) and predicted ($\hat{y}_s$) stage similarity,

$$\mathcal{L}(y_s, \hat{y}_s) = -y_s log(\hat{y}_s) + (1 - y_s) log(1 - \hat{y}_s). \tag{1}$$

For inference, staging predictions from the two network branches were averaged.

Despite the relatively small number of trainable parameters, the network continued to overfit to the training set. Since embryos show little change before and after stage onset, many image augmentations could change the stage label. To challenge the network during training without adding unwanted label noise, mixup augmentation [21] was used to blend image samples and smooth the corresponding ground truth labels by the same factor. A portion of images and labels from one synergic network branch were blended with a random selection from those fed to the other branch.

3.2   Temporal Learning

As mentioned, an embryo's appearance while progressing through morula and blastocyst stages can revert to that of a previous stage. Instead of sampling random images in each batch, images were analyzed in sequences to incorporate short-range temporal dependency during training. An LSTM layer was added after each synergic CNN branch output layer. Classification loss (categorical cross-entropy) was measured between the actual (y_t) and predicted $(\hat{y}_y)$ stages and backpropagated both before and after the LSTM layer for each frame.

$$\mathcal{L}(y_t, \hat{y}_t) = -y_t log(\hat{y}_t) \tag{2}$$

The convolutional feature extractors, stage classifiers, and LSTM layers were trained together in an end-to-end manner. The starting index for each full embryo sequence was randomly chosen (between 0 and batch size) every epoch to sample different batches of frames. The final network diagram in Fig. 2 shows how embryo stage is predicted for each image individually with the CNN-based classifier and is refined using sequence temporal context in the LSTM layer.

Since each stage onset was only sampled once per sequence, only 2 batches per sequence contained stage transitions. To train on more complex sequences containing transitions, extra sequence batches were sampled near stage onsets. Four extra batches were added per stage onset (sequence length permitting).

3.3   Pre- and Post-processing

Image pre-processing was performed by cropping to a circle centered around the centroid of Canny-detected edges. Circle radius was set as constant instead of matching the embryo boundary so embryo size variation was noticeable. Since embryo size changes during development, it can help for staging classification.

Embryo stage is clinically defined to be monotonic non-decreasing across a time-lapse sequence. The network output stage predictions can oscillate between stages and are therefore restructured to enforce monotonic non-decreasing order. Each sequence is restructured by selecting the monotonic non-decreasing series of stage classifications that has lowest error with the network output stage classification predictions. Negative label likelihood (NLL) loss [12,13] or mean absolute error (MAE, Global) [17] was computed at each frame and summed across the sequence. For a sequence of length N, monotonic non-decreasing stage classifications were formed from morula i and blastocyst j stage onset frames: $\forall(i \in [2, N-1], j \in [3, N]|j > i)$. Stage onset was then selected from the minimum restructured sequence index at which that stage was predicted.

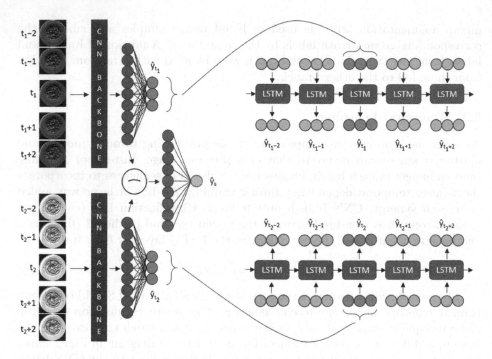

Fig. 2. Proposed network diagram for embryo staging. Two image sequence batches are fed in parallel through separate convolutional feature extractors and then classified into stages. Staging predictions are refined with an LSTM layer. Classification error (2) is computed at both LSTM layer input and output. Fully-connected layers from each classifier are concatenated and used to predict whether the input image fed through each branch are at the same stage. Synergic loss (1) is backpropagated through both classifier branches.

4   Experiments

4.1   Dataset Overview

Experiments were performed on a dataset of 117 human embryo time-lapse imaging sequences of days 1–5 in vitro collected at the Pacific Centre for Reproductive Medicine from 2017–2019. Image frames were acquired every 15 min at a single focal plane. Morula and blastocyst stage onset times were annotated by an embryologist. All frames following the morula and blastocyst stage onset were labeled as morula and blastocyst stage for classification, respectively. The dataset breakdown by stage is summarized in Table 1.

In all images, pixels corresponding to embryo well and time stamp located along the bottom rows were removed (set to zero).

Table 1. Overview of embryo sequences stage duration and onset

Stage	Total frames	Mean frames per sequence ± SD	Mean onset frame ± SD
Cleavage	32,526 (71.9%)	278.00 ± 28.82	–
Morula	8,761 (19.4%)	74.88 ± 19.71	279.00 ± 28.82
Blastocyst	3,922 (8.7%)	33.52 ± 21.89	353.88 ± 24.24

4.2 Implementation Details

Images were resized to 320×320 pixels. Standard images augmentations were used including horizontal and vertical flipping, rotation up to 360°, horizontal and vertical translation by up to 10% of the image height/width. Network training used Adam optimizer with initial learning rate chosen empirically as 1×10^{-4} for networks containing LSTM layers and 3×10^{-5} otherwise. The learning rate was reduced on plateau of 8 epochs by a factor of 0.9 and early stopping was applied after validation loss had no longer decreased for 15 epochs.

Image and label blending with mixup augmentation used scaling factor sampled from beta distribution with $\alpha = 0.2$ for each instance. A batch consisted of two 32-frame image sequences (one for each synergic branch) for LSTM networks or 32 random frames otherwise. Image sequences for LSTM network batches were sampled by iterating through embryo sequences with stride 32. When embryo sequence length was not divisible by 32, another batch was sampled containing the final $\{N-32, N-31, ..., N\}$ frames of that sequence. For onset oversampling, the 1st extra sequence batch starting index SI for stage s was randomly chosen as $SI_s \in [24, 32)$ frames before stage onset SO_s. The extra batch indices were $\{SO_s - SI_s + 8 \times k_s, SO_s - SI_s + 8 \times k_s + 1, ..., SO_s - SI_s + 8 \times k_s + 31\}$, where extra batch $k_s = \{0, 1, 2, 3\}$. In image pre-processing, circle radius was set to 190 pixels, which contained the entire embryo.

4.3 Experimental Results

Since no publicly available dataset exists for this task, embryo sequences were randomly partitioned into training, validation, and test sets using ratio 70/15/15%, respectively. To reduce performance bias in dataset splits, 5-fold cross-validation was performed by rolling the sequences in each set. Results presented were averaged across the folds.

Results of the baseline VGG-16 and fully-connected classifier and addition of synergic learning, mixup augmentation, LSTM layer, onset oversampling, and image pre-processing are presented. Classification performance was measured with precision, recall, and F1-Score for each stage in Table 2. Stage onset performance was measured as the mean absolute error in frames between predicted and actual stage onset across test set sequences in Table 3.

Table 2. Stage classification performance

Syn. loss	Mixup aug.	LSTM	Onset overspl.	Image Pre-proc.	Stage	Precision	Recall	F1-Score
		Baseline			Cleavage	0.969	0.956	0.963
					Morula	0.812	0.854	0.832
					Blastocyst	0.921	0.914	0.917
✓					Cleavage	0.968	0.961	0.965
					Morula	0.831	0.843	0.837
					Blastocyst	0.913	**0.934**	0.924
✓	✓				Cleavage	0.970	0.959	0.965
					Morula	0.825	0.866	0.845
					Blastocyst	0.929	0.924	0.926
✓	✓	✓			Cleavage	0.970	**0.981**	**0.976**
					Morula	**0.879**	0.868	0.873
					Blastocyst	**0.949**	0.884	0.915
✓	✓	✓	✓		Cleavage	0.977	0.970	0.974
					Morula	0.857	**0.896**	0.876
					Blastocyst	0.946	0.905	0.925
✓	✓	✓	✓	✓	Cleavage	**0.979**	0.973	**0.976**
					Morula	0.871	0.888	**0.880**
					Blastocyst	0.926	0.929	**0.927**

5   Discussion

The single CNN baseline is similar to the "Single Focus" method used in [10] with VGG16 backbone instead of ResNeXt101 (which was found empirically to be a stronger baseline for this staging task and dataset).

Synergic loss allowed the network to make use of stage labels and similarity between images to improve staging and morula onset error. Averaging predictions from the two synergic network branches slightly improved the results over each branch individually.

Table 3. Mean absolute stage onset error with restructured predictions

Syn. loss	Mixup aug.	LSTM	Onset overspl.	Image pre-proc.	Stage	Unproc.	NLL	MAE
		Baseline			Morula	46.32	13.96	13.64
					Blastocyst	5.21	5.17	5.17
✓					Morula	39.86	12.37	12.18
					Blastocyst	5.76	4.77	4.77
✓	✓				Morula	41.51	12.60	12.50
					Blastocyst	4.76	**4.32**	**4.32**
✓	✓	✓			Morula	20.23	11.98	11.98
					Blastocyst	5.67	5.12	5.12
✓	✓	✓	✓		Morula	24.40	11.94	11.94
					Blastocyst	5.04	4.47	4.47
✓	✓	✓	✓	✓	Morula	20.16	**11.04**	**11.04**
					Blastocyst	5.26	4.77	4.72

Adding mixup augmentation had largest improvement on blastocyst stage onset. Since blastocyst stage classification relies on size of embryo structures rather than texture of cells, blending larger blastocyst stage embryos with smaller cleavage or morula stage embryos could provide better sample variation.

Incorporating short-range temporal context with LSTM layer significantly improved morula stage classification by considering embryo stage at previous frames. While increasing sequence length may have improved performance, GPU size was a limiting factor. Since there were so few batches containing stage onsets, many blastocyst stage transitions were missed. Adding stage onset oversampling gave the network more stage transitions to train on and compensated for this blastocyst stage performance drop.

Image pre-processing ensured maximal embryo area was visible as network input, reducing pixel information loss during image resizing. Subtle cell adhesion (texture) differences could be better extracted for predicting morula stage.

Unprocessed sequences contained many incorrect morula stage predictions in frames far from the actual onset. Both temporal post-processing strategies significantly improved morula stage onset. Global MAE occasionally optimized to slightly better stage onsets, though both losses gave nearly identical solutions.

NLL loss and Global MAE are optimal for prediction restructuring if the number of mis-classifications before and after the actual onset are relatively balanced and could cancel out. In our experiments, the majority of incorrect stage predictions often occurred before or after the actual onset. Hence, higher classification performance did not guarantee lower stage onset error after restructuring.

6 Conclusions

Embryo stage classification and embryo morula and blastocyst stage onset prediction were improved by incorporating synergic and temporal learning. Image and sequence augmentations further improved each method by reducing overfitting to the training set. The proposed approach can more accurately predict morula and blastocyst stage onset to assist embryologists in their workflow of choosing high quality embryos for transfer in IVF treatment.

References

1. National Center for Chronic Disease Prevention and Health Promotion: Division of Reproductive Health: 2018 Assisted Reproductive Technology National Summary Report. https://www.cdc.gov/art/reports/2018/fertility-clinic.html. Accessed 9 Feb 2021
2. Rienzi, L., et al.: Time of morulation and trophectoderm quality are predictors of a live birth after euploid blastocyst transfer: a multicenter study. Fertil. Steril. 112(6), 1080–1093 (2019)
3. Motato, Y., de los Santos, M.J., Escriba, M.J., Ruiz, B.A., Remohí, J., Meseguer, M.: Morphokinetic analysis and embryonic prediction for blastocyst formation through an integrated time-lapse system. Fertil. Steril. 105(2), 376–384 (2016)

4. Desai, N., Ploskonka, S., Goodman, L.R., Austin, C., Goldberg, J., Falcone, T.: Analysis of embryo morphokinetics, multinucleation and cleavage anomalies using continuous time-lapse monitoring in blastocyst transfer cycles. Reprod. Biol. Endocrinol. **12**(1), 1–10 (2014)
5. Jacobs, C., et al.: Correlation between morphokinetic parameters and standard morphological assessment: what can we predict from early embryo development? a time-lapse-based experiment with 2085 blastocysts. JBRA Assist. Reprod. **24**(3), 273 (2020)
6. Basile, N., et al.: The use of morphokinetics as a predictor of implantation: A multicentric study to define and validate an algorithm for embryo selection. Hum. Reprod. **30**(2), 276–283 (2014)
7. Meseguer, M., Herrero, J., Tejera, A., Hilligsøe, K.M., Ramsing, N.B., Remohí, J.: The use of morphokinetics as a predictor of embryo implantation. Hum. Reprod. **26**(10), 2658–2671 (2011)
8. Cruz, M., Garrido, N., Herrero, J., Pérez-Cano, I., Muñoz, M., Meseguer, M.: Timing of cell division in human cleavage-stage embryos is linked with blastocyst formation and quality. Reprod. Biomed. Online **25**(4), 371–381 (2012)
9. Feyeux, M., et al.: Development of automated annotation software for human embryo morphokinetics. Hum. Reprod. **35**(3), 557–564 (2020)
10. Leahy, B.D., et al.: Automated measurements of key morphological features of human embryos for IVF. In: Martel, A.L., et al. (eds.) MICCAI 2020. LNCS, vol. 12265, pp. 25–35. Springer, Cham (2020). https://doi.org/10.1007/978-3-030-59722-1_3
11. Khan, A., Gould, S., Salzmann, M.: Deep convolutional neural networks for human embryonic cell counting. In: Hua, G., Jégou, H. (eds.) ECCV 2016. LNCS, vol. 9913, pp. 339–348. Springer, Cham (2016). https://doi.org/10.1007/978-3-319-46604-0_25
12. Ng, N.H., McAuley, J.J., Gingold, J., Desai, N., Lipton, Z.C.: Predicting embryo morphokinetics in videos with late fusion nets & dynamic decoders. In: International Conference on Learning Representations (Workshop) (2018)
13. Liu, Z., et al.: Multi-task deep learning with dynamic programming for embryo early development stage classification from time-lapse videos. IEEE Access **7**, 122153–122163 (2019)
14. Dirvanauskas, D., Maskeliunas, R., Raudonis, V., Damasevicius, R.: Embryo development stage prediction algorithm for automated time lapse incubators. Comput. Methods Programs Biomed. **177**, 161–174 (2019)
15. Raudonis, V., Paulauskaite-Taraseviciene, A., Sutiene, K., Jonaitis, D.: Towards the automation of early-stage human embryo development detection. Biomed. Eng. Online **18**(1), 1–20 (2019)
16. Lau, T., Ng, N., Gingold, J., Desai, N., McAuley, J. and Lipton, Z.C.: Embryo staging with weakly-supervised region selection and dynamically-decoded predictions. In: Machine Learning for Healthcare Conference, pp. 663–679. PMLR (2019)
17. Malmsten, J., Zaninovic, N., Zhan, Q., Rosenwaks, Z., Shan, J.: Automated cell division classification in early mouse and human embryos using convolutional neural networks. Neural Comput. Appl. **33**(7), 2217–2228 (2020). https://doi.org/10.1007/s00521-020-05127-8
18. Hochreiter, S., Schmidhuber, J.: Long short-term memory. Neural Comput. **7**, 1735–1780 (1997)

19. Simonyan, K. and Zisserman, A.: Very deep convolutional networks for large-scale image recognition. arXiv preprint arXiv:1409.1556, (2014)
20. Zhang, J., Xie, Y., Wu, Q., Xia, Y.: Medical image classification using synergic deep learning. Med. Image Anal. **54**, 10–19 (2019)
21. Zhang, H., Cisse, M., Dauphin, Y.N., Lopez-Paz, D.: mixup: Beyond empirical risk minimization. In: International Conference on Learning Representations (2018)

Deep Neural Dynamic Bayesian Networks Applied to EEG Sleep Spindles Modeling

Carlos A. Loza$^{(\boxtimes)}$ (ID) and Laura L. Colgin (ID)

Department of Neuroscience, Center for Learning and Memory,
The University of Texas at Austin, Austin, USA
carlos.loza@utexas.edu, colgin@mail.clm.utexas.edu

Abstract. We propose a generative model for single–channel EEG that incorporates the constraints experts actively enforce during visual scoring. The framework takes the form of a dynamic Bayesian network with depth in both the latent variables and the observation likelihoods—while the hidden variables control the durations, state transitions, and robustness, the observation architectures parameterize Normal–Gamma distributions. The resulting model allows for time series segmentation into local, reoccurring dynamical regimes by exploiting probabilistic models and deep learning. Unlike typical detectors, our model takes the raw data (up to resampling) without pre–processing (e.g., filtering, windowing, thresholding) or post–processing (e.g., event merging). This not only makes the model appealing to real–time applications, but it also yields interpretable hyperparameters that are analogous to known clinical criteria. We derive algorithms for exact, tractable inference as a special case of Generalized Expectation Maximization via dynamic programming and backpropagation. We validate the model on three public datasets and provide support that more complex models are able to surpass state–of–the–art detectors while being transparent, auditable, and generalizable.

1 Introduction

Sleep spindles are a hallmark of stage 2 non–REM sleep. They are the result of interactions between GABAergic reticular neurons and excitatory thalamic cells [30]. Their proposed functions include memory consolidation [29], cortical development, and potential biomarkers for psychiatric disorders [10]. Hence, proper detection and modeling using electroencephalogram (EEG) are crucial.

A sleep spindle is defined as an oscillatory burst in the range $\sim 11 - 15$ Hz (sigma band) with duration between 0.5 and 2 s., waxing–waning envelope,

Supported by NSF CAREER grant 1453756. All code and data can be accessed at https://github.com/carlosloza/DNDBN.

Electronic supplementary material The online version of this chapter (https://doi.org/10.1007/978-3-030-87240-3_53) contains supplementary material, which is available to authorized users.

M. de Bruijne et al. (Eds.): MICCAI 2021, LNCS 12905, pp. 550–560, 2021.
https://doi.org/10.1007/978-3-030-87240-3_53

and maximal in amplitude in central EEG electrodes. EEGers usually adhere to clinical manuals [13,22,28] to visually categorize multi–channel EEG traces into sleep stages. They are also trained to identify non–brain–related artifacts (e.g., eye and muscle activity). Currently, the large amount of data calls for machine learning techniques to guide principled, human–like automatic EEG scorers.

Most automatic sleep spindles detectors comprise four stages: pre–processing, decomposition, decision making, and feature extraction [1]. The first stage usually involves bandpassing and artifact rejection, while the second stage applies wavelets or other windowing techniques for quasi–stationary processes. Decision making takes the form of hard–thresholding plus cross–validation, and lastly, feature extraction characterizes the events for further analysis. A myriad of approaches use slight variations of this pipeline [4,20,24,34] (reviews in [1,33]).

Even though this methodology provides adequate results, it suffers from three main drawbacks: i) analysis based on filtered EEG, ii) lack of theoretical foundations for hyperparameter settings, and iii) reported results are often in–sample predictions, which is prone to overfitting. The first point not only differs from human–like scoring, it is also computationally expensive for online applications. The latter points are consequences of a model–less framework; that is, amplitude thresholds are data (and scorer) dependent, which severely limits generalization.

Instead, we propose a generative model with the following components: i) robustness against artifacts via observations modeled as conditional heteroscedastic non–linear generalized t likelihoods, ii) high–capacity deep learning architectures that parameterize such likelihoods, iii) reoccurring modes with distinctive dynamics that characterize non–spindle and spindle regimes, iv) semi–Markovian states that generalize the geometric regime durations of Hidden Markov Models (HMM), v) tractable exact inference via message passing routines (unlike variational approximations [7,8,14,18]), and vi) uncertainty quantification via posterior probabilities. The result is a model that combines probabilistic modeling with deep learning in an effort to mimic the constraints EEGers actively enforce when visually scoring EEG: a Deep Neural Dynamic Bayesian Network.

We validate the model on three public datasets and achieve comparable or better performance than the state–of–the–art. The results open the door to more complex models and sophisticated inference in the future (e.g., real–time detection). The paper continues as follows: Sect. 2 details the model, Sect. 3 presents the results and some discussion, and Sect. 4 concludes the paper.

2   Generative Model for Single–channel EEG

Let $\mathbf{y} = \{y_n\}_{n=1}^{N}$ be an ordered collection of N univariate random variables over time. For our case, $\mathbf{y}$ is a single–channel, single–trial EEG trace embedded into a dynamical system with K discrete hidden states, labels, or modes (non–spindle and spindle). Such states are represented by K–dimensional multinomials $\mathbf{z_n}$, while their corresponding durations are kept track via $\mathbf{d_n}$, D–dimensional multinomials that act as counter variables. Given a mode k, we pose each y_n as a

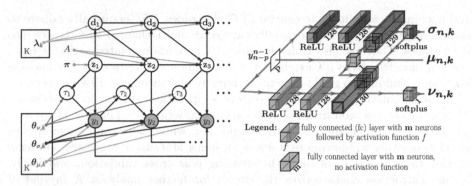

Fig. 1. Left. Generative model for EEG. Observations are shaded. **Right.** Deep networks for Normal–Gamma models (colors paired with graphical model).

Normal random variable with location, $\mu_{n,k}$, and scale, $\sigma_{n,k}$, both parameterized by functions (linear or non–linear) of the previous p samples, y_{n-p}^{n-1}, where p is the autoregressive order. To account for robustness, $\sigma_{n,k}$ is also modulated by a conditionally Gamma distributed hidden variable, τ_n. In similar fashion, $\nu_{n,k}$, the main parameter of said Gamma distribution (i.e. the degrees of freedom), is a function of y_{n-p}^{n-1}. In short, given a state k, τ_n is marginally Gamma and y_n is conditionally Normal given τ_n; this results in y_n being marginally generalized t:

$$\tau_n | \mathbf{z_n}, y_{n-p}^{n-1}; \{\boldsymbol{\theta}_{\nu,k}\}_{k=1}^{K} \sim \mathrm{Gamma}\left(\frac{\nu_{n,k}}{2}, \frac{\nu_{n,k}}{2}\right) \qquad (1)$$

$$y_n | \tau_n, \mathbf{z_n}, y_{n-p}^{n-1}; \{\boldsymbol{\theta}_{\mu,k}\}_{k=1}^{K}, \{\boldsymbol{\theta}_{\sigma,k}\}_{k=1}^{K} \sim \mathcal{N}\left(\mu_{n,k}, \frac{\sigma_{n,k}}{\sqrt{\tau_n}}\right) \qquad (2)$$

where $\boldsymbol{\theta}_{\mu,k}$, $\boldsymbol{\theta}_{\sigma,k}$, and $\boldsymbol{\theta}_{\nu,k}$ parameterize a deep neural network (DNN) with dedicated models, $F_{\boldsymbol{\theta}_{\mu,k}}(y_{n-p}^{n-1})$, $G_{\boldsymbol{\theta}_{\sigma,k}}(y_{n-p}^{n-1})$, $H_{\boldsymbol{\theta}_{\nu,k}}(y_{n-p}^{n-1})$, for $\mu_{n,k}$, $\sigma_{n,k}$, and $\nu_{n,k}$, respectively (Fig. 1). The deep observation models allow for rich representations, whereas the deep 3–level hidden state accounts for robustness and regime durations beyond the geometric paradigm of HMM. Lastly, the autoregressive structure is key to capture local temporal dependencies (e.g., sleep spindles). The result is the Deep Neural Dynamic Bayesian Network (DNDBN) of Fig. 1.

We use a 1-of-K notation for $\mathbf{z_n}$ and $\mathbf{d_n}$ (e.g., $z_{n,k}$ means the k–th entry of $\mathbf{z_n}$ is 1). $\mathbf{z_1}$ is parameterized by $\boldsymbol{\pi}$, while $\{\mathbf{z_n}\}_{n=2}^{N}$ obey semi–Markovian dynamics:

$$p(\mathbf{z_n} | \mathbf{z_{n-1}}, \mathbf{d_{n-1}}; A) = \begin{cases} \delta(\mathbf{z_n}, \mathbf{z_{n-1}}) & , d_{n-1,1} > 1 \\ p(\mathbf{z_n} | \mathbf{z_{n-1}}; A) & , d_{n-1,1} = 1 \end{cases} \qquad (3)$$

where A is a $K \times K$ stochastic matrix, and $\delta(\cdot, \cdot)$ is the Kronecker delta function. In words, the hidden label, $\mathbf{z_n}$, transitions to a different regime only when the counter variable is equal to 1; otherwise, it remains in the same mode.

Similarly, $\mathbf{d_n}$ samples a new duration only after a regime ends; otherwise, it decreases by one (i.e., explicit duration hidden semi–Markov model [9,35]).

$$p(\mathbf{d_n}|\mathbf{z_n},\mathbf{d_{n-1}};\{\boldsymbol{\lambda}_k\}_{k=1}^K) = \begin{cases} \delta(d_{n,j-1},d_{n-1,j}) & ,d_{n-1,1} > 1 \\ p(\mathbf{d_n}|z_{n,k};\{\boldsymbol{\lambda}_k\}_{k=1}^K) = p(\mathbf{d_n};\boldsymbol{\lambda}_k) & ,d_{n-1,1} = 1 \end{cases} \quad (4)$$

where $\{\boldsymbol{\lambda}_k\}_{k=1}^K$ is a collection of parameters for each regime durations.

Consequently, we can write the complete data log–likelihood over parameters, $\boldsymbol{\theta} = \{\{\boldsymbol{\lambda}_k\}_{k=1}^K, A, \boldsymbol{\pi}, \{\boldsymbol{\theta}_{\mu,k}\}_{k=1}^K, \{\boldsymbol{\theta}_{\sigma,k}\}_{k=1}^K, \{\boldsymbol{\theta}_{\nu,k}\}_{k=1}^K\}$, as (ignoring constants):

$$\log p(\mathbf{y},\boldsymbol{\tau},\mathbf{z},\mathbf{d}|\boldsymbol{\theta}) \propto \sum_{k=1}^K z_{1,k} \log \pi_k + \sum_{i=1}^D \sum_{k=1}^K z_{1,k} d_{1,i} \log p(\mathbf{d_1};\boldsymbol{\lambda}_k) +$$

$$\sum_{n=2}^N \sum_{k=1}^K \sum_{j=1}^K z_{n-1,j} z_{n,k} d_{n-1,1} \log A_{j,k} + \sum_{n=2}^N \sum_{k=1}^K z_{n,k} d_{n-1,1} \log p(\mathbf{d_n};\boldsymbol{\lambda}_k) + \quad (5)$$

$$\sum_{n=1}^N \sum_{k=1}^K z_{n,k} \left\{ \frac{\Gamma(\frac{\nu_{n,k}+1}{2})}{\Gamma(\frac{\nu_{n,k}}{2})\sqrt{\pi \nu_{n,k}}\sigma_{n,k}} \left(1 + \frac{1}{\nu_{n,k}} \left(\frac{y_n - \mu_{n,k}}{\sigma_{n,k}} \right)^2 \right)^{-\left(\frac{\nu_{n,k}+1}{2}\right)} \right\}$$

where $\Gamma(\cdot)$ is the gamma function, $A_{j,k}$ is the j–th row, k–th column entry of A, and $\mathbf{z} = \{\mathbf{z_n}\}_{n=1}^N$ (similar for $\mathbf{d}$ and $\boldsymbol{\tau}$). Three tasks are key: a) given $\boldsymbol{\theta}$, calculate the marginal log–likelihood (LL) of the observations (i.e., $p(\mathbf{y};\boldsymbol{\theta})$), b) given $\mathbf{y}$ and $\boldsymbol{\theta}$, estimate the most likely hidden state sequence, and c) given $\mathbf{y}$, estimate $\boldsymbol{\theta}$. We refer to the tasks as LL calculation, inference, and learning, respectively.

In terms of interpretability, the location networks, $\{F_{\boldsymbol{\theta}_{\mu,k}}(y_{n-p}^{n-1})\}_{k=1}^K$, are reduced to single fully–connected layers with no non–linearities. This design choice not only renders linear autoregressive models (i.e., infinite impulse response filters), but it also helps with validation while keeping the model more interpretable without sacrificing performance (as shown in the next sections).

2.1 Learning Model Parameters

Maximizing Eq. (5) analytically is intractable; rather, we exploit a variation of the Generalized Expectation–Maximization (GEM) algorithm for HMM [27]. We implement coordinate ascent on the log–likelihood: first, $\boldsymbol{\theta}$ is fixed and we obtain expectations of $\log p(\mathbf{y},\boldsymbol{\tau},\mathbf{z},\mathbf{d}|\boldsymbol{\theta})$ under the posterior distribution of the latent variables given $\mathbf{y}$. Then, $\boldsymbol{\theta}$ is optimized (either globally or locally) keeping the expectations fixed. This constitutes one GEM iteration and guarantees an increase of the marginal log–likelihood of the training data. GEM then continues until convergence to local optima (tracked by successive log–likelihoods).

The factorizations induced by the model yield closed–form solutions of the expectations without appealing to variational approximations that usually require recognition networks to compute local potentials [7,8,14,16,18]. Hence, our proposed learning routine does not suffer from the so–called amortization gap [2] nor "posterior collapse" [5,12]. Next, for simplicity, we outline the GEM algorithm for the case of a single training sequence; yet, the batch case is straight-forward.

E–Step. Let $\alpha_{\mathbf{z_n}, \mathbf{d_n}} = p(y_1^n, \mathbf{z_n}, \mathbf{d_n})$ be the joint probability of the first n observations and the n–th hidden variables, and similarly $\beta_{\mathbf{z_n}, \mathbf{d_n}} = p(y_{n+1}^N | \mathbf{z_n}, \mathbf{d_n})$. Next, by induction and d–separation [17] (details in supplementary material):

$$\alpha_{\mathbf{z_n}, \mathbf{d_n}} = p(y_n | \mathbf{z_n}) \sum_{\mathbf{d_{n-1}}} p(\mathbf{d_n} | \mathbf{z_n}, \mathbf{d_{n-1}}) \sum_{\mathbf{z_{n-1}}} p(\mathbf{z_n} | \mathbf{z_{n-1}}, \mathbf{d_{n-1}}) \alpha_{\mathbf{z_{n-1}}, \mathbf{d_{n-1}}} \qquad (6)$$

$$\beta_{\mathbf{z_n}, \mathbf{d_n}} = \sum_{\mathbf{z_{n+1}}} p(y_{n+1} | \mathbf{z_{n+1}}) p(\mathbf{z_{n+1}} | \mathbf{z_n}, \mathbf{d_n}) \sum_{\mathbf{d_{n+1}}} p(\mathbf{d_{n+1}} | \mathbf{z_{n+1}}, \mathbf{d_n}) \beta_{\mathbf{z_{n+1}}, \mathbf{d_{n+1}}} \qquad (7)$$

These expressions are key for all tasks: LL calculation reduces to $p(\mathbf{y}; \boldsymbol{\theta}) = \sum_{\mathbf{z_N}} \sum_{\mathbf{d_N}} \alpha_{\mathbf{z_N}, \mathbf{d_N}}$, while inference replaces sums by "max" operators plus backtracking in Eq. (6) (i.e., Viterbi Algorithm [31] or maximum a posteriori (MAP) estimate). Next, once again by d–separation (supplementary material):

$$\eta(\mathbf{z_n}, \mathbf{d_n}) = p(\mathbf{z_n}, \mathbf{d_n} | \mathbf{y}) = \frac{\alpha_{\mathbf{z_n}, \mathbf{d_n}} \beta_{\mathbf{z_n}, \mathbf{d_n}}}{p(\mathbf{y}; \boldsymbol{\theta})} \quad (8) \qquad \gamma(\mathbf{z_n}) = p(\mathbf{z_n} | \mathbf{y}) = \sum_{\mathbf{d_n}} \eta(\mathbf{z_n}, \mathbf{d_n}) \quad (9)$$

$$\xi(z_{n,k}, \mathbf{z_{n-1}}, \mathbf{d_n}) = p(z_{n,k}, \mathbf{z_{n-1}}, \mathbf{d_n}, d_{n-1,1} | \mathbf{y}; \{\boldsymbol{\lambda}\}_{k=1}^K, A) = \qquad (10)$$

$$\frac{\alpha_{\mathbf{z_{n-1}}, d_{n-1,1}} p(y_n | z_{n,k}) p(\mathbf{d_n}; \boldsymbol{\lambda}_k) p(z_{n,k} | \mathbf{z_{n-1}}; A) \beta_{\mathbf{z_n}, \mathbf{d_n}}}{p(\mathbf{y}; \boldsymbol{\theta})}$$

where $p(y_n | z_{n,k})$ is given by a conditional generalized t distribution. By construction, the entries of $\mathbf{z_n}$ and $\mathbf{d_n}$ are binary; hence, their expectations are merely the probabilities of taking the value 1.

M–step We plug in the previous expectations into Eq. (5) and optimize each $\boldsymbol{\theta}$ component. The optimal parameters for the discrete hidden variables are simple global maximizers due to convexity (i.e., partial derivatives equal to 0):

$$\pi_k^{(t)} \propto \gamma(z_{1,k}) \quad (11) \qquad A_{j,k}^{(t)} \propto \sum_{n=2}^N \sum_{\mathbf{d_n}} \xi(z_{n,k}, z_{n-1,j}, \mathbf{d_n}) \quad (12)$$

$$\lambda_{k,i}^{(t)} \propto \eta(z_{1,k}, d_{1,i}) + \sum_{n=2}^N \sum_{\mathbf{z_{n-1}}} \xi(z_{n,k}, \mathbf{z_{n-1}}, d_{n,i}) \quad (13)$$

where (t) denotes estimates for the t–th GEM iteration. $\gamma(z_{n,k}) = \mathbb{E}\{z_{n,k}\}$ (analogous for η and ξ) and $\lambda_{k,i}^{(t)}$ is the probability of duration i under regime k.

On the other hand, the DNN coefficients ($\{\boldsymbol{\theta}_{\mu,k}\}_{k=1}^K$, $\{\boldsymbol{\theta}_{\sigma,k}\}_{k=1}^K$, $\{\boldsymbol{\theta}_{\nu,k}\}_{k=1}^K$) parameterize a highly non–linear, non–convex function, which is optimized via backpropagation and stochastic gradient descent (i.e. local optima of the last line of Eq. (5)). In short, part of the M–step subroutine of one GEM iteration involves updating the DNN parameters by fitting the location, scale, and degrees of freedom of each regime given the current expectations. For more details and full derivations, refer to the supplementary material.

3 Results

3.1 Validation with Access to Ground Truth

The DREAMS Sleep Spindles dataset [3] was used to validate the proposed model. 30–min–long, single–channel (CZ-A1 or C3-A1) recordings from 8 subjects with corresponding sleep spindles expert scores (2 experts) are available. All traces were resampled 50 Hz and z–scored prior to feeding them to the model. No bandpass filtering, artifact rejection, nor windowing were implemented. The data was then partitioned into 8 folds of training and test sets (i.e., 7 subjects for training, 1 for testing). The performance measure is the by–sample Matthews Correlation Coefficient (MCC) between MAP output and expert scores (on test set). We detail several model variants due to the modularity of the framework:

a) Supervised setting (S, E1, E2) where the expert labels (either experts union, expert 1, or expert 2, resp.) are used to fit the model (i.e., no EM needed).
b) Unsupervised setting (U) where no labels are used (i.e., EM required).
c) Hidden Markov (HM) or hidden semi–Markov (HSM) latent state dynamics (i.e., setting $D = 1$ and allowing self–transitions reduces the model to HM).
d) "Linear + 2 Non–Linear" or "3 Non–Linear" depending whether the location models, $\{F_{\theta_{\mu,k}}(y_{n-p}^{n-1})\}_{k=1}^K$, are linear (Fig. 1) or non–linear, respectively.

We set $K = 2$, $p = 5$, and $D = 1$ (HM) or $D = 50 \times 15$ (HSM). We allowed self–transitions for non–spindles to accommodate long spindle–free intervals. For the deep models, we used stochastic gradient descent (batch size $= 32$), early–stopping (75/25 split, patience $= 3$, max epochs $= 10$), and Adam [15] ($\alpha = 0.001$, $\beta_1 = 0.9$, $\beta_2 = 0.999$) on Python's tensorflow probability [6]. In each fold, a subject was held–out for validation and the best model out of 3 was kept for testing (according to MCC or LL for supervised or unsupervised cases, respectively).

Table 1 summarizes MAP results (without any post–processing) compared to three previous efforts [20,24,34]. The supervised–HSM setting has the highest average MCC, while the unsupervised setting outperforms two of the state–of–the–art methods. It is worth noting that the authors in [24] use window–based spectral methods to suppress artifacts in a pre–processing stage and all other methods in Table 1 bandpass the signal. Therefore, DNDBN not only performs feature extraction and detection, but it also handles outliers in a principled manner. All DNDBN types of Table 1 are "Linear + 2 Non–Linear". Incidentally, the DNDBN(S, HSM) "3 Non–Linear" variant achieves the same MCC of 0.459. As a side note, the approach in [26] achieves an average MCC of 0.071.

For the unsupervised setting, we set sensible initial conditions: i) $\boldsymbol{\pi} = [1, 0]^\top$, ii) $A = \left(\begin{smallmatrix} 0.5 & 0.5 \\ 1 & 0 \end{smallmatrix}\right)$ (no self–transitions for sleep spindles), iii) uniform durations for non–spindles and $\mathcal{N}(1\text{ s.}, 0.15\text{ s.})$ for spindles, and iv) initial posterior marginals, $\gamma(\mathbf{z_n})$, equal to the average of 8 detectors implemented in the package *wonambi* (https://github.com/wonambi-python/wonambi): six state–of–the–art [10,19–21,23,32] and two native to *wonambi*. We refer to this mixture of experts type of detector as MixDetectors.

Table 1. MCC metrics (DREAMS Sleep Spindles). Best results marked in bold.

Algorithm	Subject									
	S1	S2	S3	S4	S5	S6	S7	S8	Mean	SD
Wendt et al. [34]	0.49	0.40	0.39	0.20	0.49	0.54	0.13	0.230	0.362	0.154
Martin et al.. [20]	0.43	0.54	0.40	0.21	0.51	0.58	0.22	0.25	0.399	0.149
Parekh et al. [24]	0.50	0.57	**0.54**	0.24	**0.57**	**0.62**	0.25	**0.26**	0.447	0.163
DNDBN(U, HSM)	0.56	0.48	0.45	0.26	0.54	0.61	0.19	0.23	0.419	0.163
DNDBN(S, HSM)	**0.58**	**0.60**	0.48	**0.29**	0.57	0.61	**0.25**	0.25	**0.459**	0.162

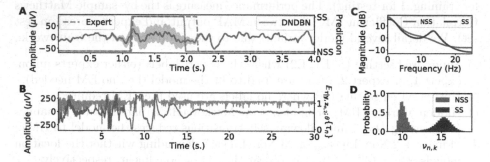

Fig. 2. A. Example of MAP (red) and posterior marginal detections (shaded red, larger shaded areas imply larger probabilities of a sleep spindle). **B.** Example of expected value of τ_n under the posterior for EEG with and without artifacts. **C.** Log frequency response of learned location coefficients, $\theta_{\mu,k}$. **D.** Histogram of learned degrees of freedom, $\nu_{n,k}$. SS: sleep spindle, NSS: non–sleep spindle.

Figure 2 validates the model while illustrating its advantages over classic detectors. DNDBN is able to not only provide binary labels (MAP estimates), but it also quantifies the uncertainty via the posterior marginal, $\gamma(\mathbf{z_n})$. The model also provides a proxy for robustness via the expected value of the posterior τ_n. In particular, by conjugacy, τ_n remains Gamma with updated means dependent of the observation parameters [25]. The first 15 s of Fig. 2B show EEG plagued with artifacts, while the rest is clean EEG. Likewise, the posterior τ_n is able to track such changes while remaining bounded. Figure 2C shows clear spectral differences between modes: the non–spindle regime depicts the well–known $1/f$ spectral distribution [11] while its counterpart has evident oscillatory activity in the sigma band. Lastly, the distributions of $\{\nu_{n,k}\}_{k=1}^2$ confirm the expert opinion that most artifacts appeared during non–spindle epochs (i.e., smaller values imply longer tails of the generalized t distribution). These results highlight the fact that detection is only one of the many attributes of the model.

Different local solutions can be compared also. Table 2 details predictive negative log–likelihoods and shows that the parameters set by expert 2 have stronger predictive power and follow the EEG dynamics closer (under the model in question). Also unsurprisingly, the "3 Non–Linear" models outperform the "Linear + 2 Non–Linear" counterparts. However, if we are interested in detection alone, "Linear + 2 Non–Linear" might suffice, as confirmed by the average MCC.

Table 2. Average predictive negative log–likelihood (NLL). HSM variants.

Observation model	Linear + 2 Non–Linear				3 Non–Linear			
Type	S	E1	E2	U	S	E1	E2	U
NLL	39876	40151	35312	39751	38553	38754	34192	38861

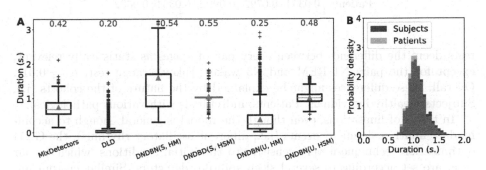

Fig. 3. Distribution of sleep spindles durations. **A.** DREAMS Sleep Spindles dataset. Mean values marked with triangles. Areas under the precision–recall curves in red. Out–of–sample results with the exception of MixDetectors case. **B.** DREAMS Subjects and Patients datasets (N2 stage events only)

Figure 3A compares several models in terms of sleep spindles durations and the area under the precision–recall curve (AUC) computed based on posterior marginals, $\gamma(\mathbf{z_n})$. An additional fully supervised deep learning detector (DLD) was implemented as well for benchmarking (6 128 fc layers + ReLU). Figure 3A illustrates the interplay between interpretability, performance, and complexity. For instance, DLD performs poorly in both AUC and durations (no underlying probabilistic model). MixDetectors performs well, as expected. The HM variants seem to miss the mark when it comes to durations (see mean values). Conversely, the HSM counterparts seem appropriate for both supervised and unsupervised cases, which suggests that a hidden semi–Markov state is indeed principled.

3.2 Sleep Spindles Density Varies Throughout Sleep Stages

The DREAMS Subjects and Patients datasets [3] provide additional validation. They consist of 20 and 27 whole–night recordings ($\approx$ 8–9 h), respectively. We train the model (DNDBN(U, HSM), "Linear + 2 Non–Linear") on the DREAMS Sleep Spindles dataset and perform inference on both Subjects and Patients datasets (one central channel resampled at 50 Hz). The learned duration distributions (Fig. 3B) resemble previous large–scale studies [33], while at the same time, constitute a proof of concept of the learning and inference subroutines.

Table 3 summarizes the sleep spindle rate (by–sample proportions) of each sleep stage (scoring provided in the datasets). As expected, N2 stages have relatively larger sleep spindles densities. One–way ANOVAs per dataset reveal statistical significance across sleep stages ($p < 0.01$), while multiple comparisons

Table 3. Average density (by–sample proportion) of putative sleep spindles.

Dataset	Sleep stage				
	Wake	N1	N2	N3	REM
Subjects	0.0426	0.0701	0.1354	0.0865	0.0623
Patients	0.0354	0.0792	0.0992	0.0364	0.0632

tests deem the difference between every par of stages as statistically relevant, except for the pairs N1–REM and N3–wake (Tukey's range test, $\alpha = 0.01$). Overall, these differences might be explained by the nature of the cohorts (i.e. Subjects: healthy individuals, Patients: individuals with various pathologies).

In terms of limitations, even though the model is general enough to accommodate multi–scale micro–events (i.e., different oscillatory rhythms), the GEM optimization technique is very sensitive to the initial conditions, which, in our case, are set according to several sleep spindles detectors. Similar custom initializations might be needed to handle additional events, such as K-complexes. A Bayesian approach might also benefit the modeling by introducing principled priors; however, the learning and inference routines would become more convoluted. Lastly, the proposed framework models single–channel EEG only; in the future, multi–channel recordings can be integrated via multivariate likelihoods.

4   Conclusion

Probabilistic models over sequences are principled frameworks for robust detection and parametrization of sleep spindles. Future work will go beyond the sigma band to characterize and detect high–frequency oscillations in other structures where ground truth is unavailable, e.g. hippocampal gamma rhythms.

References

1. Coppieterst Wallant, D., Maquet, P., Phillips, C.: Sleep spindles as an electrographic element: description and automatic detection methods. Neural Plast. **2016**, 6783812 (2016)
2. Cremer, C., Li, X., Duvenaud, D.: Inference suboptimality in variational autoencoders. In: International Conference on Machine Learning, pp. 1078–1086. PMLR (2018)
3. Devuyst, S.: The DREAMS databases and assessment algorithm (January 2005). https://doi.org/10.5281/zenodo.2650142
4. Devuyst, S., Dutoit, T., Stenuit, P., Kerkhofs, M.: Automatic sleep spindles detection overview and development of a standard proposal assessment method. In: 2011 Annual International Conference of the IEEE Engineering in Medicine and Biology Society, pp. 1713–1716. IEEE (2011)
5. Dieng, A.B., Kim, Y., Rush, A.M., Blei, D.M.: Avoiding latent variable collapse with generative skip models. In: The 22nd International Conference on Artificial Intelligence and Statistics, pp. 2397–2405. PMLR (2019)

6. Dillon, J.V., et al.: Tensorflow distributions. arXiv preprint arXiv:1711.10604 (2017)
7. Dong, Z., Seybold, B., Murphy, K., Bui, H.: Collapsed amortized variational inference for switching nonlinear dynamical systems. In: International Conference on Machine Learning, pp. 2638–2647. PMLR (2020)
8. Ebbers, J., Heymann, J., Drude, L., Glarner, T., Haeb-Umbach, R., Raj, B.: Hidden Markov model variational autoencoder for acoustic unit discovery. In: InterSpeech, pp. 488–492 (2017)
9. Ferguson, J.: Variable duration models for speech. In: Proceedings of Symposium on the Application of Hidden Markov Models to Text and Speech (1980)
10. Ferrarelli, F., et al.: Reduced sleep spindle activity in schizophrenia patients. Am. J. Psychiatry **164**(3), 483–492 (2007)
11. Freeman, W., Quiroga, R.Q.: Imaging Brain Function with EEG: Advanced Temporal and Spatial Analysis of Electroencephalographic Signals. Springer, New York (2012) https://doi.org/10.1007/978-1-4614-4984-3
12. He, J., Spokoyny, D., Neubig, G., Berg-Kirkpatrick, T.: Lagging inference networks and posterior collapse in variational autoencoders. arXiv preprint arXiv:1901.05534 (2019)
13. Iber, C.: The AASM manual for the scoring of sleep and associated events: Rules. Terminology and Technical Specification (2007)
14. Johnson, M.J., Duvenaud, D., Wiltschko, A.B., Datta, S.R., Adams, R.P.: Composing graphical models with neural networks for structured representations and fast inference. In: Proceedings of the 30th International Conference on Neural Information Processing Systems, pp. 2954–2962 (2016)
15. Kingma, D.P., Ba, J.: Adam: A method for stochastic optimization. arXiv preprint arXiv:1412.6980 (2014)
16. Kingma, D.P., Welling, M.: Auto-encoding variational bayes. arXiv preprint arXiv:1312.6114 (2013)
17. Koller, D., Friedman, N.: Probabilistic Graphical Models: Principles and Techniques. MIT Press, Cambridge (2009)
18. Krishnan, R., Shalit, U., Sontag, D.: Structured inference networks for nonlinear state space models. In: Proceedings of the AAAI Conference on Artificial Intelligence, vol. 31 (2017)
19. Leclercq, Y., Schrouff, J., Noirhomme, Q., Maquet, P., Phillips, C.: fMRI artefact rejection and sleep scoring toolbox. Comput. Intell. Neurosci. **2011**, 598206 (2011)
20. Martin, N., et al.: Topography of age-related changes in sleep spindles. Neurobiol. Aging **34**(2), 468–476 (2013)
21. Mölle, M., Bergmann, T.O., Marshall, L., Born, J.: Fast and slow spindles during the sleep slow oscillation: disparate coalescence and engagement in memory processing. Sleep **34**(10), 1411–1421 (2011)
22. Niedermeyer, E., da Silva, F.L.: Electroencephalography: basic principles, clinical applications, and related fields. Lippincott Williams & Wilkins (2005)
23. Nir, Y., et al.: Regional slow waves and spindles in human sleep. Neuron **70**(1), 153–169 (2011)
24. Parekh, A., Selesnick, I.W., Rapoport, D.M., Ayappa, I.: Sleep spindle detection using time-frequency sparsity. In: 2014 IEEE Signal Processing in Medicine and Biology Symposium (SPMB), pp. 1–6. IEEE (2014)
25. Peel, D., McLachlan, G.J.: Robust mixture modelling using the t distribution. Stat. Comput. **10**(4), 339–348 (2000)
26. Penny, W.D., Roberts, S.J.: Dynamic models for nonstationary signal segmentation. Comput. Biomed. Res. **32**(6), 483–502 (1999)

27. Rabiner, L.R.: A tutorial on hidden Markov models and selected applications in speech recognition. Proc. IEEE **77**(2), 257–286 (1989)
28. Rechtschaffen, A., Kales, A.: A Manual of Standardized Terminology. Techniques and Scoring System for Sleep Stages of Human Subjects, Brain Information Service/Brain Research Institute (1968)
29. Schabus, M., et al.: Sleep spindles and their significance for declarative memory consolidation. Sleep **27**(8), 1479–1485 (2004)
30. Steriade, M., McCormick, D.A., Sejnowski, T.J.: Thalamocortical oscillations in the sleeping and aroused brain. Science **262**(5134), 679–685 (1993)
31. Viterbi, A.: Error bounds for convolutional codes and an asymptotically optimum decoding algorithm. IEEE Trans. Inf. Theory **13**(2), 260–269 (1967)
32. Wamsley, E.J., et al.: Reduced sleep spindles and spindle coherence in schizophrenia: mechanisms of impaired memory consolidation? Biol. Psychiatry **71**(2), 154–161 (2012)
33. Warby, S.C., et al.: Sleep-spindle detection: crowdsourcing and evaluating performance of experts, non-experts and automated methods. Nat. Methods **11**(4), 385 (2014)
34. Wendt, S.L., Christensen, J.A., Kempfner, J., Leonthin, H.L., Jennum, P., Sorensen, H.B.: Validation of a novel automatic sleep spindle detector with high performance during sleep in middle aged subjects. In: 2012 Annual International Conference of the IEEE Engineering in Medicine and Biology Society, pp. 4250–4253. IEEE (2012)
35. Yu, S.Z.: Hidden semi-Markov models. Artif. Intell. **174**(2), 215–243 (2010)

Few Trust Data Guided Annotation Refinement for Upper Gastrointestinal Anatomy Recognition

Yan Li[✉], Kai Lan, Xiaoyi Chen, Li Quan, and Ni Zhang

NEC Laboratories, Beijing, China
{li-yan,lan_kai,chen_xiaoyi,quan_li,zhangni_nlc}@nec.cn

Abstract. The performance of anatomy site recognition is critical for computer-aided diagnosis systems such as the quality evaluation of endoscopic examinations and the automatic generating of electronic medical records. To achieve an accurate recognition model, it requires extensive training samples and precise annotations from human experts, especially for deep learning based methods. However, due to the similar appearance of gastrointestinal (GI) anatomy sites, it is hard to annotate accurately and is expensive to acquire such high quality dataset at a large scale. Therefore, to balance the cost-performance trade-offs, in this work we propose an effective annotation refinement approach which leverages a small amount of trust data that is accurately labelled by experts to further improve the training performance on a large amount of noisy label data. In particular, we adopt noise robust training on noisy dataset with additional constraints of adaptively assigned sample weights that are learned from trust data. Controlled experiments on synthetic datasets with generated noisy annotations validate the effectiveness of our proposed method. For practical use, we design a manual process to come up with a small amount of trust data with reliable annotations from a noisy upper GI dataset. Experimental evaluations validates that leveraging a small amount of trust data can effectively rectify incorrect labels and improve testing accuracy on noisy datasets. Our proposed annotation refinement approach provides a cost effective solution to acquire high quality annotations for medical image recognition tasks.

Keywords: Data refinement · Endoscopic image · Noisy label · Meta learning

1 Introduction

Anatomy site recognition of upper gastrointestinal endoscopic images extracts semantic information for key gastric sites. This technique can be used in computer-aided diagnosis systems such as quality evaluation of esophagogastroduodenoscopy (EGD) examinations [1] and automatic generating of electronic medical records. In order to ensure that all examinations can be evaluated correctly, or to avoid missing key sites in automatic generation of diagnostic results, it is critical to achieve accurate recognition models. Due to the complexity and similarity in appearances of different gastrointestinal (GI) anatomy sites, it is expensive and time-consuming to acquire a high quality dataset which has adequate clean annotations and a large amount of training samples [2]. It

© Springer Nature Switzerland AG 2021
M. de Bruijne et al. (Eds.): MICCAI 2021, LNCS 12905, pp. 561–570, 2021.
https://doi.org/10.1007/978-3-030-87240-3_54

is relatively easier to obtain a large number of data with coarse annotations. However, to some extent training with noise data may lead models to under-performed accuracy. An efficient labeling framework for similar tasks should have high noise tolerances and can be enhanced by extra unbiased data. Here we propose a cost-performance trade-off solution to tackle noise data refinement problem with a noise labeled real-world upper GI dataset and a small amount of clean annotations.

Earlier studies propose to make use of meta parameters such as confusion matrices or example weights which are extracted from clean annotations to prevent model training from fitting to noise. As a instance, [3] proposes to learn sample weights from clean annotations and guide model learning by the weighted loss in each iteration. [8] builds a confusion matrix of clean data to re-adjust the model loss. [9] uses clean data to train a teacher network to distill useful information for noisy label learning networks. Some other methods try to unsupervisely identify noise samples via prediction errors. Han, B. et al. [7] train two networks to filter out possible noise samples for each other according to training losses. In these approaches, information of image samples that are identified as noise data is not fully used, which might cause poor learning efficiency. Some recent works focus on using the combination of multiple noisy label learning modules such as noise robust losses, pseudo labeling, noise modeling, etc., and have achieved expressive successes. [4] and [5] adopt noise robust loss with alternate optimizations of network parameters and pseudo labels to overcome the noise fitting issue. Another hybrid noisy label learning approach [6] reports state-of-the-art results by a large margin to other approaches through introducing MixMach [10] semi-supervised learner and GMM based noisy label identifier. These methods have achieved good results without using any known clean data.

In this work, we build an upper GI dataset which consists of large scale manual labeled endoscopic images with imperfect annotations and a small number of precise labels. We target at proposing an effective noisy label learning approach to refine our real-world dataset with better accuracy and recovery performance than existing works. To achieve our goal, we employ the proposed method based on a state-of-the-art noisy label learning work, i.e. DivideMix [6] with additional constraints of a training loss with adaptively assigned sample weights that are learned from trust data. In this work, we define the annotations of trust data as *reliable annotations*. We report a manual selection process to pick out reliable annotations from a large scale noise dataset. Our proposed annotation refinement approach provides a cost effective solution on acquiring high quality annotations for medical image recognition tasks.

2   Methods

We define our task as the image classification. In this paper, we use noisy label learning techniques with the regularization of adaptively weighted loss to rectify annotations of training datasets for followed-up anatomy recognition tasks. To decouple and extract useful information from noisy labeled upper GI dataset we propose to build our framework based on a state-of-the-art noisy label learning approach [6], which is implemented based on two mutually supported models with noise robust learning modules, i.e. noise identifications [11], pseudo label estimators [10] and mixup [12] augmentations. In addition, to make full use of manually selected standard annotations that are built for this project, we add a weighted sample loss based on the meta-learning method [3]. In this work, this loss adaptively adjusts the learning importance of both possible noise and possible clean data with learned gradients on trust data.

2.1 Adaptive Weighted Loss

As mentioned, we adopt a regularization loss with sample-wise weights that are learned from a small set of clean samples to further constrain the model from fitting to noise. In order to estimate the learning weights w of each sample, which impact differently to the noise-robust training, following [3] we adopt a meta-learning process in each iteration. The idea is to use an additional gradient descent step to calculate gradients on a clean validation set with respect to a set of zero initialized weights and use it to update the weights for each noise sample during training.

We present the noisy label learning model (with probabilistic outputs) by $f(\theta, x)$, where θ indicates network parameters and x is input image. To achieve adaptive weighted loss, in particular, in each iteration, first we make a copy of the main network $f(\cdot)$ as meta-network $f^{m}(\cdot)$ with the same network parameters θ in each iteration. The meta-network is updated separately for computation of example weights.

We conduct one step inference with initial weights on a batch of N samples (x, y) which probably contain both possible clean and noise data. The loss calculated in this step will be used to update the parameters of meta-network. For each iteration, we establish the gradient graph:

$$\theta' = \theta - \alpha \, \Delta \left(\frac{1}{N} \sum_{i=1}^{N} \epsilon_i L\big(y_i, f^{m}(\theta, x_i)\big) \right) \Bigg|_{\theta}, \tag{1}$$

where $L(\cdot)$ indicates cross entropy loss between network predictions and given ground truths y. The α is the pre-defined learning rate. We can use automatic differentiation to calculate the gradient at $\epsilon_i = 0$. For ground truths y, to integrate example weights as regularizations, we calculate loss on gradually updated pseudo labels instead of original noise labels. Next we calculate the cross entropy loss C on a batch of M trust samples (x^t, y^t) with updated meta-network by:

$$C = \frac{1}{M} \sum_{i=1}^{M} L\big(y_i^t, f^{m}(\theta', x_i^t)\big). \tag{2}$$

The gradients g_i of initial weights can be calculated by taking derivative of C with respect to ϵ_i:

$$g_i = -\frac{\delta}{\delta \epsilon_i}(C) \Bigg|_{\epsilon_i=0}. \tag{3}$$

Finally, the example weight w_i for the loss of noise example x_i is updated by clamping and normalizing the gradient g_i, as follows:

$$w_i' = \max(g_i, 0), \tag{4}$$

$$w_i = \frac{w_i'}{\sum_{j=1}^{N} w_i'}, \tag{5}$$

if $\sum_{j=1}^{N} w_i' = 0$ then we remain $w_i = w_i'$.

We use the estimated example weights w_i to form an adaptive weighted cross-entropy loss L_w, which is formulated as:

$$L_w = \frac{1}{N} \sum_{i=1}^{N} w_i L(y_i, f(\theta, x_i)). \tag{6}$$

Intuitively, the adaptive weighted loss guides the model fitting to correct labels by giving more importance to the examples which can achieve similar gradient as on trust data.

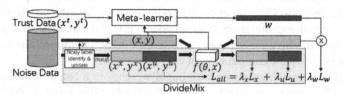

Fig. 1. Illustration map of our proposed trust data guided data refinement framework. Based on DivideMix, we adopt a weighted loss which are learned from trust data to guide model learning

2.2 Trust Data Guided Annotation Refinement Framework

We illustrate our framework in Fig. 1. Following [6], our framework classifies training data to possible clean and possible noise samples by GMM based noise identifier [11]. And, it updates labels of clean data and generates pseudo labels for noise data by averaging multiple predictions. The method adopts mixup [12] to add linear interpolations to the data. In this work, the total loss is composed of 1) cross-entropy loss L_x on possible clean data, 2) L2 loss L_u on possible noise data, and 3) aforementioned adaptive weighted loss L_w. We use pre-defined parameters λ_x, λ_u and λ_w to adjust the importance among three losses. The overall loss L_{all} is formulated as follows:

$$L_{all} = \lambda_x \frac{1}{N^x} \sum_{i=1}^{N^x} y_i^x \log(f(\theta, x_i^x)) + \lambda_u \frac{1}{N^u} \sum_{j=1}^{N^u} \|y_j^u - f(\theta, x_j^u)\| + \lambda_w L_w, \tag{7}$$

where x^x and x^u are mixed samples of possible clean and noise data, respectively, y^x and y^u are gradually updated ground truth labels.

To obtain refined annotations, we adopt the averaged predictions with multiple networks and augmentation on all training data until models are fully trained. We use hard coded labels as the final data refinement results.

3 Experiments

We firstly conduct controlled experiments of our noisy label learning method on a clean dataset with generated annotations to explore the effectiveness of different modules and demonstrate performance comparisons with related works. Then we report details of data refinement results on our real-world dataset, which is used for upper gastrointestinal anatomy recognition.

3.1 Controlled Experiments

Experiment Details. The experiments are conducted using CIFAR-10 with synthetic noise labels. The original dataset consists of 50,000 training samples and 10,000 testing samples with clean labels of 10 classes. We follow [7] to use symmetric flipping and asymmetric methods to generate noise labels for training data and keep testing data unchanged. We verify our method on two types of noise labels with different noise ratios (specifically, 0.2, 0.5 and 0.8 for symmetric flipping noise, denoted as "Sym0.2", "Sym0.5" and "Sym0.8", 0.4 and 0.8 for asymmetric noise, denoted as "ASym0.4" and "ASym0.8"). Compared with the symmetric flipping method, asymmetric noise is more reasonable for simulating real-world situations [5]. To generate trust data, we randomly select 10 images for each class from training data and assign them with clean labels. As a result, there are totally 49,900 noise data for training, 10,000 clean data for testing, and 100 trust data in our controlled experiments.

We use ResNet18 with pre-activations [13] as backbone network and adopt exactly the same hyper-parameters with [6] (Concretely: SGD training with totally 300 epochs, first 10 epochs for warmup training, with batch size 128, a stepped learning rate starts at 0.02 and decays by factor 0.1 at epoch 150) except lambdas of three losses. The detailed settings of weights of losses used in our experiments are listed in Table 1. The parameters are empirically determined on verification results. In the controlled experiments we set the batch size of trust data as 100. All of the models are trained on a single Titan RTX GPU with PyTorch [15] framework.

Table 1. Weights of losses for CIFAR-10 controlled experiments.

	Sym0.2	Sym0.5	Sym0.8	ASym0.4	ASym0.8
λ_x	5.0	5.0	5.0	5.0	1.0
λ_u	0.0	25.0	25.0	0.0	25.0
λ_w	1.0	1.0	1.0	1.0	50.0

Results. We compare our adaptive weighted loss guided DivideMix (denoted as "DMAWL") with standard cross entropy baseline (denoted as "CE") and original DivideMix (denoted as "DM") on top one accuracy of simulated datasets. For experiments of DivedeMix we report results from their paper and from our re-implementations.

In addition, to demonstrate the effectiveness of our proposed trust data guided learning scheme, we run experiments with trust data by directly fine-tuning (denoted as "with FT") on trained baseline models and original DivideMix models. we use standard cross entropy to train additional 1,000 epochs on trust data with learning rate 0.0002 and batch size 100. For DevideMix we use the models with the best performance and fine-tuned on both of the two ensemble models. In addition, we report results of only using trust data guided loss (denoted as "L2R").

The experimental results are reported in Table 2. It shows that our framework achieves significant accuracy improvements on most of the noisy situations. Especially for the case of severe noise (i.e. Asym0.8), both the baseline and DivideMix cannot achieve satisfied results. They cannot handle the situation when data is miss-labeled to one single incorrect category with a noise ratio larger than 0.5. Through our proposed adaptive weighted loss and trust data, the accuracy on severe noise is improved from lower than 60% to 82.92% for the best results. Directly fine-tuning on trust data can hardly obtain performance gains for baseline models and DivdeMix models except on Asym0.8. From the results we find that, the smaller the noise is, the less effects are brought by trust data. For data with 20% symmetric noise, our method cannot make further improvements because the performance is close to the upper limit of networks trained on totally clean set. The meta-learning module contributes especially in the case that model degrades severely in the presence of high asymmetric noise.

Table 2. Testing accuracy of different methods w/ or w/o trust data on CIFAR-10 controlled experiments (%). We report results of the best model and the last model. (*Results of re-implementations, values in brackets are results reported in their paper.)

	Trust data	Sym0.2 Best l Last	Sym0.5 Best l Last	Sym0.8 Best l Last	ASym0.4 Best l Last	ASym0.8 Best l Last
CE	w/o	86.54 l 83.80	80.20 l 58.91	59.12 l 26.83	84.84 l 77.54	57.79 l 54.57
DM*	w/o	96.21 l 96.02 (96.10 l 95.70)	94.43 l 94.08 (94.60 l 94.40)	92.23 l 92.03 (93.20 l 92.90)	93.35 l 92.83 (93.40 l 92.10)	49.91 l 49.29 (-)
CE with FT	w/	87.97 l 87.85	81.79 l 81.65	62.06 l 61.92	85.90 l 85.43	71.64 l 71.33
L2R	w/	88.08 l 86.95	82.54 l 76.27	64.13 l 50.08	85.76 l 84.64	79.51 l 76.39
DM with FT	w/	96.22 l **96.21**	94.45 l 94.30	92.31 l 92.22	93.39 l 93.16	62.92 l 62.62
DMAWL (ours)	w/	**96.35** l 96.16	**95.55** l **95.11**	**93.35** l **93.08**	**94.64** l **94.43**	**82.92** l **78.87**

Table 3. Comparison of recovery rate on CIFAR-10 controlled experiments (%).

	Sym0.2	Sym0.5	Sym0.8	ASym0.4	ASym0.8
DM	98.03	94.79	91.56	92.80	49.07
DMAWL (ours)	**98.16**	**96.43**	**92.81**	**94.79**	**77.01**

In addition, we report the recovery rate of noise labels brought by our method and original DivideMix in Table 3. Note that all of the data refinement results are conducted via models of the last epoch. As shown in the table, our approach can achieve improvements on different noise ratios and noise types, especially in the case of asymmetric generated noise. The results of controlled experiment verify that the adaptive weighted loss and trust data can further improve the noise tolerance.

3.2 Experiments on Upper GI Dataset

Upper Gastrointestinal Anatomy Dataset. For practical use, we have a large scale upper GI dataset which is coarsely labeled. In particular, we collect over 70,000 upper GI anatomy images through clinical endoscopic examinations of over 1,500 different cases. Image which are not related to standard endoscopic examination, such as surgery, vitro, or defacement etc. are filtered out. We ask doctors with years of EGD Examination experiences to manually annotate them to 30 kinds of predefined anatomy sites. In total we have 67,851 coarsely labeled samples in training data. In addition, we have 3,448 fine annotated images from about 160 different cases for testing purpose. Figure 2 presents the distribution of our training data.

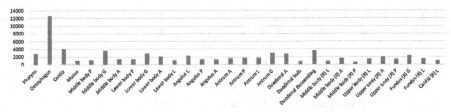

Fig. 2. Data distribution of our upper gastrointestinal anatomy dataset.

Collection of Reliable Annotations. To pick out samples with reliable annotations as trust data from the existing noise dataset, we design an iterative selection process with the assistance of 10 to 20 doctors. In particular, we randomly pick 5 samples from each class. We ask multiple doctors to point out incorrect samples and assign tags of them by voting. Then we replace the unqualified samples with the newly picked samples in a random manner. This process is repeated for three times in our work till we finally obtain a small set of samples with high quality annotations. There are totally 150 reliable annotations in our experiments, which has a significant gap with the amount of training data. We coarsely estimate that the overall noise rate of our GI dataset is about 30%, According to the consensus of multiple doctors.

Experiment Details. We follow the settings reported in the controlled experiments. We use ResNet18 with pre-activations as backbone network with 10 epochs of warmup and totally 300 epochs for training of data refinement networks. We use 128 as batch size of trust data which is the same with the batch size of training data. For weights of different losses, we define $\lambda_w = 1.0$, $\lambda_x = 5.0$ and $\lambda_u = 0$.

To obtain a refined dataset, we use the hard coded predictions (transferring probabilistic outputs to a single category number by argmax) of the last epoch as rectified annotations for possible noise data and remain possible clean data unchanged. For our real-world anatomy recognition task, considering speed-accuracy trade off, we have a separate training framework which is deployed in SEResNet34 [14] with standard training schemes which are developed by Caffe [16]. Therefore, we report the final recognition accuracy through training on this model with the annotations that are refined by the proposed approach.

Results. We report the final recognition accuracy of models that are trained on our refined data (denoted as "DMAWL Refined"), directly trained on original labels (denoted as "original") and trained on refined labels by standard dividemix (denoted as "DM refined").

Table 4 summaries the final recognition accuracies that are retrained on our real-world model via data refinement approaches mentioned in this paper. Our proposed data refinement approach outperforms standard training by a large margin on top one accuracy (from 73.70% to 78.28%). Compared with original noisy label learning approach, the usage of reliable annotations and our proposed adaptive weighted loss can bring 2 ~ 3% improvements.

Table 4. Testing accuracy on our upper GI dataset (%). All of the models are trained with same parameters on differently refined labels or fine-tuned on trust data.

Labels	FT	Trust data	Top1 Accuracy
Original	N	w/o	73.70
DM refined	N	w/o	75.85
Original	Y	w/	73.77
DM refined	Y	w/	75.86
DMAWL refined (ours)	N	w/	**78.28**

In addition, we illustrate the detailed evaluation results with a confusion matrix of recall for totally 30 categories in Fig. 3. The more concentrated of high values on the diagonals, the better recall it indicates. From the result of confusion matrix, we find that through training with our refined annotations, the number of true positive predictions of many upper GI categories are increased, which significantly improves our quality evaluation system for EGD examinations.

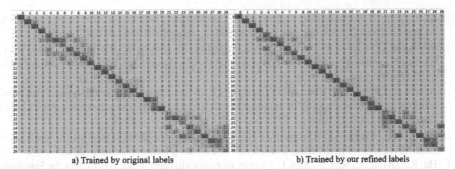

a) Trained by original labels   b) Trained by our refined labels

Fig. 3. Confusion matrixes of the 30-category upper GI anatomy recognition results that are trained on data with original and refined labels.

4 Conclusion

In this work, we target at refining the annotation of a large scale upper GI data with mostly noise and few reliable annotations. To solve this problem, we propose to combine a state-of-the-art noisy label learning scheme with an adaptive weighted loss. Our method is quantify evaluated on simulated datasets and verified on our real-world dataset. Moreover, for our upper GI anatomy recognition task, it is worthwhile to use extra effort on collecting a small number of trust data with reliable annotations under the cost-performance trade-off. For future work, we are considering to extend the ideas of noisy label learning techniques to more complicated tasks, such as refining the training data for lesion detections and segmentations, in order to solve more problems related to efficient medical image labeling.

References

1. Wu, L., et al.: Randomised controlled trial of WISENSE, a real-time quality improving system for monitoring blind spots during esophagogastroduodenoscopy. Gut 68, 2161–2169 (2019)
2. Quan, L., Li, Y., Chen, X., Zhang, N.: An Effective data refinement approach for upper gastrointestinal anatomy recognition. In: Martel, A.L., et al. (eds.) MICCAI 2020. LNCS, vol. 12261, pp. 43–52. Springer, Cham (2020). https://doi.org/10.1007/978-3-030-59710-8_5
3. Ren, M., Zeng, W., Yang, B., Urtasun, R.: Learning to reweight examples for robust deep learning. In: International Conference on Machine Learning, pp. 4334–4343 (2018)
4. Tanaka, D., Ikami, D., Yamasaki, T., Aizawa, K.: Joint optimization framework for learning with noisy labels, In: Proceedings of IEEE Conference on Computer Vision and Pattern Recognition, pp. 5552–5560 (2018)
5. Yi, K., Wu, J.: Probabilistic end-to-end noise correction for learning with noisy labels. In: Proceedings of the IEEE Conference on Computer Vision and Pattern Recognition, pp. 7017–7025 (2019)
6. Li, J., Socher, R., Hoi, S.C.H.: Dividemix: learning with noisy labels as semi-supervised learning. arXiv:2002.07394 (2020)
7. Han, B., et al.: Coteaching: robust training of deep neural networks with extremely noisy labels. Adv. Neural Inf. Process. Syst. 31, 8527–8537 (2018)
8. Hendrycks, D., Mazeika, M., Wilson, D., Gimpel, K.: Using trusted data to train deep networks on labels corrupted by severe noise. Adv. Neural Inf. Process. Syst. 31, 10456–10465 (2018)

9. Li, Y., Yang, J., Song, Y., Cao, L., Luo, J., Li, L.J.: Learning from noisy labels with distillation. In: International Conference on Computer Vision, pp. 1928–1936 (2017)

10. Berthelot, D., Carlini, N., Goodfellow, I., Papernot, N., Oliver, A., Raffel, C.A.: MixMatch: a holistic approach to semi-supervised learning. Adv. Neural Inf. Process. Syst. **32**, 5049–5059 (2019)

11. Arazo, E., Ortego, D., Albert, P., O'Connor, N.E., Mcguinness, K.: Unsupervised label noise modeling and loss correction. In: International Conference on Machine Learning, pp. 312–321 (2019)

12. Zhang, H., Cisse, M., Dauphin, Y.N., Lopez-Paz, D.: Mixup: beyond empirical risk minimization. In: International Conference on Learning Representations (2018)

13. He, K., Zhang, X., Ren, S., Sun, J.: Identity mappings in deep residual networks. In: European Conference on Computer Vision, pp. 630–645 (2016)

14. Hu, J., Shen, L., Albanie, S., Sun, G., Wu, E.: Squeeze-and-excitation networks. IEEE Trans. Pattern Anal. Mach. Intell. **42**, 2011–2023 (2020)

15. Paszke, A., Gross, S., Massa, F., Lerer, A., Chintala, S.: PyTorch: an imperative style, high-performance deep learning library. Adv. Neural Inf. Process. Syst. **32**, 8026–8037 (2019)

16. Jia, Y., et al.: Caffe: convolutional architecture for fast feature embedding. In: Proceedings of the 22nd ACM International Conference on Multimedia, pp. 675–678 (2014)

Asymmetric 3D Context Fusion
for Universal Lesion Detection

Jiancheng Yang[1,2], Yi He[2], Kaiming Kuang[2], Zudi Lin[3], Hanspeter Pfister[3],
and Bingbing Ni[1(✉)]

[1] Shanghai Jiao Tong University, Shanghai, China
nibingbing@sjtu.edu.cn
[2] Dianei Technology, Shanghai, China
[3] Harvard University, Cambridge, MA, USA

Abstract. Modeling 3D context is essential for high-performance 3D medical image analysis. Although 2D networks benefit from large-scale 2D supervised pretraining, it is weak in capturing 3D context. 3D networks are strong in 3D context yet lack supervised pretraining. As an emerging technique, *3D context fusion operator*, which enables conversion from 2D pretrained networks, leverages the advantages of both and has achieved great success. Existing 3D context fusion operators are designed to be spatially symmetric, *i.e.*, performing identical operations on each 2D slice like convolutions. However, these operators are not truly equivariant to translation, especially when only a few 3D slices are used as inputs. In this paper, we propose a novel asymmetric 3D context fusion operator (A3D), which uses different weights to fuse 3D context from different 2D slices. Notably, A3D is NOT translation-equivariant while it significantly outperforms existing symmetric context fusion operators without introducing large computational overhead. We validate the effectiveness of the proposed method by extensive experiments on DeepLesion benchmark, a large-scale public dataset for universal lesion detection from computed tomography (CT). The proposed A3D consistently outperforms symmetric context fusion operators by considerable margins, and establishes a new *state of the art* on DeepLesion. To facilitate open research, our code and model in PyTorch is available at https://github.com/M3DV/AlignShift.

Keywords: 3D context · Universal lesion detection · DeepLesion · A3D

1 Introduction

Computer vision for medical image analysis has been dominated by deep learning [11,15], thanks to the availability of large-scale open datasets [1,6,18,24]

J. Yang and Y. He—These authors have contributed equally.

© Springer Nature Switzerland AG 2021
M. de Bruijne et al. (Eds.): MICCAI 2021, LNCS 12905, pp. 571–580, 2021.
https://doi.org/10.1007/978-3-030-87240-3_55

and powerful infrastructure. In this study, we focus on 3D medical image analysis, *e.g.*, computed tomography (CT) and magnetic resonance imaging (MRI). Spatial information from 3D voxel grids can be effectively learned by convolutional neural networks (CNNs), while 3D context modeling is still essential for high-performance models. There have been considerable debates over 2D and 3D representation learning on 3D medical images; 2D networks benefit from large-scale 2D pretraining [3], whereas the 2D representation is fundamentally weak in large 3D context. 3D networks learn 3D representations; However, few publicly available 3D medical datasets are large enough for 3D pretraining.

Recently, there have been a family of techniques that enable building 3D networks with 2D pretraining [2,9,13,22,23], we refer to it as *3D context fusion operators*. See Sect. 2.1 for a review of existing techniques. These operators learn 3D representations while their (partial) learnable weights can be initialized from 2D convolutional kernels. Existing 3D context fusion operators are convolution-like, *i.e.*, either axial convolutions to fuse slice-wise information [2,13,23] or shifting adjacent slices [9,22]. Therefore, these operators are designed to be spatially **symmetric**: each 2D slice is operated identically. However, convolution-like operations are not truly translation-equivariant [12], due to padding and limited effective receptive fields. In many 3D medical image applications, only a few 2D slices are used as inputs to models due to the memory constraints. It may be meaningless to pursue translation-equivariance in these cases.

In this study, we propose a novel **asymmetric** 3D context fusion operator (A3D). See Sect. 2.2 for the methodology details. Basically, given D slices of 3D input features, A3D uses different weights to fuse the input D slices for each output slice. Therefore, the A3D is **NOT** translation-equivariant. However, it significantly outperforms existing symmetric context fusion operators without introducing large computational overhead in terms of both parameters and FLOPs. We validate the effectiveness of the proposed method by extensive experiments on DeepLesion benchmark [21], a large-scale public dataset for universal lesion detection from computed tomography (CT). As described in Sect. 3, the proposed A3D consistently outperforms symmetric context fusion operators by considerable margins, and establishes a new *state of the art* on DeepLesion.

2   Methods

2.1   Preliminary: 3D Context Fusion Operators with 2D Pretraining

In this section, we briefly review the 3D context fusion operators that enable 2D pretraining, including (a) no fusion, (b) I3D [2], (c) P3D [13], (d) ACS [23] and (e) Shift [9,22]. As an emerging technique, 3D context fusion operator leverages advantages of both 2D pretraining and 3D context modeling.

Given a 3D input feature $\boldsymbol{X_i} \in \mathbb{R}^{C_i \times D \times H \times W}$, we would like to obtain a transformed 3D output $\boldsymbol{X_o} \in \mathbb{R}^{C_o \times D \times H \times W}$ with a (pretrained) 2D convolutional kernel $\boldsymbol{W_{2D}} \in \mathbb{R}^{C_i \times C_o \times K \times K}$, where $D \times H \times W$ is the spatial size of 3D features, C_i and C_o are the input and output channels, and K denotes the kernel size. For simplicity, only cases with same padding are considered here. Apart from

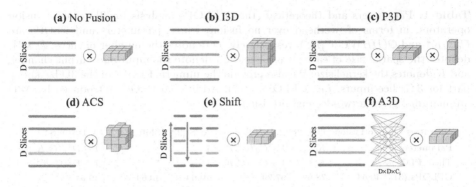

Fig. 1. Illustration of various 3D context fusion operators: (a) no fusion, (b) I3D [2], (c) P3D [13], (d) ACS [23], (e) Shift [9,22] and (f) the proposed A3D. In each sub-figure, left: D slices of C_i-channel 3D features as inputs; middle: $\otimes$ means convolution; right: illustration of convolutional kernels.

convolutions, we simply convert 2D pooling and normalization into 3D [22]. We then introduce each operator as follows:

(a) no fusion. We run 2D convolutions on each 2D slice, which is equivalent to 3D convolutions with $\boldsymbol{W_{3D}} \in \mathbb{R}^{C_i \times C_o \times 1 \times K \times K}$ converted from the 2D kernel.

(b) I3D [2]. I3D is basically an initialization technique for 3D convolution, $\boldsymbol{W_{I3D}} \in \mathbb{R}^{C_i \times C_o \times K \times K \times K}$ is initialized with K repeats of $\boldsymbol{W_{2D}}/K$, so that the distribution expectation of 3D features is the same as that of 2D features.

(c) P3D [13]. P3D convolution is a $1 \times K \times K$ 3D convolution followed by a $K \times 1 \times 1$ 3D convolution, where the first convolutional kernel is converted from a 2D kernel (same as *no fusion*), and the second is initialized as $[0, ..., 1, ..., 0]$ (*e.g.*, $[0, 1, 0]$ if $K = 3$) to make it as no fusion before training.

(d) ACS [23]. ACS runs 2D-like (3D) convolutions in three views of 3D volumes, by splitting the 2D kernel into three 3D kernels: $\boldsymbol{W_a} \in \mathbb{R}^{C_i \times C_o^{(a)} \times 1 \times K \times K}$, $\boldsymbol{W_c} \in \mathbb{R}^{C_i \times C_o^{(c)} \times 1 \times K \times K}$ and $\boldsymbol{W_s} \in \mathbb{R}^{C_i \times C_o^{(s)} \times 1 \times K \times K}$ $(C_o^{(a)} + C_o^{(c)} + C_o^{(s)} = C_o)$. 3D context is fused with layer-by-layer ACS transformation without introducing computational cost compared to no fusion.

(e) Shift [9,22]. Shift is a family of techniques that fuse 3D context by shifting adjacent 2D slices. Take TSM [9] as an example. It first splits the input feature $\boldsymbol{X_i} \in \mathbb{R}^{C_i \times D \times H \times W}$ by channel into 3 parts: $\boldsymbol{X_i^+} \in \mathbb{R}^{C_i^+ \times D \times H \times W}$, $\boldsymbol{X_i^-} \in \mathbb{R}^{C_i^- \times D \times H \times W}$ and $\boldsymbol{X_i^=} \in \mathbb{R}^{C_i^= \times D \times H \times W}$ $(C_i^+ + C_i^- + C_i^= = C_i)$. $\boldsymbol{X_i^+}$, $\boldsymbol{X_i^-}$ and $\boldsymbol{X_i^=}$ are then shifted up, shifted down and kept among the axial axis (D dimension), respectively. Finally, a 3D convolution with $\boldsymbol{W_{3D}} \in \mathbb{R}^{C_i \times C_o \times 1 \times K \times K}$ (as in no fusion) can fuse 3D context with a single slice. AlignShift [22] is a shift

Table 1. Parameters and theoretical (theo.) FLOPs analysis for 3D context fusion operators, in terms of overhead over no fusion, whose parameters and FLOPs are $C_o C_i K^2$ and $\mathcal{O}(DHWC_oC_iK^2)$, respectively. D denotes the number of slices, $D \times W$ denotes the spatial size of each slice, C_i and C_o denote the input and output channel, and K denotes the kernel size. We also provide the numeric FLOPs of the 3D backbone part for 3/7-slice inputs, $i.e.$, GFLOPs (3/7). Additional FLOPs introduced by A3D are marginal given a two-decimal precision.

Operators	No Fusion	I3D [2]	P3D [13]	ACS [23]	Shift [9,22]	A3D (Ours)
Parameters	1	K	$1 + C_o/(C_iK)$	1	1	$1 + D^2/(C_oK^2)$
Theo. FLOPs	1	K	$1 + C_o/(C_iK)$	1	1	$1 + D/(C_oK^2)$
GFLOPs (3)	40.64	78.69	67.79	40.64	40.64	40.64
GFLOPs (7)	94.83	183.61	158.18	94.83	94.83	94.83

Algorithm 1: Asymmetric 3D Context Fusion (A3D)

> **Input:** 3D input feature $X_i \in \mathbb{R}^{C_i \times D \times H \times W}$.
> **Parameter:** asymmetric fusion weight $P \in \mathbb{R}^{D \times D \times C_i}$,
>   2D (pretrained) convolutional kernel $W_{2D} \in \mathbb{R}^{C_i \times C_o \times K \times K}$.
> **Output:** 3D output feature $X_o \in \mathbb{R}^{C_o \times D \times H \times W}$.
> **1** $W_{3D} = unsqueeze(W_{2D}, dim = 2) \in \mathbb{R}^{C_i \times C_o \times 1 \times K \times K}$,
> **2** $X = einsum(\text{``}cdhw, dkc \rightarrow ckhw\text{''}, [X_i, P]) \in \mathbb{R}^{C_i \times D \times H \times W}$,
> **3** $X_o = Conv3D(X, kernel = W_{3D})$.

---

operator adaptive to medical imaging thickness, thus improves the performance of TSM on mixed-thickness data ($e.g.$, a mix of thin- and thick-slice CT scans).

Figure 1 illustrates these operators, and Table 1 summarizes the computational overhead over no fusion, in terms of parameters and FLOPs. Apart from theoretical FLOPs, we also provide the numeric FLOPs for 3/7-slice inputs to better understand the algorithm complexity in practice. To fairly compare these methods, only FLOPs in 3D backbone are counted, those in 3D-to-2D feature layer and detection heads on 2D feature maps are ignored. Interestingly, additional FLOPs introduced by A3D are marginal given a two-decimal precision.

2.2   Asymmetric 3D Context Fusion (A3D)

The 3D context fusion operators above are designed to be spatially symmetric, $i.e.$, each 2D slice is transformed identically to ensure these convolution-like operations to be translation-equivariant. However, in many medical imaging applications, only a few slices are used as model inputs because of memory constraints ($D = 3$ or 7 in this study). In this case, padding (zero or others) on the axial axis induces a significant distribution shift near top and bottom slices. Moreover, convolution-like operations are not truly translation-equivariant [12] due to limited effective receptive fields. It is not necessary to use spatially symmetric operators in pursuit of translation-equivariance for 3D context fusion.

To address this issue, we propose a novel asymmetric 3D context fusion operator (A3D), which uses different weights to fuse 3D context for each slice. Mathematically, given a 3D input feature $X_i \in \mathbb{R}^{C_i \times D \times H \times W}$, A3D fuses features from different slices by creating dense linear connections within the slice dimension for each channel separately. We introduce a trainable asymmetric fusion weight $P \in \mathbb{R}^{D \times D \times C}$, then

$$X^{(c)} = P^{(c)} \cdot X_i^{(c)} \in \mathbb{R}^{D \times H \times W}, c \in \{1, ..., C\}, \qquad (1)$$

where $P^{(c)} \in \mathbb{R}^{D \times D}$ and $X_i^{(c)} \in \mathbb{R}^{D \times H \times W}$ denotes the channel c of P and X_i, respectively, $\cdot$ denotes matrix multiplication. The output X denotes the 3D features after 3D context fusion, it is then transformed by a 3D convolution with $W_{3D} \in \mathbb{R}^{C_i \times C_o \times 1 \times K \times K}$ (as in no fusion). A3D can be implemented using Einstein summation and 3D convolution in lines of code. Einstein summation saves up extra memories occupied by intermediate results of operations such as transposing, therefore makes A3D faster and more memory-efficient. We depict a PyTorch-fashion pseudo-code of A3D in Algorithm 1. Batch dimension is ignored for simplicity, while the algorithm is easily batched by changing "$cdhw, dkc \rightarrow ckhw$" into "$bcdhw, dkc \rightarrow bckhw$". A3D is a simple operator that can be plugged into any 3D image model with ease.

To facilitate stable training and faster convergence, the convolution kernels in A3D operation can be initialized with ImageNet [3] pretrained weights to take advantage of supervised pretraining. Furthermore, we initialize each channel of asymmetric fusion weight $P^{(c)}$ with a identity matrix $I \in \mathbb{R}^{D \times D}$ added with a random perturbation following uniform distribution in $[-0.1, 0.1]$, *i.e.*, the A3D is initialized to be like no fusion before training.

Compared to symmetric 3D context fusion operators, A3D uses dense linear connections to gather global contextual information along the axial axis (illustrated in Fig. 1 (f)), thus avoids the padding issue around the top and bottom slices. Besides, as depicted in Table 1, A3D introduces negligible computational overhead in terms of both parameters and FLOPs compared with no fusion. Since D is typically much smaller than C_o, A3D is more lightweight than I3D [2] and P3D [13]. Moreover, as A3D can be implemented with natively supported *einsum*, it is faster than ACS [23] and Shift [9,22] with channel splitting in actual running time. Note that the A3D is NOT translation-equivariant, as it uses different weights for each output slice to fuse the 3D context from input D slices. However, it significantly outperforms existing symmetric context fusion operators with negligible computational overhead.

2.3   Network Structure for Universal Lesion Detection

We develop a universal lesion detection model following Mask R-CNN [4]. An overview of our network is depicted in Fig. 2. The network consists of a DenseNet-121 [5] based 3D backbone with 3D context fusion operators (the proposed A3D or others) plugged in and 2D detection heads. The network backbone takes a gray-scale 3D tensor in shape of $1 \times D \times H \times W$ as input, where D is the

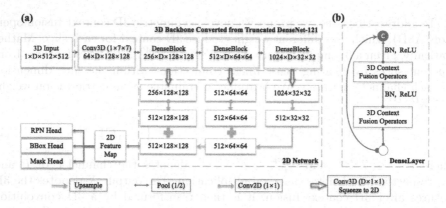

Fig. 2. Universal lesion detection model on DeepLesion [21]. The 3D backbone derived from DenseNet-121 [5,20] takes a grey-scale 3D input of $D \times 512 \times 512$, where D is the number of slices ($D \in \{3, 7\}$ in this study). Features from different scales are collected and fused together in a feature pyramid [10]. Detection is based on instance segmentation framework using Mask R-CNN [4,22].

number of slices included in each sample ($D \in \{3, 7\}$ in this study). Three dense blocks gradually downsample feature maps and increase number of channels while the depth dimension stays at D. After spatial and channel-wise unification by upsampling and $D \times 1 \times 1$ convolution, 3D features output by three dense blocks are added together and squeezed to 2D by a $D \times 1 \times 1$ convolution. Finally, the 2D feature maps are used for lesion detection on key slices.

3 Experiments

3.1 Dataset and Experiment Settings

DeepLesion dataset [21] includes 32,120 axial CT sclies extracted from 10,594 studies of 4,427 patients. There are 32,735 lesions labelled in various organs in total. Each slice contains 1 to 3 lesions, sizes of which range from 0.21 to 342.5mm. RECIST diameter coordinates and bounding boxes are annotated in key slices. Adjacent slices within the range of ±15mm from the key slice are given as contextual information.

Our experiments are based on the official code of AlignShift [22], and A3D code is merged into the same code repository. Since DeepLesion does not contain pixel-wise segmentation labels, we use GrabCut [14] to generate weak segmentation labels from RECIST annotations following [22,26]. Input CT Hounsfield units are clipped to $[-1024, 2050]$ and then normalized to $[-50, 205]$. For Align-Shift [22], we process the inputs as in its official code since it uses imaging thickness as inputs. For A3D and other counterparts, we normalize the axial thickness of all data to 2mm and resize each slice to 512×512. In terms of data augmentation, we apply random horizontal flip, shift, rescaling and rotation during the training stage. No test-time augmentation is adopted. We follow the

Table 2. Performance evaluated on the large-scale DeepLesion benchmark [21] of the proposed A3D versus other 3D context fusion operators, in terms of sensitivities (%) at various false positives (FPs) per image.

Methods	Slices	0.5	1	2	4	8	16	Avg. [0.5,1,2,4]
No Fusion	×3	72.57	79.89	86.80	91.04	94.24	96.32	82.58
I3D [2]	×3	72.01	80.09	86.54	91.29	93.91	95.68	82.48
P3D [13]	×3	62.13	73.21	82.14	88.6	92.37	94.95	76.52
ACS [23]	×3	72.82	81.15	87.40	91.35	94.69	96.42	83.18
TSM [9]	×3	71.80	80.11	86.97	91.10	93.75	95.56	82.50
AlignShift [22]	×3	73.00	81.17	87.05	91.78	94.63	95.48	83.25
A3D (Ours)	×3	**74.10**	**81.81**	**87.87**	**92.13**	**94.60**	**96.50**	**83.98**
No Fusion	×7	73.66	82.15	87.72	91.38	93.86	95.98	83.73
I3D [2]	×7	75.37	83.43	88.68	92.20	94.52	96.07	84.92
P3D [13]	×7	74.84	82.17	87.57	91.72	94.90	96.23	84.07
ACS [23]	×7	78.38	85.39	90.07	93.19	95.18	96.75	86.76
TSM [9]	×7	75.98	83.65	88.44	92.14	94.89	96.50	85.05
AlignShift [22]	×7	78.68	85.69	90.37	93.49	95.48	97.05	87.06
A3D (Ours)	×7	**80.27**	**86.73**	**91.33**	**94.12**	**96.15**	**97.33**	**88.11**

official data split of 70%/15%/15% for training, validation and test, respectively. As per [8,20,26], the proposed method and its counterparts are evaluated on the test set using sensitivities at various false positive levels (*i.e.*, FROC analysis). We also implement the mentioned 3D context fusion operators to validate the effectiveness of the proposed A3D.

3.2 Performance Analysis

We compare A3D with a variety of 3D context fusion operators (see Sect. 2.1) on the DeepLesion dataset. Table 2 gives the detailed performances of A3D and all its counterparts on 3 and 7 slices. A3D delivers superior performances compared with all counterparts on both 3 slices and 7 slices. We attribute this performance boost to A3D's ability of gathering information among globally along the axial axis by creating dense connections among slices, which can be empirically validated by the observation that A3D has a higher performance boost on 7 slices that on 3 slices when compared with the previous *state-of-the-art* AlignShift [22] (+1.05 vs. +0.73) since 7 slices provide more contextual information. Moreover, A3D introduces no padding along the axial axis, this advantage also leads to the performance boost compared to other operators. Note that AlignShift-based model is adaptive to imaging thickness, which is an orthogonal contribution to this study. The asymmetric operation-based methods could be potentially improved by adapting imaging thickness.

Table 3 shows a performance comparison of A3D and previous *State of the Art*. Without heavy engineering and data augmentations, our proposed method outperforms the previous *state-of-the-art* AlignShift [22] on both 3 slices and 7 slices by considerable margin. It is worth noting that A3D with image only

Table 3. Performance evaluated on the large-scale DeepLesion benchmark [21] of the proposed A3D versus previous *state-of-the-art*, in terms of sensitivities (%) at various false positives (FPs) per image.

Methods	Venue	Slices	0.5	1	2	4	8	16	Avg. [0.5,1,2,4]
3DCE [19]	MICCAI'18	×27	62.48	73.37	80.70	85.65	89.09	91.06	75.55
ULDor [16]	ISBI'19	×1	52.86	64.80	74.84	84.38	87.17	91.80	69.22
V.Attn [17]	MICCAI'19	×3	69.10	77.90	83.80	–	–	–	–
Retina. [26]	MICCAI'19	×3	72.15	80.07	86.40	90.77	94.09	96.32	82.35
MVP [8]	MICCAI'19	×3	70.01	78.77	84.71	89.03	–	–	80.63
MVP [8]	MICCAI'19	×9	73.83	81.82	87.60	91.30	–	–	83.64
MULAN [20]	MICCAI'19	×9	76.12	83.69	88.76	92.30	94.71	95.64	85.22
Bou.Maps [7]	MICCAI'20	×3	73.32	81.24	86.75	90.71	–	–	83.01
MP3D [25]	MICCAI'20	×9	79.60	85.29	89.61	92.45	–	–	86.74
AlignShift [22]	MICCAI'20	×3	73.00	81.17	87.05	91.78	94.63	95.48	83.25
AlignShift [22]	MICCAI'20	×7	78.68	85.69	90.37	93.49	95.48	97.05	87.06
ACS [23]	JBHI'21	×3	72.82	81.15	87.40	91.35	94.69	96.42	83.18
ACS [23]	JBHI'21	×7	78.38	85.39	90.07	93.19	95.18	96.75	86.76
A3D	Ours	×3	74.10	81.81	87.87	92.13	94.60	96.50	83.98
A3D	Ours	×7	**80.27**	**86.73**	**91.33**	**94.12**	**96.15**	**97.33**	**88.11**

surpasses MULAN [20] by nearly 3% even though it takes less slices and no additional information apart from CT images such as medical report tags and demographic information as inputs (Fig. 3).

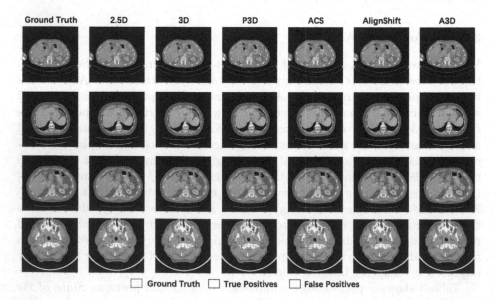

Fig. 3. Visualization of DeepLesion slices highlighted with ground truth and predictions generated by different 3D context fusion operators.

4   Conclusion

In this study, we focus on 3D context fusion operators that enable 2D pretraining, which is an emerging technique that leverages advantages of both 2D pretraining and 3D context modeling. We analyze the unnecessary pursuit of translation-equivariance in existing spatially symmetric 3D context fusion operators especially when only a few 2D slices are used as model inputs. To this end, we further propose a novel asymmetric 3D context fusion operator (A3D) that is translation-equivariant. The A3D significantly outperforms existing symmetric context fusion operators without introducing large computational overhead. Extensive experiments on DeepLesion benchmark validate the effectiveness of the proposed method, and we establish a new *state of the art* that surpasses prior arts by considerable margins.

Acknowledgment. This work was supported by National Science Foundation of China (U20B2072, 61976137).

References

1. Antonelli, M., Reinke, A., Bakas, S., et al.: The medical segmentation decathlon. arXiv preprint arXiv:2106.05735 (2021)
2. Carreira, J., Zisserman, A.: Quo vadis, action recognition? a new model and the kinetics dataset. In: CVPR, pp. 6299–6308 (2017)
3. Deng, J., Dong, W., Socher, R., Li, L.J., Li, K., Fei-Fei, L.: Imagenet: a large-scale hierarchical image database. In: CVPR, pp. 248–255 (2009)
4. He, K., Gkioxari, G., Dollár, P., Girshick, R.B.: Mask r-cnn. In: ICCV, pp. 2980–2988 (2017)
5. Huang, G., Liu, Z., Van Der Maaten, L., Weinberger, K.Q.: Densely connected convolutional networks. In: CVPR, vol. 1, p. 3 (2017)
6. Jin, L., et al.: Deep-learning-assisted detection and segmentation of rib fractures from CT scans: development and validation of fracnet. EBioMedicine **62**, 103106 (2020)
7. Li, H., Han, H., Zhou, S.K.: Bounding maps for universal lesion detection. In: Martel, A.L., et al. (eds.) MICCAI 2020. LNCS, vol. 12264, pp. 417–428. Springer, Cham (2020). https://doi.org/10.1007/978-3-030-59719-1_41
8. Li, Z., Zhang, S., Zhang, J., Huang, K., Wang, Y., Yu, Y.: MVP-Net: multi-view FPN with position-aware attention for deep universal lesion detection. In: Shen, D., et al. (eds.) MICCAI 2019. LNCS, vol. 11769, pp. 13–21. Springer, Cham (2019). https://doi.org/10.1007/978-3-030-32226-7_2
9. Lin, J., Gan, C., Han, S.: Tsm: temporal shift module for efficient video understanding. In: ICCV, pp. 7083–7093 (2019)
10. Lin, T.Y., Dollár, P., Girshick, R.B., He, K., Hariharan, B., Belongie, S.J.: Feature pyramid networks for object detection. In: VPR, pp. 936–944 (2016)
11. Litjens, G., et al.: A survey on deep learning in medical image analysis. Med. Image Anal. **42**, 60–88 (2017)
12. Luo, W., Li, Y., Urtasun, R., Zemel, R.S.: Understanding the effective receptive field in deep convolutional neural networks. In: NIPS (2016)

13. Qiu, Z., Yao, T., Mei, T.: Learning spatio-temporal representation with pseudo-3d residual networks. In: ICCV, pp. 5533–5541 (2017)
14. Rother, C., Kolmogorov, V., Blake, A.: GrabCut interactive foreground extraction using iterated graph cuts. ACM Trans. Graph. (TOG) **23**(3), 309–314 (2004)
15. Shen, D., Wu, G., Suk, H.I.: Deep learning in medical image analysis. Annu. Rev. Biomed. Eng. **19**, 221–248 (2017)
16. Tang, Y.B., Yan, K., Tang, Y.X., Liu, J., Xiao, J., Summers, R.M.: Uldor: a universal lesion detector for CT scans with pseudo masks and hard negative example mining. In: ISBI, pp. 833–836. IEEE (2019)
17. Wang, X., Han, S., Chen, Y., Gao, D., Vasconcelos, N.: Volumetric attention for 3d medical image segmentation and detection. In: Shen, D., et al. (eds.) MICCAI 2019. LNCS, vol. 11769, pp. 175–184. Springer, Cham (2019). https://doi.org/10.1007/978-3-030-32226-7_20
18. Wei, D., et al.: MitoEM dataset: large-scale 3D mitochondria instance segmentation from EM images. In: Martel, A.L., et al. (eds.) MICCAI 2020. LNCS, vol. 12265, pp. 66–76. Springer, Cham (2020). https://doi.org/10.1007/978-3-030-59722-1_7
19. Yan, K., Bagheri, M., Summers, R.M.: 3D context enhanced region-based convolutional neural network for end-to-end lesion detection. In: Frangi, A.F., Schnabel, J.A., Davatzikos, C., Alberola-López, C., Fichtinger, G. (eds.) MICCAI 2018. LNCS, vol. 11070, pp. 511–519. Springer, Cham (2018). https://doi.org/10.1007/978-3-030-00928-1_58
20. Yan, K., et al.: Mulan: Multitask universal lesion analysis network for joint lesion detection, tagging, and segmentation. In: MICCAI (2019)
21. Yan, K., et al.: Deep lesion graphs in the wild: relationship learning and organization of significant radiology image findings in a diverse large-scale lesion database. In: CVPR, pp. 9261–9270 (2018)
22. Yang, J., et al.: *AlignShift*: bridging the gap of imaging thickness in 3D anisotropic volumes. In: Martel, A.L., et al. (eds.) MICCAI 2020. LNCS, vol. 12264, pp. 562–572. Springer, Cham (2020). https://doi.org/10.1007/978-3-030-59719-1_55
23. Yang, J., et al.: Reinventing 2d convolutions for 3d images. IEEE J. Biomed. Health Inf. (2021)
24. Yang, J., Shi, R., Ni, B.: Medmnist classification decathlon: a lightweight automl benchmark for medical image analysis. In: ISBI (2021)
25. Zhang, S., et al.: Revisiting 3D context modeling with supervised pre-training for universal lesion detection in CT slices. In: Martel, A.L., et al. (eds.) MICCAI 2020. LNCS, vol. 12264, pp. 542–551. Springer, Cham (2020). https://doi.org/10.1007/978-3-030-59719-1_53
26. Zlocha, M., Dou, Q., Glocker, B.: Improving RetinaNet for CT lesion detection with dense masks from weak RECIST labels. In: Shen, D., et al. (eds.) MICCAI 2019. LNCS, vol. 11769, pp. 402–410. Springer, Cham (2019). https://doi.org/10.1007/978-3-030-32226-7_45

Detecting Outliers with Poisson Image Interpolation

Jeremy Tan[1]($\boxtimes$), Benjamin Hou[1], Thomas Day[2], John Simpson[2], Daniel Rueckert[1], and Bernhard Kainz[1,3]

[1] Imperial College London, SW7 2AZ London, UK
j.tan17@imperial.ac.uk
[2] King's College London, St Thomas' Hospital, SE1 7EH London, UK
[3] Friedrich–Alexander University Erlangen–Nürnberg, Erlangen, Germany

Abstract. Supervised learning of every possible pathology is unrealistic for many primary care applications like health screening. Image anomaly detection methods that learn normal appearance from only healthy data have shown promising results recently. We propose an alternative to image reconstruction-based and image embedding-based methods and propose a new self-supervised method to tackle pathological anomaly detection. Our approach originates in the foreign patch interpolation (FPI) strategy that has shown superior performance on brain MRI and abdominal CT data. We propose to use a better patch interpolation strategy, Poisson image interpolation (PII), which makes our method suitable for applications in challenging data regimes. PII outperforms state-of-the-art methods by a good margin when tested on surrogate tasks like identifying common lung anomalies in chest X-rays or hypoplastic left heart syndrome in prenatal, fetal cardiac ultrasound images. Code available at https://github.com/jemtan/PII.

Keywords: Outlier detection · Self-supervised learning

1 Introduction

Doctors such as radiologists and cardiologists, along with allied imaging specialists such as sonographers shoulder the heavy responsibility of making complex diagnoses. Their decisions often determine patient treatment. Unfortunately, diagnostic errors lead to death or disability almost twice as often as any other medical error [26]. In spite of this, the medical imaging workload has continued to increase over the last 15 years [3]. For instance, on-call radiology, which can involve high-stress and time-sensitive emergency scenarios, has seen a 4-fold increase in workload [3].

One of the major goals of anomaly detection in medical images is to find abnormalities that radiologists might miss due to excessive workload or inattention blindness [7]. Most of the existing, automated methods are only suitable for

© Springer Nature Switzerland AG 2021
M. de Bruijne et al. (Eds.): MICCAI 2021, LNCS 12905, pp. 581–591, 2021.
https://doi.org/10.1007/978-3-030-87240-3_56

detecting gross differences that are highly visible, even to observers without medical training. This undermines their usefulness in routine applications. Detecting anomalies at the level of medical experts typically requires supervised learning. This has been achieved for specific applications such as breast cancer [29] or retinal disease [6]. However, detecting arbitrary irregularities, without having any predefined target classes, remains an unsolved problem.

Recently, self-supervised methods have proven effective for unsupervised learning [11,17]. Some of these methods use a self-supervised task that closely approximates the target task [4,10] (albeit without labels). There are also self-supervised methods that closely approximate the task of outlier detection. For example, foreign patch interpolation (FPI) trains a model to detect foreign patterns in an image [25]. The self-supervised task used for training takes a patch from one sample and inserts it into another sample by linearly interpolating pixel intensities. This creates training samples with irregularities that range from subtle to more pronounced. But for data with poor alignment and varying brightness, FPI's linear interpolation will lead to patches that are clearly incongruous with the rest of the image. This makes the self-supervised task too easy and reduces the usefulness of the learned features.

Contribution: We propose Poisson image interpolation (PII), a self-supervised method that trains a model to detect subtle irregularities introduced via Poisson image editing [18]. We demonstrate the usefulness of PII for anomaly detection in chest X-ray and fetal ultrasound data. Both of these are challenging datasets for conventional anomaly detection methods because the normal data has high variation and outliers are subtle in appearance.

Related Work: Reconstruction-based outlier detection approaches can use autoencoders, variational autoencoders (VAEs) [33], adversarial autoencoders (AAEs) [5], vector quantised variational autoencoders (VQ-VAE) [20], or generative adversarial networks (GANs) [22]. Some generative models are also used for pseudo-healthy image generation [30]. Reconstruction can be performed at the image [2], patch [28], or pixel [1] level. In each case, the goal is to replicate test samples as closely as possible using only features from the distribution of normal samples [2]. Abnormality is then measured as intensity differences between test samples and their reconstructions. Unfortunately, raw pixel differences lack specificity, making semantic distinctions more difficult.

Disease classifiers specialize in making fine semantic distinctions. They do this by learning very specific features and ignoring irrelevant variations [13]. To harness the qualities that make classifiers so successful, some methods compare samples as embeddings within a learned representation. For example, deep support vector data description (SVDD) learns to map normal samples to a compact hypersphere [21]. Abnormality is then measured as distance from the center of the hypersphere. Other methods, such as [15], learn a latent representation using a VQ-VAE and exploit the autoregressive component to estimate the likelihood of a sample. Furthermore, components of the latent code with low likelihood can be replaced with samples from the learned prior. This helps to prevent the model from reconstructing anomalous features. A similar approach has also been proposed using transformers [19]. Overall, comparing samples in a learned representation space can allow for more semantic distinctions. But with only normal training examples, the learned representation may emphasize irrelevant features, *i.e.*, those pertaining to variations *within* the normal class. This often requires careful calibration for applications.

Self-supervised methods aim to learn more relevant representations by training on proxy tasks. One of the most effective strategies is to train a classifier to recognize geometric transformations of normal samples [9, 24]. By classifying transformations, the network learns prominent features that can act as reliable landmarks. Outliers that lack these key features will be harder to correctly classify when transformed. The anomaly score is thus inversely proportional to the classification accuracy. This works well for natural images, but in medical applications, disease appearance can be subtle. Many outliers still contain all of the major anatomical landmarks.

To target more subtle abnormalities, some methods use a localized self-supervised task. For example, FPI synthesizes subtle defects within random patches in an image [25]. The corresponding pixel-level labels help the network to learn which regions are abnormal given the surrounding context. This approach showed good performance for spatially aligned brain MRI and abdominal CT data [34]. CutPaste [14] also synthesizes defects by translating patches within an image. This is effective for detecting damage or manufacturing defects in natural images [14]. However, unlike cracks or scratches seen in manufacturing, many medical anomalies do not have sharp discontinuities. Overfitting to obvious differences between the altered patch and its surroundings can limit generalization to more organic and subtle outliers. We propose to resolve this issue using Poisson image editing [18]. This helps to create more subtle defects (for training) which in turn improves generalization to real abnormalities.

2   Method

To begin, we provide a brief description of FPI. Consider two normal training samples, x_i and x_j, of dimension $N \times N$, as well as a random patch h, and a random interpolation factor $\alpha \in [0, 1]$. FPI replaces the pixels in patch h with a convex combination of x_i and x_j, to produce a training image $\widetilde{x}_i$ (Eq. 1). Note that $\widetilde{x}_i = x_i$ outside of h. For a given training image $\widetilde{x}_i$, the corresponding label is $\widetilde{y}_i$, as specified by Eq. 2.

$$\widetilde{x}_{i_p} = (1 - \alpha)x_{i_p} + \alpha x_{j_p} , \ \forall \, p \in h \tag{1}$$

$$\widetilde{y}_{i_p} = \begin{cases} \alpha & \text{if } p \in h \\ 0 & \text{otherwise} \end{cases} \tag{2}$$

This approach has similarities to mixup [32], a data augmentation method that generates convex combinations of images and their respective labels. In the case of FPI, the training data only contains normal samples. Without having any class labels, FPI calculates its own labels as convex combinations of self (0 for y_i) and non-self (1 for y_j) as shown in Eq. 2.

If x_i and x_j have vastly different intensity levels or structures, the interpolated patch will be inconsistent with the rest of the image. These differences are easy to spot and provide no incentive for the model to learn features that constitute "normal" (a much harder task). To create more challenging cases, we use a technique for seamless image blending. Poisson image editing [18] blends the content of a source image (x_j) into the context of a destination image (x_i). Rather than taking the raw intensity values from the source, we extract the relative intensity differences across the image, *i.e.* the image gradient. Combining the gradient with Dirichlet boundary conditions (at the edge of the patch) makes it possible to calculate the absolute intensities within the patch. This is illustrated in Fig. 1.

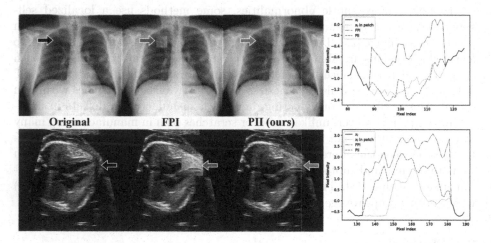

Fig. 1. Examples of patches altered by FPI (convex combination) and PII (Poisson blending). Arrows in the images indicate the location of the line plotted on the right. Altering patches can simulate subtle (top) or dramatic (bottom) changes to anatomical structures. In both cases, PII blends the changes into the image more naturally.

More formally, let f_{in} be a scalar function representing the intensity values within the patch h. The goal is to find intensity values of f_{in} that will:

1. match the surrounding values, f_{out}, of the destination image, along the border of the patch (∂h), and
2. follow the relative changes (image gradient), $\mathbf{v}$, of the source image.

$$\min_{f_{in}} \iint_h |\nabla f_{in} - \mathbf{v}|^2 \quad \text{with} \quad f_{in}\Big|_{\partial h} = f_{out}\Big|_{\partial h} \tag{3}$$

$$\Delta f_{in} = \operatorname{div} \mathbf{v} \text{ over } h, \quad \text{with} \quad f_{in}\Big|_{\partial h} = f_{out}\Big|_{\partial h} \tag{4}$$

These conditions are specified in Eq. 3 [18] and its solution is the Poisson equation (Eq. 4). Intuitively, the Laplacian, $(\Delta \cdot = \frac{\partial^2 \cdot}{\partial x^2} + \frac{\partial^2 \cdot}{\partial y^2})$, should be close to zero in regions that vary smoothly and have a larger magnitude in areas where the gradient ($\mathbf{v}$) changes quickly.

To find f_{in} for discrete pixels, a finite difference discretization can be used. Let p represent a pixel in h and let $q \in N_p$ represent the four directly adjacent neighbours of p. The solution should satisfy Eq. 5 (or Eq. 6 if any neighbouring pixels q overlap with the patch boundary ∂h) [18].

$$|N_p| f_{in_p} - \sum_{q \in N_p} f_{in_q} = \sum_{q \in N_p} \mathbf{v}_{pq} \tag{5}$$

$$\sum_{q \in N_p \cap h} \left(f_{in_p} - f_{in_q} \right) = \sum_{q \in N_p \cap \partial h} f_{out_q} + \sum_{q \in N_p} \mathbf{v}_{pq} \tag{6}$$

In our case the image gradient comes from finite differences in the source image x_j, i.e. $\mathbf{v}_{pq} = x_{j_p} - x_{j_q}$. Meanwhile, the boundary values, f_{out_q}, come directly from the destination image, x_{i_q}. This system can be solved for all $p \in h$ using an iterative solver. In some cases, this interpolation can cause a smearing effect. For example, when there is a large difference between the boundary values at opposite ends of the patch, but the gradient within the patch (from x_j) is very low. To compensate for this, Perez et al. suggest using the original gradient (from x_i) if it is larger than the gradient from x_j (Eq. 7) [18]. We modify this to introduce the interpolation factor, α, that FPI uses to control the contribution of x_j to the convex combination. In this case, α controls which image gradients take precedence (Eq. 8). This creates more variety in training samples, i.e. more ways in which two patches can be combined. It also helps create a self-supervised task with varying degrees of difficulty, ranging from very subtle to more prominent structural differences. Figure 1 demonstrates that this formulation can blend patches seamlessly.

$$\mathbf{v}_{pq} = \begin{cases} x_{i_p} - x_{i_q} & \text{if } |x_{i_p} - x_{i_q}| > |x_{j_p} - x_{j_q}| \\ x_{j_p} - x_{j_q} & \text{otherwise} \end{cases} \tag{7}$$

$$\mathbf{v}_{pq} = \begin{cases} (1 - \alpha)(x_{i_p} - x_{i_q}) & \text{if } |(1 - \alpha)(x_{i_p} - x_{i_q})| > |\alpha(x_{j_p} - x_{j_q})| \\ \alpha(x_{j_p} - x_{j_q}) & \text{otherwise} \end{cases} \tag{8}$$

PII uses the same loss as FPI, which is essentially a pixel-wise regression of the interpolation factor α [25]. The loss is given in Eq. 9:

$$\mathcal{L}_{bce} = -\tilde{y}_{i_p} log A_s(\tilde{x}_{i_p}) - (1 - \tilde{y}_{i_p}) \log(1 - A_s(\tilde{x}_{i_p})) \tag{9}$$

The inputs, $\tilde{x}_i$, are training samples that contain a random patch with values f_{in}, computed via Poisson blending as described above. The output of the model is used directly as an anomaly score, A_s. A diagram of the setup is given in Fig. 2.

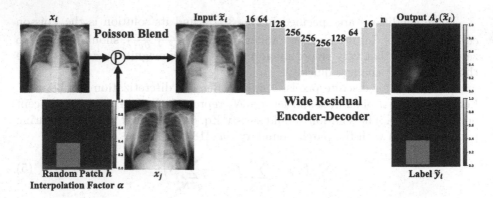

Fig. 2. Illustration of PII self-supervised training. The network architecture starts with a single convolutional layer (gray), followed by residual blocks (blue/green). Values above each block indicate the number of feature channels in the convolutional layers. In all experiments we use a single output channel, i.e. $n = 1$. (Color figure online)

Architecture and Specifications:
We follow the same network architecture as FPI, a wide residual encoder-decoder [25]. The encoder is a standard wide residual network [31] with a width of 4 and depth of 14. The decoder has the same structure but in reverse. The output has the same shape as the input and uses a sigmoid activation. Training is done using Adam [12] with a learning rate of 10^{-3} for 50 epochs. The self-supervised task uses patches, h, that are randomly generated with size $h_s \sim U(0.1N, 0.4N)$, and center coordinates $h_c \sim U_2(0.1N, 0.9N)$. Each patch is also given a random interpolation factor $\alpha \sim U(0.05, 0.95)$.

Implementation: We use TensorFlowV1.15 and train on a Nvidia TITAN Xp GPU. Training on our largest dataset for 50 epochs takes about 11 h. PII solves partial differential equations on the fly to generate training samples dynamically. To achieve this we use multiprocessing [16] to generate samples in parallel. The code is available at https://github.com/jemtan/PII.

3   Evaluation and Results

To evaluate the performance of PII, we compare with an embedding-based method, a reconstruction-based method, and a self-supervised method. For the embedding-based method, we use Deep SVDD [21] with a 6 layer convolutional neural network. Meanwhile, a vector-quantized variational autoencoder (VQ-VAE2) [20] is used as a reconstruction-based method. The VQ-VAE2 is trained using the same wide residual encoder-decoder architecture as PII, except the decoder is given the same capacity as the encoder to help produce better reconstructions. FPI [25], which our method builds upon, is used as a self-supervised benchmark method. To compare each method, we calculate the average precision (AP) for each of the datasets described below. Average precision is a scalar metric for the area under the precision-recall curve.

Data: Our first dataset is ChestX-ray14 [27], a public chest X-ray dataset with 108,948 images from 32,717 patients showing 14 pathological classes as well as a normal class. From this large dataset, we extract 43,322 posteroanterior (PA) views of adult patients (over 18) and split them into male (σ) and female (φ) partitions. All X-ray images are resized to 256×256 (down from 1024×1024) and normalized to have zero mean and unit standard deviation. The training/test split is summarized in Table 1.

The second dataset consists of a total of 13380 frames from 108 patients acquired during routine fetal ultrasound screening. This is an application where automated anomaly detection in screening services would be of most use. We use cardiac standard view planes [8], specifically 4-chamber heart (4CH) and 3-vessel and trachea (3VT) views, from a private and de-identified dataset of ultrasound videos. For each selected standard plane, 20 consecutive frames (10 before and 9 after) are extracted from the ultrasound videos. Images are 224×288 and are normalized to zero mean, unit standard deviation. Normal samples consist of healthy images from a single view (4CH/3VT) and anomalous images are composed of alternate views (3VT/4CH) as well as pathological hearts of the same view (4CH/3VT). For pathology we use cases of hypoplastic left heart syndrome (HLHS), a condition that affects the development of the left side of the heart [23]. The training/test split is outlined in Table 1. The scans are of volunteers at 18–24 weeks gestation (Ethics: *anonymous during review*), in a fetal cardiology clinic, where patients are referred to from primary screening and secondary care sites. Video clips have been acquired on Toshiba Aplio i700, i800 and Philips EPIQ V7 G devices.

Results: We compare our method with recent state-of-the-art anomaly detection methods in Table 1.

Example test images are shown in Fig. 3. The VQ-VAE2 reconstruction error indicates that sharp edges are difficult to reproduce accurately. Meanwhile FPI is sensitive to sharp edges because of the patch artifacts produced during training. In contrast, PII is sensitive to specific areas that appear unusual.

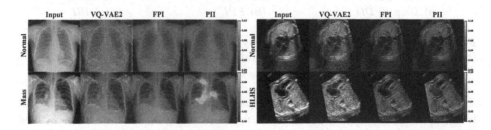

Fig. 3. Examples of test X-ray (left) and ultrasound (right) images with pixel-wise anomaly scores from each method. Note that the VQ-VAE2 reconstruction error is scaled down by a factor of 10.

Table 1. Each dataset is presented in one column. The train-test split is shown for each partition (top). Note that ultrasound images are extracted from videos as 20 frame clips. Average precision is also listed for each method (bottom).

Dataset	Chest X-ray		Fetal US	
	♂PA	♀PA	4CH	3VT
	Number of images			
Normal Train	17852	14720	283 × 20	225 × 20
Normal Test	2634	2002	34 × 20	35 × 20
Anomalous Test	3366	2748	54 × 20	38 × 20
	Average precision			
Deep SVDD	0.565	0.556	0.685	0.893
VQ-VAE2	0.503	0.516	0.617	0.578
FPI	0.533	0.586	0.658	0.710
PII	**0.690**	**0.703**	**0.723**	**0.929**

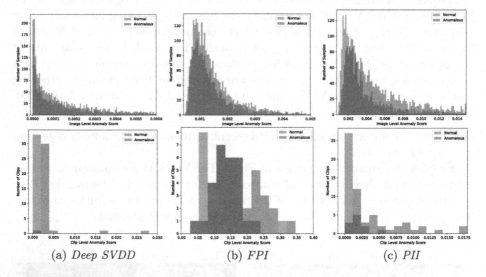

(a) *Deep SVDD*   (b) *FPI*   (c) *PII*

Fig. 4. Histograms of image level anomaly scores for Chest X-ray Female PA data (top) and clip level anomaly scores for Fetal US 3VT data (bottom).

To see each method's ability to separate normal from anomalous, we plot histograms of anomaly scores for each method in Fig. 4. The difficulty of these datasets is reflected in the fact that existing methods have almost no discriminative ability. On average, PII gives anomalous samples slightly higher scores than normal samples. The unusually high performance in the 3VT dataset is partly due to the small size of the dataset as seen in Fig. 4.

Discussion: We have shown that our method is suitable to detect pathologies when they are considered anomalies compared to a training set that contains only healthy subjects. Training from only normal data is an important aspect in our field since a) data from healthy volunteers is usually available more easily, b) prevalence for certain conditions is low, thus collecting a well balanced training set is challenging and c) supervised methods would require in the ideal case equally many samples from every possible disease they are meant to detect. The latter is particularly a problem in rare diseases where the number of patients are very low in the global population.

4   Conclusion

In this work we have discussed an alternative to reconstruction-based anomaly detection methods. We base our method on the recently introduced FPI method, which formulates a self-supervised task through patch-interpolation based on normal data only. We advance this idea by introducing Poisson Image Interpolation, which mitigates interpolation issues for challenging data like chest X-Rays and fetal ultrasound examinations of the cardio-vascular system. In future work we will explore spatio-temporal support for PII, which is in particular relevant for ultrasound imaging.

Acknowledgements. Support from Wellcome Trust IEH Award iFind project [102431] and UK Research and Innovation London Medical Imaging and Artificial Intelligence Centre for Value Based Healthcare. JT was supported by the ICL President's Scholarship.

References

1. Alaverdyan, Z., Jung, J., Bouet, R., Lartizien, C.: Regularized siamese neural network for unsupervised outlier detection on brain multiparametric magnetic resonance imaging: application to epilepsy lesion screening. Med. Image Anal. **60**, 101618 (2020)
2. Baur, C., Denner, S., Wiestler, B., Navab, N., Albarqouni, S.: Autoencoders for unsupervised anomaly segmentation in brain MR images: a comparative study. Med. Image Anal. **69**, 101952 (2021)
3. Bruls, R., Kwee, R.: Workload for radiologists during on-call hours: dramatic increase in the past 15 years. Insights into Imaging **11**(1), 1–7 (2020)
4. Chen, T., Kornblith, S., Norouzi, M., Hinton, G.: A simple framework for contrastive learning of visual representations. In: International Conference on Machine Learning, pp. 1597–1607. PMLR (2020)
5. Chen, X., Konukoglu, E.: Unsupervised detection of lesions in brain MRI using constrained adversarial auto-encoders. In: MIDL Conference book, MIDL (2018)
6. De Fauw, J., et al.: Clinically applicable deep learning for diagnosis and referral in retinal disease. Nat. Med. **24**(9), 1342–1350 (2018)
7. Drew, T., Võ, M., Wolfe, J.: The invisible gorilla strikes again: sustained inattentional blindness in expert observers. Psychol. Sci. **24**(9), 1848–1853 (2013)

8. Fasp, N.: NHS Fetal Anomaly Screening Programme Handbook Valid from August 2018. Technical Report (2018). www.facebook.com/PublicHealthEngland
9. Golan, I., El-Yaniv, R.: Deep anomaly detection using geometric transformations. In: Advances in Neural Information Processing Systems, pp. 9758–9769 (2018)
10. He, K., Fan, H., Wu, Y., Xie, S., Girshick, R.: Momentum contrast for unsupervised visual representation learning. In: Proceedings of the IEEE/CVF Conference on Computer Vision and Pattern Recognition, pp. 9729–9738 (2020)
11. Hénaff, O.J., et al.: Data-efficient image recognition with contrastive predictive coding. arXiv preprint arXiv:1905.09272 (2019)
12. Kingma, D.P., Ba, J.: Adam: A method for stochastic optimization. arXiv preprint arXiv:1412.6980 (2014)
13. LeCun, Y., Bengio, Y., Hinton, G.: Deep learning. Nature **521**(7553), 436–444 (2015)
14. Li, C.L., Sohn, K., Yoon, J., Pfister, T.: Cutpaste: self-supervised learning for anomaly detection and localization. In: Proceedings of the IEEE/CVF Conference on Computer Vision and Pattern Recognition, pp. 9664–9674 (2021)
15. Marimont, S.N., Tarroni, G.: Anomaly detection through latent space restoration using vector quantized variational autoencoders. In: 2021 IEEE 18th International Symposium on Biomedical Imaging (ISBI), pp. 1764–1767. IEEE (2021)
16. McKerns, M.M., Strand, L., Sullivan, T., Fang, A., Aivazis, M.A.: Building a framework for predictive science. arXiv preprint arXiv:1202.1056 (2012)
17. Oord, A.v.d., Li, Y., Vinyals, O.: Representation learning with contrastive predictive coding. arXiv preprint arXiv:1807.03748 (2018)
18. Pérez, P., Gangnet, M., Blake, A.: Poisson image editing. In: ACM SIGGRAPH 2003 Papers, pp. 313–318 (2003)
19. Pinaya, W.H.L., et al.: Unsupervised brain anomaly detection and segmentation with transformers. arXiv preprint arXiv:2102.11650 (2021)
20. Razavi, A., Oord, A.V.D., Vinyals, O.: Generating diverse high-fidelity images with vq-vae-2. In: ICLR Workshop DeepGenStruct (2019)
21. Ruff, L., et al.: Deep one-class classification. In: International Conference on Machine Learning, pp. 4393–4402. PMLR (2018)
22. Schlegl, T., Seebōck, P., Waldstein, S.M., Langs, G., Schmidt-Erfurth, U.: f-anogan: fast unsupervised anomaly detection with generative adversarial networks. Med. Image Anal. **54**, 30–44 (2019)
23. Simpson, J.: Hypoplastic left heart syndrome. Ultrasound Obstet. Gynecol. Official J. Int. Soc. Ultrasound Obstet. Gynecol. **15**(4), 271–278 (2000)
24. Tack, J., Mo, S., Jeong, J., Shin, J.: Csi: novelty detection via contrastive learning on distributionally shifted instances. arXiv preprint arXiv:2007.08176 (2020)
25. Tan, J., Hou, B., Batten, J., Qiu, H., Kainz, B.: Detecting outliers with foreign patch interpolation. arXiv preprint arXiv:2011.04197 (2020)
26. Tehrani, A.S.S., et al.: 25-year summary of us malpractice claims for diagnostic errors 1986–2010: an analysis from the national practitioner data bank. BMJ Qual. Saf. **22**(8), 672–680 (2013)
27. Wang, X., Peng, Y., Lu, L., Lu, Z., Bagheri, M., Summers, R.M.: Chestx-ray8: hospital-scale chest x-ray database and benchmarks on weakly-supervised classification and localization of common thorax diseases. In: Proceedings of the IEEE Conference on Computer Vision and Pattern Recognition, pp. 2097–2106 (2017)
28. Wei, Q., Ren, Y., Hou, R., Shi, B., Lo, J.Y., Carin, L.: Anomaly detection for medical images based on a one-class classification. In: Medical Imaging 2018: Computer-Aided Diagnosis, vol. 10575, p. 105751M. International Society for Optics and Photonics (2018)

29. Wu, N., et al.: Deep neural networks improve radiologists' performance in breast cancer screening. IEEE Trans. Med. Imaging **39**(4), 1184–1194 (2019)
30. Xia, T., Chartsias, A., Tsaftaris, S.A.: Pseudo-healthy synthesis with pathology disentanglement and adversarial learning. Med. Image Anal. **64**, 101719 (2020)
31. Zagoruyko, S., Komodakis, N.: Wide residual networks. In: Richard C. Wilson, E.R.H., Smith, W.A.P. (eds.) Proceedings of the British Machine Vision Conference (BMVC), pp. 87.1-87.12. BMVA Press (September 2016). https://doi.org/10.5244/C.30.87, https://dx.doi.org/10.5244/C.30.87
32. Zhang, H., Cisse, M., Dauphin, Y.N., Lopez-Paz, D.: mixup: Beyond empirical risk minimization. International Conference on Learning Representations (2018)
33. Zimmerer, D., Kohl, S., Petersen, J., Isensee, F., Maier-Hein, K.: Context-encoding variational autoencoder for unsupervised anomaly detection. In: International Conference on Medical Imaging with Deep Learning-Extended Abstract Track (2019)
34. Zimmerer, D., et al.: Medical out-of-distribution analysis challenge (2020)

MG-NET: Leveraging Pseudo-imaging for Multi-modal Metagenome Analysis

Sathyanarayanan N. Aakur[1]([✉]), Sai Narayanan[2], Vineela Indla[1],
Arunkumar Bagavathi[1], Vishalini Laguduva Ramnath[1],
and Akhilesh Ramachandran[2]

[1] Department of Computer Science, Oklahoma State University, Stillwater, OK, USA
{saakurn,vindla,abagava,vlagudu}@okstate.edu
[2] Oklahoma Animal Disease Diagnostic Laboratory, College of Veterinary Medicine,
Oklahoma State University, Stillwater, OK, USA
{ssankar,rakhile}@okstate.edu

Abstract. The emergence of novel pathogens and zoonotic diseases like
the SARS-CoV-2 have underlined the need for developing novel diagnosis
and intervention pipelines that can learn rapidly from small amounts of
labeled data. Combined with technological advances in next-generation
sequencing, metagenome-based diagnostic tools hold much promise to
revolutionize rapid point-of-care diagnosis. However, there are significant challenges in developing such an approach, the chief among which
is to learn self-supervised representations that can help detect novel
pathogen signatures with very low amounts of labeled data. This is particularly a difficult task given that closely related pathogens can share
more than 90% of their genome structure. In this work, we address these
challenges by proposing MG-Net, a self-supervised representation learning framework that leverages multi-modal context using pseudo-imaging
data derived from clinical metagenome sequences. We show that the proposed framework can learn robust representations from *unlabeled data*
that can be used for downstream tasks such as metagenome sequence
classification with limited access to labeled data. Extensive experiments
show that the learned features outperform current baseline metagenome
representations, given only 1000 samples per class.

Keywords: Metagenome analysis · Automatic diagnosis with
metagenomics · Multi-modal disease intervention

1 Introduction

Advances in DNA sequencing technologies [21,22] have made possible the determination of whole-genome sequences of simple unicellular (e.g., bacteria) and

Electronic supplementary material The online version of this chapter (https://
doi.org/10.1007/978-3-030-87240-3_57) contains supplementary material, which is
available to authorized users.

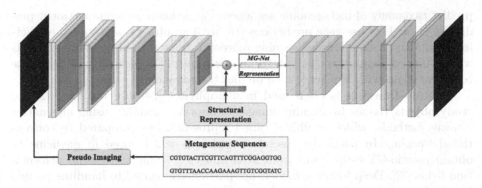

Fig. 1. Overall Architecture. Our approach, MG-Net, is illustrated here. There are three major components (i) a global structural feature, (ii) a pseudo-imaging module to generate multi-modal representation, and (iii) integrated structural reasoning with attention for learning robust features with self-supervision.

complex multicellular (e.g., human) organisms at a cheaper, faster, and larger scale. The abundance of collected genome sequences require reliable and scalable frameworks to detect novel pathogens and study further mutations of such pathogens to mitigate threatening disease transmissions. Zoonotic diseases, like SARS-CoV-2, are a prime example for the need for rapid learning from noisy and limited data, due to their ability to mutate and cause pandemic situations. To this end, DNA sequencing-based approaches, such as metagenomics, have been explored by several researchers [7,11] for plant and animal disease diagnostics. Metagenome-based diagnostics are pathogen agnostic and theoretically have unlimited multiplexing capability. Unlike traditional methods, metagenome-based diagnostics can also provide information on the host's genetic makeup that can aid in personalized medicine [10]. However, metagenome diagnostics encounter the problem of long-tail distribution of pathogen sequences in the data. The problem aggravates for pathogen detection tasks when we consider pathogens from the same genus. For example, *Mannheimia haemolytica* and *Pasteurella multocida* share as much as 95.5% of their genome. In this work, we consider the Bovine Respiratory Disease Complex (BRD) as a model and aim to detect the presence of six associated bacterial pathogens, namely *Mannheimia haemolytica, Pasteurella multocida, Bibersteinia trehalosi, Histophilus somni, Mycoplasma bovis,* and *Trueperella pyogenes.*

One of the major challenges in metagenome-based diagnostics is the need for specialized bioinformatics pipelines for analysis of the enormous amounts of DNA sequences for detecting disease markers [7,11]. Machine learning research offers several opportunities to analyze DNA sequences collected directly from environment samples [6]. Deep learning models, in particular, have been explored for representation learning from metagenome sequences for many associated tasks including, but not limited to: capturing simple nucleotide representations with reverse complement CNNs and LSTMs [4], depth-wise separable convolutions to predict taxonomy of metagenome sequences [5], genomic sub-compartment prediction [1] and disease gene predictions [13] with graph representations, predicting taxonomy of sequences by learning representations with bidirectional LSTMs with k-mer embedding and self attention mechanism [18], learning metagenome representations with ResNet [12] to predict the taxonomy.

Pseudo-Imaging is often used in astrophysics [24] and medicine [27] to study objects/tissues by forming images in another modality using alternative sensing methods, which exhibit detailed representations compared to conventional imaging. In particular, pseudo imaging is widely used in medicine to obtain pseudo-CT estimations from MRI images [17] and ultrasound deformation fields [32]. Deep learning models are particularly suited to handling pseudo image data, due to the success of convolutional neural networks in computer vision research [12,30]. However, there have been very few methods used for metagenome analysis. For example, Self-Organizing Maps (SOM) [26] and Growing Self-Organizing Maps (GSOM) [25] have been used to represent metagenome sequences as images and a shallow CNN model was used for disease prediction. A matrix representation of a polygenetic tree has been used with CNN to predict host phenotype of metagenome sequences [28].

2   MG-NET: Leveraging Pseudo-imaging for Multimodal Metagenome Analysis

In this section, we introduce our MG-NET framework for extracting robust, self-supervised representations from metagenome data. Our approach has three major components: (i) capturing a global structural prior for each metagenome sequence conditioned on the metagenome structure, (ii) extracting local structural features from pseudo-images generated from metagenome sequences, and (iii) integrate local and global structural features in an integrated, attention-driven structural reasoning module for multi-modal feature extraction. The overall approach is illustrated in Fig. 1. We jointly model the global and local structural properties in a unified framework, which is trained in a self-supervised manner, *without labels*, to capture robust representations aimed for metagenome classification with limited and unbalanced labeled data.

2.1   Capturing the Global Structure with Graph Representations

First, we construct a global graph representation of the *entire* metagenome sample, i.e., the graph provides a structural representation of the sequenced clinical sample. We take inspiration from the success of De Bruijn graphs for genome analysis [20,23] and use a modified version to represent the metagenome sample. Given a metagenome sample $\mathcal{X}_j$ with sequence reads $X_0, X_1, \ldots X_n \in \mathcal{X}_j$, we construct a weighted, directed graph whose nodes are populated by k-mers x_j such that $x_0, x_1, \ldots x_l \in X_i$. Each k-mer is a subsequence from a genome read X_i of length k, extracted using a sliding window of length k and stride s. Each edge direction is determined by order of occurrence of each observed k-mer

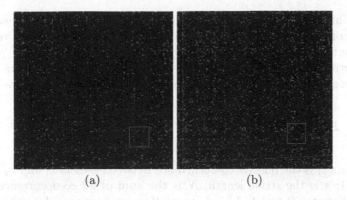

(a)   (b)

Fig. 2. Pseudo-Images generated from our approach. Highly distinct patterns returned from attention maps for the species *Mannheimia haemolytica* from different clinical samples are shown with a red bounding box. (Color figure online)

in the sliding window. The edge weights are iteratively updated based on the observation of the co-occurrence of the nodes and are a function of the frequency of co-occurrence of the k-mers. The updated weights are given by

$$\Psi(x_i, x_j) = f_s\left(\frac{e_{i,j}}{(\|e_{i,j} - e'_{i,j}\|_2)}\right) \tag{1}$$

where $e_{i,j}$ is the current weight between the k-mer nodes x_i and x_j and $e'_{i,j}$ is the new weight to be updated; $f_s(\cdot)$ is a weighted update function that bounds the new weight within a given range. In our experiments, we bound the edge weights to be between -2 and 2 and hence set $f_s(q) = 2\sqrt{max(q-1,1)} + (min(q-2,2)+2)$ to capture the *relative* increase in frequency to highlight structures that emerge through repeated co-occurrence while suppressing spurious links. The edge weights are initially set to 1. Given this structural representation, we extract features for each k-mer using *node2vec* [9] to capture the "community" or neighborhood structure of a k-mer to reject clutter due to observation noise [16]. The resulting representation x_i^{st} for each k-mer x_i captures its neighborhood within a sequence and provides a structural prior over the metagenome structure. The global structural representation for a sequence X_i is the average-pooled (AP) representations of each k-mer given by $X_i^g = AP(\{x_1^{st}, x_2^{st}, \ldots x_n^{st}\})$.

2.2   Pseudo-Imaging for Local Structural Properties

The second step in our framework is to generate a pseudo-image I_r for each metagenome sequence read $X_r \in \mathcal{X}$. The key intuition behind generating a pseudo-image is to represent and learn recurring patterns (or "fingerprints" [31]) in metagenome sequence reads that belong to the same pathogen species automatically. For example, in Fig. 2, it can be seen that two sequences from the same pathogen *Mannheimia haemolytica* have recurring patterns across clinical samples. Inspired by the success of Gray Level Co-occurrence Matrix (GLCM) [2,3], we use a histogram-based formulation to provide a visual representation of a metagenome sequence. Instead of binary co-occurrence, we use *relative* co-occurrence to generate images with varying intensity. Each "*pixel*" $p_{i,j} \in I_r$ is representative of the frequency of co-occurrence between the k-mers x_i and x_j in a sequence read. The resulting pseudo-image representation is given by

$$I_r(i,j) = \sum_{i=1}^{4^k} \sum_{j=i+s}^{4^k} \begin{cases} 255 * f(i,j)/N, & \text{if } f(x_i, x_j) > \lambda_{min} \\ 0, & \text{otherwise} \end{cases} \tag{2}$$

where $f(x_i, x_j)$ is the relative co-occurrence between k-mers x_i and x_j computed using Eq. 1; s is the stride length; N is the sum of all co-occurrences to scale the value between 0 and 1; λ_{min} is a cutoff parameter to reduce the impact of noise introduced due to any read errors [16]. Note that the $e(i,j)$ is computed *per sequence read* and not at the sample level as done in Sect. 2.1. This allows us to model sequence-level patterns and hence capture species-specific patterns. The depth of the image is set to be 3 and the pixel values are duplicated to simlate an RGB image and hence allow us to leverage advances in deep neural networks to extract automated features. In our experiments, we set $\lambda_{min} = 0$.

2.3   Structural Reasoning with Attention

The third and final step in our framework is to use the global representations (Sect. 2.1) to help learn local structural properties from pseudo-images (Sect. 2.2) using attention as a structural reasoning mechanism. Specifically, we use a convolutional neural network (CNN) to extract local structural features from a given pseudo-image. As can be seen from Fig. 1, we use the intermediate (4th convolutional block) layer of the CNN as a local feature representation X_i^{lc} of a sequence read X_i. We obtain a robust representation by using attention-based reasoning mechanism given by

$$X_i^{MG} = GAP(softmax(X_i^g \odot X_i^{lc}) \odot X_i^g) \tag{3}$$

where GAP refers to the Global Average Pooling function [19] and X_i^g is global structural feature representation provided by the global graph representation from Sect. 2.1. We train the network end-to-end by having a decoder block (a mirrored network of deconvolutional operations) to reconstruct the input pseudo-image from this structural representation. Note that our goal is to learn robust representations from limited labeled metagenome data rather than reconstruction or segmentation. Hence, adding skip connections like U-Net [29] will allow the network to "cheat" and not learn robust, "compressed" representations. We augment the features from the CNN with structural features from the node2vec representations using Eq. 3, where we flatten the feature maps so that they match the dimensions for element-wise multiplication in the attention mechanism. We use L2-norm between the reconstructed and the actual pseudo-image as the objective function to train the network in a self-supervised manner. Formally, we define the loss function as $\mathcal{L}_{recons} = \|I_i' - I_i\|^2$, where I_i' and I_i refer to the reconstructed and actual pseudo-images, respectively. We show empirically (Sect. 3) that the integrated reasoning during training enhances the performance as opposed to mere concatenation of auto-encoder and global features.

2.4 Implementation Details

We use a 4-layer convolutional neural network based on VGG-16 [30] as our feature extractor in Sect. 2.2. We mirror the network to have a 4-layer decoder network to reconstruct the pseudo-image. The network is trained end-to-end for 25 epochs with a batch size of 64 and converges in about 30 min on a server with an NVIDIA Titan RTX and a 32-core AMD ThreadRipper CPU. The extracted representations are then finetuned for 10 epochs with a 3-layer deep neural network for pathogen classification. The learning rate is set to 1×10^{-4} for both stages and optimized using the standard gradient descent optimizer. Empirically, we find that having a k-mer length of 5 and stride of 10 provides the best results and present other variations in the ablation study (Sect. 3) for completeness. Default parameters were used for node2vec representations. All networks are trained from scratch during the pre-training phase.

Table 1. Recognition results. Performance of the proposed MG-NET with varying number of training samples the recognition task on clinical metagenome data. Precision and Recall are reported for each class. Note that *P. Multocida* and *B. Trehalosi* have maximum of 21 and 17 samples across all settings, respectively.

Max training sequences/ Class	Host		H. somni.		M. bovis.		M. haemo.		P. multo.		T. pyoge.		B. treha.	
	Prec.	Rec.	Prec.	Rec.	Prec.	Rec.	Prec.	Rec.	Prec.	Rec.	Prec.	Rec.	Prec.	Rec.
0	0.97	0.23	0.26	0.61	0.60	0.50	0.01	0.08	0.01	0.16	0.02	0.93	0.00	0.00
25	0.96	0.23	0.13	0.02	0.48	0.15	0.02	0.33	0.02	0.13	0.00	0.07	0.00	0.03
100	0.87	0.27	0.07	0.05	0.48	0.74	0.03	0.13	0.00	0.01	0.00	0.00	0.00	0.00
250	0.90	0.34	0.16	0.09	0.42	0.90	0.03	0.14	0.00	0.01	0.01	0.87	0.00	0.00
500	0.97	0.85	0.66	0.46	0.59	0.96	0.33	0.67	0.03	0.01	0.10	1.00	0.00	0.00
1000	0.98	0.89	0.70	0.57	0.68	0.96	0.34	0.61	0.04	0.01	0.14	1.00	0.00	0.10
MG-Net (All)	**0.99**	**0.98**	**0.84**	**0.74**	**0.88**	**0.95**	**0.59**	**0.69**	**0.42**	**0.18**	**0.94**	**1.00**	**0.02**	**0.33**
Node2Vec [23] (All)	0.87	0.65	0.16	0.39	0.19	0.14	0.04	0.24	0.05	0.01	0.02	0.13	0.00	0.00

3   Experimental Evaluation

Data Collection. For constructing the dataset for the training and evaluation of automated metagenome-based pathogen detection, we collected metagenome sequences from 13 Bovine Respiratory Disease Complex (BRDC) lung specimens at a local (name redacted to preserve anonymity) diagnostic laboratory using the DNeasy Blood and Tissue Kit (Qiagen, Hilden, Germany). Sequencing libraries are prepared from the extracted DNA using the Ligation Sequencing Kit and the Rapid Barcoding Kit. Prepared libraries are sequenced using MinION (R9.4

Flow cells), and sequences with an average Q-score of more than 7 are used in the final genome. RScript MinIONQC [15] was used for quality assessment.

Annotation and Quality Control. We used the MiFi platform [8][1] for labeling metagenome sequence data. This platform is based on the modified version of the bioinformatics pipeline discussed by Stobbe *et al.* [31]. Using MiFi, unique signature sequences referred to as *e-probes* were developed for the pathogen of interest. These e-probes were then used to identify and label pathogen specific sequences in the metagenome reads, and differentiate them from other sequences (host, commensals, and other pathogen sequences). Clinical metagenome samples from 7 patients were used for training, 1 for validation, while sequences from 5 patients were used for evaluation.

Table 2. Comparison with other representations. Performance evaluation of machine learning baselines using other metagenome representations. Average F1 scores are reported across pathogen and host classes.

Classifier	Node2Vec [23]		SPK [23]		WLK [23]		GSK [23]		Seq2Vec		MG-Net	
	Host	Path.	Host	Path.	Host	Path.	Host	Path.	Host	Path.	Host	Path.
Linear	0.05	0.07	0.02	0.04	0.04	0.05	0.03	0.04	–	–	0.86	0.38
LR	0.82	0.04	0.86	0.13	0.87	0.10	0.87	0.02	–	–	0.97	0.53
SVM	0.81	0.07	0.86	0.11	0.86	0.10	0.86	0.05	–	–	0.97	0.53
MLP	0.85	0.09	0.86	0.08	0.88	0.07	0.84	0.00	–	–	**0.98**	0.54
DL	0.74	0.10	0.78	0.09	0.81	0.13	0.79	0.07	0.758	0.362	**0.98**	**0.63**

Metrics and Baselines. To quantitatively evaluate our approach, we use precision, recall, and the F-score for each class. We do not use accuracy as a metric since real-life metagenomes can be highly skewed towards host sequences. Precision and recall, on the other hand, allow us to quantify the false alarm rates and provide more precise detection accuracy. We compare against other representation learning frameworks for metagenome analysis proposed, such as graph-based approaches [23], and a deep learning model termed Seq2Vec, an end-to-end sequence-based learning model based on DeePAC [4]. For classification, we consider both traditional baselines such as logistic regression (LR), support vector machines (SVM), and multi-layer perceptron (MLP), as well as a deep neural network (DL). The MLP baseline has two hidden layers with 256 neurons each, while the deep learning baseline has 3 hidden layers with 256, 512 and 1024 neurons each with a ReLU activation function. We choose the hyperparameters for each of the baselines using an automated grid search and the best performing models from the validation set were taken for evaluation on the test set.

3.1 Quantitative Evaluation

We evaluate our approach and0 report the quantitative results in Table 1 and Table 2. We evaluate under different settings to assess the robustness of the proposed framework under limited data and limited supervision. We also compare

---

[1] https://bioinfo.okstate.edu/.

against comparable representation learning approaches to highlight the importance of attention-based reasoning to integrate global and local structural information in a unified framework. We significantly outperform all baselines when training with the entire training data and offer competitive performance when fine-tuned with only 500 labeled samples per class.

Effect of Limited Labels. Since our representations are learned in a self-supervised manner, we also evaluate its metagenome recognition performance with limited labeled data and summarize results in Table 1. First, we evaluate when there is no labeled data using k-means clustering to segment the features into groups and align the predicted clusters with the ground-truth using the Hungarian method, following prior works [14]. It can be seen that we perform reasonably well considering we do not use any labels from the ground-truth to train. As expected, the performance gets better with the use of increasing amounts of labeled data. It is interesting to note that we match the performance of fully supervised models like sequence-level graph-based representations [23] and end-to-end deep learning models like Seq2Vec with as little as 500 samples and outperform their performance with as little as 1000 labeled samples per class. Given the performance of the linear classifier (Table 2), we can see that our approach learns robust representations with limited data.

Table 3. Ablation Studies. Performance evaluation of different variations to evaluate the effect of each design choice in the overall framework. * indicates final model.

Approach	Host			Pathogen		
	Precision	Recall	F1	Precision	Recall	F1
Autoencoder Only	0.950	0.800	0.869	0.242	0.453	0.315
Autoencoder + Structural Priors	0.980	0.970	0.975	0.523	0.543	0.533
Structural Priors Only	0.990	0.960	0.975	0.522	0.615	0.565
MG-Net (k = 3, s = 5)	0.980	0.980	0.980	0.598	0.628	0.613
MG-Net (k = 3, s = 10)	0.99	0.980	0.984	0.595	0.640	0.617
MG-Net (k = 4, s = 5)	0.980	0.980	0.980	0.610	0.598	0.604
MG-Net (k = 4, s = 10)	0.980	0.980	0.980	0.580	0.581	0.581
MG-Net (k = 5, s = 5)	0.980	0.980	0.980	0.573	0.588	0.581
MG-Net* (k-5, s = 10)	**0.990**	**0.980**	**0.984**	**0.615**	**0.648**	**0.631**

Comparison with Other Representations. We compare our representations with other baselines and summarize the results in Table 2. We use a mix of traditional approaches (MLP, SVM, and LR) and a deep neural network (DL). We also train a linear classifier on top of each of the representations to assess their robustness. As can be seen, the representations from MG-Net outperform all other baselines by a significant margin. In fact, a linear classifier outperforms all other baseline representations that use deep neural networks. Our MG-Net features with a deep learning classifier achieve an average pathogen F-score of 63% and a host F-score of 98%, which are significantly higher than baseline representations. Evaluation with 5-fold cross validation (see supplementary material) corroborate the results. It is interesting to note that both the graph kernels and Seq2Vec use sequence read-level features, and our approach with only an autoencoder from Table 3 has comparable performance indicating that the global structural features have a significant impact on the performance.

Ablation Studies. Finally, we systematically evaluate each component of the framework independently to identify their contribution. Specifically, we provide ablations of our approach with using only image features (autoencoder only), image+structural features without the MG-Net architecture (Autoencoder + Structural Priors), and only structural features (from node2vec). From Table 3, we can see that the use of structural priors greatly improves the performance. We remove the structural prior and use an autoencoder trained on only the pseudo-images (*Autoencoder Only*) and use a late fusion strategy (*Autoencoder + Structural Priors*) to evaluate the structural reasoning module. While the performance is better than other baselines, the final MG-Net architecture outperforms all variations. Finally, we also vary the length of k-mer sequences and stride lengths and see that the performance increases with an increase in stride lengths while reducing the k-mer length reduces the performance.

4   Conclusion and Future Work

In this work, we presented MG-Net, one of the first efforts to offer a multi-modal perspective to metagenome analysis using the idea of co-occurrence statistics to construct pseudo-images. A novel, attention-based structural reasoning framework is introduced to perform multi-modal feature fusion, allowing for joint optimization over multiple modalities. Extensive real-world clinical data experiments show that the learned representations outperform existing baselines by a significant margin and offer a way forward for metagenome classification under limited resources. We aim to leverage these results to build automated diagnosis and intervention pipelines for novel pathogen diseases with limited supervision.

Acknowledgement. This research was supported in part by the US Department of Agriculture (USDA) grants AP20VSD and B000C011.

We thank Dr. Kitty Cardwell and Dr. Andres Espindola (Institute of Biosecurity and Microbial Forensics, Oklahoma State University) for providing access and assisting with use of the MiFi platform.

References

1. Ashoor, H., et al.: Graph embedding and unsupervised learning predict genomic sub-compartments from hic chromatin interaction data. Nature Commun. **11**(1), 1–11 (2020)
2. Bagari, A., Kumar, A., Kori, A., Khened, M., Krishnamurthi, G.: A combined radio-histological approach for classification of low grade gliomas. In: Crimi, A., Bakas, S., Kuijf, H., Keyvan, F., Reyes, M., van Walsum, T. (eds.) BrainLes 2018. LNCS, vol. 11383, pp. 416–427. Springer, Cham (2019). https://doi.org/10.1007/978-3-030-11723-8_42

3. Ballerini, L., Li, X., Fisher, R.B., Rees, J.: A query-by-example content-based image retrieval system of non-melanoma skin lesions. In: Caputo, B., Müller, H., Syeda-Mahmood, T., Duncan, J.S., Wang, F., Kalpathy-Cramer, J. (eds.) MCBR-CDS 2009. LNCS, vol. 5853, pp. 31–38. Springer, Heidelberg (2010). https://doi.org/10.1007/978-3-642-11769-5_3

4. Bartoszewicz, J.M., Seidel, A., Rentzsch, R., Renard, B.Y.: Deepac: predicting pathogenic potential of novel dna with reverse-complement neural networks. Bioinformatics **36**(1), 81–89 (2020)

5. Busia, A., et al.: A deep learning approach to pattern recognition for short dna sequences. BioRxiv p. 353474 (2019)

6. Ching, T., et al.: Opportunities and obstacles for deep learning in biology and medicine. J. R. Soc. Interface **15**(141), 20170387 (2018)

7. Chiu, C.Y., Miller, S.A.: Clinical metagenomics. Nat. Rev. Genet. **20**(6), 341–355 (2019)

8. Espindola, A.S., Cardwell, K.F.: Microbe finder (mifi®): implementation of an interactive pathogen detection tool in metagenomic sequence data. Plants **10**(2), 250 (2021)

9. Grover, A., Leskovec, J.: node2vec: Scalable feature learning for networks. In: Proceedings of the 22nd ACM SIGKDD International Conference on Knowledge Discovery and Data Mining, pp. 855–864 (2016)

10. Hamburg, M.A., Collins, F.S.: The path to personalized medicine. N. Engl. J. Med. **363**(4), 301–304 (2010)

11. Hasan, M.R., et al.: A metagenomics-based diagnostic approach for central nervous system infections in hospital acute care setting. Sci. Rep. **10**(1), 1–11 (2020)

12. He, K., Zhang, X., Ren, S., Sun, J.: Deep residual learning for image recognition. In: Proceedings of the IEEE Conference on Computer Vision and Pattern Recognition (CVPR) (June 2016)

13. Hwang, S., et al.: Humannet v2: human gene networks for disease research. Nucleic Acids Res. **47**(D1), D573–D580 (2019)

14. Ji, X., Henriques, J.F., Vedaldi, A.: Invariant information clustering for unsupervised image classification and segmentation. In: Proceedings of the IEEE/CVF International Conference on Computer Vision, pp. 9865–9874 (2019)

15. Lanfear, R., Schalamun, M., Kainer, D., Wang, W., Schwessinger, B.: Minionqc: fast and simple quality control for minion sequencing data. Bioinformatics **35**(3), 523–525 (2019)

16. Laver, T., et al.: Assessing the performance of the oxford nanopore technologies minion. Biomol. Detect. Quantification **3**, 1–8 (2015)

17. Leu, S.C., Huang, Z., Lin, Z.: Generation of pseudo-CT using high-degree polynomial regression on dual-contrast pelvic MRI data. Sci. Rep. **10**(1), 1–11 (2020)

18. Liang, Q., Bible, P.W., Liu, Y., Zou, B., Wei, L.: Deepmicrobes: taxonomic classification for metagenomics with deep learning. NAR Genomics Bioinform. **2**(1), lqaa009 (2020)

19. Lin, M., Chen, Q., Yan, S.: Network in network. arXiv preprint arXiv:1312.4400 (2013)

20. Lin, Y., Yuan, J., Kolmogorov, M., Shen, M.W., Chaisson, M., Pevzner, P.A.: Assembly of long error-prone reads using de bruijn graphs. Proc. Nat. Acad. Sci. **113**(52), E8396–E8405 (2016)

21. Metzker, M.L.: Sequencing technologies–the next generation. Nat. Rev. Genet. **11**(1), 31–46 (2010)

22. Mikheyev, A.S., Tin, M.M.: A first look at the oxford nanopore minion sequencer. Mol. Ecol. Resour. **14**(6), 1097–1102 (2014)

23. Narayanan, S., Ramachandran, A., Aakur, S.N., Bagavathi, A.: Gradl: a framework for animal genome sequence classification with graph representations and deep learning. In: 2020 19th IEEE International Conference on Machine Learning and Applications (ICMLA), pp. 1297–1303. IEEE (2020)

24. Nelson, R.J., Mooney, J.M., Ewing, W.S.: Pseudo imaging. In: Algorithms and Technologies for Multispectral, Hyperspectral, and Ultraspectral Imagery XII, vol. 6233, p. 62330M. International Society for Optics and Photonics (2006)

25. Nguyen, H.T., et al.: Growing self-organizing maps for metagenomic visualizations supporting disease classification. In: Dang, T.K., Küng, J., Takizawa, M., Chung, T.M. (eds.) FDSE 2020. LNCS, vol. 12466, pp. 151–166. Springer, Cham (2020). https://doi.org/10.1007/978-3-030-63924-2_9

26. Nguyen, T.H.: Metagenome-based disease classification with deep learning and visualizations based on self-organizing maps. In: Dang, T.K., Küng, J., Takizawa, M., Bui, S.H. (eds.) FDSE 2019. LNCS, vol. 11814, pp. 307–319. Springer, Cham (2019). https://doi.org/10.1007/978-3-030-35653-8_20

27. Pennec, X., Cachier, P., Ayache, N.: Tracking brain deformations in time sequences of 3d us images. Pattern Recogn. Lett. **24**(4–5), 801–813 (2003)

28. Reiman, D., Metwally, A.A., Sun, J., Dai, Y.: Popphy-cnn: a phylogenetic tree embedded architecture for convolutional neural networks to predict host phenotype from metagenomic data. IEEE J. Biomed. Health Inf. **24**(10), 2993–3001 (2020)

29. Ronneberger, O., Fischer, P., Brox, T.: U-Net: convolutional networks for biomedical image segmentation. In: Navab, N., Hornegger, J., Wells, W.M., Frangi, A.F. (eds.) MICCAI 2015. LNCS, vol. 9351, pp. 234–241. Springer, Cham (2015). https://doi.org/10.1007/978-3-319-24574-4_28

30. Simonyan, K., Zisserman, A.: Very deep convolutional networks for large-scale image recognition. arXiv preprint arXiv:1409.1556 (2014)

31. E-probe diagnostic nucleic acid analysis (edna): a theoretical approach for handling of next generation sequencing data for diagnostics. J. Microbiol. Methods **94**(3), 356–366 (2013)

32. Sun, H., Xie, K., Gao, L., Sui, J., Lin, T., Ni, X.: Research on pseudo-ct imaging technique based on an ultrasound deformation field with binary mask in radiotherapy. Medicine **97**(38), e12532 (2018)

Multimodal Multitask Deep Learning for X-Ray Image Retrieval

Yang Yu[1], Peng Hu[2], Jie Lin[1], and Pavitra Krishnaswamy[1(✉)]

[1] Institute for Infocomm Research, A*STAR, Singapore, Singapore
{yu_yang,pavitrak}@i2r.a-star.edu.sg
[2] College of Computer Science, Sichuan University, Chengdu, Sichuan, China

Abstract. Content-based image retrieval (CBIR) is of increasing interest for clinical applications spanning differential diagnosis, prognostication, and indexing of electronic radiology databases. However, meaningful CBIR for radiology applications requires capabilities to address the semantic gap and assess similarity based on fine-grained image features. We observe that images in radiology databases are often accompanied by free-text radiologist reports containing rich semantic information. Therefore, we propose a Multimodal Multitask Deep Learning (MMDL) approach for CBIR on radiology images. Our proposed approach employs multimodal database inputs for training, learns semantic feature representations for each modality, and maps these representations into a common subspace. During testing, we use representations from the common subspace to rank similarities between the query and database. To enhance our framework for fine-grained image retrieval, we provide extensions employing deep descriptors and ranking loss optimization. We performed extensive evaluations on the MIMIC Chest X-ray (MIMIC-CXR) dataset with images and reports from 227,835 studies. Our results demonstrate performance gains over a typical unimodal CBIR strategy. Further, we show that the performance gains of our approach are robust even in scenarios where only a subset of database images are paired with free-text radiologist reports. Our work has implications for next-generation medical image indexing and retrieval systems.

Keywords: Content-based image retrieval · Multimodal representation learning · Fine-grained retrieval · Diagnostic radiographs.

1 Introduction

The growth in large-scale electronic radiology databases presents rich opportunities for clinical decision support. In particular, multimedia databases such as

Electronic supplementary material The online version of this chapter (https://doi.org/10.1007/978-3-030-87240-3_58) contains supplementary material, which is available to authorized users.

the Radiology Information System (RIS) and the Picture Archiving and Communication Systems (PACS) store anatomical, pathological and functional information for millions of patients. As such, there is increasing interest in content-based image retrieval (CBIR) capabilities to search these databases and retrieve relevant cases for a range of diagnostic, research and educational purposes [2,3,15,22]. Specific use cases of interest could include differential diagnosis, enhanced assessment of rare conditions, severity assessment, and prognostication based on actualized outcomes in patients with similar images in the database [3]. As manual CBIR is often infeasible, automated CBIR methods are desirable.

Automated CBIR solutions have had substantive successes for forensics, retail, mobile and photo archival applications [25]. However, their uptake in medicine has been limited [2,3]. While some studies have showed applicability of CBIR for histopathology images [10], there have been far fewer demonstrations of CBIR in the radiology domain. In part, this is because radiology images are often generated by non-light based contrast mechanisms and do not contain color information. Further, meaningful CBIR in radiology requires capabilities to extend beyond gross pixel-based features and assess similarity of semantic content based on domain knowledge or on fine-grained details in the image.

We note that images in radiology databases are often accompanied by textual reports detailing radiologist interpretation and observations. Therefore, we propose a multimodal CBIR framework to leverage the rich semantic information in textual reports alongside visual (and semantic) information in the images. We build on deep learning based multimodal retrieval approaches developed for computer vision applications [12,13,30,31] and adapt these for training with radiology images and text reports. Our framework is amenable to addressing both image queries and multimodal queries, and can be customized to address the fine-grained nature of CBIR in radiology [7,8,10,11,29]. Our main contributions are as follows:

1. We introduce a customizable deep learning framework that leverages multimodal databases comprising radiology images and free-text reports for content-based image retrieval in radiology.
2. To enhance our framework for fine-grained image retrieval, we present extensions that (a) learn descriptors of abnormal regions in the image and (b) employ triplet loss for metric learning.
3. In extensive experiments on MIMIC-CXR, a real-world radiology multimodal dataset, we demonstrate good performance gains over standard baselines. We further show that these gains remain robust in scenarios where a subset of images lack associated textual reports.

2   Related Work

CBIR in Radiology: Early CBIR efforts in radiology employed handcrafted features [19,20,23]. Other early efforts focused on using regions of interest defined by clinical users to identify database images with visually similar patterns [1]. Recent efforts have explored the more generalizable deep learning based CBIR techniques for retrieval based on modality and/or body part similarity [2,4,24, 27], and for retrieval on chest X-Rays [5,8]. However, such methods are yet to be demonstrated for challenging clinical retrieval use cases with limited user input, and are not set up to fully leverage the multimedia information available in radiology databases.

Multimodal Retrieval: There has been longstanding interest in integrating multimodal information to inform medical image retrieval [2,3,15,22]. One study proposed a probabilistic latent semantic analysis method to combine images with short textual descriptors for a modality and body-part similarity retrieval task [4]. However, this approach is limited to short descriptors for the textual modality, focused on single task learning and is not suitable for use cases like differential diagnosis which only allow unimodal query inputs. In contrast, the computer vision literature has advanced deep learning techniques to effectively learn semantic representations from a variety of multimodal data types to enhance CBIR [12,13,21,30,31]. While these advanced deep learning methods have been demonstrated on natural scene image datasets, they have yet to be translated to domain-specific tasks.

Fine-Grained Image Retrieval: While typical image retrieval approaches focus on macro-level similarities between query and database, fine-grained retrieval seeks to rank images based on similarities in more subtle features. To address this challenge, prior works have employed either attention-based mechanisms or ranking loss terms. Attention-based mechanisms (e.g., Selective Convolutional Descriptor Aggregation (SCDA) [29]) localize the objects of interest by discarding the noisy background and keeping useful deep descriptors. Ranking loss approaches employ triplets of samples to learn relative distances between samples and optimize metric learning [7,9,11], and can be beneficial for fine-grained image retrieval [28]. A recent study [8] employed an Attention-based Triplet Hashing (ATH) approach that uses deep triplet loss and a fine-grained attention mechanism for chest X-ray retrieval. However, these approaches have yet to be demonstrated in domain-specific tasks and/or in multimodal retrieval settings.

3   Methods

Problem Formulation: We consider data inputs for the image and text modalities. We denote the set of n_i samples of the i-th modality as $\mathcal{X}_i = \{x_1^i, x_2^i, \ldots, x_{n_i}^i\}$, where x_j^i denotes the j-th data input for the i-th modality. As each data input can be associated with a multiplicity of class labels, the corresponding label matrix for i-th modality is denoted as $\mathcal{Y}_i = [y_1^i, y_2^i, \ldots, y_{n_i}^i]$. Given a multimodal database $[\mathcal{X}^D, \mathcal{Y}^D]$ and a unimodal image query x^Q, the retrieval task is to identify a ranked list of the most similar images from the database.

Multimodal Multitask Deep Learning (MMDL) Framework: Figure 1 illustrates the proposed MMDL framework. The representation learning task is to learn, for each modality, modality-specific transformation functions. We

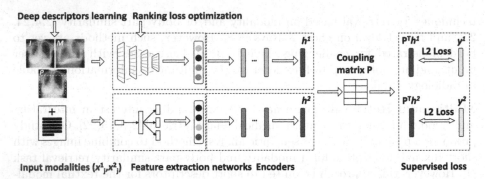

Fig. 1. Proposed Multimodal Multitask Deep Learning (MMDL) Framework. The training framework extracts features from the inputs, reduces dimensionality and projects the multimodal information into a common subspace for semantic representation learning. The encoders and coupling matrix define the common subspace and are learnt during network training. The projected representation is fed into a linear classifier for supervised learning of the multimodal transformation functions. During testing, the query image is transformed into the common subspace so as to rank similar database images for retrieval.

accomplish this with two feature extraction and encoder networks to project the data inputs from each modality into a common subspace where the intra-class variation is minimized and inter-class variation is maximized.

For the j-th data input, we denote the output of the i-th encoder network as h_j^i. We denote learnable parameters of the i-th modality-specific transformation function as Θ_i. Then, the objective function for the i-th encoder network is:

$$\mathcal{L}^i(x_j^i) = \frac{1}{n_i} \sum_{j=1}^{n_i} [||P^T h_j^i - y_j^i||_2], \tag{1}$$

where P is a learnable matrix used to define the common subspace and $\mathcal{L}^i(x_j^i)$ is the supervised loss. The rows of P correspond to the dimensionality of the encoder output while columns of P correspond to the semantic categories. For each modality, the supervised loss $\mathcal{L}$ leverages the labels to enhance the discriminative power of the predefined common space.

To optimize the transform functions for all m modalities, the system needs to simultaneously minimize the total loss across the m modality-specific networks:

$$\min \sum_{i=1}^{m} \mathcal{L}^i. \tag{2}$$

Since the trainable parameters are shared amongst all modality-specific networks, the MMDL learns weights Θ_i and P simultaneously in a joint manner. With this learning strategy, MMDL is able to bridge the heterogeneity gap between various modalities and learn common representations by simultaneously preserving the semantic discrimination and modality invariance. Moreover, the optimized networks trained from multimodal learning could further enhance the performance in the multitask implementation where only one of the modalities is available during querying phase.

During testing, we use the encoder outputs to compute the cosine distance between image query x^Q and each sample in the multimodal database $[\mathcal{X}^D, \mathcal{Y}^D]$. We then rank database entries by the distance to obtain the retrieval output.

Learning Fine-Grained Image Features: To enhance performance of our multimodal multitask deep learning framework for the fine-grained nature of the content in the images, we present two extensions. First, we employ the Selective Convolutional Descriptor Aggregation (SCDA) method [29] to localize the abnormal regions in the images, and learn the deep descriptors for these abnormal regions. As shown in Fig. 1, we employ SCDA on image I to generate attention masks M and then threshold using these masks to generate the image descriptors or patches P, which then serve as the input for the multimodal multitask retrieval. Second, we employ the triplet loss [11] to optimize the metric learning for better distinctions based on subtle image features. Specifically, the ranking loss optimization is applied to enhance the extracted feature representations for the image modality. Together, these enhancements serve to highlight the fine-grained features in the images for more precise retrieval.

4   Experiment Setup

Datasets: We evaluated our multimodal retrieval framework on the MIMIC Chest X-ray (MIMIC-CXR) Database v2.0.0. This dataset contains 377,110 images and paired reports corresponding to 64,579 patients with their annotations categorized into 14 classes [17]. We removed the lateral images and only considered frontal images. From the remaining dataset, we randomly sampled 189,036 image-text pairs for training and 26,750 image-text pairs for validation. From the remainder of the dataset, we selected 4,075 image-text pairs for testing. For the test set selection, we used stratified sampling to represent the class distributions of the native dataset. We tested retrieval performance using each test image as a query against the remainder of the test samples as the database. We generated inputs for multimodal retrieval tasks by first extracting the corresponding features from images and their paired reports. We generated the labels for the images and reports using the CheXpert Natural Language Processing (NLP) labeler with the "U-Zero" setting [16].

Feature Representation Learning: For each image (I), we generate a mask (M)-derived patch (P) using the VGG16 model [26] pre-trained on CheXpert [16] with the SCDA approach. The image patches are further represented by a 1,024-dimensional convolutional neural network (CNN) feature extracted from the final fully connected layer of the DenseNet121 model [14] pre-trained on CheXpert using ranking loss optimization. For reports, we used the free-text from "Findings" and "Impression" sections to extract a 300-dimensional text representation via Doc2Vec model [18] pre-trained on the MIMIC-CXR.

Implementation Details: For each modality, the proposed network includes three fully-connected layers with each layer following a Rectified Linear Unit (ReLU) activation function. The three fully-connected layers have 1024, 1024, 512 hidden units and share weights. We randomly initialized P. We used four Nvidia RTX 2080 Super GPUs for training the proposed MMDL model with PyTorch. Hyperparameters are in Supplement Table S1.

Baselines: We compared our approach to the conventional unimodal retrieval baseline that applies a fine-tuned DenseNet121 network on the image query (Image D121). For this, we first trained the network on ImageNet [6] and fine-tuned on the CheXpert dataset, then extracted features from average pooling layers, and computed similarity using cosine distance to retrieve relevant database images. For more direct and rigorous comparison with our proposed method, we also implemented a unimodal version of our proposed method using the image query where the network is trained only with image modality (Image MMDL).

To further compare with other state-of-the-art (SOTA) unimodal retrieval approaches, we also implemented the ATH method [8]. Specifically, we adapted the ATH pipeline for multi-label inputs and retrained the model on the entire MIMIC-CXR dataset (14 classes) to evaluate performance (ATH). Further, we integrated the ATH method within our MMDL framework and repeated the previous evaluation to obtain a more rigorous comparison of performance (ATH-MMDL).

Experiments: In a first experiment, we focused on training with a database comprising complete image and text pairs, and evaluating with unimodal image queries. For this, we performed retrieval using our proposed method and the above baselines. In a second experiment, we simulated the scenario where not all radiology images in the database would have paired textual reports. We considered scenarios where only 25%, 50%, 75%, or 90% of the images had associated reports, and compared the performance against the completely paired image-text experiment. For these experiments, we first trained the networks with the image-text pairs and then utilized the remaining images to further train the network.

Evaluation Metrics: In each experiment, we evaluated the mean Average Precision (mAP) score for all testing samples and the Average Precision (AP) for each individual class for the Top 10 retrieved images. Retrieval is considered accurate if any of the positive labels in the retrieved image overlap with the query image. We also computed the mAP for the Top 1, 5, 20, and k (where k is number of test queries) retrieved images and the Top 1 and Top 5 Accuracy (Acc-1 and Acc-5 with sigmoid probability of the top 1 and top 5 predicted labels).

5   Results

Fig. 2 shows results for the two aforementioned multimodal retrieval experiments on the MIMIC-CXR dataset. The first experiment focuses on retrieval of database images for a given image query, and evaluates our multimodal multitask framework against the standard baseline unimodal image retrieval approach (Panel A). The second experiment focuses on comparing the effectiveness of retrieval models trained in scenarios where not all database images have accompanying text reports (Panel B).

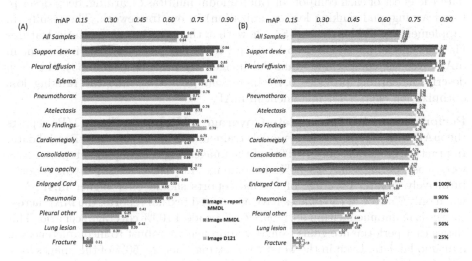

Fig. 2. Performance comparison for proposed MMDL method and baselines: A) All samples mAP and individual class AP for unimodal and multimodal retrieval tasks. (B) All samples mAP and individual class AP for multimodal retrieval in scenarios with incomplete data coverage, as indicated by percentage of images paired with textual reports.

Performance of Multimodal vs. Unimodal Retrieval: Figure 2 (A) reports mAP of our proposed multimodal method and two unimodal image retrieval baselines (with DensetNet121 and with our MMDL framework). For the unimodal retrieval task, our MMDL framework (Image-MMDL) performs comparably to the standard approach of learning features with DenseNet121 (Image-D121). We note that inclusion of semantic features from the textual reports with our proposed multimodal approach provides a 4% boost in mAP over the conventional DenseNet121-based unimodal retrieval baseline. Supplement Table S2 details additional performance metrics for varying numbers of the top retrieved images. Our results suggest that the weight sharing strategy in our multimodal multitask learning approach is able to enhance semantic consistency amongst the modalities.

Comparison with SOTA Unimodal Baselines: We benchmarked our results against the recent unimodal ATH retrieval baseline and found that the ATH baseline achieves all samples mAP of 0.515. When integrated within our MMDL framework, the ATH-MMDL baseline achieves all samples mAP of 0.541. In both cases,the ATH results are lower than our DenseNet121-based unimodal retrieval implementations (0.643 for Image-D121, and 0.650 for Image-MMDL). Supplement Table S2 also details additional performance metrics for the ATH baseline implementations using varying numbers of top retrieved images.

Ablation Studies: Next, we performed ablation studies to systematically evaluate the effect of each component (multimodal multitask learning, deep descriptors learning, and ranking loss optimization) in our framework. The results, in supplementary Tables S3 and S4, show that the injection of semantic features via multimodal multitask deep learning (Table S4) provides 3.0% increment in mAP over the comparable unimodal image-only baseline (Table S3); the deep descriptor learning provides 0.4% increment in mAP; while the ranking loss optimization offers 1.1% increment in mAP.

Performance with Incomplete Coverage of Reports: Figure 2 (B) reports the mAP with MMDL retrieval models trained with incomplete multimodal data. In practical scenarios, only 75–90% of the images might have paired reports. As a worst case, perhaps only 25–50% of the images may have accompanying reports. Intuitively, greater coverage of textual reports should offer higher mAP. Yet, when only 75–90% of the images have associated textual reports, the mAP across classes is maintained within 98.1–99.7% of the ideal 100% coverage scenario. This shows that performance gains from our approach are robust to incompleteness in multimodal data. Even in the worst case setting when 25–50% of the images have associated textual reports, our MMDL approach still offers a 1.9–2.5% boost in mAP over the conventional DenseNet121 unimodal retrieval baseline.

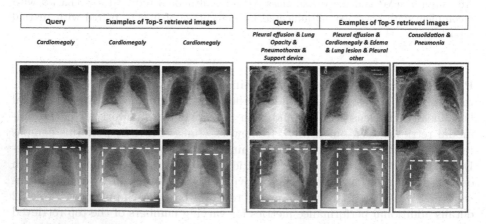

Fig. 3. Exemplar query images with two of the top 5 retrieved cases. In each case, original images are overlaid with attention masks and image patches (white dotted box) generated by the deep descriptor learning step. The correctly retrieved results are marked with green box while incorrectly retrieved results are marked with red box. (Color figure online)

Visualization of Retrieved Images: For further characterization, we present exemplar results on 2 queries from the testing set in Fig. 3. For each query, the test image and associated ground truth labels are shown, alongside two of the top 5 retrieved cases and their labels. We observe that the returned images have good overlap in labels with respect to the query images when correctly retrieved. Further, there is good overlap in the visual features between query and returned images despite inherent inter-subject variations. For the second query, although the pleural effusion is inconspicuous, the correctly retrieved example also corresponds to a mild pleural effusion. Furthermore, even for the wrongly retrieved example, there is a sensible semantic match - specifically the query corresponds to a lung opacity but the returned image corresponds to consolidation and pneumonia which are sub-types of lung opacities. Finally, we note that the full text report for this case mentioned an effusion, although the CheXpert labeler did not infer pleural effusion as it was based on the impression section only [16]. These illustrations highlight the potential of the proposed method to identify semantically similar X-Ray images, especially when accurate retrieval relies on fine-grained features.

6   Conclusion

We proposed a deep learning framework that learns common semantic representations for CBIR based on a combination of radiology images and textual reports. Comprehensive experimental results on a public Chest X-Ray dataset demonstrate the effectiveness of our framework to boost retrieval performance. Future work to improve report-based feature extraction could offer additional performance improvements. Our work establishes a new baseline for content-based medical image retrieval and has implications for practical clinical use cases in differential diagnosis, cohort selection and prognostication.

Acknowledgements. Research efforts were supported by funding and infrastructure for deep learning and medical imaging research from the Institute for Infocomm Research, Science and Engineering Research Council, A*STAR, Singapore. We thank Victor Getty, Vijay Chandrasekhar and Ivan Ho Mien from the Institute for Infocomm Research, A*STAR for their valuable inputs. We also acknowledge insightful discussions with Jayashree Kalpathy-Cramer at the Massachusetts General Hospital, Boston, USA.

References

1. https://contextflow.com/
2. Ahmad, J., Sajjad, M., Mehmood, I., Baik, S.W.: Sinc: saliency-injected neural codes for representation and efficient retrieval of medical radiographs. PLoS ONE **12**(8), e0181707 (2017)
3. Akgül, C.B., Rubin, D.L., Napel, S., Beaulieu, C.F., Greenspan, H., Acar, B.: Content-based image retrieval in radiology: current status and future directions. J. Digit. Imaging **24**(2), 208–222 (2011)

4. Cao, Y., et al.: Medical image retrieval: a multimodal approach. Cancer Inf. **13**, CIN-S14053 (2014)
5. Chen, Z., Cai, R., Lu, J., Feng, J., Zhou, J.: Order-sensitive deep hashing for multimorbidity medical image retrieval. In: Frangi, A.F., Schnabel, J.A., Davatzikos, C., Alberola-López, C., Fichtinger, G. (eds.) MICCAI 2018. LNCS, vol. 11070, pp. 620–628. Springer, Cham (2018). https://doi.org/10.1007/978-3-030-00928-1_70
6. Deng, J., Dong, W., Socher, R., Li, L.J., Li, K., Fei-Fei, L.: Imagenet: a large-scale hierarchical image database. In: 2009 IEEE Conference on Computer Vision and Pattern Recognition, pp. 248–255. IEEE (2009)
7. Dodds, E., Nguyen, H., Herdade, S., Culpepper, J., Kae, A., Garrigues, P.: Learning embeddings for product visual search with triplet loss and online sampling. arXiv preprint arXiv:1810.04652 (2018)
8. Fang, J., Fu, H., Liu, J.: Deep triplet hashing network for case-based medical image retrieval. Med. Image Anal. **69**, 101981 (2021)
9. Gómez, R.: Understanding ranking loss, contrastive loss, margin loss, triplet loss, hinge loss and all those confusing names. Raúl Gómez blog (2019)
10. Hegde, N., et al.: Similar image search for histopathology: smily. NPJ Digital Med. **2**(1), 1–9 (2019)
11. Hoffer, E., Ailon, N.: Deep metric learning using triplet network. In: Feragen, A., Pelillo, M., Loog, M. (eds.) SIMBAD 2015. LNCS, vol. 9370, pp. 84–92. Springer, Cham (2015). https://doi.org/10.1007/978-3-319-24261-3_7
12. Hu, P., Peng, D., Wang, X., Xiang, Y.: Multimodal adversarial network for cross-modal retrieval. Knowl.-Based Syst. **180**, 38–50 (2019)
13. Hu, P., Zhen, L., Peng, D., Liu, P.: Scalable deep multimodal learning for cross-modal retrieval. In: Proceedings of the 42nd International ACM SIGIR Conference on Research and Development in Information retrieval, pp. 635–644 (2019)
14. Huang, G., Liu, Z., Van Der Maaten, L., Weinberger, K.Q.: Densely connected convolutional networks. In: Proceedings of the IEEE Conference on Computer Vision and Pattern Recognition, pp. 4700–4708 (2017)
15. Hwang, K.H., Lee, H., Choi, D.: Medical image retrieval: past and present. Healthc. Inf. Res. **18**(1), 3 (2012)
16. Irvin, J., et al.: Chexpert: a large chest radiograph dataset with uncertainty labels and expert comparison. In: Proceedings of the AAAI Conference on Artificial Intelligence, pp. 590–597 (2019)
17. Johnson, A.E., et al.: Mimic-cxr-jpg, a large publicly available database of labeled chest radiographs. arXiv preprint arXiv:1901.07042 (2019)
18. Le, Q., Mikolov, T.: Distributed representations of sentences and documents. In: International Conference on Machine Learning, pp. 1188–1196. PMLR (2014)
19. Liu, Y., Rothfus, W.E., Kanade, T.: Content-based 3d neuroradiologic image retrieval: preliminary results. In: Proceedings 1998 IEEE International Workshop on Content-Based Access of Image and Video Database, pp. 91–100. IEEE (1998)
20. Ma, L., Liu, X., Gao, Y., Zhao, Y., Zhao, X., Zhou, C.: A new method of content based medical image retrieval and its applications to ct imaging sign retrieval. J. Biomed. Inf. **66**, 148–158 (2017)
21. Mourão, A., Martins, F., Magalhaes, J.: Multimodal medical information retrieval with unsupervised rank fusion. Comput. Med. Imaging Graph. **39**, 35–45 (2015)
22. Müller, H., Michoux, N., Bandon, D., Geissbuhler, A.: A review of content-based image retrieval systems in medical applications-clinical benefits and future directions. Int. J. Med. Informatics **73**(1), 1–23 (2004)
23. Pilevar, A.H.: Cbmir: content-based image retrieval algorithm for medical image databases. J. Med. Signals Sens. **1**(1), 12 (2011)

24. Qayyum, A., Anwar, S.M., Awais, M., Majid, M.: Medical image retrieval using deep convolutional neural network. Neurocomputing **266**, 8–20 (2017)
25. Schaer, R., Otálora, S., Jimenez-del Toro, O., Atzori, M., Müller, H.: Deep learning-based retrieval system for gigapixel histopathology cases and the open access literature. J. Pathol. Inf. **10** (2019)
26. Simonyan, K., Zisserman, A.: Very deep convolutional networks for large-scale image recognition. arXiv preprint arXiv:1409.1556 (2014)
27. Sklan, J.E., Plassard, A.J., Fabbri, D., Landman, B.A.: Toward content-based image retrieval with deep convolutional neural networks. In: Medical Imaging 2015: Biomedical Applications in Molecular, Structural, and Functional Imaging, vol. 9417, p. 94172C. International Society for Optics and Photonics (2015)
28. Wang, J., et al.: Learning fine-grained image similarity with deep ranking. In: Proceedings of the IEEE Conference on Computer Vision and Pattern Recognition, pp. 1386–1393 (2014)
29. Wei, X.S., Luo, J.H., Wu, J., Zhou, Z.H.: Selective convolutional descriptor aggregation for fine-grained image retrieval. IEEE Trans. Image Process. **26**(6), 2868–2881 (2017)
30. Zhen, L., Hu, P., Peng, X., Goh, R.S.M., Zhou, J.T.: Deep multimodal transfer learning for cross-modal retrieval. IEEE Trans. Neural Netw. Learn. Syst., 1–13 (2020)
31. Zhen, L., Hu, P., Wang, X., Peng, D.: Deep supervised cross-modal retrieval. In: Proceedings of the IEEE/CVF Conference on Computer Vision and Pattern Recognition, pp. 10394–10403 (2019)

Linear Prediction Residual for Efficient Diagnosis of Parkinson's Disease from Gait

Shanmukh Alle$^{(\boxtimes)}$ (ID) and U. Deva Priyakumar (ID)

Center for Computational Natural Sciences and Bioinformatics, IIIT Hyderabad,
Hyderabad, India
shanmukh.alle@research.iiit.ac.in, deva@iiit.ac.in

Abstract. Parkinson's Disease (PD) is a chronic and progressive neurological disorder that results in rigidity, tremors and postural instability. There is no definite medical test to diagnose PD and diagnosis is mostly a clinical exercise. Although guidelines exist, about 10–30% of the patients are wrongly diagnosed with PD. Hence, there is a need for an accurate, unbiased and fast method for diagnosis. In this study, we propose LPGNet, a fast and accurate method to diagnose PD from gait. LPGNet uses Linear Prediction Residuals (LPR) to extract discriminating patterns from gait recordings and then uses a 1D convolution neural network with depth-wise separable convolutions to perform diagnosis. LPGNet achieves an AUC of 0.91 with a 21 times speedup and about 99% lesser parameters in the model compared to the state of the art. We also undertake an analysis of various cross-validation strategies used in literature in PD diagnosis from gait and find that most methods are affected by some form of data leakage between various folds which leads to unnecessarily large models and inflated performance due to overfitting. The analysis clears the path for future works in correctly evaluating their methods.

Keywords: Parkinson's diagnosis · Gait · Model evaluation · Convolutional neural networks · Linear prediction analysis · Signal processing

1 Introduction

Parkinson's Disease (PD) is a neurological disorder that affects neurons in the brain responsible for motor control which leads to tremors, bradykinesia (slowed movement), limb rigidity, balance and gait problems. It is the second most common neurological disorder after Alzheimer's, affects about 10 million people worldwide [4] and is considered a chronic disease. Although the condition is

Electronic supplementary material The online version of this chapter (https://doi.org/10.1007/978-3-030-87240-3_59) contains supplementary material, which is available to authorized users.

not fatal, disease complications due to symptoms can be serious as they start gradually and develop over time. The Centers for Disease Control and Prevention (CDC) rates complications from PD as the 14th largest cause of death in the United States of America [16]. Unfortunately, the cause of the condition is not yet known and there is no known cure for treating the condition.

To this day there is no definite medical test to diagnose PD and diagnosis is still a clinical exercise [2,14] where an expert draws a conclusion from medical history, symptoms observed, and a neurological examination of a subject. Slow and gradual onset of symptoms and a possibility of human error make diagnosis inaccurate. About 10–30% of the patients initially diagnosed with PD are later diagnosed differently [2,15]. Although there is no known cure for treating the disorder, several therapies [10,14] exist that have shown promise to improve the quality of life of affected patients and reduce severe complications. Development of a fast, standardized, and accurate method for diagnosis of PD is expected to help the lives of affected patients.

Machine learning and statistical methods have shown promise in diagnosing various medical conditions in recent years. Several attempts have been made to use machine learning to build models to diagnose PD from various modalities like speech [11,20], handwriting patterns [7,17], gait patterns [12,19], etc. Although speech and handwriting based models perform well, they have a problem of large variability between demographics as they are not as universal as gait. Hence we believe that gait is the modality to look forward to, to build generalizable models for Parkinson's diagnosis. Few recent works using deep learning methods show good performance in diagnosis Parkinson's from gait. Zhao et al. [23] use a hybrid CNN, LSTM model to achieve 98.6% accuracy, Maachi et al. [5] use a 1D CNN to achieve 98.7% accuracy, Xia et al. [21] use a deep attention based neural network to achieve 99.07% accuracy. Although all the works mentioned use some form of K-fold cross-validation to evaluate the performance of the model they build, there are a few significant differences in the methods used to generate cross-validation folds, making it difficult to compare performance reported in different studies.

We propose LPGNet (Linear Prediction residual based Gait classification Network) a deep learning based model that diagnoses PD from gait with good accuracy while being fast and small enough to be used in embedded systems to enable the method to be cheap and widely accessible. The secondary contribution is that we analyze and compare various methods for obtaining cross-validation folds (train-test splits) used in current literature to check them for data leakage and to gain clarity on how to correctly evaluate the model we build. The code for all the experiments discussed is made public to ensure reproducibility for future research[1].

---

[1] https://github.com/devalab/parkinsonsfromgait.

2 Materials and Methods

2.1 Dataset

We use a publicly available dataset [8] that is a collection of data collected from three different studies [6,9,22]. The dataset consists of 306 gait recordings from 93 patients with PD and 73 healthy control subjects. Each recording is a two-minute long measurement of Vertical Ground Reaction Forces (VGRF) measured under each foot, sampled at a rate 100 Hz as a subject walks at their usual pace at ground level. Each recording includes 18 time series signals where 16 of them are VGRFs measured at 8 points under each foot and the remaining two represent the total VGRF under each foot. The database includes multiple recordings for some subjects where they were made to perform an additional task of solving arithmetic problems while walking.

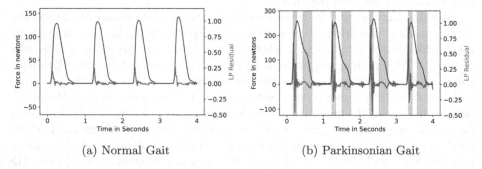

(a) Normal Gait   (b) Parkinsonian Gait

Fig. 1. VGRFs measured from the sensor under the rear part of the right foot for a period of 4 s along with the aligned LPR along with anomalous regions highlighted in the case of the Parkinsonian gait recording.

2.2 Data-Leakage Experiment

To evaluate different validation strategies mentioned in literature, we create a holdout test set that is used to measure performance of the best models obtained from each validation strategy to understand the presence of data leakage. The test set is made up of all VGRF recordings originating from 20% of the subjects. A subject level separation is necessary as recordings originating from the same subject generally are similar. All the recordings are normalized to unit variance. Each recording is then split into windows of length 100 samples with a 50% overlap. Each window inherits the class of the source recording and is considered a separate sample. We then split the windows from the remaining 80% subjects available into the train and validation sets maintaining a 90:10 split with the following strategies.

- **Window Level:** Random 10% of all the available windows make up the validation set and the remaining 90% make up the train set. This method represents the validation strategies [1,21,23].

- **Within Recording:** Random 10% of windows from each recording make up the validation set and the remaining 90% windows in each recording make up the train set. This method represents the validation strategy used by Maachi et al. [5]
- **Subject Level:** Windows belonging to 10% of the subjects make up the validation set and windows belonging to the remaining 90% make up the train set.

We use a Convolution Neural Network (CNN) with three 1D convolution layers with RELU activation followed by a fully connected layer with sigmoid activation function to evaluate the differences between different validation strategies. The model is trained from scratch for each validation method with binary cross-entropy loss and L2 regularization. We use early stopping while training the model and choose the parameters from the epoch that gives the best validation accuracy to evaluate on the test set. We then examine the differences between the validation and test performance of the model representing each validation method to draw conclusions. Where ever necessary we use stratified splits to maintain the ratio of PD and control samples in the train, validation, and holdout test sets.

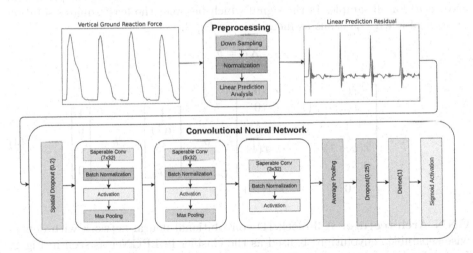

Fig. 2. LPGNet (Linear Prediction residual Gait classification Network) pipeline for Parkinson's Diagnosis

2.3   Model Pipeline

The prediction pipeline in LPGNet involves 3 main steps which include preprocessing, generating the LPR, and performing diagnosis with a CNN as shown in Fig. 2. The following sections give further details.

Preprocessing: Each time series signal (VGRF measured at a point under the foot) is downsampled 50 Hz and then normalized to unit variance. To avoid artifacts while downsampling, the raw signal is filtered with a moving average filter of order 2.

Linear Prediction Residual: Linear Prediction (LP) is a mathematical operation where future samples in a time series are estimated as a linear combination of p past samples [13] as shown in Eq. 1. The coefficients $a(i)$ of a linear predictor model information about the source of the time series. Linear Prediction Residual (LPR) is the prediction error $e(n)$ (Eq. 2) that holds information specific to the time series that LP does not capture. This gives us the ability to separate normal gait patterns from the VGRF recordings and distill information specific to Parkinsonian gait in the residual.

$$\widehat{x}(n) = -\sum_{i=1}^{p} a(i)x(n-i) \tag{1}$$

$$e(n) = x(n) - \widehat{x}(n) \tag{2}$$

The coefficients of the linear predictor are obtained by minimizing the prediction error $e(n)$ for all samples in the signal which becomes the least-squares solution of a set of linear equations as mentioned in Eq. 3.

$$Xa = b \tag{3}$$

$$X = \begin{bmatrix} x(1) & 0 & \cdots & 0 \\ x(2) & x(1) & \cdots & 0 \\ \vdots & \vdots & \cdots & \vdots \\ x(p+1) & \cdots & \cdots & x(1) \\ \vdots & \vdots & \cdots & \vdots \\ 0 & \cdots & 0 & x(m) \end{bmatrix}, \quad a = \begin{bmatrix} 1 \\ a(1) \\ \vdots \\ a(p) \end{bmatrix}, \quad b = \begin{bmatrix} 1 \\ 0 \\ \vdots \\ 0 \end{bmatrix} \tag{4}$$

CNN Architecture: A 3-block 1D convolutional neural network with depthwise separable convolutions is used as the clasifier in LPGNet as seen in Fig. 2. In depth-wise separable convolutions [3], a normal convolution is replaced with channel-wise and point-wise convolutions in succession which reduces the computational and parametric complexity of the network while maintaining its predictive power. They are also less susceptible to overfitting compared to a normal CNN because of the lesser number of learnable parameters.

Each convolution block contains a separable convolution layer, batch normalization layer, ELU activation followed by a max-pooling operation. The three blocks are followed by a global average pooling operation which computes the average over the time dimension resulting in a fixed dimensional vector for the final fully connected layer with sigmoid activation. Using global average pooling at this level enables the model to accept time series of varying lengths which in turn enables us to perform one step inference without resorting to padding/windowing.

Training: Training is done in two steps: Training the linear predictors to find optimal coefficients to generate the LPR followed by training a 1D-CNN to perform diagnosis.

We train a separate linear predictor for each of the 18 time series in a gait recording. To find the LP coefficients, we concatenate all control gait recordings across time with padding equal to the order of the LP and then find the least-squares solution for each time series. Padding is added to prevent different recordings from affecting each other in the optimization process. Once we obtain the coefficients for the linear predictors, we generate LPRs for each of the 306 gait recordings which are then used to train a CNN. LPRs are used to train the CNN as normal gait characteristics in a subject's recording are removed when the predicted time series is subtracted from the original recording and the resulting LPR is more discriminating for classifying between normal and PD. We use the signal processing toolbox available in Matlab to generate LPRs.

Training the CNN is also done in two steps, the generated LPRs are divided into windows representing 2 s maintaining a 50% overlap between successive windows. Each window is assigned the label of the source recording. A CNN is then trained to classify these windows. Once the network converges keeping the weights of the convolution backbone of the network frozen, the fully connected layer is retrained to classify each two-minute recording at once. The second training step is necessary to calibrate the last fully connected layer to changes that might come up due to average pooling over the entire recording. The ADAM optimizer with its default parameters was used to train the network. Binary cross-entropy with label smoothing was used as the objective function.

Considering the small size of the dataset, different forms of regularization are used to control overfitting. We use spatial dropout [18] at the input to the CNN, followed by dropout at the input to the final logistic layer. L2 regularization on all learnable parameters along with gradient clipping was used to aid stability of the training process. We use TensorFlow library to implement and train the model.

2.4   Evaluation

We use stratified 10-fold cross-validation while maintaining a subject level separation between the folds to evaluate the performance of the models considered. We report accuracy, AUC, and F1 scores with the mean and standard deviation measured over the 10 folds. We also report the number of trainable parameters and inference time to classify a two-minute recording on a single thread on an Intel Xeon E5-2640 v4 processor. The inference times reported are an average of 1000 runs. To enforce the use of a single thread we use the capabilities of the SLURM workload manager.

3   Results and Discussion

Table 1 summarizes the train, validation, and test performance of models achieving the best validation performance with each strategy considered (as seen in Sect. 2.2). We also report the absolute difference between each model's performance on the validation and test sets to evaluate the degree of overfitting that in turn provides insight into the extent of data leakage in the validation strategy used.

Table 1. Data leakage experiment: Train, Validation and Test set performance of a baseline CNN for each validation strategy, expressed in the form of percentage accuracy (loss) to understand the presence of data leakage.

Split strategy	Train	Validation	Test	Difference (validation-test)
Within recording	**95.9 (0.285)**	**95.9 (0.284)**	74.9 (0.637)	21.0 (0.353)
Window level	94.6 (0.301)	94.1 (0.308)	74.3 (0.661)	19.8 (0.353)
Subject level	88.7 (0.387)	74.7 (0.572)	**78.8 (0.580)**	**4.1 (0.008)**

The window level and within recording split strategies show very good validation performance but perform poorly on the hold-out test set where the subject level split strategy performs the best. The within recording and window level split strategies show a large drop in performance between validation and test sets compared to the subject level split strategy. This signifies that these strategies are not good validation methods as it indicates heavy data leakage between the train and validation sets. Hence a subject level separation between the train and test folds should be maintained for correctly measuring the performance of a model when a holdout test set is not available. This explains the extremely good performance seen in works of Zhao et al. [23], Xia et al. [21], Maachi et al. [5] despite their models being relatively large for the number of training recordings available as large models overfit easily and generally perform very well in conditions where data leakage exists between the train and test sets.

Table 2. Comparison of various baseline and ablation models

Method	AUC	F1 Score	Accuracy	Inference time (ms)	Parameters
LPGNet	**91.7 ± 9.4**	**93.2 ± 3.6**	**90.3 ± 5.8**	**9.3 ms**	4933
Ablation	90.4 ± 8.1	91.2 ± 4.9	87.6 ± 6.7	13.4 ms	**4735**
Baseline	87.6 ± 11.4	88.7 ± 6.9	83.6 ± 9.7	20.6 ms	16001
1D-ConvNet [5]	86.7 ± 10.3	88.2 ± 6.8	82.5 ± 10.1	195.1 ms	445841

As a baseline, we use the same model used in the data leakage experiment to represent the performance of a simple CNN. We average the probabilities predicted at the window level to get the prediction probabilities for a gait recording.

We also compare with the 1D ConvNet model proposed by Maachi et al. [5] when evaluated with a subject level 10 fold CV used in this work. To analyze the effect of the LPR we perform an ablation study where we train the proposed model with normalized VGRF signals. Table 2 summarizes the performances of various methods considered in this study. The proposed LPGNet performs the best with an AUC of 91.7. It is also the fastest model with an inference time of 9.1 ms (3.7 ms for generating LPR and 5.6 ms for classification), the model is also considerably smaller than others with just 4,933 parameters (198 coefficients in 18 linear predictors of order 11 and 4,735 parameters in the 1D CNN). The large reduction in model size can be attributed to the corrected validation strategy used to evaluate the model while tuning the model architecture. We believe that the models built by Avasarano et al. [1] with 1.5 million parameters and Maachi et al. [5] with 445,841 parameters are relatively large as their model validation strategies were biased towards large models due to data leakage between the train and validation folds. Additional reduction in model size and inference time can be attributed to the use of depth-wise separable convolutions that are more efficient. Faster inference speed can be attributed to the model's ability to classify the entire sequence at once which removes the need for breaking the VGRF recordings into windows and averaging of window level predictions.

Ablation Study: A reduction in performance across the metrics was seen when using normalized VGRF signals 100 Hz to train the model. Apart from worse performance, an increase in variation (standard deviation) in accuracy and F1 score between folds is observed. This shows the role of the LPR in improving the performance and stability of the model. When using the LPR no loss in performance is observed when the VGRFs are sampled 50 Hz, this contributes to the faster inference compared to the ablation experiment despite LPR generation taking up additional time.

LPR also provides a level of interpretability into how the model arrives at a decision as it is the error between modeled normal gait and real gait. A strong deviation from zero in the LPR signifies a deviation of the subject's gait from normal which indicates a higher chance of positive PD diagnosis. It can be seen in Fig. 1b that the deviations are higher in the case of Parkinsonian gait compared to normal gait. Since LPR has the same temporal resolution as the VGRF signal, we can identify parts of the gait cycle where a PD subject's gait deviates heavily from normal gait. The highlighted areas in the Parkinsonian gait in Fig. 1b point to such areas.

The ability of the proposed LPGNet to identify Parkinsonian gait accurately while being small and fast opens new avenues for it to be deployed in embedded systems with limited memory and compute resources which would go a long way in early diagnosis of Parkinson's in the developing countries of the world where clinical experts are not abundant.

4 Conclusion

In this work, we present a novel method LPGNet that uses linear prediction residual with a 1D CNN to efficiently diagnose Parkinson's from VGRF recordings. The proposed method achieves good discriminative performance with an accuracy of 90.3% and an F1 score of 93.2%. The model proposed is also orders of magnitude smaller and faster than methods described in literature. The proposed linear prediction residual aids in improving the interpretability of the method by pointing to the positions in gait patterns that deviate from normal. We also evaluate different validation strategies used in literature and identify the presence of data leakage and show that a subject level separation is necessary for correct evaluation of a method. This experiment clears the path for future works in correctly evaluating their methods by identifying sub-optimal strategies that are susceptible to data leakage.

Acknowledgements. This study was supported by funding from IHub-Data and IIIT Hyderabad. We would also like to thank Dr. K Sudarsana Reddy for the discussions we had regarding the theoretical correctness of the method presented.

References

1. Aversano, L., Bernardi, M.L., Cimitile, M., Pecori, R.: Early detection of parkinson disease using deep neural networks on gait dynamics. In: 2020 International Joint Conference on Neural Networks (IJCNN), pp. 1–8. IEEE (2020)
2. Berardelli, A., et al.: Efns/mds-es recommendations for the diagnosis of parkinson's disease. Eur. J. Neurol. **20**(1), 16–34 (2013)
3. Chollet, F.: Xception: Deep learning with depthwise separable convolutions. CoRR abs/1610.02357 (2016). http://arxiv.org/abs/1610.02357
4. Dorsey, E.R., et al.: Projected number of people with parkinson disease in the most populous nations, 2005 through 2030. Neurology **68**(5), 384–386 (2007)
5. El Maachi, I., Bilodeau, G.A., Bouachir, W.: Deep 1d-convnet for accurate parkinson disease detection and severity prediction from gait. Expert Syst. Appl. **143**, 113075 (2020)
6. Frenkel-Toledo, S., Giladi, N., Peretz, C., Herman, T., Gruendlinger, L., Hausdorff, J.M.: Treadmill walking as an external pacemaker to improve gait rhythm and stability in parkinson's disease. Movement Disorders: Official J. Movement Disorder Soc. **20**(9), 1109–1114 (2005)
7. Gil-Martín, M., Montero, J.M., San-Segundo, R.: Parkinson's disease detection from drawing movements using convolutional neural networks. Electronics **8**(8), 907 (2019)
8. Goldberger, A.L., et al.: PhysioBank, PhysioToolkit, and PhysioNet: components of a new research resource for complex physiologic signals. Circulation **101**(23), e215–e220 (2000). circulation Electronic Pages: http://circ.ahajournals.org/content/101/23/e215.full PMID:1085218; https://doi.org/10.1161/01.CIR.101.23.e215
9. Hausdorff, J.M., Lowenthal, J., Herman, T., Gruendlinger, L., Peretz, C., Giladi, N.: Rhythmic auditory stimulation modulates gait variability in parkinson's disease. Eur. J. Neurosci. **26**(8), 2369–2375 (2007)

10. Jankovic, J., Poewe, W.: Therapies in parkinson's disease. Curr. Opin. Neurol. **25**(4), 433–447 (2012)
11. Karabayir, I., Goldman, S.M., Pappu, S., Akbilgic, O.: Gradient boosting for parkinson's disease diagnosis from voice recordings. BMC Med. Inform. Decis. Mak. **20**(1), 1–7 (2020)
12. Lei, H., Huang, Z., Zhang, J., Yang, Z., Tan, E.L., Zhou, F., Lei, B.: Joint detection and clinical score prediction in parkinson's disease via multi-modal sparse learning. Expert Syst. Appl. **80**, 284–296 (2017)
13. Makhoul, J.: Linear prediction: a tutorial review. Proc. IEEE **63**(4), 561–580 (1975)
14. Massano, J., Bhatia, K.P.: Clinical approach to parkinson's disease: features, diagnosis, and principles of management. Cold Spring Harbor Perspectives Med. **2**(6), a008870 (2012)
15. News, G., Media: Quarter of parkinson's sufferers were wrongly diagnosed, says charity. https://www.theguardian.com/society/2019/dec/30/quarter-of-parkinsons-sufferers-were-wrongly-diagnosed-says-charity, December 2019
16. Sherry L. Murphy, J.X., Kenneth D. Kochanek, E.A., Tejada-Vera, B.: Deaths: Final data for 2018. National Vital Statistics Reports 69(14) (2021)
17. Thomas, M., Lenka, A., Kumar Pal, P.: Handwriting analysis in parkinson's disease: current status and future directions. Movement Disorders Clinical Practice **4**(6), 806–818 (2017)
18. Tompson, J., Goroshin, R., Jain, A., LeCun, Y., Bregler, C.: Efficient object localization using convolutional networks. CoRR abs/1411.4280 (2014). http://arxiv.org/abs/1411.4280
19. Wahid, F., Begg, R.K., Hass, C.J., Halgamuge, S., Ackland, D.C.: Classification of parkinson's disease gait using spatial-temporal gait features. IEEE J. Biomed. Health Inform. **19**(6), 1794–1802 (2015)
20. Wroge, T.J., Özkanca, Y., Demiroglu, C., Si, D., Atkins, D.C., Ghomi, R.H.: Parkinson's disease diagnosis using machine learning and voice. In: 2018 IEEE Signal Processing in Medicine and Biology Symposium (SPMB), pp. 1–7. IEEE (2018)
21. Xia, Y., Yao, Z., Ye, Q., Cheng, N.: A dual-modal attention-enhanced deep learning network for quantification of parkinson's disease characteristics. IEEE Trans. Neural Syst. Rehabil. Eng. **28**(1), 42–51 (2019)
22. Yogev, G., Giladi, N., Peretz, C., Springer, S., Simon, E.S., Hausdorff, J.M.: Dual tasking, gait rhythmicity, and parkinson's disease: which aspects of gait are attention demanding? Eur. J. Neurosci. **22**(5), 1248–1256 (2005)
23. Zhao, A., Qi, L., Li, J., Dong, J., Yu, H.: A hybrid spatio-temporal model for detection and severity rating of parkinson's disease from gait data. Neurocomputing **315**, 1–8 (2018)

Primary Tumor and Inter-Organ Augmentations for Supervised Lymph Node Colon Adenocarcinoma Metastasis Detection

Apostolia Tsirikoglou[1]([✉])[iD], Karin Stacke[1,3][iD], Gabriel Eilertsen[1,2][iD], and Jonas Unger[1,2][iD]

[1] Department of Science and Technology, Linkoping University, Linköping, Sweden
apostolia.tsirikoglou@liu.se
[2] Center for Medical Image Science and Visualization, Linkoping University, Linköping, Sweden
[3] Sectra AB, Linköping, Sweden

Abstract. The scarcity of labeled data is a major bottleneck for developing accurate and robust deep learning-based models for histopathology applications. The problem is notably prominent for the task of metastasis detection in lymph nodes, due to the tissue's low tumor-to-non-tumor ratio, resulting in labor- and time-intensive annotation processes for the pathologists. This work explores alternatives on how to augment the training data for colon carcinoma metastasis detection when there is limited or no representation of the target domain. Through an exhaustive study of cross-validated experiments with limited training data availability, we evaluate both an inter-organ approach utilizing already available data for other tissues, and an intra-organ approach, utilizing the primary tumor. Both these approaches result in little to no extra annotation effort. Our results show that these data augmentation strategies can be an efficient way of increasing accuracy on metastasis detection, but fore-most increase robustness.

Keywords: Computer aided diagnosis · Computational pathology · Domain adaptation · Inter-organ · Colon cancer metastasis

1 Introduction

Colon cancer is the third most common cancer type in the world, where the majority of the cases are classified as adenocarcinoma [26]. Along with grading the primary tumor, assessment of the spread of the tumor to regional lymph nodes is an important prognostic factor [5]. The pathologist is therefore required

Electronic supplementary material The online version of this chapter (https://doi.org/10.1007/978-3-030-87240-3_60) contains supplementary material, which is available to authorized users.

to not only assess the primary tumor but in high resolution, scan multiple lymph node sections, a task that is both challenging and time-consuming. Deep learning-based methods could be of use in assisting the pathologist, as they have shown great success for other histopathology applications [21]. However, a significant challenge is the need for large, annotated datasets, which in the case of lymph node metastasis detection is heightened due to the low tumor-to-non-tumor ratio in the tissue, and the annotation expertise needed. In this paper, we study how data with lower acquisition and annotation cost can be used to augment the training dataset, thus reducing the need for a large cohort size of the target lymph node metastasis data. We explore this using *inter-organ augmentations*, i.e., utilizing data from different organs from existing public datasets. Leveraging the uniformity across staining, scanning, and annotation protocols, we investigate how potential similarities and differences across tissue and cancer types can be useful for the target application. In addition to the inter-organ augmentations, we also consider *intra-organ augmentations*, by using data from the primary tumor. Gathering labeled data from the primary tumor requires little extra work, as tissue samples of it typically are acquired in conjunction with the lymph node sections, and the high tumor-to-non-tumor ratio allows for faster annotations. Furthermore, we investigate three different data availability scenarios based on annotation cost (in terms of time and effort).

In summary, we present the following set of contributions:

- The first large-scale study on inter-organ data augmentations in digital pathology for metastasis detection. This includes a rigorous experimental setup of different combinations of inter- and intra-organ training data. We test both direct augmentation between organs, as well as transformed data by means of Cycle-GAN [29] in order to align the source images with the target domain.
- We measure the impact on performance of lymph node colon tumor metastasis detection in three different scenarios, each representing a different effort/cost in terms of gathering and annotating the data.
- In addition to inter-organ augmentation, we show how intra-organ augmentation, using data from the primary tumor, can increase robustness for detection on lymph node data.

The results point to how inter-organ data augmentations can be an important tool for boosting accuracy, but fore-mostly for increasing the robustness of deep pathology applications. How to make the best use of source organ data depends on its similarity to the target organ, where more similar data results in no or detrimental impact on performance together with Cycle-GAN transformed images, whereas the opposite is true for dissimilar data. Finally, we highlight the importance of making use of data from the primary tumor, as a low-effort strategy for increasing the robustness of lymph node metastasis detection.

2   Related Work

A number of deep learning-based methods have previously been presented for metastases detection, primarily facilitated by the CAMELYON16 and -17

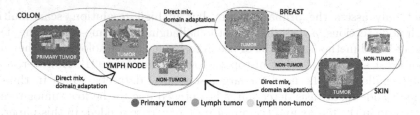

Fig. 1. Same-distribution (colon primary tumor) as target, and inter-organ (breast and skin) augmentations. Both alternatives are explored either as data direct mix, or image synthesis as domain adaptations to the target distribution (lymph node colon adenocarcinoma metastasis).

challenges [2,6], where large datasets of whole-slide images of sentinel lymph node sections for breast cancer metastases were collected and made publicly available. As these types of large datasets are not available for all tissue and cancer types, different approaches have been taken to harness the data in other domains. These can be divided in to two, in many cases orthogonal categories: manipulation of the model, such as transfer learning [12,27] and domain adaptation [7,20], and manipulation of the data, which is the focus of this paper.

Examples of augmentations that have shown successful for histopathology applications are geometric transformations (rotation, flipping, scaling), color jittering [22,25], and elastic deformations [11]. Furthermore, methods using generative adversarial networks (GANs [9]) to synthetically generate data have proven useful [4,10,13,14]. In this work, we omit the step of generating synthetic data, and instead, investigate the possibility to augment the target dataset with 1) already existing publicly available datasets of other tissue types, and 2) same-distribution data with lower annotation cost.

Using the primary tumor for metastasis detection has been done in Zhou et al. [28] for preoperative investigation using ultrasound imaging, and in Lu et al. [17], where metastatic tumor cells were used to find the primary source. To our knowledge, this is the first time the efficiency of using primary tumor data for metastatic cancer detection is investigated in histopathology.

3 Method

To provide a deeper understanding of the impact of inter- and intra-organ augmentation strategies, we set up an experimental framework that evaluates different perspectives in terms of data availability and augmentation protocols. As illustrated in Fig. 1, we propose to leverage the readiness of the primary colon cancer tumor, as well as already existing carcinoma datasets for different organs tissue (breast and skin). We evaluate different training data compositions for three different data availability scenarios of the target domain (colon lymph node metastasis), as illustrated in Fig. 2. In what follows we outline the datasets, target scenarios, augmentation techniques, as well as evaluation protocol. For details on the experimental setup, we refer to the supplementary material.

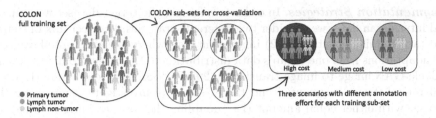

Fig. 2. Colon training set and annotation cost scenarios overview. The full colon train-ing set consists of all tissue types (left). Cross-validation of limited data access simula-tion is possible by dividing the full dataset into four sub-sets (middle). For each sub-set, three scenarios for annotation effort are created (right): high (only lymph node tissue), medium (primary and very little lymph node), and low (primary tumor).

Datasets. In the conducted experiments, the target colon adenocarcinoma dataset consists of data from 37 anonymized patients, where data from 5 patients were used as the test set, and the rest used for training and validation [19]. The dataset contains images from both primary and lymph node tumor samples, as well as lymph node non-tumor tissue. For the inter-organ augmentations, we selected breast and skin carcinoma, driven by their high clinical occurrence, existing datasets availability, and cancer type/similarity compared to colon ade-nocarcinoma. The breast dataset [16] consists of whole slide images of sentinel breast lymph node sections, containing breast cancer metastasis. This cancer type, originating from epithelial cells, is similar to colon cancer. On the other hand, the skin cancer dataset [15,24] consists of abnormal findings identified as basal cell carcinoma, squamous cell carcinoma, and squamous cell carcinoma in situ. These tissue and cancer types are more different from regional colon lymph nodes and colon adenocarcinoma. The whole-slide images of all three datasets were sampled to extract patches. The data were extracted at a resolution of 0.5 microns per pixel, with a size of 256×256. All three datasets are publicly available for use in legal and ethical medical diagnostics research.

Target Scenarios. To simulate limited access of target domain training data, but also cross-validate the experiments' performance and the outcome obser-vations, the available full colon dataset was divided into four subsets, ensuring balance between the different tissue types, as well as number of patients. Each sub-set has non-tumor lymph node tissue data from at least five patients, tumor lymph node samples from at least six patients, and primary tumor samples from at least four patients. Considering the different costs of annotation effort of the primary tumor and lymph node tissue we identify three baseline experiments: 1) the **high** cost scenario including only lymph node tissue data, 2) the **medium** cost including primary tumor data along with lymph node tissue from only two patients on average, and 3) the **low** cost case including only primary tumor (i.e., no target tumor representation) and lymph node non-tumor tissue from just two patients on average (Fig. 2). The inter-organ augmentations do not charge the baseline experiments with extra annotation effort for the experts, since they utilize already available annotated data.

Augmentation Strategies. In order to augment the target dataset with intra- and inter-organ data, we consider two different strategies: 1) direct mix of source and target training data, and 2) image synthesis where the augmented samples are adaptations from one organ's data distribution to the target domain through a Cycle-GAN image-to-image-translation [29]. Furthermore, to evaluate the optimal ratio between augmented and target data, we investigate augmenting the dataset with either equal amount (i.e., doubling the total training set size), or half the amount.

While the direct mix allows for understanding of how data from a different organ impact the target domain, the domain-adaptation of images demonstrates if there are features in the source domain that can be utilized if the data distribution is aligned with the target domain. Although there are other strategies for aligning the domains, such as stain normalization [18], the Cycle-GAN provides us with a representative method for investigating the performance of transformed source data. Note that since the target domain is formulated in three different scenarios, the different inter- and intra-organ augmentations are evaluated in three separate experiments, i.e., Cycle-GANs need to be trained separately for each target scenario.

Evaluation Protocol. To evaluate task performance for the different combinations of training data, we train a deep classifier for tumor detection and evaluate it on the lymph node colon adenocarcinoma metastasis test data. The network consists of three convolutional and two fully connected layers with dropout and batch-normalization for regularization. We employ standard geometric and color jittering augmentations. The networks are trained with Adam optimizer for 50 epochs, out of which the best model is selected. For the GAN augmentation, we used a vanilla Cycle-GAN[1], trained for 250,000 iterations, using colon data defined per experiment and the organ's entire dataset.

We cross-validate each target scenario between the four colon dataset subsets. Moreover, we run each experiment's convolutional network five times to ensure adequate statistical variation. This work puts special emphasis on formulating augmentation schemes that lead to stable and robust results, highlighting that they are at least equally important as factors like training data size and downstream task reported performance.

In addition to the evaluation in terms of classification performance we also include a measure of the representation shift between source and target data [22, 23]. This measure takes the distributions of layer activations over a dataset in a classifier, and compares this between the source and target domains, capturing the model-perceived similarity between the datasets.

4   Results

The experimental setup with different scenarios, data, augmentation strategies, and amount of augmented data, lead to a total of 50 individual experiments.

---

[1] https://github.com/vanhuyz/CycleGAN-TensorFlow.

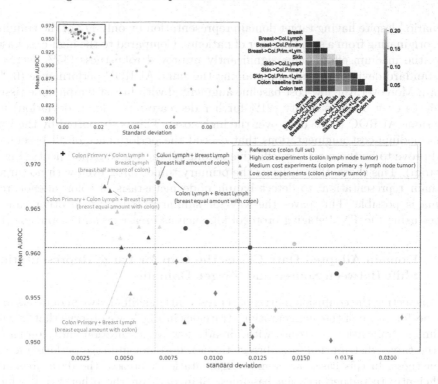

Fig. 3. Mean AUROC to standard deviation for all the experiments (top left), and detailed view for the best performing ones with respect to the corresponding baselines (bottom). The representation shift (top right) is measured between different datasets using a classifier trained for the baseline medium effort scenario.

Figure 3 plots the mean AUROC (Area Under the Receiver Operating Characteristic) curve against the standard deviation, computed over the 4 sub-sets' performance for 5 trainings per sub-set, for the different cost scenarios, as well as the representation shift between different datasets. Baselines are noted with dashed line crosses and the different augmentation experiments for each scenario are color and marker coded. The best performing setups are reported in Table 1, where each column corresponds to a different scenario (marked with different colors in Fig. 3 top left and bottom). For an exhaustive presentation of the numbers and legends for all the experiments, we refer to the supplementary material. In what follows, we will focus on the most interesting observations made from the results in Fig. 3 and Table 1.

4.1 Primary Tumor Is an Inexpensive Training Data Source for Lymph Node Colon Cancer Metastasis Detection

Inspecting the baseline performances of the three scenarios in Fig. 3 and Table 1, it is clear that the most robust approach is the medium annotation effort scenario, despite having target domain representation by only 1/3 of the training set, originating from a small number of patients. Compared to the high-cost baseline, the medium effort offers significantly improved robustness (45% decrease in standard deviation) while maintaining the same AUROC performance (0.2% drop). Moreover, the low-cost baseline augmented with breast lymph node tissue achieves even better stability (21% further decrease in standard deviation) for the same AUROC (0.1% drop over the high cost). This is supported by the various medium-cost augmentations that exceed the performance of all baselines, and prove to be the most cost-effective among all the tested experiments (Fig. 3 bottom). This shows that utilizing the primary tumor data, even with no target domain representation, to detect lymph node metastasis of colon adenocarcinoma is possible. This paves the way for similar possibilities in other cancer types using the TNM staging protocol [3], such as breast [8] and esophagus [1].

4.2   Domain Adapted Data Closes the Gap for Large Representation Shift Between Source and Target Domain

By inspecting the confusion matrix of representation shifts between datasets and the performance of the augmentation strategies in Fig. 3, we observe that image-to-image translation of tissues with already low representation shift compared to the target distribution (e.g., breast tissue), does not improve performance or robustness. In this case, direct mix augmentation increases the data diversity, sufficiently to outperform the baselines. Skin data on the other hand, which exhibits a much larger representation shift from the colon lymph node metastasis samples, show a significant avail from data adaptation. Direct mix augmentation with such a dataset drastically decreases performance, indicating that the added diversity does not contribute to convergence. Domain adaptation closes the gap in the representation shift, leading improved performance as compared to the baseline.

Table 1. Mean AUROC for the best performing augmentation strategies for all three annotation effort scenarios. Bre./Skin→Col. and Prim.→Lym. denote data domain transformation using Cycle-GAN, and (equal/half am.) is the amount of added images in relation to the size of the available baseline training set.

Augmentation	Mean AUROC± stddev		
	High	Medium	Low
	0.9607 ± 0.01206	0.9589 ± 0.0067	0.9538 ± 0.0114
+Breast(equal am.)	0.9684 ± 0.0076	0.9671 ± 0.0042	0.9598 ± 0.0053
+Breast(half am.)	0.9680 ± 0.0069	0.9676 ± 0.0039	0.9591 ± 0.0079
+Bre.→Col.(equal am.)	0.9577 ± 0.0171	0.9521 ± 0.0117	0.9502 ± 0.0167
+Bre.→Col.(half am.)	0.9635 ± 0.0096	0.9523 ± 0.0084	0.9518 ± 0.0122
+Skin→Col.(equal am.)	0.9589 ± 0.0173	0.9555 ± 0.0054	0.9528 ± 0.0208
+Skin→Col.(half am.)	0.9611 ± 0.0146	0.9635 ± 0.0043	0.9536 ± 0.0135
+Prim.→Lym.	–	0.9630 ± 0.0070	–

4.3   Robustness Is Improved by Out-of-Domain Data

One of the central observations from the experiments regards the differences in robustness for different augmentation scenarios, with the robustness measured from the standard deviation over multiple trainings. Classical image augmentation is the most critical component for increasing both generalization performance and robustness, and is applied in all of our different experiments. In addition to this, both primary tumor data, as well as inter-organ augmentations using breast and skin data, can provide an additional boost in terms of performance. However, analyzing the relations between AUROC and standard deviation in Fig. 3, we can see a more pronounced impact on robustness.

As discussed above, the medium-effort scenario is on pair with the high-effort scenario in terms of AUROC, and gives overall lower variance. For the inter-organ augmentations, breast data do not benefit from adaptation by means of the Cycle-GAN, while this is essential for reaping the benefits of the skin data. These results point to how the out-of-domain data (e.g., primary tumor, or other organ tissue) can have a regularizing effect on the optimization, which has a significant impact on robustness. This means that in certain situations it is better not to perform data adaptation since this will decrease the regularizing effect (e.g., primary tumor data, or breast data with low representation shift). However, if the data is widely different (e.g., skin data, with large representation shift), it is necessary to perform adaptation in order to benefit from augmentation.

5   Conclusion

This paper presented a systematic study on the impact of inter- and intra-organ augmentations under different training data availability scenarios for lymph node colon adenocarcinoma metastasis classification. The results show that accuracy can be boosted by utilizing data from different organs, or from the primary tumor, but most importantly how this has an overall positive effect on the robustness of a model trained on the combined dataset.

One of the important aspects when incorporating data from a different domain is the strategy used for performing augmentation. For a source dataset that more closely resembles the target data, adaptation of the image content can have a detrimental effect, while for different data image adaptation is a necessity. For future work, it would be of interest to closer define when to adapt and when not to. This could potentially be quantified with the help of measures that aim at comparing model-specific differences between datasets, such as the representation shift used in these experiments. Moreover, there are other types of data and augmentation strategies that could be explored, as well as model-specific domain adaptation and transfer-learning techniques. We believe that utilization of inter-organ data formulations will be an important tool in future machine learning-based medical diagnostics.

Acknowledgments. We would like to thank Martin Lindvall for the interesting discussions and insights into the use of cancer type-specific primary tumor data for lymph node metastasis detection, and Panagiotis Tsirikoglou for the suggestions in results analysis. This work was partially supported by the Wallenberg AI, Autonomous Systems and Software Program (WASP) funded by the Knut and Alice Wallenberg Foundation, the strategic research environment ELLIIT, and the VINNOVA grant 2017-02447 for the Analytic Imaging Diagnostics Arena (AIDA).

References

1. Ajani, J.A., D'Amico, T.A., Bentrem, D.J., Chao, J., Corvera, C., et al.: Esophageal and Esophagogastric Junction Cancers, Version 2.2019, NCCN Clinical Practice Guidelines in Oncology. J. National Comprehensive Cancer Network **17**(7), 855–883 (2019)
2. Bándi, P., Geessink, O., Manson, Q., Dijk, M.V., Balkenhol, M., et al.: From detection of individual metastases to classification of lymph node status at the patient level: the CAMELYON17 Challenge. IEEE Trans. Med. Imaging **38**(2), 550–560 (2019)
3. Brierley, J., Gospodarowicz, M., Wittekind, C. (eds.): UICC TNM Classification of Malignant Tumours, 8th edn. Wiley-Blackwell, Chichester, November 2016
4. Brieu, N., Meier, A., Kapil, A., Schoenmeyer, R., Gavriel, C.G., et al.: Domain Adaptation-based Augmentation for Weakly Supervised Nuclei Detection. arXiv preprint arXiv:1907.04681 (2019)
5. Compton, C.C., Fielding, L.P., Burgart, L.J., Conley, B., Cooper, H.S., et al.: Prognostic factors in colorectal cancer. Arch. Pathol. Lab. Med. **124**, 16 (2000)
6. Ehteshami Bejnordi, B., Veta, M., Johannes van Diest, P., van Ginneken, B., Karssemeijer, N., et al.: Diagnostic assessment of deep learning algorithms for detection of lymph node metastases in women with breast cancer. JAMA **318**(22), 2199–2210 (2017)
7. Figueira, G., Wang, Y., Sun, L., Zhou, H., Zhang, Q.: Adversarial-based domain adaptation networks for unsupervised tumour detection in histopathology. In: 2020 IEEE 17th International Symposium on Biomedical Imaging (ISBI), pp. 1284–1288, April 2020
8. Fitzgibbons, P.L., Page, D.L., Weaver, D., Thor, A.D., Allred, D.C., et al.: Prognostic factors in breast cancer. College of American Pathologists Consensus Statement 1999. Archives Pathol. Laboratory Med. **124**(7), 966–978 (2000)
9. Goodfellow, I.J., Pouget-Abadie, J., Mirza, M., Xu, B., Warde-Farley, D., et al.: Generative adversarial nets. In: Advances in Neural Information Processing Systems 27: Annual Conference on Neural Information Processing Systems 2014, 8–13 December 2014, Montreal, Quebec, Canada, pp. 2672–2680 (2014)
10. Hou, L., Agarwal, A., Samaras, D., Kurc, T.M., Gupta, R.R., et al.: Robust histopathology image analysis: to label or to synthesize? In: 2019 IEEE/CVF Conference on Computer Vision and Pattern Recognition (CVPR), pp. 8525–8534 (2019)
11. Karimi, D., Nir, G., Fazli, L., Black, P.C., Goldenberg, L., et al.: Deep learning-based gleason grading of prostate cancer from histopathology images-role of multi-scale decision aggregation and data augmentation. IEEE J. Biomed. Health Inform. **24**(5), 1413–1426 (2020)
12. Khan, U.A.H., et al.: Improving prostate cancer detection with breast histopathology images. In: Reyes-Aldasoro, C.C., Janowczyk, A., Veta, M., Bankhead, P., Sirinukunwattana, K. (eds.) ECDP 2019. LNCS, vol. 11435, pp. 91–99. Springer, Cham (2019). https://doi.org/10.1007/978-3-030-23937-4_11

13. Krause, J., Grabsch, H., Kloor, M., Jendrusch, M., Echle, A., et al.: Deep learning detects genetic alterations in cancer histology generated by adversarial networks. J. Pathol. **254**, 70–79 (2021)
14. Levine, A.B., Peng, J., Farnell, D., Nursey, M., Wang, Y., et al.: Synthesis of diagnostic quality cancer pathology images by generative adversarial networks. J. Pathol. **252**(2), 178–188 (2020)
15. Lindman, K., Rose, J.F., Lindvall, M., Stadler, C.B.: Skin data from the visual sweden project DROID (2019). https://doi.org/10.23698/aida/drsk
16. Litjens, G., Bandi, P., Ehteshami Bejnordi, B., Geessink, O., Balkenhol, M., et al.: 1399 H&E-stained sentinel lymph node sections of breast cancer patients: the CAMELYON dataset. GigaScience **7**(6), June 2018
17. Lu, M.Y., Zhao, M., Shady, M., Lipkova, J., Chen, T.Y., et al.: Deep Learning-based Computational Pathology Predicts Origins for Cancers of Unknown Primary. arXiv:2006.13932 [cs, q-bio], June 2020
18. Macenko, M., Niethammer, M., Marron, J.S., Borland, D., Woosley, J.T., et al.: A method for normalizing histology slides for quantitative analysis. In: 2009 IEEE International Symposium on Biomedical Imaging: From Nano to Macro, pp. 1107–1110. IEEE, Boston, June 2009
19. Maras, G., Lindvall, M., Lundstrom, C.: Regional lymph node metastasis in colon adenocarcinoma (2019). https://doi.org/10.23698/aida/lnco
20. Ren, J., Hacihaliloglu, I., Singer, E.A., Foran, D.J., Qi, X.: Unsupervised domain adaptation for classification of histopathology whole-slide images. Frontiers Bioeng. Biotechnol. **7**, 102 (2019)
21. Serag, A., et al.: Translational ai and deep learning in diagnostic pathology. Front. Med. **6**, 185 (2019)
22. Stacke, K., Eilertsen, G., Unger, J., Lundström, C.: Measuring domain shift for deep learning in histopathology. IEEE J. Biomed. Health Inform. **25**(2), 325–336 (2021)
23. Stacke, K., Eilertsen, G., Unger, J., Lundström, C.: A Closer Look at Domain Shift for Deep Learning in Histopathology. arXiv preprint arXiv:1909.11575 (2019)
24. Stadler, C.B., Lindvall, M., Lundström, C., Bodén, A., Lindman, K., et al.: Proactive construction of an annotated imaging database for artificial intelligence training. J. Digit. Imaging **34**, 105–115 (2021)
25. Tellez, D., Litjens, G., Bándi, P., Bulten, W., Bokhorst, J.M., et al.: Quantifying the effects of data augmentation and stain color normalization in convolutional neural networks for computational pathology. Med. Image Anal. **58**, 101544 (2019)
26. Wild, C., Weiderpass, E., Stewart, B. (eds.): World Cancer Report: Cancer Research for Cancer Prevention. International Agency for Research on Cancer, Lyon, France (2020)
27. Xia, T., Kumar, A., Feng, D., Kim, J.: Patch-level tumor classification in digital histopathology images with domain adapted deep learning. In: 2018 40th Annual International Conference of the IEEE Engineering in Medicine and Biology Society (EMBC), pp. 644–647 (2018)
28. Zhou, L.Q., Wu, X.L., Huang, S.Y., Wu, G.G., Ye, H.R., et al.: Lymph node metastasis prediction from primary breast cancer US images using deep learning. Radiology **294**(1), 19–28 (2020)
29. Zhu, J., Park, T., Isola, P., Efros, A.A.: Unpaired image-to-image translation using cycle-consistent adversarial networks. In: 2017 IEEE International Conference on Computer Vision (ICCV), pp. 2242–2251 (2017)

Radiomics-Informed Deep Curriculum Learning for Breast Cancer Diagnosis

Giacomo Nebbia[1], Saba Dadsetan[1], Dooman Arefan[2], Margarita L. Zuley[2,3],
Jules H. Sumkin[2,3], Heng Huang[4], and Shandong Wu[1,2,5(✉)]

[1] Intelligent Systems Program, School of Computing and Information, University of Pittsburgh, Pittsburgh, PA, USA
{gin2,sad149}@pitt.edu, wus3@upmc.edu

[2] Department of Radiology, University of Pittsburgh School of Medicine, Pittsburgh, PA, USA
doa14@pitt.edu, {zuleyml,sumkjh}@upmc.edu

[3] Magee-Womens Hospital, University of Pittsburgh Medical Center, Pittsburgh, PA, USA

[4] Department of Electrical and Computer Engineering, Swanson School of Engineering, University of Pittsburgh, Pittsburgh, PA, USA
heng.huang@pitt.edu

[5] Department of Biomedical Informatics and Department of Bioengineering, University of Pittsburgh, Pittsburgh, PA, USA

Abstract. Convolutional Neural Networks (CNNs) are traditionally trained solely using the given imaging dataset. Additional clinical information is often available along with imaging data but is mostly ignored in the current practice of data-driven deep learning modeling. In this work, we propose a novel deep curriculum learning method that utilizes radiomics information as a source of additional knowledge to guide training using customized curriculums. Specifically, we define a new measure, termed radiomics score, to capture the difficulty of classifying a set of samples. We use the radiomics score to enable a newly designed curriculum-based training scheme. In this scheme, the loss function component is weighted and initialized by the corresponding radiomics score of each sample, and furthermore, the weights are continuously updated in the course of training based on our customized curriculums to enable curriculum learning. We implement and evaluate our methods on a typical computer-aided diagnosis of breast cancer. Our experiment results show benefits of the proposed method when compared to a direct use of radiomics model, a baseline CNN without using any knowledge, the standard curriculum learning using data resampling, an existing difficulty score from self-teaching, and previous methods that use radiomics features as additional input to CNN models.

Keywords: Breast cancer · Classification · Deep learning · Radiomics · Curriculum learning

1 Introduction

The traditional way of training Convolutional Neural Networks (CNNs) makes use of images as the sole source of data. In medical imaging applications, additional information or clinical knowledge are often available along with the data, such as pre-assessment

© Springer Nature Switzerland AG 2021
M. de Bruijne et al. (Eds.): MICCAI 2021, LNCS 12905, pp. 634–643, 2021.
https://doi.org/10.1007/978-3-030-87240-3_61

made by clinicians, auxiliary tasks in relation to a target task, qualitative clinical experience, etc. These sources of information can be useful for a target task but are mostly ignored in the current practice of data-driven deep learning modeling.

Incorporating domain knowledge into machine learning is a well-studied topic for traditional machine learning [1]. Previous works explicitly embed knowledge in a model's structure, by, for example, defining a kernel for Support Vector Machine [2], by building a basic building block for Bayesian networks [3], or by weighing different features in the loss function [4]. These methods are challenging for deep learning models because it is harder to adapt these methods to manipulate the features automatically learned in deep learning. This poses a need to investigate new strategies to incorporate domain knowledge (especially medical contexts) into deep learning [5, 6].

Radiomics features are pre-defined and hand-crafted imaging features that are often extracted from segmented areas or full images to capture microstructural information. Such features have successfully been applied to and shown promising effects in numerous medical imaging tasks [7–10]. Researchers have explored to directly incorporate radiomics features into deep learning for joint modeling [11, 12]. It remains a challenging but important question to develop new strategies and methodologies to leverage radiomics to improve deep learning modeling.

In this work, we propose a novel strategy to harness radiomics features as a form of knowledge to improve CNN-based deep learning. We first define a new measure, termed radiomics score, to capture the difficulty of classifying a set of samples. Then we design a novel radiomics score-guided curriculum training scheme. In this scheme, the radiomics score is used to weigh each image's loss function component; and furthermore, in the course of training, the weights (initialized by the radiomics scores) are continuously updated based on our customized curriculums to enable curriculum learning. We evaluate our methods by comparing it to (1) the radiomics model, (2) a baseline CNN without using any knowledge, (2) standard curriculum learning using data resampling, (3) an existing difficulty score from self-teaching, and (4) previous methods that directly use radiomics features as additional input to CNN models. Experiments on a breast cancer imaging dataset for the computer-aided diagnosis (malignant vs benign) task show outperforming results of our method.

Our main contributions include: (I) we create radiomics score as a source of knowledge to augment deep learning training, (II) we propose a novel training scheme guided the radiomics score and customized curriculum strategies to control the training loss, and (III) we show the applicability and generalizability of the proposed training scheme on a different type of difficulty score from the radiomics score.

2   Methods

Figure 1 illustrates the framework of our proposed method. The first step is to extract radiomics features and compute a difficulty score from these radiomics features to characterize each image. For imaging data with segmentations available, one can apply a standard radiomics pipeline [16] to extract a comprehensive set of radiomics features, as usually done in the radiomics community. Typical radiomics features (see [16] for more details on the feature definitions) include shape, textures, first-order features, grey-level based features, etc. After extracting radiomics features, we normalize all the radiomics features by subtracting the mean and dividing by the standard deviation. Then, we train a logistic regression classifier on the target task with L1 regularization, from which we can calculate a radiomics score as the predicted probability of the ground truth label, as shown in Eq. 1.

$$RadScore_n = \mathbb{1}(I_n \text{ is positive})p_n + \mathbb{1}(I_n \text{ is negative})(1 - p_n) \qquad (1)$$

In Eq. 1, $\mathbb{1}$ represents the indicator function returning 1 if the condition between parentheses is met and 0 otherwise; I_n is the n-th image, and p_n is the probability of I_n being classified as positive. The hypothesis is that the higher the radiomics score is, the easier the sample is for classification, and this hypothesis provides a form of knowledge that may help for the image classification tasks.

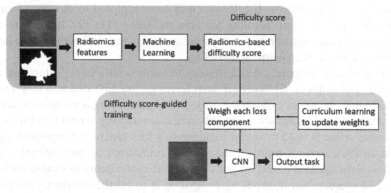

Fig. 1. Flowchart depicting our proposed approach. Top, we extract radiomics features, build a machine learning classifier, and obtain a difficulty score. Bottom, the score is used to guide curriculum training by weighing the loss function using customized curriculums.

Once we obtain the radiomics-based difficulty score, we use it to implement a difficulty score-guided curriculum training of CNN models. To do so, we choose to use the following approach to adjust each image's loss component:

$$L = \sum_{n=1}^{N} w_n L_n, \qquad (2)$$

where L denotes the overall loss value, N is the total number of images, L_n is the loss component associated with image n, and w_n is the weight assigned to image n's loss component. At the beginning, w_n is initialized as the corresponding radiomics score, i.e., $w_n = RadScore_n$. Here, the weight w_n is not fixed during the training. Instead, we propose to continuously update the weight w_n using customized curriculums to enable curriculum learning (i.e., easier cases are emphasized first, harder cases second).

Following the previous hypothesis, i.e., easier cases have higher radiomics scores, the initial weights of loss are set to the corresponding radiomics scores for each sample. To leverage curriculum learning for improved training effects, we propose to apply customized curriculums to control and continuously update the weights (i.e., w_n) as training progresses. By dynamically adjusting the weights, we guide the model training to focus the learning more on a subset of samples of different interests (e.g., easier/hardier cases) in the course of the training.

We can customize various forms of curriculums to achieve the goal of dynamically adjusting the weights. Here we define two curriculums (Fig. 2), as an example, to implement our method. In Curriculum 1 (Fig. 2, left), we decrease the weight of easy samples (i.e., those with $RadScore_n > 0.5$) and increase that of hard samples until they all reach the same value of 0.5. The rationale behind Curriculum 1 is to encourage the model to first learn more from easy cases and then have it deal with the learning equivalently for both easy and hard cases (when the weights are all the same). In Curriculum 2 (Fig. 2, right), we apply the same strategy of changing the weights at the initial phase, but then we keep increasing the weight of hard samples and decreasing that of easy samples until an upper and lower bound are reached. Here the rationale is to start the training in the same way but encourage the model to focus almost solely on hard cases toward the end of the training, after features descriptive of easy cases should have already been learned. The two proposed curriculums represent and also extend the intuitive implementations of the "easy-to-hard" curriculum learning. It should be pointed out that more and different forms of curriculums can be customized and used directly in our proposed method.

We set the time in training when the weights stop changing to half and two-thirds of the training time for Curriculum 1 and 2, respectively. The upper and lower bounds for Curriculum 2 are chosen based on the maximum and minimum score values in the training set (0.05 and 1 in our dataset). This choice makes the selection of these two hyperparameters data-driven and less arbitrary.

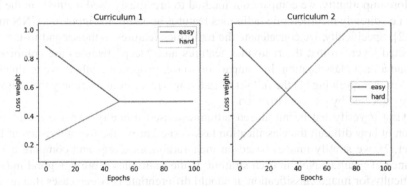

Fig. 2. The customized curriculums to guide dynamic updating of the weights in adjusting the training loss. Left: Curriculum 1: weights for easy cases decrease and weights for hard cases increase until they both reach the same value of 0.5. Right: Curriculum 2: weights for easy cases decrease until they reach a lower bound and weights for hard cases increase until they reach an upper bound.

3 Evaluation

3.1 Evaluation Settings of the Proposed Method

We perform an evaluation of the proposed method with multiple settings. First, we compare our method to two baselines, i.e., the direct classification results of the radiomics model and a baseline CNN model directly trained using the imaging data alone (without using any additional knowledge or curriculum learning).

Second, we compare our method to a classic curriculum learning [13] strategy that relies on data resampling, where the radiomics-based difficulty scores are directly used to determine the training order of the images. Specifically, in every epoch, each image is sampled (without replacement) with probability proportional to its difficulty score, so that easier samples are more likely to be seen at the beginning of each epoch. This comparison is to reveal the benefit of the customized curriculums applied to the radiomics difficulty scores.

In addition, in order to further assess the generalizable effects of the proposed curriculum-guided training, we apply it to an existing difficulty score (not using the radiomics score) obtained from self-teaching [14]. With self-teaching, a model is trained in the standard approach (i.e., no prior difficulty score used) and its predictions on each image are recorded. These predictions are then used to define a self-taught score to guide re-training of the model [14]. We plug the self-teaching score to replace the radiomics scores in the training process described in Sect. 2. This experiment is to show that the proposed curriculum-guided training can base on different forms of the difficult scores, and this will have a substantial indication of value for scenarios where segmentation and thus radiomics-based difficulty score is not available.

More importantly, we compare our method to previously used methods in the literature, i.e., directly incorporating radiomics features as additional input to a CNN model [11, 12]. Specifically, we concatenate the radiomics features to the second-to-last fully connected layer, so that the radiomics features and "deep" features are combined to generate a final classification. In comparison to our proposed method, we evaluate the effects of incorporating (1) the full set of radiomics features and (2) only the radiomics features selected by L1 regularization.

At last, to verify that the introduced radiomics-based difficulty score is a good approximation of how difficult the classification task is (i.e., prove the hypothesis about Eq. 1 in Sect. 2), we stratify images based on their radiomics scores and compute the performance of vanilla CNNs in each stratum. If the radiomics score is a good indicator of difficulty for image classification, it should differentiate between cases that are easy to classify and cases that are not, when using vanilla CNNs for the classification. For this reason, we expect the performance of the vanilla CNNs to be higher for easy cases than it is for hard ones. Specifically, we stratify the images in the validation set (not the training/testing set) by the radiomics-based difficulty score into "easy", "intermediate", and "hard" three levels. We compute the first and third quartile of the radiomics scores to find two thresholds to separate the images into the three strata. We chose these thresholds so that the majority of the scores are considered intermediate, and fewer scores are considered hard or easy.

3.2 Dataset, Experimental Setup, and Performance Metrics

We use CBIS-DDSM [15], a publicly available dataset of digitized film mammogram images for experiments. This dataset contains 1,462 images of mass lesions that are either benign (N = 722) or malignant (N = 740). The target task is to perform a computer-aided diagnosis of malignant cases from benign cases. The segmentations acquired by radiologists are available for all lesions, and we use these segmentations to extract radiomics features. This dataset has a prespecified training/test split available (1,135 images in the training set and 327 images in the testing set) and we stick to this prespecified split in our experiments. We further set aside 20% of the training set as a validation set.

The input of CNN models is cropped images for considerations of computational efficiency and maintain the high resolution of the lesion regions. We crop the images around the lesion segmentation and resize them to 300x300 pixels if needed (if the lesion segmentation fits in a bounding box smaller than 300x300, then resizing is not needed).

We use pyradiomics v3.0.1 [16] to extract 102 typical radiomics features from the full resolution images and we use scikit-learn v0.23.2 to train the logistic regression classifier with L1 regularization (optimal regularization strength found using grid search on the validation set). We experiment with three CNN architectures: DenseNet [17], GoogleNet [18], and ResNet18 [19]. For all three, we use ImageNet [20] pre-trained weights. The top, fully connected layers for these models are replaced with two fully connected layers with output sizes 100 and 1, initialized using a Xavier normal [21]. We apply typical data augmentation, including random horizontal and vertical flip, translation, rotation, and shear.

We run experiments using pytorch 1.5.1 and torchvision 0.6.0, and we train our models using the Adam optimizer [22] and the binary cross-entropy loss for 100 epochs. We experiment with learning rates of 10^{-4}, 10^{-5}, 10^{-6}, and batch sizes of 10 and 32. During the training of each model, we select the weights that achieve the lowest validation loss. For the proposed difficulty score-guided training, we consider the weights that achieve the lowest validation loss after all scores are constant (i.e., they reach the value of 0.5 or one of the two bounds, as shown in Fig. 2).

We report performance as Area Under the Receiver Operating Characteristics Curve (AUC) [23]. For CBIS-DDSM, we compare test AUCs using the DeLong's method [24].

4 Experiment Results

Table 1 shows the classification results and various comparisons. We performed a detailed interpretation of the results in the next paragraph.

Table 1. Percentage test AUC values. The asterisk indicates the statistical significance of improvement over the baseline: $p < 0.05$.

	Test AUC		
Radiomics Model	67.25		
Deep Learning models	DenseNet	GoogleNet	ResNet18
Baseline	71.12	70.44	69.23
Radiomics as input (all features) [11, 12]	72.51	67.83	66.58
Radiomics as input (L1 regularization) [11, 12]	74.35	65.72	67.05
Radiomics/ Resampling	71.36	66.20	71.61
Radiomics/ Curriculum 1	75.92*	70.49	74.18*
Radiomics/ Curriculum 2	**77.12***	**73.79**	73.35
Self-taught/ Resampling	73.12	66.46	68.35
Self-taught/ Curriculum 1	74.30	71.90	73.19
Self-taught/ Curriculum 2	72.32	68.55	**74.56***

As can be seen from Table 1, almost all the deep learning models (including baseline CNNs) outperform the radiomics model. More importantly, the radiomics-based curriculum learning outperforms the baseline CNNs, and the direct use of radiomic scores for data resampling, and Curriculum 2 seems to work better than Curriculum 1. When it comes to the use of the self-taught score, we see a similar trend that the proposed curriculums perform better than the direct resampling and the baseline CNNs. In addition, when compared to the previous methods that directly incorporate radiomics features into deep learning modeling, our models largely outperform those methods. These results reported in Table 1 indicate that (1) our methods of incorporating knowledge boosts the classification performance, (2) our proposed difficulty score-guided curriculum training makes better use of the score than the resampling strategy does, and thus the observed higher classification effects, and (3) our curriculum-guided training strategy uses the information captured by radiomics more effectively than simply concatenating the radiomics features into deep learning.

In addition, if we compare the two different scores across training strategies, the radiomics score generally outperforms the self-taught score. These results show the customized curriculums can enhance the training regardless of the type of the scores.

While there are some variations among the three different tested CNN models, they collectively show the overall benefits of the proposed methods. Noticeably, the performance improvement reached by DenseNet is always statistically significant when using the radiomics score coupled with our proposed curriculums.

Table 2 shows the test performance stratified by the three difficulty levels (i.e., "easy", "intermediate", and "hard") as captured by the radiomics score. Here, we observe the expected test AUC stratification, i.e., higher test AUC is achieved for easy cases, and lower AUC is reached for hard cases. This verifies our hypothesis from Sect. 2 that the radiomics score is a good indicator of the image classification difficulty, and thus provides support to the proposed curriculum training method.

Table 2. Stratified test AUC values on the best performing baseline CNN (i.e., DenseNet).

Difficulty	Test AUC
Hard	44.74
Intermediate	74.78
Easy	91.67

Finally, Table 3 reports the test AUC values stratified by the radiomics score of the samples in terms of three levels. As can be seen, both the two customized curriculums increase model performance in every difficulty level, with the exception on Curriculum 2 on easy cases. This exception indicates the potential cost under the corresponding settings in order to archive the observed overall benefits of our method. In particular, Curriculum 2 improves performance on the hard cases more than Curriculum 1 at the cost of a decrease in performance for easy cases. This is compatible with the design of the two curriculums, where Curriculum 2 is meant to emphasize hard cases a lot more than easy ones toward the end of training.

Table 3. Percentage test AUCs for DenseNet stratified by radiomics-based difficulty levels.

Difficulty Stratification	Baseline	Curriculum 1	Curriculum 2
Hard	44.74	52.50	54.80
Intermediate	74.78	79.38	80.76
Easy	91.67	92.69	90.93
Overall	71.12	75.92	77.12

5 Conclusions

In this paper, we propose a novel technical method to improve deep learning-based breast cancer diagnosis by leveraging radiomics as a source of knowledge. An important motivation of our study is to take advantage of radiomics, which has been extensively studied and shown promising effects in recent years, to inform and enhance deep learning for image classification. The experiment results and comparisons show the merits of our proposed method. Our study shows the potential values of radiomics to inform improved deep learning for more scenarios/tasks, and moreover, we show that other types of difficulty score can also benefit from our proposed curriculum training scheme. It should be pointed out that depending on the applications, image segmentation may be needed for our method in order to compute radiomics on regions of interest. Future work includes developing additional and more sophisticated curriculums (specific or agnostic to datasets/tasks), experimenting with different implementations of the radiomics pipeline, more comprehensive radiomics feature sets (including incorporation of clinical variables), more datasets, and additional model architectures, as well as evaluating other types of difficulty scores.

Acknowledgements. This work was supported by National Institutes of Health grants (1R01CA193603, 3R01CA193603-03S1, and 1R01CA218405), the UPMC Hillman Cancer Center Developmental Pilot Program, and an Amazon Machine Learning Research Award. This work used the Extreme Science and Engineering Discovery Environment (XSEDE), which is supported by National Science Foundation (NSF) grant number ACI-1548562. Specifically, it used the Bridges-2 system, which is supported by NSF award number ACI-1928147, at the Pittsburgh Supercomputing Center (PSC).

References

1. Jan, T., Debenham, J.: Incorporating prior domain knowledge into inductive machine learning. J. Mach. Learn., 1–42 (2007)
2. Lauer, F., Bloch, G.: Incorporating prior knowledge in support vector machines for classification: a review. Neurocomputing **71**, 1578–1594 (2008). https://doi.org/10.1016/j.neucom.2007.04.010
3. Langseth, H., Nielsen, T.D.: Fusion of domain knowledge with data for structural learning in object oriented domains. J. Mach. Learn. Res. **4**, 339–368 (2003). https://doi.org/10.1162/153244304773633870
4. Culos, A., et al.: Integration of mechanistic immunological knowledge into a machine learning pipeline improves predictions. Nat. Mach. Intell. **2**, 619–628 (2020). https://doi.org/10.1038/s42256-020-00232-8
5. Jiménez-Sánchez, A., et al.: Medical-based deep curriculum learning for improved fracture classification. In: Shen, D., et al. (eds.) MICCAI 2019. LNCS, vol. 11769, pp. 694–702. Springer, Cham (2019). https://doi.org/10.1007/978-3-030-32226-7_77
6. Tang, Y., Wang, X., Harrison, A.P., Lu, L., Xiao, J., Summers, R.M.: Attention-guided curriculum learning for weakly supervised classification and localization of thoracic diseases on chest radiographs. In: Shi, Y., Suk, H.-I., Liu, M. (eds.) MLMI 2018. LNCS, vol. 11046, pp. 249–258. Springer, Cham (2018). https://doi.org/10.1007/978-3-030-00919-9_29
7. Gillies, R.J., Kinahan, P.E., Hricak, H.: Radiomics: images are more than pictures, they are data. Radiology **278**, 563–577 (2016). https://doi.org/10.1148/radiol.2015151169
8. Wibmer, A., et al.: Haralick texture analysis of prostate MRI: utility for differentiating noncancerous prostate from prostate cancer and differentiating prostate cancers with different Gleason scores. Eur. Radiol. **25**(10), 2840–2850 (2015). https://doi.org/10.1007/s00330-015-3701-8

9. Li, H., et al.: Quantitative MRI radiomics in the prediction of molecular classifications of breast cancer subtypes in the TCGA/TCIA data set. NPJ Breast Cancer **2**, 1 (2016). https://doi.org/10.1038/npjbcancer.2016.12

10. Nebbia, G., Zhang, Q., Arefan, D., Zhao, X., Wu, S.: Pre-operative microvascular invasion prediction using multi-parametric liver MRI radiomics. J. Digit. Imag. **33**(6), 1376–1386 (2020). https://doi.org/10.1007/s10278-020-00353-x

11. Wang, X., et al.: Deep learning combined with radiomics may optimize the prediction in differentiating high-grade lung adenocarcinomas in ground glass opacity lesions on CT scans. Eur. J. Radiol. **129**, 109150 (2020). https://doi.org/10.1016/j.ejrad.2020.109150

12. Zhang, Y., Lobo-Mueller, E.M., Karanicolas, P., Gallinger, S., Haider, M.A., Khalvati, F.: Improving prognostic performance in resectable pancreatic ductal adenocarcinoma using radiomics and deep learning features fusion in CT images. Sci. Rep. **11**, 1–11 (2021). https://doi.org/10.1038/s41598-021-80998-y

13. Bengio, Y., Louradour, J., Collobert, R., Weston, J.: Curriculum learning. In: Proceedings of the 26th Annual International Conference on Machine Learning, pp. 41–48 (2009). https://doi.org/10.1145/1553374.1553380

14. Hacohen, G., Weinshall, D.: On the power of curriculum learning in training deep networks. In: 36th International Conference on Machine Learning, ICML 2019, pp. 4483–4496, June 2019

15. Lee, R.S., Gimenez, F., Hoogi, A., Miyake, K.K., Gorovoy, M., Rubin, D.L.: Data descriptor: a curated mammography data set for use in computer-aided detection and diagnosis research. Sci. Data **4**, 1–9 (2017). https://doi.org/10.1038/sdata.2017.177

16. Van Griethuysen, J.J.M., et al.: Computational radiomics system to decode the radiographic phenotype. Cancer Res. **77**, e104–e107 (2017). https://doi.org/10.1158/0008-5472.CAN-17-0339

17. Huang, G., Liu, Z., Van Der Maaten, L., Weinberger, K.Q.: Densely connected convolutional networks. In: Proceedings of the IEEE Conference on Computer Vision and Pattern Recognition, pp. 4700–4708 (2017). https://doi.org/10.1016/j.midw.2011.06.009

18. Szegedy, C., et al.: Going deeper with convolutions. In: Proceedings of the IEEE Conference on Computer Vision and Pattern Recognition, pp. 1–9 (2015). https://doi.org/10.1089/pop.2014.0089.

19. He, K., Zhang, X., Ren, S., Sun, J.: Deep residual learning for image recognition. In: Proceedings of the IEEE Conference on Computer Vision and Pattern Recognition, pp. 770–778 (2016). https://doi.org/10.1016/0141-0229(95)00188-3

20. Deng, J., Dong, W., Socher, R., Li, L.-J., Kai, L., Li, F.-F.: ImageNet: a large-scale hierarchical image database. In: 2009 IEEE Conference on Computing Vision and Pattern Recognition, pp. 248–255 (2010). https://doi.org/10.1109/cvpr.2009.5206848

21. Glorot, X., Bengio, Y.: Understanding the difficulty of training deep feedforward neural networks. J. Mach. Learn. Res. **9**, 249–256 (2010)

22. Kingma, D.P., Ba, J.L.: Adam: a method for stochastic optimization. In: 3rd International Conference Learning Representations, ICLR 2015 - Conference Track Proceeding, pp. 1–15 (2015)

23. Fawcett, T.: An introduction to ROC analysis. Pattern Recogn. Lett. **27**, 861–874 (2006). https://doi.org/10.1016/J.PATREC.2005.10.010

24. DeLong, E.R., DeLong, D.M., Clarke-Pearson, D.L.: Comparing the areas under two or more correlated receiver operating characteristic curves: a nonparametric approach. Biometrics **44**(3), 837–845 (1988)

Integration of Imaging
with Non-Imaging Biomarkers

Lung Cancer Risk Estimation with Incomplete Data: A Joint Missing Imputation Perspective

Riqiang Gao[1]([✉]), Yucheng Tang[1], Kaiwen Xu[1], Ho Hin Lee[1], Steve Deppen[2], Kim Sandler[2], Pierre Massion[2], Thomas A. Lasko[1,2], Yuankai Huo[1], and Bennett A. Landman[1]

[1] EECS, Vanderbilt University, Nashville, TN 37235, USA
riqiang.gao@vanderbilt.edu
[2] Vanderbilt University Medical Center, Nashville, TN 37235, USA

Abstract. Data from multi-modality provide complementary information in clinical prediction, but missing data in clinical cohorts limits the number of subjects in multi-modal learning context. Multi-modal missing imputation is challenging with existing methods when 1) the missing data span across heterogeneous modalities (e.g., image vs. non-image); or 2) one modality is largely missing. In this paper, we address imputation of missing data by modeling the joint distribution of multi-modal data. Motivated by partial bidirectional generative adversarial net (PBiGAN), we propose a new Conditional PBiGAN (C-PBiGAN) method that imputes one modality combining the conditional knowledge from another modality. Specifically, C-PBiGAN introduces a conditional latent space in a missing imputation framework that jointly encodes the available multi-modal data, along with a class regularization loss on imputed data to recover discriminative information. To our knowledge, it is the first generative adversarial model that addresses multi-modal missing imputation by modeling the joint distribution of image and non-image data. We validate our model with both the national lung screening trial (NLST) dataset and an external clinical validation cohort. The proposed C-PBiGAN achieves significant improvements in lung cancer risk estimation compared with representative imputation methods (e.g., AUC values increase in both NLST (+2.9%) and in-house dataset (+4.3%) compared with PBiGAN, p < 0.05).

Keywords: Missing data · Multi-modal · GAN · Lung cancer

1 Introduction

Lung cancer has the highest cancer death rate [1] and early diagnosis with low-dose computed tomography (CT) can reduce the risk of dying from lung cancer by 20% [2,3]. Risk factors (e.g., age and nodule size) are widely used in machine learning and established prediction models [4–7]. With deep learning

© Springer Nature Switzerland AG 2021
M. de Bruijne et al. (Eds.): MICCAI 2021, LNCS 12905, pp. 647–656, 2021.
https://doi.org/10.1007/978-3-030-87240-3_62

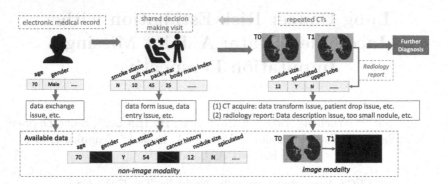

Fig. 1. Missing data in multiple modalities. The upper panel shows a general screening process. In practice, missing data can happen at different phases (as red text). Lower panel shows that patient may miss risk factors or/and follow-up CT scans.

techniques, CT image features can be automatically extracted at the nodule-level [8], scan-level [9], or patient-level with longitudinal scans [10]. Previous studies demonstrated that CT image features and risk factors provide complementary information, which is combined to improve lung cancer risk estimation [11].

In the clinical screening process (Fig. 1), patients' demographic information (e.g., age and gender) is captured in electronic medical records (EMR). In the shared decision-making (SDM) visit, lung cancer risk factors (e.g., smoke status) are collected to determine if a chest CT is necessary. For each performed CT scan, a radiology report is created. Then, such a process might recur according to clinical guidelines. Extensive efforts have been made to collect comprehensive information for patients. However, data can be missing due to multiple issues from data entry, data exchange, data description, et cetera.

Missing data mechanisms were categorized into three types [12]: 1) missing completely at random (MCAR): the missing has no dependency on data, 2) missing at random (MAR): the missing only depends on observed variables, 3) missing not at random (MNAR): the missing may be affected by unobserved variables. To address missing data problems, various imputation approaches were proposed to "make-up" missing data for downstream analyses [13–18]. Mean imputation is widely used to fill missing data with population averages. Last observation carried forward (LOCF) [13] takes the last observation as a replacement for missing data, which has been used in clinical serial trials. Soft-imputer [14] provides a convex algorithm for minimizing the reconstruction error corresponding to a bound on the nuclear norm. Recently, deep learning based imputation methods have been developed using generative models [17,18] (e.g., variants of variational auto-encoder (VAE) [19] and generative adversarial net (GAN) [20]). The partial bi-directional GAN (PBiGAN) [18], an encoder-decoder imputation framework, has been validated as a state-of-the-art performance of imputations. However, majority methods have limited imputation within a single modality, which can lead to two challenges in multi-modal context: 1) it is hard to integrate data spanning across heterogeneous modalities (e.g., image vs. non-image) into a single-modal imputation framework, 2) recovering discriminative information is unattainable when data are largely missing in target modality (limiting case: data are completely missing).

We posit that essential information missed in one modality can be maintained in another. In this paper, we propose the Conditional PBiGAN (C-PBiGAN) to model the joint distribution across modalities by introducing 1) a conditional latent space in multi-modal missing imputation context; 2) a class regularization loss to capture discriminative information during imputation. Herein, we focus on lung cancer risk estimation, where risk factors and serial CT scans are two essential modalities for rendering clinical decisions. C-PBiGAN achieves superior predicting performance of downstream multi-modal learning tasks in three broad settings: 1) missing data in image modality, 2) missing data in non-image modality, and 3) both modalities have missing data. With C-PBiGAN, we validate that 1) CT images are conducive to impute missed factors for better risk estimation, and 2) lung nodules with malignancy phenotype can be imputed conditioned on risk factors.

Our contributions are three folds: (**1**) To our knowledge, we are the first to impute missing data by modeling joint distribution of image and non-image data with adversarial training; (**2**) Our model can impute visually realistic data and recover discriminative information, even when the target data in target modality are completely missing; (**3**) Our model achieves superior downstream predicting performance compared with benchmarks with simulated missing (MCAR) and missing in practice (MNAR).

2   Theory

Encoder-Decoder and PBiGAN Framework. PBiGAN [18] is a recently proposed imputation method with encoder-decoder framework based on bidirectional GAN (BiGAN) [21]. Our conditional PBiGAN (C-PBiGAN) is shown in Fig. 2, where the PBiGAN [18] is consist of "black text" components. Note that PBiGAN only deals with a single modality (i.e., modality A in Fig. 2).

The generator of PBiGAN includes a separate encoder and decoder. The decoder g^A transforms a latent code z into a complete data space X^A, where z is a feature space (e.g., z_o^A) or sampled from a simple distribution (e.g., Gaussian). The encoder $q^A(z_o^A|x^A, m)$, denoted as q^A for simplification, maps the missing distribution p_m of an incomplete data (x^A, m) into a latent vector z_o^A, where $x^A \in \mathbb{R}^n$ denotes complete data, and $m \in \{0, 1\}^n$ is a missing indicator with same dimension of x^A that determines which entries in x^A are missing (i.e., 1 for observed, 0 for missing).

The discriminator D of PBiGAN takes the observed data $[x^A, m]$ and its corresponding latent code z_o^A as the "real" tuple in adversarial training. The "fake" tuple $(\hat{x}^A, \hat{m}, \hat{z})$ is comprised of 1) a random latent code $\hat{z}$ sampled from a simple distribution $p_{\hat{z}}$ (e.g., Gaussian), 2) missing indices $\hat{m}$ from a missing distribution $p_{\hat{m}}$, and 3) the generated data $\hat{x}^A$ based on random latent code $\hat{z}$.

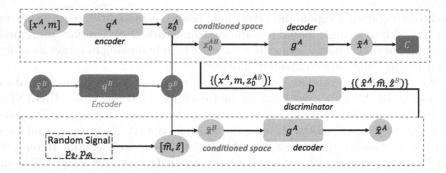

Fig. 2. Structure of the proposed C-PBiGAN. The orange and green characters high-light our contributions compared with PBiGAN [18]. m is the missing index of target modality A and z is the corresponding latent space. $\tilde{x}^B$ is the complete data of condi-tional modality B, which can be fully observed or imputed. $\tilde{x}^A$ is the imputed data of A based on observed data $[x^A, m]$ and $\tilde{x}^B$. $\hat{x}^A$ is the generated data of A based on $\tilde{x}^B$ and noise distributions of $p_{\hat{z}}$ and $p_{\hat{m}}$. C is a classifying module along with cross-entropy loss regularizing the generator for keeping the identities of imputed data. (Color figure online)

The loss function of PBiGAN is defined as follows, which is minimax optimized:

$$L\left(D, g^A, q^A\right) = \mathbb{E}_{(x^A, m) \sim p_m} \mathbb{E}_{z_o^A \sim q^A(z_o^A | x^A, m)}[\log D(x^A, m, z_o^A)] \\ + \mathbb{E}_{(\cdot, \hat{m}) \sim p_{\hat{m}}} \mathbb{E}_{\hat{z} \sim p_{\hat{z}}}[\log(1 - D(g^A(\hat{z}, \hat{m}), \hat{m}, \hat{z}))] \tag{1}$$

The Proposed Conditional PBiGAN. The original PBiGAN [18] imputes data within a single modality, which does not utilize complementary information from multiple modalities. Herein, we propose C-PBiGAN to impute one modality conditioned on another, and a cross-entropy loss is optimized during generator training to effectively preserve discrimination for imputed data.

As Fig. 2, when imputing A (target modality), the conditional data $\tilde{x}^B$ is complete, either fully observed or imputed. Two encoders q^A and q^B are used to map data space to latent space for modality A and B, respectively. The GAN loss of our method $L_G\left(D, g^A, q^A, q^B\right)$, also denoted as L_G, is written as follows:

$$L_G = \mathbb{E}_{(x^A, m) \sim p_m} \mathbb{E}_{z_o^{AB} \sim [q^A(z_o^A | x^A, m), q^B(\hat{z}^B | \tilde{x}^B)]}[\log D(x^A, m, z_o^{AB})] \\ + \mathbb{E}_{(\cdot, \hat{m}) \sim p_{\hat{m}}} \mathbb{E}_{\hat{z}^B \sim [p_{\hat{z}}, q^B(\hat{z}^B | \tilde{x}^B)]}[\log(1 - D(g^A(\hat{z}^B, \hat{m}), \hat{m}, \hat{z}^B))] \tag{2}$$

Different from Eq. (1) of PBiGAN focusing on single modality A, the latent space z_o^{AB} in Eq. (2) includes the knowledge from two modalities.

To enforce the imputed $\tilde{x}^A$ or generated $\hat{x}^A$ having the same identity with x^A even when data are largely missing, we further introduce a feature extraction net C along with cross-entropy loss (the second term in Eq. 3) when training the generator. Specifically, C-PBiGAN is optimized with:

$$\min_{g^A, q^A} (\max_D (L_G(D, g^A, q^A, q^B)) - \mathbb{E}_{\tilde{x}^A \sim g^A(\cdot)}[\log p(y | C(\tilde{x}^A))]) \tag{3}$$

where y is class label and $p(y|C(\tilde{x}^A))$ is the prediction from C. Modules q^B, C can be pretrained or trained with g^A, q^A simultaneously.

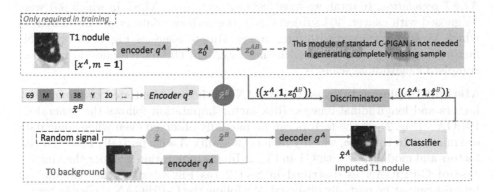

Fig. 3. An instantiation of limiting C-PBiGAN: imputing TP1 nodule in longitudinal context. $\tilde{x}^B$ is the imputed risk factor of TP1. $[x_A, m = 1]$ is complete TP1 data only used in training, as the upper dashed box. "TP0 background" is the observed TP0 (or TP1 in training phase) image with center masked, which is fed to q^A to make the imputed TP1 with a similar background as TP0. A comparable setting C-PBiGAN$^{\#}$ is fed with TP0 without masking center.

Different from conditional GAN [22], 1) our model can utilize the partially observed data in the imputation context, and 2) a module C along with cross-entropy loss is introduced to highlight identity preservation of imputed data.

A limiting case of C-PBiGAN is to impute data that is completely missing (i.e., $m = 0$). In this case, complete data for training (i.e., $m = 1$) are needed, and it is the generated $\hat{x}^A$, rather than $\tilde{x}^A$ as in Fig. 2, that used for downstream task. In Eq. (3), the $\tilde{x}^A$ is replaced with $\hat{x}^A$. One of our tasks imputing nodules belongs to this limiting case, as Fig. 3 (details in Sect. 3).

3   Experiment Designs and Results

Datasets. We consider two longitudinal CTs (TP0 for previous, TP1 for current) as the complete data for image modality. The non-image modality includes the following 14 risk factors: *age, sex, education, body mass index, race, quit smoke time, smoke status, pack-year, chronic obstructive pulmonary disease, personal cancer history, family lung cancer history, nodule size, spiculation, upper lobe of nodule*. The first two, the middle nine, and the last three factors come from EMR, SDM visit, and radiology report (Fig. 1), respectively.

Two datasets are studied, 1) the national lung screening trail (NLST) [3] and 2) an in-house screening dataset from Vanderbilt Lung Screening Program (VLSP, https://www.vumc.org/radiology/lung). Patients in NLST are selected if 1) they have 14 selected risk factors available, 2) have a tissue-based diagnosis, and 3) the diagnosis happened within 2 years of the last scan if it is a cancer case. *Note that selected subjects are all high-risk patients (all received biopsies), the distinction between cancer/non-cancer in our cohort is hard than in the whole NLST population.* In total, we have 3889 subjects from NLST in which 601 were diagnosed with cancer. 404 subjects from the in-house dataset are evaluated, in which 45 were diagnosed with lung cancer. Due to issues as Fig. 1, the available factors have an average of 32% missing rate, and 60% of patients do not have complete longitudinal scans.

Method Implementations. C-PBiGAN has been instantiated to impute risk factors and longitudinal images. Risk factor imputation follows the general C-PBiGAN (Fig. 2), as the factors can be partially observed even when some data are missing. In this case, we only replace modality A with partially observed risk factors and modality B with CT in Fig. 2. Image imputation is under the limiting case of C-PBiGAN as described in Sect. 2 (as Fig. 3), since the "nodule" of interest cannot be partially observed. We follow the C-PBiGAN theory in Sect. 2 for image imputation, and we also utilize information from longitudinal context in practice. We assume the background of a nodule would not substantially change between TP0 and TP1. Thus, motivated by masking strategies of [24, 25], nodule background is borrowed from observed CT (i.e., TP0 image) of the same patient by masking its center when generating the target time point (i.e., TP1 image), see "TP0 background" in Fig. 3. In brief, we target at the problem of missing whole image, while the implementation is kind of central inpainting based on our assumption. We have reconstruction regularization motivated by PBiGAN and UNet [23] skip connections in image-modality implementation.

Given a CT scan, we follow Liao's pipeline [9] to preprocess the data and detect the top five confidence nodule regions for downstream work. Rather than imputing a whole 3D CT scan, we focus on imputing the nodule areas of interest in 2D context with axial/coronal/sagittal directions as 3 channels (i.e., $3 \times 128 \times 128$). Considering 1) radiographic reports regarding TP0 are rarely available, and 2) TP1 plays a more important role in lung cancer risk estimation [10], we focus on the imputation on TP1 of image modality in this study. The TP0 image is copied with the TP1 image when TP1 is observed and TP0 is missing.

Networks. The structures of encoder, decoder, and discriminator are 1) adapted from face example in PBiGAN [18] for image modality, and 2) separately comprised of four dense layers for non-image modality. A unified multi-modal longitudinal model (MLM), including an image path and a non-image path, is used for lung cancer risk estimation to evaluate the effectiveness of imputations. The image path includes a backbone of ResNeTP18 [26] to extract image features and a LSTM [27] to integrate longitudinal images (from TP0 and TP1). Risk factor features are extracted by a module with four dense layers. The image path and non-image path in the MLM are validated to be effective by comparing with representative prediction models (i.e., AUC in NLST: image-path model (0.875) vs. Liao et al. [9] (0.872) with image data only, non-image path model

Table 1. AUC results (%) of the test set of NLST (upper, a case of MCAR mechanism) and external in-house set (lower, a case of MNAR mechanism). Generally, each row or each column represents an imputation option for image-missing or risk-factor-missing, respectively. "Image-only" or "factor-only" represents predicting only use imputed longitudinal-images or factors, respectively.

Method	image-only	Mean-imput	Soft-imputer	PBiGAN	C-PBiGAN	fully-observed
test set of longitudinal NLST (30% factors, 50% TP1 image are missing, MCAR)						
factor-only	N/A	79.73	79.46	79.14	83.04	86.24
LOCF	73.45	83.76	83.80	83.79	84.00	86.21
PBiGAN	76.54	83.02	83.82	83.29	83.51	85.90
C-PBiGAN#	82.70	85.00	85.62	85.17	85.87	86.72
C-PBiGAN	84.15	85.72	85.90	85.91	**86.20**	88.27
fully-observed	87.48	88.23	88.40	88.44	88.46	89.57
external test of in-house dataset (MNAR)						
factor-only	N/A	75.17	83.46	84.40	86.56	N/A
LOCF	75.52	82.83	87.11	86.99	87.63	N/A
PBiGAN	73.44	80.85	84.43	84.88	85.86	N/A
C-PBiGAN#	80.59	83.87	86.57	87.19	87.69	N/A
C-PBiGAN	82.61	85.29	88.11	88.49	**89.19**	N/A

(0.883) vs. Mayo clinical model [7] (0.829)). The image and non-image features are combined for the final prediction.

Settings and Evaluations. The NLST is randomly split into train/validation/test sets with 2340/758/791 subjects. The in-house dataset of 404 subjects is externally tested when training is finished in NLST. We follow the experimental setup of PBiGAN opensource code [18] when training C-PBiGAN, e.g., use Adam optimizer with a learning rate of 1e-4. The max number of training epochs is set to 200. Our experiments are performed with Python 3.7 and PyTorch 1.5 on GTX Titan X. The mask size of "TP0 background" is 64×64. The area under the receiver operating characteristic (AUC) [28] for lung cancer risk estimation is used to quantitatively evaluate the effectiveness of imputations.

Imputation Baselines. Representative imputations (introduced in Sec. 1) of image (i.e., LOCF [13] and PBiGAN [18]) and non-image (i.e., mean imputation, soft-imputer [14] and PBiGAN [18]) are combined for comparison as in Table 1. As a comparable setting of ours, C-PBiGAN# denotes feeding TP0 nodule without masking the center, rather than "TP0 background" in Fig. 3.

Results and Discussion. Table 1 shows 1) tests of NLST (upper) with 30% of missing in risk factors and 50% of missing in longitudinal TP1 and 2) external tests of in-house data with missing in practice. The C-PBiGAN combination (**bold** in Table 1) significantly improves all imputation combinations without C-PBiGAN across the image and non-image modalities (p<0.05, bootstrapped two-tailed test, n = 2000 [29]) in both NLST and external clinical dataset (e.g., C-PBiGAN increases 4.3% AUC on PBiGAN in the external cohort). Those indicate our model effectively imputes data when missing in both modalities for cancer risk estimation.

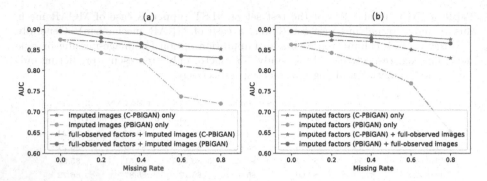

Fig. 4. (a) AUCs of various TP1-image missing rates when factors are fully observed in NLST, and (b) AUCs of various factor missing rates when images are fully observed in NLST. The left start point is under condition that data are not missing.

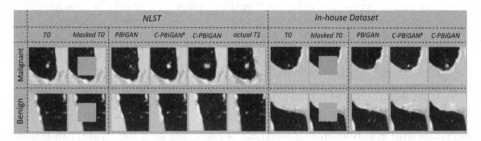

Fig. 5. Qualitative results of imputed TP1 nodules (upper: malignant, bottom: benign). Malignant/benign cases from C-PBiGAN are most distinguishable.

Figure 4 compares proposed C-PBiGAN with PBiGAN in terms of the lung cancer predicting performance in NLST under (a) various TP1 missing rates when factors are fully observed, (b) various factor missing rates when longitudinal images are fully observed. Our model outperforms PBiGAN in the image-missing and factor-missing contexts of different rates. A more obvious superiority can be found when only using the imputed modality for prediction (e.g., C-PBiGAN: 0.830 vs. PBiGAN: 0.652 when risk factors have missing rate of 80%), and the imputed factors conditioned on images can even achieve higher AUC than the fully observed factors at some missing rates. Those indicate the information from conditional modality in C-PBiGAN does help the imputation.

Figure 5 shows malignant and benign cases from NLST and in-house dataset. Both PBiGAN and proposed C-PBiGAN can reconstruct visually realistic images, while malignant and bengin cases from PBiGAN are harder to distinguish.

As a comparable setting, C-PBiGAN$^\#$ is less effective than C-PBiGAN (Table 1, Fig. 5) given the current setting and network structure. It is probably because when feeding TP0 without masking center to provide nodule background (i.e., C-PBiGAN$^\#$), the central nodule region of imputed TP1 can be fit to the center of TP0, just like the nodule background of imputed TP1 is designed to fit TP0 nodule background. This limits the discrimination of imputed TP1, as the examples in Fig. 5. Thus, it is essential to separate "background" and "nodule" during learning, since we want the "background" of imputed TP1 to be close to observed TP0 while the "nodule" of imputed TP1 should mainly be conditioned on risk factors. Motivated by strategies in [24,25], our C-PBiGAN is fed with TP0 background masking the center when imputing the TP1 (in Fig. 3).

4 Conclusion

We propose a novel deep learning based missing imputation model for multi-modal data. By modeling the joint distribution of multiple modalities, the proposed C-PBiGAN can effectively impute the missing data across image and non-image modalities. We validate our method on a large-scale NLST dataset (MCAR) and an external clinical cohort (MNAR). Given no restriction on data type, our model can be readily extended to other multi-modal missing contexts.

Acknowledgement. This research was supported by NSF CAREER 1452485, R01 EB017230 and R01 CA253923. This study was supported in part by U01 CA196405 to Massion. This project was supported in part by the National Center for Research Resources, Grant UL1 RR024975-01, and is now at the National Center for Advancing Translational Sciences, Grant 2 UL1 TR000445-06. This study was funded in part by the Martineau Innovation Fund Grant through the Vanderbilt-Ingram Cancer Center Thoracic Working Group and NCI Early Detection Research Network 2U01CA152662 to PPM.

References

1. Siegel, R.L., Miller, K.D., Jemal, A.: Cancer statistics, 2019. CA Cancer J. Clin. **69**, 7–34 (2019)
2. Aberle, D.R., et al.: Reduced lung-cancer mortality with low-dose computed tomographic screening. N. Engl. J. Med. **365**, 395–409 (2011)
3. National Lung Screening Trial Research Team, et al.: The national lung screening trial: overview and study design. Radiology **258**, 243–253 (2011)
4. Huang, P., et al.: Prediction of lung cancer risk at follow-up screening with low-dose CT: a training and validation study of a deep learning method. Lancet Digit. Heal. **1**, e353–e362 (2019)
5. Tammemägi, M.C., et al.: Selection criteria for lung-cancer screening. N. Engl. J. Med. **368**, 728–736 (2013)
6. Swensen, S.J.: The probability of malignancy in solitary pulmonary nodules. Arch. Intern. Med. **157**, 849 (1997)
7. McWilliams, A., et al.: Probability of cancer in pulmonary nodules detected on first screening CT. N. Engl. J. Med. **369**, 910–919 (2013)
8. Liu, L., Dou, Q., Chen, H., Qin, J., Heng, P.A.: Multi-task deep model with margin ranking loss for lung nodule analysis. IEEE Trans. Med. Imaging **39**, 718–728 (2020)
9. Liao, F., Liang, M., Li, Z., Hu, X., Song, S.: Evaluate the malignancy of pulmonary nodules using the 3-D deep leaky noisy-or network. IEEE Trans. Neural Networks Learn. Syst. **2019**, 1–12 (2019)

10. Gao, R., et al.: Time-distanced gates in long short-term memory networks. Med. Image Anal. **65**, 101785 (2020)
11. Gao, R. et al.: Deep Multi-path Network Integrating Incomplete Biomarker and Chest CT Data for Evaluating Lung Cancer Risk. arXiv:2010.09524 (2021)
12. Rubin, D.B.: Inference and missing data. Biometrika **63**, 581–592 (1976). https://doi.org/10.1093/biomet/63.3.581
13. Van Buuren, S.: Flexible imputation of missing data. CRC Press (2018)
14. Mazumder, R., Hastie, T., Edu, H., Tibshirani, R., Edu, T., Jaakkola, T.: Spectral regularization algorithms for learning large incomplete matrices. J. Mach. Learn. Res. **11**, 2287–2322 (2010)
15. Yoon, J., Jordon, J., Van Der Schaar, M.: GAIN: missing data imputation using generative adversarial nets. In: International Conference on Machine Learning, pp. 9042–9051. International Machine Learning Society (IMLS) (2018)
16. Stekhoven, D.J., Bühlmann, P.: Missforest-Non-parametric missing value imputation for mixed-type data. Bioinformatics **28**, 112–118 (2012)
17. Mattei, P.A., Freiisen, J.: Miwae: deep generative modelling and imputation of incomplete data sets. In: 36th International Conference on Machine Learning, ICML 2019, pp. 7762–7772 (2019)
18. Cheng, S., Li, -Xian, Marlin, B.M.: Learning from irregularly-sampled time series: a missing data perspective. In: International Conference Machine Learning (2020)
19. Kingma, D.P., Welling, M.: Auto-encoding variational bayes. In: International Conference on Learning Representations, ICLR (2014)
20. Goodfellow, I.J., et al.: Generative adversarial nets. In: Advances in Neural Information Processing Systems, pp. 2672–2680 (2014)
21. Donahue, J., Krähenbühl, P., Darrell, T.: Adversarial Feature Learning (2016)
22. Mirza, M., Osindero, S.: Conditional Generative Adversarial Nets. arXiv Prepr. arXiv:1411.1784 (2014)
23. Ronneberger, O., Fischer, P., Brox, T.: U-Net: convolutional networks for biomedical image segmentation. In: Navab, N., Hornegger, J., Wells, W.M., Frangi, A.F. (eds.) MICCAI 2015. LNCS, vol. 9351, pp. 234–241. Springer, Cham (2015). https://doi.org/10.1007/978-3-319-24574-4_28
24. Jin, D., Xu, Z., Tang, Y., Harrison, A.P., Mollura, D.J.: CT-realistic lung nodule simulation from 3D conditional generative adversarial networks for robust lung segmentation. In: Frangi, A.F., Schnabel, J.A., Davatzikos, C., Alberola-López, C., Fichtinger, G. (eds.) MICCAI 2018. LNCS, vol. 11071, pp. 732–740. Springer, Cham (2018). https://doi.org/10.1007/978-3-030-00934-2_81
25. Mirsky, Y., Mahler, T., Shelef, I., Elovici, Y.: CT-GAN: malicious tampering of 3D medical imagery using deep learning. In: Proceedings of the 28th USENIX Security Symposium, pp. 461–478 (2019)
26. He, K., Zhang, X., Ren, S., Sun, J.: Deep residual learning for image recognition. In: IEEE Computer Society Conference on Computer Vision and Pattern Recognition, pp. 770–778 (2016)
27. Hochreiter, S., Schmidhuber, J.: Long short-term memory. Neural Comput. **9**, 1735–1780 (1997). https://doi.org/10.1162/neco.1997.9.8.1735
28. Fawcett, T.: An introduction to ROC analysis. Pattern Recognit. Lett. **27**, 861–874 (2006). https://doi.org/10.1016/j.patrec.2005.10.010
29. Mateuszbuda: Statistical functions based on bootstrapping for computing confidence intervals and p-values comparing machine learning models and human readers. https://github.com/mateuszbuda/ml-stat-util. Accessed 27 Feb 2021

Co-graph Attention Reasoning Based Imaging and Clinical Features Integration for Lymph Node Metastasis Prediction

Hui Cui[1] , Ping Xuan[2] , Qiangguo Jin[3] , Mingjun Ding[4], Butuo Li[4], Bing Zou[4], Yiyue Xu[4], Bingjie Fan[4], Wanlong Li[4], Jinming Yu[4], Linlin Wang[4](✉), and Been-Lirn Duh[1]

[1] Department of Computer Science and Information Technology, La Trobe University, Melbourne, Australia
[2] Department of Computer Science and Technology, Heilongjiang University, Harbin, China
[3] College of Intelligence and Computing, Tianjin University, Tianjin, China
[4] Department of Radiation Oncology, Shandong Cancer Hospital and Institute, Shandong First Medical University and Shandong Academy of Medical Sciences, Jinan, China

Abstract. Lymph node metastasis (LNM) is the most critical prognosis factor in esophageal squamous cell carcinoma (ESCC). Effective and adaptive integration of preoperative CT images and multi-sourced non-imaging clinical factors is a challenging issue. In this work, we propose a graph-based reasoning model to learn new representations from multi-categorical clinical parameters for LNM prediction. Given CT, general, diagnostic, pathological, and hematological clinical information, we firstly propose a graph construction strategy with category-wise contextual attention to embed multi-categorical features as graph node attributes. Secondly, we introduce a co-graph attention layer composed of a conventional graph attention network (con-GAT) and a correlation-based GAT (corr-GAT) to learn new representations. Corr-GAT complements con-GAT by difference-based correlations across image regions in global spectral space. Experimental results of ablation studies and comparison with others over 924 lymph nodes demonstrated improved performance and contributions of our major innovations. Our model has the potential to foster early prognosis and personalized surgery or radiotherapy planning in ESCC patients.

Keywords: Graph attention network · Cross-modal fusion · Lymph node metastasis

1 Introduction

Esophageal cancer (EC) is the seventh most common cancer leading to death in the United States [1]. The overall 5-year survival rate between 2010 and 2016 is 19.9%. [2] Diagnosed at early stage 1, EC patients' chance of surviving 5 years can be increased to 47.1%. [2] An independent factor in EC prognosis is lymph node metastasis (LNM), which is most common in the significant histological EC subtype, esophageal squamous

© Springer Nature Switzerland AG 2021
M. de Bruijne et al. (Eds.): MICCAI 2021, LNCS 12905, pp. 657–666, 2021.
https://doi.org/10.1007/978-3-030-87240-3_63

cell carcinoma (ESCC). [3] Thus, predicting LNM preoperatively is critical in clinical treatment decisions and planning for radiotherapy and surgery.

Computed tomography (CT) is widely used for preoperative LNM assessment [4]. A clinical criterion of LNM detection is based on the size of lymph nodes (LN), such as short-axis diameter [5, 6]. The reported sensitivity on preoperative CT is only 37.3–67.2% [4] when using LN size alone. Since quantitative image texture analysis provides rich information reflecting tumor heterogeneity, radiomics models attracted intensive interests in clinical research. For instance, a multivariable logistic regression model [7] has been widely used for CT based cancer staging and prognosis. Image features in radiomics are usually handcrafted, which limits the generalization to broad applications. Deep learning provides an end-to-end approach for thorough multi-level multi-scale feature extraction. The radiomics model proposed by [6] fused hand-crafted and deep features for LNM prediction and performed feature selection and correlation analysis for dimension reduction. The effective and adaptive integration of imaging features and non-imaging clinical parameters for LNM prediction is a critical yet challenging task.

Existing models for heterogeneous data integration can be divided into two categories. The first category joints features extracted from single modality or sourced data via a full connection at a late stage. For instance, X-Ray images and radiology reports are firstly sent to two separate encoders and then fused by two fully connected layers [8]. In [9], parallel CNNs are performed where each of them takes an image and transformed non-imaging features as input. The outputs are fused by a fully connected layer for final classification. The second type of methods is inspired by the mapping between language expressions and images. For instance, cross-modal semantic content correlations between images and captions are embedded by [10] for sentiment analysis. Ye *et al.* [11] performs convolutional LSTM over text phrases and incorporates attention mechanism to map linguistic features to image regions for segmentation. Even though the second type of methods enables the interaction between images and other modalities, the interactions are limited by the rigid-like structure of imaging feature space.

Graph models are unique in node connections and topology. Since graph edge can propagate information across different nodes, graph-based reasoning supports long-range node-wise relation modelling across different image regions in an irregular domain. [12, 13] Graph attention network (GAT) [14] computes the hidden representations of graph nodes by self-attention learning from node-neighbors, which has also been used in image recognition tasks. For instance, S. Mo *et al.* [15] proposed mutual information based GAT to leverage multi-modality MRI sequences for liver segmentation.

This work proposes a graph-based reasoning model to learn new representations from imaging and non-imaging multi-categorical clinical parameters for LNM prediction. Given CT and clinical parameters, we firstly propose a graph construction strategy to embed multi-categorical features as graph node attributes. Considering multi-categorical or multi-sourced data may have various decision-making contributions, we design a category-wise contextual attention mechanism for node attributes. Secondly, a co-graph attention network is proposed to learn new representations by connection modeling and context-based reasoning between local and long-range regions in graph feature space.

Our contributions are summarized below. The first innovation is a category-wise attention module to adaptively adjust the weights of imaging, general, diagnostic, pathological and hematological clinical factors when constructing graph node attributes. The second contribution is a co-graph attention layer which is composed of a conventional GAT (con-GAT) [14] and correlation-based GAT (corr-GAT). Con-GAT transforms input features using attention weights which are obtained by summarizing neighbouring nodes. Corr-GAT focus on difference-based correlations across image regions in global spectral space. Co-graph associates complementary information as the final new node representation.

2 Dataset

397 ESCC patients from Shandong Cancer Hospital were enrolled in this retrospective study. Inclusion criteria were as follows: (a) patients over 18 years old who took Contrast-Enhanced CT within 15 days before surgery between October 2013 and November 2018; (b) patients with pathologically confirmed LN status after surgery. Exclusion criteria included: (a) patients who received prior treatment before surgery; (b) patients who developed adenocarcinoma or other tumours. Each patient has 1 to 6 LNs. LNs of four patients are shown in Fig. 1. In total, there are 924 (798 negative and 126 positive) LNs. All the LNs have pathologically con-firmed status after surgery.

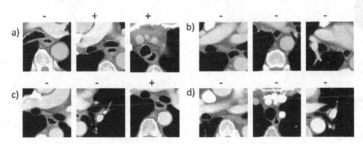

Fig. 1. Examples of four patients with manually delineated LNs by red color. Each patient has various numbers and locations of LNs with different class labels. − and + denote negative and positive classes (Color figure online).

CT images were reconstructed using a matrix of 512×512 pixels. The pixel spacing varies between different scans within a range of 0.5859 mm $\times$ 0.5859 mm and 0.9766 mm $\times$ 0.9766 mm. The slice thickness is 5 mm. The number of slices in different patients' thorax CT scans is between 58 and 117. Manual delineations of LNs were performed and verified by experienced radiation oncologists on CT.

There are 20 non-imaging clinical parameters following into four categories. The first type is general information of age (62.69 ± 7.37) and gender (83.12% Male). The second type includes histories of smoking (60.00% Yes), drinking (57.79% Yes), hypertension (24.6% Yes), diabetes (7.8% Yes), heart disease (5.79% Yes), cerebral infarction (2.26% Yes), and gastritis (17.18% Yes). Third type pathologic features include preoperative tumor markers of CEA (3.06 ± 1.97), NSE (13.94 ± 4.45), CA19–9 (13.27 ± 9.82),

CA125 (11.90 ± 5.92), CA72–4 (2.63 ± 4.45), CA15–3 (15.64 ± 8.66), Cyfra21–1 (3.26 ± 2.20), and SCC (1.55 ± 1.03). Last category is hematological parameters of neutrophil count (4.29 ± 2.00), lymphocyte count (1.83 ± 0.67), and platelet count (252.90 ± 67.48).

3   Method

The proposed model is given in Fig. 2. Given volumetric CT patches at LN level, and clinical parameters, we firstly extract image features and non-imaging features. In this work, we exploit 3D U-Net [16, 17] encoder as an image feature extractor and design a clinical encoder based on 1D CNN. Then a graph is constructed where the attribute vector of each graph node is composed of image features and clinical features with category-wise attention. Afterwards, the proposed co-graph attention layer formulates new node attribute representations in graph space by learning from cross-modal connections and neighbouring and long-range contextual correlations across image regions. Finally, the transformed features by co-graph attention layer are fused with category-wise attention graph attributes for final classification.

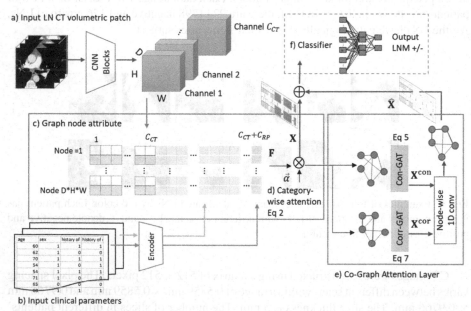

Fig. 2. Overview of the propose co-graph attention network for LNM prediction from preoperative CT and non-imaging clinical parameters. (a) LN volumetric patches and (b) clinical parameters are firstly fed into encoders for feature extraction. The extracted features are adaptively embedded as (c) graph node attributes with (d) category-wise attention. (e) Co-graph attention layer generates new node representations which contain complementary con-GAT and corr-GAT outputs. (f) New node representations and category-wise attentional attributes are fused for final positive or negative LNM classification.

3.1 Graph Node Attributes Construction with Category-Wise Attention

We propose a fully connected graph $G = (V, E, \mathbf{X})$ with category-wise attention to adaptively embed image and clinical features, where V, E are graph nodes and edges, and $\mathbf{X}$ represents category-wise attentional node attribute matrix.

Graph Nodes and Attributes. Image feature extractor consists of five blocks where each block has two 3D convolutional (conv3D) layers and one 3D max pooling layer with kernel size $2 \times 2 \times 2$. Each conv3D layer has a sequence of $3 \times 3 \times 1$ kernel, $1 \times 1 \times 0$ padding, $1 \times 1 \times 1$ stride, Batch Normalization, and ReLU operations. The output from the fifth block is denoted by $\mathbf{F}_{CT} \in \mathbb{R}^{H \times W \times D \times C_{CT}}$, where H, W, D are the height, width, and depth. C_{CT} denotes the number of channels in features extracted from CT. The clinical parameter encoder is composed of 1D convolution, batch normalization, and 1D average pooling. The output clinical representation is denoted by $\vec{f}_{RP} \in \mathbb{R}^{1 \times C_{RP}}$.

To map multi-modal features $\mathbf{F}_{CT}$ and $\vec{f}_{RP}$ to graph node, we firstly reshape $\mathbf{F}_{CT}$ to $\mathbb{R}^{N_V \times C_{CT}}$ where $N_V = H \times W \times D$ is the number of graph nodes as shown in Fig. 2 (c). Based on the nature of the convolutional operation, a graph node v_i can be considered as corresponding to a particular region in the input image. Secondly, $\vec{f}_{RP}$ is repeated N_V times column-wise resulting in $\mathbf{F}_{RP} \in \mathbb{R}^{N_V \times C_{RP}}$. Concatenated feature matrix $\mathbf{F}$ is obtained by $\mathbf{F} = \begin{bmatrix} \mathbf{F}_{CT} \| \mathbf{F}_{RP} \end{bmatrix} \in \mathbb{R}^{N_V \times N_F}$, $N_F = C_{CT} + C_{RP}$, where $\|$ represents concatenation.

Category-wise Contextual Attention. Considering that multi-category or multi-sourced data may have various clinical decision-making contributions, we propose a category-wise contextual attention mechanism which is inspired by [18]. Given M different categories of features, we firstly calculate the informative score $\mathbf{s}_m$ of m-th category as

$$\mathbf{s}_m = \tanh\left(\mathbf{W}_m^{\text{node}}\mathbf{F}_m + \vec{b}_m\right) \tag{1}$$

where $\mathbf{s}_m \in \mathbb{R}^{N_V \times N_m}$, $\mathbf{F}_m \in \mathbb{R}^{N_V \times N_m}$ is the attribute matrix of m-th category where N_m represents the number of features. $\mathbf{W}_m^{\text{node}} \in \mathbb{R}^{N_m \times N_m}$ and $\vec{b}_m$ are trainable weight matrix and bias vector. $\mathbf{W}_m^{\text{node}}$ and $\vec{b}_m$ are initialized by Xavier normalization. Secondly, attention score $\vec{\alpha}_m$ for m-th category is obtained by performing SoftMax normalization as

$$\vec{\alpha}_m = \frac{\exp\left(\mathbf{S}_m \vec{h}_m\right)}{\sum_{m=1}^{M} \exp\left(\mathbf{S}_m \vec{h}_m\right)} \tag{2}$$

where $\vec{\alpha}_m$ is a $N_V \times 1$ column vector, $\vec{h}_m$ a learnable $N_m \times 1$ row vector is to capture contextual relations. $\vec{h}_m$ is initialized by Xavier normalization and learnt during the training process.

Given attention score $\vec{\alpha}_m$, attention enhanced feature vector of m-th category is obtained as:

$$\mathbf{X}_m = \vec{\alpha}_m \circ \mathbf{F}_m \tag{3}$$

Finally, $\mathbf{X}$ is obtained by concatenating $\mathbf{X}_m, m \in [1, M]$ row-wise. In this work, there are $M = 5$ categories of data which are CT imaging features and four categories of clinical parameters. The attribute vector of node v_i is denoted by $\vec{x}_i$, which is the i-th row of $\mathbf{X}$.

3.2 Co-graph Attention Based Reasoning

We propose a co-graph attention layer to infer the new representation $\vec{x}_i'$ of node v_i from local and long-range regions in graph space. The co-graph layer consists of conventional GAT (con-GAT) [14] where a node's new representation is obtained by weighted averaging its neighbouring nodes, and a corr-GAT where weight score is measured by feature difference based correlation.

Con-GAT. Given input node attribute vectors $\vec{x}_i$ and $\vec{x}_j$ of v_i and v_j, the importance of node j's attribute vector to node i is calculated by [14] as

$$e_{ij}^{con} = \frac{\exp\left(LeakyReLU\left(\vec{\beta}^T\left[\mathbf{W}^{con}\vec{x}_i \| \mathbf{W}^{con}\vec{x}_j\right]\right)\right)}{\sum_{k=1}^{K}\exp\left(LeakyReLU\left(\vec{\beta}^T\left[\mathbf{W}^{con}\vec{x}_i \| \mathbf{W}^{con}\vec{x}_k\right]\right)\right)} \tag{4}$$

where $\vec{\beta} \in \mathbb{R}^{2N_F}$ is weight vector of a single layer feedforward neural network (NN), $\mathbf{W}^{con}$ is a trainable weight matrix. $\vec{\beta}$ and $\mathbf{W}^{con}$ are initialized by Xavier initialization. $\|$ represents concatenation. K is the number of nodes connected to v_i, which means the node j 's importance score is normalized by considering all other nodes connected to v_i. The output new representation of node v_i from the con-GAT layer is obtained as

$$\vec{x}_i^{con} = \sum_{j=1}^{K} e_{ij}^{con}\mathbf{W}^{con}\vec{x}_i \tag{5}$$

Corr-GAT. Inspired by [19], corr-GAT computes the importance of node j to node i by measuring feature difference based correlation. This is different from the concatenation operation in Eq. (5) which averagely groups neighbouring node. The importance of v_j to v_i in corr-GAT is calculated by applying a sequence of dot product, linear transformation, LeakyReLU nonlinearity as

$$d_{ij} = LeakyReLU\left(\vec{\gamma}^T\vec{x}_i^T\mathbf{W}_\theta^{corT}\mathbf{W}_\varnothing^{cor}\vec{x}_j\right) \tag{6}$$

where $\vec{\gamma}$ is weight vector of a single layer feedforward NN. $\mathbf{W}_\varnothing, \mathbf{W}_\theta$ are learnable weights of two 1D convolutional layers that are initialized by Xavier normalization. The normalized new representation of node v_i after corr-GAT layer is obtained as

$$\vec{x}_i^{cor} = \sum_{j=1}^{K} e_{ij}^{cor}\mathbf{W}_\varphi^{cor}\vec{x}_i \tag{7}$$

where $e_{ij}^{cor} = \exp\left(d_{ij}\right)/\sum_{k=1}^{K}\exp\left(d_{ik}\right)$ is SoftMax normalized d_{ij}.

Co-graph Integration and Final Classification. Given new node attribute representations $\mathbf{X}^{con} = [\vec{x}_i^{con}]$ and $\mathbf{X}^{cor} = [\vec{x}_i^{cor}]$ from con-GAT and corr-GAT, we perform node-wise 1D convolution to integrate two graphs, as shown in Fig. 2(e). The output is denoted by $\hat{\mathbf{X}}$. Afterwards, $\hat{\mathbf{X}}$ is added with $\mathbf{X}$ before sending to a final classifier.

The classifier in Fig. 2(f) consists of a sequence of dropout, a fully connected layer with 72 units, ReLU, a fully connected layer with six units, and a fully connected layer with two units for output.

4 Experiments

Experimental Settings. The model was implemented by PyTorch on a Nvidia GeForce GTX 2080 Super graphic card with 8G memory. U-Net encoder was implemented by [20]. GAT-related layers were implemented based on pyGAT [14]. We use cross-entropy loss and Adam optimization. The initial learning rate was 0.01 and scaled by 0.5 every 25 epochs. The network is trained for 200 epochs. The batch size was 20.

Data Pre-processing. All the CT scans are resampled to voxel size of $1 \times 1 \times 1$ mm3 for further processing as there are different voxel sizes. Volumetric patches located at the centre of manually segmented LNs are extracted from resampled CT volumes. The volumetric patch size is defined as $200 \times 200 \times 130$ mm^3 empirically. Categorical features are pre-processed by one-hot encoding. Numerical features are normalized to [0, 1].

60%, 20% and 20% of LNs are randomly selected by stratified sampling for training, validation, and testing. Data augmentations include shifting along x, y and z axis, and rotation of $[-5, +5]$ are performed during the training process, generating 1914 negative and 1600 positive LN samples.

Ablation Study and Comparison Results. Since there are imbalanced LNM class distributions (6.3 negative:1 positive) in the whole patient study and the test dataset, evaluation measures include *sensitivity* = *truepositive/totalpositive*, *specificity* = *truenegative/totalnegative*, the area under Receiver Operating Characteristic Curve (AUC), and Precision-Recall Area Under Curve (PR AUC). All the reported results are at LN level and calculated by averaging three training sessions.

We firstly perform an ablation study to evaluate the contributions of co-graph attention layer. The results using CT images alone are given in Table 1. As shown, con-GAT improved the backbone classification to an AUC of 0.726, sensitivity of 0.685, specificity of 0.795, and PR AUC of 0.765. The corr-GAT achieved better AUC, sensitivity, and specificity than con-GAT, while PR AUC slightly decreased by 0.7%. When using the co-graph attention layer, the image-based classification improved the results of a single graph by 2.3%, 0.5%, 1.2% and 1% concerning four evaluation measures, respectively.

The second ablation study in Table 2 further considered four categories of clinical parameters by category-wise attention and co-GAT. Our first finding is that overall, clinical parameters contributed to improved LNM prediction results when compared with Table 1. Secondly, adaptive fusion by category-wise attention mechanism improved the

Table 1. Ablation study results using CT images alone.

Con-GAT	Corr-GAT	AUC	Sensitivity	Specificity	PR AUC
		0.718	0.650	0.790	0.753
√		0.726	0.685	0.795	0.765
	√	0.759	0.709	0.824	0.758
√	√	0.782	0.714	0.836	0.775

performance of those methods without the attention module. For instance, category-wise attention improved the con-GAT results by 2.7%, 0.3%, 3.4% and 1.4% with respect to AUC, sensitivity, specificity, and PR AUC. The third finding is that corr-GAT contributed to the model's ability to screen positive samples as reflected by sensitivity, while con-GAT showed its power in classifying negative LNM as indicated by specificity and PR AUC. We explain the above findings by corr-GAT's focus on differences between neighboring node attribute vectors while con-GAT performs information summarization based on group averaging. Co-graph complements con-GAT and corr-GAT, which achieved better results. Our final model achieved the best results with AUC of 0.823, sensitivity of 0.793, specificity of 0.852 and PR AUC of 0.840.

Table 2. Ablation study results using both CT and clinical parameters.

Category-wise attention	Con-GAT	Corr-GAT	AUC	Sensitivity	Specificity	PR AUC
	√		0.794	0.724	0.814	0.818
		√	0.808	0.727	0.786	0.817
	√	√	0.810	0.731	0.810	0.828
√	√		0.821	0.747	0.840	0.832
√		√	0.815	0.782	0.843	0.822
√	√	√	0.823	0.793	0.852	0.840

Our model uses a graph neural network to derive new representations in an irregular domain. To demonstrate the effectiveness of novel feature learning in non-rigid like feature space, we compared related models with public code, including a widely used radiomics benchmark model based on multivariable logistic regression [7], 3D UNet encoder with gated attention [20], and a method which transforms non-image data to an image for CNN [9]. Table 3 shows the experimental results when comparing to other methods. The first finding is that our model achieved the best results, followed by models [9] and [7] which combined both images and clinical data. Our second finding is that recall the results in Table 1 and Table 2, the new feature representations learnt by GAT in graph feature space outperformed pure CNN layers when validated by our dataset. The average time taken to train our model is about 5 h. Future work includes investigating the impact of LN patch sizes on the results, and pre-train our image encoder using public CT LN dataset.

Table 3. Comparison with other methods

Method		AUC	Sensitivity	Specificity	PR AUC
Multivariable logistic regression [7]	CT	0.688	0.689	0.816	0.739
	CT + clinical	0.713	0.739	0.779	0.775
Attention gated network [20]		0.639	0.646	0.817	0.721
Deep Insight [9]		0.739	0.703	0.791	0.821
Ours		0.823	0.793	0.852	0.840

5 Conclusion

We propose a new graph node attribute and co-graph attention layer to integrate imaging and multi-categorical clinical factors for LNM prediction. Category-wise attention adaptively fused multi-sourced data with learned attention scores. Co-graph layer performs correlation reasoning by learning from neighbouring regions without the limitation of rigid-structure feature space. The improved performance suggested that our model has the potential to predict LNM from preoperative CT for personalized treatment planning in ESCC patients.

References

1. Lee, H.N., Kim, J.I., Shin, S.Y., Kim, D.H., Kim, C., Hong, I.K.: Combined CT texture analysis and nodal axial ratio for detection of nodal metastasis in esophageal cancer. Br. J. Radiol. **93**, 20190827 (2020)
2. Cancer Stat Facts: Esophageal Cancer. National Cancer Institute. https://seer.cancer.gov/statfacts/html/esoph.html
3. Lee, J.Y., et al.: Improved detection of metastatic lymph nodes in oesophageal squamous cell carcinoma by combined interpretation of fluorine-18-fluorodeoxyglucose positron-emission tomography/computed tomography. Cancer Imaging **19**, 40 (2019)
4. Liu, J., Wang, Z., Shao, H., Qu, D., Liu, J., Yao, L.: Improving CT detection sensitivity for nodal metastases in oesophageal cancer with combination of smaller size and lymph node axial ratio. Eur. Radiol. **28**(1), 188–195 (2017). https://doi.org/10.1007/s00330-017-4935-4
5. Foley, K., Christian, A., Fielding, P., Lewis, W., Roberts, S.: Accuracy of contemporary oesophageal cancer lymph node staging with radiological-pathological correlation. J Clin. Radiol. **72**, 693-e691 (2017)
6. Wu, L., et al.: Multiple level CT radiomics features preoperatively predict lymph node metastasis in esophageal cancer: a multicentre retrospective study. Front Oncol **9**, 1548 (2019)

7. Vallieres, M., Freeman, C.R., Skamene, S.R., El Naqa, I.: A radiomics model from joint FDG-PET and MRI texture features for the prediction of lung metastases in soft-tissue sarcomas of the extremities. Phys Med Biol **60**, 5471–5496 (2015)

8. Chauhan, G., et al.: Joint modeling of chest radiographs and radiology reports for pulmonary edema assessment. In: Martel, A.L., et al. (eds.) MICCAI 2020. LNCS, vol. 12262, pp. 529–539. Springer, Cham (2020). https://doi.org/10.1007/978-3-030-59713-9_51

9. Sharma, A., Vans, E., Shigemizu, D., Boroevich, K.A., Tsunoda, T.: DeepInsight: a methodology to transform a non-image data to an image for convolution neural network architecture. Sci. Rep. **9**, 11399 (2019)

10. Zhang, K., Zhu, Y., Zhang, W., Zhu, Y.: Cross-modal image sentiment analysis via deep correlation of textual semantic. Knowl.-Based Syst. **216** (2021)

11. Ye, L., Liu, Z., Wang, Y.: Dual convolutional LSTM network for referring image segmentation. IEEE Trans. Multimedia **22**, 3224–3235 (2020)

12. Wu, Z., Pan, S., Chen, F., Long, G., Zhang, C., Philip, S.Y.: A comprehensive survey on graph neural networks. IEEE Trans. Neural Netw. Learn. Syst. **32**, 4–24 (2020)

13. Xuan, P., Zhang, Y., Cui, H., Zhang, T., Guo, M., Nakaguchi, T.: Integrating multi-scale neighbouring topologies and cross-modal similarities for drug–protein interaction prediction. Briefings in Bioinf. **119** (2021)

14. Veličković, P., Cucurull, G., Casanova, A., Romero, A., Lio, P., Bengio, Y.: Graph attention networks. Int. Conf. Learn. Rep. (2018)

15. Mo, S., et al.: Multimodal priors guided segmentation of liver lesions in MRI using mutual information based graph co-attention networks. In: Martel, A.L., et al. (eds.) MICCAI 2020. LNCS, vol. 12264, pp. 429–438. Springer, Cham (2020). https://doi.org/10.1007/978-3-030-59719-1_42

16. Isensee, F., Jaeger, P.F., Kohl, S.A.A., Petersen, J., Maier-Hein, K.H.: nnU-Net: a self-configuring method for deep learning-based biomedical image segmentation. Nat. Methods **18**, 203–211 (2021)

17. Jin, Q., Meng, Z., Sun, C., Cui, H., Su, R.: RA-UNet: a hybrid deep attention-aware network to extract liver and tumor in CT scans. Front. Bioeng. Biotechnol. **8**, 1–15 (2020)

18. Sheng, N., Cui, H., Zhang, T., Xuan, P.: Attentional multi-level representation encoding based on convolutional and variance autoencoders for lncRNA-disease association prediction. Briefings Bioinf. **22** (2020)

19. Wang, T., Wang, G., Tan, K.E., Tan, D.: Spectral Pyramid Graph Attention Network for Hyperspectral Image Classification. arXiv preprint arXiv:2001.07108 (2020)

20. Schlemper, J., et al.: Attention gated networks: learning to leverage salient regions in medical images. Med Image Anal **53**, 197–207 (2019)

Deep Orthogonal Fusion: Multimodal Prognostic Biomarker Discovery Integrating Radiology, Pathology, Genomic, and Clinical Data

Nathaniel Braman[✉], Jacob W. H. Gordon, Emery T. Goossens, Caleb Willis, Martin C. Stumpe, and Jagadish Venkataraman

Tempus Labs, Inc., Chicago, IL, USA
{nathaniel.braman,jagadish.venkataraman}@tempus.com
https://www.tempus.com/

Abstract. Clinical decision-making in oncology involves multimodal data such as radiology scans, molecular profiling, histopathology slides, and clinical factors. Despite the importance of these modalities individually, no deep learning framework to date has combined them all to predict patient prognosis. Here, we predict the overall survival (OS) of glioma patients from diverse multimodal data with a Deep Orthogonal Fusion (DOF) model. The model learns to combine information from multiparametric MRI exams, biopsy-based modalities (such as H&E slide images and/or DNA sequencing), and clinical variables into a comprehensive multimodal risk score. Prognostic embeddings from each modality are learned and combined via attention-gated tensor fusion. To maximize the information gleaned from each modality, we introduce a multimodal orthogonalization (MMO) loss term that increases model performance by incentivizing constituent embeddings to be more complementary. DOF predicts OS in glioma patients with a median C-index of 0.788 ± 0.067, significantly outperforming ($p = 0.023$) the best performing unimodal model with a median C-index of 0.718 ± 0.064. The prognostic model significantly stratifies glioma patients by OS within clinical subsets, adding further granularity to prognostic clinical grading and molecular subtyping.

1 Introduction

Cancer diagnosis and treatment plans are guided by multiple streams of data acquired from several modalities, such as radiology scans, molecular profiling, histology slides, and clinical variables. Each characterizes unique aspects of tumor biology and, collectively, they help clinicians understand patient prognosis and assess therapeutic options. Advances in molecular profiling techniques have

Electronic supplementary material The online version of this chapter (https://doi.org/10.1007/978-3-030-87240-3_64) contains supplementary material, which is available to authorized users.

© Springer Nature Switzerland AG 2021
M. de Bruijne et al. (Eds.): MICCAI 2021, LNCS 12905, pp. 667–677, 2021.
https://doi.org/10.1007/978-3-030-87240-3_64

enabled the discovery of prognostic gene signatures, bringing precision medicine to the forefront of clinical practice [1]. More recently, computational techniques in the field of radiology have identified potential imaging-based phenotypes of treatment response and patient survival. Such approaches leverage large sets of explicitly designed image features (commonly known as radiomics [2]) or entail the novel discovery of image patterns by optimizing highly parameterized deep learning models such as convolutional neural networks (CNN) [3] for prediction. Along similar lines, the digitization of histopathology slides has opened new avenues for tissue-based assays that can stratify patients by risk from H&E slide images alone [4]. Given the complementary nature of these various modalities in comprehensive clinical assessment, we hypothesize that their combination in a rigorous machine learning framework may predict patient outcomes more robustly than qualitative clinical assessment or unimodal strategies.

Glioma is an intuitive candidate for deep learning-based multimodal biomarkers owing to the presence of well-characterized prognostic information across modalities [5], as well as its severity [6]. Gliomas can be subdivided by their malignancy into histological grades II-IV [5]. Grades differ in their morphologic and molecular heterogeneity [7], which correspond to treatment resistance and short-term recurrence [8,9]. Quantitative analysis of glioma [10] and its tumor habitat [11] on MRI has demonstrated strong prognostic potential, as well as complex interactions with genotype [12] and clinical variables [13].

Most deep multimodal prediction strategies to date have focused on the fusion of biopsy-based modalities [14–16]. For instance, previous work integrating molecular data with pathology analysis via CNN or graph convolutional neural networks (GCN) has shown that a deep, multimodal approach improves prognosis prediction in glioma patients [14,15]. Likewise, Cheerla et al. integrated histology, clinical, and sequencing data across cancer types by condensing each to a correlated prognostic feature representation [16]. Multimodal research involving radiology has been predominantly correlative in nature [12,13]. Some have explored late-stage fusion approaches combining feature-based representations from radiology with similar pathology [17] or genomic features [18] to predict recurrence. While promising, these strategies rely on hand-crafted feature sets and simple multimodal classifiers that likely limit their ability to learn complex prognostic interactions between modalities and realize the full additive benefit of integrating diverse clinical modalities.

To our knowledge, no study to date has combined radiology, pathology, and genomic data within a single deep learning framework for outcome prediction or patient stratification. Doing so requires overcoming several challenges. First, owing to the difficulty of assembling multimodal datasets with corresponding outcome data in large quantities, fusion schemes must be highly data efficient in learning complex multimodal interactions. Second, the presence of strongly correlated prognostic signals between modalities [16] can create redundancy and hinder model performance.

In this paper, we introduce a deep learning framework that combines radiologic, histologic, genomic, and clinical data into a fused prognostic risk score.

Using a novel technique referred to as Deep Orthogonal Fusion (DOF), we train models using a Multimodal Orthogonalization (MMO) loss function to maximize the independent contribution of each data modality, effectively improving predictive performance. Our approach, depicted in Fig. 1, first trains unimodal embeddings for overall survival (OS) prediction through a Cox partial likelihood loss function. Next, these embeddings are combined through an attention-gated tensor fusion to capture all possible interactions between each data modality. Fusion models are trained simultaneously to predict OS and minimize the correlation between unimodal embeddings. We emphasize the following contributions:

Deep Fusion of Radiology, Pathology, and Omics Data: We present a powerful, data-efficient framework for combining oncologic data across modalities. Our approach enabled a previously unexplored deep integration of radiology with tissue-based modalities and clinical variables for patient risk stratification. This fusion model significantly improved upon unimodal deep learning models. In particular, we found that integrating radiology into deep multimodal models, which is under-explored in previous prognostic studies, conferred the single greatest performance increase. This finding suggests the presence of independent, complementary prognostic information between radiology and biopsy-based modalities and warrants their combination in future prognostic studies.

MMO: To mitigate the effect of inherent correlations between data modalities, we present an MMO loss function that penalizes correlation between unimodal embeddings and encourages each to provide independent prognostic information. We find that this training scheme, which we call DOF, improves prediction by learning and fusing disentangled, prognostic representations from each modality. DOF was also found to outperform a fusion scheme that enforces correlated representations between modalities [16], emphasizing that the dissimilarity of these clinical data streams is crucial to their collective strength.

Multi-parametric Radiology FeatureNet: A neural network architecture that can fuse CNN-extracted deep features from local tumor regions on multiple image sequences (e.g., Gd-T1w and T2w-FLAIR scans) with global hand-crafted radiomics features extracted across the full 3D region-of-interest.

Independent Prognostic Biomarker of OS in Glioma Patients: Using 15-fold Monte Carlo cross-validation with a 20% holdout test set, we evaluate deep fusion models to predict glioma prognosis. We compare this multimodal risk score with existing prognostic clinical subsets and biomarkers (grade, *IDH* status) and investigate its prognostic value within these outcome-associated groups.

2   Methodology

Let X be a training minibatch of data for N patients, each containing M modalities such that $X = [x_1, x_2, ..., x_M]$. For each modality m, x_m includes data from

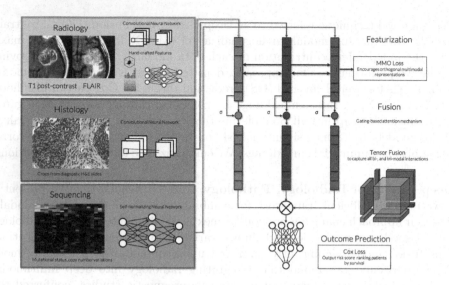

Fig. 1. DOF model architecture and training.

for N patients. Φ_m denotes a trainable unimodal network, which accepts x_m and generates a deep embedding $h_m = \Phi_m(x_m) \in \mathbb{R}^{l_1 x N}$.

When $M > 1$, we combine embeddings from each modality in a multimodal fusion network. For each h_m, an attention mechanism is applied to control its expressiveness based on information from the other modalities. An additional fully connected layer results in h_m^S of length l_2. Attention weights of length l_2 are obtained through a bilinear transformation of h_m with all other embeddings (denoted as $H_{\bar{m}}$), then applied to h_m^S to yield the attention-gated embedding:

$$h_m^* = a_m * h_m^S = \sigma(h_m^T * W_A * H_{\bar{m}}) * h_m^S. \tag{1}$$

To capture all possible interactions between modalities, we combine attention-weighted embeddings through an outer product between modalities, known as tensor fusion [19]. A value of 1 is also included in each vector, allowing for partial interactions between modalities and for the constituent unimodal embeddings to be retained. The output matrix

$$F = \begin{bmatrix} 1 \\ h_1^* \end{bmatrix} \otimes \begin{bmatrix} 1 \\ h_2^* \end{bmatrix} \otimes \dots \otimes \begin{bmatrix} 1 \\ h_M^* \end{bmatrix} \tag{2}$$

is an M-dimensional hypercube of all multimodal interactions with sides of length $l_2 + 1$. Figure 1 depicts F for the fusion of radiology, pathology, and genomic data. It contains subregions corresponding to unaltered unimodal embeddings, pairwise fusions between 2 modalities, and trilinear fusion between all three of the modalities. A final set of fully connected layers, denoted by Φ_F, is applied to tensor fusion features for a final fused embedding $h_F = \Phi_F(F)$.

2.1   Loss Functions

Networks are trained using L which is a linear combination of the terms

$$L = L_{pl} + \gamma L_{MMO} \tag{3}$$

where γ is a scalar weighting the contribution of MMO loss relative to Cox partial likelihood loss. When training unimodal networks, γ is always zero. Performance for various values of γ are included in the Table S4.

MMO Loss: To address the shortcoming of multimodal models converging to correlated predictors, we introduce MMO loss. Inspired by Orthogonal Low-rank Embedding [20], we stipulate that unimodal embeddings preceding fusion should be orthogonal. This criterion enforces that each modality introduced contributes unique information to outcome prediction, rather than relying on signal redundancy between modalities. Each Φ_m is updated through MMO loss to yield embeddings that better complement other modalities. Let $H \in \mathbb{R}^{l_1 x M * N}$ be the set of embeddings from all modalities. MMO loss is computed as

$$L_{MMO} = \frac{1}{M * N} \sum_{m=1}^{M} max(1, ||h_m||_*) - ||H||_* \tag{4}$$

where $|| \cdot ||_*$ denotes the matrix nuclear norm (i.e., the sum of the matrix singular values). This loss is the difference between the sum of nuclear norms per embedding and the nuclear norm of all embeddings combined. It penalizes the scenario where the variance of two modalities separately is decreased when combined and minimized when all unimodal embeddings are fully orthogonal. The per-modality norm is bounded to a minimum of 1 to prevent the collapse of embedding features to zero.

Cox Partial Likelihood Loss: The final layer of each network, parameterized by β, is a fully connected layer with a single unit. This output functions as a Cox proportional hazards model using the deep embedding from the previous layer, h, as its covariates. This final layer's output, θ, is the log hazard ratio, which is used as a risk score. The log hazard ratio for patient i is $\theta_i = h_i^T * \beta$.

We define the negative log likelihood L_{pl} as our cost function

$$L_{pl} = - \sum_{i:E_i=1} \left(\theta_i - log \sum_{j:t_i \geq t_j} e^{\theta_j} \right) \tag{5}$$

where $t \in \mathbb{R}^{Nx1}$ indicates the time to date of last follow up. The event vector, $E \in \{0,1\}^{Nx1}$, equals 1 if death was observed or 0 if a patient was censored (still alive) at last follow up. Each patient i with an observed event is compared against all patients whose observation time was greater than or equal to t_i.

2.2 Modality-Specific Networks for Outcome Prediction

Radiology: A multiple-input CNN was designed to incorporate multiparametric MRI data and global lesion measurements, shown in Fig. S1. The backbone of the network is a VGG-19 CNN [21] with batch normalization, substituting the final max pooling layer with a 4×4 adaptive average pooling. Two pre-trained [22] CNN branches separately extract features from Gd-T1w and T2w-FLAIR images, which are then concatenated and passed through a fully connected layer. A third branch passes hand-crafted features (described in Sect. 3) through a similar fully connected layer. Concatenated embeddings from all branches are fed to 2 additional fully connected layers. All fully connected layers preceding the final embedding layer have 128 units.

Histology, Genomic, and Clinical Data: We reused the models proposed in [14] - a pre-trained VGG-19 CNN with pretrained convolutional layers for Histology and a Self-Normalizing Neural Network (SNN) for genomic data. We also use this SNN for analysis of clinical data, which was not explored in [14].

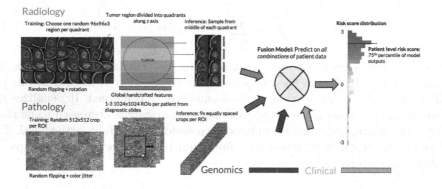

Fig. 2. Sampling multiple radiology & pathology images for patient level risk scores.

3 Experimental Details

Radiology: 176 patients (see patient selection in Fig. S2) with Gd-T1w and T2w-FLAIR scans from the TCGA-GBM [23] and TCGA-LGG [24] studies were obtained from TCIA [25] and annotated by 7 radiologists to delineate the enhancing lesion and edema region. Volumes were registered to the MNI-ICBM standardized brain atlas with 1 mm isotropic resolution, processed with N4 bias correction, and intensity normalized. $96 \times 96 \times 3$ patches were generated from matching regions of Gd-T1w and T2w-FLAIR images within the enhancing lesion. For each patient, 4 samples were generated from four even quadrants of the tumor along the z-axis. Patch slice position was randomized in unimodal training and fixed to the middles of quadrants during inference and fusion network training. Nine features including size, shape, and intensity measures were extracted separately from Gd-T1w and T2w-FLAIR images, and summarized in three fashions for a total of 56 handcrafted features, listed in Table S1.

Pathology and Genomics: We obtained 1024×1024 normalized regions-of-interest (ROIs) and DNA sequencing data curated by [15]. Each patient had 1–3 ROIs from diagnostic H&E slides, totaling 372 images. DNA data consisted of 80 features including mutational status and copy number variation (Table S2).

Clinical Information: 14 clinical features were included into an SNN for the prediction of prognosis. The feature set included demographic and treatment details, as well as subjective histological subtype (see Table S3).

Implementation Details: The embedding size for unimodal networks, l_1, was set to 32. Pre-fusion scaled embedding size, l_2, was 32 for $M = 2$, 16 for $M = 3$, and 8 for $M = 4$. Post-fusion fully connected layers consisted of 128 units each. The final layer of each network had a single unit with sigmoid activation, but its outputs were rescaled between -3 and 3 to function as a prognostic risk score. Unimodal networks were trained for 50 epochs with linear learning rate decay, while multimodal networks were trained for 30 epochs with learning rate decay beginning at the 10th epoch. When training multimodal networks, the unimodal embedding layers were frozen for 5 epochs to train the fusion layers only, then unfrozen for joint training of embeddings and fusion layers.

Statistical Analysis: All models were trained via 15-fold Monte Carlo cross-validation with 20% holdout using the patient-level splits provided in [15]. The primary performance metric was the median observed concordance index (C-index) across folds, a global metric of prognostic model discriminant power. We evaluated all possible combinations of a patient's data (see sampling strategy in Fig. 2) and used the 75th percentile of predicted risk score as their overall prediction. C-indexes of the best-performing unimodal model and the DOF multimodal model were compared with a Mann-Whitney U test [26]. Binary low/high-risk groups were derived from the risk scores, where a risk score >0 corresponded to high risk. For Kaplan-Meier (KM) curves and calculation of hazard ratio (HR), patient-level risk scores were pooled across validation folds.

Table 1. Median C-index of unimodal and fusion models with and without MMO loss.

Group	Model	Cox loss only	With MMO loss
Unimodal	Rad	0.718 ± 0.064	–
	Path	0.715 ± 0.054	–
	Gen	0.716 ± 0.063	–
	Clin	0.702 ± 0.049	–
Pairwise fusion	Path+Gen	0.711 ± 0.055	0.752 ± 0.072
	Gen+Clin	0.702 ± 0.053	0.703 ± 0.052
	Rad+Gen	0.761 ± 0.071	0.766 ± 0.067
	Rad+Path	0.742 ± 0.067	0.752 ± 0.072
	Rad+Clin	0.746 ± 0.068	0.736 ± 0.067
	Path+Clin	0.696 ± 0.051	0.690 ± 0.043
Triple fusion	Path+Gen+Clin	0.704 ± 0.059	0.720 ± 0.056
	Rad+Path+Clin	0.748 ± 0.067	0.741 ± 0.067
	Rad+Gen+Clin	0.754 ± 0.066	0.755 ± 0.067
	Rad+Path+Gen	0.764 ± 0.062	**0.788 ± 0.067**
Full Fusion	Rad+Path+Gen+Clin	**0.775 ± 0.061**	0.785 ± 0.077

4   Results and Discussion

Genomic- and pathology-only model performance metrics are practically similar
(Table 1). However, the CNN-only (C-index = 0.687 × 0.067) and feature-only
(C-index = 0.653 × 0.057) configurations of the radiology model underperform
relative to the aforementioned unimodal models. Combining the radiology CNN
features with the handcrafted features results in the strongest unimodal model.
In contrast, clinical features are the least prognostic unimodal model.

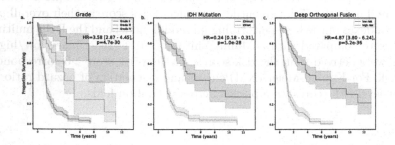

Fig. 3. Stratification by (a) grade, (b) IDH mutation status, and (c) DOF risk groups.

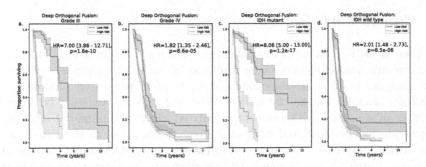

Fig. 4. DOF risk groups stratify patients by OS within (a, b) grade & (c, d) IDH subsets.

Deep fusion models integrating radiology outperform individual unimodal models, naive ensembles of unimodal models, as well as fusions of only clinical and/or biopsy-derived modalities. The full fusion model (C-index = 0.775 ± 0.061) achieves the best performance when trained with Cox loss [27] alone, second only to the Rad+Path+Gen model trained with MMO loss. Naive late fusion ensembles (i.e., averaging unimodal risk scores) exhibit inferior performance for Rad+Path+Gen with (C-index = 0.735 ± 0.063) and without (C-index = 0.739 ± 0.062) clinical features, confirming the benefits of deep fusion.

The addition of MMO loss when training these deep fusion models consistently improves their performance at five different weightings (Table S4), with best performance for both at $\gamma = .5$. When all fusion models are trained at this weighting, 8 of 11 improve in performance. DOF combining radiology, pathology, and genomic data predicts glioma survival best overall with a median C-index of 0.788 ± 0.067, a significant increase over the best unimodal model (p = 0.023).

An ablation study was conducted to investigate the contributions of components of the fusion module (modality attention-gating and tensor fusion). We found that a configuration including both yields the best performance, but that strong results can also be achieved with a simplified fusion module (Table S5).

In Fig. 3, KM plots show that the stratification of patients by OS in risk groups derived from this model perform comparably to established clinical markers. In Fig. 4, risk groups further stratify OS within grade and *IDH* status groups. In sum, these results suggest that the DOF model provides useful prognostic value beyond existing clinical subsets and/or individual biomarkers.

To further benchmark our approach, we implemented the fusion scheme of [16], who combined pathology images, DNA, miRNA, and clinical data, which we further modified to also include radiology data. The network and learning approach is described in-depth in Table S6. In contrast to DOF, [16] instead seeks to maximize the correlation between modality embeddings prior to prediction. A model combining radiology, pathology, and genomic data achieved C-index = 0.730 ± 0.05, while a model excluding the added radiology arm stratified patients by OS with C-index = 0.715 ± 0.05.

5 Conclusions

We present DOF, a data efficient scheme for the novel fusion of radiology, histology, genomic, and clinical data for multimodal prognostic biomarker discovery. The integration of multi-dimensional data from biopsy-based modalities and radiology strongly boosts the ability to stratify glioma patients by OS. The addition of a novel MMO loss component, which forces unimodal embeddings to provide independent and complementary information to the fused prediction, further improves prognostic performance. Our DOF model incorporating radiology, histology, and genomic data significantly stratifies glioma patients by OS within outcome-associated subsets, offering additional granularity to routine clinical markers. DOF can be applied to any number of cancer domains, modality combinations, or new clinical endpoints including treatment response.

References

1. El-Deiry, W.S., et al.: The current state of molecular testing in the treatment of patients with solid tumors, 2019. CA Cancer J. Clin. **69**(4), 305–343 (2019)
2. Gillies, R.J., Kinahan, P.E., Hricak, H.: Radiomics: images are more than pictures, they are data. Radiology **278**(2), 563–577 (2015)
3. Saba, L., et al.: The present and future of deep learning in radiology. Eur. J. Radiol. **114**, 14–24 (2019)
4. Skrede, O.-J., et al.: Deep learning for prediction of colorectal cancer outcome: a discovery and validation study. Lancet (London England) **395**(10221), 350–360 (2020)
5. Louis, D.N., et al.: The 2016 world health organization classification of tumors of the central nervous system: a summary. Acta Neuropathologica **131**(6), 803–820 (2016)
6. Siegel, R.L., Miller, K.D., Jemal, A.: Cancer statistics, 2017. CA Cancer J. Clin. **67**(1), 7–30 (2017)
7. Olar, A., Aldape, K.D.: Using the molecular classification of glioblastoma to inform personalized treatment. J. Pathol. **232**(2), 165–177 (2014)
8. Stupp, R., et al.: Radiotherapy plus concomitant and adjuvant temozolomide for glioblastoma. New Engl. J. Med. **352**(10), 987–996 (2005)
9. Parker, N.R., et al.: Intratumoral heterogeneity identified at the epigenetic, genetic and transcriptional level in glioblastoma. Sci. Rep. **6**, 22477 (2016)
10. Bae, S., et al.: Radiomic MRI phenotyping of glioblastoma: improving survival prediction. Radiology **289**(3), 797–806 (2018)
11. Prasanna, P., et al.: Radiomic features from the peritumoral brain parenchyma on treatment-Naïve multi-parametric MR imaging predict long versus short-term survival in glioblastoma multiforme: preliminary findings. Eur. Radiol. **27**(10), 4188–4197 (2017)
12. Beig, N., et al.: Radiogenomic-based survival risk stratification of tumor habitat on Gd-T1w MRI is associated with biological processes in glioblastoma. Clin. Cancer Res. **26**(8), 1866–1876 (2020)
13. Beig, N., et al.: Sexually dimorphic radiogenomic models identify distinct imaging and biological pathways that are prognostic of overall survival in glioblastoma. Neuro-Oncology **23**(2), 251–263 (2021)

14. Chen, R.J., et al.: Pathomic fusion: an integrated framework for fusing histopathology and genomic features for cancer diagnosis and prognosis. arXiv:1912.08937 [cs, q-bio]. version: 1, 18 December 2019
15. Mobadersany, P., et al.: Predicting cancer outcomes from histology and genomics using convolutional networks. Proc. Natl. Acad. Sci. **115**(13), E2970–E2979 (2018)
16. Cheerla, A., Gevaert, O.: Deep learning with multimodal representation for pan-cancer prognosis prediction. Bioinformatics **35**(14), i446–i454 (2019)
17. Vaidya, P., et al.: RaPtomics: integrating radiomic and pathomic features for predicting recurrence in early stage lung cancer. In: Medical Imaging 2018: Digital Pathology, vol. 10581, p. 105810M. International Society for Optics and Photonics, 6 March 2018
18. Subramanian, V., Do, M.N., Syeda-Mahmood, T.: Multimodal fusion of imaging and genomics for lung cancer recurrence prediction. arXiv:2002.01982 [cs, eess, q-bio], 5 February 2020
19. Zadeh, A., et al.: Tensor fusion network for multimodal sentiment analysis. arXiv:1707.07250 [cs], 23 July 2017
20. Lezama, J., et al.: O \ Ln'E: orthogonal low-rank embedding, a plug and play geometric loss for deep learning. arXiv:1712.01727 [cs, stat], 5 December 2017
21. Simonyan, K., Zisserman, A.: Very deep convolutional networks for large-scale image recognition. arXiv:1409.1556 [cs], 10 April 2015
22. Deng, J., et al.: ImageNet: a large-scale hierarchical image database. In: 2009 IEEE Conference on Computer Vision and Pattern Recognition, June 2009, pp. 248–255 (2009). ISSN: 1063–6919
23. Scarpace, L., et al.: Radiology Data from The Cancer Genome Atlas Glioblas-toma Multiforme [TCGA-GBM] collection. In collab. with TCIA Team. type: dataset (2016)
24. Pedano, N., et al.: Radiology Data from The Cancer Genome Atlas Low Grade Glioma [TCGA-LGG] collection. In collab. with TCIA Team. type: dataset (2016)
25. Clark, K., et al.: The cancer imaging archive (TCIA): maintaining and operating a public information repository. J. Digit. Imaging **26**(6), 1045–1057 (2013)
26. Mann, H.B., Whitney, D.R.: On a test of whether one of two random variables is stochastically larger than the other. Ann. Math. Stat. **18**(1), 50–60 (1947)
27. Ching, T., Zhu, X., Garmire, L.X.: Cox-nnet: an artificial neural network method for prognosis prediction of high-throughput omics data. PLOS Comput. Biol. **14**(4), e1006076 (2018)

A Novel Bayesian Semi-parametric Model for Learning Heritable Imaging Traits

Yize Zhao[1], Xiwen Zhao[1], Mansu Kim[2], Jingxuan Bao[2], and Li Shen[2]([✉])

[1] Department of Biostatistics, Yale University School of Public Health,
New Haven, NJ, USA
yize.zhao@yale.edu

[2] Department of Biostatistics, Epidemiology, and Informatics,
University of Pennsylvania Perelman School of Medicine, Philadelphia, PA, USA
li.shen@pennmedicine.upenn.edu

Abstract. Heritability analysis is an important research topic in brain imaging genetics. Its primary motivation is to identify highly heritable imaging quantitative traits (QTs) for subsequent in-depth imaging genetic analyses. Most existing studies perform heritability analyses on regional imaging QTs using predefined brain parcellation schemes such as the AAL atlas. However, the power to dissect genetic underpinnings under QTs defined in such an unsupervised fashion is largely deteriorate with inner partition noise and signal dilution. To bridge the gap, we propose a new semi-parametric Bayesian heritability estimation model to construct highly heritable imaging QTs. Our method leverages the aggregate of genetic signals to imaging QT construction by developing a new brain parcellation driven by voxel-level heritability. To ensure biological plausibility and clinical interpretability of the resulting brain heritability parcellations, hierarchical sparsity and smoothness, coupled with structural connectivity of the brain, are properly imposed on genetic effects to induce spatial contiguity of heritable imaging QTs. Using the ADNI imaging genetic data, we demonstrate the strength of our proposed method, in comparison with the standard GCTA method, in identifying highly heritable and biologically meaningful new imaging QTs.

Keywords: Imaging genetics · Heritability estimation · Bayesian semi-parametric modeling

1 Introduction

Brain imaging genetics is an emerging and rapidly growing data science field that arises with the recent advances in acquiring multimodal neuroimaging data and high throughput genotyping and sequencing data [9,14,16,23]. To characterize

Supported in part by NIH RF1 AG068191, R01 LM013463, and R01 AG071470. Data used in this study were obtained from the Alzheimer's Disease Neuroimaging Initiative database (adni.loni.usc.edu), which was funded by NIH U01 AG024904.

M. de Bruijne et al. (Eds.): MICCAI 2021, LNCS 12905, pp. 678–687, 2021.
https://doi.org/10.1007/978-3-030-87240-3_65

genetic contributions on heritable neuroimaging quantitative traits (QTs), we are able to gain new insights into the pathobiological mechanism from genetics to brain structure and function, and their impact on behaviors and diseases.

The concept of heritability [19] has thus emerged under imaging genetics paradigm to describe the proportion of the total imaging phenotypic variance that is explained by the aggregated genetic effect captured by pedigree information [17] or all the single nucleotide polymorphisms (SNPs) on a genotyping array [24]. Under existing heritability studies, atlas-based brain parcellations like automated anatomical labeling (AAL) [18] are routinely used to define imaging traits based on certain imaging modality. However, within each region of interest (ROI) defined under such an unsupervised brain parcellation, some areas may be impacted marginally by SNPs, leading to a dilution of power to dissect genetic contribution. Thus, there is an urgent need to construct a heritability map at voxel level to accurately provide cartography for the truly heritable brain areas.

It is a challenging task to accurately construct a biological interpretable heritability map over whole brain voxel-wise neuroimaging measurements. Most of the existing heritability modeling [5,22], including the widely used genome-wide complex trait analysis (GCTA) [22], can only handle univariate phenotype without an efficient way to accommodate the phenotypic correlation. A few recent attempts [6,10,25] start to explore heritability analysis for multivariate or large-scale phenotypes especially given the highly correlated collections from neuroimaging data. Those methods, though providing promising results under their applications, are either unable to handle high-dimensional phenotypes like the voxel-wise traits due to a direct inverse of the phenotypic covariance matrix, or fail to incorporate biologically plausible assumptions like the smoothness over brain topology for the heritability estimates.

In this paper, we propose a new Bayesian joint voxel-wise heritability analysis to construct highly heritable imaging QTs based on the estimated heritability map. From the analytical perspective, this requires an efficient and meaningful variance component selection under high dimensional imaging responses. Despite there is a broad literature on Bayesian sparsity and shrinkage, almost none of them deals with the selection and estimation on the variance components. Under neuroimaging studies with unique spatial correlation across voxels and structural/functional interactions among ROIs, we also need to properly consider the underlying biological information, which otherwise will cause a power loss in heritable traits detection and implausible interpretation.

To address all the above limitations and challenges, we propose a new semi-parametric Bayesian heritability estimation model, and apply it to the imaging genetic data from the Alzheimer's Disease Neuroimaging Initiative (ADNI) [12, 15,20] to construct highly heritable and biologically meaningful imaging QTs. Our major contributions are summarized as follows:

– We create a brain heritability map under a novel Bayesian integrative heritability analysis for high dimensional voxel-wise imaging phenotypes. We jointly incorporate the brain connectivity information and spatial correlation among voxels to enhance analytical power and biological interpretation.

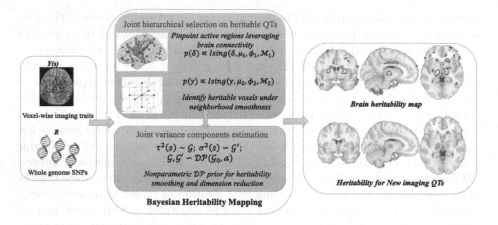

Fig. 1. The illustration of our proposed method Brain Heritability Mapping (BHM).

– We make the very first attempt to construct heritable imaging QTs with stronger genetic dissection power by removing the constrain from the traditional unsupervised brain atlas for the QT definition. These new QTs will provide a great potential to uncover in-depth genetic underpinnings.
– We demonstrate the effectiveness of our method in an empirical study to construct heritable imaging QTs using the structural magnetic resonance imaging (MRI) and genome-wide genotyping data from the ADNI cohort [12, 15, 20]. These novel imaging QTs are highly heritable in comparison with the heritable AAL-based QTs discovered by the standard GCTA method.

2   Method

Our overarching goal is to construct more powerful neuroimaging endophenotypes with strong genetic dissection power based on an innovative "brain heritability map". We propose a Bayesian semi-parametric model to jointly estimate voxel-specific heritability over whole brain imaging traits. Within the Bayesian paradigm, a hierarchical Ising-Spike-and-Slab prior is used to simultaneously impose sparsity on heritabilities at 1) brain regions while accounting for correlations induced by brain structural connectivity; 2) voxels while considering the dependency among adjacent voxels. To enhance biological insight and reduce parameter space, we further assign a Dirichlet process ($\mathcal{DP}$) prior on the genetic and environmental variance components, so that each of them are identical within a contiguous brain area. Based on the result, a brain heritability map can be constructed directly with the new imaging phenotype defined under the "active" subregions; and the heritability for each of them is also estimated. Please see Fig. 1 for the schematic design of our method.

Brain Heritability Mapping. To estimate the additive genetic heritability for imaging traits $\{y(s)\}_{s=1}^{S}$ over S voxels adjusting for clinical covariate $\mathbf{X}$, we build the following mixed effect models

$$y(s) = \mathbf{X}\boldsymbol{\beta}(s) + g(s) + e(s), \quad s = 1, \ldots, S. \tag{1}$$

At each voxel s, $g(s)$ is the genetic random effect with $g(s) \sim \mathrm{N}(0, \mathbf{R}\tau^2(s))$, $\mathbf{R}$ is the empirical genetic relationship matrix among subjects that is calculated directly from the SNP data, and $\tau^2(s)$ is the variance explained by the genetics. The residual error $e(s) \sim \mathrm{N}(0, \mathbf{I}\sigma^2(s))$ with $\sigma^2(s)$ denotes the variance explained by the environmental effects. Based on the two variance components, the voxel-specific heritability can be calculated by $h(s) = \frac{\tau^2(s)}{\tau^2(s)+\sigma^2(s)}$ for $s = 1, \ldots, S$.

The assembly of $\{h(s)\}_{s=1}^S$ will provide a genetics cartographic map over the human brain. To impose a structural driven sparsity, we introduce a regional binary indicate set $\boldsymbol{\delta} = (\delta_1, \ldots, \delta_K)$ for the K ROIs and a voxel-level set $\boldsymbol{\gamma} = (\gamma_1, \ldots, \gamma_S)$ to jointly distinguish brain locations with and without active genetic impact. Based on them, we impose the following sparse group spike-and-slab prior for each genetic variance $\tau^2(s)$:

$$\tau^2(s) \sim (1 - \gamma(s)\delta(k))I_0 + \gamma(s)\delta(k)\mathcal{G}, \text{ with voxel } s \text{ belonging to region } k, \tag{2}$$

where I_0 is a point mass at zero, and $\mathcal{G}$ is a probability function where the nonzero component of $\tau^2(s)$ samples from. It is straightforward to see $\tau^2(s)$ (and $h(s)$) is nonzero only if $\gamma(s) = \delta(k) = 1$. Therefore, we can effectively narrow down heritable brain traits by stochastically excluding the regions with negligible genetic effect and only locate fine scale signals for the heritable regions.

To further induce a biologically plausible coupling of selection status in light of brain connectivity and spatial correlation, we resort to the Ising model for each of the indicator set

$$p(\boldsymbol{\delta}) \propto \mathrm{Ising}(\boldsymbol{\delta}, \mu_1, \phi_1, \mathcal{M}_1); \quad p(\boldsymbol{\gamma}) \propto \mathrm{Ising}(\boldsymbol{\gamma}, \mu_2, \phi_2, \mathcal{M}_2) \tag{3}$$

where graphs $\mathcal{M}_1$ and $\mathcal{M}_2$ summarize the region and voxel level structural information, and $\mu_1, \mu_2, \phi_1, \phi_2$ are the sparsity and smoothness parameters. Under hyper-prior (3), "connected" trait units have a higher possibility to be jointly included or excluded from the model, aligning with the biological expectation.

Imaging Endophenotype and Heritability Estimation. In terms of the environmental variance, we assume $\sigma^2(s) \sim \mathcal{G}'$ with a prior probability function $\mathcal{G}'$. Under such a fine scale voxel-level imaging phenotypes, it is biologically meaningful to assume the spatially contiguous voxels share similar heritability. To impose such smoothness of $\{h(s)\}$ over the identified contiguous hertiable brain areas, we assume the joint distribution of $\mathcal{G}$ and $\mathcal{G}'$ follows a nonparametric $\mathcal{DP}$ prior with scalar parameter α

$$\mathcal{G}, \mathcal{G}' \sim \mathcal{DP}(\mathcal{G}_0, \alpha), \quad \mathcal{G}_0 = \text{Inverse Gamma} \times \text{Inverse Gamma}', \tag{4}$$

where $\mathcal{G}_0$ is the base measure defined by a joint of two independent Inverse Gamma (IG) distributions. The nonparametric and discrete nature of (4) can be clearly seen under the following sticking-breaking representation [13]

$$\mathcal{G} = \sum_{k=1}^{\infty} \pi_k I_{\theta_k}; \quad \mathcal{G}' = \sum_{k=1}^{\infty} \pi_k I_{\theta'_k}; \quad \theta_k, \theta'_k \sim \mathcal{G}_0; \tag{5}$$

with $\pi_k = \pi'_k \prod_{h=1}^{k-1}(1 - \pi'_h)$ and $\pi'_k \sim \mathrm{Beta}(1, \alpha)$. This allows us to more robustly accommodate the potential irregular distribution of variance components, while inducing a clustering effect of $\{h(s)\}$ with each contiguous area sharing the identical heritability estimate. Meanwhile, given the estimation unit of variance components moves from voxel to brain area, the risk of overfitting could be dramatically reduced with much less unknown parameters, and facilitate a more accurate heritability mapping estimation with meaningful smoothness effect.

Combing all the model specifications, we name our model Bayesian Heritability Mapping (BHM) which is semi-parametric; and develop a Markov chain Monte Carlo (MCMC) algorithm to conduct posterior inference. We rely on Gibbs samplers with data augmentation to obtain posterior draws embedded with a truncated stick-breaking process to approximate the $\mathcal{DP}$ representation. The eventual heritability map is captured by the median probability model [2] under the posterior inclusion probabilities of $\{\delta\}$ and $\{\gamma\}$. Simultaneously, we could obtain the heritability of each identified imaging QTs using the posterior median of $\{h(s)\}$. Given the defined QTs are the brain areas with active correspondence with genetics, we anticipate higher heritabilities of them than those under the traditional AAL defined regional imaging traits. Meanwhile, given the sparse nature of our method, we also expect a number of unwarranted regional heritable signals to be excluded from the result.

3   Experiments and Results

Data and Materials. The neuroimaging and genotypinig data used in this work were obtained from the Alzheimer's Disease Neuroimaging Initiative (ADNI) database (adni.loni.usc.edu) [12,15,20]. The up-to-date information about the ADNI is available at www.adni-info.org. The participants ($N = 1,472$) include 341 cognitively normal (CN), 85 significant memory concern (SMC), 265 early mild cognitive impairment (EMCI), 495 late MCI (LMCI), and 286 AD subjects at the ADNI-GO/2 baseline. See Table 1 for characteristics of these participants.

Structural MRI scans were processed with voxel-based morphometry (VBM) using the Statistical Parametric Mapping (SPM) software tool [1]. All scans were aligned to a T1-weighted template image, segmented into gray matter (GM), white matter (WM) and cerebrospinal fluid (CSF) maps, normalized to the standard Montreal Neurological Institute (MNI) space as $2 \times 2 \times 2\,\mathrm{mm}^3$ voxels. The GM maps were extracted and smoothed with an 8mm FWHM kernel, and analyzed in this study. A total of 144,999 voxels, covering cortical, sub-cortical, and cerebellar regions and measuring GM density, were studied in this work as voxel-level imaging traits. Based on the AAL atlas [18], 116 ROI-level traits were also obtained by averaging all the voxel-level measures within each ROI.

Table 1. Participant characteristics. Age and sex are used as covariates in our study.

Diagnosis	CN	SMC	EMCI	LMCI	AD	Overall
Number	341	85	265	495	286	1472
Age (mean ± sd)	75.1 ± 5.4	72.4 ± 5.7	71.2 ± 7.1	73.9 ± 7.6	75.1 ± 8.0	73.9 ± 7.2
Sex (M/F)	182/159	36/49	147/118	306/189	162/124	833/639
Education(mean ± sd)	16.3 ± 2.6	16.7 ± 2.6	16.1 ± 2.6	16.0 ± 2.9	15.3 ± 3.0	16.0 ± 2.8
APOE ε4 present	24.9%	34.1%	36.2%	41.6%	46.5%	37.3%

For the genotyping data, we performed quality control using the following criteria: genotyping call rate >95 %, minor allele frequency >5%, and Hardy Weinberg Equilibrium >1e−6. A total of 565,373 SNPs were used for estimating heritability. The structural connectivity computed from diffusion MRI (dMRI) was used as connectivity information of the BHM model [8]. The preprocessed dMRI data of 291 participants were obtained from the human connectome project database, and the FSL software was used to construct the structural connectivity [21]. The distance-dependent consensus thresholding method was applied to generate group-level connectivity and to avoid overestimating short-range connections. This was used as the connectivity information in our analyses [3].

Implementation and Evaluation. We applied our proposed BHM model on the VBM and genetics data adjusting for age, sex and the first ten genetic principal components. The tuning parameters in the Ising priors were determined by the auxiliary method [11], the shape and scale parameters in the Inverse Gamma distributions were set to be 0.1 to provide non-informative support, and we assigned α to be a noninformative Gamma distribution $G(1,1)$. We started with multiple chains with 10,000 iterations 5000 burn-in under random initials (Matlab2020b implementation, 2.4 GHz CPU, 64 GB Memory, Windows System). Each run took ∼6 h to finish, and both trace plots and GR method [7] were used to confirm the posterior convergence. We also implemented the GCTA model to calculate the marginal heritability of the AAL-based regional QTs using their provided pipeline in PLINK format. Eventually, we summarized the heritability for each ROI, our defined QT within the corresponding ROI, and the size in voxels of each heritable sub-region (Table 2).

ADNI Results. Table 2 shows the heritability estimation results of comparing the proposed BHM model and the traditional GCTA model, including 47 BHM-identified ROIs. For most of these ROIs, BHM was able to identify new imaging QTs (i.e., the subregion of each ROI with size indicated by N_{Voxels}) with higher heritability than the GCTA-estimated heritability for the entire ROI based average measure. For instance, the BHM heritability estimates in bilateral superior frontal gyri (0.532, 0.409) are higher than GCTA results (0.124, 0.119). BHM also successfully captured working memory related regions better than GCTA, and these regions were known to be significantly heritable, including inferior, middle, and superior frontal gyri [4]. All these observations demonstrate the promise of the BHM method in identifying new highly heritable imaging QTs,

Table 2. Performance comparison. The GCTA column includes the heritability of the average VBM measure in the ROI estimated by the conventional GCTA method. The BHM column includes the BHM-estimated heritability of the identified imaging QT (i.e., the subregion of each ROI with size indicated by N_{Voxels}).

Region	Left hemisphere			Right hemisphere		
	GCTA	BHM	N_{Voxels}	GCTA	BHM	N_{Voxels}
Precentral				0.063	0.443	281
Frontal_Sup	0.124	0.532	352	0.119	0.409	429
Frontal_Sup_Orb	0.031	0.533	82	0.000	0.507	42
Frontal_Mid	0.109	0.321	331	0.003	0.446	259
Frontal_Inf_Oper	0.210	0.403	148	0.098	0.539	174
Frontal_Inf_Tri	0.014	0.390	194	0.223	0.452	564
Frontal_Inf_Orb				0.261	0.613	118
Rolandic_Oper				0.333	0.442	75
Frontal_Sup_Medial	0.311	0.323	264	0.265	0.421	227
Rectus	0.184	0.400	32	0.154	0.321	55
Cingulum_Ant				0.328	0.473	113
Cingulum_Mid	0.470	0.353	108			
Cingulum_Post				0.398	0.678	55
ParaHippocampal				0.240	0.487	41
Occipital_Inf				0.244	0.343	35
Postcentral	0.090	0.485	289	0.123	0.226	235
Parietal_Inf				0.336	0.554	66
SupraMarginal				0.172	0.428	62
Angular				0.028	0.391	56
Caudate	1.000	0.433	117	1.000	0.509	273
Putamen	0.000	0.288	312	0.000	0.377	356
Pallidum				0.068	0.419	75
Thalamus	0.720	0.445	446	0.613	0.391	324
Temporal_Pole_Sup	0.196	0.324	80	0.257	0.424	330
Temporal_Mid	0.091	0.340	123	0.085	0.478	298
Temporal_Pole_Mid	0.052	0.288	203	0.026	0.482	228
Temporal_Inf	0.000	0.323	101			
Cerebelum_8	0.322	0.450	409	0.456	0.369	150
Cerebelum_9	0.000	0.265	283	0.111	0.382	121
Cerebelum_10	0.285	0.132	58			

which can be used for subsequent in-depth brain imaging genetic analysis. Of note, some identified ROIs are relatively small and warrant replication in inde-

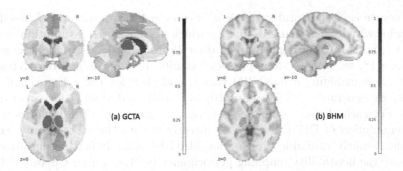

Fig. 2. Heritability maps estimated by (a) the conventional GCTA method and (b) the proposed BHM method. In the GCTA map, the entire ROI is painted with the estimated heritability. In the BHM map, only the identified voxels forming new heritable imaging QTs are painted with the estimated heritability.

Table 3. Simulation results under BHM, GCTA and MEGHA: RMSE for heritability estimation and AUC for heritability mapping. The Monte Carlo standard deviation is included in the parentheses.

Scenario	Method	IG(5, 10)		IG(0.5, 1)	
		RMSE	AUC	RMSE	AUC
$\mathcal{DP}$	GCTA	0.021 (6.211e−04)	0.966	0.043 (2.216e−05)	0.886
	MEGHA	0.013 (5.788e−05)	0.988	0.044 (6.824e−05)	0.960
	BHM	0.002 (3.578e−05)	0.999	0.002 (1.213e−04)	0.992
IG	GCTA	0.018 (3.431e−05)	0.914	0.032 (8.343e−05)	0.784
	MEGHA	0.005 (8.102e−06)	0.908	0.026 (3.877e−05)	0.792
	BHM	0.004 (8.354e−05)	0.995	0.004 (1.088e−04)	0.897

pendent cohorts. In addition, the GCTA method also identified some heritable traits that are worth detailed imaging genetic analysis.

Figure 2 shows the heritability maps estimated by (a) the conventional GCTA method and (b) the proposed BHM method. GCTA estimates the heritability of the average voxel measure of each ROI. Given that the AAL ROIs are quite large, the map looks nonsparse. However, the map covers part of the white matter region. This appears counter-intuitive, since VBM only measures gray matter density. In contrast, the BHM map identifies heritable voxels only in gray matter region, which appears to be biologically more precise and meaningful.

Simulation Results. We also perform simulations to evaluate the perform of BHM model compared with GCTA and massively expedited genome-wide heritability analysis (MEGHA) [5] in heritability mapping and estimation. We consider phenotypes measured over a 100×100 square with 10,000 voxels which is partitioned into 16 equally sized squared regions. Across the regions, we generate a scale-free connectivity network as prior information. We consider two scenarios to generate environmental variance σ^2–Scenario 1 is to generate σ^2 based on a $\mathcal{DP}$ prior with base measure $IG(0.5, 1)$, which fits our model assumption (see also Eq. (4)); Scenario 2 is to generate σ^2 directly from $IG(0.5, 1)$. As for the genetic variance τ^2, we first set four pieces of significantly heritable areas lying over two regions including around 100 voxels, and then for the active signal index set $\mathcal{R}$, we generate $\tau^2(\mathcal{R}) \sim IG(5, 10)$ or $\tau^2(\mathcal{R}) \sim IG(0.5, 1)$ for two variance cases. For each of these simulated settings, we generate 100 MC datasets. The implementation of BHM and GCTA directly follows the ADNI study, and we use the publicly available pipeline for MEGHA with 10,000 permutation. We evaluate the heritability mapping performance by Area under Curve (AUC) for identifying the heritable voxels, and the heritability estimation by root-mean-square error (RMSE) for $h(s)$. All the results are summarized in Table 3.

In general, the proposed BHM model considerably outperforms GCTA and MEGHA in all the simulated settings for both heritability mapping and estimation. A higher variance in σ^2 expectedly deteriorates the model performance for all the methods. In addition, when the model assumption for BHM is not satisfied, we see minor decrease of AUC and increase of RMSE based on the proposed method, indicating the robustness of our approach.

4   Conclusion

We have proposed a new semi-parametric Bayesian heritability estimation model to construct highly heritable and biologically meaningful imaging quantitative traits (QTs). Our method leverages the aggregate of genetic signals to imaging QT construction by developing a new brain parcellation driven by voxel-level heritability. To ensure biological plausibility and clinical interpretablity of the resulting brain heritability parcellations, hierarchical sparsity and smoothness, coupled with structural connectivity of the brain, have been properly imposed on genetic effects to induce spatial contiguity of heritable imaging QTs. Using the ADNI imaging genetic data, we have demonstrated the strength of our proposed method, in comparison with the standard GCTA method, in identifying highly heritable and biologically meaningful new imaging QTs.

References

1. Ashburner, J., Friston, K.J.: Voxel-based morphometry-the methods. Neuroimage **11**(6), 805–21 (2000)
2. Barbieri, M.M., Berger, J.O., et al.: Optimal predictive model selection. Ann. Stat. **32**(3), 870–897 (2004)
3. Betzel, R.F., Griffa, A., Hagmann, P., Mišić, B.: Distance-dependent consensus thresholds for generating group-representative structural brain networks. Netw. Neurosci. **3**(2), 475–496 (2019)
4. Blokland, G.A., McMahon, K.L., Thompson, P.M., Martin, N.G., de Zubicaray, G.I., Wright, M.J.: Heritability of working memory brain activation. J. Neurosci. **31**(30), 10882–10890 (2011)

5. Ge, Y., et al.: Massively expedited genome-wide heritability analysis (megha). Proc. Natl. Acad. Sci. **112**(8), 2479–2484 (2015)
6. Ge, T., et al.: Multidimensional heritability analysis of neuroanatomical shape. Nat. Commun. **7**, e13291 (2016)
7. Gelman, A., Meng, X.L.: Simulating normalizing constants: from importance sampling to bridge sampling to path sampling. Stat. Sci. **13**, 163–185 (1998)
8. Glasser, M.F., et al.: The minimal preprocessing pipelines for the human connectome project. Neuroimage **80**, 105–124 (2013)
9. Hibar, D.P., et al.: Voxelwise gene-wide association study (vGeneWAS): multivariate gene-based association testing in 731 elderly subjects. Neuroimage **56**(4), 1875–91 (2011)
10. Luo, S., Song, R., Styner, M., Gilmore, J., Zhu, H.: FSEM: Functional structural equation models for twin functional data. J. Am. Stat. Assoc. **114**(525), 344–357 (2019)
11. Møller, J., Pettitt, A.N., Reeves, R., Berthelsen, K.K.: An efficient Markov chain Monte Carlo method for distributions with intractable normalising constants. Biometrika **93**(2), 451–458 (2006)
12. Saykin, A.J., et al.: Genetic studies of quantitative MCI and AD phenotypes in ADNI: progress, opportunities, and plans. Alzheimers Dement. **11**(7), 792–814 (2015)
13. Sethuraman, J.: A constructive definition of dirichlet priors. Stat. Sin. **4**, 639–650 (1994)
14. Shen, L., Thompson, P.M.: Brain imaging genomics: integrated analysis and machine learning. Proc. IEEE Inst. Electr. Electron. Eng. **108**(1), 125–162 (2020)
15. Shen, L., et al.: Genetic analysis of quantitative phenotypes in AD and MCI: imaging, cognition and biomarkers. Brain Imaging Behav. **8**(2), 183–207 (2014)
16. Stein, J.L., et al.: Voxelwise genome-wide association study (vGWAS). Neuroimage **53**(3), 1160–74 (2010)
17. Thompson, P.M., et al.: Genetic influences on brain structure. Nat. Neurosci. **4**(12), 1253–8 (2001)
18. Tzourio-Mazoyer, N., et al.: Automated anatomical labeling of activations in SPM using a macroscopic anatomical parcellation of the MNI MRI single-subject brain. Neuroimage **15**(1), 273–89 (2002)
19. Visscher, P.M., Hill, W.G., Wray, N.R.: Heritability in the genomics era-concepts and misconceptions. Nat. Rev. Genet. **9**(4), 255–66 (2008)
20. Weiner, M.W., et al.: Recent publications from the Alzheimer's disease neuroimaging initiative: reviewing progress toward improved ad clinical trials. Alzheimer's Dementia **13**(4), e1–e85 (2017)
21. Woolrich, M.W., et al.: Bayesian analysis of neuroimaging data in FSL. Neuroimage **45**(1), S173–S186 (2009)
22. Yang, J., Lee, S.H., Goddard, M.E., Visscher, P.M.: GCTA: a tool for genome-wide complex trait analysis. Am. J. Hum. Genet. **88**(1), 76–82 (2011)
23. Yao, X., et al.: Mining regional imaging genetic associations via voxel-wise enrichment analysis. In: 2019 IEEE EMBS International Conference on Biomedical & Health Informatics (BHI), pp. 1–4 (2019)
24. Zhao, B., et al.: Heritability of regional brain volumes in large-scale neuroimaging and genetic studies. Cereb. Cortex **29**(7), 2904–2914 (2019)
25. Zhao, Y., Li, T., Zhu, H.: Bayesian sparse heritability analysis with high-dimensional neuroimaging phenotypes. Biostatistics, kxaa035 (2020). https://doi.org/10.1093/biostatistics/kxaa035

Combining 3D Image and Tabular Data via the Dynamic Affine Feature Map Transform

Sebastian Pölsterl$^{(\boxtimes)}$ 🆔, Tom Nuno Wolf, and Christian Wachinger 🆔

Artificial Intelligence in Medical Imaging (AI-Med), Department of Child and Adolescent Psychiatry, Ludwig-Maximilians-Universität, Munich, Germany
sebastian.poelsterl@med.uni-muenchen.de

Abstract. Prior work on diagnosing Alzheimer's disease from magnetic resonance images of the brain established that convolutional neural networks (CNNs) can leverage the high-dimensional image information for classifying patients. However, little research focused on how these models can utilize the usually low-dimensional tabular information, such as patient demographics or laboratory measurements. We introduce the Dynamic Affine Feature Map Transform (DAFT), a general-purpose module for CNNs that dynamically rescales and shifts the feature maps of a convolutional layer, conditional on a patient's tabular clinical information. We show that DAFT is highly effective in combining 3D image and tabular information for diagnosis and time-to-dementia prediction, where it outperforms competing CNNs with a mean balanced accuracy of 0.622 and mean c-index of 0.748, respectively. Our extensive ablation study provides valuable insights into the architectural properties of DAFT. Our implementation is available at https://github.com/ai-med/DAFT.

1 Introduction

In recent years, deep convolutional neural networks (CNNs) have become the standard for classification of Alzheimer's disease (AD) from magnetic resonance images (MRI) of the brain (see e.g. [4, 29] for an overview). CNNs excel at extracting high-level information about the neuroanatomy from MRI. However, brain MRI only offers a partial view on the underlying changes causing cognitive decline. Therefore, clinicians and researchers often rely on tabular data such as patient demographics, family history, or laboratory measurements from cerebrospinal fluid for diagnosis. In contrast to image information, tabular data is typically low-dimensional and individual variables capture rich clinical knowledge.

S. Pölsterl and T.N. Wolf—These authors contributed equally to this work.

Electronic supplementary material The online version of this chapter (https://doi.org/10.1007/978-3-030-87240-3_66) contains supplementary material, which is available to authorized users.

© Springer Nature Switzerland AG 2021
M. de Bruijne et al. (Eds.): MICCAI 2021, LNCS 12905, pp. 688–698, 2021.
https://doi.org/10.1007/978-3-030-87240-3_66

Due to image and tabular data being complementary to each other, it is desirable to amalgamate both sources of information in a single neural network such that one source of information can inform the other. The effective integration is challenging, because the dimensionality mismatch between image and tabular data necessitates an architecture where the capacity required to summarize the image is several orders of magnitude higher than the one required to summarize the tabular data. This imbalance in turn implicitly encourages the network to focus on image-related parameters during training, which ultimately can result in a model that is only marginally better than a CNN using the image data alone [24]. Most existing deep learning approaches integrate image and tabular data naïvely by concatenating the latent image representation with the tabular data in the final layers of the network [6, 10, 18, 21, 23, 26]. In such networks, the image and tabular parts have only minimal interaction and are limited in the way one part can inform the other. To enable the network to truly view image information in the context of the tabular information, and vice versa, it is necessary to increase the network's capacity and interweave both sources of information.

We propose to increase a CNN's capacity to fuse information from a patient's 3D brain MRI and tabular data via the Dynamic Affine Feature Map Transform (DAFT). DAFT is a generic module that can be integrated into any CNN architecture that establishes a two-way exchange of information between high-level concepts learned from the 3D image and the tabular biomarkers. DAFT uses an auxiliary neural network to dynamically incite or repress each feature map of a convolutional layer conditional on both image *and* tabular information. In our experiments on AD diagnosis and time-to-dementia prediction, we show that DAFT leads to superior predictive performance than using image or tabular data alone, and that it outperforms previous approaches that combine image and tabular data in a single neural network by a large margin.

2   Related Work

A naïve approach to combine image and tabular data is to first train a CNN on the image data, and use its prediction (or latent representation) together with tabular data in a second, usually linear, model. This way, the authors of [19] combined regions of interest extracted from brain MRI with routine clinical markers to predict progression to AD. Since image descriptors are learned independently of the clinical markers, descriptors can capture redundant information, such as a patient's age, instead of complementing it. This is alleviated when using a single network that concatenates the clinical information with the latent image representation prior to the last fully connected (FC) layer, which has been done in [10] with histopathology images, genomic data, and demographics for survival prediction, and in [18, 26] with hippocampus shape and clinical markers for time-to-dementia prediction. The disadvantage of this approach is that tabular data only contributes to the final prediction linearly. If concatenation is followed by a multilayer perceptron (MLP), rather than a single FC layer, non-linear relationships between image and tabular data can be captured. This was applied by the authors of [23] to learn from digital pathology images and genomic data, and of [6,21] to learn from brain MRI and clinical markers for AD diagnosis. Closely related to the above, the authors of [5,20,27] use an MLP on the tabular data before concatenation, and on the combined representation after concatenation. However, both approaches are restricted to interactions between the global image descriptor and tabular data and do not support fine-grained interactions.

In contrast, Duanmu et al. [3] fused information in a multiplicative manner for predicting response to chemotherapy. They use an auxiliary network that takes the tabular data and outputs a scalar weight for each feature map of every other convolutional layer of their CNN. Thus, a patient's tabular data can amplify or repress the contribution of image-derived latent representations at multiple levels. The downside of their approach is that the number of weights in the auxiliary network scales quadratically with the depth of the CNN, which quickly becomes impracticable. The Feature-wise Linear Modulation (FiLM) layer, used in visual question-answering, is the most similar to our approach [25]. FiLM has an auxiliary network that takes the text of the question and outputs an affine transformation to scale and shift each feature map of a convolutional layer. In the medical domain, the only approach based on FiLM is for image segmentation to account for lesion size or cardiac cycle phase [14]. In contrast to the above, we focus on disease prediction and utilize tabular information that is complementary to the image information, rather than describing image contents or semantics. Moreover, our proposed DAFT scales and shifts feature maps of a convolutional layer conditional on both image *and* tabular data.

3   Methods

We are seeking a CNN that utilizes high-dimensional 3D image information and seamlessly accounts for complementary low-dimensional tabular information in its predictions. We use a ResNet [11] architecture and achieve tight integration of both sources of information by dynamically scaling and shifting the feature maps of a 3D convolutional layer, conditional on a patient's image and clinical tabular information. Since tabular information often comprises demographics and summary measures that describe the patient's state as a whole, we require a level exchange of information between tabular and image data. Therefore, we propose to affinely transform the output of a convolutional layer in the last residual block, which is able to describe the image in terms of high-level concepts rather than primitive concepts, such as edges. Figure 1 summarizes our network.

For the i-th instance in the dataset, we let $\mathbf{x}_i \in \mathbb{R}^P$ denote the tabular clinical information, and $\mathbf{F}_{i,c} \in \mathbb{R}^{D \times H \times W}$ denote the c-th output (feature map) of a convolutional layer based on the i-th volumetric image ($c \in \{1,\ldots,C\}$). We propose to incite or repress high-level concepts learned from the image, conditional on the image and tabular data. To this end, we learn the Dynamic Affine Feature Map Transform (DAFT), with scale $\alpha_{i,c}$ and offset $\beta_{i,c}$:

$$\mathbf{F}'_{i,c} = \alpha_{i,c}\mathbf{F}_{i,c} + \beta_{i,c}, \qquad \alpha_{i,c} = f_c(\mathbf{F}_{i,c}, \mathbf{x}_i), \qquad \beta_{i,c} = g_c(\mathbf{F}_{i,c}, \mathbf{x}_i), \qquad (1)$$

where f_c, g_c are arbitrary functions that map the image and tabular data to a scalar. We model f_c, g_c by a single auxiliary neural network h_c that outputs

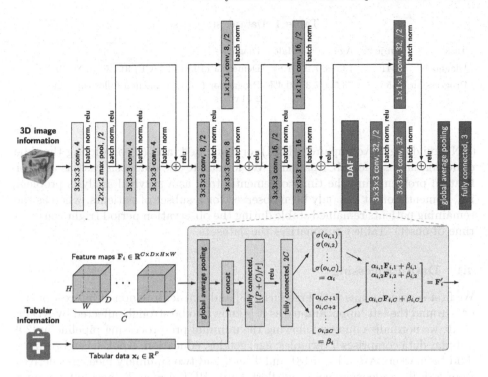

Fig. 1. Our proposed network architecture with the Dynamic Affine Feature Map Transform (DAFT) in the last residual block. DAFT combines $\mathbf{F}_i$, a $C \times D \times H \times W$ tensor of C feature maps from a convolutional layer, and $\mathbf{x}_i \in \mathbb{R}^P$, a vector of tabular data, to affinely transform the C feature maps via scales $\boldsymbol{\alpha}_i$ and shifts $\boldsymbol{\beta}_i$, where $r = 7$.

one $\boldsymbol{\alpha}$, $\boldsymbol{\beta}$ pair, which we refer to as DAFT (see Fig. 1). DAFT first creates a bottleneck by global average pooling of the image feature map, concatenating the tabular data, and squeezing the combined vector by a factor r via an FC layer. Next, both vectors are concatenated and fed to an MLP that outputs vectors $\boldsymbol{\alpha}_i$ and $\boldsymbol{\beta}_i$; following [13], all FC layers do not have bias terms. Note that we only apply an activation function $\sigma(\cdot)$ to the scale $\boldsymbol{\alpha}$, but not the offset $\boldsymbol{\beta}$ – we explore linear, sigmoid, and tanh activations in our experiments. The proposed DAFT is computationally efficient, because it does not depend on the number of instances in the dataset, nor the spatial resolution of the feature maps. DAFT can dynamically produce a scale factor $\alpha_{i,c}$ and offset $\beta_{i,c}$ conditional on a patient's specific image and tabular information due to parameter sharing. Finally, note that our principal idea of DAFT is not restricted to the CNN in Fig. 1, but can be integrated in any type of CNN.

4 Experiments

We evaluate DAFT on two tasks using T1 brain MRI from the Alzheimer's Disease Neuroimaging Initiative [15]: (i) diagnosing patients as cognitively normal

Table 1. Dataset statistics.

Task	Subjects	Age	Male	Diagnosis
Diagnosis	1341	73.9 ± 7.2	51.8%	Dementia (19.6%), MCI (40.1%), CN (40.3%)
Progression	755	73.5 ± 7.3	60.4%	Progressor (37.4%), median follow-up time 2.01 years

(CN), mild cognitive impaired (MCI), or demented, and (ii) predicting the time of dementia onset for MCI patients. We formulate the diagnosis task as a classification problem, and the time-to-dementia task as a survival analysis problem, i.e., dementia onset has only been observed for a subset of patients, whereas the remaining patients remained stable during the observation period (right censored time of onset). Table 1 summarizes the datasets.

4.1   Data Processing

We first segment scans with FreeSurfer [8] and extract a region of interest of size 64^3 around the left hippocampus, as it is known to be strongly affected by AD [9]. Next, we normalize images following the minimal pre-processing pipeline in [29]. Tabular data comprises 9 variables: age, gender, education, ApoE4, cerebrospinal fluid biomarkers $A\beta_{42}$, P-tau181 and T-tau, and two summary measures derived from 18F-fluorodeoxyglucose and florbetapir PET scans. To account for missing values, we adopt an approach similar to [17] by appending binary variables indicating missingness for all features, except age, gender, and education which are always present. This allows the network to use incomplete data and learn from missingness patterns. In total, tabular data comprises $P = 15$ features. To avoid data leakage due to confounding effects of age and sex [29], data is split into 5 non-overlapping folds using only baseline visits such that diagnosis, age and sex are balanced across folds [12]. We use one fold as test set and combine the remaining folds such that 80% of it comprise the training set and 20% the validation set. For the diagnosis task, we extend the training set, but not validation or test, by including each patient's longitudinal data (3.49 ± 2.56 visits per patient). For the time-to-dementia task, we only include patients with MCI at baseline and exclude patients with bi-directional change in diagnosis such that all patients remain MCI or progress to dementia.

4.2   Evaluation Scheme

We consider two unimodal baselines: (i) a ResNet [11] that only uses the image information based on the architecture in Fig. 1, but without DAFT, and (ii) a linear model using only the tabular information.[1] Moreover, we fit a separate linear model where the latent image representation from the aforementioned ResNet is combined with the tabular data (Linear model/w ResNet features). The linear

---

[1] We explored gradient boosted models too, but did not observe any advantage.

Table 2. Predictive performance for the diagnosis task (columns 4–5) and time-to-dementia task (columns 6–7). Values are mean and standard deviation across 5 folds. Higher values are better. I indicates the use of image data, T of tabular data, with L/NL denoting a linear/non-linear model.

	I	T	Balanced accuracy		Concordance index	
			Validation	Testing	Validation	Testing
Linear model	✗	L	0.571 ± 0.024	0.552 ± 0.020	0.726 ± 0.040	0.719 ± 0.077
ResNet	✓	–	0.568 ± 0.015	0.504 ± 0.016	0.669 ± 0.032	0.599 ± 0.054
Linear model /w ResNet features	✓	L	0.585 ± 0.050	0.559 ± 0.053	0.743 ± 0.026	0.693 ± 0.044
Concat-1FC	✓	L	0.630 ± 0.043	0.587 ± 0.045	0.755 ± 0.025	0.729 ± 0.086
Concat-2FC	✓	NL	0.633 ± 0.036	0.576 ± 0.036	0.769 ± 0.026	0.725 ± 0.039
1FC-Concat-1FC	✓	NL	0.632 ± 0.020	0.591 ± 0.024	0.759 ± 0.035	0.723 ± 0.056
Duanmu et al. [3]	✓	NL	0.634 ± 0.015	0.578 ± 0.019	0.733 ± 0.031	0.706 ± 0.086
FiLM [25]	✓	NL	0.652 ± 0.033	0.601 ± 0.036	0.750 ± 0.025	0.712 ± 0.060
DAFT	✓	NL	0.642 ± 0.012	**0.622 ± 0.044**	0.753 ± 0.024	**0.748 ± 0.045**

Table 3. Training time for one epoch using a NVIDIA GeForce GTX 1080 Ti GPU.

ResNet	Concat-1FC	Concat-2FC	1FC-Concat-1FC	Duanmu et al. [3]	FiLM [25]	DAFT
8.9 s	8.9 s	8.9 s	8.9 s	9.0 s	8.7 s	9.0 s

model is a logistic regression model for diagnosis, and Cox's model for time-to-dementia prediction [2]. As baselines, we evaluate three concatenation-based networks with the same ResNet backbone as in Fig. 1. In Concat-1FC, the tabular data is concatenated with the latent image feature vector and fed directly to the final classification layer. Thus, it models tabular data linearly – identical to the linear model baseline – but additionally learns an image descriptor, as in [10,18,26]. In Concat-2FC, the concatenated vector is fed to an FC bottleneck layer prior to the classification layer, similar to [6,21]. Inspired by [27], 1FC-Concat-1FC feeds the tabular information to an FC bottleneck layer before concatenating it with the latent image representation, as in Concat-1FC. In addition, we evaluate the network introduced by Duanmu et al. [3], and the FiLM layer [25], originally proposed for visual question answering, in place of DAFT. Our implementation of the DAFT and competing methods is available at https://github.com/ai-med/DAFT.

The networks Concat-2FC, 1FC-Concat-1FC, FiLM and DAFT, contain a bottleneck layer, which we set to 4 dimensions, which is roughly one fourth of the number of tabular features. For FiLM and DAFT, we use the identity function $\sigma(x) = x$ in the auxiliary network for the scale $\alpha_{i,c}$. In the diagnosis task, we minimize the cross-entropy loss. For progression analysis, we account for right censored progression times by minimizing the negative partial log likelihood of Cox's model [7]. We use the AdamW optimizer for both tasks [22]. We train for 30 and 80 epochs in the diagnosis and progression task respectively, and shrink the initial learning rate by 10 when 60% has been completed and by 20 when 90% has been completed. For each network, we optimize learning rate and weight decay on the validation set using grid search with a total of 5×3 configurations.[2] We report the performance on the test set with respect to the best performing model on the validation set. For diagnosis, we use the balanced accuracy (bACC [1]) to account for class imbalance, and for time-to-dementia analysis, we use an inverse probability of censoring weighted estimator of the concordance index (c-index [28]), which is identical to the area under the receiver operating characteristics curve if the outcome is binary and no censoring is present.

5   Results

Predictive Performance. Table 2 summarizes the predictive performance of all models. For both tasks, we observe that the linear model using only the tabular data outperforms the ResNet using only the image data. This is expected as tabular data comprises amyloid-specific measures derived from cerebrospinal fluid and PET imaging that are known to become abnormal before changes in MRI are visible [16]. Moreover, when learning image descriptors independently of the tabular clinical data and combining both subsequently, the predictive performance does not increase significantly (third row). This suggests that the learned image descriptor is not complementing the clinical information. In the diagnosis task, all Concat networks were successful in extracting complementary image information, leading to an increase in the average bACC by at least 0.024. In the time-to-dementia task, the improvement in c-index is at most 0.01 over the linear model, and when accounting for the variance, the improvement must be considered insignificant. The network by Duanmu et al. [3] generally performs worse than the Concat networks and is outperformed by the linear model on the time-to-dementia task too (0.013 higher mean c-index). The FiLM-based network has a marginal lead over all Concat networks on the diagnosis task (0.011 higher mean bACC), but falls behind on the time-to-dementia task (0.013 lower mean c-index). These results clearly demonstrate that Concat approaches cannot achieve the level of integration to fully utilize the complementary nature of image and tabular information, and that integrating tabular data with both low and high-level descriptors of the image, as done by Duanmu et al. [3], can severely deteriorate performance. Our proposed DAFT network is the only approach that excels at integrating image and tabular data for both tasks by outperforming competing methods by a large margin (0.021 higher bACC, 0.019 higher c-index). Finally, Table 3 summarizes the training times of networks, which shows that the runtime increase due to DAFT is negligible.

Ablation Study. To better understand under which settings DAFT can best integrate tabular data, we perform an ablation study with respect to (i) its location within the last ResBlock, (ii) the activation function σ for the scale $\alpha_{i,c}$, and (iii) whether one of scale and offset is sufficient. Following [25], we turn the

---

[2] Learning rate $\in \{0.03, 0.013, 0.0055, 0.0023, 10^{-3}\}$, weight decay $\in \{0, 10^{-4}, 10^{-2}\}$.

Table 4. Test set performance for different configurations. The proposed configuration (last row) uses DAFT before the first convolution with shift, scale, and $\sigma(x) = x$.

Configuration	Balanced accuracy	Concordance index
Before Last ResBlock	0.598 ± 0.038	0.749 ± 0.052
Before Identity-Conv	0.616 ± 0.018	0.745 ± 0.036
Before 1st ReLU	0.622 ± 0.024	0.713 ± 0.085
Before 2nd Conv	0.612 ± 0.034	0.759 ± 0.052
$\alpha_i = 1$	0.581 ± 0.053	0.743 ± 0.015
$\beta_i = 0$	0.609 ± 0.024	0.746 ± 0.057
$\sigma(x) = \text{sigmoid}(x)$	0.600 ± 0.025	0.756 ± 0.064
$\sigma(x) = \tanh(x)$	0.600 ± 0.025	0.770 ± 0.047
Proposed	0.622 ± 0.044	0.748 ± 0.045

parameters of batch normalization layers that immediately precede the DAFT off. From Table 4, we observe that DAFT is relatively robust to the choice of location. Importantly, DAFT outperforms all Concat networks on the diagnosis task, irrespective of its location. For the progression task, only the location before the first ReLU leads to a performance loss. Regarding the type of transformation, the results for diagnosing show that scaling is more essential than shifting, but removing any of them comes with a decisive performance drop of at least 0.013 bACC. For progression analysis, the DAFT's capacity seems to be sufficient if one of them is present. Finally, leaving the scale parameter unconstrained, as in the proposed configuration, is clearly beneficial for diagnosis, but for progression analysis constraining the scale leads to an increase in mean c-index. Only two configurations of DAFT are outperformed by Concat baselines, which highlights its robustness. Moreover, the optimal configuration differs between tasks, hence, further gains are possible when optimizing these choices for a given task.

Impact of α, β. To compare the impact of α and β, we run test time ablations on the fully trained models from the diagnosis task by modifying α or β during inference. Figure 2 (left) shows that the range of α and β is higher for DAFT than for FiLM, which uses α and β values close to zero. DAFT expresses a more dynamic behavior than FiLM, which could explain its performance gain. Next, we remove the conditioning information from either α or β by replacing it with its mean across the training set (Fig. 2, center). The larger difference when removing conditioning from β suggests that DAFT is more effective in integrating tabular information to shift feature maps, whereas FiLM is overall less effective and depends on scaling and shifting. This is also supported by our third test time ablation, where we add Gaussian noise to α or β (Fig. 2, right). For DAFT, the performance loss is larger when distorting β, whereas it is equal for FiLM. In addition, DAFT seems in general more robust to inaccurate α or β than FiLM.

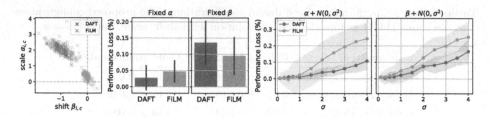

Fig. 2. Left: Scatter plot of $\alpha_{i,c}$ and $\beta_{i,c}$ for one feature map. Middle: Performance loss when setting $\boldsymbol{\alpha}$ or $\boldsymbol{\beta}$ to its mean. Right: Performance loss when distorting $\boldsymbol{\alpha}$ or $\boldsymbol{\beta}$.

6   Conclusion

Brain MRI can only capture a facet of the underlying dementia-causing changes and other sources of information such as patient demographics, laboratory measurements, and genetics are required to see the MRI in the right context. Previous methods often focus on extracting image information via deep neural networks, but naïvely account for other sources in the form of tabular data via concatenation, which results in minimal exchange of information between image- and tabular-related parts of the network. We proposed the Dynamic Affine Feature Map Transform (DAFT) to incite or repress high-level concepts learned from a 3D image, conditional on both image and tabular information. Our experiments on Alzheimer's disease diagnosis and time-to-dementia prediction showed that DAFT outperforms previous deep learning approaches that combine image and tabular data. Overall, our results support the case that DAFT is a versatile approach to integrating image and tabular data that is likely applicable to many medical data analysis tasks outside of dementia too.

Acknowledgements. This research was supported by the Bavarian State Ministry of Science and the Arts and coordinated by the Bavarian Research Institute for Digital Transformation, and the Federal Ministry of Education and Research in the call for Computational Life Sciences (DeepMentia, 031L0200A).

References

1. Brodersen, K.H., Ong, C.S., Stephan, K.E., Buhmann, J.M.: The balanced accuracy and its posterior distribution. In: 20th International Conference on Pattern Recognition, pp. 3121–3124 (2010)
2. Cox, D.R.: Regression models and life tables (with discussion). J. R. Stat. Soc. Ser. B (Stat. Methodol.) **34**, 187–220 (1972)
3. Duanmu, H., et al.: Prediction of pathological complete response to neoadjuvant chemotherapy in breast cancer using deep learning with integrative imaging, molecular and demographic data. In: Martel, A.L., et al. (eds.) MICCAI 2020. LNCS, vol. 12262, pp. 242–252. Springer, Cham (2020). https://doi.org/10.1007/978-3-030-59713-9_24

4. Ebrahimighahnavieh, M.A., Luo, S., Chiong, R.: Deep learning to detect Alzheimer's disease from neuroimaging: a systematic literature review. Comput. Meth. Program. Biomed. **187**, 105242 (2020)
5. El-Sappagh, S., Abuhmed, T., Islam, S.M.R., Kwak, K.S.: Multimodal multitask deep learning model for Alzheimer's disease progression detection based on time series data. Neurocomputing **412**, 197–215 (2020)
6. Esmaeilzadeh, S., Belivanis, D.I., Pohl, K.M., Adeli, E.: End-to-end Alzheimer's disease diagnosis and biomarker identification. In: Machine Learning in Medical Imaging, pp. 337–345 (2018)
7. Faraggi, D., Simon, R.: A neural network model for survival data. Stat. Med. **14**(1), 73–82 (1995)
8. Fischl, B.: FreeSurfer. Neuroimage **62**(2), 774–781 (2012)
9. Frisoni, G.B., Ganzola, R., Canu, E., Rub, U., Pizzini, F.B., et al.: Mapping local hippocampal changes in Alzheimer's disease and normal ageing with MRI at 3 Tesla. Brain **131**(12), 3266–3276 (2008)
10. Hao, J., Kosaraju, S.C., Tsaku, N.Z., Song, D.H., Kang, M.: PAGE-Net: interpretable and integrative deep learning for survival analysis using histopathological images and genomic data. Biocomputing **2020**, 355–366 (2019)
11. He, K., Zhang, X., Ren, S., Sun, J.: Deep residual learning for image recognition. In: CVPR, pp. 770–778 (2016)
12. Ho, D.E., Imai, K., King, G., Stuart, E.A.: Matching as nonparametric preprocessing for reducing model dependence in parametric causal inference. Polit. Anal. **15**(3), 199–236 (2007)
13. Hu, J., Shen, L., Albanie, S., Sun, G., Wu, E.: Squeeze-and-excitation networks. IEEE Trans. Pattern Anal. Mach. Intell. **42**(8), 2011 2023 (2020)
14. Jacenków, G., O'Neil, A.Q., Mohr, B., Tsaftaris, S.A.: INSIDE: steering spatial attention with non-imaging information in CNNs. In: Martel, A.L., et al. (eds.) MICCAI 2020. LNCS, vol. 12264, pp. 385–395. Springer, Cham (2020). https://doi.org/10.1007/978-3-030-59719-1_38
15. Jack, C.R., et al.: The Alzheimer's disease neuroimaging initiative (ADNI): MRI methods. J. Magn. Reson. Imaging **27**(4), 685–691 (2008)
16. Jack, C.R., Knopman, D.S., Jagust, W.J., Petersen, R.C., Weiner, M.W., et al.: Tracking pathophysiological processes in Alzheimer's disease: an updated hypothetical model of dynamic biomarkers. Lancet Neurol. **12**(2), 207–216 (2013)
17. Jarrett, D., Yoon, J., van der Schaar, M.: Dynamic prediction in clinical survival analysis using temporal convolutional networks. IEEE J. Biomed. Health Inform. **24**(2), 424–436 (2020)
18. Kopper, P., Pölsterl, S., Wachinger, C., Bischl, B., Bender, A., Rügamer, D.: Semi-structured deep piecewise exponential models. In: Proceedings of AAAI Spring Symposium on Survival Prediction - Algorithms, Challenges, and Applications 2021, vol. 146, pp. 40–53 (2021)
19. Li, H., Habes, M., Wolk, D.A., Fan, Y.: A deep learning model for early prediction of Alzheimer's disease dementia based on hippocampal magnetic resonance imaging data. Alzheimer's Dementia **15**(8), 1059–1070 (2019)
20. Li, S., Shi, H., Sui, D., Hao, A., Qin, H.: A novel pathological images and genomic data fusion framework for breast cancer survival prediction. In: International Conference of the IEEE Engineering in Medicine & Biology Society (EMBC), pp. 1384–1387 (2020)
21. Liu, M., Zhang, J., Adeli, E., Shen, D.: Joint classification and regression via deep multi-task multi-channel learning for Alzheimer's disease diagnosis. IEEE Trans. Biomed. Eng. **66**(5), 1195–1206 (2019)

22. Loshchilov, I., Hutter, F.: Decoupled weight decay regularization. In: 7th International Conference on Learning Representations (2019)
23. Mobadersany, P., Yousefi, S., Amgad, M., Gutman, D.A., et al.: Predicting cancer outcomes from histology and genomics using convolutional networks. Proc. Natl. Acad. Sci. **115**(13), E2970–E2979 (2018)
24. Pelka, O., Friedrich, C.M., Nensa, F., Mönninghoff, C., Bloch, L., et al.: Sociodemographic data and APOE-e4 augmentation for MRI-based detection of amnestic mild cognitive impairment using deep learning systems. PLOS ONE **15**(9), e0236868 (2020)
25. Perez, E., Strub, F., de Vries, H., Dumoulin, V., Courville, A.: FiLM: visual reasoning with a general conditioning layer. In: AAAI, vol. 32 (2018)
26. Pölsterl, S., Sarasua, I., Gutiérrez-Becker, B., Wachinger, C.: A wide and deep neural network for survival analysis from anatomical shape and tabular clinical data. In: Machine Learning and Knowledge Discovery in Databases, pp. 453–464 (2020)
27. Spasov, S., Passamonti, L., Duggento, A., Liò, P., Toschi, N.: A parameter-efficient deep learning approach to predict conversion from mild cognitive impairment to Alzheimer's disease. Neuroimage **189**, 276–287 (2019)
28. Uno, H., Cai, T., Pencina, M.J., D'Agostino, R.B., Wei, L.J.: On the C-statistics for evaluating overall adequacy of risk prediction procedures with censored survival data. Stat. Med. **30**(10), 1105–1117 (2011)
29. Wen, J., Thibeau-Sutre, E., Diaz-Melo, M., Samper-González, J., et al.: Convolutional neural networks for classification of Alzheimer's disease: overview and reproducible evaluation. Med. Image Anal. **63**, 101694 (2020)

Image-Derived Phenotype Extraction for Genetic Discovery via Unsupervised Deep Learning in CMR Images

Rodrigo Bonazzola[1,2]([✉]), Nishant Ravikumar[1,2], Rahman Attar[3],
Enzo Ferrante[4], Tanveer Syeda-Mahmood[5], and Alejandro F. Frangi[1,2]

[1] Centre for Computational Imaging and Simulation Technologies in Biomedicine
(CISTIB), Schools of Computing and Medicine, University of Leeds, Leeds, UK
[2] Leeds Institute of Cardiovascular and Metabolic Medicine, School of Medicine,
University of Leeds, Leeds, UK
scrb@leeds.ac.uk
[3] Department of Bioengineering, Faculty of Engineering, Imperial College London,
London, UK
[4] Research Institute for Signals, Systems and Computational Intelligence, sinc(i),
FICH-UNL/CONICET, Santa Fe, Argentina
[5] IBM Almaden Research Center, San Jose, USA

Abstract. Prospective studies with linked image and genetic data, such
as the UK Biobank (UKB), provide an unprecedented opportunity to
extract knowledge on the genetic basis of image-derived phenotypes.
However, the extent of phenotypes tested within so-called genome-wide
association studies (GWAS) is usually limited to handcrafted features,
where the main limitation to proceed otherwise is the high dimensionality of both the imaging and genetic data. Here, we propose an approach where the phenotyping is performed in an unsupervised manner,
via autoencoders that operate on image-derived 3D meshes. Therefore,
the latent variables produced by the encoder condense the information
related to the geometry of the biologic structure of interest. The network's training proceeds in two steps: the first is genotype-agnostic and
the second enforces an association with a set of genetic markers selected
via GWAS on the intermediate latent representation. This genotype-
dependent optimisation procedure allows the refinement of the phenotypes produced by the autoencoder to better understand the effect of
the genetic markers encountered. We tested and validated our proposed
method on left-ventricular meshes derived from cardiovascular magnetic
resonance images from the UKB, leading to the discovery of novel genetic
associations that, to the best of our knowledge, had not been yet reported
in the literature on cardiac phenotypes.

1 Introduction

The emergence of population-scale prospective studies with linked imaging and
genetic data, such as the UK Biobank [7], has enabled research into the genetics

© Springer Nature Switzerland AG 2021
M. de Bruijne et al. (Eds.): MICCAI 2021, LNCS 12905, pp. 699–708, 2021.
https://doi.org/10.1007/978-3-030-87240-3_67

of image-derived phenotypes, a field called imaging genetics. One of the main challenges of this field is the high dimensionality of both the imaging and genetic datasets. This problem is usually addressed on the imaging side by deriving hand-crafted features from the images, based on prior expert knowledge supporting their clinical relevance. For example, in the case of cardiac images, these phenotypes could be the volumes of the different cardiac chambers, the myocardial mass, or functional parameters such as the ejection fraction.

In this work, we propose a different approach, based on unsupervised learning, to extract phenotypes from image-derived 3D meshes which describe the organs of interest's geometry. This approach is outlined in Fig. 1.

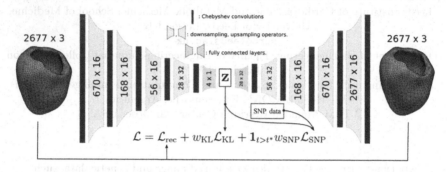

Fig. 1. Scheme of the encoder-decoder network implemented in this work. $\mathbf{1}$ is the indicator function. The numbers $V \times C$ within the downsampling and upsampling operators represent the dimension of their output, where V is the number of vertices and C is the number of channels.

Its input consists of a set of 3D meshes representing the geometry of the organs of interest, obtained by previous segmentation of the images. A graph-convolutional autoencoder is trained for t^* epochs to perform dimensionality reduction on the 3D meshes, without any genetic data being input to the network. Each component of the latent representation found by the autoencoder, which will encode the modes of variation of the set of meshes, are employed as a phenotype in a genome-wide association study (GWAS), where a reduced set of candidate loci are chosen based on a significance criterion. Then, the autoencoder is further trained adding to the cost function a term that enforces an association between each of the chosen loci and the corresponding latent variables; this step aims to produce a fine-tuning of the phenotype. Finally, the latent representation is tested in a GWAS using an independent set. We applied this approach to cardiovascular magnetic resonance (CMR) images from the UK Biobank, where the left ventricle (LV) at end-diastole was our object of study.

Our contributions are, therefore, two-fold: on one hand, we propose an approach to study the genetic basis of image-derived phenotypes, based on unsupervised deep learning; on the other hand, we discover novel a genetic association that has not been previously reported in the literature on cardiac phenotypes, to the best of our knowledge.

1.1   Related Work

Other studies using unsupervised approaches to derive phenotypes have been performed. Studies on the genetic basis of 3D-mesh-derived facial features have been published [8], which used hierarchical clustering to obtain distinctive regions of the face. Each of these regions was projected onto a linear space using principal component analysis (PCA), and canonical correlation analysis (CCA) was used to find linear combinations of region-specific PCs that are maximally correlated with each of the single nucleotide polymorphisms (SNPs) tested.

On the other hand, several studies have been published in the recent years in the field of cardiac imaging genetics using CMR data, which focused on hand-crafted phenotypes. In [3], the authors investigate left-ventricular wall thickness at end-diastole, performing and association test with a set of genetic variants in a vertex-by-vertex fashion. Other studies ([2] and [12]) perform GWAS on global LV phenotypes: chamber volume at end-diastole and end-systole, stroke volume, ejection fraction and myocardial mass.

Also, work has been performed on CMR images to extract biomarkers using unsupervised dimensionality reduction approaches [4]. Still, this work has not relied on an intermediate 3D mesh representation of the cardiac chambers, nor has been used for the purpose of genetic discovery.

2   Methods

2.1   Description of the Data

All the data used for this work comes from the UK Biobank project, data accession number 11350.

Cardiovascular Magnetic Resonance (CMR) Data. The CMR imaging protocol used to obtain the raw imaging data is described elsewhere [11]. For a given individual and time point, this data consists of a stack of 10–12 short-axis view slices (SAX) along with three long-axis view (LAX) slices. The cardiac segmentation algorithm utilised is described in detail in [1]. This algorithm produces as output a set of registered meshes, i.e. meshes with the same number of vertices and the same connectivity. For LV, the meshes encompass the endocardial and epicardial surfaces. As mentioned before, in this work we studied only end-diastole.

The LV mesh for subject i, $i = 1, ..., N$, can then be represented as pairs $(\mathbf{S}_i, A)$, where $\mathbf{S}_i = \left[\, x_{i1}\, y_{i1}\, z_{i1} \mid ... \mid x_{iM}\, y_{iM}\, z_{iM} \,\right] \in \mathbb{R}^{M \times 3}$ is the shape and A is the adjacency matrix of the mesh. The number of individuals and mesh vertices are $N = 29051$ and $M = 2677$, respectively.

Genotype Data. SNP microarray data is available for all the individuals in the UK Biobank cohort. This microarray covers $\sim$800k genetic variants including SNPs and short indels. The design of this microarray has been described in detail in [6]. An augmented set of ~9.5M variants was obtained from these genotyped markers through imputation, after filtering by a minor allele frequency (MAF) threshold of 1%, a Hardy-Weinberg equilibrium p-value threshold of 10^{-5} and an imputation info score of 0.3. Also, only the autosomes (chromosomes 1 through 22) were used. .

2.2   Graph-Convolutional Autoencoder

To perform dimensionality reduction, we propose using an encoder-decoder approach. The encoder E consists of convolutional and pooling layers, whereas D consists of unpooling layers. To leverage the topology of the mesh, we utilise graph-convolutional layers. Since the vertices are not in a rectangular grid, the usual convolution, pooling and unpooling operations defined for such geometry are not adequate for this task and need to be suitably generalised. There are several methods to achieve this, but they all can be classified into two large groups: spatial or spectral [14]. In this work we applied a method belonging to the latter category, which relies on expressing the features in the Fourier basis of the graph, as will be explained below after providing some background of spectral graph theory.

The Laplace-Beltrami operator of a graph with adjacency matrix A is defined as $L = D - A$, where D is the degree matrix, i.e. a diagonal matrix where $D_{ii} = \sum_j A_{ij}$ is the number of edges connected to vertex i. The Fourier basis of the graph can be obtained by diagonalising the Laplace operator, $L = U^t \Lambda U$. The columns of U constitute the Fourier basis, and the operation of convolution $\star$ for a graph can be defined in the following manner

$$x \star y = U(U^t x \odot U^t y), \tag{1}$$

where $\odot$ is the element-wise product (also known as Hadamard product). All spectral methods for convolution rely on this definition, and differ from one another in the form of the kernel. In this work, a parameterisation proposed in [9] was used. The said method is based on the Chebyshev family of polynomials $\{T_i\}$. The kernel g_ξ is defined as:

$$g_\xi(L) = \sum_{i=1}^{K} \xi_i T_i(L), \tag{2}$$

where K is the highest degree of the polynomials considered (in this work $K = 6$). Chebyshev polynomials have the advantage of being computable recursively through the relation $T_i(x) = xT_{i-1}(x) - T_{i-2}(x)$ and the base cases $T_1(x) = 1$ and $T_2(x) = x$. It is also worth mentioning that the filter described by Eq. 2, despite its spectral formulation, has the characteristic of being local.

Following [13], each of the three spatial coordinates of each vertex are input as a separate channel of the autoencoder. Downsampling and upsampling operations used in this study are based on a surface simplification algorithm proposed in [10]. These operations are defined before training each layer, using a single template shape. Here we utilise the mean shape $\bar{\mathbf{S}} = (1/N)\sum_{i=1}^{N}\mathbf{S_i}$ as a template.

2.3   GWAS

According to the traditional GWAS scheme [5], we tested each genetic variant l, with dosage $X_l \in [0,2]$, for association with each of the LV latent features z_j through a univariate linear model $z_k = \beta_{lk}X_l + \epsilon_{lk}$, where ϵ_{lk} is the component not explained by the genotype, which we model as a normal random variable. The null hypothesis tested is that $\beta_{lk} = 0$. From linear regression, one obtains an estimate $\hat{\beta}_{lk}$ of the effect size β_{lk}, along with the standard error of this estimate, $\text{se}(\hat{\beta}_{lk})$. Finally, the p-value for the association can be computed from these values.

Before GWAS, the phenotypes (i.e. latent variables) were adjusted for a set of covariates: height, BMI, age, sex, diastolic and systolic blood pressure. For this, a multilinear model is used, and the new phenotypes are the residues obtained from there. Also, rank-based inverse normalisation is performed on these phenotypes so that the usual closed-form formulas for hypothesis testing can be utilised.

To avoid issues related to population stratification, only individuals with British ancestry were utilised, leading to the aforementioned sample size of $N = 29051$. No filtering was performed based on pathologies.

2.4   Proposed Algorithm

The proposed method is described in Algorithm 1.

Loss Function. The loss function $\mathcal{L}_1$ for the first training stage consists of two terms:

$$\mathcal{L}_1 = \mathcal{L}_{\text{rec}} + w_{\text{KL}}\mathcal{L}_{\text{KL}}, \tag{3}$$

where $\mathcal{L}_{\text{rec}}$ is the reconstruction loss and $\mathcal{L}_{\text{KL}}$ is the variational regularisation term, computed as the Kullback-Leibler divergence of the latent representation $\mathbf{z}$ with an isotropic normal distribution. For the second training stage, another term is added, $\mathcal{L}_{\text{SNP}}$, which encourages a stronger association between each of the latent variables k and a set of SNPs S_k:

$$\mathcal{L}_2 = \mathcal{L}_1 + w_{\text{SNP}}\mathcal{L}_{\text{SNP}} \tag{4}$$

$$\mathcal{L}_{\text{SNP}} = -\sum_k \sum_{l_k \in S_k} \text{Corr}(X_{l_k}, z_k) \tag{5}$$

During training, the correlation in Eq. 5 is computed as the sample correlation for each batch. The reference and alternative alleles are chosen so that the correlation is positive within the training set[1].

---

[1] Swapping the alleles corresponds to performing the transformation $X_l \mapsto 2 - X_l$.

Data: 3D meshes S_i and linked genotype dosages X_{il}
Result: Network weights, GWAS summary statistics.
Hyperparameters: network architecture, $w_{KL} > 0$, $w_{SNP} > 0$.
ProcrustesAlignment(S_i);
PartitionDataset(S_i, X_{il});
InitialiseWeights();
while *Stop criterion is not met* **do**
$\quad\mid$ Perform optimisation step with loss $\mathcal{L}_1$;
end
Select the best epoch within the validation set;
for *each latent variable k* **do**
$\quad\mid$ Perform GWAS within the training set;
$\quad\mid$ Extract set of genetic markers S_k based on a significance criterion;
$\quad\mid$ **for** *each SNP l_k in S_k* **do**
$\quad\quad\mid$ **if** $Corr(X_{l_k}, z_k) < 0$ **then**
$\quad\quad\quad\mid$ $X_{l_k} \mapsto 2 - X_{l_k}$
$\quad\quad\mid$ **end**
$\quad\mid$ **end**
end
while *Stop criterion is not met* **do**
$\quad\mid$ Perform optimisation step with loss $\mathcal{L}_2$;
end
Select the best epoch within the validation set;
Perform GWAS within the test set;

Algorithm 1: Workflow of the proposed method.

In this work, our criterion for SNP selection was passing the Bonferroni significance threshold (taken as the usual genome-wide threshold of 5×10^{-8} divided by the number of latent variables tested). As we explain below, this criterion led to a single term in the summation.

2.5   Implementation Details

The architecture of the network is detailed in Fig. 1. After each convolutional layer, a ReLU activation function was applied. Importantly, the latent representation's size $\dim(\mathbf{z})$ was chosen as 4. We found this number enough to capture the most salient global features explaining the variability in LV shape. The reconstruction loss employed was the vertex-wise mean squared error (MSE), averaged across the vertices of each mesh. An ablation study was performed to assess the impact of different parameters: learning rate γ, w_{KL} and w_{SNP}. γ was chosen as 10^{-3}, whereas for the first training stage $w_{KL} = 10^{-3}$. Batch size was 100. Also, in the second training stage and for each parameter configuration, experiments with different seeds were performed. These seeds controlled the partition of the full dataset into training, validation and test sets. The sizes of these sets were 5000, 1000 and 23051, respectively.

3   Results and Discussion

After the first training stage, GWAS was performed on each of the 4 components of $\mathbf{z}$. Only one of them, z_1, yielded a Bonferroni-significant association. We mention here that after applying shape PCA on the meshes, no significant association was found.

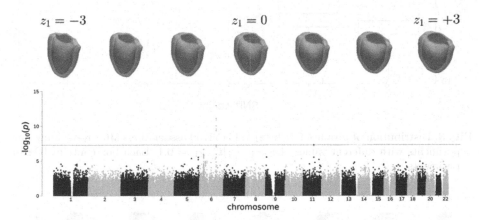

Fig. 2. Effect of latent variable z_1 (before re-training) and corresponding Manhattan plot.

Figure 2 shows both the morphological impact of changes in this variable, and the Manhattan plot displaying the GWAS p-values. The association lies on chromosome 6, and we mapped it to gene PLN with high confidence, based on the literature on the genetics of cardiac phenotypes.

As can also be determined visually, latent variable z_1 was found to correlate with LV sphericity index s. Spearman correlation between s and z_1 is 0.432 (computed across real meshes). The index s was calculated as $s = A_{\text{sph}}(V_{\text{CH}})/A_{\text{CH}}$, where $A_{\text{sph}}(V) = (36\pi V^2)^{\frac{1}{3}}$ is the surface area of a sphere of volume V, whereas A_{CH} and V_{CH} are the surface area of and the volume enclosed by the convex hull of the LV, respectively. To the best of our knowledge, this association between PLN and LV sphericity had not been previously reported.

For the second training stage, the effect of the $\mathcal{L}_{\text{SNP}}$ term on the strength of the genetic association was studied. $\mathcal{L}_{\text{SNP}}$ consists of a single term, corresponding to the leading SNP in the PLN locus, rs11153730. Results are displayed in Fig. 3. Each observation corresponds to a different experiment. In each experiment, the sets of training and testing samples varies according to the random seed, but their size is fixed. We confirmed that the additional training is beneficial for obtaining a better phenotype, i.e. a phenotype that yields a stronger association on the GWAS. It is worth noting that the effectiveness of this procedure was not obvious in advance, since individual SNPs' effect sizes on complex traits are generally small.

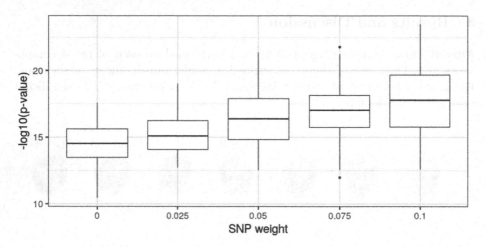

Fig. 3. Distribution of p-values for the z_1-rs11153730 associations after re-training, for experiments with different values of w_{SNP} with $w_{KL} = 0.1$. Each box contains 60 ± 10 experiments.

Fig. 4. Morphologic effect of latent variable z_1, after re-training. Color represents the deviations in shape with respect to the meshes from Fig. 2. (Color figure online)

We studied the change in shape modeled by the z_1 latent variable after retraining with $w_{SNP} = w_{KL} = 0.1$, for one particular experiment. This change is shown in Fig. 4. The w_{KL} coefficient was increased with respect to the first training stage to avoid overfitting of the $\mathcal{L}_{SNP}$ term.

4   Conclusions

We have proposed and validated an approach based on graph-convolutional autoencoders to extract phenotypes from 3D meshes of biological structures, for genetic discovery through GWAS; and applied it to CMR-derived left-ventricular meshes in the end-diastolic phase.

In particular, a genetic association was found between a genetic locus linked to PLN gene and a latent variable that correlates with LV sphericity. To the best of our knowledge, this had not been reported before, even though mutations PLN have been linked to dilated cardiomyopathy, a disease characterised by an increase in LV sphericity. Furthermore, we have shown that an additional stage of training that aims to refine the phenotype effectively improves the genetic association, and is therefore recommended.

On the other hand, we acknowledge that our method as presented here results in a loss of information as compared to the original CMR images, since it only captures the LV myocardial surface's geometry. However, this is not a drawback of the method itself, since texture information can be added seamlessly as node features (in addition to spatial coordinates). Besides, diseases could be investigated in this context by encouraging the latent representation to be discriminative of disease status, whenever diagnosis information is available. Finally, UKB provides images for the whole cardiac cycle, from which dynamic patterns could be extracted. These ideas are left as future work.

Acknowledgments. This work was funded by the following institutions: The Royal Academy of Engineering [grant number CiET1819\19] and EPSRC [TUSCA EP/V04799X/1] (R.B., N.R. and A.F.F.), The Royal Society through the International Exchanges 2020 Round 2 scheme [grant number IES\R2\202165] (R.B., E.F. and A.F.F). E.F. was also funded by ANPCyT [grant number PICT2018-3907] and UNL [grant numbers CAI+D 50220140100-084LI, 50620190100-145LI].

References

1. Attar, R., et al.: Quantitative CMR population imaging on 20,000 subjects of the UK Biobank imaging study: LV/RV quantification pipeline and its evaluation. Med. Image Anal. **56**, 26–42 (2019). https://doi.org/10.1016/j.media.2019.05.006
2. Aung, N., et al.: Genome-wide analysis of left ventricular image-derived phenotypes identifies fourteen loci associated with cardiac morphogenesis and heart failure development. Circulation **140**(16), 1318–1330 (2019). https://doi.org/10.1161/CIRCULATIONAHA.119.041161
3. Biffi, C., et al.: Three-dimensional cardiovascular imaging-genetics: a mass univariate framework. Bioinformatics **34**(1), 97–103 (2018). https://doi.org/10.1093/bioinformatics/btx552
4. Biffi, C., et al.: Learning interpretable anatomical features through deep generative models: application to cardiac remodeling. In: Frangi, A.F., Schnabel, J.A., Davatzikos, C., Alberola-López, C., Fichtinger, G. (eds.) MICCAI 2018. LNCS, vol. 11071, pp. 464–471. Springer, Cham (2018). https://doi.org/10.1007/978-3-030-00934-2_52
5. Bush, W.S., Moore, J.H.: Chapter 11: genome-wide association studies. PLOS Comput. Biol. **8**(12), 1–11 (2012). https://doi.org/10.1371/journal.pcbi.1002822
6. Bycroft, C., et al.: Genome-wide genetic data on ∼500,000 UK Biobank participants. bioRxiv (2017). https://doi.org/10.1101/166298
7. Bycroft, C., et al.: The UK Biobank resource with deep phenotyping and genomic data. Nature **562**(7726), 203–209 (2018). https://doi.org/10.1038/s41586-018-0579-z
8. Claes, P., et al.: Genome-wide mapping of global-to-local genetic effects on human facial shape. Nat. Genet. **50**(3), 414–423 (2018)
9. Defferrard, M., Bresson, X., Vandergheynst, P.: Convolutional neural networks on graphs with fast localized spectral filtering. In: Advances in Neural Information Processing Systems (2016). https://arxiv.org/abs/1606.09375

10. Garland, M., Heckbert, P.S.: Surface simplification using quadric error metrics. In: Proceedings of the 24th annual conference on Computer graphics and interactive techniques - SIGGRAPH 1997, pp. 209–216. ACM Press (1997). https://doi.org/10.1145/258734.258849

11. Petersen, S.E., et al.: Imaging in population science: cardiovascular magnetic resonance in 100,000 participants of UK Biobank - rationale, challenges and approaches. J. Cardiovasc. Magn. Reson. **15**(1), 46 (2013). https://doi.org/10.1186/1532-429X-15-46

12. Pirruccello, J.P., et al.: Analysis of cardiac magnetic resonance imaging in 36,000 individuals yields genetic insights into dilated cardiomyopathy. Nature Commun. **11**(1) (December 2020). https://doi.org/10.1038/s41467-020-15823-7

13. Ranjan, A., Bolkart, T., Sanyal, S., Black, M.J.: Generating 3D Faces Using Convolutional Mesh Autoencoders. In: Ferrari, V., Hebert, M., Sminchisescu, C., Weiss, Y. (eds.) Computer Vision - ECCV 2018, vol. 11207, pp. 725–741. Springer, Cham (2018). https://doi.org/10.1007/978-3-030-01219-9_43

14. Zhou, J., et al.: Graph neural networks: a review of methods and applications. arXiv preprint arXiv:1812.08434 (2018)

GKD: Semi-supervised Graph Knowledge Distillation for Graph-Independent Inference

Mahsa Ghorbani[1,2]([✉]), Mojtaba Bahrami[1,2], Anees Kazi[1],
Mahdieh Soleymani Baghshah[2], Hamid R. Rabiee[2], and Nassir Navab[1,3]

[1] Computer Aided Medical Procedures, Technical University of Munich,
Munich, Germany
mahsa.ghorbani@tum.de
[2] Sharif University of Technology, Tehran, Iran
mghorbani@ce.sharif.edu
[3] Whiting School of Engineering, Johns Hopkins University, Baltimore, USA

Abstract. The increased amount of multi-modal medical data has opened the opportunities to simultaneously process various modalities such as imaging and non-imaging data to gain a comprehensive insight into the disease prediction domain. Recent studies using Graph Convolutional Networks (GCNs) provide novel semi-supervised approaches for integrating heterogeneous modalities while investigating the patients' associations for disease prediction. However, when the meta-data used for graph construction is not available at inference time (e.g., coming from a distinct population), the conventional methods exhibit poor performance. To address this issue, we propose a novel semi-supervised approach named GKD based on the knowledge distillation. We train a teacher component that employs the label-propagation algorithm besides a deep neural network to benefit from the graph and non-graph modalities only in the training phase. The teacher component embeds all the available information into the soft pseudo-labels. The soft pseudo-labels are then used to train a deep student network for disease prediction of unseen test data for which the graph modality is unavailable. We perform our experiments on two public datasets for diagnosing Autism spectrum disorder, and Alzheimer's disease, along with a thorough analysis on synthetic multi-modal datasets. According to these experiments, GKD outperforms the previous graph-based deep learning methods in terms of accuracy, AUC, and Macro F1.

Keywords: Population-based disease prediction · Semi-supervised graph neural networks · Knowledge distillation

Electronic supplementary material The online version of this chapter (https://doi.org/10.1007/978-3-030-87240-3_68) contains supplementary material, which is available to authorized users.

M. de Bruijne et al. (Eds.): MICCAI 2021, LNCS 12905, pp. 709–718, 2021.
https://doi.org/10.1007/978-3-030-87240-3_68

1 Introduction

Acquisition and exploitation of the wide range of genetic, phenotypic, and behavioral information along with the imaging data increase the necessity of an integrated framework for analyzing medical imaging and non-imaging modalities [2,3,5,14]. On the other hand, with the success of conventional deep neural networks, there has been an extensive effort on utilizing them for multi-modal datasets [11,19,26]. Considering the relationships between the patients based on one or multiple important modalities is beneficial as it helps to analyze and study the similar cohort of patients together. By viewing the patients as nodes and their associations as edges, graphs provide a natural way of representing the interactions among a population [21,28]. Meanwhile, Graph Convolutional Networks (GCNs) provide deep network architectures for integrating node features and graph modalities of data and recently have been used for such population-level analysis [24]. In the non-medical domain, ChebyNet [7] was one of the first methods which extended the theory of graph signal processing into deep learning. Further, Kipf et al. [18] simplified the ChebyNet by introducing a method named GCN to perform semi-supervised classification on citation datasets. Parisot et al. [24] exploited GCN for the disease prediction problem with multi-modal datasets. Then, GCN is adopted in different forms for medical applications including disease classification [10,15,16], brain analysis [20], and mammogram analysis [9].

In practice, all the graph modalities are not always available for new or unseen patients. This can be due to missing observations or costly and time-consuming data collection for graph construction. Moreover, the test samples may not belong to the primary population with the same set of measured modalities. In such a setting, where the new coming samples are isolated nodes with no edges connecting them to the rest of the population graph, the conventional graph-based models might fail to generalize to new isolated samples. This is due to the reason that GCN-based methods smooth each node's features by averaging with its neighbors in each layer of the network, which makes the features of adjacent nodes close to each other and dissimilar to the farther ones [29]. This also makes it hard for the model, learned over the connected components, to perform as well on isolated or low-degree nodes with unsmoothed high-variance features. To address this limitation, a comprehensive approach is required to make the most use of the extra modality information among the training samples.

In order to overcome this challenge, inspired by the knowledge distillation technique [4,13], we propose a novel student-teacher method to leverage all modalities at training and operate independently from the graph at inference time. Our proposed method encapsulates all the graph-related data using Label-Propagation Algorithm (LPA) [30] into pseudo-labels produced by a so-called Teacher model. The pseudo-labels are then used to train a network (Student) that learns the mapping from high variance and noisy original input space to the output without filtering the features throughout the network. By doing so, the student network makes predictions without needing the presence of the graph modality on the future unseen data.

2 Knowledge Distillation Framework

Complex models or an ensemble of separately trained models can extract complicated structures from data but are not efficient for deployment. The knowledge of
a trained model is saved in its parameters, and transferring the learned weights
is not straightforward while changing the structure, form, and size of the target
model. However, a model's knowledge can be viewed as a mapping from its input
to the output. In a classification task, the relative magnitude of class probabilities of a trained model's output is a rich source of information and represents
the similarity between classes. This similarity measure gained by a teacher network can be used as a source of knowledge to train a smaller and more efficient
network (student). Caruana et al. [4] followed by Hinton et al. [13] show that the
student model generalizes in the same way as the teacher and is superior compared to being trained by hard targets from scratch. Inspired by this framework,
we utilize all the available modalities only during the learning procedure of the
teacher and then transfer the obtained knowledge into a student that does not
have any assumption regarding the graph's availability at inference time. Just
to note, the existing distillation methods are aimed at model compression, while
our method takes advantage of the teacher-student framework to exclude the
graph modality in test time that is a novel direction in literature.

3 Methods

We provide a detailed description of our method by explaining the problem
definition and notations, the teacher structure, and the student network that is
the final classifier. An overview of the method is available in Fig. 1.

Problem Definition and Notations: Assume that N patients are given. We
represent the subject interactions graph as $G(V, E, A)$, where V is the set of
patient vertices ($|V| = N$), $E \in (V \times V)$ is the set of edges and $A \in \mathbb{R}^{N \times N}$
shows the adjacency matrix. Let D be the diagonal degree matrix for A. We also
define the F-dimensional node features as $X \in \mathbb{R}^{N \times F}$. The state of patients is
described by the one-hot labels $Y \in \mathbb{R}^{N \times C}$ where C denotes the total number of
states (e.g., diseased or normal). According to our semi-supervised setting, the
training samples consist of two mutually exclusive sets of labeled samples (V_L)
and unlabeled ones (V_U) where $V = V_L \cup V_U$. The ground-truth labels Y_L are
only available for V_L. Finally, our objective is to train a teacher to predict the
class probabilities for unlabeled training nodes (V_U) and use them together with
labeled training nodes as Y^T to train a student network with cross-entropy loss
function and apply it on unseen test samples.

Teacher for Knowledge Integration: Here, we develop a mechanism to attain
a teacher using both the node features and graph interactions. To this end, we
first try to extract the available knowledge from the node features by training a
deep neural network $\mathcal{D}_\theta^T$, where θ shows the network's parameters. The teacher
network is trained by the labeled training pairs (X_L, Y_L). Afterward, the trained

network is applied to the rest of the unlabeled data (X_U) and predicts the best possible soft pseudo-labels $\hat{Y}_U$ for them. Together with the ground-truth labels Y_L, we initialize the graph G with $Y_L \cup \hat{Y}_U$ as node labels. Up to this point, each patient's node labels are predicted independently of their neighbors in the graph. To add the graph information to the node labels, we employ the well-known LPA to distribute the label information all around the graph. However, applying the original LPA might forget the primary predictions for each node and result in the over-emphasizing on the graph. We try to compensate for this issue by adding a Remembrance Term (RT) to LPA, which avoids forgetting the initial predictions. The k-th iteration of our modified LPA is as follow:

$$Y^{T^{(k)}} = (1 - \alpha)D^{-1}AY^{T^{(k-1)}} + \alpha \underbrace{Y^{T^0}}_{\substack{\text{Remembrance} \\ \text{Term (RT)}}} \; ; \; Y^{T^0} = Y_L \cup \hat{Y}_U, \tag{1}$$

$$Y_L^{T^{(k)}} = Y_L,$$

where Y^{T^0} is the initial predictions by the teacher network, and Y^{T^k} denotes the set of labels at k-th iteration. At each iteration, the labels of training nodes are replaced with their true labels (Y_L). The output of the last iteration is known as teacher soft pseudo-labels (Y^T). The effect of graph neighborhood and the remembrance term is adjustable by the coefficient α.

Student as Final Classifier for Inference Time: As described in distillation framework [13], the student network is trained to mimic the input-output mapping learned by the teacher network. Specifically, the student network is trained using the class-probabilities produced by the teacher (Y^T). For this purpose, we use another neural network $\mathcal{D}_\phi^S$ with parameters ϕ called student network and train its parameters with the cross-entropy loss function between Y^S and Y^T in which Y^S is the output predictions of the student network $\mathcal{D}_\phi^S$. The intuition behind minimizing the cross-entropy between Y^S and Y^T is that it is equivalent to minimizing the KL-divergence distance between the distribution of the Y^S and Y^T random variables, enforcing the student network to imitate the predictions of the teacher as a rich model.

4   Experiments and Results

In this section, we perform experimental evaluations to explore the performance of the proposed method for disease prediction compared with state-of-the-art methods.

Experimental Setup: We show our results on two multi-modal medical datasets and analyze the behavior of the methods. Finally, a set of multi-modal synthetic datasets are created and further analysis on the stability and robustness of the proposed method are carried out on them. We choose EV-GCN [15], and the introduced framework by Parisot et al. [23] as two graph-based approaches incorporating GCN architecture designed for multi-modal disease

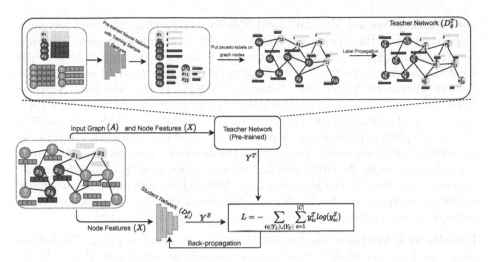

Fig. 1. Overview of the proposed method. The training set comprises of labeled and unlabeled nodes. Unlabeled nodes are in grey and other colors depicts the node labels. The upper part: teacher network. The lower part: student network. The student is trained by cross-entropy loss between teacher and student outputs to inject the teachers knowledge into the student's parameters.

prediction, DNN-JFC [27] as a multi-modal method, fully connected deep neural network (DNN) and the introduced method by Abraham et al. [1] as unimodal baselines. For GCN-based methods, when the graph modality is not available during the inference time, we assume that nodes are isolated, and the identity matrix with appropriate size is used as the adjacency matrix. Since DNN-JFC uses the features of all modalities for classification, at the inference time, when one modality (graph modality) is not available, mean imputation is employed to fill the values for this method. All the results are obtained from 5 different initialization seeds. Fully-connected neural network is the selected structure for both teacher and student in the experiments. For all methods, we examined networks with one, two and three hidden layers with a set of different number of units and checked the set of $\{5e-3, 1e-2\}$ for learning rate and $\{0.1, 0.3, 0.5\}$ for dropout. Adam optimizer is utilized for training of the networks [17]. The results are reported on the test set using the best configuration (selected based on the validation set) for each method per dataset.

4.1 Autism Spectrum Disorder Diagnosis on the ABIDE Dataset

The Autism Brain Imaging Data Exchange (ABIDE) [6,8] database provides the neuroimaging (functional MRI) and phenotypic data of 1112 patients with binary labels indicating the presence of diagnosed Autism Spectrum Disorder (ASD). To have a fair comparison with other methods, we choose the same 871 patients comprising 468 normal and 403 ASD patients. We also follow the same pre-processing steps of Parisot et al. [23] to select 2000 features from fMRI for node features. We use 55%, 10%, and 35% for train, validation, and test split, respectively, and the labels are available for 40% of training samples. For graph construction, first, we discard the phenotypic features that are only available for ASD patients including Autism Diagnostic Interview-Revised scores (ADIs), Autism Diagnostic Observation Schedule (ADOS) scores, Vineland Adaptive Behavior scores (VINELAND), Wechsler Intelligence Scale of Children (WISC), to prevent label leakage, since they are mostly available for ASD patients. The rest of the phenotypic features of patients are used for training a simple auto-encoder to both reconstruct the input via the decoder and classify the ASD state via a classifier using the latent low-dimensional representation of data (encoder's output). The auto-encoder is trained with the weighted sum of the mean squared errors as unsupervised (for labeled and unlabeled training samples) and cross-entropy as supervised (for labeled training samples) loss functions.

Results and Analysis: The boxplot of results are shown in Fig. 2. GKD shows higher performance in all metrics. Improvement in accuracy and AUC indicates that GKD 's performance is not sensitive to the decision threshold and learns a more reliable latent representation than its competitors. Moreover the results of GKD compared to its version without the remembrance term (GKD w/o RT) shows the effectiveness and necessity of this term when using the GKD. On the contrary, DNN shows appropriate accuracy, but it fails in AUC. This means that the embedding space of classes learned by DNN is not differentiable and selecting the right threshold is so effective. On the other hand, it has been previously suggested that graphs are good choices for modeling multi-modal datasets [15, 23]. However, in the current setting where the graph modality is not available, Parisot et al. and EV-GCN's performance drop which indicates that the graph availability in inference time has an important role in their architecture.

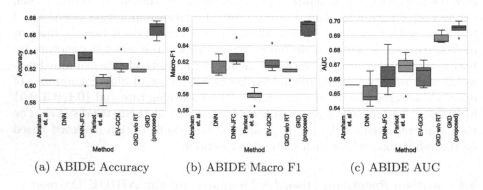

(a) ABIDE Accuracy   (b) ABIDE Macro F1   (c) ABIDE AUC

Fig. 2. Boxplot results of compared methods on the ABIDE dataset

4.2   Alzheimer's Progression Prediction on the TADPOLE Dataset

The Alzheimer's Disease Prediction Of Longitudinal Evolution (TADPOLE) [22] is a subset of Alzheimer's Disease Neuroimaging Initiative (ADNI) data which is introduced in the TADPOLE challenge. The dataset has binary labels indicating if the patient's Alzheimer's status will be progressed in the next six months. The dataset contains 151 instances with diagnosis or progression in Alzheimer's disease and 1878 who keep their status. The 15 biomarkers suggested by challenge organizers which are known to be informative in addition to the current status of patients are chosen as features. These biomarkers contains cognitive tests, MRI measures, PET measures, Cerebrospinal fluid (CSF) measures, APOE and age risk factors. We choose the biomarkers which are missing for more than 50% of patients for constructing a graph based on their available values instead of imputing the missing ones. $A\beta$, Tau and $pTau$ (CSF measures) and FDG and $AV45$ (PET measures) are the sparse biomarkers. We follow the steps described in [23] for graph construction. For each sparse biomarker, we connect every pair of nodes with the absolute distance less than a threshold which is chosen empirically based on the training data. Then the union of constructed graphs are chosen as the final graph. We use 65%, 10%, and 25% for train, validation, and test split, respectively, and the labels are available for 10% of training samples.

Results and Analysis: The results of Alzheimer's prediction are provided in Fig. 3. GKD is superior to its competitors in all metrics. TADPOLE is a highly imbalanced dataset and enhancing accuracy and Macro F1 means efficiency of the proposed method in diagnosis of both classes. Graph-based methods (Parisot et al. and EV-GCN) show more stability and competitive results on the TADPOLE dataset. Hence, it can be concluded that the constructed graph is more effective in the ABIDE dataset and its absence at inference time leads to inadequate performance. Despite the ABIDE dataset, DNN-JFC demonstrates unstable results on the TADPOLE. Comparing DNN-JFC and DNN with graph-based methods indicates that using graph-modality features directly results in more instability than using them as graphs.

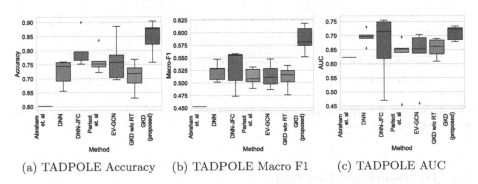

(a) TADPOLE Accuracy   (b) TADPOLE Macro F1   (c) TADPOLE AUC

Fig. 3. Boxplot results of compared methods on the TADPOLE dataset

4.3 Synthetic Dataset

To investigate the effect of graph modality in the training phase on the results of DNN-JFC, GCN, EV-GCN, and GKD as multi-modal approaches, we create a set of synthetic datasets with different rates of missing values in the graph-modality. We generate 2000 samples, distributed equally in two classes, with 128-dimensional node features and 4-dimensional graph-modality features. The algorithm of node-feature generation is adopted from [12], which is based on creating class points normally distributed around vertices on a hypercube [25]. The features of the graph modality are drawn from a 4-dimensional standard normal distribution with mean μ_{c_1} and μ_{c_2} for each class, respectively. Then we randomly remove the graph feature values by $P_{missing}$ rate and follow the steps of the TADPOLE graph construction.

Results and Analysis: The results of the experiment are illustrated in Fig. 4. This experiment also supports the claim that GCN-based architectures rely on the neighborhood in the graph, and inaccessibility to the graph deteriorates the efficiency of methods. As shown in Fig. 4, increasing the rate of missing values in training degrades the trained classifier, but with increasing the missing rate, the similarity between the training and testing data overcomes and the classifier improves. It should be noted that the shortcoming of GCN-based methods in comparison with DNN-JFC is another reason showing the weakness of GCN architecture in taking advantage of graph modality. It is also worth noting that GKD benefits from both features and graph and consistently outperforms in all metrics.

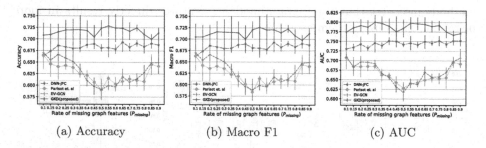

| (a) Accuracy | (b) Macro F1 | (c) AUC |

Fig. 4. Results of multi-modal methods by changing the rate of missing value in the graph-features at training

4.4 Discussion and Conclusion

This paper proposes a semi-supervised method to integrate distinct modalities (including imaging and non-imaging) into a unified predictive model that benefits from the graph information during the training phase and operates independently from graph modality during the inference time.

Unlike previous graph convolutional-based methods that rely on graph metadata, our method shows superior performance when the graph modality is missing among unseen samples. As a proof of concept, we evaluated our method for the task of diagnosing the Autism spectrum disorder on the ABIDE dataset and predicting Alzheimer's disease progress using the TADPOLE dataset. Extensive experimental results show that GKD manifests a consistent enhancement in terms of accuracy, AUC, and Macro F1 compared to the SOTA methods. Additionally, the ability of the method on a set of semi-labeled synthetic datasets is examined. We believe that this work opens the path to learn the classifier as it utilizes all the available information during the training time without being concerned about the availability of one or multiple modalities while diagnosing a patient's status.

References

1. Abraham, A., Milham, M.P., Di Martino, A., Craddock, R.C., Samaras, D., Thirion, B., Varoquaux, G.: Deriving reproducible biomarkers from multi-site resting-state data: An autism-based example. Neuroimage **147**, 736–745 (2017)
2. Abrol, A., Fu, Z., Du, Y., Calhoun, V.D.: Multimodal data fusion of deep learning and dynamic functional connectivity features to predict alzheimer's disease progression. In: 2019 41st Annual International Conference of the IEEE Engineering in Medicine and Biology Society (EMBC), pp. 4409–4413. IEEE (2019)
3. Bi, X.a., Cai, R., Wang, Y., Liu, Y.: Effective diagnosis of alzheimer's disease via multimodal fusion analysis framework. Frontiers Genetics **10**, 976 (2019)
4. Buciluǎ, C., Caruana, R., Niculescu-Mizil, A.: Model compression. In: Proceedings of the 12th ACM SIGKDD International Conference on Knowledge Discovery and Data Mining, pp. 535–541 (2006)
5. Cai, Q., Wang, H., Li, Z., Liu, X.: A survey on multimodal data-driven smart healthcare systems: approaches and applications. IEEE Access **7**, 133583–133599 (2019)
6. Craddock, C., et al.: The neuro bureau preprocessing initiative: open sharing of preprocessed neuroimaging data and derivatives. Frontiers in Neuroinformatics 7 (2013)
7. Defferrard, M., Bresson, X., Vandergheynst, P.: Convolutional neural networks on graphs with fast localized spectral filtering. arXiv preprint arXiv:1606.09375 (2016)
8. Di Martino, A., Yan, C.G., Li, Q., Denio, E., Castellanos, F.X., Alaerts, K., Anderson, J.S., Assaf, M., Bookheimer, S.Y., Dapretto, M., et al.: The autism brain imaging data exchange: towards a large-scale evaluation of the intrinsic brain architecture in autism. Mol. Psychiatry **19**(6), 659–667 (2014)
9. Du, H., Feng, J., Feng, M.: Zoom in to where it matters: a hierarchical graph based model for mammogram analysis. arXiv preprint arXiv:1912.07517 (2019)
10. Ghorbani, M., Kazi, A., Baghshah, M.S., Rabiee, H.R., Navab, N.: Ra-gcn: Graph convolutional network for disease prediction problems with imbalanced data. arXiv preprint arXiv:2103.00221 (2021)
11. Guo, Z., Li, X., Huang, H., Guo, N., Li, Q.: Deep learning-based image segmentation on multimodal medical imaging. IEEE Trans. Radiation Plasma Med. Sci. **3**(2), 162–169 (2019)
12. Guyon, I.: Design of experiments of the nips 2003 variable selection benchmark. In: NIPS 2003Workshop on Feature Extraction and Feature Selection, vol. 253 (2003)
13. Hinton, G., Vinyals, O., Dean, J.: Distilling the knowledge in a neural network. arXiv preprint arXiv:1503.02531 (2015)

14. Huang, S.C., Pareek, A., Seyyedi, S., Banerjee, I., Lungren, M.P.: Fusion of medical imaging and electronic health records using deep learning: a systematic review and implementation guidelines. NPJ digital Med. **3**(1), 1–9 (2020)
15. Huang, Y., Chung, A.C.S.: Edge-variational graph convolutional networks for uncertainty-aware disease prediction. In: Martel, A.L., Abolmaesumi, P., Stoyanov, D., Mateus, D., Zuluaga, M.A., Zhou, S.K., Racoceanu, D., Joskowicz, L. (eds.) MICCAI 2020. LNCS, vol. 12267, pp. 562–572. Springer, Cham (2020). https://doi.org/10.1007/978-3-030-59728-3_55
16. Kazi, A., Shekarforoush, S., Arvind Krishna, S., Burwinkel, H., Vivar, G., Kortüm, K., Ahmadi, S.-A., Albarqouni, S., Navab, N.: InceptionGCN: receptive field aware graph convolutional network for disease prediction. In: Chung, A.C.S., Gee, J.C., Yushkevich, P.A., Bao, S. (eds.) IPMI 2019. LNCS, vol. 11492, pp. 73–85. Springer, Cham (2019). https://doi.org/10.1007/978-3-030-20351-1_6
17. Kingma, D.P., Ba, J.: Adam: A method for stochastic optimization. arXiv preprint arXiv:1412.6980 (2014)
18. Kipf, T.N., Welling, M.: Semi-supervised classification with graph convolutional networks. arXiv preprint arXiv:1609.02907 (2016)
19. Lee, G., Nho, K., Kang, B., Sohn, K.A., Kim, D.: Predicting alzheimer's disease progression using multi-modal deep learning approach. Sci. Rep. **9**(1), 1–12 (2019)
20. Li, X., Duncan, J.: Braingnn: Interpretable brain graph neural network for fmri analysis. bioRxiv (2020)
21. Liu, J., Tan, G., Lan, W., Wang, J.: Identification of early mild cognitive impairment using multi-modal data and graph convolutional networks. BMC Bioinformatics **21**(6), 1–12 (2020)
22. Marinescu, R.V., et al.: Tadpole challenge: Prediction of longitudinal evolution in alzheimer's disease. arXiv preprint arXiv:1805.03909 (2018)
23. Parisot, S., et al.: Disease prediction using graph convolutional networks: application to autism spectrum disorder and alzheimer's disease. Med. Image Anal. **48**, 117–130 (2018)
24. Parisot, S., Ktena, S.I., Ferrante, E., Lee, M., Moreno, R.G., Glocker, B., Rueckert, D.: Spectral Graph Convolutions for Population-Based Disease Prediction. In: Descoteaux, M., Maier-Hein, L., Franz, A., Jannin, P., Collins, D.L., Duchesne, S. (eds.) MICCAI 2017. LNCS, vol. 10435, pp. 177–185. Springer, Cham (2017). https://doi.org/10.1007/978-3-319-66179-7_21
25. Pedregosa, F., et al.: Scikit-learn: Machine learning in python. J. Mach. Learn. Res.**12**, 2825–2830 (2011)
26. Venugopalan, J., Tong, L., Hassanzadeh, H.R., Wang, M.D.: Multimodal deep learning models for early detection of alzheimer's disease stage. Sci. Rep. **11**(1), 1–13 (2021)
27. Xu, T., Zhang, H., Huang, X., Zhang, S., Metaxas, D.N.: Multimodal deep learning for cervical dysplasia diagnosis. In: Ourselin, S., Joskowicz, L., Sabuncu, M.R., Unal, G., Wells, W. (eds.) MICCAI 2016. LNCS, vol. 9901, pp. 115–123. Springer, Cham (2016). https://doi.org/10.1007/978-3-319-46723-8_14
28. Yang, H., et al.: Interpretable multimodality embedding of cerebral cortex using attention graph network for identifying bipolar disorder. In: Shen, D., et al. (eds.) MICCAI 2019. LNCS, vol. 11766, pp. 799–807. Springer, Cham (2019). https://doi.org/10.1007/978-3-030-32248-9_89
29. Zhang, S., Tong, H., Xu, J., Maciejewski, R.: Graph convolutional networks: a comprehensive review. Comput. Soc. Networks **6**(1), 1–23 (2019)
30. Zhu, X., Ghahramani, Z.: Learning from labeled and unlabeled data with label propagation (2002)

Outcome/Disease Prediction

Predicting Esophageal Fistula Risks Using a Multimodal Self-attention Network

Yulu Guan[1], Hui Cui[1] (ID), Yiyue Xu[2], Qiangguo Jin[3] (ID), Tian Feng[4] (ID), Huawei Tu[1] (ID), Ping Xuan[5] (ID), Wanlong Li[2], Linlin Wang[2(✉)], and Been-Lirn Duh[1] (ID)

[1] Department of Computer Science and Information Technology, La Trobe University, Melbourne, Australia
[2] Department of Radiation Oncology, Shandong Cancer Hospital and Institute, Shandong First Medical University and Shandong Academy of Medical Sciences, Jinan, China
[3] College of Intelligence and Computing, Tianjin University, Tianjin, China
[4] School of Software Technology, Zhejiang University, Hangzhou, China
[5] Department of Computer Science and Technology, Heilongjiang University, Harbin, China

Abstract. Radiotherapy plays a vital role in treating patients with esophageal cancer (EC), whereas potential complications such as esophageal fistula (EF) can be devastating and even life-threatening. Therefore, predicting EF risks prior to radiotherapies for EC patients is crucial for their clinical treatment and quality of life. We propose a novel method of combining thoracic Computerized Tomography (CT) scans and clinical tabular data to improve the prediction of EF risks in EC patients. The multimodal network includes encoders to extract salient features from images and clinical data, respectively. In addition, we devise a self-attention module, named VisText, to uncover the complex relationships and correlations among different features. The associated multimodal features are integrated with clinical features by aggregation to further enhance prediction accuracy. Experimental results indicate that our method classifies EF status for EC patients with an accuracy of 0.8366, F1 score of 0.7337, specificity of 0.9312 and AUC of 0.9119, outperforming other methods in comparison.

Keywords: Esophageal fistula prediction · Self attention · Multimodal attention

1 Introduction

Esophageal cancer (EC) is the 6th most common cause of cancer-related death. To treat patients with unresectable locally advanced esophageal squamous cell carcinoma (SCC), chemotherapy and/or radiotherapy have demonstrated effectiveness and received considerable attention [1, 2]. Unfortunately, esophageal fistula (EF) is one of the complications resulting from these treatments [3]. Around 4.8–22.1% of the EC patients developed EF due to chemoradiotherapy [5]; this drastically reduces life expectancy to a rate of two to three months.

Electronic supplementary material The online version of this chapter (https://doi.org/10.1007/978-3-030-87240-3_69) contains supplementary material, which is available to authorized users.

M. de Bruijne et al. (Eds.): MICCAI 2021, LNCS 12905, pp. 721–730, 2021.
https://doi.org/10.1007/978-3-030-87240-3_69

Identifying risk factors during patient selection in radiotherapy-related treatments is crucial as it helps cancer management strategies, resulting in an improvement in the patients' quality of life. Various risk factors have been previously identified in many studies, including cancer anatomy, smoking history, age and so on [3, 4]. Researchers attempted to pinpoint the risk factors with statistical tools in clinical studies, such as meta-analysis, univariate, and multivariate logistic regression analyses [3–6]. Studies that focus solely on clinical variables neglect the imperative features found in the radiographic scans. The rise of deep learning in medical image classification has enabled the use of 2D Convolutional Neural Networks (CNNs) [7] to classify Computerized Tomography (CT) scans of EC patients. However, the method in [7] requires extensive user input and prior knowledge in selecting and pre-processing 3D CT images into applicable 2D visual data. Besides, 3D-spatial information will be lost while breaking the 3D CT scans into 2D images [8].

Leveraging the recent success of computer vision and text recognition in deep learning, multimodal CNNs can be the key to a more promising prognostic EF prediction. CNNs have gained tremendous success in image classification, segmentation, and synthesis throughout the years [9–11]. We also observed that multimodal data has been explored in image segmentation tasks that integrate techniques in natural language processing (NLP) and computer vision. A popular approach in this area is to employ CNNs or Recurrent Neural Network (RNNs) to decode text and images separately and then perform concatenation to yield a pixel-wise image segmentation result [12–14]. This fosters the idea of combining tabular data and image data for classification tasks. Frey *et al.* proposed a novel method to transform non-image genomic data into well-organized image-like forms [15]. Luis *et al.* fused genomic data, microscopic images and clinical information in an end-to-end deep learning method for automatic patient risk predictions for 33 types of pancreatic cancers [16]. Chauhan *et al.* implemented separate encoders and classifiers for each respective modality and a downstream joint loss function to combine all multimodal information to improve the performance of pulmonary oedema assessments [17]. Yap *et al.* employed a late fusion technique to combine multimodal data [18]. The common shortcoming of these current multimodal methods lies in the incapability of uncovering multimodal interdependencies and correlations. Xu *et al.* introduced a multimodal method with handcrafted clinical features and cervigram images to predict cervical dysplasia [19]. However, the major drawback with handcrafted features lies in the heavy domain-driven design that may not be generally applied to others.

To fill these gaps, we propose an end-to-end multimodal network to predict EF risks from heterogeneous data (i.e., CT scans and clinical data). The contribution of this study can be concluded as: (1) introducing a self-attention module, named VisText, to uncover the correlations between visual and clinical data; (2) clinical parameters are embedded through a text encoder, and transformed into an image-like high-dimensional vector to facilitate visual-text concatenation on which the VisText self-attention module operates.

2   Dataset

The retrospective study includes data relating to EC patients collected from 2014–2019. Among the total 553 eligible patients, 186 patients who had developed EF were assigned to the case group (positive group); the remaining 367 patients who had not developed EF were assigned to the control group (negative group). Negative cases were then matched with positive cases to the ratio of 2:1 by the diagnosis time of EC, marriage, gender, and race, which has been commonly adopted in existent clinical research. The data collected for this study includes clinical and image data. The development of EF from time of scan ranges in [0, 1401] days. All data, excluding those on treatment, were collected before treatment.

The clinical dataset was collected using a standardized questionnaire, including general, diagnostic, treatment, and hematological data. There are 7 numerical and 27 categorical variables. Categorical variables were further processed to one-hot encoding matrices for each patient (n = 74). Chemoradiotherapy (CRT) regimen is available in this retrospective study. Reason for this is integrating CRT information in a prediction model can assist clinicians to devise treatment plans and estimate the probability of developing fistula. Numerical attributes underwent a min-max normalization to [0, 1]. Clinical data variable details are provided in the supplementary materials.

CT images were reconstructed by applying a matrix of 512×512 pixels with a voxel size of $0.8027 \times 0.8027 \times 5$ mm^3. The 3D CT thoracic scans consisted of various numbers of 2D-slices, ranging from 71 to 114. We further constructed a $200 \times 200 \times 130$ mm^3 cube for the volume of interest located at the center of the tumor region.

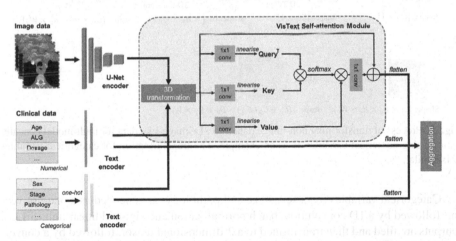

Fig. 1. Pipeline of the multimodal network with the VisText self-attention module. Visual features and salient clinical features are extracted by U-Net encoder and text encoder, respectively. VisText self-attention module extracts underlying connections between images and clinical features to enhance the integrated multimodal features. Self-attention enhanced multimodal features are aggregated with clinical data for final classification. Details of 3D transformation module is given in Fig. 2.

3 Multimodal Network with the VisText Self-attention Module

Given the input 3D CT scans and clinical data, we propose a multimodal network to predict EF as positive or negative. Its major components include CNN blocks for extracting visual features, a text encoder for extracting salient clinical text features, and a Vis-Text self-attention module for uncovering visual-text multimodal dependencies. Self-attention enhanced multimodal features and clinical features are aggregated for final prediction. The network is illustrated in Fig. 1.

3.1 Multimodal Feature Extraction

Our multimodal network takes a 3D image V, and non-image clinical data with categorical variables, O, and numerical variables, M, as input. We employ the U-Net's encoder [10] as the network's CNN backbone for extracting visual features. Specifically, the global visual feature (GVF) map is the output from the encoder's 5th layer, represented as $GVF \in \{\mathbb{R}^{C_V \times D_V \times W_V \times H_V}\}$, where C_V, D_V, H_V and W_V are the dimensions of visual feature channel, depth, height and width.

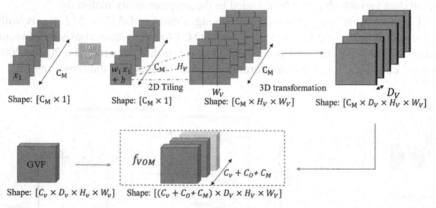

Fig. 2. Process of transforming non-image data into 3D image-like data for multimodal concatenation using numerical variables M as an example. The transformation for categorical variables O is similar.

Categorical variables (e.g., sex, stage and pathology) are encoded as a one-hot vector, followed by a 1D convolution, batch normalization and sigmoid linear unit (SiLU). Outputs are tiled and then transformed to a 2-dimensional tensor, followed by a conversion into a 3-dimensional tensor, represented as $O_{3D} \in \{\mathbb{R}^{C_O \times D_O \times H_O \times W_O}\}$, where C_O is the total number of one-hot features, and D_O, W_O and H_O are identical to D_V, H_V and W_V. This process is illustrated in Fig. 2.

On top of the categorical features embedded in the one-hot vectors, numerical features of the clinical data are also essential for predicting EF risks in EC patients. They are normalized to [0, 1], and experience the same operation and transformation as the categorical features. The processed numerical features are represented as $M_{3D} \in \{\mathbb{R}^{C_M \times D_M \times W_M \times H_M}\}$.

Given GVF, O_{3D} and M_{3D}, we can construct a joint multimodal feature representation by concatenating them along the channel/feature dimension. The shape of this feature tensor f_{VOM} is $(C_v + C_O + C_M) \times D_V \times H_V \times W_V$ as shown in Fig. 2.

3.2 VisText Self-attention Module for Multimodal Dependency Extraction

Inspired by the success of the self-attention mechanism [20, 21], we introduced VisText, a self-attention module to learn the correlations between images and clinical data. Given a multimodal feature vector f_{VOM}, the module first produces a set of query-key-value triad by a 1×1 convolution operation and linear transformation to $q_{VOM} = W_q f_{VOM}$, $k_{VOM} = W_k f_{VOM}$, and $v_{VOM} = W_v f_{VOM}$, where $\{W_q, W_k . W_v\}$ are trainable parameters. Different from [14], each query-key-value triad maintains the dimension of the original input feature vector. We compute the VisText self-attention map $\hat{f}_{VOM}$ as follows,

$$\hat{f}_{VOM} = \sum_{V'} \sum_{O'} \sum_{M'} \alpha_{V,O,M,V',O',M'} \times v_{V'O'M'},$$

$$\alpha_{V,O,M,V',O',M'} = Softmax(q^T_{V'O'M'} \times k_{VOM}), \tag{1}$$

where $\alpha_{V,O,M,V',O',M'}$ is the attention score considering the correlation between all features from all V, O, M modalities. V', O', M' represent any one feature from visual, categorical, and numerical modality. We notice that the visual-clinical correlation extracted by the VisText self-attention module is irrelevant to the order of the collected clinical attributes.

The following step converts $\hat{f}_{VOM}$ to the same shape as f_{VOM} via a linear layer with ReLU activation function. It is then added element-wise with f_{VOM} to yield a final residual connection, as follows,

$$f_{VisText} = \hat{f}_{VOM} + f_{VOM}. \tag{2}$$

The output feature $f_{VisText}$ is flattened and summed along the channel dimension before aggregation.

3.3 Aggregation and Optimization

During aggregation, we concatenate the flattened O_{3D} and M_{3D} with the VisText self-attention feature vector along the channel axis via a linear layer. Our network is trained using the cross-entropy loss via the Adam optimizer, with an initial learning rate of 0.001, batch size of 10 and epochs of 200. Table 1 describes the details of our network.

Table 1. Detailed architecture for VisText self-attention model.

	Name	Hyperparameters
Image encoder	Conv blocks 1–4	Each block: Conv = {Conv3d, $3 \times 3 \times 1$, stride 1, padding [1,1,0], batch normalization and ReLU} Conv = {Conv3d, $3 \times 3 \times 1$, stride 1, padding [1,1,0], batch normalization and ReLU} Maxpool3D, $2 \times 2 \times 2$, stride 2

(continued)

Table 1. (*continued*)

Name		Hyperparameters
	Conv block 5	Conv = {Conv3d, $3 \times 3 \times 1$, stride 1, padding [1,1,0], batch normalization and ReLU} Conv = {Conv3d, $3 \times 3 \times 1$, stride 1, padding [1,1,0], batch normalization and ReLU}
Text encoder	Categorical	1D convolution operation (kernel = 1, stride = 1), batch normalization and SiLU
	Numerical	1D convolution operation (kernel = 1, stride = 1), batch normalization and SiLU
VisText self-attention	q_{VOM}, k_{VOM}, v_{VOM}	3D convolution operation (kernel = $1 \times 1 \times 1$, stride = 1), ReLU
	$\hat{f}_{VOM}$	3D convolution operation (kernel = $1 \times 1 \times 1$, stride = 1), ReLU
Aggregation		Linear operation

4 Experiments and Results

4.1 Experimental Settings and Pre-processing

Our network was implemented with PyTorch and trained on an Intel® Core™ i5–4460 Processor paired with a NVIDIA GEFORCE GTX 960 GPU. On average, the training duration lasted 3 h and 47 min.

All the input images were resized to $48 \times 48 \times 27$ and normalized to [0, 1]. 75% of the patients' data was used as the training set and remaining 25% was used for testing. Due to an imbalanced class issue in the dataset, data augmentation, such as pixel shifting and rotation, was performed across all CT scans in training set, leading to 1,251 positive and 1,100 negative samples in training set. CT scans in test set has 47 positive and 92 negative samples.

4.2 Comparison Methods and Evaluation Measures

To evaluate the performance of the proposed model in predicting EF risks, we compared against existing methods, including: (a) a gated attention model [22], which was initially proposed for CT classification; (2) logistic regression risk prediction model for predicting EF risks in EC with only clinical data [5]; (c) a late fusion technique for fusing multimodal data [18], which was initially designed for skin lesion classification; (d) a joint loss function for multimodal representation learning [17], which was introduced for pulmonary oedema assessments from chest radiographs; and only the joint loss function from [17] was applied to our dataset for comparison evaluation. (e) the cross-modal self-attention (CMSA) module from [14] which was originally proposed for image segmentation. Following [14], multimodal features (unencoded clinical data and encoded image data) were fed to CMSA at three levels before final fusion.

Evaluation measures include accuracy, specificity, F1 score, and area under the receiver operating curve (ROC) (AUC). F1 score $= 2 \times$ precision $\times$ recall / (precision + recall). We ran each experiment three times and reported the average and standard deviation for both comparative and ablative studies.

4.3 Comparison with Others and Ablation Study Results

Ablation study was performed to evaluate the contributions of each component in our network. As shown by the results in Table 2, the baseline model is image only using U-Net encoder, which achieved an accuracy of 0.6978, F1 score of 0.5067, specificity of 0.8217 and AUC of 0.7341. If we embed original clinical data with images by VisText self-attention module excluding text encoder, the results were improved to an accuracy of 0.7770, F1 score of 0.6632, specificity of 0.8406 and AUC of 0.8273. When combining clinical data with text encoder but without self-attention module, the results were lifted to an accuracy of 0.8225, F1 score of 0.7062, specificity of 0.9143 and AUC of 0.8953. Finally, with the addition of VisText self-attention module, the results were further increased to an accuracy of 0.8366, F1 score of 0.7337, specificity of 0.9312 and AUC of 0.9119. The results demonstrated the contributions of transforming original clinical data into an image-like high-dimensional vector to facilitate visual-text concatenation.

Table 2. Ablation study results.

Module			Results			
Text data	Text encoder	VisText self attn	Accuracy	F1 score	Specificity	AUC
			0.6978 ± 0.0176	0.5067 ± 0.0396	0.8188 ± 0.0136	0.7341 ± 0.0291
✓		✓	0.7770 ± 0.0305	0.6632 ± .0522	0.8406 ± 0.0136	0.8273 ± .0254
✓	✓		0.8225 ± 0.0068	0.7062 ± 0.0125	0.9203 ± 0.0103	0.8953 ± 0.0096
✓	✓	✓	**0.8366 ± 0.0179**	**0.7337 ± 0.0416**	**0.9312 ± 0.0051**	**0.9119 ± 0.0122**

The comparison results with others are given in Table 3. All evaluation measures show that our model is superior when compared to the unimodal and other multimodal approaches, followed by the second best [14] and third best [18] multimodal models. The improved performance over unimodality analyses [5, 22] demonstrated that combining two modalities can effectively improve model performance. The three other multimodal approaches did not perform as well as the proposed model because our VisText attention module further enhanced the performance by uncovering multimodal intra-correlations which the other multimodal methods lack. We also draw classification activation map (CAM) using the output from VisText self-attention module to visually interpret the learnt discriminative regions by the self-attention module. As shown by the two positive and two negative cases in Fig. 3, the module learnt to assign higher importance to the tumor regions during the prediction process.

Table 3. Comparison with other architectures by accuracy, F1 score, Specificity and AUC.

Method		Accuracy	F1 score	Specificity	AUC
Text only	Xu et al. 2019 [5]	0.7589	0.6667	0.7766	0.8196
Image only	Schlemper et al. 2019 [22]	0.7218 ± 0.0271	0.5403 ± 0.0246	0.8442 ± 0.0455	0.7350 ± 0.0326
Text + Image	Yap et al. 2018 [18]	0.7578 ± 0.0265	0.6337 ± 0.051	0.8261 ± 0.0089	0.8123 ± 0.0144
	Chauhan et al. 2020 [17]	0.6739 ± 0.0244	0.5193 ± 0.0532	0.7500 ± 0.0089	0.6885 ± 0.0363
	Ye et al. 2019 [14]	0.7770 ± 0.0204	0.6417 ± 0.0177	0.8732 ± 0.0359	0.7736 ± 0.013
	Proposed method	**0.8366 ± 0.0179**	**0.7337 ± 0.0416**	**0.9312 ± 0.0051**	**0.9119 ± 0.0122**

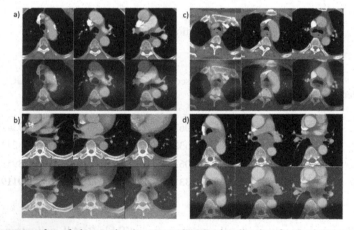

Fig. 3. Four examples of class activation map (CAM) visualization for the interpretation of the network's self-attention module. Two negative class cases are given in a) and b), two positive cases are given in c) and d).

5   Conclusion

In this paper, we proposed a novel multimodal network for EF predictions for EC patients. We incorporated a text encoder to produce salient clinical features and VisText self-attention module to extract visual-clinical multimodal correlations which further improved EF prediction. Comparative analysis and ablation study demonstrated that the proposed method outperformed other state-of-the-art multimodal approaches in the field.

References

1. Hirano, H., Boku, N.: The current status of multimodality treatment for unresectable locally advanced esophageal squamous cell carcinoma. Asia Pac J Clin Oncol **14**, 291–299 (2018)
2. Borggreve, A.S., et al.: Surgical treatment of esophageal cancer in the era of multimodality management. Ann. New York Acad. Sci. (2018)
3. Tsushima, T., et al.: Risk factors for esophageal fistula associated with chemoradiotherapy for locally advanced unresectable esophageal cancer: a supplementary analysis of JCOG0303. Medicine **95** (2016)
4. Zhang, Y., Li, Z., Zhang, W., Chen, W., Song, Y.: Risk factors for esophageal fistula in patients with locally advanced esophageal carcinoma receiving chemoradiotherapy. Onco Targets Ther **11**, 2311–2317 (2018)
5. Xu, Y., et al.: Development and validation of a risk prediction model for radiotherapy-related esophageal fistula in esophageal cancer. Radiat. Oncol. **14**, 181 (2019)
6. Zhu, C., Wang, S., You, Y., Nie, K., Ji, Y.: Risk factors for esophageal fistula in esophageal cancer patients treated with radiotherapy: a systematic review and meta-analysis. Oncol Res Treat **43**, 34–41 (2020)
7. Cui, H., Xu, Y., Li, W., Wang, L., Duh, H.: Collaborative learning of cross-channel clinical attention for radiotherapy-related esophageal fistula prediction from CT. In: Martel, A.L., et al. (eds.) MICCAI 2020. LNCS, vol. 12261, pp. 212–220. Springer, Cham (2020). https://doi.org/10.1007/978-3-030-59710-8_21
8. Starke, S., et al.: 2D and 3D convolutional neural networks for outcome modelling of locally advanced head and neck squamous cell carcinoma. Sci. Rep. **10**, 1–13 (2020)
9. Wu, Z., Shen, C., Van Den Hengel, A.: Wider or deeper: revisiting the resnet model for visual recognition. Pattern Recogn. **90**, 119–133 (2019)
10. Jin, Q., Meng, Z., Sun, C., Cui, H., Su, R.: RA-UNet: a hybrid deep attention-aware network to extract liver and tumor in CT scans. Front. Bioengi. Biotechnol. **8**, 1–15 (2020)
11. Jin, Q., Cui, H., Sun, C., Meng, Z., Su, R.: Free-form tumor synthesis in computed tomography images via richer generative adversarial network. Knowl.-Based Syst. **218**, 106753 (2021)
12. Li, R., et al.: Referring image segmentation via recurrent refinement networks. In: IEEE Conference on Computer Vision and Pattern Recognition (CVPR), pp. 5745–5753 (2018)
13. Shi, H., Li, H., Meng, F., Wu, Q.: Key-word-aware network for referring expression image segmentation. In: Ferrari, V., Hebert, M., Sminchisescu, C., Weiss, Y. (eds.) ECCV 2018. LNCS, vol. 11210, pp. 38–54. Springer, Cham (2018). https://doi.org/10.1007/978-3-030-01231-1_3
14. Ye, L., Rochan, M., Liu, Z., Wang, Y.: Cross-modal self-attention network for referring image segmentation. In: IEEE conference on computer vision and pattern recognition (CVPR), pp. 10502–10511 (2019)
15. Sharma, A., Vans, E., Shigemizu, D., Boroevich, K.A., Tsunoda, T.: DeepInsight: a methodology to transform a non-image data to an image for convolution neural network architecture. Sci. Rep. **9**, 1–7 (2019)
16. Silva, L.A.V., Rohr, K.: Pan-cancer prognosis prediction using multimodal deep learning. In: International Symposium on Biomedical Imaging (ISBI), pp. 568–571. IEEE (2020)
17. Chauhan, G., et al.: Joint modeling of chest radiographs and radiology reports for pulmonary edema assessment. In: Martel, A.L., et al. (eds.) MICCAI 2020. LNCS, vol. 12262, pp. 529–539. Springer, Cham (2020). https://doi.org/10.1007/978-3-030-59713-9_51
18. Yap, J., Yolland, W., Tschandl, P.: Multimodal skin lesion classification using deep learning. Exp. Dermatol. **27**, 1261–1267 (2018)
19. Xu, T., Zhang, H., Huang, X., Zhang, S., Metaxas, D.N.: Multimodal deep learning for cervical dysplasia diagnosis. In: Ourselin, S., Joskowicz, L., Sabuncu, M.R., Unal, G., Wells, W. (eds.)

MICCAI 2016. LNCS, vol. 9901, pp. 115–123. Springer, Cham (2016). https://doi.org/10.1007/978-3-319-46723-8_14

20. Vaswani, A., et al.: Attention is all you need. arXiv preprint arXiv:1706.03762 (2017)
21. Wang, X., Girshick, R., Gupta, A., He, K.: Non-local neural networks. In: IEEE Conference on Computer Vision and Pattern Recognition (CVPR), pp. 7794–7803 (2018)
22. Schlemper, J., et al.: Attention gated networks: learning to leverage salient regions in medical images. Med. Image Anal. 53, 197–207 (2019)

Hybrid Aggregation Network for Survival Analysis from Whole Slide Histopathological Images

Jia-Ren Chang[1], Ching-Yi Lee[1], Chi-Chung Chen[1], Joachim Reischl[2], Talha Qaiser[2], and Chao-Yuan Yeh[1(✉)]

[1] aetherAI, Taipei, Taiwan
{JiaRenChang,cylee,chenchc,joeyeh}@aetherai.com
[2] AstraZeneca, London, UK
{joachim.reischl,talha.qaiser1}@astrazeneca.com

Abstract. Understanding of prognosis and mortality is crucial for evaluating the treatment plans for patients. Recent developments of digital pathology and deep learning bring the possibility of predicting survival time using histopathology whole slide images (WSIs). However, most prevalent methods usually rely on a small set of patches sampled from a WSI and are unable to directly learn from an entire WSI. We argue that a small patch set cannot fully represent patients' survival risks due to the heterogeneity of tumors; moreover, multiple WSIs from one patient need to be evaluated together. In this paper, we propose a Hybrid Aggregation Network (HANet) to adaptively aggregate information from multiple WSIs of one patient for survival analysis. Specifically, we first extract features from WSIs using a convolutional neural network trained in a self-supervised manner, and further aggregate feature maps using two proposed aggregation modules. The self-aggregation module propagates informative features to the entire WSI, and further abstract features to region representations. The WSI-aggregation module fuses all the region representations from different WSIs of one patient to predict patient-level survival risk. We conduct experiments on two WSI datasets that have accompanying survival data, *i.e.*, NLST and TCGA-LUSC. The proposed method achieves state-of-the-art performances with concordance indices of 0.734 for NLST and 0.668 for TCGA-LUSC, outperforming existing approaches.

Keywords: Survival analysis · WSIs · Prognosis

1 Introduction

Survival analysis is becoming a popular field in healthcare research. The purpose of survival analysis is to examine how specified factors (e.g. smoking, age, treat-

Electronic supplementary material The online version of this chapter (https://doi.org/10.1007/978-3-030-87240-3_70) contains supplementary material, which is available to authorized users.

M. de Bruijne et al. (Eds.): MICCAI 2021, LNCS 12905, pp. 731–740, 2021.
https://doi.org/10.1007/978-3-030-87240-3_70

ment, etc.) affect the occurrence probability of a particular event (e.g. death, recurrence of the disease, etc.) at a certain time point. Clinicians can exploit survival analysis to evaluate the significance of prognostic variables and subsequently make an early decision among treatment options.

The abilities of deep neural networks to learn complex and nonlinear relations between variables and an individual's survival risk have drawn increasing attention in the survival analysis field. Recently, Hao *et al.* proposed a Cox-PASNet for prognosis by integrating genomic data and clinical data [7]. In addition to genomic and clinical data, image-based survival prediction has become an active field due to the rise of digital medical imaging. These image data include histopathological images [20], CT images [5], MR images [12], etc. One of the most challenging problems in image-based healthcare research is that a whole slide histopathological image (WSI) has a very high resolution (i.e. typically have $10^6 \times 10^6$ pixels), compared to natural images from ImageNet [3] that usually have $10^3 \times 10^3$ pixels. Previous studies [2, 13, 20, 21, 23, 24] have demonstrated that pathological images can be utilized as additional information in survival analysis using deep learning. Yao *et al.* localized and classified cells in the WSIs, and calculated quantitative features of these cells to predict survival outcomes which can achieve better results than traditional imaging biomarker. Zhu *et al.* [23] proposed DeepConvSurv that adopts image patches from regions of interest (RoIs) to train a deep neural network for survival analysis.

The disadvantage of patch-based methods relying on RoIs is that the labeling of RoIs requires highly trained experts to ensure the correctness. To lower the burden of RoI annotations, [2] and [21] exploited the multiple instance learning [4] to discover relevant features from WSIs. Furthermore, Zhu *et al.* [24] extended DeepConvSurv to analyze WSIs without RoIs by clustering similar image patterns. Li *et al.* [13] integrated the local patch features from WSIs by using graph models to learn the relations of local patches. These studies suggested that learning survival risk by integrating image patches from WSIs can utilize discriminative patterns to achieve better patients' survival prediction. However, features learnt from local patches lack global topological representations of WSIs for evaluating survival risks. Moreover, we argue that the WSIs from different parts of tissue from one patient should be considered together, but previous studies [2, 13, 21, 23] lack patient-level integration.

In this paper, we introduce the Hybrid Aggregation Network (HANet) which consists of the self-aggregation module and WSI-aggregation module to merge information from multiple WSIs of one patient for survival analysis. The overall framework of the proposed approach is illustrated in Fig. 1. The WSIs are first fed into a self-supervised pre-trained model to extract representative feature maps. The HANet then performs self-aggregation and WSI-aggregation on the input feature maps to predict patient-level survival risks. As shown in Fig. 2a, the self-aggregation module propagates global information to local features, and then abstract features to region representations via an attentive bilinear pooling. The WSI-aggregation module fuses all the region representations from multiple WSIs into patient-level features to predict survival risk, as shown in Fig. 2b. The experimental results demonstrated the HANet can achieve superior performance

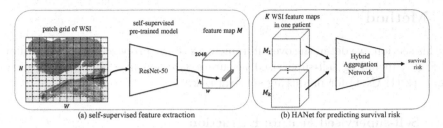

Fig. 1. The overall framework of the proposed method, including (a) self-supervised feature extraction and (b) HANet for predicting survival risk. An input WSI is fed into a unsupervised pre-trained model to derive its own feature map M. We further make a patient-level survival risk prediction by aggregating K feature maps of one patient.

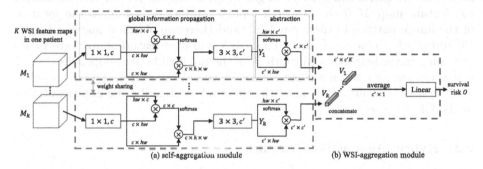

Fig. 2. The proposed HANet consists of (a) self-aggregation module and (b) WSI-aggregation module. The feature map M_k first passes through self-aggregation module to obtain a compact region representation V_k. The self-information module shares weights for all K WSI feature maps. The region representation V_k can be recognized as a descriptor of the k-th WSI of a patient. We further fuse these representations to predict the patient-level survival risk O using WSI-aggregation information module. The orange boxes denote the convolutional/linear layers with the kernel size and the number of channels.

on several publicly available datasets. We also offer adequate ablation studies for analyzing the effectiveness of the proposed HANet.

To summarise, the proposed method makes the contributions as follows:

- We present a novel Hybrid Aggregation Network (HANet) which can predict patient-level survival risk from Whole Slide histopathological Images (WSIs).
- We propose self-aggregation module and WSI-aggregation module for integrating information from multiple WSIs of one patient.
- The proposed HANet achieves superior performance on several public datasets, such as NLST and TCGA-LUSC.

2   Method

In this section, we detail the overall framework of the proposed method, as shown in Fig. 1, including self-supervised feature extraction and Hybrid Aggregation Network (HANet) for predicting survival risk.

2.1   Self-supervised Feature Extraction

In our framework as shown in Fig. 1, the WSIs first pass through feature extraction to derive compact feature maps. The input WSI is divided to 224×224 pixel patches at $10\times$ magnification without overlap. Each image patch is further fed into a feature extraction network to get a representation vector. We reconstruct the feature map M from the patch representations according to the location of the image patches in the input WSI, and therefore obtain a matrix of 2,048 (features) $\times h \times w$ for each slide. The size h and w of feature matrix are varied according to the height H and width W of the input slide. We adopt a ResNet-50 [10] as our encoder which was pre-trained in an self-supervised manner on public WSI dataset [1] using MoCo algorithm [8], and we describe the implementation in the following section.

2.2   Hybrid Aggregation Network

We suppose that there are K WSIs of one patient for making a survival prediction; therefore, we can obtain K feature maps after feature extraction. The K feature maps are further fed into HANet for predicting the patient-level survival risk. As shown in Fig. 2, the proposed HANet is consisted of two modules, including the self-aggregation module and the WSI-aggregation module. We detail the two proposed module in the following subsections.

Self-aggregation Module. The self-aggregation module consists of two parts: *global information propagation* and *region representation abstraction*. The core idea of the proposed global information propagation is to gather information from the entire hispathologicial slide into a compact set and distribute them to each location adaptively. Thus the output feature maps can encode the relations between locations without large receptive fields. The *region representation abstraction* is applied to handle the varied h and w in different WSIs, which could project the input feature map into fixed-size region representations. Therefore, the region representation abstraction produces a low-dimensional but highly discriminative image representation.

The input feature map is denoted $M \in \mathbb{R}^{2048 \times h \times w}$, where h and w indicate the height and width of the input. Note that the h and w are varied depending on the input WSI. The input feature map is first fed into a 1×1 convolutional layer $\phi(\cdot)$ to derive output $X \in \mathbb{R}^{c \times h \times w}$, *i.e.*, $X = \phi(M)$, where c is the number of channels. Then we reshape X into $X \in \mathbb{R}^{c \times hw}$, and the output of global information propagation Y is computed as $Y = \psi(WX)$, where $W \in \mathbb{R}^{c \times c}$

denotes the relation between channels and $\psi(\cdot)$ is a 3×3 convolutional layer with c' number of channels. The relation matrix W is obtained by:

$$W_{ij} = \frac{\exp(-(XX^\top)_{ij})}{\sum_{u=1}^{c} \exp(-(XX^\top)_{iu})}. \tag{1}$$

Unlike [18] performs non-local operation in spatial dimension, the global information propagation is performed in the channel dimension, and thus reduces computational costs.

The region representation abstraction is further applied to abstract the output Y into a compact region representation $V \in \mathbb{R}^{c' \times c'}$ by calculating $V = Y\varphi(Y^\top)$, where $\varphi(\cdot)$ is a softmax function on spatial dimension. The projection matrix $\varphi(Y^\top)$ can be understood as the attention maps of Y which reveals the important regions of the input WSI. Therefore, we can adopt arbitrary spatial sizes of input WSI feature maps into compact region representations in the same size $(c' \times c')$ via the region representation abstraction.

WSI-Aggregation Module. We suppose that there are K WSIs of a patient for survival risk prediction, and thus we can acquire K region representations $(V_1, ..., V_k)$ via previous introduced self-aggregation module. Note that the self-aggregation module are weight-shared among the WSIs of a patient, i.e., K WSI feature maps are processed using the same module. These compact region representations are first concatenated into a matrix $S \in \mathbb{R}^{c' \times c'K}$, where $S = \text{concat}(V_1, V_2, ..., V_k)$. We further aggregate these region representations via averaging:

$$C_i = \frac{1}{c'K} \sum_{j=1}^{c'K} S_{ij}, \tag{2}$$

where $C \in \mathbb{R}^{c'}$. The final survival risk prediction O is achieved by a linear layer τ; that is, $O = \tau(C)$.

2.3   Loss

The HANet aggregates intra-slide and inter-slide information to make a final survival prediction. For i-th patient WSIs passing through the proposed model, the output survival score of this patient is denoted as O_i. We assume that the label of the i-th patient as (t_i, δ_i) where t_i is the observed or censored time for the patient i and δ_i indicates whether the survival time of the patient i is censored $(\delta_i = 0)$ or observed $(\delta_i = 1)$. The negative log partial likelihood loss [23,24] is denoted as:

$$\text{loss} = -\sum_{i=1}^{N} \delta_i(O_i - \log \sum_{j:t_j \geq t_i} \exp(O_j)), \tag{3}$$

where j is from the set whose survival time is equal or larger than t_i and N is batch size. The loss function penalizes the discordance between patients' survival time and predicted survival scores.

3   Experiment

3.1   Dataset

Two public survival analysis datasets which provide WSIs, National Lung Screening Trial - lung squamous cell carcinoma (NLST-LUSC) [16] and The Cancer Genome Atlas (TCGA) [11], are used in this study.

The NLST randomized 53,454 adults of age 55 to 74 with at least 30-year smoking history as high risk group for lung cancer screening. Please note that these WSIs in one patient are sectioned from the tissue resected as part of lung cancer during a treatment; that is, the WSIs in one patient are resected in the same period. This fact suggests the importance of aggregating WSIs for survival risk prediction.

3.2   Implementation Details

Our method is implemented using PyTorch [14]. We chose a ResNet-50 [10] as our feature extraction model. For training self-supervised feature extraction model, we adopt MoCo [8] for learning visual representations. The model was trained with 300 epochs using 245,458 512×512 pathological image patches from [1]. The data augmentations are similar to MoCo [8], including random resize crop to 224×224, random color jitter, random blur, random horizontal flip and random vertical flip. The moment coefficient is set to 0.999 and memory bank size is set to 16,384. We employ batch size of 128 and initial learning rate 0.3 which is dropped by a factor of 0.1 at 150 and 225 epoch. The optimizer is SGD with weight decay 0.0001 for self-supervised learning. The feature extraction model takes five days for training on 4 NVIDIA 1080ti GPU.

For the proposed HANet, we set the c to 64 and c' to 32. We train HANet with 80 epochs and batch size was set to 32 on both datasets. The optimizer is RMSprop with weight decay 0.0003. The initial learning rate is set to 0.01 and dropped by a factor of 0.1 at 20 and 40 epoch. We perform five fold cross-validation on NLST for ablation study. The TCGA-LUSC is randomly split to train and test set, and the setting of experiment is exactly the same with NLST without hyperparameter tuning. The experiments were conducted on a single NVIDIA 1080ti GPU and the training of a single model takes around 20 min.

3.3   Evaluation Metric

To compare the performances in survival risk prediction, we take the standard evaluation metric in survival analysis [13,15,21,23,24] concordance index (C-index) as evaluation metric. The C-index is defined as follows:

$$\text{C-index} = \frac{\# \text{ of ordered pairs}}{\# \text{ of all possible ranking pairs}}. \tag{4}$$

Equation 4 indicates the fraction of all pairs of subjects whose survival times are correctly ordered among all possible ranking pairs. The value of 0 means the worst performance that the ranking order are all inverted, 0.5 is a random guess and 1 is the perfect ranking performance.

3.4 Ablation Study

We conduct ablation study for demonstrating the effectiveness of the proposed aggregation module. For studying the effectiveness of spatial-level aggregation module, we compare three methods that can be employed to handle arbitrary size of input, that is, spatial pyramid pooling [9], the compact bilinear pooling [6], and the proposed module. For studying the effectiveness of WSI-level aggregation module, we compare the adopted average pooling and the weighted summation (the weight is achieved by the softmax function). The results of ablation study are shown in Table 1. The average pooling in WSI aggregation consistently achieves better performance in three spatial aggregation type. The best combination of aggregation type is the proposed HANet, and this demonstrates the effectiveness of the proposed modules.

In addition, we compare the feature maps extracted from ImageNet pre-trained ResNet-50 (C-index $= 0.570 \pm 0.088$) with histopathological images pre-trained ResNet-50 using MoCo [8] (C-index $= 0.734 \pm 0.077$). The result significantly showed the feature maps extracted from histopathological images pre-trained model can achieve superior performance than ImageNet pre-trained model. This suggests that our feature extraction model can capture meaningful representations of WSI images and thus benefit survival analysis. Moreover, we remove the WSI-aggregation module in our model and achieve C-index $= 0.640 \pm 0.102$ which is shown in Table 1. The result suggests that the WSI aggregation is important for survival analysis.

Table 1. The ablation study on different types of aggregation. We report the mean C-index and 95% confidence interval (CI) of five fold cross-validation on NLST. The results of ablation study demonstrate the effectiveness of the proposed method. The none type of WSI-aggregation indicates that the model predicts survival risk using only one slide without aggregation.

Self-aggregation type	WSI-aggregation type	Mean C-index/95% CI
Spatial pyramid pooling [9]	Weighted sum	0.707 ± 0.055
Spatial pyramid pooling [9]	Averaging	0.718 ± 0.057
Compact bilinear pooling [6]	Weighted sum	0.670 ± 0.058
Compact bilinear pooling [6]	Averaging	0.692 ± 0.025
The proposed module	None	0.640 ± 0.102
The proposed module	Weighted sum	0.686 ± 0.090
The proposed module	Averaging	$\mathbf{0.734 \pm 0.077}$

Table 2. The C-index comparison with other state-of-the-art methods on two public datasets. The proposed method has superior performance on survival prediction.

Model	NLST	TCGA-LUSC
LASSO-Cox [17] (ROI-based)	0.474	0.528
DeepConvSurv [23] (ROI-based)	0.629	–
DLS [19]	–	0.621
WSISA [24]	0.654	0.638
DeepAttnMISL [22]	0.696	–
DeepGraphSurv [13]	0.707	0.660
HANet (ours)	**0.734**	**0.668**

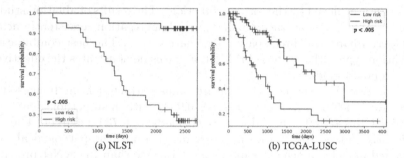

(a) NLST   (b) TCGA-LUSC

Fig. 3. The Kaplan-Meier curves of lung cancer patients in (a) NLST and (b) TCGA-LUSC validation set. We split low and high-risk groups according to the median of predicted survival risk. We perform a log-rank test to validate the significance between two groups ($p < .005$).

3.5   Results

Table 2 presents the C-index from state-of-the-art studies on two datasets. The RoI-based methods, such as LASSO-Cox [17] and DeepConvSurv [23], only utilize limited local information from RoIs which have worse performance. The WSISA [24] and DeepGraphSurv [13] can utilize whole slide information by clustering and selecting image patches from the entire WSI. However, the WSISA [24] uses small model for feature extraction as well as the DeepGraphSurv [13] uses Imagenet pre-trained VGG-16 without finetuning. It might drops performances by the inferior feature extraction model.

Instead of using ImageNet pre-trained VGG-16 as in [13], the proposed feature extraction model is self-supervised trained on pathological image patches, and thus can find representative visual pattern on WSIs. Moreover, the proposed HANet learns to aggregate intra-WSI and inter-WSI information. This led the proposed method to achieve stat-of-the-art results of C-index **0.734** on NLST and **0.668** on TCGA-LUSC.

Given the trained survival models, we can use the predicted survival risk to split patients into low or high-risk group for further clinical assessments.

Two groups are split by the *median* of predicted survival risk. Patients with short survival time should be classified into high-risk group and vice versa. The proposed model was evaluated if it can correctly classify patients into two groups. The Kaplan-Meier curves are shown in Fig. 3(a) NLST and (b) TCGA-LUSC. We conduct log-rank test to gauge the difference between low and high-risk groups, and it suggests that how well the model can classify patients into two groups. It is shown that the proposed method can achieve statistical significant results ($p < .005$) on classifying low-risk and high-risk groups. This suggests that the proposed method can be a recommendation system for personalized treatment in the future.

4   Conclusion

In this paper, we have presented a Hybrid Aggregation Network (HANet) for survival risk prediction. Specifically, we introduce a self-aggregation and a WSI-aggregation module to capture survival-related information in multiple WSIs. The experiments show that the proposed HANet can achieve superior performance consistently on two survival analysis datasets, i.e. NLST and TCGA-LUSC. In the future, we would like to apply HANet to many other survival risk prediction tasks of various cancer subtypes.

Acknowledgement. We are grateful to Taiwan's National Center for High-performance Computing for providing computing resources. We also thank Szu-Hua Chen, M.D. (aetherAI) for valuable advice.

References

1. Bejnordi, B.E., et al.: Diagnostic assessment of deep learning algorithms for detection of lymph node metastases in women with breast cancer. JAMA **318**(22), 2199–2210 (2017)
2. Courtiol, P., et al.: Deep learning-based classification of mesothelioma improves prediction of patient outcome. Nat. Med. **25**(10), 1519–1525 (2019)
3. Deng, J., Dong, W., Socher, R., Li, L.J., Li, K., Fei-Fei, L.: Imagenet: a large-scale hierarchical image database. In: 2009 IEEE Conference on Computer Vision and Pattern Recognition, pp. 248–255. IEEE (2009)
4. Dietterich, T.G., Lathrop, R.H., Lozano-Pérez, T.: Solving the multiple instance problem with axis-parallel rectangles. Artif. Intell. **89**(1–2), 31–71 (1997)
5. Ganeshan, B., Miles, K.A., Young, R.C., Chatwin, C.R.: Hepatic enhancement in colorectal cancer: texture analysis correlates with hepatic hemodynamics and patient survival. Acad. Radiol. **14**(12), 1520–1530 (2007)
6. Gao, Y., Beijbom, O., Zhang, N., Darrell, T.: Compact bilinear pooling. In: Proceedings of the IEEE Conference on Computer Vision and Pattern Recognition, pp. 317–326 (2016)
7. Hao, J., Kim, Y., Mallavarapu, T., Oh, J.H., Kang, M.: Interpretable deep neural network for cancer survival analysis by integrating genomic and clinical data. BMC Med. Genomics **12**(10), 1–13 (2019)

8. He, K., Fan, H., Wu, Y., Xie, S., Girshick, R.: Momentum contrast for unsupervised visual representation learning. arXiv preprint arXiv:1911.05722 (2019)
9. He, K., Zhang, X., Ren, S., Sun, J.: Spatial pyramid pooling in deep convolutional networks for visual recognition. IEEE Trans. Pattern Anal. Mach. Intell. **37**(9), 1904–1916 (2015)
10. He, K., Zhang, X., Ren, S., Sun, J.: Deep residual learning for image recognition. In: Proceedings of the IEEE Conference on Computer Vision and Pattern Recognition, pp. 770–778 (2016)
11. Kandoth, C., et al.: Mutational landscape and significance across 12 major cancer types. Nature **502**(7471), 333–339 (2013)
12. Kim, J.H., et al.: Breast cancer heterogeneity: Mr imaging texture analysis and survival outcomes. Radiology **282**(3), 665–675 (2017)
13. Li, R., Yao, J., Zhu, X., Li, Y., Huang, J.: Graph CNN for survival analysis on whole slide pathological images. In: Frangi, A.F., Schnabel, J.A., Davatzikos, C., Alberola-López, C., Fichtinger, G. (eds.) MICCAI 2018. LNCS, vol. 11071, pp. 174–182. Springer, Cham (2018). https://doi.org/10.1007/978-3-030-00934-2_20
14. Paszke, A., et al.: Pytorch: an imperative style, high-performance deep learning library. In: Advances in Neural Information Processing Systems, pp. 8024–8035 (2019)
15. Steck, H., Krishnapuram, B., Dehing-Oberije, C., Lambin, P., Raykar, V.C.: On ranking in survival analysis: Bounds on the concordance index. In: Advances in Neural Information Processing Systems, pp. 1209–1216 (2008)
16. Team, N.L.S.T.R.: Reduced lung-cancer mortality with low-dose computed tomographic screening. New England J. Med. **365**(5), 395–409 (2011)
17. Tibshirani, R.: The lasso method for variable selection in the cox model. Stat. Med. **16**(4), 385–395 (1997)
18. Wang, X., Girshick, R., Gupta, A., He, K.: Non-local neural networks. In: Proceedings of the IEEE Conference on Computer Vision and Pattern Recognition, pp. 7794–7803 (2018)
19. Wulczyn, E., et al.: Deep learning-based survival prediction for multiple cancer types using histopathology images. PLoS One **15**(6), e0233678 (2020)
20. Yao, J., Wang, S., Zhu, X., Huang, J.: Imaging biomarker discovery for lung cancer survival prediction. In: Ourselin, S., Joskowicz, L., Sabuncu, M.R., Unal, G., Wells, W. (eds.) MICCAI 2016. LNCS, vol. 9901, pp. 649–657. Springer, Cham (2016). https://doi.org/10.1007/978-3-319-46723-8_75
21. Yao, J., Zhu, X., Huang, J.: Deep multi-instance learning for survival prediction from whole slide images. In: Shen, D., Liu, T., Peters, T.M., Staib, L.H., Essert, C., Zhou, S., Yap, P.-T., Khan, A. (eds.) MICCAI 2019. LNCS, vol. 11764, pp. 496–504. Springer, Cham (2019). https://doi.org/10.1007/978-3-030-32239-7_55
22. Yao, J., Zhu, X., Jonnagaddala, J., Hawkins, N., Huang, J.: Whole slide images based cancer survival prediction using attention guided deep multiple instance learning networks. Medical Image Analysis p. 101789, July 2020. https://doi.org/10.1016/j.media.2020.101789. https://linkinghub.elsevier.com/retrieve/pii/S1361841520301535
23. Zhu, X., Yao, J., Huang, J.: Deep convolutional neural network for survival analysis with pathological images. In: 2016 IEEE International Conference on Bioinformatics and Biomedicine (BIBM), pp. 544–547. IEEE (2016)
24. Zhu, X., Yao, J., Zhu, F., Huang, J.: Wsisa: making survival prediction from whole slide histopathological images. In: Proceedings of the IEEE Conference on Computer Vision and Pattern Recognition, pp. 7234–7242 (2017)

Intracerebral Haemorrhage Growth Prediction Based on Displacement Vector Field and Clinical Metadata

Ting Xiao[1]([✉]), Han Zheng[2], Xiaoning Wang[2], Xinghan Chen[2], Jianbo Chang[3], Jianhua Yao[2], Hong Shang[2], and Peng Liu[1]

[1] Harbin Institute of Technology, Harbin, China
xiaoting1@hit.edu.cn
[2] Tencent Healthcare, Shenzhen, China
[3] Peking Union Medical College Hospital, Beijing, China

Abstract. Intracerebral hemorrhage (ICH) is the deadliest type of stroke. Early prediction of stroke lesion growth is crucial in assisting physicians towards better stroke assessments. Existing stroke lesion prediction methods are mainly for ischemic stroke. In ICH, most methods only focus on whether the hematoma will expand but not how it will develop. This paper explored a new, unknown topic of predicting ICH growth at the image-level based on the baseline non-contrast computerized tomography (NCCT) image and its hematoma mask. We propose a novel end-to-end prediction framework based on the displacement vector fields (DVF) with the following advantages. 1) It can simultaneously predict CT image and hematoma mask at follow-up, providing more clinical assessment references and surgery indication. 2) The DVF regularization enforces a smooth spatial deformation, limiting the degree of the stroke lesion changes and lowering the requirement of large data. 3) A multimodal fusion module learns high-level associations between global clinical features and spatial image features. Experiments on a multi-center dataset demonstrate improved performance compared to several strong baselines. Detailed ablation experiments are conducted to highlight the contributions of various components.

Keywords: Hemorrhage growth prediction · Displacement vector field · Clinical metadata · Multi-modal fusion · DNN · Stroke

1 Introduction

Stroke was the second largest cause of death globally (5.5 million deaths) after ischaemic heart disease, with 80 million prevalent cases globally in 2016 [10],

---

T. Xiao and H. Zheng—Equal contribution.

Electronic supplementary material The online version of this chapter (https://doi.org/10.1007/978-3-030-87240-3_71) contains supplementary material, which is available to authorized users.

M. de Bruijne et al. (Eds.): MICCAI 2021, LNCS 12905, pp. 741–751, 2021.
https://doi.org/10.1007/978-3-030-87240-3_71

among which 16% cases were Intracerebral Haemorrhage (ICH), but with greater morbidity and mortality than ischaemic stroke [2,12]. Treatment decision, including whether a surgery is necessary for hematoma evacuation, usually requires evaluating the current status of hematomas, such as their volume and location, and the risk of further hematoma expansion (HE) [8]. The former is usually observed from non-contrast computerized tomography (NCCT), the routine, first-line of choice, imaging method for emergent ICH evaluation worldwide. The latter, HE, as a result of active bleeding after symptom onset, leads to neurological deterioration and increases the risk of poor functional outcome and death [8,14]. The current practices of HE prediction are formulated as a classification problem. HE is defined positive if hematoma volume growth between the baseline and the follow-up CT is more than 6mL or relative volume growth is more than 33%, otherwise negative. Existing solutions for HE include visually observing the presence of several radiological signs from NCCT or CT-angiography (CTA) and radiomics based automatic methods [6,15,23].

We argue that predicting hematoma growth in such a binarization form provides quite limited information for clinical decision-making, and its cut-off value being controversial. Alternatively, directly predicting the shape of hematoma growth at the image-level, if possible, would provide more comprehensive information, such as the volume, shape, and location of hematoma, and thus could better assist clinical decision making. Similar works have been explored to predict infarct region growth for ischemic stroke [11,20,21]. However, such image-level lesion growth prediction has not been studied for ICH. Therefore, we explored this unknown topic for the first time in this work. More specifically, we defined the task as predicting the follow-up NCCT image and the hematoma mask on it, with the input of the baseline NCCT image and the hematoma mask on it.

Existing solutions of infarct region growth prediction for ischemic stroke can be categorized into two groups. The first group of methods applied machine learning models to predict each voxel independently disregarding spatial context, such as using multivariate linear regression [11,20,21] and decision tree [5,16]. The second group used deep CNN based segmentation methods, usually, U-Net architecture [18,22,24,26,27], which usually outperform the first group by taking into account the spatial context and alleviating the need for post-processing to close the structural holes in the resulting mask. However, it is still limited in that such prediction task is intrinsically different from segmentation task, and the lesion growth mechanism cannot be leveraged by segmentation approaches. Additionally, clinical metadata, such as treatment parameters, TICI score, were used in some methods to improve performance [17,19]. However, these methods fuse images and clinical metadata through simply feature concatenation without considering their correlations that can be important for prediction.

Unlike previous segmentation alike Deep CNN, we proposed a novel network designed specifically for lesion growth prediction, with the assumption that the follow-up lesion shape is a spatially transformed version of the initial lesion shape, which can be formulated as displacement vector fields (DVF). DVF is commonly used in medical image registration [4,13,25], however, we introduced DVF into the task of lesion growth prediction for the first time. More specifically,

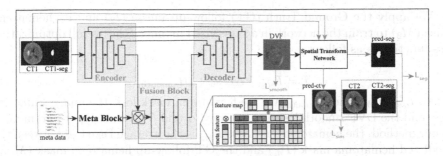

Fig. 1. Overview of the framework. The red box represents the fusion operation. (Color figure online)

instead of learning region growth shape directly, as segmentation approaches do, our method learns the DVF using a CNN, which can be used to warp the initial NCCT image and lesion mask into the predicted follow-up image and its lesion mask respectively via a spatial transformer network (STN) [9]. Additionally, to overcome the shortcomings of current methods in correlation mining between images and clinical features, we design a multi-modal fusion module to learn high-level associations between imaging and clinical metadata and integrate them for joint prediction. Experimental results show that the performance has improved compared with several strong baselines. We also conduct detailed ablation experiments to highlight the contributions of various components.

2 Method

Our method mainly consists of two functional blocks: DVF based network and clinical metadata fusion. The main idea is that the follow-up hematomas are based on the initial hematomas' spatial displacements. Therefore, we introduce the concept of DVF to perform hematoma growth prediction. Clinical metadata and radiological images provide complementary information. An intuitive idea is to incorporate clinical metadata into our DVF based framework to relate imaging with clinical information. Unlike other methods [17,19], our clinical metadata consider enriched information about patient condition, such as the patient's age, sex, time from symptom onset to baseline NCCT, etc.

2.1 The DVF Based Framework

The DVF based framework consists of encoder, decoder, and STN, as illustrated in Fig. 1. Encoder, decoder, and skip-connections are similar to that of 3D U-Net [7]. The encoder takes initial image (I_1) and its hematoma mask (M_1) as input to capture multi-resolution image features. The decoder receives the features of the encoder's corresponding layer to predict a DVF (F). And then, the STN applies the DVF to warp the initial NCCT image and its hematoma mask into the predicted follow-up image (P_i) and its hematoma mask (P_m).

We apply the Ground Truth (GT) follow-up image (I_2) and its hematoma mask (M_2) to train the network in a supervised manner. The overall optimization loss function is as follows:

$$L = L_{\text{seg}} + \alpha L_{\text{sim}} + \beta L_{\text{smooth}}, \tag{1}$$

where L_{seg}, L_{sim} and L_{smooth} are the hematoma mask loss, the image similarity loss, and the DVF smooth loss, respectively. α and β are two hyper-parameters. In our method, the hematoma mask loss L_{seg} is the average Dice loss between the predicted hematoma mask (P_m) and the GT follow-up hematoma mask (M_2),

$$L_{\text{seg}} = 1 - 2 * |M_2 \cap P_m| / (|M_2| + |P_m|). \tag{2}$$

The image similarity loss L_{sim} is the intensity mean squared error between the predicted follow-up image (P_i) and the GT follow-up image (I_2),

$$L_{\text{sim}} = 1/|\Omega| * \sum_{v \in \Omega} [I_2(v) - P_i(v)]^2 \tag{3}$$

where v is a voxel and Ω is the 3-D image spatial domain. Minimizing L_{sim} and L_{seg} will learn a DVF F and encourage the network to generate predicted image and hematoma mask approximate to their GTs. To avoid the unrealistic images generated by excessive deformation, we impose a smooth DVF regularization to constrain the displacement changes in three spatial directions. Usually, this regularization is modeled as a function of the spatial gradient of the DVF,

$$L_{\text{smooth}} = \sum_{v \in \Omega} \|\Delta F(v)\|^2. \tag{4}$$

2.2   Fusing Clinical Metadata

Clinical metadata fusion was implemented by the meta block and the fusion block, as shown in the second input branch of Fig. 1. The meta block is a fully connected network to extract global meta-features. Image features are local features due to the receptive field of the convolution kernel. The fusion block relates the local imaging with the global clinical information for high-level associations. We design a special fusion operator denoted as $\otimes$. The schematic diagram of the fusion operation is shown in the red box in Fig. 1. For the sake of intuition, we simplify the 3-D feature map to 2-D and take one channel of a patient as an example. Assuming that the patient's meta feature is a 4×1 vector, the imaging channel has only 3 feature maps, and their size is 2×2. The first step of the fusion operation is to take the value of the corresponding pixel (voxel) in the feature map and reshape them into a 1×3 vector. Repeat the above process for each pixel (voxel) of the feature maps and obtain four 1×3 image feature vectors. The clinical feature vector is then multiplied by each image feature vector to obtain four 4×3 fused feature maps. After this operation, each image feature map carries clinical information, and the two modalities can complement each other. Then the fused features are passed through the fusion block's subsequent convolutional layer to learn more high-level fusion features. After fusing the clinical information, the decoder receives the fused features and features of the encoder's corresponding layer to predict a new DVF.

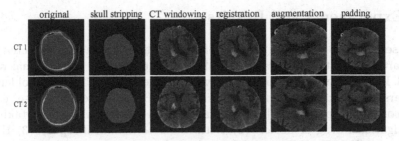

Fig. 2. Examples of one patient's two CT going the main preprocessing steps.

2.3 Data Preprocessing and Augmentation

Image processing pipeline is shown in Fig. 2, starting from stripping skull and extracting brain using [1]. Then, we set each image to a regular stroke window [0,80] HU. Afterwards, the follow-up CT2 was co-registered to the baseline CT1 (two rigid registrations followed by two affine registrations, the threshold of Dice for successful registration was empirically set to 0.94) using Advanced Normalization Tools (ANTs) [3], followed by foreground cropping and resizing (with padding) to 128×128 in the x-y plane. In the z-axis, we perform equal-distance random sampling to a size of 64. More specifically, for CT z-spacing <2 mm, we averagely group slices and resample to 64 at equal intervals. For CT z-spacing > 3 mm, we first duplicate slices 1–2 times before resampling them to 64. For more sampling details, please see supplementary material. For data augmentation, we implemented random intensity shift with an offset of 20 and probability of 0.5, random crop with the size of $32 \times 32 \times 4$ and random rotation within $5°$ in x, y-axis, and $60°$ in the z-axis. Finally, intensity of each scan was normalized to $[-1, 1]$. For 21-dimensional metadata, we rescale continuous feature to $(0,1)$, and the binary feature is encoded as follows: 1 for the positive value, -1 for the negative value. 0 for missing value in both continuous and binary features.

2.4 Implementation Details

Both encoder and decoder have four resolution blocks. In the encoder path, each block contains two convolutions with $3 \times 3 \times 3$ kernels followed by ReLU, and then a $2 \times 2 \times 2$ max pooling with strides of two. This three-layer block is repeated to capture multi-resolution features. The meta block consists of two fully connected layers. The fusion block comprises a fusion operation and three 3-D convolutional layers with a kernel size of 1. In the decoder path, each layer consists of an up-convolution of $2 \times 2 \times 2$ by strides of two, followed by two $3 \times 3 \times 3$ convolutions each followed by a ReLU. Skip connections of equal resolution in the encoder provide the essential high-resolution features to the decoder. DVF is obtained by a three-channel convolution without activation function. The network was implemented in PyTorch with Adam optimization and trained in NVIDIA Tesla P40 for 300 epochs with batch size 8. The learning rate was set to 0.001. Hyperparameters α and β were all set to 1.

3   Experiments and Results

Datasets. The 3-D NCCT dataset of hemorrhagic stroke was collected from our partner hospitals. Ethical approval was granted by the institutional review board. Inclusion and exclusion criteria were as follows: (1) ICH patient older than 18 years with Intraparenchymal hemorrhage (IPH) only, excluding cases caused by secondary hemorrhages like trauma, cerebral infarction, or undertaking an investigational drug or device. (2) With access to a clinical report. (3) Having at least two NCCTs (CT1 and CT2), where the acquisition time of CT1 should be within 24 h of stroke onset, and the interval between CT2 and CT1 should be between 8–72 h (if multiple CT2 is available, use the latest). The CT1's hematoma volume should not be less than 5 ml. CT2 should be registered to CT1 successfully, and both CTs do not have severe visual artifacts. 410 patients meet the above criteria.

For determining hematoma mask, we first applied a model [7] to automatically segment and then manually check and repair it using software platform ITK-SNAP 3.2. The studies were labeled independently by two trained assistants and confirmed by the third senior doctor. All annotation staff were blinded to patients' clinical status. We randomly split our dataset into 320, 20, and 70 volumes for train, validation, and test, respectively. The Dice coefficient (Dice) and its absolute change ratio ($\text{Doc} = \frac{|P_m - M_2| \cap |M_2 - M_1|}{|P_m - M_2| + |M_2 - M_1|}$) are used to measure the performance of predicted area. Dice measures the spatial overlap of hematoma, Doc focuses on the change of hematoma. The average prediction error of hematoma volume ($\text{AEV} = P_m - M_2$) and the absolute average error of hematoma volume ($\text{AAEV} = |P_m - M_2|$) are used to evaluate the performance of predicting volume. All results are the average of four random runs.

Table 1. The results compared with baselines.

Method	Dice	Doc	AEV	AAEV
Benchmark	0.686 ± 0.02	0	-12.9 ± 0.9	12.9 ± 0.9
3D U-Net	0.692 ± 0.017	0.453 ± 0.012	6.2 ± 0.7	12.2 ± 1.7
Ours	$\mathbf{0.735 \pm 0.001}$	$\mathbf{0.483 \pm 0.002}$	$\mathbf{4.8 \pm 0.6}$	$\mathbf{10.5 \pm 0.5}$

Results Comparison with Baselines. We compared with two baselines: CT1's hematoma mask as the predicted mask (benchmark) and a standard 3D U-Net [7] for segmentation. Taking CT1's hematoma as the predicted hematoma is a strong baseline, as for most cases the hematoma of CT2 have only slight changes compared to CT1's hematoma, thus CT2 appear the same as CT1. The standard 3D U-Net has the same architecture with our encoder/decoder portion, except for the last output layer of the decoder, its output is a segmented mask. Both CT1 and CT2 are used to train the standard 3D U-Net.

Results are reported in Table 1. Our method achieves the best results in both area and volume prediction. The benchmark achieves 68.6% on Dice and 12.9ml on AAEV, but it cannot predict the hematoma's change, and the AEV is 12.9ml less than the GT volume. The 3D U-Net is only 0.6% higher on Dice and 0.7ml less on AAEV than the benchmark. Comparing the Dice and Doc, our method is 4.3% and 3.0% higher than that of 3D U-Net, and the standard deviation is reduced by order of magnitude, which demonstrates that our DVF-based method is significantly better on both prediction performance and stability than the segmentation-based method. The hematoma volume predicted by our method and the 3D U-Net is larger than the GT. While, our method still achieves 1.4ml and 1.7ml improvements on AEV and AAEV than the 3D U-net, respectively.

Table 2. Results of ablation experiments.

Method	DVF	Aug	Meta	Img	Dice	Doc	AEV	AAEV
Ours	√				0.721 ± 0.005	0.467 ± 0.002	6.1 ± 2.6	12.4 ± 1.3
	√	√			0.717 ± 0.005	0.473 ± 0.003	6.1 ± 0.7	12.1 ± 0.7
	√	√	√		0.726 ± 0.005	0.479 ± 0.004	6.0 ± 0.6	11.4 ± 0.7
	√	√	√	√	0.735 ± 0.001	0.483 ± 0.002	4.8 ± 0.6	10.5 ± 0.5

Ablation Study. We conduct four ablation studies. a) Basic DVF scheme only with mask prediction (DVF); b) The DVF with data augmentation (Aug); c) The DVF with both data augmentation and clinical metadata (Meta); d) The DVF with data augmentation, clinical metadata and CT image prediction (Img). The results are shown in Table 2, significance test result is in supplementary material. Several findings can be drawn. (1) Although data augmentation does not significantly improve the Dice of hematoma prediction, it allows the model to focus more on hematomas' changes and has a lower volume difference, which is a positive feedback for our prediction network. (2) Fusing structured clinical data improves the Dice and Doc values simultaneously and reduces the volume prediction error. It demonstrates that combining clinical data can further enhance the model's capabilities. (3) Adding image prediction achieves the best average performance on all metrics, indicating that adding image prediction task can significantly improve performance. (4) All methods have a relatively small standard deviation, which shows that our DVF-based method is stable.

Case Study. Figure 3 shows two cases. The first three columns are networks' input, and the fourth column is the GT hematoma mask. The middle two columns are networks' output. The last column is the overlap of CT2 and hematoma masks of CT1, CT2, and the predicted, where the white contour is the hematoma border in the CT1; the green area is the CT2's GT hematoma mask; the red area is FP prediction, meaning that the area is not a hematoma, but the model predicts it as. The blue area represents the FN prediction, meaning that it is a true hematoma, but the model did not predict. Comparing the

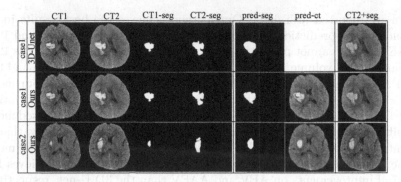

Fig. 3. Examples of the predicted image and mask.

first row and the second row, there are two findings. 1) 3D U-Net can only predict the hematoma mask, while our method can predict both the hematoma mask and CT2. 2) Our method has less FP (red area on the left and top of hematoma) and FN (blue area on the bottom of hematoma), indicating that our method performs better than the 3D U-Net. From the second and third rows, it can observe that our predicted CT can predict that the tissue around the hematoma squeezes the ventricle, providing more assessment references and surgery indication.

Results on the External Test Set. Our data set was mainly collected from 5 centers. To verify our method's robustness among different centers, we conduct cross-validation using data from 4 centers for training and data from the last center as an external test set. Because the data in center 3 accounts for about 50% of the total data. If it is used as an external test set, the remaining training data is too little so that the model will overfit. Therefore, Table 3 only shows the results among centers 1, 2, 4, and 5. The results vary among different centers. This is due to the different CT acquisition parameters between different centers and the inconsistent CT thickness. Our model performs the worst on center 5, mainly caused by the large proportion of CT with thicker slices in center 5. Our method tends to serve better on thin CTs because thin-slice CTs contain more detailed information compared to thick-slice CTs.

Table 3. Results on the external test set.

Test	Dice	Doc	AEV	AAEV
C1	0.716 ± 0.009	0.439 ± 0.007	10.6 ± 2.1	19.9 ± 1.4
C2	0.716 ± 0.002	0.475 ± 0.006	-0.9 ± 1.4	17.9 ± 0.2
C4	0.733 ± 0.007	0.475 ± 0.004	2.5 ± 2.5	18.4 ± 1.4
C5	0.619 ± 0.005	0.355 ± 0.001	-1.7 ± 1.3	19.3 ± 0.7

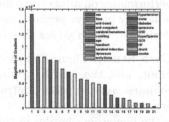

Fig. 4. Ranking result of the importance of clinical features.

The Importance Analysis on Clinical Features. We further explore which clinical feature plays a leading role. Each element's derivative of the data relative to the loss function is a scalar, and its magnitude indicates the loss change absolute value caused by 1 unit element change. Thus, given a learned model, the magnitude of the derivative of the input element relative to the loss defines the corresponding element's importance. We calculate the gradient of clinical meta-data relative to the trained model on the training set and test set. Figure 4 shows the rank results of clinical features.

In our study, sex, time from symptom onset to baseline NCCT, antiboard, anti-coagulant, and cerebral-hematoma are the top 5 most important factors for predicting stroke lesion. This study is basically in accordance with published literatures in hematoma expansion [6,15,23]. Sex is the top factor. One possible explanation is the difference in living habits may cause that sex differences. The time interval from the onset to baseline NCCT is another important factor in lesion outcome prediction. This is because patients admitted earlier after the onset of symptoms are accompanied by a higher detection rate of hematoma growth in follow-up CT. The use of anticoagulants and anti boards and the history of cerebral hemorrhage are also key factors in stroke lesion prediction.

4   Conclusion

This work explored a new clinical problem in predicting the hematoma growth in ICH and proposed a solution based on the DVF. The framework can simultaneously predict the follow-up CT image and stroke lesion mask. Experiments on the hemorrhagic stroke NCCT dataset demonstrate that our method could significantly improve performance compared with baselines. Detailed ablation experiments show that combing clinical metadata, data augmentation, and image prediction can further enhance the performance. The results of independent external experiments show that our method tends to serve better on thin CTs.

References

1. Akkus, Z., Kostandy, P., Philbrick, K.A., Erickson, B.J.: Robust brain extraction tool for CT head images. Neurocomputing **392**, 189–195 (2020)
2. An, S.J., Kim, T.J., Yoon, B.W.: Epidemiology, risk factors, and clinical features of intracerebral hemorrhage: an update. J. Stroke **19**(1), 3 (2017)
3. Avants, B.B., Tustison, N., Song, G.: Advanced normalization tools (ants). Insight j **2**(365), 1–35 (2009)
4. Balakrishnan, G., Zhao, A., Sabuncu, M.R., Guttag, J., Dalca, A.V.: Voxelmorph: a learning framework for deformable medical image registration. IEEE Trans. Med. Imaging **38**(8), 1788–1800 (2019)
5. Bauer, S., Gratz, P.P., Gralla, J., Reyes, M., Wiest, R.: Towards automatic mri volumetry for treatment selection in acute ischemic stroke patients. In: 2014 36th Annual International Conference of the IEEE Engineering in Medicine and Biology Society, pp. 1521–1524. IEEE (2014)

6. Chen, Q., et al.: Clinical-radiomics nomogram for risk estimation of early hematoma expansion after acute intracerebral hemorrhage. Academic radiology (2020)
7. Çiçek, Ö., Abdulkadir, A., Lienkamp, S.S., Brox, T., Ronneberger, O.: 3D U-Net: learning dense volumetric segmentation from sparse annotation. In: Ourselin, S., Joskowicz, L., Sabuncu, M.R., Unal, G., Wells, W. (eds.) MICCAI 2016. LNCS, vol. 9901, pp. 424–432. Springer, Cham (2016). https://doi.org/10.1007/978-3-319-46723-8_49
8. Hemphill, J.C., III., et al.: Guidelines for the management of spontaneous intracerebral hemorrhage: a guideline for healthcare professionals from the american heart association/american stroke association. Stroke **46**(7), 2032–2060 (2015)
9. Jaderberg, M., Simonyan, K., Zisserman, A., Kavukcuoglu, K.: Spatial transformer networks. In: Proceedings of the 28th International Conference on Neural Information Processing Systems-Volume 2, pp. 2017–2025 (2015)
10. Johnson, C.O., et al.: Global, regional, and national burden of stroke, 1990–2016: a systematic analysis for the global burden of disease study 2016. The Lancet Neurology **18**(5), 439–458 (2019)
11. Kemmling, A., et al.: Multivariate dynamic prediction of ischemic infarction and tissue salvage as a function of time and degree of recanalization. J. Cerebral Blood Flow Metabolism **35**(9), 1397–1405 (2015)
12. Krishnamurthi, R.V., et al.: Global and regional burden of first-ever ischaemic and haemorrhagic stroke during 1990–2010: findings from the global burden of disease study 2010. Lancet Global Health **1**(5), e259–e281 (2013)
13. Li, H., Fan, Y.: Non-rigid image registration using self-supervised fully convolutional networks without training data. In: 2018 IEEE 15th International Symposium on Biomedical Imaging (ISBI 2018), pp. 1075–1078. IEEE (2018)
14. Li, Z., et al.: Hematoma expansion in intracerebral hemorrhage: an update on prediction and treatment. Frontiers Neurol. **11**, 702 (2020)
15. Liu, J., et al.: Prediction of hematoma expansion in spontaneous intracerebral hemorrhage using support vector machine. EBioMedicine **43**, 454–459 (2019)
16. McKinley, R., et al.: Fully automated stroke tissue estimation using random forest classifiers (faster). J. Cerebral Blood Flow Metabolism **37**(8), 2728–2741 (2017)
17. Pinto, A., Mckinley, R., Alves, V., Wiest, R., Silva, C.A., Reyes, M.: Stroke lesion outcome prediction based on MRI imaging combined with clinical information. Frontiers Neurol. **9**, 1060 (2018)
18. Pinto, A., Pereira, S., Meier, R., Wiest, R., Alves, V., Reyes, M., Silva, C.A.: Combining unsupervised and supervised learning for predicting the final stroke lesion. Med. Image Anal. **69**, 101888 (2021)
19. Robben, D., et al.: Prediction of final infarct volume from native CT perfusion and treatment parameters using deep learning. Med. Image Anal. **59**, 101589 (2020)
20. Rose, S.E., et al.: Mri based diffusion and perfusion predictive model to estimate stroke evolution. Magnetic Resonance Imaging **19**(8), 1043–1053 (2001)
21. Scalzo, F., Hao, Q., Alger, J.R., Hu, X., Liebeskind, D.S.: Regional prediction of tissue fate in acute ischemic stroke. Ann. Biomed. Eng. **40**(10), 2177–2187 (2012)
22. Soltanpour, M., Greiner, R., Boulanger, P., Buck, B.: Ischemic stroke lesion prediction in CT perfusion scans using multiple parallel u-nets following by a pixel-level classifier. In: 2019 IEEE 19th International Conference on Bioinformatics and Bioengineering (BIBE), pp. 957–963. IEEE (2019)
23. Song, Z., et al.: Noncontrast computed tomography-based radiomics analysis in discriminating early hematoma expansion after spontaneous intracerebral hemorrhage. Korean J. Radiol. **21** (2020)

24. Stier, N., Vincent, N., Liebeskind, D., Scalzo, F.: Deep learning of tissue fate features in acute ischemic stroke. In: 2015 IEEE International Conference on Bioinformatics and Biomedicine (BIBM), pp. 1316–1321. IEEE (2015)

25. de Vos, B.D., Berendsen, F.F., Viergever, M.A., Staring, M., Išgum, I.: End-to-end unsupervised deformable image registration with a convolutional neural network. In: Cardoso, M.J., et al. (eds.) DLMIA/ML-CDS -2017. LNCS, vol. 10553, pp. 204–212. Springer, Cham (2017). https://doi.org/10.1007/978-3-319-67558-9_24

26. Winzeck, S., et al.: Isles 2016 and 2017-benchmarking ischemic stroke lesion outcome prediction based on multispectral MRI. Frontiers Neurol. 9, 679 (2018)

27. Yu, Y., et al.: Use of deep learning to predict final ischemic stroke lesions from initial magnetic resonance imaging. JAMA Network Open 3(3), e200772–e200772 (2020)

AMINN: Autoencoder-Based Multiple Instance Neural Network Improves Outcome Prediction in Multifocal Liver Metastases

Jianan Chen[1,3]([✉]), Helen M. C. Cheung[2,3], Laurent Milot[4], and Anne L. Martel[1,3]

[1] Department of Medical Biophysics, University of Toronto, Toronto, ON, Canada
geoff.chen@mail.utoronto.ca
[2] Department of Medical Imaging, University of Toronto, Toronto, ON, Canada
[3] Sunnybrook Research Institute, Toronto, ON, Canada
[4] Centre Hospitalier Universitaire de Lyon, Lyon, France

Abstract. Colorectal cancer is one of the most common and lethal cancers and colorectal cancer liver metastases (CRLM) is the major cause of death in patients with colorectal cancer. Multifocality occurs frequently in CRLM, but is relatively unexplored in CRLM outcome prediction. Most existing clinical and imaging biomarkers do not take the imaging features of all multifocal lesions into account. In this paper, we present an end-to-end autoencoder-based multiple instance neural network (AMINN) for the prediction of survival outcomes in multifocal CRLM patients using radiomic features extracted from contrast-enhanced MRIs. Specifically, we jointly train an autoencoder to reconstruct input features and a multiple instance network to make predictions by aggregating information from all tumour lesions of a patient. Also, we incorporate a two-step normalization technique to improve the training of deep neural networks, built on the observation that the distributions of radiomic features are almost always severely skewed. Experimental results empirically validated our hypothesis that incorporating imaging features of all lesions improves outcome prediction for multifocal cancer. The proposed AMINN framework achieved an area under the ROC curve (AUC) of 0.70, which is 11.4% higher than the best baseline method. A risk score based on the outputs of AMINN achieved superior prediction in our multifocal CRLM cohort. The effectiveness of incorporating all lesions and applying two-step normalization is demonstrated by a series of ablation studies. A Keras implementation of AMINN is released (https://github.com/martellab-sri/AMINN).

Electronic supplementary material The online version of this chapter (https://doi.org/10.1007/978-3-030-87240-3_72) contains supplementary material, which is available to authorized users.

M. de Bruijne et al. (Eds.): MICCAI 2021, LNCS 12905, pp. 752–761, 2021.
https://doi.org/10.1007/978-3-030-87240-3_72

1   Introduction

Colorectal cancer (CRC) is the fourth most common non-skin cancer and the second leading cause of cancer deaths in the United States [16]. 50–70% of CRC patients eventually develop liver metastases, which become the predominant cause of death [18]. Hepatic resection is the only potential cure for colorectal cancer liver metastases (CRLM), providing up to 58% 5-year overall survival, compared to 11% in patients with chemotherapy alone [4,5]. Stratifying CRLM patients can lead to better treatment selection and improvements in prognosis.

Main-stream clinical risk scores utilize clinical factors for CRLM (number of metastases>1, lymph node status, *etc.*) for prognosis and treatment planning, but none are strong predictors of long-term survival [13]. A variety of medical imaging biomarkers have also been developed based on texture and intensity features extracted from the tumors and/or liver tissue in routinely collected Magnetic Resonance Imaging (MRI) to predict CRLM patient outcome [2,3,11].

Most medical-imaging-based prognostic models (not limited to CRLM) focus solely on the largest lesion when dealing with multifocality. Incorporating features from multiple lesions in modeling is difficult because it's unknown how each tumor contributes to patient outcome. Although 90% of CRLM patients present with multifocal metastases, there is currently no medical-imaging based prognostic biomarker designed specifically for multifocal CRLM [16]. In this paper, we address this gap with a radiomics-based multiple instance learning framework.

We hypothesize that using features from all lesions of multifocal cancers improves outcome prediction. We propose an end-to-end autoencoder-based multiple instance neural network (AMINN) that predicts multifocal CRLM patient outcome based on radiomic features extracted from contrast-enhanced 3D MRIs (Fig. 1). This model is jointly trained to reconstruct input radiomic features with an autoencoder branch and to make predictions with a multiple instance branch. This two-branch structure encourages the learning of prognosis-relevant representations of tumors. The multiple instance neural network aggregates the representations of all tumors of a patient for survival prediction. In addition, we propose a feature transformation technique tailored to radiomic features to suit the need of deep learning models. Ablation experiments were performed to show the effect of all the components of our framework.

2   Methods

Multiple Instance Learning (MIL): MIL is a class of machine learning algorithms that learn from a bag of instances, where labels are available at the bag-level but not at instance level [10]. MIL has been widely adopted in medical image and video analysis as it performs well in weakly-supervised situations [12]. In most cases, detection or prediction using medical images is modeled as a MIL problem where the whole image (or 3d scan) is a bag and image patches (or cubes) cropped from the image are instances. Our work differs from previous

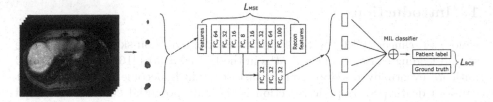

Fig. 1. An overview of the proposed autoencoder-based multiple instance learning network. The network structure between the curly brackets is shared for each tumor of the same patient.

MIL studies in treating multifocal tumors as instances. This allows us to generate more accurate instance representations (as image patches can sometimes contain redundant or irrelevant information) and also simplify the MIL problem by reducing the number of instances.

We formulated the task of patient outcome prediction as a MIL problem as follows:

$$p_i = \mathcal{M}\left(f\left(\Phi\left(x_{ij|j=1...k}\right)\right)\right) \tag{1}$$

where p_i is the probability of event $y_i \in \{0, 1\}$ (3-year overall survival) for patient i and $X_i = \{x_{i1}, ..., x_{ik}\}$ is a bag of instances representing a varying number of k tumors in patient i. Φ represents the encoding function which maps input features x_{ij} to latent representations h_{ij}. f represents the transformation function parameterized by the neural network for extracting information from instance-level representations and $\mathcal{M}$ is a permutation invariant multiple instance pooling function that aggregates instance representations to obtain bag representations and bag labels.

Radiomics Features: Radiomics is an emerging field where mathematical summarizations of tumors are extracted from medical images for use in modeling [7]. Radiomic approaches have shown promising results in various tasks and imaging modalities. For the analysis of multifocal cancers, radiomic models usually use features from the largest tumor, resulting in the loss of information [22]. Deep learning-based workflows have been investigated as an alternative to conventional radiomics, however, building a generalizable deep learning-based model requires huge datasets that are currently impractical in medical imaging and radiomics is still the generally preferred approach [1].

We extract radiomic features from preoperative 10-minute delayed T1 MRI scans using pyradiomics (v2.0.0) [19]. The 3D MRI scans were resampled to [1.5, 1.5, 1.5] isotropic spacing using B-spline interpolation. Images were normalized using Z-score normalization with 99th percentile suppression, rescaled to range [0, 100] and discretized with a bin size of 5. Tumors are delineated slice by slice by a radiologist with 6 years of experience in abdominal imaging. For each tumor lesion, we extract 959 features (including 99 original features, 172 Laplacian of Gaussian filtered features and 688 wavelet features) from three categories to mathematically summarize its intensity, shape and texture properties. However, we only kept the 99 original features as the models trained with original features performed the best. A list of the final features is included in **Table S1**.

Feature Normalization: Normalization is a common technique in machine learning for bias correction and training acceleration. Ideally, for deep learning frameworks, input data should be scaled, centered at zero and relatively symmetrically distributed around the mean to alleviate oscillating gradients and avoid vanishing (due to packed points) or exploding (due to extreme values) gradients in gradient descent [15]. We observe that the distribution of most of our radiomic features are severely skewed, calling for normalization. However, Z-score normalization, one of the most commonly-used techniques for radiomic feature normalization, is less informative for data that is not approximately normally distributed. As an example, after Z-score normalization of tumor volume, above 90% of samples are packed in the range $[-0.5, 0]$, while the largest lesions have Z-value up to 8 (Fig. 2). This is undesirable as tumor volume is correlated with many radiomic features and is an important predictor of patient outcome [21].

We utilize a simple two-step normalization algorithm that combines logarithmic and Z-score transformation to tackle this problem:

$$l_{ij} = \log(f_{ij} - 2\min(F_j) + \text{median}(F_j)) \tag{2}$$

$$Z_{ij} = \frac{l_{ij} - \mu}{\sigma} \tag{3}$$

where $f_{ij} \in F_j = \{f_{1j}, f_{2j}, ...f_{nj}\}$ is the jth input feature of the ith sample. A logarithmic transformation is first applied to reduce skewness. To handle negative and zero values, we incorporate a feature-specific constant, $-2\min(F_j) + \text{median}(F_j)$. It is a combination of a first $-\min(F_j)$ term, that ensures values are larger than or equal to zero, and a second term $\text{median}(F_j) - \min(F_j)$, that aims to avoid log operations on extremely small numbers. This is followed by a standard Z-score transformation (Fig. 2).

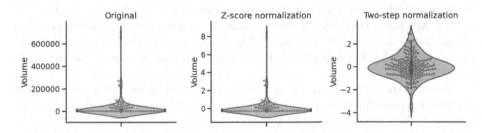

Fig. 2. Volume of 181 lesions from 50 patients with Z-score normalization and two-step normalization.

Network Structure: The autoencoder applied to reduce the dimensionality of input radiomic features is composed of a four-layer encoder and a four-layer decoder. All layers are fully-connected layers with ReLU activation except the last layer, which uses sigmoid activation. While autoencoders can be employed to select radiomic features by learning their low-dimensional representations in an unsupervised manner [2], we wish to directly extract the most relevant latent representations for patient prognosis. Thus the bottleneck layer of the autoencoder is connected to the multiple instance prediction network as input in order to select latent features that are informative of patient outcome.

The MIL network consists of 3 fully connected layers followed by a MIL pooling layer that integrates instance representations to predict patient label. Each fully connected layer consists of 32 hidden units. Various pooling methods are tested with the same backbone network [20, 23]:

$$
\mathcal{M} = \begin{cases} max : p = \max_{j=1,\ldots k} f_j \\ average : p = \frac{1}{k} \sum_{j=1}^{k} f_j \\ lse : p = r \log \left[\frac{1}{k} \sum_{j=1}^{k} \exp (r f_j) \right] \\ att : p = \sum_{j=1}^{k} a_j f_j \end{cases}
$$

The selection of MIL pooling functions usually depends on the specific task. Since it is unclear how multiple tumors collectively determine clinical outcome, in addition to *max*, *average* and *lse* (log-sum-exp, a smooth approximation of *max*), we also test attention-based pooling (*att*) [8]. The *att* model incorporates a trainable pooling module that learns the importance of each instance, and can be more flexible and interpretable compared to standard pooling functions [23].

3   Experiments

We used a retrospective cohort of 108 colorectal cancer liver metastases (CRLM) patients that were eligible for hepatic resection and were treated at our institution. Informed consent was waived by the institutional review board. All patients received Gadobutrol-enhanced MRI prior to hepatic resection. 50 patients had multifocal CRLM, and each had 2–17 lesions (mean 3.6), leading to a total of 181 lesions. Median patient overall survival was 30 months and was right-censored to obtain 3-year survival (19/50 events). The cohort has been described in detail in previous work [3]. Demographics of the multifocal cohort is available in **Table S2**.

We used the unifocal patient subset of our dataset as an independent cohort for hyper-parameter tuning using grid search in 10 repeated 3-fold cross validation (CV). Using the selected hyper-parameters, we trained AMINN on the multifocal patient subset over 10 repeated 3-fold CV with splits stratified by patient outcome, each in 100 epochs without dropout. The model was optimized using Adam optimization with an initial learning rate of 0.0001, $\beta_1 = 0.9$ and $\beta_2 = 0.999$, on a compound loss – mean squared error loss for reconstruction of the autoencoder plus binary cross-entropy loss from the patient-level prediction:

$$
\textbf{Loss} = \mathbf{L}_{\text{MSE}}(x_{recon}, x) + \alpha \mathbf{L}_{\text{BCE}}(\hat{y}, y) \tag{4}
$$

where α is a hyperparameter for the trade-off between reconstruction error and prediction error. In our experiments, $\alpha = 1$ provides robust results and given the small sample size of the dataset, we didn't attempt to tune this parameter.

Models were evaluated using area under the ROC curve (AUC) with mean and 95% confidence interval, and accuracy (ACC) with mean and standard deviation, estimated across 10 repeats of 3-fold cross validations for predicting 3-year survival. We chose AUC as the primary evaluation criteria because our dataset is slightly imbalanced. Commonly-used models in radiomic analysis: support vector machine (SVM), random forest (RF) and logistic regression (LR), combined with LASSO-selected features of the largest lesion were built as baselines [17]. We also built an abridged version of AMINN trained only on the largest lesions of multifocal patients for comparison. Further, we performed a series of ablation experiments to evaluate the effect of each component in our network.

We predicted the outcome for each fold based on a model trained on the other two folds, and combined the outputs of each test fold to obtain the survival probability predictions for all patients. We derived a binary risk score by median dichotimizing the model outputs. We compared our binary score against other clinical and imaging biomarkers for CRLM, including the Fong risk score, a clinical risk score for treatment planning, the tumor burden score (TBS), a recent risk score calculated from tumor counts and size, and target tumor enhancement (TTE), a biomarker specifically designed for 10-minutes delayed gadobutrol-enhanced MRIs [3,6,14]. All risk scores were evaluated using concordance-index (c-index) and hazard ratio (HR) in univariate cox proportional hazard models. In addition, a multivariable cox model incorporating all risk scores as predictors was used to evaluate the adjusted prognostic value of our risk score.

4   Results

The proposed autoencoder-based multiple instance neural network (AMINN) outperforms baseline machine learning algorithms by a large margin (Table 1).

Table 1. Comparison of machine learning models for predicting multifocal CLRM patient outcome in 10 repeated runs of 3-fold cross validation.

Model	Setting	AUC (95%CI)	Accuracy
Largest lesion			
SVM	Largest	0.58 (0.54–0.62)	0.62 ± 0.08
RF	Largest	0.58 (0.55–0.62)	0.60 ± 0.10
LR	Largest	0.58 (0.55–0.62)	0.62 ± 0.07
AMINN-unifocal	Largest	0.58 (0.57–0.59)	0.61 ± 0.08
Multiple instances			
AMINN	Max	0.67 (0.65–0.69)	0.65 ± 0.11
AMINN	LSE	0.67 (0.66–0.68)	0.67 ± 0.09
AMINN	Average	0.70 (0.67–0.73)	0.68 ± 0.09
AMINN	Attention [8]	0.65 (0.62-0.68)	0.65 ± 0.10

AMINN with average pooling has the best performance in both area under the ROC curve (AUC) and accuracy (ACC). Compared to the best performing baseline method (logistic regression with LASSO feature selection based on the largest tumor), AMINN with average pooling achieves a 11.4% increase in AUC and a 5.7% increase in ACC. The unifocal version of AMINN, which has the same structure as AMINN but only uses features from the largest tumor, has similar performance as other baselines. Attention-based pooling doesn't show superior performance over other methods, suggesting that more data are required for the proposed framework to learn the lethality (weights) of each lesion.

We then tested the performance gain from each component of AMINN, namely incorporating all lesions into outcome prediction (**multi**), two-step feature transformation (**log**), and feature reduction with an autoencoder (**ae**) (Table 2). Incorporating all lesions consistently improved AUC, by 3.0%, 4.1%, 7.9% and 8.3% compared to AMINN with the baseline setting (using features from the largest lesion, z-score transformation and fully connected network), with **log**, with **ae**, and with **log+ae**, respectively. Notably, adding two-step normalization also boosts performance by 4.4% to 6.5% from AMINN using only Z-score normalization. In contrast, incorporating two-step normalization results in similar performance compared to Z-score normalization in baseline non-neural network models (for conciseness, only results for logistic regression are shown). This supports the hypothesis that two-step normalization improves the training of neural networks with radiomic features as inputs. The autoencoder component of AMINN only improves model performance when multiple lesions are considered. One possible explanation is that the autoencoder overfits quickly in the abridged models.

Table 2. Ablation studies with different components of AMINN.

Model	AUC (95%CI)	Accuracy
$LogisticRegression$	0.58 (0.54–0.62)	0.62 ± 0.07
$LogisticRegression_{log}$	0.59 (0.54–0.63)	0.63 ± 0.09
$AMINN_{baseline}$	0.58 (0.57–0.59)	0.61 ± 0.08
$AMINN_{log}$	0.63 (0.63–0.64)	0.61 ± 0.05
$AMINN_{ae}$	0.57 (0.56–0.59)	0.57 ± 0.08
$AMINN_{log+ae}$	0.62 (0.59–0.64)	0.58 ± 0.07
$AMINN_{multi}$	0.61 (0.59–0.63)	0.62 ± 0.11
$AMINN_{multi+log}$	0.68 (0.66–0.69)	0.64 ± 0.14
$AMINN_{multi+ae}$	0.65 (0.62–0.68)	0.62 ± 0.08
$AMINN_{multi+log+ae}$	0.70 (0.67–0.73)	0.68 ± 0.09

When compared to clinical and imaging biomarkers for CRLM, our risk score is the only one that achieved predictive value in univariate cox regression modeling for our cohort of multifocal patients (Table 3), with c-index of 0.63 and HR of

2.88 (95%CI: 1.12-7.74). In multivariable analysis where all four biomarkers are included as predictors, our risk score remained the only one showing predictive value.

Interestingly, performance of all models improved when applied to the entire cohort of 58 unifocal and 50 multifocal cases (Table 4). We observe that although both TBS and the Fong score take into account the number of tumor lesions, they fail to further stratify within multifocal patients as all patients have elevated risk from multifocality. Similarly, although TTE incorporates intensity features of the two largest lesions when there are multiple tumours, it has less predictive value when all patients have two or more tumors. The gap in predictive power of these biomarkers between predicting on the full cohort and the multifocal subset highlights the need for developing multifocality-aware cancer prediction models.

Table 3. Uni- and multi-variable cox regression models using different biomarkers on multifocal cohort. Q-values derived from p-values with Benjamini–Hochberg correction.

Method	Univariate analysis			Multivariable analysis		
	HR (95%CI)	C-index	Q-value	HR (95%CI)	C-index	Q-value
Fong [6]	1.44 (0.60–3.44)	0.55	0.41	1.19 (0.45–3.15)	0.72	0.47
TBS [14]	1.73 (0.77–3.88)	0.58	0.18	2.59 (0.96–7.01)		0.06
TTE [3]	2.29 (0.86–6.11)	0.59	0.10	1.72 (0.63–4.76)		0.29
AMINN	2.88 (1.12–7.44)	0.63	0.03*	4.22 (1.38–12.88)		0.01*

Table 4. Existing biomarkers have higher predictive value on the full cohort (unifocal subset plus multifocal subset)

Method	HR (95% CI)	C-index	p-value
Fong score [6]	1.96 (0.94–4.08)	0.57	0.07
TBS [14]	1.72 (1.02–2.89)	0.61	0.04
TTE [3]	2.82 (1.38–5.77)	0.65	<0.01

5   Conclusion and Discussion

In conclusion, we propose an end-to-end autoencoder-based multiple instance neural network (AMINN) to predict outcomes of multifocal CRLM patients from radiomic features extracted from contrast-enhanced MRIs. By incorporating information from all lesions for patient outcome prediction, our model achieves a 11.4% increase in AUC and 5.7% increase in accuracy compared to commonly-used methods. The risk score built from the outputs of our network outperforms other clinical and imaging biomarkers and remains the only one with predictive value in our multifocal CRLM cohort. We also demonstrate the ability of our two-step normalization technique to improve the performance of our radiomic-feature-based neural network.

The primary limitation of our study is its small sample size. Preoperative multifocal CRLM scans are difficult to acquire because most multifocal patients are not eligible for surgery and have poor survival [18]. To the best of our knowledge, there is no public imaging dataset curated for outcome prediction of multifocal patients in any tumor type. One of the goals of this study is to draw attention to multifocal diseases in the medical image computing community.

Ideally, we would like to use an automatic segmentation model, for instance, the nnUNet [9] to perform tumor segmentation in order to avoid inter- / intra-observer variability and to automate the analysis pipeline. However, although we were able to obtain high accuracy in segmenting the whole liver using deep learning, we found that the results for liver tumor segmentation were unsatisfactory. We therefore opted to use manual segmentation by a radiologist in order to evaluate our methodology. The development of a more accurate detection and segmentation network for MRI liver lesions is the subject of future work. Similarly, we did not apply whole image-based deep learning because the features learned can be biased if the network cannot detect or segment lesions accurately.

Acknowledgements. The authors would like to thank The Natural Sciences and Engineering Research Council of Canada (NSERC) for funding, and acknowledge the contribution of Drs. Karanicolas, Law and Coburn in helping to create the patient cohort for this study.

References

1. Afshar, P., Mohammadi, A., Plataniotis, K.N., Oikonomou, A., Benali, H.: From handcrafted to deep-learning-based cancer radiomics: challenges and opportunities. IEEE Signal Process. Mag. **36**(4), 132–160 (2019)
2. Chen, J., Milot, L., Cheung, H.M.C., Martel, A.L.: Unsupervised clustering of quantitative imaging phenotypes using autoencoder and gaussian mixture model. In: Shen, D., et al. (eds.) MICCAI 2019. LNCS, vol. 11767, pp. 575–582. Springer, Cham (2019). https://doi.org/10.1007/978-3-030-32251-9_63
3. Cheung, H.M., et al.: Late gadolinium enhancement of colorectal liver metastases post-chemotherapy is associated with tumour fibrosis and overall survival post-hepatectomy. European radiology, pp. 1–8 (2018)
4. Fernandez, F.G., Drebin, J.A., Linehan, D.C., Dehdashti, F., Siegel, B.A., Strasberg, S.M.: Five-year survival after resection of hepatic metastases from colorectal cancer in patients screened by positron emission tomography with f-18 fluorodeoxyglucose (fdg-pet). Ann. Surg. **240**(3), 438 (2004)
5. Ferrarotto, R., et al.: Durable complete responses in metastatic colorectal cancer treated with chemotherapy alone. Clin. Colorectal Cancer **10**(3), 178–182 (2011)
6. Fong, Y., Fortner, J., Sun, R.L., Brennan, M.F., Blumgart, L.H.: Clinical score for predicting recurrence after hepatic resection for metastatic colorectal cancer: analysis of 1001 consecutive cases. Ann. Surg. **230**(3), 309 (1999)
7. Gillies, R.J., Kinahan, P.E., Hricak, H.: Radiomics: images are more than pictures, they are data. Radiology **278**(2), 563–577 (2016)

8. Ilse, M., Tomczak, J.M., Welling, M.: Attention-based deep multiple instance learning. In: International Conference in Machine Learning (2018)
9. Isensee, F., Jaeger, P.F., Kohl, S.A., Petersen, J., Maier-Hein, K.H.: nnu-net: a self-configuring method for deep learning-based biomedical image segmentation. Nat. Methods **18**(2), 203–211 (2021)
10. Maron, O., Lozano-Pérez, T.: A framework for multiple-instance learning. In: Advances in Neural Information Processing Systems, pp. 570–576 (1998)
11. Nakai, Y., et al.: Mri findings of liver parenchyma peripheral to colorectal liver metastasis: A potential predictor of long-term prognosis. Radiology, p. 202367 (2020)
12. Quellec, G., Cazuguel, G., Cochener, B., Lamard, M.: Multiple-instance learning for medical image and video analysis. IEEE Rev. Biomed. Eng. **10**, 213–234 (2017)
13. Roberts, K., et al.: Performance of prognostic scores in predicting long-term outcome following resection of colorectal liver metastases. Br. J. Surg. **101**(7), 856–866 (2014)
14. Sasaki, K., et al.: The tumor burden score: a new "metro-ticket" prognostic tool for colorectal liver metastases based on tumor size and number of tumors. Ann. Surg. **267**(1), 132–141 (2018)
15. Shi, J.J.: Reducing prediction error by transforming input data for neural networks. J. Comput. Civ. Eng. **14**(2), 109–116 (2000)
16. Siegel, R.L., Miller, K.D., Jemal, A.: Cancer statistics. CA: Aancer J. Clinicians **69**(1), 7–34 (2019)
17. Tibshirani, R.: Regression shrinkage and selection via the lasso. J. Roy. Stat. Soc.: Ser. B (Methodol.) **58**(1), 267–288 (1996)
18. Valderrama-Treviño, A.I., Barrera-Mera, B., Ceballos-Villalva, J.C., Montalvo-Javé, E.E.: Hepatic metastasis from colorectal cancer. Euroasian J. Hepato-Gastroenterology **7**(2), 166 (2017)
19. Van Griethuysen, J.J., et al.: Computational radiomics system to decode the radiographic phenotype. Can. Res. **77**(21), e104–e107 (2017)
20. Wang, X., Yan, Y., Tang, P., Bai, X., Liu, W.: Revisiting multiple instance neural networks. Pattern Recogn. **74**, 15–24 (2018)
21. Welch, M.L., et al.: Vulnerabilities of radiomic signature development: the need for safeguards. Radiother. Oncol. **130**, 2–9 (2019)
22. Yip, S.S., Aerts, H.J.: Applications and limitations of radiomics. Phys. Med. Biol. **61**(13), R150 (2016)
23. Zhou, S.K., Rueckert, D., Fichtinger, G.: Handbook of medical image computing and computer assisted intervention. Academic Press (2019)

Survival Prediction Based on Histopathology Imaging and Clinical Data: A Novel, Whole Slide CNN Approach

Saloni Agarwal[✉], Mohamedelfatih Eltigani Osman Abaker,
and Ovidiu Daescu

Department of Computer Science, University of Texas at Dallas, Richardson,
TX 75080, USA
{saloni.agarwal1,m.abaker,daescu}@utdallas.edu

Abstract. Current methods for whole slide image (WSI) histopathology subregion classification and survival prediction rely on phenotype clustering from randomly sampled image tiles or on analyzing key tiles selected by experts from the much larger in size WSIs. These approaches do not capture the whole tissue region present in a histopathology image, also missing the spatial distribution of features that could be critical for good survival predictors. We propose a novel method that extracts a whole slide feature map (WSFM) in the first step and then uses it to train the survival prediction model. Specifically, we partition the WSI into tiles, and for each tile extract InceptionV3 features followed by PCA dimension reduction. The low dimension features of each tile are stored as the channel information in the WSFM. The resulting WSFM preserves the tile adjacency information and captures the entire tissue in the WSI. To overcome the small-size data set concern inherent to previous methods, we design a siamese survival convolutional neural network (SSCNN) that takes the WSFM and multivariate clinical features as input and predicts the survival score. We train the SSCNN using a novel loss function that combines a modified pairwise ranking loss and a bounded inverse term. The key advantages of the proposed method are that it does not require pixel-level annotations, a notorious bottleneck, and it can be easily adapted for any type of tumor without performance dependence on other parameters like the number of clusters. Experimental results demonstrate the effectiveness of the proposed SSCNN over other state-of-the-art survival analysis approaches.

Keywords: Histopathology · Siamese network · Survival analysis

1 Introduction

With large histopathology images being captured at an increasing rate and the development of deep learning models for analyzing such images, image-based

M. de Bruijne et al. (Eds.): MICCAI 2021, LNCS 12905, pp. 762–771, 2021.
https://doi.org/10.1007/978-3-030-87240-3_73

computer-aided diagnosis based on segmentation and classification has rapidly expanded. Histopathology images capture tumor growth and morphology with great details, making them highly suitable for patient risk assessment. However, there is little work on histopathology image-based survival analysis because of the challenges involved. First, in contrast to the natural images where a label is learned for a relatively smaller image (100s or 1000s of pixels) input, in histopathology studies labels are learned from extremely large inputs (WSIs might be 10^{10} pixels). Thus, patient samples with corresponding labels are usually limited, while a massive number of parameters need to be optimized when using WSIs in deep learning-based approaches. Second, histopathology images have extremely heterogeneous structures and textures, and annotating specific regions to facilitate risk prediction is laborious and often unfeasible.

Previously build histopathology image-based survival models follow a two-step approach where scores are predicted for each tile in the first step and aggregated using various strategies in the second step [19]. However, these techniques fail to capture the complexity of disease progression, reflected from a range of histopathology structures. More recent works are based on multiple instance learning and attention mechanism, but they still sample tiles to form the phenotype representation [16,17]. The performance of these models relies heavily on the choice of the number of phenotype clusters. Moreover, sampling a few tiles from the histopathology images can not ensure a complete view of the tissue, and the tile adjacency information is also lost. Graph convolutional network-based methods have also been developed, but they require prior knowledge to construct a complete graph representation [9].

The first neural network-based survival analysis model is the DeepSurv [6], which employs a fully connected network to represent nonlinear risk and optimizes the Cox partial likelihood. Most of the other deep learning-based models use this loss function [19], and the Cox partial likelihood is approximated over the training batch instead of calculating it for the entire training dataset. In [14], a censored cross-entropy loss function is used, where the time is discretized based on percentiles, and survival analysis is treated as a classification task. In [5], a rank-based loss function is proposed, defined as the sum of an extended mean squared error loss and a pairwise ranking loss based on ranking information on survival data. We suggest a new loss function for training the survival analysis models inspired by this ranking loss. The primary motivation for viewing survival analysis as a ranking problem is that clinical management decisions involving treatment and monitoring are based on the relative risk of patients.

In this paper, we propose siamese style pairwise learning for solving survival prediction as a ranking problem. The siamese style pairwise training helps overcome the small size of patient samples problem, as N training samples can form upto $O(N^2)$ pairs. A key contribution in this paper is the generation of whole slide feature maps (WSFM) by assembling tile level features extracted from the pre-trained InceptionV3 [11] model. WSFM provides a complete view of the histopathology image, preserving the tile location information. Due to the creation of WSFM, the whole histopathology image can be processed at once, and no annotations by pathologists are needed on the histopathology images, overcoming perphaps the most important and costly bottleneck. Additionally, clinical information along with histopathology images is a better indicator of survival. Hence, we propose SSCNN for survival prediction taking the multivariate clinical feature along with WSFM as input. We evaluate the model performance using the Concordance index (C-index) [2] value and compare it with other state-of-the-art methods on a publicly available dataset.

2   Methodology

In survival analysis, patient i with feature vector X_i has a pair of associated labels (t_i, o_i), where t_i is the observed time (in days/months/years) and o_i is the indicator of event occurrence, set to 1 for an uncensored instance (death observed at t_i) and to 0 for a censored instance (death not observed by t_i). The feature vector X_i is given as input to a model which outputs a survival score yp_i. The standard performance metric for survival prediction is the C index [2], which is the probability that a randomly chosen pair of subjects (i, j) is correctly ordered by the model in terms of event times. Since the effectiveness of a survival predictor is based on comparing two subjects, we train our model on subject pairs (i, j) instead of on individual subjects. We use siamese style training with batch gradient descent because siamese networks are trained on data pairs instead of individual data. This model training style overcomes the often encountered inefficacy due to overfitting, caused by optimizing large sets of deep learning parameters with limited patient samples.

In siamese architecture, the base network is replicated to form "twins", with shared weight matrices at each layer [7]. The features of a subject pair (X_i, X_j) are provided as input to the twin network that predicts (yp_i, yp_j) as the output. We build our model on permissible pairs having more than a constant difference d_{diff} in the observed time. In deep learning-based models, we want the training dataset to be clean to avoid confusion during model training. A difference of d_{diff} ensures that the features of the pair being compared are significantly different for relative survival score estimation and model weight training. The model is trained such that if t_i is greater than t_j then yp_i should be greater than yp_j and vice versa, using the proposed loss function described next.

2.1   Loss Function

A loss function L_{Total} is used for training the siamese survival predictor. This loss function is the sum of two terms, L_1 and L_2, where L_1 is an extended pairwise ranking loss and L_2 is an inverse function that constraints the survival model to predict a valid outcome.

Specifically, L_1 is a pairwise ranking loss that calculates the rank error between survival instances and the predicted values. For a pair of instances i and j, $(t_i - t_j)$ is the true difference between the occurrence times, say $true_{diff}$, and $(py_i - py_j)$ is the predicted score difference, say $pred_{diff}$, then the error is defined as $(true_{diff} - pred_{diff})^2$. If we consider the ranking space of a pair of survival instances as a sample, the error can be regarded as the square error of the sample. We have right-censored data in survival analysis, so not all data can be compared for survival score ranking. If $true_{diff} > 0$ then $pred_{diff}$ should be at least $true_{diff}$ or it will lead to an error. Similarly, if $true_{diff} < 0$ then $pred_{diff}$ should be at most $true_{diff}$ or it will lead to an error. In both the conditions, a loss is incurred either when the product of $true_{diff}$ and $pred_{diff}$ is less than 0, or when the product is greater than 0 but the absolute true difference is larger than the absolute predicted difference. Therefore, a comparability indicator for a pair of instances is defined as:

$$\text{comp}(i,j) = \begin{cases} 1 : true_{diff} \times pred_{diff} < 0 \\ 1 : true_{diff} \times pred_{diff} > 0 \text{ and } |true_{diff}| > |pred_{diff}| \\ 0 : \text{otherwise} \end{cases} \quad (1)$$

The L_1 term is then formulated as:

$$L_1(i,j) = \text{comp}_{i,j} \times (true_{diff} - pred_{diff})^2 \quad (2)$$

We train the model on only those permissible pairs that are d_{diff} days apart so we don't want the model to predict very close survival scores while training. To prevent the model from computing survival scores with a difference of less than c (a constant) and for better model convergence, we also compute another term L_2, defined as follows, where ϵ is a small constant

$$L_2(i,j) = \begin{cases} \frac{1}{|pred_{diff}|+\epsilon} : |pred_{diff}| < c \\ 0 \quad : \text{otherwise} \end{cases} \quad (3)$$

Finally, L_{Total} is defined as:

$$L_{\text{Total}}(i,j) = \alpha L_1(i,j) + \beta L_2(i,j) + \lambda_1 \|\theta\|_1 + \lambda_2 \|\theta\|_2^2 \quad (4)$$

with α and β positive constants, θ weights of the network, and λ_1 and λ_2 regularization parameter.

2.2 Whole Slide Feature Maps (WSFM)

Due to the large size of histopathology images it is not feasible to feed a whole WSI to a CNN model. Instead, a WSI is partitioned into tiles and the much smaller tile images are processed independently. Hence, we create a downscaled representation of a histopathology image as a WSFM to train a CNN. A WSFM is better than using a thumbnail image because the tile feature extraction helps capture morphological details in WSFM, but the lower objective magnification thumbnail images lack any tumor progression information. For generating the WSFM, we first tile the histopathology image at 1.25X, 5X, and 20X objective magnification to form small RGB images of size 128×128, 512×512, 2048×2048 pixels, respectively. The tiles at different magnifications are centered at the same WSI location as in [4]. These tiles are pre-processed to remove

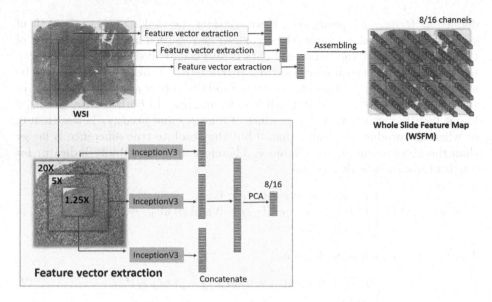

Fig. 1. Process of creating WSFM from a WSI.

white background that does not contain a tissue region. The tiles containing tissue regions are sent as input to an ImageNet pre-trained InceptionV3 model and converted to a 2048 feature vector. We concatenate the feature vectors from all the magnifications forming a vector of size 6144 for a given tile center. Further, we reduce the dimension of the extracted feature vector down from 6144 by training a PCA model on all the feature vectors extracted from a subset of 100 training histopathology images. We do this to decrease the overall WSFM dimension while capturing the essential features, and experiment with 8 and 16 PCA extracted features. The pyramidal tiles of size $128 \times 128 \times 3$, $512 \times 512 \times 3$, and $2048 \times 2048 \times 3$ (3 RGB color channels) are replaced with their corresponding 8/16 feature vectors. In most histopathology image-based methods, individual tiles are studied independently, including in [4]. Instead, we assemble all the tile features from a histopathology image in place and create the whole slide feature map (WSFM) to capture the whole tissue. The entire process of the WSFM creation is shown in Fig. 1.

2.3   Siamese Survival Convolutional Neural Network (SSCNN)

Figure 2 gives an overview of the proposed framework. The base network in SSCNN takes the WSFM representation of the histopathology image and multivariate clinical features as input and predicts a survival score. It consists of a convolution block and a fully connected block. The convolution block has two convolutional layers followed by a 2D average pooling layer. We extract all the histopathology features to be used in survival analysis as an output from this convolution block. This convolutional block structure ensures that we can extract

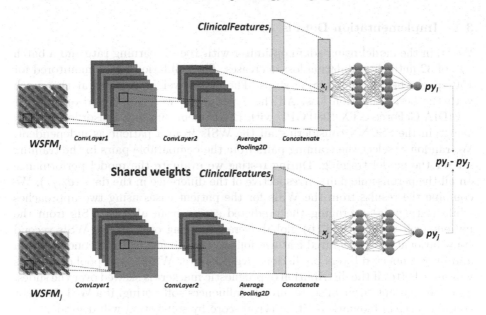

Fig. 2. Siamese survival convolutional neural network model architecture.

histopathology features from WSFM of any width and height during testing. We concatenate the histopathology features extracted from the convolution block with the multivariate clinical features. These concatenated features are input to the fully connected block. The fully connected block has two fully connected layers with dropouts between them and gives the survival score as the output.

3   Experiments

We focus on the Glioblastoma multiforme (GBM) brain cancer in our study. We downloaded the diagnostic WSIs and clinical data for GBM from The Cancer Genome Atlas (TCGA) [13] cohort. This dataset contains 860 h&e stained diagnostic histopathology slides and clinical information from 389 patients. The clinical data include patient age, gender, ethnicity, race, vital status, days to last follow up and days to death information. The clinical features used in the building of SSCNN are numerical age, and categorical gender, ethnicity and race. We use one hot-encoding for the categorical features and drop the last category. The vital status gives the patient mortality information, the occurrence indicator, o_i, is set to 1 if the vital status is dead and 0 otherwise. The observed time, t_i, is days to death for the patients who died and days to last follow up for others. We divide the whole dataset into 70% patient cases for model development and 30% patient cases for testing. We further divide the development set into 80% cases for training and 20% for validation during model building.

3.1   Implementation Details

We train the model using adam optimizer with 10e−4 learning rate and a batch size of 32 until the validation loss decreases. The validation loss is monitored for 150 epochs before the training stops. The model weight is saved at the epoch with the best validation loss. All the experiments were conducted on a single NVIDIA GeForce GTX 1080 GPU with 12 GB memory using Tensorflow library. We train the SSCNN considering all the WSIs from a patient as independent. We randomly select the training pairs from the comparable pairs in the training set for the model training. During testing we evaluate the model performance on all the permissible pairs, irrespective of the difference in the days (d_{diff}). We combine the results from the WSIs for the patient cases using two approaches while testing: (a) averaging the predicted scores from all the WSIs from the patient (soft-voting), and (b) global average pooling over all the WSIs second convolutional layer generated features followed by clinical features concatenation and fully connected layers prediction (avg-pool over WSIs). Avg-pool over WSIs will serve better if the distribution of prognostic markers is skewed over the slides. Since the patient survival score directly influences soft voting, if a WSI does not contain survival biomarkers, its survival score by soft-voting will degrade.

We also experimented by training the model using only the pairwise ranking loss term (L_1), without the inverse term (L_2), and it did not converge well. A combination of ranking loss and the inverse term led to convergence with good validation results. Further, we also analyzed the performance of the model using the average of the tile feature vectors, instead of forming WSFM, but simply averaging the PCA generated tile features for a WSI resulted in low performance. After experimenting with different settings, the best validation results were obtained by combining both techniques, the WSFM construction and the training of the Siamese network with the proposed loss function.

Comparison Models. The proposed model is compared with the state-of-the-art survival predictors including regularized Cox proportional hazards models: (LASSO-Cox) [12], and elastic net penalized Cox (EN-Cox) [15], Parametric censored regression models with Weibull [8], Random Survival Forest (RSF) [3] and Multi-Task Learning model for Survival Analysis (MTLSA) [10]. CellProfiler [1] extracted cell shape, size, texture, distribution of the pixel intensity and nuclei texture, and pixel intensity distribution serve as features for these models. WSISA [20]and DeepMISL [16] with 7 clusters are two other WSI-based survival models we compare our results with. We don't compare the model performance with [14] because they formulate survival prediction as a classification problem.

3.2   Results and Discussion

The C-index is the standard metric to analyze the performance of survival prediction. We compare the performance of different models based on their C- index values. The C-index values range from 0 to 1, and a larger value means a better survival predictor. SSCNN is trained on 1000 to 10,000 training permissible pairs with 8 and 16 PCA features, 32 first convolutional layer filters, 32 second convolutional layer filters, 32 first dense layer units, 32 second dense layer units, with a dropout of 0.2 between the dense layers. The best SSCNN is obtained by training on 3000 permissible pairs with 16 PCA features and soft-voting to get patient-level predictions. Table 1 shows the C-index values from various survival models, and the proposed SSCNN has the best performance among all models. The c-index value for SSCNN without WSI features is 0.6201 and without clinical features is 0.493. This suggests that both WSI and clinical features are important for the survival prediction. Additionally, our approach has better performance than the MR Image-based survival prediction in [18] with an 0.51 accuracy score. This comparison suggests that using histopathology images with the clinical data is more effective than using only MRI images in the survival prediction.

Table 1. Performance comparison of the proposed methods and other existing related methods using C-index values.

Models	LASSO-Cox	EN-Cox	RSF	Weibull	MTLSA	WSISA	DeepMISL	Proposed
C index	0.496	0.501	0.539	0.521	0.577	0.585	0.576	**0.6221**

Although the number of permissible pairs of WSIs in the training set is in the order of a hundred thousand, we can train the model on much fewer pairs while retaining good performance. Figure 3 shows the performance of SSCNN trained on 1000 to 10,000 training pairs for 8 and 16 PCA generated features using both soft voting and avg-pool over WSIs to get patient-level survival scores. The plot in Fig. 3 shows that SSCNN is exceptionally robust to the various model parameters, like the number of PCA features used, the strategy used to get patient-level scores, and the number of training pairs used during model building. Contrary to this, the performance of phenotype cluster-based approaches [16,17] relies heavily on the number of clusters used.

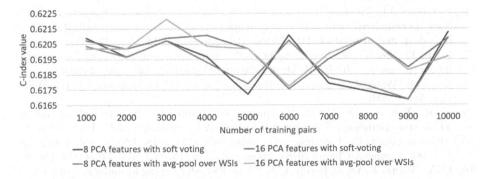

Fig. 3. Test pairs c-index values over different number of training pairs.

To compare the effect of the number of PCA features used to create the feature map on the model performance, we observe that results with 16 PCA features are mostly slightly higher than the results from 8 PCA features. This should be expected because 16 PCA features capture more variation from the original WSI. The model performances are stable with respect to the number of training pairs, ranging from 0.617 to 0.622, which is a desirable outcome.

4   Conclusion

We proposed a novel feature map generation technique from the multi-resolution WSIs that stores information from the whole tissue. We build a siamese-style CNN using this feature map and clinical data for survival prediction. This model learns the survival patterns without any annotations on the WSI. Compared to other models, this model has a superior performance because of the inclusion of both clinical and histopathology information and the correct positioning of features from the tiles in the generated feature map. Future research will focus on analyzing this model's performance for other types of cancer from various anatomic sites. We will also study methods to combine MR images and histopathology images for the survival prediction.

References

1. Carpenter, A.E., et al.: CellProfiler: image analysis software for identifying and quantifying cell phenotypes. Genome Biol. **7**(10), 1–11 (2006)
2. Harrell, F.E., Califf, R.M., Pryor, D.B., Lee, K.L., Rosati, R.A.: Evaluating the yield of medical tests. JAMA **247**(18), 2543–2546 (1982)
3. Ishwaran, H., Gerds, T.A., Kogalur, U.B., Moore, R.D., Gange, S.J., Lau, B.M.: Random survival forests for competing risks. Biostatistics **15**(4), 757–773 (2014)
4. Jaber, M.I., et al.: A deep learning image-based intrinsic molecular subtype classifier of breast tumors reveals tumor heterogeneity that may affect survival. Breast Cancer Res. **22**(1), 12 (2020)
5. Jing, B., et al.: A deep survival analysis method based on ranking. Artif. Intell. Med. **98**, 1–9 (2019)
6. Katzman, J.L., Shaham, U., Cloninger, A., Bates, J., Jiang, T., Kluger, Y.: DeepSurv: personalized treatment recommender system using a cox proportional hazards deep neural network. BMC Med. Res. Methodol. **18**(1), 1–12 (2018)
7. Koch, G., Zemel, R., Salakhutdinov, R.: Siamese neural networks for one-shot image recognition. In: ICML Deep Learning Workshop, Lille, vol. 2 (2015)
8. Lee, E.T., Wang, J.: Statistical Methods for Survival Data Analysis, vol. 476. Wiley, Hoboken (2003)
9. Li, R., Yao, J., Zhu, X., Li, Y., Huang, J.: Graph CNN for survival analysis on whole slide pathological images. In: Frangi, A.F., Schnabel, J.A., Davatzikos, C., Alberola-López, C., Fichtinger, G. (eds.) MICCAI 2018. LNCS, vol. 11071, pp. 174–182. Springer, Cham (2018). https://doi.org/10.1007/978-3-030-00934-2_20
10. Li, Y., Wang, J., Ye, J., Reddy, C.K.: A multi-task learning formulation for survival analysis. In: Proceedings of the 22nd ACM SIGKDD International Conference on Knowledge Discovery and Data Mining, pp. 1715–1724 (2016)

11. Szegedy, C., Vanhoucke, V., Ioffe, S., Shlens, J., Wojna, Z.: Rethinking the inception architecture for computer vision. In: Proceedings of the IEEE Conference on Computer Vision and Pattern Recognition, pp. 2818–2826 (2016)
12. Tibshirani, R.: The Lasso method for variable selection in the Cox model. Stat. Med. **16**(4), 385–395 (1997)
13. Tomczak, K., Czerwińska, P., Wiznerowicz, M.: The cancer genome atlas (TCGA): an immeasurable source of knowledge. Contemp. Oncol. **19**(1A), A68 (2015)
14. Wulczyn, E., et al.: Deep learning-based survival prediction for multiple cancer types using histopathology images. PLOS ONE **15**(6), e0233678 (2020)
15. Yang, Y., Zou, H.: A cocktail algorithm for solving the elastic net penalized Cox's regression in high dimensions. Stat. Interface **6**(2), 167–173 (2013)
16. Yao, J., Zhu, X., Huang, J.: Deep multi-instance learning for survival prediction from whole slide images. In: Shen, D., et al. (eds.) MICCAI 2019. LNCS, vol. 11764, pp. 496–504. Springer, Cham (2019). https://doi.org/10.1007/978-3-030-32239-7_55
17. Yao, J., Zhu, X., Jonnagaddala, J., Hawkins, N., Huang, J.: Whole slide images based cancer survival prediction using attention guided deep multiple instance learning networks. Med. Image Anal. **65**, 101789 (2020)
18. Yogananda, C.G.B., et al.: Fully automated brain tumor segmentation and survival prediction of gliomas using deep learning and MRI. In: Crimi, A., Bakas, S. (eds.) BrainLes 2019. LNCS, vol. 11993, pp. 99–112. Springer, Cham (2020). https://doi.org/10.1007/978-3-030-46643-5_10
19. Zhu, X., Yao, J., Huang, J.: Deep convolutional neural network for survival analysis with pathological images. In: 2016 IEEE International Conference on Bioinformatics and Biomedicine (BIBM), pp. 544–547. IEEE (2016)
20. Zhu, X., Yao, J., Zhu, F., Huang, J.: WSISA: making survival prediction from whole slide histopathological images. In: Proceedings of the IEEE Conference on Computer Vision and Pattern Recognition, pp. 7234–7242 (2017)

Beyond Non-maximum Suppression - Detecting Lesions in Digital Breast Tomosynthesis Volumes

Yoel Shoshan, Aviad Zlotnick, Vadim Ratner[✉], Daniel Khapun, Ella Barkan, and Flora Gilboa-Solomon

IBM Research, University of Haifa Campus, 3498825 Haifa, Israel
{yoels,aviad,vadimra,ella,flora}@il.ibm.com, daniel.khapun@ibm.com

Abstract. Detecting the specific locations of malignancy signs in a medical image is a non-trivial and time-consuming task for radiologists. A complex, 3D version of this task, was presented in the DBTex 2021 Grand Challenge on Digital Breast Tomosynthesis Lesion Detection. Teams from all over the world competed in an attempt to build AI models that predict the 3D locations that require biopsy. We describe a novel method to combine detection candidates from multiple models with minimum false positives. This method won the second place in the DBTex competition, with a very small margin from being first and a standout from the rest. We performed an ablation study to show the contribution of each one of the different new components in the proposed ensemble method, including additional performance improvements done after the competition.

Keywords: Digital Breast Tomosynthesis · Lesion Detection · Ensemble · DBTex 2021. · Deep neural networks

1 Introduction

Breast cancer (BC) is the leading cause of cancer death in women worldwide [1]. Digital Mammography (DM), a 2D X-ray based projection of the breast, is the standard of care for breast screening and several studies (e.g. [2,3]) reported impressive results of AI models trained on large-scale DM data for diagnosing BC.

Digital Breast Tomosynthesis (DBT) imaging is a new modality that offers a higher diagnostic accuracy than DM. DBT, a three-dimensional (3D) imaging modality, acquires scans by rotating an X-ray tube around the breast. The DBT volume is reconstructed from these projection views, having between 30 to 100 slices per breast side-view.

In this paper, we describe the methods to detect breast lesions and provide the location and size of a bounding box, as well as a confidence score for each detected lesion candidate.

Electronic supplementary material The online version of this chapter (https://doi.org/10.1007/978-3-030-87240-3_74) contains supplementary material, which is available to authorized users.

© Springer Nature Switzerland AG 2021
M. de Bruijne et al. (Eds.): MICCAI 2021, LNCS 12905, pp. 772–781, 2021.
https://doi.org/10.1007/978-3-030-87240-3_74

The American Association of Physicists in Medicine (AAPM), along with the SPIE (the international society for optics and photonics) and the National Cancer Institute (NCI), recently conducted a challenge called DBTex for the detection of biopsy-proven breast lesions in DBT volumes [4]. Using our methods, we reached the second place in this challenge, with a very small margin from first place.

We continued to research additional performance improvements after the challenge deadline, as the organizers allowed the submission of additional models to be tested on the competition validation set.

Our contribution is in the new methods to combine results of detectors and classifiers, and the introduction of novel filters that augment non-maximal suppression algorithms to reduce false positive candidates.

2 The DBTex Grand Challenge

The goal of DBTex Grand Challenge was to detect breast lesions that subsequently underwent biopsy and provide the location and size of a bounding box, as well as a confidence score for each detected lesion candidate.

Two performance metrics were used. Primary Metric: average sensitivity for one, two, three, and four false-positives per DBT view (3D images), only on views with a biopsied finding. Secondary Metric: average sensitivity for two false-positives per view for all views [5].

A predicted box counts as a true positive if the distance in pixels in the original image between its center point and the center of a ground truth box is less than half of the ground truth box diagonal or 100 pixels, whichever is larger. Regarding the third dimension (depth), the ground truth bounding box is assumed to span 25% of volume slices before and after the ground truth center slice and the predicted box center slice is required to be included in this range to be considered a true positive. Lesions that did not undergo biopsy did not have annotations (ground truth boxes) [4].

3 Data Sets

The DBTex grand challenge [6] made available DBT studies from 1000 women, each study with 2 to 4 volumes (total of 3592 volumes) on The Cancer Imaging Archive (TCIA) [7]. The data set contains DBT studies with breast cancers, biopsy-proven benign lesions, actionable (requiring followup) non-biopsied findings, as well as normal studies (scans without any findings). The competition organizers divided it into train (700 studies, 2596 volumes, 200 positive volumes), validation (120 studies), and test (180 studies). It contains both masses and architectural distortions (changes to the tissue which indicate potential malignancy), without calcifications. Ground truth information was provided only for the DBTex train set. We defined three sets: Ablation-Train - our inhouse dataset and one half of the DBTex train set; Ablation-Validation - the other half of the

DBTex train set; Ablation-Test - DBTex validation set, for which only the primary metric is available. There was no patient overlap between the different sets.

Model training was limited to Ablation-Train, Hyper-Parameters tuning was performed by a grid search on Ablation-Validation, and neither were done on Ablation-Test. The best performing hyper-parameters were used for testing.

Our inhouse dataset contained both MG and DBT images (1472 positive biopsy studies, 2235 negative biopsy studies, 9624 normal studies), including local annotations of over ten thousand women from both the USA and Israel. Inhouse data collection was approved by the Healthcare Networks institutional review boards (with a waiver for the need to obtain written informed consent) and was compliant with the Health Insurance Portability and Accountability Act. The data sets were fully de-identified.

4   Method

We applied several different detectors on each slice of the DBT volume, gathered all proposed candidates, used several filtering techniques to minimize false positive candidates and kept the best bounding boxes candidates. A total of nine detection models were used, and a single classifier model.

4.1   Models Architecture

We used RetinaNet-based [8] object detectors utilizing ResNet [9] and Feature Pyramid Network (FPN) [10] backbone, a proven architecture for detecting objects on different scales. For all detectors, the input was a single image slice. The output was bounding boxes at different positions, sizes and aspect ratios in that slice, with corresponding malignancy probability scores. See Fig. 1. Observing that findings in DBT slices tend to look sharp in usually a small set of adjacent slices, and very blurry in others, we hypothesized that a 2D detector will perform well as a candidates generator. For training we used finding-level annotations, namely slice number and the bounding box of the finding contour. We pretrained detector models weights first on either COCO [11] or ImageNet [12], then on 2D Digital Mammography images, and only then on DBT slices. Input intensity values were linearly scaled to the range $[-1.0, 1.0]$. Images were downsampled by a factor of 2 on xy axes, but not on slices axis.

Apart from the initialization weights, different models vary in the following aspects: Trained only on the in-house dataset or additionally fine-tuned on half of DBTex train set; Rotation augmentation - either any angle or none; Training only on images with biopsied findings or on all; ResNet34 or ResNet50 backbones. Additionally, standard image augmentations were used for all models - namely horizontal and vertical flips, and color augmentation - gamma and contrast modifications. Non Maximum-Suppression was activated on the outputs of all individual detector models, prior entering the ensemble module.

The CLS component (explained in the Ensemble section) makes use of a single slice classifier model. It is based on a fully convolutional 2D InceptionResnetV2

[13], and was trained similarly to the detectors on a slice level, but using only a single slice-level label of either containing to-be biopsied lesion or not. During training typical image augmentation was used - horizontal and vertical flips, any angle rotation, color modifications - gamma and contrast. It was pretrained on ImageNet only. Pytorch version 1.6 [14] was used for training all detection and classification models.

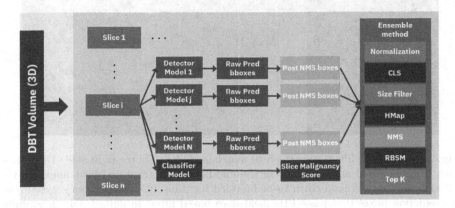

Fig. 1. An overview of the entire method. First, prediction boxes are collected per model per slice for an entire DBT volume. Then, all of the aggregated prediction boxes enter the ensemble method. Our novel ensemble components are CLS, HMap and RBSM.

4.2 Ensemble

For a given 3D mammography volume, predictions undergo multiple steps, in the following order (the overall architecture is illustrated in Fig. 1):

1. Aggregation of predictions bounding boxes: For a single volume (3D image) we collected the top 3 prediction boxes from each slice, from all detector models. All the boxes were processed disregarding the origin slice. Per 2D predicted box, we stored the slice it originated from, to be finally used in the 3D predicted box. See Fig. 1.

2. Normalization: All scores were normalized linearly per model so that the lowest bounding box score per model, over all the train data, was 0, and the highest was 1.

3. Slice classifier weighting (CLS): We added the classifier model described in the architecture section. This model computes a score per slice, and was trained without full localization information. We modified the bounding box scores originating from each slice by the score predicted on the same slice by the slice-classifier. After some experimentation, we decided to multiply the box score by (1.33+slice_score).

4. Filtering by size: We discarded bounding boxes whose diagonal was over 800 pixels, and halved the scores of bounding boxes with a diagonal smaller than 100.

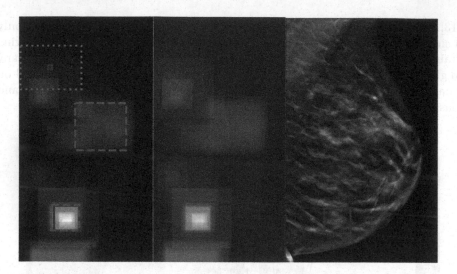

Fig. 2. An example illustrating the heat map based filter. On the right side - DBT slice, with to-be biopsied area in blue. On the middle side, a generated heat map. On the left side, in blue - ground truth to-be biopsied location. In dashed green - 2 examples of prediction boxes that passed the heat map based filter. In dotted red, an example of a box that failed the heat map based filter and was therefore rejected, including the tested central area in purple. (Color figure online)

5. Heat map based filtering (HMap): One novel component in our ensemble method involves the creation of a heat map, representing predictions of multiple detector models. See Fig. 2. Each pixel in the generated heat map was the sum of the number of bounding boxes that cover that pixel added to the sum of scores of these bounding boxes. We set a threshold proportional to the maximal value in the heat map, and then rejected any bounding box in which a rectangle in the center had at least one pixel with a value less than the threshold. We set the threshold to $1.7\sqrt{heatmap.max()}$ and the examined rectangle size to 35×35. These values were found experimentally.

6. Non-Maximal Suppression: We use the Non-Maximal Weighted algorithm (NMW) [15], with an overlap threshold of 0.26.

7. Rank based score modification (RBSM): We exponentially reduce the scores of the remaining boxes, w.r.t. their rank. We multiply each score by $\alpha^{max(0,rank-2)}$. We found $\alpha = 0.685$ to produce good results.

8. Top K: We chose the top 6 bounding boxes for evaluation.

9. 3D prediction boxes generation: Taking the boxes that "survived" all steps up to this point, we take for each box the slice origin which we stored (mentioned in Step 1), and build a 3D prediction box by extruding the 2D prediction box along the depth axis symmetrically in two directions. We use a fixed size of 25% of the total depth, since the competition evaluation metric, by design, only considers the center slice (z axis) of the predicted 3D box when deciding whether a prediction is considered a hit or not, that is unlike xy axis in which min and max x and y are all considered. The rationale for this metric is that the artifacts in DBT modality only allow to properly annotate the central slice of lesions.

For completeness, we also describe the difference between the more advanced, post competition, version of our algorithm, and the specific version used for the final test set submission in the competition. In the competition version, all boxes from all detectors on all slices were aggregated, and not only the top 3. For Non-Maximum Suppression we did not use NMW, instead we modified to following elements in the standard NMS method. (1) Unlike the standard IoU (Intersection over union), we measured overlap by intersection over the larger of the overlapping areas. We found 0.05 to be the best threshold for rejecting overlapping boxes. (2) Additionally, RBSM was not used.

5   Results

First, we present the final results in the DBTex competition [4]. Then, we analyze the contribution of the main components of our proposed method, and compare our results against prior art in the field. We report the primary metric, secondary metric and sensitivity at one false-positive per image on ablation-validation set, and the primary metric on ablation-test set (since it is the only one made available by the competition organizers).

Due to lack of direct access to the competition validation and test ground truth, we analytically estimated confidence intervals [16]:

$$CI = Z_{\alpha/2}\sqrt{\frac{p(1-p)}{n}}, \tag{1}$$

with $Z_{\alpha/2}$ being is the critical value of the Normal distribution at $\alpha/2$ (1.96 for CI of 95%), n - population size (number of positive images in a set), and p - the measured sensitivity.

5.1   Results on DBTex Dataset

The results on the final DBTex competition test set are presented in Table 1 including confidence intervals and significance of difference compared to our method's performance (two-proportion z-test), computed based on information in [5]. As the table shows, our difference from the primary and secondary metric performance of the winner is extremely small, while having a significant difference compared to the 3rd place.

5.2   Ablation Study Results

We examine the effect of our method's components on overall performance by testing different combinations in comparison to several known variants of Non Maximum Suppression.

Table 1. Competition results on DBTex Test Set

Team position	Primary metric	Second. metric
1st place	0.912 (0.867–0.957, p = 0.95)	0.912 (0.867–0.957, p = 0.82)
2nd place (ours)	0.910 (0.865–0.955)	0.904 (0.857–0.951)
3rd place	0.853 (0.797–0.909, p = 0.15)	0.868 (0.814–0.922, p = 0.35)
4th place	0.814 (0.752–0.876, p = 0.02)	0.831 (0.772–0.890, p = 0.08)
5th place	0.811 (0.749–0.873, p = 0.02)	0.801 (0.738–0.864, p = 0.02)

Table 2. Ablation study results, reported on Ablation-Validation and Ablation-Test sets. Given a base NMS variant algorithm (NMS-Variant), we show the effect of adding combinations of Classifier-based slice weighting (CLS), Heatmap-based filtering (HMap), and Rank Based Score Modification (RBSM). NMS variants include NMS, Soft NMS (sNMS) [18], Non-Maximal Weighted (NMW) [15], and Weighted Box Fusion (WBF) [20].

NMS Variant	HMap	RBSM	CLS	Ablation-Validation			Ablation-Test
				Primary (%)	Secondary (%)	1 fp Sens. (%)	Primary (%)
Baseline methods							
NMS				89 (83–94)	89 (84–95)	78 (71–85)	88 (81–95)
WBF				89 (83–94)	89 (83–94)	77 (69–84)	86 (78–93)
sNMS				93 (88–97)	94 (89–98)	83 (76–89)	89 (82–95)
NMW				93 (88–97)	93 (88–97)	83 (77–90)	89 (82–96)
Our Methods							
NMW	✓			93 (88–97)	94 (90–98)	84 (77–90)	90 (83–96)
NMW	✓	✓		94 (90–98)	91 (86–96)	88 (82–93)	90 (83–96)
NMW		✓		94 (90–98)	92 (87–97)	86 (80–92)	89 (83–96)
NMW			✓	94 (90–98)	**97 (94–100)**	83 (76–89)	90 (83–96)
NMW	✓		✓	95 (92–99)	**97 (94–100)**	86 (80–92)	91 (84–97)
NMW		✓	✓	96 (92–99)	94 (90–98)	89 (84–94)	91 (84–97)
NMS	✓	✓	✓	91 (86–96)	89 (84–95)	84 (78–91)	89 (82–96)
WBF	✓	✓	✓	94 (90–98)	92 (87–97)	86 (80–92)	**92 (86–98)**
sNMS	✓	✓	✓	96 (93–100)	94 (90–98)	92 (87–97)	**92 (86–98)**
NMW	✓	✓	✓	**97 (93–100)**	94 (90-98)	**93 (89-98)**	**92 (86-98)**

The components we analyzed are: CLS, HMap, and RBSM. Other components, such as normalization and size based filtering, as well as the bounding box aggregation method, remained constant throughout our ablation study. We give more details on combinations with NMW since it gave best results on the ablation validation set.

Results on ablation-validation and ablation-test sets, presented in Table 2, show that the highest contributing components, when used alone, are RBSM and CLS, but each of the three components has a positive impact on at least on one of the metrics, and all three together produce the highest primary score.

Further breakdown of results in Table 2 is presented in Table 1 in the appendix. There, we present results on five sub-groups of Ablation-Validation set: three ranges of lesion sizes, roughly corresponding to breast cancer stages [17]: T1 (up to 2cm), T2 (between 2 cm and 5 cm) and T3 (5 cm and larger), and two biopsy results (positive and negative).

The most significant difference, in some cases larger than the corresponding confidence intervals, appears to be between the positive and negative biopsy groups, with majority of tested methods achieving better results on the positive group on tested metrics. One possible explanation is that the decision to perform a biopsy may be based on information not present in the DBT image of a patient, resulting in a biopsy negative (or positive) label of a healthy-looking volume. We assume that this happens more often with benign lesions, resulting in the above difference. Additionally, performance seems to improve with an increase of tumor size for all tested models and metrics.

5.3 Comparison with Prior Art

A typical task in object detection is choosing a single bounding box from a collection of many bounding boxes that are assumed to cover the same detected object. These boxes may be produced by a single or multiple detector models/variants. Typical examples of the latter are an ensemble of models, or test-time augmentation.

We compare our algorithm to several NMS variants in the literature:

1. Non-Maximum Suppression (NMS) [18]: Removes low-scoring bounding boxes that overlap with others. The algorithm of choice until recently.
2. Soft-NMS [18]: Overlapping bounding boxes are not discarded but rather their scores are reduced. This avoids losing a good detection because of a small overlap with a bounding box that has a higher score. Guo et al. [19] address the same issue by keeping k top bounding boxes instead of only one.
3. Non-Maximal Weighted (NMW) [15]: Fuses the coordinates of overlapping boxes instead of just selecting the top score box. It uses a weighted average of the selected box with its overlapping boxes, where the weight is the product of the box score and its overlap with the selected box.
4. Weighted Box Fusion (WBF) [20]: Fuses both coordinates and scores of overlapping boxes using a weighted average, where the weight is the box score (ignoring the overlap). After averaging, the score is adjusted by multiplying by the number of overlapping boxes divided by the number of participating models.

Furthermore, we report the effect of using our contributions (HMap, RBSM and CLS) in conjunction with the different NMS variants. We combine the algorithms by replacing the non-maximum suppression component (NMW) used in our method with each of the variants.

Results (Table 2) show that our algorithm outperforms all baseline NMS variants (NMS, NMW, WBF, and sNMS). It is important to note that adding our proposed methods to each of the NMS variants results in improved performance on all tested metrics.

6   Discussion

We presented a DBT lesion detector that is based on an ensemble of several detector models and a classifier. A novel algorithm was used to combine the 2D models' predictions to produce 3D boxes predictions. Ablation study was performed on the contribution of the different components of the algorithm, and it was compared to state-of-the-art methods in the field. It was shown to outperform the current state-of-the art.

The DBTex dataset focuses on tumors and architectural distortions, and does not contain calcifications. This limits our analysis, since calcifications may affect radiologist's decision to perform a biopsy.

We've shown that the proposed ensemble method can be combined with other non-maximum suppression techniques, resulting in improved performance in all tested metrics.

The proposed algorithm was used in the DBTex 2021 competition, and out of 136 registered participants, reached 2nd place.

References

1. Cancer statistics, 2019 - Siegel - 2019 - CA: A Cancer Journal for Clinicians - Wiley Online Library. https://acsjournals.onlinelibrary.wiley.com/doi/full/10.3322/caac.21551. Accessed 28 Feb 2021
2. McKinney, S.M., et al.: International evaluation of an AI system for breast cancer screening. Nature **577**, 89–94 (2020). https://doi.org/10.1038/s41586-019-1799-6
3. Schaffter, T., et al.: Evaluation of combined artificial intelligence and radiologist assessment to interpret screening mammograms. JAMA Netw. Open. **3**, e200265–e200265 (2020). https://doi.org/10.1001/jamanetworkopen.2020.0265
4. SPIE-AAPM-NCI DAIR Digital Breast Tomosynthesis Lesion Detection Challenge (DBTex). http://spie-aapm-nci-dair.westus2.cloudapp.azure.com/competitions/4. Accessed 21 Feb 2021
5. Buda, M., et al.: Detection of masses and architectural distortions in digital breast tomosynthesis: a publicly available dataset of 5,060 patients and a deep learning model. ArXiv201107995 Cs Eess. (2021)
6. DBTex Challenge — SPIE Medical Imaging. https://spie.org/conferences-and-exhibitions/medical-imaging/digital-detection-challenge. Accessed 16 Feb 2021
7. Breast Cancer Screening - Digital Breast Tomosynthesis (BCS-DBT). https://wiki.cancerimagingarchive.net/pages/viewpage.action?pageId=64685580. Accessed 16 Feb 2021
8. Lin, T.-Y., Goyal, P., Girshick, R., He, K., Dollar, P.: Focal Loss for Dense Object Detection. Presented at the Proceedings of the IEEE International Conference on Computer Vision (2017)
9. He, K., Zhang, X., Ren, S., Sun, J.: Deep Residual Learning for Image Recognition. Presented at the Proceedings of the IEEE Conference on Computer Vision and Pattern Recognition (2016)
10. Lin, T.-Y., Dollar, P., Girshick, R., He, K., Hariharan, B., Belongie, S.: Feature Pyramid Networks for Object Detection. Presented at the Proceedings of the IEEE Conference on Computer Vision and Pattern Recognition (2017)

11. Lin, T.-Y., et al.: Microsoft COCO: Common Objects in Context. ArXiv14050312 Cs. (2015)
12. Deng, J., Dong, W., Socher, R., Li, L., Li, K., Li, F.-F.: ImageNet: a large-scale hierarchical image database. In: 2009 IEEE Conference on Computer Vision and Pattern Recognition, pp. 248–255 (2009). https://doi.org/10.1109/CVPR.2009. 5206848
13. Szegedy, C., Ioffe, S., Vanhoucke, V., Alemi, A.: Inception-v4, Inception-ResNet and the Impact of Residual Connections on Learning. ArXiv160207261 Cs. (2016)
14. Paszke, A., et al.: PyTorch: An Imperative Style, High-Performance Deep Learning Library. ArXiv191201703 Cs Stat. (2019)
15. Ning, C., Zhou, H., Song, Y., Tang, J.: Inception Single Shot MultiBox Detector for object detection. In: 2017 IEEE International Conference on Multimedia Expo Workshops (ICMEW), pp. 549–554 (2017). https://doi.org/10.1109/ICMEW.2017. 8026312
16. Daniel, W.W.: Biostatistics: a foundation for analysis in the health sciences. 7th edn. Wiley (1999)
17. Stages of Breast Cancer — Understand Breast Cancer Staging. https:// www.cancer.org/cancer/breast-cancer/understanding-a-breast-cancer-diagnosis/ stages-of-breast-cancer.html. Accessed 22 July 2021
18. Bodla, N., Singh, B., Chellappa, R., Davis, L.S.: Soft-NMS - improving object detection with one line of code. Presented at the Proceedings of the IEEE International Conference on Computer Vision (2017)
19. Guo, R., et al.: 2nd Place Solution in Google AI Open Images Object Detection Track 2019. ArXiv191107171 Cs. (2019)
20. Solovyov, R., Wang, W., Gabruseva, T.: Weighted boxes fusion: ensembling boxes from different object detection models. Image Vis. Comput. **107**, 104117 (2021). https://doi.org/10.1016/j.imavis.2021.104117

A Structural Causal Model for MR Images of Multiple Sclerosis

Jacob C. Reinhold[1]([✉]), Aaron Carass[2], and Jerry L. Prince[1,2]

[1] Department of Electrical and Computer Engineering, Johns Hopkins University, Baltimore, MD 21218, USA
jcreinhold@gmail.com, prince@jhu.com
[2] Department of Computer Science, Johns Hopkins University, Baltimore, MD 21218, USA
aaron_carass@jhu.com

Abstract. Precision medicine involves answering counterfactual questions such as "Would this patient respond better to treatment A or treatment B?" These types of questions are causal in nature and require the tools of causal inference to be answered, e.g., with a structural causal model (SCM). In this work, we develop an SCM that models the interaction between demographic information, disease covariates, and magnetic resonance (MR) images of the brain for people with multiple sclerosis. Inference in the SCM generates counterfactual images that show what an MR image of the brain would look like if demographic or disease covariates are changed. These images can be used for modeling disease progression or used for downstream image processing tasks where controlling for confounders is necessary.

Keywords: Causal inference · Multiple sclerosis · MRI

1 Introduction

Scientific inquiry and precision medicine involve answering causal questions, e.g., "does this medicine treat this disease?" Causal questions are asked to determine the effect of interventions on variables of interest. The main tool used to investigate these phenomena, however, is statistical inference which lacks effective methods to establish causal relationships outside of randomized control trials (RCTs).

If a scientist has a plausible model of the relationships between covariates, and proper measurements have been made (controlling for confounds), then causal questions can be answered with observational data instead of an RCT [17,18]. A directed acyclic graph (DAG) can serve as such a model, with edges pointing from cause to effect, i.e., the effect is a function of the cause. Such a DAG is known as a structural causal model (SCM), representing a generative model on which we can emulate interventions and generate counterfactuals.

© Springer Nature Switzerland AG 2021
M. de Bruijne et al. (Eds.): MICCAI 2021, LNCS 12905, pp. 782–792, 2021.
https://doi.org/10.1007/978-3-030-87240-3_75

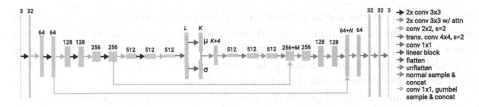

Fig. 1. VAE with hierarchical latent space: Deep neural network representing the recognition model (encoder) and generative model of the observed image (decoder). The variables $K, L, M, N \in \mathbb{N}$ are experiment-specific parameters.

Medical imaging problems like image harmonization can be viewed as asking the counterfactual question: "What would this image look like if it had been acquired with scanner X?" [2]. Inference in SCMs, however, is difficult for such high-dimensional problems. A tractable approach is to amortize inference across datasets and local variables with a neural network [8,23]. Pawlowski et al. [16] implemented an SCM with amortized inference for healthy MR images of the brain; we extended their model to the clinical and radiological phenotype of multiple sclerosis (MS) [20]. MS presents as hyper-intense lesions in T_2-weighted MR images, which are readily apparent in white matter (WM).

We propose an SCM that encodes a causal functional relationship with MR images of the brain for people with MS (PwMS). With this we can answer counterfactual questions like: "What would this brain look like if it did not have lesions?" Counterfactual questions improve our understanding of the relationship between clinical phenotype and disease presentation in MR. This is of interest in MS because of the "clinico-radiological paradox," related to the moderate correlation between lesion load and cognitive performance. Our contributions are: 1) an extension of the SCM model [16] to include MS-related covariates and 2) a novel method to generate realistic high-resolution (HR) counterfactual images with a variational autoencoder (VAE) embedded in the SCM.

2 Related Work

Machine learning (ML) has recently been used to create data-driven disease models, e.g., by training a network to show the MR progression of Alzheimer's [19]. The images may appear convincing, but methods that don't make assumptions about causal structure are fundamentally limited. A deep network trained to predict disease progression may use spurious correlations that fit the data but do not reflect the true causal effect [26]. RCTs are traditionally used to establish causal effect because observational data contains potential confounders. An SCM, by contrast, makes explicit causal assumptions and can control for confounders, providing a more robust model of disease.

Our work is also related to WM lesion filling—where lesions are made to appear as healthy WM. This is useful because many image processing methods are developed on lesion-free images. Our method can be used for lesion filling by intervening to set the lesion volume to 0. However, our method is more general as we can generate an image for any covariate change in the SCM. It is also more robust because it controls for confounders. For example, lesion filling with a CycleGAN, where one domain is healthy and the other is MS, may result in undesired side effects as noted by Cohen et al. [4].

Finally, our work is related to disentangled representations [3] which is where a subset of the latent variables control a factor of variation. These factors are generally constructed to be marginally independent which is unrealistic. Our SCM provides a systematic way to intervene on factors while accounting for dependency among factors. It also efficiently encodes inductive biases necessary to disentangle factors [15].

3   Methods

3.1   Background on Structural Causal Models

An SCM [17] is a tuple $\mathbf{M} = \langle \mathbf{U}, \mathbf{V}, \mathbf{F}, P(\mathbf{u}) \rangle$, where $\mathbf{U}$ are unobserved background variables, $\mathbf{V}$ are explicitly modeled variables (called *endogenous* variables), $\mathbf{F}$ are the functional relationships between $v_i \in \mathbf{V}$ (i.e., $v_i = f_i(\mathrm{pa}_i, u_i)$ where $\mathrm{pa}_i \subset \mathbf{V}$ are the parents of v_i), and $P(\mathbf{u})$ is the probability distribution over $\mathbf{U}$.

Each element of $\mathbf{U}$ is assumed to be independent. This assumption is known as *causal sufficiency* and also implies $\mathbf{V}$ includes all common causes between pairs in $\mathbf{V}$ [29]. We also assume independence between cause and mechanism; that is, if $\mathrm{pa}_i \to v_i$, then the distribution of pa_i and the function f mapping pa_i to v_i are independent of one another [5]. This implies f_i does not change if the distribution of a parent changes. We also assume f_j is invariant to changes to f_k, where $k \neq j$.

Counterfactual inference is accomplished with three steps:

Abduction	Predict $u_i, \forall u_i \in \mathbf{U}$ given $f_i \in \mathbf{F}$ and $v_i \in \mathbf{V}$.
Action	Modify $\mathbf{M} \to \mathbf{M}_A$ (edge removal) for intervention $\mathrm{do}(A)$.
Prediction	Use $\mathbf{M}_A$ and predicted u_i's, to recompute $\tilde{v}_i$ under f_i's.

Inference on medical images is challenging for SCMs for numerous reasons. One issue is $f_i \in \mathbf{F}$ must be an invertible function so the u_i's can be computed. Invertibility, however, is computationally costly and prohibitive with HR images. As a result, we decompose $\mathbf{u_x} \in \mathbf{U}$ (associated with image $\mathbf{x}$) into invertible and non-invertible terms $\varepsilon_\mathbf{x}$ and $\mathbf{z_x}$, respectively. Specifically, $\mathbf{z_x}$ is computed explicitly with a VAE, see [16] for details.

3.2   High-Resolution Image Generation in Variational Autoencoders

Pawlowski et al. [16] use a simple VAE architecture as a proof-of-concept; they only use the middle slices of an MR image volume downsampled to 64×64 pixels. In our work, we do two experiments using more slices and images of size

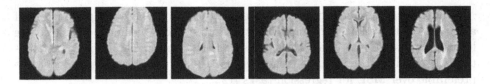

Fig. 2. Samples from the generative model: The above six images were generated by (unconditionally) ancestrally sampling the generative model trained on 128×128 images.

128×128 and 224×224 pixels. HR image generation with VAEs is a topic of increasing interest [32]. We introduce a novel means of generating HR images with a VAE by adding a hierarchical binary latent space (BLS) z_0 and z_1—in addition to the normal latent space z_2—that preserves structural information. Although inspired by [6], we use a hierarchy of latent spaces and a KL term on the BLS. Our VAE architecture is shown in Fig. 1 and a non-counterfactual unconditional sample from our SCM/VAE is shown in Fig. 2.

The BLS consists of a straight-through relaxed Bernoulli distribution using the Gumbel reparameterization [11] on two levels of the feature space. With z_0 of size $N \times 64 \times 64$ and z_1 is $M \times 16 \times 16$, where $M, N \in \mathbb{N}$ are experiment-specific parameters. The Bernoulli distribution has a prior of 0.5 (for both levels); the encoder balances minimizing the latent spaces KL divergence and image reconstruction. The BLSs (not shown) are very noisy which encourages the network to use all the latent spaces.

We use a conditional VAE, conditioned on $\mathbf{c} = (n, v, b, l)^\top$, where n is slice number, b is brain volume, v is ventricle volume, and l is lesion volume. We concatenate these values in the normal latent space z_2 to make $\mathbf{z}_c = [\mathbf{z}_2, \mathbf{c}]$, which is then input to the lowest level of our generative model. To condition on $\mathbf{c}$ when also using BLSs, we used squeeze-and-excite (channel-wise) attention [9] on the convolutional layers of the generative model, where the input to the squeeze-and-excite layers is $\mathbf{c}$.

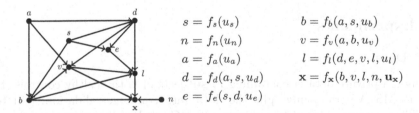

$$s = f_s(u_s) \qquad b = f_b(a, s, u_b)$$
$$n = f_n(u_n) \qquad v = f_v(a, b, u_v)$$
$$a = f_a(u_a) \qquad l = f_l(d, e, v, l, u_l)$$
$$d = f_d(a, s, u_d) \qquad \mathbf{x} = f_{\mathbf{x}}(b, v, l, n, \mathbf{u_x})$$
$$e = f_e(s, d, u_e)$$

Fig. 3. Proposed structural causal model. (Only $\mathbf{V}$ are shown in the graph.) a is age, d is the duration of MS symptoms, l is the lesion volume of the subject, n is the slice number, $\mathbf{x}$ is the image, b is the brain volume, s is biological sex, e is the expanded disability severity score (EDSS), and v is the ventricle volume. The right-hand side shows the functional relationships $\mathbf{F}$ associated with $\mathbf{V}$ and $\mathbf{U}$ of the SCM.

The generative model outputs image-sized location and scale parameters of a Laplace distribution. For simplicity, we assume pixels are independent so the location and scale images are the same size as the image. The Laplace distribution was used instead of a Gaussian because the L_1 loss has been noted to be preferable over MSE for regression tasks [12].

We trained the VAE for 2000 epochs using a one-cycle learning rate schedule [27] with the Adam optimizer [13] starting at 2×10^{-5}, increasing to 5×10^{-4} for the first 10% of training epochs, and decaying to 5×10^{-8}. We also used a schedule for scaling the KL terms [28]; specifically, we started the KL scaling parameter for the normal distribution, λ_2, at 0, and linearly increased it 1 over 600 epochs. The KL terms for the BLSs (λ_0 and λ_1) started at 1 and linearly increased them to different levels which were experiment-specific. We used weight normalization [24] on all convolutional layers, and we clipped the norm of the gradients at 100. Finally, we used one inverse autoregressive affine normalizing flow on the posterior [14].

3.3   Learning and Inference in the SCM and VAE

The functions $f_i \in \mathbf{F}$, which take causal parents pa_i as input and outputs $v_i \in \mathbf{V}$, are represented by normalizing flows (NFs) and neural networks. All f_i in our SCM are linear rational spline NFs [7], except the functions for the image $f_\mathbf{x}$, biological sex f_s, and slice number f_n. $f_\mathbf{x}$ is the generative model part of the VAE (see Sect. 3.2). The s is sampled from a Bernoulli distribution (0 for male, 1 for female). The n is sampled from a uniform distribution (the minimum to the maximum of slice number in training). The base distributions for the NFs are normal distributions with the log of the empirical mean and variance of the training data as the location and scale parameters, respectively. During training, the networks associated with the NFs and the VAE are jointly updated with backpropagation using the negative evidence lower bound as a loss function. At inference, the learned $\mathbf{F}$ are fixed. A counterfactual image is generated using the single-world intervention graph formalism [22]. The code and a full listing of hyperparameters can be found at the link in the footnote[1].

4   Experiments

4.1   Data

We used a private dataset containing 77 subjects of both healthy controls (HC) and PwMS. We randomly split the data into training, validation, and testing. Training consisted of 68 subjects (38 HC, 30 PwMS) which included 171 unique scans (47 HC, 124 PwMS). The validation set had 7 subjects (4 HC, 3 MS) which included 16 unique scans (5 HC, 11 MS). The test set had 2 subjects (1 HC, 1 MS) with 4 unique scans (2 HC, 2 MS). Each scan had both an MPRAGE and FLAIR image which were bias field-corrected with N4 [31] and

---

[1] https://github.com/jcreinhold/counterfactualms.

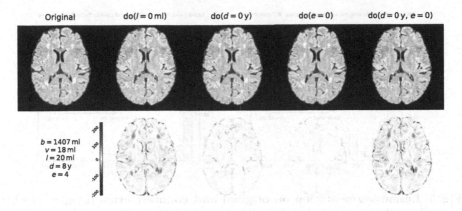

Fig. 4. Example counterfactual images: The first image is the original FLAIR image of a PwMS, the remaining images show counterfactual images. From left-to-right the interventions set the 1) lesion volume to 0 mL, 2) duration of symptoms to 0 years, 3) EDSS to 0, and 4) duration and EDSS to 0.

the MPRAGE was registered to the MNI ICBM152 brain atlas with ANTs [1]. The FLAIR image was super-resolved [34] and then registered to its corresponding MPRAGE in the MNI space. The WM mean was then normalized to 1 using the fuzzy c-means estimated WM mean [21]. Brain volume was measured based on ROBEX [10]. Lesion volume was measured based on Zhang et al. [33], trained on a private dataset. Ventricle volume was computed using Shao et al. [25] on the MPRAGE. The ventricle segmentation occasionally segmented lesions as ventricles for PwMS; the lesion mask was used to remove the incorrect voxels and the ventricle volume was updated accordingly. Axial slices of each image were converted to PNGs by extracting three adjacent slices, clipping the intensity at the 99.5^{th} percentile, then rescaling and quantizing to $[0, 255]$. In all experiments, uniform random noise is added to discrete random variables [30].

4.2 SCM for MS

The SCM we use for all experiments is shown as a DAG in Fig. 3. Note that we only show the endogenous variables $\mathbf{V}$ in Fig. 3. The SCM can alternatively be represented as a system of equations relating $\mathbf{V}$ and $\mathbf{U}$ via the equations $\mathbf{F}$ as shown on the right-hand side of Fig. 3. The u_i are the background variables associated with an endogenous variable v_i. Note that $\mathbf{u_x} = (\mathbf{z_x}, \varepsilon_{\mathbf{x}})$ where $\mathbf{z_x} = (\mathbf{z_0}, \mathbf{z_1}, \mathbf{z_2})$ are the non-invertible latent space terms estimated with the recognition model of the VAE and $\varepsilon_{\mathbf{x}}$ is the invertible term. The edges in this SCM were determined by starting from the SCM described in Pawlowski et al. [16], densely adding reasonable edges to the new variables (d, e, and l), and pruning the edges over many experiments based on the quality of the counterfactual images.

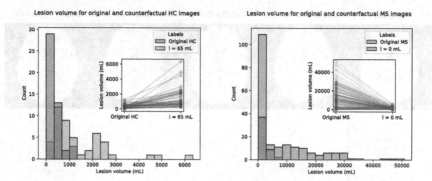

Fig. 5. Lesion segmentation on original and counterfactual images: The left figure shows the histogram of segmented lesion volumes for original and counterfactual HC images, where lesion volume was set to 65 mL. The right figure shows a similar plot for original and counterfactual MS images, where lesion volume was set to 0 mL. The inset plots show the same data with the lesion volume for original on the left and counterfactual on the right.

4.3   Small Images, Large Range

In this experiment, we randomly cropped the PNG images to 224×224 pixels and downsampled to 128×128 with linear interpolation. We used the middle 60 axial slices of the image volume. The KL schedule for z_0 and z_1 started at $\lambda_0 = \lambda_1 = 1$ and ended at $\lambda_0 = 4.4$ and $\lambda_1 = 1.1$. We used batches of 342 for training. Latent space size parameters are $K = 100$, $L = 8192$, $M = 4$, and $N = 1$.

We show an example set of counterfactual images in Fig. 4. These counterfactuals are generated from a subject in the training set; the quality of counterfactuals on the validation and test set were worse. See the discussion section for details.

To quantify the effect of our counterfactuals on lesion volume, we used a lesion segmentation method on the original and counterfactual images. The results are summarized in Fig. 5. Setting the lesion volume to 0 mL for PwMS consistently moves the lesion volume to near zero; however, for HCs, setting the lesion volume to 65 mL does not result in a consistent move to 65 mL. This is likely due to processing slices individually instead of as a volume; using 2D images to generate counterfactuals with statistics computed from 3D volumes is sub-optimal.

4.4   Large Images, Small Range

In this experiment, we cropped the images to 224×224 and did not downsample the cropped patches. We used the middle 10 axial slices of the image volume as input for training, validation, and testing. The KL warm-up is similar to the previous experiment except for the KL schedule for z_1 started at $\lambda_1 = 0.5$. We used a batch size of 128 for training. The latent space size parameters are

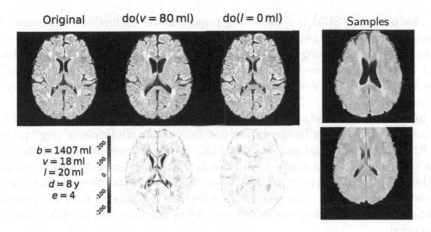

Fig. 6. Large image counterfactuals and samples: The images on the left show an intervention setting 1) the ventricle volume to 80 mL, 2) the lesion volume to 0 mL. The two images on the right are samples from the generative model, similar to Fig. 2.

$K = 120$, $L = 25088$, $M = 8$, and $N = 2$. An example intervention and samples from the generative model are shown in Fig. 6.

5   Discussion

In this work, we proposed an SCM to generate counterfactual images with application to MS. We showed that we can produce believable counterfactual images given a limited data set of healthy and MS subjects.

This work has many limitations. A noted problem with counterfactuals is that they are usually unverifiable. Counterfactuals imagine what some variable would look like in a parallel universe where all but the intervened on variables and their descendants were the same. Since counterfactual images cannot be validated, they should not be used for diagnosis or prognosis. The tools of counterfactuals, however, give researchers better ways to control for known confounders—potentially improving the performance of and trust in an ML system.

Another limitation is that our model created poor counterfactual images for images outside the training set. However, causal inference is usually about retrospective analysis; generalization to new samples is not a requirement of a successful causal model. A statistical learning view of ML is that the aim of a predictor is to minimize the loss function under the true data distribution (i.e., the risk). Causal inference, however, is concerned with estimating the causal effect of a change in a covariate. Regardless, optimal hyperparameters and more data would likely help the SCM work better on unseen data.

Alternatives to the VAE should be further investigated because the causal effect of variables on one another is likely unidentifiable in the presence of the latent variables $\mathbf{z}$; this means that the functions f_i do not necessarily reflect the true causal effect for a change in the parents.

The proposed SCM is a launching point but needs further refinement. For example, the type of treatment used should be included as variables and would make for more interesting counterfactuals (e.g., "what would this patient's lesion load be if they had received interferon beta instead of glatiramer acetate?"). The fact that we do not use this variable in our model can only hamper its performance.

In spite of the limitations, our SCM framework provides a principled way to generate a variety of images of interest for various applications (e.g., pseudo-healthy image synthesis) and analysis (e.g., counterfactual analysis of the effect of a medicine on lesion load). Such a framework can augment normal scientific analysis by providing a principled basis on which to test the effect of interventions outside of an RCT, and the generated counterfactual images have the potential to improve the performance of image processing pipelines and precision medicine in general.

References

1. Avants, B.B., Tustison, N., Song, G.: Advanced normalization tools (ANTS). Insight J. **2**(365), 1–35 (2009)
2. Castro, D.C., Walker, I., Glocker, B.: Causality matters in medical imaging. Nat. Commun. **11**(1), 1–10 (2020)
3. Chen, X., et al.: InfoGAN: interpretable representation learning by information maximizing generative adversarial nets. In: 30th International Conference on Neural Information Processing Systems, pp. 2180–2188 (2016)
4. Cohen, J.P., Luck, M., Honari, S.: Distribution matching losses can hallucinate features in medical image translation. In: Frangi, A.F., Schnabel, J.A., Davatzikos, C., Alberola-López, C., Fichtinger, G. (eds.) MICCAI 2018. LNCS, vol. 11070, pp. 529–536. Springer, Cham (2018). https://doi.org/10.1007/978-3-030-00928-1_60
5. Daniusis, P., et al.: Inferring deterministic causal relations. arXiv preprint arXiv:1203.3475 (2012)
6. Dewey, B.E., Zuo, L., Carass, A., He, Y., Liu, Y., Mowry, E.M., Newsome, S., Oh, J., Calabresi, P.A., Prince, J.L.: A disentangled latent space for cross-site MRI harmonization. In: Martel, A.L., Abolmaesumi, P., Stoyanov, D., Mateus, D., Zuluaga, M.A., Zhou, S.K., Racoceanu, D., Joskowicz, L. (eds.) MICCAI 2020. LNCS, vol. 12267, pp. 720–729. Springer, Cham (2020). https://doi.org/10.1007/978-3-030-59728-3_70
7. Dolatabadi, H.M., et al.: Invertible generative modeling using linear rational splines. In: International Conference on Artificial Intelligence and Statistics, pp. 4236–4246. PMLR (2020)
8. Gershman, S., Goodman, N.: Amortized inference in probabilistic reasoning. In: Proceedings of the Annual Meeting of the Cognitive Science Society, vol. 36 (2014)
9. Hu, J., Shen, L., Sun, G.: Squeeze-and-excitation networks. In: IEEE Conference on Computer Vision and Pattern Recognition, pp. 7132–7141 (2018)
10. Iglesias, J.E., et al.: Robust brain extraction across datasets and comparison with publicly available methods. IEEE Trans. Med. Imag. **30**(9), 1617–1634 (2011)
11. Jang, E., Gu, S., Poole, B.: Categorical reparameterization with gumbel-softmax. In: International Conference on Learning Representations (2017)

12. Kendall, A., Gal, Y.: What uncertainties do we need in bayesian deep learning for computer vision? In: Advances in Neural Information Processing Systems, pp. 5574–5584 (2017)

13. Kingma, D.P., Ba, J.L.: Adam: a method for stochastic gradient descent. In: ICLR: International Conference on Learning Representations, pp. 1–15 (2015)

14. Kingma, D.P., et al.: Improved variational inference with inverse autoregressive flow. Adv. Neural. Inf. Process. Syst. **29**, 4743–4751 (2016)

15. Locatello, F., et al.: A sober look at the unsupervised learning of disentangled representations and their evaluation. J. Mach. Learn. Res. **21**(209), 1–62 (2020)

16. Pawlowski, N., et al.: Deep structural causal models for tractable counterfactual inference. In: Advances in Neural Information Processing Systems (2020)

17. Pearl, J.: Causality. Cambridge University Press (2009)

18. Peters, J., et al.: Elements of Causal Inference. MIT Press, Cambridge (2017)

19. Ravi, D., Alexander, D.C., Oxtoby, N.P.: Degenerative Adversarial NeuroImage nets: generating images that mimic disease progression. In: Shen, D., Liu, T., Peters, T.M., Staib, L.H., Essert, C., Zhou, S., Yap, P.-T., Khan, A. (eds.) MICCAI 2019. LNCS, vol. 11766, pp. 164–172. Springer, Cham (2019). https://doi.org/10.1007/978-3-030-32248-9_19

20. Reich, D.S., Lucchinetti, C.F., Calabresi, P.A.: Multiple sclerosis. N. Engl. J. Med. **378**(2), 169–180 (2018)

21. Reinhold, J.C., et al.: Evaluating the impact of intensity normalization on MR image synthesis. In: Medical Imaging 2019: Image Processing, vol. 10949, p. 109493H. International Society for Optics and Photonics (2019)

22. Richardson, T.S., Robins, J.M.: Single world intervention graphs (SWIGs): a unification of the counterfactual and graphical approaches to causality. Center for the Statistics and the Social Sciences, University of Washington Series. Working Paper **128**(30), 2013 (2013)

23. Ritchie, D., Horsfall, P., Goodman, N.D.: Deep amortized inference for probabilistic programs. arXiv preprint arXiv:1610.05735 (2016)

24. Salimans, T., Kingma, D.P.: Weight normalization: a simple reparameterization to accelerate training of deep neural networks. In: 30th International Conference on Neural Information Processing Systems, pp. 901–909 (2016)

25. Shao, M., et al.: Brain ventricle parcellation using a deep neural network: application to patients with ventriculomegaly. NeuroImage: Clinical **23**, 101871 (2019)

26. Shpitser, I., Pearl, J.: Complete identification methods for the causal hierarchy. J. Mach. Learn. Res. **9**, 1941–1979 (2008)

27. Smith, L.N.: Cyclical learning rates for training neural networks. In: 2017 IEEE Winter Conference on Applications of Computer Vision (WACV), pp. 464–472. IEEE (2017)

28. Sønderby, C.K., et al.: Ladder variational autoencoders. In: 30th International Conference on Neural Information Processing Systems, pp. 3745–3753 (2016)

29. Spirtes, P.: Introduction to causal inference. J. Mach. Learn. Res. **11**(5), 1643–1662 (2010)

30. Theis, L., et al.: A note on the evaluation of generative models. In: ICLR: International Conference on Learning Representations, pp. 1–10 (2016)

31. Tustison, N.J., et al.: N4ITK: improved N3 bias correction. IEEE Trans. Med. Imag. **29**(6), 1310–1320 (2010)

32. Vahdat, A., Kautz, J.: NVAE: a deep hierarchical variational autoencoder. In: Advances in Neural Information Processing Systems, vol. 33 (2020)

33. Zhang, H., et al.: Multiple sclerosis lesion segmentation with tiramisu and 2.5D stacked slices. In: Shen, D., Liu, T., Peters, T.M., Staib, L.H., Essert, C., Zhou, S., Yap, P.-T., Khan, A. (eds.) MICCAI 2019. LNCS, vol. 11766, pp. 338–346. Springer, Cham (2019). https://doi.org/10.1007/978-3-030-32248-9_38
34. Zhao, C., et al.: SMORE: a self-supervised anti-aliasing and super-resolution algorithm for MRI using deep learning. IEEE Trans. Med. Imag. **40**(3), 805–817 (2021)

EMA: Auditing Data Removal from Trained Models

Yangsibo Huang[1], Xiaoxiao Li[1,2](✉), and Kai Li[1](✉)

[1] Princeton University, Princeton, NJ, USA
{yangsibo,xl132}@princeton.edu, li@cs.princeton.edu
[2] The University of British Columbia, Vancouver, BC, Canada

Abstract. Data auditing is a process to verify whether certain data have been removed from a trained model. A recently proposed method [10] uses Kolmogorov-Smirnov (KS) distance for such data auditing. However, it fails under certain practical conditions. In this paper, we propose a new method called Ensembled Membership Auditing (EMA) for auditing data removal to overcome these limitations. We compare both methods using benchmark datasets (MNIST and SVHN) and Chest X-ray datasets with multi-layer perceptrons (MLP) and convolutional neural networks (CNN). Our experiments show that EMA is robust under various conditions, including the failure cases of the previously proposed method. Our code is available at: https://github.com/Hazelsuko07/EMA.

Keywords: Privacy · Machine learning · Auditing

1 Introduction

An important aspect of protecting privacy for machine learning is to verify if certain data are used in the training of a machine learning model, i.e., data auditing. Regulations such as GDPR [18] and HIPPA [1] require institutions to allow individuals to revoke previous authorizations for the use of their data. In this case, such data should be removed not only from storage systems, but also from trained models.

Previous work focuses on data removal instead of data auditing. Some investigate how training data can be memorized in model parameters or outputs [3,20] so as to show the importance of data removal. Others study data removal methods from trained models, especially those that does not require retraining the model [2,4]. However, independent of how data is removed, in order to meet the compliance of data privacy regulations, it is important, especially for healthcare applications such as medical imaging analysis, to have a robust data auditing process to verify if certain data are used in a trained model.

Electronic supplementary material The online version of this chapter (https://doi.org/10.1007/978-3-030-87240-3_76) contains supplementary material, which is available to authorized users.

© Springer Nature Switzerland AG 2021
M. de Bruijne et al. (Eds.): MICCAI 2021, LNCS 12905, pp. 793–803, 2021.
https://doi.org/10.1007/978-3-030-87240-3_76

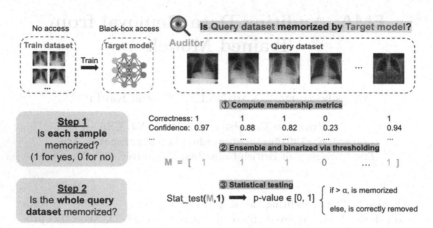

Fig. 1. EMA method consists of two steps: 1) the auditor first infers if each sample in the query set is memorized by the target model; 2) then it ensembles the results and see if the whole query set is memorized.

The data auditing problem is an under-studied area. The closely related work is by Liu et al. [10] who proposed an auditing method to verify if a query dataset is removed, based on Kolmogorov-Smirnov (KS) distance and a calibration dataset. However, the method may fail under certain practical conditions, such as when the query dataset is similar to the training dataset or when the calibration dataset is not of high quality.

To overcome these limitations, we propose an Ensembled Membership Auditing (EMA) method, inspired by membership inference attacks [16], to audit data removal from a trained model (see Fig. 1). It is a 2-step procedure which ensembles multiple metrics and statistical tools to audit data removal. To verify if a trained model memorizes a query dataset, first, EMA auditor infers whether the model memorizes each sample of the query dataset based on *various metrics*. Second, EMA ensembles multiple membership metrics and utilizes statistical tools to aggregate the sample-wise results and obtain a final auditing score.

Our contributions are summarized as follows:

1. We propose Ensembled Membership Auditing (EMA), an effective method to measure if certain data are memorized by the trained model.
2. EMA method improves the cost-efficiency of the previous approach [10], as it does not need to train a model on the query dataset.
3. Our experiments on benchmark datasets and Chest X-ray datasets demonstrate that our approach is robust under various practical settings, including the conditions that the previous method fails.

2   Preliminary

2.1   Problem Formulation

Our formulation of the data auditing problem is similar to that proposed by Liu et al. [10]. Suppose the dataset D is sampled from a given distribution $\mathbb{D} \subset \mathbb{R}^d$, where d denotes the input dimension. A machine learning model $f_D : \mathbb{R}^d \rightarrow \mathcal{C}$ is trained on D to learn the mapping from an input to a label in the output space $\mathcal{C}$. We denote the inference with a data point $x \in \mathbb{R}^d$ as $f_D(x)$. The auditory institution (or the auditor) aims to tell if a query dataset D_q is memorized by the trained model f_D.

In real applications, most machine learning models for healthcare are provided as Application Programming Interface (APIs). Users only have access to the model outputs rather than model parameters, referred to as a black-box access. Hence, similar to [10], we assume a black-box setting for data auditing: the auditor has access to 1) the algorithm to train f_D, and 2) $f_D(D_q)$, probability outputs of the query data D_q on f_D. The auditor does **not** have access to the training dataset, nor the network parameters of f_D.

2.2 Previous Method

Let us use D to denote the training dataset and D_{cal} to denote the calibration dataset. Liu et al. proposes an auditing method [10] that uses Kolmogorov-Smirnov (KS) distance to compare the distance between $f_D(D_q)$ and $f_{D_q}(D_q)$ and that between $f_{D_{cal}}(D_q)$ and $f_{D_q}(D_q)$, where D and D_{cal} are drawn from the same domain. The criteria is defined as:

$$\rho_{KS} = KS(f_D(D_q), f_{D_q}(D_q))/KS(f_{D_{cal}}(D_q), f_{D_q}(D_q)), \tag{1}$$

where $\rho_{KS} \geq 1$ indicates the query dataset D_q has been forgotten by f_D. However, the ρ_{KS} formula may fail in the following scenarios:

- when the query dataset is very similar to the original training dataset, the numerator is small, which will lead to a false negative result;
- when the calibration set is of low quality, the denominator is small, which will lead to a false positive result.

Section 4 provides experimental results of the above limitations of using ρ_{KS}.

3 The Proposed Method

This section presents Ensembled Membership Auditing (EMA), a 2-step procedure to audit data removal from a trained model.

3.1 Membership Inference Attack

The key idea of our approach is inspired by the Membership Inference Attack (MIA) [16], which shares a same black-box setting as that of auditing data removal. A black-box MIA attacker aims to identify if a **single** datapoint is a member of a machine learning model's training dataset. Formally, given an example x and a target trained model f_D, MIA formulates a decision rule h and computse $h(x; f_D) \in [0, 1]$, the probability of x being a member of f_D's training

---

Algorithm 1. Ensembled Membership Auditing (EMA)

Input: A, the training algorithm; f_D, the target model; D_q, the query dataset; D_{cal}, the calibration dataset;

   $g_1, \cdots, g_m$, m different metrics for membership testing.

Output: $\rho_{EMA} \in [0,1]$, the possibility that D_q is memorized by f_D

1: **procedure** ENSEMBLEDMEMBERSHIPAUDITING

2:     $\tau_1, \cdots, \tau_m \leftarrow$ INFERMEMBERSHIPTHRESHOLDS$(A, D_{cal}, g_1, \cdots, g_m)$

3:     $\mathbf{M} \leftarrow \mathbf{0}$ ▷ $\mathbf{M} \in \{0,1\}^{|D_q|}$, the inferred membership of each sample in D_q

4:     **for** $(x_i, y_i) \in D_q$ **do**

5:         $\mathbf{M}_i \leftarrow \mathbf{1}\{g_1(f_D, (x_i, y_i)) \geq \tau_1\} \cup \mathbf{1}\{g_2(f_D, (x_i, y_i)) \geq \tau_2\} \cup \cdots \cup \mathbf{1}\{g_m(f_D, (x_i, y_i)) \geq \tau_m\}$

6:     **end for**

7:     $\rho_{EMA} \leftarrow$ 2SAMP-PVALUE$(\mathbf{M}, \mathbf{1})$ ▷ 2SAMP-PVALUE() returns the p-value of a two-sample statistical test, which determines if two populations are from the same distribution

8:     **return** ρ_{EMA}

9: **end procedure**

Table 1. Comparing EMA with the method by Liu et al. [10].

	Calibration model	Query model	High quality calibration set	Query data similar to training data
EMA	Need to train	No need to train	No need	Robust
Liu et al.	Need to train	Need to train	Need	Not robust

dataset. The final results are binarized by a threshold, and 1 indicates the membership. To formulate the decision rule h, in addition to knowing trained model outputs, MIA requires knowing another set of data (we refer to as calibration data), which is assumed to be similar to the training dataset. Previous work suggests that the decision rule h can either be a machine learning model that is trained on the calibration data [11,15,16], or be thresholds of certain metrics that are computed using the calibration data [17]. Motivated by recent successes of MIA on single data points, we propose a framework that adapts MIA to audit whether a set of data points is removed.

3.2 Ensembled Membership Auditing (EMA)

We propose Ensembled Membership Auditing (EMA), a 2-step auditing scheme for data removal (see Algorithm 1): to verify if a query dataset is memorized by a trained model, the auditor first infers if each sample is memorized based on certain metrics, and then utilizes some statistical tools to aggregate the sample-wise results and to infer the probability that the query dataset is memorized. We name this probability as the EMA score and denote it by ρ_{EMA}.

Algorithm 2. Infer Membership Thresholds [17]

Input: A, the training algorithm; D_{cal}, the calibration dataset; $g_1, \cdots, g_m$, m different metrics for membership testing.
Output: $\tau_1, \cdots, \tau_m$, thresholds for m different metrics for membership inference.

1: **procedure** INFERMEMBERSHIPTHRESHOLDS
2:     Split D_{cal} into a training dataset D_{cal}^{train} and a test set D_{cal}^{test}
3:     $f_{D_{cal}} \leftarrow A(D_{cal}^{train})$ ▷ Train the calibration model
4:     **for** $i \in [m]$ **do**
5:         $\mathbf{V}_{train} \leftarrow \{g_i(f_{D_{cal}}, s) | s \in D_{cal}^{train}\}$ ▷ Compute metrics for training dataset
6:         $\mathbf{V}_{test} \leftarrow \{g_i(f_{D_{cal}}, s) | s \in D_{cal}^{test}\}$ ▷ Compute metrics for test dataset
7:         $\tau_i \leftarrow \arg\max_{\tau \in [\mathbf{V}_{train}, \mathbf{V}_{test}]}(BA(\tau))$ ▷ Infer the threshold based on Eq 2
8:     **end for**
9:     **return** $\tau_1, \cdots, \tau_m$
10: **end procedure**

Step 1: Infer if Each Sample Is Memorized. Given the target model f_D, which is trained with training dataset D, and query dataset D_q, the first step infers if each sample in D_q is memorized by f_D (see Algorithm 1, line 2 to line 6). The auditor first computes $\tau_1, \cdots, \tau_m$, thresholds for m different metrics by running a standard membership inference pipeline [17] on the calibration set. To select thresholds to identify training data, we define balanced accuracy on calibration data based on the balanced accuracy regarding True Positive Rate (TPR) and True Negative Rate (TNR):

$$BA(\tau) = \frac{TPR(\tau) + TNR(\tau)}{2} \tag{2}$$

where given a threshold τ, $TPR(\tau) = \sum_{s \in D_{cal}^{train}} \mathbf{1}\{g_i(s) > \tau\} / |D_{cal}^{train}|$, and $TNR(\tau) = \sum_{s \in D_{cal}^{test}} \mathbf{1}\{y_i(s) \geq \tau\} / |D_{cal}^{test}|$. The best threshold is selected to maximize the balanced accuracy (see Algorithm 2). For each sample in D_q, it will be inferred as a member or memorized by the target model, if it gets a membership score higher than the threshold for at least one metric (Algorithm 1, line 3 to 6). The auditor stores the membership results in $\mathbf{M} \in \{0,1\}^{|D_q|}$: $\mathbf{M}_i = 1$ indicates that the i-th sample in D_q is inferred as memorized by f_D, and $\mathbf{M}_i = 0$ indicates otherwise.

Our scheme uses the following 3 metrics for membership inference [17]:

- *Correctness:* $g_{corr}(f, (x, y)) = \mathbf{1}\{\arg\max_i f(x)_i = y\}$
- *Confidence:* $g_{conf}(f, (x, y)) = f(x)_y$
- *Negative entropy:* $g_{entr}(f, (x, y)) = \sum_i f(x)_i \log(f(x)_i)$

Step 2: Aggregate Sample-Wise Auditing Results. Given $\mathbf{M}$, the sample-wise auditing results from step 1, the auditor infers if the whole query dataset is memorized. A simple approach is to perform majority voting on $\mathbf{M}$, however, the state-of-the-art MIA approaches[17] achieve only $\sim$70% accuracy with benchmark datasets. Majority voting may not achieve reliable results.

The unreliability of a single entry in $\mathbf{M}$ motivates us to consider using the distribution of $\mathbf{M}$: ideally, if a query dataset D_q^* is memorized, it should give $\mathbf{M}_{D_q^*} = 1$. Thus, we run a two-sample statistical test: we fix one sample to be $\mathbf{1}$ (an all-one vector), and use $\mathbf{M}$ as the second sample. We set the null hypothesis to be that *2 samples are drawn from the same distribution* (i.e., $\mathbf{M}$ is the sample-wise auditing results for a memorized query dataset). The test will return a p-value, which is the final output of our EMA scheme, and we denote it as ρ_{EMA}. We interpret ρ_{EMA} as follow: if $\rho_{\text{EMA}} \leq \alpha$, the auditor can reject the null hypothesis, and conclude that the query dataset is not memorized. Here, α is the threshold for statistical significance, and is set to 0.1 by default.

Comparison with the Previous Method. Table 1 lists the differences between our method and Liu et al.'s [10]. As shown, our approach is more cost-efficient since it does not require training a model on the query dataset. It also addresses limitations of the previous method by avoiding possible false-positive (due to low quality calibration data) and false-negative cases (due to similar query data to training data), which we are going to show in the next section.

4   Experiments

We conduct two experiments to validate EMA and compare it with the method by Liu et al. [10]. The first (see Sect. 4.1) uses benchmark datasets (MNIST and SVHN) and the second (see Sect. 4.2) uses Chest X-ray datasets. Both methods are implemented in Pytorch framework [13]. We present the main results by using t-test as the statistical aggregation step of EMA. Appendix B provides the results of EMA using different statistical tests, and more results under various constraints of the query dataset.

4.1   Benchmark Datasets (MNIST and SVHN)

We start with verifying the feasibility of EMA and explaining the experiment setting on benchmark datasets for the ease of understanding.[1] MNIST dataset [9] contains 60,000 images with image size 28×28. SVHN dataset [12] contains 73,257 images in natural scenes with image size 32×32. We generate the training dataset, the calibration set, and the query dataset as follow.

Training Dataset. We randomly sample 10,000 images from MNIST as the training dataset and split it equally to 5 non-overlapping folds. Each fold contains 2,000 images.

Calibration Set. We sample 1,000 images from MNIST (disjoint with the training dataset) as the calibration set. To simulate a low-quality calibration set in practice, we keep $k\%$ of the original images, add random Gaussian noise to $(100 - k)/2\%$ of the images, and randomly rotate the other $(100 - k)/2\%$ of the images. We vary k in our evaluation.

---

[1] The real medical data experiment follows the similar setting.

Table 2. Auditing scores of both methods on **benchmark datasets**. Each column corresponds to a query dataset, and each row corresponds to a calibration set with quality controlled by k. False positive results are in red, while false negative results are in *blue*.

k	M1	M2	M3	M4	M5	M6	S
100	0.91	0.90	0.90	0.89	0.90	*0.82*	2.34
90	0.96	0.95	0.96	0.95	0.95	*0.86*	2.59
80	0.98	0.95	0.97	0.95	0.96	*0.87*	3.75
70	0.98	0.96	0.97	0.96	0.96	*0.88*	1.08
60	1.02	0.99	1.01	1.00	0.99	*0.90*	4.74
50	1.07	1.04	1.06	1.05	1.05	*0.94*	2.62

(a) ρ_{KS} scores of method by Liu et al.

k	M1	M2	M3	M4	M5	M6	S
100	1.00	1.00	1.00	1.00	1.00	0.00	0.00
90	1.00	1.00	1.00	1.00	1.00	0.00	0.00
80	1.00	1.00	1.00	1.00	1.00	0.00	0.00
70	1.00	1.00	1.00	1.00	1.00	0.00	0.00
60	1.00	1.00	1.00	1.00	1.00	0.00	0.00
50	1.00	1.00	1.00	1.00	1.00	0.00	0.00

(b) ρ_{EMA} scores of EMA (t-test)

Query Dataset. We design the following three kinds of query dataset:

- $\{M1, M2, M3, M4, M5\}$: 5 folds of MNIST images used in training, each with 2,000 images;
- **M6**: 2,000 images randomly selected from the MNIST dataset (disjoint with the training and the calibration set);
- **S**: 2,000 images randomly selected from the SVHN dataset.

Target Model. The target model is a three-layer multi-layer perceptron of hidden size (256, 256). Its training uses SGD optimizer [14] with learning rate 0.05 run for 50 epochs. The learning rate decay is set to 10^{-4}.

Results and Discussion. Fig. 2(a) shows that the distribution of metrics on M1 (memorized) is clearly distinguishable from those of M6 and S (not memorized). This validate that EMA can be used to infer whether a query dataset is memorized by the target model.

EMA gives correct auditing results even when the calibration set is of low quality, while the method by Liu et al. [10] may fail. Note that for method by Liu et al., $\rho_{KS} \leq 1$ indicates that the dataset is removed. For EMA, $\rho_{EMA} \leq \alpha$ ($\alpha = 0.1$) indicates that the dataset is removed. Table 2(a) shows when a query dataset is included in the training dataset of the target model (columns 'M1' to 'M5') and the calibration set's quality the calibration set's quality drops to $k = 60$, Liu et al. [10] returns false positive results on M1, M3, and M4 (i.e. $\rho_{KS} \geq 1$) On the contrary, EMA returns correct EMA scores despite variations in the quality of the calibration set.

Liu et al.'s method fails when the query dataset is similar to but not included in the training dataset (shown in column 'M6' in Table 2(a)). By contrast, EMA is robust for such a scenario. Both methods give correct answers for query dataset 'S' from SVHN whose appearance is significantly different from that of MNIST.

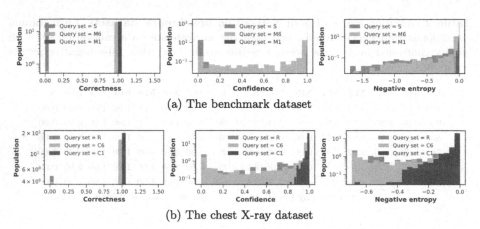

(a) The benchmark dataset

(b) The chest X-ray dataset

Fig. 2. Distribution of correctness, confidence, and negative entropy scores of query datasets M1 (memorized), M6 and S (not memorized) for benchmark dataset; of query datasets C1 (memorized), C6 and R (not memorized) for chest X-ray dataset.

4.2 Chest X-Ray Datasets

We further evaluate EMA on medical image analysis. We use two Chest X-ray datasets, including COVIDx [19], a recent public medical image dataset which contains 15,173 Chest X-ray images, and the Childx dataset [7], which contains 5,232 Chest X-ray images from children. We perform pneumonia/normal classification on both datasets. Appendix A.1 provides details and sample images of both datasets. We describe the training dataset, the calibration set, and the query dataset as follow.

Training Dataset. We randomly sample 4,000 images from COVIDx as the training dataset and split it equally to 5 non-overlapping folds. Each fold contains 800 images.

Calibration Set. We generate the calibration set using a subset of the COVIDx dataset, which is disjoint with the training dataset and contains 4,000 images as well. To simulate a potentially low-quality calibration set, we keep $k\%$ of the original images, and add random Gaussian noise to $(100 - k)\%$ of the images.

Query Dataset. We evaluate with different query datasets, including

- $\{C1, C2, C3, C4, C5\}$, 5 folds of COVIDx images used in training, each with 800 images;
- $C6$, 800 images randomly selected from the COVIDx dataset (disjoint with the training and the calibration set);
- R, 800 images randomly selected from the Childx dataset.

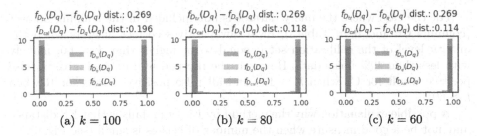

Fig. 3. Visualization of $f_{D_{tr}}(D_q)$, $f_{D_{cal}}(D_q)$, and $f_{D_q}(D_q)$ of the Chest X-ray datasets, with the query dataset is included in the training dataset. $f_{D_{tr}}(D_q)$ highly overlaps with $f_{D_q}(D_q)$, but the KS distance between them is larger than the KS distance between $f_{D_{cal}}(D_q)$ and $f_{D_q}(D_q)$. This suggests KS distance may not be a good measure of distributions of prediction outputs.

Table 3. Auditing scores of both methods on **Chest X-ray datasets.** Each column corresponds to a query dataset, and each row corresponds to a calibration set with quality controlled by k. False positive results are in red while false negative results are in *blue*.

k	C1	C2	C3	C4	C5	C6	R
100	1.19	1.20	1.23	1.20	1.21	*0.99*	*0.53*
90	1.25	1.28	1.23	1.24	1.25	1.01	*0.71*
80	1.26	1.27	1.2	1.22	1.21	1.00	*0.96*
70	1.28	1.26	1.24	1.21	1.24	1.02	*0.73*
60	1.32	1.32	1.31	1.26	1.31	1.09	1.06

(a) ρ_{KS} scores of method by Liu et al.

k	C1	C2	C3	C4	C5	C6	R
100	1.00	1.00	1.00	1.00	1.00	0.00	0.00
90	1.00	1.00	1.00	1.00	1.00	0.00	0.00
80	1.00	1.00	1.00	1.00	1.00	0.00	0.00
70	1.00	1.00	1.00	1.00	1.00	0.00	0.00
60	1.00	1.00	1.00	1.00	1.00	0.00	0.00

(b) ρ_{EMA} scores of EMA (t-test)

Target Model. The target model is ResNet-18 [5]. We use the Adam optimizer [8] with learning rate 2×10^{-5} and run for 30 epochs (weight decay is set to 10^{-7}).

Results and Discussion. The results further validate EMA can be used to infer whether a query dataset is memorized by the target model. As shown in Fig. 2(b), the distribution of membership metrics on C1 (memorized) is clearly distinguishable from those of C6 and R (not memorized); however, the difference between distributions of metrics for memorized and not-memorized query datasets is smaller when compared to that on benchmark datasets. One potential rationale for this difference is that we perform a 10-way classification on benchmark datasets, but only do a binary classification for Chest X-ray datasets. Thus, the auditor may get less information from the final prediction of the target model on Chest X-ray datasets, as the final prediction has fewer classes.

When the query dataset is a subset of the training dataset (columns 'C1' to 'C5' in Table 3), EMA correctly indicates that the query dataset is memorized ($\rho_{EMA} = 1$), whereas the results of the method by Liu et al. [10] are all false positive.

For the case where the query dataset is not included in the training dataset (columns C6 and R in Table 3), EMA always gives correct answers when the quality level of the calibration set is equal to or higher than $k = 60$, namely with less than 40% noisy data. However, the method by Liu et al. gives a false positive result for C6 when $k = 100$ and all false positive results for R when $k > 60$.

A possible explanation why the method by Liu et al. fails is that KS distance may not be a good measure when the number of classes is small (see Fig. 3).

5   Conclusion

This paper presents EMA, a 2-step robust data auditing procedure to verify if certain data are used in a trained model or if certain data has been removed from a trained model. By examining if each data point of a query set is memorized by a target model and then aggregating sample-wise auditing results, this method not only overcomes two main limitations of the state-of-the-art, but also improves efficiency. Our experimental results show that EMA is robust for medical images, comparing with the state-of-the-art, under practical settings, such as lower-quality calibration dataset and statistically overlapping data sources.

Future work includes testing EMA with more medical imaging tasks, and more factors that may affect the algorithm's robustness, such as the requirements of the calibration data, different training strategies and models, and other aggregation methods.

Acknowledgement. This project is supported in part by Princeton University fellowship and Amazon Web Services (AWS) Machine Learning Research Awards. The authors would like to thank Liwei Song and Dr. Quanzheng Li for helpful discussions.

References

1. Act, A.: Health insurance portability and accountability act of 1996. Public Law **104**, 191 (1996)
2. Bourtoule, L., et al.: Machine unlearning. arXiv preprint arXiv:1912.03817 (2019)
3. Carlini, N., Liu, C., Erlingsson, Ú., Kos, J., Song, D.: The secret sharer: evaluating and testing unintended memorization in neural networks. In: 28th USENIX Security Symposium (USENIX Security 19), Santa Clara, CA, pp. 267–284. USENIX Association, August 2019. https://www.usenix.org/conference/usenixsecurity19/presentation/carlini
4. Guo, C., Goldstein, T., Hannun, A., Maaten, L.v.d.: Certified data removal from machine learning models. arXiv preprint arXiv:1911.03030 (2019)
5. He, K., Zhang, X., Ren, S., Sun, J.: Deep residual learning for image recognition. In: Proceedings of the IEEE Conference on Computer Vision and Pattern Recognition, pp. 770–778 (2016)
6. Kermany, D., Zhang, K., Goldbaum, M., et al.: Labeled optical coherence tomography (OCT) and chest X-ray images for classification. Mendeley Data **2**(2) (2018)
7. Kermany, D.S., et al.: Identifying medical diagnoses and treatable diseases by image-based deep learning. Cell **172**(5), 1122–1131 (2018)

8. Kingma, D.P., Ba, J.: Adam: a method for stochastic optimization. arXiv preprint arXiv:1412.6980 (2014)
9. LeCun, Y., Cortes, C.: MNIST handwritten digit database (2010). http://yann.lecun.com/exdb/mnist/
10. Liu, X., Tsaftaris, S.A.: Have you forgotten? A method to assess if machine learning models have forgotten data. In: Martel, A.L., et al. (eds.) MICCAI 2020. LNCS, vol. 12261, pp. 95–105. Springer, Cham (2020). https://doi.org/10.1007/978-3-030-59710-8_10
11. Nasr, M., Shokri, R., Houmansadr, A.: Machine learning with membership privacy using adversarial regularization. In: Proceedings of the 2018 ACM SIGSAC Conference on Computer and Communications Security, pp. 634–646 (2018)
12. Netzer, Y., Wang, T., Coates, A., Bissacco, A., Wu, B., Ng, A.Y.: Reading digits in natural images with unsupervised feature learning (2011)
13. Paszke, A., et al.: PyTorch: an imperative style, high-performance deep learning library. arXiv preprint arXiv:1912.01703 (2019)
14. Ruder, S.: An overview of gradient descent optimization algorithms. arXiv preprint arXiv:1609.04747 (2016)
15. Salem, A., Zhang, Y., Humbert, M., Berrang, P., Fritz, M., Backes, M.: ML-leaks: model and data independent membership inference attacks and defenses on machine learning models. arXiv preprint arXiv:1806.01246 (2018)
16. Shokri, R., Stronati, M., Song, C., Shmatikov, V.: Membership inference attacks against machine learning models. In: 2017 IEEE Symposium on Security and Privacy (SP), pp. 3–18. IEEE (2017)
17. Song, L., Mittal, P.: Systematic evaluation of privacy risks of machine learning models. arXiv preprint arXiv:2003.10595 (2020)
18. Voigt, P., Von dem Bussche, A.: The EU general data protection regulation (GDPR). Intersoft consulting (2018)
19. Wang, L., Lin, Z.Q., Wong, A.: COVID-net: a tailored deep convolutional neural network design for detection of COVID-19 cases from chest X-ray images. Sci. Rep. 10(1), 1–12 (2020)
20. Zhang, Y., Jia, R., Pei, H., Wang, W., Li, B., Song, D.: The secret revealer: generative model-inversion attacks against deep neural networks. In: Proceedings of the IEEE/CVF Conference on Computer Vision and Pattern Recognition, pp. 253–261 (2020)

AnaXNet: Anatomy Aware Multi-label Finding Classification in Chest X-Ray

Nkechinyere N. Agu[1(✉)], Joy T. Wu[2], Hanqing Chao[1], Ismini Lourentzou[3], Arjun Sharma[2], Mehdi Moradi[2], Pingkun Yan[1], and James Hendler[1]

[1] Rensselaer Polytechnic Institute, Troy, NY 12180, USA
agun@rpi.edu
[2] IBM Research, Almaden Research Center, San Jose, CA 95120, USA
[3] Virginia Tech, Blacksburg, VA 24061, USA

Abstract. Radiologists usually observe anatomical regions of chest X-ray images as well as the overall image before making a decision. However, most existing deep learning models only look at the entire X-ray image for classification, failing to utilize important anatomical information. In this paper, we propose a novel multi-label chest X-ray classification model that accurately classifies the image finding and also localizes the findings to their correct anatomical regions. Specifically, our model consists of two modules, the detection module and the anatomical dependency module. The latter utilizes graph convolutional networks, which enable our model to learn not only the label dependency but also the relationship between the anatomical regions in the chest X-ray. We further utilize a method to efficiently create an adjacency matrix for the anatomical regions using the correlation of the label across the different regions. Detailed experiments and analysis of our results show the effectiveness of our method when compared to the current state-of-the-art multi-label chest X-ray image classification methods while also providing accurate location information.

Keywords: Graph convolutional networks · Multi-label chest X-ray image classification · Graph representation

1 Introduction

Interpreting a radiology imaging exam is a complex reasoning task, where radiologists are able to integrate patient history and image features from different anatomical locations to generate the most likely diagnoses. Convolutional Neural Networks (CNNs) have been widely applied in earlier works in automatic Chest X-ray (CXR) interpretation, one of the most commonly requested medical imaging modality. Many of these works have framed the problem either as a multi-label abnormality classification problem [18,27], an abnormality detection and localization problem [5,22,29], or an image-to-text report generation problem [14,28]. However, these models fail to capture inter-dependencies between

© Springer Nature Switzerland AG 2021
M. de Bruijne et al. (Eds.): MICCAI 2021, LNCS 12905, pp. 804–813, 2021.
https://doi.org/10.1007/978-3-030-87240-3_77

features or labels. Leveraging such contextual information that encodes relational information among pathologies is crucial in improving interpretability and reasoning in clinical diagnosis.

To this end, Graph Neural Networks (GNN) have surfaced as a viable solution in modeling disease co-occurrence across images. Graph Neural Networks (GNNs) learn representations of the nodes based on the graph structure and have been widely explored, from graph embedding methods [7,23], generative models [25,32] to attention-based or recurrent models [15,24], among others. For a comprehensive review on model architectures, we refer the reader to a recent survey [31]. In particular, Graph Convolutional Networks (GCNs) [13] utilize *graph convolution* operations to learn representations by aggregating information from the neighborhood of a node, and have been successfully applied to CXR image classification. For example, the multi-relational ImageGCN model learns image representations that leverage additional information from related images [17], while CheXGCN and DD-GCN incorporate label co-occurrence GCN modules to capture the correlations between labels [2,16]. To mitigate the issues with noise originating from background regions in related images, recent work utilizes attention mechanisms [1,34] or auxiliary tasks such as lung segmentation [3,6]. However, none of these works consider modeling correlations among anatomical regions and findings, *e.g.*, output the anatomical location semantics for each finding.

We propose a novel model that captures the dependencies between the anatomical regions of a chest X-ray for classification of the pathological findings, termed **Ana**tomy-aware **X**-ray Network (AnaXNet). We first extract the features of the anatomical regions using an object detection model. We develop a method to accurately capture the correlations between the various anatomical regions and learn their dependencies with a GCN model. Finally, we combine the localized region features via attention weights computed with a non-local operation [26] that resembles self-attention.

The main contributions of this paper are summarized as follows: 1) we propose a novel multi-label CXR findings classification framework that integrates both global and local anatomical visual features and outputs accurate localization of clinically relevant anatomical regional levels for CXR findings, 2) we propose a method to automatically learn the correlation between the findings and the anatomical regions and 3) we conduct in-depth experimental analysis to demonstrate that our proposed AnaXNet model outperforms previous baselines and state-of-the-art models.

2   Methodology

We first describe our proposed framework for multi-label chest X-ray classification. Let CXR image collection comprised of a set of N chest-X ray images $\mathcal{C} = \{x_1, \ldots, x_N\}$, where each image x_i is associated with a set of M labels $\mathcal{Y}_i = \{y_i^1, \ldots, y_i^M\}$, with $y_i^m \in \{0, 1\}$ indicating whether the label for pathology M appears in image x_i or not. Then the goal is to design a model that predicts the label set for an unseen image as accurately as possible, by utilizing

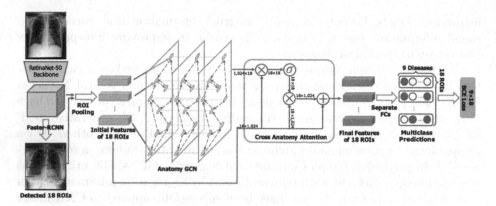

Fig. 1. Model overview. We extract anatomical regions of interest (ROIs) and their corresponding features, feed their vectors to a Graph Convolutional Network that learns their inter-dependencies, and combine the output with an attention mechanism, to perform the final classification with a dense layer. Note that throughout the paper, we use the term *bounding box* instead of ROI.

the correlation among anatomical region features $R_i = f(x_i) \in \mathbb{R}^{k \times d}$, where k is the number of anatomical region embedding, each with dimensionality d. The anatomical region feature extractor f is described in Subsect. 3.3.

Given this initial set of anatomical region representations R_i, we define a normalized adjacency matrix $A \in \mathbb{R}^{k \times k}$ that captures region correlations, and utilize a GCN $Z_i = g(R_i, A) \in \mathbb{R}^{k \times d}$ to update R_i as follows:

$$R_i^{t+1} = \phi(A R_i^t W_1^t), \tag{1}$$

where $W_1^l \in \mathbb{R}^{d \times d}$ is the learned weight matrix and $\phi(.)$ denotes a non-linear operation, *e.g.*, ReLU [33] in our experiments, and t is the number of stacked GCN layers. To construct the adjacency matrix A, we extract co-occurrence patterns between anatomical regions for label pairs. More specifically, the label co-occurrence matrix can be computed based on Jaccard similarity:

$$J(r_i, r_j) = \frac{1}{M} \sum_{m=1}^{M} \frac{|\mathcal{Y}_i^m \cap \mathcal{Y}_j^m|}{|\mathcal{Y}_i^m \cup \mathcal{Y}_j^m|}, \tag{2}$$

where r_i and r_i represent anatomical regions, $\mathcal{Y}_i$ is the set for region r_i and label m across all images and $\cap, \cup$ denote the intersection and union over multisets. However, this label co-occurrence construction may overfit the training data due to incorporating noisy rare occurrences. To mitigate such issues, we use a filtering threshold τ, *i.e.*,

$$A_{ij} = \begin{cases} 1 & \text{if } J(R_i, R_j) \geq \tau \\ 0 & \text{if } J(R_i, R_j) < \tau \end{cases}, \tag{3}$$

where A is the final adjacency matrix.

Table 1. Dataset characteristics. # Images (number of images) and # Bboxes (number of bounding boxes) labeled with L1-L9. There are a total of 217,417 images in the dataset, of which 153,333 have at least one of the L1-L9 labels globally. Of these images, 3,877,010 bounding boxes were extracted automatically and 720,098 of them have at least one or more of the 9 labels.

Label ID	Description	# Images (1)	# Bboxes
L1	Lung opacity	132,981	584,638
L2	Pleural effusion	68,363	244,005
L3	Atelectasis	76,868	240,074
L4	Enlarged cardiac silhouette	55,187	58,929
L5	Pulmonary edema/hazy opacity	33,441	145,965
L6	Pneumothorax	9,341	22,906
L7	Consolidation	16,855	53,364
L8	Fluid overload/heart failure	6,317	18,066
L9	Pneumonia	32,042	95,215
All 9 labels	**Positive/total**	153,333/217,417	720,098/3,877,010

To capture both global and local dependencies between anatomical regions, we leverage a non-local operation that resembles self-attention [26]:

$$Q_i = \text{softmax}\left(R_i Z_i^T\right) R_i, \tag{4}$$

where $Q_i \in \mathbb{R}^{k \times d}$. The final prediction is computed via

$$\hat{y} = [R_i; Q_i] W_2^T, \tag{5}$$

where $W_2 \in \mathbb{R}^{2d \times M}$ is a fully connected layer to obtain the label predictions. The network is trained with a multi-label cross-entropy classification loss

$$L = \frac{1}{N} \sum_{i=1}^{N} \sum_{m=1}^{M} y_i^m log(\sigma(\hat{y}_i^m)) + (1 - y_i^m) log(1 - \sigma(\hat{y}_i^m)), \tag{6}$$

where σ is the Sigmoid function and $\{\hat{y}_i^m, y_i^m\} \in \mathbb{R}^M$ are the model prediction and ground truth for example x_i, respectively. The model architecture is summarized in Fig. 1.

3 Experiments

We describe experimental details, *i.e.*, evaluation dataset, metrics, *etc..*, and present quantitative and qualitative results, comparing AnaXNet with several baselines.

Table 2. Intersection over Union scores (IoU) are calculated between the automatically extracted anatomical bounding box (Bbox) regions and a set of single manual ground truth bounding boxes for 1000 CXR images. Average precision and recall across 9 CXR pathologies are shown for the NLP derived labels at: right lung (RL), right apical zone (RAZ), right upper lung zone (RULZ), right mid lung zone (RMLZ), right lower lung zone (RLLZ), right costophrenic angle (RCA), left lung (LL), left apicl zone (LAZ), left upper lung zone (LULZ), left mid lung zone (LMLZ), left lower lung zone (LLLZ), left costophrenic angle (LCA), mediastinum (Med), upper mediastinum (UMed), cardiac silhouette (CS) and trachea (Trach).

Bbox abbreviation	RL	RAZ	RULZ	RMLZ	RLLZ	RHS	RCA	LL	LAZ
Bbox IoU	0.994	0.995	0.995	0.989	0.984	0.989	0.974	0.989	0.995
NLP precision	0.944	0.762	0.857	0.841	0.942	0.897	0.871	0.943	0.800
NLP recall	0.98	0.889	0.857	0.746	0.873	0.955	0.808	0.982	1.00
Bbox abbreviation	LULZ	LMLZ	LLLZ	LHS	LCA	Med	UMed	CS	Trach
Bbox IoU	0.995	0.986	0.979	0.985	0.950	0.972	0.993	0.967	0.983
NLP precision	0.714	0.921	0.936	0.888	0.899	N/A	N/A	0.969	N/A
NLP recall	0.938	0.972	0.928	0.830	0.776	N/A	N/A	0.933	N/A

3.1 Dataset

Existing annotations of large-scale CXR datasets [11,12,27] are either weak global labels for 14 common CXR findings extracted from reports with Natural Language Processing (NLP) methods [11], or are manually annotated with bounding boxes for a smaller subset of images and for a limited number of labels [4,21]. None of these annotated datasets describe the anatomical location for different CXR pathologies. However, localizing pathologies to anatomy is a key aspect of radiologists' reasoning and reporting process, where knowledge of correlation between image findings and anatomy can help narrow down potential diagnoses.

The Chest ImaGenome dataset builds on the works of [29,30] to fill this gap by using a combination of rule-based text-analysis and atlas-based bounding box extraction techniques to structure the anatomies and the related pathologies from 217,417 report texts and frontal images (AP or PA view) from the MIMIC-CXR dataset [12]. In summary, the text pipeline [30] first sections the report and retains only the finding and impression sentences. Then it uses a prior curated CXR concept dictionary (lexicons) to identify and detect the context (negated or not) for name entities required for labeling the 18 anatomical regions and 9 CXR pathology labels from each retained sentence. The pathology labels are associated with the anatomical region described in the same sentence with a natural language parser, SpaCy [9], and clinical heuristics provided by a radiologist was used to correct for obvious pathology-to-anatomy assignment errors (e.g. lung opacity wrongly assigned to mediastinum). Finally the pathology label(s) for each of the 18 anatomical regions from repeated sentences are grouped to the exam level. A separate anatomy atlas-based bounding box pipeline extracts the coordinates from each frontal images for the 18 anatomical regions [29].

Table 3. Comparison of our approach against baselines (AUC score).

Method	L1	L2	L3	L4	L5	L6	L7	L8	L9	**AVG**
Faster R-CNN	0.84	0.89	0.77	0.85	0.87	0.77	0.75	0.81	0.71	0.80
GlobalView	**0.91**	0.94	0.86	0.92	0.92	**0.93**	0.86	0.87	0.84	0.89
CheXGCN	0.86	0.90	0.91	0.94	**0.95**	0.75	**0.89**	**0.98**	0.88	0.90
AnaXNet (ours)	0.88	**0.96**	**0.92**	**0.99**	**0.95**	0.80	**0.89**	**0.98**	**0.97**	**0.93**

Table 1 shows high-level statistics of the generated dataset. Dual annotations for 500 random reports (disagreement resolved via consensus) were curated at sentence level by a clinician and a radiologist, who also annotated the bounding boxes for 1000 frontal CXRs (single annotation). For the 9 pathology, the overall NLP average precision and recall without considering localization are 0.9819 and 0.9875, respectively. More detailed results by anatomical regions are shown in Table 2.

3.2   Baselines

The anatomical region feature extractor $f(x_i) \in \mathbb{R}^{k \times d}$ is a Faster R-CNN with with ResNet-50 [8] as base model. Additional implementation details, e.g., hyper-parameters, are provided later on (Subsect. 3.3). We perform comprehensive analysis on the Chest ImaGenome dataset. We compare our AnaXNet model against: 1) **GlobalView** we implement a DenseNet169 [10] model as a baseline method to contrast the effectiveness of location-aware AnaXNet versus a global view of the image, 2) **Faster R-CNN** [19] followed by a fully-connected layer, i.e., without the GCN and attention modules, to establish a baseline accuracy for the classification task using the extracted anatomical features, and 3) **CheXGCN** We re-implement the state-of-the-art model CheXGCN [2] that utilizes GCNs to learn the label dependencies between pathologies in the X-ray images. The model uses a CNN for feature extraction and a GCN to learn the relationship between the labels via word embeddings. We replace the overall CNN with Faster R-CNN for a fair comparison with our model, but retain their label co-occurrence learning module.

3.3   Implementation Details

We train the detection model to detect the 18 anatomical regions. To obtain the anatomical features, we take the final output of the model and perform non-maximum suppression for each object class using an IoU threshold. We select all the regions where any class probability exceeds our confidence threshold. We use a value of 0.5 for τ. For each region, we extract a 1024 dimension convolutional feature vector. For multiple predictions of the same anatomical region, we select the prediction with the highest confidence score and drop the duplicates. When the model fails to detect a bounding box, we use a vector of zeros to represent

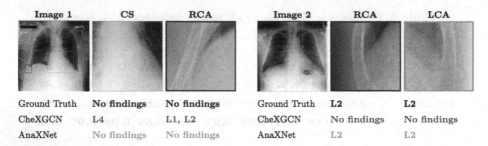

Image 1	CS	RCA	Image 2	RCA	LCA

Ground Truth	**No findings**	**No findings**	Ground Truth	**L2**	**L2**
CheXGCN	L4	L1, L2	CheXGCN	No findings	No findings
AnaXNet	No findings	No findings	AnaXNet	L2	L2

Fig. 2. Examples of the results obtained by our best two models. The overall chest X-ray image is shown alongside two anatomical regions. The predictions from best performing models are compared against the ground-truth labels.

the anatomical features of the region within the GCN. We use detectron2[1] to train Faster R-CNN to extract anatomical regions and their features. Our GCN model is made up of two GCN layers with output dimensionality of 512 and 1024 respectively. We train with Adam optimizer, and 10^{-4} learning rate for 25 epochs in total.

3.4   Results and Evaluation

Results are summarized in Table 3. The evaluation metric is Area Under the Curve (AUC). Note that the baseline GlobalView is in fact a global classifier and does not produce a localized label. The remaining rows in Table 3 show localized label accuracy. For the localized methods, the reported numbers represent the average AUC of the model for each label over the various anatomical regions. If a finding is detected at the wrong anatomical location, it counts as false detection. For fair comparison, we use the same 70/10/20 train/validation/testing split across patients to train each model. AnaXNet model obtains improvements over the previous methods while also localizing the diseases in their correct anatomical region. The GlobalView is most likely limited because it focuses on the entire image instead of a specific region.

The CheXGCN model outperforms the other two baselines but is also limited because it focuses on one section and uses label dependencies to learn the relationship between the labels, while ignoring the relationships between the anatomical regions of the chest X-ray image. In Table 2, we visualize the output from both the CheXGCN model and our AnaXNet model. The CheXGCN model had difficulty predicting small anatomical regions like the costophrenic angles, while our model had additional information from the remaining anatomical regions, which helped in its prediction. Also the CheXGCN model struggled with enlarged cardiac silhouette label because information from the surrounding labels is needed in order to accurately tell if the heart is enlarged.

In Fig. 3 we also visualize the output of Grad-CAM [20] method on the GlobalView model to highlight the importance of the localization, while the prediction

---

[1] https://github.com/facebookresearch/detectron2.

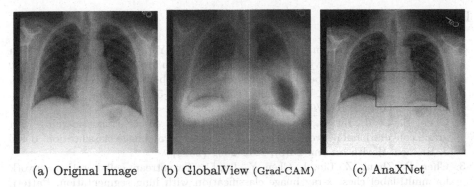

(a) Original Image   (b) GlobalView (Grad-CAM)   (c) AnaXNet

Fig. 3. An example image from our dataset with enlarged cardiac silhouette. The GlobalView network detects this label correctly, but as the Grad-CAM activation map shows (b), the attention of the network is not on the cardiac region. Our method detects the finding in the correct bounding box (c).

of Enlarged Cardiac Silhouette was correct, the GlobalView model was focused on the lungs. Our method was able to provide accurate localization information as well as the finding.

4   Conclusion

We described a methodology for localized detection of diseases in chest X-ray images. Both the algorithmic framework of this work, and the dataset of images labeled for pathologies in the semantically labeled bounding boxes are important contributions. For our AnaXNet design, a Faster R-CNN architecture detects the bounding boxes and embeds them. The resulting embedded vectors are then used as input to a GCN and an attention block that learn representations by aggregating information from the neighboring regions.

This approach accurately detects any of the nine studied abnormalities and places it in the correct bounding box in the image. The 18 pre-specified bounding boxes are devised to map to the anatomical areas often described by radiologists in chest X-ray reports. As a result, our method provides all the necessary components for composing a structured report. Our vision is that the output of our trained model, subject to expansion of the number and variety of findings, will provide both the finding and the anatomical location information for both downstream report generation and other reasoning tasks. Despite the difficulty of localized disease detection, our method outperforms a global classifier. As our data shows (See Fig. 3), global classification can be unreliable even when the label is correct as the classifier might find the correct label for the wrong reason at an irrelevant spot.

Acknowledgements. This work was supported by the Rensselaer-IBM AI Research Collaboration, part of the IBM AI Horizons Network.

References

1. Cai, J., Lu, L., Harrison, A.P., Shi, X., Chen, P., Yang, L.: Iterative attention mining for weakly supervised thoracic disease pattern localization in chest X-rays. In: Frangi, A.F., Schnabel, J.A., Davatzikos, C., Alberola-López, C., Fichtinger, G. (eds.) MICCAI 2018. LNCS, vol. 11071, pp. 589–598. Springer, Cham (2018). https://doi.org/10.1007/978-3-030-00934-2_66
2. Chen, B., Li, J., Lu, G., Yu, H., Zhang, D.: Label co-occurrence learning with graph convolutional networks for multi-label chest x-ray image classification. IEEE J. Biomed. Health Inform. **24**(8), 2292–2302 (2020)
3. Chen, B., Zhang, Z., Lin, J., Chen, Y., Lu, G.: Two-stream collaborative network for multi-label chest x-ray image classification with lung segmentation. Pattern Recogn. Lett. **135**, 221–227 (2020)
4. Filice, R.W., et al.: Crowdsourcing pneumothorax annotations using machine learning annotations on the NIH chest X-ray dataset. J. Digit. Imaging **33**(2), 490–496 (2020)
5. Gabruseva, T., Poplavskiy, D., Kalinin, A.: Deep learning for automatic pneumonia detection. In: Proceedings of the IEEE/CVF Conference on Computer Vision and Pattern Recognition Workshops, pp. 350–351 (2020)
6. Gordienko, Y., et al.: Deep learning with lung segmentation and bone shadow exclusion techniques for chest X-ray analysis of lung cancer. In: Hu, Z., Petoukhov, S., Dychka, I., He, M. (eds.) ICCSEEA 2018. AISC, vol. 754, pp. 638–647. Springer, Cham (2019). https://doi.org/10.1007/978-3-319-91008-6_63
7. Grover, A., Leskovec, J.: node2vec: scalable feature learning for networks. In: Proceedings of the 22nd ACM SIGKDD International Conference on Knowledge Discovery and Data Mining, pp. 855–864 (2016)
8. He, K., Zhang, X., Ren, S., Sun, J.: Deep residual learning for image recognition. In: Proceedings of the IEEE Conference on Computer Vision and Pattern Recognition, pp. 770–778 (2016)
9. Honnibal, M., Montani, I., Van Landeghem, S., Boyd, A.: spaCy: Industrial-strength Natural Language Processing in Python. Zenodo (2020)
10. Huang, G., Liu, Z., Van Der Maaten, L., Weinberger, K.Q.: Densely connected convolutional networks. In: Proceedings of the IEEE Conference on Computer Vision and Pattern Recognition, pp. 4700–4708 (2017)
11. Irvin, J., et al.: CheXpert: a large chest radiograph dataset with uncertainty labels and expert comparison. In: Proceedings of the AAAI Conference on Artificial Intelligence, vol. 33, pp. 590–597 (2019)
12. Johnson, A.E., et al.: MIMIC-CXR, a de-identified publicly available database of chest radiographs with free-text reports. Sci. Data **6**, 1–8 (2019)
13. Kipf, T.N., Welling, M.: Semi-supervised classification with graph convolutional networks. In: International Conference on Learning Representations (2017)
14. Li, C.Y., Liang, X., Hu, Z., Xing, E.P.: Hybrid retrieval-generation reinforced agent for medical image report generation. arXiv:1805.08298 (2018)
15. Li, Y., Tarlow, D., Brockschmidt, M., Zemel, R.: Gated graph sequence neural networks. In: International Conference on Learning Representations (2016)
16. Liu, D., Xu, S., Zhou, P., He, K., Wei, W., Xu, Z.: Dynamic graph correlation learning for disease diagnosis with incomplete labels. arXiv:2002.11629 (2020)
17. Mao, C., Yao, L., Luo, Y.: ImageGCN: multi-relational image graph convolutional networks for disease identification with chest X-rays. arXiv:1904.00325 (2019)

18. Rajpurkar, P., et al.: CheXNet: radiologist-level pneumonia detection on chest X-rays with deep learning. arXiv:1711.05225 (2017)
19. Ren, S., He, K., Girshick, R., Sun, J.: Faster R-CNN: towards real-time object detection with region proposal networks. arXiv:1506.01497 (2015)
20. Selvaraju, R.R., Cogswell, M., Das, A., Vedantam, R., Parikh, D., Batra, D.: Grad-CAM: visual explanations from deep networks via gradient-based localization. In: Proceedings of the IEEE International Conference on Computer Vision (2017)
21. Shih, G., et al.: Augmenting the national institutes of health chest radiograph dataset with expert annotations of possible pneumonia. Radiol. Artif. Intell. (2019)
22. Sirazitdinov, I., Kholiavchenko, M., Mustafaev, T., Yixuan, Y., Kuleev, R., Ibragimov, B.: Deep neural network ensemble for pneumonia localization from a large-scale chest X-ray database. Comput. Electr. Eng. **78**, 388–399 (2019)
23. Tang, J., Qu, M., Wang, M., Zhang, M., Yan, J., Mei, Q.: LINE: large-scale information network embedding. In: Proceedings of the 24th International Conference on World Wide Web, pp. 1067–1077 (2015)
24. Veličković, P., Cucurull, G., Casanova, A., Romero, A., Lio, P., Bengio, Y.: Graph attention networks. In: International Conference on Learning Representations (2018)
25. Wang, H., et al.: GraphGAN: graph representation learning with generative adversarial nets. In: Proceedings of the AAAI Conference on Artificial Intelligence, vol. 32 (2018)
26. Wang, X., Girshick, R., Gupta, A., He, K.: Non-local neural networks. In: Proceedings of the IEEE Conference on Computer Vision and Pattern Recognition, pp. 7794–7803 (2018)
27. Wang, X., Peng, Y., Lu, L., Lu, Z., Bagheri, M., Summers, R.M.: ChestX-ray8: hospital-scale chest X-ray database and benchmarks on weakly-supervised classification and localization of common thorax diseases. In: Proceedings of the IEEE Conference on Computer Vision and Pattern Recognition, pp. 2097–2106 (2017)
28. Wang, X., Peng, Y., Lu, L., Lu, Z., Summers, R.M.: TieNet: text-image embedding network for common thorax disease classification and reporting in chest X-rays. In: Proceedings of the IEEE Conference on Computer Vision and Pattern Recognition, pp. 9049–9058 (2018)
29. Wu, J., et al.: Automatic bounding box annotation of chest X-ray data for localization of abnormalities. In: 2020 IEEE 17th International Symposium on Biomedical Imaging (ISBI), pp. 799–803. IEEE (2020)
30. Wu, J.T., Syed, A., Ahmad, H., et al.: AI accelerated human-in-the-loop structuring of radiology reports. In: Americal Medical Informatics Association (AMIA) Annual Symposium (2020)
31. Wu, Z., Pan, S., Chen, F., Long, G., Zhang, C., Philip, S.Y.: A comprehensive survey on graph neural networks. IEEE Trans. Neural Netw. Learn. Syst. **32**, 4–24 (2020)
32. You, J., Ying, R., Ren, X., Hamilton, W., Leskovec, J.: GraphRNN: generating realistic graphs with deep auto-regressive models. In: International Conference on Machine Learning, pp. 5708–5717. PMLR (2018)
33. Zeiler, M.D., et al.: On rectified linear units for speech processing. In: 2013 IEEE International Conference on Acoustics, Speech and Signal Processing, pp. 3517–3521. IEEE (2013)
34. Zhou, Y., Zhou, T., Zhou, T., Fu, H., Liu, J., Shao, L.: Contrast-attentive thoracic disease recognition with dual-weighting graph reasoning. IEEE Trans. Med. Imaging **40**, 1196–1206 (2021)

Projection-Wise Disentangling for Fair and Interpretable Representation Learning: Application to 3D Facial Shape Analysis

Xianjing Liu[1], Bo Li[1], Esther E. Bron[1], Wiro J. Niessen[1,2], Eppo B. Wolvius[1], and Gennady V. Roshchupkin[1(✉)]

[1] Erasmus MC, Rotterdam, The Netherlands
g.roshchupkin@erasmusmc.nl
[2] Delft University of Technology, Delft, The Netherlands

Abstract. Confounding bias is a crucial problem when applying machine learning to practice, especially in clinical practice. We consider the problem of learning representations independent to multiple biases. In literature, this is mostly solved by purging the bias information from learned representations. We however expect this strategy to harm the diversity of information in the representation, and thus limiting its prospective usage (e.g., interpretation). Therefore, we propose to mitigate the bias while keeping almost all information in the latent representations, which enables us to observe and interpret them as well. To achieve this, we project latent features onto a learned vector direction, and enforce the independence between biases and projected features rather than all learned features. To interpret the mapping between projected features and input data, we propose projection-wise disentangling: a sampling and reconstruction along the learned vector direction. The proposed method was evaluated on the analysis of 3D facial shape and patient characteristics (N = 5011). Experiments showed that this conceptually simple method achieved state-of-the-art fair prediction performance and interpretability, showing its great potential for clinical applications.

Keywords: Fair representation learning · Disentangled representation learning · Interpretability · 3D shape analysis

1 Introduction

Machine learning techniques, especially deep learning, have emerged as a powerful tool in many domains. However, its susceptibility to bias present in training datasets and tasks poses, brings a new challenge for the practical applicability, i.e., spurious performance in the training and evaluation stage with limited generalizability to application in new conditions [1]. To mitigate (confounding) bias in data analysis, traditional statistic methods use special techniques such as control-matching [2] and stratification [3]. However,

Electronic supplementary material The online version of this chapter (https://doi.org/10.1007/978-3-030-87240-3_78) contains supplementary material, which is available to authorized users.

M. de Bruijne et al. (Eds.): MICCAI 2021, LNCS 12905, pp. 814–823, 2021.
https://doi.org/10.1007/978-3-030-87240-3_78

due to the end-to-end training scheme and the need for large-size training data, these techniques are no longer favored by the machine learning field.

In *"Representation Learning"* one tries to find representations (i.e., learned features, Z) of the data that are related to specific attributes (i.e., the learning target, t). Especially, a *fair representation* means it contains no information of sensitive attributes (i.e., bias, s) [8]. Existing methods for fair representation learning can be categorized into two types: 1) adversarial training, in which methods are trained to predict the bias from the representation, and subsequently minimize the performance of the adversary to remove bias information from the representation [6, 7], and 2) variational autoencoder (VAE) - based methods [8–11], which minimize the dependency between the latent representation and the bias using Mutual Information (MI) or Maximum Mean Discrepancy (MMD) metrics [8, 10]. Despite their potential to facilitate fair representation, these models are not interpretable, which can limit their applicability in clinical practice [13]. Moreover, the fairness of these methods is approached by purging all bias information from the learned representations (i.e., $MI(Z, s) \rightarrow 0$). This strict strategy can reduce the diversity of information in Z (Fig. S3 in supplementary).

To address these issues, we propose a novel projection-wise disentangling strategy for auto-encoder-based fair and interpretable representation learning. We construct z_p as a linear projection of latent features onto a vector direction, and learn the fair representation by minimizing the correlation between z_p and s, i.e., $MI(z_p, s) \rightarrow 0$. Compared with existing strategies of global-constraint $MI(Z, s) \rightarrow 0$, the proposed conditional-constraint strategy can maintain the diversity of information in Z, and thus 1) obtains an **optimal trade-off** between reconstruction quality and fairness, fitting the proposed method into **semi-supervised extensions**; Also, 2) we propose projection-wise disentangling to **interpret** the disentanglement of correlated attributes; 3) Our method can easily handle **multiple and continuous biases**.

In this paper, we applied the proposed method to clinical applications of 3D facial shape analysis. For epidemiological studies the bias (confounding) problem is crucial, because it can create association that is not true, or association that is true, but misleading, thus leading to wrong diagnosis or therapy strategy. The bias problem becomes significant for AI since it has been used more and more for medical data analysis. While

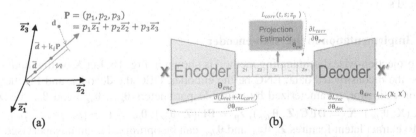

(a)   (b)

Fig. 1. a) $P = [p_1, p_2, \ldots p_n]$ represents a vector in latent space ($n = 3$). d is the latent representation of a datapoint. $\overline{d}$ and $\overline{d} + k_i P$ show the sampling along P in Eq. 6. When sampling along P, z_p changes and this change is correlated to t ($|Corr(z_p, t)|$ is maximized) while independent to s ($|Corr(z_p, s)|$ is minimized); b) Framework of the proposed method.

we applied our framework to 3D facial images due to our clinical interest, the approach can be generally used for other type of data.

2 Methods

2.1 Projecting Features onto a Direction in the Latent Space

A latent space can be viewed as a vector subspace of $\mathbb{R}^n$ (n latent features) with basis vectors[1] $\overrightarrow{z}_i$ ($i = 1, 2 \ldots n$). Some recent studies show that most attributes (target or bias) of the input data have a vector direction that predominantly captures its variability [4, 5]. We aim to couple a vector direction with the target while being independent to the bias. A vector direction in the latent space can be represented by $\mathbf{P} = [p_1, p_2, ..., p_n]$ as a linear combination of the basis vectors (Fig. 1a). Let $\mathbf{D}^{d \times n} = [\mathbf{z}_1^d, \mathbf{z}_2^d, \ldots, \mathbf{z}_n^d]$ (d datapoints) be the latent representations of the input data. For each datapoint, $\mathbf{d} = [z_1, z_2, \ldots, z_n]$ is its latent representations, and $z_p = \mathbf{d} * \mathbf{P}/\|\mathbf{P}\| = (p_1 z_1 + p_2 z_2 + \cdots + p_n z_n)/\|\mathbf{P}\|$ can be viewed as a scalar projection of $\mathbf{d}$ onto vector $\mathbf{P}$ (Fig. 1a).

The canonical correlation analysis theory suggests that a linear combination (projection) of multiple variables can have a maximum or minimum linear relationship with specific attributes [14]. We thus formulate the linear relationship between $\mathbf{z_p}$ and attributes as follow:

$$\mathbf{z_p} = \beta_0 + \beta_s \mathbf{s} + \beta_t \mathbf{t} + \varepsilon, \tag{1}$$

where $\mathbf{z_p}^d$ is the projections of d datapoints, $\mathbf{s}^d$ denotes a bias and $\mathbf{t}^d$ denotes a target with d datapoints, ε is the error term, β_0 is a constant, β_s and β_t denote the coefficients of $\mathbf{s}$ and $\mathbf{t}$.

Subsequently, we link a vector direction $\mathbf{P}$ with the target and bias, with the goal to estimate a $\mathbf{P}$ that minimizes $|\beta_s|$ (equivalent to minimizing $|Corr(\mathbf{z_p}, \mathbf{s})|$) while maximizing $|\beta_t|$ (equivalent to maximizing $|Corr(\mathbf{z_p}, \mathbf{t})|$). Here $Corr(., .)$ is the Pearson correlation coefficient which ranges between $[-1, 1]$.

To promote a linear relation between $\mathbf{z_p}$ and $(\mathbf{t}, \mathbf{s})$, which is required for effective bias minimization (Eq. 1), a correlation loss (Eq. 3) is further proposed. The correlation loss not only estimates $\mathbf{P}$, but also encourages the encoder to generate $\mathbf{z_p}$ linearly correlated to $\mathbf{t}$ and $\mathbf{s}$.

2.2 Implementation in the Auto-encoder

Our projection-wise disentangling method is shown in Fig. 1b. Let $\mathbf{X}$ be the input data, $\mathbf{X}'$ be the reconstructed data; ENC be the encoder, DEC the decoder, and PE the projection estimator, parameterized by trainable parameters θ_{enc}, θ_{dec}, and θ_{pe}, i.e.:$\mathbf{Z} = \text{ENC}(\mathbf{X}; \theta_{enc}); \mathbf{X}' = \text{DEC}(\mathbf{Z}; \theta_{dec}); \mathbf{z_p} = \text{PE}(\mathbf{Z}; \theta_{pe}); \theta_{pe} = \mathbf{P} = [p_1, p_2, ..., p_n]$.

To extract latent features $\mathbf{Z}$, θ_{enc} and θ_{dec} can be optimized in an unsupervised way using a reconstruction loss (L_{rec}), as quantified by the mean squared error between the input data and the reconstructed data:

$$L_{rec}(\mathbf{X}; \mathbf{X}') = \|\mathbf{X} - \mathbf{X}'\|^2. \tag{2}$$

---

[1] Non-orthogonal basis vectors.

Based on Eq. 1, to estimate θ_{pe}, as well as to optimize θ_{enc} and thereby promote the linear correlation between z_p and (s, t), we propose a correlation loss (L_{corr}):

$$L_{corr}(t, s; z_p) = |Corr(z_p, s)| - \eta |Corr(z_p, t)|, \tag{3}$$

where η can be considered as a Lagrange multiplier to balance the correlations.

L_{corr} can handle binary and continuous biases. In case s is categorical, it can be converted into dummy variables. Besides, it can be easily extended to handle tasks with multiple biases $s_i^d (i = 1, 2, \ldots m)$:

$$L_{corr}(t, s_1, s_2, \ldots, s_m; z_p) = |Corr(z_p, s_1)| + \ldots + |Corr(z_p, s_m)| - \eta |Corr(z_p, t)|. \tag{4}$$

Combining Eq. 2 and 3, we optimize the proposed framework using a multi-task loss function (L_{joint}):

$$\theta_{enc}, \theta_{dec}, \theta_{pe} \leftarrow argmin \, L_{joint} = L_{rec} + \lambda L_{corr}, \tag{5}$$

where λ balances the magnitude of the reconstruction quality and the fairness terms.

In addition, for applications where (target and bias) attributes are only available for part of the data, we provide a semi-supervised implementation of the method to fully exploit the data. In particular, for each training batch, we update the parameters in two steps: 1) update by L_{rec} based on the unlabelled data (half batch), and 2) update by L_{joint} based on the labelled data (half batch). A detailed implementation is provided in the supplementary file (Algorithm 1).

2.3 Projection-Wise Disentangling for Interpretation

When θ_{enc}, θ_{dec} and θ_{pe} are determined, for each given input, its fair projection z_p can be estimated and used for fair prediction of t with logistic or linear regression (LR), i.e., $\hat{t} = LR(z_p)$. To provide insight into the effect of fair representation on feature extraction, we visualize the reconstructed images and establish its correspondence with the target attribute $\hat{t}_i$:

$$X_i' = DEC(\overline{d} + k_i P; \theta_{dec}), \text{ and}$$
$$\hat{t}_i = LR(PE(\overline{d} + k_i P; \theta_{pe})), i = 1, 2 \ldots h, \tag{6}$$

where X'_i is the reconstructed image sampled along the direction $P = [p_1, p_2, \ldots, p_n]$ in the latent space (Fig. 1a), namely the projection-wise disentangling. $\overline{d} = [\overline{d}_1, \overline{d}_2, \ldots, \overline{d}_n]$ denotes the mean latent representations of datapoints over the testing set, representing a reference point for the sampling. k_i is a self-defined parameter to control the step of the sampling.

3 Experiments

3.1 Dataset and Tasks

We applied the proposed method to analyzing how the facial shape is related to patient characteristics and clinical parameters.

The data used in this work is from a multi-ethnic population-based cohort study [12]. We included 9-year-old children that underwent raw 3D facial shape imaging with a 3dMD camera system [15]. We built the raw data into a template-based dataset following Booth's procedures [16]. Additionally, extensive phenotyping was performed regarding gender, BMI, height, ethnicity, low to moderate maternal alcohol exposure (i.e., drinking during pregnancy), maternal age and maternal smoking (during pregnancy). The binary phenotypes were digitalized. Gender: 1 for female and 0 for male; Ethnicity: 1 for Western and 0 for non-Western; Maternal alcohol exposure: 1 for exposed and 0 for non-exposed; Maternal smoking: 1 for exposed and 0 for non-exposed.

The first experiment was designed to extract features only related to the target and independent to defined biases. We therefore investigated the relation between basic characteristics (gender, height and BMI) [22]. When predicting one of the attributes, the other two were considered as biases. The number of labelled samples for this experiment was 4992.

In the second experiment we evaluate the applicability of the method in a clinical practice setting. Low to moderate maternal alcohol consumption during pregnancy could have effects on children's facial shape [17]. We aim to predict if a child was exposed or not, and to explain which part of the face is affected. As suggested by [17], gender, ethnicity, BMI, maternal age and smoking were considered as biases in this task. For this experiment we had access to 1515 labelled samples (760 non-exposed and 755 exposed) and 3496 missing-label samples. The missing-label samples were only involved in the semi-supervised learning settings of our method.

Details about the data characteristic of the two experiments can be found in Fig. S4 and Table S1 in the supplementary file.

3.2 Implementation Details

For this 3D facial morphology analysis, we implemented the proposed method using a 3D graph convolution network based on Gong's work [18], which was originally designed for unsupervised 3D shape reconstruction. Correlation (Eq. 3) was computed on batch level. For the correlation term (Eq. 3), to minimize $|Corr(\mathbf{z_p}, \mathbf{s})|$ we set η to 0.5; for the joint loss function (Eq. 5), L_{rec} and L_{corr} were equally weighted ($\lambda = 1$). The training stopped after 600 epochs. For all experiments, the batch size was 64. The number of latent features was 32 for the first experiment, and 64 for the second.

The proposed method was compared with the following state-of-the-art models:

- 3D graph autoencoders (3dAE) [18]: An unsupervised model for 3D shape reconstruction, which serves as a baseline reconstruction method without any restriction on latent features, i.e., does not support fair representation learning.
- VAE-regression (VAE-reg) [19]: A supervised VAE model for interpretable prediction, but unable to handle the bias during training. After training, it is able to reconstruct a list of images by taking as input a list of corresponding t.
- VFAE-MI [8]: A supervised VAE-based method for fair representation using MMD loss. Since MMD is not applicable to multiple or continuous biases, we replaced it by MI loss [21], which is considered to be equivalent or better than the MMD loss in some applications [11].

- BR-Net [7]: A supervised adversarial-training-based method, which uses statistical correlation metric as the adversarial loss rather than the commonly used cross-entropy or MSE losses. The original method was designed to handle single bias only. As the authors suggested [20], we adapted the method to handle multiple biases by adding one more BP network for each additional bias.

3.3 Evaluation Metrics

The proposed method was evaluated in terms of fair prediction performance and interpretability. In the context of this paper, a fair prediction means the prediction $\hat{t}$ is unbiased by s; The fairness in disentanglement (interpretability) is to disentangle facial features related to the target but not confounded by bias.

Prediction results were evaluated on two aspects: the prediction accuracy and fairness. For binary prediction, the accuracy was quantified by the area under the receiver operating characteristic curve (AUC); for prediction of continuous variables, the root mean square error (R-MSE) was measured. The fairness of the prediction was quantified by $|Corr(\hat{t}, s_i)|$ (simplified as $|Corr(s_i)|$) in the range of [0, 1], which measures to what extent the prediction is biased by the bias. In Table 1 and 2 we added '+' or '-' for $|Corr(s_i)|$, in order to tell if there is an overestimation (+) or underestimation (-) in the prediction when given a larger s_i. For autoencoder-based methods, the reconstruction error (Rec error) was quantified by mean L_1 distance. All results were based on 5-fold cross-validation.

Since only VAE-reg [19] is interpretable, the proposed method is compared with it in terms of interpretability. To qualitatively evaluate the interpretation performance, we provide ten frames of faces reconstructed by Eq. 6 (Fig. S2), and the corresponding difference heatmap between the first and the last frame. For each task, k_i in Eq. 6 was adjusted to control the range of $\hat{t}_i$ to be the same as that of VAE-reg [19].

4   Results

4.1   Phenotype Prediction for Gender, BMI, and Height

Fair Prediction. Compared to other fair methods, our method overall obtained the best fair prediction accuracy (R-MSE and AUC), while controlling the biases information at the lowest level ($|Corr(s_i)|$) (Table 1). Compared with VFAE-MI, the proposed method showed much better reconstruction quality, with similar performance to 3dAE. In addition, we observed a more robust training procedure of the proposed method than the compared VAE- and adversarial-based methods (Fig. S1).

Table 1. Fair prediction of gender, BMI, and height. ' $+$ ' and '$-$' means there was respectively an overestimation and underestimation in the prediction when given a larger s_i. The best result for each row among 'Methods with fairness' is in bold.

	Methods without fairness		Methods with fairness		
	3dAE	VAE-reg	VFAE-MI	BR-Net	Ours
Gender prediction					
AUC ↑	–	0.866	0.807	0.802	**0.840**
|Corr (BMI)| ↓	–	−0.061	−0.037	−0.035	**−0.035**
|Corr (height)| ↓	–	−0.133	−0.072	−0.073	**−0.047**
Rec error ↓	0.273	0.275	0.671	–	**0.295**
BMI prediction					
R-MSE ↓	–	1.786	2.706	2.504	**2.373**
|Corr (height)| ↓	–	+0.338	+0.056	+0.047	**+ 0.023**
|Corr (gender)| ↓	–	+0.027	+0.034	+0.021	**+ 0.013**
Rec error ↓	0.273	0.276	0.681	–	**0.278**
Height prediction					
R-MSE (cm) ↓	–	5.303	6.617	6.480	**6.222**
|Corr (BMI)| ↓	–	+0.424	+0.030	+0.026	**+0.013**
|Corr (gender)| ↓	–	−0.201	−0.049	−0.038	**−0.028**
Rec error ↓	0.273	0.277	0.738	–	**0.278**

Interpretation. Figure 2 provides visualizations of the facial features that are used by the methods for prediction tasks. The baseline model (VAE-reg) [19] captured all features to boost the prediction, whereas our model captured the features that only related to the target and independent to biases. For gender prediction, our result is similar to that of VAE-reg because gender is nearly unbiased by height and BMI in the dataset (Fig. S4). However, BMI and height are positively correlated in our dataset. The similar heatmaps for the BMI and height prediction of the VAE-reg indicate that it captured common facial features for the two tasks, and thus failed to disentangle the confounding bias. In contrast, our model learned a target-specific representation, showing a strong correlation to the target attribute and without being confounded by other (bias) attributes.

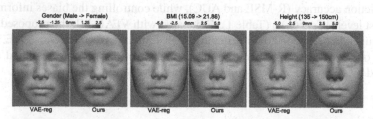

Fig. 2. Interpretation of facial features extracted for prediction. VAE-reg: results of VAE-reg [19]; Ours: results of the proposed method. Red and blue areas refer to inner and outer facial changes towards the geometric center of the 3D face, respectively. (Color figure online)

4.2 Prediction on Maternal Alcohol Consumption During Pregnancy

Fair Prediction. In the second experiment, our method achieved a similar prediction accuracy as the other fair methods while controlling the bias information at the lowest level (Table 2). Compared with the results of 'Ours', the results of 'Ours-SSL' showed that the semi-supervised strategy further improved the prediction accuracy and reconstruction quality by additionally including missing-label training data, when controlling the bias information at a similar level.

Table 2. Fair prediction results. Ours-SSL refers to the semi-supervised settings of our method.

	Methods without fairness		Methods with fairness			
	3dAE	VAE-reg	VFAE-MI	BR-Net	Ours	Ours-SSL
AUC ↑	–	0.768	0.572	0.563	0.579	**0.587**
ICorr (ethnicity)I ↓	–	+0.430	+0.076	**+0.024**	+0.040	+0.037
ICorr (maternal smoking)I ↓	–	+0.125	+0.050	+0.044	**+0.032**	+0.039
ICorr (maternal age)I ↓	–	+0.257	+0.054	+0.044	**+0.031**	+0.034
ICorr (BMI)I↓	–	−0.418	−0.079	−0.055	−0.030	**-0.019**
ICorr (gender) I↓	–	+0.078	+0.070	+0.069	+0.044	**+0.036**
Rec error ↓	0.276	0.331	0.656	–	0.344	**0.316**

Interpretation. We compared the interpretation results of the VAE-reg and our methods (Fig. 3), and further explained the gap between the baseline and our results, by visualizing our results with gradually increased fairness (Fig. 4): from left to right, in the first figure L_{corr} corrected for no bias in the training; in the last figure L_{corr} corrected for all bias in the training (Eq. 4). Our results suggest low to moderate maternal alcohol exposure during pregnancy affected children's facial shape. Affected regions were shown by our methods in Fig. 3, which is consistent with existing findings [17].

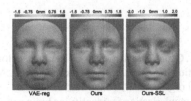

Fig. 3. Heatmaps show how the facial shape changes from non-exposed to exposed.

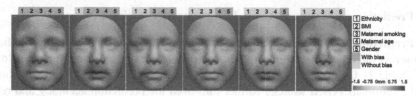

Fig. 4. Explanation about how the target features were disentangled from biases.

5 Discussion and Conclusion

In this paper, we proposed a projection-wise disentangling method and applied it to 3D facial shape analysis. For evaluated tasks (BMI, height, gender and maternal alcohol exposure prediction), we achieved the best prediction accuracy while controlling the bias information at the lowest level. Also, we provide a solution to interpret prediction results, which improved mechanistic understanding of the 3D facial shape. In addition, given the shortage of labelled data in many domains, we expect that the proposed method with its semi-supervised extension can serve as an important tool to fully exploit available data for fair representation learning.

Beyond the presented application, the proposed method is widely applicable to prediction tasks, especially to clinical analysis where confounding bias is a common challenge. For future work, we plan to disentangle aging effects from pathology of neurodegenerative diseases.

The improvement of our method mainly comes from the projection strategy. Previous methods learn fair features Z by forcing $MI(Z, s) \rightarrow 0$, which encourages a global independence between Z and s, i.e., **any linear or non-linear combination of Z** contains no information of s. This strong restriction leads to a decrease in diversity of learned features (Fig. S3), resulting in huge reconstruction error in VFAE-MI (Tables 1 and 2). This loss of diversity also explains why the prediction accuracy of baselines were limited although using all features in Z for prediction. In contrast, our strategy can be viewed as a conditional independence between Z and s, i.e., **only a linear combination of Z** (the projection z_p) being independent to s. This strategy allows that most of the information can be kept in the latent space, and thus minimizing the conflicts between reconstruction quality and fairness in auto-encoder models. This is crucial especially when the biases contain much more information of the input than the target does, e.g., the second experiment.

References

1. Tommasi, T., Patricia, N., Caputo, B., Tuytelaars, T.: A Deeper look at dataset bias. In: Csurka, G. (ed.) Domain Adaptation in Computer Vision Applications. ACVPR, pp. 37–55. Springer, Cham (2017). https://doi.org/10.1007/978-3-319-58347-1_2
2. Adeli, E., et al.: Chained regularization for identifying brain patterns specific to HIV infection. Neuroimage **183**, 425–437 (2018)
3. Pourhoseingholi, M.A., Baghestani, A.R., Vahedi, M.: How to control confounding effects by statistical analysis. Gastroenterol. Hepatol. Bed Bench **5**, 79–83 (2012)

4. Zhou, B., Bau, D., Oliva, A., Torralba, A.: Interpreting deep visual representations via network dissection. IEEE Trans. Pattern Anal. Mach. Intell. **41**(9), 2131–2145 (2019)
5. Balakrishnan, G., Xiong, Y., Xia, W., Perona, P.: Towards causal benchmarking of bias in face analysis algorithms. In: Vedaldi, A., Bischof, H., Brox, T., Frahm, J.-M. (eds.) ECCV 2020. LNCS, vol. 12363, pp. 547–563. Springer, Cham (2020). https://doi.org/10.1007/978-3-030-58523-5_32
6. Xie, Q., Dai, Z., Du, Y., Hovy, E., Neubig, G.: Controllable invariance through adversarial feature learning. In: NIPS (2017)
7. Adeli, E., et al.: Representation learning with statistical independence to mitigate bias. In: WACV (2021)
8. Louizos, et al.: The variational fair autoencoder. In: ICLR (2016)
9. Creager, E., et al.: Flexibly fair representation learning by disentanglement. In: PMLR (2019)
10. Alemi, A.A., Fischer, I., Dillon, J.V., Murphy, K.: Deep variational information bottleneck. In: ICLR (2017)
11. Botros, P., Tomczak, J.M.: Hierarchical vampprior variational fair auto-encoder. arXiv preprint arXiv: 1806.09918 (2018)
12. Jaddoe, V.W., Mackenbach, J.P., Moll, H.A., Steegers, E.A., Tiemeier, H., Verhulst, F.C., et al.: The generation R study: study design and cohort profile. Eur. J. Epidemiol. **21**, 475–484 (2006)
13. Vellido, A.: The importance of interpretability and visualization in machine learning for applications in medicine and health care. Neural Comput. Appl. **32**(24), 18069–18083 (2019). https://doi.org/10.1007/s00521-019-04051-w
14. Härdle, W.K., Simar, L.: Canonical correlation analysis. In: Applied Multivariate Statistical Analysis, pp. 361-372. Springer, Heidelberg (2003).https://doi.org/10.1007/978-3-540-72244-1_14
15. 3dMD. https://3dmd.com/. Accessed Feb 3 2021
16. Booth, J., Roussos, A., Ponniah, A., Dunaway, D., Zafeiriou, S.: Large scale 3D morphable models. Int. J. Comput. Vis. **126**, 233–254 (2017)
17. Muggli, E., Matthews, H., Penington, A., et al.: Association between prenatal alcohol exposure and craniofacial shape of children at 12 months of age. JAMA Pediatr. **171**(8), 771–780 (2017)
18. Gong, S., Chen, L., Bronstein, M., Zafeiriou, S.: SpiralNet++: a fast and highly efficient mesh convolution operator. In: ICCVW (2019)
19. Zhao, Q., Adeli, E., Honnorat, N., Leng, T., Pohl, K.M.: Variational autoencoder for regression: application to brain aging analysis. In: Shen, D., et al. (eds.) MICCAI 2019. LNCS, vol. 11765, pp. 823–831. Springer, Cham (2019). https://doi.org/10.1007/978-3-030-32245-8_91
20. Zhao, Q., Adeli, E., Pohl, K.M.: Training confounder-free deep learning models for medical applications. Nat. Commun. **11**, 6010 (2020)
21. Belghazi, M.I., et al.: Mutual information neural estimation. In: ICML (2018)
22. Tobias, M., et al.: Cross-ethnic assessment of body weight and height on the basis of faces. Pers. Individ. Differ. **55**(4), 356–360 (2013)

Attention-Based Multi-scale Gated Recurrent Encoder with Novel Correlation Loss for COVID-19 Progression Prediction

Aishik Konwer[1], Joseph Bae[2], Gagandeep Singh[3], Rishabh Gattu[3], Syed Ali[3], Jeremy Green[3], Tej Phatak[3], and Prateek Prasanna[2(✉)]

[1] Department of Computer Science, Stony Brook University, Stony Brook, NY, USA
akonwer@cs.stonybrook.edu
[2] Department of Biomedical Informatics, Stony Brook University,
Stony Brook, NY, USA
prateek.prasanna@stonybrook.edu
[3] Department of Radiology, Newark Beth Israel Medical Center, Newark, NJ, USA

Abstract. COVID-19 image analysis has mostly focused on diagnostic tasks using single timepoint scans acquired upon disease presentation or admission. We present a deep learning-based approach to predict lung infiltrate progression from serial chest radiographs (CXRs) of COVID-19 patients. Our method first utilizes convolutional neural networks (CNNs) for feature extraction from patches within the concerned lung zone, and also from neighboring and remote boundary regions. The framework further incorporates a multi-scale Gated Recurrent Unit (GRU) with a correlation module for effective predictions. The GRU accepts CNN feature vectors from three different areas as input and generates a fused representation. The correlation module attempts to minimize the correlation loss between hidden representations of concerned and neighboring area feature vectors, while maximizing the loss between the same from concerned and remote regions. Further, we employ an attention module over the output hidden states of each encoder timepoint to generate a context vector. This vector is used as an input to a decoder module to predict patch severity grades at a future timepoint. Finally, we ensemble the patch classification scores to calculate patient-wise grades. Specifically, our framework predicts zone-wise disease severity for a patient on a given day by learning representations from the previous temporal CXRs. Our novel multi-institutional dataset comprises sequential CXR scans from N = 93 patients. Our approach outperforms transfer learning and radiomic feature-based baseline approaches on this dataset.

Keywords: COVID-19 · Correlation · Attention · Gated recurrent unit · Transfer learning

Electronic supplementary material The online version of this chapter (https://doi.org/10.1007/978-3-030-87240-3_79) contains supplementary material, which is available to authorized users.

1 Introduction

Coronavirus disease 2019 (COVID-19) remains at the forefront of threats to public health. As a result, there continues to be a critical need to further understand the progression of the disease process. In the United States, chest radiographs (CXRs) are the most commonly used imaging modality for the monitoring of COVID-19. On CXR, COVID-19 infection has been found to manifest as opacities within lung regions. Previous studies have demonstrated that the location, extent, and temporal evolution of these findings can be correlated to disease progression [12]. Studies have shown that COVID-19 infection frequently results in bilateral lower lung opacities on CXR and that these opacities may migrate to other lung regions throughout the disease's clinical course [7,12]. This suggests that COVID-19 progression may be appreciable on CXR via examination of the spatial spread of radiographic findings across multiple timepoints.

Despite the many studies analyzing the use of CXRs in COVID-19, machine learning applications have been limited to diagnostic tasks including differentiating COVID-19 from viral pneumonia and predicting clinical outcomes such as mortality and mechanical ventilation requirement [4,6]. Many of these studies have reported high sensitivities and specificities for the studied outcomes, but they remain constrained due to deficiencies in publicly available datasets [8]. Furthermore, none have attempted to computationally model the temporal progression of COVID-19 from an imaging perspective. Significantly, most studies have also not explicitly taken into account the spatial evolution of CXR imaging patterns within lung regions that have been demonstrated to correlate with disease severity and progression [7,12]. In this study we take advantage of a unique longitudinal COVID-19 CXR dataset and propose a novel deep learning (DL) approach that exploits the spatial and temporal dependencies of CXR findings in COVID-19 to predict disease progression (Fig. 1).

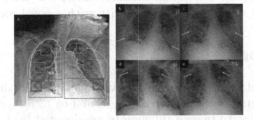

Fig. 1. (a) depicts a CXR in which lung fields have been divided into three equal zones. Disease information in patches from primary zone (Pp) are more similar to those from neighboring zone (Np) than the remote zones (Rp). (b)–(e) depict serial CXRs taken for one patient over several days of COVID-19 infection. We note a progression of imaging findings beginning with lower lobe involvement in (b) with spread to middle lung involvement in (c) and upper lung region involvement in (d) and (e).

Previous deep learning (DL) based COVID-19 studies have mainly considered single timepoint CXRs [1,10]. Unlike these studies, we analyze CXRs from multiple timepoints to capture lung infiltrate progression. Recurrent neural networks (RNNs) have been widely employed for time series prediction tasks in computer vision problems. Recently, RNNs have also found success in analyzing tumor evolution [18] and treatment response from serial medical images [14,16]. A Gated Recurrent Unit (GRU) is an RNN which controls information flow using two gates - a Reset gate and an Update gate. Thus, relevant information from past timepoints are forwarded to future timepoints in the form of hidden states. GRUs have been used extensively to predict disease progression [9].

In this work, we aim to explore how the different zones of an image are correlated to each other. Many studies have demonstrated the spatial progression of COVID-19 seen on CXR imaging with lung opacities generally being noted in lower lung regions in earlier disease stages before gradually spreading to involve other areas such as the middle and upper lung [7,12,15]. Therefore, two neighboring lung zones should have a higher similarity measure than two far-apart zones. Unlike previous approaches, we propose a multi-scale GRU [17] which can accept three distinct inputs at the same timepoint. Apart from primary patches Pp of concerned zone, patches from Neighbor Np and Remote areas Rp are also used as inputs to a GRU cell at a certain timepoint. We include a Correlation module to maximize the correlation measure between Pp and Np, while minimizing the correlation between Pp and Rp. Finally, an attention layer is applied over hidden states to obtain patch weights and give relative importance to patches collected from multiple timepoints. The major contributions of this paper are the following: (1) Our work uses a multi-scale GRU framework to model the progression of lung infiltrates over multiple timepoints to predict the severity of imaging infiltrates at a later stage. (2) Disease patterns in adjacent regions tend to be spatially related to each other. COVID-19 imaging infiltrates exhibit similar patterns of correlation across lung regions on CXRs. We are the first to use a dedicated correlation module within our temporal encoder that exploits this latent state inter-zone similarity with a novel correlation loss.

2   Methodology

Varying numbers of temporal images are available for each patient. The number of timepoints is equal to d which may vary from 4 to 13 for a given patient. The images corresponding to these d timepoints are denoted by $I_{t_1}, I_{t_2}, \ldots I_{t_{d-1}}, I_{t_d}$. The left and right lung masks are generated from these images using a residual U-Net model [1]. These masks are each further subdivided into 3 lung zones - Upper (L_1, R_1), Middle (L_2, R_2) and Lower (L_3, R_3) zones. Our collaborating radiologists assigned severity grades to each of the 6 zones as $g_0 = 0$, $g_1 = 1$ or $g_2 = 2$ depending on the zonal infiltrate severity. This procedure mirrors the formulation of other scoring systems [6]. We train 6 different models for each of the six zones - M_{L_1}, M_{L_2}, M_{L_3}, M_{R_1}, M_{R_2}, and M_{R_3}. We adopt this zone-wise granular approach to overcome the need of image registration.

2.1 Overview

We implement an Encoder-Decoder framework based on seq2seq model [2] in order to learn sequence representations. Specifically, our framework (Fig. 2) includes two recurrent neural networks: a multi-scale encoder and a decoder. The training of the multi-scale encoder involves fusion of three input patches each from Pp, Np and Rp - concerned (current zone of interest), neighboring, and remote zones, at each of the timepoints, to generate a joint feature vector. The attention weighted context vector that we obtain from the encoder is finally used as input to the decoder. The decoder at its first timepoint attempts to classify this encoder context vector into the 3 severity labels. The multi-scale encoder is trained with the help of a correlation module to retain only relevant information from each of the patches of three distinct zones.

2.2 Patch Extraction

Each image zone is divided into sixteen square grids. These grids are resized to dimension 128×128 and used as primary patches Pp for the concerned zone. Now, for each zone, we also consider 8 patches from the boundary of two adjoining neighbor zones. For example, in the case of L_1 zone, we use 4 patch grids from R_1 boundary and 4 patch grids from L_2 boundary. Similarly, in the case of middle zone L_2, we use 4 patch grids each from nearest L_1 and L_3 boundaries. Thus we build a pool of 8 neighboring Np patches for each concerned lung zone. Additionally, we also create a cluster of 8 Rp patches coming from the far-away boundaries of remote zones. e.g. Rp patches for L_1 is collected from boundaries of L_3.

2.3 Feature Extraction

For a particular model, say M_{L_1}, each Pp patch from zone L_1 is fed as input to a Convolutional neural network (CNN) to predict the severity scores at a given timepoint. Similarly, one random patch from each of Np and Rp, are also passed into the same CNN. As an output of the CNN, we obtain three 1×256 dimensional feature vectors. The CNN network configuration contains five convolutional layers, each associated with an operation of max-pooling. The network terminates with a fully connected layer.

2.4 Encoder

Multi-scale GRU. The GRU module used here is a multi-scale extension of the standard GRU. It houses different gating units - the reset gate and the update gate which control the flow of relevant information in a GRU. The GRU module takes P_t, N_t, and R_t as inputs (denoted by X_t^i, where i = 1, 2, 3) at time step t and monitors four latent variables, namely the joint representation h_t, and input-specific representations h_t^1, h_t^2 and h_t^3. The fused representation h_t is actually treated as a single descriptor for the multi-input data that helps in learning the temporal context of our data over multiple timepoints. The input-specific representations h_t^1, h_t^2 and h_t^3 constitute the projections of three distinct inputs. They are used to calculate two correlation measures among them in the GRU module. The computation within this module may be formally expressed as follows:

$$r_t^i = \sigma(W_r^i X_t^i + U_r h_{t-1} + b_r^i) \tag{1}$$

$$z_t^i = \sigma(W_z^i X_t^i + U_z h_{t-1} + b_z^i) \tag{2}$$

$$\tilde{h}_t^i = \varphi(W_h^i X_t^i + U_h(r_t^i \odot h_{t-1}) + b_h^i), i = 1, 2, 3 \tag{3}$$

$$r_t = \sigma(\sum_{i=1}^{3} w_t^i(W_r^i X_t^i + b_r^i) + U_r h_{t-1}) \tag{4}$$

$$z_t = \sigma(\sum_{i=1}^{3} w_t^i(W_z^i X_t^i + b_z^i) + U_z h_{t-1}) \tag{5}$$

$$\tilde{h}_t = \varphi(\sum_{i=1}^{3} w_t^i(W_h^i X_h^i + b_h^i) + U_h(r_t \odot h_{t-1}) \tag{6}$$

$$h_t^i = (1 - z_t^i) \odot h_{t-1} + z_t^i \odot \tilde{h}_t^i, i = 1, 2, 3 \tag{7}$$

$$h_t = (1 - z_t) \odot h_{t-1} + z_t \odot \tilde{h}_t \tag{8}$$

where σ is the logistic sigmoid function and φ is the hyperbolic tangent function, r and z are the input to the reset and update gates, and h and $\tilde{h}$ represent the activation and candidate activation, respectively, of the standard GRU [3]. W_r, W_z, W_h, U_r, U_z and U_h are the weight parameters learned during training. w_t^i (i = 1, 2, 3) are also learned parameters. b_r, b_z and b_h are the biases. X_t^i (i = 1, 2, 3) are the CNN feature vectors of patches from the three zones - P_p, N_p and R_p.

Correlation Module. In order to obtain a better joint representation for temporal learning, we introduce an important component into the multi-scale encoder, one that explicitly captures the correlation between the three distinct inputs. Our model explicitly applies a correlation-based loss term in the fusion process. The principle of our model is to maximize the correlation between features from Pp and Np, and to minimize the correlation between features from Pp and Rp. Pearson coefficient has been used to compute the correlation. Hence this module computes the correlation between the projections h_t^1, h_t^2 and also between h_t^1, h_t^3 obtained from the GRU module. We denote the correlation-based loss function as

$$L_{corr} = max[corr(h_t^1, h_t^2)] + min[corr(h_t^1, h_t^3)] \tag{9}$$

For all patients, independently for each patch from Pp and Np zones, we maximized the correlation function. Similarly we minimized the correlation function for each patch from Pp and Rp zones.

Attention Module. The hidden state from each GRU cell is passed through an attention network. The attention weights $\alpha_1, \alpha_2, \ldots, \alpha_{d-1}$ are computed for each timepoint. These scores are then fed to a softmax layer to obtain the probability weight distribution, such that the summation of all attention weights covering the available $d-1$ timepoints of the encoder equals to 1. We compute a weighted summation of these attention weights and the GRU hidden states' vectors to construct a holistic context vector for the encoder output.

2.5 Decoder

The attention weighted context vector obtained from $d-1$ timepoints of the encoder is used as an input to the decoder. A linear classifier and softmax layer is applied on the GRU decoder's hidden state to obtain three severity scores - g_0, g_1, and g_2. For each patient and zone, we predict 16 such patch classification scores for the $I_{t_d}^{th}$ image. We employ majority voting as an ensemble procedure on these scores to obtain the final patient-wise grade.

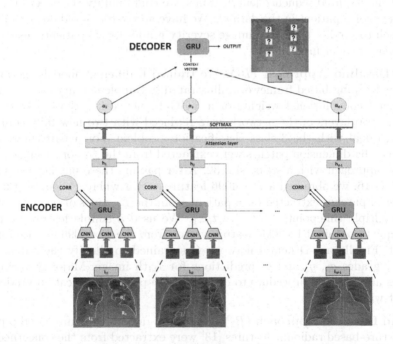

Fig. 2. Architecture of the proposed approach. We show here model M_{L_1} which deals with patches from L_1 zone. At each timepoint, 3 patches each from Pp, Np and Rp are inputs to CNN network. The generated CNN features are passed into a GRU cell. Fused hidden state GRU output h_t is used to calculate attention weights. Attention weighted summation of multiple such hidden states form the context vector for decoding purpose.

3   Experimental Design

3.1   Dataset Description

Our multi-institutional dataset, COVIDProg [5], contains 621 antero-posterior CXR scans from 93 COVID-19 patients, collected from multiple days. 23 cases were obtained from *Newark Beth Israel Medical Center*. The remaining 70 cases curated from *Stony Brook University Hospital*. All the CXRs were of dimension 3470×4234. Additional details can be found in Supplementary Sect. 3.

3.2   Implementation Details

For training the CNN and GRU, a cross entropy loss function was used along with the designed correlation loss discussed earlier. Optimization of the network was done using Adam Optimizer. Each of the 6 models is trained once for 300 iterations with a batch size of 30 and a learning rate of 0.001. The total number of epochs is 20. We used pack padded sequence to mask out all losses that surpassed the required sequence length. Thus, we could nullify the effect of missing timesteps for a patient in the dataset. We have adopted a 5-fold cross validation approach to predict the I_{t_d} th image severity grades for 93 patients, using $d-1$ images as encoder input.

First Baseline Approach (B_1). We trained 6 different models based on a transfer learning based framework, illustrated in Supplementary Sect. 4. All the pretrained convolutional weights of a VGG-16 network [11] were kept same. The last two layers of the network were replaced with two new fully connected layers to deal with the 3-class classification problem. For a particular model, M_{L_1}, 64×64 dimension patches were extracted from the L_1 zone using a sliding window approach with a stride size 32. After passing these patches as input to our VGG-16, we obtained a $P \times 4096$ feature vector where P denotes the total number of patches extracted for a patient from the L_1 zones of images collected from multiple timepoints $t_1, t_2, \ldots, t_{d-1}$. We used a simple feature averaging technique to obtain a 1×4096 feature vector from the $P \times 4096$ feature for each patient. Finally a 1-D neural network was trained to classify the patches into severity grades g_0, g_1 and g_2 predictions for I_{td}th image. Majority voting was used as an ensemble procedure to convert these patch classification grades to a patient-wise grade.

Second Baseline Approach (B_2). We built a radiomic feature based pipeline. 445 texture-based radiomic features [13] were extracted from the concerned lung zone. These features were similarly averaged into a single feature vector and classified using random forest classifier.

4   Results

Averaged results are presented after 5 runs of model-testing. Accuracy is computed for each of the 6 lung zones, while the precision and recall are measured

Table 1. Quantitative results on left lung zones

Methods	Left lung upper							Left lung middle							Left lung lower						
	Acc (%)	*Pre*			*Rec*			*Acc*(%)	*Pre*			*Rec*			*Acc* (%)	*Pre*			*Rec*		
		0	**1**	**2**	**0**	**1**	**2**		**0**	**1**	**2**	**0**	**1**	**2**		**0**	**1**	**2**	**0**	**1**	**2**
Baseline-1	60.21	0.53	0.72	0.53	0.43	0.51	**0.77**	62.36	0.55	0.69	0.71	**0.65**	0.63	0.60	59.13	0.53	0.67	0.72	0.56	0.69	0.80
Baseline-2	56.98	0.54	0.60	0.52	0.63	0.56	0.66	54.83	0.58	0.74	0.67	0.50	0.71	0.61	60.21	**0.74**	0.69	0.76	0.45	0.69	0.73
Variant-1	66.66	0.54	0.68	**0.72**	0.67	0.73	0.56	67.74	0.61	0.69	0.63	0.54	0.69	0.64	63.44	0.65	0.72	0.79	0.67	0.72	0.75
Variant-2	73.11	0.59	0.71	0.66	0.66	0.77	0.69	70.96	0.63	0.74	0.69	0.56	0.67	0.59	68.89	0.61	0.76	0.81	**0.70**	**0.74**	0.67
Our approach	**75.26**	**0.69**	**0.73**	0.58	**0.68**	**0.81**	0.75	**72.04**	**0.72**	**0.82**	**0.77**	0.63	**0.84**	**0.65**	**73.11**	0.66	**0.78**	**0.83**	0.69	0.72	**0.85**

Table 2. Quantitative results on right lung zones

Methods	Right lung upper							Right lung middle							Right lung lower						
	Acc (%)	*Pre*			*Rec*			*Acc* (%)	*Pre*			*Rec*			*Acc*(%)	*Pre*			*Rec*		
		0	**1**	**2**	**0**	**1**	**2**		**0**	**1**	**2**	**0**	**1**	**2**		**0**	**1**	**2**	**0**	**1**	**2**
Baseline-1	61.29	0.54	0.62	0.54	0.59	0.64	0.63	62.36	0.60	0.63	**0.67**	0.61	0.64	0.59	58.06	0.71	0.58	0.64	0.52	0.63	0.73
Baseline-2	55.91	0.50	0.66	0.57	0.61	0.58	0.66	59.13	0.63	0.60	0.65	0.64	0.55	0.56	53.76	0.74	0.51	0.58	0.57	0.66	0.70
Variant-1	69.89	0.55	0.64	**0.65**	0.60	**0.73**	0.71	**67.74**	0.68	0.63	0.58	0.68	0.74	0.68	63.44	0.71	0.63	0.57	**0.58**	0.73	0.77
Variant-2	73.11	0.62	0.69	0.64	0.63	0.71	0.75	64.51	0.67	**0.66**	0.61	0.65	**0.76**	0.70	65.59	0.73	0.68	**0.77**	0.54	0.72	0.80
Our approach	**76.34**	**0.67**	**0.72**	0.56	**0.71**	0.70	**0.78**	64.51	**0.69**	**0.66**	0.64	**0.68**	0.73	**0.72**	**69.89**	**0.74**	**0.79**	0.76	0.57	**0.75**	**0.83**

for each of the severity grades, g_0, g_1 and g_2. The results using our approach and the two baseline methods are illustrated in Tables 1 and 2 for the left and the right lung zones, respectively. In all the zones, except R_3, our method performed significantly better than both baseline approaches. For example, in left lung upper zone, we achieved an accuracy of 75.26%. The baseline accuracies were 60.21% and 56.98% for B_1 and B_2, respectively.

Ablation Study. In order to capture the gradual improvement of our framework through different stages, we conducted a serial ablation study and built two sub-variants of our frameworks. 1) Variant-1: This variant uses only multi-scale GRU cells which concatenate the inputs from two distinct patches - Pp, Np to generate the fused representation. Both the Correlation module and the Attention module were removed from our framework. Though neighboring patches were taken into consideration, we do not exploit the explicit correlation between Pp, Np and between Pp, Rp. Also, the encoder output vector does not consider the relative importance of hidden states generated for multiple timepoints. 2) Variant-2: This variant consists of the Correlation module. However, the Attention module is neglected and equal importance is assigned to all the zone patches collected from multiple timepoint' images. Thus we gradually zeroed into our framework which outperforms the sub-variants by a large margin in most zones. The results in Tables 1 and 2 suggest that exploiting the correlation between the nearby zones and remote zone patches leads to an increase in prediction performance. Moreover, the use of an attention layer to provide individual patch importance further boosts the accuracy. As an example, it can seen that for the left lung middle zone, our M_{L_1} accuracy is 72.04% while for Variant-1 and Variant-2 it is 67.74% and 70.96%, respectively.

Testing with $d - 2$ Timepoints as Encoder Input. We designed an experimental setup to analyze how the framework performs when patches from only first $d - 2$ images are used as input to our GRU encoder. However, the task is still to predict the severity scores of I_{t_d} image. From Supplementary Table 1, we can observe that even in this experimental setting, our model achieves highest accuracies for L_1, L_2, R_1 and R_3 - 64.51, 62.36, 66.67 and 59.13, while achieving competitive scores for the other two zones. This suggests that our framework can perform well even if we have fewer number of timepoints as encoder input.

5   Conclusion

COVID-19 CXRs reveal varied spatial correlations among the lung infiltrates across different zones. Adjacent zones are generally found to be more correlated than two distant regions. We build a multi-scale GRU based encoder-decoder framework which accepts multiple inputs from different lung zones at a single timepoint. Unlike generative approaches, our model does not require registration between images from different timepoints. A novel two component correlation loss is introduced to explore the spatial correlations within nearby and distant lung fields in latent representation. Finally we use an attention layer to judge the relative importance of the images from available timepoints for computing the disease severity score at a future timepoint.

Acknowledgments. Reported research was supported by the OVPR and IEDM seed grants, 2020 at Stony Brook University, NIGMS T32GM008444, and NIH 75N92020D00021 (subcontract). The content is solely the responsibility of the authors and does not necessarily represent the official views of the National Institutes of Health.

References

1. Bae, J., et al.: Predicting mechanical ventilation requirement and mortality in COVID-19 using radiomics and deep learning on chest radiographs: a multi-institutional study. arXiv preprint arXiv:2007.08028 (2020)
2. Bahdanau, D., Cho, K., Bengio, Y.: Neural machine translation by jointly learning to align and translate. arXiv preprint arXiv:1409.0473 (2014)
3. Cho, K., et al.: Learning phrase representations using rnn encoder-decoder for statistical machine translation. In: Proceedings of the 2014 Conference on Empirical Methods in Natural Language Processing, pp. 1724–1734 (October 2014)
4. Hu, Q., Drukker, K., Giger, M.L.: Role of standard and soft tissue chest radiography images in COVID-19 diagnosis using deep learning. In: Medical Imaging 2021: Computer-Aided Diagnosis, vol. 11597, p. 1159704. International Society for Optics and Photonics (February 2021)
5. Konwer, A., et al.: Predicting COVID-19 lung infiltrate progression on chest radiographs using spatio-temporal LSTM based encoder-decoder network. Med. Imag. Deep Learn. (MIDL) **143**, 384–398 (2021)
6. Kwon, Y.J.F., et al.: Combining initial radiographs and clinical variables improves deep learning prognostication in patients with COVID-19 from the emergency department. Radiol. Artif. Intell. **3**(2), e200098 (2020)
7. Litmanovich, D.E., Chung, M., Kirkbride, R.R., Kicska, G., Kanne, J.P.: Review of chest radiograph findings of COVID-19 pneumonia and suggested reporting language. J. Thorac. Imaging **35**(6), 354–360 (2020)

8. López-Cabrera, J.D., Orozco-Morales, R., Portal-Diaz, J.A., Lovelle-Enríquez, O., Pérez-Díaz, M.: Current limitations to identify COVID-19 using artificial intelligence with chest X-ray imaging. Heal. Technol. **11**(2), 411–424 (2021)

9. Pavithra, M., Saruladha, K., Sathyabama, K.: GRU based deep learning model for prognosis prediction of disease progression. In: 3rd International Conference on Computing Methodologies and Communication (ICCMC), pp. 840–844 (2019)

10. Shi, F., et al.: Review of artificial intelligence techniques in imaging data acquisition, segmentation and diagnosis for COVID-19. arXiv arXiv:2004.02731 [cs, eess, q-bio] (April 2020)

11. Simonyan, K., Zisserman, A.: Very deep convolutional networks for large-scale image recognition. arXiv preprint arXiv:1409.1556 (2014)

12. Toussie, D., et al.: Clinical and chest radiography features determine patient outcomes in young and middle age adults with COVID-19. Radiology **297**, E197–E206 (2020)

13. Van Griethuysen, J.J.M., et al.: Computational radiomics system to decode the radiographic phenotype. Cancer Res. **77**(21), e104–e107 (2017)

14. Wang, C., et al..: Toward predicting the evolution of lung tumors during radiotherapy observed on a longitudinal MR imaging study via a deep learning algorithm. Med. Phys. **46**, 4699–4707 (2019)

15. Wong, H.Y.F., et al.: Frequency and distribution of chest radiographic findings in COVID-19 positive patients. Radiology **296**, E72–E78 (2020)

16. Xu, Y., et al.: Deep learning predicts lung cancer treatment response from serial medical imaging. Clin. Cancer Res. **25**, 3266–3275 (2019)

17. Yang, X., Ramesh, P., Chitta, R., Madhvanath, S., Bernal, E.A., Luo, J.: Deep multimodal representation learning from temporal data. In: Proceedings of the IEEE Conference on Computer Vision and Pattern Recognition, pp. 5447–5455 (2017)

18. Zhang, L., Lu, L., Summers, R., Kebebew, E., Yao, J.: Convolutional invasion and expansion networks for tumor growth prediction. IEEE Trans. Med. Imaging **37**, 638–648 (2018)

Correction to: Transfer Learning of Deep Spatiotemporal Networks to Model Arbitrarily Long Videos of Seizures

Fernando Pérez-García (iD), Catherine Scott, Rachel Sparks (iD),
Beate Diehl (iD), and Sébastien Ourselin (iD)

Correction to:
Chapter "Transfer Learning of Deep Spatiotemporal
Networks to Model Arbitrarily Long Videos of Seizures"
in: M. de Bruijne et al. (Eds.): *Medical Image Computing*
***and Computer Assisted Intervention – MICCAI 2021*,**
LNCS 12905, https://doi.org/10.1007/978-3-030-87240-3_32

In the first line of Sect. 2.2, "A neurophysiologist (A.A.)" has been replaced by "A neurophysiologist (C.S.)". In addition, Equation (1) and one mathematical expression two paragraphs above this had been rasterized into images. This has been remedied.

The updated version of this chapter can be found at
https://doi.org/10.1007/978-3-030-87240-3_32

Author Index

Printed in the United States
by Baker & Taylor Publisher Services